D0076158

Circuit Analysis

Circuit Analysis

SECOND EDITION

David R. Cunningham
University of Missouri—*Rolla*

John A. Stuller
University of Missouri—*Rolla*

JOHN WILEY & SONS, INC.
New York • Chichester • Brisbane • Toronto • Singapore

To Our Students

Cover Design DesignHeads, Boston
Cover Image Jose Alexanco, ''Espacios de separacior III'' 1981
Photo Credits page xxv, portrait of Volta courtesy of Institution of Electrical Engineers; page xxvi, portrait of Ampere from Historical Pictures Service, Chicago; portrait of Faraday painted by Thomas Phillips from *Ten Founding Fathers of Electrical Science,* by Ben Dibner, Burndy Library publication no. 11, 1954; portraits of Ohm and Fourier from Historical Pictures Service, Chicago; page xxvii, portrait of Laplace from Historical Pictures Service, Chicago; portrait of Henry from Smithsonian Archives, Record Unit 95, Photograph Collection, 1880s– ; portrait of Kirchhoff from Historical Pictures Service, Chicago; portrait of Siemens courtesy of Institution of Electrical Engineers.

Previously published by Houghton Mifflin Company.

ISBN: 0 471 12484 2

Printed in the Unites States of America
10 9 8 7 6 5 4

Brief Contents

Contents

Preface

Circuit Analysis is written for the two-semester electric circuits course typically offered as the first course in the standard electrical engineering curriculum. Knowledge of first-year calculus is assumed. A high school physics course or a knowledge of college physics is desirable but not necessary. This book does not assume a familiarity with complex arithmetic, matrices, or differential equations. It is designed for flexibility of coverage and to accommodate students with differing backgrounds.

Our objective in writing the second edition remains the same as in the first: to enable students to excel in circuit analysis while developing a solid foundation for later courses in systems theory. The second edition is substantially revised to be more effective for the student. A great deal of effort has been made to make this edition more efficient and easier to learn and teach from, while retaining the technical strengths and integrity of the first edition. The revisions include the following:

- Text, examples, and problems have been streamlined for increased pedagogical effectiveness. Reference and advanced material not found in standard curricula has been eliminated.
- Chapters 8 and 9 have been rewritten to conform to more popular coverage of differential equations and dynamic circuit response. First-order circuits are covered before second-order ones, and the mathematical sophistication required of the student has been reduced.
- Laplace transforms have been moved forward to Chapter 14.
- Coupled coils and mutual inductance are covered as a separate topic in Chapter 17.
- All introductory PSpice material is now contained in the PSpice manual that accompanies this text. The text and manual are still keyed for cross referencing.

Coverage and Organization

To help students recognize the interrelation of major ideas, we divided the topics into five major parts: Part One, Fundamentals; Part Two, Time-Domain Circuit Analysis;

Part Three, Frequency-Domain Circuit Analysis; Part Four, Series and Transform Methods; and Part Five, Selected Topics. Some chapters can be covered in an alternative sequence. The flowchart on page xix shows the prerequisites for each chapter.

Part One, Fundamentals, begins with an introduction to electrical current, voltage, and power. It then presents Kirchhoff's current law (KCL) and Kirchhoff's voltage law (KVL), giving examples of each law before introducing the network components. We then introduce sources, resistance R, capacitance C, and inductance L.

Circuits composed of only sources and Rs are strongly emphasized in the early development. However, a very few examples of simple circuits containing Ls or Cs are also shown to the student from an early point. This teaching strategy effectively prevents the mental block that too often develops in students when this topic is postponed. No calculus is needed other than the ability to write down derivatives and integrals and to evaluate them for elementary functions like $i(t) = 3t$, e^t and $\cos 2t$.

Part Two, Time-Domain Circuit Analysis, contains all the fundamental techniques of linear circuit analysis, except graph theory, beginning with node-voltage analysis. Here, and throughout the book, we present new material in the context of preceding work. For example, in Chapter 4, we present circuit analysis by node voltages as a generalization of the analysis of a parallel circuit. Similarly, we present mesh-current analysis as a generalization of the analysis of a series circuit. We conclude Part Two with a standard treatment for solving differential equations and finding the response of RC, RL, and RLC circuits for constant and exponential inputs.

Part Three, Frequency-Domain Circuit Analysis, develops phasor circuit analysis beginning with basic definitions and concluding with poles, zeros, and Bode plots. We describe the mathematical transformations from time domain to frequency domain in greater detail and depth than many texts do. The treatment is designed to provide students with a particularly strong foundation for future courses in linear systems.

Part Four, Series and Transform Methods, begins with the Laplace Transform to emphasize its link to the s-domain circuit analysis techniques of Chapter 13. Fourier series and transforms follow. All techniques in Part Four are presented as straightforward extensions of the techniques covered previously.

The material in Part Five, Selected Topics, can be used to tailor the book to different curriculum requirements or to feed into subsequent courses. These topics can be covered in any of the sequences shown in the flowchart on page xix.

Concluding appendixes provide introductions to elementary matrix manipulations, the algebra of complex numbers, and the second-order differential equation.

The topic of circuit analysis necessarily precedes that of circuit design. We emphasize circuit analysis throughout this text. However, we do include a limited number of design examples and problems where appropriate to the students' level of proficiency.

PSpice

A complete, new manual entitled *Introduction to PSpice with Student Exercise Disk* accompanies this text. The manual contains a student exercise disk and gives students direct, hands-on instruction with PSpice. The manual teaches students how to use PSpice as a tool to learn more about circuits and to see circuit design principles in action.

The manual begins with installation of PSpice. Students then work through an introductory hands-on tutorial and eight worked-out examples of PSpice programs. A separate chapter containing dozens of problems, graduated in difficulty from simple to complex, directly extends the presentation of circuit design principles provided in the

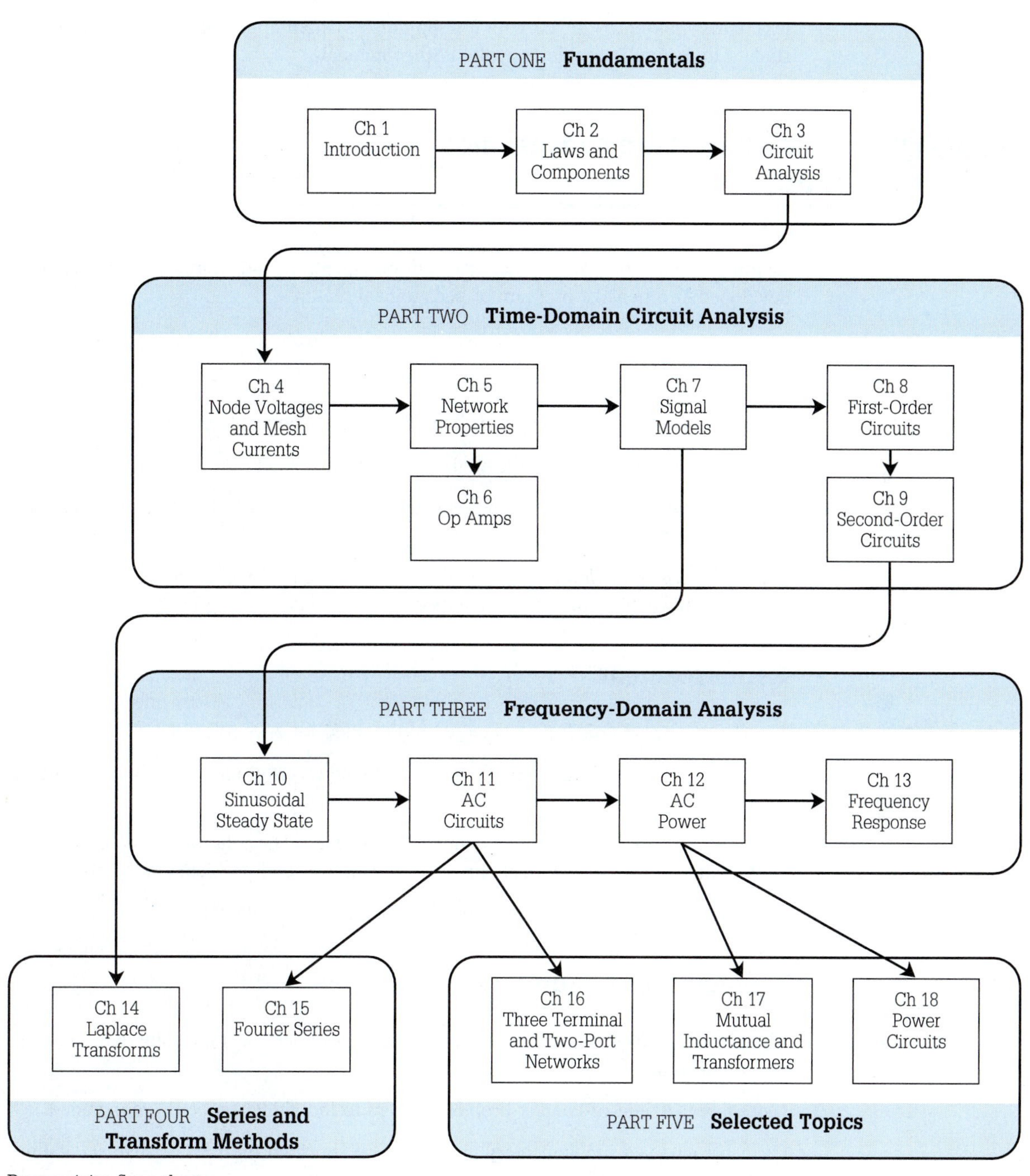

Prerequisite flow chart

text. Cross references keyed with a floppy disk symbol appear in the text to refer the student to particular problems in the PSpice manual.

Pedagogy: Functional Aids to Enhance Student Learning

Learning comes through practice and reinforcement. To aid in this process, we have developed an extensive program of pedagogical features.

Perspective

- *Chapter introductions* highlight the content and basic concepts presented.
- *New material* is presented in the context of preceding work.
- *Frequent reference to practical systems and devices* motivates students and helps them to relate theory to real problems.
- *A brief history of the development of electrical science and engineering* conveys appreciation for the achievements of many people over hundreds of years.

Reinforcement

- *Key definitions* are concisely stated and highlighted for easy reference.
- *Remember statements* that identify principal or noteworthy ideas conclude individual sections.
- More than 185 carefully chosen examples are included. The solutions to these examples are organized to serve as models for solving exercises and problems in the book.
- A *Summary* section and a *Key Facts* table conclude each chapter.

Problems and Exercises

- More than 390 exercises appear within chapters, designed for immediate reinforcement of the material.
- More than 620 end-of-chapter problems keyed to individual sections are included.
- Over 220 unkeyed supplementary problems also appear at the ends of chapters, calling on the student to combine skills learned in various chapters.
- Answers to all exercises appear after each exercise for easy checking.
- Answers to many problems appear at the end of the book.

Symbols

- A marginal floppy disk ( ) and reference refers students to particular problems in the PSpice manual.

- A floppy disk and problem number (5.3) following a text problem refers students to problems in the PSpice manual.

Acknowledgments

We thank the University of Missouri-Rolla for providing an ideal teaching environment in which we could develop this book. This text benefited immeasurably from the feedback we received from instructors and students here. We thank many of our colleagues for their suggestions.

D. R. C.
J. A. S.

The Beginnings

The development of modern electrical engineering has been intimately intertwined with the scientific discovery and understanding of electric and magnetic phenomena. The history of this development is a fascinating story. In some instances, the application of observed phenomena encouraged a quest for understanding; in other instances, the search for understanding pointed the way to new phenomena and applications.

Surely the first practical, although uncontrolled, use of electricity occurred when a fire started by lightning was used for heating or cooking. The first reported application of magnetism was by Emperor Huan-ti of China, who used a magnetic compass as early as 2637 B.C. Static electricity, developed by friction on amber (the Greek name for amber is *electron*), is said to have been first observed over twenty centuries later by Thales of Miletus, the Greek philosopher.

Dr. William Gilbert, court physician to Queen Elizabeth of England, published six books in 1600 containing the results of seventeen years of intense research on electricity and magnetism. His second book is considered the first published work on electricity, and for this work he is called the founding father of modern electricity.

In 1660 Otto von Guericke, a German physicist, constructed the first machine for generating static electricity. A sulfur sphere revolved around its axis while a cloth was held against the sphere.

The Leyden jar, a form of capacitor that could store electrical energy, was developed in 1745. Although several people claimed the development, Heinrich Bernt Wilhelm von Kleist published his discovery first. As early as 1767, Joseph Bozolus proposed using Leyden jars and long wires to transmit coded messages by creating sparks.

Alessandro Volta

The quantitative study of electricity began with Charles A. Coulomb, who was the founder of electrostatics and the school of experimental physics in France. In 1785, with the aid of a small torsional balance electroscope, he established Coulomb's law by showing that the force of attraction or repulsion between two charged spheres varied inversely with the square of the distance between their centers.

Alessandro Volta developed the voltaic pile, an early form of the electric battery in 1796, and the publication of his work in 1800 provided the greatest boost for electrical experimentation since the Leyden jar. The battery provided the first continuous supply of electrical energy.

The next truly great discovery occurred in 1820 when the Danish physicist and experimenter Hans Christian Oersted found that a wire connecting two battery terminals

André Marie Ampère

would influence the needle of a compass placed under it. This demonstrated that an electric current creates a magnetic field.

Scientists were extremely active over the next few years. Perhaps the most significant work was performed by André Marie Ampère, who discovered that a force was exerted between wires carrying current. This helped establish the elementary laws of electrodynamics.

In England in 1821, Michael Faraday showed that electricity can produce continuous mechanical motion. His greatest discovery occurred in 1831 when he verified his prediction that magnetism could induce electric current. This opened the door for practical applications of electricity, and electrical engineering was born.

Michael Faraday

The German mathematician and physicist Georg Simon Ohm published *The Galvanic Theory of Electricity* in Berlin in 1827 and established Ohm's law: The current through a conductor is proportional to the voltage across it. Not to be overlooked are the contributions of two famous experimentalists and mathematicians of the late eighteenth and early nineteenth centuries: Jean Baptiste Joseph Fourier and Marquis Pierre Simon de Laplace, who developed mathematical techniques now used in circuit analysis.

Joseph Henry, a professor in Albany, New York, conducted many experiments on electromagnetism and constructed an electric motor in 1829. While others, such as the German mathematician Karl Friedrich Gauss, contributed to electromagnetic theory, James Clerk Maxwell unified this theory with the concept of displacement current. This completed what are now known as Maxwell's equations and contributed to the 1864 prediction of electromagnetic wave propagation.

By the end of 1831 Hypolite Pixii of Paris demonstrated a rotating electromagnetic generator that could supply a constant voltage and current. In 1839 Professor Jacobi of Russia demonstrated a battery-powered boat, and in 1840 Robert Davidson, a Scot, equipped a car to run on rails using a solenoid-type motor.

Georg Simon Ohm

Gustav Robert Kirchhoff, a Prussian working in Germany, published some of his work between 1845 and 1882. This work gave us Kirchhoff's voltage law and Kirchhoff's current law, which form the basis of electric circuit analysis.

Following the development of the Morse code by Samuel F. B. Morse and Alfred Vail, a public telegraph was opened in the United States in 1844 and in England in 1845; the modern era of communications began. In 1861 the New York-to-San Francisco telegraph line was completed and established high-speed trans-continental communications. A major breakthrough occurred in 1875 when Alexander Graham Bell developed the telephone.

Comparable advances were being made in electric power. By 1851 a few electromechanical generators or dynamos were in use for commercial electroplating, and other new applications of electricity evolved rapidly. The South Foreland lighthouse in England first shined an arc lamp to sea in 1858 and established electric lighting as practical. In 1879 Thomas A. Edison invented the incandescent light, which opened the way to electric lights for the home. The world was introduced to electrically powered transportation by an electric train with a 3-horsepower motor, which was constructed by the Siemens and Halske Company in 1879 for a trade fair in Berlin.

Jean Baptiste Joseph Fourier

In the post-Civil War period, static electricity was taught as science in the physics department, while galvanic electricity* was taught in the chemistry department. The telegraph as well as the dynamo and the incandescent lamp created a greater demand for trained electrical workers than these educational programs could provide. This demand was recognized prior to 1880, but the first formal electrical engineering programs were

* Electricity supplied at constant voltage was called galvanic electricity, because it was first generated by galvanic action in a battery.

Marquis Pierre Simon de Laplace

Joseph Henry

Gustav Robert Kirchhoff

William Siemens

introduced at the Massachusetts Institute of Technology and Cornell in 1882. By the end of the decade most universities had introduced a formal program in electrical science or engineering, and in 1893 Ohio State University graduated Bertha Lamme, the first woman electrical engineer. The programs differed, but all were built on a strong foundation of mathematics, physics, and mechanical engineering.

With many people trained in electrical science and engineering, progress was rapid. Nikola Tesla produced a two-phase alternator in 1887 and soon developed the first practical ac motor. This established the feasibility of ac power, which is the basis of our power systems.

In 1894 Marchese Guglielmo Marconi successfully transmitted feeble electrical signals through space. This introduced wireless communication, which was well established by the beginning of World War One.

The products developed since 1918 that rely on electrical phenomena are too numerous to mention. Some developments could best be described as revolutionary, whereas others were more evolutionary. Developments of both types were required for practical products.

More recent developments include television broadcasting (1936), radar (World War Two), and the electronic digital computer (1945). In 1948 John Bardeen, Walter Brattain, and William Shockley invented the transistor, which made the modern era of communications possible. The first communications satellite was orbited in 1970.

The pocket calculator and desktop computer are perhaps the most conspicuous products of the modern era, but many less visible advances have been made. Navigation satellites guide shipping, solar cells supply power to remote weather stations, and the newer home appliances are more efficient.

This brief history is obviously incomplete, but we hope that it will stimulate you to read more about this fascinating subject. As for future developments, well, these are partly up to you.

1

Introduction

One can almost picture the dismay of Dr. William Gilbert when his first book on electricity and magnetism, published in 1600, was criticized by the famous scientist Sir Francis Bacon as having little substance. Nevertheless, the publication of his nearly two decades of experimental work had a significant impact on science, for he encouraged others to repeat and expand on his experiments. Progress seemed slow in the following two centuries, and the most significant electrical devices developed were electrostatic generators and the Leyden jar, which were often used for entertainment in homes of the wealthy. Although scientific investigation continued, it was more qualitative than quantitative. Experimenters observed that certain materials can be electrically charged to attract or repel each other, and that magnets attract some materials and not others. Typically these phenomena were neither precisely measured nor described mathematically.

A scant 15 years before the nineteenth century began, Charles A. Coulomb used careful measurements to establish the mathematical relationship for the forces between charged objects, and laid the cornerstone for the quantitative theory of *electrostatics.* What was still lacking was a continuous source of electrical energy to replace the fleeting sparks of the electrostatic generator. As the nineteenth century began, Alexander Volta satisfied this need when he published his work on the voltaic pile. This primitive battery provoked an explosion in scientific experimentation, and permitted H. C. Oersted to establish the link between electricity and magnetism in 1820.

André Marie Ampère was so inspired by Oersted that only two years later he published his paper on *electrodynamics* that enunciated two properties of electricity: electrical tension (voltage) and electrical intensity (current). He also used careful measurements to establish the mathematical relationship for the forces between wires that carry current, and thus did for electrodynamics what Coulomb did for electrostatics.

The pace of scientific investigation quickened, but the lack of a standard set of electrical units hindered progress. Scientists could share the mechanical dimensions of their apparatus, but they could not compare electrical measurements. Standards slowly evolved, and more was accomplished by Michael Faraday and others in the first half of the nineteenth century than in the previous two centuries. By the end of the American Civil War, James Clerk Maxwell had essentially completed the *theory* of electricity and magnetism, and established the electromagnetic theory of light. This ushered in the modern era of electrical science and the growth in *technology* that accompanied it. Success of this new technology required *both* the standardization of units and the availability of *practical* instrumentation. In this chapter we introduce the units that evolved and relate these to practical measurements of voltage and current.

1.1 Units

The International System of Units, abbreviated SI (from the French name for the system, *Système International des Unites*) and popularly called the metric system, is universally used in circuit analysis. The system defines six fundamental units, as shown in Table 1.1. We will not need the units for temperature or luminous intensity. The remaining four fundamental units are the basis for all units used in circuit analysis. For example, force is the product of mass and acceleration. Therefore, the unit of force has

TABLE 1.1
International system of units

Quantity	*Unit*	*Symbol*
Length	meter	m
Mass	kilogram	kg
Time	second	s
Electric current	ampere	A
Temperature	degree kelvin	K
Luminous intensity	candela	cd

TABLE 1.2
Prefixes

Prefix	*Abbreviation*	*Multiplier*
pico	p	10^{-12}
nano	n	10^{-9}
micro	μ	10^{-6}
milli	m	10^{-3}
kilo	k	10^{3}
mega	M	10^{6}
giga	G	10^{9}
tera	T	10^{12}

the derived dimension of kilogram-meter per second squared, or newton (N). A force acting over a distance is work (energy) and has the dimension of newton-meter or joule (J). Power is work per unit time. The unit of power is the joule per second or watt (W). *Electric power* is universally measured in watts.

In the SI system we also attach prefixes to the unit to specify powers of 10. The prefixes most commonly used in circuit analysis are listed in Table 1.2. For example, the following are equivalent values of power in watts (W).

$$5000 \text{ nW} \qquad 5 \ \mu\text{W} \qquad 0.005 \text{ mW} \qquad 0.000005 \text{ W}$$

We will use an uppercase letter to denote a quantity that *cannot be* a function of time. As an example, average power is not a time function, so we use an uppercase P to represent the average power supplied to the 100-W lamp in your room:

$$P = 100 \text{ W} \tag{1.1}$$

A lowercase letter implies that a quantity *can be, but is not necessarily,* a function of time. The instantaneous power supplied to the 100-W lamp in your room is a function of time:

$$\begin{aligned} p &= p(t) \\ &= 100[1 + \cos 2\pi 120t] \text{ W} \end{aligned} \tag{1.2}$$

so we use a lowercase p.

The instantaneous power supplied to a lamp in your car is constant:

$$\begin{aligned} p &= p(t) \\ &= 6 \text{ W} \end{aligned} \tag{1.3}$$

Although the instantaneous power was constant in this case, we have used a lowercase p, because instantaneous power *can be* a function of time. We will often omit the explicit dependence of a quantity on time unless it is needed for emphasis or clarity.

Remember

The fundamental electrical unit in the SI system is the ampere (A).

EXERCISES

1. Write the following quantities in terms of prefixed values that contain no more than three digits. (For example, 360,000 m is equivalent to 360 km).
   (a) 0.002 m *answer:* 2 mm (b) 15,000 m *answer:* 15 km
   (c) 0.05 s *answer:* 50 ms (d) 0.0004 ms *answer:* 400 ns
   (e) 27,000 kW *answer:* 27 MW (f) 0.007 s *answer:* 7 ms

2. With voltages given in volts (V) and currents in amperes (A), which of the following are consistent with the notation used in Eqs. (1.1), (1.2), and (1.3)?
   (a) $p = 120\sqrt{2}\cos 2\pi 60t$ *answer:* Consistent
   (b) $I = 5t^2$ *answer:* Inconsistent
   (c) $p = 12$ *answer:* Consistent
   (d) $I = 12 \sin 2t$ *answer:* Inconsistent

1.2 Current and Charge

All electric phenomena are manifestations of *electric charge.* If you have removed a sweater and had it stick to your shirt, you have observed the force of *static* electric charge. Many devices, such as motors, rely on the magnetic force caused by *moving* electric charge, which we call *electric current.* The SI system uses this magnetic force to define the unit of electric current, the *ampere* (A). Current rather than charge is selected as the fundamental electrical unit, because a precise standard for current is easier to establish.

A simple relationship exists between current and charge. Refer to Fig. 1.1. A current i of one ampere in the direction of the arrow will cause a net positive charge of one coulomb (C) per second to pass through the surface in the direction of the arrow. Therefore a coulomb has the units of ampere-second. As indicated by the small arrows on two charges in Fig. 1.1, a positive charge that passes through the surface in the direction of the current arrow, and a negative charge that passes through the surface in the opposite direction, will both yield a positive contribution to current i.

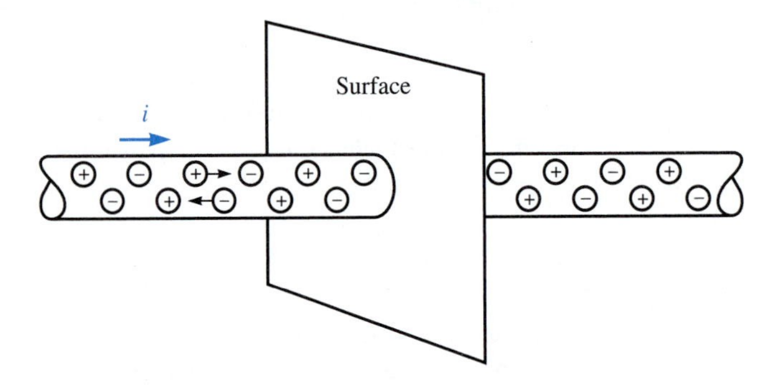

FIGURE 1.1 Electric current due to charge flow in a fluorescent lamp

We can determine the charge that passes through the surface in the time interval $-\infty$ to t if we integrate current with respect to time.†

DEFINITION Charge

$$q(t) = \int_{-\infty}^{t} i(\lambda)d\lambda \tag{1.4}$$

where i is current in amperes
q is charge in coulombs
t is time in seconds

† We use lowercase i and q to imply that these quantities are functions of time. The variable of integration is changed to λ because t appears in the upper limit of the integral.

The smallest quantities of charge are carried by the proton (1.602×10^{-19} C) and by the electron (-1.602×10^{-19} C). The quantity of charge involved in simple electrical processes can be very large. For example, 3,036 C of charge is required to electroplate 1 gram (g) of copper.

Charges of opposite sign attract each other, and charges of like sign repel each other. Two 1-C charges separated by 1 m would exert a force on each other of

$$F = 8.99 \times 10^9 \text{ newton (N)}$$

which is approximately 1 million tons. Fortunately, an object normally has approximately equal quantities of positive and negative charge, whose effects cancel each other with respect to forces exerted on other objects.

Differentiation of Eq. (1.4) yields current in terms of charge:

Current

$$i = \frac{dq}{dt} \tag{1.5}$$

In practice, we measure current with an ammeter, as shown in Fig. 1.2a. The meter will indicate the current through the meter, and device *A*, in the direction from 1 to 2. For convenience we simplify the circuit drawing as shown in Fig. 1.2b.

The symbol *i* represents the ammeter reading, and the arrow represents the location and relative connection of the meter. By convention the arrow points from the point where the (+) terminal of the meter is connected toward the point where the (−)

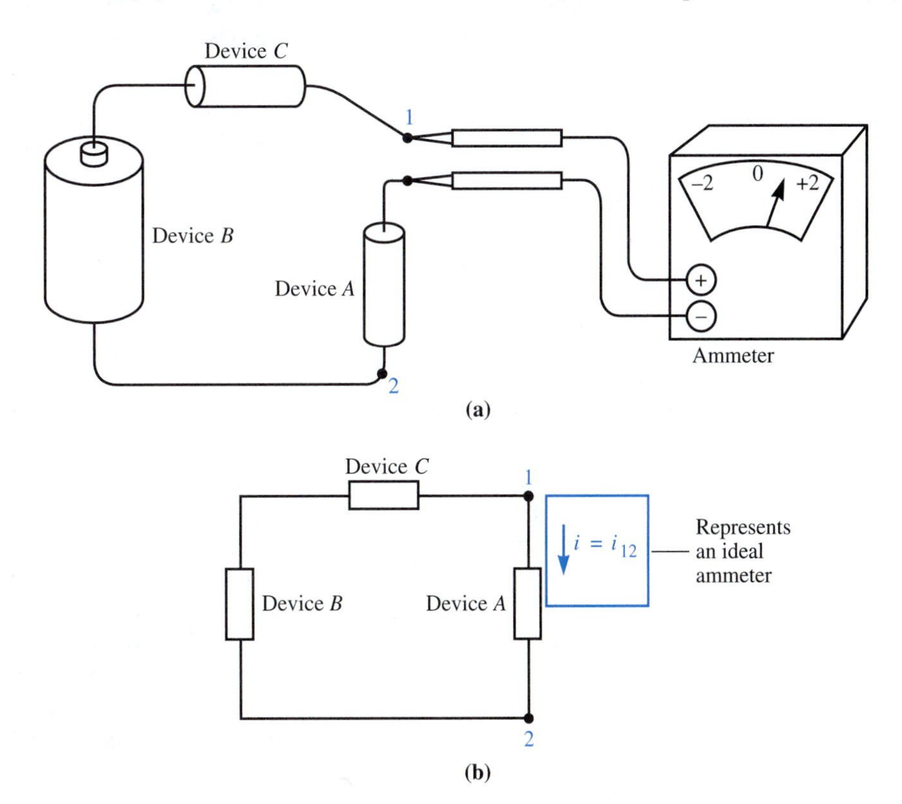

FIGURE 1.2 Current measurement. (*a*) Measuring current i_{12} with an ammeter; (*b*) symbolic representation of current measurement (box shown to emphasize that the letter *i* and the arrow correspond to the meter reading and terminal connections of the ammeter)

terminal of the meter is connected, as shown in Fig. 1.2. The value of i may be positive or negative. The representation of Fig. 1.2b implies that an ideal current measurement has no effect on the circuit.

We often use subscripts to indicate the current reference direction. The equivalent reference arrow points from the location of the first subscript toward the location of the second subscript. For current i as shown in Fig. 1.2b,

$$i = i_{12} = -i_{21} \tag{1.6}$$

We read i_{12} as i one-two, not i twelve.

Remember

A current symbol i represents the current measured by an ideal ammeter. The reference arrow indicates the relative meter connections. The arrow points from the point where the positive (+) terminal of the meter is connected toward the point where the negative (−) terminal of the meter is connected. With the double-subscript notation for current (i_{12}), the reference arrow points from the point labeled with the first subscript (1) toward the point labeled with the second subscript (2). We must always include *both a current symbol and a reference arrow (or double subscripts) on a circuit diagram to define a current.*

EXERCISES

3. The starter for an automobile requires 200 A when operated. What charge, in coulombs, passes through the starter of your car if it takes 5 s to start?
   answer: 1 kC
4. A current of i_{12} amperes passes through a wire from point 1 to point 2. Sketch current $i_{12}(t)$ and the charge $q(t)$ that passes through the wire between time 0 and t, for the following currents. *Hint:* Use Eq. (1.4).
   (a) $i_{12} = 10e^{-2t}$ — *answer:* $5(1 - e^{-2t})$ C
   (b) $i_{12} = 10 \cos 2\pi t$ — *answer:* $(5/\pi) \sin 2\pi t$ C
5. Charge q passes through a wire from point 1 to point 2. Sketch $q(t)$ and $i_{12}(t)$ for $0 \le t \le 2$ s, if $q(t)$, in coulombs, is defined by the following functions. *Hint:* Use Eq. (1.5).
   (a) $q = 10e^{-2t}$ — *answer:* $-20e^{-2t}$ A
   (b) $q = 10 \cos 2\pi t$ — *answer:* $-20\pi \sin 2\pi t$ A

1.3 Voltage

Have you ever felt a small shock when you walked across a carpet and touched a doorknob? You may also have heard a faint click, and possibly have seen a small flash of light. If so, you observed the effects of a small electric current passing through air. The current was forced through the air by the *voltage* of your hand with respect to the doorknob. In passing through the air, the charge performed work. This is the source of the light and sound. The work per coulomb performed by positive charge moving from point 1 to point 2 is the voltage of point 1 with respect to point 2. We often call this voltage the *voltage drop* going from point 1 to point 2. The SI unit of voltage is the *volt* (V), which has the equivalent units of joule per coulomb (J/C). This definition gives us the following equation:

DEFINITION Voltage

$$v = \frac{dw}{dq} \tag{1.7}$$

where v is voltage in volts
w is energy in joules
q is charge in coulombs

In practice, we measure the voltage of one point with respect to another with a voltmeter, as shown in Fig. 1.3a. The voltmeter indicates the voltage *across* device *A* going from point 1 to point 2. For convenience we simplify the drawing as shown in Fig. 1.3b. The symbol v represents the voltmeter reading. The (+) and (−) signs correspond to the points where the similarly marked meter terminals are connected. The quantity v can be positive or negative. The representation of Fig. 1.3b implies an ideal voltage measurement that does not affect any part of the circuit.

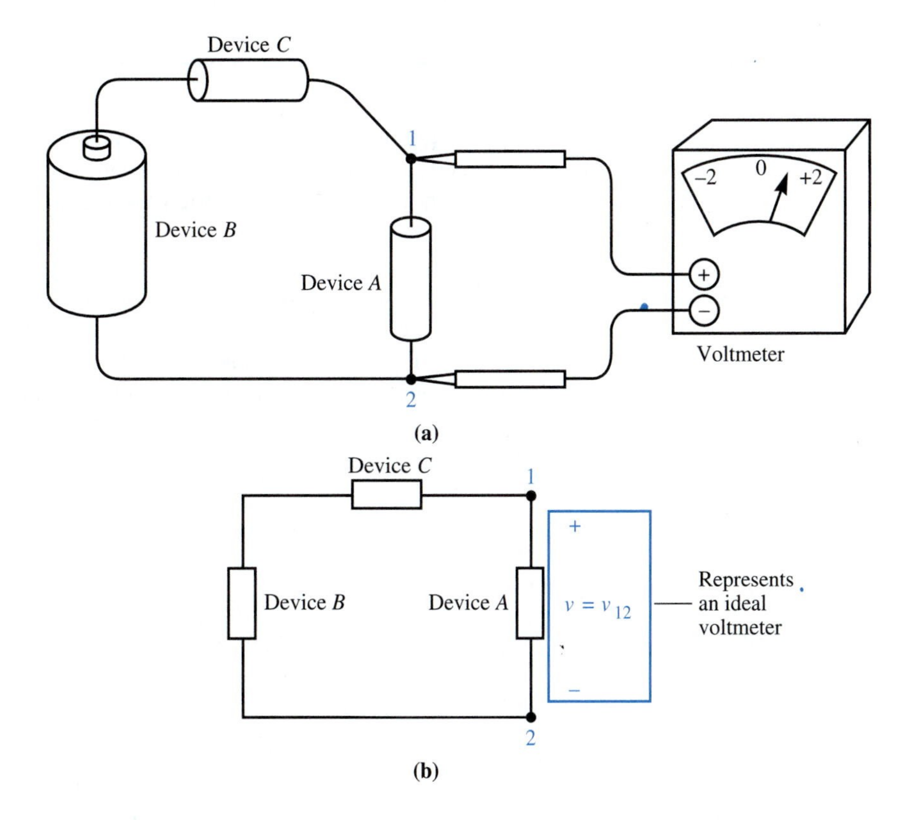

FIGURE 1.3 Voltage measurement. (*a*) Measuring voltage v_{12} with a voltmeter; (*b*) symbolic representation of voltage measurement (box shown to emphasize that the letter v and the polarity marks correspond to the meter reading and terminal connections of the voltmeter)

We often use double subscripts to indicate the meter connections. The first subscript indicates where the (+) terminal is connected, and the second subscript indicates where the (−) terminal is connected. For the voltage v as shown in Fig. 1.3b,

$$v = v_{12} = -v_{21} \tag{1.8}$$

Double subscript notation is widely used in electric power engineering.

Remember

A voltage symbol v represents the voltage measured by an ideal voltmeter. The location of the (+) and (−) signs corresponds to the points where the like-marked terminals of the meter are connected. With the double-subscript notation for voltage (v_{12}), the first subscript indicates the point in the circuit where the (+) terminal of the voltmeter is connected, and the second subscript indicates the point where the (−) terminal is connected. We must always include *both a symbol and polarity marks (or double subscripts) on a circuit diagram to define a voltage.*

EXERCISES

6. A device converts all electrical energy delivered to it directly into heat. Temperature measurements indicate that the energy input was 50 J when a charge of 4 C passed through the device. If the voltage across the device was constant, what was its value? *answer:* 12.5 V

7. In Exercise 3 you calculated that 1000 C of charge passed through your starter to start the engine. If the voltage across your starter was 10 V, how much energy was input to the starter? *answer:* 10 kJ

1.4 Power

Electric power requirements for personal and industrial use vary from a fraction of a milliwatt for an electric watch to several megawatts for large motors used in industry. We will now see how to calculate electric power from voltage and current measurements.

Voltage v, as shown in Fig. 1.4, is the energy in joules per coulomb absorbed by device D as positive charge passes through from the (+)-marked end to the (−)-marked end of device D. Current i is the positive charge in coulombs per second that moves in the same direction. Therefore the product of voltage v and current i is the instantaneous electric power, p, absorbed by (delivered to) device D in joules per second, or watts (W):

$$p = \frac{dw}{dt} = \left(\frac{dw}{dq}\right)\left(\frac{dq}{dt}\right) = vi \tag{1.9}$$

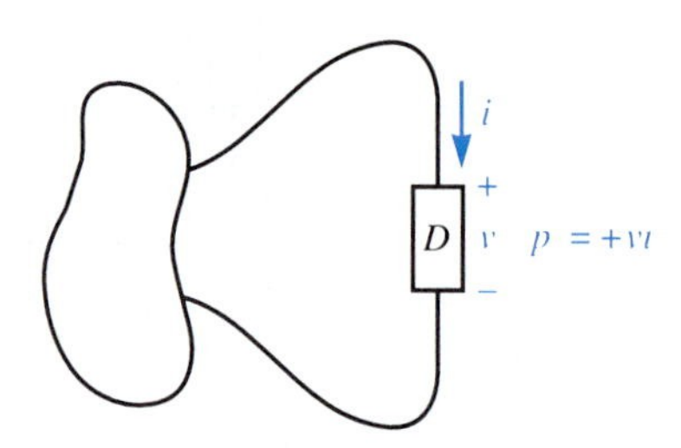

FIGURE 1.4 Relative reference directions for the passive sign convention

To simplify discussion, we call the relative reference directions for v and i shown in Fig. 1.4—the reference arrow for i entering the end of the device with the (+) reference mark and leaving the end with the (−) reference mark—the *passive sign convention.* This gives the following equation to calculate the power absorbed:

Power Absorbed—Passive Sign Convention

$$p = +vi \tag{1.10}$$

When we calculate the instantaneous power absorbed, p, it may be a positive or a negative quantity. If we calculate a negative power, the device supplies power to the rest of the circuit.

When working problems we will also encounter voltage and current references oriented according to the *active sign convention* as shown in Fig. 1.5. For these reference directions Eq. (1.10) gives

$$p = (-v)i = -vi \tag{1.11}$$

This yields a convenient rule to determine the sign on the product vi used to calculate the power absorbed. If the current reference arrow passes through the device from (+) to (−), the power absorbed is $+vi$. If the current reference passes through the device from (−) to (+), the power absorbed is $-vi$. We use the passive sign convention unless otherwise noted.

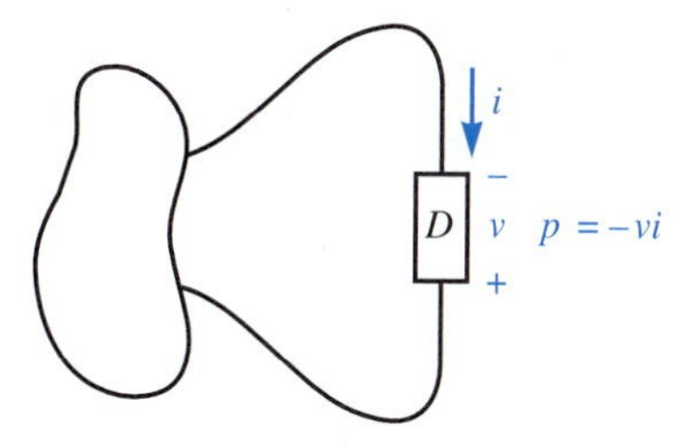

FIGURE 1.5 Relative reference directions for the active sign convention

If we use the passive sign convention, as in Fig. 1.4, the electrical energy absorbed by device D in the interval t_1 to t_2 is given by the following equation:

Energy Absorbed

$$W = \int_{t_1}^{t_2} vi\, dt \tag{1.12}$$

In general, the *instantaneous power* p is a function of time. The *average power* absorbed by a device in the interval t_1 to t_2 is a constant.

Average Power

$$P = \frac{1}{t_2 - t_1} \int_{t_1}^{t_2} p\, dt \tag{1.13}$$

Conservation of energy requires that the power absorbed by an element be supplied from some other source: perhaps chemical energy from a battery or mechanical energy driving a generator. This observation gives us the power balance equation:

Power Balance Equation

The algebraic sum of the instantaneous electric power absorbed by all components in a network is zero:

$$\sum p(t)_{\text{absorbed}} = 0 \tag{1.14}$$

where the sum is over all components in the network, including the sources.

Integration of Eq. (1.14) gives the result that the algebraic sum of the *average* electric power absorbed by all components in a network also must be zero.

EXAMPLE 1.1 The voltage and current for the device shown in Fig. 1.4 are $v = 12t$ V and $i = 0.5t$ A, for $0 \leq t \leq 10$ s.

(a) Find the instantaneous power absorbed by the device D.

(b) Determine the average power absorbed in the interval $0 \leq t \leq 10$ s.

Solution The voltage and current references in Fig. 1.4 are assigned according to the passive sign convention. Therefore:

(a)
$$p = vi = (12t)(0.5t) = 6t^2 \text{ W}$$

(b)
$$P = \frac{1}{t_2 - t_1}\int_{t_1}^{t_2} p\, dt = \frac{1}{10}\int_0^{10} 6t^2 dt = 200 \text{ W}$$

EXAMPLE 1.2 Given the system shown in Fig. 1.6, determine the power p_B absorbed by component B at some instant in time, if the power absorbed by the other elements is $p_1 = -1$ W, $p_2 = 2$ W, $p_3 = 3$ W, and $p_4 = 4$ W at the same instant in time.

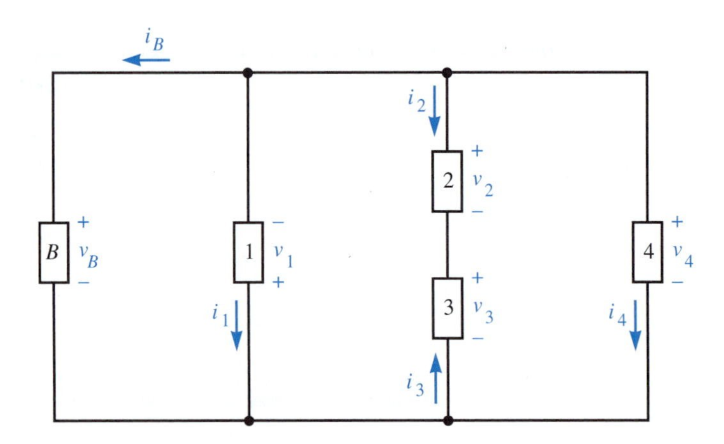

FIGURE 1.6 Circuit for Example 1.2

Solution From Eq. (1.14),

$$\sum p_{\text{absorbed}} = 0$$
$$p_B + p_1 + p_2 + p_3 + p_4 = 0$$
$$p_B + (-1) + (2) + (3) + (4) = 0$$
$$p_B = -8 \text{ W}$$

Component B absorbs −8 W or supplies +8 W.

Remember

The relative reference directions for v and i shown in Fig. 1.4—the reference arrow for current i entering the end of the device with the (+) reference mark for v and leaving at the end with the (−) mark—is called the passive sign convention. For the passive sign convention, the product of v and i gives the power absorbed by the device. Unless otherwise noted we use the passive sign convention. The sum of all electric power absorbed in a network is zero.

EXERCISES

8. Determine which of the devices shown in Fig. 1.6 have the voltage and current references selected according to the passive sign convention.
   (a) Device 1 *answer:* active (b) Device 2 *answer:* passive
   (c) Device 3 *answer:* active (d) Device 4 *answer:* passive
   (e) Device B *answer:* passive

9. The power absorbed by each device in the circuit of Fig. 1.6 is given in Example 1.2. Determine the specified current if the device voltage is as stated.
   (a) $v_1 = -12$ V; determine i_1. *answer:* $-\frac{1}{12}$ A
   (b) $v_2 = \frac{24}{5}$ V; determine i_2. *answer:* $\frac{5}{12}$ A
   (c) $i_3 = -\frac{5}{12}$ A; determine v_3. *answer:* $\frac{36}{5}$ V
   (d) $v_4 = 12$ V; determine i_4. *answer:* $\frac{1}{3}$ A
   (e) $v_B = 12$ V; determine i_B. *answer:* $-\frac{2}{3}$ A

10. An electrical system with five devices is shown in the following diagram. Determine the power absorbed by the specified device.
   (a) Device A *answer:* −96 W (b) Device B *answer:* −24 W
   (c) Device C *answer:* 480 W (d) Device D *answer:* −480 W
   (e) Device E *answer:* 120 W
   (f) Sum the powers absorbed by the devices. Is the power balance equation satisfied? *answer:* yes

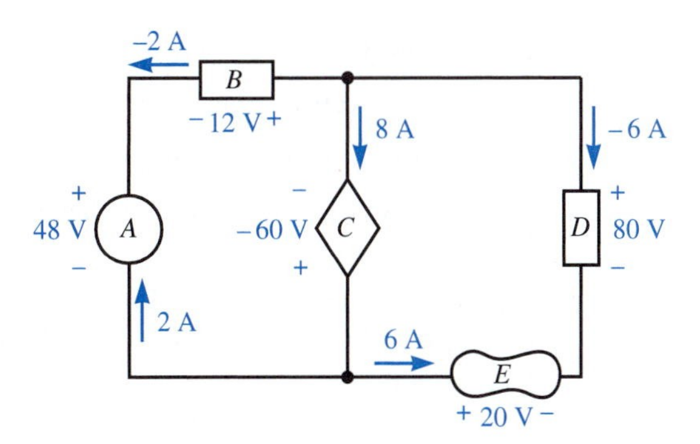

11. When actuated, the voltage across an automobile starter is 10 V, and the current through it is 200 A. What is the instantaneous power absorbed by the starter when in operation? *answer:* 2 kW

12. The starter described in Problem 11 is engaged for 5 s, left idle for 5 s, and then actuated for an additional 10 s before the engine starts. What is the average power absorbed by the starter during this 20-s time interval? *answer:* 1.5 kW

1.5 Summary

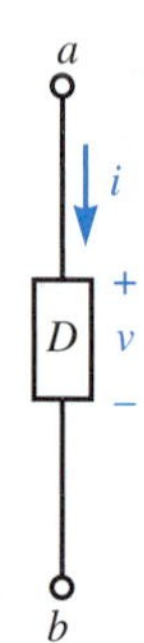

In this chapter we introduced the fundamental units of the SI system and two derived electrical units, charge and voltage. Current i, in amperes, through a circuit component from end a to end b ($i = i_{ab}$) is the net rate of positive charge, in coulombs per second, passing through the device from end a to end b. We can interpret the numerical value of i_{ab} as the current measured by an ideal ammeter when the current i_{ab} enters the (+)-marked terminal of the ammeter and exits at the (−)-marked terminal. Voltage v, in volts, of point a with respect to point b ($v = v_{ab}$) is defined as the energy in joules per coulomb lost by a positive charge as it moves from point a to point b in a circuit. The numerical value of v_{ab} is the voltage indicated by an ideal voltmeter with the (+) terminal connected to the point indicated by the first subscript (a) and the (−) terminal connected to the point indicated by the second subscript (b).

The passive sign convention implies that the current reference arrow passes through a device from the (+) to the (−) voltage reference, as shown in the illustration to the left. For the passive sign convention, the power absorbed by (delivered to) the device is given by the product of voltage and current ($p = vi$). If the current reference arrow passes through a device from the (−) to the (+) voltage reference, the power absorbed by the device is given by $p = -vi$.

Conservation of energy requires that the sum of the instantaneous electric power absorbed by all components in a network is zero. This gives us the power balance equation ($\Sigma p = 0$). The power balance equation also applies to average power.

KEY FACTS Concept	Equation	Section	Page
❑ We use the SI system of units.		1.1	2
❑ The unit of energy, the joule (J), is a newton-meter.		1.1	3
❑ The unit of power, the watt (W), is a joule per second.		1.1	3
❑ A coulomb (C) is an ampere-second.	(1.4)	1.2	4
❑ An ampere (A) is a coulomb per second.	(1.5)	1.2	5
❑ A volt (V) is a joule per coulomb	(1.7)	1.3	7
❑ Electric power in watts (W) is voltage times current.	(1.9)	1.4	8
❑ The passive sign convention implies that the current reference arrow passes through the device from the (+) to the (−) voltage reference. For the passive sign convention, the power absorbed by the device is the product of voltage and current ($p = vi$).	(1.10)	1.4	9
❑ Conservation of energy gives the power balance equation: The sum of the instantaneous powers absorbed by all components in a circuit is zero ($\Sigma p = 0$).	(1.14)	1.4	10
❑ The power balance equation is also valid for average power.		1.4	10

PROBLEMS

Section 1.2

1. A current i passes through a component from point 1 to point 2. The current is zero for $t < 0$, and defined by the following equations for $t \geq 0$. The charge that passes through the device from point 1 to point 2 is q. Sketch $i(t)$ and $q(t)$ for $-1 < t < 2$ s.

   (a) $i = 4$ A (b) $i = 10t$ A

   (c) $i = 5e^{-t}$ A (d) $i = 20\pi \sin 4\pi t$ A

2. A charge q passes through a device from point 1 to point 2. The charge is zero for $t < 0$, and defined by the following equations for $t \geq 0$. The current that passes through the device from point 1 to point 2 is i. Sketch $q(t)$ and $i(t)$ for $-1 < t < 2$ s.

   (a) $q = 5t$ C (b) $q = 5\,t^2$ C

   (c) $q = 10(1 - e^{-t})$ C (d) $q = 10\pi \sin 2\pi t$ C

3. A current of 1 μA flows through a device from terminal 1 to terminal 2. Assume that all current is caused by electron flow. How many electrons enter device terminal 1 in a time interval of 1 ns? This is a relatively small current and a short time interval. Explain why we can usually ignore the discrete nature of charge in circuit analysis.

Section 1.3

4. Temperature measurements indicate that a heater has provided 1200 J of energy to a container of water in a time interval of 1 min. The charge that has passed through the heater in this time interval is 100 C. What is the constant voltage across the heater?

5. An electrical system consists of three parts, A, B, and C, as indicated in the following illustration. It is known that 2 J is absorbed by device A for each coulomb of positive charge that moves through part A from terminal a to terminal b. At the same time, if 1 C of positive charge moves through part B from terminal b to terminal c, 4 J is absorbed by element B.

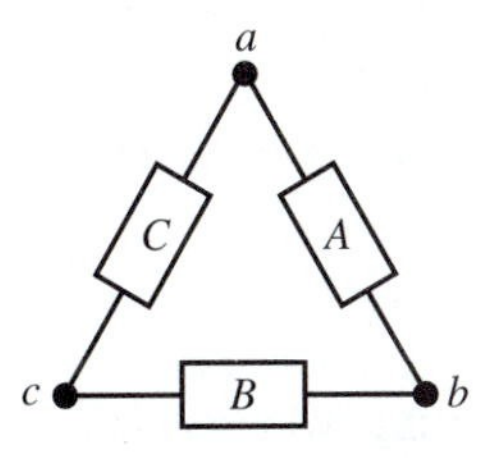

   (a) What is v_{ab}?

   (b) What is v_{bc}?

   (c) Based on the fact that charge will not accumulate inside a device and on conservation of energy, what would be the energy absorbed by device C for each coulomb of charge that passes through device C from c to a?

   (d) What is v_{ca}?

   (e) Find the sum of v_{ab}, v_{bc}, and v_{ca}.

Section 1.4

6. Given meter connections that follow the passive sign convention, find and plot the instantaneous power absorbed by device D as a function of time and the total energy absorbed by D from time 0 to t, and calculate the average power absorbed by D over the interval $0 < t < \infty$. The voltage and current are in volts and amperes, respectively.

   (a) $v = 10$ and $i = 5$, $t > 0$

   (b) $v = 10$ and $i = 5e^{-4t}$, $t > 0$

   (c) $v = 10e^{-4t}$ and $i = 5e^{-4t}$, $t > 0$

   (d) $v = 10$ and $i = 20 \cos 2\pi t$, $t > 0$

   (e) $v = 10 \cos 2\pi t$ and $i = 20 \cos 2\pi t$, $t > 0$

   (f) $v = 10 \cos 2\pi t$ and $i = 20 \sin 2\pi t$, $t > 0$

   (g) $v = 10 \cos 2\pi t$ and $i = 20 \cos (2\pi t + \pi/4)$, $t > 0$

   (h) $v = 10 \cos 2\pi t$ and $i = 20 \cos 4\pi t$, $t > 0$

7. Determine the power absorbed by each device in the circuit below. Check your result with the power balance equation.

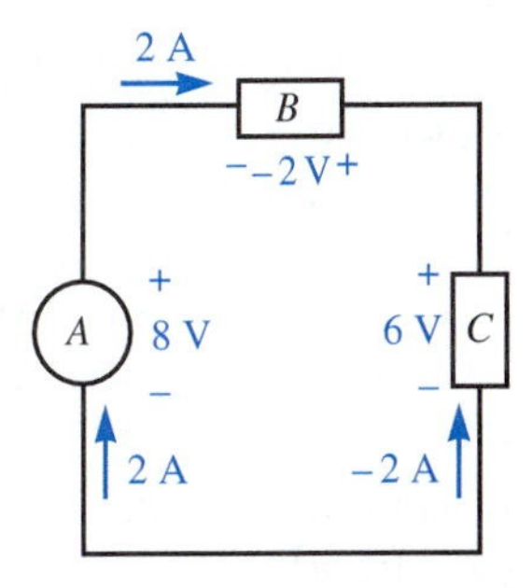

8. Determine the power absorbed by each device in the circuit on page 14. Check your result with the power balance equation.

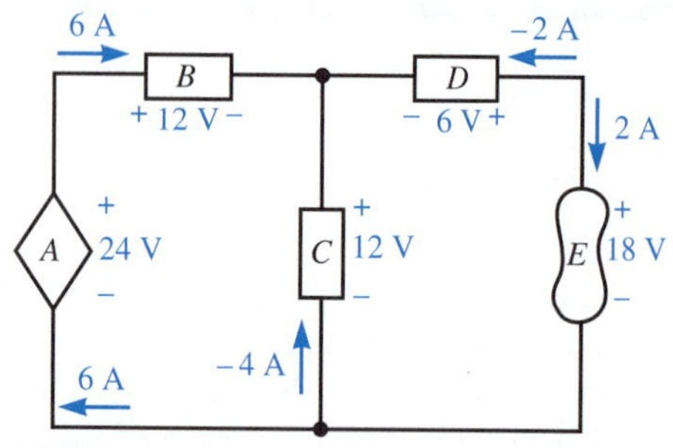

9. Calculate the power absorbed by the specified device in the circuit below if the following voltages and currents are known: $v_{12} = -5$ V, $i_{12} = -6$ A, $v_{23} = -6$ V, $i_{32} = 6$ A, $v_{30} = 21$ V, $i_{30} = -4$ A, $v_{10} = 10$ V, $i_{01} = -6$ A, $v_{43} = 7$ V, $i_{43} = 2$ A, $v_{45} = -9$ V, $i_{54} = 2$ A, $v_{50} = 37$ V, and $i_{05} = 2$ A.
   (a) Device A
   (b) Device B
   (c) Device C
   (d) Device D
   (e) Device E
   (f) Device F
   (g) Device G

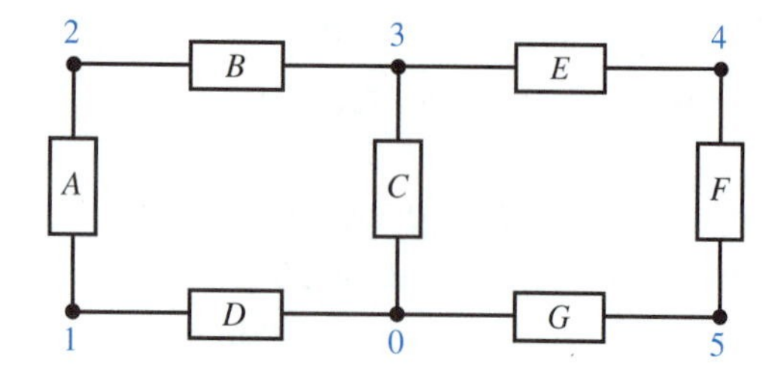

Supplementary Problems

10. A photodetector releases an average of one electron for every ten photons that strike its surface. If 5×10^{12} photons per second strike the surface, how many coulombs of charge are released in one second? What is the average current through the photodetector?

11. A power supply must be designed to electroplate 20 g of copper when the current is constant during a time period of 30 min. What current must the power supply provide? (Copper is a bivalent element and requires two electrons per atom of plated material.) *Hint:* See the paragraph that follows Eq. (1.4).

12. Professor Cox converted a Volkswagen Rabbit to electric power. After a typical day's commute, the battery charger supplies a current of

$$i = 30\left(1 - \frac{t}{21600}\right) \text{ A} \qquad 0 < t < 21600 \text{ s}$$

   (a) What charge passes through the battery in this time interval?
   (b) The battery has a terminal voltage of 85 V. Determine the instantaneous power input to the battery for $0 < t < 21600$ s.
   (c) What is the energy absorbed by the battery in this time interval?
   (d) What is the average power absorbed by the battery in this time interval?
   (e) If electricity costs 12 cents per kilowatt-hour (kWh), what are the electrical costs for the day's commute? (A kWh is the energy supplied when the battery absorbs 1 kW for 1 h.)

13. You plan to purchase a wind-driven generator to charge the battery of your electric car. The generator is rated to provide a maximum output power of 10 kW in a 60-km/h wind. The maximum power output of a wind-driven generator is proportional to the cube of the wind velocity. The car requires an energy input of 5 kWh to charge the battery after a trip to school and back. (One kWh is the energy supplied when the battery absorbs 1 kW for 1 h.) How many hours does it take to charge the battery after a round trip to school if (a) the wind velocity is 20 km/h; and (b) the wind velocity is 15 km/h?

14. The instantaneous power p absorbed by a two-terminal component is $60e^{-4t}$ W for $t \geq 0$. If $v_{ab}(t) = 3e^{-2t}$ V, $t > 0$, calculate the total charge input to terminal a over the interval $0 \leq t < \infty$.

15. The famous bicycle racer Eddy Merckz once held the 1-h closed-course distance record of just over 30 mi. It has been estimated that during this hour his power output was approximately seven-tenths of one horsepower. (One horsepower is approximately 745.7 W.)
   (a) What was his power output in watts?
   (b) What was his energy output in joules?
   (c) If his pedals had been connected to an electrical generating system with an overall efficiency of 80 percent and a terminal voltage of 12 V, what would be the electric power output of the system?
   (d) What would be the output current of the generator?

16. A family of energy wasters lives in a house with six rooms and two baths. Each room and bath has 150 W of lighting, which is left on an average of

6 h per day. If electricity costs 12 cents per kWh, how much a week does it cost for the energy used by the lights?

17. An automobile battery has a terminal voltage of approximately 12 V when the engine is not running and the starter motor is not engaged. A car is parked at a picnic with the stereo playing, which requires 60 W, and some lights on, which require 120 W. With a load of this type, the battery will supply approximately 1.2 MJ of energy before it will no longer start the car.
    (a) What power must the battery supply?
    (b) What current must the battery supply?
    (c) What is the approximate amount of time that the car can remain parked with the lights on and the stereo playing and still start?

18. You wish to design an electric teapot for a motor home. It must heat 1 liter (L) of water, and raise the temperature from 20°C to 100°C in 10 min. If heat losses are neglected, what must be the power rating of the heater? The motor home has a 12-V electrical system. What current is required? (It requires approximately 4.2×10^3 J to raise the temperature of 1 L of water one Celsius degree.)

19. Silver is a univalent element, so one electron is required to electroplate each atom of silver. There are 6.023×10^{23} atoms in a gram atomic weight (Avogadro's number). The atomic weight of silver is 107.88, and the charge of an electron is -1.602×10^{-19} C.
    (a) How many coulombs are required to electroplate a gram atomic weight of silver? (This is the quantity of electric charge often called a faraday.)
    (b) How many coulombs are required to electroplate 1 g of silver?

20. A copper bar with a square cross section of 1 cm $\times$ 1 cm is conducting 400 A. The current is entirely due to the motion of free electrons. In copper the free electron density is approximately 10^{29} electrons/m^3.
    (a) What is the average velocity (in the opposite direction to the current) of the electrons due to this current?
    (b) How long would it take an electron to travel 1 km at this velocity?
    (c) Does this contradict the common statement, "Electricity travels at the speed of light"?

21. You must design a power supply for an electroplating system that deposits 5 g of copper per hour.
    (a) Determine the output current and power if the output voltage must be 6 V. *Hint:* See the paragraph that follows Eq. (1.4).
    (b) If the power supply has an efficiency of 70 percent, what is the input power to the power supply?

22. When energized, an automobile starter motor requires 200 A. This causes the terminal voltage of the battery to drop from 12 V to 10 V. If the terminal voltage of the starter is 8 V, what power is dissipated in the wires that connect the starter motor to the battery.

23. On the high-speed position the fan motor for a car heater draws 10 A at 12 V.
    (a) If the motor is 60 percent efficient, what is the output power in watts and in horsepower (1 hp = 745.7 W). [percent efficiency = 100 (mechanical power out)/(electrical power in)]
    (b) The motor speed is directly proportional to the voltage. What voltage is required for the motor to run at one-half speed?
    (c) The power requirements for the fan are proportional to the cube of the motor speed. If the motor efficiency is independent of speed, what current is required to run the motor at one-half speed?

24. A direct lightning stroke to the Cathedral of Learning at the University of Pittsburgh was recorded on June 10, 1939.
    (a) The current from the cloud to ground is approximated by $i = 0$ for $t < 0$ and $i = -20{,}000e^{-5000t}$ A for $t \geq 0$. What charge was transferred from the cloud to ground?
    (b) Just before a lightning stroke, the electric field between the base of the cloud and ground is 10 kV/m. If the base of the cloud was 3 km above ground, determine the voltage V_m of the cloud with respect to ground ($V_m < 0$), just before the lightning stroke.
    (c) If the voltage of the cloud with respect to ground is $v = V_m e^{-5000t}$ V for $t \geq 0$, determine the instantaneous power dissipated in the lightning stroke.
    (d) What was the total energy dissipated in the lightning stroke?

2

Network Laws and Components

As engineers we must predict the performance of *systems* containing electrical *devices*. We will base our analysis on a *circuit diagram* that represents a mathematical model of the system. The circuit diagram consists of an interconnection of network *components*. Each network component models the *idealized* properties of an electrical device by the use of a *component equation*. These components model sources of electrical energy, the conversion of electrical energy into heat, and the storage of energy in electric and magnetic fields. We assume that all electrical energy is confined to the network components. Conservation of charge and energy then lets us describe the interaction of the components by two axioms: *Kirchhoff's current law* (KCL) and *Kirchhoff's voltage law* (KVL).

2.1 Kirchhoff's Current Law

In Chapter 1, we commented that the net charge of an object is ordinarily zero, and experiments show that electric charge is neither created nor destroyed in a network. Because of this, if we were to visualize a closed surface, such as an imaginary soap bubble, enclosing any portion of a circuit we would expect that any charge entering the closed surface would be matched by a like quantity of charge leaving along some other path. This statement of the conservation of charge is formalized in Kirchhoff's current law (KCL), which is one of the two axioms of circuit theory.

AXIOM KCL

Kirchhoff's current law states that, at any instant in time, the algebraic sum of all currents leaving any closed surface is zero:

$$i_1 + i_2 + i_3 + \cdots + i_N = 0 \tag{2.1a}$$

or in abbreviated notation

$$\sum_{k=1}^{N} i_k = 0 \tag{2.1b}$$

where i_k is the kth current of the N currents leaving the closed surface.

When applying KCL, we must treat a current leaving the closed surface along a path as the negative of a current entering along that path. We can then rephrase KCL to state that *the sum of the currents entering a closed surface must be equal to the sum of the currents leaving the closed surface.*

We assume that electrical energy is confined to the devices and interconnecting wires of the physical system. From the standpoint of our network diagram, this implies that all currents flow through the symbols for the network components and the connecting line segments. The symbol for a network component and the associated line segments form a *branch,* and the termination of a branch is a *node.*† For example, in Fig. 2.1 a single branch (branch E) connects node 2 to node 3, and current i_x flows through branch E.

EXAMPLE 2.1 For the circuit shown in Fig. 2.1, determine current i_x.

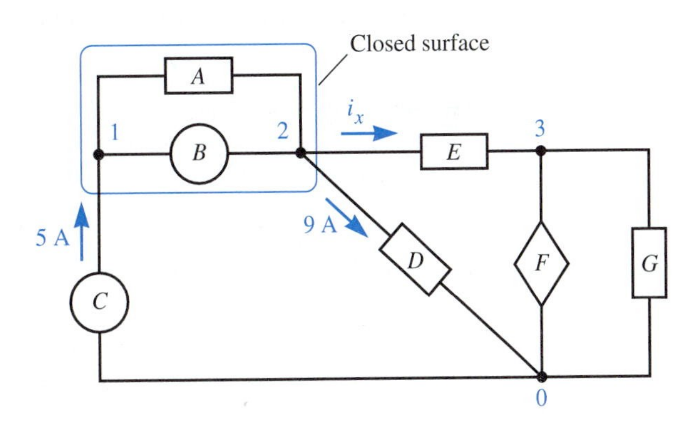

FIGURE 2.1 Network for Example 2.1

† Some books define a node as a point where two or more branches terminate.

Solution If we select the closed surface as shown, the only unknown in the corresponding KCL equation is current i_x. The sum of the currents leaving the closed surface is

$$-(5) + (9) + i_x = 0$$

The minus sign appears on the 5-A current because the reference arrow for the 5-A current *enters* the closed surface. Equation 2.2 gives

$$i_x = -4 \text{ A}$$

This implies that an ammeter connected as indicated by the reference arrow for i_x would indicate −4 A.

In the preceding example we saw that we only needed to know the current through the branches (C, D, and E) *cut* by our selected closed surface. These branches form a *cutset* (the removal of these branches *cuts* our network into two separate parts, and the replacement of any one of these branches reconnects the parts).† A wise choice for the cutset often simplifies a problem.

EXAMPLE 2.2 Determine current i_a, if $i_b = 2$ A, $i_c = -3$ A, $i_d = 5$ A, and $i_e = -12$ A in Fig. 2.2.

FIGURE 2.2
Application of KCL to a closed surface

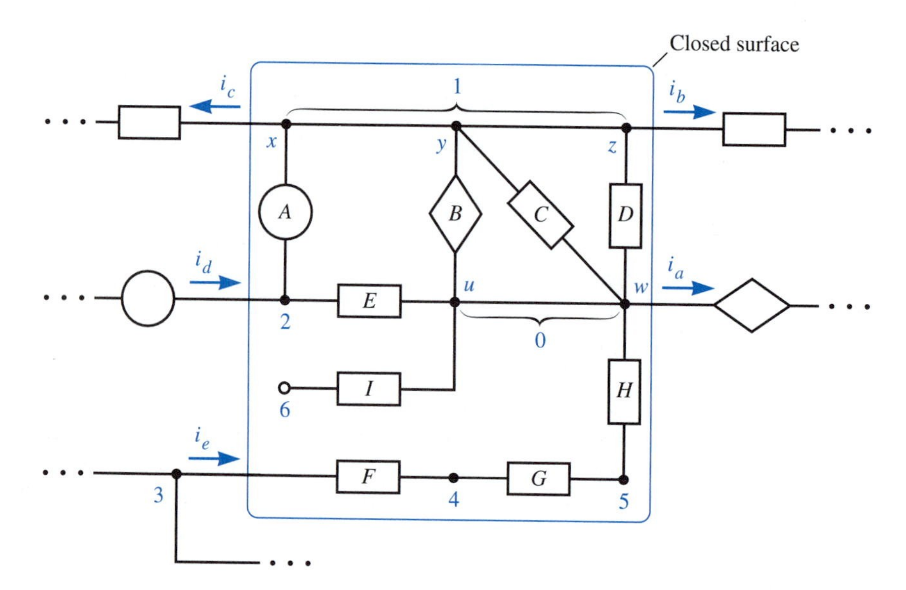

Solution If we select the closed surface as shown, the only unknown cutset current is i_a. From KCL the sum of the currents leaving the closed surface is zero:

$$i_a + i_b + i_c - i_d - i_e = 0$$

A minus sign appears on i_d and i_e in the above equation, because the associated reference arrows *enter* the closed surface. Substitution of the given current values into this KCL equation gives

$$i_a + (2) + (-3) - (5) - (-12) = 0$$

We can easily solve this to obtain

$$i_a = -6 \text{ A}$$

We will now introduce some definitions to simplify future discussion.

† We need the concept of a cutset if we use the computer program PSpice® for network analysis.

Nodes, Short Circuits, and Open Circuits

The termination of one or more branches is a node. Thus points 2 through 6 form nodes in the circuit of Fig. 2.2. The straight line segment between points u and w represents an ideal wire connecting these two points. This implies that the voltage v_{uw} is zero regardless of the value of the current i_{uw}. We say that there is a *short circuit* between points u and w. We consider these points as a single node (labeled node 0). In general, all points connected to each other by short circuits form a single node. Therefore, points x, y, and z form a single node (labeled node 1).

In Fig. 2.2, no current can flow directly from node 6 to node 2, because no branch connects these two nodes. A path through which no current can flow, regardless of the voltage across it, is an *open circuit.*

Parallel and Series Connections and Junctions

Branches B, C, and D of Fig. 2.2 share two common nodes (nodes 1 and 2) and are therefore in *parallel.* The same voltage (in this case v_{12}) must appear across all components in parallel.

A *junction* is a node where three or more branches terminate. In Fig. 2.2, nodes 0 through 3 are also junctions. Branches F, G, and H connect nodes 3 and 0 with no junctions at the intervening nodes (nodes 4 and 5) and are therefore in *series.*† Application of KCL will quickly establish that the same current (in this case i_{30}) passes through components in series.

Remember

The sum of all currents leaving any closed surface is zero. The currents through those branches that lie totally inside or totally outside a closed surface play no part in the writing of the KCL equation for that surface. We should select our closed surfaces wisely to make the problem as simple as possible.

EXERCISES

1. Refer to the network shown below.

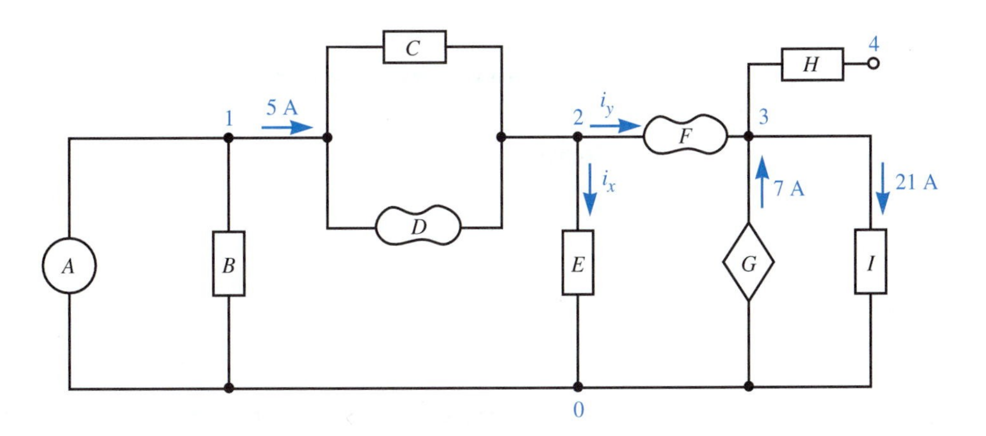

† Some books define a combination of series branches, such as F, G, and H, to be a single branch.

(a) How many nodes and junctions are in this network? *answer:* 5, 4

(b) Identify all parallel combinations of components in this network. *answer:* A and B, C and D, G and I

(c) Determine currents i_x and i_y. *answer:* -9 A, 14 A

(d) Determine currents i_{32} and i_{34}. *answer:* -14 A, 0 A

2. For the following network, $i_1 = 3$ A, $i_2 = -4$ A, and $i_3 = 5$ A.

(a) How many nodes and junctions are in this network? *answer:* 7, 6

(b) Identify any series or parallel combinations of components in this network. *answer:* The two left-hand components are in series, and the two right-hand components are in parallel.

(c) Determine i_4 from a single KCL equation. *answer:* 6 A

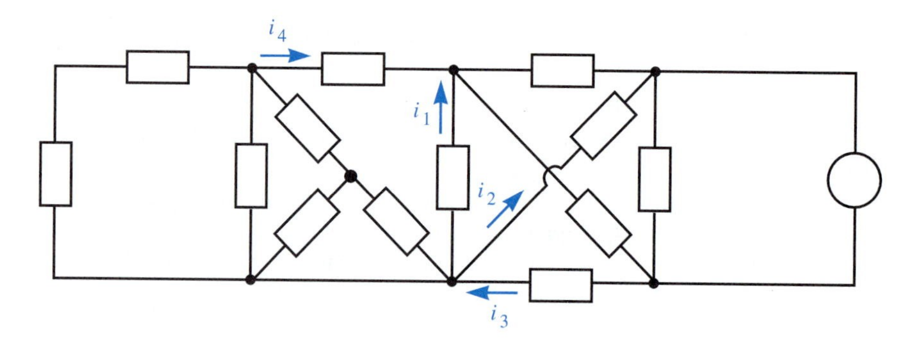

2.2 Kirchhoff's Voltage Law

Voltage v_{12} is the work in joules per coulomb *performed* by the charge moving along any path from point 1 to point 2. An energy source must supply the same energy per coulomb to move charge from point 1 to point 2 along any path. This statement of the conservation of energy is formalized in Kirchhoff's voltage law (KVL), which is the second and last axiom of circuit theory.

AXIOM KVL

Kirchhoff's voltage law states that, at any instant in time, the algebraic sum of all voltage drops taken around any closed path is zero:

$$v_1 + v_2 + v_3 + \cdots + v_N = 0 \tag{2.2a}$$

or in abbreviated notation

$$\sum_{k=1}^{N} v_k = 0 \tag{2.2b}$$

where v_k is the voltage drop, taken in the direction of the path, along the kth segment of the N segments in the closed path.

When writing a KVL equation, we will use $+v_k$ if the path takes us from the (+) to the (−) reference mark for v_k. If the path is from the (−) to the (+) reference mark, we will use the negative of the value of v_k. A restatement of KVL is *the voltage difference of one point with respect to another is independent of the path.* We will now apply KVL to an example circuit.

We can apply KVL to any closed path that is convenient. The following example demonstrates that a wise choice can often save considerable effort when analyzing a circuit.

EXAMPLE 2.3 For the circuit shown in Fig. 2.3, determine voltage v_x.

FIGURE 2.3
Circuit for Example 2.3

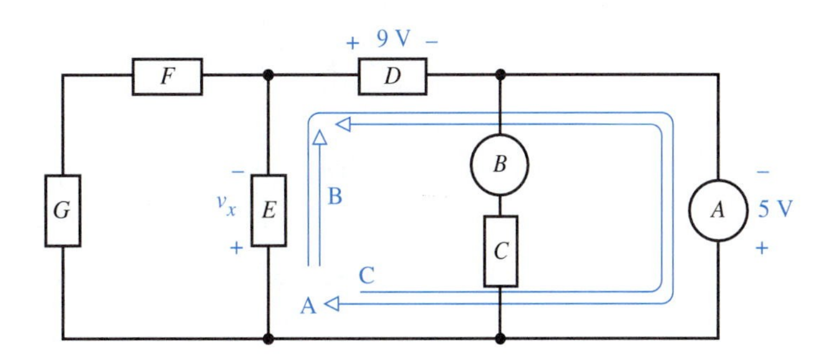

Solution Select a closed path *A* as indicated in Fig. 2.3. The sum of the voltages is

$$v_x + 9 - 5 = 0$$

The minus sign appears on 5 because the chosen path enters the end of the component with the (−) reference. Solving the preceding equation yields

$$v_x = -4 \text{ V}$$

We can see that there was no need to know the voltage across component *B* or *C*. Our selected path did not follow these branches. For the same reason, the voltages across components *F* and *G* did not enter into our KVL equation.

We could skip a bit of algebra by using the alternative statement of KVL: The voltage difference between two points is independent of the path. The voltage drop along path B must equal the sum of the voltage drops along path C:

$$v_x = 5 - 9 = -4 \text{ V}$$

EXAMPLE 2.4 Determine v_a in the network shown in Fig. 2.4 if $v_b = 2$ V, $v_c = -3$ V, $v_d = 5$ V, and $v_e = -12$ V.

Solution If we select the closed path as shown in Fig. 2.4, the only unknown voltage in the path is v_a. Application of KVL gives

$$v_a + v_b + v_c - v_d - v_e = 0$$

A minus sign appears on v_d and v_e because, for the chosen direction of the path, we go from the (−) to the (+) reference for these two voltages. Substitution of the given voltage values into this KVL equation gives

$$v_a + (2) + (-3) - (5) - (-12) = 0$$

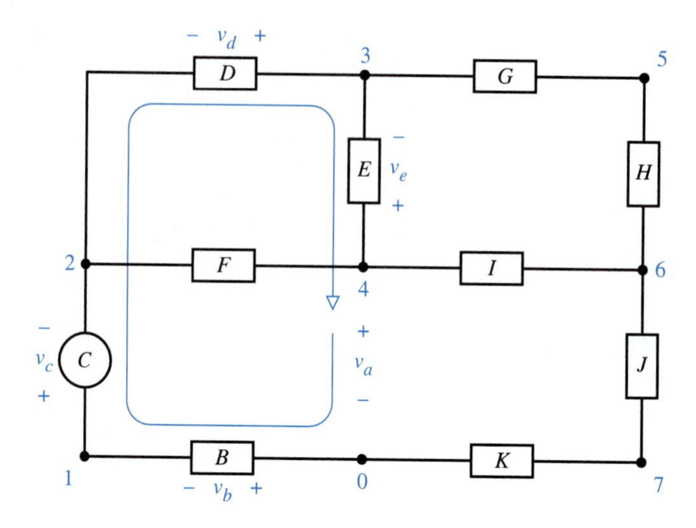

FIGURE 2.4 Application of KVL to a closed path

We can easily solve this to obtain

$$v_a = -6 \text{ V}$$

We now introduce a few additional definitions that will be useful in our work.

Loops, Meshes, and Planar Networks

Observe that the closed path we chose in the preceding example *did not* follow along a path of branches as we went from node 4 to node 0. The voltage v_a is simply voltage v_{40} and not a branch voltage. This closed path is not a loop. A *loop* is a closed path of branches that does not pass through the same node twice. Paths A, B, C, and D shown in Fig. 2.5 are loops.

A loop that does not enclose another loop, such as paths A, B, and C in Fig. 2.5, is a *mesh.* Loop D is not a mesh because it encloses meshes B and C. We define meshes only for a planar network. We can always draw a *planar network* so that lines cross only at points of connection. If we cannot draw the circuit diagram so that no unconnected lines cross, the circuit is *nonplanar.*†

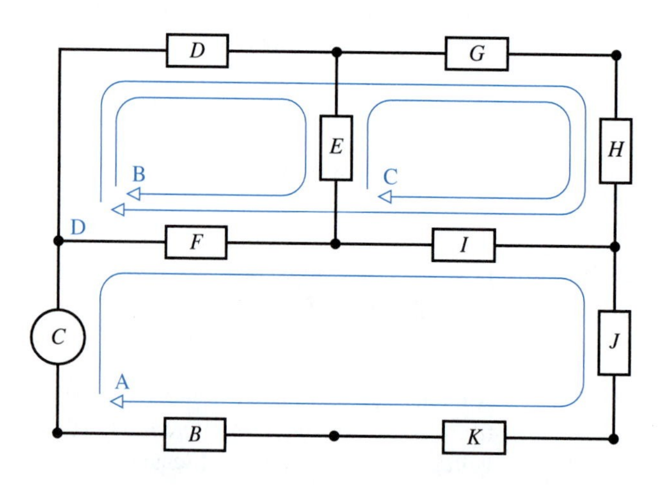

FIGURE 2.5 Some loops shown for the network of Fig. 2.4

† See Problem 3 at the end of this chapter for an example of a nonplanar network.

Remember The sum of the voltage drops around any closed path is zero. Voltages that are not on a closed path play no part in writing a KVL equation for that path. We should select our closed path wisely to make the problem as simple as possible.

EXERCISES

3. The network shown below has voltage v_{10} equal to 5 V.
   (a) How many nodes and junctions are in the network? *answer:* 5, 3
   (b) Identify series and parallel combinations of components in the network.
   answer: The two left-hand components are in series, and two series components connect node 1 to node 0.
   (c) How many meshes are in the network? *answer:* 3
   (d) How many loops can be found in the network? *answer:* 6
   (e) Determine voltage v_x and voltage v_y. *answer:* 9 V, 14 V

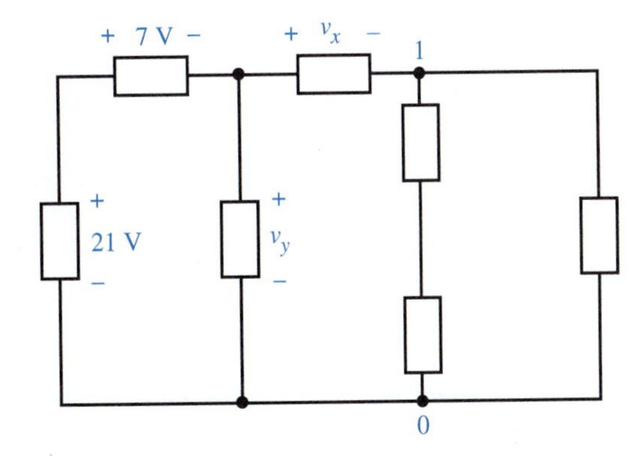

4. For the circuit shown below, $v_{10} = 5$ V, $v_{20} = 17$ V, and $v_{30} = 21$ V.
   (a) Write a KVL equation around the indicated closed path and solve for v_{12}.
   answer: −12 V
   (b) Use the fact that KVL says that the voltage difference is independent of the path, and equate the voltage difference between node 1 and node 2 along path *A* to the sum of the differences along path *B*. This gives v_{12} directly.
   answer: −12 V
   (c) Use KVL to find v_{13}. *answer:* −16 V

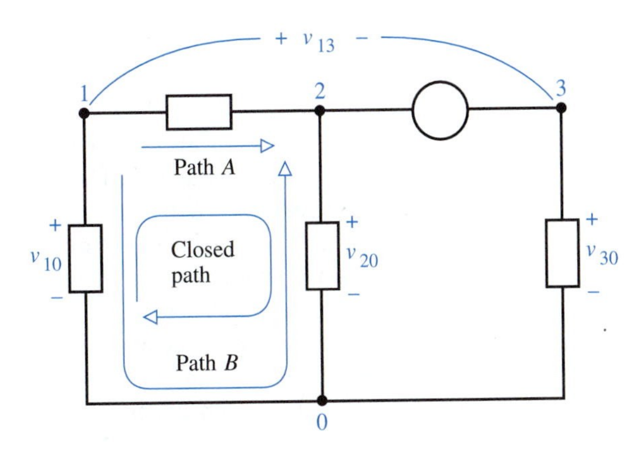

5. For the circuit in Fig. 2.4, $v_{10} = -2$ V, $v_{20} = 1$ V, $v_{30} = 18$ V, $v_{40} = 6$ V, $v_{50} = 16$ V, $v_{60} = 22$ V, and $v_{70} = 29$ V. Find v_a, v_{21}, v_{12}, v_{53}, v_{16}, v_{67}, v_e, and v_{14}.

   answer: 6 V, 3 V, −3 V, −2 V, −24 V, −7 V, −12 V, −8 V

2.3 Independent Sources

An electrical system can perform no useful function without a source of electrical energy. Sources of electrical energy include generators or alternators that convert mechanical energy into electrical energy, but other devices are also used. For example, a flashlight relies on a battery to convert chemical energy into electrical energy, and a solar cell converts light (electromagnetic energy) directly into a conveniently used form of electrical energy. In each case we can model the availability of electrical energy with network components called *sources.*

Voltage Source

Under normal operating conditions a practical energy source, such as the battery in an automobile, can maintain a voltage across its terminals that is relatively independent of the current required by the accessories. We idealize such a device with a network component called an *independent voltage source,* or more simply a *voltage source.*

DEFINITION Voltage Source

An **independent voltage source** is a two-terminal network component with terminal voltage v_{ab} specified by a time function $v(t)$ that is independent of the terminal current i_{ab}. The source equation is

$$v_{ab} = v(t) \tag{2.3}$$

and the component symbol is shown in Fig. 2.6.

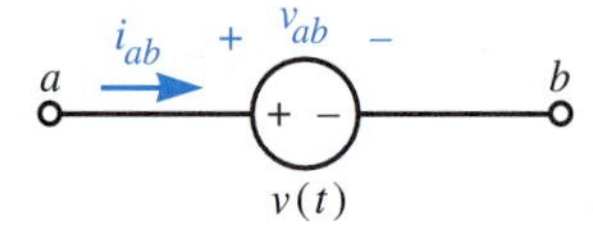

FIGURE 2.6
Voltage source symbol

The (+) and (−) marks inside the circle of the voltage source symbol (Fig. 2.6) identify the component as a voltage source. The current i_{ab} is determined by what is connected to the voltage source. Observe that current i_{ab} has no effect on voltage v_{ab} as given by Eq. (2.3). In Fig. 2.7 we graphically depict the component equation for an independent voltage source with voltage $v_{ab} = 12$ V to emphasize that the terminal voltage is independent of the terminal current. The definition of a voltage source implies that a *short circuit* can be considered to be a voltage source of value zero. Because an independent voltage source represents an idealization of a practical device, such as a battery, many engineers refer to an independent voltage source as an *ideal voltage source.*

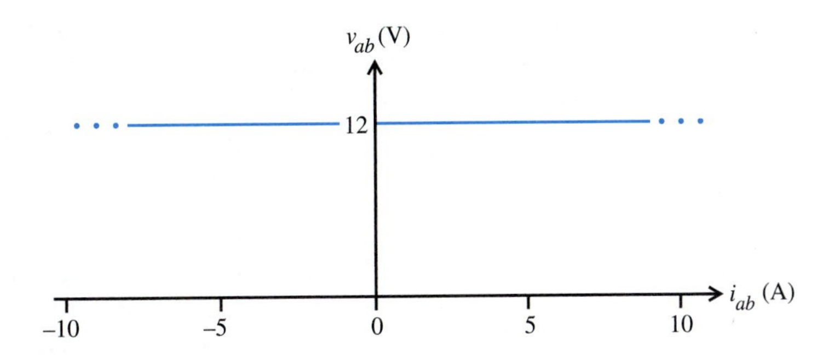

FIGURE 2.7 Terminal characteristics of an independent voltage source

Current Source

A few practical devices generate a current that is relatively independent of the terminal voltage over the normal range of operating voltages. A constant-current transformer, once widely used to supply power to incandescent street lights, is a good example. Another example is an automobile alternator, which provides a current relatively independent of terminal voltage under certain operating conditions. We idealize such a device with a network component called an *independent current source,* or more simply, a *current source.*

DEFINITION Current Source

An **independent current source** is a two-terminal network component with terminal current i_{ab} specified by a time function $i(t)$ that is independent of the terminal voltage v_{ab}. The source equation is

$$i_{ab} = i(t) \tag{2.4}$$

and the component symbol is shown in Fig. 2.8.

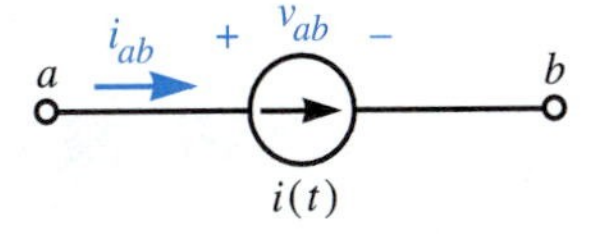

FIGURE 2.8 Current source symbol

The arrow inside the circle of the current source symbol identifies the component as a current source. Voltage v_{ab} is determined by what is connected to the current source. Observe that voltage v_{ab} has no effect on current i_{ab} as given by Eq. (2.4).

In Fig. 2.9 we graphically depict the component equation of an independent current

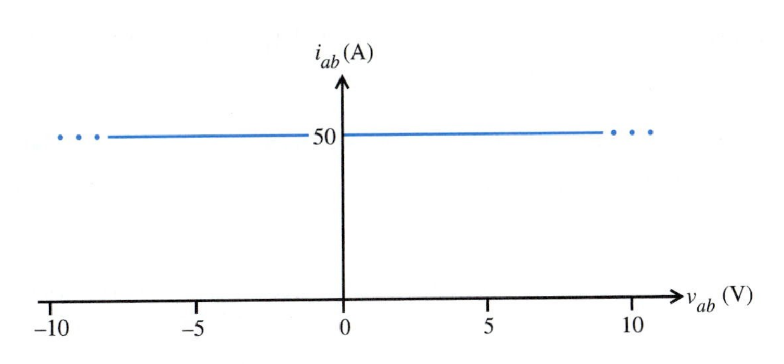

FIGURE 2.9 Terminal characteristics of an independent current source

source with current $i_{ab} = 50$ A to emphasize that the terminal current is independent of the terminal voltage. Engineers frequently refer to an independent current source as an *ideal current source,* because it is an idealization of a practical device. The definition of a current source implies that an *open circuit* can be considered to be a current source of value zero.

We can now begin to form a model for a practical electrical system.

EXAMPLE 2.5 A relatively crude model of an automobile electrical system is shown in Fig. 2.10. The voltage source represents the lead-acid storage battery, and the current source represents the alternator.† We have lumped the automobile accessories together in one box, called the *load,* which requires a current of $i_L = 30$ A. Determine the load voltage v_L, the battery current i_B, and the power absorbed by the battery.

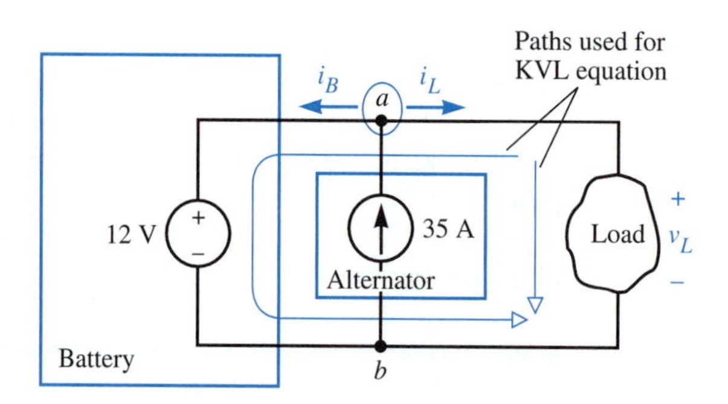

FIGURE 2.10 A crude model of an automobile electrical system

Solution Application of KVL to the paths indicated by arrows gives the load voltage:

$$v_L = 12 \text{ V}$$

We can obtain the battery current by summing the currents leaving the indicated closed surface:

$$i_B - 35 + 30 = 0$$

which gives

$$i_B = 5 \text{ A}$$

The power absorbed by the battery is

$$p = 12i_B = 12(5) = 60 \text{ W}$$

Therefore we are storing chemical energy in the battery that will be returned in the form of electrical energy at a later time. For this reason we say that the battery is being "charged."

We will require more types of network components to refine this model.

Remember The voltage across a voltage source is independent of the current through it, and the current through a current source is independent of the voltage across it.

EXERCISES

6. Determine voltages v_{10}, v_x, v_{13}, v_{20}, and v_y in the following network.
answer: 15 V, 15 V, 12 V, −5 V, −3 V

† The alternating current (ac) generator (alternator) on an automobile includes rectifiers to convert the ac to dc (direct current). Rectifiers are described in books on electronics.

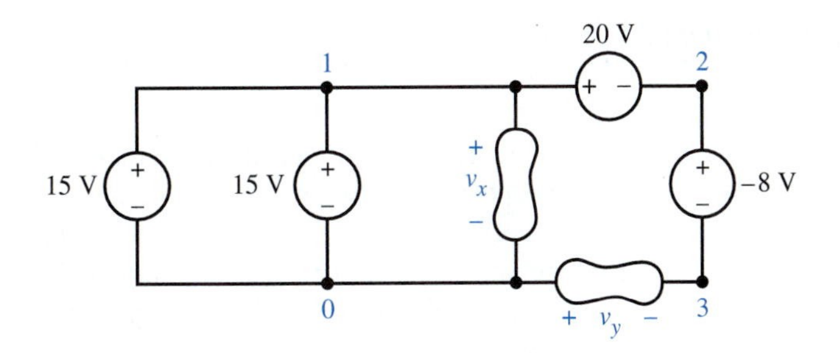

7. Determine currents i_{10}, i_{23}, i_x, and i_y in the following circuit.
answer: -15 A, 15 A, 35 A, -27 A

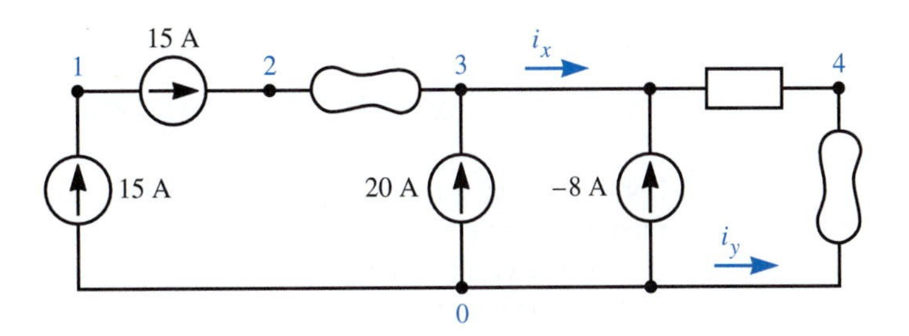

8. For the networks shown, find i_x, v_{ab}, and the power absorbed by the right-hand source.

(a) (b)

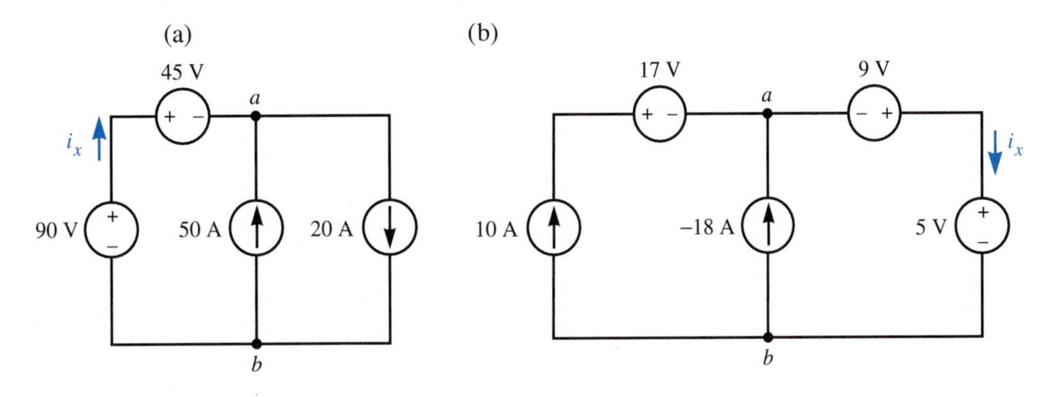

answer: -30 A, 45 V, 900 W

answer: -8 A, -4 V, -40 W

(c) (d)

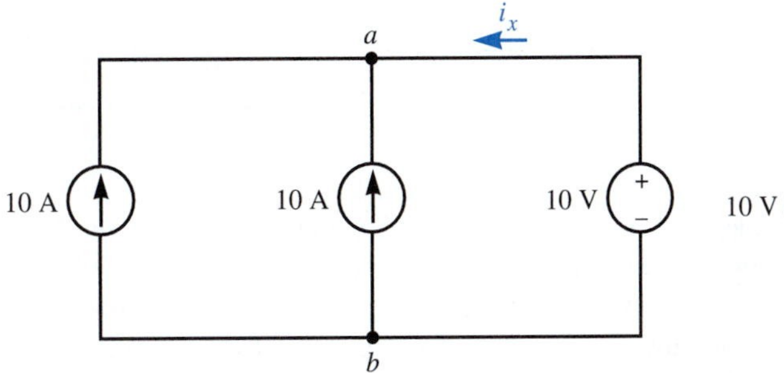

answer: -20 A, 10 V, 200 W

answer: -10 A, 20 V, -100 W

2.4 Dependent Sources

Many electrical systems take a small amount of electric power from one source and use this to control the delivery of a larger amount of electric power from another source. A common example is an amplifier used in a tape player. The magnetized tape moving past the tape head, which consists of magnetic material and a small coil of wire, generates a very small voltage signal that delivers a small fraction of a watt to the amplifier. This low-power signal ultimately controls the delivery of tens of watts to the speaker system. The power to the speaker is obtained from the amplifier power source: a car battery or an electrical outlet. We model this ability to control the delivery of power by the introduction of four-terminal network components called *dependent sources* or *controlled sources.*

Voltage-Controlled Voltage Source

Some practical voltage amplifiers generate a large voltage across two terminals that is intended to be proportional to a smaller voltage established across two other terminals. A tape player amplifier is one such device. We idealize a practical voltage amplifier with a four-terminal network component called a *voltage-controlled voltage source* (VCVS).

DEFINITION VCVS

A **voltage-controlled voltage source** is a four-terminal network component that establishes a voltage v_{cd} between two points c and d in the circuit that is proportional to a voltage v_{ab} between two points a and b. The control equation is

$$v_{cd} = \mu v_{ab} \tag{2.5}$$

and the component symbol is shown in Fig. 2.11.

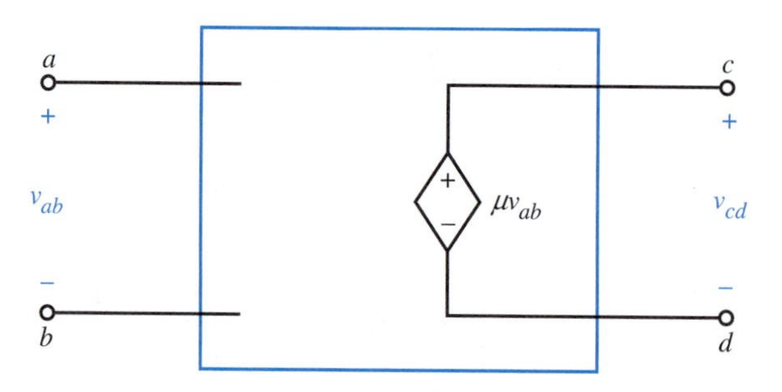

FIGURE 2.11 VCVS symbol

The (+) and (−) marks inside the diamond of the component symbol (Fig. 2.11) identify the component as a voltage source.

Observe that voltage v_{cd} depends only on the constant μ, a dimensionless constant called the *voltage gain,* and the *control voltage* v_{ab}. Current i_{cd} can affect voltage v_{cd} only if it affects voltage v_{ab}. Although we show the VCVS symbol with four terminals, *the control voltage v_{ab} is typically the voltage across some other network component,* as shown in Fig. 2.12.

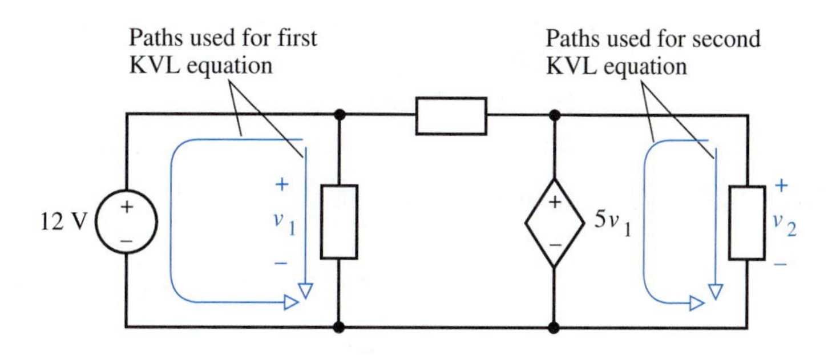

FIGURE 2.12
Example with a VCVS

EXAMPLE 2.6 Determine voltage v_2 in the circuit of Fig. 2.12.

Solution Kirchhoff's voltage law easily establishes that

$$v_1 = 12 \text{ V}$$

and

$$v_2 = 5v_1 = 5(12) = 60 \text{ V}$$

Voltage-Controlled Current Source

Some devices permit the control of a current directly with a voltage. One such electronic device is a field-effect transistor. We idealize such a device with a four-terminal network component called a *voltage-controlled current source* (VCCS).

DEFINITION VCCS

A **voltage-controlled current source** is a four-terminal network component that establishes a current i_{cd} in a branch of the circuit that is proportional to the voltage v_{ab} between two points a and b. The control equation is

$$i_{cd} = gv_{ab} \tag{2.6}$$

and the component symbol is shown in Fig. 2.13.

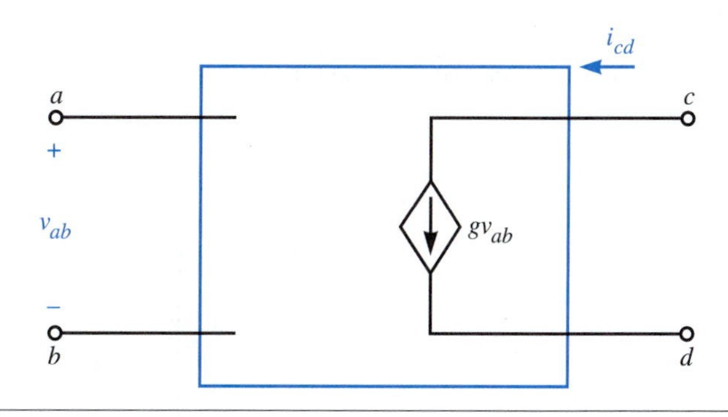

FIGURE 2.13
VCCS symbol

The arrow inside the diamond of the component symbol (Fig. 2.13) identifies the component as a current source.

Observe that current i_{cd} depends only on the control voltage v_{ab} and the constant g, called the *transconductance* or mutual conductance. Constant g has dimensions of ampere per volt, or siemens (S).

EXAMPLE 2.7 Determine current i_x in the circuit of Fig. 2.14.

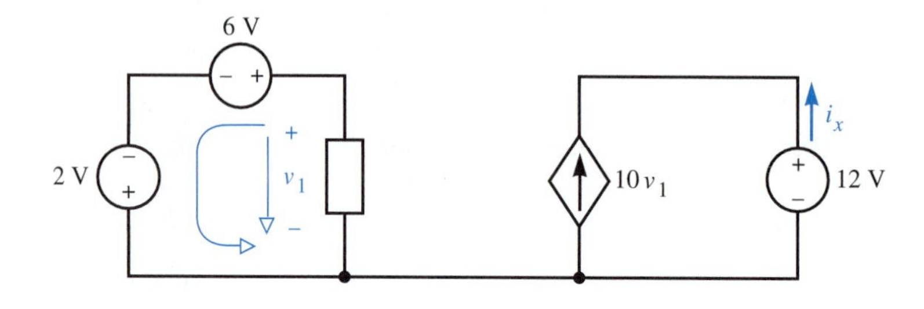

FIGURE 2.14
Example with a VCCS

Solution We can use KVL to show that

$$v_1 = 6 - 2 = 4 \text{ V}$$

and KCL to determine that

$$i_x = -10v_1 = -10(4) = -40 \text{ A}$$

Current-Controlled Voltage Source

Electronic devices that generate a voltage that is almost proportional to a current can be constructed. We idealize such a device with a four-terminal network component called a *current-controlled voltage source* (CCVS).

DEFINITION CCVS

A **current-controlled voltage source** is a four-terminal network component that establishes a voltage v_{cd} between two points c and d in the circuit that is proportional to current i_{ab} in some branch of the circuit. The control equation is

$$v_{cd} = ri_{ab} \tag{2.7}$$

and the component symbol is shown in Fig. 2.15.

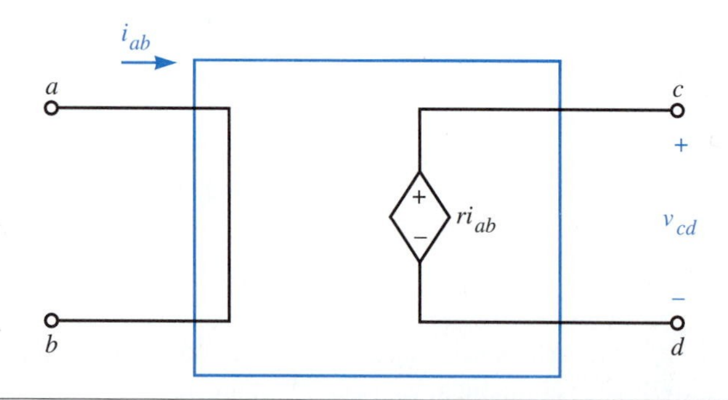

FIGURE 2.15
CCVS symbol

The (+) and (−) marks inside the diamond of the component symbol (Fig. 2.15) identify the component as a voltage source.

Observe that voltage v_{cd} depends only on the *control current* i_{ab} and the constant r, called the *transresistance* or mutual resistance. Constant r has the dimension of volt per ampere, or ohm (Ω). Although we show the CCVS symbol with four terminals, terminals a and b and the lines joining them are not usually shown explicitly. *The control current* i_{ab} *is typically the current through some other network component,* as shown in Fig. 2.16.

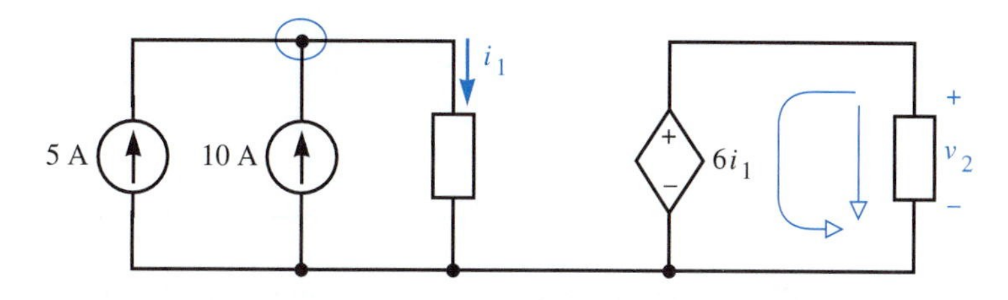

FIGURE 2.16
Example with a CCVS

EXAMPLE 2.8 Determine voltage v_2 for the network shown in Fig. 2.16.

Solution We can use KCL to establish that

$$i_1 = 15 \text{ A}$$

and KVL gives us

$$v_2 = 6i_1 = 6(15) = 90 \text{ V}$$

Current-Controlled Current Source

Some devices permit the control of one current with another. An example is the electronic device called a bipolar junction transistor. We idealize a practical current amplifier with a four-terminal network component called a *current-controlled current source* (CCCS).

DEFINITION CCCS

A **current-controlled current source** is a four-terminal network component that establishes a current i_{cd} in one branch of a circuit that is proportional to current i_{ab} in some branch of the network. The control equation is

$$i_{cd} = \beta i_{ab} \tag{2.8}$$

and the component symbol is shown in Fig. 2.17.

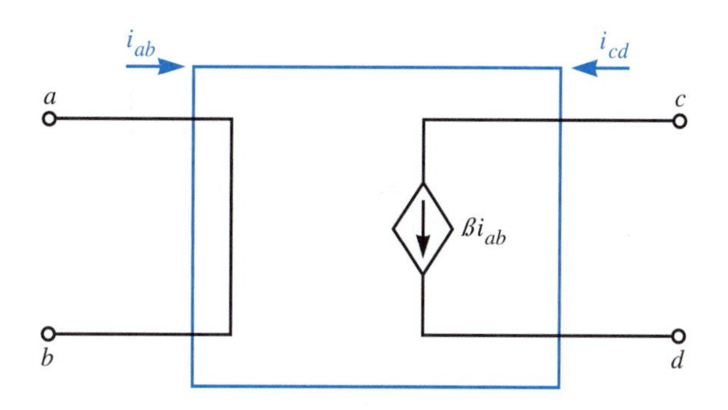

FIGURE 2.17
CCCS symbol

The arrow inside the diamond of the component symbol in Fig. 2.17 identifies the component as a current source. Observe that current i_{cd} depends only on the control current i_{ab} and the dimensionless constant β, called the *current gain.*

EXAMPLE 2.9 Determine current i_2 for the circuit shown in Fig. 2.18.

FIGURE 2.18
Example with a CCCS

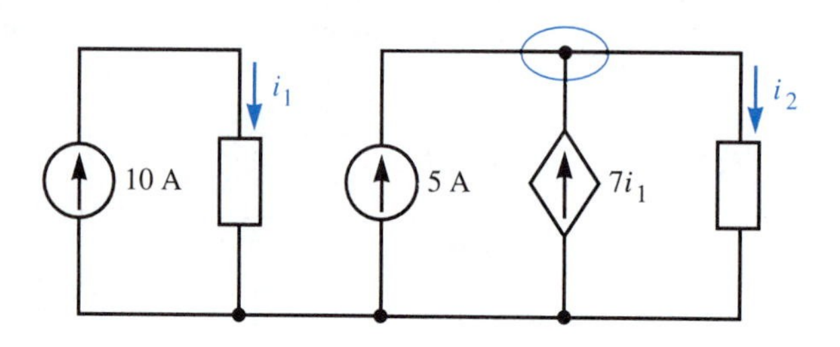

Solution We can obtain the solution from two KCL equations,

$$i_1 = 10 \text{ A}$$

and

$$i_2 = 5 + 7i_1 = 5 + 7(10) = 75 \text{ A}$$

A network component that can supply positive average power (absorb negative average power) is an *active component.* Independent and dependent sources are active components. In the following sections we will introduce three components that cannot supply positive average power (the average absorbed power is never negative). These are *passive components.* A network that contains one or more active components is an *active network.* A *passive network* contains only passive components.

In addition to providing sources of electrical energy, physical devices, unavoidably or by design, convert electrical energy into heat or store energy in electric or magnetic fields. We model these properties of devices with network components called *elements.* In the next section, we will consider the conversion of electrical energy into heat.

Remember

Dependent sources model the control of a voltage or current by another voltage or current. The voltage across a dependent voltage source is determined by the control equation, so this voltage depends *only* on the control variable and the gain constant. Similarly, the current through a dependent current source is determined by its control equation, so this current depends only on the control variable and the gain constant.

EXERCISES 9. For the following networks, determine v_{ab} and i_y.

(a)

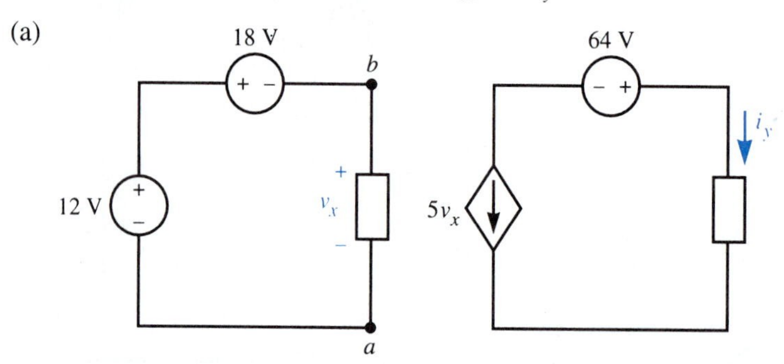

answer: 6 V, 30 A

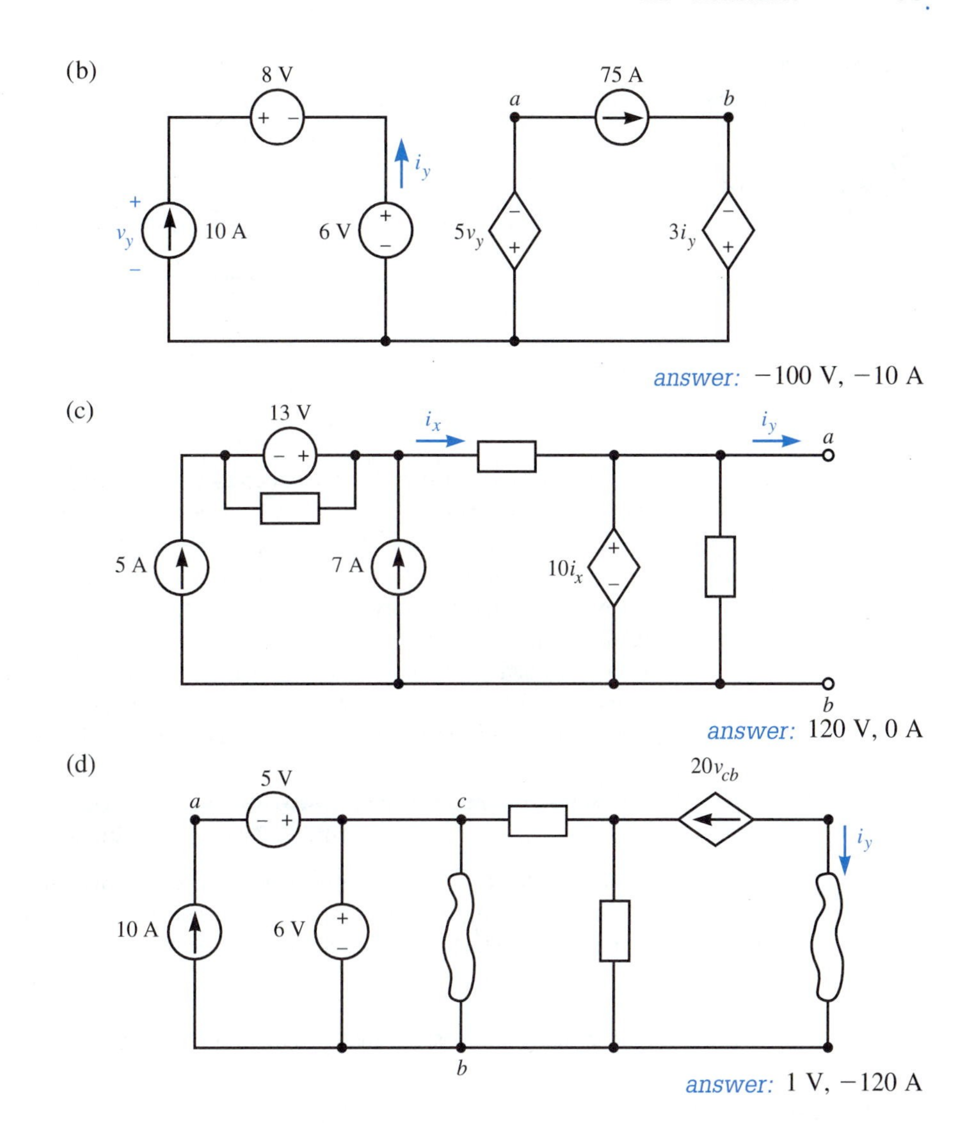

2.5 Resistance

We are all familiar with devices, such as an electric furnace, that are intentionally designed to convert electrical energy into heat. Other devices, such as a television receiver, unavoidably convert some fraction of the electrical energy they absorb into heat. We call a device that converts electrical energy into heat a *dissipative device,* because it dissipates electrical energy. Even a simple piece of wire exhibits this characteristic. Perhaps you have noticed that the cord to an electric toaster was warm when you unplugged it. Electric charge moving through a material such as metal causes the molecules of the material to vibrate and thus converts electrical energy into heat.

We call a physical device designed to convert electrical energy into heat a *resistor.* If we were to experimentally measure the relationship between the voltage across and the current through a physical resistor, we might obtain a characteristic such as that given by curve *a* or curve *b* in Fig. 2.19. In general the voltage is a function of current:

$$v_{ab} = f(i_{ab}) \tag{2.9}$$

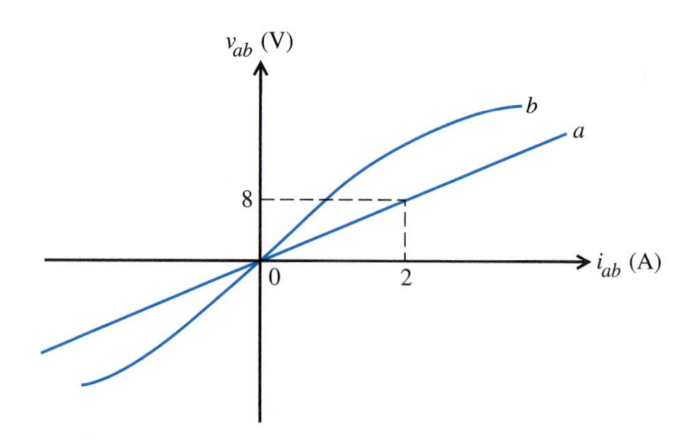

FIGURE 2.19 Measured volt-ampere characteristics of physical resistors: (*a*) linear; (*b*) nonlinear

A resistor constructed of a semiconducting material, such as silicon carbide, could have a characteristic such as curve b of Fig. 2.19. We say that the device exhibits *nonlinear resistance* and refer to the device as a *nonlinear resistor.*

Experiments show that for a metallic conductor at constant temperature, the voltage across the conductor is directly proportional to the current through it. This proportional relationship is referred to as *Ohm's law.* A resistor constructed of metal would obey Ohm's law and would have a characteristic that follows a straight line, as given by curve a of Fig. 2.19. We say that the device exhibits *linear resistance* and is a *linear resistor.* A linear resistor is the basis for the network component that we call a *resistance* or *ideal resistor.*†

DEFINITION Resistance

A **resistance** is a two-terminal network component with terminal voltage v_{ab} directly proportional to current i_{ab}. The constant of proportionality is also called resistance. The terminal equation is

$$v_{ab} = Ri_{ab} \tag{2.10}$$

and the component symbol is shown in Fig. 2.20.

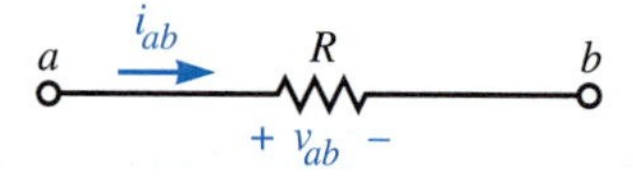

FIGURE 2.20 Resistance symbol

Resistance R has the dimension of volt per ampere, or *ohm* (Ω). Observe that Eq. (2.10) implies *the passive sign convention.* That is, the current arrow for $i = i_{ab}$ enters the end of the resistance with the (+) reference mark for $v = v_{ab}$.

We must always observe the relative reference directions for the voltage and current when we apply Ohm's law. For example, if we use current i_{ba} to calculate voltage v_{ab}, the current reference arrow passes through the device from (−) to (+) (*the active sign convention*) and Eq. (2.10) gives us

$$v_{ab} = Ri_{ab} = R(-i_{ba}) = -Ri_{ba} \tag{2.11}$$

We often work with the reciprocal of resistance, which is called *conductance.*

† Some texts refer to both the physical device and the network component as a resistor and use the word resistance to denote only the constant of proportionality R.

DEFINITION Conductance

Conductance G is the reciprocal of resistance:

$$G = \frac{1}{R} \quad (2.12)$$

and

$$i_{ab} = Gv_{ab} \quad (2.13)$$

Conductance G has the dimension of ampere per volt or *siemens* (S). [An archaic term for siemens is the *mho* (℧), which is ohm spelled backward.] We refer to Eqs. (2.10) and (2.13) as Ohm's law equations.

EXAMPLE 2.10 Determine the resistance of the resistor with terminal characteristics given by curve a of Fig. 2.19.

Solution The slope of line a in Fig. 2.19 is

$$R = \frac{v_{ab}}{i_{ab}} = \frac{8}{2} = 4\ \Omega$$

Within the measured current range, we can describe the device by the equation

$$v_{ab} = 4i_{ab}$$

and we would model the device by a 4-Ω resistance.

We can easily calculate the power absorbed by a resistance if the component voltage and current are known.

$$p = v_{ab}i_{ab} \quad (2.14)$$

where the voltage and current references satisfy the passive sign convention. Substitution from Ohm's law gives

$$p = (Ri_{ab})i_{ab} = Ri_{ab}^2 \quad (2.15)$$

or

$$p = (v_{ab})\frac{1}{R}v_{ab} = \frac{1}{R}v_{ab}^2 \quad (2.16)$$

We can see from either of these two equations that the power absorbed by a resistance is never negative:

$$p \geq 0 \quad (2.17)$$

Therefore resistance is a passive component.

We show an example of the relation between the voltage across, the current through, and the power and energy absorbed by a 2-Ω resistance in Fig. 2.21. Note that the voltage and current waveforms have exactly the same shape, and that the power and energy absorbed are never negative.

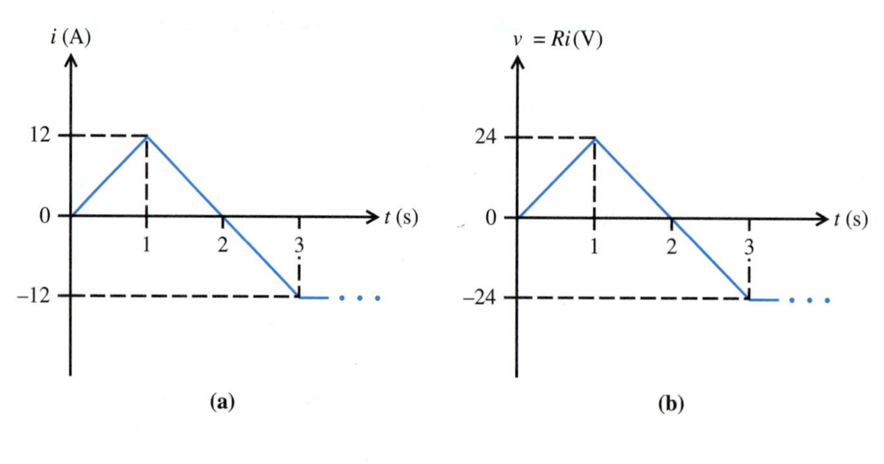

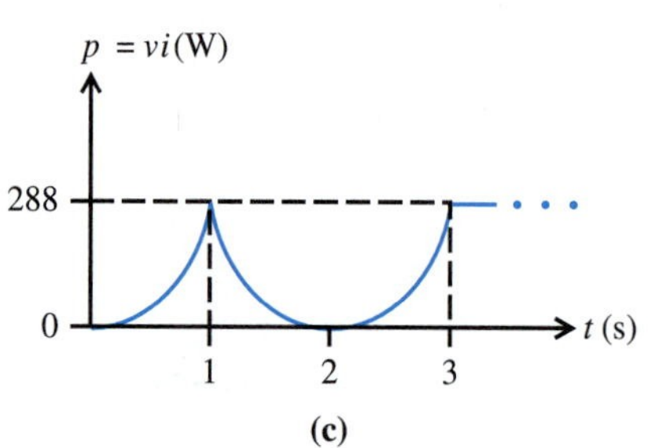

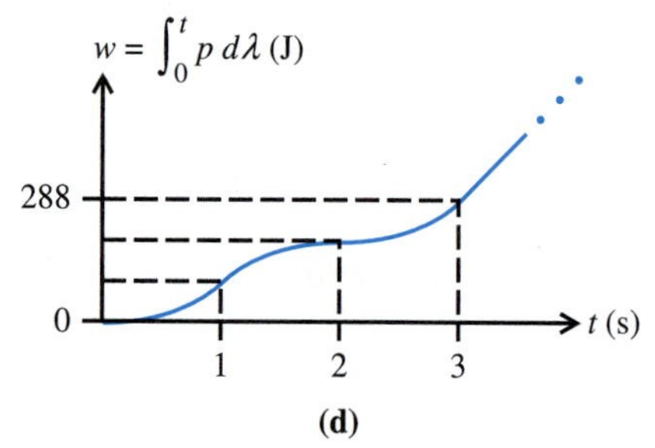

FIGURE 2.21 An example of the relation between current, voltage, power, and energy absorbed by a 2-Ω resistance. (*a*) current; (*b*) voltage; (*c*) power; (*d*) energy

EXAMPLE 2.11 We can improve the model we used for a storage battery in Example 2.5 by including a resistance in series with the voltage source, as shown in Fig. 2.22. This resistance is a model for the *internal resistance* of the battery. Assume that the load current remains unchanged. Determine the load voltage v_{ab}.

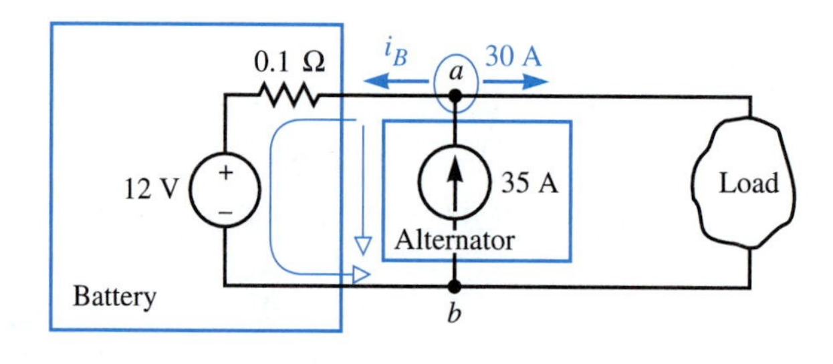

FIGURE 2.22 An improved model of an automobile electrical system

Solution We obtain current i_B by the use of KCL:

$$i_B - 35 + 30 = 0$$

which gives

$$i_B = 5 \text{ A}$$

We can now obtain voltage v_{ab} by the use of KVL and Ohm's law:

$$v_{ab} = 0.1i_B + 12 = 0.1(5) + 12 = 12.5 \text{ V}$$

The alternator is supplying more current than is required by the load, so the battery terminal voltage is greater than 12 V.

Remember Resistance is the network model that accounts for the conversion of electrical energy into heat and for voltage that is proportional to current.

EXERCISES

10. The volt-ampere measurements of a device are shown in the following figure. What value of resistance would have this terminal equation? answer: 120 Ω

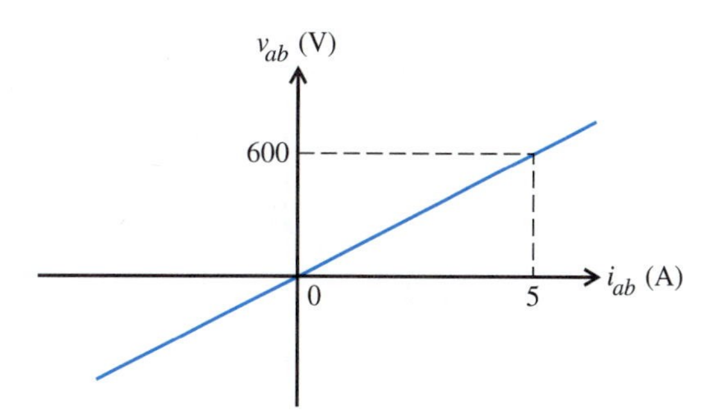

11. The terminal characteristics of a *practical voltage source* were measured and found to be as shown below. Determine a series connection of a resistance and an independent voltage source that would have the same terminal characteristics.
answer: 10 V, 2 Ω

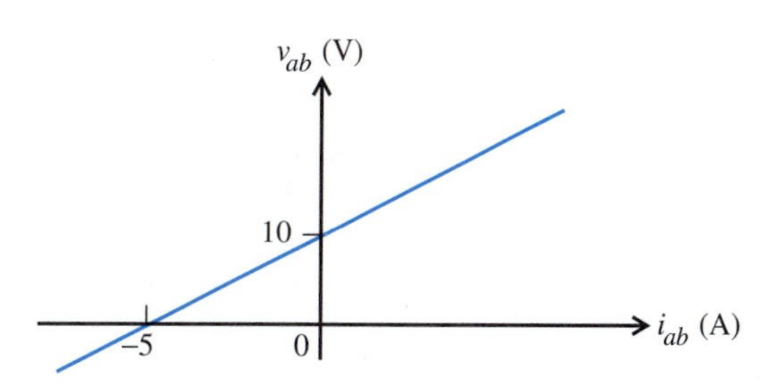

12. For the following nine circuits, find v_{ab}, i_{ab}, and the power absorbed by the resistance.

(a)

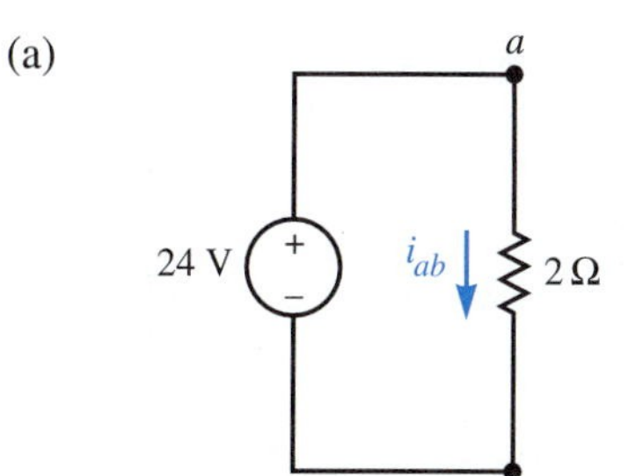

answer: 24 V, 12 A, 288 W

(b)

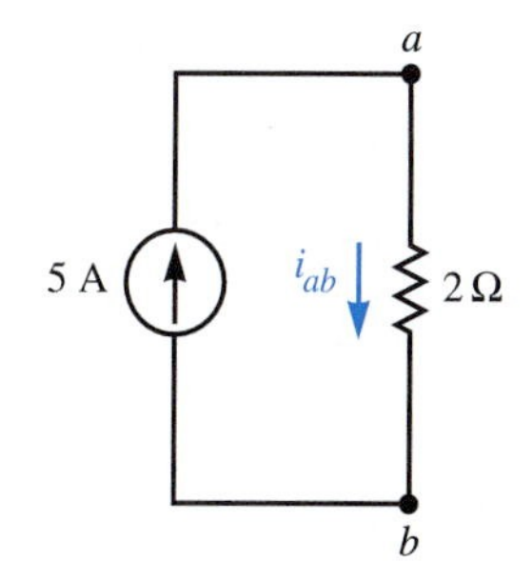

answer: 10 V, 5 A, 50 W

(c)

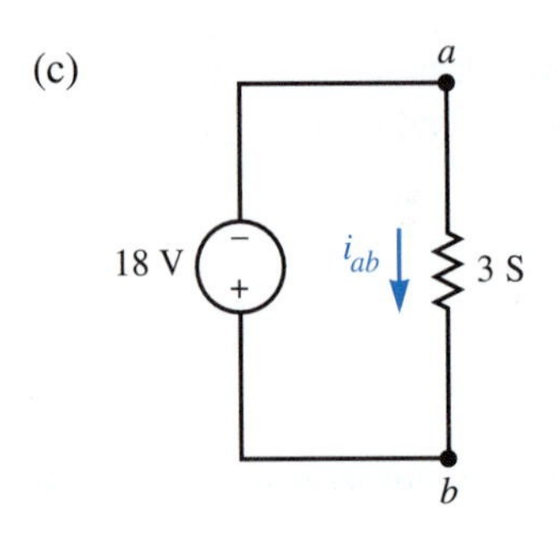

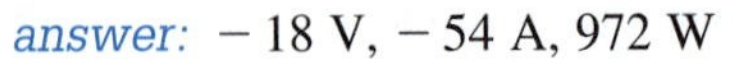

answer: − 18 V, − 54 A, 972 W

(d)

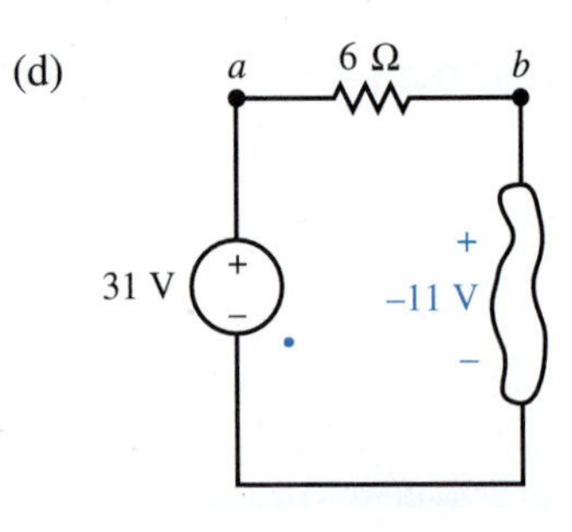

answer: 42 V, 7 A, 294 W

(e)

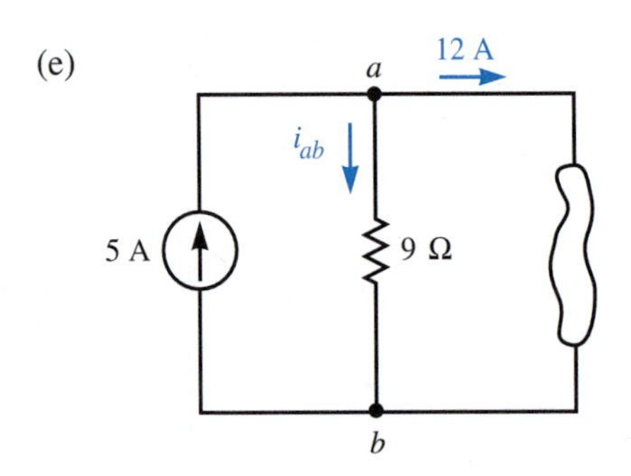

answer: − 63 V, − 7 A, 441 W

(f)

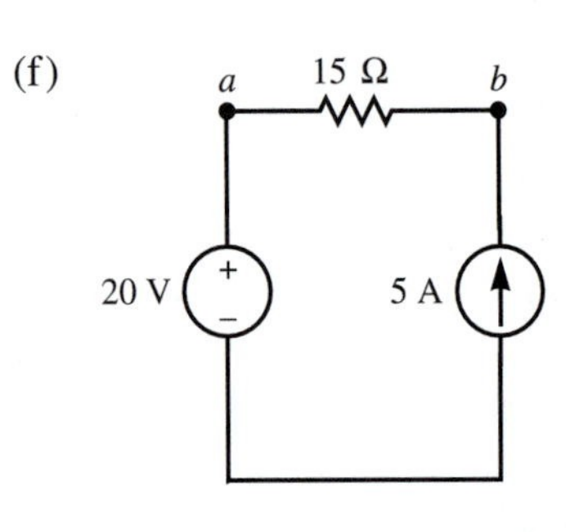

answer: − 75 V, − 5 A, 375 W

(g)

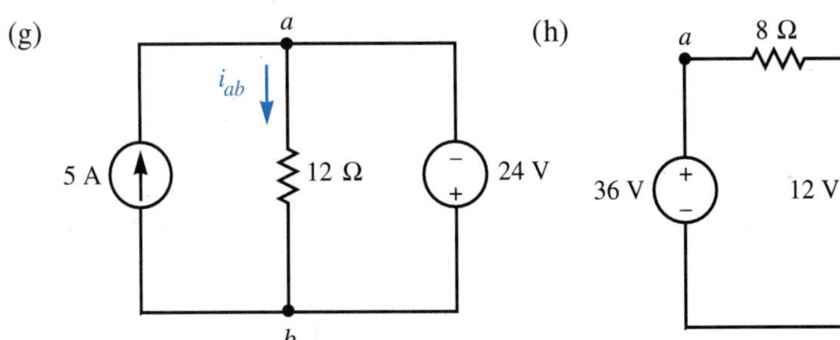

answer: − 24 V, − 2 A, 48 W

(h)

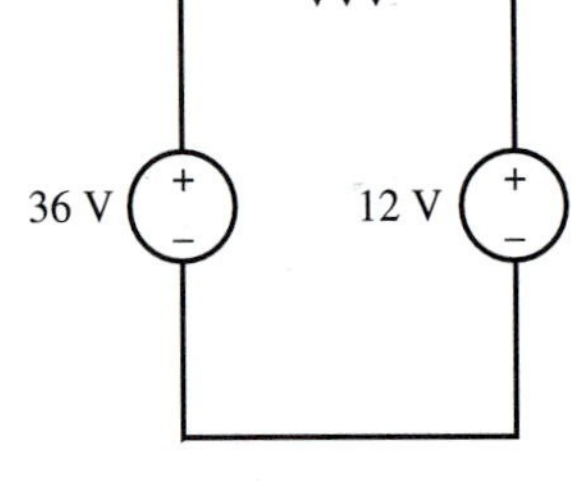

answer: 24 V, 3 A, 72 W

(i)

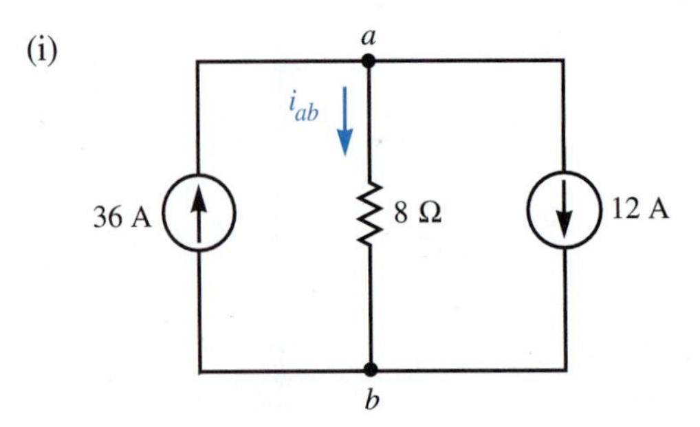

answer: 192 V, 24 A, 4608 W

13. Determine current i_2 in the following circuit.

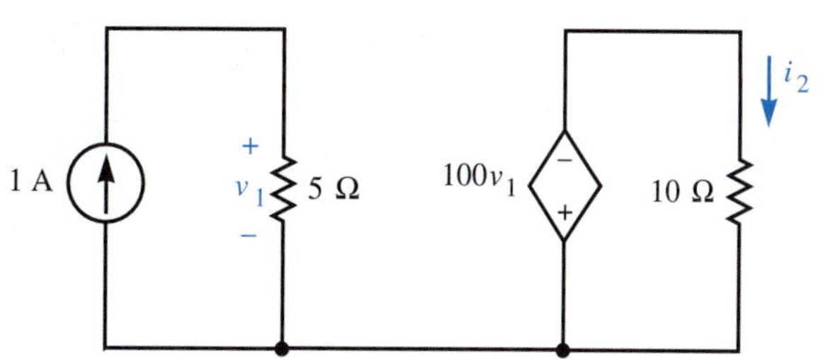

answer: −50 A

14. Determine voltage v_2 in the following circuit.

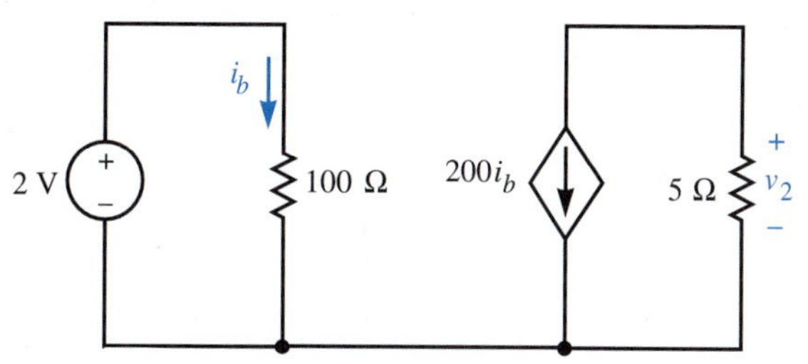

answer: −20 V

15. The alternator modeled as the current source in Example 2.11 is no longer charging, so the current source becomes zero (an open circuit). What is the battery terminal voltage v_{ab} if $i_L = 30$ A? *answer:* 9 V

2.6 Capacitance

Removal of a wool sweater often establishes a charge on the sweater and an opposite charge on the shirt. The resulting electric field will pull the sweater back toward the shirt. This demonstrates that energy is stored in the electric field. The electric field around the power line to your school also stores electrical energy. Electric fields always contain energy.

A physical device that is intentionally designed to store energy in an electric field is called a *capacitor,* or occasionally a condenser. We make extensive use of these devices in electronic and power systems. For instance, a radio tuner uses a capacitor of adjustable value. A capacitor is physically made from two conducting plates (often aluminum foil) separated by a dielectric (nonconducting or insulating) medium, such as plastic, waxed paper, or even air. The physical device is depicted as connected to an electrical system in Fig. 2.23.

Charge of equal magnitude but opposite sign accumulates on the opposing capacitor plates. When the voltage v is not zero, the resulting electric field causes an *electric flux* ψ to exist in the region between the two plates. The total electric flux ψ will equal the excess charge q on the (+) plate. As long as ψ is constant, i will be zero, but as v changes, ψ will change and i will be nonzero.

Conduction current

$$i = \frac{d}{dt} q \tag{2.18}$$

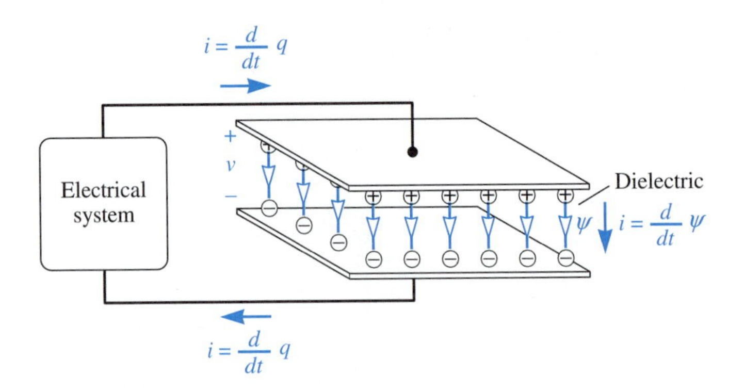

FIGURE 2.23 A parallel-plate capacitor with air dielectric

in the wire is the result of moving charge. Although no charge passes through the insulating region between the two plates, the changing electric flux causes a *displacement current*

$$i = \frac{d}{dt} \psi \tag{2.19}$$

in the dielectric region. This displacement current precisely equals the conduction current, so KCL is not violated.

The ratio of flux to voltage is approximately constant for most dielectrics. Because flux ψ is equal to charge q, the charge-to-voltage ratio is also a constant C:

$$C = \frac{q}{v} \tag{2.20}$$

This lets us write current i in terms of voltage v:

$$\begin{aligned} i &= \frac{d}{dt} q = \frac{d}{dt} \left[\left(\frac{q}{v} \right) v \right] \\ &= \frac{d}{dt} [Cv] \\ &= C \frac{d}{dt} v \end{aligned} \tag{2.21}$$

The constant of proportionality C is the *capacitance* of the plates.

We model energy storage in an electric field with a network component called a *capacitance* or *ideal capacitor*.† All currents induced by changing voltage are assumed to flow through such components.

DEFINITION Capacitance

A **capacitance** is a two-terminal network component with current i_{ab} directly proportional to the time derivative of the voltage v_{ab}. The constant of proportionality C is also called the capacitance. The terminal equation is

† Some texts refer to both the practical device and the network component as a capacitor and use the word capacitance to denote only the constant of proportionality C.

$$i_{ab} = C \frac{d}{dt} v_{ab} \tag{2.22}$$

and the component symbol is shown in Fig. 2.24.

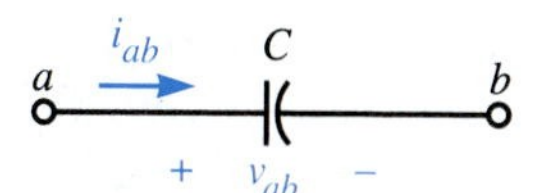

FIGURE 2.24
Capacitance symbol

Capacitance C has the dimension of coulomb per volt, or farad (F). Observe that Eq. (2.22) implies the passive sign convention. With the *active sign convention,* Eq. (2.22) gives us

$$i_{ab} = C \frac{d}{dt} v_{ab} = C \frac{d}{dt} (-v_{ba}) = -C \frac{d}{dt} v_{ba} \tag{2.23}$$

EXAMPLE 2.12 Voltages and currents are assigned according to the passive sign convention. The voltage across a 3-μF capacitance is

$$v(t) = 24e^{40t} \text{ V}$$

Calculate the capacitance current i.

Solution

$$\begin{aligned} i &= C \frac{d}{dt} v = 3 \times 10^{-6} \frac{d}{dt} (24e^{40t}) \\ &= 0.00288e^{40t} \text{ A} \end{aligned}$$

Integration of Eq. (2.22) with respect to t gives†

$$\int_{v_{ab}(-\infty)}^{v_{ab}(t)} dv_{ab} = \frac{1}{C} \int_{-\infty}^{t} i_{ab}(\lambda)\, d\lambda \tag{2.24}$$

Completion of the integration on the left-hand side gives us

$$v_{ab}(t) - v_{ab}(-\infty) = \frac{1}{C} \int_{-\infty}^{t} i_{ab}\, d\lambda \tag{2.25}$$

With the assumption that the capacitance voltage is zero at time minus infinity,

$$v_{ab}(t) = \frac{1}{C} \int_{-\infty}^{t} i_{ab}\, d\lambda \tag{2.26}$$

We can write Eq. (2.26) as

$$v_{ab}(t) = \frac{1}{C} \int_{-\infty}^{t_0} i_{ab}\, d\lambda + \frac{1}{C} \int_{t_0}^{t} i_{ab}\, d\lambda \tag{2.27}$$

† The variable of integration is changed from t to λ because t appears in the upper limit.

which is of the form

$$v_{ab}(t) = v_{ab}(t_0) + \frac{1}{C}\int_{t_0}^{t} i_{ab}\, d\lambda \tag{2.28}$$

We often call the voltage

$$v_{ab}(t_0) = \frac{1}{C}\int_{-\infty}^{t_0} i_{ab}\, d\lambda \tag{2.29}$$

the *initial condition* at time t_0.

EXAMPLE 2.13 Voltages and currents are assigned according to the passive sign convention. The voltage across a 3-μF capacitance at $t = 0$ is $v(0) = 4$ V. The current is

$$i(t) = 24e^{-40t} \text{ mA}$$

for $t \geq 0$. Determine $v(t)$ for $t \geq 0$.

Solution

$$\begin{aligned} v(t) &= v(0) + \frac{1}{C}\int_0^t i\, d\lambda \\ &= 4 + \frac{1}{3 \times 10^{-6}}\int_0^t 24 \times 10^{-3}\, e^{-40\lambda}\, d\lambda \\ &= 204 - 200e^{-40t} \text{ V} \end{aligned}$$

A capacitance does not dissipate electrical energy in the form of heat as does a resistance. It *stores energy* in an electric field as the magnitude of the voltage increases. It returns this energy to the rest of the circuit as the magnitude of the voltage decreases. The stored energy in joules (J) is

$$\begin{aligned} w(t) &= \int_{-\infty}^{t} p(\lambda)\, d\lambda = \int_{-\infty}^{t} v(\lambda)i(\lambda)\, d\lambda \\ &= \int_{-\infty}^{t} v(\lambda)C\left[\frac{d}{d\lambda}v(\lambda)\right] d\lambda \\ &= \int_{v(-\infty)}^{v(t)} Cv(\lambda)\, dv(\lambda) = \frac{1}{2}Cv^2(\lambda)\Bigg|_{v(-\infty)=0}^{v(t)} \end{aligned} \tag{2.30}$$

which gives

$$w(t) = \frac{1}{2}Cv^2(t) \tag{2.31}$$

Because C is greater than zero, it follows that

$$w \geq 0 \tag{2.32}$$

Therefore capacitance is a passive component.

An example of the relation between the current through, the voltage across, the power supplied to, and the energy stored in a 2-F capacitance is shown in Fig. 2.25.

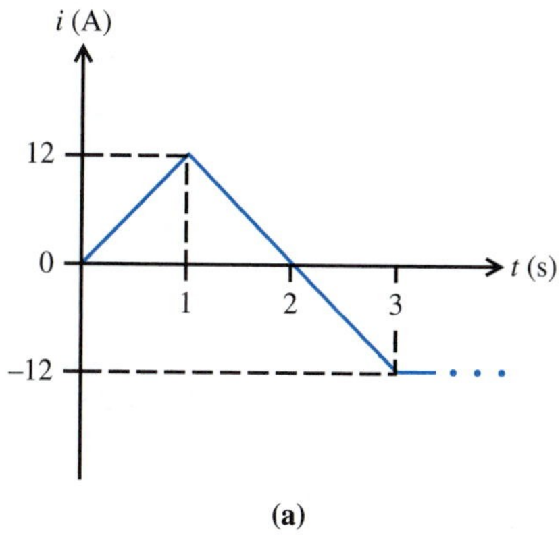

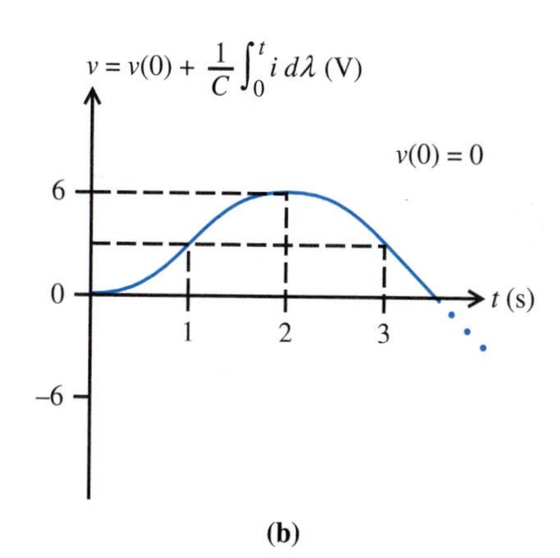

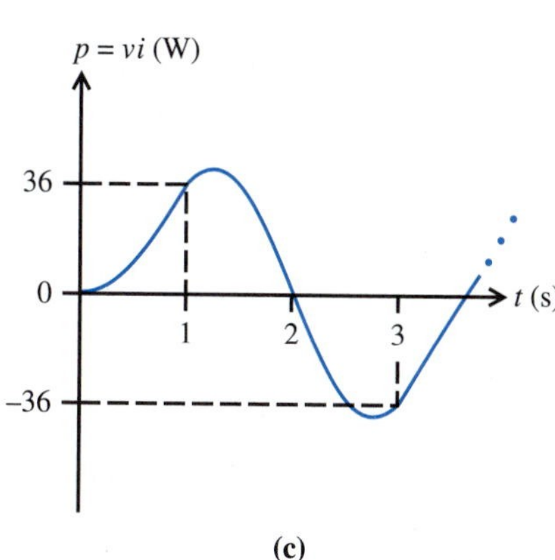

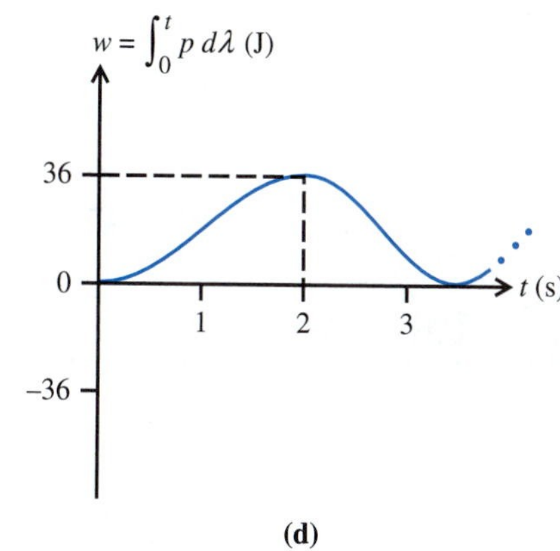

FIGURE 2.25
An example of the relation between current, voltage, power supplied to, and energy stored by a 2-F capacitance. (*a*) current; (*b*) voltage; (*c*) power; (*d*) energy

(This figure assumes that voltage v equals zero when t equals zero.) Note that the voltage and current waveforms for the capacitance do not have the same shape. The instantaneous power supplied to the capacitance can be either positive or negative. Power supplied is negative when the capacitance returns stored energy to the rest of the circuit. The energy stored can never be negative.

Capacitors with values of capacitance between a few picofarads and a few thousand microfarads are regularly encountered in electronic systems. Most capacitors have a capacitance that is very nearly independent of voltage and current. These find wide use in electronics and power systems. Some practical devices make use of voltage-dependent capacitance. One such device, the varactor diode, can be used for electronic tuning in radios, where it replaces a mechanically adjusted capacitor.

Remember

Capacitance is our network model that accounts for energy stored in an electric field and for current induced by changing voltage.

EXERCISES

16. Assume the passive sign convention for the relation between v and i. If $C = 0.01$ F and $v(t) = 100 \cos 2\pi 60t$ V, determine and sketch the following quantities (the quantity $2\pi 60t$ is in radians): $i(t)$, $p(t)$, and $w(t)$.
answer: $-120\ \pi \sin 2\pi 60t$ A, $-18{,}850 \sin 2\pi 120t$ W, $25 + 25 \cos 2\pi 120t$ J

17. For the network shown on the next page, $v_a = 24e^{-3t}$ V and $v_b = 2e^{-3t}$ V. Determine the following quantities for $t > 0$: v_1, i_1, i_2, and i_3.

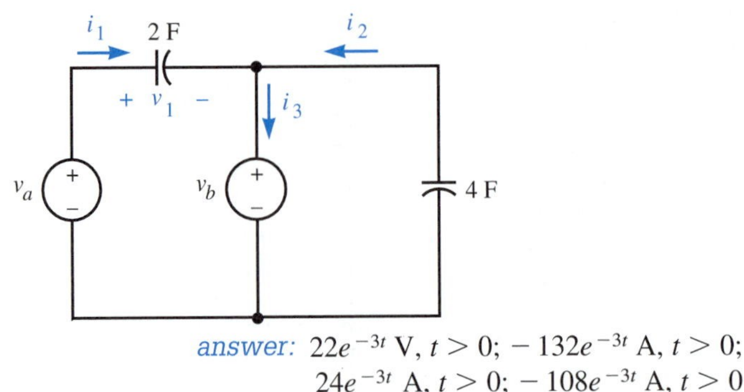

answer: $22e^{-3t}$ V, $t > 0$; $-132e^{-3t}$ A, $t > 0$; $24e^{-3t}$ A, $t > 0$; $-108e^{-3t}$ A, $t > 0$

18. Show that for the active sign convention,

$$v_{ab}(t) = v_{ab}(t_0) - \frac{1}{C}\int_{t_0}^{t} i_{ba}\, d\lambda$$

answer: Integrate both sides of Eq. (2.23)

2.7 Inductance

When we pick up a paper clip with a permanent magnet, we use the energy stored in the magnetic field to lift the paper clip. Current flowing through a wire also creates a magnetic field, and therefore magnetic flux, ϕ, around the wire. Magnetic fields always contain energy.

A physical device designed to store energy in a magnetic field is called an *inductor,* a choke, or occasionally a coil. We make extensive use of these devices in such diverse equipment as transformers, radios, and radar. An automobile ignition coil, for example, uses an inductor. We can physically make an inductor by winding a coil of wire around some supporting structure or form, such as plastic or iron. The physical device is depicted as connected to an electrical system in Fig. 2.26. Current i causes magnetic flux to link (pass through) each turn of the coil. As indicated in the figure, not all flux will link every turn. With resistance neglected, Faraday's law gives

$$v = N\frac{d}{dt}\phi \tag{2.33}$$

where ϕ is the average magnetic flux linking a turn and N is the number of turns. For coil forms constructed of linear magnetic materials, such as plastic, the ratio of current i to magnetic flux ϕ is constant. We define the constant L:

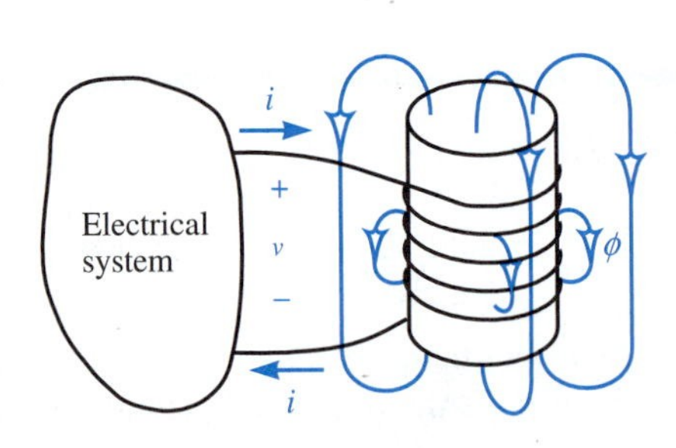

FIGURE 2.26
An N-turn inductor or coil

$$L = \frac{N\phi}{i} \tag{2.34}$$

This permits us to write voltage v in terms of current i:

$$v = N\frac{d}{dt}\phi = \frac{d}{dt}\left[\frac{N\phi}{i}i\right]$$

$$= \frac{d}{dt}(Li)$$

$$= L\frac{d}{dt}i \tag{2.35}$$

The constant of proportionality L is the *inductance* of the coil.

We model energy storage in a magnetic field with a network component called an *inductance* or *ideal inductor*.† All voltages created by a changing current are assumed to appear across such components.

DEFINITION Inductance

An **inductance** is a two-terminal network component with voltage v_{ab} proportional to the time derivative of the current i_{ab}. The constant of proportionality L is also called the inductance. The terminal equation is

$$v_{ab} = L\frac{d}{dt}i_{ab} \tag{2.36}$$

and the component symbol is shown in Fig. 2.27.

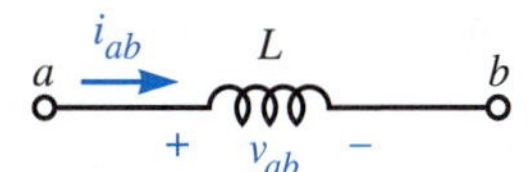

FIGURE 2.27 Inductance symbol

Inductance L has the dimension of volt-second per ampere, or henry (H). Observe that Eq. (2.36) implies the passive sign convention. With the active sign convention, Eq. (2.36) gives us

$$v_{ab} = L\frac{d}{dt}i_{ab} = L\frac{d}{dt}(-i_{ba}) = -L\frac{d}{dt}i_{ba} \tag{2.37}$$

EXAMPLE 2.14 Voltages and currents are assigned according to the passive sign convention. The current through a 3-mH inductance is

$$i(t) = 24 \cos 40t \text{ A}$$

Calculate the inductance voltage v.

† Some texts refer to both the practical device and the network component as an inductor and use the word inductance to denote only the constant of proportionality L.

Solution

$$v = L\frac{d}{dt}i = 3 \times 10^{-3}\frac{d}{dt}24\cos 40t$$
$$= -2.88 \sin 40t \text{ V}$$

Integration of Eq. (2.36) with respect to t gives

$$\int_{i_{ab}(-\infty)}^{i_{ab}(t)} di_{ab} = \frac{1}{L}\int_{-\infty}^{t} v_{ab}\, d\lambda \tag{2.38}$$

Completion of the integration on the left-hand side gives us

$$i_{ab}(t) - i_{ab}(-\infty) = \frac{1}{L}\int_{-\infty}^{t} v_{ab}\, d\lambda \tag{2.39}$$

With the assumption that inductance current is zero at time minus infinity,

$$i_{ab}(t) = \frac{1}{L}\int_{-\infty}^{t} v_{ab}\, d\lambda \tag{2.40}$$

We can write Eq. (2.40) as

$$i_{ab}(t) = \frac{1}{L}\int_{-\infty}^{t_0} v_{ab}\, d\lambda + \frac{1}{L}\int_{t_0}^{t} v_{ab}\, d\lambda \tag{2.41}$$

which is of the form

$$i_{ab}(t) = i_{ab}(t_0) + \frac{1}{L}\int_{t_0}^{t} v_{ab}\, d\lambda \tag{2.42}$$

We often call the current

$$i_{ab}(t_0) = \frac{1}{L}\int_{-\infty}^{t_0} v_{ab}\, d\lambda \tag{2.43}$$

the *initial condition* at time t_0.

EXAMPLE 2.15 Voltages and currents are assigned according to the passive sign convention. The current through a 3-mH inductance at $t = 0$ is $i(0) = 4$ A. The inductance voltage is

$$v(t) = 24\cos 4000t \text{ V} \qquad \text{for } t \geq 0$$

Determine $i(t)$ for $t \geq 0$.

Solution

$$i(t) = i(0) + \frac{1}{L}\int_{0}^{t} v\, d\lambda$$
$$= 4 + \frac{1}{3 \times 10^{-3}}\int_{0}^{t} 24\cos 4000\lambda\, d\lambda$$
$$= 4 + 2\sin 4000t \text{ A} \qquad \text{for } t \geq 0$$

As contrasted with capacitance, which accounts for energy stored in an electric field, inductance is our network model that accounts for energy stored in a magnetic field. An inductance does not dissipate electrical energy in the form of heat as does a resistance. It only stores energy in a magnetic field as the magnitude of the current increases. It returns this energy to the rest of the circuit as the magnitude of the current decreases. The energy stored is

$$\begin{aligned} w(t) &= \int_{-\infty}^{t} p(\lambda)\, d\lambda = \int_{-\infty}^{t} i(\lambda)v(\lambda)\, d\lambda \\ &= \int_{-\infty}^{t} i(\lambda)L\left[\frac{d}{d\lambda} i(\lambda)\right] d\lambda \\ &= \int_{i(-\infty)}^{i(t)} Li(\lambda)\, di(\lambda) = \frac{1}{2} Li^2(\lambda)\Bigg|_{i(-\infty)=0}^{i(t)} \end{aligned} \tag{2.44}$$

which gives

$$w(t) = \frac{1}{2} Li^2(t) \tag{2.45}$$

Because inductance L is greater than zero, it follows that

$$w \geq 0 \tag{2.46}$$

Therefore inductance is a passive component.

An example of the relation between the voltage across, the current through, the power supplied to, and the energy stored in a 2-H inductance is shown in Fig. 2.28. Note

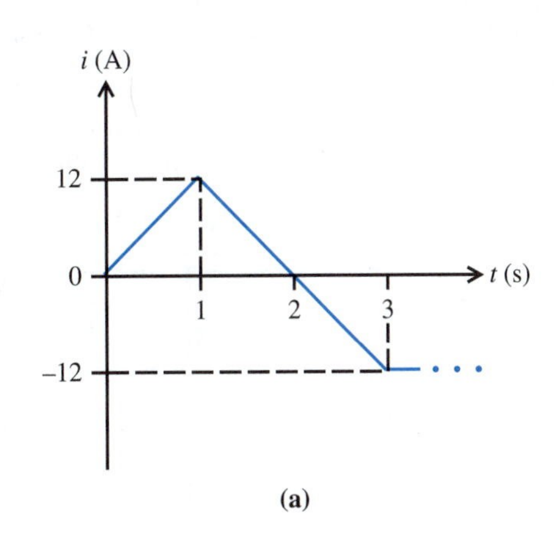

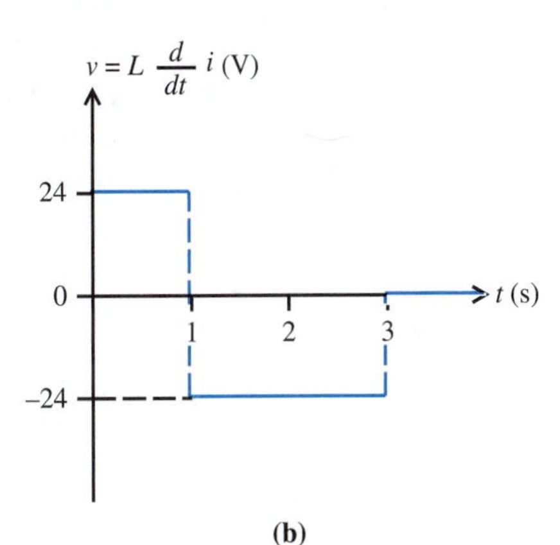

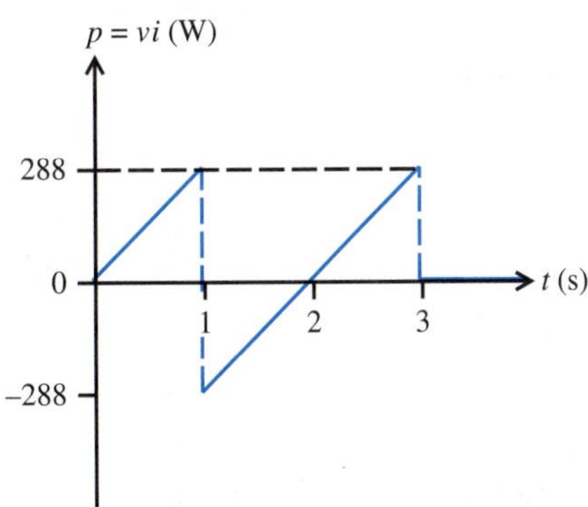

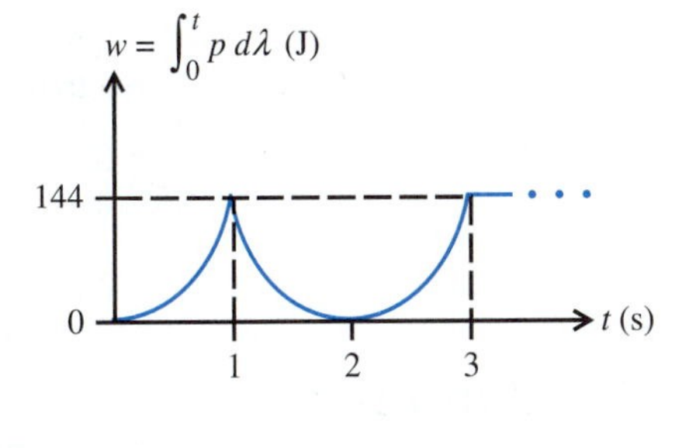

FIGURE 2.28
An example of the relation between current, voltage, power supplied to, and energy stored by a 2-H inductance. (*a*) current; (*b*) voltage; (*c*) power; (*d*) energy

that the voltage and current waveforms for an inductance do not have the same shape. The power supplied to the inductance can be either positive or negative. Power supplied is negative when the inductance returns stored energy to the rest of the circuit. The energy stored can never be negative.

Inductors with inductance values between a few microhenries and several henries are regularly found in physical systems. Inductors used in communication equipment, such as radios and television, are often wound on plastic forms. We can accurately model such devices with an inductance that is independent of current.

We cannot accurately model a coil wound on ferromagnetic material, such as iron, by a fixed inductance. Nevertheless, a constant inductance model often provides a useful approximation for iron-cored transformers and some other devices.

Examples of devices that use nonlinear magnetic material are the iron-core transformer and the electric motor. Devices that intentionally exploit the nonlinear magnetization of ferromagnetic material are the magnetic core memory of a computer and a device called a magnetic amplifier. Although they were once common, magnetic core memories and magnetic amplifiers are seldom encountered in modern equipment, having been replaced by less expensive semiconductor devices such as transistors and thyristors.

Remember Inductance is our network model that accounts for energy stored in a magnetic field and for voltage induced by a changing current.

EXERCISES

19. Assume the passive sign convention for the relation between v and i. If $L = 10$ mH and $i(t) = 100 \cos 2\pi 60t$ A, determine the following quantities. (The quantity $2\pi 60t$ is in radians.): $v(t)$, $p(t)$, and $w(t)$.
    answer: $-120\ \pi \sin 2\pi 60t$ V, $-18{,}850 \sin 2\pi 120t$ W, $25 + 25 \cos 2\pi 120t$ J
20. For the network shown below, $i_a = 24e^{-3t}$ A and $i_b = 2e^{-3t}$ A for $t \geq 0$. Determine the following quantities for $t > 0$: i_1, i_2, v_1, and v_2.
    answer: $22e^{-3t}$ V, $t > 0$; $2e^{-3t}$ A, $t > 0$; $-132e^{-3t}$ V, $t > 0$; $24e^{-3t}$ V, $t > 0$

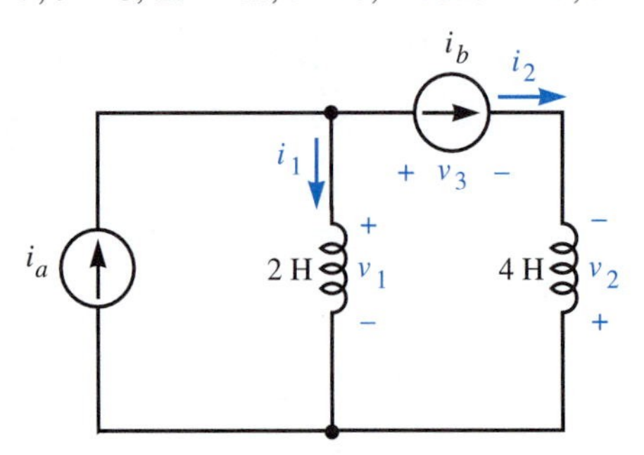

21. Show that for the active sign convention,

$$i_{ab}(t) = i_{ab}(t_0) - \frac{1}{L}\int_{t_0}^{t} v_{ba}\, d\lambda$$

answer: Integrate both sides of Eq. (2.37).

2.8 Summary

We introduced Kirchhoff's current law and Kirchhoff's voltage law as the two axioms of network theory:

TABLE 2.1
Physical properties modeled by network components

Network Component	Physical Property Modeled
Independent source	Source of electrical energy
Dependent source	Control of electrical energy
Resistance	Conversion of electrical energy to heat
Capacitance	Storage of energy in an electric field
Inductance	Storage of energy in a magnetic field

KCL: The sum of the currents leaving any closed surface is zero.
KVL: The sum of the voltage drops around any closed path is zero.

These axioms are based on the assumption that the network does not radiate electromagnetic energy.

We presented definitions relating to network theory and introduced the network components as ideal models for physical properties of practical devices. Table 2.1 contains physical properties modeled by network components. The equations describing these components, in conjunction with KCL and KVL, are the basis of network analysis (we will introduce three additional models in later chapters: the ideal op amp in Chapter 6 and mutual inductance and ideal transformers in Chapter 18).

The following Key Facts provides a convenient list of many definitions used in circuit analysis. While you need not memorize the list, you must read it *carefully* to be sure that you understand these definitions.

KEY FACTS

	Concept	Equation	Section	Page
❑	*KCL:* The sum of the currents leaving any closed surface is zero.	(2.1)	2.1	17
❑	A network component and its associated line segments is a *branch.*		2.1	18
❑	The termination of one or more branches is a *node.*		2.1	19
❑	A line segment connected between two points is a *short circuit.* The voltage across a short circuit is always zero.		2.1	19
❑	All points connected to each other by short circuits form a single node.		2.1	19
❑	*Parallel* branches share two nodes and have the same voltage.		2.1	19
❑	A path that does not follow a branch or a line is an *open circuit.* The current through an open circuit is always zero.		2.1	19
❑	A node that terminates three or more branches is a *junction.*		2.1	19
❑	*Series* branches are connected end to end with no junctions at the connecting nodes and have the same current.		2.1	19

KEY FACTS

Concept	Equation	Section	Page
❑ *KVL:* The sum of the voltage drops around any closed path is zero.	(2.2)	2.2	22
❑ A *loop* is a closed path of branches that does not pass through the same node twice.		2.2	22
❑ We can draw the circuit diagram of a *planar* network so that lines cross only where connected.		2.2	22
❑ A *mesh* is a loop that does not enclose another loop. We define meshes only for planar networks.		2.2	22
❑ Voltage across an *independent voltage source* does not depend on the current through it.	(2.3)	2.3	24
❑ A zero-volt voltage source is equivalent to a *short circuit.*		2.3	24
❑ Current through an *independent current source* does not depend on the voltage across it.	(2.4)	2.3	25
❑ A zero-ampere current source is equivalent to an *open circuit.*		2.3	26
❑ Voltage across a *dependent voltage source* is determined by the control equation.	(2.5) (2.7)	2.4	28
❑ Current through a *dependent current source* is determined by the control equation.	(2.6) (2.8)	2.4	29 31
❑ *Resistance* voltage is proportional to current.	(2.10)	2.5	34
❑ *Conductance* is the reciprocal of resistance.	(2.12)	2.5	35
❑ *Conductance* current is proportional to voltage.	(2.13)	2.5	35
❑ Power absorbed by resistance is proportional to current squared.	(2.15)	2.5	35
❑ Power absorbed by resistance is proportional to voltage squared.	(2.16)	2.5	35
❑ *Capacitance* current is proportional to the derivative of voltage.	(2.22)	2.6	40
❑ *Capacitance* voltage is proportional to the integral of current.	(2.26)	2.6	41
❑ The energy stored in capacitance is proportional to the voltage squared.	(2.31)	2.6	42
❑ *Inductance* voltage is proportional to the derivative of current.	(2.36)	2.7	45
❑ *Inductance* current is proportional to the integral of voltage.	(2.40)	2.7	46
❑ Energy stored in inductance is proportional to the current squared.	(2.45)	2.7	47

PROBLEMS

Section 2.1

1. For the network shown below, $i_1 = 2$ A, $i_2 = 8$ A, and $i_3 = 5$ A. Determine currents i_{12}, i_{21}, i_{32}, i_{24} and i_{05}.

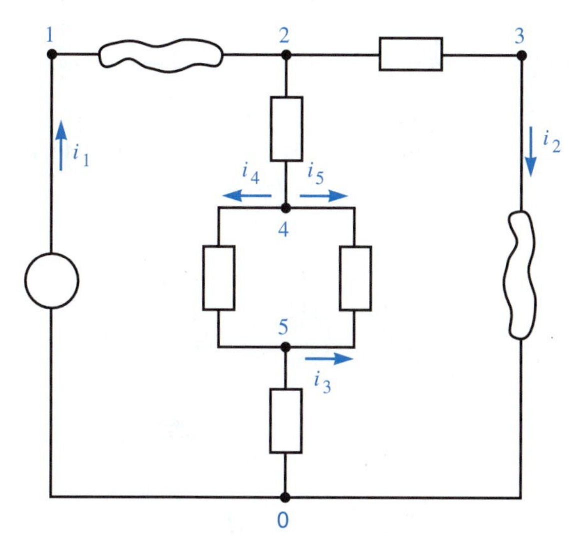

2. Determine currents i_4 and i_5 for the network described in Problem 1.

3. (a) For the network shown below, determine currents i_x and i_y.

   (b) Determine currents i_{30}, i_{14}, and i_{40}.

   (c) Is the network planar?

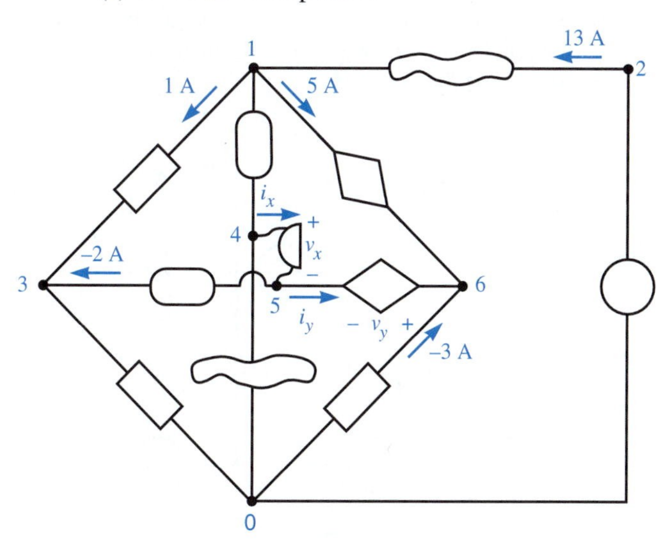

4. Determine currents i_x, i_y, and i_z for the following circuit.

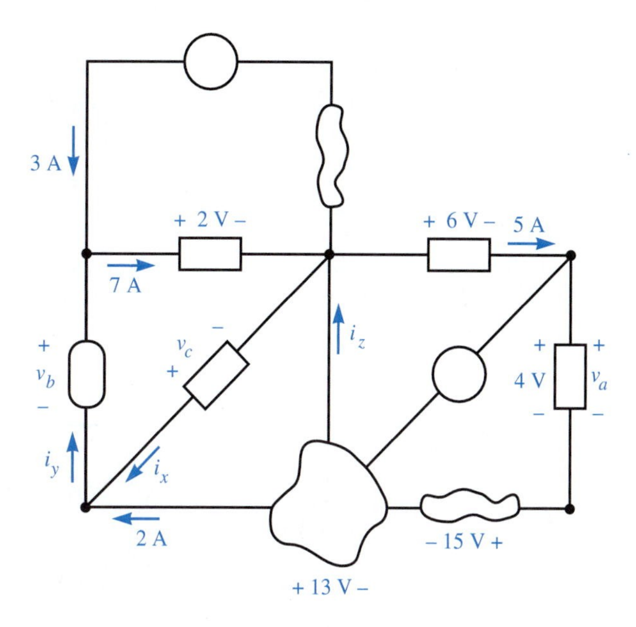

Section 2.2

5. For the network shown below, $v_{10} = 1$ V, $v_{20} = 2$ V, $v_{30} = 3$ V, $v_{40} = 4$ V, $v_{50} = 5$ V, and $v_{60} = 6$ V. Determine the specified voltages: v_{12}, v_{21}, v_{34}, v_{43}, v_{41}, v_{15}, v_{45}, v_{16}, v_{62}, v_{53}, v_{64} and v_{03}.

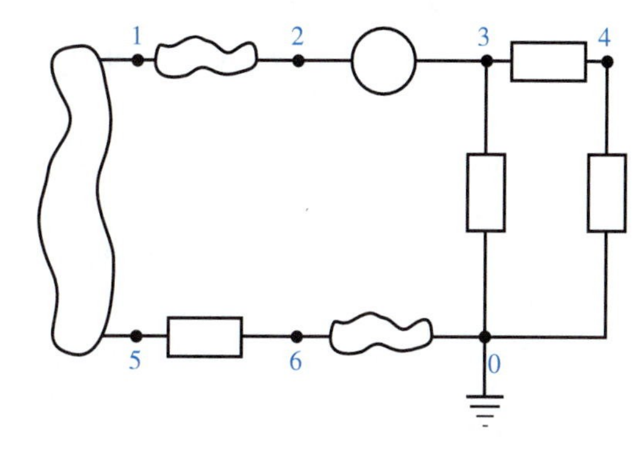

6. For the network used in Problem 4, find v_a, v_b, and v_c.

7. For the network used in Problem 3 if $v_{10} = 6$ V, $v_{20} = 8$ V, $v_{30} = 11$ V, $v_{40} = 15$ V, $v_{50} = 20$ V, and $v_{60} = 26$ V, determine voltages v_{13}, v_{15}, v_{52}, v_{24}, v_x, and v_y.

Section 2.4

8. For the following network, find i_x, v_{ab}, i_{ab}, and the power absorbed by the 15-V voltage source.

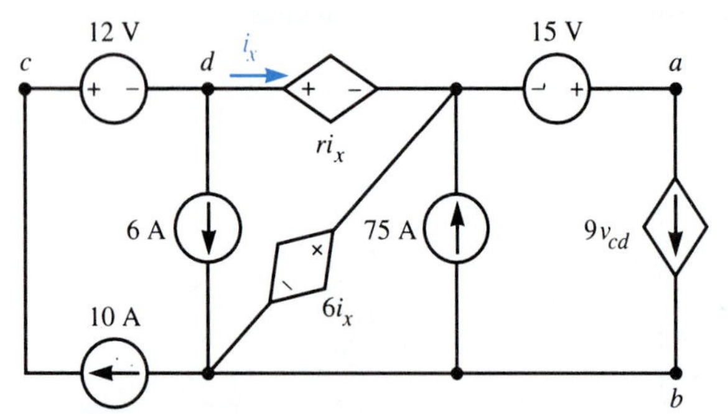

9. For the following network, determine i_1, v_1, i_x, v_x, and v_{ab}, if no other components are connected to terminals a and b.

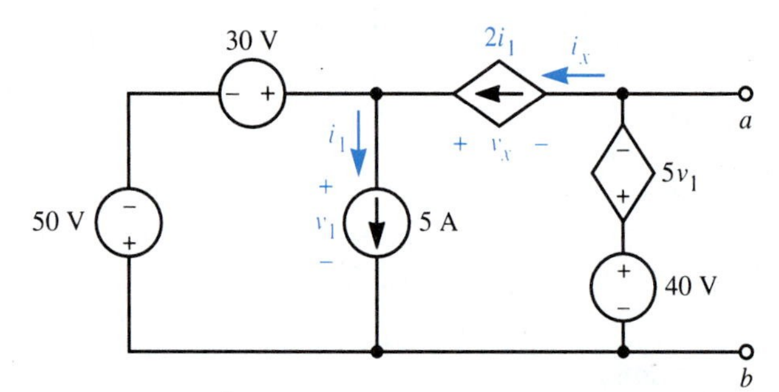

10. Determine the output voltage v_o as a function of the input voltage v_i in the following amplifier circuit. What is the output voltage if the VCVS has a gain A of 100,000? Do you see why this circuit is called a voltage follower?

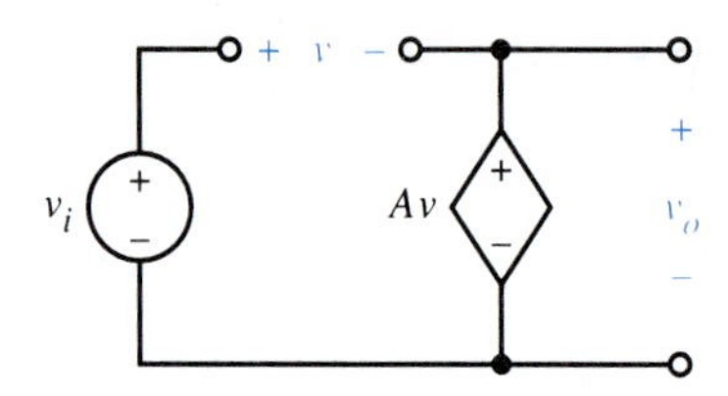

Section 2.5

11. For the network below, find the total energy absorbed by the load when $i = 0$ for $t < 0$, $i = 5e^{-2t}$ A for $t \geq 0$, and $v_s = 10$ V.

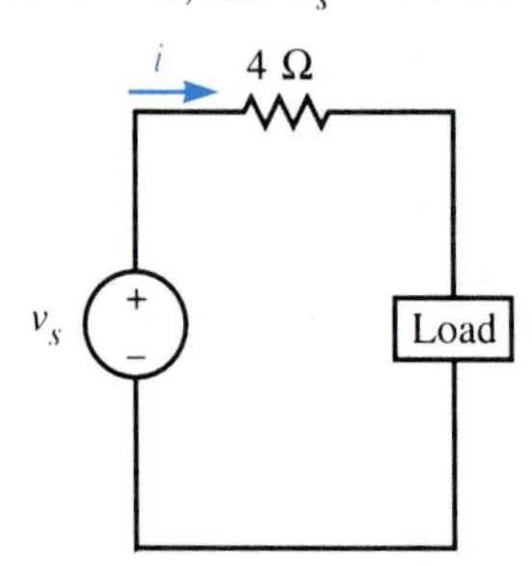

12. Given the following network, find i_x, v_1, i_{ba}, and the power absorbed by the $-$ 30-V source.

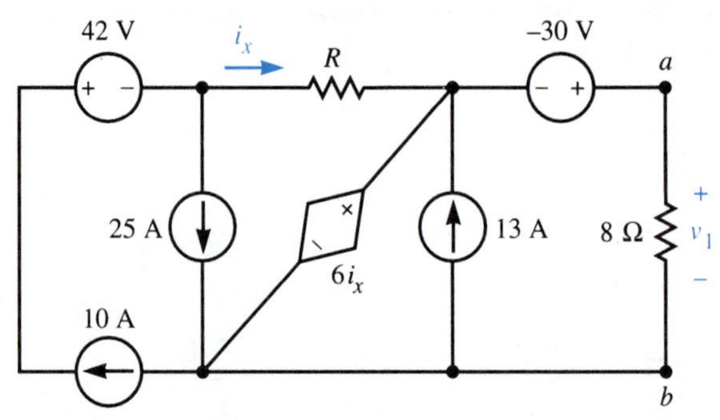

13. The voltage $v_x = 9$ V and the current $i_y = -1$ A have been measured for the network below. Find v_{10}, v_{20}, v_{30}, v_{40}, i_1, i_2, i_3, and i_4.

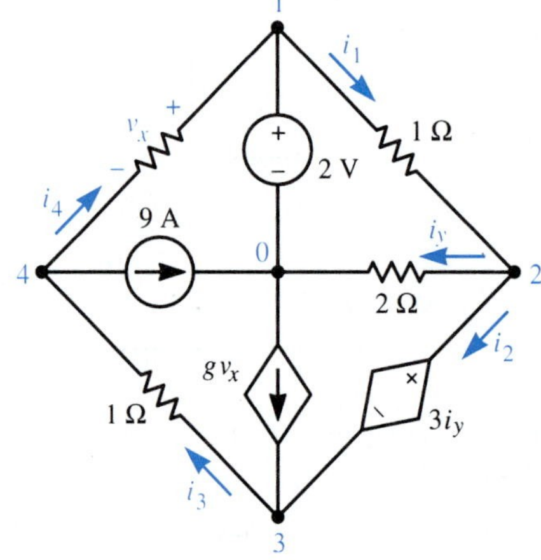

14. For the network above, currents $i_1 = -2$ A, $i_2 = -4$ A, $i_3 = -1$ A, and $i_4 = -10$ A have been measured. Find v_{10}, v_{20}, v_{30}, and v_{40}.

15. Part of a network is shown below. In the most convenient order for you, determine (a) v_x; (b) v_z; (c) i_y; (d) i_a; and (e) R.

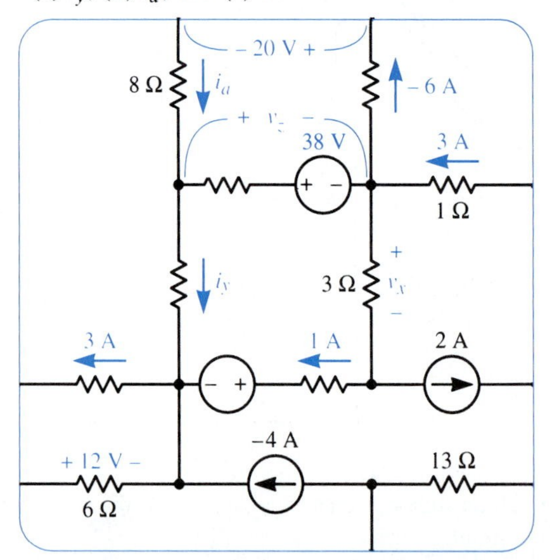

16. Part of a network is shown below. In the most convenient order for you, determine (a) v_z; (b) v_x; (c) i_y; (d) i_a; and (e) R.

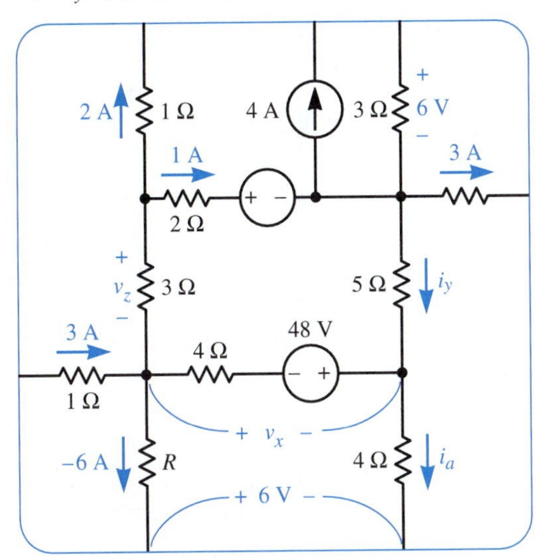

Section 2.6

17. Determine the capacitance current i_{ab} for $t > 0$ if the capacitance is 5 μF and the voltage v_{ab} in volts is as given below for $t \geq 0$.

(a) 12 (b) $12e^{-4000t}$ (c) 50 cos $300t$

18. Determine the capacitance voltage v_{ab} for $t \geq 0$ if the capacitance is 6 μF, $v_{ab}(0)$ is 6 V, and current i_{ab} in amperes is as given below for $t \geq 0$.

(a) 12 (b) $0.36e^{-8000t}$ (c) 0.48 cos $2000t$

19. Capacitance current i_{ab} for a $\frac{1}{2}$-F capacitance is as shown below. If $v(0) = -4$ V, plot v_{ab} for $0 \leq t \leq 4$ s.

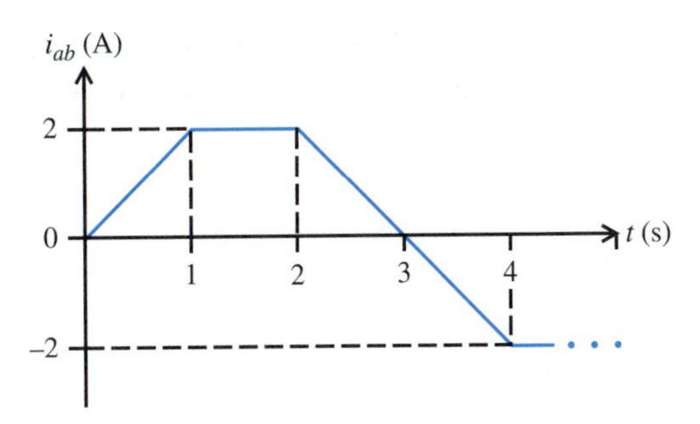

20. If the ordinate in the above graph is changed to v_{ab} in volts, where v_{ab} is the voltage across a $\frac{1}{2}$-F capacitance, plot the current i_{ab} through the capacitance for $0 \leq t \leq 4$ s.

21. For the following network, determine current i_{ab} and voltage v_{cd} for $t > 0$. The source voltage is $v_s = 12$ cos $200t$ V for $t \geq 0$, and $v_{cd}(0)$ is 3 V.

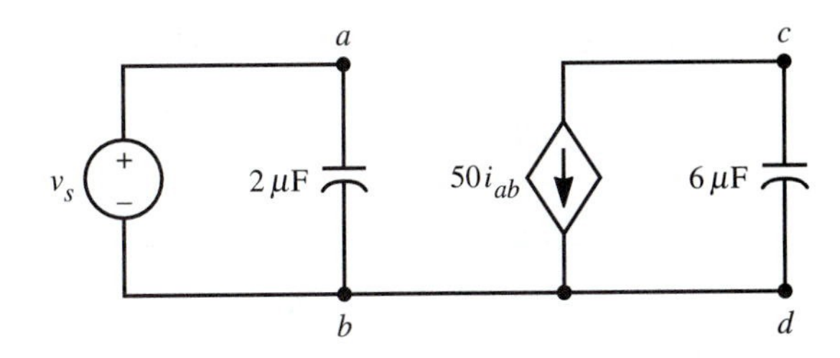

Section 2.7

22. Determine the inductance voltage v_{ab} for $t > 0$ if the inductance is 5 mH and the current i_{ab} in amperes is as given below for $t \geq 0$.

(a) 12 (b) $13e^{-200t}$ (c) 20 cos $(2\pi 60t)$

23. Determine the inductance current i_{ab} for $t \geq 0$ if the inductance is 18 mH, $i_{ab}(0)$ is 2 A, and voltage v_{ab} in volts is as given below for $t \geq 0$.

(a) 72 (b) $18e^{-400t}$ (c) 40 cos $90t$

24. The current i_{ab} through a 2-H inductance is as given in Problem 19. Plot the voltage v_{ab} across the inductance for $0 \leq t \leq 4$ s.

25. Change the label on the ordinate of the graph shown in Problem 19 to to v_{ab} volts. If v_{ab} is the voltage across a 2-H inductance and $i_{ab}(0) =$ 2 A, plot i_{ab} for $0 \leq t \leq 4$ s.

26. For the network shown below, determine voltage v_{ab} and current i_{cd} for $t > 0$. The source current is $i_s = 12e^{-2t}$ A for $t \geq 0$, and $i_{cd}(0)$ is 2 A.

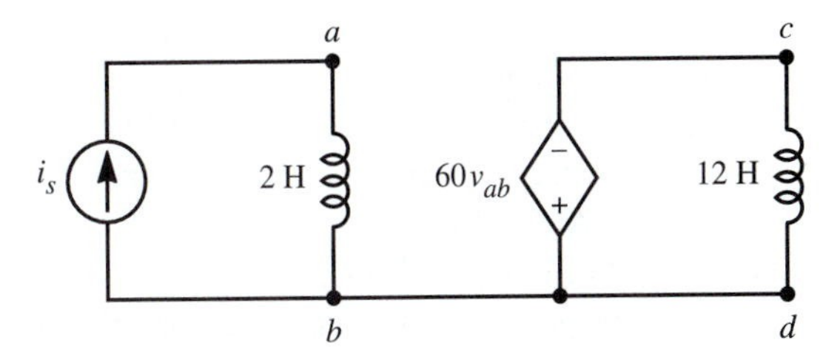

Supplementary Problems

27. Determine voltages v_1 and v_2 in the following circuit.

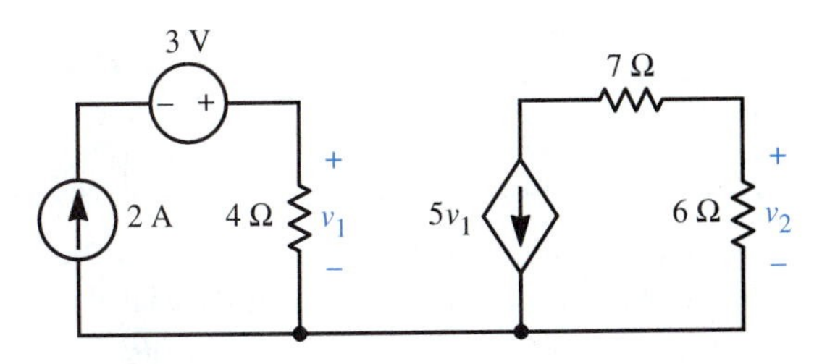

28. Select the current gain β in the following amplifier circuit so that 16 W of output power is delivered to the 4-Ω resistance. Calculate power deliv-

ered to the circuit by the 2-V source. The ratio of this output power to the input power is the *power gain* of the amplifier.

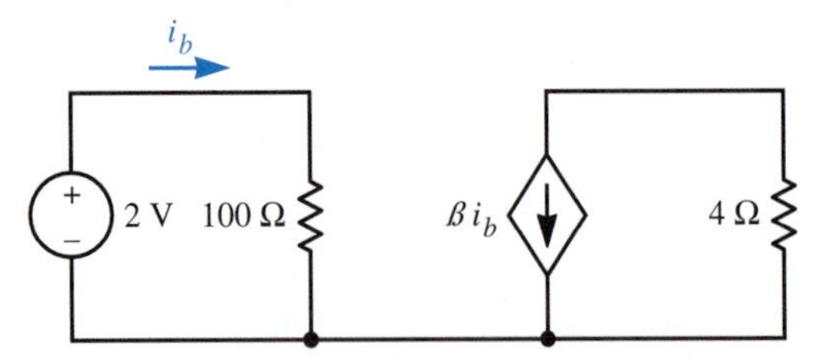

29. Measurements on a two-terminal network device indicate that

$$v_{ab} = 10 + 5i_{ab}$$

(a) Construct a network model for the device. Use two network components in series.

(b) Construct a network model with two network components in parallel that has the terminal equation given.

30. A four-terminal device has the characteristics $i_{ab} = 0$, $v_{bd} = 0$, and $v_{cd} = -10v_{ab} + 2i_{cd}$. Use network components to construct a model.

31. An electric water heater must raise the temperature of 30 gal of water from 10°C to 70°C in 1 h. The supply voltage to the heater is 240 V. Neglect all heat losses and assume that the heater is a resistance. Determine the power rating of the heater and the resistance of the heating element. What is the electric current through the heater? At 12 cents per kWh, how much does it cost to heat the 30 gal of water? (Approximately 15,800 J is required to raise the temperature of one gallon of water one degree Celsius.)

32. The unknown devices shown in the figure below absorb the indicated powers. Determine the unknown current i_x.

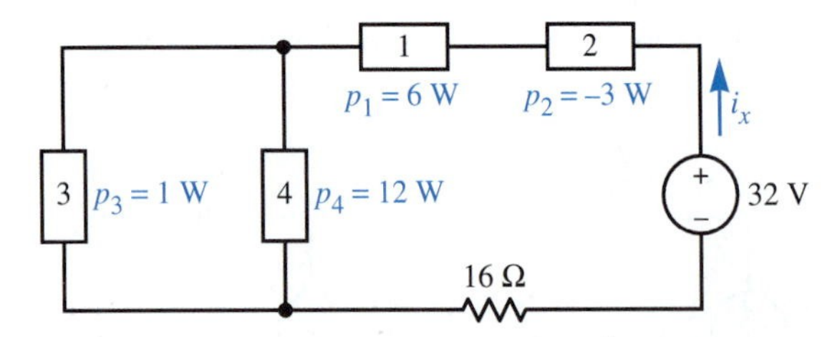

33. In the preceding circuit current i_x is known to be 1 A. What is the value of the resistance labeled 3?

34. Determine the output voltage v_o as a function of the input voltage v_i in the following amplifier circuit. What is the output voltage if A is 100,000? What is the output voltage if A is 10,000? Does the value of A have much effect on the output voltage as long as A is very large?

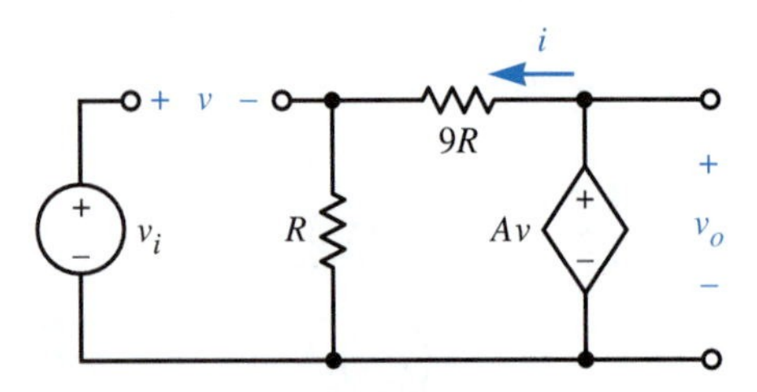

35. In the amplifier circuit of the preceding problem, replace the resistance of value $9R$ with a resistance of value αR. Assume that $\alpha \ll A$. Select α so that a 10-mV input voltage v_i gives an output voltage $v_o \cong 1$ V.

36. For the network below, determine i_{45}, i_{30}, i_{01}, i_{12}, i_{23}, v_{10}, v_{12}, v_{20}, v_{43}, and v_{53} for $t > 0$. For $t \geq 0$ the source values in amperes are $i_a = 10t^2$, $i_b = 5t^2$, and $i_c = 2t^2$. The initial capacitance voltage is $v_{12} = 5$ V.

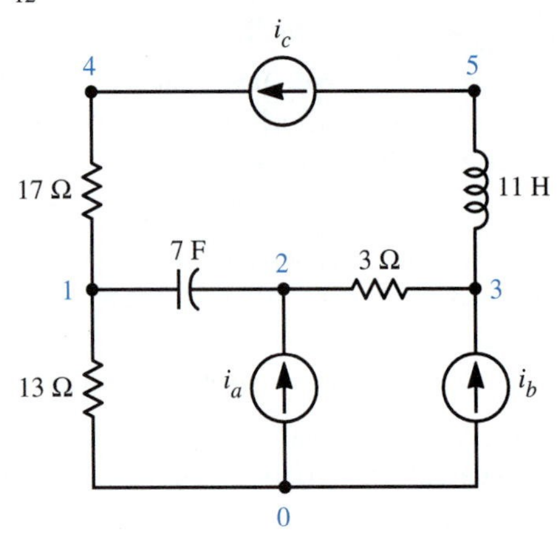

37. For the network shown below, determine v_{ab}, v_1, i_1, and the power absorbed by the dependent voltage source.

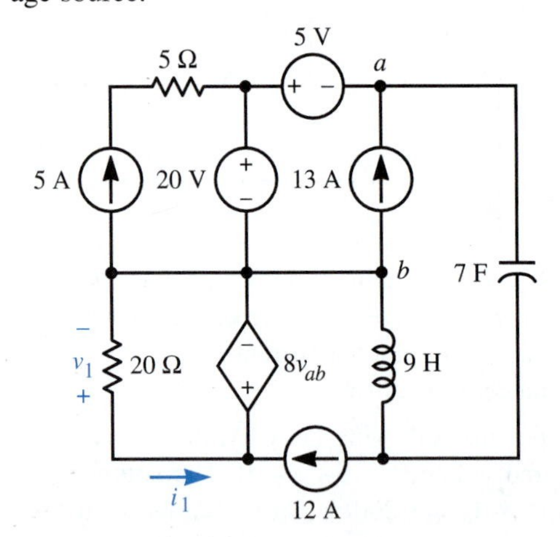

38. For the network below, determine v_{10}, v_{20}, v_{30}, v_{12}, v_{23}, v_{34}, i_x, i_y, i_z, and i_w for $t > 0$. The initial condition is $i_z(0) = 5$ A. The source voltages for $t \geq 0$ are $v_a = 2e^{-2t}$ V, $v_b = 7e^{-2t}$ V, and $v_c = 3e^{-2t}$ V.

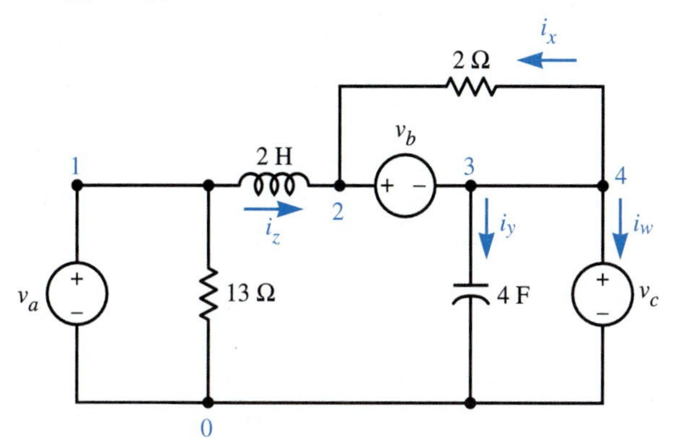

39. Determine i_{10}, i_{20}, i_{23}, v_{20}, v_{12}, v_{30}, and v_{34} for $t > 0$ for the network shown below. The sources in amperes for $t \geq 0$ are $i_a = 6e^{-2t}$, $i_b = 16e^{-2t}$, and $i_c = 7e^{-2t}$. The initial capacitance voltage is $v_c(0) = 10$ V.

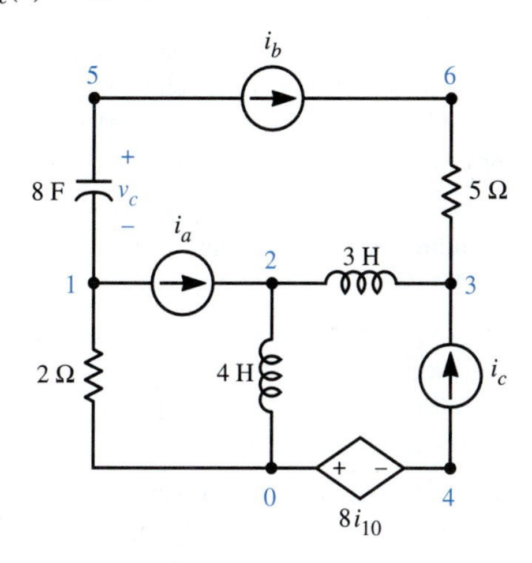

40. A two-terminal device is found to have the volt-ampere characteristics

$$v_{ab} = 5i_{ab} + 10\frac{d}{dt}i_{ab}$$

Construct a network model for the device. Use two network components in series.

41. A two-terminal device has the volt-ampere characteristics

$$i_{12} = 0.1v_{12} + 5\frac{d}{dt}v_{12}$$

Construct a network model for this device that consists of two components in parallel.

42. In the following circuit write a KVL equation in terms of voltage v and current i. This gives an equation of the form $v = R_{eq}i$. What is the equivalent resistance R_{eq}?

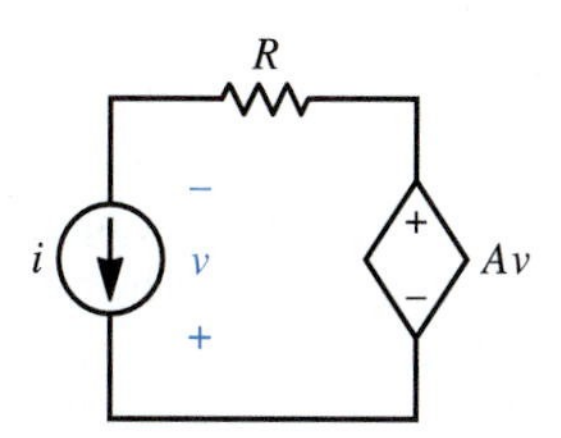

43. Replace the resistance in the preceding circuit with a capacitance of value C. Write a KVL equation in terms of voltage v and the integral of current i. Differentiate this equation to obtain an equation of the form $i = C_{eq}\, dv/dt$. What is the equivalent capacitance C_{eq}? This ability of an amplifier to increase the apparent value of a capacitance is called the *Miller effect.*

44. Replace the resistance in the circuit of Problem 42 with an inductance of value L. Use KVL to obtain an equation of the form $v = L_{eq}\, di/dt$ and give the value of the equivalent inductance L_{eq}.

45. The two dependent sources in the following circuit form one model for an *ideal transformer* (other models are discussed in Chapter 17). Calculate current i_a as a function of v_a, R, and n_1/n_2. The ratio v_a/i_a is the equivalent resistance that is connected to the independent source.

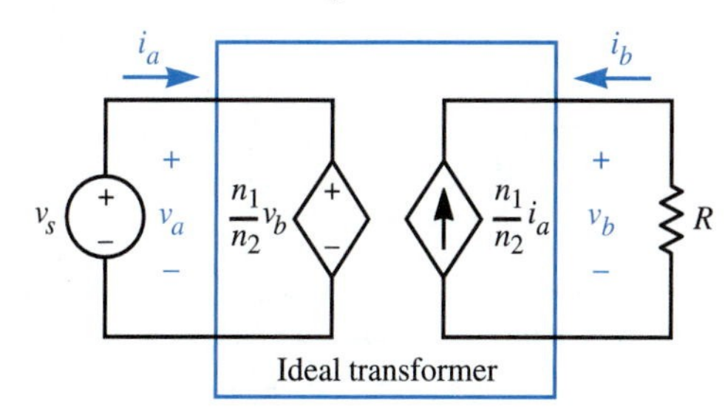

46. When energized, an automobile starter motor requires 200 A. This causes the terminal voltage of the battery to drop from 12 V to 10 V, and the voltage across the starter motor is 8 V.

(a) The battery can be modeled as a 12 V source in series with an internal resistance of value R_S. Determine R_S.

(b) What is the resistance of the wires that connect the battery to the starter motor?

(c) If the starter motor is 50 percent efficient, what is the output horsepower of the starter? (1hp = 745.7 W)

47. The open-circuit voltage of a transistor-radio battery is 9 V, and the battery terminal voltage is 8 V with a 40-Ω resistive load. Construct a network model for the battery that consists of a voltage source and a series resistance.

48. The audio amplifier for a public address system is modeled as a voltage-controlled voltage source (VCVS) that is controlled by the voltage across a 50-Ω resistance. The microphone supplies an average power of 50 μW to this resistance, and the VCVS must supply 25 W of average power to the 4-Ω speaker. What voltage gain μ is required for the VCVS?

49. A scanning electron microscope uses an electrostatically deflected electron beam. The voltage across the deflection plates must increase linearly from -150 V to $+150$ V in a time of 1 ms. The capacitance of the deflection plates is 6 pF. What current must be supplied to the deflection plates during the 1-ms scan interval?

50. A television receiver uses a cathode-ray tube (picture tube) with a magnetically deflected electron beam. The horizontal deflection coil is modeled as a 2-mH inductance.

    (a) The inductance current must increase linearly from -1 A to $+1$ A in a time of 53.5 μs. What must be the inductance voltage in this time interval?

    (b) For the retrace (return of the beam to the starting point), the current must decrease from $+1$ A to -1 A in 10 μs. What must be the average voltage across the inductance during retrace?

    (c) If the inductance voltage on retrace is not constant but varies as the first half of one cycle of a sine, what is the peak voltage?

51. A capacitor stores the energy required to generate one spark in an electronic ignition system for an automobile. The energy stored in the capacitor must be 0.32 J at 400 V. Assume that the capacitor is ideal. What capacitance must it have?

52. We must design a system to detonate the explosive lens used to form a missile nose cone. The system must store 50,000 J of energy in an inductor that conducts 100 A. What must be the inductance of the coil?

53. An old-style ignition system for a motorcycle uses a mechanical switch (ignition points) to interrupt the connection between the battery and the ignition coil. When the points are open, the coil is in series with a capacitor. Assume that the coil is an ideal inductance and that the capacitor is ideal with a capacitance of 0.2 μF. If all energy stored in the inductance is transferred to the capacitance and the capacitance reaches a maximum voltage of 400 V when the initial inductance current is 4 A, what is the value of the inductance?

54. A practical power system that relies heavily on solar cells must store large quantities of energy for use at night. A new mutual fund, Bay Area Research Fund, is asked to invest in a corporation that plans to store energy in large inductors made of superconducting wire. Each storage unit has a coil with an inductance of 1.44 kH and a maximum inductor current of 10 kA.

    (a) What is the maximum energy stored in the inductance in joules and in megawatt-hours? (1 MWh = 3.6×10^9 J)

    (b) How many hours can the inductor supply power to 250 residences if each residence has an average power consumption of 4 kW during the nighttime hours?

    (c) On the average, 50 percent of the maximum energy-storage capability of a unit will be sold to customers each day at a net yield to the mutual fund of one cent per kWh. What is the maximum amount that the mutual fund can afford to invest per unit if a yield of 10 percent per year is required?

3

Introduction to Circuit Analysis

In this chapter we use KCL, KVL, and the component terminal equations to determine the voltages and currents in some simple, but highly important, circuits. We first apply KCL to the analysis of the general *two-node* or *parallel circuit.* This analysis provides us with our first two-terminal *equivalent circuits.* The analysis also yields the *current divider* equation, which shows how the sum of currents through parallel conductances divides proportionally to the individual conductances. We next use KVL to analyze the general *single-loop* or *series circuit.* Analysis of the series circuit yields additional two-terminal equivalent circuits. This analysis also shows that the voltage across series resistances divides proportionally to the individual resistances. This is the *voltage divider* relation. We use the equivalent circuits we have developed to simplify and analyze circuits that initially appear complicated.

3.1 Two-Node or Parallel Circuits

Most equipment for industry, appliances for the house, and automotive accessories are designed to run on a specified voltage. For instance, automotive accessories are designed so that approximately 12 V should appear across each unit. Ideally they should be connected in *parallel.* An example of parallel-connected components is shown in Fig. 3.1.

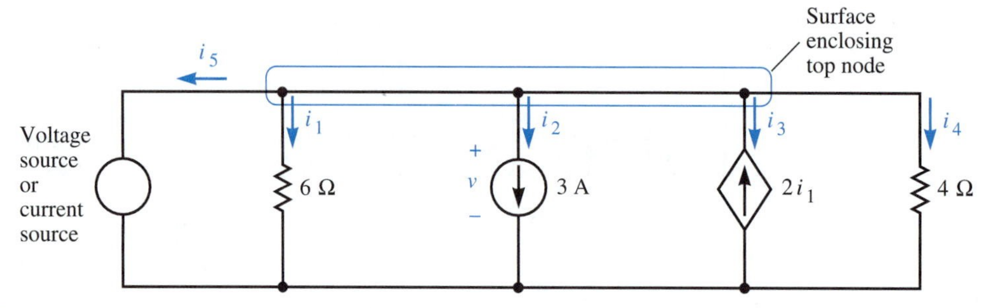

FIGURE 3.1
A parallel circuit

DEFINITION Parallel Components

Two or more components are in parallel if their terminals are connected to two common (shared) nodes.

As an alternative definition, *a set of components is connected in parallel if every combination of two components in the set forms a loop.* Application of KVL to a loop of any two of the components in parallel gives the following result:

At any specified time the same voltage v must appear across each component connected in parallel.

The *general procedure* for solving a two-node or parallel circuit problem is to *first find the voltage v between the nodes.* If this is specified by a voltage source, the problem is simple. If the voltage is not known, apply KCL to a surface enclosing one node, write all unknown currents in terms of the voltage v with the use of terminal or control equations, and solve for v. Once v is known, all currents are easily calculated. We will now demonstrate the procedure with three simple examples.

EXAMPLE 3.1 For the network of Fig. 3.1, the left-hand source is a 12-V voltage source with the (+) reference mark at the top of the figure. Determine the indicated currents.

Solution The first four component currents are

$$i_1 = \frac{1}{6} v = \frac{1}{6}(12) = 2 \text{ A}$$
$$i_2 = 3 \text{ A}$$
$$i_3 = -2i_1 = -2(2) = -4 \text{ A}$$
$$i_4 = \frac{1}{4} v = \frac{1}{4}(12) = 3 \text{ A}$$

We can now find the current through the voltage source from a KCL equation for a surface enclosing the top node (or the bottom node).

$$i_5 + i_1 + i_2 + i_3 + i_4 = 0$$

We can write this KCL equation directly in terms of v with the use of the component equations.

$$i_5 + \frac{1}{6} v + 3 - 2\left(\frac{1}{6} v\right) + \frac{1}{4} v = 0$$

With $v = 12$ V,

$$i_5 = -4 \text{ A}$$

If the voltage between the nodes is not specified by a voltage source, we can readily find the voltage by the use of KCL and terminal, source, and control equations.

EXAMPLE 3.2 For the circuit of Fig. 3.1, the left-hand source is a 12-A current source (reference arrow pointing up). Determine the voltage v.

Solution Use the component equations to write a KCL equation in terms of the voltage v for a surface enclosing the top node. This is the same as in the preceding example, except that current i_5 is now known and voltage v is unknown:

$$-12 + \frac{1}{6} v + 3 - 2\left(\frac{1}{6} v\right) + \frac{1}{4} v = 0$$

This gives

$$v = 108 \text{ V}$$

Once voltage v is known, the currents not already specified by the independent current sources are easily found.

It is no more difficult to write the necessary equations for a two-node circuit with resistance, capacitance, and inductance than it is for the resistive case.

EXAMPLE 3.3 Find the indicated currents for the network shown in Fig. 3.2, on page 60, if (a) the left-hand source is a voltage source; (b) the left-hand source is a current source.

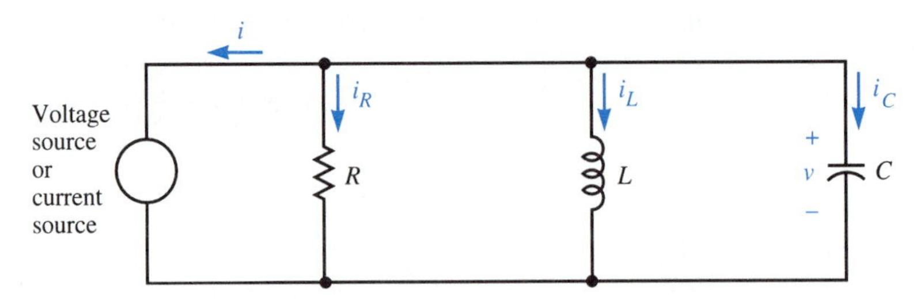

FIGURE 3.2
A parallel *RLC* circuit

Solution (a) The element currents are given by the terminal equations (the dependence of v and i on t is implicit):

$$i_R = \frac{1}{R} v$$

$$i_L = \frac{1}{L} \int_{-\infty}^{t} v \, d\lambda$$

$$i_C = C \frac{d}{dt} v$$

where voltage v is specified by the voltage source. Current i through the source can now be found from a KCL equation for a surface enclosing the top node:

$$i + i_R + i_L + i_C = 0$$

Substitution from the component equations gives us

$$i + \frac{1}{R} v + \frac{1}{L} \int_{-\infty}^{t} v \, d\lambda + C \frac{d}{dt} v = 0$$

where voltage v is known (with a little practice, we can write this last equation directly).
(b) If the voltage source in the network of Fig. 3.2 is replaced by a current source, we cannot find the element currents immediately. The KCL equation for a surface enclosing the top node, as given in part (a), is an *integro-differential equation* (includes both integration and differentiation) and must first be solved for v in terms of i. The solution of equations of this type is described in Chapters 8 and 9.

Remember The same voltage v appears across each component in parallel. If the circuit is driven by a voltage source, voltage v is known. If the circuit is driven by a current source, write a KCL equation in terms of v and solve for this voltage.

EXERCISES

1. For the circuit shown below, find currents i_1, i_2, i_3, i_4, and i_5, and the powers p_1, p_2, p_3, p_4, and p_5 absorbed by each component.
   answer: 12 A, 8 A, 24 A, −5 A, −39 A, 1440 W, 960 W, 2880 W, −600 W, −4680 W

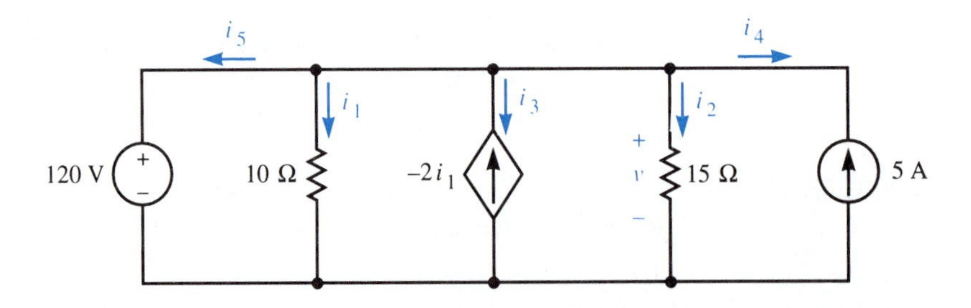

2. Replace the voltage source in the preceding network with a current source, with the reference arrow pointing toward the bottom of the page, of value 49 A. Find voltage v and currents i_1, i_2, i_3, i_4, and i_5.

   answer: -120 V, -12 A, -8 A, -24 A, -5 A, 49 A

3. For the circuit shown below, find currents i_1, i_2, i_3, i_4, i_5, and i_6 for $t > 0$ if $i_3(0) = -5$ A.

   answer: $2 \cos 2t$ A, $-320 \sin 2t$ A, $-5 + 320 \sin 2t$ A, $-\cos 2t$ A, $5 - \cos 2t$ A, -5 A

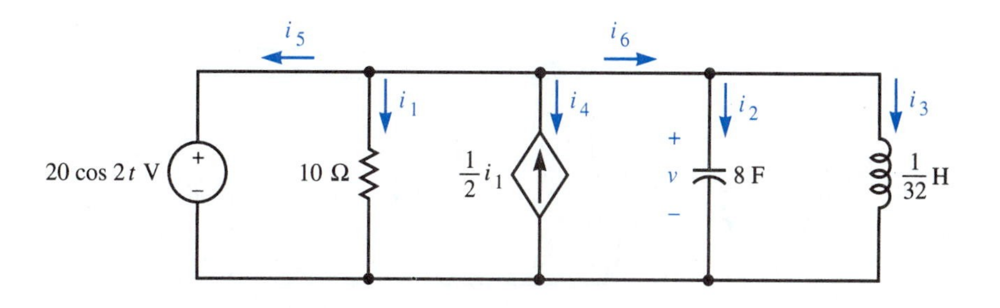

4. Replace the voltage source in the circuit shown above with a current source (reference arrow pointing up) of value $i_s = 2 \cos 2t$ A for $t \geq 0$. Write a KCL equation that gives an intergro-differential equation that can be solved for voltage v. The initial conditions are $v(0) = 20$ V and $i_3(0) = 10$ A. Do not solve for voltage v.

   answer: $8 \frac{d}{dt} v + \frac{1}{20} v + 10 + 32 \int_0^t v \, d\lambda = 2 \cos 2t,\ t > 0$

3.2 Parallel Elements

The analysis of a network containing parallel elements can often be simplified if we replace the parallel elements with an *equivalent element.* Consider the N parallel resistances of Fig. 3.3a on page 62. We can relate the current i to the individual element currents by a KCL equation:

$$i = i_1 + i_2 + \cdots + i_N \tag{3.1}$$

Substitution of the terminal equations for resistance into Eq. (3.1) yields

$$\begin{aligned} i &= \frac{1}{R_1} v + \frac{1}{R_2} v + \cdots + \frac{1}{R_N} v \\ &= \left(\frac{1}{R_1} + \frac{1}{R_2} + \cdots + \frac{1}{R_N} \right) v \\ &= \frac{1}{R_p} v \end{aligned} \tag{3.2}$$

where the *equivalent resistance R_p for N resistances connected in parallel* is

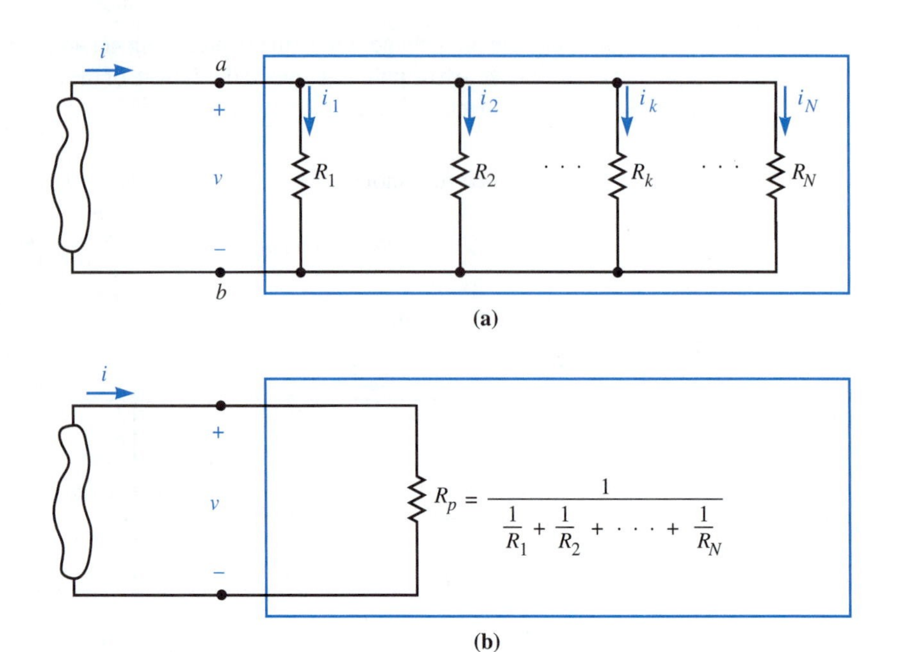

FIGURE 3.3 The equivalent resistance of N resistances in parallel: (*a*) parallel resistances; (*b*) the equivalent resistance

Resistances in Parallel

$$R_p = \frac{1}{\frac{1}{R_1} + \frac{1}{R_2} + \cdots + \frac{1}{R_N}} \tag{3.3}$$

Thus the *reciprocals of resistances in parallel are added* to give the *reciprocal* of the equivalent resistance. By defining the conductances $G_k = 1/R_k$, we can see that *conductances in parallel are added.*

Conductances in Parallel

$$G_p = G_1 + G_2 + \cdots + G_N \tag{3.4}$$

If the parallel elements of Fig. 3.3 were inductances, substitution of the corresponding terminal equations into Eq. (3.1) would yield

$$\begin{aligned} i &= \frac{1}{L_1}\int_{-\infty}^{t} v\,d\lambda + \frac{1}{L_2}\int_{-\infty}^{t} v\,d\lambda + \cdots + \frac{1}{L_N}\int_{-\infty}^{t} v\,d\lambda \\ &= \left(\frac{1}{L_1} + \frac{1}{L_2} + \cdots + \frac{1}{L_N}\right)\int_{-\infty}^{t} v\,d\lambda \\ &= \frac{1}{L_p}\int_{-\infty}^{t} v\,d\lambda \end{aligned} \tag{3.5}$$

where the *equivalent inductance* L_p *of N inductances in parallel* is

Inductances in Parallel

$$L_p = \frac{1}{\dfrac{1}{L_1} + \dfrac{1}{L_2} + \cdots + \dfrac{1}{L_N}} \tag{3.6}$$

Thus the *reciprocals of inductances in parallel are added* to give the *reciprocal* of the equivalent inductance.

Similarly, if the parallel elements of Fig. 3.3 were capacitances, substitution of the corresponding terminal equations into Eq. (3.1) would yield

$$\begin{aligned} i &= C_1 \frac{d}{dt} v + C_2 \frac{d}{dt} v + \cdots + C_N \frac{d}{dt} v \\ &= (C_1 + C_2 + \cdots + C_N) \frac{d}{dt} v \\ &= C_p \frac{d}{dt} v \end{aligned} \tag{3.7}$$

where the *equivalent capacitance for N capacitances connected in parallel* is

Capacitances in Parallel

$$C_p = C_1 + C_2 + \cdots + C_N \tag{3.8}$$

Thus *capacitances in parallel are added* to give the equivalent capacitance.

Equations (3.3), (3.6), and (3.8) give us our first two-terminal *equivalent circuits.* The circuits are equivalent in the sense that *the relationship between the terminal voltage v and the terminal current i is the same for both the original parallel elements and the equivalent parallel element.* Equivalent networks play an important role in the design and analysis of all types of electric circuits. We will now use an equivalent parallel resistance to analyze a simple circuit.

EXAMPLE 3.4 We model three lights on a boat by three parallel resistances, as shown in Fig. 3.4a on the next page. Find the equivalent resistance and the current i for the parallel combination of resistances. The voltage $v = 12$ V is given.

Solution

$$R_p = \frac{1}{(1/4) + (1/6) + (1/12)} = \frac{12}{3 + 2 + 1} = 2\ \Omega$$

$$i = \frac{1}{R_p} v = \frac{12}{2} = 6 \text{ A}$$

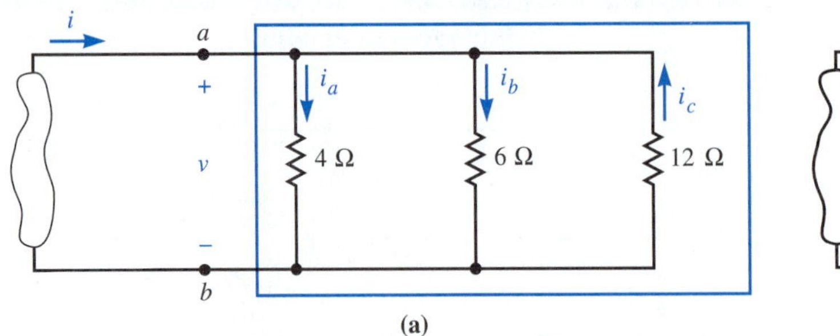

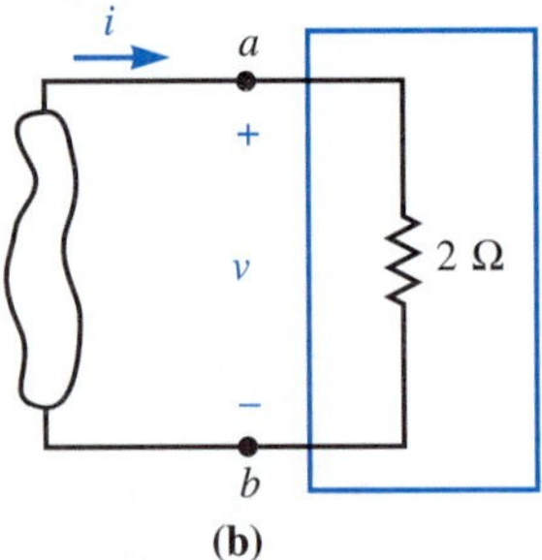

FIGURE 3.4
Equivalent parallel resistance example: (*a*) parallel resistances; (*b*) equivalent parallel resistance

Note that it really saves us no time to combine resistances to find R_p if the desired result is i rather than R_p. It is just as easy to write KCL for the top node to obtain i directly:

$$i = \frac{12}{4} + \frac{12}{6} + \frac{12}{12} = 6 \text{ A}$$

Further note that the identities of all resistances and individual resistance currents, i_a, i_b, and i_c, in the original circuit are lost in the equivalent circuit of Fig. 3.4b. Nevertheless the use of equivalent parallel resistances is often a valuable aid in the simplification of a complex network.

Remember

We add reciprocals of resistances in parallel to obtain the reciprocal of the equivalent resistance (conductances in parallel are added). We add reciprocals of inductances in parallel to obtain the reciprocal of the equivalent inductance. We add capacitances in parallel to obtain the equivalent capacitance.

EXERCISES

5. Replace the network to the right of terminals a and b in the following four circuits by a single equivalent element. Use this equivalent value to calculate the voltage v_{ab} if the current i into terminal a is e^{2t} A.

(a)

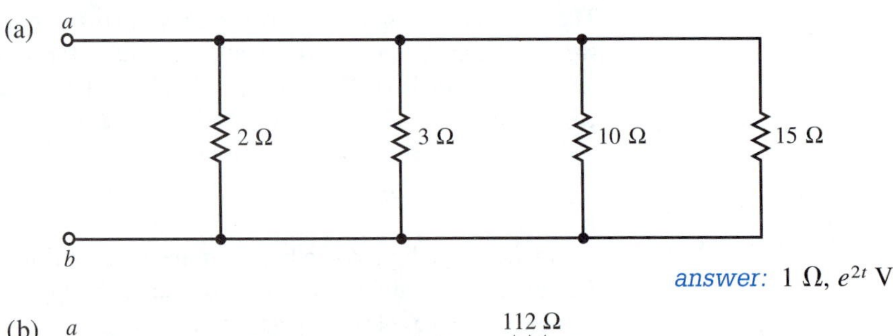

answer: 1 Ω, e^{2t} V

(b)

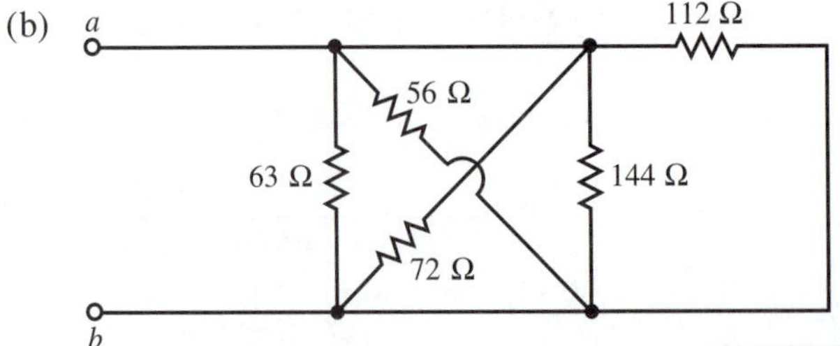

answer: 15.75 Ω, $15.75e^{2t}$ V

(c)

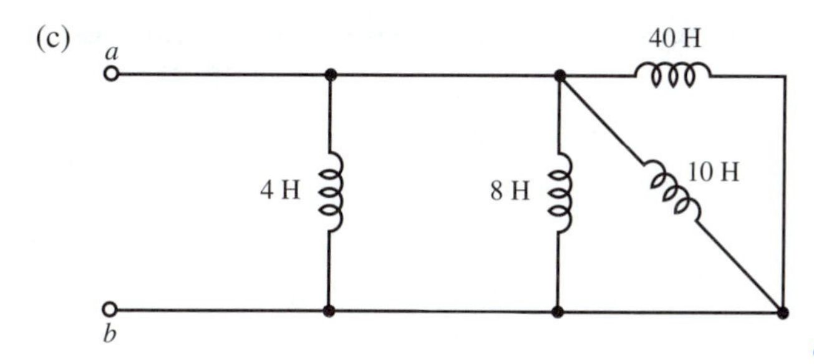

answer: 2 H, $4e^{2t}$ V

(d)

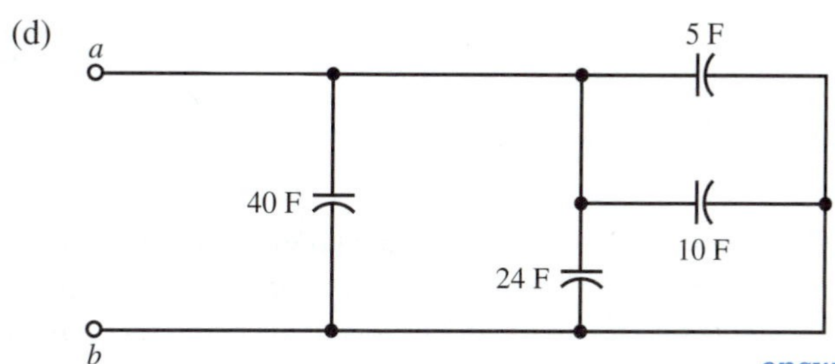

answer: 79 F, $6.329e^{2t}$ mV

6. Replace the network to the right of terminals a and b in the following circuit with an equivalent circuit that consists of three components in parallel. Use this equivalent circuit to determine the current into terminal a, if v is $360e^{2t}$ V.

   answer: 6 Ω, 6 H, 25 F, $18{,}090e^{2t}$ A

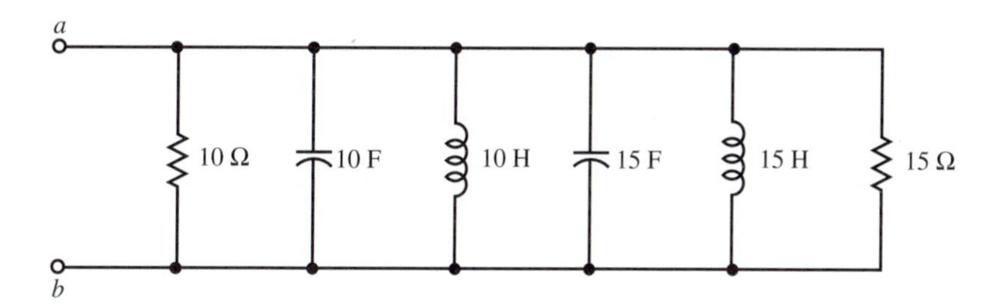

3.3 Current Divider

The same voltage appears across all components in parallel, and the current through a resistance is inversely proportional to the resistance. This implies that the current through a parallel combination of resistances divides proportionally to the inverse of each resistance. We can express this mathematically by using the equivalent parallel resistance R_p as given by Eq. (3.3) to write the current i_k through resistance R_k of Fig. 3.3 in terms of the current i entering node a.

$$i_k = \frac{1}{R_k} v = \frac{1}{R_k}(R_p i) = \frac{\dfrac{1}{R_k}}{\dfrac{1}{R_p}} i \tag{3.9}$$

Substitution for R_p from Eq. (3.3) gives us the following:

Current Divider

$$i_k = \frac{\frac{1}{R_k}}{\frac{1}{R_1} + \frac{1}{R_2} + \cdots + \frac{1}{R_N}} i \qquad (3.10)$$

or

$$i_k = \frac{G_k}{G_1 + G_2 + \cdots + G_N} i \qquad (3.11)$$

Equations (3.10) and (3.11) are known as the *current divider relation.* This says that *the current through a parallel combination of conductances divides proportionally to the conductance.* Although this relation is not of the same fundamental importance as KVL and KCL, it is frequently a time-saver and provides a thought pattern that is very useful in electronic circuit design. We must use care in applying the current divider relation. The resistances must be in parallel, and if the current i enters the node, current i_k must leave the node or the sign on the equation will change. This is illustrated by the following example.

EXAMPLE 3.5 For the example of parallel resistances shown in Fig. 3.4, suppose that voltage v is unknown, but current i has been measured and is 6 A. Determine the values of i_b and i_c.

Solution The current divider relation gives

$$i_b = \frac{1/6}{(1/4) + (1/6) + (1/12)} 6 = 2 \text{ A}$$

and

$$i_c = -\frac{1/12}{(1/4) + (1/6) + (1/12)} 6 = -1 \text{ A}$$

The current divider relation should become a part of an engineer's analysis tools, since it is a valuable aid in analysis of more complex networks.

Remember

The current through a parallel combination of conductances divides proportionally to the individual conductances.

EXERCISES

7. For the following networks, determine currents i_x and i_y with the use of the current divider relation. (Be careful with the current reference directions.)

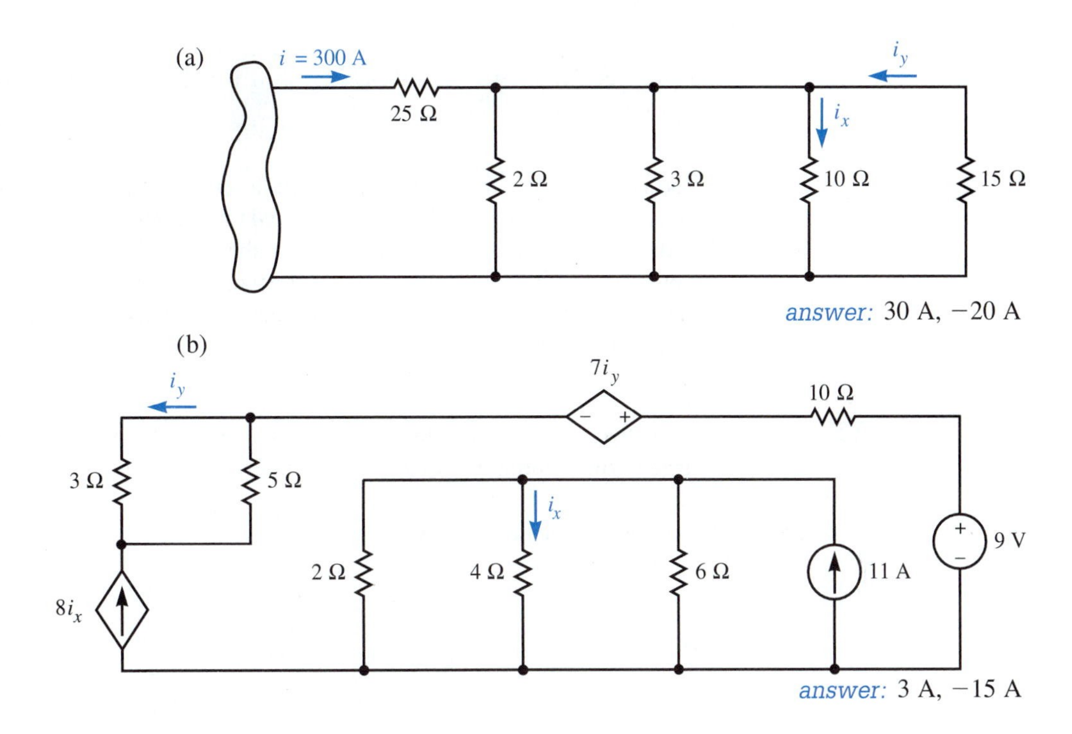

answer: 30 A, −20 A

answer: 3 A, −15 A

3.4 Single-Loop or Series Circuits

Although most electrical equipment is connected to the power source in parallel, internally there are often many devices connected in *series,* or devices modeled by components in series. An example of series-connected components is shown in Fig. 3.5.

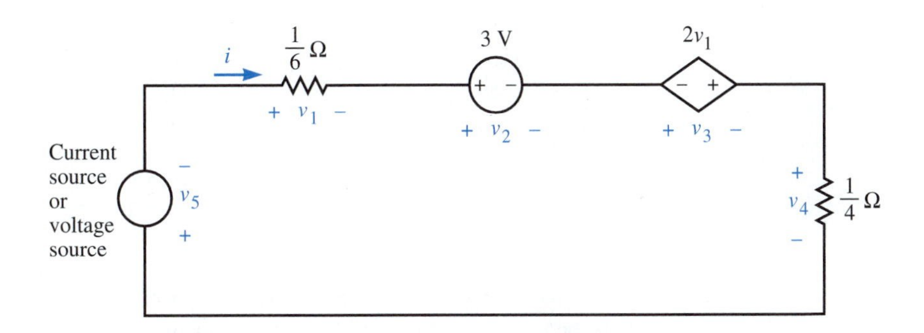

FIGURE 3.5
A series circuit

DEFINITION Series Components

Two or more components are in **series** if they are connected end to end with only two components connected at each of the intervening nodes.

Application of KCL to a closed surface that cuts any two branches of a series circuit gives the following result:

> At any specified time the same current flows through each component connected in series.

The *general procedure* to solve a single-loop circuit problem is to first find the current i through the elements. If this is specified by a current source, current i is known. If the current is not known, apply KVL around the loop. Write all unknown voltages in terms of the current i with the use of terminal or control equations and solve for i. Once i is found, the individual component voltages are easily calculated. The following three examples demonstrate the procedure.

EXAMPLE 3.6 For the network of Fig. 3.5, the left-hand source is a 12-A current source with the reference arrow pointing up. Find the indicated voltages.

Solution The first four component voltages are

$$v_1 = \frac{1}{6} i = \frac{1}{6}(12) = 2 \text{ V}$$

$$v_2 = 3 \text{ V}$$

$$v_3 = -2v_1 = -2(2) = -4 \text{ V}$$

$$v_4 = \frac{1}{4} i = \frac{1}{4}(12) = 3 \text{ V}$$

We can now find the voltage across the current source from a KVL equation written around the loop:

$$v_5 + v_1 + v_2 + v_3 + v_4 = 0$$

We can write this KVL equation directly in terms of i with the use of the component equations.

$$v_5 + \frac{1}{6} i + 3 - 2\left(\frac{1}{6} i\right) + \frac{1}{4} i = 0$$

With $i = 12$ A,

$$v_5 = -\frac{12}{6} - 3 + 2\left(\frac{12}{6}\right) - \frac{12}{4} = -4 \text{ V}$$

If the current through the loop is not specified by a current source, we can easily find the current by the use of KVL and terminal, source, and control equations.

EXAMPLE 3.7 For the circuit of Fig. 3.5, the left-hand source is a 12-V source with the (+) reference mark at the top of the page, so $v_5 = -12$ V. Determine current i.

Solution Use the component equations to write KVL around the loop in terms of current i. This is the same as in the preceding example, except that voltage v_5 is now known and current i is unknown:

$$-12 + \frac{1}{6} i + 3 - 2\left(\frac{1}{6} i\right) + \frac{1}{4} i = 0$$

This gives

$$i = 108 \text{ A}$$

Now that we know current i, the voltages not already specified by voltage sources are easily found.

It is no more difficult to write the necessary equations for a single-loop circuit with resistance, inductance, and capacitance than it is for the resistive case.

EXAMPLE 3.8 Find the indicated voltages for the circuit shown in Fig. 3.6 if (a) the left-hand source is a current source, and (b) the left-hand source is a voltage source. (This would be a model for an unloaded power transmission line. R and L are the resistance and inductance of the wire, and C is the capacitance between wires.)

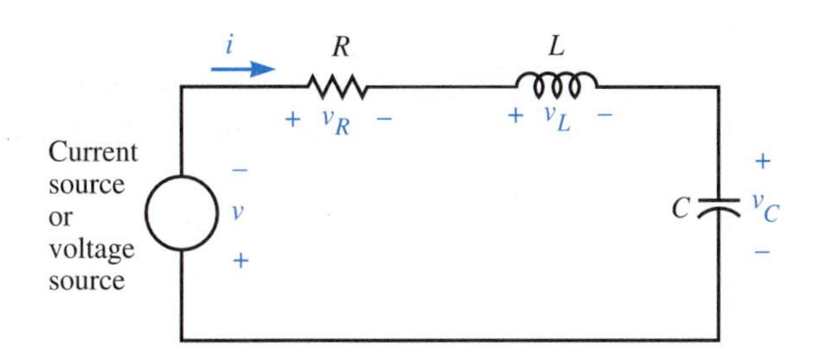

FIGURE 3.6
A series *RLC* circuit

Solution (a) The terminal equations give the element voltages:

$$v_R = Ri$$

$$v_L = L\frac{d}{dt}i$$

$$v_C = \frac{1}{C}\int_{-\infty}^{t} i \, d\lambda$$

where current i is specified by the current source. We can now find the voltage v across the current source from a KVL equation around the loop:

$$v + v_R + v_L + v_C = 0$$

Substitution from the component equations gives us

$$v + Ri + L\frac{d}{dt}i + \frac{1}{C}\int_{-\infty}^{t} i \, d\lambda = 0$$

(With a little practice, we can write the KVL equation directly in terms of i.) With current i known, this KVL equation gives voltage v.

(b) If the current source in the network is replaced by a voltage source, the element voltages cannot be found immediately. The KVL equation given in part (a) is an integro-differential equation and must be solved for current i in terms of voltage v, which is specified by the voltage source.

Remember

The same current i flows through each component in series. If the circuit is driven by a current source, current i is known. If the circuit is driven by a voltage source, write a KVL equation in terms of i and solve for this current.

EXERCISES

8. For the network shown below, determine voltages v_1, v_2, v_3, v_4, and v_5, and the powers p_1, p_2, p_3, p_4, and p_5 absorbed by each component.
answer: 12 V, 24 V, 8 V, −5 V, −39 V, 1440 W, 2880 W, 960 W, −600 W, −4680 W

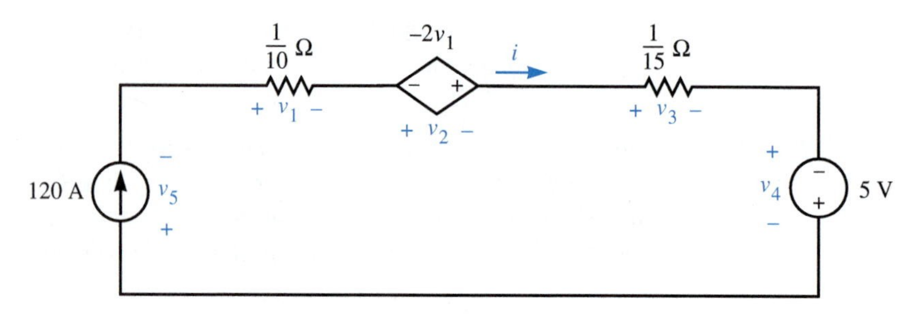

9. Replace the current source in the circuit above with a voltage source, (+) reference mark at the bottom, of value 49 V. Find current i and voltages v_1, v_2, v_3, v_4, and v_5.
answer: −120 A, −12 V, −24 V, −8 V, −5 V, 49 V

10. For the circuit shown below, find voltages v_1, v_2, v_3, v_4, v_5, and v_6 for $t > 0$, if $v_3(0) = -5$ V.
answer: 2 cos 2t V, −320 sin 2t V, −5 + 320 sin 2t V, −cos 2t V, −cos 2t + 5 V, −5 V

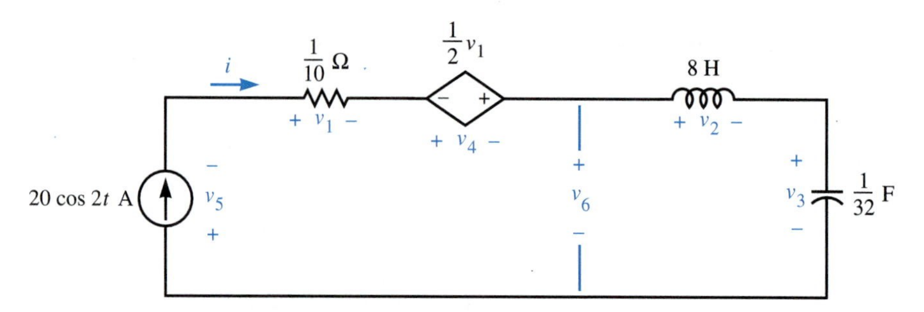

11. Replace the current source in the network above with a voltage source, (+) reference mark at the top, of value $v_s = 2\cos 2t$ V for $t \geq 0$. Write a KVL equation that gives a differential equation that can be solved for i. The initial conditions are $i(0) = 20$ A and $v_3(0) = 10$ V. Do not solve for current i.

answer: $$8\frac{d}{dt}i + \frac{1}{20}i + 10 + 32\int_0^t i\,d\lambda = 2\cos 2t,\ t > 0$$

3.5 Series Elements

The analysis of a network containing series elements can often be simplified if we replace the series elements with an equivalent element. Consider the N series resistances of Fig. 3.7a. We can relate the voltage v to the individual element voltages by a KVL equation:

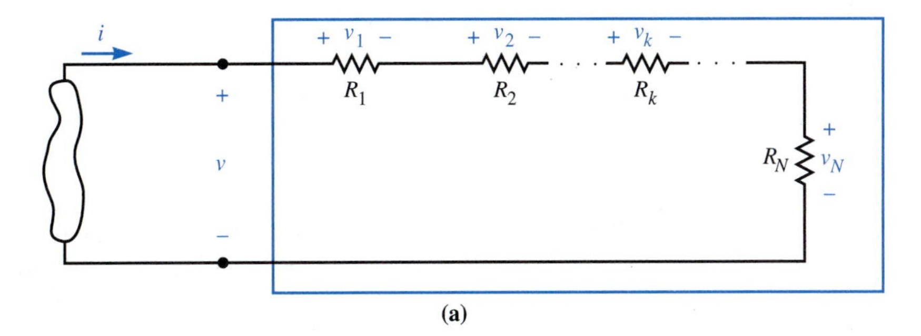

(a)

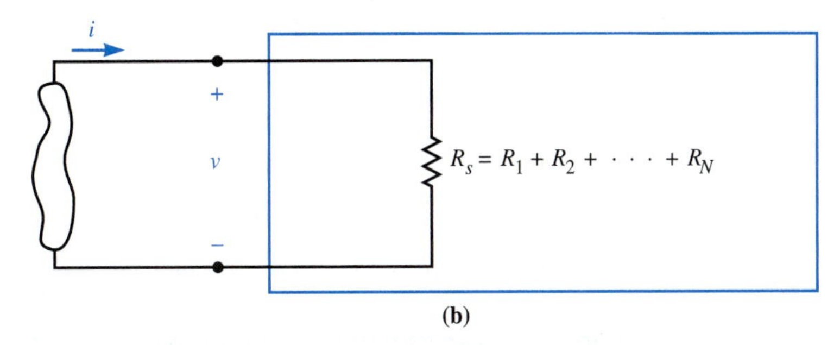

(b)

FIGURE 3.7 The equivalent resistance of N resistances in series: (*a*) series resistances; (*b*) the equivalent resistance

$$v = v_1 + v_2 + \cdots + v_N \tag{3.12}$$

Substitution of the terminal equations for resistance into Eq. (3.12) yields

$$\begin{aligned} v &= R_1 i + R_2 i + \cdots + R_N i \\ &= (R_1 + R_2 + \cdots + R_N) i \\ &= R_s i \end{aligned} \tag{3.13}$$

where the *equivalent resistance R_s for N resistances connected in series* is

Resistances in Series

$$R_s = R_1 + R_2 + \cdots + R_N \tag{3.14}$$

We see that *resistances in series are added* to give the equivalent resistance.

If the series elements of Fig. 3.7 were inductances, substitution of the corresponding terminal equations into Eq. (3.12) would yield

$$\begin{aligned} v &= L_1 \frac{d}{dt} i + L_2 \frac{d}{dt} i + \cdots + L_N \frac{d}{dt} i \\ &= (L_1 + L_2 + \cdots + L_N) \frac{d}{dt} i \\ &= L_s \frac{d}{dt} i \end{aligned} \tag{3.15}$$

where the *equivalent inductance of N inductances connected in series* is

Inductances in Series

$$L_s = L_1 + L_2 + \cdots + L_N \tag{3.16}$$

Thus *inductances in series are added* to give the equivalent inductance.

Similarly, if the series elements of Fig. 3.7 were capacitances, substitution of the corresponding terminal equations into Eq. (3.12) would yield

$$\begin{aligned} v &= \frac{1}{C_1}\int_{-\infty}^{t} i\,d\lambda + \frac{1}{C_2}\int_{-\infty}^{t} i\,d\lambda + \cdots + \frac{1}{C_N}\int_{-\infty}^{t} i\,d\lambda \\ &= \left(\frac{1}{C_1} + \frac{1}{C_2} + \cdots + \frac{1}{C_N}\right)\int_{-\infty}^{t} i\,d\lambda \\ &= \frac{1}{C_s}\int_{-\infty}^{t} i\,d\lambda \end{aligned} \tag{3.17}$$

where the *equivalent capacitance of N capacitances connected in series* is

Capacitances in Series

$$C_s = \frac{1}{\dfrac{1}{C_1} + \dfrac{1}{C_2} + \cdots + \dfrac{1}{C_N}} \tag{3.18}$$

Thus the *reciprocals of capacitances in series are added* to give the *reciprocal* of the equivalent capacitance.

As with parallel circuits, the equivalence is with respect to the terminal voltage and current, v and i only. The individual element voltages do not appear in the equivalent circuit.

EXAMPLE 3.9 Find the equivalent resistance for the series combination of resistances shown in Fig. 3.8a. Then determine voltage v. Current $i = 2$ A is known.

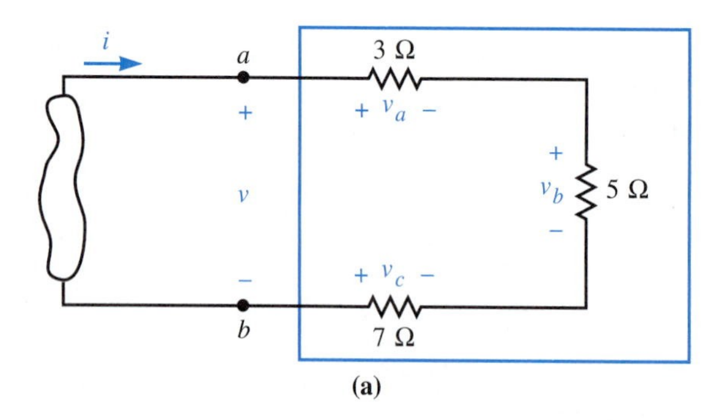

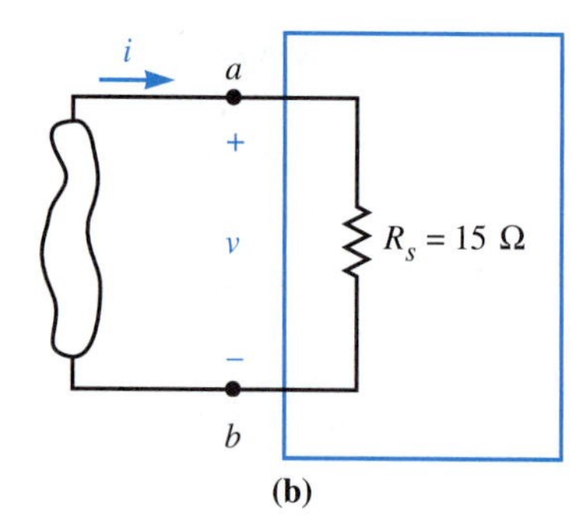

FIGURE 3.8
Equivalent series resistance example: (*a*) series resistances; (*b*) equivalent resistance

Solution

$$R_s = 3 + 5 + 7 = 15\ \Omega$$

$$v = R_s i = 15(2) = 30\ \text{V}$$

We should note that it really saved no time to find the equivalent series resistance since the desired result was to find v rather than R_s. The voltage v is just as easily found from KVL as follows:

$$v = 3(2) + 5(2) + 7(2) = 30\ \text{V}$$

Further note that the identities of all the original resistances and the original resistance voltages, v_a, v_b, and v_c, are lost in the equivalent circuit of Fig. 3.8b. Nevertheless the use of an equivalent series resistance is often a valuable aid in the simplification of more complex networks.

Remember We add resistances in series to obtain the equivalent resistance. We add inductances in series to obtain the equivalent inductance. We add reciprocals of capacitances in series to obtain the reciprocal of the equivalent capacitance.

EXERCISES

12. What is the equivalent element looking into terminals a and b of the following four networks? Use this equivalent value to calculate the current i if v_{ab} is $148e^{3t}$ V.

(a)

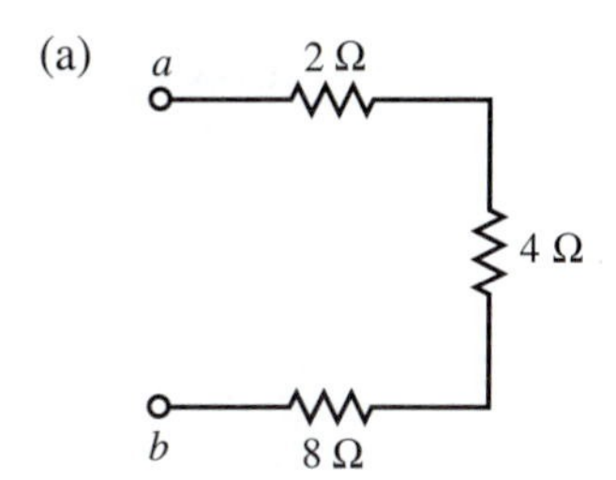

answer: 14 Ω, $\frac{74}{7}e^{3t}$ A

(b)

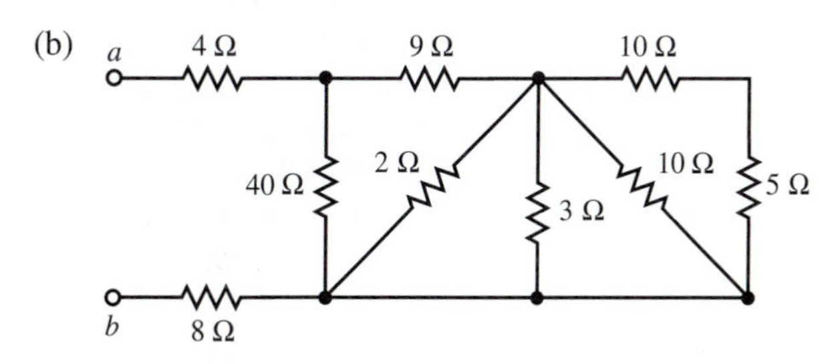

answer: 20 Ω, $\frac{37}{5}e^{3t}$ A

(c)

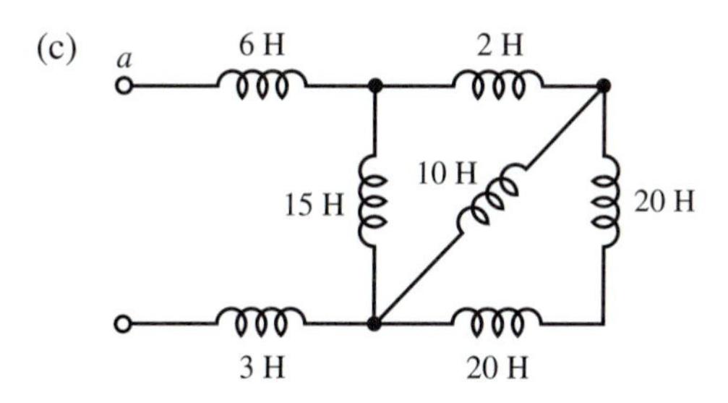

answer: 15 H, $\frac{148}{45}e^{3t}$ A

(d)

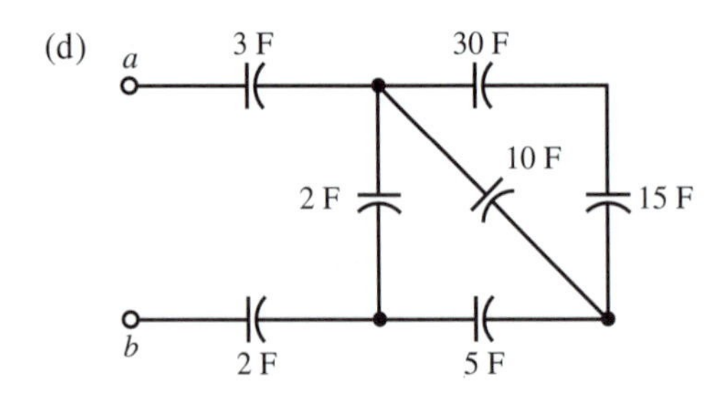

answer: 1 F, 444 e^{3t} A

3.6 Voltage Divider

The current is the same through all components in series, and the voltage across a resistance is directly proportional to the resistance. This implies that the voltage across a series connection of resistances divides proportionally to each resistance. The equivalent series resistance R_s given by Eq. (3.14) can be used to give the voltage v_k across resistance R_k of Fig. 3.7 in terms of the voltage v across the series combination:

$$v_k = R_k i = R_k \left(\frac{1}{R_s} v \right) \tag{3.19}$$

Substitution for R_s from Eq. (3.14) gives us

Voltage Divider

$$v_k = \frac{R_k}{R_1 + R_2 + \cdots + R_N} v \tag{3.20}$$

The latter equation is known as the *voltage divider relation.* This says that *the voltage across a series combination of resistances divides proportionally to the resistance.* Although this relation is not of the same fundamental importance as KVL and KCL, it is frequently a time-saver and provides a useful way for us to think about circuit design. We must use care in applying the voltage divider relation. The resistances must be in series, and the path around a loop containing voltages v and v_k must go into opposite polarity marks (+ or −) for v and v_k, or the sign on Eq. (3.20) will change.

EXAMPLE 3.10 For the previous example of series resistances shown in Fig. 3.8, suppose that current i is unknown, but voltage v is measured to be 30 V. Determine the value of v_b and v_c.

Solution From the voltage divider relation,

$$v_b = \frac{5}{3 + 5 + 7} 30 = 10 \text{ V}$$

and

$$v_c = -\frac{7}{3 + 5 + 7} 30 = -14 \text{ V}$$

The voltage divider relation should become a part of an engineer's analysis tools, since it is a valuable aid in analysis of more complex networks.

Remember The voltage across a series combination of resistances divides proportionally to the resistance.

EXERCISES

13. The storage battery for a boat is modeled as a 12-V voltage source in series with a 0.1-Ω resistance. We will use a resistance of 0.2 Ω as a crude model for the boat's starter motor. Use the voltage divider relation to determine the voltage across the starter motor when it is starting the engine. (The two resistances and the source are in series.) *answer:* 8 V

14. For the following networks calculate voltages v_x and v_y.

(a)

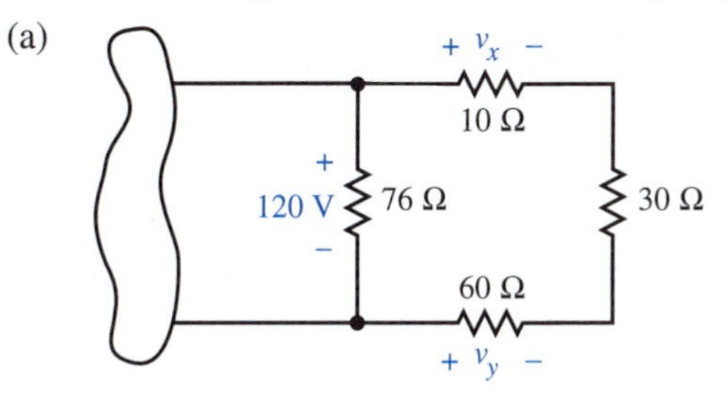

answer: 12 V, − 72 V

(b)

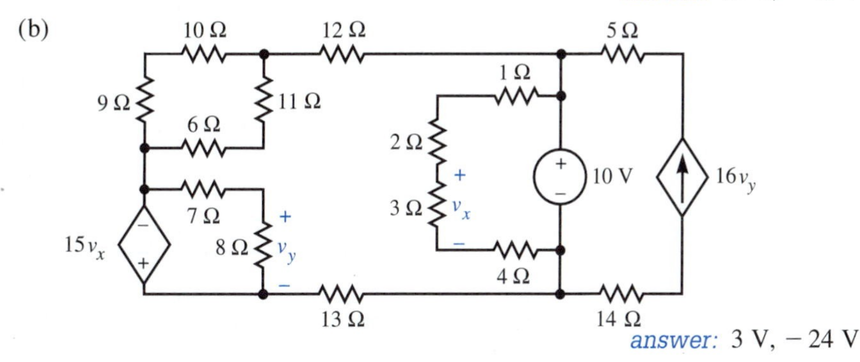

answer: 3 V, − 24 V

3.7 Network Analysis by Simplification

We have seen how to replace a group of parallel or series elements of the same type by a single equivalent element without affecting the remainder of the circuit. The equivalent element *simplifies* the circuit. Repeated replacement of series or parallel elements by an equivalent element may permit us to greatly simplify a complicated circuit and solve for some voltage or current value we need. The following example demonstrates network analysis by simplification.

EXAMPLE 3.11 The circuit shown in Fig. 3.9 on page 76 is a model for the lighting system on a small boat. Determine the terminal voltage v_b of the battery and the voltage v_L for the 10-Ω lamp.

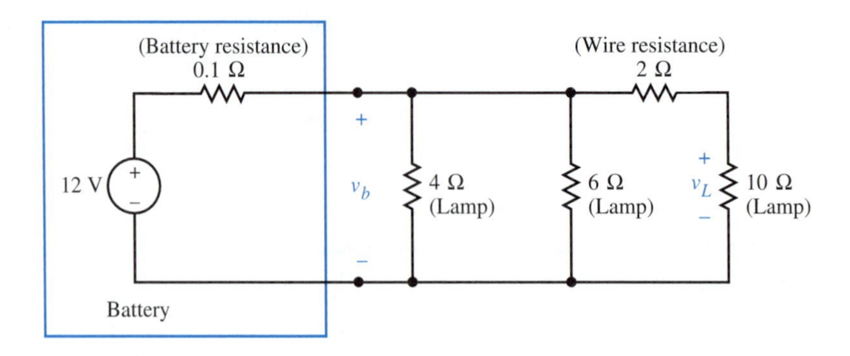

FIGURE 3.9 Lighting system for a small boat

Solution The equivalent resistance for the series combination of the 2-Ω and 10-Ω resistances is

$$R_s = 2 + 10 = 12\ \Omega$$

This equivalent resistance is in parallel with the 4-Ω and 6-Ω resistances. The equivalent resistance for this parallel combination is

$$R_p = \frac{1}{(1/4) + (1/6) + (1/12)} = 2\ \Omega$$

This gives us the simple equivalent circuit (Fig. 3.10).

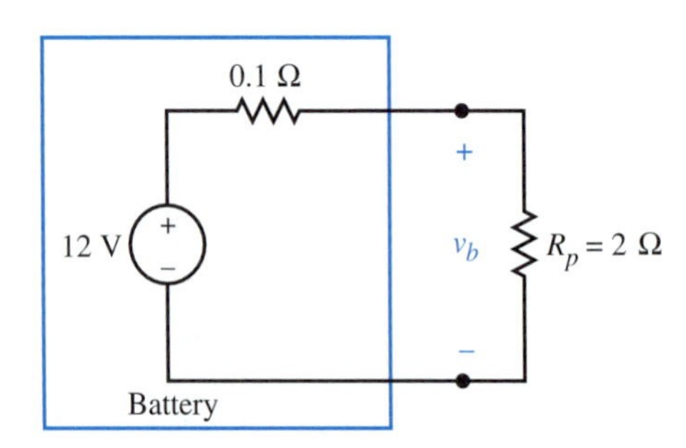

FIGURE 3.10 An equivalent lighting circuit

We can obtain voltage v_b from the voltage divider relation:

$$v_b = \frac{2}{2 + 0.1} 12 = 11.43\ \text{V}$$

Because the battery resistance is much lower than the load resistance, which consists of the lights, most of the 12-V internal voltage of the battery appears across the battery terminals.

Observe that the voltage v_L does not appear in the equivalent circuit. We use the original circuit and voltage v_b calculated from the equivalent circuit to calculate voltage v_L. Use of the voltage divider relation with the original circuit gives

$$v_L = \frac{10}{10 + 2} v_b = \frac{10}{12}(11.43) = 9.52\ \text{V}$$

We can see that the 2-Ω resistance of the wire further reduces the voltage for the 10-Ω lamp.

A physical system is only approximated by a network model. For example, the circuit of Fig. 3.9 neglected the resistance of the wire between the 4-Ω and 6-Ω lamps.

For the network shown in Fig. 3.10, the voltage source and series resistance to the left of terminals *a* and *b* represent a *practical voltage source.* The inclusion of the 0.1-Ω resistance gives a more practical model for a physical voltage source than would an ideal voltage source, because the output voltage v_b would decrease as the output current increased. The 2-Ω resistance R_s to the right of terminals *a* and *b* represents the *equivalent load* on the practical source. (The load absorbs power from the source.)

Caution must be exercised in the use of parallel and series equivalents. Be sure the components are really in parallel or series. Some components may not be in parallel or series with any other component. For example, see Fig. 3.11. The resistances in these circuits are neither in series nor in parallel. In Chapter 2, we defined a network such as that in Fig. 3.11a, which can be drawn so that no lines cross, to be a *planar network.* The network of Fig. 3.11b cannot be drawn so that no lines cross and is a *nonplanar network.*

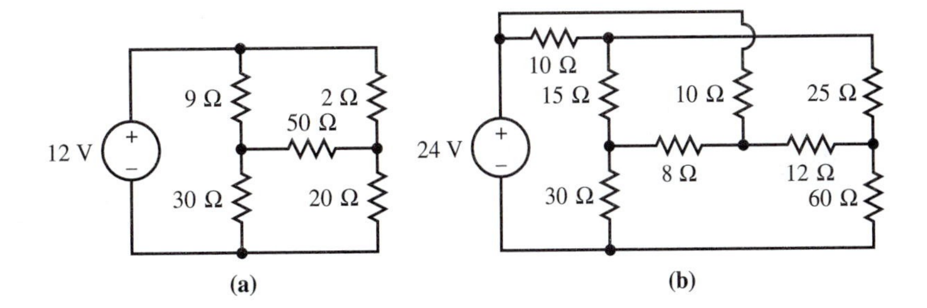

FIGURE 3.11 Networks with no series or parallel components: (*a*) planar network; (*b*) nonplanar network

Remember

Check to be sure that components are in parallel before calculation of a parallel equivalent or use of the current divider equation. Be sure that components are in series before calculation of a series equivalent or use of the voltage divider equation.

EXERCISES

15. Calculate v_x and v_y for the following three networks.

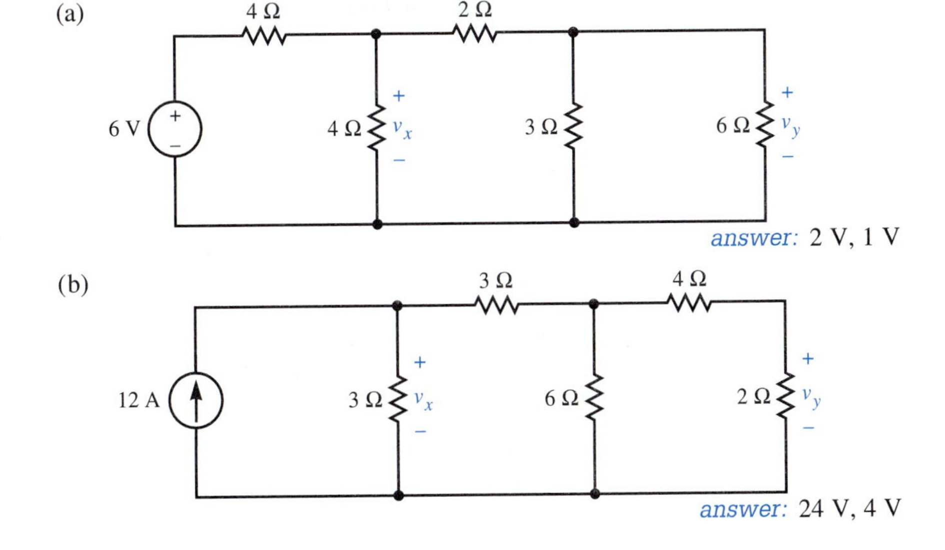

answer: 2 V, 1 V

answer: 24 V, 4 V

(c)

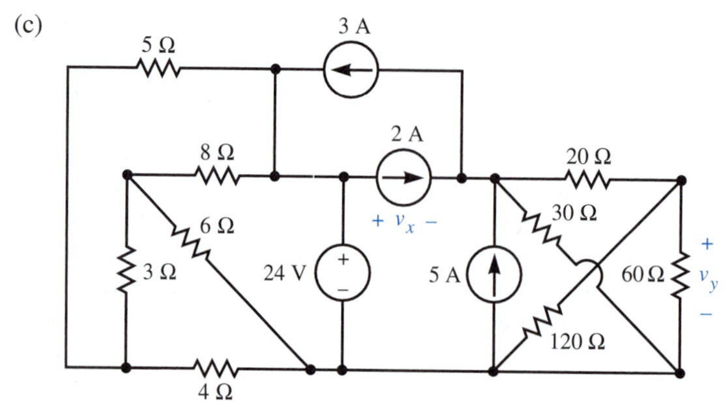

answer: -56 V, $\frac{160}{3}$ V

3.8 Summary

We began this chapter by analyzing circuits composed of components that were all in parallel. This analysis led to the concept of an equivalent component or circuit, and to the current divider relation. A similar analysis of series circuits yielded additional equivalent components and the voltage divider relation. We used these equivalent components to reduce a more complicated network to one that was easily analyzed.

KEY FACTS Concept	Equation	Section	Page
❑ Branches that share two common nodes are in *parallel.*		3.1	58
❑ The same voltage v appears across all components in parallel (Example 3.1).		3.1	58
❑ If a parallel circuit is driven by a voltage source, the voltage v is known. The current through this voltage source is obtained from a KCL equation (Example 3.1).		3.1	58
❑ If a parallel circuit is driven by a current source, write a KCL equation in terms of v and solve for this voltage (Example 3.2).		3.1	59
❑ The reciprocals of resistances in parallel are added to give the reciprocal of the equivalent resistance.	(3.3)	3.2	62
❑ Conductances in parallel are added to give the equivalent conductance.	(3.4)	3.2	62

KEY FACTS	**Concept**	**Equation**	**Section**	**Page**
❑	The reciprocals of inductances in parallel are added to give the reciprocal of the equivalent inductance.	(3.6)	3.2	63
❑	Capacitances in parallel are added to give the equivalent capacitance.	(3.8)	3.2	63
❑	For parallel resistances, the current divides proportionally to the reciprocal of resistance (conductance).	(3.10) (3.11)	3.3	66
❑	Branches that are connected end to end with no junctions at these connecting nodes are in *series.*		3.4	67
❑	The same current i passes through all components in series (Example 3.6).		3.4	68
❑	If a series circuit is driven by a current source, the current i is known. The voltage across this current source is obtained from a KVL equation (Example 3.6).		3.4	68
❑	If a series circuit is driven by a voltage source, write a KVL equation in terms of i and solve for this current (Example 3.7).		3.4	68
❑	Resistances in series are added to give the equivalent resistance.	(3.14)	3.5	71
❑	Inductances in series are added to give the equivalent inductance.	(3.16)	3.5	72
❑	The reciprocal of capacitances in series are added to give the reciprocal of the equivalent capacitance.	(3.18)	3.5	72
❑	For series resistances, the voltage divides proportionally to the resistance.	(3.20)	3.6	74
❑	Series and parallel combinations of components often provide a convenient method to solve for a voltage or current in a complicated circuit (Example 3.11).		3.7	75
❑	Individual component voltages and currents will not all appear in an equivalent circuit. Use the voltages and currents calculated for the equivalent circuit to determine the voltages and currents in the original circuit.		3.7	76
❑	Be sure that the components are in series or parallel before you calculate an equivalent series or parallel component.		3.7	77

PROBLEMS

Section 3.1

1. For the circuit below v_s is 60 V. (a) Determine voltage v_{ab}; the open-circuit voltage v_{oc}; currents i_a, i_b, and i_c; and the instantaneous power absorbed by each component. (b) Short-circuit terminals c and d, then calculate voltage v_{ab} and the short-circuit current i_{cd}.

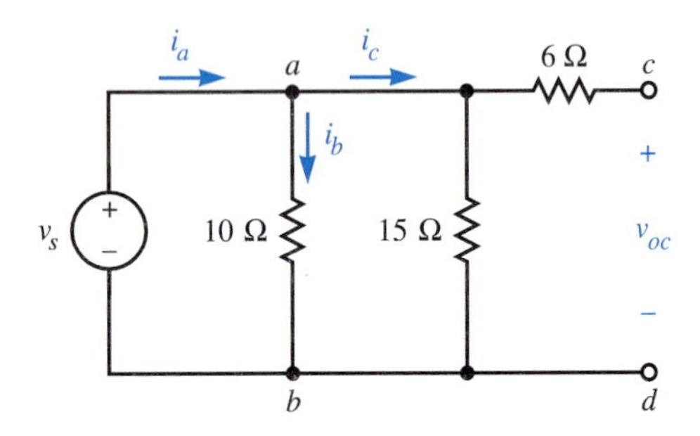

2. Replace the voltage source in the preceding circuit with a current source of 10 A that has the reference arrow pointing up, then repeat Problem 1.

3. Repeat Problem 1 for the following circuit.

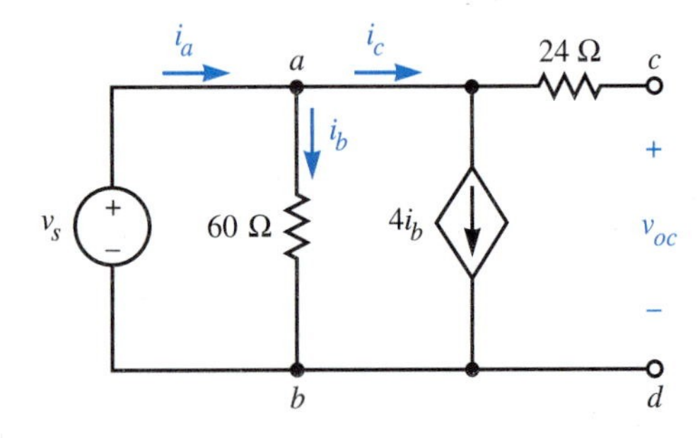

4. Repeat Problem 2 for the preceding circuit.

5. Calculate the power absorbed by each component in the following circuit if v_s is 30 cos 2t V. Is the ratio of v_{ab}/i_a a resistance?

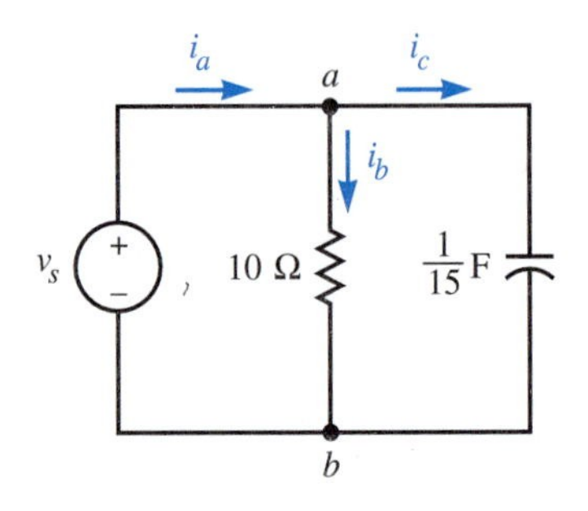

6. Calculate voltages v_{ab}, v_{cb}, and v_{ac} and currents i_1, i_2, and i_3 for the network shown below.

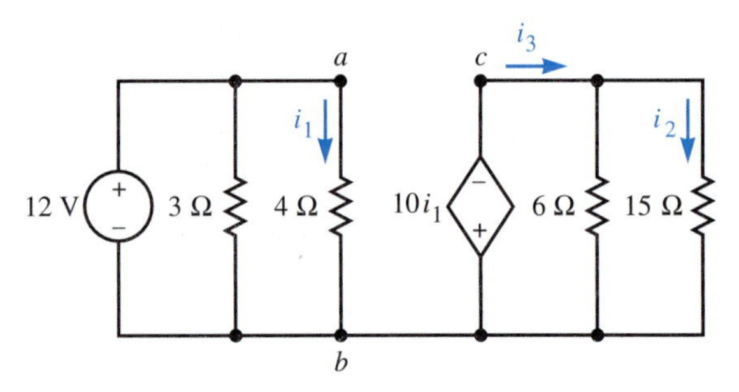

7. Calculate voltages v_{ab} and v_{cd} and currents i_1 and i_2 for the network shown below.

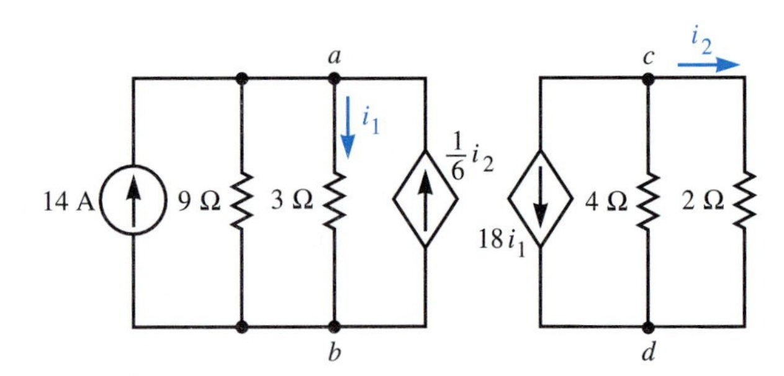

Section 3.2

8. Use Eq. (3.3) and algebra to show that the equivalent resistance of two resistances R_1 and R_2 in parallel is given by

$$R_p = \frac{R_1 R_2}{R_1 + R_2}$$

From Eq. (3.3), develop a relation similar to the above equation for three resistances in parallel, then for four resistances in parallel.

9. Determine a single equivalent component with respect to terminals a and b for each of the following networks.

(a)

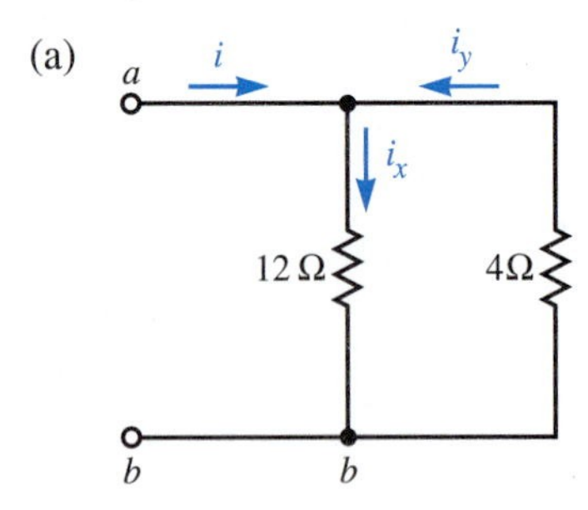

(b)

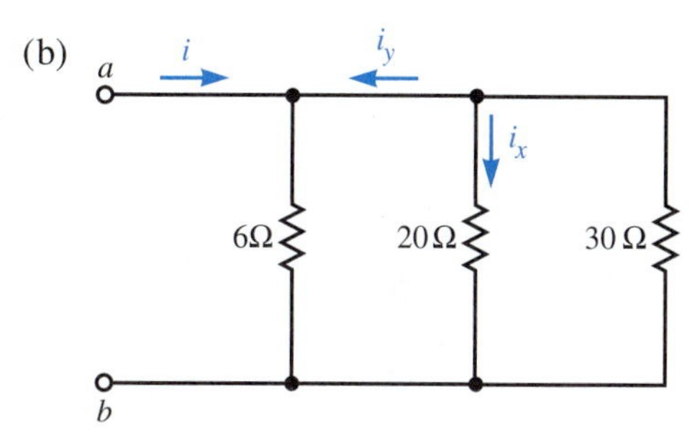

(c)

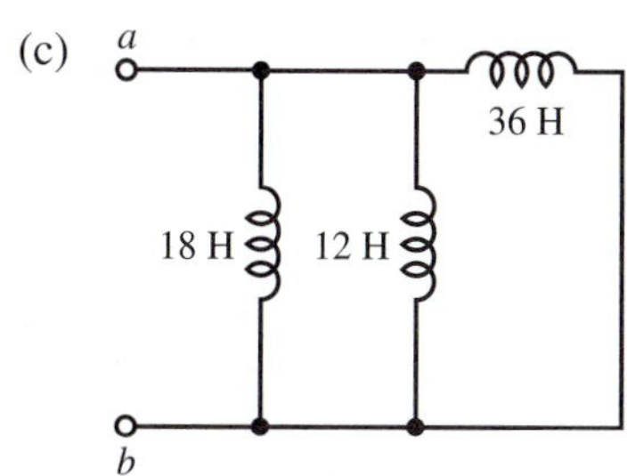

(d)

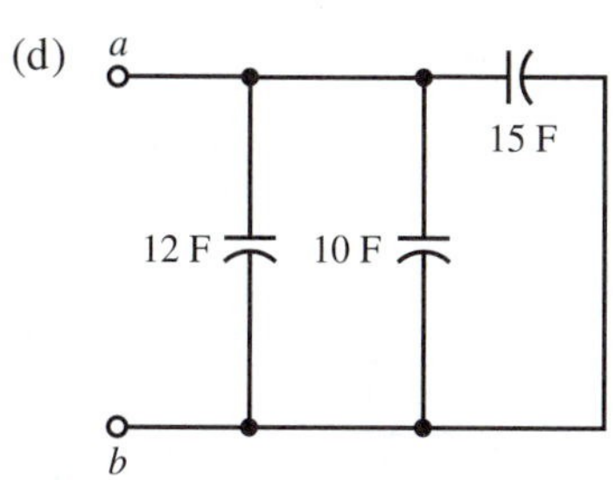

10. Replace the network to the right of terminals a and b in the circuit below with a single resistance, inductance, and capacitance.

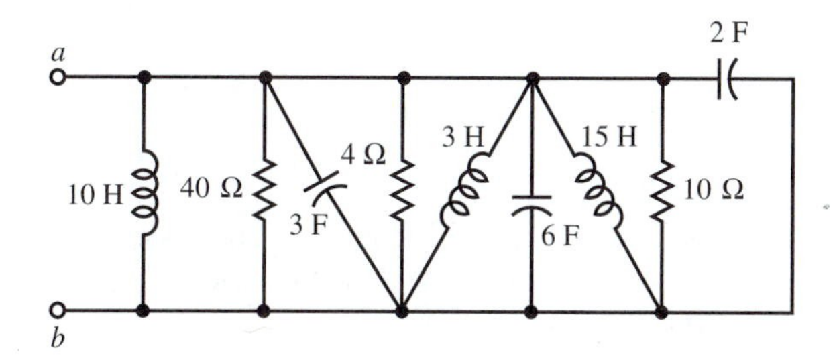

Section 3.3

11. Use the current divider equation to determine currents i_b and i_c in the circuit of Problem 2.

12. If current $i = 24$ A, determine currents i_x and i_y in the network, shown in the specified problem.

    (a) Problem 9a (b) Problem 9b

13. You have an ammeter that reaches full scale when the current through it is 1 mA. The meter resistance is known to be 50 Ω. What resistance must you place in parallel with this meter so that you can use the meter with this *shunt resistance* in an application that requires a meter that measures 1 A at full scale? Refer to the illustration below.

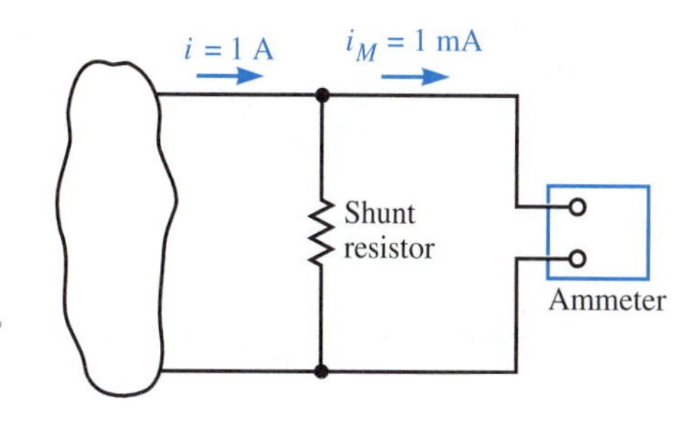

14. Determine voltage v for the network shown below.

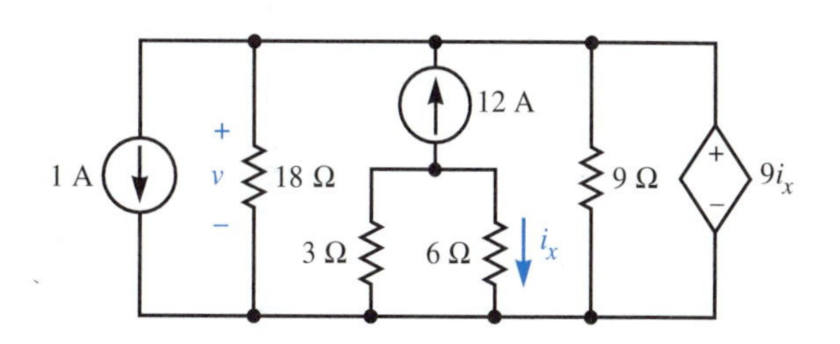

15. For the dependent voltage source shown above, substitute a dependent current source with the reference arrow pointing up and a control equation giving this current a value of $2i_x$. Determine v.

Section 3.4

16. For the circuit below i_s is 6 A. (a) Determine current i; voltages v_{ab}, v_{bc}, v_{ac}, and the open-circuit voltage v_{oc}. Also calculate the power absorbed by each component. (b) Short-circuit terminals d and e, then calculate current i, voltage v_{ac}, and the short-circuit current i_{de}.

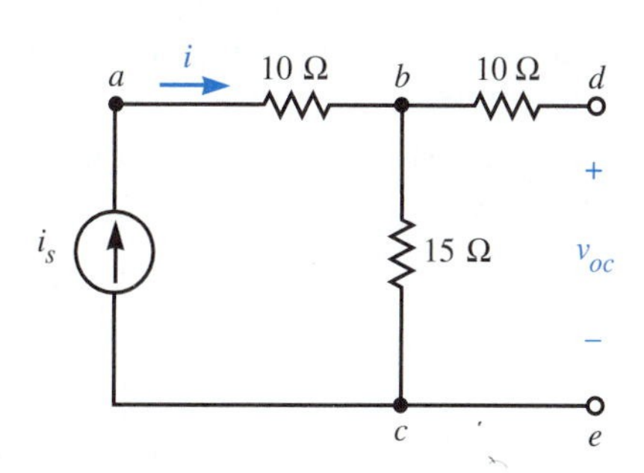

17. Replace the current source in the preceding circuit with a voltage source of 400 V that has the (+) reference at the top, then repeat Problem 16.

18. Repeat Problem 16 for the following circuit.

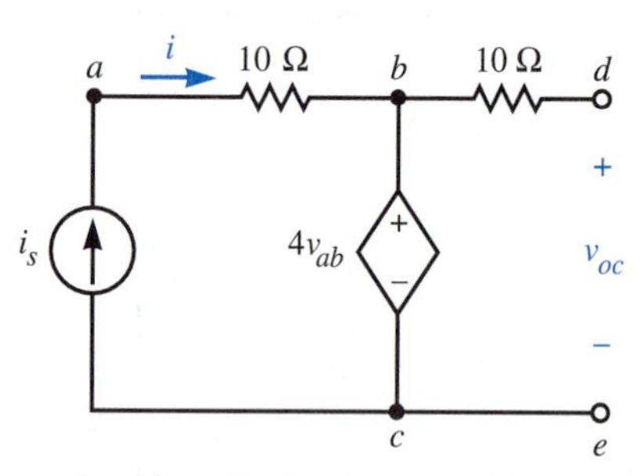

19. Repeat Problem 17 for the preceding circuit.

20. Calculate the power absorbed by each component in the following circuit if i_s is $50 \cos 2t$ A. Is the ratio of v_{ac}/i a resistance?

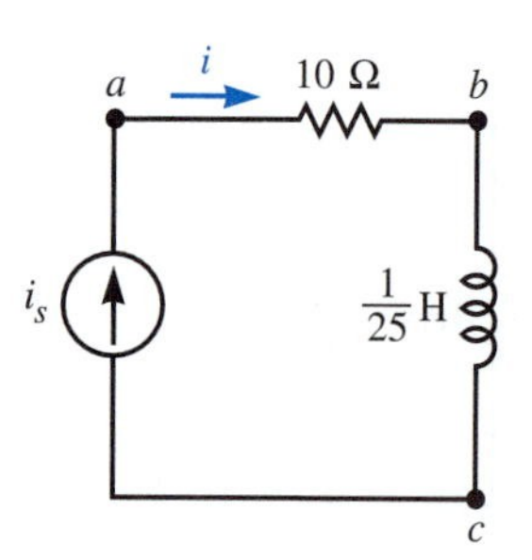

21. Find the voltage v, the current i, and the power absorbed by each component of the network given below.

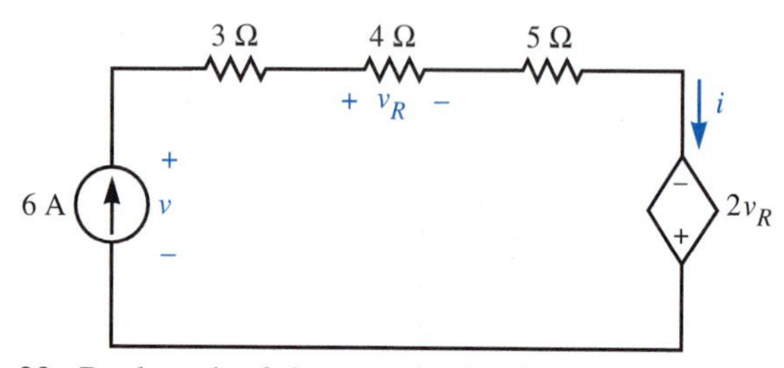

22. Replace the 6-A source in the circuit above with a 24-V voltage source having the (+) reference mark toward the top of the page. Determine voltage v and current i.

23. Calculate voltages v_1, v_2, and v_3 in the network shown below.

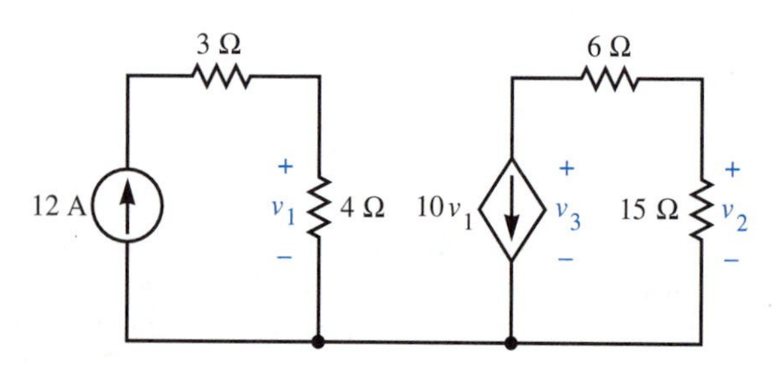

24. For the networks shown below, find v_o or i_o.

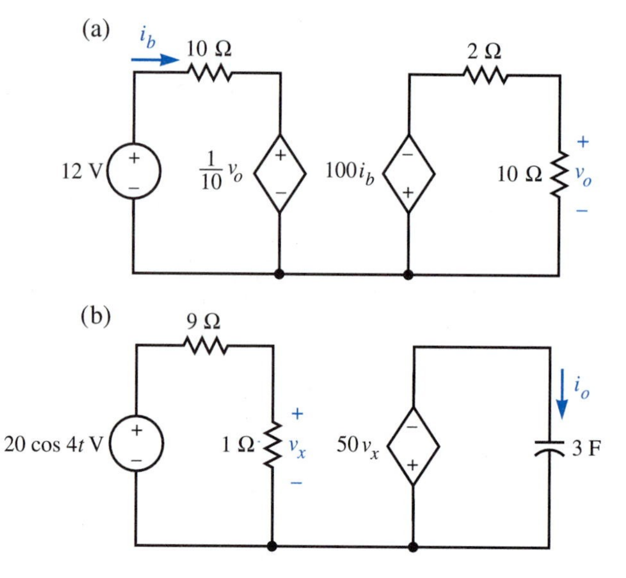

25. Determine v_1, v_2, v_3, v_4, and v_5 for the circuit shown below, if $v_3(0) = 0$ and (*a*) $i_s = 20 \cos 2t$ A; (*b*) $i_s = 20 \cos 4t$ A.

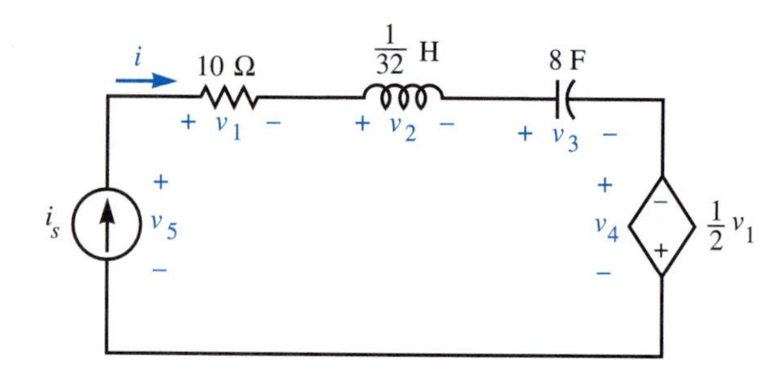

Section 3.5

26. Determine a single equivalent component with respect to terminals a and b for each of the following three networks.

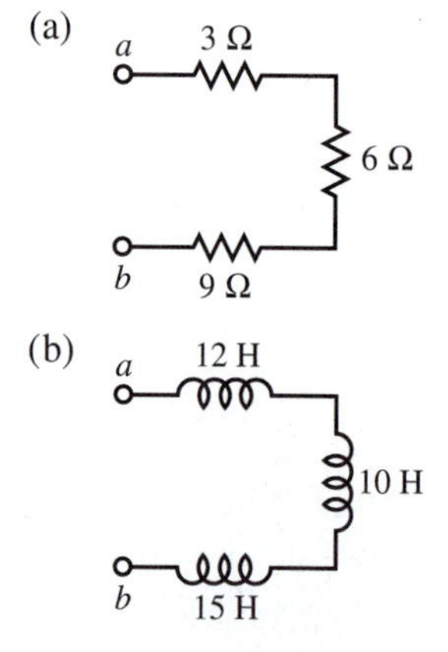

(c)

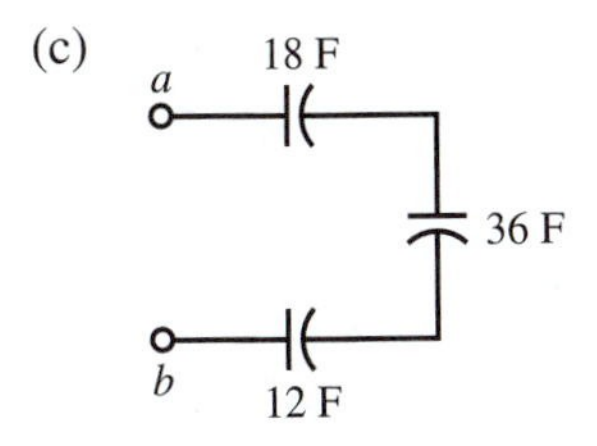

27. Replace the network to the right of terminals a and b in the network shown below by an equivalent series resistance R_s, inductance L_s, and capacitance C_s.

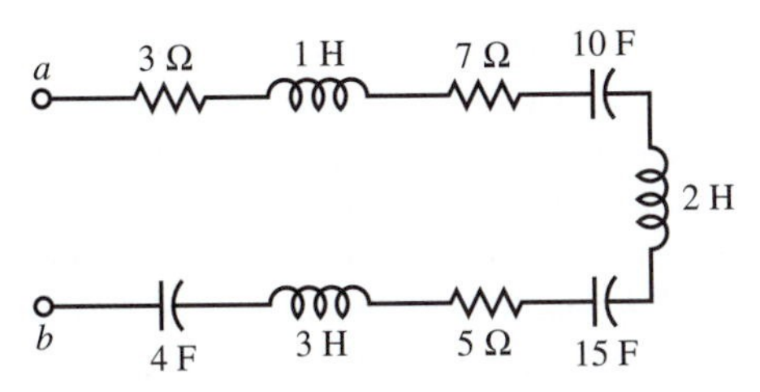

Section 3.6

28. Use the voltage divider equation to calculate voltages v_{ab} and v_{cb} in the network of Problem 16.

29. Determine current i for the network shown below.

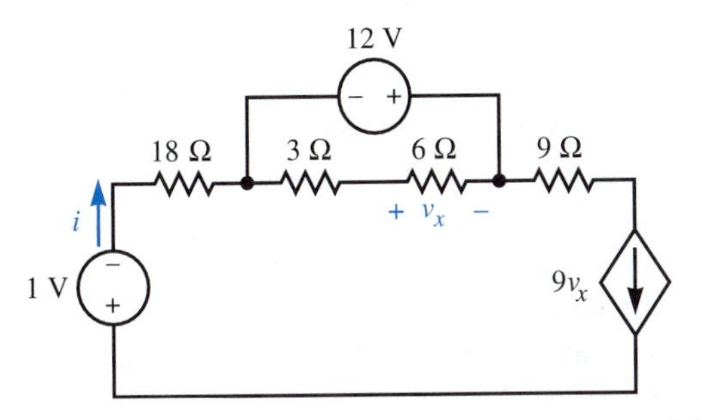

30. An unknown resistance R_x can be measured in terms of known resistances R_a, R_b, and R_c with an instrument called a Wheatstone bridge. Refer to the illustration below. We say that the bridge is balanced when $v_{12} = 0$. For the balanced condition, find R_x as a function of R_a, R_b, and R_c.

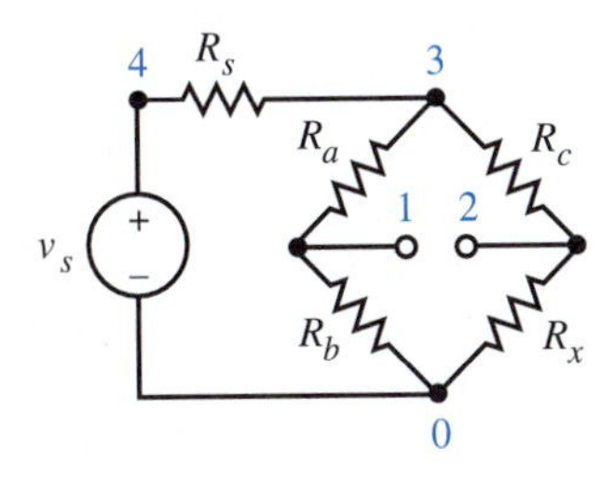

Section 3.7

31. Find the voltage v across the current source for the following network.

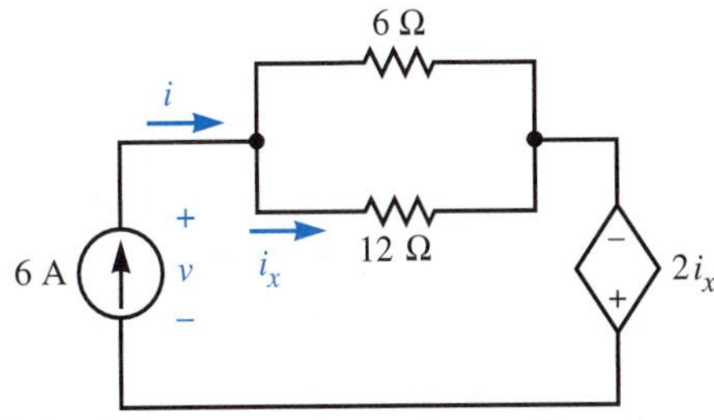

32. If the 6-A current source shown above is replaced by a 100-V voltage source, find the power absorbed (a) by the independent voltage source, and (b) by the dependent voltage source.

33. Determine the voltage v and the power absorbed by each component in the following networks.

(a)

(b)

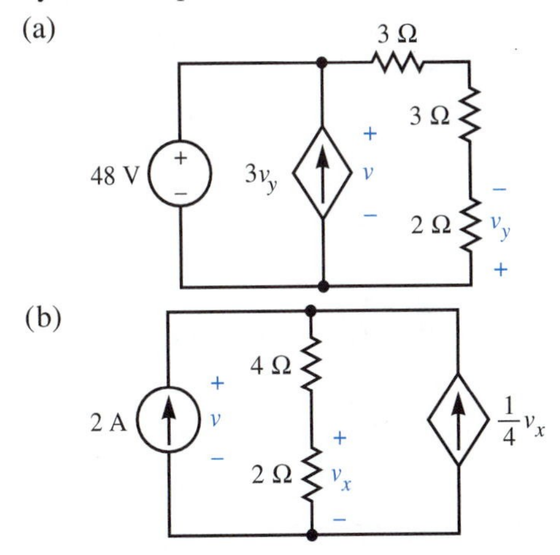

34. For the network that follows, determine voltage v, current i, and the power absorbed by the independent source.

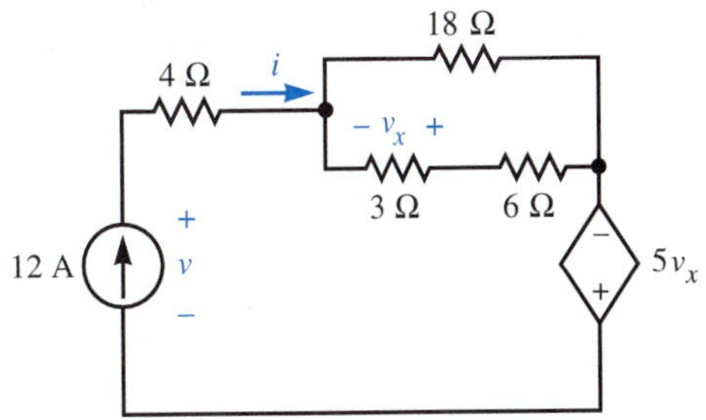

35. Find the voltage v and the current i in the network given below.

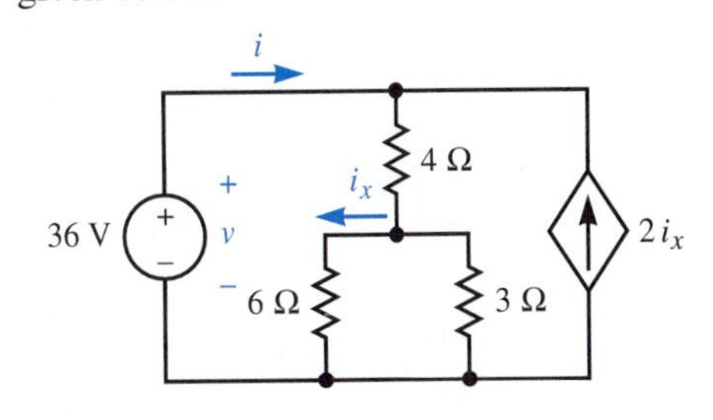

36. For the network shown below, find current i.

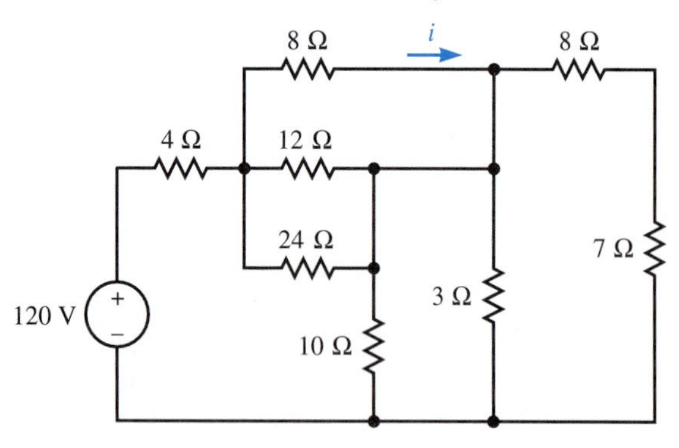

Supplementary Problems

37. Find the equivalent resistance when connections to the network shown below are made to the terminals specified.
    (a) a and b (b) c and d
    (c) e and f (d) e and d
    (e) c and e (f) b and d
    (g) a and c

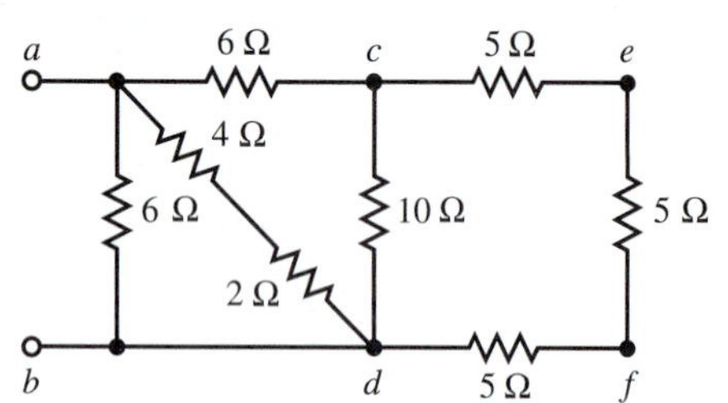

38. For the following networks determine the equivalent resistance connected between terminals a and b.
    (a)

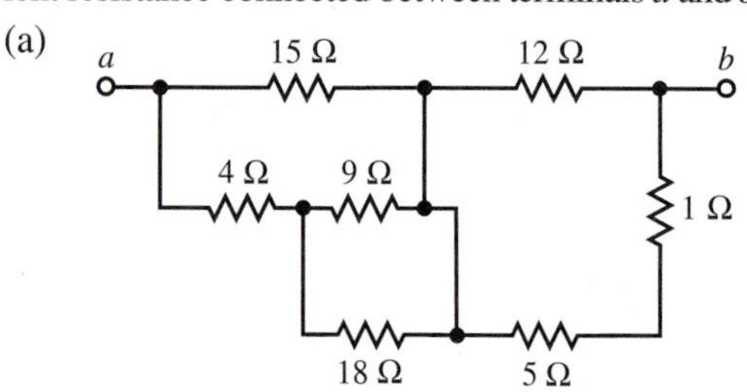

(b)

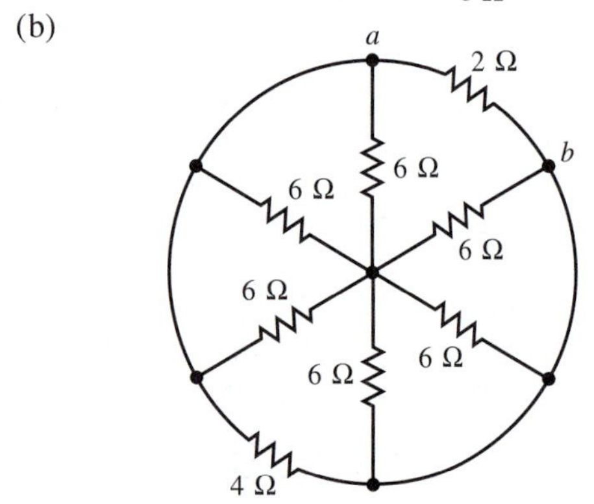

39. Use symmetry to calculate the resistance between terminals a and b in the following circuit.

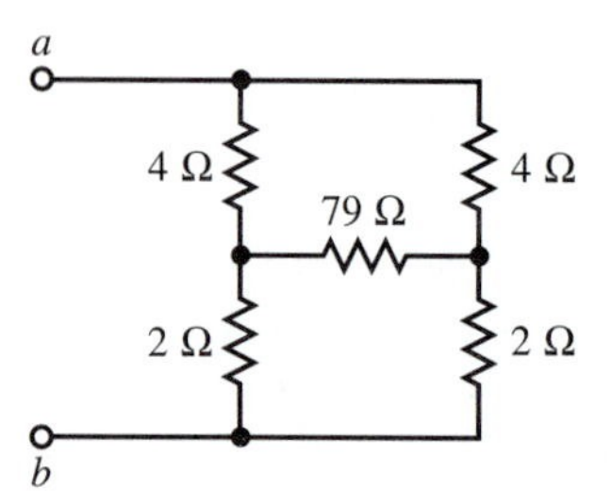

40. A practical ammeter can often be modeled as a resistance. The meter reading is proportional to the current through the resistance. A meter has a full-scale reading of 10 A when the voltage across the meter is 50 mV. What resistance must be placed in parallel with the meter as a *shunt* so that the combination will yield a full-scale reading of 100 A?

41. A practical voltmeter can often be modeled as a resistance, and the meter reading, although calibrated in volts, is proportional to the current the voltage causes to flow through this resistance. A 10-V meter requires 1 mA of current for a full-scale reading. What resistance must be placed in series with the voltmeter so that 100 V is required for a full-scale reading?

42. An ammeter with a full-scale reading of 50 μA has a resistance of $R_M = 1$ kΩ. What resistance R_s must we place in series with this meter so that we can use it as a voltmeter with the specified full-scale range V_{max}.
    (a) 3 V
    (b) 10 V
    (c) 30 V
    (d) 100 V
    (e) Determine the ratio $(R_M + R_s)/V_{max}$ in each case. This ohms-per-volt parameter is often given on the front of multirange voltmeters.

43. A practical voltmeter does not have infinite resistance and will *load* the circuit. This can cause significant errors in some instances. Many multirange voltmeters are specified to have a resistance of 20 kΩ/V. This means that the meter resistance in ohms is 20,000 times the full-scale value for the range selected. For the following circuit calculate the indicated voltages.
    (a) The actual voltage v_x.

(b) The meter reading if the voltage v_x is measured with the meter on the 3-V scale.

(c) The meter reading if the voltage is read on the 10-V scale.

(d) The voltage calculated in part c is closer to the actual value than that calculated in part b. In practice the result might not be better. The maximum meter error is usually given as a percentage of the full-scale value of the selected range. If the maximum meter error is given as ± 5 percent of full scale, what would be the range of voltages that you might actually measure in parts b and c? Which range should you use?

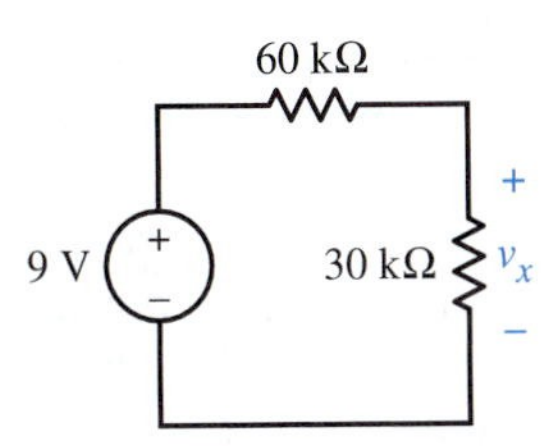

44. One standard for a family of ammeters is that the voltage across the meter is 50 mV when the meter is reading the full-scale value I_{max}.

(a) Determine the meter resistance R_M as a function of I_{max}.

(b) An ammeter is connected in series with a voltage source of value v_s and resistance of value R as shown in the following illustration. Without the meter, the current is v_s/R. Determine the minimum value of R as a function of I_{max} so that the meter resistance introduces an error of less than 10 percent in the current.

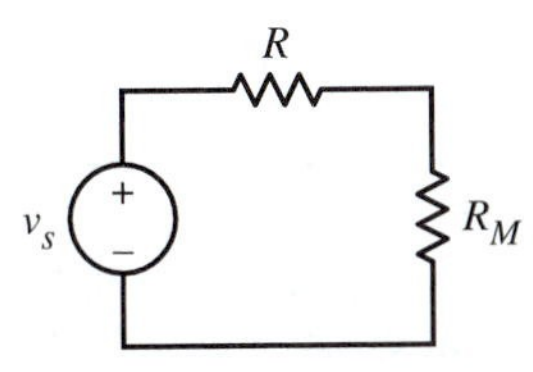

45. A low-current ammeter is used to measure resistance in an instrument called an ohmmeter. One design is shown in the following figure. We can simplify this circuit model by use of the equivalent series resistance $R_2 = R_z + R_M$. The scale of the ohmmeter is determined by R_1, and R_z is adjusted so that when terminals a and b are short-circuited, the meter reads a full-scale value. The ammeter reads a full-scale value of 50 μA when the voltage across the meter is 1 V.

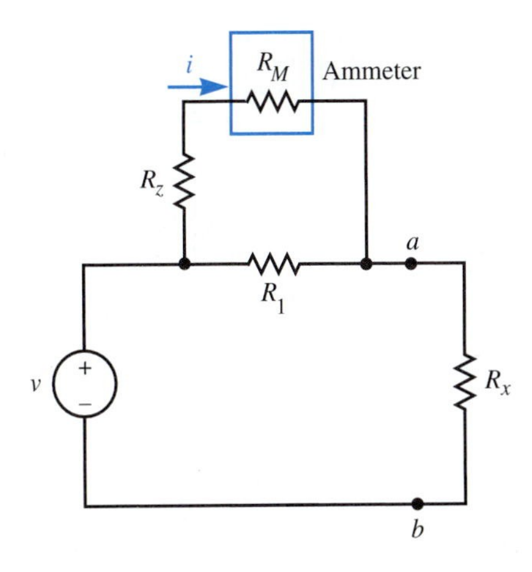

(a) What is the resistance R_M of the ammeter?

(b) Resistance R_2 is the series combination of R_M and R_z. Find an equation that gives i as a function of R_1, R_2, R_x, and v.

(c) Find an equation that gives R_x as a function of R_1, R_2, v, and i.

(d) With voltage $v = 3$ V, the ammeter reading must be 25 μA (one-half scale) when $R_x = 30\ \Omega$ and 50 μA when $R_x = 0$. What must be the value of R_1, R_2, and R_z?

(e) If v changes and the meter is set to read full scale for $R_x = 0$ by adjusting R_z, will the half-scale reading of 25 μA still correspond to $R_x = 30\ \Omega$?

(f) A resistance R_s is included in series with the voltage source v to yield a more accurate model. With R_1 as previously calculated, will adjustment of R_z for a full-scale reading when $R_x = 0$ yield a proper reading when $R_x = 30\ \Omega$?

46. A small-signal equivalent circuit for a field-effect transistor (FET) amplifier is shown on the next page. Terminals G, S, and D represent the gate, source, and drain, respectively.

(a) Determine v_{GS}, i, i_1, i_2, and v_o.

(b) You wish to deliver the same current to both R_1 and R_2. What resistance would you place in series with R_2?

(c) You wish to deliver the same power to both R_1 and R_2. What resistance would you place in series with R_2?

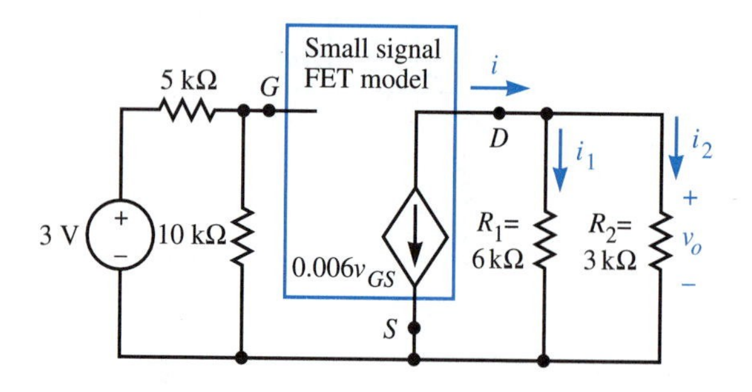

47. Semiconductor devices, such as a field-effect transistor (FET), operate properly only within a certain range of voltages and must be *biased* to operate within that range. The equivalent circuit shown below represents the bias problem for an FET. The large-signal transconductance is $g = 0.001$ S when the drain current i_D is in the range $0.8 < i_D < 1.2$ mA.
   (a) Use the value of $R_2 = 100$ kΩ, and select R_1 so that the current i_D is 1 mA.
   (b) Select R_1 and R_2 so that the series combination is 100 kΩ and v_{S0} is 1 V.

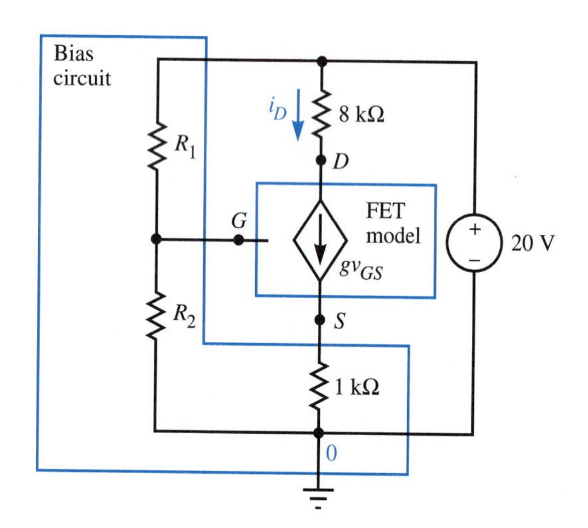

48. Semiconductor devices operate properly only within a certain range of voltages and currents and must be *biased* to operate within that range. The following figure represents the bias problem for a bipolar transistor, where terminals *B, C,* and *E* are the base, collector, and emitter connections, respectively. You must bias for a collector current i_C of 1 mA. Determine (a) the currents i_B and i_E; (b) the value of v_{CO}, v_{EO}, and v_{BO}; (c) the value of i_1 and i_2 if $R_1 = 10$ kΩ; and (d) the value of R_2.

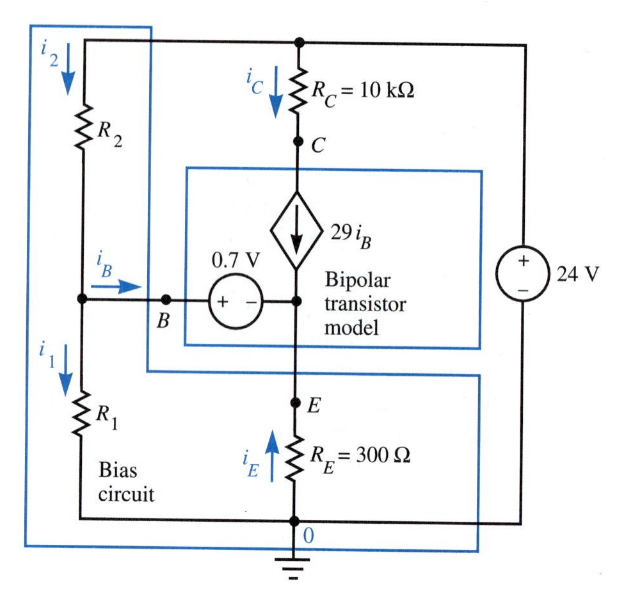

49. For each of the following two networks, replace the sources with a single equivalent source that will have the same effect on the load that is connected to terminals *a* and *b*.

(a)

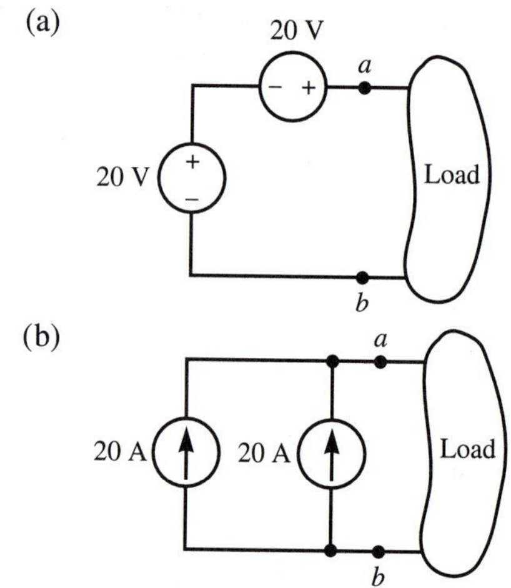

50. You want to design a 16-W electric coffee warmer for your car. The warmer is connected to a part of the electrical system that is modeled as a 12-V source in series with a 2-Ω resistance. The warmer can be modeled as a resistance. What two values of resistance can the warmer have? Which value should you use? Why? What is the voltage across the warmer for the value of resistance selected? How much power is absorbed by the 2-Ω system resistance?

51. You plan to convert your classic Volkswagen Beetle from a 6-V to a modern 12-V electrical system. You wish to retain some of the 6-V instrument lights and accessories. Two 6-V devices,

one modeled as a resistance of 4 Ω and the other modeled as a resistance of 12 Ω, are to be operated in parallel from the 12-V system. A resistance is inserted in series with the parallel combination as shown in the following illustration.

(a) What value of R should you use to supply 6 V to the two devices?

(b) Determine the power absorbed by each component.

(c) The device modeled as a 12-Ω resistance fails as an open circuit. What is the power absorbed by the 4-Ω device?

(d) The device modeled as a 4-Ω resistance fails as an open circuit. What is the power absorbed by the 12-Ω device?

(e) Can you see any problems with this design? Can you propose a better design?

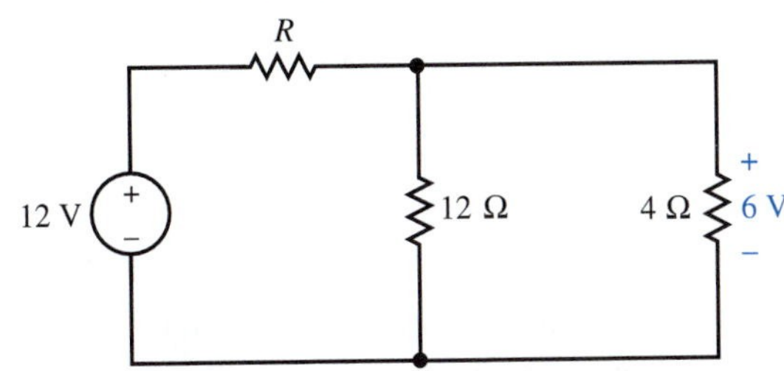

52. The heater fan on an automobile is operated by a 12-V dc motor with a permanent-magnet field. We can model the motor as a resistance of 2 Ω in series with a dependent voltage source with a value dependent on motor speed n, as shown in the following illustration. The power absorbed by this dependent source is equal to the mechanical power developed by the motor.

(a) Calculate the current i and the average mechanical power P developed as a function of speed n, where n is in revolutions per minute.

(b) The system has been designed to run at a speed n_1 that corresponds to the maximum average mechanical power P_1 that can be developed. Determine the speed n_1 and the current i_1 at power P_1.

(c) We must design a two-speed controller for the motor. For the slow speed, we will insert a resistance of value R in series with the motor. The mechanical power required for the fan is proportional to the cube of the motor speed. What value of series resistance R is required if P_2 is to be one-half of P_1, and what is the speed n_2 and current i_2 at power P_2?

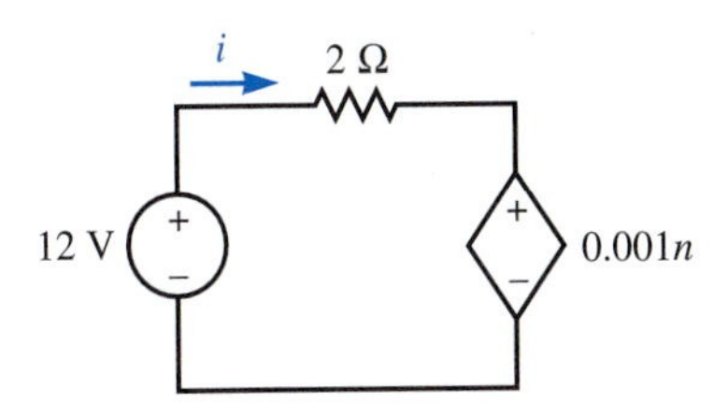

53. In electronics we often need to *match* the resistance R_S of a source to the resistance R_L of the load. (One reason for this is a topic in electromagnetic field theory.) A simple way to do this is with a resistive L network (R_1 and R_2) as shown in the following circuit. We need to make the equivalent resistance for the network to the right of terminals a and b (with R_L connected) equal to R_S. We must also make the equivalent resistance for the network to the left of terminals c and d (with R_S and v_S connected, but with v_S set equal to zero) equal to R_L. For $R_S > R_L$, use equivalent parallel and series resistance combinations and some tedious algebra to show that we need to have

(a) $R_1 = \beta R_S$ where $\beta = \sqrt{1 - R_L/R_S}$

(b) $R_2 = R_L/\beta$

(c) Use the voltage divider relation and a parallel equivalent resistance to show that the *voltage gain* is

$$G_v = v_2/v_1 = 1 - \beta$$

The gain is less than one, so the L network is an *attenuator.*

(d) The power gain is the ratio of the power delivered to R_L divided by the power delivered to the network to the right of terminals a and b. Show that

$$G_P = (1 - \beta)^2(R_S/R_L)$$

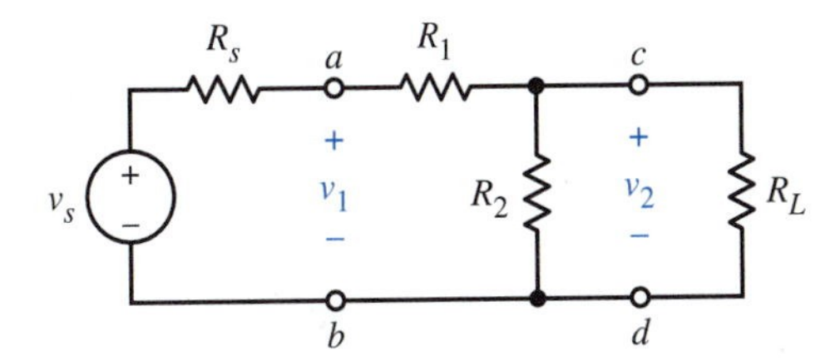

54. Refer to the preceding problem. Design a resistive L matching network for $R_S = 100$ Ω and $R_L = 36$ Ω. Determine the voltage gain and the power gain.

55. A resistive L network can also be used to match a source resistance R_S to a load resistance R_L when $R_S < R_L$. (Refer to Problem 53 for the defini-

tions.) This configuration is shown in the following circuit. Show that

(a) $R_1 = \alpha R_L$ where $\alpha = \sqrt{1 - R_S/R_L}$

(b) $R_2 = R_S/\alpha$

(c) the voltage gain is

$$G_V = 1/(1 + \alpha)$$

(d) and the power gain is

$$G_P = [1/(\alpha + 1)]^2(R_S/R_L)$$

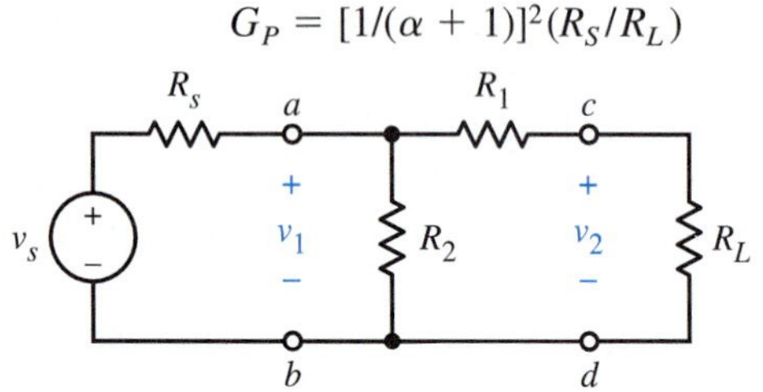

56. Refer to the preceding problem. Design a resistive L matching network for $R_S = 36\ \Omega$ and $R_L = 100\ \Omega$. Determine the voltage gain and the power gain.

57. The resistive L in the circuit of Problem 53 is often used as an attenuator when we need to match to the source resistance R_S but not to the load resistance R_L. That is, we specify the voltage gain $G_v = v_2/v_1$ and only require that the equivalent resistance of the network to the right of terminals a and b must equal R_S. For $R_S > R_L$, derive the following results. (Refer to Problem 53 for the definitions.)

(a) $R_1 = (1 - G_v)R_S$

(b) $R_2 = G_v R_S R_L/(R_L - G_v R_S)$

(c) $0 < G_v < R_L/R_S < 1$

(d) $G_P = G_v^2 R_S/R_L$

58. Refer to Problem 57. Design a resistive L attenuator with a voltage gain $G_v = 1/4$ that is matched to a source with 600-Ω resistance when the load is 300 Ω. What is the resistance seen by the 300-Ω load (the resistance of the circuit to the left of terminals c and d in the network of Problem 53) when $v_S = 0$?

59. A symmetrical pi (Π) attenuator is a useful resistive network. This can provide any specified gain $G_v = v_2/v_1$ between zero and one, and at the same time match a source of resistance R to a load of resistance R. That is, the resistance of the network to the right of terminals a and b is R with the load connected. Similarly, the resistance to the left of terminals c and d is R with the source and series resistance connected and with $v_S = 0$. Use equivalent series and parallel resistance values to establish the following two results.

(a) $G_v = R_2R/(R_1R_2 + R_1R + R_2R)$

(b) $R = R_2(R_1R_2 + R_1R + R_2R)/(R_1R_2 + R_2^2 + R_1R + 2R_2R)$

(c) Substitute the equation from part a into the equation in part b to obtain

$$R_2 = [(1 + G_v)/(1 - G_v)]R$$

(d) Substitute the result of part c into the equation in part a to obtain

$$R_1 = [(1 - G_v^2)/2G_v]R$$

(e) Show that the power gain (the ratio of the power delivered to the load resistance to the power delivered to the network to the right of terminals a and b) is

$$G_P = G_v^2$$

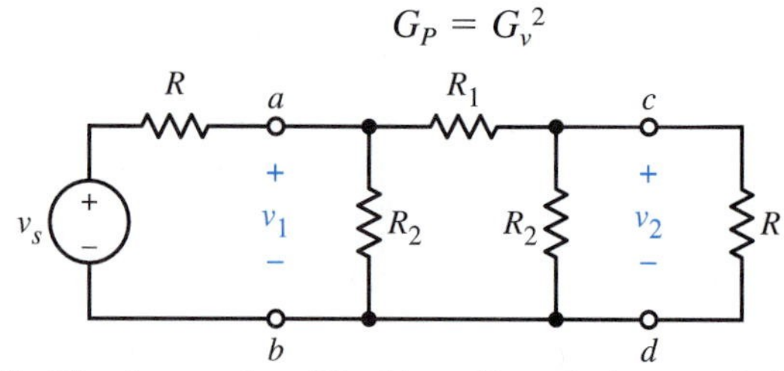

60. Use the results of Problem 59 to design a resistive pi attenuator with a voltage gain of 1/2 when the source and load resistances are 300 Ω.

61. Use the results of Problem 59 to develop the design equations for the balanced attenuator shown below. Express R_a and R_b as a function of G_v.

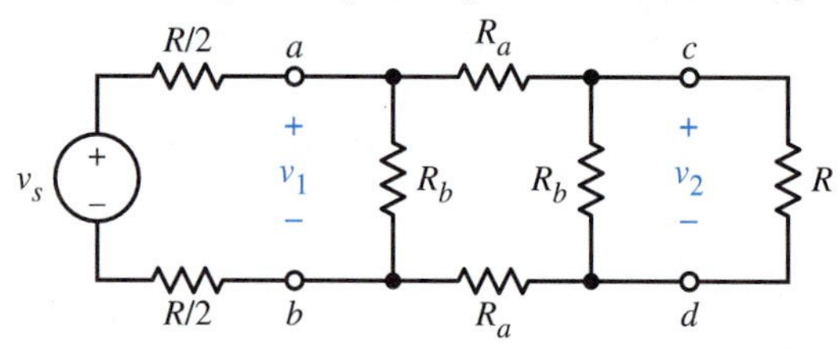

62. Twelve resistances, each with the same value R, are connected to form the edges of a cube as shown in the following illustration. What resistance would be measured between the indicated terminals?

(a) a and b

(b) b and c

(c) a and d

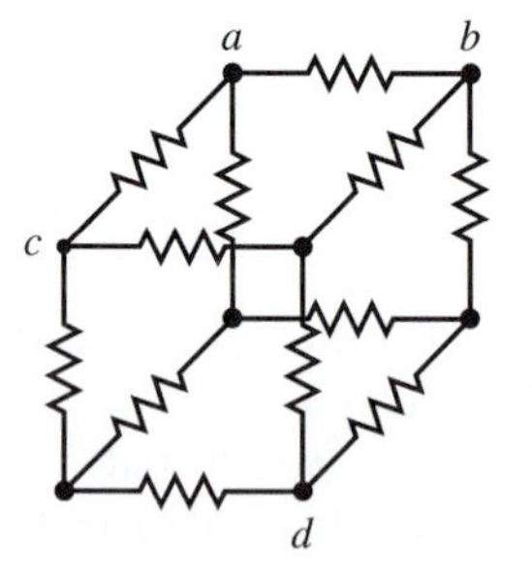

4

Node Voltages and Mesh Currents

In this chapter we develop two systematic approaches to network analysis. The first method, called node-voltage analysis, is applicable to both planar and nonplanar electrical networks. Node-voltage analysis is a generalization of the method we used in Chapter 3 to analyze simple parallel circuits. The second method, called mesh-current analysis, is applicable to only *planar* networks. Mesh-current analysis is a generalization of the method we used in Chapter 3 to analyze simple series circuits. Both methods yield a set of linearly independent equations that we can solve to obtain the *node voltages* or *mesh currents*.

These systematic approaches efficiently combine KCL, KVL, and the component equations to arrive at a complete specification of all component voltages and currents in the network. In addition, the methods described here are also useful to establish network theorems presented in later chapters.

4.1 Node Voltages

A standard method to measure device voltages is to connect one terminal of a voltmeter or oscilloscope to a node of an electrical system and measure the voltages of all other nodes with respect to this *reference node*. This method is illustrated in Fig. 4.1.

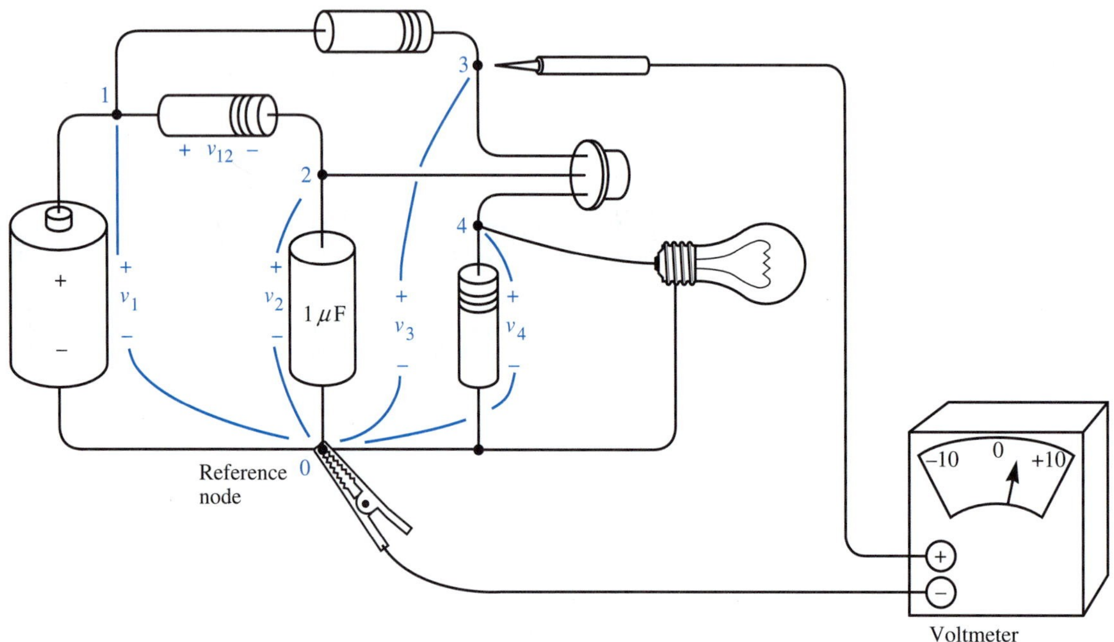

FIGURE 4.1
Measuring node voltages v_1, v_2, v_3, and v_4

Five nodes are accessible in the system illustrated. Therefore, four node voltages can be measured with respect to the reference node. We define node voltage v_k as the voltage measured with the (+) terminal of the voltmeter connected to node k and the (−) terminal of the voltmeter connected to the reference node, node 0. That is, for an $(N + 1)$-node circuit (see Fig. 4.1 for $N + 1 = 5$),

DEFINITION Node Voltages

$$v_k = v_{k0}, \qquad k = 0, 1, 2, \ldots, N \tag{4.1}$$

Once we have measured all node voltages, we know the voltage across any device that is connected directly to the reference node. Moreover, we can calculate the voltage across any device that is not directly connected to the reference node as the difference of two node voltages. For example, in Fig. 4.1, KVL gives us

$$v_{12} = v_1 - v_2 \tag{4.2}$$

In general, for a connected circuit with $N + 1$ nodes, the voltage across any branch is

Branch Voltages

$$v_{kj} = v_k - v_j \qquad k, j = 0, 1, 2, \ldots, N \tag{4.3}$$

We can readily adapt this experimental procedure to the analysis of a network model. The first step is to draw the network diagram and select node 0, the *reference node*. The choice of a reference node is completely arbitrary, but the node with the largest number of components or voltage sources connected to it is usually most convenient.

After we choose a reference node, the next step is to number the remaining nodes 1, 2, . . . , N. The choice of a numbering scheme is arbitrary.

We next write the *node-voltage equations* to solve for the node voltages. This procedure is *easily* and best learned by studying examples.

We begin with a circuit for which the node-voltage equations are all KVL equations. The analysis is very similar to that of a parallel circuit driven by a voltage source.

Node-Voltage Equations That Are KVL Equations

We must determine only one node voltage for a parallel or two-node circuit. If the circuit is driven by a voltage source, a KVL equation gives the node voltage immediately. In a more complicated circuit, a KVL equation† also gives us the node voltage for any node connected to the reference node by a path of voltage sources. We will demonstrate this with the following example.

EXAMPLE 4.1 Write a set of node-voltage equations for the network shown in Fig. 4.2.

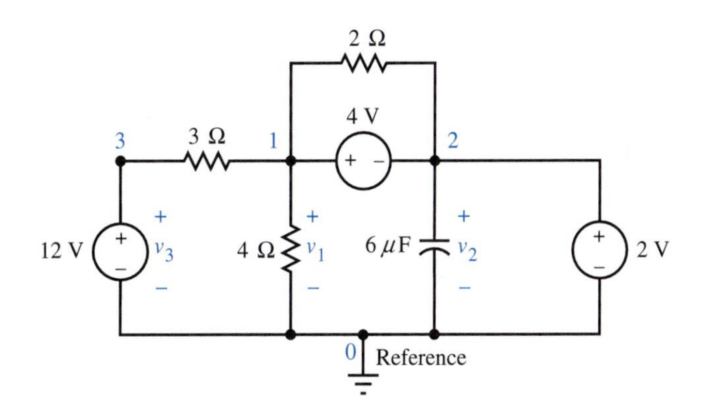

FIGURE 4.2 Network with node-voltage equations that are KVL equations

Solution Since there are four nodes and three voltage sources in the network of Fig. 4.2, we must write three KVL equations to obtain the node voltages. The node-voltage equation for node 1 is found from KVL by equating voltages along the paths indicated in the following illustration. (Only the components that influence the equations are shown explicitly; the other components are represented by dotted lines in the illustrations.)

† These KVL equations are often called *constraint equations*, because voltages between nodes are constrained by the voltage sources.

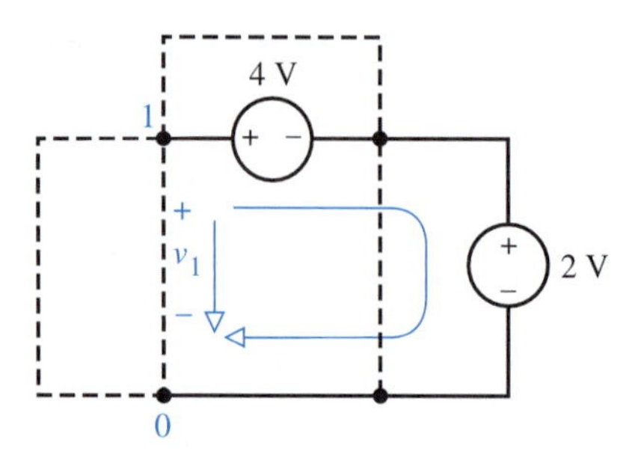

$$v_1 = 4 + 2 = 6 \text{ V}$$

We quickly obtain the node-voltage equations for nodes 2 and 3 from KVL:

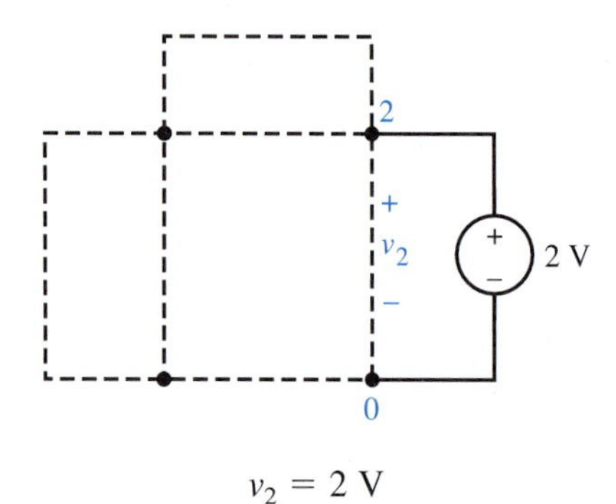

$$v_2 = 2 \text{ V}$$

and

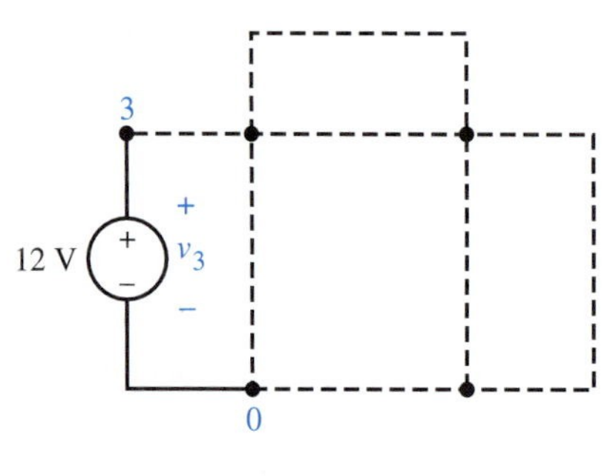

$$v_3 = 12 \text{ V}$$

We can collect these three node-voltage equations in matrix form to yield

$$\begin{bmatrix} v_1 \\ v_2 \\ v_3 \end{bmatrix} = \begin{bmatrix} 6 \\ 2 \\ 12 \end{bmatrix} \text{V}$$

Notice that the values of resistance and capacitance have no effect on the node voltages in this circuit. The element currents are obtained directly from the terminal equations, but we must use KCL equations to find the currents through the voltage sources.

In this section we examined a circuit for which all node-voltage equations were KVL equations. In the next section we consider a circuit for which all node-voltage equations are KCL equations. The analysis is very similar to that of a parallel circuit driven by a current source.

Remember We will write a KVL equation for each node connected to the reference node by a path of voltage sources. The KVL equation is a node-voltage equation and must be written in terms of known source values and node voltages.

EXERCISES

1. Write a set of node-voltage equations for the network shown below. Use KVL to determine the value of voltage v_{12}, and calculate current i_b.

   answer: −12 V, 36 V, −48 V, 8 A

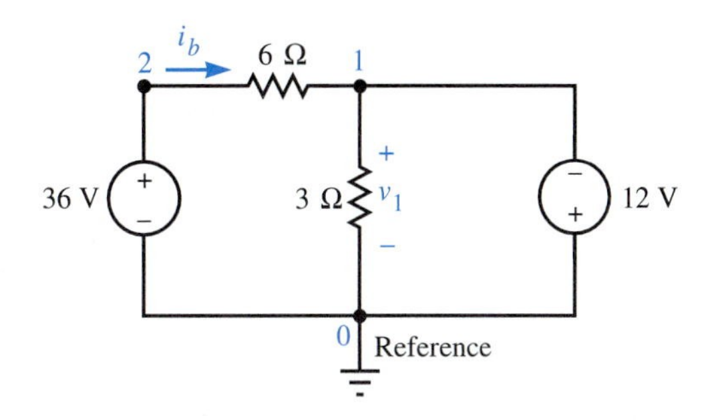

2. Replace the 12-V source in the circuit above with a dependent voltage source [(+) reference mark at the top]. The control equation is $v = 6i_b$. Then repeat Exercise 1. Write i_b in terms of the node voltages v_1 and v_2. The unknown current i_b must not appear in the final node-voltage equations. (Control variables must always be written in terms of node voltages or known voltages and currents.)

   answer: 18 V, 36 V, −18 V, 3 A

Node-Voltage Equations That Are KCL Equations

We obtain the node voltage for a parallel circuit driven by a current source by writing and solving a single KCL equation. The following example shows how we use the same approach for finding the node voltages for a more complicated circuit not excited by voltage sources.

EXAMPLE 4.2 The node numbers for the network of Fig. 4.3 are arbitrarily assigned. (a) Write a set of node-voltage equations for the network for Fig. 4.3 and solve for all node voltages. (b) Replace the 6-Ω resistance with a 6-F capacitance and the 10-Ω resistance with a 10-H inductance. Write the node-voltage equations.

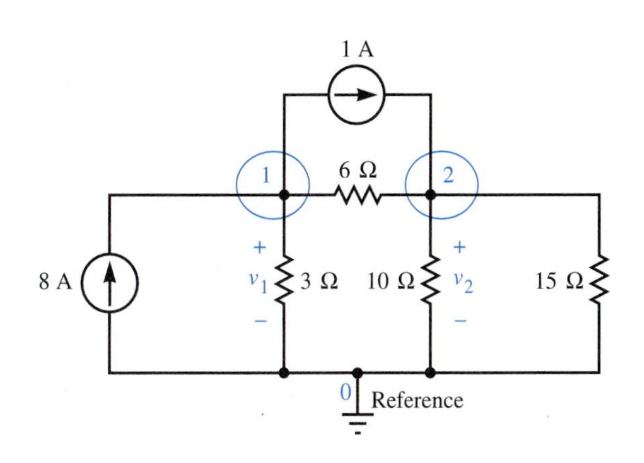

FIGURE 4.3 Network with node-voltage equations that are KCL equations

Solution (a) We obtain the node-voltage equations by the use of KCL. The KCL equation for a surface enclosing node 1 is

$$-8 + 1 + i_{10} + i_{12} = 0$$

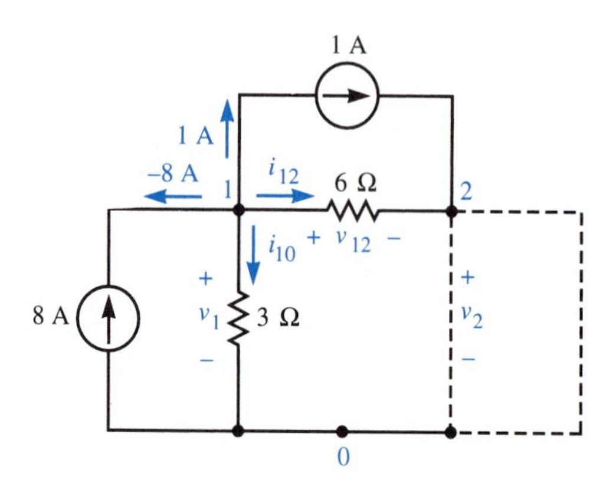

The terminal equation for the 3-Ω resistance gives

$$i_{10} = \frac{1}{3} v_1$$

and (with the use of KVL) the terminal equation for the 6-Ω resistance gives

$$i_{12} = \frac{1}{6} v_{12} = \frac{1}{6} (v_1 - v_2)$$

Substitution from these terminal equations lets us write our KCL equation as

$$-8 + 1 + \frac{1}{3} v_1 + \frac{1}{6} (v_1 - v_2) = 0$$

The first three equations could have been bypassed and this last equation written directly. Just remember that the current i_{kj} through an element is obtained by substitution of $v_k - v_j$ for v_{kj} in the terminal equation.

We can rearrange the preceding equation to give

$$\frac{1}{2} v_1 - \frac{1}{6} v_2 = 7$$

This is the first node-voltage equation. We obtain the second node-voltage equation from a KCL equation for a surface enclosing node 2:

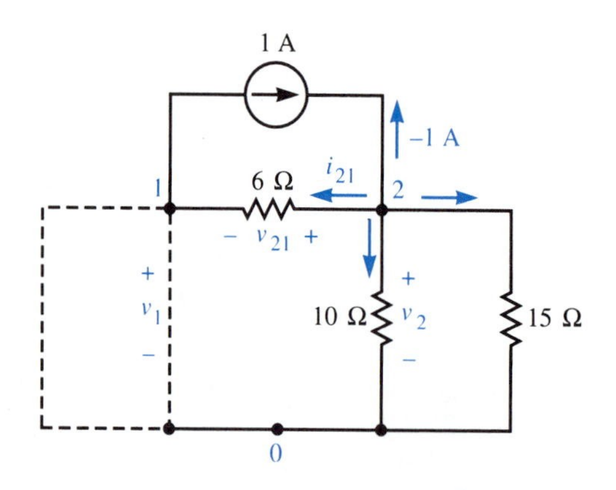

$$-1 + \frac{1}{6}(v_2 - v_1) + \frac{1}{10}v_2 + \frac{1}{15}v_2 = 0$$

which we can rearrange to give

$$-\frac{1}{6}v_1 + \frac{1}{3}v_2 = 1$$

These two node-voltage equations written as a single matrix equation are

$$\begin{bmatrix} \frac{1}{2} & -\frac{1}{6} \\ -\frac{1}{6} & \frac{1}{3} \end{bmatrix} \begin{bmatrix} v_1 \\ v_2 \end{bmatrix} = \begin{bmatrix} 7 \\ 1 \end{bmatrix}$$

Note that the only unknowns that appear in this equation are node voltages.

We can solve this matrix equation by Cramer's rule† (or with a calculator) to give the node voltages:

$$\Delta = \begin{vmatrix} \frac{1}{2} & -\frac{1}{6} \\ -\frac{1}{6} & \frac{1}{3} \end{vmatrix} = \left(\frac{1}{2}\right)\left(\frac{1}{3}\right) - \left(-\frac{1}{6}\right)\left(-\frac{1}{6}\right) = \frac{5}{36}$$

$$v_1 = \frac{\begin{vmatrix} 7 & -\frac{1}{6} \\ 1 & \frac{1}{3} \end{vmatrix}}{\Delta} = \frac{(7)(1/3) - (1)(-1/6)}{5/36} = 18 \text{ V}$$

$$v_2 = \frac{\begin{vmatrix} \frac{1}{2} & 7 \\ -\frac{1}{6} & 1 \end{vmatrix}}{\Delta} = \frac{(1/2)(1) - (7)(-1/6)}{5/36} = 12 \text{ V}$$

The voltage across any component is given by either a node voltage or the difference in node voltages. For example,

$$v_{12} = v_1 - v_2 = 18 - 12 = 6 \text{ V}$$

Thus

$$i_{12} = \frac{1}{6}v_{12} = \frac{1}{6}(18 - 12) = 1 \text{ A}$$

(b) With the two resistances replaced by a capacitance and inductance as requested, the first KCL equation is

$$-8 + 1 + \frac{1}{3}v_1 + 6\frac{d}{dt}(v_1 - v_2) = 0$$

and the second KCL equation is

$$-1 + 6\frac{d}{dt}(v_2 - v_1) + \frac{1}{10}\int_{-\infty}^{t} v_2 \, d\lambda + \frac{1}{15}v_2 = 0$$

† See Appendix A.

The inclusion of inductance and capacitance does not significantly complicate writing the two node-voltage equations, but the resulting integro-differential equations are more difficult to solve than algebraic equations. We will postpone the solution of such equations until Chapters 8 and 9.

This section has dealt with a circuit with node-voltage equations that are all KCL equations. We will introduce the concept of supernodes in the next section. These require both KCL and KVL equations.

Remember

We will write a KCL equation for each node (except the reference node) that has no voltage sources connected to it. The KCL equation is a node-voltage equation and must be written in terms of node voltages and known source values.

EXERCISES

3. (a) Write a set of node-voltage equations for the network shown below, and solve for the node voltages. Determine the value of v_{12}.

   answer: 70 V, − 105 V, 175 V

   (b) Write a KCL equation for a surface enclosing the reference node. Observe that this is the negative of the sum of the two KCL equations written in part (a), and is thus linearly dependent on them. This is why we will not use a KCL equation for the reference node as a node-voltage equation.

   answer: $-\frac{1}{4}v_1 - \frac{1}{2}v_2 = 35$

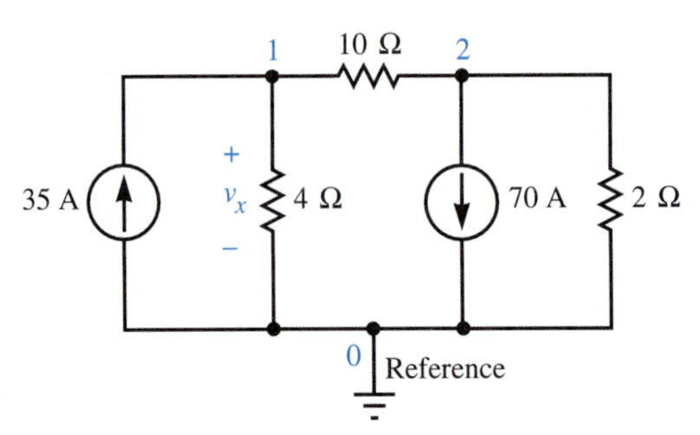

4. Replace the 70-A source in the above network with a dependent current source (reference arrow pointing up). The control equation is $i = \frac{1}{4}v_x$. Repeat Exercise 3(a). Write v_x in terms of the node voltage. The unknown voltage v_x must not appear in the final equations.

   answer: 120 V, 70 V, 50 V

5. Replace the 10-Ω resistance in the above figure with a 10-H inductance and the 2-Ω resistance with a 2-F capacitance. Write the two node-voltage equations, but do not solve them.

answer:

$$\begin{bmatrix} \frac{1}{4} + \frac{1}{10}\int_{-\infty}^{t} d\lambda & -\frac{1}{10}\int_{-\infty}^{t} d\lambda \\ -\frac{1}{10}\int_{-\infty}^{t} d\lambda & 2\frac{d}{dt} + \frac{1}{10}\int_{-\infty}^{t} d\lambda \end{bmatrix} \begin{bmatrix} v_1 \\ v_2 \end{bmatrix} = \begin{bmatrix} 35 \\ -70 \end{bmatrix}$$

Supernodes Require KCL and KVL Equations

To facilitate discussion, we call nodes that are connected to each other by voltage sources, but not to the reference node by a path of voltage sources, a *supernode*. We include in the supernode all components connected in parallel with these sources. A supernode requires one equation for each node contained within the supernode. We must write a KCL equation for a surface enclosing the supernode. The remaining equations are KVL equations.

For a shortcut, we can select one node within the supernode as a *principal node*, and use KVL to write the other node voltages in the supernode as a sum of this principal node voltage and voltage source values. We then write a KCL equation for a surface enclosing the supernode, just as we would for a node of a parallel circuit. The following example demonstrates the procedure.

EXAMPLE 4.3 Write a set of node-voltage equations for the network shown in Fig. 4.4.

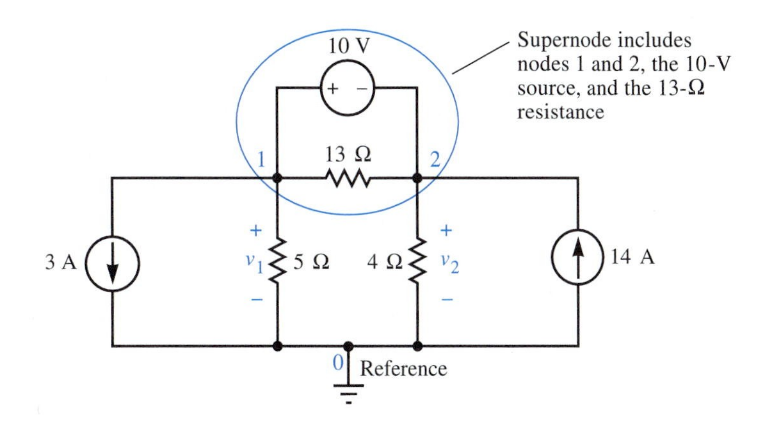

FIGURE 4.4 Network with one supernode that requires both a KCL and a KVL equation

Solution With the nodes numbered as in Fig. 4.4, nodes 1 and 2, together with the connecting voltage source and 13-Ω resistance, form one supernode. A KCL equation for a surface enclosing the supernode is

$$3 + \frac{1}{5}v_1 + \frac{1}{4}v_2 - 14 = 0$$

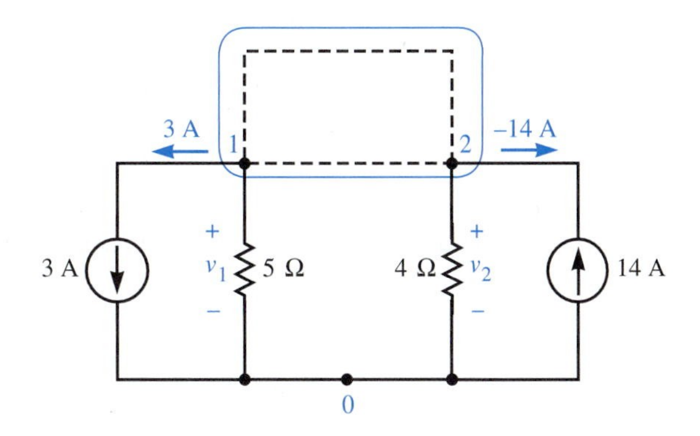

which is the first node-voltage equation. Use of KVL to equate the voltages along the paths indicated in the following illustration gives

$$v_1 - v_2 = 10$$

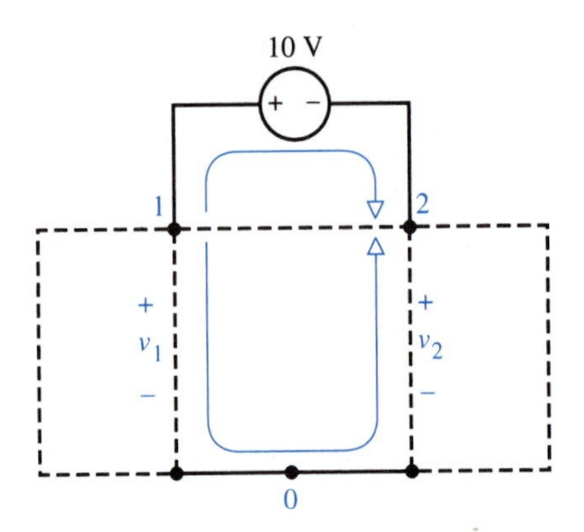

which is the second node-voltage equation, and relates the voltage source within the supernode to the node voltages. We can write these two equations as

$$\begin{matrix}\text{From KCL} \\ \\ \text{From KVL}\end{matrix}\left\{\begin{bmatrix}\frac{1}{5} & \frac{1}{4} \\ 1 & -1\end{bmatrix}\begin{bmatrix}v_1 \\ v_2\end{bmatrix} = \begin{bmatrix}11 \\ 10\end{bmatrix}\right\}\begin{matrix}\text{Current due to current source} \\ \\ \text{Voltage due to voltage source}\end{matrix}$$

and solve for the node voltages.

The labor necessary to write and solve the node-voltage equations is reduced if we use KVL to write

$$v_2 = v_1 - 10$$

directly on the network diagram before we write the KCL equation. Because we have written v_2 in terms of v_1, we call node 1 a *principal node*.

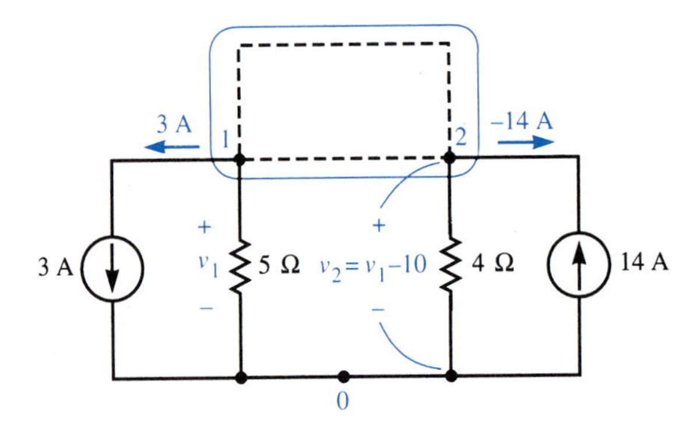

We can now write the KCL equation for a surface enclosing the supernode as

$$3 + \frac{1}{5}v_1 + \frac{1}{4}(v_1 - 10) - 14 = 0$$

which immediately yields

$$v_1 = 30 \text{ V}$$

Observe how this shortcut reduced the analysis procedure to one similar to that for a simple parallel circuit. The second node voltage is given by the KVL equation as

$$v_2 = 30 - 10 = 20 \text{ V}$$

This section has investigated a group of nodes we call a supernode. In the following section we will examine the general circuit.

Remember

Nodes within a supernode are connected by a path of voltage sources, but are not connected to the reference node by a path of voltage sources. A supernode requires one node-voltage equation that is a KCL equation. The remaining node-voltage equations are KVL equations.

EXERCISES

6. (a) Write a set of node-voltage equations for the network shown in the circuit below (v_x must be written in terms of the node voltages). First, write four node-voltage equations (two KVL and two KCL equations). Next, use the KVL equation to write v_3 and v_4 in terms of v_1 and v_2 directly on the network diagram. Then write a KCL equation for each supernode. The resulting two node-voltage equations should contain v_1 and v_2 as the only unknowns. Solve for all node voltages.

answer:

$$\begin{bmatrix} \frac{3}{5} & -\frac{1}{2} & \frac{11}{20} & -\frac{1}{2} \\ -\frac{1}{2} & -\frac{3}{2} & \frac{3}{2} & 1 \\ 1 & 0 & -1 & 0 \\ 0 & 1 & 0 & -1 \end{bmatrix} \begin{bmatrix} v_1 \\ v_2 \\ v_3 \\ v_4 \end{bmatrix} = \begin{bmatrix} 0 \\ 0 \\ 10 \\ 40 \end{bmatrix},$$

$$\begin{bmatrix} \frac{23}{20} & -1 \\ 1 & -\frac{1}{2} \end{bmatrix} \begin{bmatrix} v_1 \\ v_2 \end{bmatrix} = \begin{bmatrix} -\frac{29}{2} \\ 55 \end{bmatrix},$$

146.47 V, 182.94 V, 136.47 V, 142.94 V

(b) Use the node voltages found in part (a) to solve for i_{10} and i_{14}. Now write a KCL equation for a surface enclosing node 1 and solve for i_{13}. Observe that we could not write i_{13} in terms of v_1 and v_3 alone, but that $i_{13} = i_{30} + i_{32}$ can be written in terms of v_2 and v_3. This is why we wrote the node-voltage equations for surfaces enclosing the supernodes. answer: 14.65 A, 1.765 A, −16.41A

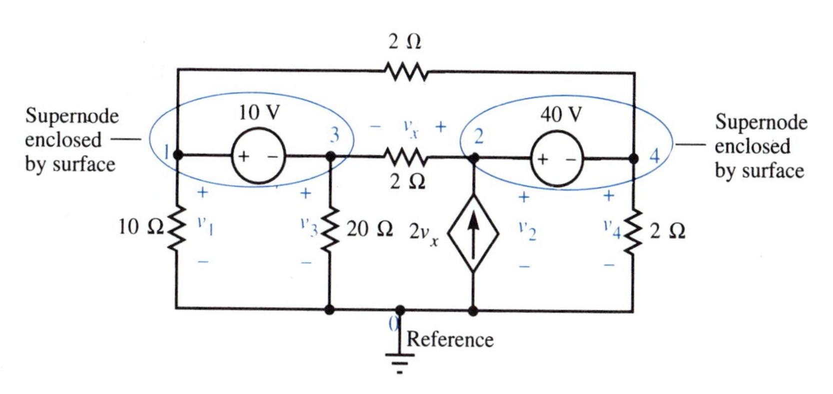

4.2 Network Analysis by Node Voltages

In many networks, some nodes are connected to the reference node by a path of voltage sources, some nodes are not connected to voltage sources, and some nodes form supernodes. We shall see that the methods we developed in Section 4.1 still apply.

EXAMPLE 4.4 Use the method of node voltages to analyze the circuit shown in Fig. 4.5.

FIGURE 4.5
A more general node-voltage example

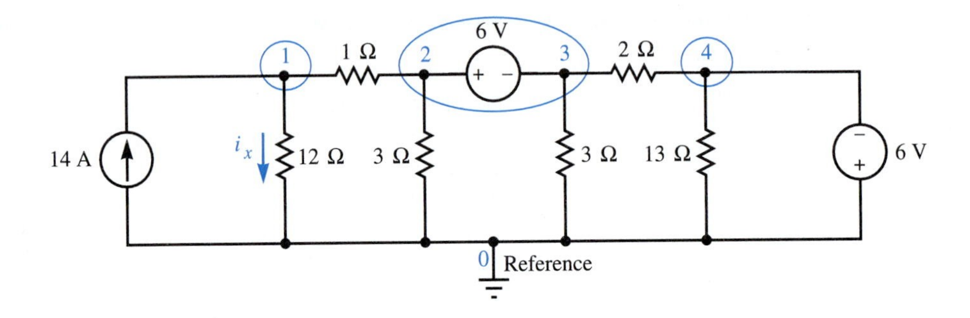

Solution Select the reference node as shown. Node 1 requires a KCL equation. Nodes 2 and 3 form a supernode and require both KCL and KVL equations. Node 4 requires only a KVL equation. Write a KCL equation for a surface that encloses node 1:

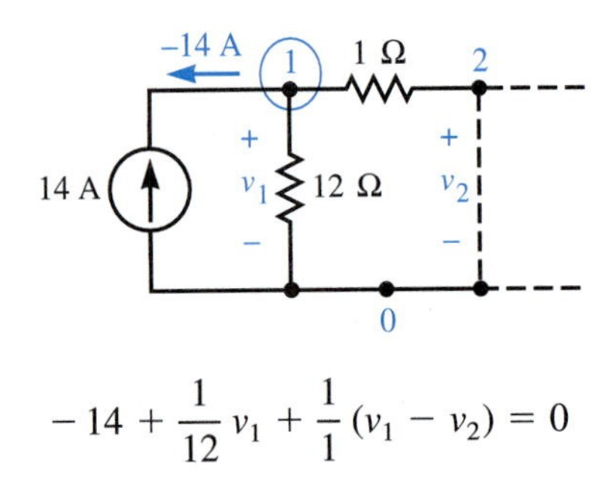

$$-14 + \frac{1}{12}v_1 + \frac{1}{1}(v_1 - v_2) = 0$$

Write a KCL equation for a surface enclosing the supernode formed by nodes 2 and 3:

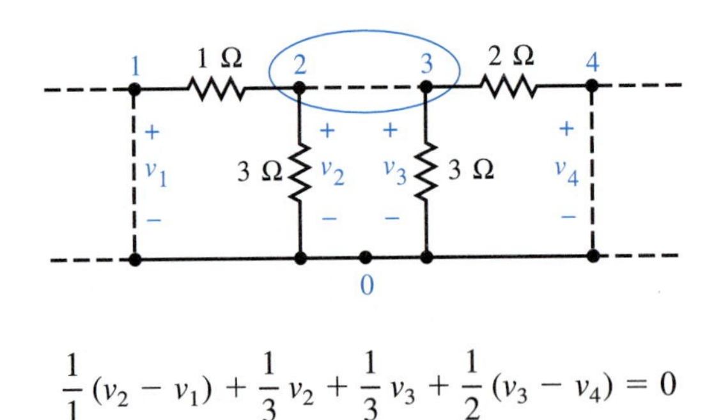

$$\frac{1}{1}(v_2 - v_1) + \frac{1}{3}v_2 + \frac{1}{3}v_3 + \frac{1}{2}(v_3 - v_4) = 0$$

The third and fourth node-voltage equations are obtained from KVL:

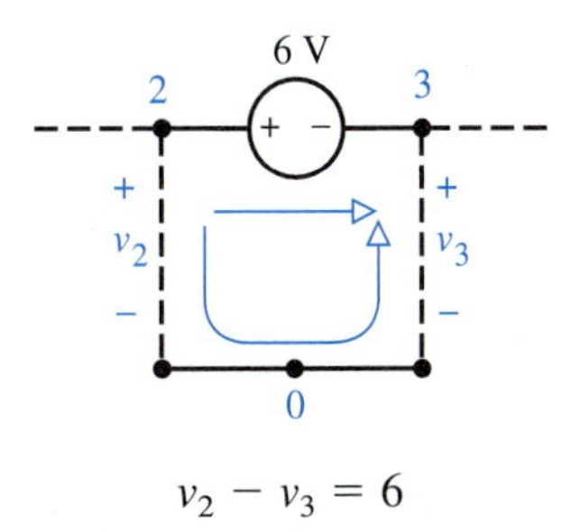

$$v_2 - v_3 = 6$$

and

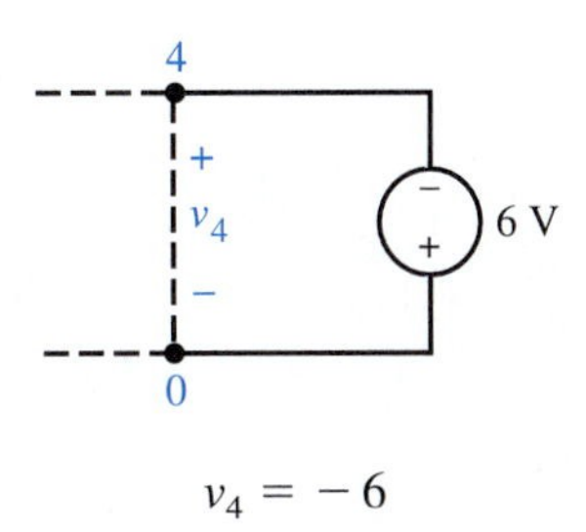

$$v_4 = -6$$

These four node-voltage equations form a single matrix equation:

$$\begin{matrix}\text{From KCL}\\ \\ \text{From KVL}\end{matrix}\begin{bmatrix} \frac{13}{12} & -1 & 0 & 0 \\ -1 & \frac{4}{3} & \frac{5}{6} & -\frac{1}{2} \\ 0 & 1 & -1 & 0 \\ 0 & 0 & 0 & 1 \end{bmatrix}\begin{bmatrix} v_1 \\ v_2 \\ v_3 \\ v_4 \end{bmatrix} = \begin{bmatrix} 14 \\ 0 \\ 6 \\ -6 \end{bmatrix}\begin{matrix}\text{Currents due to current sources}\\ \\ \text{Voltages due to voltage sources}\end{matrix}$$

We call the left-hand matrix the *node transformation matrix*.

We can reduce the labor involved in writing and solving the node-voltage equations by writing

$$v_3 = v_2 - 6$$

and

$$v_4 = -6$$

directly on the network diagram before writing the KCL equations. The first KCL equation remains the same, and the second KCL equation becomes

$$\frac{1}{1}(v_2 - v_1) + \frac{1}{3}v_2 + \frac{1}{3}(v_2 - 6) + \frac{1}{2}[(v_2 - 6) - (-6)] = 0$$

in which case the node-voltage equations reduce to

$$\begin{bmatrix} \frac{13}{12} & -1 \\ -1 & \frac{13}{6} \end{bmatrix}\begin{bmatrix} v_1 \\ v_2 \end{bmatrix} = \begin{bmatrix} 14 \\ 2 \end{bmatrix}$$

The left-hand matrix is the *reduced node transformation matrix*. We can solve this matrix equation for the *principal-node* voltages, v_1 and v_2:

$$\begin{bmatrix} v_1 \\ v_2 \end{bmatrix} = \begin{bmatrix} 24 \\ 12 \end{bmatrix} \text{V}$$

We obtain the two remaining node voltages v_3 and v_4 from our two KVL equations.

A summary of the shortcut procedure we have developed follows. We present the list for easy reference, not for memorization.

Shortcut Method ➤

Analysis of Node Voltages

1. Draw the network diagram and select the reference node. Usually the node with the largest number of voltage sources or components connected to it is the most convenient choice for the reference node.
2. Number the remaining nodes in sequence.
3. Write the node-voltage equations as follows:
   (a) Identify each node that is connected to the reference node by a path of voltage sources. Use KVL to write each of these node voltages in terms of the source voltages. Write these relationships on the network diagram.
   (b) Identify each group of nodes and components that form a supernode. Select one node within each group as a principal node. Use KVL to write the voltage of the other nodes within this group as the sum of the principal-node voltage and the source voltages. Write these relationships on the network diagram.
   (c) The remaining nodes, other than the reference node, are also principal nodes.
   (d) Now write a KCL equation for surfaces that enclose each node identified in (c) and each supernode.
   Remember that all unknowns must be written in terms of principal-node voltages.
4. Solve the KCL equations for the principal-node voltages. Each remaining node voltage is then obtained from a KVL equation.

We will now apply the shortcut procedure to a five-node network.

EXAMPLE 4.5 Use the shortcut procedure to write a set of node-voltage equations for the network shown in Fig. 4.6.

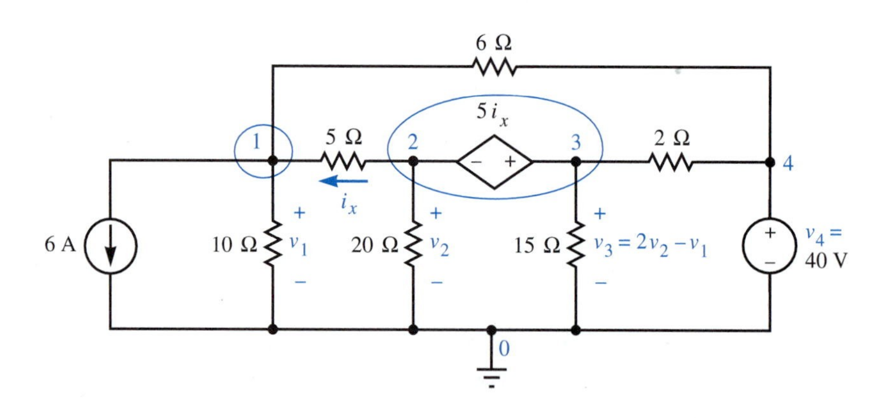

FIGURE 4.6 Example circuit for the shortcut procedure for node-voltage analysis

Solution

1. Select the reference node as shown.
2. Number the remaining nodes 1, 2, 3, and 4 as indicated.
3. Write the node-voltage equations.
   (a) Node 4 is connected to the reference node by a voltage source. Application of KVL gives us

$$v_4 = 40 \text{ V}$$

   Write $v_4 = 40$ V on the network diagram as shown.
   (b) Nodes 2 and 3 form a supernode. Select node 2 as the principal node. We again use KVL and Ohm's law to obtain

$$v_3 = 5i_x + v_2 = 5\left[\frac{1}{5}(v_2 - v_1)\right] + v_2 = 2v_2 - v_1$$

   Write $v_3 = 2v_2 - v_1$ on the network diagram as shown.
   (c) Node 1 is also a principal node, because no voltage sources are connected to it.
   (d) Write a KCL equation for a surface enclosing node 1:

$$6 + \frac{1}{10}v_1 + \frac{1}{5}(v_1 - v_2) + \frac{1}{6}(v_1 - 40) = 0$$

   Next write a KCL equation for a surface enclosing the supernode:

$$\frac{1}{5}(v_2 - v_1) + \frac{1}{20}v_2 + \frac{1}{15}(2v_2 - v_1) + \frac{1}{2}[(2v_2 - v_1) - 40] = 0$$

4. These two KCL equations, written as a single matrix equation, give us

$$\begin{bmatrix} \frac{7}{15} & -\frac{1}{5} \\ -\frac{23}{30} & \frac{83}{60} \end{bmatrix} \begin{bmatrix} v_1 \\ v_2 \end{bmatrix} = \begin{bmatrix} \frac{2}{3} \\ 20 \end{bmatrix}$$

   We can easily solve this matrix equation for $v_1 = 10$ V and $v_2 = 20$ V. From our two KVL equations, $v_3 = 30$ V and $v_4 = 40$ V.

The following example illustrates how easy it is to apply the shortcut procedure to even a complicated-looking network.

EXAMPLE 4.6 Use the shortcut procedure to write a set of node-voltage equations for the network shown in Fig. 4.7 on the next page.

Solution

1. Select the reference node as shown.
2. Number the remaining nodes 1, 2, 3, and 4.
3. Write the node-voltage equations.
   (a) Node 4 is connected to the reference node by the dependent voltage source. From KVL,

$$v_4 = -8i_x = -8\left(\frac{1}{2}v_1\right) = -4v_1$$

   Write this on the network diagram.

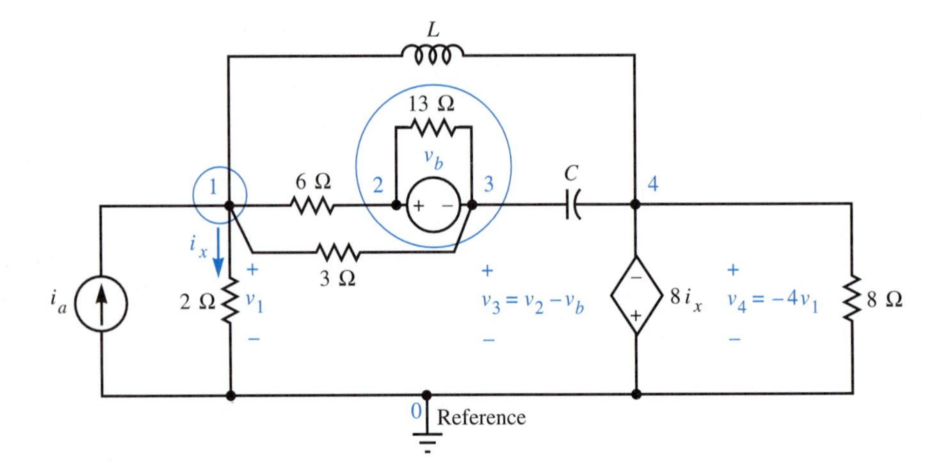

FIGURE 4.7 Node-voltage example with inductance and capacitance

(b) Nodes 2 and 3 form a supernode. Select node 2 as the principal node. From KVL,

$$v_3 = v_2 - v_b$$

Write this on the network diagram.

(c) Node 1 will also be a principal node, because no voltage sources are connected to it.

(d) Write a KCL equation for a surface enclosing node 1:

$$-i_a + \frac{1}{2} v_1 + \frac{1}{3} [v_1 - (v_2 - v_b)] + \frac{1}{6} (v_1 - v_2) + \frac{1}{L} \int_{-\infty}^{t} [v_1 - (-4v_1)] \, d\lambda = 0$$

Next apply KCL to a surface enclosing the supernode:

$$\frac{1}{6} (v_2 - v_1) + \frac{1}{3} [(v_2 - v_b) - v_1] + C \frac{d}{dt} [(v_2 - v_b) - (-4v_1)] = 0$$

4. We can write the two KCL equations as a single matrix differential equation:

$$\begin{bmatrix} 1 + \dfrac{5}{L} \displaystyle\int_{-\infty}^{t} d\lambda & -\dfrac{1}{2} \\ 4C \dfrac{d}{dt} - \dfrac{1}{2} & C \dfrac{d}{dt} + \dfrac{1}{2} \end{bmatrix} \begin{bmatrix} v_1 \\ v_2 \end{bmatrix} = \begin{bmatrix} i_a - \dfrac{1}{3} v_b \\ C \dfrac{d}{dt} v_b + \dfrac{1}{3} v_b \end{bmatrix}$$

where we use the notation

$$\int_{-\infty}^{t} v \, d\lambda = \int_{-\infty}^{t} d\lambda \cdot v$$

in writing the matrix equation. The solution of differential equations is postponed until Chapters 8 and 9, but once the principal-node voltages, v_1 and v_2, are determined, v_3 and v_4 are easily obtained from the two KVL equations.

The networks of Figs. 4.2 through 4.7 are all one-part (connected) networks. Occasionally the need arises to analyze a network like that shown in Fig. 4.8. This is an unconnected network consisting of *two separate parts*. To analyze this network by the method of node voltages, one can select a reference node for each part, as shown. Notice that $v_{00'}$, and thus v_{12}, is indeterminant, because no line connects the two parts of this circuit. More generally, a network with S separate parts will require S reference nodes, one for each part.

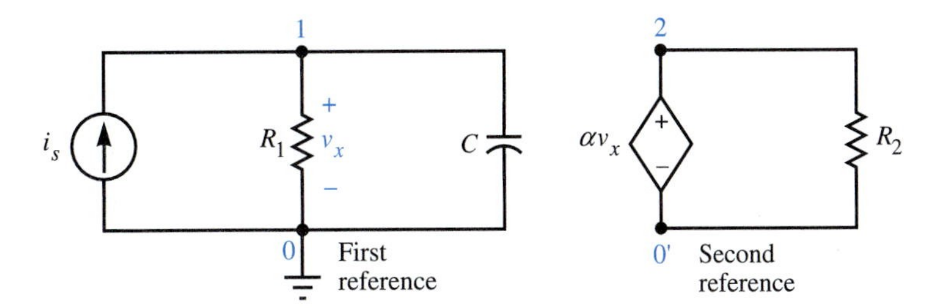

FIGURE 4.8 Two-part network with two reference nodes

An alternative procedure to analyze a network with separate parts is to connect one node in each part to a common reference node. For example, connect nodes 0 and 0′ in the circuit of Fig. 4.8. (This is the procedure to use for network analysis by the computer program PSpice.) When you do this, remember that voltage v_{12} is indeterminate in the original circuit.

Remember

The number of node-voltage equations required for a network is equal to the number of nodes minus the number of separate parts. If there are no loops of voltage sources, the number of node-voltage equations that are KVL equations is equal to the number of voltage sources. The remainder of the node-voltage equations are KCL equations. The effort required to solve the node-voltage equations is largely determined by the number of KCL equations.

EXERCISES

7. Use the method of analysis by node voltages to calculate the node voltages of the network shown in the following circuit. Then calculate the power absorbed by the dependent source. *answer:* − 100 V, 10 V, 20 V, − 220 W

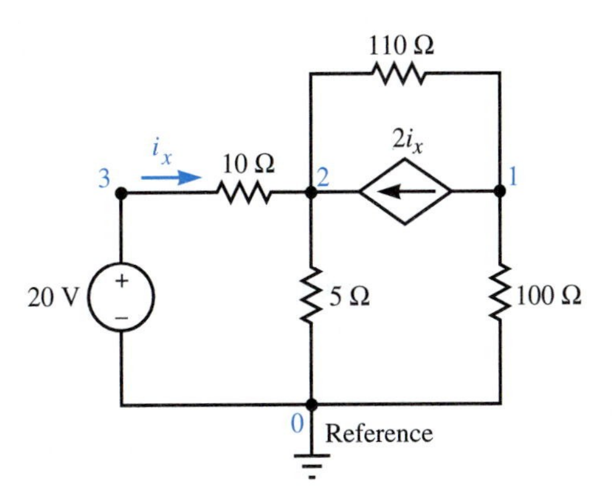

8. Replace the 20-V source in the above circuit with a short circuit and the 5-Ω resistance with the 20-V source. Keep the (+) reference mark at the top. Then repeat Exercise 7. *answer:* 219 V, 20 V, 0 V, − 796 W

9. Replace the dependent current source of the circuit used in Exercise 7 with a dependent voltage source [(+) reference to the right] having a control equation of $v = 110i_x$. Then repeat Exercise 7. *answer:* 230 V, − 1 V, 20 V, − 1016.4 W

10. (a) Replace the right-hand 2-Ω resistance of the circuit used in Exercise 6 with a voltage source of value 100 V [the (+) reference mark at the top]. Write the four node-voltage equations.

answer:
$$\begin{bmatrix} \frac{3}{5} & -\frac{1}{2} & \frac{11}{20} & -\frac{1}{2} \\ 1 & 0 & -1 & 0 \\ 0 & 1 & 0 & 0 \\ 0 & 0 & 0 & 1 \end{bmatrix} \begin{bmatrix} v_1 \\ v_2 \\ v_3 \\ v_4 \end{bmatrix} = \begin{bmatrix} 0 \\ 10 \\ 140 \\ 100 \end{bmatrix}$$

(b) Use the shortcut procedure and repeat part (a). This should give one KCL equation that contains v_1 as the only unknown. Solve for all node voltages.

answer: 109.13 V, 140 V, 99.13 V, 100 V

11. Write the node-voltage equations for the network of Fig. 4.8 on page 105.

answer:
$$\begin{bmatrix} C\frac{d}{dt} + \frac{1}{R} & 0 \\ -\alpha & 1 \end{bmatrix} \begin{bmatrix} v_1 \\ v_2 \end{bmatrix} = \begin{bmatrix} i_s \\ 0 \end{bmatrix}$$

12. Replace the upper voltage source of Fig. 4.5 on page 100 with a dependent voltage source [(+) reference mark on the left] having a control equation of $v = 3i_x$.

(a) Write the four node-voltage equations and arrange in matrix form. You should have two KCL equations and two KVL equations.

answer:
$$\begin{bmatrix} \frac{13}{12} & -1 & 0 & 0 \\ -1 & \frac{4}{3} & \frac{5}{6} & -\frac{1}{2} \\ \frac{1}{4} & -1 & 1 & 0 \\ 0 & 0 & 0 & 1 \end{bmatrix} \begin{bmatrix} v_1 \\ v_2 \\ v_3 \\ v_4 \end{bmatrix} = \begin{bmatrix} 14 \\ 0 \\ 0 \\ -6 \end{bmatrix}$$

(b) Use the two KVL equations written in part (a) to write v_3 and v_4 in terms of the specified voltage and current sources and node voltages v_1 and v_2. Write these on the network diagram. Directly from the network diagram, write two KCL equations that contain v_1 and v_2 as the only unknowns.

answer:
$$\begin{bmatrix} \frac{13}{12} & -1 \\ -\frac{29}{24} & \frac{13}{6} \end{bmatrix} \begin{bmatrix} v_1 \\ v_2 \end{bmatrix} = \begin{bmatrix} 14 \\ -3 \end{bmatrix}$$

(c) Solve for all node voltages and the power absorbed by the dependent voltage source. *answer:* 24 V, 12 V, 6 V, − 6 V, 48 W

4.3 Mesh Currents

We can determine the state of an electrical system by calculating or measuring each device current. There is a method that permits a solution for all device currents in terms of a smaller number of currents, called *mesh currents*. The method of analysis by mesh currents applies only to planar networks, that is, networks that can be drawn so that no lines cross.† Circuit analysis by mesh currents is very popular among students and some electrical engineers, even though it has no experimental equivalent.

We calculate the mesh currents by writing and solving a set of KVL and KCL equations, called *mesh-current equations*.

We can understand the definition of mesh currents by considering the system of electrical devices shown in Fig. 4.9. We assign currents i_1, i_2, and i_3 clockwise around each mesh as if it were an isolated single loop or series circuit. These are the mesh currents. A mesh current can be measured by a single ammeter reading only if that mesh current passes through a branch in the outer loop, as do mesh currents i_1 and i_2. We can see that mesh current i_3 does not constitute the total current through any single device.

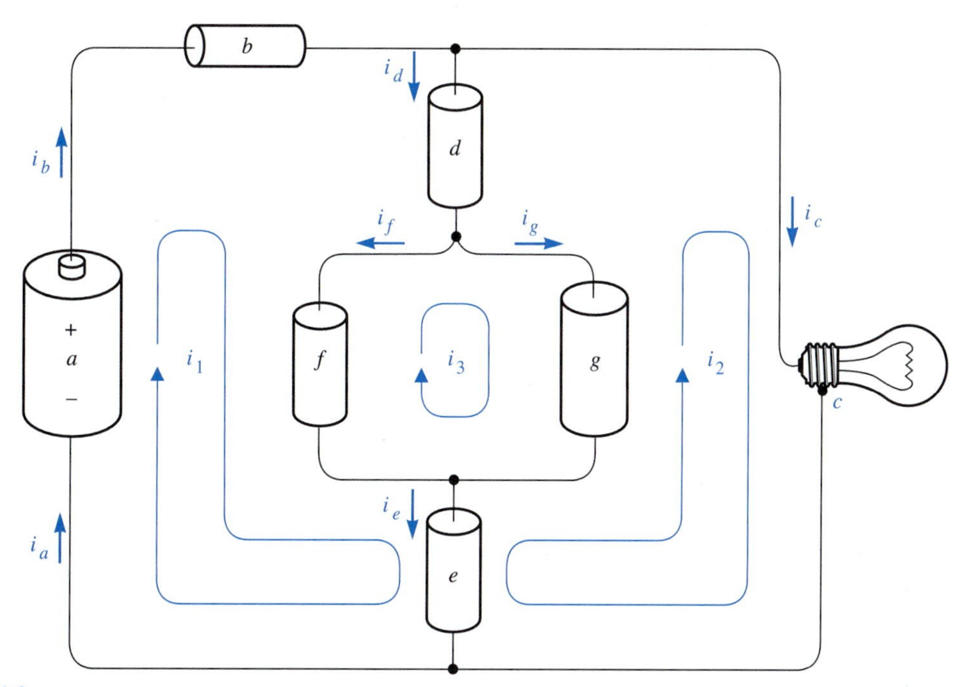

FIGURE 4.9
Circulating currents i_1, i_2, and i_3 are mesh currents

There are seven device currents to find. KCL directly implies $i_a = i_b$, and a KCL equation for a surface enclosing the top node yields $i_d = i_b - i_c = i_a - i_c$. It is not too hard to see that we can represent the device currents by a smaller set of currents with the aid of KCL. A clever way to do this is to write all currents in terms of mesh currents. Clearly

$$i_a = i_1 \tag{4.4}$$

$$i_b = i_1 \tag{4.5}$$

$$i_c = i_2 \tag{4.6}$$

and in general

† A generalization, called *link-current analysis* or *loop analysis*, applies to any network. Link-current analysis is not discussed in this chapter.

Mesh Currents

The current i_x through any device x in the outer loop is

$$i_x = i_j \tag{4.7}$$

where current i_j is the mesh current through the device x in the same direction as current i_x.

Application of KCL to a surface enclosing the top node yields

$$i_d = i_1 - i_2 \tag{4.8}$$

Another way to look at this latter equation is that device current i_d is composed of two components: i_1 in the direction of i_d, and i_2 in the opposite direction. (This is analogous to representing the voltage across a device connecting two nodes as the difference of two node voltages.) From this line of reasoning, device current i_e is

$$i_e = i_1 - i_2 \tag{4.9}$$

We can easily verify this by applying KCL to a surface enclosing the bottom node of the network. Repeated application of KCL will verify that the remaining device currents are represented in terms of the mesh currents as

$$i_f = i_1 - i_3 \tag{4.10}$$

and

$$i_g = i_3 - i_2 \tag{4.11}$$

In general,

Branch Currents

The current i_x through any device x not in the outer loop is

$$i_x = i_j - i_k \tag{4.12}$$

where current i_j is the mesh current through device x in the direction of i_x, and current i_k is the mesh current through device x in the opposite direction. (This assumes that all mesh currents are assigned in a clockwise or all in a counter-clockwise direction.)

The first step in the analysis by mesh currents is to draw the network diagram so that no unconnected lines cross. (If this is not possible, the circuit is not planar, so we cannot use this method.)

The second step is to assign a mesh current and number to each mesh. (Mesh currents are usually assigned clockwise, but this is not a necessity.) Any numbering scheme will work.

We next write the mesh-current equations to solve for the mesh currents. Like analysis by node voltages, the procedure is easily and best learned from examples.

We begin with a circuit for which mesh-current equations are all KCL equations. The analysis is very similar to that of a series circuit driven by a current source.

Mesh-Current Equations That Are KCL Equations

A series circuit has only one mesh and one mesh current. Kirchhoff's current law gives this current if the circuit is driven by a current source. In some more complicated networks, mesh currents can also be determined from KCL equations.† We use KCL if we can go from the interior of a mesh to the exterior of the circuit along a path that cuts across only branches with current sources (see the dashed paths A, B, and C shown on Fig. 4.10). We will demonstrate the procedure by example.

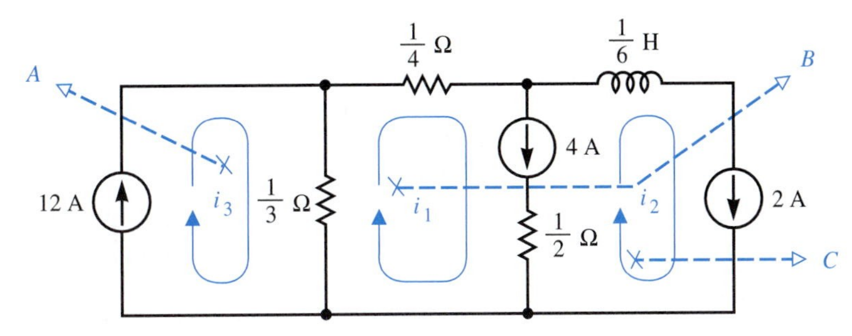

FIGURE 4.10 Network with mesh-current equations that are KCL equations

EXAMPLE 4.7 Write a set of mesh-current equations for the network shown in Fig. 4.10, and solve for all mesh currents.

Solution Since there are three meshes in the network of Fig. 4.10, three mesh currents must be found. We can always write the current through any current source in terms of at most two mesh currents. Thus from KCL (only the components that influence the equation are shown explicitly in the following illustrations),

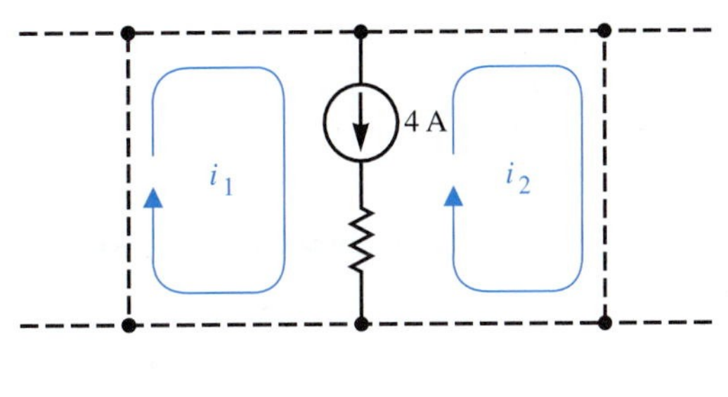

$$i_1 - i_2 = 4 \text{ A}$$

and KCL also gives

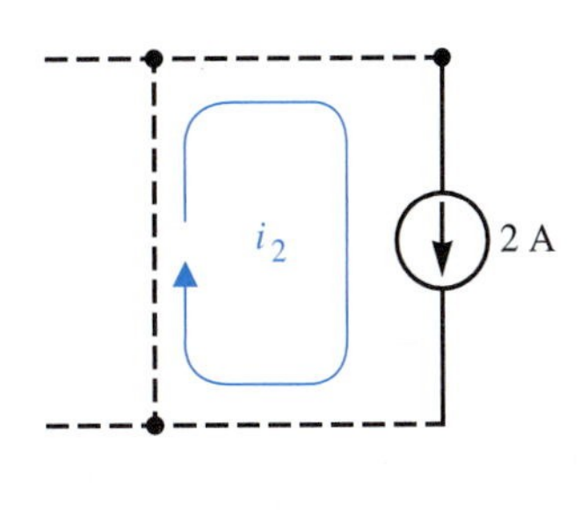

$$i_2 = 2 \text{ A}$$

† These KCL equations are often called *constraint equations*, because a mesh current or the difference between two mesh currents is constrained by a current source.

and

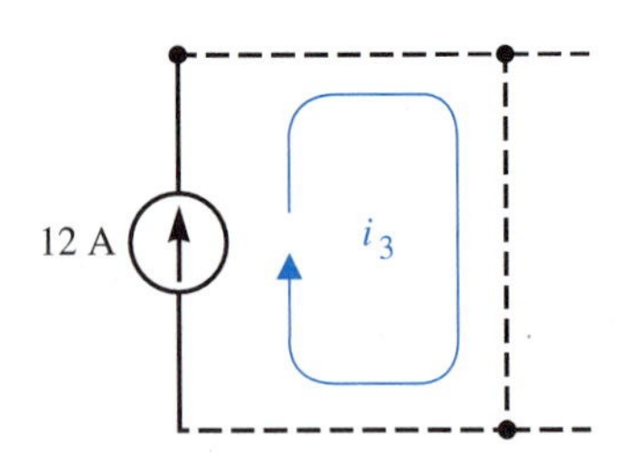

$$i_3 = 12 \text{ A}$$

The three mesh-current equations yield

$$\begin{bmatrix} i_1 \\ i_2 \\ i_3 \end{bmatrix} = \begin{bmatrix} 6 \\ 2 \\ 12 \end{bmatrix} \text{A}$$

The element voltages are easily found by use of the terminal equations, but KVL equations are required to find the voltage across a current source.

In this section we have investigated a circuit for which all mesh-current equations are KCL equations. In the next section we examine a circuit for which all mesh-current equations are KVL equations. The analysis is very similar to that of a series circuit driven by a voltage source.

Remember

We will write a KCL equation for a mesh if we can go from the interior of the mesh to the exterior of the circuit along a path that cuts across only branches with current sources. This KCL equation is a mesh-current equation and must be written in terms of known sources and mesh currents.

EXERCISES

13. Write a set of mesh-current equations for the network shown below. Use KCL to determine the value of i_{12} and v_b. *answer:* −12 A, 36 A, −48 A, 8V

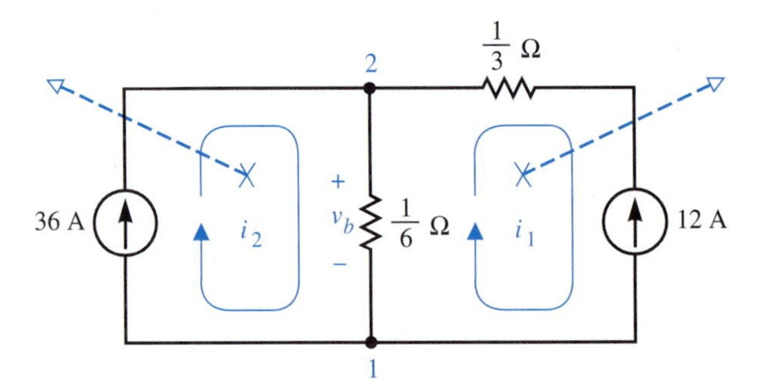

14. Replace the 12-A source in the circuit above with a dependent current source having the reference arrow pointing down. The control equation is $i = 6v_b$. Repeat Exercise 13. Write v_b in terms of mesh currents i_1 and i_2. The unknown voltage v_b must not appear in the mesh-current equations.
answer: 18 A, 36 A, −18 A, 3 V

Mesh-Current Equations That Are KVL Equations

A KVL equation is used to determine the mesh current for a series circuit without a current source. We shall show that this same technique gives us the mesh currents for more complicated networks not excited by current sources.

EXAMPLE 4.8 (a) Write a set of mesh-current equations for the network shown in Fig. 4.11, and solve for all mesh currents. (b) Replace the $\frac{1}{6}$-Ω resistance with a $\frac{1}{6}$-H inductance and the $\frac{1}{10}$-Ω resistance with a $\frac{1}{10}$-F capacitance. Write the mesh-current equations.

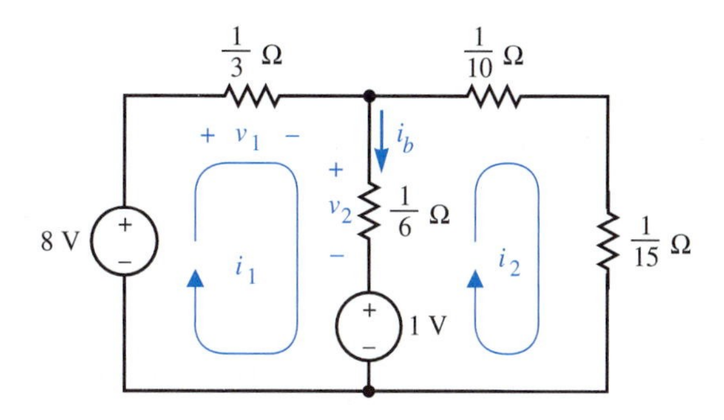

FIGURE 4.11 Network with mesh-current equations that are KVL equations

Solution (a) Select the mesh currents as shown in the network of Fig. 4.11. The mesh-current equations are obtained from KVL. Application of KVL to the closed path around mesh 1 gives

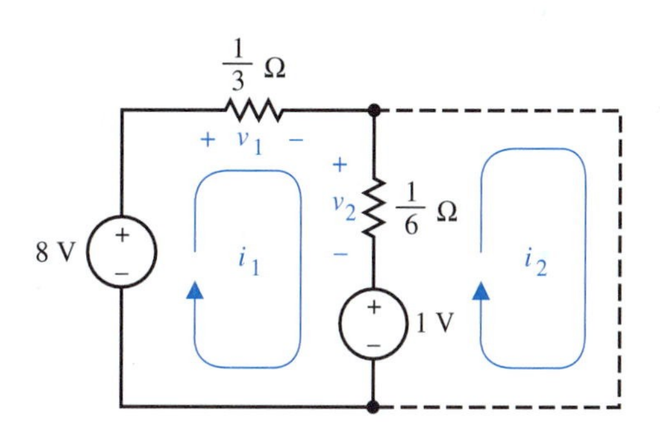

$$-8 + v_1 + v_2 + 1 = 0$$

The element terminal equations are

$$v_1 = \frac{1}{3} i_1$$

and

$$v_2 = \frac{1}{6}(i_1 - i_2)$$

Substitution from these terminal equations into the KVL equation gives the first mesh-current equation:

$$-8 + \frac{1}{3} i_1 + \frac{1}{6}(i_1 - i_2) + 1 = 0$$

We can bypass the first three steps and write this last equation directly. Just remember that the total current through a branch, which is in common with mesh k and mesh j, in the direction of mesh current i_k is $i = i_k - i_j$.

The second mesh-current equation is obtained from KVL applied around mesh 2.

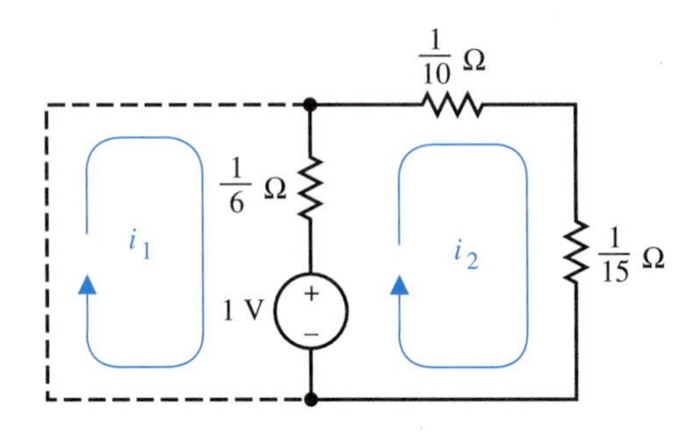

$$-1 + \frac{1}{6}(i_2 - i_1) + \frac{1}{10}i_2 + \frac{1}{15}i_2 = 0$$

We can write these two mesh-current equations as a single matrix equation:

$$\begin{bmatrix} \frac{1}{2} & -\frac{1}{6} \\ -\frac{1}{6} & \frac{1}{3} \end{bmatrix} \begin{bmatrix} i_1 \\ i_2 \end{bmatrix} = \begin{bmatrix} 7 \\ 1 \end{bmatrix}$$

Note that the only unknowns that appear in this equation are mesh currents.

We can easily solve these equations by Cramer's rule or substitution to give

$$\begin{bmatrix} i_1 \\ i_2 \end{bmatrix} = \begin{bmatrix} 18 \\ 12 \end{bmatrix} \text{A}$$

The current through any component is given by either a mesh current or the difference between two mesh currents. For example,

$$\begin{aligned} i_b &= i_1 - i_2 \\ &= 18 - 12 = 6 \text{ A} \end{aligned}$$

Thus

$$v_2 = \frac{1}{6}i_b = \frac{1}{6}(6) = 1 \text{ V}$$

(b) With the two resistances replaced by an inductance and a capacitance, the first KVL equation becomes

$$-8 + \frac{1}{3}i_1 + \frac{1}{6}\frac{d}{dt}(i_1 - i_2) + 1 = 0$$

and the second KVL equation is

$$-1 + \frac{1}{6}\frac{d}{dt}(i_2 - i_1) + 10\int_{-\infty}^{t} i_2 \, d\lambda + \frac{1}{15}i_2 = 0$$

The solution of integro-differential equations will be covered in Chapters 8 and 9.

We have considered a circuit for which all mesh-current equations are KVL equations. We introduce the concept of a supermesh in the next section. This requires both KVL and KCL equations.

Remember

We write a KVL equation for each mesh that has no current source in any branch. This KVL equation is a mesh-current equation and must be written in terms of mesh currents and known source values.

EXERCISES

15. (a) Write a set of mesh-current equations for the network shown below and solve for the mesh currents. Determine the value of i_{12}.

answer: 70 A, − 105 A, 175 A

(b) Write a KVL equation around the outer loop of the following circuit. Observe that this is the sum of the two KVL equations written in part (a). This is why we will not use a KVL equation around loops that are not meshes when we write mesh-current equations.

answer: $\frac{1}{4}i_1 + \frac{1}{2}i_2 = -35$

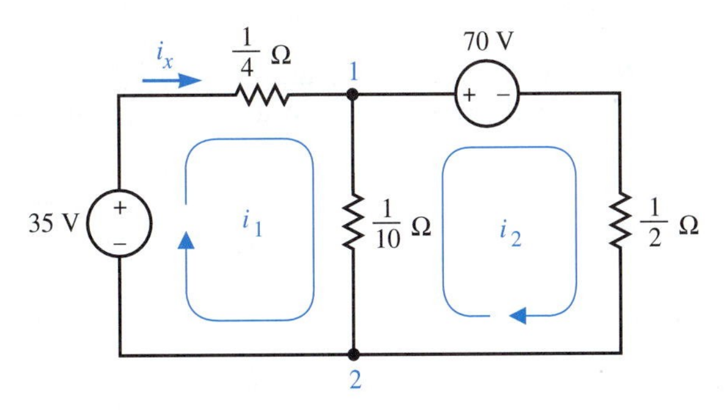

16. Replace the 70-V source in the network shown above with a dependent voltage source having the (+) reference mark at the right-hand side. The control equation is $v = \frac{1}{4}i_x$. Repeat Exercise 15(a). Write i_x in terms of the mesh currents. The unknown current i_x must not appear in the mesh-current equations.

answer: 120 A, 70 A, 50 A

17. For the network used in Exercise 15, replace the $\frac{1}{10}$-Ω resistance with a $\frac{1}{10}$-F capacitance and the $\frac{1}{2}$-Ω resistance with a $\frac{1}{2}$-H inductance. Write the two mesh-current equations, but do not solve them.

answer:
$$\begin{bmatrix} \frac{1}{4} + 10\int_{-\infty}^{t} d\lambda & -10\int_{-\infty}^{t} d\lambda \\ -10\int_{-\infty}^{t} d\lambda & \frac{1}{2}\frac{d}{dt} + 10\int_{-\infty}^{t} d\lambda \end{bmatrix} \begin{bmatrix} i_1 \\ i_2 \end{bmatrix} = \begin{bmatrix} 35 \\ -70 \end{bmatrix}$$

Supermeshes Require KVL and KCL Equations

Meshes that share a current source with other meshes, none of which contains a current source in the outer loop, form a *supermesh.* A path around a supermesh does not pass through a current source. A path around each mesh contained within a supermesh passes through a current source. The total number of equations required for a supermesh is equal to the number of meshes contained in the supermesh. We must write a KVL equation around the supermesh. The remaining equations are KCL equations.

We also demonstrate a shortcut. We designate one mesh within a supermesh as a *principal mesh,* and use KCL to write the other mesh currents in the supermesh as a sum of this principal-mesh current and current source values. We then write a KVL equation for the loop around the supermesh. This last step is similar to the analysis of a series circuit that does not have a current source.

EXAMPLE 4.9 Write a set of mesh-current equations for the network shown in Fig. 4.12.

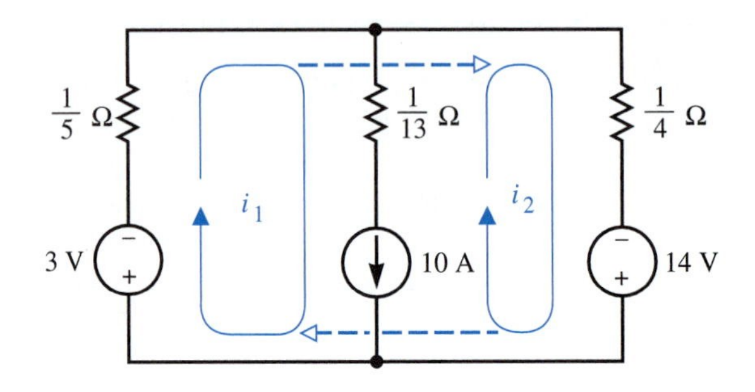

FIGURE 4.12 Network with one supermesh that requires both a KVL and a KCL equation

Solution With the network as drawn in Fig. 4.12, meshes 1 and 2, together with the shared current source and $\frac{1}{13}$-Ω resistance, form one supermesh.

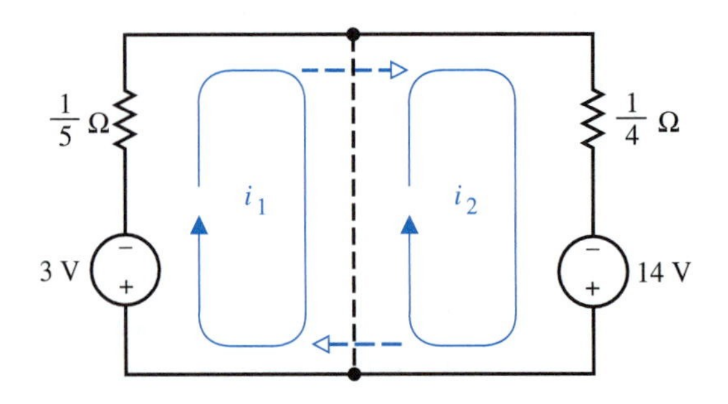

A KVL equation around the supermesh is

$$3 + \frac{1}{5}i_1 + \frac{1}{4}i_2 - 14 = 0$$

which is the first mesh-current equation. The use of KCL permits us to write the 10-A current source current in terms of mesh-current components:

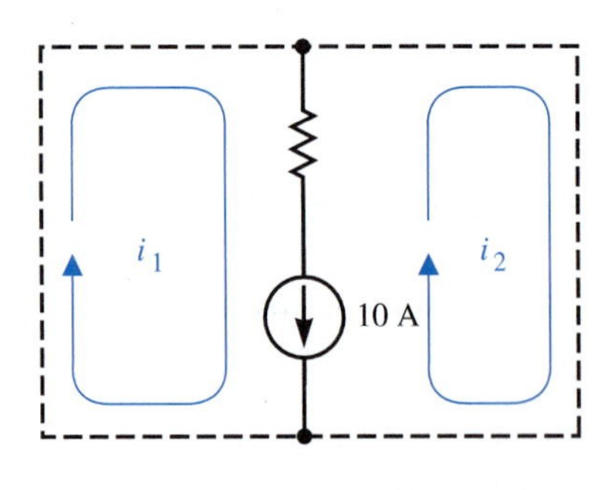

$$i_1 - i_2 = 10$$

This is the second mesh-current equation. We can write the two mesh-current equations

$$\begin{matrix}\text{From KCL} \\ \text{From KVL}\end{matrix}\left\{\begin{bmatrix} \frac{1}{5} & \frac{1}{4} \\ 1 & -1 \end{bmatrix}\begin{bmatrix} i_1 \\ i_2 \end{bmatrix} = \begin{bmatrix} 11 \\ 10 \end{bmatrix}\right\}\begin{matrix}\text{Voltage due to voltage sources} \\ \text{Current due to current source}\end{matrix}$$

and solve for the mesh currents.

The labor needed to write and solve the mesh-current equations can be reduced if we use KCL to write

$$i_2 = i_1 - 10$$

directly on the network diagram before writing the KVL equation. Because we have written i_2 in terms of i_1, we call mesh 1 a *principal mesh.*

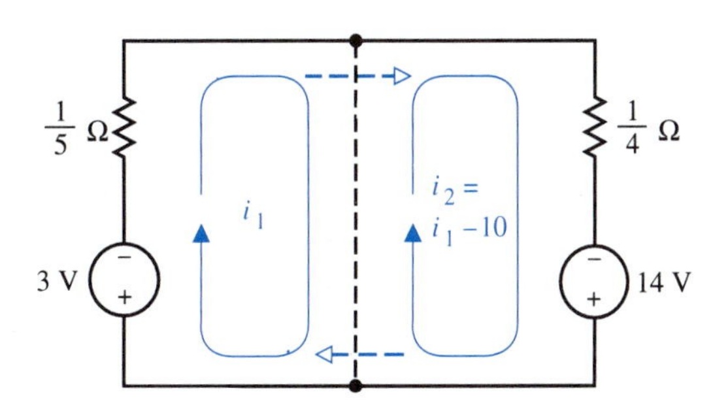

The KVL equation around the supermesh can be written as

$$3 + \frac{1}{5} i_1 + \frac{1}{4}(i_1 - 10) - 14 = 0$$

which immediately yields

$$i_1 = 30 \text{ A}$$

Observe how this shortcut reduced the analysis procedure to one similar to that for a simple series circuit. The second mesh current is given by the KCL equation as

$$i_2 = 30 - 10 = 20 \text{ A}$$

Remember

We can go from the interior of any mesh within a supermesh to the interior of any other mesh within the same supermesh by cutting across branches that contain current sources. You cannot reach the exterior of the network by cutting across a branch containing a current source (see the dashed paths in Fig. 4.12). A supermesh requires one mesh-current equation that is a KVL equation. The remaining mesh-current equations are KCL equations.

EXERCISES

18. (a) Develop a set of mesh-current equations for the network shown below. First, write four mesh-current equations (two KVL and two KCL equations). Next, express i_3 and i_4 in terms of i_1 and i_2, and write them directly on the network diagram. Then write a KVL equation for each supermesh. The resulting two

mesh-current equations should contain i_1 and i_2 as the only unknowns. Solve for all four mesh currents.

answer:
$$\begin{bmatrix} \frac{3}{5} & -\frac{1}{2} & \frac{11}{20} & -\frac{1}{2} \\ -\frac{1}{2} & -\frac{3}{2} & \frac{3}{2} & 1 \\ 1 & 0 & -1 & 0 \\ 0 & 1 & 0 & -1 \end{bmatrix} \begin{bmatrix} i_1 \\ i_2 \\ i_3 \\ i_4 \end{bmatrix} = \begin{bmatrix} 0 \\ 0 \\ 10 \\ 40 \end{bmatrix},$$

$$\begin{bmatrix} \frac{23}{20} & -1 \\ 1 & -\frac{1}{2} \end{bmatrix} \begin{bmatrix} i_1 \\ i_2 \end{bmatrix} = \begin{bmatrix} -\frac{29}{2} \\ 55 \end{bmatrix},$$

146.47 A, 182.94 A, 136.47 A, 142.94 A

(b) Use the mesh currents calculated in part (a) to solve for v_1 and v_6 shown on the network diagram. Now write a KVL equation around mesh 1 and solve for the voltage on the left end of the 10-A source with respect to the right end. Observe that we could not write this voltage in terms of i_1 and i_3, but that this voltage is $v_2 + v_5$, which can be written in terms of i_2 and i_3. This is why we wrote the KVL equations around the supermeshes.

answer: 14.647 V, 1.7647 V, −16.412 V

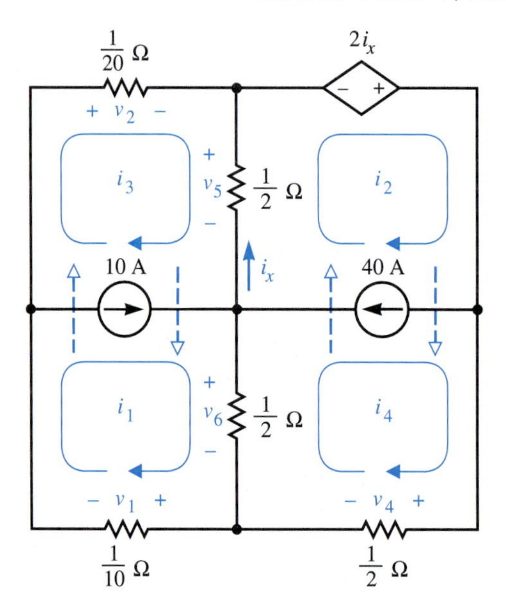

4.4 Network Analysis by Mesh Currents

A more complex network may have current sources in the outer loop, meshes without current sources, and meshes that form a supermesh. The following example demonstrates that the analysis methods we developed still apply.

EXAMPLE 4.10 Analyze the circuit of Fig. 4.13 by the method of mesh currents.

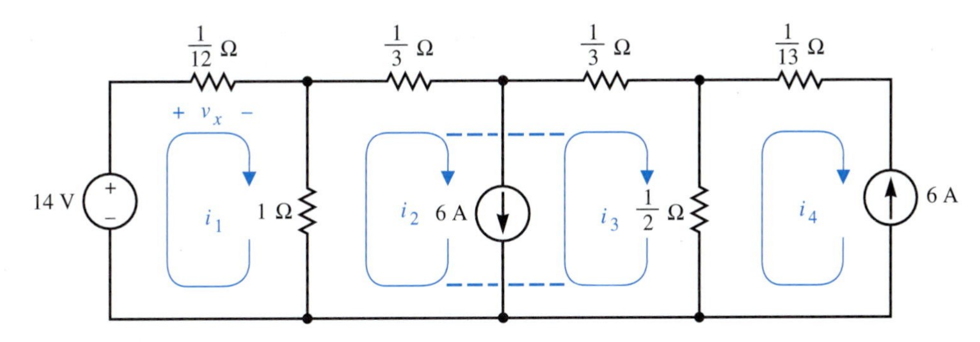

FIGURE 4.13
A more general mesh-current example

Solution Assign the mesh currents as shown in Fig. 4.13. Mesh 1 requires a KVL equation. Meshes 2 and 3 form a supermesh that requires one KVL equation and one KCL equation. Mesh 4 requires only a KCL equation. Write a KVL equation around mesh 1:

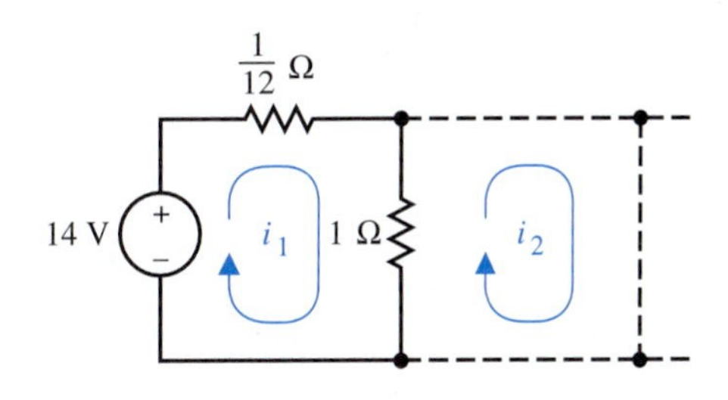

$$-14 + \frac{1}{12} i_1 + 1(i_1 - i_2) = 0$$

Write a KVL equation around meshes 2 and 3 (a supermesh):

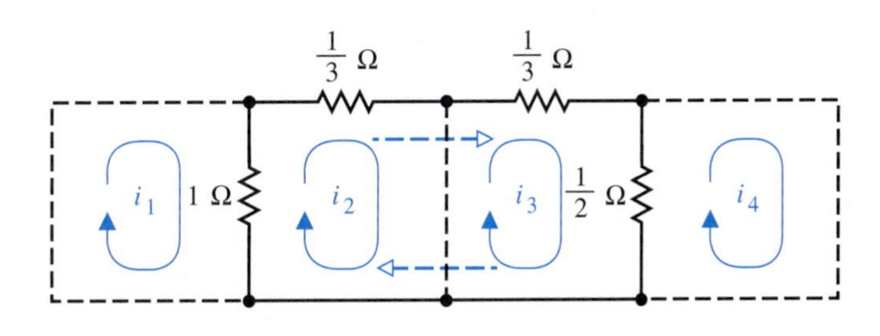

$$1(i_2 - i_1) + \frac{1}{3} i_2 + \frac{1}{3} i_3 + \frac{1}{2} (i_3 - i_4) = 0$$

The third and fourth mesh equations are KCL equations:

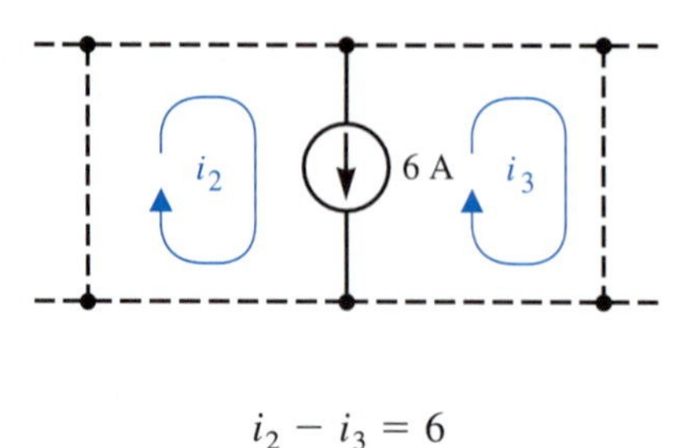

$$i_2 - i_3 = 6$$

and

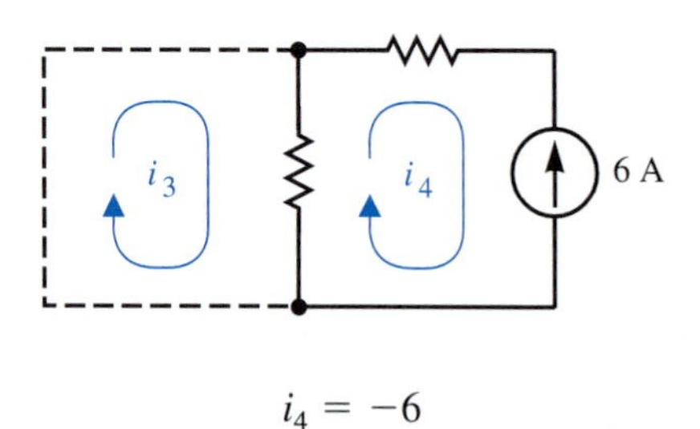

$$i_4 = -6$$

We can write these four mesh-current equations as a single matrix equation:

$$\begin{bmatrix} \frac{13}{12} & -1 & 0 & 0 \\ -1 & \frac{4}{3} & \frac{5}{6} & -\frac{1}{2} \\ 0 & 1 & -1 & 0 \\ 0 & 0 & 0 & 1 \end{bmatrix} \begin{bmatrix} i_1 \\ i_2 \\ i_3 \\ i_4 \end{bmatrix} = \begin{bmatrix} 14 \\ 0 \\ 6 \\ -6 \end{bmatrix}$$

From KVL (first two rows) — Voltages due to voltage sources

From KCL (last two rows) — Currents due to current sources

The left-hand matrix is the *mesh transformation matrix.*

We can reduce the labor involved in writing and solving the mesh-current equations by writing

$$i_3 = i_2 - 6$$

and

$$i_4 = -6$$

directly on the network diagram before writing the KVL equations. The first KVL equation remains the same, and the second KVL equation is

$$1(i_2 - i_1) + \frac{1}{3} i_2 + \frac{1}{3}(i_2 - 6) + \frac{1}{2}[(i_2 - 6) - (-6)] = 0$$

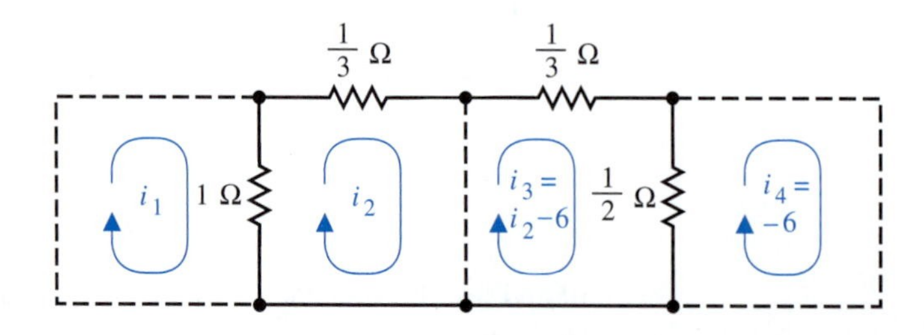

in which case the mesh-current equations reduce to

$$\begin{bmatrix} \frac{13}{12} & -1 \\ -1 & \frac{13}{6} \end{bmatrix} \begin{bmatrix} i_1 \\ i_2 \end{bmatrix} = \begin{bmatrix} 14 \\ 2 \end{bmatrix}$$

We refer to the left-hand matrix as the *reduced mesh transformation matrix.* When we solve for the *principal-mesh* currents, i_1 and i_2, we get

$$\begin{bmatrix} i_1 \\ i_2 \end{bmatrix} = \begin{bmatrix} 24 \\ 12 \end{bmatrix} \text{A}$$

We obtain currents i_3 and i_4 from the two KCL equations.

A summary of the shortcut procedure we have developed follows. The list is presented for easy reference, not for memorization.

Shortcut Method ➤

Analysis of Mesh Currents

1. Given a network diagram, draw the network so that no lines cross without being connected. If this cannot be done, the network is not planar and the method of mesh currents is not applicable.
2. Number each mesh beginning with 1, and assign a mesh current clockwise to each mesh.
3. Write the mesh-current equations as follows:
   (a) Identify each mesh that has its mesh current defined by current sources. Use KCL to write each of these mesh currents in terms of source currents. Write these relationships on the network diagram.
   (b) Identify each group of meshes that form a supermesh. Select one mesh within each supermesh as a principal mesh. Use KCL to write each other mesh current within the group in terms of the principal-mesh current and source currents. Write these relationships on the network diagram.
   (c) The remaining meshes are also principal meshes.
   (d) Now write a KVL equation around each mesh identified in (c) and around each supermesh. Remember that all unknowns must be written in terms of principal-mesh currents.
4. Solve the KVL equations for the principal-mesh currents. Each remaining mesh current is then obtained from a KCL equation.

We will now apply the shortcut procedure to a simple network.

EXAMPLE 4.11 Use the shortcut procedure to write a set of mesh-current equations for the network shown in Fig. 4.14.

FIGURE 4.14
Example circuit for the shortcut procedure for mesh-current analysis

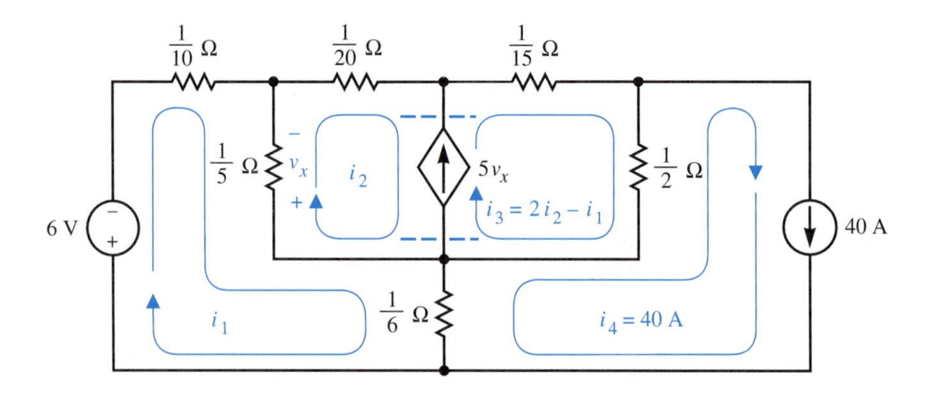

Solution

1. The network is drawn so that no lines cross without being connected.
2. Assign the mesh currents as shown.
3. Write a set of mesh-current equations.
   (a) Mesh current i_4 is determined by the current source. From KCL,

$$i_4 = 40 \text{ A}$$

   (b) Meshes 2 and 3 form a supermesh. Select mesh 2 as the principal mesh. From KCL,

$$i_3 - i_2 = 5v_x = 5\left[\frac{1}{5}(i_2 - i_1)\right]$$

   This gives

$$i_3 = 2i_2 - i_1$$

   Write this on the network diagram.
   (c) Mesh 1 is also a principal mesh, because it contains no current sources.
   (d) Write a KVL equation around mesh 1:

$$6 + \frac{1}{10}i_1 + \frac{1}{5}(i_1 - i_2) + \frac{1}{6}(i_1 - 40) = 0$$

   Next apply KVL around the supermesh:

$$\frac{1}{5}(i_2 - i_1) + \frac{1}{20}i_2 + \frac{1}{15}(2i_2 - i_1) + \frac{1}{2}[(2i_2 - i_1) - 40] = 0$$

4. We can write these two KVL equations as a single matrix equation,

$$\begin{bmatrix} \frac{7}{15} & -\frac{1}{5} \\ -\frac{23}{30} & \frac{83}{60} \end{bmatrix} \begin{bmatrix} i_1 \\ i_2 \end{bmatrix} = \begin{bmatrix} \frac{2}{3} \\ 20 \end{bmatrix}$$

that is easily solved to give $i_1 = 10$ A and $i_2 = 20$ A. The two KCL equations give $i_3 = 30$ A and $i_4 = 40$ A.

The following example illustrates how easy it is to apply the shortcut procedure to even a complicated-looking network.

EXAMPLE 4.12 Use the shortcut procedure to write a set of mesh-current equations for the network shown in Fig. 4.15.

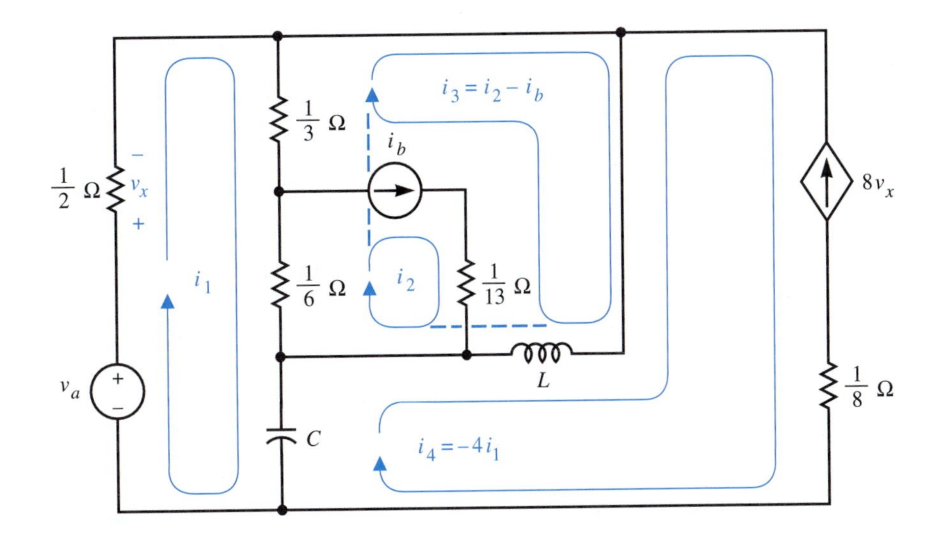

FIGURE 4.15 Mesh-current example with inductance and capacitance

Solution

1. The network is drawn so that no lines cross without being connected.
2. Assign the mesh currents as shown.
3. Write a set of mesh-current equations.
   (a) Mesh current i_4 is determined by the dependent current source. From KCL,

$$i_4 = -8v_x = -8\left(\frac{1}{2}i_1\right) = -4i_1$$

   Write this on the network diagram.

   (b) Meshes 2 and 3 form a supermesh. Select mesh 2 as the principal mesh. From KCL,

$$i_b = i_2 - i_3$$

   which gives

$$i_3 = i_2 - i_b$$

   Write this on the network diagram.

   (c) Mesh 1 is also a principal mesh, because it contains no current sources.

   (d) Write a KVL equation around mesh 1:

$$-v_a + \frac{1}{2}i_1 + \frac{1}{3}[i_1 - (i_2 - i_b)] + \frac{1}{6}(i_1 - i_2) + \frac{1}{C}\int_{-\infty}^{t}[i_1 - (-4i_1)]\,d\lambda = 0$$

   Next apply KVL around the supermesh (meshes 2 and 3):

$$\frac{1}{6}(i_2 - i_1) + \frac{1}{3}[(i_2 - i_b) - i_1] + L\frac{d}{dt}[(i_2 - i_b) - (-4i_1)] = 0$$

4. We can write these two KVL equations as a single matrix equation:

$$\begin{bmatrix} 1 + \dfrac{5}{C}\displaystyle\int_{-\infty}^{t} d\lambda & -\dfrac{1}{2} \\ 4L\dfrac{d}{dt} - \dfrac{1}{2} & L\dfrac{d}{dt} + \dfrac{1}{2} \end{bmatrix} \begin{bmatrix} i_1 \\ i_2 \end{bmatrix} = \begin{bmatrix} v_a - \dfrac{1}{3}i_b \\ L\dfrac{d}{dt}i_b + \dfrac{1}{3}i_b \end{bmatrix}$$

The solution of a matrix differential equation of this type is postponed until Chapter 9, but once the principal-mesh currents, i_1 and i_2, are determined, i_3 and i_4 are easily obtained from these two KCL equations.

Mesh equations are also suitable for planar networks with two or more separate parts, as in Fig. 4.8. Each separate part has an outer loop, but this introduces no added complication.

Remember

One mesh-current equation is required for each mesh. If there are no cutsets of current sources, the number of mesh-current equations that are KVL equations is equal to the number of meshes minus the number of current sources. The remainder of the mesh-current equations are KCL equations. The effort required to solve the mesh-current equations is largely determined by the number of KVL equations.

EXERCISES

19. Use the method of analysis by mesh currents to calculate the mesh currents for the network in the following figure. Then calculate the power absorbed by the dependent source. *answer:* −100 A, 10 A, 20 A, −220 W

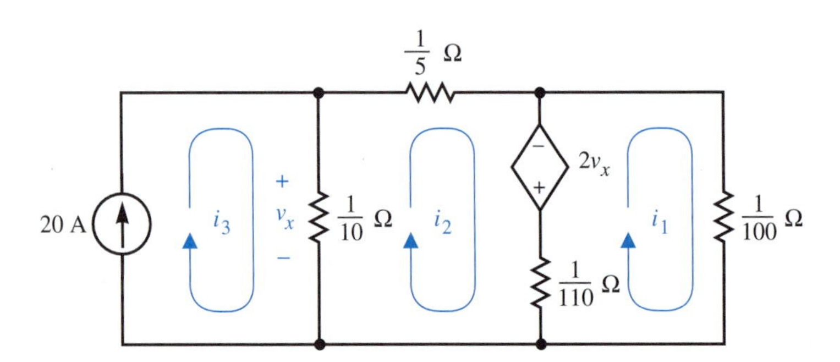

20. Replace the 20-A source in the network shown above with an open circuit and the $\frac{1}{5}$-Ω resistance with a 20-A source (reference arrow pointing to the right). Then repeat Exercise 19. *answer:* 219.05 A, 20 A, −796.2 W

21. Replace the dependent voltage source of the preceding network with a dependent current source (current reference arrow pointing down) with a control equation of $i = 110v_x$. Then repeat Exercise 19. *answer:* −100 A, 10 A, 20 A, −220 W

22. Replace the bottom $\frac{1}{2}$-Ω resistance of the circuit used in Exercise 18 on page 116 with a current source of value 100 A (reference arrow pointing toward the left). Use the shortcut procedure to write the mesh-current equations. This would leave one KVL equation that contains i_1. Solve for all mesh currents.
answer: 109.3 A, 140 A, 99.13 A, 100 A

23. Replace the center current source of Fig. 4.13 with a dependent current source, reference arrow pointing down, with a control equation $i = 3v_x$.
    (a) Write the four mesh-current equations, and arrange in matrix form. You should have two KVL equations and two KCL equations.

answer:
$$\begin{bmatrix} \frac{13}{12} & -1 & 0 & 0 \\ -1 & \frac{4}{3} & \frac{5}{6} & -\frac{1}{2} \\ \frac{1}{4} & -1 & 1 & 0 \\ 0 & 0 & 0 & 1 \end{bmatrix} \begin{bmatrix} i_1 \\ i_2 \\ i_3 \\ i_4 \end{bmatrix} = \begin{bmatrix} 14 \\ 0 \\ 0 \\ -6 \end{bmatrix}$$

(b) Use the two KCL equations to write i_3 and i_4 in terms of i_1 and i_2. Write these on the network diagram. Directly from the network diagram, write two KVL equations that contain i_1 and i_2 as the only unknowns.

answer:
$$\begin{bmatrix} \frac{13}{12} & -1 \\ -\frac{29}{24} & \frac{13}{16} \end{bmatrix} \begin{bmatrix} i_1 \\ i_2 \end{bmatrix} = \begin{bmatrix} 14 \\ -3 \end{bmatrix}$$

(c) Solve for all mesh currents and the power absorbed by the dependent source.

answer: 24 A, 12 A, 6 A, −6 A, 48 W

4.5 The Best Method

The best network analysis method to use depends not only on the network to be analyzed but also on the information required. If we require the voltage across and the current through each component, direct application of the method of node voltages or mesh currents may be the best method. This is especially true when a computer program is available to solve the resulting equations. However, even when a computer is available, it is wise to pick the method that results in the smallest set of equations. The set of node-voltage equations can easily be reduced to the number of nodes minus the number of reference nodes, minus the number of voltage sources. The set of mesh-current equations can easily be reduced to the number of meshes minus the number of current sources.†

Often, only a subset of the component voltages and currents is required. Here considerable simplification of the analysis can result from the use of equivalent (series, parallel) circuits for portions of the network, as described in Chapter 3. Try to eliminate nodes if the method of node voltages is used. Try to eliminate meshes if the method of mesh currents is used.

Although an impractical situation, we are occasionally confronted by a circuit diagram that contains a cutset of current sources, as shown in Fig. 4.16a. A cutset of current sources overconstrains the currents, so the voltages across these current sources are indeterminate. Other component voltages and currents are unique, so we can solve for the mesh currents in such a circuit, but not the node voltages. We can adapt analysis by node voltages to this circuit in two ways:

† This assumes that all cutsets of current sources and loops of voltage sources have been eliminated as described later in this section.

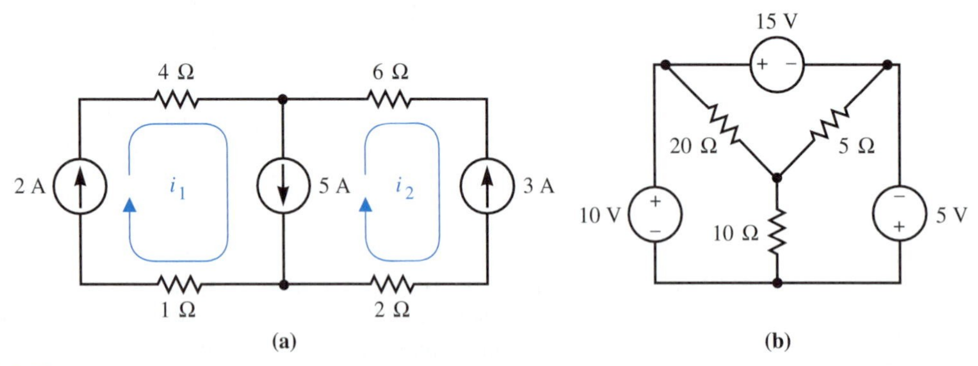

FIGURE 4.16
Circuits with (a) a cutset of current sources, and (b) a loop of voltage sources

1. Draw a closed surface that defines the cutset of current sources. Select one reference node for the part of the circuit inside the closed surface and a second reference node for the part of the circuit outside the closed surface, then continue the analysis as if the circuit had two separate parts.
2. Replace one current source in the cutset with a short circuit, then continue with conventional analysis by node voltages. Remember that the voltages across the current sources are indeterminate in the original circuit.

A second impractical condition that we occasionally encounter is a loop of voltage sources, as shown in Fig. 4.16b. A loop of voltages overconstrains the voltages, so the currents through these sources are indeterminate. Other component currents and voltages are unique, so we can solve for the node voltages, but not the mesh currents. Analysis by mesh currents can be adapted to this circuit in the following way: Replace one voltage source in the loop of voltage sources with an open circuit, then continue with conventional analysis by mesh currents. Remember that the currents through the voltage sources in the original circuit are indeterminate.

It is a good idea to eliminate all cutsets of current sources and loops of voltage sources, as described above, before analysis by any method. (This is necessary for computer-aided analysis by PSpice.)

The inclusion of inductance and capacitance does not increase the difficulty of writing network equations, but it may significantly increase the difficulty of solution, as will be seen in Chapters 8 and 9. However, in certain very important circumstances, the increased complexity is only that of going from the algebra of real numbers to that of complex numbers, as will be shown in Chapter 10.

Remember

Before analyzing a circuit it is best to eliminate all cutsets of current sources and all loops of voltage sources.

We can reduce the effort required to analyze a circuit by node voltages if we can eliminate nodes by the use of equivalent series networks. For a connected network, the number of node-voltage equations that are KCL equations is one less than the number of nodes minus the number of voltage sources. The effort required to solve the node-voltage equations is largely determined by the number of KCL equations.

To reduce the effort required to analyze a circuit by mesh currents, eliminate meshes by the use of equivalent parallel networks. The number of mesh-current equations that are KVL equations is equal to the number of meshes minus the number of current sources. The effort required to solve the mesh-current equations is largely determined by the number of KVL equations.

If a network is not planar, we analyze the circuit by the method of node voltages. For a planar circuit, we can analyze the circuit by the use of node voltages or mesh currents. If the number of mesh equations that are KVL equations is less than the number of node-voltage equations that are KCL equations, the mesh-current equations will be easier to solve.

EXERCISES

24. For the following circuit, determine: (a) the number of node voltage equations that are KCL equations, and; (b) the number of mesh equations that are KVL equations. (c) Which method is most suitable to analyze this circuit?

    answer: (a) 2 (A series resistor combination and the voltage divider relation reduces this to 1.), (b) 5, (c) node voltages

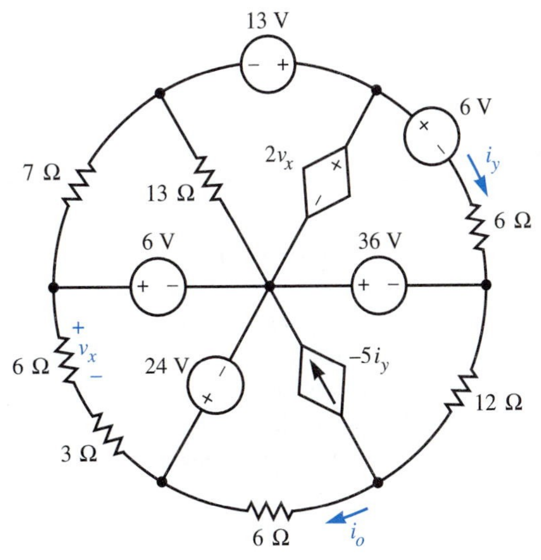

25. For the following circuit determine: (a) the number of node-voltage equations that are KCL equations, and; (b) the number of mesh current equations that are KVL equations. (c) Which method is most suitable to analyze this circuit?

    answer: (a) 5, (b) 2, (c) mesh currents

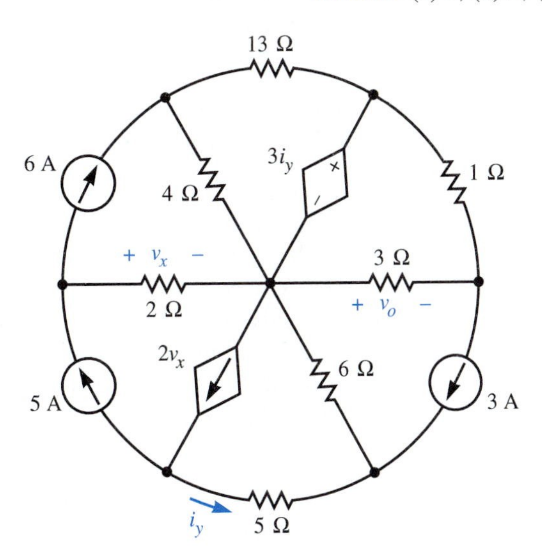

26. Determine the mesh currents in the network of Fig. 4.16a on page 124. Next replace one current source in the cutset of current sources with a short circuit, and solve for the node voltages. Use these node voltages to solve for the current through the 4-Ω resistance. This should equal mesh current i_1.

    answer: 2 A, −3 A. Replace the 5-A source with a short circuit. With the center node as a reference, the node voltages are 8 V, −2 V, 18 V, −6 V.

27. Determine the node voltages in the circuit of Fig. 4.16b on page 124. Next replace one voltage source with an open circuit, and solve for the mesh currents. Use these mesh currents to solve for the voltage across the 10-Ω resistance. This should be the same as determined from the node voltages.

    answer: 10 V, −5 V, −10/7 V. Replace the 15-V source with an open circuit. The mesh currents are 4/7 A and 5/7 A.

4.6 Summary

In this chapter we developed two systematic methods of circuit analysis. The first method, node-voltage analysis, was obtained through generalization of the methods of Chapter 3 to analyze a parallel circuit. This method is applicable to both planar and nonplanar electrical networks. We next generalized the method used to analyze a series circuit to obtain the second systematic method, called mesh-current analysis. This method is applicable only to planar networks. We followed with a brief discussion of which analysis technique is best for a given circuit.

KEY FACTS	Concept	Equation	Section	Page
❑	We can use node voltages to analyze planar or nonplanar networks.		4.1	90
❑	A node connected to the reference node by a path of voltage sources requires a KVL equation (Example 4.1).		4.1	91
❑	A node with no voltage sources connected requires a KCL equation (Example 4.2).		4.1	93
❑	Nodes connected together by a path of voltage sources, but not connected to the reference node by a path of voltage sources, form a supernode. The number of equations required is equal to the number of nodes in the supernode. We write a KCL equation for a surface enclosing the supernode. The remaining equations are KVL equations (Example 4.3).		4.1	96
❑	We require one node-voltage equation for every node except the reference node (Examples 4.4, 4.5 and 4.6).		4.2	100

KEY FACTS	Concept	Equation	Section	Page
❑	When we use network simplification with node-voltage analysis, we try to eliminate nodes.		4.5	124
❑	We use mesh-current analysis only with planar networks.		4.3	106
❑	If we can go from the inside of a mesh to the outside of the circuit by crossing only current sources, the mesh requires a KCL equation (Example 4.7).		4.3	109
❑	A mesh that does not contain a current source requires a KVL equation (Example 4.8).		4.3	111
❑	If meshes contain current sources in common branches and none of these meshes have a current source in the outer loop, the meshes form a supermesh. The number of equations required is equal to the number of meshes in the supermesh. We write a KVL equation around the supermesh. The remaining equations are KCL equations (Example 4.9).		4.3	113
❑	We require one mesh-current equation for each mesh (Examples 4.10, 4.11 and 4.12).		4.4	116
❑	When we use network simplification with mesh-current analysis, we try to eliminate meshes.		4.5	124

PROBLEMS

Section 4.1

1. Determine the node voltages in the following circuit and solve for the indicated currents if $v_a = 12e^t$ V and $v_b = 36e^t$ V.

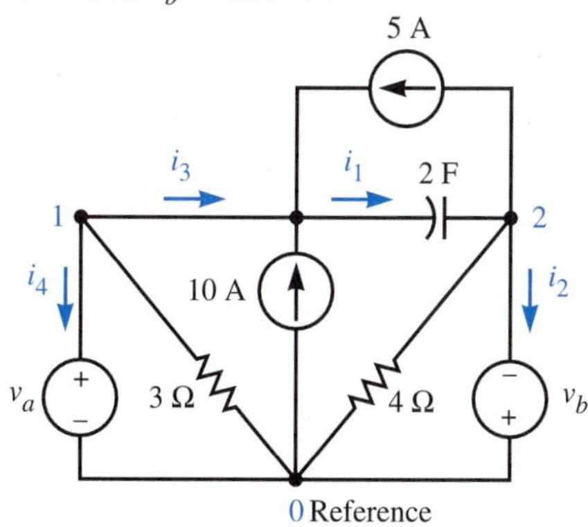

2. For the following circuit, determine the node voltages and solve for currents i_1, i_2, and i_3. 4.4

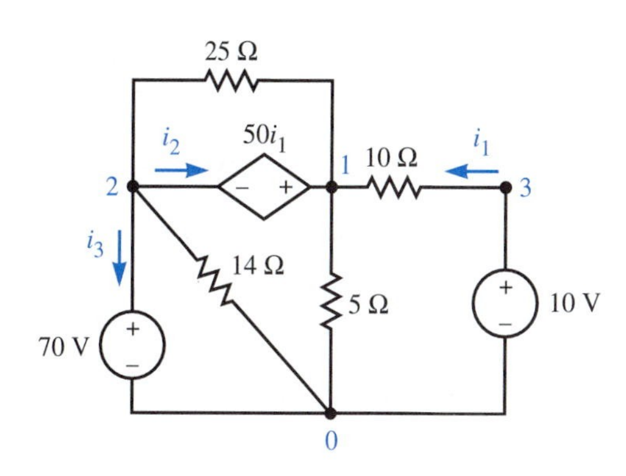

3. Use the method of node voltages to calculate voltage v_x in the following circuit. 4.1

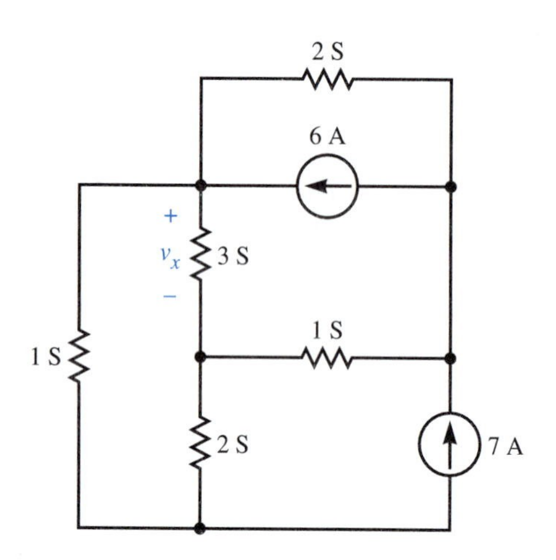

4. Determine the node voltages in the following circuit and calculate current i_x. 4.2

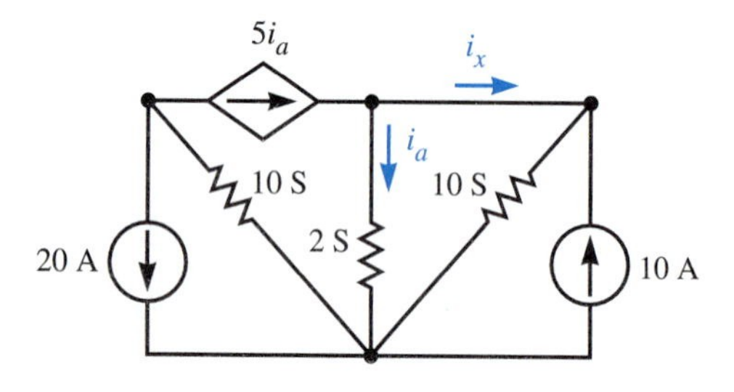

5. Write a set of node-voltage equations for the following circuit. 4.3

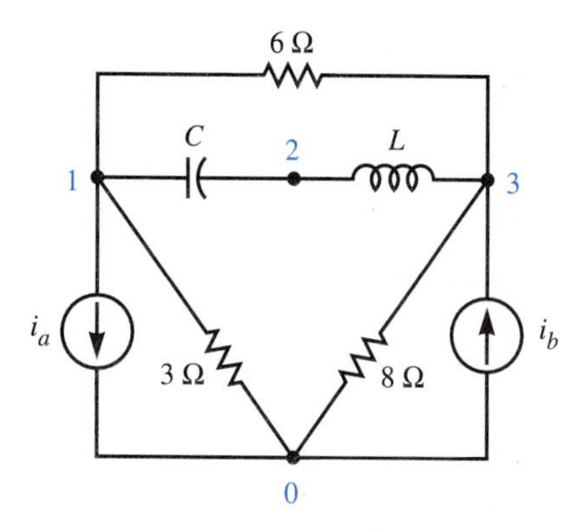

6. Determine the node voltages for the following circuit and calculate current i_x. 4.5

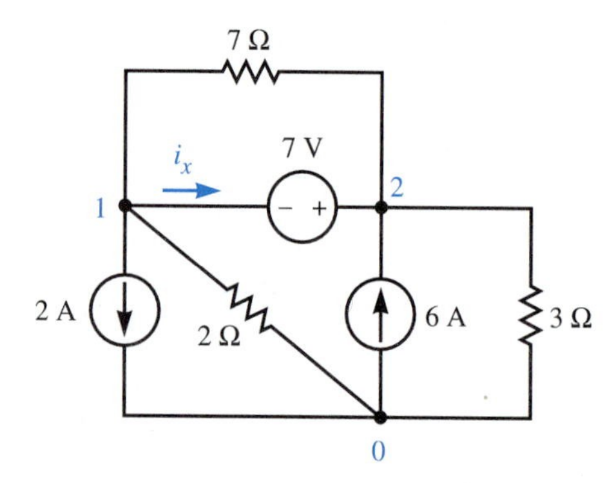

7. Replace the voltage source in the preceding network with a dependent voltage source [(+) reference mark at the right] that has a control equation $v = 1.5v_1$. Solve for the node voltages.

8. (a) Write a set of node-voltage equations in matrix form for the following circuit. You should have one KCL equation and two KVL equations.

   (b) Use the KVL equations obtained in part (a) to solve for v_2 and v_3 in terms of v_1. Write these on the network diagram. Use these to write a single KCL equation directly from the circuit diagram that contains v_1 as the only unknown. This is a shortcut. Now solve for all node voltages.

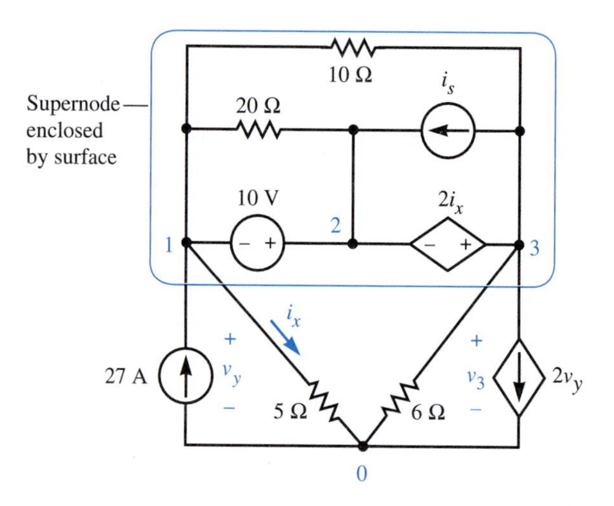

Section 4.2

9. (a) Write a set of node-voltage equations in matrix form for the following circuit. 4.6

   (b) Reduce this set of equations to two KCL equations by substitution.

(c) Use the shortcut procedure to solve for the node voltages in this circuit.

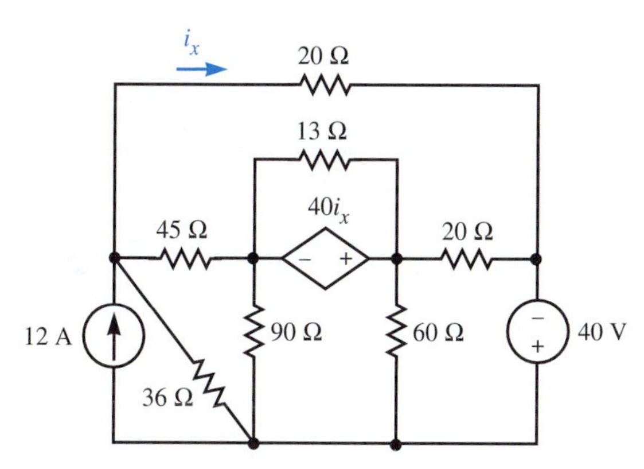

10. (a) Write a set of node-voltage equations for the following network.
 (b) Solve for the node voltages.
 (c) Find i_{12} and i_{43}.

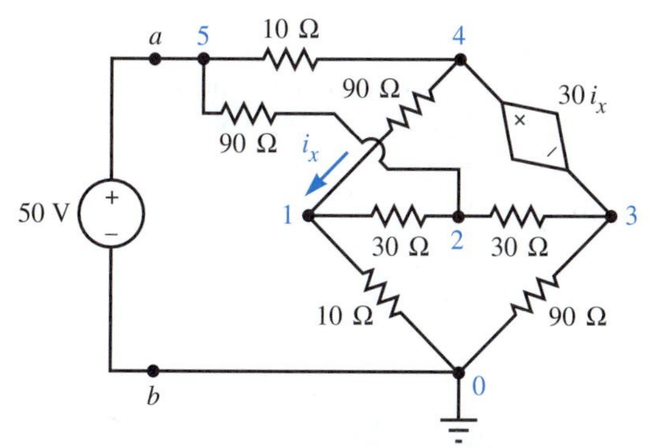

11. Use the shortcut procedure to solve for the node voltages in the following amplifier circuit. Find v_e/v_s and v_c/v_s. These are *voltage gains.* Find the *current gains* i_e/i_b and i_c/i_b. Also find the *power gains* $v_e i_e/v_s i_b$ and $-v_c i_c/v_s i_b$. Find the ratio v_s/i_b, which is the *input resistance* seen by the independent voltage source, because $i_b = 0$ when $v_s = 0$. This is a simplified small-signal equivalent circuit for a one-transistor amplifier. 4.7

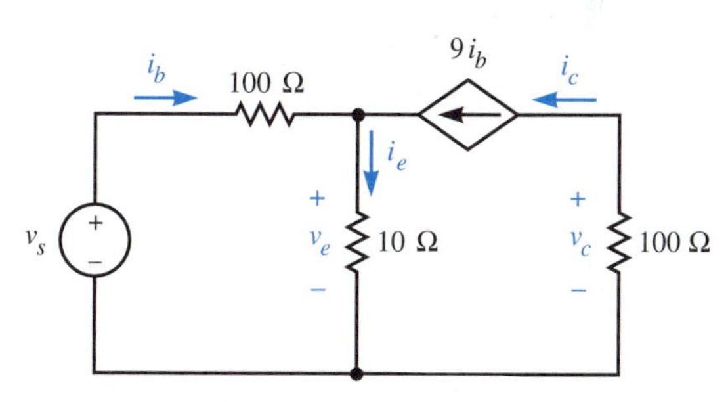

12. The network shown below represents a simplified model of the small-signal equivalent circuit of a field-effect transistor amplifier in the grounded-source configuration.
 (a) Use the method of analysis by node voltages to calculate v_o and the voltage gain v_o/v_i of the amplifier.
 (b) Connect a 100-Ω resistance between terminals G and D and repeat (a). Also calculate i_i. The ratio v_i/i_i is the input resistance of the amplifier.

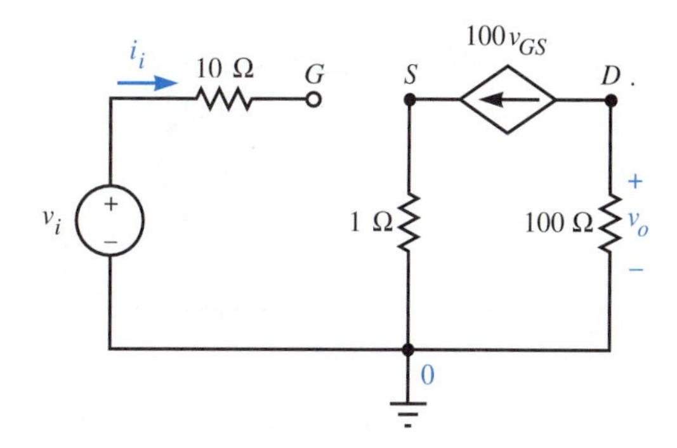

13. The network shown below represents a simplified model of the small-signal equivalent circuit of a field-effect transistor amplifier in the grounded-gate configuration.
 (a) Use the method of analysis by node voltages to calculate v_o and the voltage gain v_o/v_i of the amplifier.
 (b) Connect a 100-Ω resistance between terminals G and D and repeat (a). Also calculate i_i. The ratio v_i/i_i is the input resistance of the amplifier.

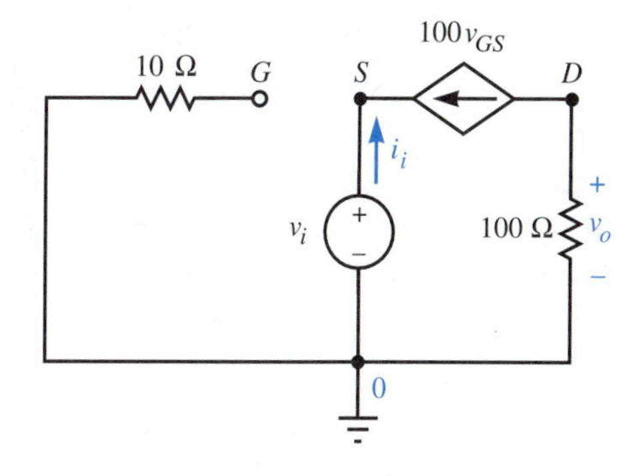

14. Transistors are often used in the Darlington connection (compound amplifier). A simplified

small-signal equivalent circuit for this connection is shown below. Analyze the circuit by the method of node voltages and solve for the gains v_e/v_b, v_c/v_b, i_e/i_b, and i_c/i_b. Calculate the ratio v_b/i_b, which is the input resistance seen by the voltage source because $i_b = 0$ when $v_b = 0$.

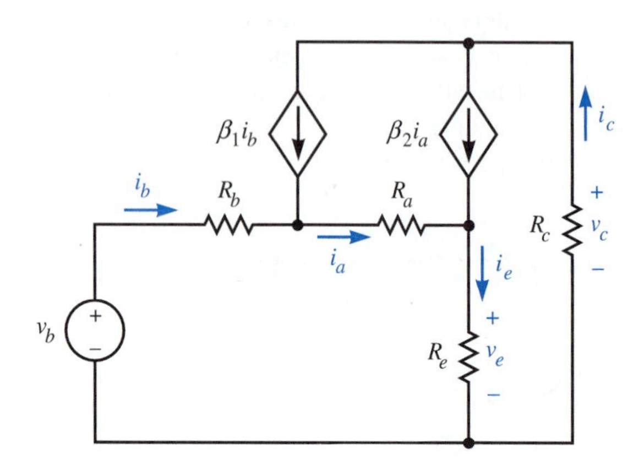

15. A simplified small-signal model for a cascode-connected two-transistor amplifier is shown below.
    (a) Use the method of node voltages to find the voltage gain, v_L/v_b. Calculate the ratio v_b/i_b. This is the input resistance seen by the voltage source, because $i_b = 0$ when $v_b = 0$.
    (b) Repeat part (a) if the short circuit between terminals a and 0 is replaced by a resistance of value R_e.

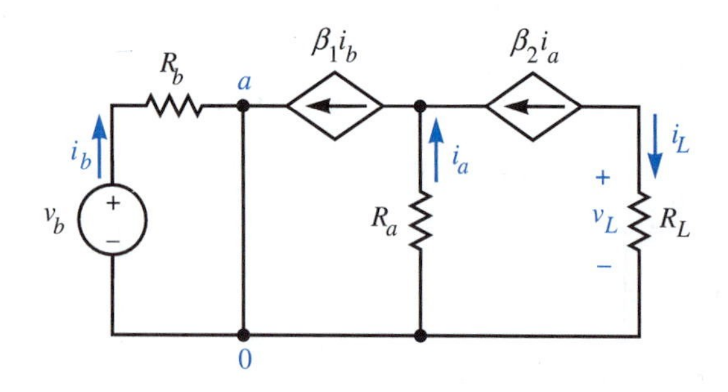

16. (a) Use the node numbers given in the following circuit to write a matrix node-voltage equation that includes one KCL equation and two KVL equations.
    (b) Use the shortcut procedure to write one KCL equation that can be solved for node voltage v_1.

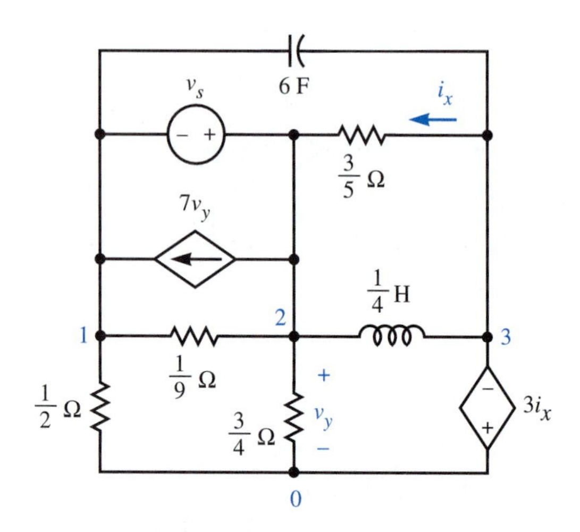

17. (a) Use the node numbers given in the following network to write a matrix node-voltage equation. This will include both KCL and KVL equations.
    (b) Use the shortcut procedure to write a matrix node-voltage equation that does not include KVL equations.

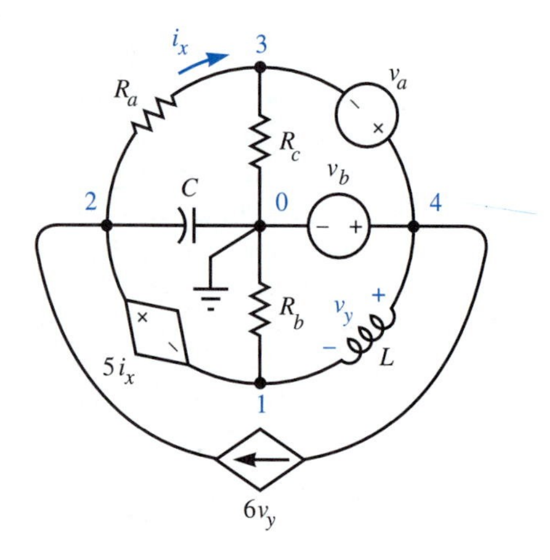

Section 4.3

18. Determine the mesh currents in the following circuit and solve for the indicated voltages if $i_a = 12e^t$ A and $i_b = 36e^t$ A.

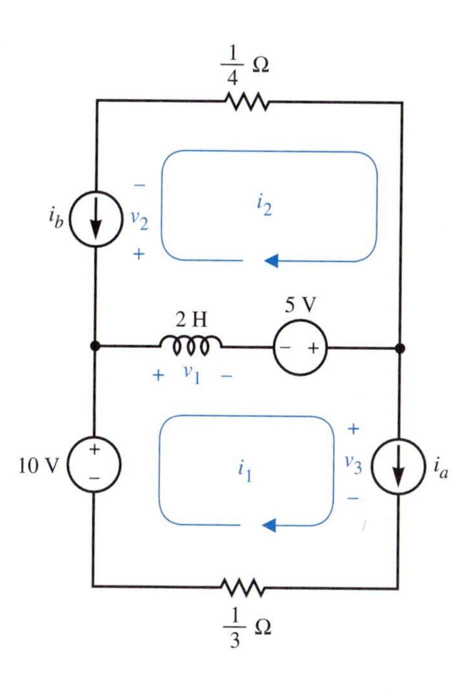

19. For the following circuit, determine the mesh currents and solve for voltages v_1, v_2, and v_3.

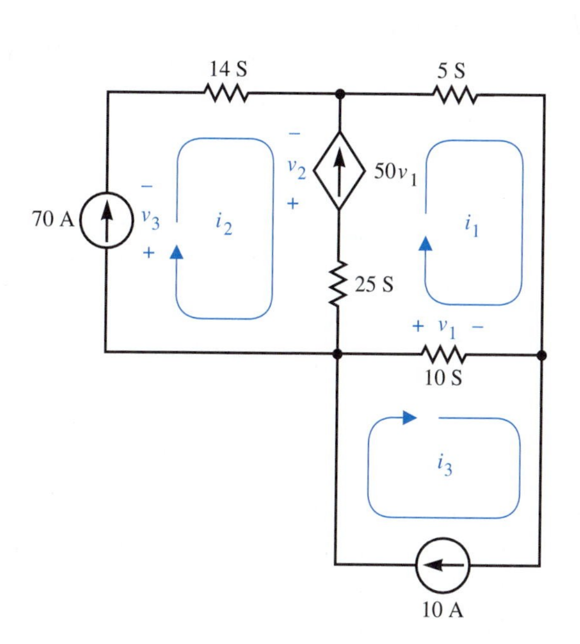

20. Use the method of mesh currents and calculate current i_x in the following circuit.

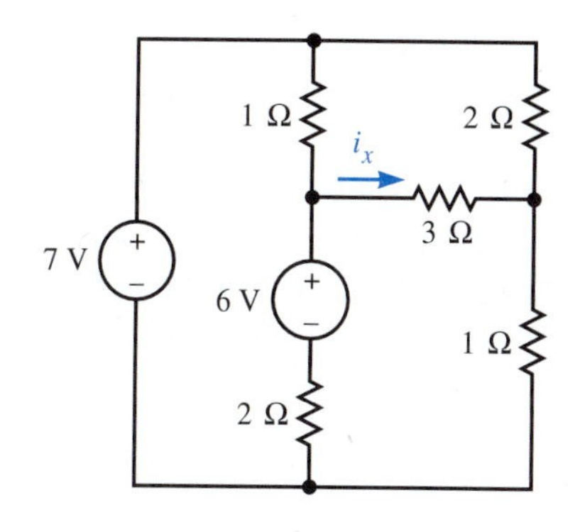

21. Determine the mesh currents in the following circuit and calculate current i_x.

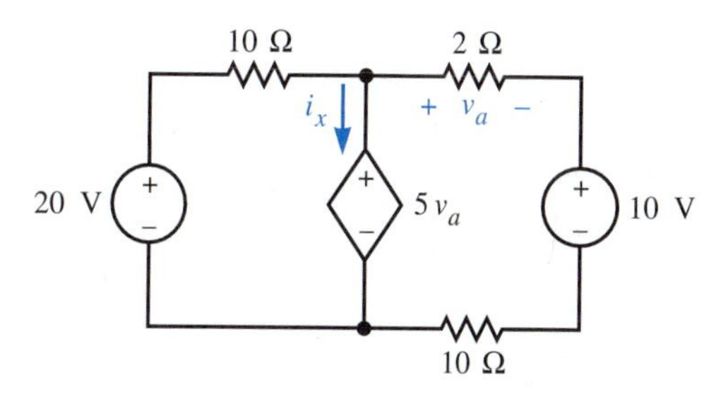

22. Write mesh-current equations for the following network.

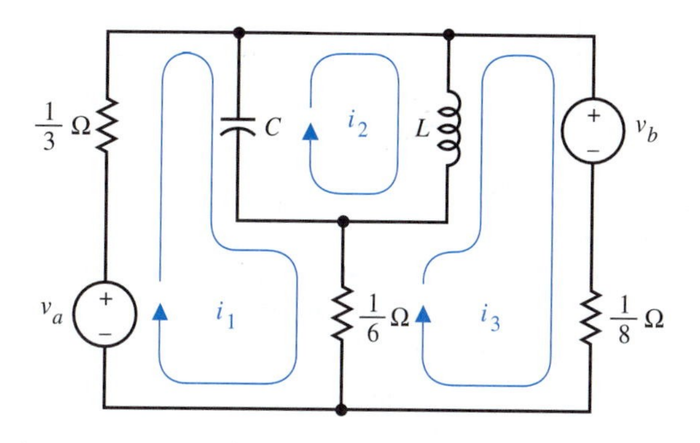

23. Solve for the mesh currents and then calculate voltage v_x for the following network.

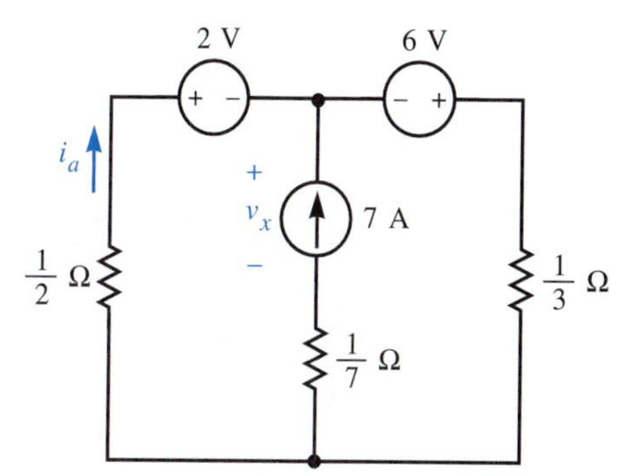

24. Replace the current source in the preceding network with a dependent current source (reference arrow pointing up) that has a control equation $i = 1.5i_a$. Solve for the mesh currents.

25. Write a set of mesh-current equations for the following network. You should have one KVL equation and two KCL equations. 4.8

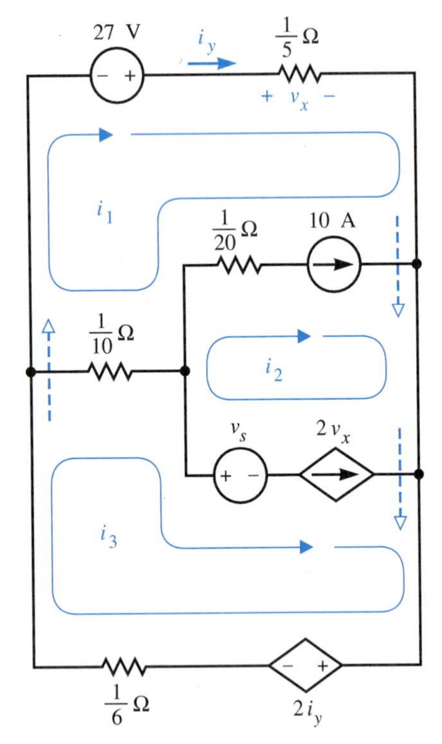

Section 4.4

26. Write a set of mesh-current equations in matrix form for the following circuit. Reduce this set of equations to two KVL equations by substitution, and solve for the mesh currents with the use of determinants. 4.9

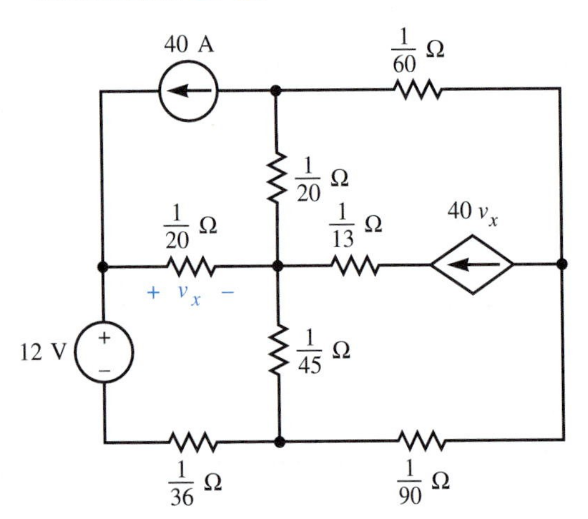

27. (a) Write a set of mesh-current equations in matrix form for the following network.
(b) Solve for the mesh currents if $v_a = -50$ V and $v_b = 150$ V.
(c) Find i_y and v_{ab}.

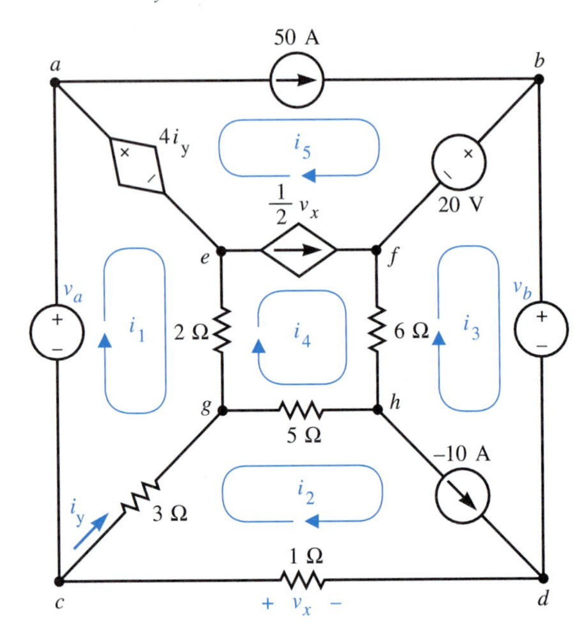

28. The following circuit is the small-signal equivalent circuit for a one-transistor amplifier driven by a current source. Solve for the mesh currents. Determine the ratio of i_y/i_s. This is the current gain of the amplifier.

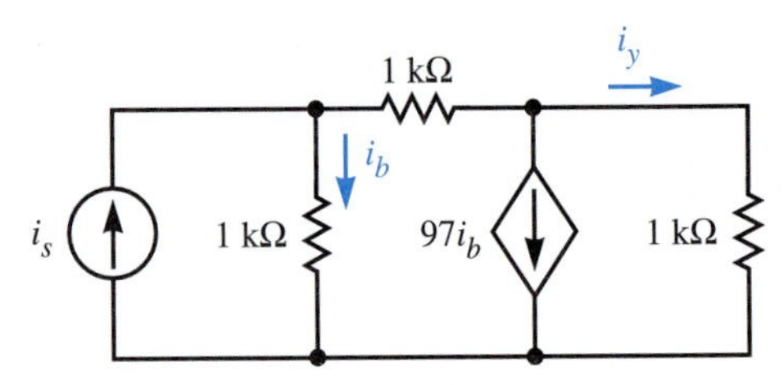

29. Use mesh currents to analyze the amplifier circuit given in Problem 11. Calculate the gains and input resistance requested in Problem 11.

30. Work Problem 14 by the method of mesh currents.

31. Work Problem 15 by the method of mesh currents.

32. For the following network, use the indicated mesh currents and write a matrix mesh-current equation that includes both KVL and KCL equations. Then use the shortcut procedure to write a matrix

mesh-current equation that does not contain any KCL equations.

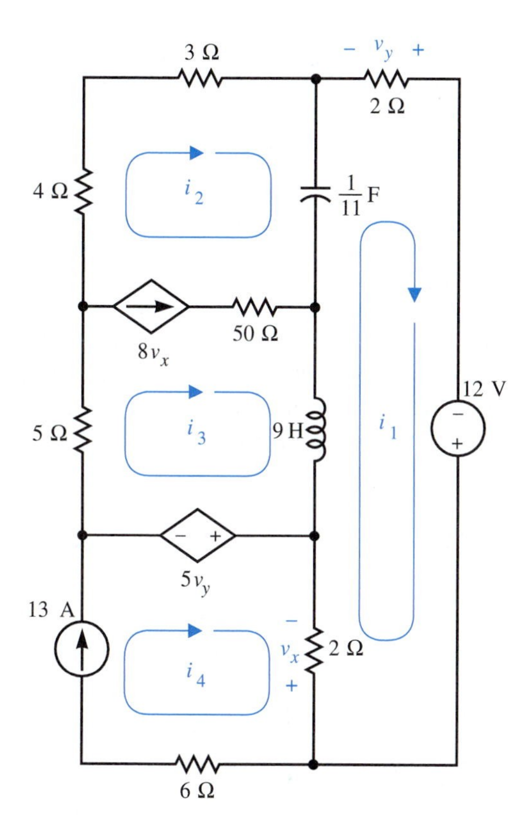

33. Write five mesh-current equations for the network of Problem 17.

Section 4.5

34. Indicate which method of analysis, node voltages or mesh currents, would be most appropriate for the networks in the indicated problems. Explain why.

    (a) Problem 1 (b) Problem 3
    (c) Problem 8 (d) Problem 10
    (e) Problem 18 (f) Problem 25
    (g) Problem 27 (h) Problem 32

Supplementary Problems

35. Draw a circuit diagram for a network with the constraints stated.

    (a) Seven nodes and the following set of mesh-current equations (x_1, x_2, and x_3 are mesh currents).

$$\begin{bmatrix} 6 & -2 & -3 \\ -2 & 11 & -4 \\ -3 & -4 & 12 \end{bmatrix} \begin{bmatrix} x_1 \\ x_2 \\ x_3 \end{bmatrix} = \begin{bmatrix} 10 \\ -20 \\ 30 \end{bmatrix}$$

    (b) Six meshes and the preceding set of node-voltage equations (x_1, x_2, and x_3 are node voltages).

36. The following is a small-signal equivalent circuit for a one-transistor amplifier. (The bottom line would be continuous on the actual circuit.) Determine v_2 and i_1 by the stated method.

    (a) The use of node-voltage equations.

    (b) The use of mesh-current equations.

    (c) The use of node-voltage equations for the left-hand part of the circuit and a mesh equation for the right-hand part of the circuit.

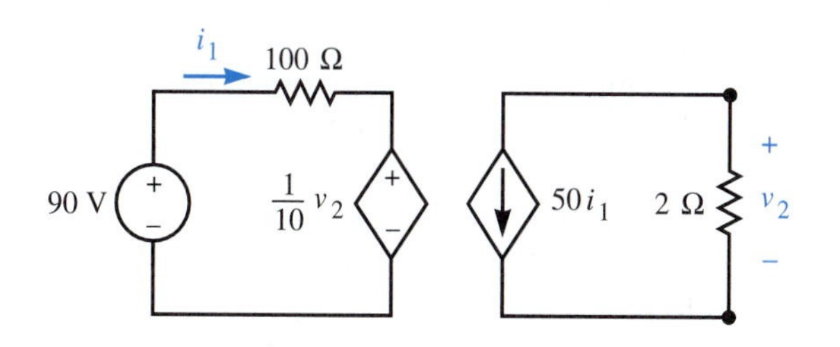

37. Determine i_o for the network shown below.

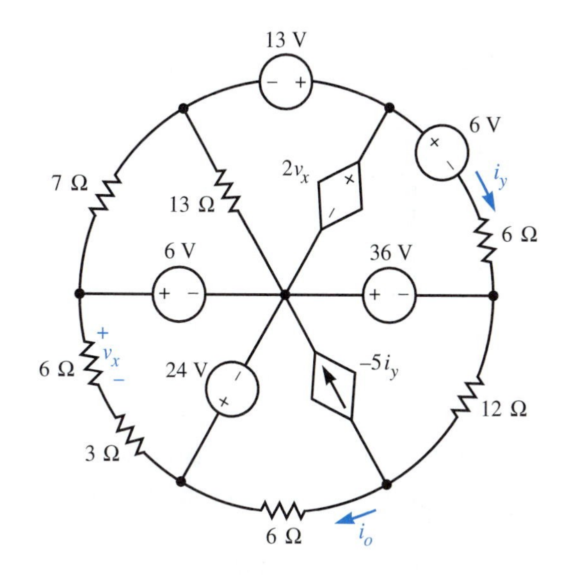

38. Calculate v_o for the following network.

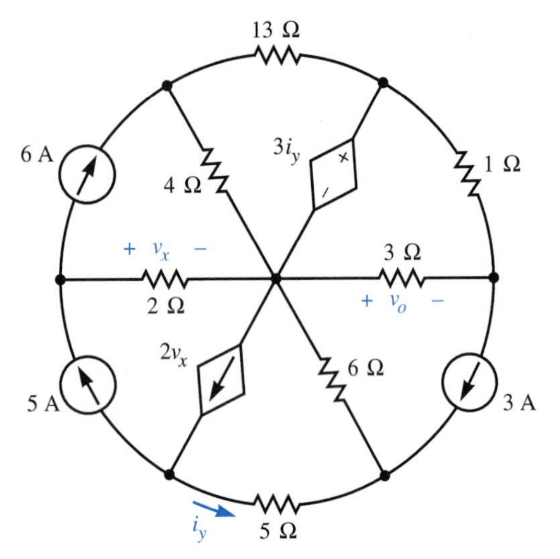

39. Power is most often supplied with a sinusoidally varying voltage (ac), but some World War II–era ships used a dc (constant-voltage) power system. The configuration shown below supplies power at nominal voltages of 120 and 240 V. This system has a slight safety advantage, because no point in the system has a voltage greater than 120 V with respect to node 0, which is connected to the ship hull. Determine currents i_1, i_2, and i_3. Calculate the voltages across the 9.5-, 20.5-, and 10-Ω load resistances. If the 1-Ω neutral wire is broken, determine the load voltages. Do you see a potential problem?

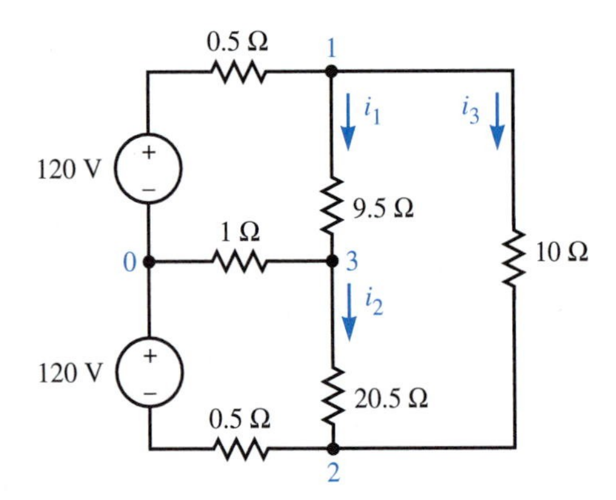

40. We have designed a transistor amplifier with a voltage gain $G_v = v_o/v_s$. The small-signal equivalent circuit follows.

    (a) We must select a transistor with a current gain β so that $9.5 < |G_v| \leq 10$. What range of values of β satisfies the gain requirements if $R_E = 1\ \text{k}\Omega$?

    (b) We know that $\beta > 30$. What is the minimum value of R_E so that we know $|G_v| < 10$?

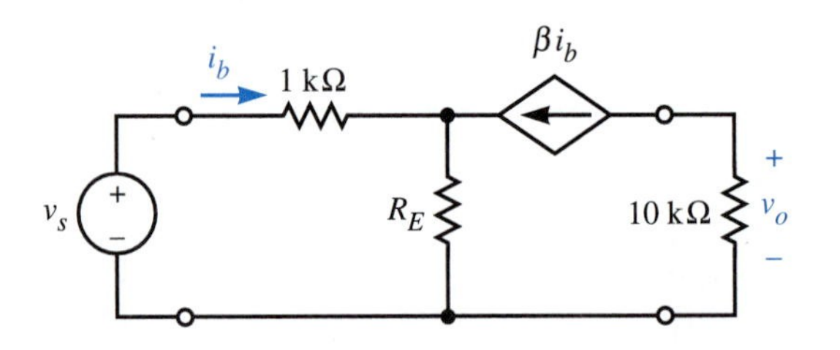

41. For the following circuit use the reference node and node numbers given to write a matrix node-voltage equation that includes both the KCL and KVL equations. Next, use the shortcut procedure to write a matrix node-voltage equation that does not contain any KVL equations.

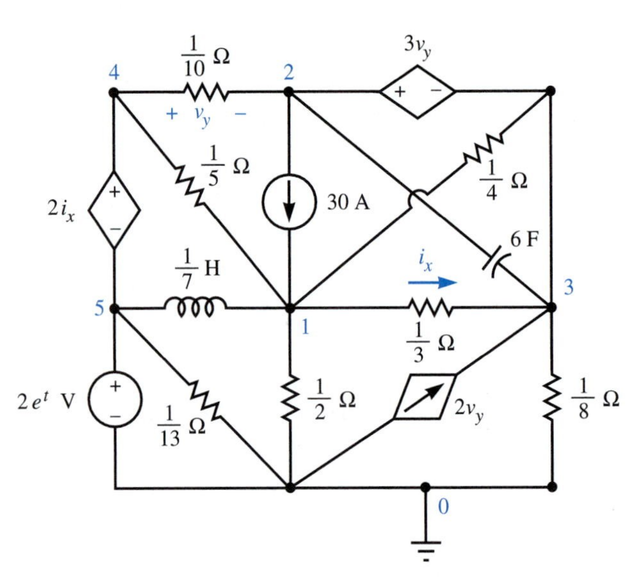

42. Analyze the three contrived networks that follow.

    (a)

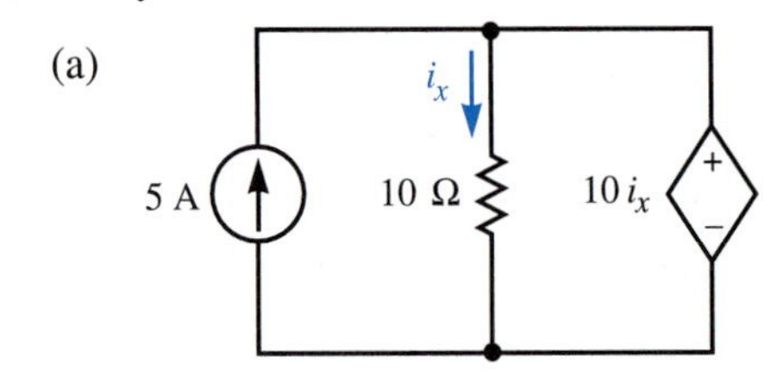

    (b)

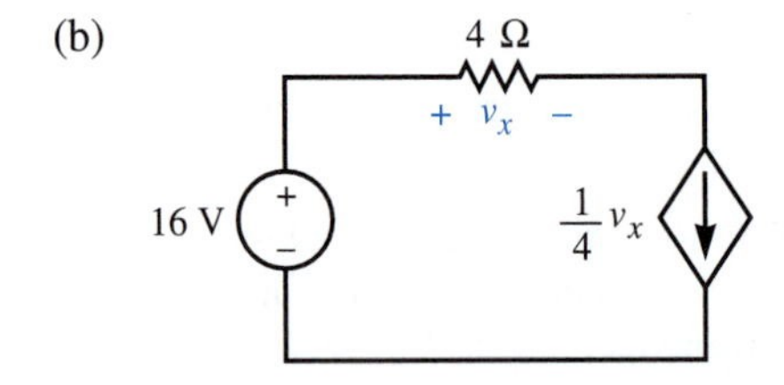

(c)

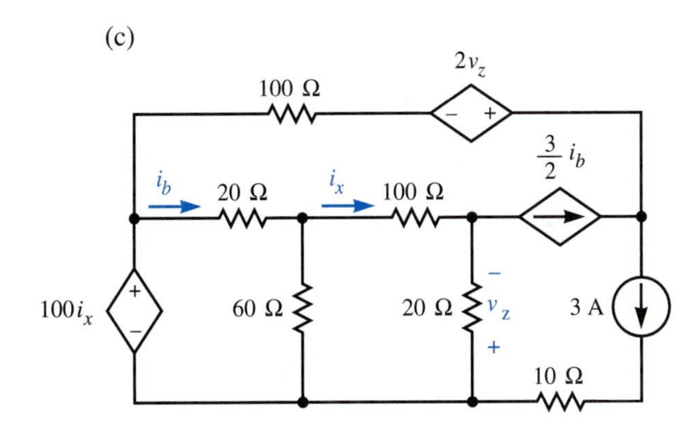

43. Use the method of your choice to calculate current i_x in the circuit in Fig. 4.17.

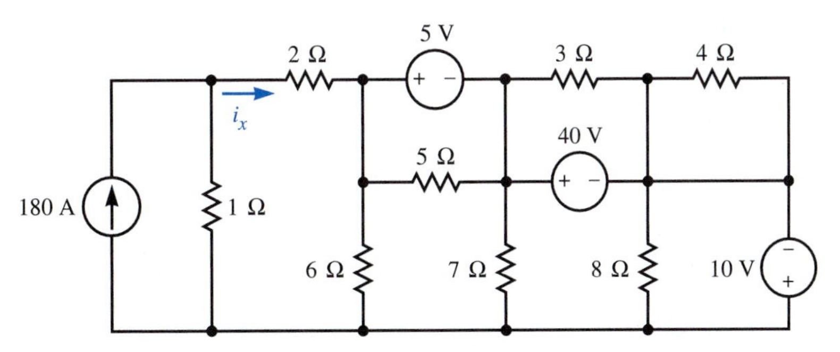

FIGURE 4.17

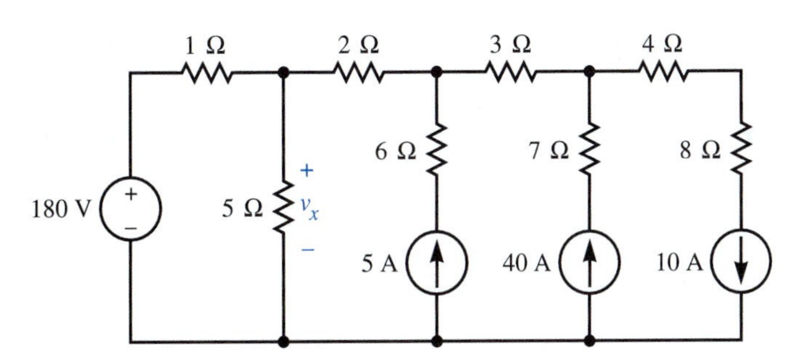

FIGURE 4.18

44. Use the most convenient method to determine voltage v_x in the network in Fig. 4.18.

45. Use some of the analysis tools introduced in Chapter 3 in conjunction with the method of node voltages to find i_x in the network in Fig. 4.19. 4.10

46. Use some of the analysis tools introduced in Chapter 3 in conjunction with the method of mesh currents to find v_x in the network in Fig. 4.20.

47. Analyze the network in Fig. 4.21 by the method of node voltages. Solve for current i_{23}.

48. Analyze the network in Fig. 4.22 by the method of mesh currents. Solve for voltage v_{23}.

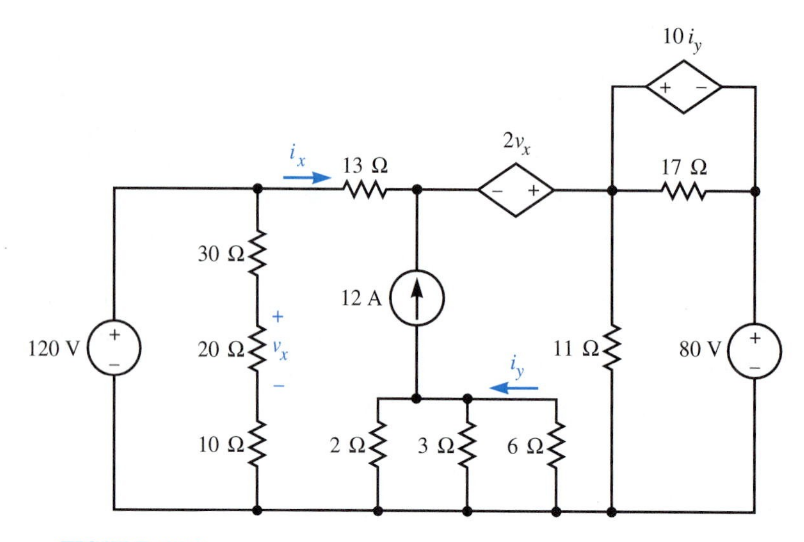

FIGURE 4.19

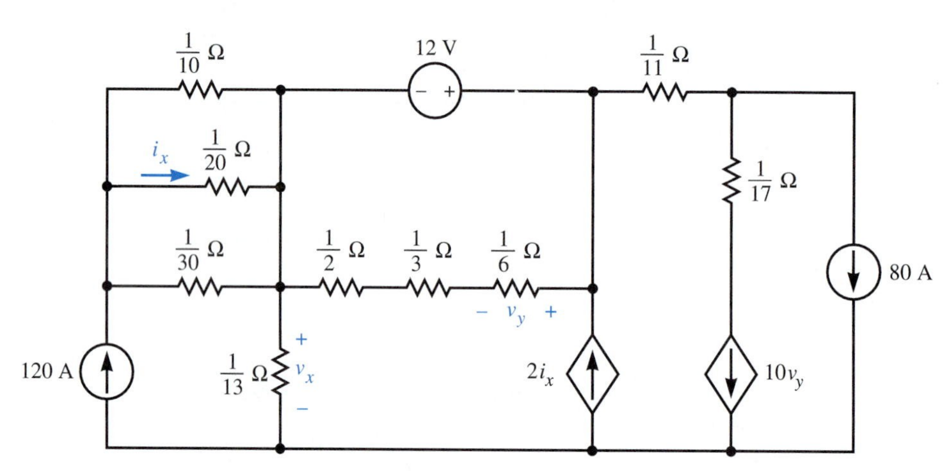

FIGURE 4.20

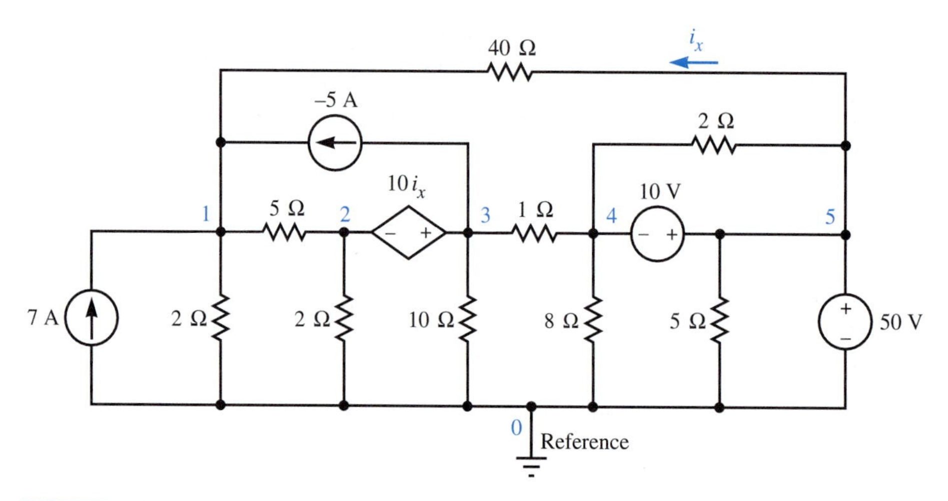

FIGURE 4.21

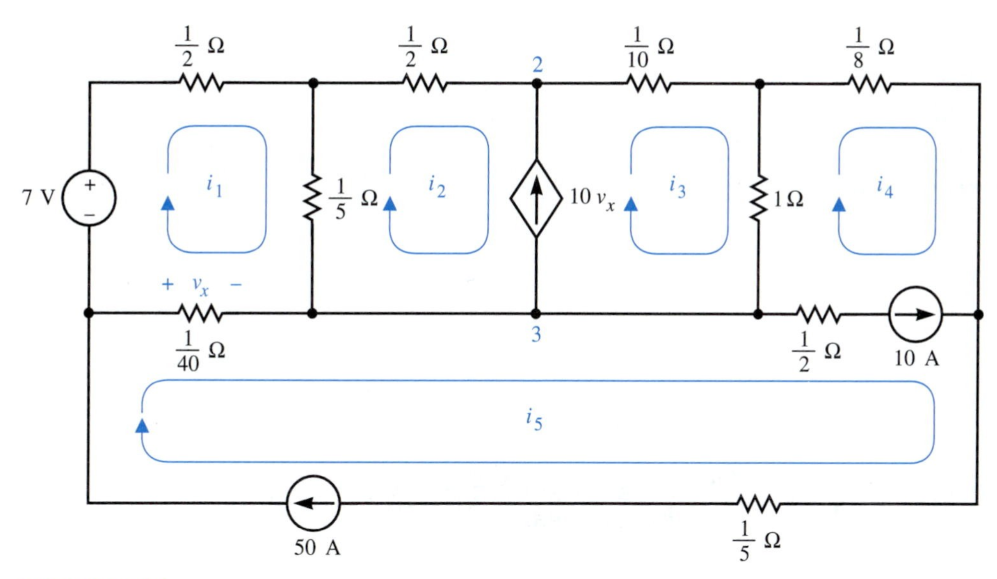

$\frac{1}{2}\ \Omega$
$\frac{1}{2}\ \Omega$
2
$\frac{1}{10}\ \Omega$
$\frac{1}{8}\ \Omega$
7 V
+
−
i_1
$\frac{1}{5}\ \Omega$
i_2
$10\ v_x$
i_3
1 Ω
i_4
+ v_x −
$\frac{1}{40}\ \Omega$
3
$\frac{1}{2}\ \Omega$
10 A
i_5
50 A
$\frac{1}{5}\ \Omega$

FIGURE 4.22

5

Network Properties

The networks considered in this book have important general properties, known as time invariance and linearity. Time invariance implies that a delay in an input (a source value) just delays the response to the source by the same amount of time. Linearity implies that if an input is multiplied by a constant, then the response is multiplied by the same constant. Linearity further implies that the response to two or more sources applied simultaneously is simply the sum of the responses for each source taken separately. In the remainder of the book, we will develop analysis techniques that exploit time invariance and linearity.

5.1 Linearity and Superposition

In this text, we consider only circuits with constant values of resistance (R), inductance (L), capacitance (C), and dependent sources with constant parameters (μ, g, β, and r). Because these values are constant, the components have two important properties:

1. The component is *time-invariant.* If we delay the input (v or i) to a time-invariant component by a time t_0, we delay the response (i or v) by the same time.
2. The component is *linear.* (a) If an input (v or i) to a linear component is multiplied by a constant a, the response (i or v) is multiplied by the same constant. (b) If the input (v_1 or i_1) to a linear component causes a response (i_1 or v_1), and a second input (v_2 or i_2) causes a second response (i_2 or v_2), then an input that is the sum of two inputs ($v = v_1 + v_2$ or $i = i_1 + i_2$) causes a response that is the sum of the two responses ($i = i_1 + i_2$ or $v = v_1 + v_2$).

A network composed of linear time-invariant components is a *linear time-invariant network* and will have the same two properties as a linear time-invariant component. We are particularly interested in Property 2b, which we call *superposition.* For a linear network we state this as follows:

THEOREM 1 Superposition Theorem

The voltage and current response of a linear network to a number of independent sources is the *sum* of the responses obtained by applying each independent source once with other independent sources set equal to zero.

We will find it useful to observe that a 0-V voltage source is equivalent to a short circuit, and a 0-A current source is equivalent to an open circuit. To apply superposition, *dependent sources are left intact and are not set equal to zero.* We can use any analysis technique to calculate each response.

A pictorial interpretation of the superposition theorem for a network with three independent sources is shown in Fig. 5.1. The values of a voltage v and a current i are to be calculated for the linear network depicted by Fig. 5.1a. The voltage and current are calculated by use of the superposition theorem as follows:

1. Set all independent sources except the voltage source of value v_a equal to zero. (A 0-V voltage source is equivalent to a short circuit and a 0-A current source is equivalent to an open circuit.) This yields the network depicted in Fig. 5.1b. The responses v' and i' due to v_a acting alone are now calculated by any convenient method.
2. Set all independent sources except the current source of value i_b equal to zero. This yields the network depicted in Fig. 5.1c. The responses v'' and i'' due to i_b acting alone are now calculated by any convenient method.
3. Set all independent sources except the voltage source of value v_c equal to zero. This yields the network depicted in Fig. 5.1d. The responses v''' and i''' due to v_c acting alone are now calculated by any convenient method.

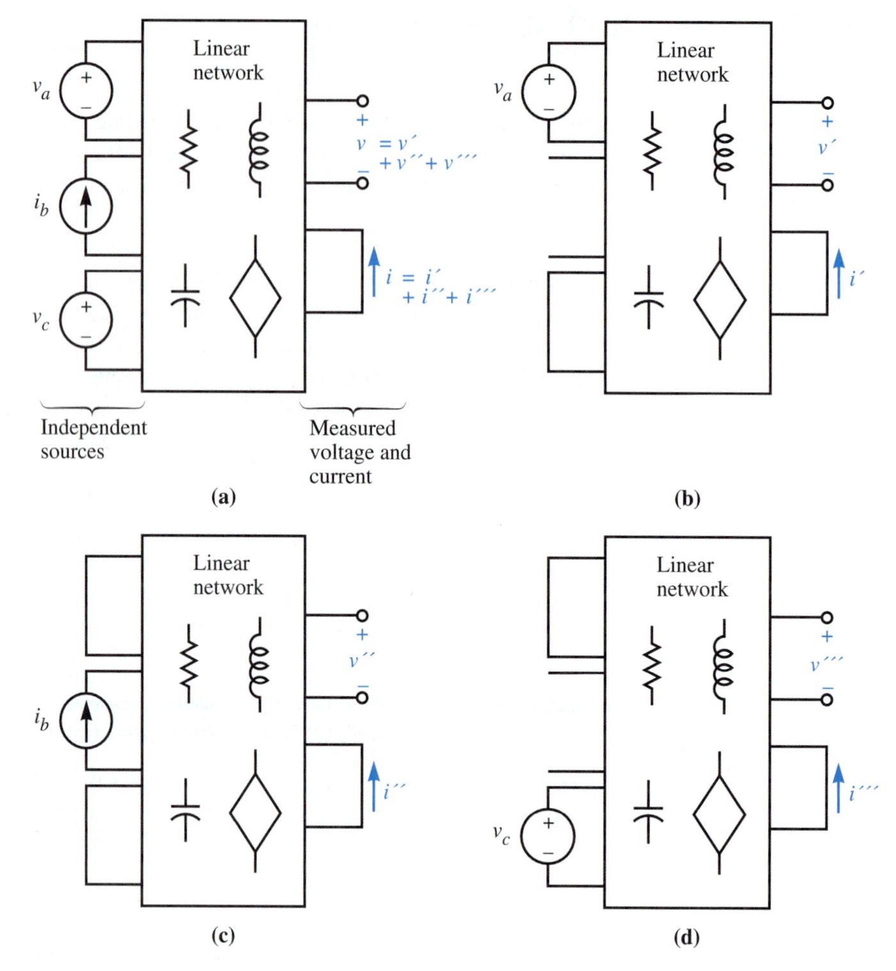

FIGURE 5.1 Pictorial interpretation of the superposition theorem for a network with three independent sources. (*a*) Original network with three independent sources; (*b*) network with all independent sources except v_a set equal to zero; (*c*) network with all independent sources except i_b set equal to zero; (*d*) network with all independent sources except v_c set equal to zero

4. The responses v and i due to all sources acting simultaneously in the original circuit of Fig. 5.1a are found as the sum of the responses due to each source acting alone:

$$v = v' + v'' + v''' \tag{5.1}$$

and

$$i = i' + i'' + i''' \tag{5.2}$$

The following two examples illustrate the use of the superposition theorem for a network containing three independent sources as in Fig. 5.1a.

EXAMPLE 5.1 Use superposition to determine voltage v_x in the network of Fig. 5.2.

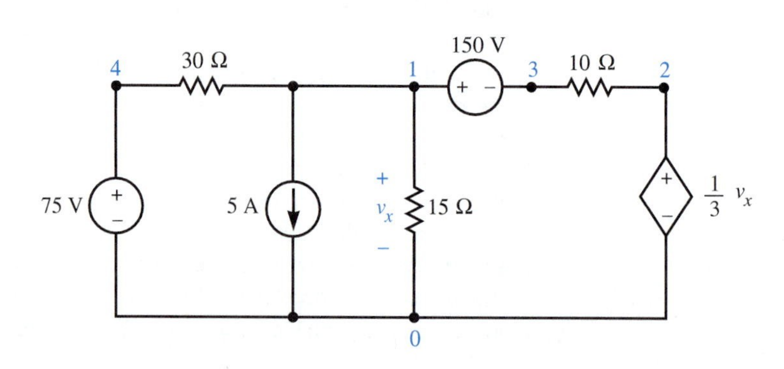

FIGURE 5.2 Original network with three independent sources

Solution Since we have three independent sources, we should use them one at a time, find the corresponding component of v_x, and add the three components.

First set all independent sources except the 75-V voltage source equal to zero. The result is shown in Fig. 5.3. Observe that the dependent source of value $\frac{1}{3}v_x$ is *not* set equal to zero.

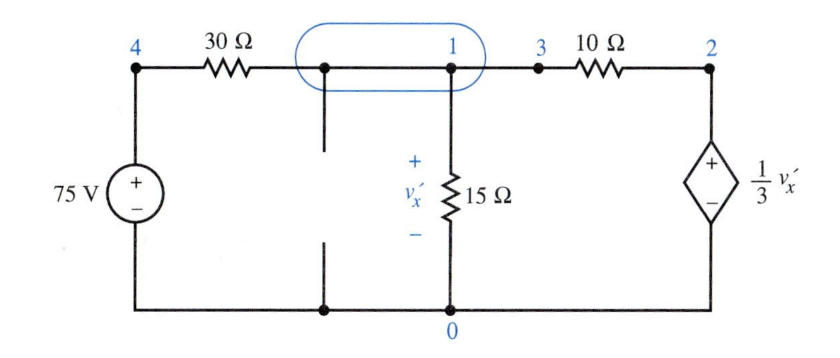

FIGURE 5.3
All independent sources except the 75-V voltage source set equal to zero

The KCL equation for the closed surface indicated in Fig. 5.3 is

$$\frac{1}{30}(v_x' - 75) + \frac{1}{15}v_x' + \frac{1}{10}\left(v_x' - \frac{1}{3}v_x'\right) = 0$$

which gives the component of v_x that corresponds to the 75-V source as

$$v_x' = 15 \text{ V}$$

Next, set all independent sources except the 5-A current source equal to zero. The result is shown in Fig. 5.4.

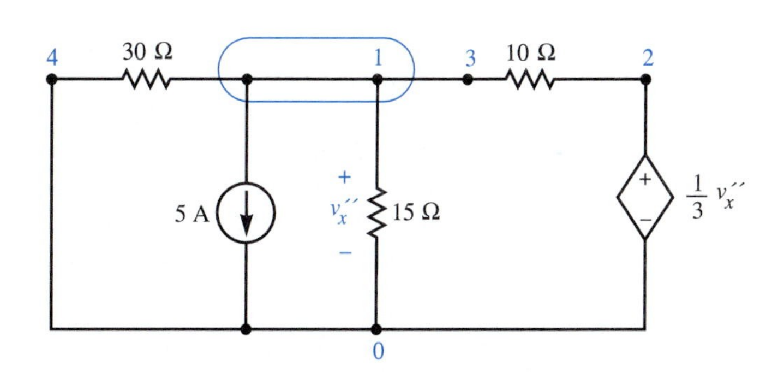

FIGURE 5.4
All independent sources except the 5-A current source set equal to zero

The KCL equation for the indicated closed surface is

$$\frac{1}{30}v_x'' + 5 + \frac{1}{15}v_x'' + \frac{1}{10}\left(v_x'' - \frac{1}{3}v_x''\right) = 0$$

which gives the component of v_x that corresponds to the 5-A source:

$$v_x'' = -30 \text{ V}$$

Now retain the last independent source, as shown in Fig. 5.5. The equivalent resistance for the 30-Ω and 15-Ω parallel resistances is

$$R_p = \frac{1}{(1/30) + (1/15)} = 10\ \Omega$$

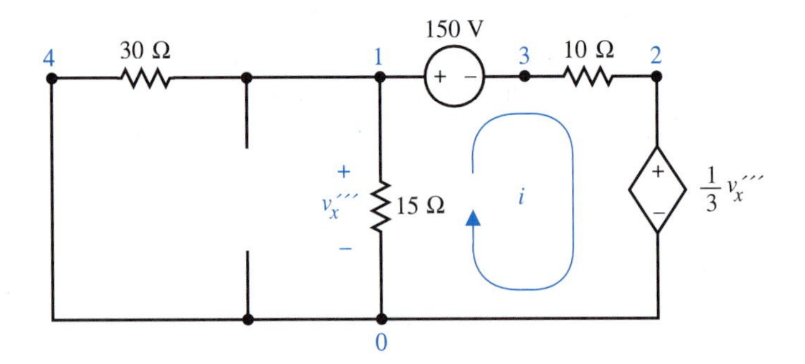

FIGURE 5.5 All independent sources except the 150-V voltage source set equal to zero

A KVL equation around the indicated loop can then be written as

$$10i + 150 + 10i + \frac{1}{3}(-10i) = 0$$

which gives

$$i = -9 \text{ A}$$

From Ohm's law, the component of v_x that corresponds to the 150-V source is

$$v_x''' = -10(-9) = 90 \text{ V}$$

Addition of the three components of v_x as required by the superposition theorem gives us

$$v_x = v_x' + v_x'' + v_x''' = 15 - 30 + 90 = 75 \text{ V}$$

The rule is that we use each independent source once. Therefore we could have set the 150-V source equal to zero and solved for the component of v_x due to both the 75-V source and the 5-A source. We would then set both the 75-V source and the 5-A source to zero and solve for the component of v_x due to the 150-V source. Voltage v_x would be the sum of these two voltage components.

Although we used superposition to calculate a voltage in the preceding example, we could just as easily have used superposition to calculate currents.

> Be sure that the reference for a voltage or current component has the same orientation as that of the original voltage or current.

If we violate this rule, the voltage or current component must be subtracted.

As we will see, in many parts of this text, the concept of superposition provides an exceedingly useful way of thinking about network analysis and design.

Remember

Superposition applies to linear circuits. Each independent source must be used once while the unused independent sources are set equal to zero. A voltage source set equal to zero is equivalent to a short circuit. A current source set equal to zero is equivalent to an open circuit. In all cases, dependent sources are left in the circuit and not changed.

EXERCISES

1. Use superposition to calculate voltage v_x in the following circuit; (a) with terminals x and y not connected and (b) with terminals x and y connected.

answer: 18 V, 18 V

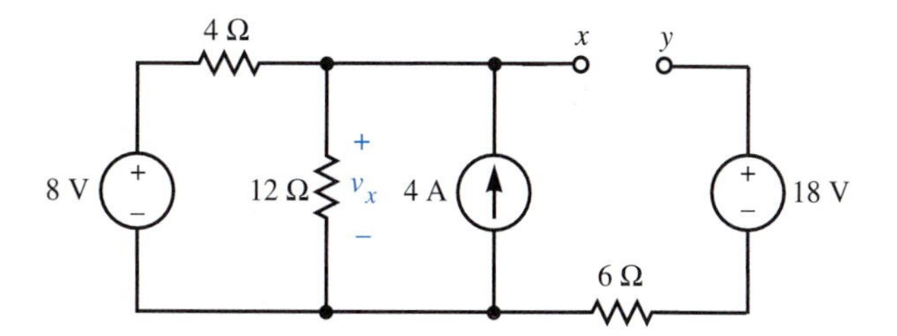

2. Use superposition to calculate current i_x in the following circuit.

answer: −47 A

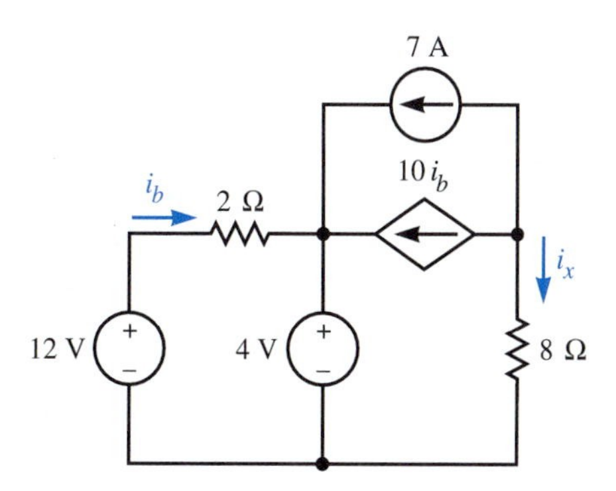

5.2 Equivalent Sources

In this section we introduce a new technique for network simplification. Many *practical sources* of electrical energy, such as a storage battery, are adequately modeled for some applications by an ideal voltage source and series resistance,† as shown in Fig. 5.6a on the following page. A KVL equation quickly provides the terminal equation given in Fig. 5.6a. This is the equation for a straight line. For an open circuit, the load current i is zero. Therefore the voltage across the internal resistance R_s is zero, and the open-circuit voltage is $v_{oc} = v = v_s$. This gives the intercept of the volt-ampere line with the voltage axis, as shown in Fig. 5.6a. With a short circuit, load voltage v is zero, and Ohm's law gives the short-circuit current $i_{sc} = i = v_s/R_s$. This is the intercept of the volt-ampere line with the current axis, as shown in Fig. 5.6a.

A *practical* current source is modeled as an ideal current source in parallel with a resistance,‡ as shown in Fig. 5.6b. A KCL equation gives the terminal equation shown

† An ideal voltage source with zero series resistance can supply infinite current and thus infinite power. This is physically impossible.

‡ An ideal current source without a parallel resistance can supply current to an infinite resistance and thus supply infinite power. This is physically impossible.

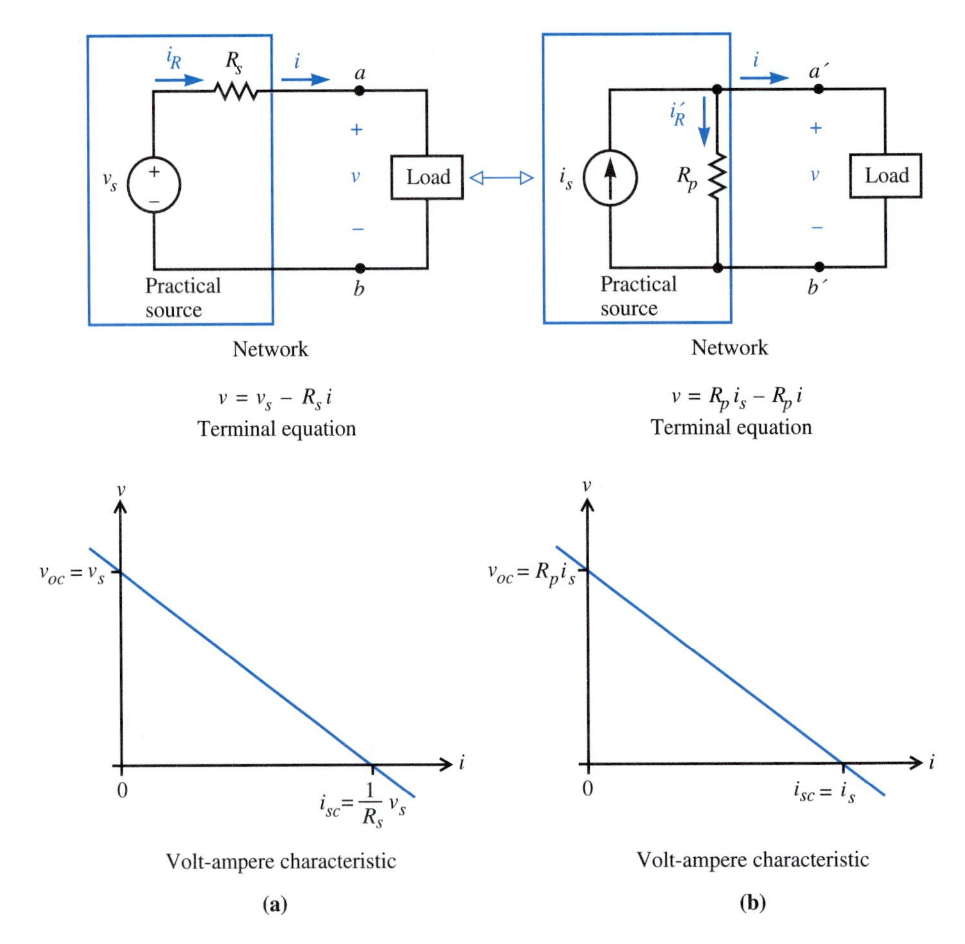

FIGURE 5.6
Practical sources and their terminal characteristics. (*a*) Practical voltage source; (*b*) practical current source

in the figure. This is the equation for a straight line. For a short circuit, the load voltage v is zero. Therefore the current through the internal resistance R_p is zero and the short-circuit current is $i_{sc} = i = i_s$. This gives the intercept of the volt-ampere line with the current axis, as shown in Fig. 5.6b. For an open-circuit, load current i is zero, and Ohm's law gives the open-circuit voltage $v_{oc} = R_p i_s$. This is the intercept of the volt-ampere line with the voltage axis, as shown in Fig. 5.6b.

Comparison reveals that the two terminal equations and volt-ampere lines shown in Fig. 5.6 are identical if

$$R_p = R_s \tag{5.3}$$

and R_s, v_s, and i_s are related by Ohm's law, that is,

$$v_s = R_s i_s = R_p i_s \tag{5.4}$$

The conclusion is

THEOREM 2 Source Transformations

A voltage source of value v_s in series with a resistance of value R_s and a current source of value i_s in parallel with a resistance of value R_p are equivalent if

$$R = R_s = R_p \tag{5.5}$$

and $v_s = v_{oc}$, $i_s = i_{sc}$, and R are related by Ohm's law:

$$v_s = Ri_s \tag{5.6}$$

The *equivalence is only with respect to the terminal characteristics* that relate v to i. The currents through R_s and R_p are only equal when the load current is $i_{sc}/2$, which gives a load voltage of $v_{oc}/2$.

A voltage source with zero series resistance has a terminal voltage that is independent of the output current and therefore cannot be replaced with an equivalent current source and a parallel resistance. A current source with zero parallel conductance (infinite parallel resistance) has an output current that is independent of the terminal voltage and therefore cannot be replaced with an equivalent voltage source and series resistance.

The following gives a numerical example of source transformations.

EXAMPLE 5.2 (a) Develop an equivalent source, with respect to terminals a and b, for the network shown in Fig. 5.7a. (b) Perform a source transformation on the network of Fig. 5.7b.

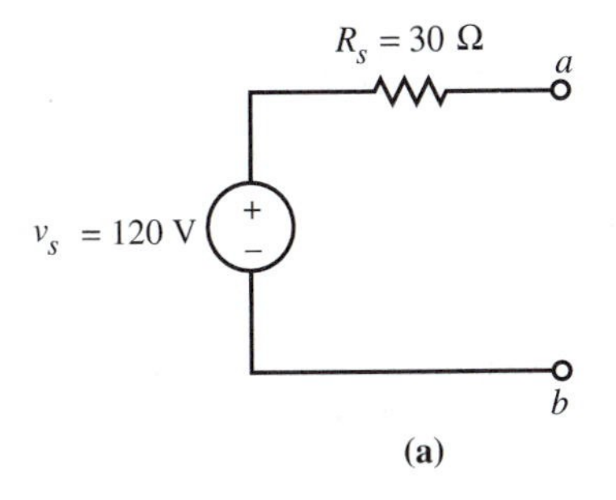

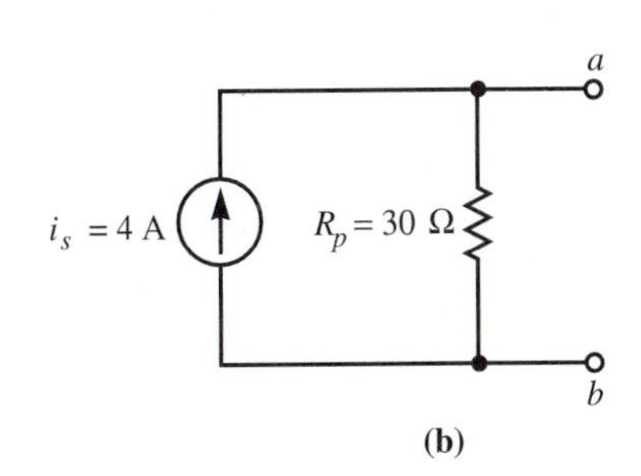

FIGURE 5.7 Equivalent source example

Solution (a) The resistance in the equivalent circuit must be the same as in the original circuit:

$$R_p = R_s = 30\ \Omega$$

The current source required for equivalence can be obtained from Ohm's law:

$$i_s = i_{sc} = i_{ab}\Big|_{v_{ab}=0} = \frac{1}{R_s}v_s = \frac{1}{30}(120) = 4\ \text{A}$$

The equivalent source is as shown in Fig. 5.7b.

(b) The resistance of the equivalent source must be the same as in the original circuit:

$$R_s = R_p = 30\ \Omega$$

The voltage source required for equivalence is given by Ohm's law:

$$v_s = v_{oc} = v_{ab}\Big|_{i_{ab}=0} = R_p i_s = 30(4) = 120 \text{ V}$$

The equivalent source is as shown in Fig. 5.7a.

Source transformations are useful to simplify a circuit for analysis. Conversion from a practical voltage source to a practical current source eliminates a node. Conversion from a practical current source to a practical voltage source eliminates a mesh. A source transformation may also permit further simplification by series and parallel equivalents. This is illustrated in the following example.

EXAMPLE 5.3 Find voltage v shown in Fig. 5.8a with the aid of repeated source transformations.

Solution Begin at the left of the circuit in Fig. 5.8a, and make repeated source transformations until only a two-node circuit remains.

Transformation of the 250-V voltage source and 10-Ω series resistance gives a current source

$$i_s = \frac{250}{10} = 25 \text{ A}$$

with a 10-Ω parallel resistance, as shown in Fig. 5.8b. (Remember that the current through this 10-Ω resistance is not the same as that through the original 10-Ω resistance.) Now find an equivalent resistance for the 10-Ω resistance in parallel with the 15-Ω resistance:

$$R_p = \frac{1}{(1/10) + (1/15)} = \frac{150}{25} = 6 \ \Omega$$

Next, make a source transformation on the 25-A current source with the equivalent 6-Ω parallel resistance:

$$v_s = 25 \times 6 = 150 \text{ V}$$

The equivalent circuit is shown in Fig. 5.8c. Find an equivalent resistance for the series combination of the 6-Ω and 4-Ω resistances:

$$R_s = 6 + 4 = 10 \ \Omega$$

This yields the circuit shown in Fig. 5.8d. Finally, transform the 150-V voltage source and the 10-Ω equivalent series resistance into a current source with a 10-Ω parallel resistance:

$$i_s = \frac{150}{10} = 15 \text{ A}$$

which yields the equivalent circuit of Fig. 5.8e. At this point, v can easily be found from a single KCL equation written for a surface enclosing the top node:

$$-15 + \frac{1}{10}v + \frac{1}{15}v = 0$$

or

$$v = 90 \text{ V}$$

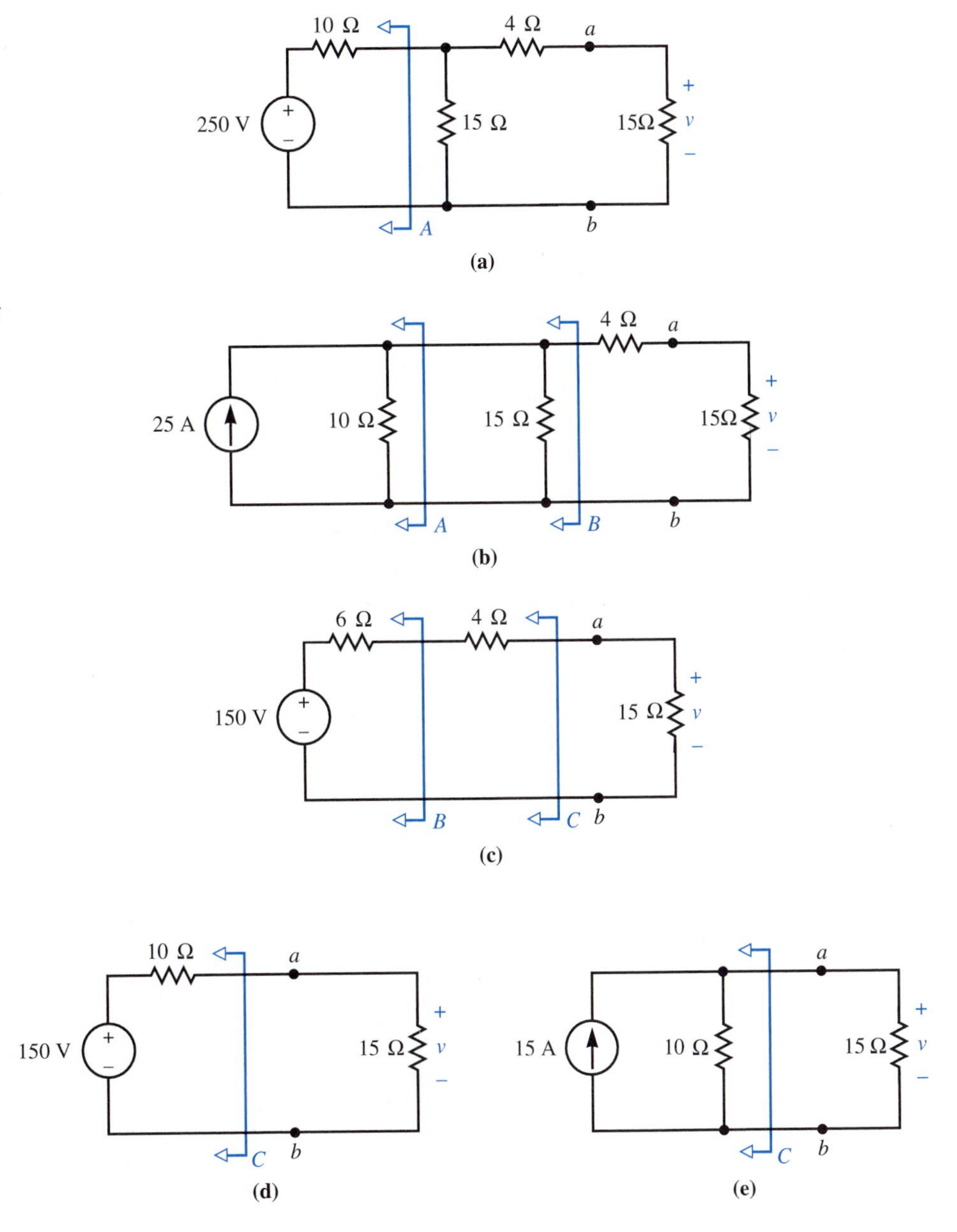

FIGURE 5.8 An example in which source transformations are convenient. (*a*) The original circuit; (*b*) after the first source transformation; (*c*) after a resistance combination and a second source transformation; (*d*) after a resistance combination; (*e*) after a third source transformation

A terminal equivalent for a voltage source and parallel resistance is just a voltage source, as shown in Fig. 5.9a. Similarly, a terminal equivalent for a current source and

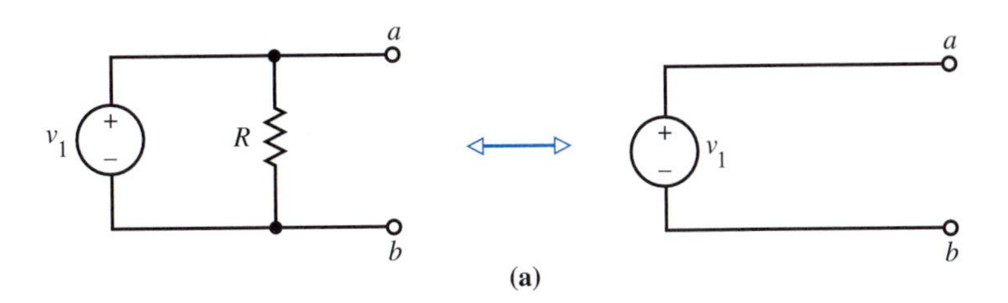

FIGURE 5.9 Sources with equivalent terminal characteristics. (*a*) Voltage source and parallel resistance; (*b*) current source and series resistance

FIGURE 5.9
Continued

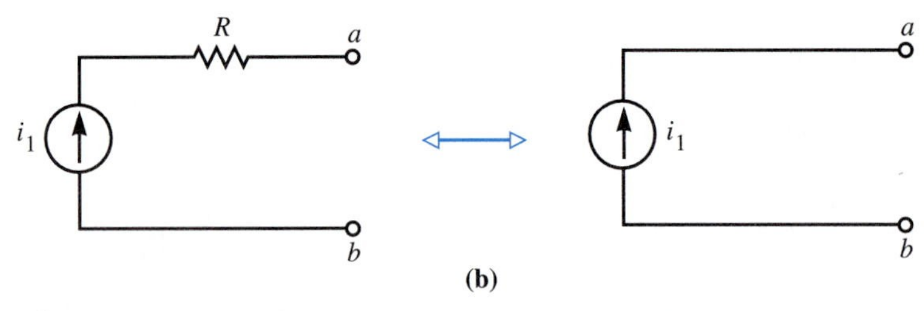

(b)

series resistance is just a current source, as shown in Fig. 5.9b.

Remember

Conversion from a practical voltage source to a practical current source eliminates a node. Conversion from a practical current source to a practical voltage source eliminates a mesh. Either conversion typically alters the current through the source resistance. The equivalence is only with respect to the terminal characteristics.

EXERCISES

3. Use repeated source transformations, and combine parallel and series resistances to find voltage v in the network below by writing a single KCL equation.

   answer: 4 V

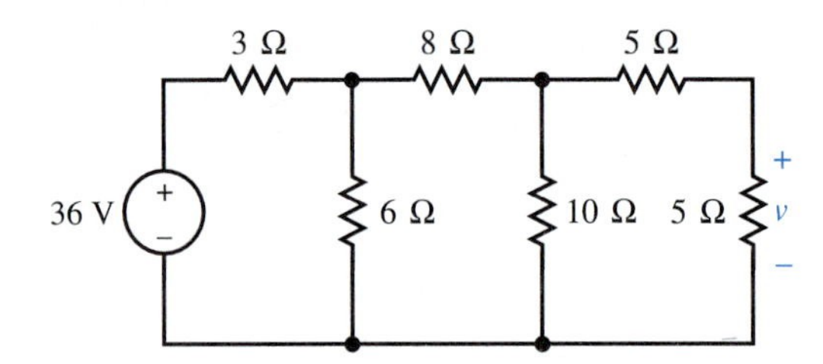

4. Use source transformations to determine currents i_x and i_y for the network that follows.

   answer: 25 A, 35 A

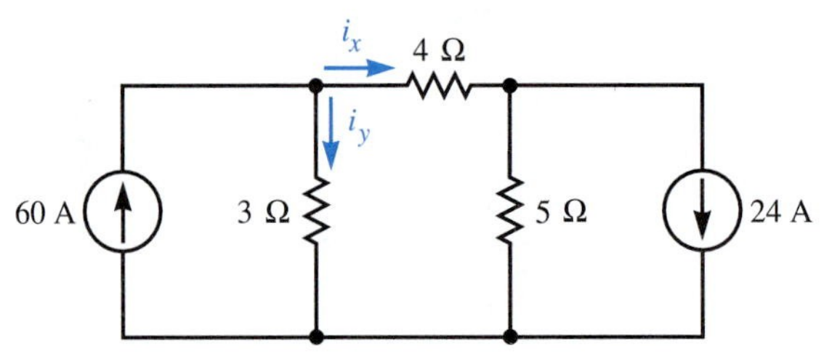

5. Use source transformations to eliminate two nodes in the network on the next page and write node-voltage equations for the remaining two nodes. This is a useful technique.

   answer:

$$\begin{bmatrix} C_1\dfrac{d}{dt}+\dfrac{1}{R_1}+\dfrac{1}{L}\displaystyle\int_{-\infty}^{t} d\lambda & -\dfrac{1}{L}\displaystyle\int_{-\infty}^{t} d\lambda \\ -\dfrac{1}{L}\displaystyle\int_{-\infty}^{t} d\lambda & C_2\dfrac{d}{dt}+\dfrac{1}{R_2}+\dfrac{1}{L}\displaystyle\int_{-\infty}^{t} d\lambda \end{bmatrix}\begin{bmatrix} v_1 \\ v_2 \end{bmatrix} = \begin{bmatrix} \dfrac{1}{R_1}v_a \\ \dfrac{1}{R_2}v_b \end{bmatrix}$$

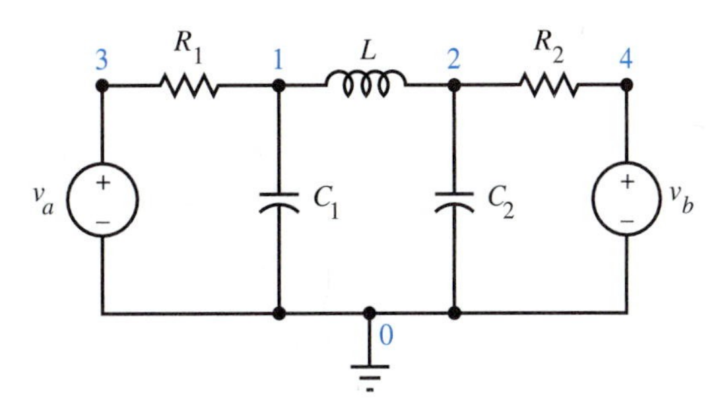

6. Use source transformations to find two-component networks that are equivalent with respect to the relation between v_{ab} and i_{ab} for the portion of the network shown to the left of terminals a and b in the following two circuits.

(a)

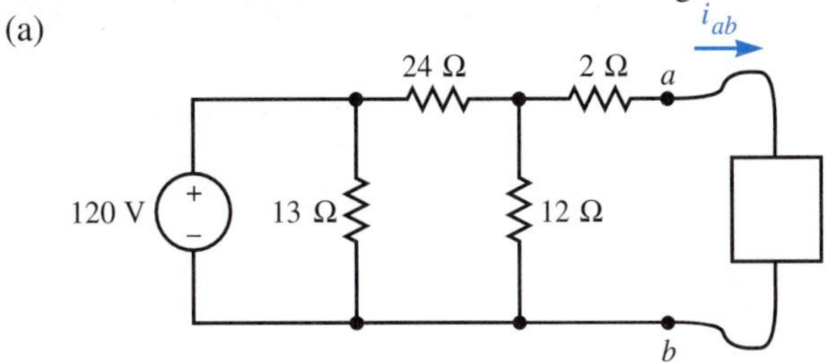

answer: A 40-V source in series with 10 Ω or a 4-A source in parallel with 10 Ω

(b)

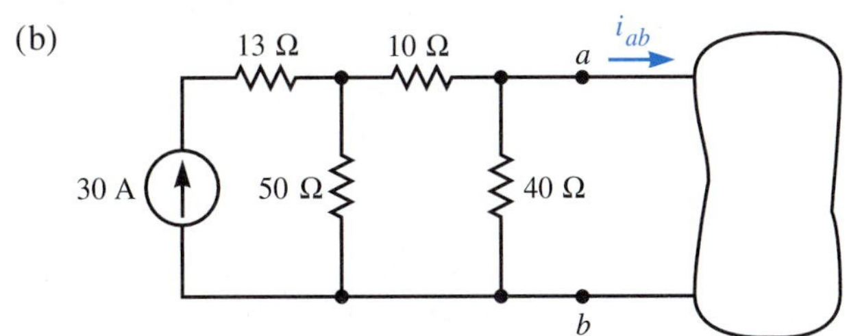

answer: A 25-A source in parallel with 24 Ω or a 600-V source in series with 24 Ω

7. A voltage source with a series resistance of value $R_s = R$ is connected to a load resistance of value R_L. An *equivalent* current source with parallel resistance $R_p = R$ is also connected to a load resistance of value R_L. For what value of R_L will the current through R_s be the same as the current through R_p? answer: $R_L = R$

5.3 Thévenin's and Norton's Theorems

Frequently we need to connect two terminals of one network, such as an audio amplifier, to two terminals of another network, such as a speaker. Here the pertinent electrical quantities are the speaker voltage and current, not the various voltages and currents internal to the amplifier. We can use superposition to substantially generalize the concept of equivalent sources to apply to this problem.

The problem is depicted in Fig. 5.10. Network A is a linear network with a bounded open-circuit voltage and contains only independent sources, resistances, and dependent sources. The control voltages and currents for dependent sources in one network are not in the other network, but no other restrictions are placed on network B.

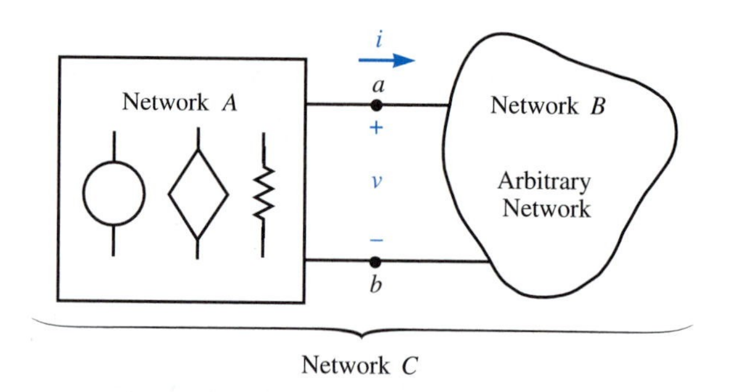

FIGURE 5.10 Network A of resistances and independent and dependent sources connected to an arbitrary network B

First calculate or measure current i and replace network B with a current source of this value so that we can apply superposition. Now set this current source to zero. Voltage v will be the open-circuit voltage:

$$v_{oc} = v|_{i=0} \tag{5.7}$$

We next set all independent sources in network A equal to zero and reactivate the current source. Linearity of network A assures us that voltage v is proportional to current i:

$$v = -R_{Th} i \tag{5.8}$$

The constant of proportionality R_{Th} does not depend on i and is the equivalent resistance looking into terminals a and b of network A when all independent sources in network A are set equal to zero.

When we return the independent sources in network A to their original value, Eqs. (5.7) and (5.8) and superposition give us

$$v = v_{oc} - R_{Th} i \tag{5.9}$$

This is the terminal equation for a voltage source and series resistance connected to network B, as shown in Fig. 5.11. This equivalent circuit for network A is completely general, because no restriction was placed on the value of current i required by network B, and R_{Th} does not depend on the value of i. We formalize this result in the following theorem.

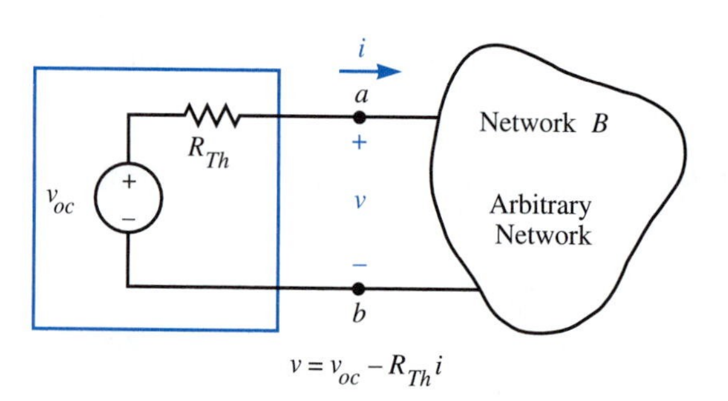

FIGURE 5.11 Network A of Fig. 5.10 replaced by a voltage source and series resistance (the Thévenin equivalent circuit)

THEOREM 3 Thévenin's Theorem

Two networks A and B are connected at only two terminals a and b as shown in Fig. 5.10. If linear network A has

1. An open-circuit voltage $|v_{oc}| < \infty$,
2. Only independent sources, resistances, and dependent sources,
3. A resistance of value R_{Th} that is measured between terminals a and b of network A when all independent sources in network A are set equal to zero and network B is disconnected from network A, and
4. The control voltages and currents for dependent sources in one network are not in the other network,

we can make a Thévenin equivalent for network A that consists of an independent voltage source of value v_{oc} and series resistance of value R_{Th} as shown in Fig. 5.11.

A source transformation yields the following theorem:

THEOREM 4 Norton's Theorem

If the magnitude of the short-circuit current is $|i_{sc}| < \infty$ for network A of Fig. 5.10, we can replace network A with a Norton equivalent circuit that consists of a current source of value i_{sc} in parallel with a resistance $R_N = R_{Th}$ as shown in Fig. 5.12. The restriction $|v_{oc}| < \infty$ is not required.

FIGURE 5.12 Network A of Fig. 5.10 replaced by a current source and parallel resistance (the Norton equivalent circuit)

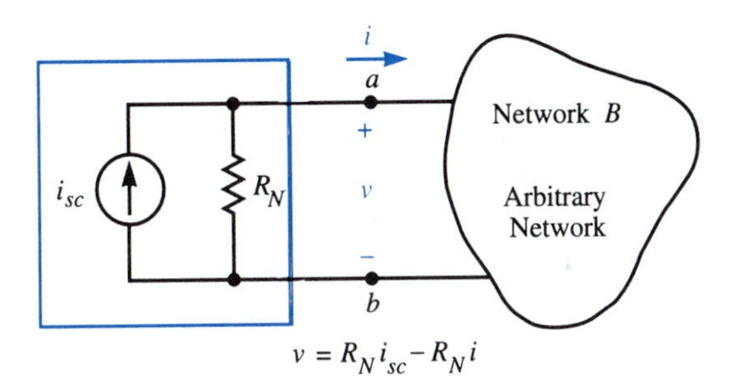

Both the Thévenin and Norton equivalent circuits are equivalent to the original circuit only with respect to the terminal equation that relates voltage v and current i.

If

$$|v_{oc}|, |i_{sc}| < \infty \tag{5.10}$$

either Thévenin's or Norton's equivalent network can be used. Application of KVL and Ohm's law to the network of Fig. 5.11 gives

$$R_{Th} = \frac{v_{oc}}{i_{sc}} \tag{5.11}$$

which provides an alternative method to calculate or measure the Thévenin or Norton equivalent resistance.

> Obviously any two of the three parameters v_{oc}, i_{sc}, and R_{Th} can be measured and the third calculated from Ohm's law. Use the method that requires the least work.

We use Thévenin's theorem if we wish to minimize the number of meshes and Norton's theorem if we wish to minimize the number of nodes.

The procedure will now be clarified by several examples where we determine Thévenin and Norton equivalent circuits.

EXAMPLE 5.4 Determine the Thévenin equivalent circuit for the network to the left of terminals a and b in Fig. 5.13 by calculating v_{oc} and R_{Th}.

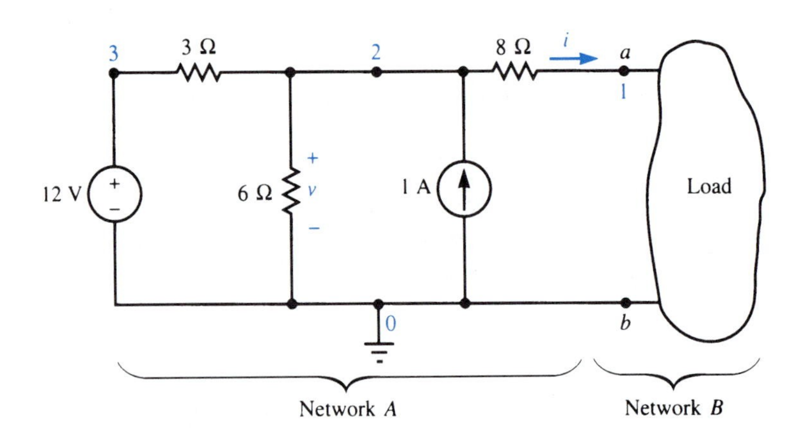

FIGURE 5.13 Network for Thévenin equivalent, Example 5.4

Solution First find the open-circuit voltage. Apply KCL to a surface enclosing the nodes indicated in Fig. 5.14:

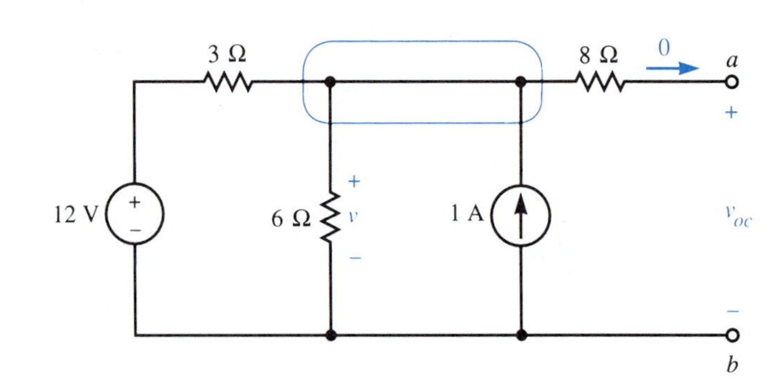

FIGURE 5.14 Diagram for calculating v_{oc} for the network of Fig. 5.13

$$\frac{1}{3}(v - 12) + \frac{1}{6}v - 1 + 0 = 0$$

which gives

$$v = 10 \text{ V}$$

A simple KVL equation gives

$$v_{oc} = 8(0) + v = 10 \text{ V}$$

Next set all independent sources to zero, as shown in Fig. 5.15.

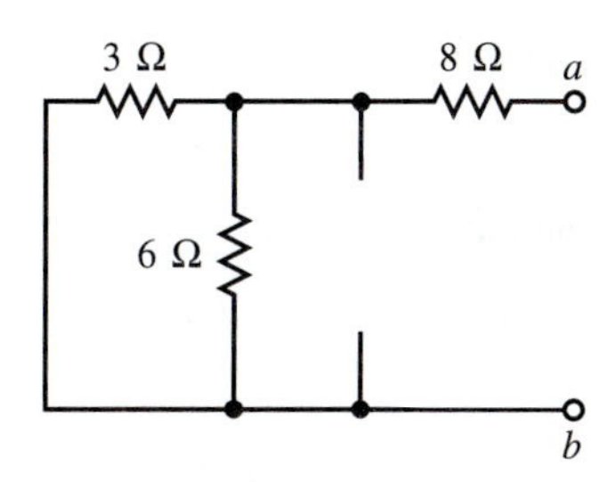

FIGURE 5.15 Diagram for calculating the Thévenin resistance with respect to terminals *a* and *b* of the network of Fig. 5.13

The 8-Ω resistance is in series with the parallel combination of the 3-Ω and 6-Ω resistances:

$$R_{Th} = 8 + \frac{1}{(1/3) + (1/6)} = 10\ \Omega$$

This gives the Thévenin equivalent shown in Fig. 5.16.

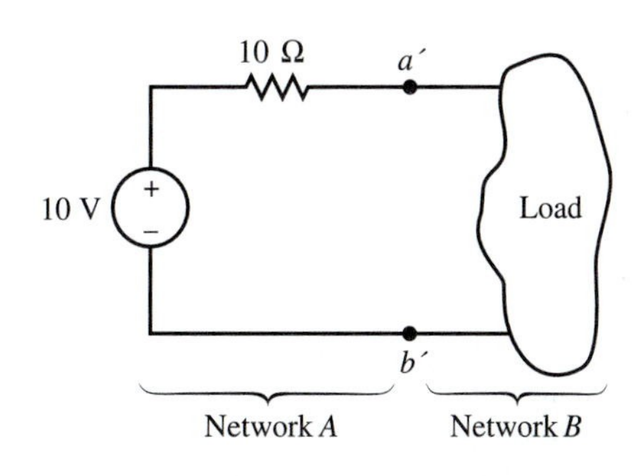

FIGURE 5.16 Network *A* of Fig. 5.13 replaced by its Thévenin equivalent

We can apply a source transformation to the network of Fig. 5.16 to obtain the Norton equivalent circuit for the network to the left of terminals *a* and *b* in Fig. 5.13, but for practice we will apply Norton's theorem directly to the circuit of Fig. 5.13.

EXAMPLE 5.5 Determine the Norton equivalent for the network to the left of terminals *a* and *b* in Fig. 5.13 by calculating i_{sc} and R_N.

Solution Determine the short-circuit current as indicated in Fig. 5.17. Application of KCL to the indicated closed surface yields

$$\frac{1}{3}(v - 12) + \frac{1}{6}v - 1 + \frac{1}{8}v = 0$$

which gives

$$v = 8\ \text{V}$$

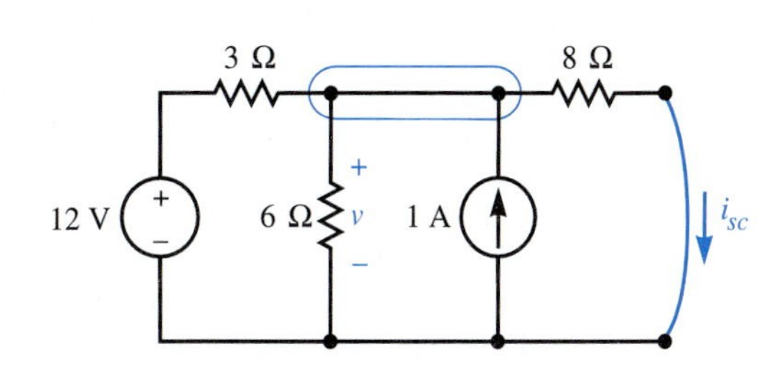

FIGURE 5.17 Diagram for calculating the short-circuit current of the network of Fig. 5.13

From KVL,

$$-v + 8i_{sc} = 0$$

which gives

$$i_{sc} = 1 \text{ A}$$

The Norton resistance can be found as in Example 5.4:

$$R_N = R_{Th} = 10\ \Omega$$

or calculated as the ratio of open-circuit voltage to short-circuit current, as shown below and in Fig. 5.18:

$$R_N = v_{oc}/i_{sc} = 10/1 = 10\ \Omega$$

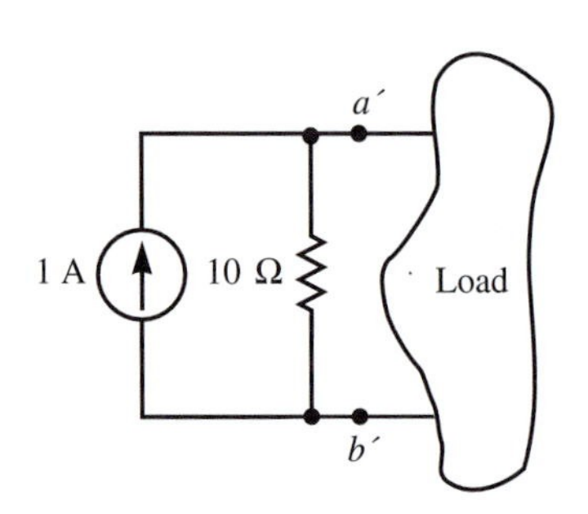

FIGURE 5.18 Network A of Fig. 5.13 replaced by its Norton equivalent

We will now examine the input characteristics of a simple amplifier.

EXAMPLE 5.6 Figure 5.19 represents a signal source (for example, a microphone) driving an emitter-follower amplifier, which in turn drives the input of a power amplifier (the load), modeled as a 10-Ω resistance. The amplifier model is somewhat simplified, and the numerical values are chosen to simplify the arithmetic. Determine the Thévenin equivalent to the right of terminals a and b with the 10-Ω load connected. (This Thévenin resistance, R_{ab}, is the *input resistance* of the amplifier with the load connected.)

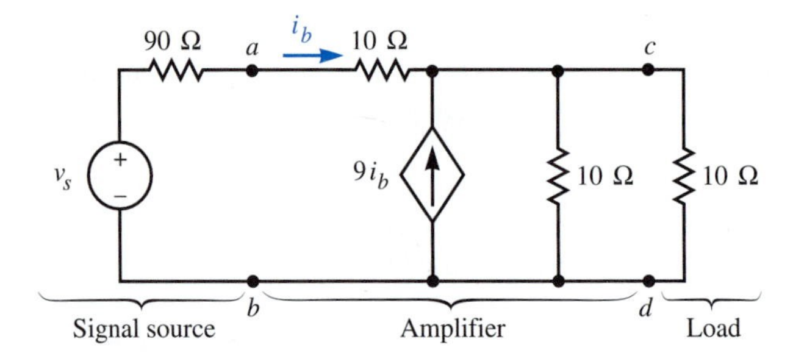

FIGURE 5.19 A signal source and amplifier supplying power to a load

Solution Disconnect the signal source from terminals a and b. Node-voltage or mesh-current analysis will verify that both the open-circuit voltage and the short-circuit current for the amplifier input terminals are zero:

$$v_{oc} = v_{ab}\Big|_{i_{ab}=0} = 0$$

and

$$i_{sc} = -i_b\Big|_{v_{ab}=0} = 0$$

Open-circuit voltage and short-circuit current are zero for any resistive network that does not contain independent sources. We cannot calculate R_{Th} from v_{oc} and i_{sc} in this case. The Thévenin resistance can be found if we connect a source of power to terminals a and b, as shown in Fig. 5.20. Although specification of the type of source (voltage or current) is not required, it is often easiest to specify it as a 1-V or 1-A source. For this example, let the source remain unspecified. Apply KCL to the indicated node.

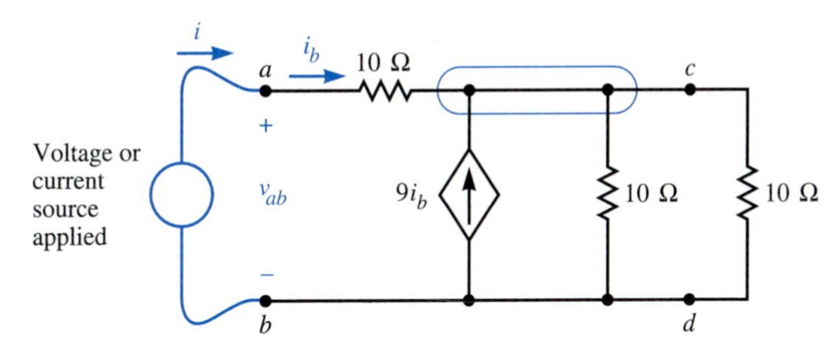

FIGURE 5.20 Measuring the input resistance of the amplifier of Fig. 5.19

$$\frac{1}{10}(v_{cd} - v_{ab}) - 9\left[\frac{1}{10}(v_{ab} - v_{cd})\right] + \frac{1}{10}v_{cd} + \frac{1}{10}v_{cd} = 0$$

This gives

$$v_{cd} = \frac{5}{6}v_{ab}$$

and the input current is

$$i = \frac{1}{10}(v_{ab} - v_{cd}) = \frac{1}{10}\left(v_{ab} - \frac{5}{6}v_{ab}\right) = \frac{1}{60}v_{ab}$$

Ohm's law yields the input resistance

$$R_{ab} = \frac{v_{ab}}{i} = \frac{v_{ab}}{v_{ab}/60} = 60\ \Omega$$

The signal source is connected to an equivalent load of 60 Ω, as shown in Fig. 5.21.

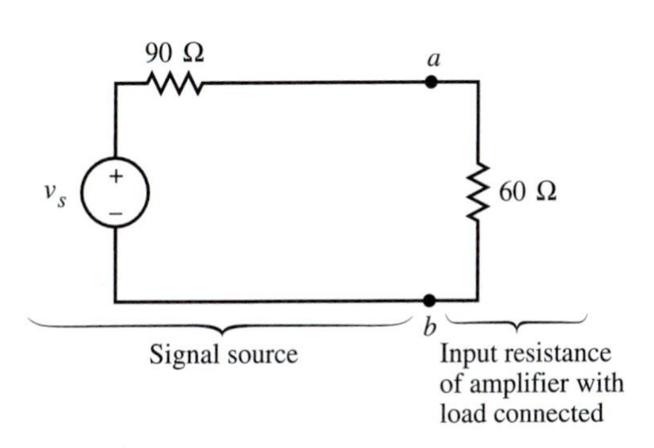

FIGURE 5.21 Equivalent load on the signal source obtained by Thévenin's theorem

We frequently call the Thévenin resistance measured between two terminals (terminals a and b in the preceding example) the *driving-point resistance* looking into terminals a and b, because this is the resistance a source must drive when connected between the two terminals.

We will now examine the output characteristics of the amplifier of Fig. 5.19.

EXAMPLE 5.7 Determine the Thévenin equivalent circuit to the left of terminals c and d in Fig. 5.19 with the source connected. (The Thévenin resistance, R_{cd}, is the *output resistance* of the amplifier with the source connected.) Calculate v_{oc} and then set the independent source v_s to zero to calculate the Thévenin resistance.

Solution Open circuit terminals c and d as shown in Fig. 5.22. KCL applied to the indicated closed surface yields

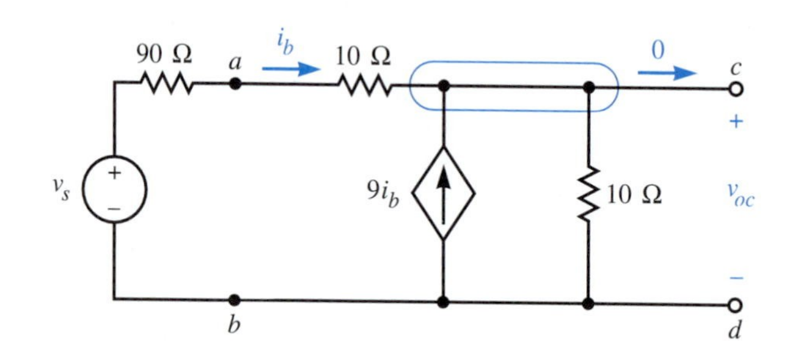

FIGURE 5.22 Signal source and amplifier with terminals c and d open-circuited

$$\frac{1}{100}(v_{oc} - v_s) - 9\left[\frac{1}{100}(v_s - v_{oc})\right] + \frac{1}{10}v_{oc} = 0$$

which gives

$$v_{oc} = \frac{1}{2}v_s$$

which is the Thévenin equivalent voltage looking into terminals c and d.

We can find the *output resistance,* or the Thévenin resistance looking into terminals c and d, with the source connected by setting all *independent sources* to zero and applying a source to terminals c and d, as shown in Fig. 5.23. KCL applied to the indicated node yields

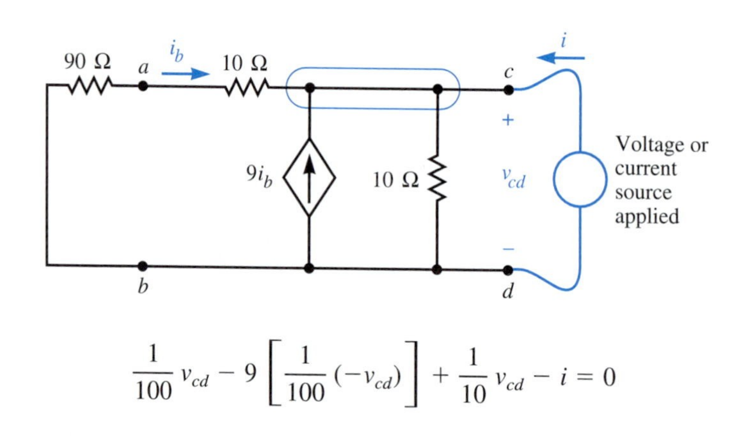

FIGURE 5.23 Source voltage v_s set equal to zero so the output resistance of the amplifier of Fig. 5.19 can be measured

$$\frac{1}{100}v_{cd} - 9\left[\frac{1}{100}(-v_{cd})\right] + \frac{1}{10}v_{cd} - i = 0$$

or

$$i = \frac{1}{5}v_{cd}$$

From Ohm's law, the Thévenin resistance looking into terminal c and d is

$$R_{cd} = \frac{v_{cd}}{v_{cd}/5} = 5\ \Omega$$

The load is connected to or ''sees'' an equivalent voltage source and series resistance as shown in Fig. 5.24. The equivalent circuits shown in Figs. 5.21 and 5.24 easily yield the current required from the source and the current delivered to the load.

FIGURE 5.24 Equivalent source seen by the load

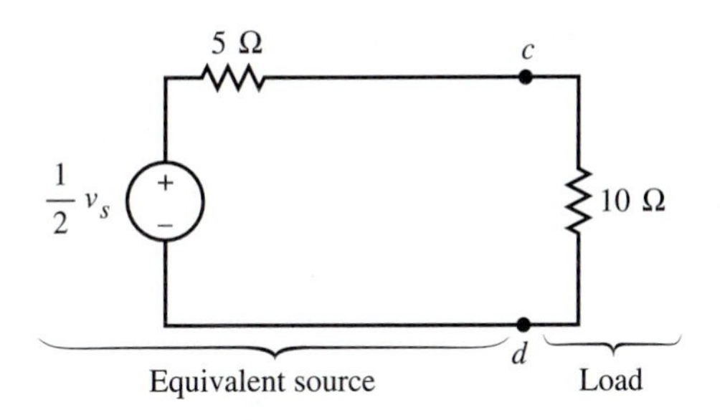

For practice we will also use another method to calculate the output resistance.

EXAMPLE 5.8 Use the open-circuit voltage and the short-circuit current to calculate the output resistance for the network of Fig. 5.25.

FIGURE 5.25 Signal source and amplifier with terminals c and d short-circuited

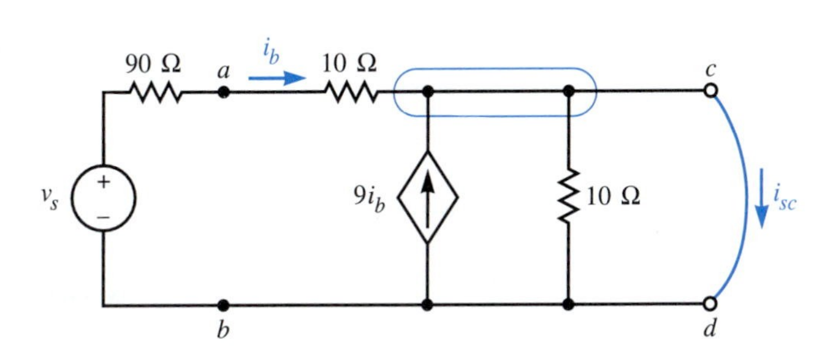

Solution The open-circuit voltage was calculated in Example 5.7:

$$v_{oc} = \frac{1}{2} v_s$$

Short-circuit terminals c and d as shown in Fig. 5.25. Application of KCL to the indicated closed surface gives

$$-\frac{1}{100} v_s - 9\left(\frac{1}{100} v_s\right) + i_{sc} = 0$$

This gives the short-circuit current:

$$i_{sc} = \frac{1}{10} v_s$$

The output resistance is

$$R_{cd} = \frac{v_{oc}}{i_{sc}} = \frac{[(1/2)v_s]}{[(1/10)v_s]}$$
$$= 5\ \Omega$$

This is the same result obtained in Example 5.7, as expected.

Remember

Thévenin's and Norton's theorems apply to circuits with independent sources, linear resistances, and linear dependent sources. This restriction does not apply to the load.

The Thévenin voltage is the open-circuit voltage. The Norton current is the short-circuit current. If the network is without an independent source, the open-circuit voltage and short-circuit current are zero.

The ratio of open-circuit voltage to short-circuit current is the Thévenin resistance (also the Norton resistance).

The Thévenin resistance can also be measured by setting all independent sources to zero (dependent sources are left as they are) and applying a voltage or current to the two terminals considered. The ratio of the terminal voltage to the input current is the Thévenin resistance.

EXERCISES

8. The following circuit from Example 3.11 represents the lighting system for a small boat. Determine the Thévenin and Norton equivalent circuits seen by the specified devices.
   (a) The battery — *answer:* 2 Ω
   (b) The 10-Ω lamp — *answer:* 2.096 Ω, 11.52 V, 5.496 A
   (c) The 4-Ω lamp — *answer:* 0.0976 Ω, 11.707 V, 120 A
   (d) The 6-Ω lamp — *answer:* 0.0968 Ω, 11.613 V, 120 A

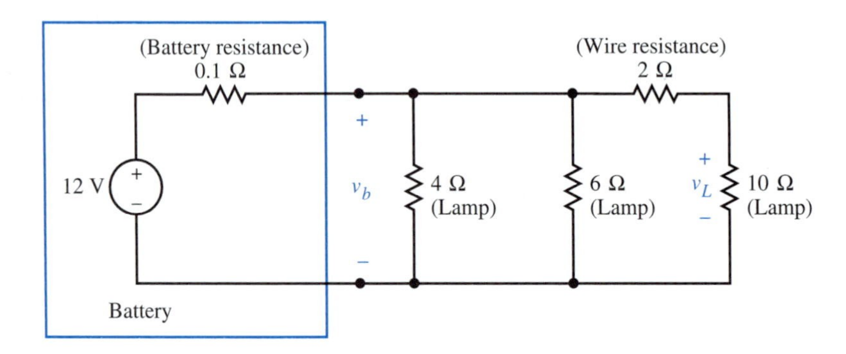

9. Determine the Thévenin and Norton equivalent circuits with respect to terminals a and b for the following two networks (The use of repeated source transformations is suggested).

(a)

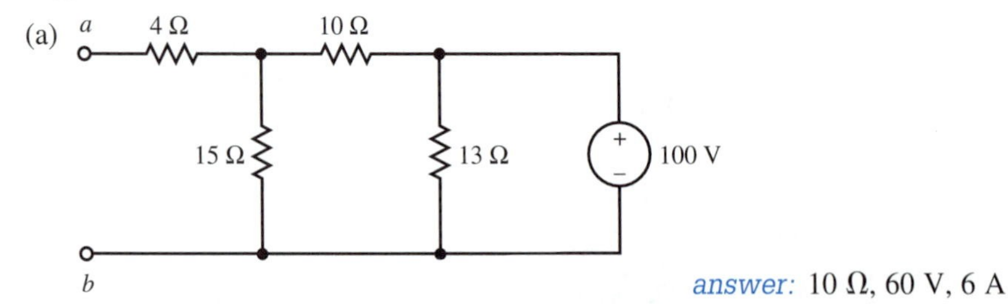

answer: 10 Ω, 60 V, 6 A

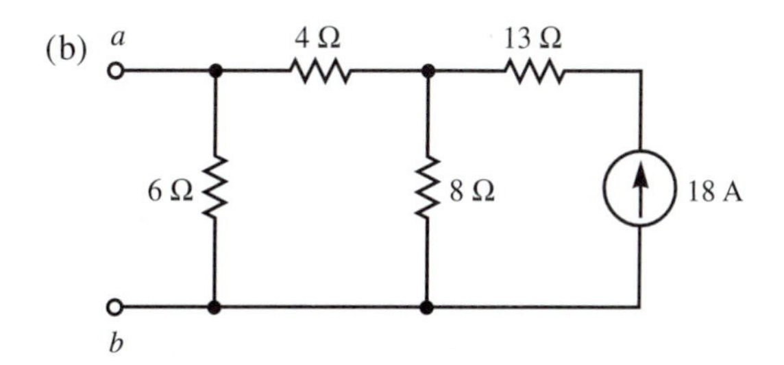

answer: 4 Ω, 48 V, 12 A

10. Determine the Thévenin and Norton equivalent circuits with respect to terminals a and b for the following network. *answer:* 10 Ω, −120 V, −12 A

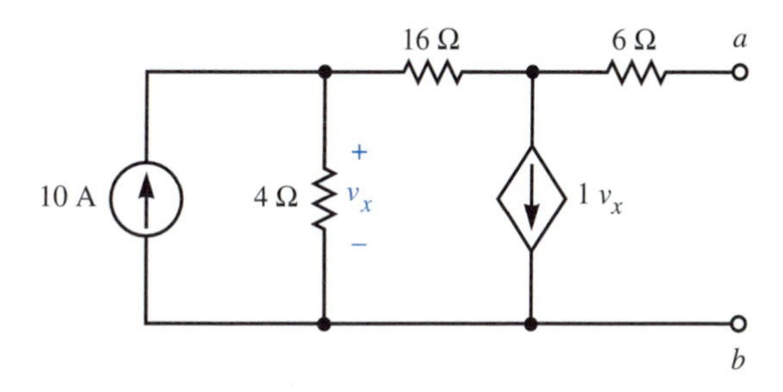

5.4 Maximum Power Transfer†

A principal function of most electrical systems is to deliver power to a load. The basic criterion for an electrical utility is to minimize the cost of providing power to the customer. Typically the utility generates electrical energy by burning fossil fuel, and high electrical and mechanical efficiency are important. In this section we will consider a different problem. For solar panels, as used on solar-powered race cars, the solar energy is free, so we try to maximize the power delivered by the array for the available illumination and accept the electrical efficiency this entails. This same problem also occurs in electronics when we try to obtain the maximum power from an amplifier without regard for efficiency.

One possibility is that the load resistance R_L and open-circuit voltage v_s, as shown in Fig. 5.26a, are fixed and the output resistance R_s of the driving system can be selected. The current is

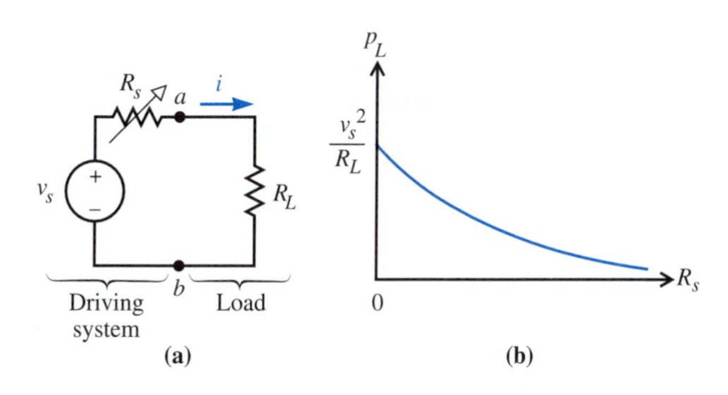

FIGURE 5.26 Load power when the source resistance R_s is adjustable. (*a*) Circuit with adjustable source resistance; (*b*) load power

† This section can be omitted without loss of continuity.

$$i = \frac{v_s}{R_s + R_L} \tag{5.12}$$

and the power absorbed by the load is

$$p_L = R_L i^2 = \frac{R_L}{(R_s + R_L)^2} v_s^2 \tag{5.13}$$

Load power p_L as a function of source resistance R_s is sketched in Fig. 5.26b. The load power p_L is obviously maximized by making the source resistance as small as possible, with the absolute maximum occurring for

$$R_s = 0 \tag{5.14}$$

and R_L and v_s fixed.

The second, more interesting, case occurs when the open-circuit voltage v_s and output resistance R_s of the driving network are specified, and the designer is free to choose the load resistance R_L, as indicated in Fig. 5.27a. As before, the power absorbed by the load is given by Eq. (5.13).

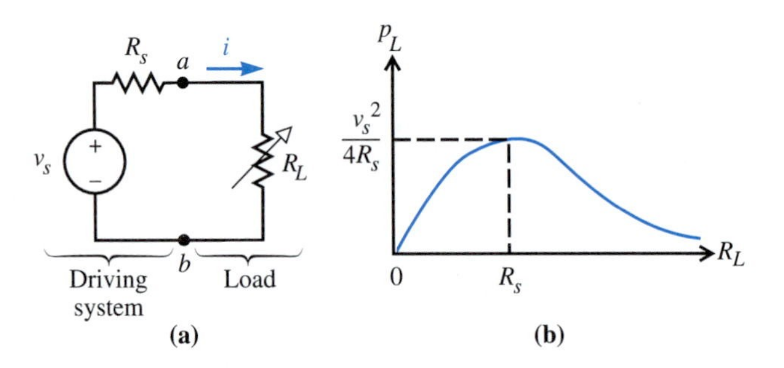

FIGURE 5.27 Load power when the load resistance is adjustable. (*a*) Circuit with adjustable load; (*b*) load power

A maximum for p_L obviously occurs interior to the interval $0 \le R_L \le \infty$ and can be found if we set the derivative of p_L with respect to R_L equal to zero. Differentiation of the equation for load power gives

$$\frac{dp_L}{dR_L} = \frac{(R_s + R_L)^2 - 2R_L(R_s + R_L)}{(R_s + R_L)^4} v_s^2 = 0 \tag{5.15}$$

This equation is satisfied by

$$R_L = R_s \tag{5.16}$$

This result is stated in the following theorem.

THEOREM 5 Maximum Power Transfer Theorem

When the source resistance is fixed and the load resistance can be selected, maximum power is absorbed by the load when the source and load resistance are equal or *matched.*

Substitution of $R_L = R_s$ into the equation for load power gives the load power maximized with respect to R_L for R_s and v_s fixed:

$$\max_{R_L} \{p_L\} = \frac{v_s^2}{4R_s} \tag{5.17}$$

When a matched load is connected to a network of resistances and voltage sources, only half of the additional power required from the sources is delivered to the load. The remainder is consumed by the network resistances. These added internal losses are equal to the power absorbed by the Thévenin resistance R_s of Fig. 5.27. This relationship is not typically true for networks with current sources or dependent sources (see Problem 36).

Remember

If you are free to select the source resistance without changing the source voltage, make the source resistance as small as possible to maximize the power delivered to the load. If you are free to select only the load resistance, make it equal to the source resistance to maximize the power delivered to the load. This latter case causes the power absorbed by the Thévenin resistance of the source to be equal to the power delivered to the load.

EXERCISES

11. For the network shown below, select R_1 so that the maximum power is absorbed by R_2. Find P_1 and P_2. *answer:* $R_1 = 0, \frac{1}{5}v_s^2$

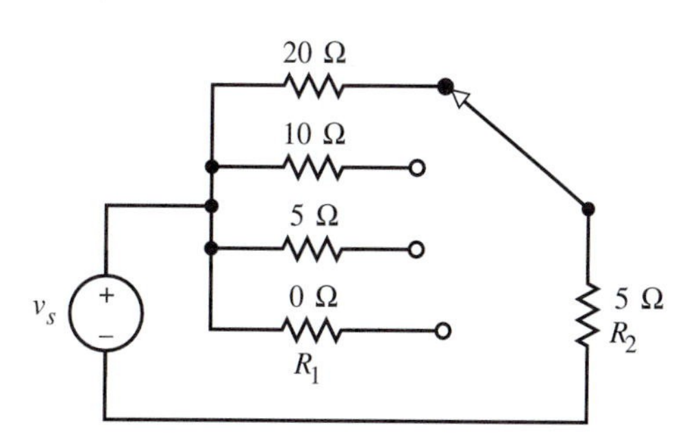

12. For the network shown above, select R_1 so that the maximum power is absorbed by R_1. Find P_1 and P_2. *answer:* $R_1 = 5\ \Omega, \frac{1}{20}v_s^2, \frac{1}{20}v_s^2$

13. Select R_L so that the power absorbed by R_L is a maximum, and find P_L for the following circuit. (Use Thévenin's theorem.) *answer:* 4 Ω, 36 W

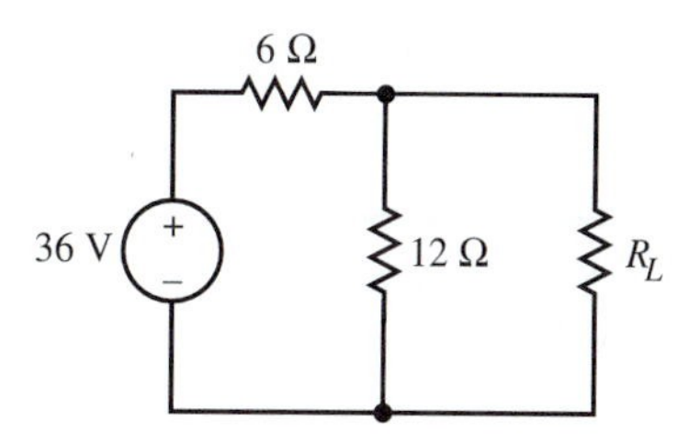

14. Find a load resistance R_L that, when connected to terminals a and b of the network described in Exercise 10, will absorb the maximum power. *answer:* 10 Ω

5.5 Duality†

Perhaps you noticed that the node-voltage equations for many examples in Sections 4.1 and 4.2 were similar in form to the mesh-current equations for examples in Sections 4.3 and 4.4, but with current and voltage interchanged. For two circuits, if the node-voltage equations for each circuit can be found when we interchange voltage and current in the mesh-current equations for the other, we say the circuits are *duals*. The examples in Chapter 4 were intentionally selected to illustrate the concept of *duality*. Duality is a consequence of the similarity in component equations used in KCL and KVL equations. This is illustrated in Fig. 5.28. Some additional dual relationships are given in Table 5.1. These are mutual relationships, and the columns in which the terms of a dual pair appear can be interchanged.

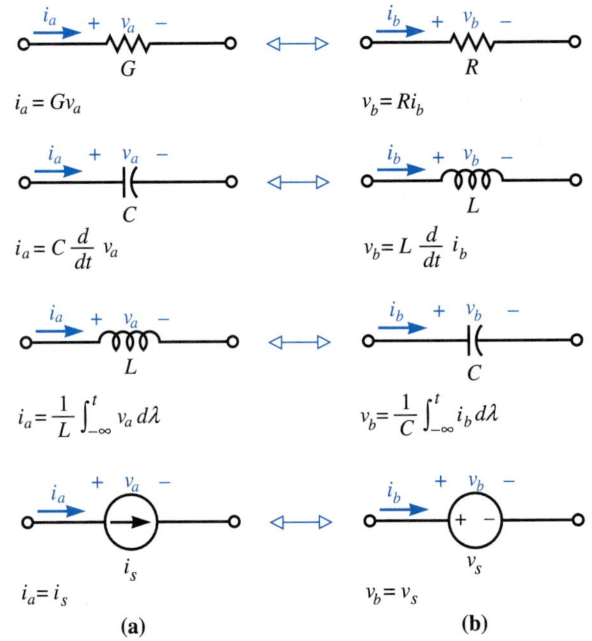

FIGURE 5.28 The dual nature of some network components: (*a*) used in KCL equations; (*b*) used in KVL equations

$$i_a = Gv_a \quad \longleftrightarrow \quad v_b = Ri_b$$

$$i_a = C\frac{d}{dt}v_a \quad \longleftrightarrow \quad v_b = L\frac{d}{dt}i_b$$

$$i_a = \frac{1}{L}\int_{-\infty}^{t} v_a\, d\lambda \quad \longleftrightarrow \quad v_b = \frac{1}{C}\int_{-\infty}^{t} i_b\, d\lambda$$

$$i_a = i_s \quad \longleftrightarrow \quad v_b = v_s$$

(a) (b)

The principal application of duality is that once you have the solution to a problem, you also know the solution to the dual problem. This will be enlightening when we examine parallel and series *RLC* circuits in Chapter 9. Duality is also useful in standardized network design. If you have a table of networks with the desired properties and the network is driven by a voltage source, the dual of that network has the dual properties when driven by a current source.

TABLE 5.1 Some dual pairs

Current-controlled current source	Voltage-controlled voltage source
Voltage-controlled current source	Current-controlled voltage source
Node	Mesh
Node voltage	Mesh current

† This section can be omitted with negligible loss of continuity.

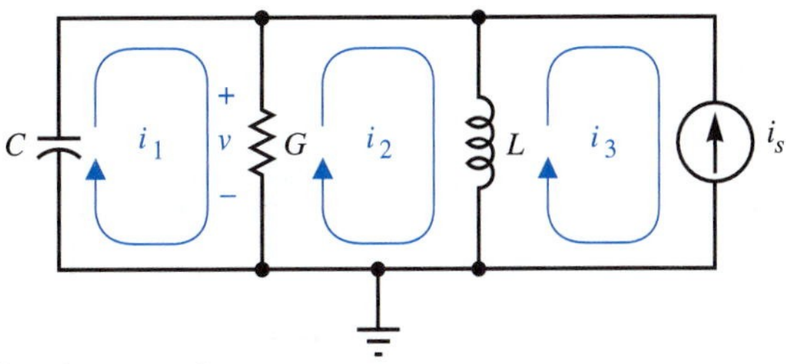

Node voltage equation

$$C\frac{d}{dt}v + Gv + \frac{1}{L}\int_{-\infty}^{t} v\, d\lambda = i_s$$

Mesh current equations

$$\begin{bmatrix} \frac{1}{C}\int_{-\infty}^{t} d\lambda + \frac{1}{G} & -\frac{1}{G} & 0 \\ -\frac{1}{G} & \frac{1}{G} + L\frac{d}{dt} & -L\frac{d}{dt} \\ 0 & 0 & 1 \end{bmatrix} \bullet \begin{bmatrix} i_1 \\ i_2 \\ i_3 \end{bmatrix} = \begin{bmatrix} 0 \\ 0 \\ -i_s \end{bmatrix}$$

(a)

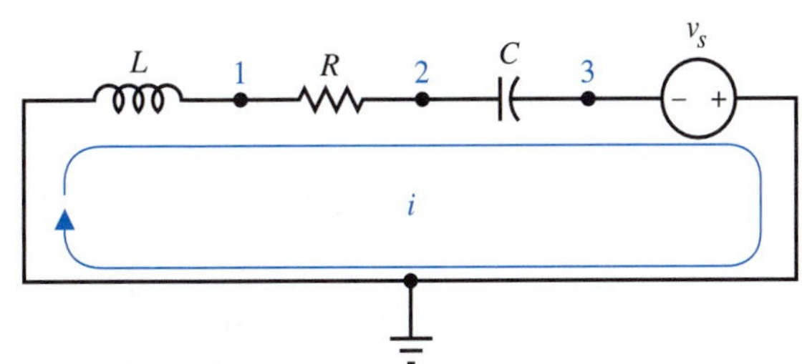

Mesh current equation

$$L\frac{d}{dt}i + Ri + \frac{1}{C}\int_{-\infty}^{t} i\, d\lambda = v_s$$

Node voltage equations

$$\begin{bmatrix} \frac{1}{L}\int_{-\infty}^{t} d\lambda + \frac{1}{R} & -\frac{1}{R} & 0 \\ -\frac{1}{R} & \frac{1}{R} + C\frac{d}{dt} & -C\frac{d}{dt} \\ 0 & 0 & 1 \end{bmatrix} \bullet \begin{bmatrix} v_1 \\ v_2 \\ v_3 \end{bmatrix} = \begin{bmatrix} 0 \\ 0 \\ -v_s \end{bmatrix}$$

(b)

FIGURE 5.29 Duality of parallel and series circuits: (*a*) a parallel circuit; (*b*) a series circuit

A planar circuit with $N + 1$ nodes and M meshes will have a *dual* circuit with $M + 1$ nodes and N meshes. Each component in the first network is replaced by its dual component in the second network.† The node-voltage equation for one circuit will have the same form as the mesh equations for the other, so the analysis of a circuit will also yield the performance of the dual network. A simple example of dual circuits is shown in Fig. 5.29.

As suggested in the preceding paragraph, the dual of a network can be constructed from a given circuit by inspection of the mesh-current and node-voltage equations. A

† Mutual inductance, introduced in Chapter 17, does not have a dual. For a more thorough discussion of duality, refer to Ernst A. Guilleman, *Introductory Circuit Theory*. New York: John Wiley & Sons, 1953.

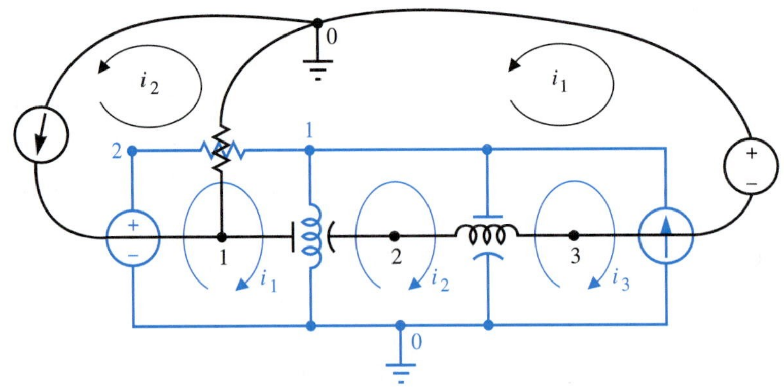

FIGURE 5.30 Graphical relationship between dual networks

convenient graphical method is to superimpose the dual network diagram on the original network diagram (refer to Fig. 5.30 as you read this paragraph). A node for the dual network is placed inside each mesh of the original network, and the reference node for the dual network is placed exterior to the original network. Each component that appears in a single mesh (or outer loop) of the original circuit becomes the dual component from the new node inside the mesh to the new reference node. Each new component is conveniently drawn across the original component. A component shared by two meshes in the original network yields a dual component shared by the two new nodes placed inside the meshes. This dual component is drawn across the original component. Source polarities are assigned so that if the source tends to force node k in one network positive, the dual source in the other network tends to force mesh current k positive.

To ensure that both mesh and node equations exhibit the dual relation, components connected to the dual reference node must be drawn so that the original reference node is not enclosed. The method requires the counterclockwise assignment of mesh currents in the dual to maintain the signs for the mesh-current equations. We can correct this if desired by redrawing the dual as a mirror image, or with the top and bottom interchanged.

Remember Once we have analyzed a network, we also have the analysis of the dual network.

EXERCISES

15. Construct the dual of the network shown below.

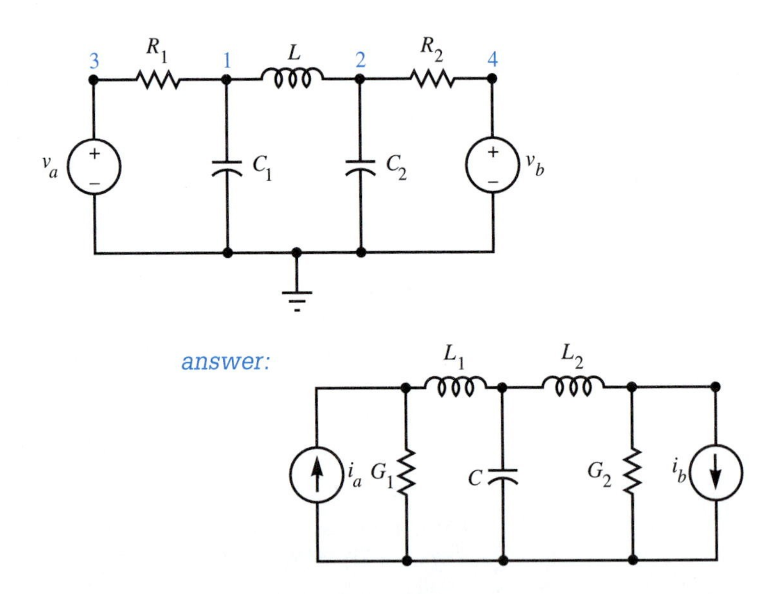

5.6 Summary

In this chapter, we presented a variety of concepts that help the engineer simplify the analysis of a circuit. We began with the definitions of linearity, time invariance, and superposition, and introduced source transformations, which provide an alternative method to solve for a specified response in some circuits. We next extended the idea of an equivalent source by the use of Thévenin's and Norton's theorems. We used these equivalent circuits to solve maximum power transfer problems. We concluded the chapter with optional sections on driving-point resistance and duality.

KEY FACTS	Concept	Equation	Section	Page
❑	Superposition applies to voltages and currents in linear circuits.		5.1	139
❑	The superposition theorem states that the total response is equal to the sum of the responses due to each independent source acting alone, with all other independent sources set equal to zero. The dependent sources are retained in each case.		5.1	139
❑	A 0-V voltage source is equivalent to a short circuit, and a 0-A current source is equivalent to an open circuit.		5.1	139
❑	To apply superposition be sure to use each independent source exactly once.		5.1	142
❑	A voltage source of value v_s in series with a resistance of value R has the same terminal characteristics as a current source of value i_s in parallel with a resistance of the same value R, if $v_s = Ri_s$.	(5.5) (5.6)	5.2	145
❑	We can use source transformations to eliminate nodes and meshes (Example 5.3).		5.2	146
❑	Thévenin's and Norton's theorems apply to linear circuits with resistance, dependent sources, and independent sources.		5.3	150
❑	The voltage source in the Thévenin equivalent circuit has a value equal to the open-circuit voltage v_{oc}.		5.3	151
❑	The current source in the Norton equivalent circuit has a value equal to the short-circuit current i_{sc}.		5.3	154
❑	The Thévenin and Norton equivalents for a circuit will have the same value of resistance R.		5.3	151
❑	The resistance R in the Thévenin and Norton equivalent circuits is related to the open-circuit voltage and short-circuit current by Ohm's law: $R = v_{oc}/i_{sc}$.	(5.11)	5.3	151

KEY FACTS	***Concept***	***Equation***	***Section***	***Page***
❑	Both the open-circuit voltage and short-circuit current are zero for a circuit with no independent sources (Example 5.6).		5.3	154
❑	We can also obtain the value of the resistance R, if we set all independent sources in the circuit to zero (dependent sources are retained), and apply a source to the two terminals. The ratio of terminal voltage to terminal current is the Thévenin resistance looking into these terminals. (This is Ohm's law.)		5.3	155
❑	If the open-circuit voltage of a circuit is fixed, we can maximize the power delivered to any fixed resistive load by minimizing the circuit output resistance (Thévenin resistance).	(5.14)	5.4	160
❑	If the open-circuit voltage and output resistance of a circuit are fixed, we can maximize the power delivered to a resistive load by selecting the load resistance equal to the output resistance.	(5.16)	5.4	160
❑	Important dual pairs are *R-G, C-L, v-i,* node-mesh, and series-parallel.		5.5	162

PROBLEMS

Section 5.1

1. Use superposition to calculate current i_y or voltage v_x as indicated, for the following three networks.

(a) 5.1

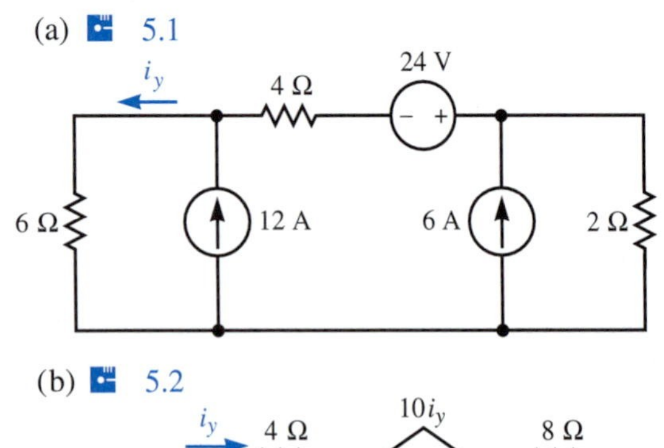

(b) 5.2

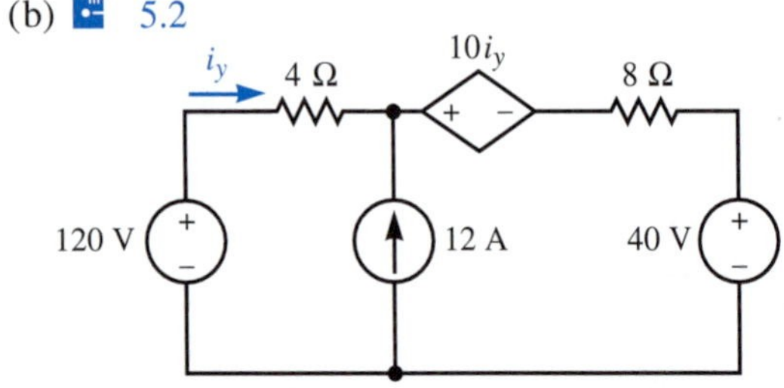

(c)

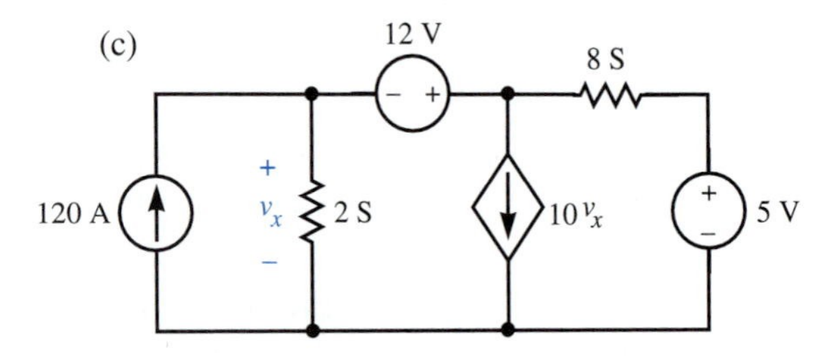

2. The following circuit is a small-signal equivalent circuit for a *difference amplifier.* Use superposition to determine the output voltage v_o.

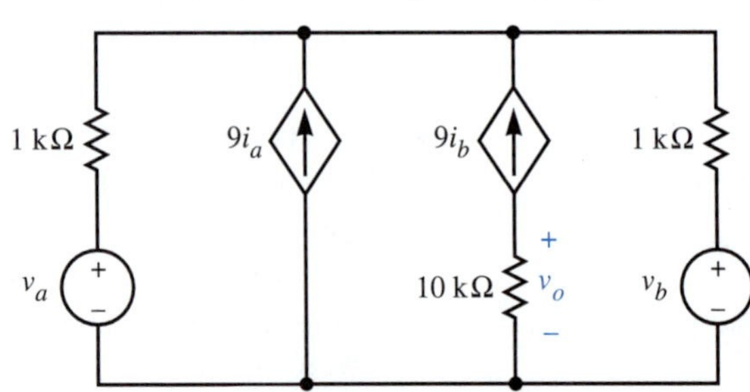

3. The following circuit is a small-signal equivalent circuit for an *audio mixer* that adds the signal from microphones *A* and *B*. 5.3
   (a) Use superposition to determine the output voltage v_o.
   (b) We must redesign the amplifier so that the voltage gain v_o/v_a is twice that of the voltage gain v_o/v_b. What value of resistance R_a will satisfy this design requirement?

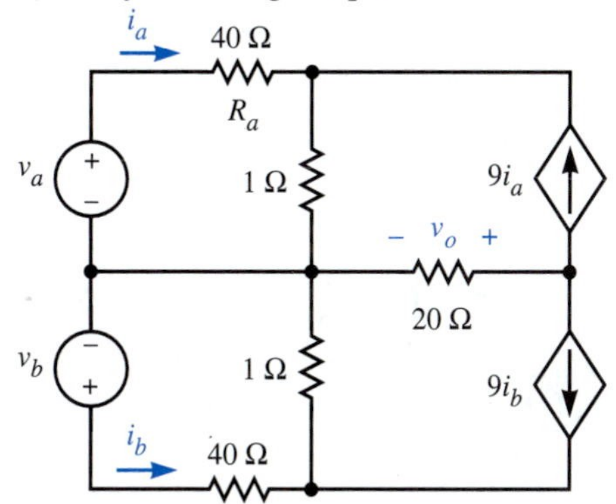

4. Use the principle of superposition to write three differential equations that can be solved for the component of current due to each independent voltage source in the following circuit. (Do not solve the equations.) We will see in later chapters that this type of problem is one of the most important applications of superposition.

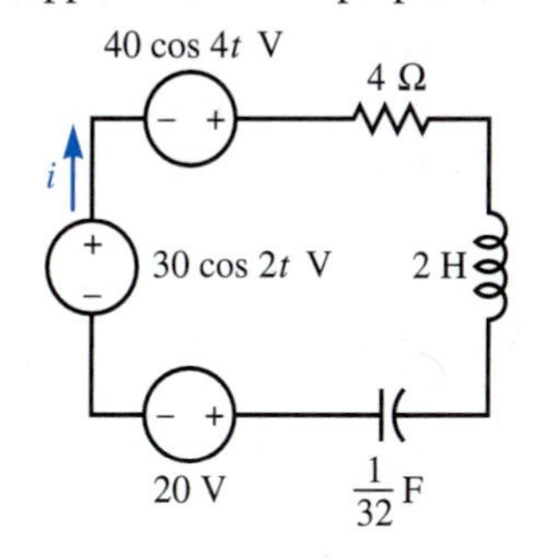

Section 5.2

5. For the following three circuits use source transformations to determine an equivalent circuit that consists of a voltage source and series resistance connected between terminals *a* and *b*.

(a)

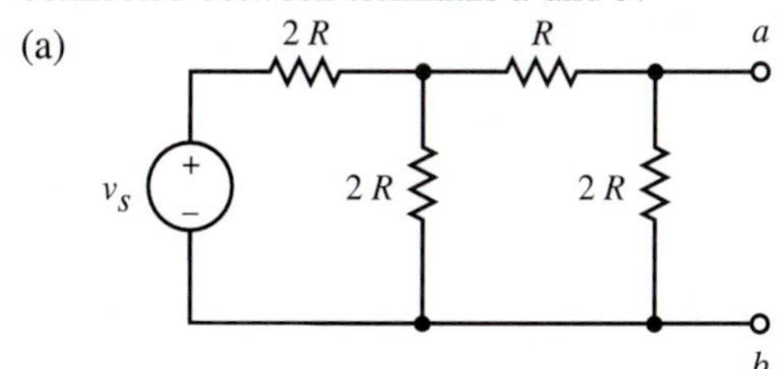

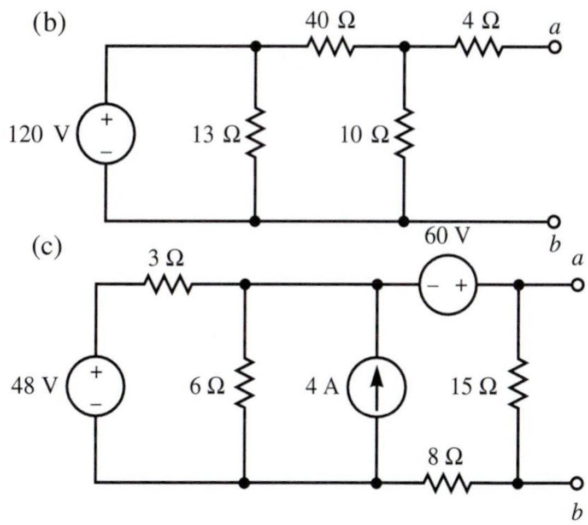

6. Use source transformations to eliminate the two current sources from the network that follows. Solve for the two remaining mesh currents. Also solve for voltages v_1, v_2, and v_3. 5.4

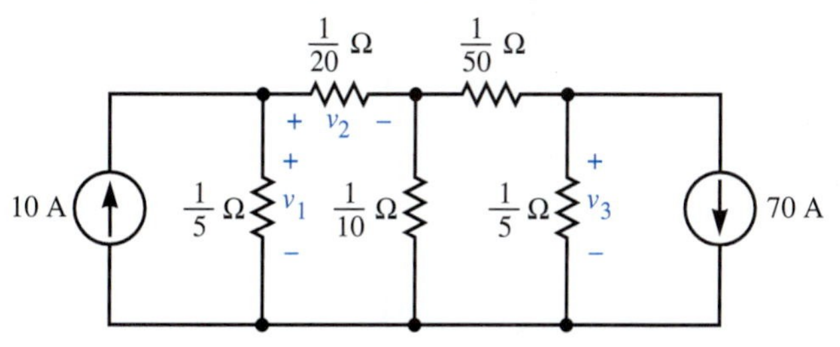

7. Use source transformations to eliminate the two voltage sources from the network that follows. Then solve for the two remaining node voltages. Also solve for currents i_1, i_2, and i_3.

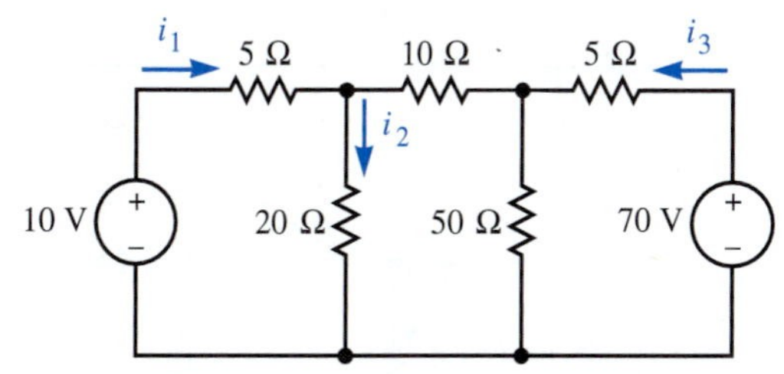

8. Use source transformations to determine a current source and parallel resistance that has the same terminal characteristics, relative to terminals *a* and *b*, as the following circuit.

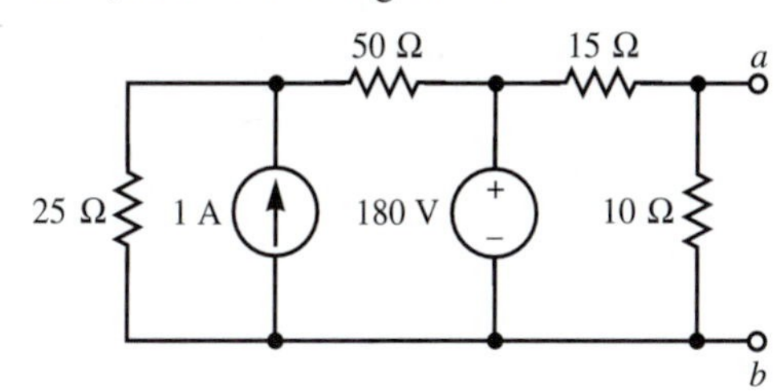

Section 5.3

9. Determine both the Thévenin and Norton equivalent circuits with respect to terminals a and b in the following six circuits.

(a)

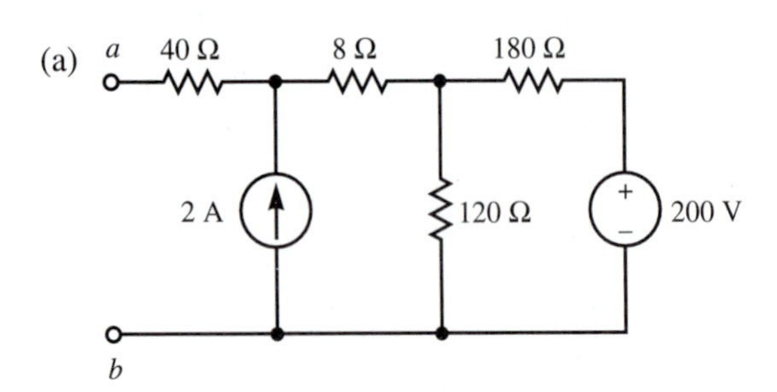

(b)

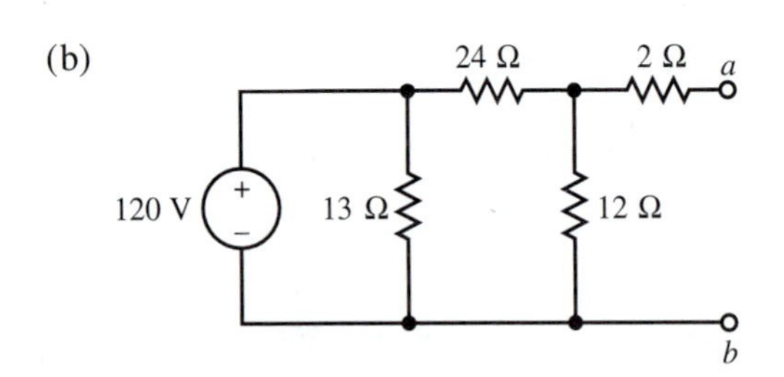

(c) 5.5

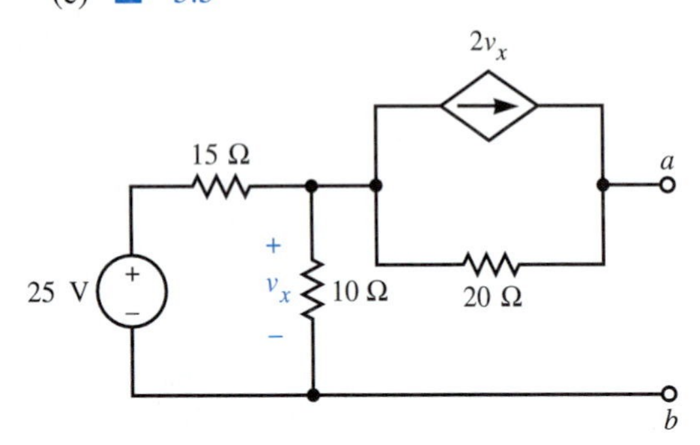

(d)

(e)

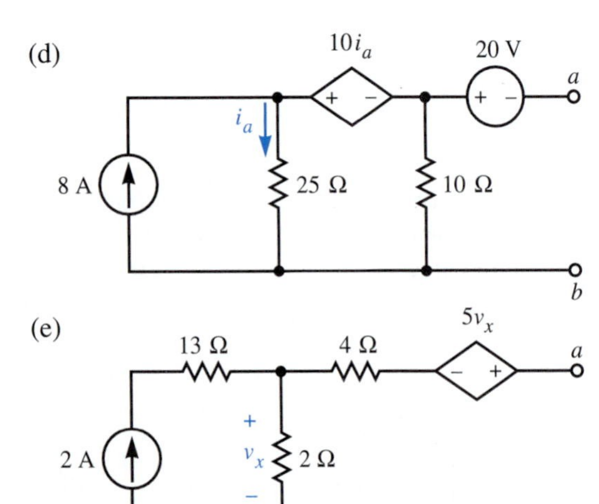

(f)

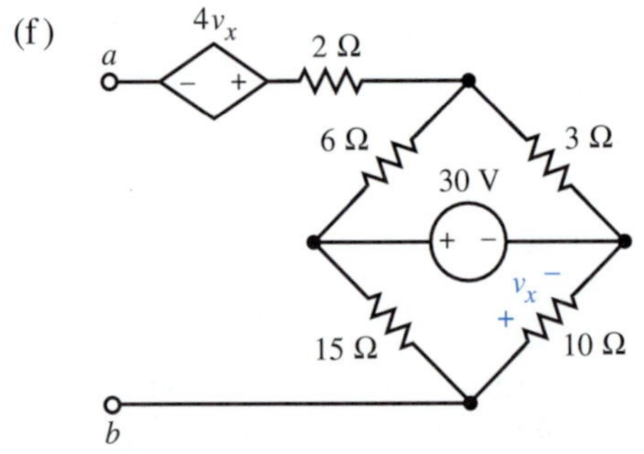

10. For the following circuit, calculate the Thévenin equivalent circuit seen by the specified source.

    (a) The 120-A source (b) The 12-V source

    (c) The 5-V source

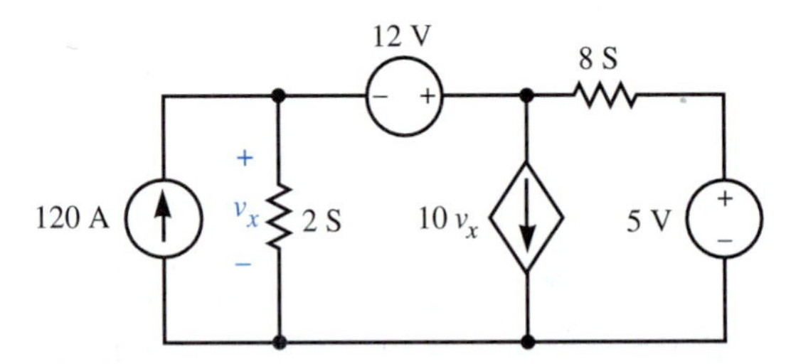

11. A 2-Ω speaker load is connected to the output of an amplifier driven by a signal generator. The output voltage v_{ab} is measured and found to be $4 \cos 2000t$ V. The speaker load is changed to 8 Ω with the signal generator unchanged, and v_{ab} is measured to be $8 \cos 2000t$ V. Sketch a block diagram that shows the connection of the three devices. What is the Thévenin equivalent circuit model for the amplifier driven by the signal generator?

12. A resistive load is connected to a power supply, and the output voltage v_{ab} is measured to be 140 V when the load current $i_{ab} = 2$ A. The load is changed, and v_{ab} is measured to be 20 V when i_{ab} is 8 A. What is the Thévenin equivalent circuit model for the power supply?

13. Solve for the current through the nonlinear load shown below. The terminal equation for the load is $v = 2i^2$.

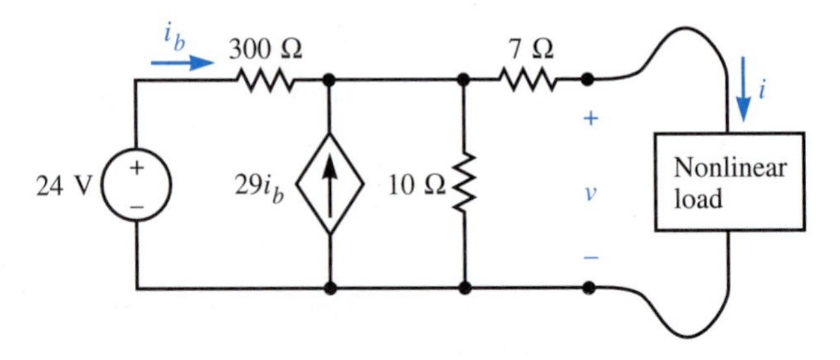

Section 5.4

14. Determine the value of resistance that will absorb the maximum power when connected to terminals a and b of the following two circuits.

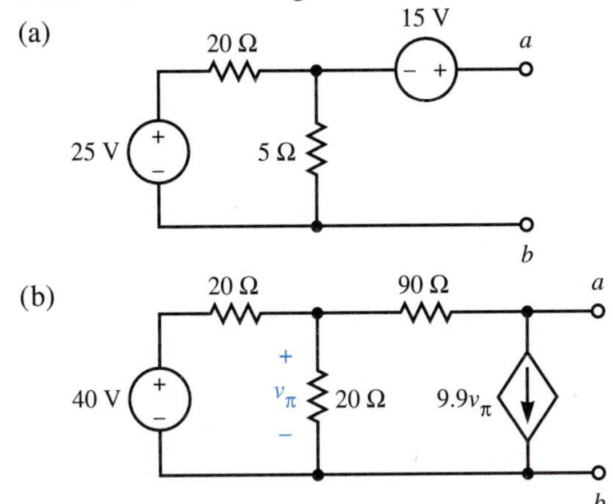

15. In the following circuit $R_1 = 6\ \Omega$. Determine the value of R_2 that maximizes the power delivered to the load.

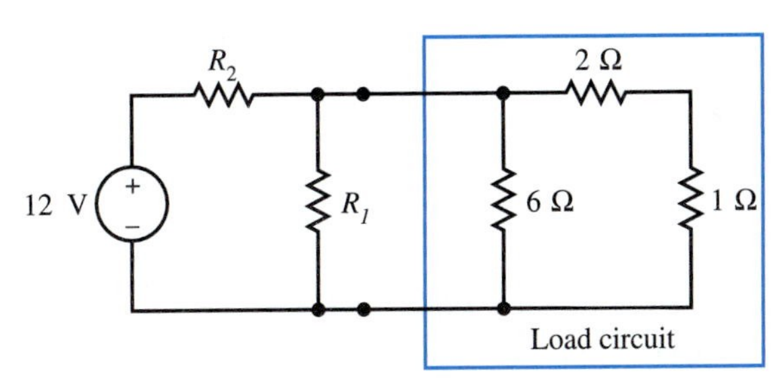

16. In the preceding circuit $R_2 = 3\ \Omega$. Determine the value of R_1 that maximizes the power delivered to the load.

17. For the following circuit, determine the value of r that maximizes the power delivered to the load. 5.6

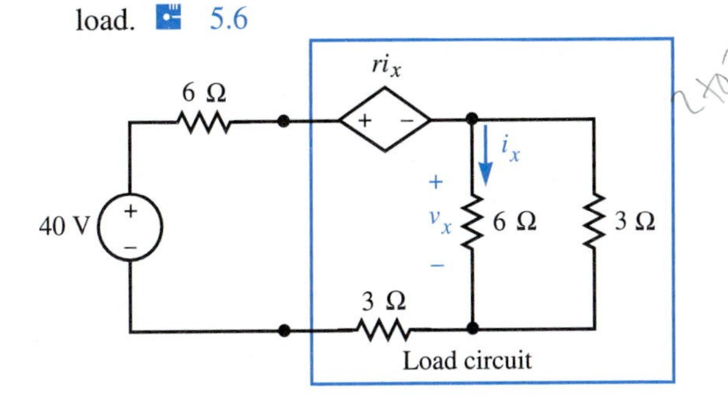

Section 5.5

18. For the following circuits:

(i) Select the lower node as the reference node for the circuit and write the node-voltage equations.

(ii) Select the left-hand mesh as mesh 1 and write the mesh-current equations.

(iii) Use the graphical procedure described in Section 5.5 to draw the dual of the circuit.

(iv) Write the mesh-current equations for the circuit constructed in part (iii). Compare these to the node-voltage equations from part (i). Do these equations demonstrate duality as expected?

(v) Write the node-voltage equations for the network constructed in part (iii). Compare these to the mesh-current equations from part (ii). Do these equations demonstrate duality as expected?

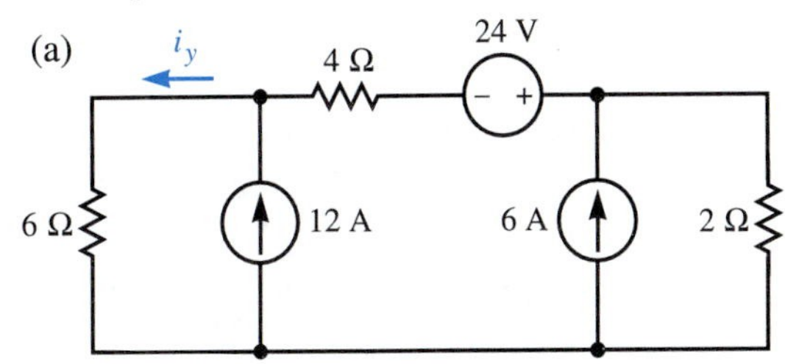

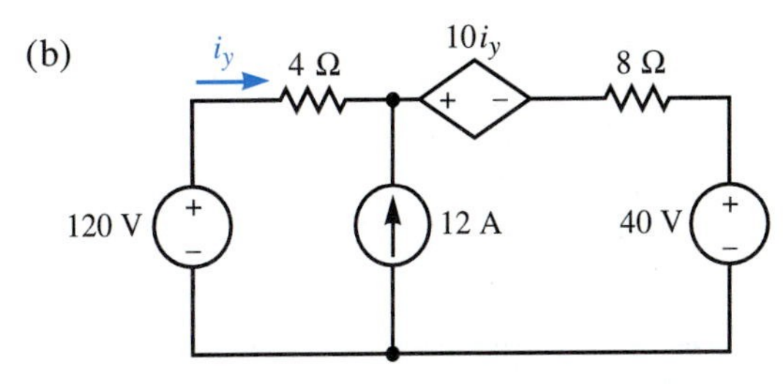

19. Use the graphical procedure described in Section 5.5 to draw the dual for the following network.

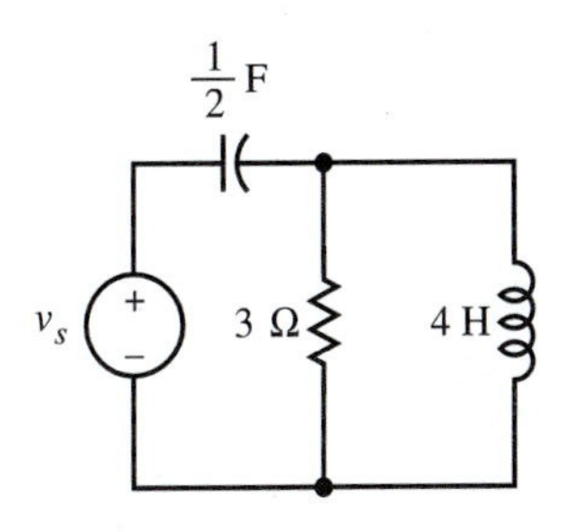

Supplementary Problems

20. Use superposition to calculate currents i_a, i_b, and i_c for the following network. What effect does voltage v_a have on current i_b, and what effect does voltage v_b have on current i_a?

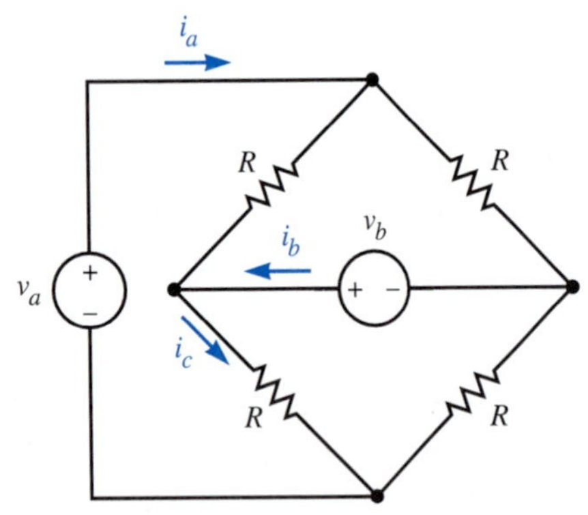

21. The small-signal model for a common collector transistor amplifier is shown in the following figure. It must have an output resistance R_o looking into terminals E and C of 10 Ω or less. What is the minimum value of β that will satisfy this requirement?

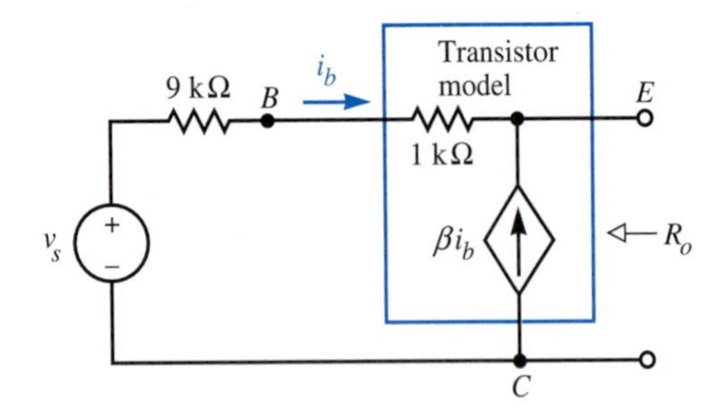

22. Determine the Thévenin and Norton equivalent circuits with respect to terminals a and b for the following network.

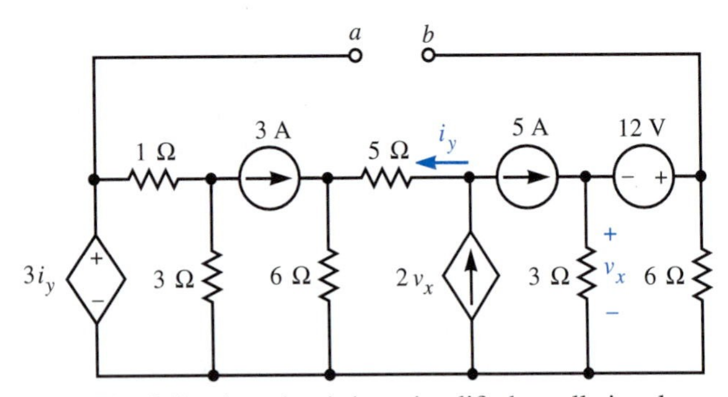

23. The following circuit is a simplified small-signal equivalent circuit for a one-transistor amplifier with *feedback*. The voltage gain v_o/v_s and output resistance (Thévenin resistance looking into the output terminals) are controlled by the feedback resistance R_f.
   (a) Select the value of R_f to design for an output resistance of 100 Ω. What is the voltage gain for this value of R_f?
   (b) Select the value of R_f to design for a voltage gain of −4.05. What is the output resistance for this value of R_f?

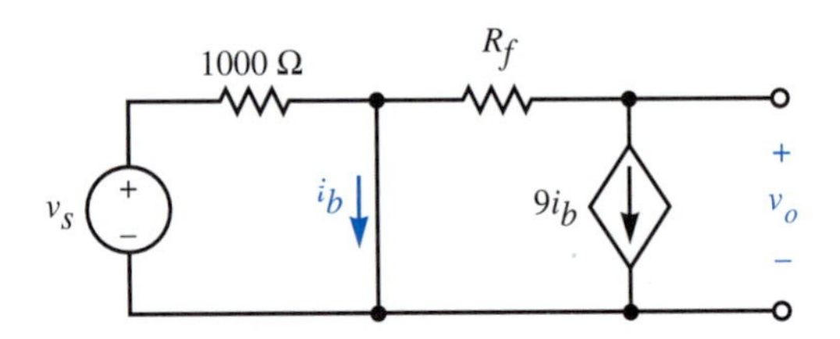

24. A simplified equivalent circuit for a grounded-base amplifier is shown below. Calculate the input resistance seen by the practical source, the voltage gain v_{cb}/v_{eb}, the current gain i_c/i_e, and the power gain $-v_{cb}i_c/v_{eb}i_e$, all with the 1000-Ω load connected.

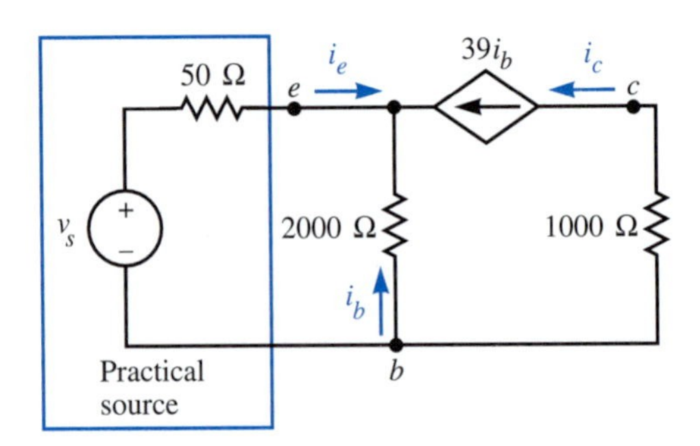

25. Construct a Thévenin equivalent circuit with respect to terminals a and b for the following network. Next, connect a 20-Ω feedback resistance between terminals a and c and construct the Thévenin equivalent with respect to terminals a and b. Compare your results. This is an example of the effect of feedback on an electronic amplifier. For the original network, an input resistance of 3 Ω is seen by the 12-V source with a 6-Ω series resistance. What is the input resistance with the 20-Ω feedback resistance connected? (Some of the output is fed back through this resistance and subtracted from the 12-V input.) We will investigate feedback circuits in Chapter 6.

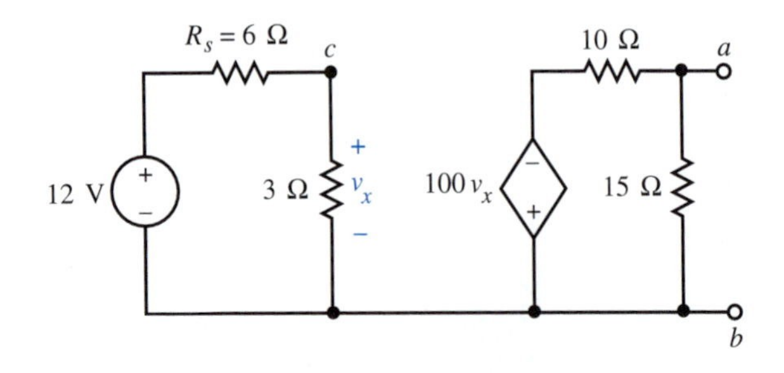

26. Calculate the voltage gain and output resistance looking into terminals a and b of the following circuit if the resistance R is replaced by an open circuit.

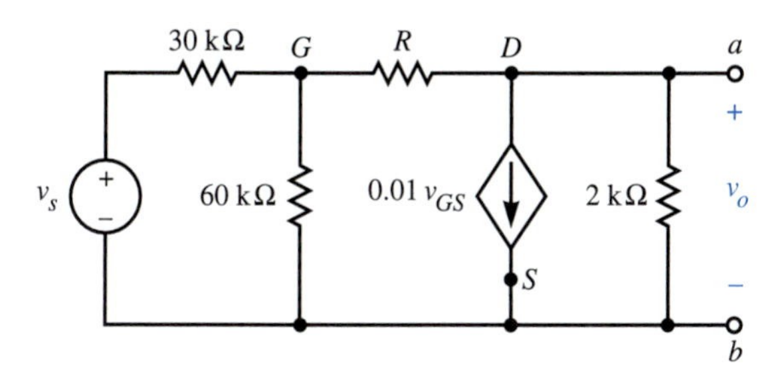

27. The preceding circuit is a simplified model of an FET amplifier. Both the voltage gain and the output resistance R_o (looking into terminals S and D) are a function of the feedback resistance R. Select R so that the output resistance is 1 kΩ. What is the voltage gain v_o/v_s for this value of R?

28. A small-signal equivalent circuit for a field-effect transistor (FET) amplifier is shown below. Terminals G, S, and D are the gate, source, and drain terminals, respectively. 5.7
    (a) Determine the value of R_2 that will absorb the maximum power.
    (b) What power is absorbed by the value of R_2 found in (a)?

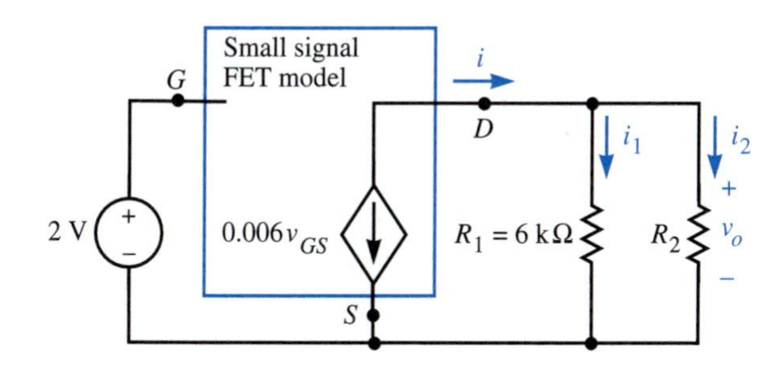

29. Transistors are often used in the *Darlington connection* (compound amplifier). A simplified small-signal equivalent circuit for this connection is shown in the following figure. Determine the Thévenin or Norton equivalent circuit for this network as seen by:
    (a) The voltage source
    (b) The resistance R_e
    (c) The resistance R_c

    if $R_a = R_b = 10\ \Omega$, $R_c = 13\ \Omega$, $R_e = 1\ \Omega$, and $\beta = 9$.

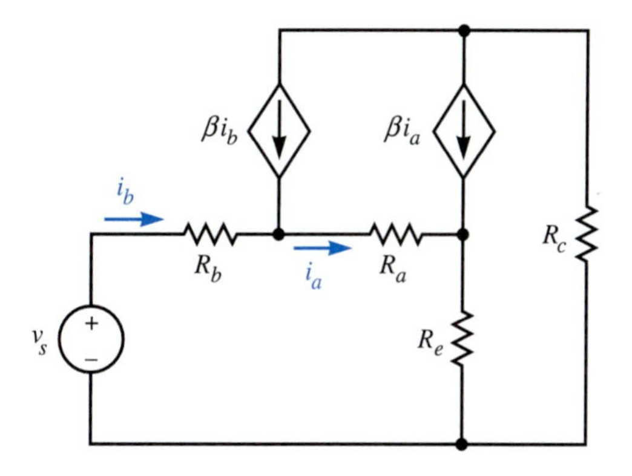

30. Semiconductor devices are linear only within certain voltage and current ranges and must be *biased* to operate within the proper range. You must design the *bias circuit* for a transistor amplifier with the large-signal equivalent circuit shown below. The collector current i_C must be 1 mA.
    (a) Calculate the input resistance looking into terminals B and 0 with the connection between terminals A and B removed.
    (b) Select resistances R_1 and R_2 so that the resistance looking into terminals A and 0 is four times the resistance looking into terminals B and 0, and so that collector current i_C is 1 mA when terminals A and B are connected. Refer to the first two steps of Problem 48 of Chapter 3 for help.

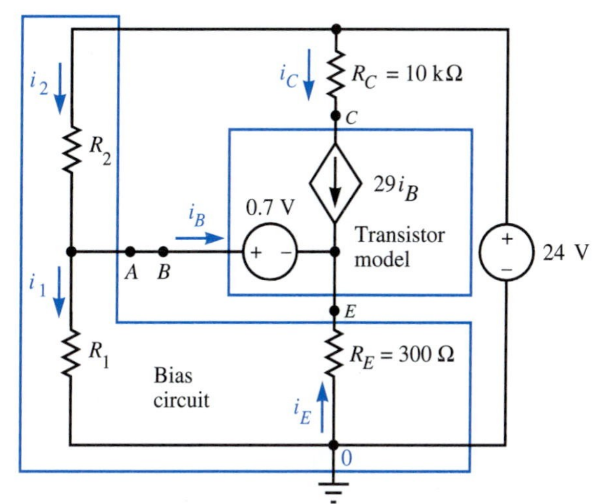

31. One solar panel for the school solar-powered racer, Sol Survivor, has an open-circuit voltage of 45 V and a short-circuit current of 3 A in noonday sun. Assume that the solar panel is a linear circuit. (This is not a very close approximation.)

(a) Determine the Thévenin equivalent circuit for the solar panel.

(b) An electronic circuit, called a maximum-power tracker, matches each panel in the solar array to the battery for maximum charging power. In noonday sun, what current should the power tracker require *from the solar panel*? (The current supplied by the solar panel will be different than the current supplied to the battery.)

(c) If the maximum-power tracker is 98 percent efficient, what is the output power of the power tracker?

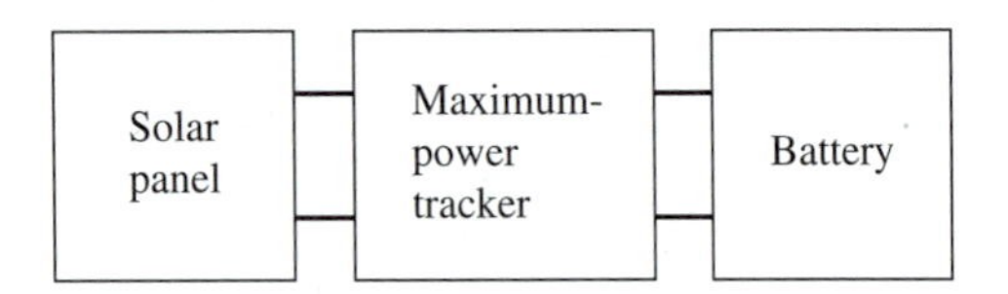

32. The terminal characteristics of a two-terminal device are experimentally measured and plotted on the following graph. Construct a Thévenin equivalent network as a model for the device. (Current i_{ab} is the current into terminal a of the device.)

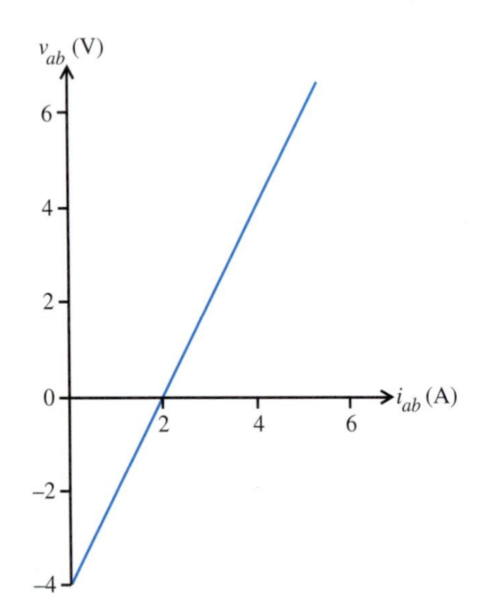

33. Analysis of electronic circuits is often facilitated by the performance of *source transformations on dependent sources.* Show that the dependent voltage source and series resistance that follows is equivalent, with respect to its effect on network B, to the dependent current source and parallel resistance shown, if $R' = R$ and $k' = k/R$.

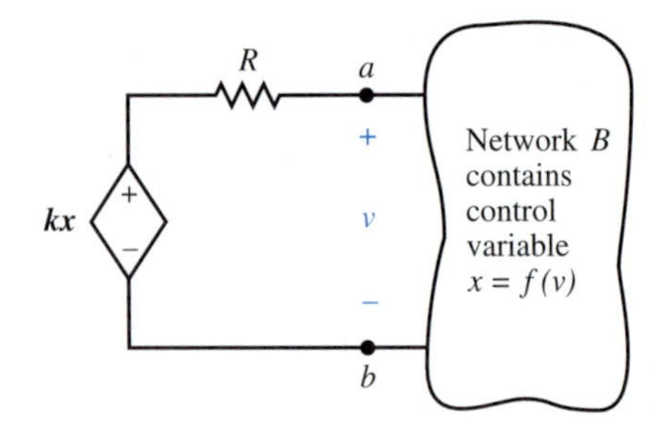

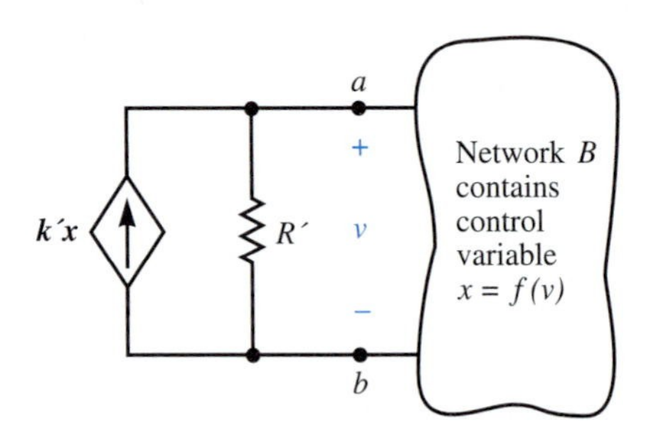

34. With the aid of the results obtained in Problem 33, convert the following network into a network that is equivalent with respect to terminals a and b, and consists of an independent voltage source, dependent voltage source, and series resistances.

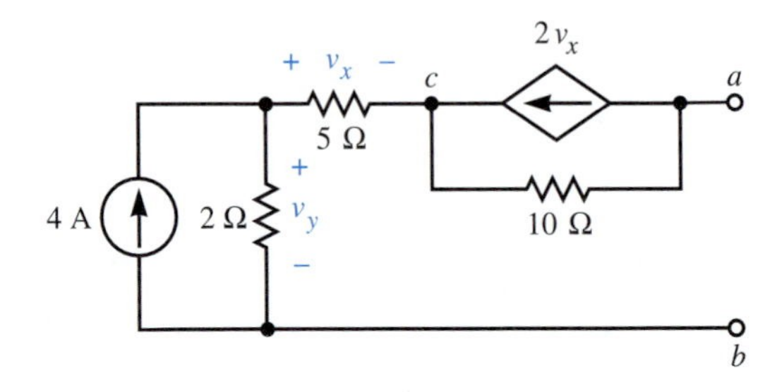

35. In the preceding network, replace the dependent current source with a dependent current source of value $2v_y$, reference arrow pointing toward the left, and repeat Problem 34.

36. This problem examines the efficiency of power delivery to a matched load. For each of the following circuits, calculate

(i) the power p_1 supplied by the sources when the network is unloaded,

(ii) the load resistance that will absorb the maximum power when connected to terminals a and b,

(iii) the power p_L absorbed by this load resistance,

(iv) the power p_2 supplied by the source when the load is connected,

(v) the change in power $p_2 - p_1$ required from the sources when the load is connected, and

(vi) the efficiency $E = [p_L/(p_2 - p_1)]100$ percent.

(vii) Is the power supplied by the voltage source in the Thévenin equivalent circuit equal to $p_2 - p_1$?

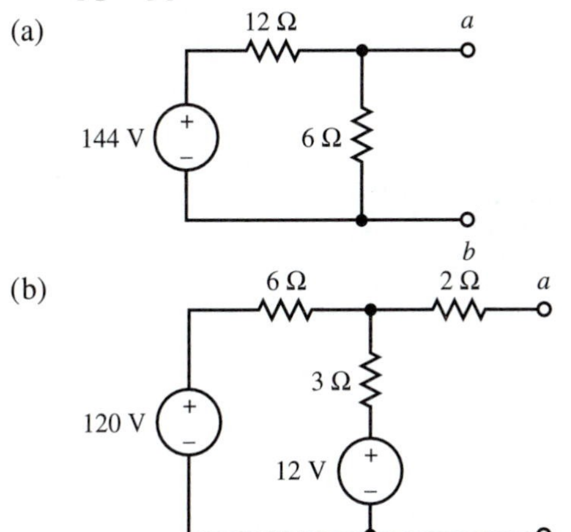

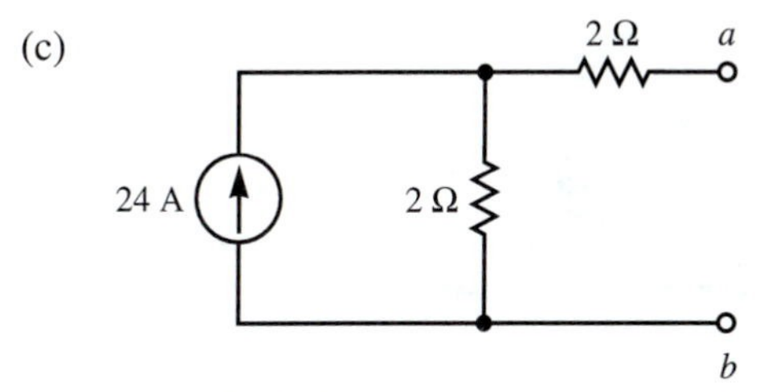

37. Determine the Thévenin and Norton equivalent circuits with respect to terminals a and b for the following network.

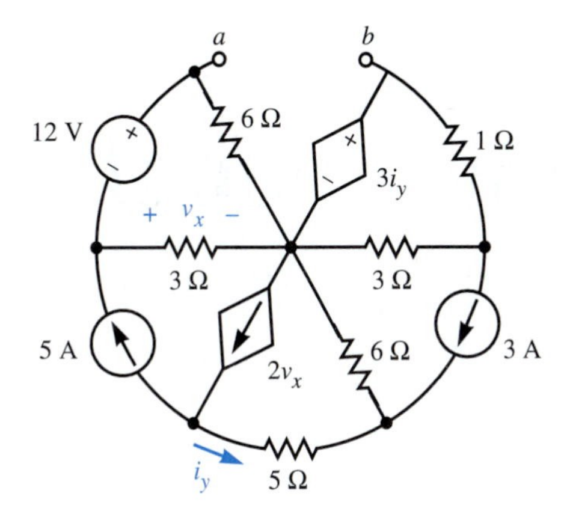

6

Operational Amplifiers

We all know that the integrated circuit, or semiconductor chip, has revolutionized digital circuit design and given us the personal computer. Although less widely publicized, integrated circuits have also revolutionized linear circuit design by providing us with small building blocks containing many resistors and transistors. Perhaps the most versatile of these devices is the *operational amplifier,* or *op amp.*

Operational amplifiers are characterized by a high input resistance, low output resistance, and large voltage gain. Op amps are used in audio amplifier systems; they also appear in many instrumentation applications, such as digital multimeters, where they amplify small voltages.

A detailed analysis of the voltages and currents internal to the op amp is seldom necessary. Usually we need consider only the terminal characteristics of the op amp. We often further simplify the analysis by introducing a four-terminal network component called the *ideal op amp.* As a tool for understanding op-amp circuits, we also introduce the concept of *negative feedback.*

6.1 Op-Amp Models

An op amp, although once constructed as an interconnection of discrete transistors and resistors, is now built as an integrated circuit on a few square millimeters of silicon and costs less than one dollar. Op amps come in several types of packages, typically with eight or more terminals, some of which may have no internal connection. The dual-in-line package shown in Fig. 6.1 is quite popular. We shall consider only the external characteristics of the op amp. In the simplest case, only five connections are made to the op amp, as shown in Fig. 6.2.

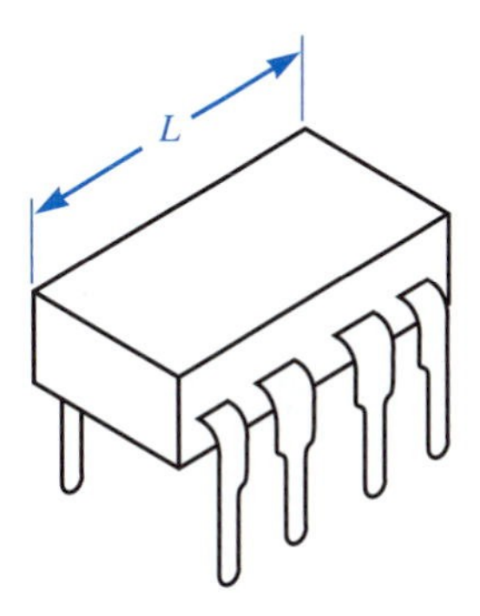

FIGURE 6.1 A dual-in-line package often used for operational amplifiers. The length L is less than $\frac{1}{2}$ inch.

Constant voltages of value V^+ and V^-, shown in Fig. 6.2, supply the electric power necessary for operation of the amplifier. (Typical values for V^+ and V^- are $+15$ V and -15 V, respectively.) The reference symbol (⏚) shown in Fig. 6.2 represents the reference from which the constant voltages are measured. Physically, the reference is some common point, such as the power-supply case, to which the circuit is connected. This *chassis ground* is usually connected to the earth (*earth ground*).

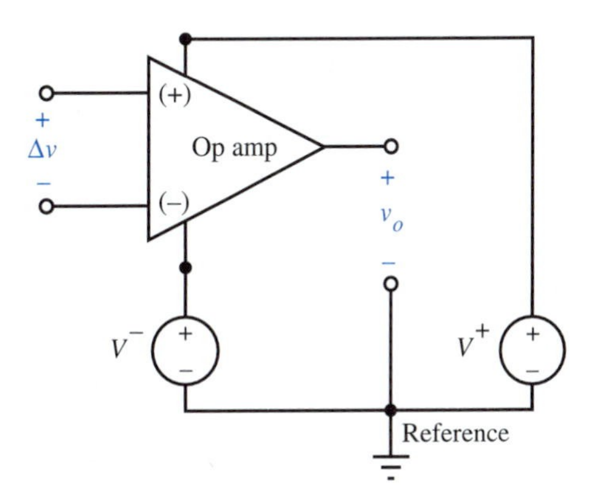

FIGURE 6.2 Op-amp symbol with power sources and reference shown

The output voltage v_o is a function of v_+, the voltage of the *noninverting* input terminal [the (+) terminal] with respect to ground, and v_-, the voltage of the *inverting* input terminal [the (−) terminal] with respect to ground. For satisfactory operation, these input-terminal voltages must be small (a few volts or less in most amplifiers). Under these conditions, the amplifier largely rejects the voltage component in common with v_+ and v_- (the *common mode voltage*) and the output voltage is essentially a function of the *error voltage* Δv:

$$\Delta v = v_+ - v_- \tag{6.1}$$

In practice, the open-circuit output voltage of an op amp is constrained to fall between V^+ and V^-, as shown in Fig. 6.3. The output current of an op amp, which is limited by the supply voltages and internal op-amp resistances, also affects the value of the saturation voltages. For linear analysis we assume that our op amps are operating in the linear region indicated in Fig. 6.3, and the open-circuit output voltage is proportional to the error voltage Δv.

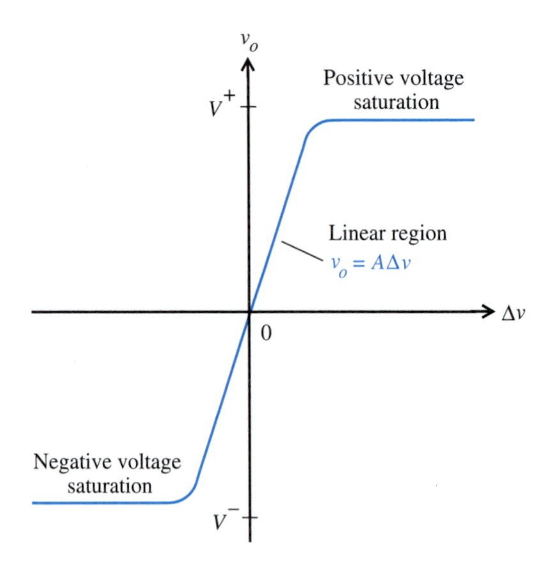

FIGURE 6.3 Open-circuit output voltage v_o as a function of error voltage Δv for an op amp

In the linear range, we model the input-terminal characteristics of an op amp by an input resistance R_i, and the output-terminal characteristics by a dependent voltage source of value $A\Delta v$ in series with an output resistance R_o, as shown in Fig. 6.4. The connection to ground and the current i_g account for the currents through the voltage sources shown in Fig. 6.2.

We significantly simplify the model of Fig. 6.4 for most circuit analysis. The

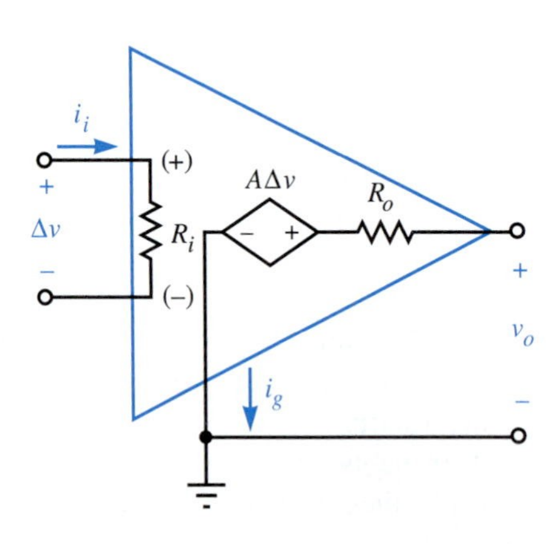

FIGURE 6.4 Op-amp model showing input resistance R_i, output resistance R_o, voltage gain A, input current i_i, and current to the reference i_g

voltage gain A is not tightly controlled by the manufacturer, but is always large (greater than 100,000 in most op amps), so we use the approximation

$$A = \infty \tag{6.2}$$

The large gain of an op amp requires *negative feedback*. Some fraction of the output voltage v_o is *fed back* and subtracted from the input voltage to form the error voltage Δv. Infinite gain drives v_o to the value required to make this error voltage zero.

$$\Delta v = v_o/A = 0 \tag{6.3}$$

The input resistance R_i for an op amp is greater than 100 kΩ (greater than 10^{12} Ω in some op amps), so we assume that the input current is zero:

$$i_i = \Delta v/R_i = 0 \tag{6.4}$$

In Section 6.5, we show that the assumption of zero input current is equivalent to an assumption of infinite input resistance:

$$R_i = \infty \tag{6.5}$$

The small output resistance R_o primarily has the effect of limiting the maximum output current for linear operation, so we assume

$$R_o = 0 \tag{6.6}$$

In terms of the model of Fig. 6.4, these approximations give us a *voltage-controlled voltage source with infinite gain and control voltage* Δv, which is the model for a new circuit component, the *ideal op amp.*

DEFINITION Ideal Op Amp

The input current i_i for an ideal op amp is zero, and negative feedback causes the output voltage v_o to assume the value required to make the error voltage Δv equal to zero:

$$i_i = 0 \tag{6.7}$$

$$\Delta v = 0 \tag{6.8}$$

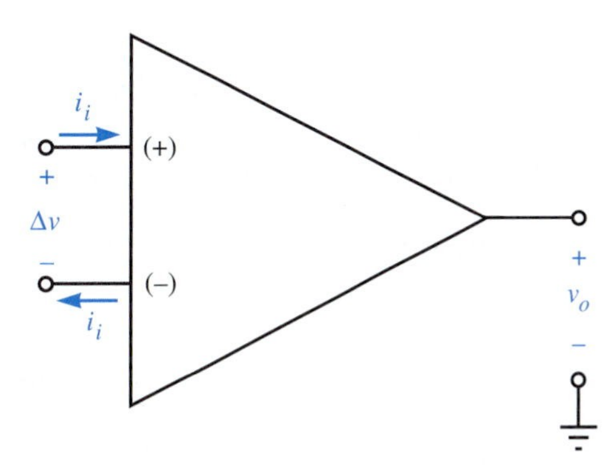

FIGURE 6.5
Ideal op-amp symbol

We must exercise care when applying KCL with the ideal op-amp symbol, because the current to ground i_g in Fig. 6.4 is not shown in Fig. 6.5. It is safest to show the ground connection if KCL is applied to a surface enclosing the op-amp symbol.

In the following section, we use the ideal op amp to analyze several practical op-amp circuits that employ negative feedback.

Remember The linear range of operation for an op amp is limited by the supply voltages and the output current. For analysis within the linear range of operation we usually use the ideal op-amp model, which is a dependent voltage source with infinite gain that is controlled by the error voltage $\Delta v = v_+ - v_-$.

EXERCISES

1. The saturation voltages of an unloaded op amp are ± 12 V, and the output resistance is $R_o = 60\ \Omega$. Use the finite-gain op-amp model (Fig. 6.4) to estimate the maximum short-circuit current available from the op amp. (In practice, the maximum output current might be listed in the manufacturer's specifications.) *answer:* 200 mA

6.2 Op-Amp Circuits

In this section we use the ideal op-amp symbol of Fig. 6.5 with Eqs. (6.7) and (6.8) to analyze some simple, but important, op-amp circuits that are used as building blocks in more complex circuits. We will see that in each op-amp circuit some fraction of the output voltage v_o is *fed back* and subtracted from the circuit input voltage v_s to obtain the *error voltage* Δv. With infinite gain, the output voltage must have the value that makes Δv zero.

The Noninverting Amplifier

In the following example we analyze an op-amp circuit that multiplies an input voltage v_s by a positive constant. This *noninverting amplifier* circuit is widely used as a building block in more complex circuits.

EXAMPLE 6.1 A noninverting voltage amplifier that uses feedback in conjunction with an operational amplifier is shown in Fig. 6.6 on the next page. Use the ideal op-amp model to determine the closed-loop voltage gain v_o/v_s.

Solution Resistances R_a and R_b are effectively in series, because $i_i = 0$, so we can use the voltage divider relation to give

See Problems 6.1 and 6.3 in the PSpice manual.

$$v_{Fb} = \frac{R_a}{R_a + R_b} v_o$$

(The subscript *Fb* designates "feedback." That is, v_{Fb} is the part of v_o *fed back to the inverting input terminal.*) The ideal op-amp assumption ($\Delta v = 0$) and KVL give

$$v_s = v_{Fb}$$

Substitution from the first equation gives

$$v_s = \frac{R_a}{R_a + R_b} v_o$$

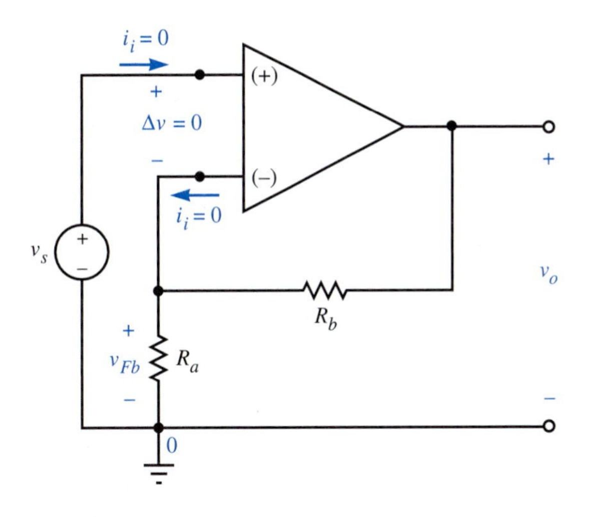

FIGURE 6.6 Noninverting voltage amplifier using an op amp with negative feedback (ideal op-amp model used)

and the *voltage gain* with feedback is

$$\frac{v_o}{v_s} = \frac{R_a + R_b}{R_a}$$

We call the voltage gain with feedback the *closed-loop voltage gain.*

From the preceding example we can see that an equivalent circuit for the noninverting op-amp circuit of Fig. 6.6 is shown in Fig. 6.7.

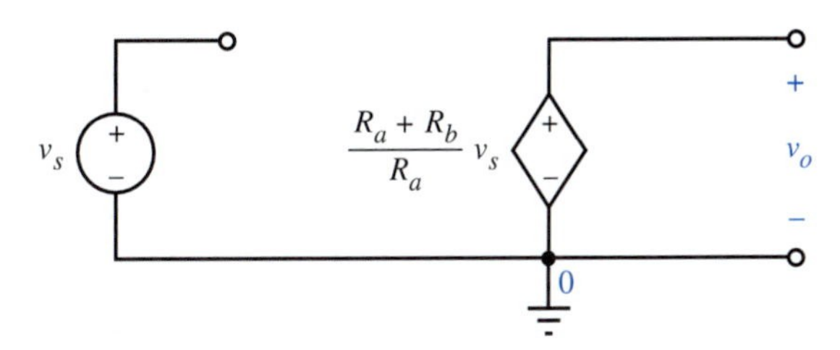

FIGURE 6.7 An equivalent circuit for the noninverting amplifier of Fig. 6.6

The open-loop voltage gain A of a practical op amp is not infinite, so the equivalent circuit of Fig. 6.7 only approximates a practical noninverting op-amp circuit. We show in Sec. 6.4 that this approximation is very good if the open-loop gain A is much greater than the closed-loop gain $v_o/v_s = (R_a + R_b)/R_a$.

The Inverting Amplifier

We will now analyze an op-amp circuit that multiplies an input signal by a negative constant. This *inverting amplifier* circuit is another of the building blocks used in more complex op-amp circuits.

EXAMPLE 6.2 Use the ideal op-amp model to determine the closed-loop voltage gain of the *inverting voltage amplifier* of Fig. 6.8. Note that the (−) terminal is at the top in this figure. We always have feedback to the inverting terminal.

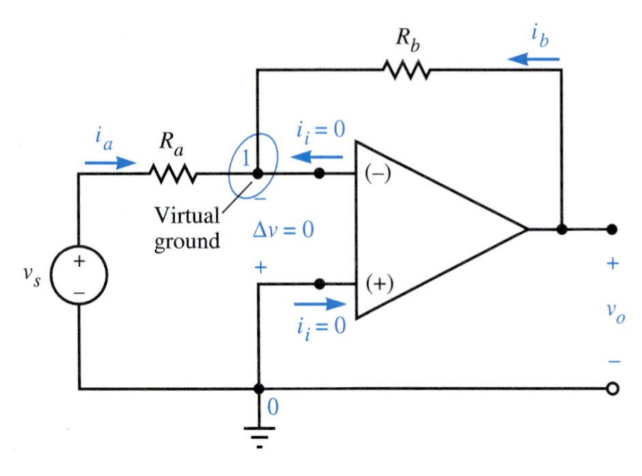

FIGURE 6.8
Inverting amplifier using an op amp with negative feedback

Solution For the ideal op-amp model,

$$v_{10} = -\Delta v = 0$$

[Node 1 is said to be a *virtual ground,* because it is at the same voltage as the reference (ground).] Thus KCL for the indicated closed surface gives

See Problems 6.2 and 6.4 in the PSpice manual.

$$-\frac{1}{R_a}v_s - \frac{1}{R_b}v_o = 0$$

and the closed-loop gain is

$$\frac{v_o}{v_s} = -\frac{R_b}{R_a}$$

From the preceding example, we can see that an equivalent circuit for the inverting op-amp circuit of Fig. 6.8 is given in Fig. 6.9. The equivalent circuit of Fig. 6.9 closely approximates a practical inverting op-amp circuit if the open-loop gain A is much greater than the magnitude of the closed-loop gain R_b/R_a.

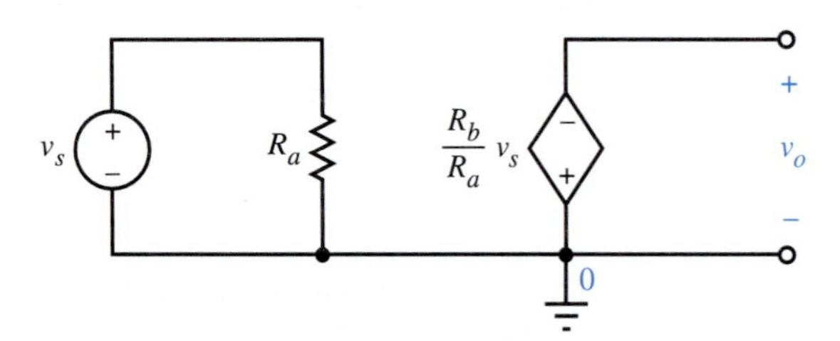

FIGURE 6.9
An equivalent circuit for the inverting amplifier of Fig. 6.8

We can replace the op-amp circuit of Fig. 6.6 or 6.8 with the equivalent circuits of Fig. 6.7 or 6.9, respectively. This can simplify the analysis of complex circuits containing several op amps.

Op-amp circuits not only perform algebraic operations, such as multiplication by a positive or negative constant (Examples 6.1 and 6.2), but they also perform integration.

An Integrator Circuit

We will now analyze an op-amp circuit with an output voltage that is the integral of the input voltage. This op-amp *integrator* is used as one component in many electronic instruments.

EXAMPLE 6.3 Determine the output voltage v_o as a function of the input voltage v_s for the circuit shown in Fig. 6.10.

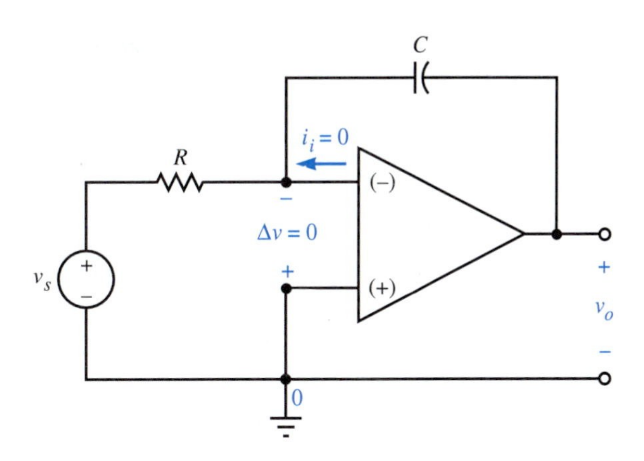

FIGURE 6.10
An integrator circuit that uses an op amp with negative feedback

Solution Application of KCL to the node at the inverting input gives

$$\frac{1}{R}(-v_s) + C\frac{d}{dt}(-v_o) = 0$$

Integration of this equation gives

$$v_o = -\frac{1}{RC}\int_{-\infty}^{t} v_s \, d\lambda$$

With the exception of the sign change, this last equation gives the output voltage v_o as $1/RC$ times the integral of the input voltage v_s. This op-amp integrator circuit is frequently used in instrumentation.

In addition to multiplication by a constant (amplification) and integration, op amps can perform other mathematical operations, such as addition (see Exercise 2) and subtraction (see Exercise 3).

Observe that in each of the preceding examples, the feedback was to the inverting terminal [the (−) terminal]. This *negative feedback* caused the output voltage to be independent of the very large, but imprecisely known, voltage gain A of the op amp. *For linear operation, we must always have feedback to the inverting terminal.* Feedback to only the noninverting terminal [the (+) terminal] always results in instability.

We typically use the ideal op amp to analyze practical op-amp circuits. We then check to see that the voltages and currents are in the linear range of operation for the op amps. If not, we must redesign the circuit. If a more accurate linear analysis is required, we can use the model of Fig. 6.4, but this is seldom necessary.

Remember

For linear operation, we *always* use op amps with feedback to the inverting terminal [the (−) terminal]. The gain of an op-amp circuit with feedback is the closed-loop gain. If the magnitude of the open-loop voltage gain A is much greater than the magnitude of the closed-loop voltage gain v_o/v_s, we can use the ideal op amp for our analysis. An amplifier has a positive closed-loop gain if the input is to the (+) terminal. If the input is to the (−) terminal, the closed-loop gain is negative. Besides amplification, we can use op amps to perform addition, subtraction (Exercises 2 and 3), and integration.

EXERCISES

2. (a) Use the ideal op-amp model ($\Delta v = 0$ and $i_i = 0$) to find an expression that relates v_o to v_1 and v_2 in the following circuit. What function does the circuit perform when $R_1 = R_2 = R_3$? *Hint:* The analysis is similar to that used in Example 6.2.

   answer: $v_o = -R_3(v_1/R_1 + v_2/R_2)$, $v_o = -(v_1 + v_2)$

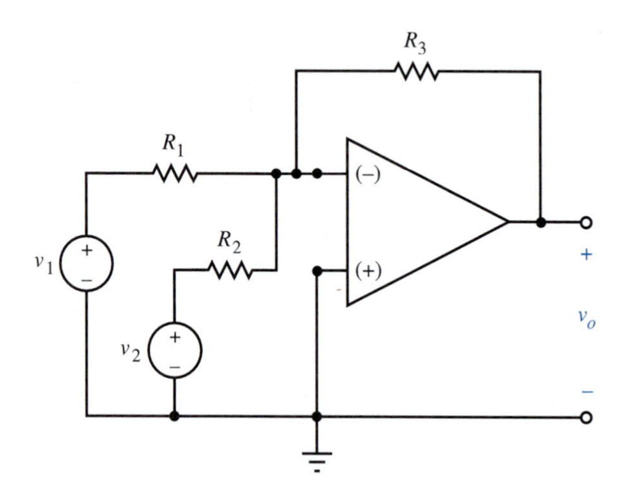

   (b) Construct an equivalent circuit for the preceding op-amp circuit, as was done in Figs. 6.7 and 6.9.

   answer: Source 1 sees a resistance of R_1. Source 2 sees a resistance of R_2. The output circuit is two VCVS in series. One has a value of $-(R_3/R_1)v_1$, and the other a value of $-(R_3/R_2)v_2$ with the (+) references toward the output terminal and the (−) references toward ground.

See Problem 6.5 in the PSpice manual.

3. (a) Use the ideal op-amp model to find an expression for the voltage v_o for the following network. What operation does the circuit perform when $R_1 = R_a$ and $R_2 = R_b$? *Hint:* Use the voltage divider relation to determine the voltage across R_b. Then proceed with an analysis similar to that used in Example 6.2.

   answer: $v_o = \frac{R_2 + R_1}{R_1} \cdot \frac{R_b}{R_a + R_b} v_2 - \frac{R_2}{R_1} v_1$, $v_o = \frac{R_2}{R_1}(v_2 - v_1)$

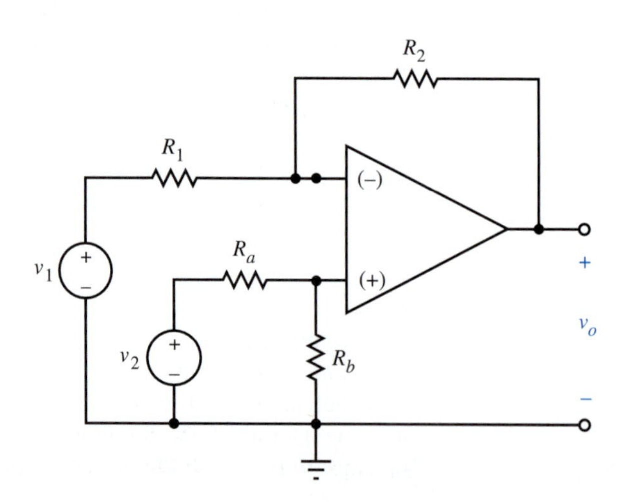

(b) Construct an equivalent circuit for the preceding op-amp circuit, as was done for Figs. 6.7 and 6.9.

answer: Source v_1 sees a resistance of R_1 in series with a VCVS of value v_b. Source v_2 sees resistances R_a and R_b in series. The voltage across R_b is v_b. The output circuit consists of two VCVS in series, one with value v_b and the other with value $-(R_2/R_1)\,v_1$. Both have (+) reference marks toward the output terminal.

4. What function does the following circuit perform? (Electronic differentiation is very sensitive to noise and is avoided if possible.) *Hint:* The analysis is very similar to that of Example 6.3.

answer: $v_o = -RC\dfrac{d}{dt}v_s$

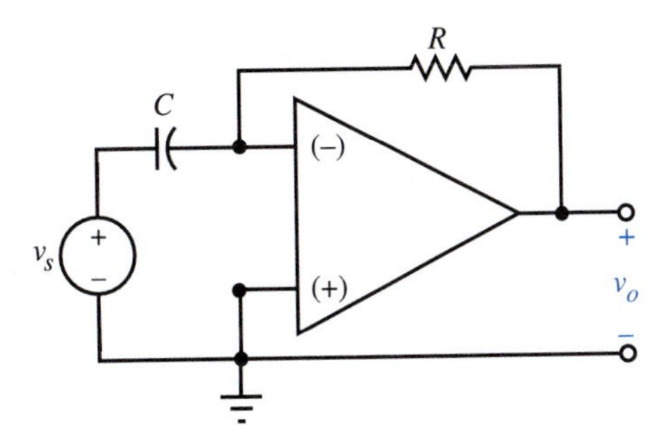

6.3 Node-Voltage Equations with Ideal Op Amps

We can easily apply the method of analysis by node voltages to networks containing op amps. If we use the op-amp model of Fig. 6.4, we simply replace the op amp by the chosen model. The network is then analyzed by our standard procedures. If we use the ideal model of Fig. 6.5, we leave the op-amp symbol in the network diagram. We adapt our method of node-voltage analysis to the ideal op-amp model by using the following information:

1. The current into each input terminal of the ideal op amp is zero.
2. The voltage difference between the two input terminals of the ideal op amp is zero.
3. The output terminal of the ideal op-amp model is connected directly to the reference node by a voltage-controlled voltage source (with infinite gain).

The method is effectively demonstrated by the following example.

EXAMPLE 6.4 Calculate the node voltages of the network shown in Fig. 6.11.

Solution First number the nodes as shown and indicate on the network diagram that the op-amp input currents and error voltage are zero. The number of node-voltage equations is five, which is one less than the number of nodes.

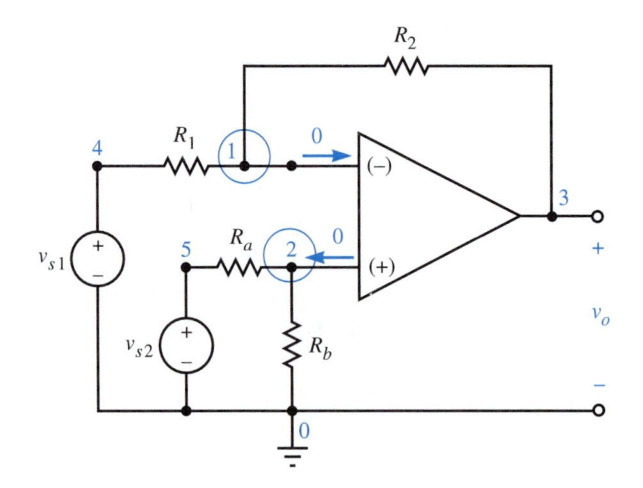

FIGURE 6.11
A difference amplifier

There is one KVL equation for each voltage source and one that includes the op-amp input voltage difference (Δv) of zero. The KVL equations are

$$v_4 = v_{s1}$$

$$v_5 = v_{s2}$$

$$v_2 = v_1 \qquad \text{because } \Delta v = 0$$

Notice that we use $\Delta v = 0$ rather than $v_3 = A\,\Delta v$ for our third KVL equation, as we would with the model of Fig. 6.4. For infinite gain A, v_3 controls Δv and makes it equal zero.

We typically make use of the three KVL equations when writing the two KCL equations (our shortcut procedure for analysis by node voltages), in which case we write the KCL equations for surfaces that enclose nodes 1 and 2 as

$$\frac{1}{R_1}(v_1 - v_{s1}) + \frac{1}{R_2}(v_1 - v_3) = 0$$

$$\frac{1}{R_a}(v_1 - v_{s2}) + \frac{1}{R_b}v_1 = 0$$

(Node 3, the output node, is connected to the reference node by a voltage-controlled voltage source, so we do not write a KCL equation for a surface enclosing this node.)

These last two equations are easily solved to give

$$v_3 = \frac{R_1 + R_2}{R_1} \cdot \frac{R_b}{R_a + R_b} v_{s2} - \frac{R_2}{R_1} v_{s1}$$

Remember

We can use node-voltage equations to analyze feedback circuits when the ideal op-amp model is assumed. The assumptions are

1. The op-amp input current is zero.
2. The error voltage is zero ($\Delta v = 0$).
3. The output terminal is connected to the reference node by a voltage-controlled voltage source with infinite gain.

EXERCISES

5. (a) Use the method of node voltages to solve for the output voltage of the following op-amp circuit.

answer: $$v_o = \frac{(R_1 + R_2)(R_3 + R_4) + R_3R_4}{R_1R_4} v_{s2} - \frac{R_2(R_3 + R_4) + R_3R_4}{R_1R_4} v_{s1}$$

See Problem 6.6 in the PSpice manual.

(b) Construct an equivalent circuit for the following op-amp circuit.

answer: Source s_1 sees R_1 in series with a VCVS of value v_{s2}. Source v_{s2} sees an open circuit. The output circuit consists of a VCVS of value $A_1 v_{s1}$ in series with a VCVS of value $A_2 v_{s2}$ where $A_1 = -\frac{R_2(R_3 + R_4) + R_3R_4}{R_1R_4}$ and

$$A_2 = +\frac{(R_1 + R_2)(R_3 + R_4) + R_3R_4}{R_1R_4}$$

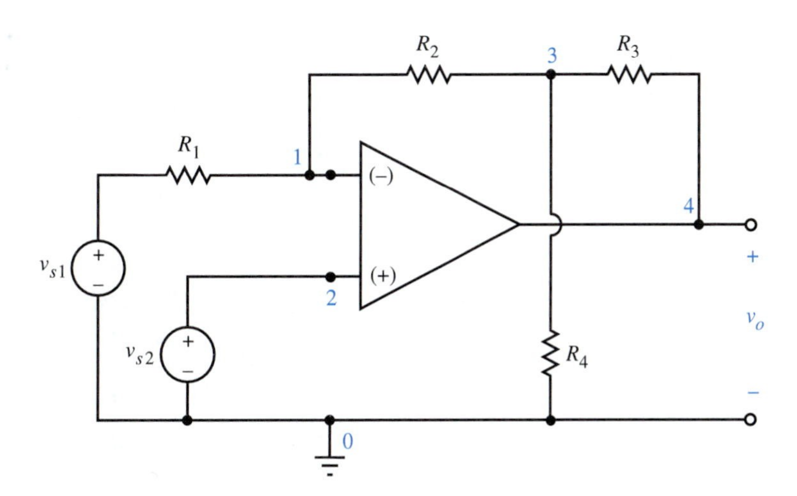

6.4 Finite Open-Loop Gain†

In this section we analyze the suitability of the infinite-gain assumption in the ideal op amp. We retain the approximations of zero output resistance

$$R_o = 0 \tag{6.9}$$

and zero input current

$$i_i = 0 \qquad (R_i = \infty) \tag{6.10}$$

† This section can be omitted without loss of continuity.

but relax the infinite gain assumption so that

$$A < \infty \tag{6.11}$$

We use this model, which is included in the circuit of Fig. 6.12, to determine the effect of finite open-loop gain A on the closed-loop gain v_o/v_s of this noninverting amplifier circuit. We express the results in terms of the *feedback ratio* β, which is the fraction of the output voltage fed back to the inverting $(-)$ terminal.

EXAMPLE 6.5 Determine the output voltage for the noninverting op-amp circuit shown in Fig. 6.12. The finite-gain op-amp model is used.

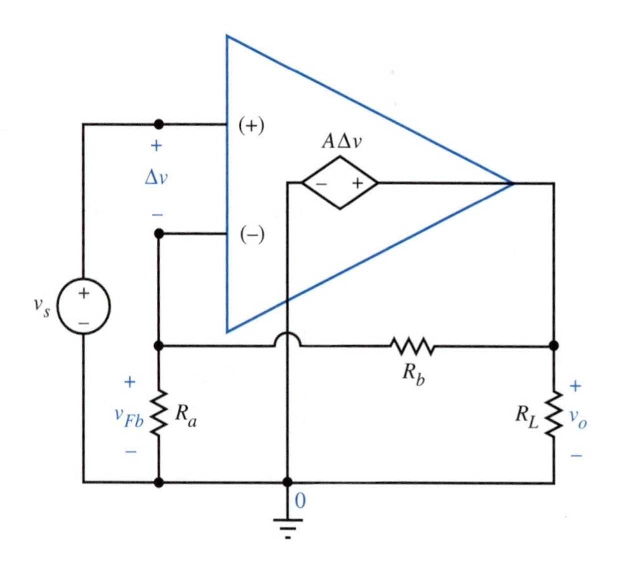

FIGURE 6.12
A noninverting amplifier circuit using an op amp with negative feedback (finite gain op-amp model shown)

Solution The voltage divider relation gives

$$v_{Fb} = \frac{R_a}{R_a + R_b} v_o = \beta v_o$$

where

$$\beta = \frac{R_a}{R_a + R_b}$$

KVL easily gives

$$\Delta v = v_s - v_{Fb} = v_s - \beta v_o$$

(This last equation clearly shows that the *feedback ratio* β is the fraction of the output voltage v_o fed back and subtracted from the input voltage v_s to provide the error voltage Δv input to the op amp.) The control equation gives

$$\Delta v = \frac{1}{A} v_o$$

Substitution of this into the preceding equation yields

$$\frac{1}{A} v_o = v_s - \beta v_o$$

which we can solve for

$$v_o = \frac{A}{A\beta + 1} v_s$$

We call $A\beta$, the product of the open-loop gain and the feedback ratio, the *loop gain.* When the loop gain becomes much greater than 1, this last equation reduces to

$$v_o = \frac{1}{\beta} v_s \tag{6.12}$$

This is the result we obtained (Example 6.1) when we used the ideal op-amp model to analyze the noninverting amplifier. We see that as long as the loop gain is much greater than one, the finite-gain op-amp model yields essentially the same result as the ideal op-amp model. Because the closed-loop gain is approximately the reciprocal of the feedback ratio (Eq. 6.12), a large loop gain implies that the open-loop gain is much greater than the closed-loop gain. Although we developed this result for a specific circuit, it is true in general.

Remember

If the open-loop gain is much greater than the magnitude of the closed-loop gain, analysis with the ideal op-amp model gives essentially the same results as analysis with the finite-gain model.

EXERCISES

6. For the following inverting amplifier circuit:
   (a) Calculate the closed-loop gain $G = v_o/v_s$.
   answer: $G = -AR_b/[(1 + A)R_a + R_b]$
   (b) Does the result obtained in part (a) reduce to that obtained in Example 6.2 as A becomes much larger than R_b/R_a? answer: Yes
   (c) Calculate the closed-loop input resistance R_i' (the resistance seen by the voltage source). answer: $R_i = [(1 + A)R_a + R_b]/(1 + A)$

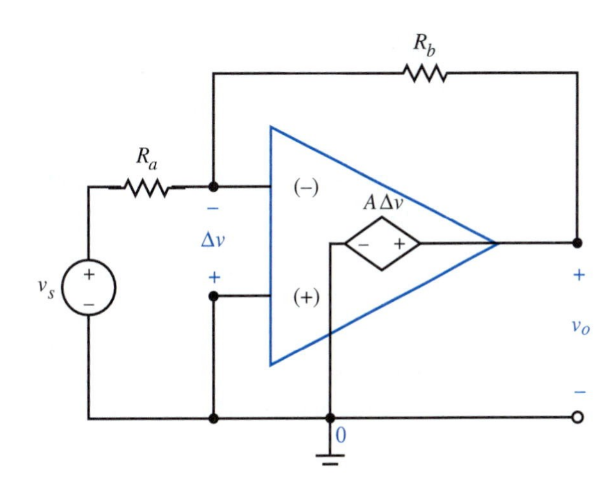

6.5 Finite Resistances†

We know that the error voltage $\Delta v = v_o/A$ is small and the input resistance R_i is large, so the assumption of zero input current is justified for the ideal op-amp. In this section we will analyze the noninverting amplifier circuit with the more precise model of Fig. 6.4 to justify the assumption of zero output resistance ($R_o = 0$) in the ideal op amp.

EXAMPLE 6.6 Calculate the output voltage of the noninverting op-amp circuit shown in Fig. 6.13.

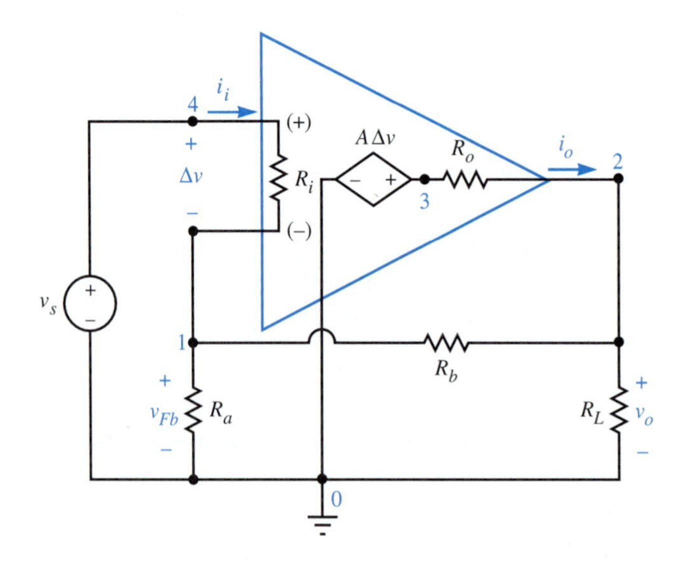

FIGURE 6.13 A noninverting amplifier circuit that uses an op amp with negative feedback (finite resistance model shown)

Solution We can easily analyze the circuit by the method of node voltages. Apply KCL to a surface enclosing node 1:

$$\frac{1}{R_a} v_{Fb} + \frac{1}{R_b}(v_{Fb} - v_o) + \frac{1}{R_i}(v_{Fb} - v_s) = 0$$

Application of KCL to a surface enclosing node 2 gives

$$\frac{1}{R_L} v_o + \frac{1}{R_b}(v_o - v_{Fb}) + \frac{1}{R_o}[v_o - A(v_s - v_{Fb})] = 0$$

The use of Cramer's rule and some complicated algebra, which we need not reproduce, gives us

$$v_o = \frac{(R_a + R_b)/R_a + (1/A)\cdot(R_o/R_i)}{1 + (1/A)\{[(R_L + R_o)/R_L][1 + (R_b/R_a) + (R_b/R_i)] + (R_o/R_a) + (R_o/R_i)\}} v_s$$

Although this equation gives the exact solution for the output voltage of the circuit in Fig. 6.13, it is cumbersome. We can reduce this result to a more manageable form with a few minor approximations. For any practical op-amp circuit,

† This section can be omitted without loss of continuity.

$$A > 10{,}000$$

and

$$R_o \ll R_a \ll R_i$$

so the output voltage is approximated by

$$v_o = \frac{\dfrac{R_a + R_b}{R_a}}{1 + \dfrac{1}{A}\left(\dfrac{R_L + R_o}{R_L}\right)\left(\dfrac{R_a + R_b}{R_a}\right)} v_s$$

$$= \frac{\dfrac{R_L}{R_L + R_o} A}{\left(\dfrac{R_L}{R_L + R_o} A\right)\left(\dfrac{R_a}{R_a + R_b}\right) + 1} v_s$$

As expected, we see that the output voltage is independent of R_i.

We will now calculate the closed-loop output resistance R_o' of the noninverting amplifier circuit. If we let $R_L = \infty$, the last equation in Example 6.6 gives the open-circuit voltage:

$$v_{oc} = v_o = \frac{A}{A\beta + 1} v_s \tag{6.13}$$

Where β is the feedback ratio, as defined in Example 6.5. When we short-circuit the output,† we can neglect the contribution to the short-circuit current of any current through R_b. Therefore the short-circuit current will be

$$i_{sc} = i_o\big|_{v_o=0} = \left.\frac{A\,\Delta v}{R_o}\right|_{v_o=0} \tag{6.14}$$

Because $R_a \ll R_i$,

$$i_{sc} \simeq \frac{Av_s}{R_o} \tag{6.15}$$

Thus the closed-loop output resistance is

$$R_o' = \frac{v_{oc}}{i_{sc}} = \frac{A}{A\beta + 1} v_s \cdot \frac{R_o}{Av_s}$$

$$= \frac{1}{A\beta + 1} R_o \tag{6.16}$$

We always design for a closed-loop gain that is much less than the open-loop gain. (This is equivalent to the statement that the loop gain $A\beta$ is much greater than one.) Therefore the closed-loop output resistance is approximately

$$R_o' \simeq \frac{1}{\beta A} R_o \tag{6.17}$$

† Short-circuiting a physical op-amp circuit will drive the op amp into the nonlinear region, but this is not a problem for our model.

We see that the closed-loop output resistance R'_o is the already small open-loop output resistance R_o divided by the loop gain. This justifies the assumption of zero output resistance for the ideal op-amp model.

Although the closed-loop output resistance R'_o is very small (a fraction of an ohm), the maximum output current is also small. The output current is limited by saturation to a magnitude significantly less than V^+/R_o, where the open-loop output resistance R_o is typically 100 Ω or less.

Remember

The closed-loop output resistance of an op-amp circuit is approximately the open-loop output resistance of the op amp divided by the loop gain. If the open-loop gain is much greater than the closed-loop gain, the closed-loop output resistance of an op-amp circuit is approximately zero. The maximum available output current is limited by the open-loop output resistance, not by the closed-loop output resistance.

EXERCISES

7. For the noninverting amplifier circuit of Fig. 6.13, $A = 100$, $R_a = 1\ \text{k}\Omega$, $R_b = 9\ \text{k}\Omega$, and $R_L = 100\ \Omega$.
   (a) Use node-voltage equations to calculate the closed-loop gain and the closed-loop output resistance seen by R_L, if $R_i = 10\ \text{k}\Omega$ and $R_o = 100\ \Omega$.
   answer: 8.203, 9.8 Ω
   (b) Use $R_i = \infty$ and $R_o = 0$ and the voltage-divider relation to calculate β. Then use Eqs. (6.13) and (6.16) to calculate the closed-loop gain and the closed-loop output resistance. Compare your results to that obtained in part (a). (The loop gain is rather small, so your approximation may not be very accurate.)
   answer: 9.091, 9.09 Ω
   (c) Repeat part (b) with the use of Eqs. (6.12) and (6.17). *answer:* 10, 10 Ω

6.6 Summary

We first described a linear model for an op amp and used the assumption of infinite open-loop gain to justify the ideal op-amp component. We analyzed four simple, but important, circuits with the aid of the ideal op amp. The assumptions that the error voltage Δv and input current i_i are zero let us adapt analysis by node voltages to circuits with ideal op amps. We closed the chapter with two sections where we investigated our ideal op-amp assumption. We concluded that the ideal op-amp assumption yields accurate results, if the loop gain is much greater than one. The loop gain is much greater than one if the open-loop gain of the op amp is much greater than the closed-loop gain of the op-amp circuit.

KEY FACTS

Concept	Equation	Section	Page
❑ An op amp is linear over a limited range of output voltages and currents. We restrict our analysis to linear operation.		6.1	176
❑ The voltage gain A of an op amp is typically greater than 100,000.		6.1	177
❑ For an ideal op amp we assume that the gain A is infinite, so the error voltage is $\Delta v = v_+ - v_- = 0$, and the input current is $i_i = 0$.	(6.7) (6.8)	6.1	177

KEY FACTS	Concept	Equation	Section	Page
❑	We always use op amps with negative feedback for linear operation.		6.2	178
❑	Negative feedback requires feedback to the inverting terminal [(−) terminal].		6.2	181
❑	We use op amps to		6.2	178
	Multiply voltages by a positive constant (Example 6.1),			179
	Multiply voltages by a negative constant (Example 6.2),			181
	Integrate voltages (Example 6.3),			182
	Add two voltages (Exercise 2), and			
	Subtract two voltages (Exercise 3).			
❑	To write the node-voltage equations for an ideal op-amp circuit, we use $\Delta v = 0$, $i_i = 0$, and the fact that the output terminal is connected to a voltage-controlled voltage source with infinite gain (Example 6.4).		6.3	183
❑	The assumption of infinite open-loop gain A is reasonable, if the open-loop gain A is much greater than the closed-loop gain v_o/v_s (Example 6.5).		6.4	186
❑	The assumption of zero output resistance is reasonable if the open-loop gain A is much greater than the closed-loop gain v_o/v_s.	(6.16)	6.5	189
❑	The maximum output current of an op amp is limited by the supply voltages and the open-loop output resistance R_o, not by the closed-loop output resistance R_o'.		6.5	190

PROBLEMS

Section 6.1

1. We must measure the parameters of an op amp to determine the model as shown in Fig. 6.4. The op amp is connected to a power source as shown in Fig. 6.2. The output voltage and current are measured to be zero when both input terminals are connected to the reference. With an error voltage Δv of 1 mV, the open-circuit output voltage v_o is 5 V, and the input current is 0.1 mA. Connection of the output terminal to the reference through a resistance value of 100 Ω reduces the output voltage to 4 V. Determine the values R_i, R_o, and A for our op-amp model. Is this a practical method to measure the op-amp parameters when the open-loop gain is 100,000 or more?

Section 6.2

2. Power is supplied to the op amp in the following noninverting amplifier circuit by $V^+ = 15$ V and

$V^- = -15$ V. If $R_a = 1$ kΩ, $R_b = 19$ kΩ and $R_L = 1$ kΩ, determine the output voltages for (a) $v_i = 0.1$ V, (b) $v_i = 0.5$ V, (c) $v_i = 1$ V and, (d) $v_i = 2$ V. (Check for saturation.)

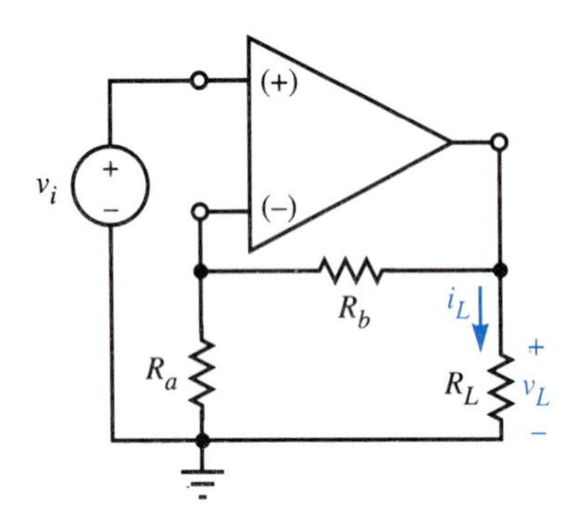

3. In the preceding op-amp circuit $R_a = 10$ kΩ, $R_b = 100$ kΩ, and $R_L = 1$ kΩ. Calculate (a) the voltage gain v_L/v_i, (b) the current gain i_L/i_i, and (c) the power gain if the op amp is ideal.

4. Power is supplied to the following inverting amplifier circuit by $V^+ = 15$ V and $V^- = -15$ V. The resistance values are $R_a = 2$ kΩ, $R_b = 20$ kΩ, and $R_L = 1$ kΩ. Determine the output voltages for (a) $v_i = 0.1$ V, (b) $v_i = 0.5$ V, (c) $v_i = 1$ V, (d) $v_i = 2$ V, and (e) $v_i = 4$ V. (Check for saturation.)

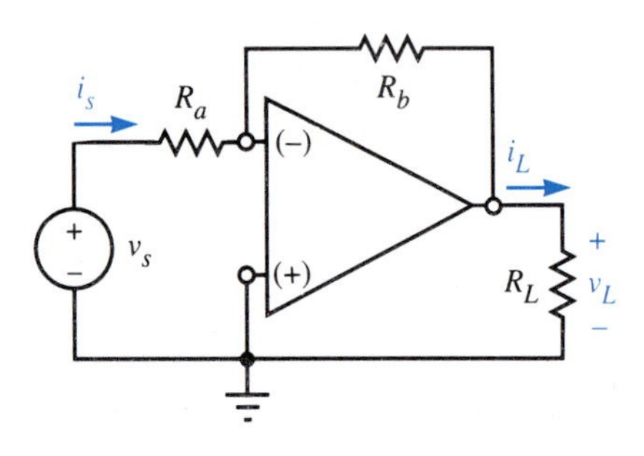

5. In the preceding op-amp circuit the resistance values are $R_a = 100$ kΩ, $R_b = 10$ kΩ, and $R_L = 1$ kΩ. Calculate (a) the voltage gain v_L/v_s, (b) the current gain i_L/i_s, and (c) the power gain of the amplifier if the op amp is ideal. Why might we be interested in an amplifier with a voltage gain less than one?

6. In the following network, replace the circuit to the left of terminals *a* and *b* with its Thévenin equivalent circuit, and calculate the power absorbed by the 2-kΩ resistance.

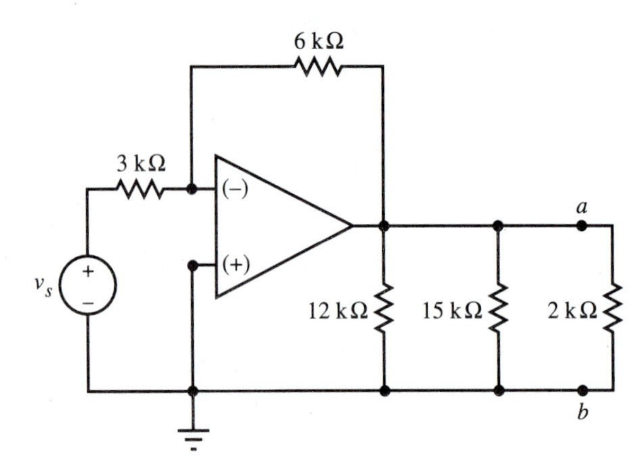

7. The inverting amplifier circuit shown below is used in an electronic instrument, called a digital multimeter, for measuring resistance. The constant voltage V_{ref} and the resistance R_{ref} are known. The voltage V_o is a measure of the unknown resistance R_x. Determine V_o as a function of V_{ref}, R_{ref}, and R_x.

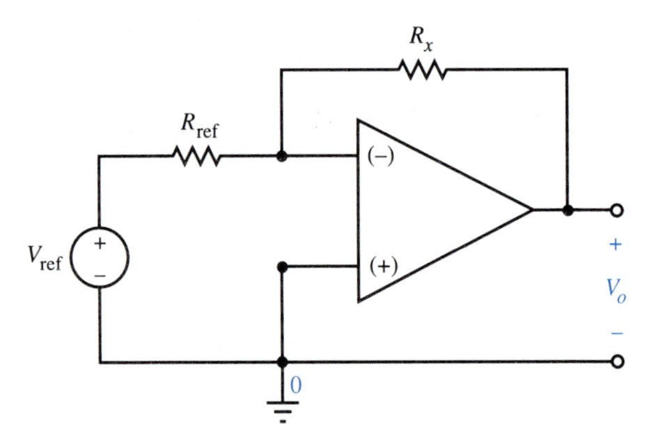

8. For the following circuit, determine the output voltage v_o as a function of the input voltage v_s.

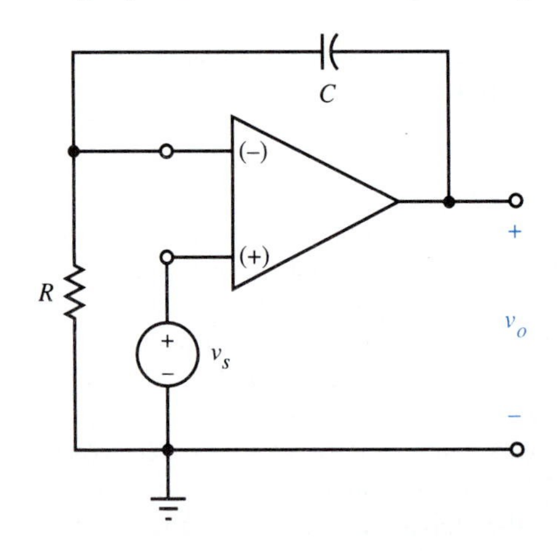

9. This problem investigates the effect of nonzero output resistance R_o on the closed-loop performance of a loaded op amp. To simulate this problem, we have connected a 100-Ω resistance to the output terminal of the ideal op amp and in front of the feedback point.
   (a) Short-circuit the 100-Ω resistance and calculate the closed-loop voltage gain v_o/v_s (assume linear operation). If the supply voltages are ±15 V, the maximum output voltage is limited by saturation. Determine an upper bound on the load voltage v_o for the 200-Ω load. What is the available short-circuit current?
   (b) Remove the short circuit that was placed across the 100-Ω resistance. Now repeat part (a) (Neglect the current through the feedback circuit when calculating the bound on the output voltage and the short-circuit current). What seems to be the principal effect of a nonzero output resistance on the closed-loop performance of an op-amp circuit?

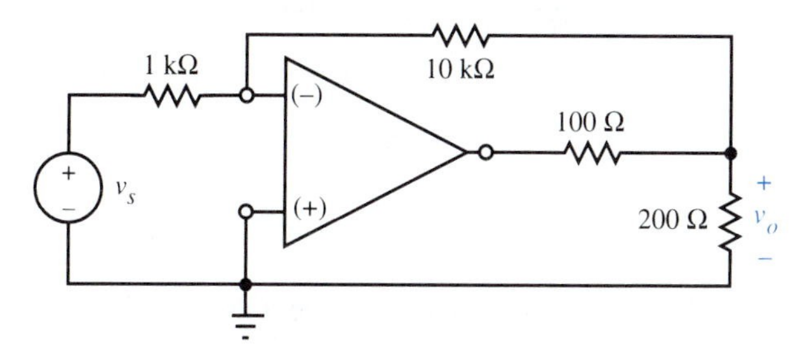

10. Use the ideal op-amp model to find the output current i for the following circuit. This is a *voltage-to-current converter.* Draw an equivalent circuit that consists of the independent source, a dependent source, a resistance, and the load device. Can you think of a practical limitation for the actual op-amp circuit?

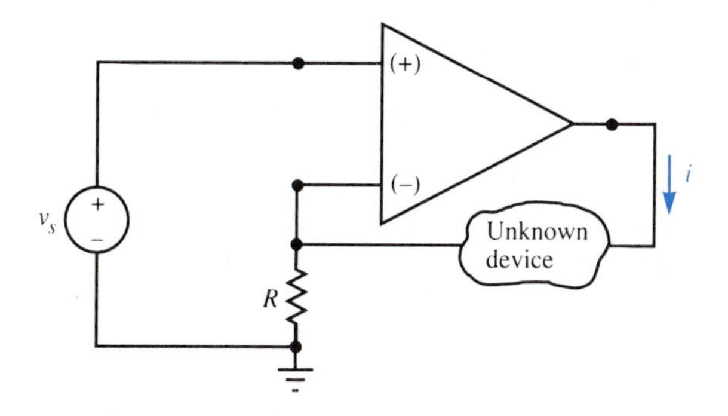

11. Use the ideal op-amp model to determine v_o as a function of v_s in the following circuit.

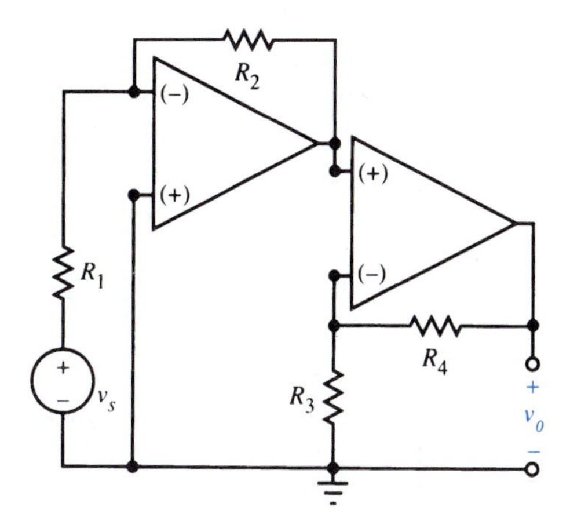

12. Use an ideal op amp and two resistances to design an amplifier with an input resistance of 1 kΩ and a voltage gain of −5. *Hint:* See Example 6.2.

13. Use an ideal op amp and two resistances to design a noninverting amplifier. The feedback circuit should draw 0.1 mA when the output voltage is 1 V, and the voltage gain must be 5. *Hint:* See Example 6.1.

14. Use an ideal op amp and resistances to design a circuit that performs the algebraic function $v_o = -5(v_1 + v_2)$. The input resistance seen by each source must be 2 kΩ. *Hint:* See Exercise 2.

15. Use an ideal op amp and resistances to design a circuit that performs the algebraic function $v_o = v_2 - v_1$. The input resistance seen by each source must be 10 kΩ. *Hint:* See Exercise 3.

16. Use an ideal op amp, one resistance, and one capacitance to design a circuit with an output that is minus one-tenth times the integral of the input. The input resistance must be 100 kΩ. *Hint:* See Example 6.3.

17. Use one ideal op amp and four resistances to design a circuit that performs the algebraic function $v_o = -(v_1 + v_2 + v_3)$. The input resistance for each source must be 20 kΩ. *Hint:* Add to the circuit of Exercise 2.

Section 6.3

18. Use the ideal op-amp model and analysis by node voltages to determine the output of the *difference amplifier* shown below.

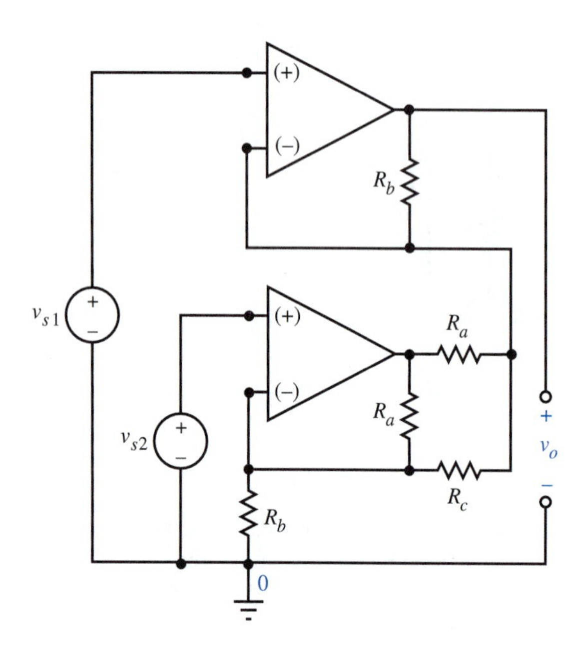

19. The following circuit is used in instrumentation. Resistances R_1 through R_4 are connected in a *bridge*. A change in value of one of the resistances changes the output voltage V_o. Use node-voltage equations to determine V_o. What relation must exist between the resistances in the bridge for the output voltage to be zero?

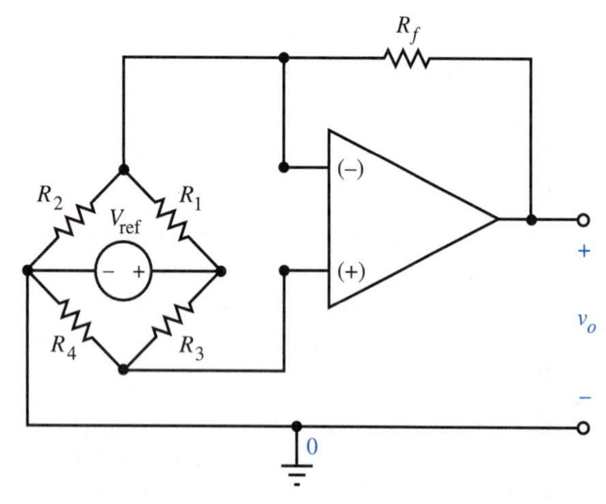

20. Write a set of node-voltage equations for the following two networks. (Do not solve the equations.)

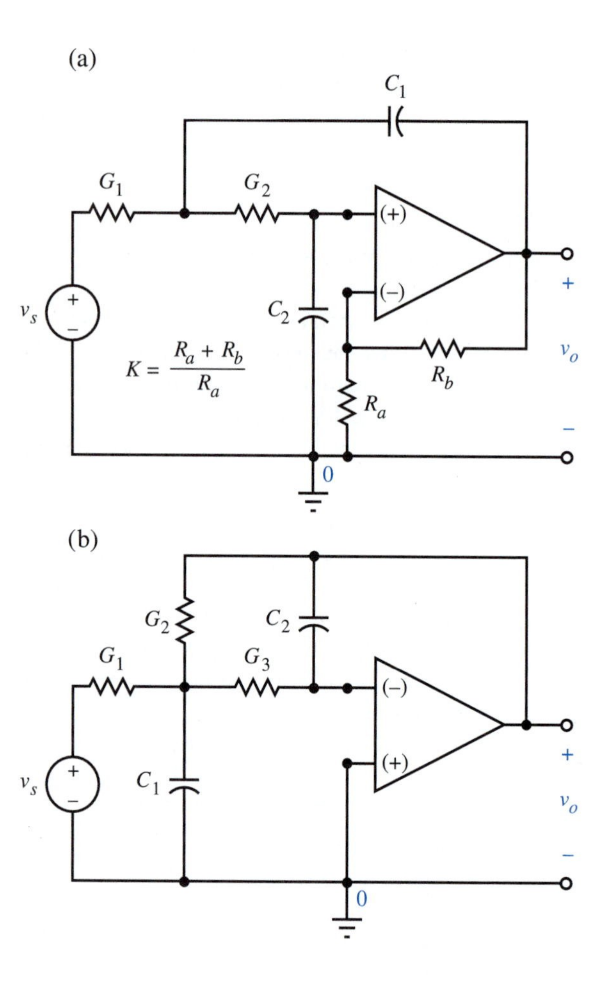

21. Solve for node voltage v_4 in the following circuit.

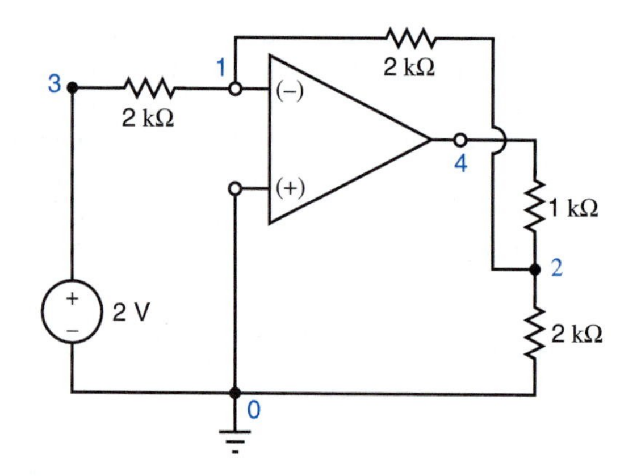

Section 6.4

22. Use the finite-gain op-amp model in the following circuit.

    (a) If component X is a capacitance of value C, write a differential equation that relates the output voltage v_o to the input voltage v_s. What function does this circuit perform as the open-loop voltage gain A becomes infinite?

(b) Repeat part (a) if component X is an inductance of value L. (This is not a practical circuit, because it is very sensitive to electrical noise.)

(c) Box X represents a resistance of value R_b in parallel with a capacitance of value C. Write a differential equation that relates v_o to v_s. Determine the differential equation as the open-loop gain A becomes infinite.

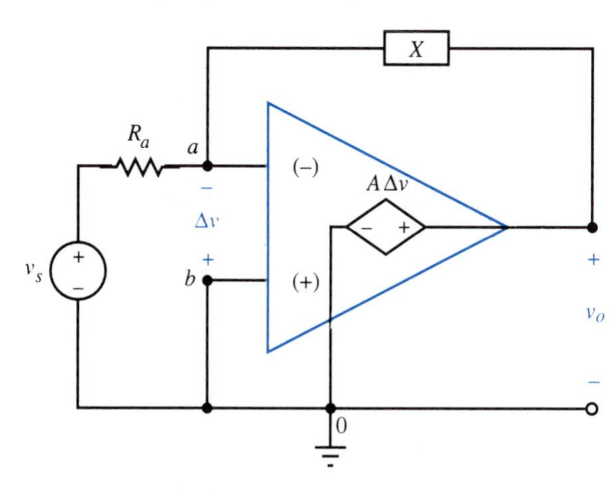

23. Use finite-gain op-amp models to determine the output voltage for the following circuit.

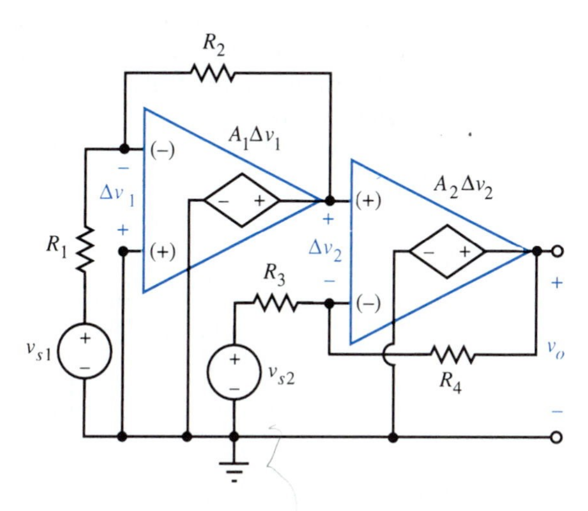

Section 6.5

24. For the inverting amplifier shown below, $R_a = 1\ \text{k}\Omega$, $R_b = 10\ \text{k}\Omega$, $R_i = 10\ \text{k}\Omega$, and $R_o = R_L = 100\ \Omega$.

(a) If $A = 100$, determine the closed-loop voltage gain $G = v_o/v_s$, the closed-loop input resistance $R_i = v_s/i_s$, and the closed-loop output resistance R_o'. Is the ideal op-amp approximation very close for this circuit?

(b) Repeat part (a) if $A = 1000$. Is the ideal op-amp approximation very close for this circuit?

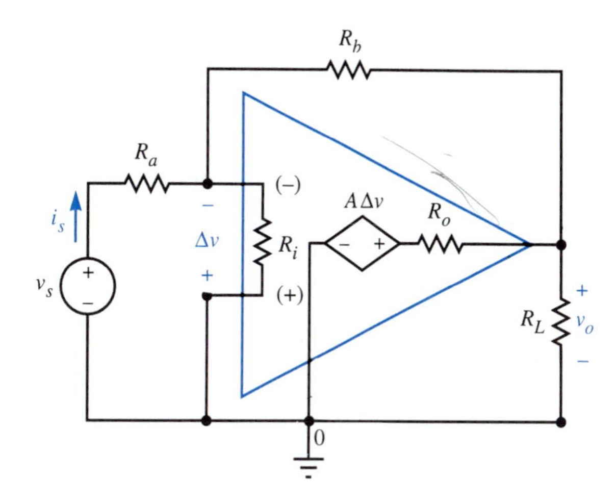

Supplementary Problems

25. Determine the output voltage in the following op-amp circuit. If the input voltage is increased to 1 V, and $V^+ = 15$ V and $V^- = -15$ V, what will be the value of v_o?

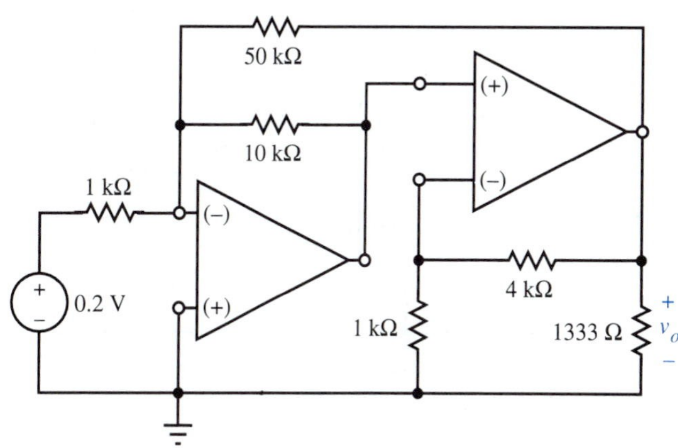

26. Design a difference amplifier to have an output of

$$v_o = 2(v_{s2} - v_{s1})$$

The resistance seen by each source must be 1 kΩ. Determine the values of R_1, R_2, R_a, and R_b. (See Example 6.4.)

27. Design a circuit with a single op amp that will have an output voltage that is directly propor-

tional to an unknown conductance G. A reference voltage of 1 V is available, and the circuit must provide an output voltage with a magnitude of 1 V (± 1 V) when the conductance has a value of 1 mS.

28. The following circuit is a current-to-voltage converter. Select the resistance R so that the output is 0.1V per μA of input current i_s.

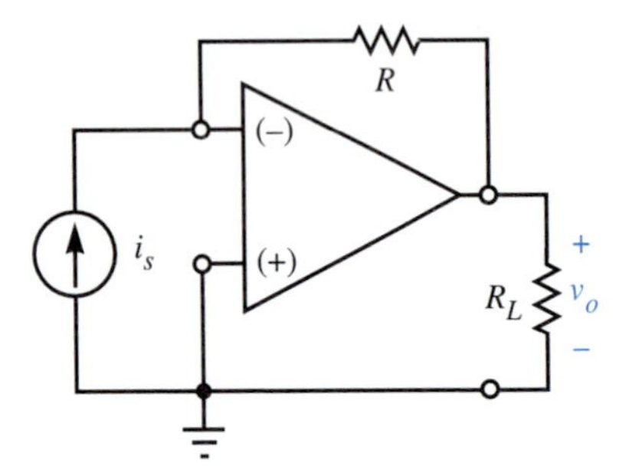

29. Apply the ideal op-amp model and determine the closed-loop gain $G = v_o/v_s$ of the following circuit. (Observe the relative location of the inverting and noninverting terminals.) Although your analysis has yielded an answer, this result is wrong. The op amp will not function properly in this mode. To obtain an insight into the problem, work the following problem.

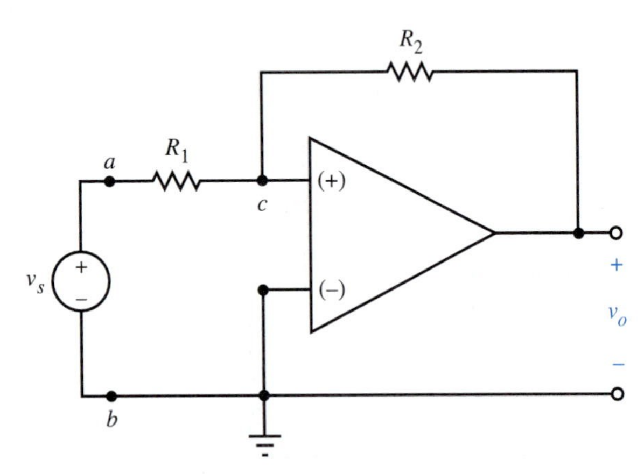

30. Use the finite-gain op-amp model of Section 6.4 to calculate the Thévenin resistance looking into terminals a and b (with the independent source disconnected) of the above circuit. Then calculate the Thévenin resistance looking into terminals c and b with the source connected. For what values of A is either of these resistances negative? We shall see in Chapter 8 that a negative resistance can lead to an unstable circuit (one that has an output that increases with time with zero input).

31. The following refers to the op-amp circuit shown below.

(a) Determine the equivalent circuit looking into terminals a and b when component X is a resistance of value R.

(b) Determine the equivalent circuit looking into terminals a and b if component X is a capacitance of value C.

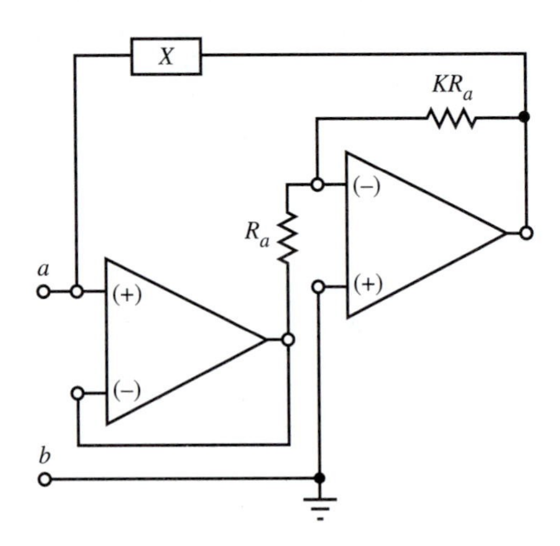

32. The following circuit can be used in coulombmetric titration. A chemical reaction takes place in the cell at a rate proportional to the current through the reaction cell. The voltage across the cell depends on the particular reaction. When these reactants are exhausted, the cell voltage rises, which stops a timer. Determine the value of R that gives a cell current i_c of 10 μA for an input voltage of -1 V.

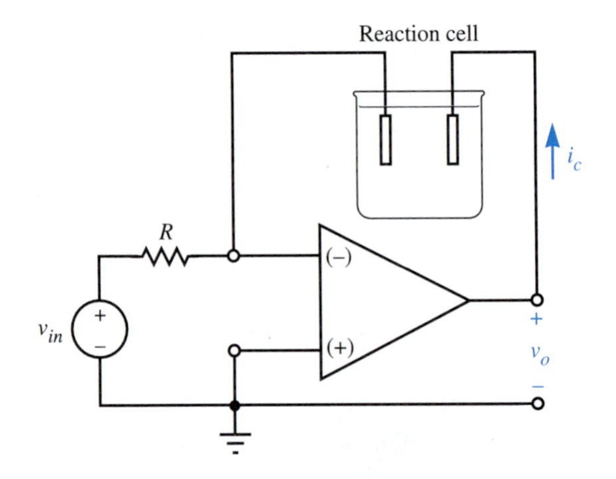

33. Determine the output voltage for the following circuit. What is the input resistance seen by each source? What function does the circuit perform?

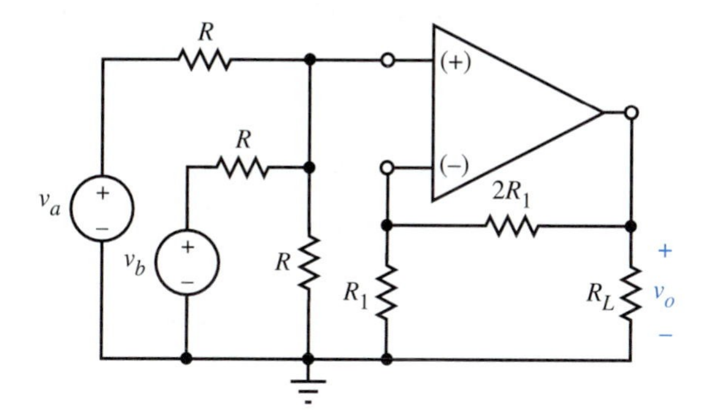

34. Use two op amps to design a circuit that meets the following specifications.
    (i) The output voltage must be +10 times the integral of the input voltage.
    (ii) The input resistance must be 10 kΩ.
    (iii) The magnitude of the output voltage of each op amp must be the same at all times, so that one op amp will not saturate before the other.
    (iv) All resistance values must be between 1 kΩ and 100 kΩ.
35. For the following circuit, show that i is approximately $-v_s/R$, if $R_L \ll R_1$. (A small value of capacitance in parallel with R_2 is needed to stabilize a practical circuit.) See Fig. 6.14.
36. For each of the following circuits, use the ideal op amp to determine the functional relationship between voltage v and current i,

$$v = f(i)$$

for the network, if the functional relationship between v_L and i_L is

$$v_L = g(i_L)$$

or
$$i_L = g^{-1}(v_L)$$

What is the equivalent circuit connected between terminals a and b if the load is
    (i) A resistance of value R?
    (ii) An inductance of value L?
    (iii) A capacitance of value C?

(a) a *negative converter*

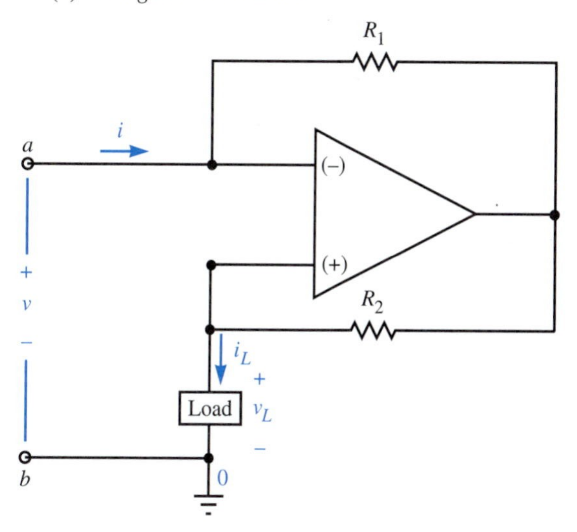

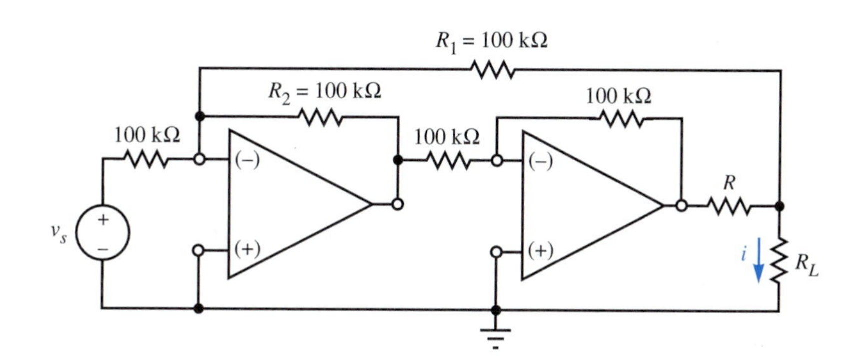

FIGURE 6.14

(b) a *gyrator*

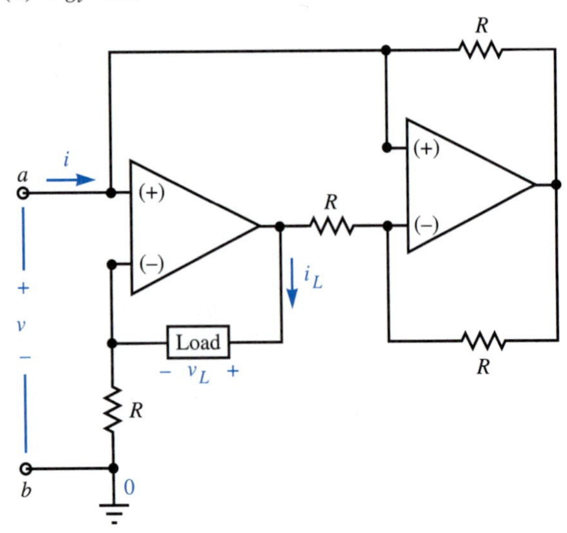

(c)

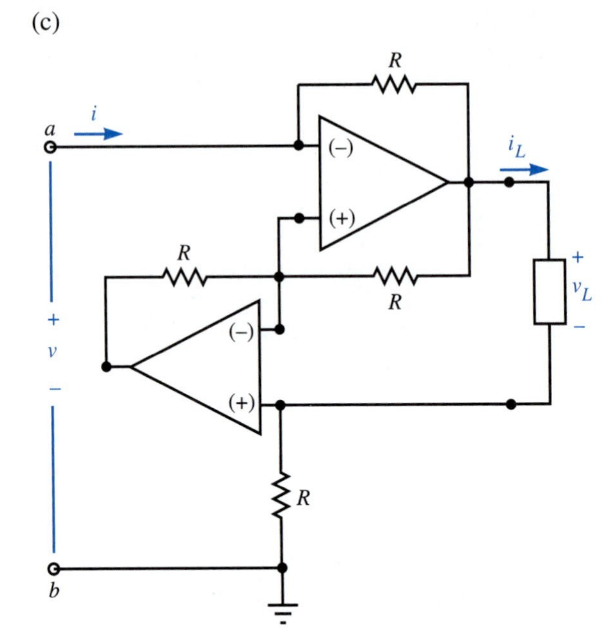

7

Signal Models

Electrical engineers often use the term *signal* or *waveform* to denote a voltage $v(t)$ or current $i(t)$ of interest in a physical circuit. Just as there is a mathematical model for the physical circuit the engineer is inspecting, so there are mathematical models for the signals in the physical circuit. We find it convenient to characterize these mathematical signal models in various ways. This chapter describes several signal models and signal characterizations that are useful in electrical engineering. We will use these signal models extensively in the remainder of the text.

7.1 DC and Related Signals

A *dc* voltage or current is one that is constant for all time, as illustrated in Fig. 7.1. (The term *dc* is applied to both voltage and current in spite of the fact that it is an abbreviation for *direct current.*)

DEFINITION DC Signal

$$v(t) = V_s \qquad -\infty < t < \infty \tag{7.1}$$

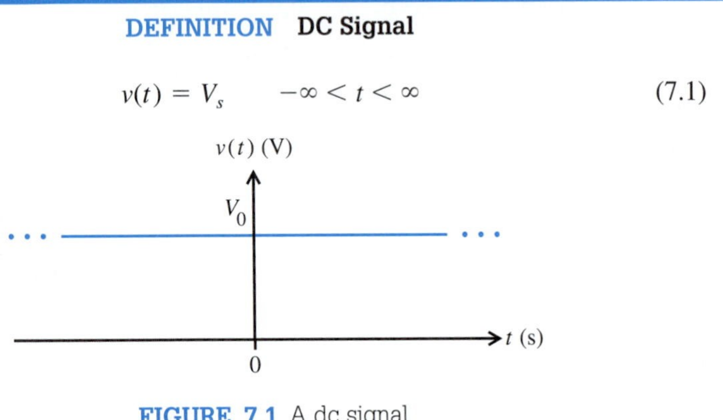

FIGURE 7.1 A dc signal

A dc voltage source of 9 V and a series resistance form an approximate model for a 9-V transistor radio battery. A dc source is obviously not an exact model for any physical device. No physical voltage or current has been constant since the beginning of time. The unit step, discussed in the following paragraph, lets us include a starting time.

The Unit Step

A waveform closely related to dc is one that is zero for $t < t_0$, and suddenly becomes some nonzero value for $t > t_0$. We describe signals of this type by use of the *unit step function,* depicted in Fig. 7.2.

DEFINITION Unit Step Function

$$u(t) = \begin{cases} 0 & t < 0 \\ 1 & t > 0 \end{cases} \tag{7.2}$$

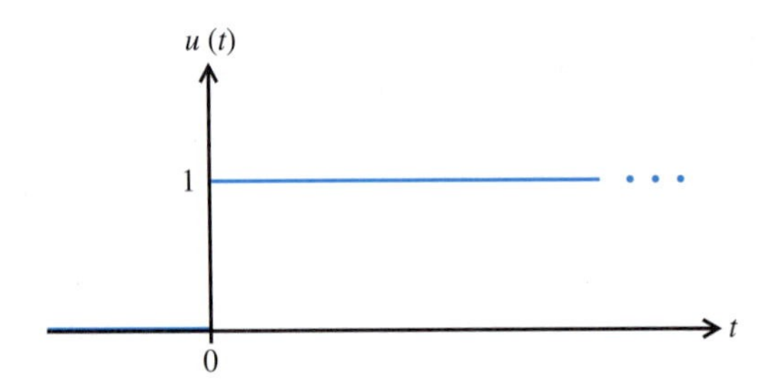

FIGURE 7.2 The unit step function

[The value of $u(0)$ is variously defined to be 0, $\frac{1}{2}$, or 1, or it is left undefined. The only restriction that we need is that $|u(0)| < \infty$. The value of $u(0)$ will not influence any results obtained in this book or the analysis of any physical system.] Observe that the unit step is zero when its argument is negative, and unity when the argument is positive. The transition, or step, occurs when the argument of the unit step changes from negative to positive.

When we use $u(t)$ to write a voltage or current, we are referring to the unit step function.

EXAMPLE 7.1 Write the signal

$$v_s(t) = \begin{cases} 0 & t < t_0 \\ V_S & t > t_0 \end{cases}$$

in terms of the unit step and sketch $v_s(t)$.

Solution The transition occurs for $t = t_0$, so the argument of the unit step must be negative for $t < t_0$, 0 for $t = t_0$, and positive for $t > 0$. Therefore

$$v_s(t) = V_s u(t - t_0) \text{ V}$$

This signal is sketched in Fig. 7.3.

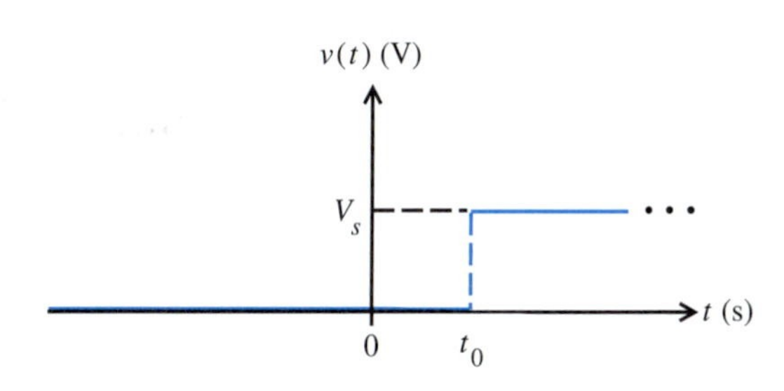

FIGURE 7.3 A time-shifted step function

EXAMPLE 7.2 The *unit pulse function*

$$\Pi(t) = \begin{cases} 1 & |t| < \dfrac{1}{2} \\ 0 & |t| > \dfrac{1}{2} \end{cases}$$

is widely used in communication systems analysis. Represent the unit pulse in terms of unit step functions.

Solution We can write the unit pulse as the superposition of two unit step functions:

$$\Pi(t) = u\left(t + \frac{1}{2}\right) - u\left(t - \frac{1}{2}\right)$$

We can visualize this result by referring to Fig. 7.4.

Observe that the unit pulse function is centered about the point where its argument is zero, and the function is zero if the magnitude of the argument is greater than one-half. For example, $\Pi[(t - t_0)/T]$ is a pulse of width T that is centered at $t = t_0$.

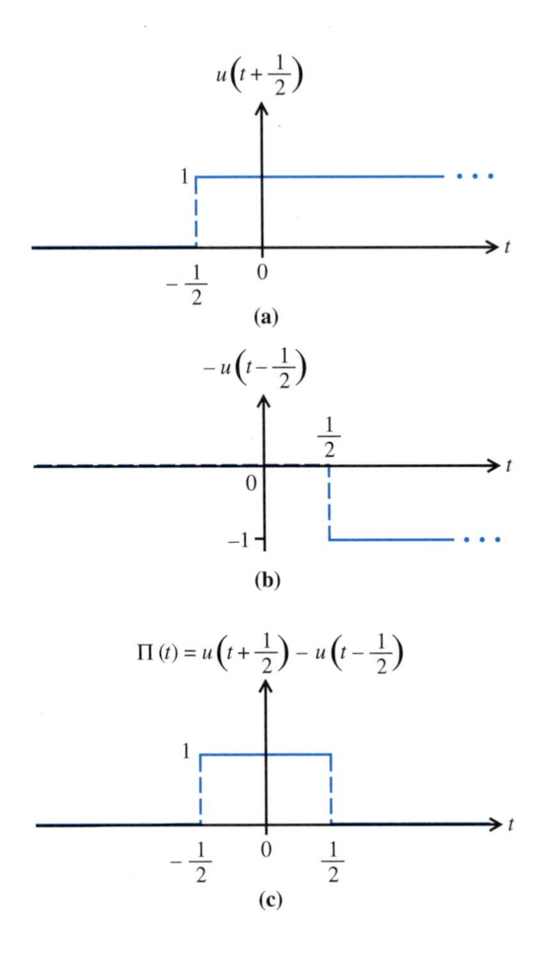

FIGURE 7.4 The unit pulse function written as a superposition of step functions. (*a*) Unit step advanced by $\frac{1}{2}$ s; (*b*) the negative of a unit step that is delayed by $\frac{1}{2}$ s; (*c*) the sum of the unit steps in *a* and *b*

We can model many other waveforms occurring in physical systems with the aid of appropriately scaled and time-shifted unit step functions.

EXAMPLE 7.3 Write the following signal with the use of the unit step function.

$$i(t) = \begin{cases} 0 & t < 0 \\ I_m \cos \omega_p t \text{ A} & t > 0 \end{cases}$$

Solution The unit step is zero for $t < 0$, and one for $t > 0$. Zero times the cosine is zero, and one times the cosine is just the cosine. Therefore,

$$i(t) = (I_m \cos \omega_p t)u(t) \text{ A}$$

[We will not concern ourselves with the value of $i(0)$ where the discontinuity occurs.]

EXAMPLE 7.4 The voltage across an inductance of L henries is given by

$$v = \begin{cases} 0 \text{ V} & t < 0 \text{ s} \\ V_s \text{ V} & t \geq 0 \text{ s} \end{cases}$$

Write the voltage v and the inductance current i in terms of the unit step (assume the passive sign convention).

Solution We can easily write voltage v in terms of the unit step:

$$v = V_s u(t) \text{ V}$$

The inductance current is the integral of the inductance voltage divided by the inductance:

$$i = \frac{1}{L}\int_{-\infty}^{t} v(\lambda)\, d\lambda = \frac{1}{L} V_s \int_{-\infty}^{t} u(\lambda)\, d\lambda$$

$$= \frac{1}{L} V_s t u(t) \text{ A}$$

We can visualize the preceding integral by sketching the unit step $u(\lambda)$ as shown in Fig. 7.5. Select some t on the λ axis. The area under the unit step to the left of time t represents the value of the integral. This area is zero for $t < 0$, and is represented by the shaded area shown in Fig. 7.5 for $t > 0$.

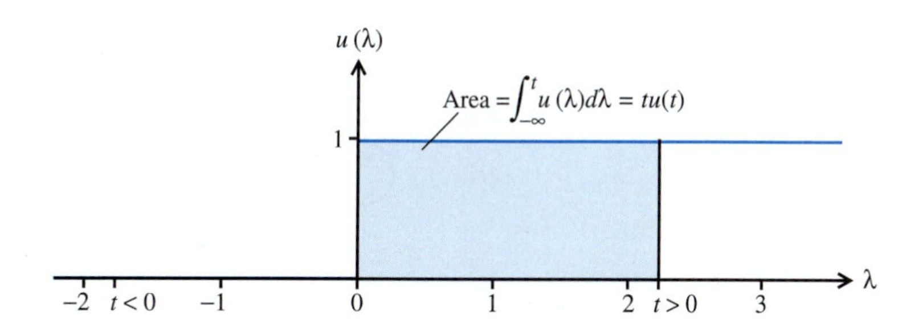

FIGURE 7.5 The integral of a unit step represented as the area under a curve

We often define the integral of the unit step to be the *unit ramp function:*

$$r(t) = tu(t) \tag{7.3}$$

The unit ramp is sketched in Fig. 7.6. The unit ramp is widely used in control system analysis, where, for example, a shaft angle can increase linearly with time as the shaft rotates.

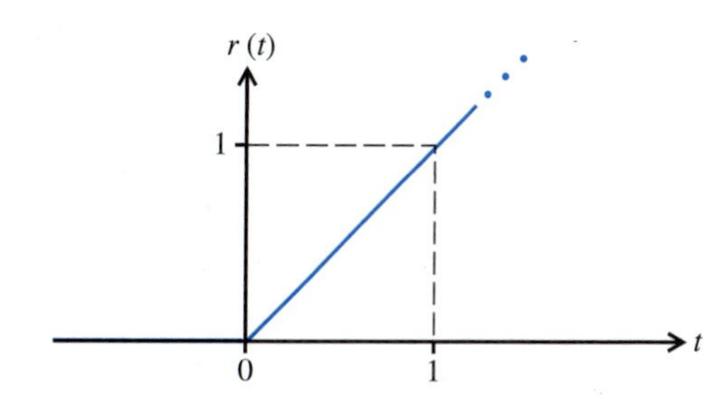

FIGURE 7.6 The unit ramp function

We often use the unit step to model a source that is switched on or off at some time t. Although this representation is useful, we should apply it with care. For example, suppose that an engineer proposed the network model of Fig. 7.8a as a model for the network of Fig. 7.7a. Although the proposal may seem accurate at first glance, it contains a trap. The Thévenin equivalent for the two networks is the same for $t > t_0$, but different for $t < t_0$.

Like our network components, the signals we have introduced *approximately*

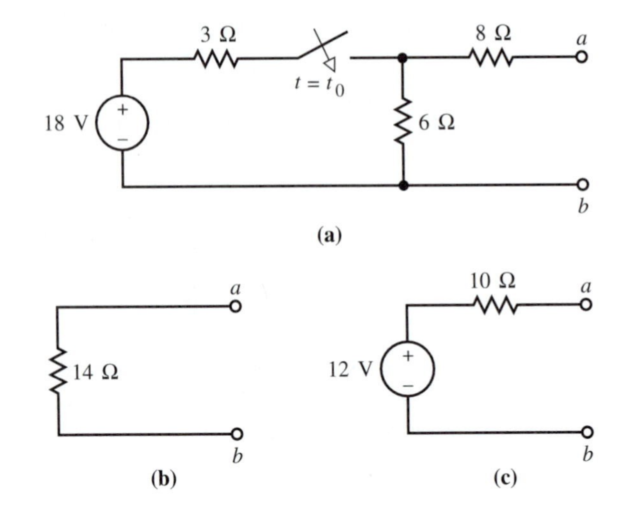

FIGURE 7.7
(*a*) Network with a switched source; (*b*) Thévenin equivalent for $t < t_0$; (*c*) Thévenin equivalent for $t > t_0$

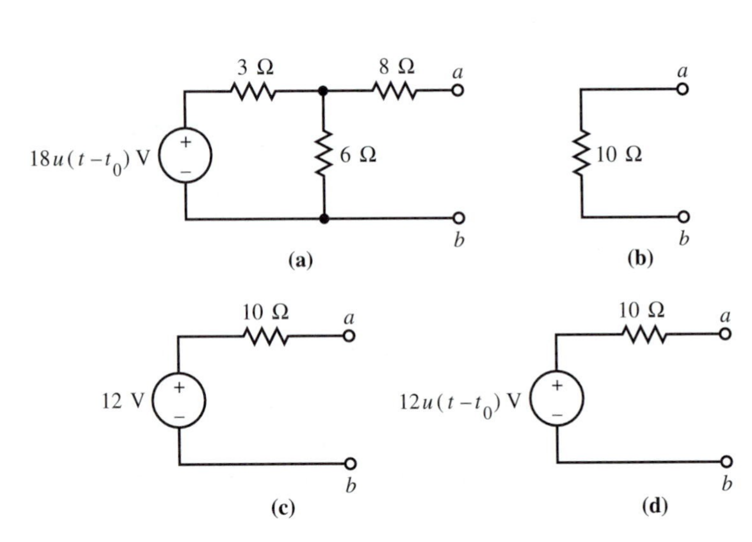

FIGURE 7.8
A unit step model for the switched source of Fig. 7.4a. (*a*) Unit step model; (*b*) Thévenin equivalent for $t < t_0$; (*c*) Thévenin equivalent for $t > t_0$; (*d*) Thévenin equivalent for all t

model physical quantities. The dc waveform is a model, because no physical signal is constant for all time. The unit step and the unit pulse are models, because no physical voltage or current can change its value instantaneously, nor can it be precisely constant for any time interval. The unit ramp is a model, because no physical voltage or current can grow precisely linearly or without bound.

Remember

A dc signal is constant for all time. A unit step is 0 when the argument is negative and has a value of 1 when the argument is positive.

EXERCISES

1. Accurately sketch the following signals and write them in terms of the unit step.

(a) $i(t) = \begin{cases} 0 & t < 0 \\ 12 \text{ A} & t > 0 \end{cases}$

answer: $12u(t)$ A

(b) $v(t) = \begin{cases} 0 & t < 0 \\ 6\text{ V} & 0 < t < 4\text{ s} \\ 0 & t > 4\text{ s} \end{cases}$ answer: $6u(t) - 6u(t-4)$ V

(c) $v(t) = \begin{cases} 0 & t < 0 \\ 4\sin t\text{ V} & t > 0 \end{cases}$ answer: $(4\sin t)u(t)$ V

(d) $i(t) = \begin{cases} 5\cos t\text{ A} & t < 0 \\ 0 & t > 0 \end{cases}$ answer: $(5\cos t)u(-t)$ A

2. Write the following signals with the use of the unit step function.

(a)

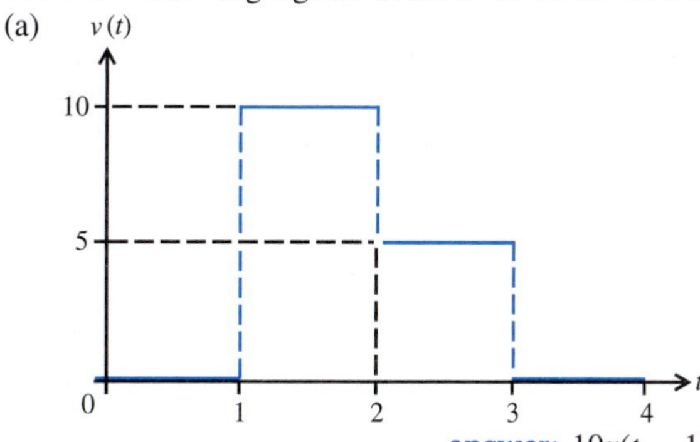

answer: $10u(t-1) - 5u(t-2) - 5u(t-3)$ V

(b)

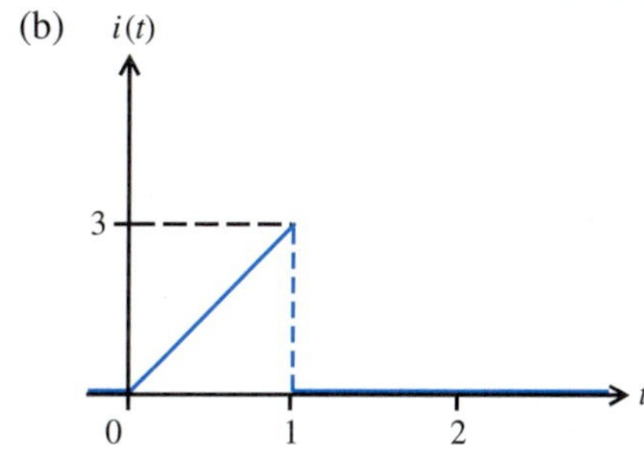

answer: $3t[u(t) - u(t-1)]$ A

3. Accurately sketch the following signals.

(a) $3u(t-2)$ (b) $2u(2-t)$

(c) $u(t) + 2u(t-1) - u(t-3)$ (d) $tu(t-1)$

(e) $2(t-1)u(t-1)$ (f) $2u(\sin \pi t)$

4. The following currents are applied to a 2-μF capacitance. If the capacitance voltage is zero for all time less than zero, write the voltage across the capacitance in terms of the unit step function, and sketch the result for $-1 \leq t \leq 4$ s.

(a) $6u(t)$ μA answer: $3tu(t)$ V

(b) $8[u(t) - u(t-1)]$ μA answer: $4tu(t) - 4(t-1)u(t-1)$ V

7.2 Delta Functions†

In later chapters we will need to consider pulses that have a duration much shorter than the natural response time of our circuit (a mechanical analogy is a hammer blow that rings a bell). We will idealize a very short pulse with the *unit impulse* or *delta function,* $\delta(t)$. We will also see that the impulse provides a convenient way to establish an initial capacitance voltage or inductance current.

† This section can be postponed until Chapter 14.

DEFINITION Unit Impulse or Delta Function

The unit impulse or delta function $\delta(t)$ is defined by the integral relation

$$\int_{-\infty}^{\infty} f(t)\delta(t)\,dt = f(0) \tag{7.4}$$

for every function $f(t)$ that is defined and continuous at $t = 0$.

The theory of the delta function is developed in a branch of mathematics known as the theory of distributions, but we can justify the properties of a delta function by treating $\delta(t)$ as a very short duration pulse of unit area. To justify the idea of the delta function being a model for a very short duration pulse of unit area we consider the unit area pulse $(1/T)\Pi(t/T)$ shown in Fig. 7.9a. It is easy to see that if T is small enough, Eq. (7.4) is approximated by

$$\int_{-\infty}^{\infty} \frac{1}{T}\Pi\left(\frac{1}{T}\right) f(t)\,dt = \int_{-T/2}^{T/2} \frac{1}{T} f(t)\,dt \tag{7.5a}$$

$$\simeq \int_{-T/2}^{T/2} \frac{1}{T} f(0)\,dt = \frac{1}{T} f(0)t \Bigg|_{-T/2}^{T/2} = f(0) \tag{7.5b}$$

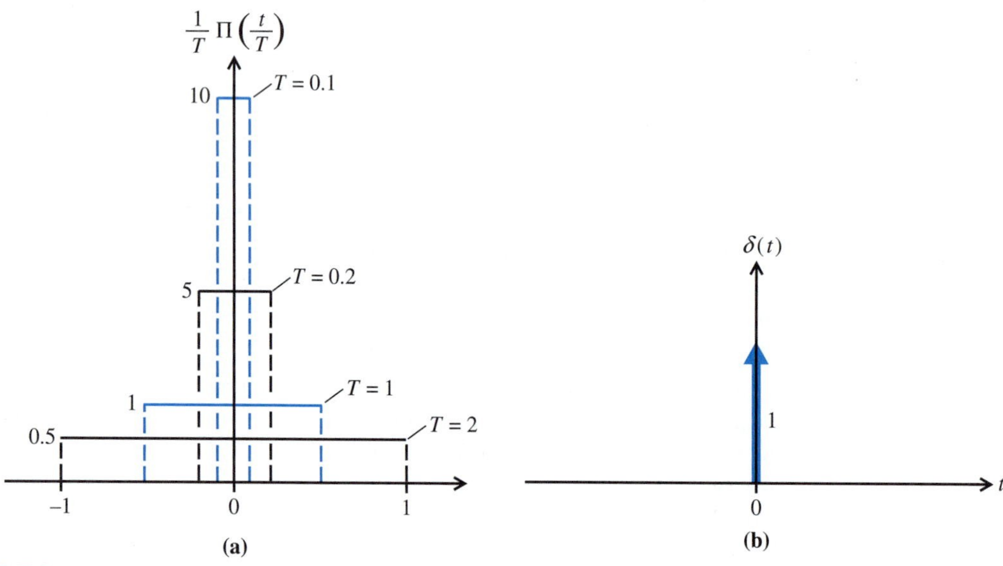

FIGURE 7.9
Approximation of the impulse or delta function by a rectangular pulse.
(*a*) Pulses of unit area; (*b*) the impulse or delta function.

Notice that the step from Eq. (7.5a) to (7.5b) is valid whenever $f(t)$ is defined and continuous at $t = 0$, because under these conditions $f(t \pm (T/2)) \simeq f(t)$ for $T \simeq 0$. As long as the pulse $(1/T)\Pi(t/T)$ is so short that $f(t)$ is approximately constant for the pulse duration, a unit area pulse approximates an impulse. In the limit $T \rightarrow 0$, the step from Eq. (7.5a) to (7.5b) holds with equality. This limiting argument gives an alternative interpretation of the delta function:

$$\int_{-\infty}^{\infty} \delta(t)\, dt = 1 \tag{7.6}$$

and

$$\delta(t) = 0 \qquad t \neq 0 \tag{7.7}$$

The delta function is represented graphically as shown in Fig. 7.6b. We will use delta functions extensively to model short-duration pulses, after we introduce transforms in Chapter 14.

Properties of the Delta Function

Several useful properties of the delta function are easily established. The delta function $\delta(t - t_0)$ is nonzero only for $t = t_0$. When we integrate the product of $\delta(t - t_0)$ and a function $f(t)$, we obtain only the value of $f(t)$ at time t_0. This *sifts out* the value of $f(t)$ at $t = t_0$ or *samples* $f(t)$ at time t_0. A simple change of variables ($\lambda = t - t_0$) lets us establish this *sifting* or *sampling property* of the unit impulse:

DEFINITION Sifting Property of the Unit Impulse

$$\int_{-\infty}^{\infty} f(t)\delta(t - t_0)\, dt = \int_{-\infty}^{\infty} f(\lambda + t_0)\delta(\lambda)\, d\lambda \tag{7.8a}$$

$$= f(t_0) \tag{7.8b}$$

It can be shown that Eq. (7.4) implies that the impulse function is an even function:

$$\delta(t_0 - t) = \delta(t - t_0) \tag{7.9}$$

The delta function is often considered to be the derivative of the unit step,

$$\delta(t) = \frac{d}{dt} u(t) \tag{7.10}$$

If we view the unit impulse as a pulse of zero width and unit area, we can easily see that

$$\int_{-\infty}^{t} \delta(\lambda)\, d\lambda = \begin{cases} 0 & t < 0 \\ 1 & t > 0 \end{cases} = u(t) \tag{7.11}$$

We will use the delta function in Chapter 14 as an idealized model for physical signals. In some problems the use of the delta function will significantly simplify our work, compared with the use of a more exact model for a pulse.

Remember We can visualize a delta function as a very narrow pulse with an area of 1.

EXERCISES

5. Evaluate the following integrals.

(a) $\int_{-\infty}^{\infty} (\cos 2\pi t)\delta(t)\, dt$ — *answer:* 1

(b) $\int_{-\infty}^{\infty} (\cos 2\pi t)\delta(t - 1)\, dt$ — *answer:* 1

(c) $\int_{-\infty}^{\infty} (\cos 2\pi\lambda)\delta(t - \lambda)\, d\lambda$ *answer:* $\cos 2\pi t$

(d) $\int_{-\infty}^{\infty} t^3\delta(t - 2)\, dt$ *answer:* 8

7.3 Exponential Signals

In Chapter 8 we will see that energy stored in the capacitance of an *RC* circuit produces voltages and currents that are exponential. That is, the voltages and currents are of the form

$$v(t) = V_0 e^{-t/\tau} \tag{7.12}$$

and

$$i(t) = I_0 e^{-t/\tau} \tag{7.13}$$

where

$$\tau = RC \tag{7.14}$$

is the *time constant.* We will also see that the voltages and currents in an *RL* circuit have a similar form, except that

$$\tau = \frac{L}{R} \tag{7.15}$$

An exponentially decaying voltage, as given by Eq. (7.12), is shown in Fig. 7.10. The value of the exponential after n time constants is

$$v(n\tau) = V_0 e^{-n} \tag{7.16}$$

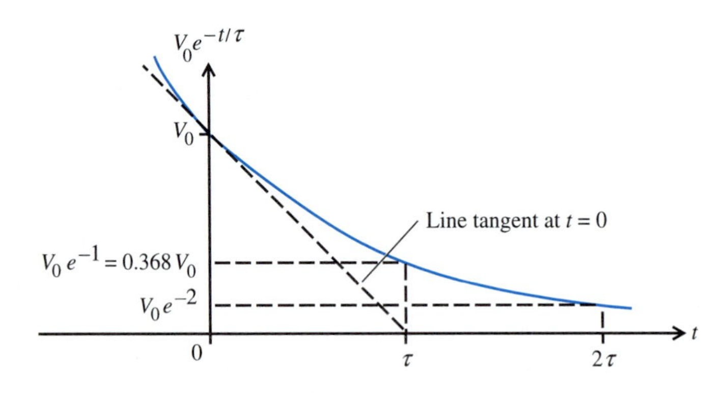

FIGURE 7.10
A decaying exponential

Thus an exponentially decaying signal decreases to 0.368 of its original value in one time constant, and to 0.0067 of its initial value in five time constants. For this reason, we often say that an exponentially decreasing signal is negligible, compared to its original value, after five time constants.

The line tangent to the exponential curve at $t = 0$ is

$$v_t(t) = v(0) + \frac{d}{dt}v\bigg|_{t=0}$$

$$= V_0 - \frac{V_0}{\tau}t \tag{7.17}$$

This tangent, shown as a dashed line in Fig. 7.10, intersects the time axis at $t = \tau$. The intersection of the tangent line with the t axis provides a method to estimate the time constant from the graph of an exponential.

We can rewrite the exponential signal of Eq. (7.12) as

$$v(t) = V_0 e^{\sigma t} \tag{7.18}$$

where σ is a real number. Equation (7.18) is sketched in Fig. 7.11. We see that the exponential decreases with time for $\sigma < 0$ and increases with time for $\sigma > 0$. We further see that *a constant is just a special case of an exponential* where $\sigma = 0$.

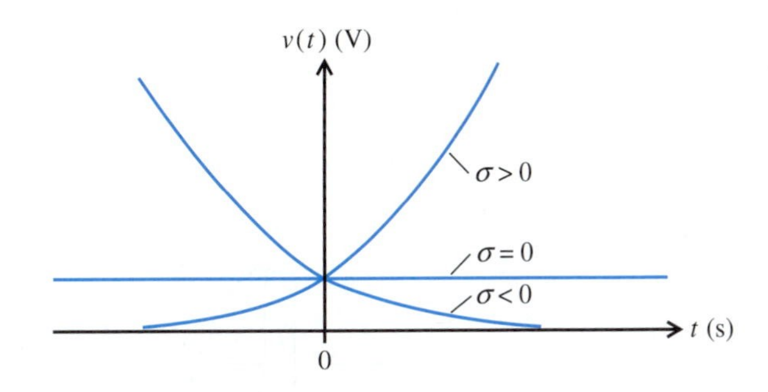

FIGURE 7.11 Exponential signals for real exponents

Not only do exponential signals occur frequently in physical systems, they are extraordinarily easy to differentiate and integrate:

$$\frac{d}{dt} V_0 e^{\sigma t} = \sigma V_0 e^{\sigma t} \tag{7.19}$$

and

$$\int V_0 e^{\sigma t}\, dt = A + \frac{1}{\sigma} V_0 e^{\sigma t} \tag{7.20}$$

where A is the constant of integration. We see that *differentiation* of the exponential $V_0 e^{\sigma t}$ with respect to t *is equivalent to multiplication* by the coefficient σ. We use this observation extensively in later chapters.

Remember

An exponential is usually considered to be negligible, compared to its original value, after five time constants. Differentiation of $e^{\sigma t}$ with respect to t is equivalent to multiplication by σ.

EXERCISES

6. Evaluate the following exponential signals at $t = 0.2$ s.
   (a) $10e^{-5t}$ — *answer:* 3.679
   (b) $8e^{-20t}$ — *answer:* 0.1465
   (c) $2e^{10t}$ — *answer:* 14.778
   (d) $\frac{d}{dt} 20e^{-4t}$ — *answer:* −35.95
7. The voltage across a 5-μF capacitance is $40e^{-2t}$ V. Determine the capacitance current. *answer:* $-0.4e^{-2t}$ mA

7.4 Sinusoidal Signals

Voltages and currents that have a sinusoidal variation with time (are sine or cosine functions of time) are very important in electrical engineering for several reasons:

1. The natural response of a circuit can contain sinusoidal terms.
2. Almost all electrical power is generated with a sinusoidal voltage waveform.
3. Linear circuits with sinusoidal sources are easily analyzed.
4. Linear circuits with nonsinusoidal sources can be analyzed by representing each source as a sum of sinusoidal sources using Fourier series or transforms, as we will see in later chapters.

A sinusoid† with arbitrary time origin and amplitude is shown in Fig. 7.12.

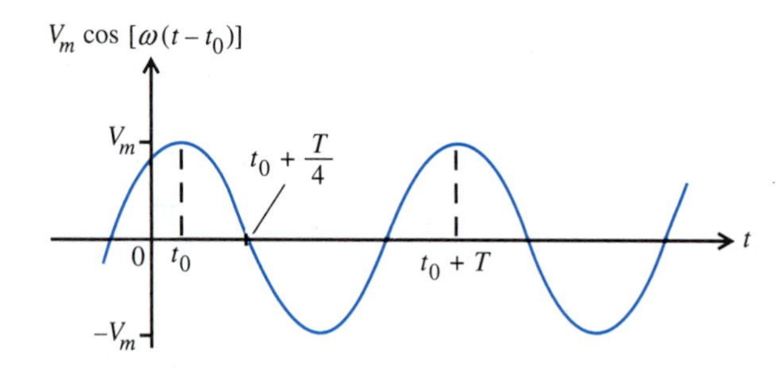

FIGURE 7.12
A sinusoid

The cosine shown in Fig. 7.12 has a time delay of t_0 s and can be written as

$$V_m \cos \omega(t - t_0) = V_m \cos(\omega t - \omega t_0) \tag{7.21a}$$

$$= V_m \cos\left(\frac{2\pi}{T} t - \frac{2\pi}{T} t_0\right) \tag{7.21b}$$

$$= V_m \cos(2\pi f t - 2\pi f t_0) \tag{7.21c}$$

$$= V_m \cos(2\pi f t + \theta) \tag{7.21d}$$

where V_m = amplitude (7.22)

ω = angular frequency, rad/s (7.23)

$f = \omega/2\pi$ = frequency, Hz (7.24)

$T = 1/f$ = period, s (7.25)

$\theta = -\omega t_0$ = phase angle, rad (7.26)

The phase angle θ is often expressed in degrees and must be converted to radians before adding θ and ωt.

In the previous section we saw that differentiation of exponential signals was equivalent to multiplication by a constant. We will now see how to represent sinusoidal signals by exponential signals so that we can take advantage of this simplicity.

We can find the Taylor series expansion of the real exponential in any calculus book:

$$e^{\sigma t} = \sum_{n=0}^{\infty} \frac{1}{n!} (\sigma t)^n \tag{7.27}$$

† The term *sinusoid* implies either the trigonometric sine or cosine function having arbitrary amplitude V_m, angular frequency ω, and phase θ.

We use this Taylor series expansion to extend the definition of the exponential to imaginary exponents:

$$e^{j\omega t} = \sum_{n=0}^{\infty} \frac{1}{n!} (j\omega t)^n \tag{7.28}$$

where

$$j^2 = -1 \tag{7.29}$$

If we separate the sum in Eq. (7.28) into two sums, the first containing even powers of j and the second containing odd powers of j, we have

$$e^{j\omega t} = \left[\sum_{n=0}^{\infty} (-1)^n \frac{1}{(2n)!} (\omega t)^{2n} \right] + j \left[\sum_{n=0}^{\infty} (-1)^n \frac{1}{(2n+1)!} (\omega t)^{2n+1} \right] \tag{7.30}$$

We can refer to any calculus book and find that the first sum in Eq. (7.30) is the Taylor series expansion for a trigonometric cosine, and the second sum is the Taylor series expansion for a sine. Therefore Eq. (7.30) yields

Euler's Identity

$$e^{j\omega t} = \cos \omega t + j \sin \omega t \tag{7.31}$$

Euler's Identity is of fundamental importance to linear circuit analysis and therefore should be memorized by all engineering students. The complex exponential is represented in the complex plane as shown in Fig. 7.13. Obviously the real part† of the complex exponential is a cosine, and the imaginary part of the complex exponential is a sine.

FIGURE 7.13
The complex exponential $e^{j\omega t}$

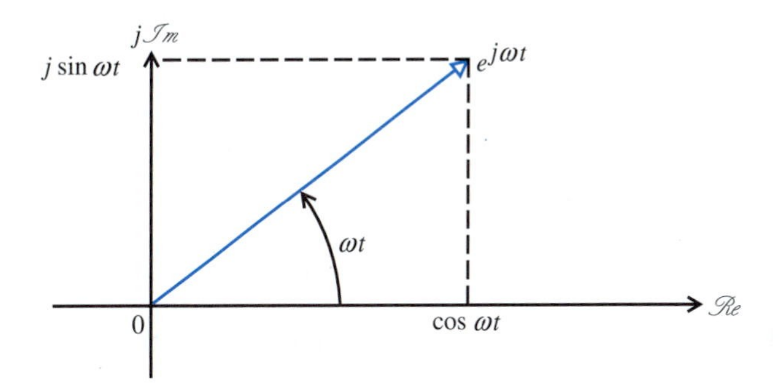

$$\cos \omega t = \mathcal{R}e\{e^{j\omega t}\} \tag{7.32}$$

$$\sin \omega t = \mathcal{I}m\{e^{j\omega t}\} \tag{7.33}$$

$$|e^{j\omega t}| = 1 \tag{7.34}$$

From the result

$$\begin{aligned} e^{-j\omega t} &= \cos(-\omega t) + j \sin(-\omega t) \\ &= \cos \omega t - j \sin \omega t \end{aligned} \tag{7.35}$$

† You should review Appendix B at this time. The real and imaginary operators used in Eqs. (7.32) and (7.33) are defined there.

the sum or difference of $\frac{1}{2}e^{j\omega t}$ and $\frac{1}{2}e^{-j\omega t}$ gives

$$\cos \omega t = \frac{e^{j\omega t} + e^{-j\omega t}}{2} \tag{7.36}$$

$$\sin \omega t = \frac{e^{j\omega t} - e^{-j\omega t}}{2j} \tag{7.37}$$

EXAMPLE 7.5 Use the complex exponential to determine the derivative of

$$v(t) = V_m \cos \omega_p t$$

Solution

$$\begin{aligned}\frac{d}{dt} V_m \cos \omega_p t &= \frac{d}{dt} \mathcal{R}e\{V_m e^{j\omega_p t}\} = \mathcal{R}e\left\{\frac{d}{dt} V_m e^{j\omega_p t}\right\} \\ &= \mathcal{R}e\{ j\omega_p V_m e^{j\omega_p t}\} = \mathcal{R}e\{\omega_p e^{j\pi/2} V_m e^{j\omega_p t}\} \\ &= \mathcal{R}e\{\omega_p V_m e^{j(\omega_p t + \pi/2)}\} \\ &= \omega_p V_m \cos(\omega_p t + \pi/2)\end{aligned}$$

Damped Sinusoids

In Chapter 9, we will see that a signal closely related to a sinusoid, the *damped sinusoid,* occurs naturally in many circuits when they are disturbed by a suddenly changed source value. A *damped cosine* with phase shift can be written as

$$\begin{aligned}V_m e^{\sigma t} \cos(\omega t + \theta) &= V_m e^{\sigma t} \mathcal{R}e\{e^{j(\omega t + \theta)}\} \\ &= \mathcal{R}e\{V_m e^{\sigma t + j(\omega t + \theta)}\} = \mathcal{R}e\{V_m e^{j\theta} e^{(\sigma + j\omega)t}\}\end{aligned} \tag{7.38}$$

The undamped sinusoid is just a special case when $\sigma = 0$.

Refer to Fig. 7.14. We have all experienced a damped sinusoidal response in mechanical systems when a weight supported by a spring is disturbed—for example, when our car hits a bump and the shock absorbers (called dampers in England) are weak.

EXAMPLE 7.6 In Chapter 9 we will encounter expressions such as

$$v = (4 + j3)e^{(-1+j2)t} + (4 - j3)e^{(-1-j2)t} \text{ V}$$

Use Euler's identity to write voltage v as a function that includes the sine and cosine.

Solution

$$\begin{aligned}v &= (4 + j3)e^{(-1+j2)t} + (4 - j3)e^{(-1-j2)t} \\ &= e^{-t}[(4 + j3)e^{j2t} + (4 - j3)e^{-j2t}] \\ &= e^{-t}[(4 + j3)(\cos 2t + j \sin 2t) + (4 - j3)(\cos 2t - j \sin 2t)] \\ &= e^{-t}\{[(4 + j3) + (4 - j3)] \cos 2t + [(4 + j3) - (4 - j3)]j \sin 2t\} \\ &= e^{-t}(8 \cos 2t - 6 \sin 2t)\end{aligned}$$

The complex exponential provides a convenient method to determine the derivative of a damped sinusoid. We use the complex variable

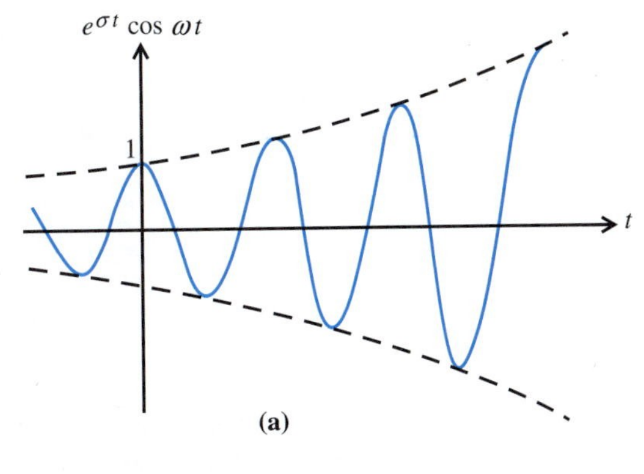

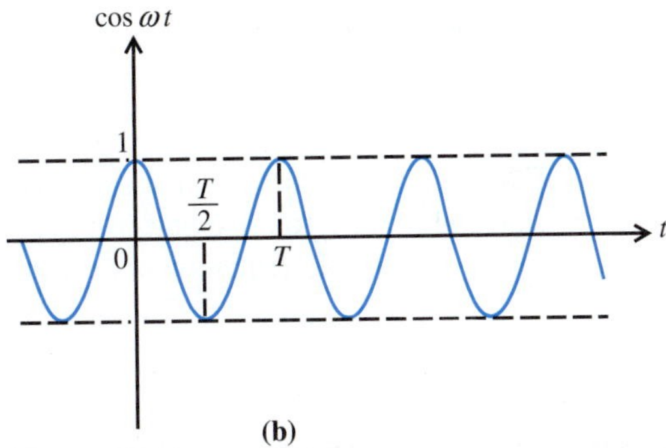

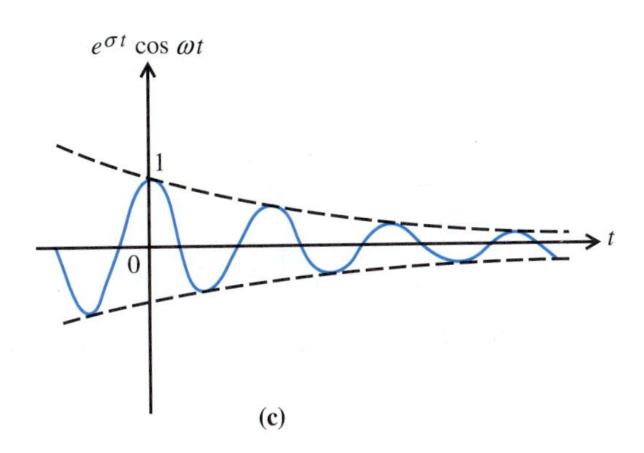

FIGURE 7.14 The cosine with exponential amplitude: (*a*) negatively damped cosine, $\sigma > 0$; (*b*) undamped cosine, $\sigma = 0$; (*c*) damped cosine, $\sigma < 0$

$$s = \sigma + j\omega \tag{7.39}$$

Then

$$\frac{d}{dt} e^{(\sigma + j\omega)t} = \frac{d}{dt} e^{st} = se^{st} \tag{7.40}$$

and we see that differentiation of the complex exponential e^{st} with respect to time is equivalent to multiplying by s.

EXAMPLE 7.7 Use the complex exponential to determine the derivative of the damped sinusoid

$$v = V_m e^{\sigma t} \cos(\omega t + \theta)$$

Solution

$$\frac{d}{dt}V_m e^{\sigma t}\cos(\omega t+\theta)=\frac{d}{dt}\mathcal{R}e\{V_m e^{\sigma t}e^{j(\omega t+\theta)}\}$$
$$=\mathcal{R}e\left\{\frac{d}{dt}e^{j\theta}V_m e^{(\sigma+j\omega)t}\right\}$$
$$=\mathcal{R}e\{(\sigma+j\omega)V_m e^{j\theta}e^{(\sigma+j\omega)t}\}$$
$$=\mathcal{R}e\{\sqrt{\sigma^2+\omega^2}\,V_m e^{\sigma t}e^{j(\omega t+\theta+\arctan\omega/\sigma)}\}$$
$$=\sqrt{\sigma^2+\omega^2}\,V_m e^{\sigma t}\cos\left(\omega t+\theta+\arctan\frac{\omega}{\sigma}\right)$$

The differentiation in the preceding example looks tedious, because it is done in general. When numerical values for σ and ω are given, the differentiation is very easy. This is especially true when we do the *rectangular-to-polar conversion* for the complex numbers with our calculator.

Remember

Euler's identity lets us write the complex exponential $e^{j\omega t}$ in terms of the trigonometric sine and cosine. The real part of the complex exponential is $\cos\omega t$ and the imaginary part is $\sin\omega t$. Differentiation of the complex exponential e^{st} is equivalent to multiplying by s.

EXERCISES

8. Evaluate the following functions at $t = 1$ s [read the note under Eq. (7.26)].
   (a) $e^{-0.5t}$ answer: 0.6065
   (b) $\cos 0.5t$ answer: 0.8776
   (c) $\cos[0.1t+(\pi/6)]$ answer: 0.812
   (d) $\cos(0.1t-30°)$ answer: 0.912
   (e) $e^{j0.5t}$ answer: $0.8776+j0.4794$
   (f) $e^{j[0.5t+(\pi/6)]}$ answer: $0.5203+j0.8540$
   (g) $e^{(-2+j0.5)t}$ answer: $0.1188+j0.06488$
9. Use the complex exponential to determine the derivative of $v(t)=e^{4t}\cos 3t$ V.
   answer: $5e^{4t}\cos(3t+36.87°)$ V/s

7.5 Average and RMS Values†

We often find that the average value of a signal over some time interval T is important. An average over a short time interval can be quite different than the average over a long time interval. For example, the average power needed for a few seconds while a large electric motor starts is much greater than the average power required by the same motor running steadily. We often model this latter case as an average over the infinite time interval $-\infty < t < \infty$. Periodic signals, such as sinusoids, repeat with a period T_0 so that $v(t+T_0)=v(t)$ for all t. In this case the average over all time is the same as the average over one period.

† This section can be postponed until Chapter 12.

Average Value or DC Component

In later chapters, we will need to calculate the average or dc value of signals. The dc component of a signal is of great importance in electronics and electrochemistry.

DEFINITION Average Value or DC Component

Over the interval t_0 to $t_0 + T$, the *dc component* of a signal is the *average value* of the signal:

$$X_{dc} = \frac{1}{T}\int_{t_0}^{t_0+T} x(t)\,dt \tag{7.41}$$

For a periodic signal with period T, the average value over the infinite time interval is the same as the average value over any period.

From Eq. (7.41), if the area between $x(t)$ and the time axis is zero (area above the axis is positive, and area below the axis is negative), the dc component is zero. Thus the dc component of a sine or cosine waveform is zero.

EXAMPLE 7.8 Calculate the average value (dc component) of the sinusoidal voltage over all time.

$$v(t) = V_m \sin \omega t \text{ V}$$

Solution The sine is periodic with period $T = 2\pi/\omega$, so we only need to calculate the average over one period.

$$\begin{aligned} V_{dc} &= \frac{1}{T}\int_0^T V_m \sin \omega t\,dt \\ &= -\frac{1}{T} V_m \frac{1}{\omega} \cos \omega t \Big|_0^T \\ &= -\frac{1}{T} V_m \frac{T}{2\pi} \cos \frac{2\pi}{T} t \Big|_0^T = 0 \text{ V} \end{aligned}$$

Average Power for DC Signals

In Chapter 2 we showed that the instantaneous power p absorbed by a resistance is given by

$$p = Ri^2 \tag{7.42}$$

or

$$p = \frac{1}{R} v^2 \tag{7.43}$$

The *average power* absorbed over the interval t_0 to $t_0 + T$ is

$$P = \frac{1}{T}\int_{t_0}^{t_0+T} p\,dt \tag{7.44}$$

For a constant (dc) voltage or current, $i(t) = I$ and $v(t) = V$, Eq. (7.44) gives the average power as

$$P = RI^2 \tag{7.45}$$

and

$$P = \frac{1}{R} V^2 \tag{7.46}$$

Effective or RMS Value

In some equipment, such as an electric heater, the average power delivered to a resistance is the important quantity. Because of this, we often specify voltages and currents of arbitrary wave shapes by their *effective values* (effective heating value). When a sinusoidal voltage is specified, it is usually in terms of its effective value. For this reason, the 60-Hz, 120-V outlet in a house has a voltage waveform described by

$$v(t) = 120\sqrt{2} \cos(2\pi 60t + \theta) \text{ V} \tag{7.47}$$

which has a maximum value of approximately 170 V. [We will soon see why the square root of 2 appears in Eq. (7.47).]

The effective value of a signal is determined from the average power delivered to a resistive load. Given an arbitrary voltage v across a resistance R, the average power absorbed by the resistance over the time interval t_0 to $t_0 + T$ is

$$P = \frac{1}{T}\int_{t_0}^{t_0+T} p\, dt \tag{7.48a}$$

$$= \frac{1}{R}\frac{1}{T}\int_{t_0}^{t_0+T} v^2\, dt \tag{7.48b}$$

If we define the effective value V_{eff} by

$$V_{\text{eff}}^2 = \frac{1}{T}\int_{t_0}^{t_0+T} v^2\, dt \tag{7.49}$$

we see that Eq. (7.48b) is of the form

$$P = \frac{1}{R} V_{\text{eff}}^2 \tag{7.50}$$

A similar development gives

$$I_{\text{eff}}^2 = \frac{1}{T}\int_{t_0}^{t_0+T} i^2\, dt \tag{7.51}$$

and

$$P = RI_{\text{eff}}^2 \tag{7.52}$$

From Eqs. (7.49) and (7.51), we see that the effective values are the square *r*oot of the average (*m*ean) of the *s*quare of the voltage or current. We will take this as our definition:

DEFINITION Root-Mean-Square (RMS) or Effective Value of a Signal†

Over the time interval t_0 to $t_0 + T$,

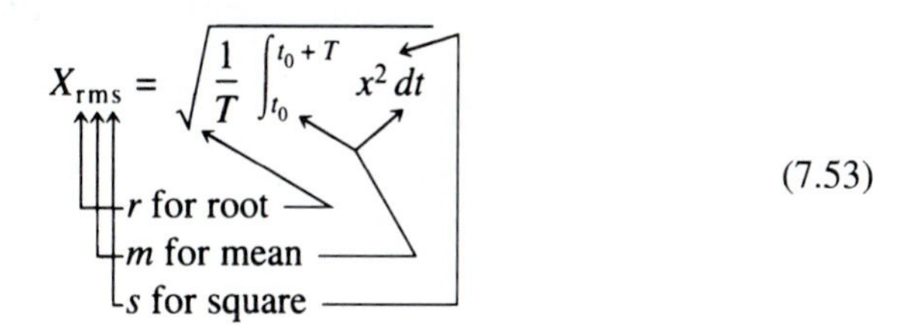

(7.53)

and

$$X_{eff} = \sqrt{\frac{1}{T}\int_{t_0}^{t_0+T} x^2\,dt} \tag{7.54}$$

For a periodic signal with period T, the *rms* or *effective value* over the infinite time interval is the same as the rms value over one period.

Thus the rms value is the same as the effective value:

$$I_{eff} \equiv I_{rms} \tag{7.55}$$

$$V_{eff} \equiv V_{rms} \tag{7.56}$$

EXAMPLE 7.9 A sinusoid

$$v(t) = V_m \cos \omega t$$

is applied to a resistance of R ohms. Calculate the effective or rms value of $v(t)$ and the power delivered to the resistance.

Solution

$$V_{rms} = \sqrt{\frac{1}{T}\int_{t_0}^{t_0+T} V_m^2 \cos^2 \omega t\,dt}$$

$$= \sqrt{\frac{1}{T}\int_{t_0}^{t_0+T} \frac{V_m^2}{2}(1 + \cos 2\omega t)\,dt}$$

$$= \frac{1}{\sqrt{2}} V_m \simeq 0.707 V_m$$

and

$$P = \frac{V_{rms}^2}{R} = \frac{1}{2}\frac{V_m^2}{R}$$

From this last example we conclude that:

† If complex signals are considered, $x^2(t)$ must be replaced by $x(t)x^*(t) = |x(t)|^2$, where $x^*(t)$ is the complex conjugate of $x(t)$.

The rms value of a sinusoid is the amplitude of the sinusoid divided by the square root of 2. The square root of 2 relation *does not apply* to the relation between maximum and rms values for most other waveforms.

EXAMPLE 7.10 Calculate the rms value of a periodic signal $v(t)$ with period T, where the signal is defined to be

$$v(t) = \begin{cases} Kt & 0 \le t < \dfrac{T}{2} \\ 0 & \dfrac{T}{2} \le t < T \end{cases}$$

for a single period, where $K = 2V_m/T$ is a positive constant and V_m is the maximum value of $v(t)$.

Solution Sketch $v(t)$ for $-2T \le t \le 2T$. This will help visualize that

$$\begin{aligned} V_{\text{rms}} &= \sqrt{\frac{1}{T}\int_0^T v^2\,dt} \\ &= \sqrt{\frac{1}{T}\left[\int_0^{T/2} (Kt)^2\,dt + \int_{T/2}^{T} (0)^2\,dt\right]} \\ &= \sqrt{\frac{1}{T}\frac{K^2T^3}{24}} = \frac{1}{2\sqrt{6}}KT = \frac{V_m}{\sqrt{6}} \end{aligned}$$

We see that for this signal the rms value is the maximum value divided by $\sqrt{6}$ and not by $\sqrt{2}$ as it is for a sinusoid.

Remember

We call the average value of a signal the dc component. The effective value and rms value of a signal are the same. The rms value is the square *r*oot of the *m*ean *s*quare value. We can use the rms value of a signal to calculate the power delivered to a resistance. The rms value of a sinusoid is the amplitude (maximum value) of the sinusoid divided by the square root of 2. This square root of 2 relation does not hold for most other signals.

EXERCISES

10. Calculate the dc component (average value) of the following signals.
    (a) $i_a = 5$ A — *answer:* 5 A
    (b) $v_b = 4\cos\omega t + 3\cos\omega t$ V — *answer:* 0
    (c) $i_c = 4 + 3\cos 2t$ A — *answer:* 4 A
    (d) $v_d = 8\cos^2\omega t$ V — *answer:* 4 V
11. Calculate the rms (effective) value of the following signals.
    (a) $i_a = 5$ A — *answer:* 5 A
    (b) $v_b = 3 + 4\cos 2t$ V — *answer:* $\sqrt{17}$ V

7.6 Superposition of Power†

In Chapter 5 we found that superposition holds for voltages and currents. That is, the voltage and current response of a linear network to a number of independent sources is the *sum* of the responses obtained by applying each independent source once with other independent sources set equal to zero. We will now investigate whether superposition applies to mean square values of voltages and currents and to power. Consider a voltage $v(t)$ that is the sum of two voltages:

$$v(t) = v_1(t) + v_2(t) \tag{7.57}$$

The mean square value of $v(t)$ is given by

$$\begin{aligned} V_{\text{rms}}^2 &= \lim_{T\to\infty} \frac{1}{T} \int_{-T/2}^{T/2} [v_1(t) + v_1(t)]^2 \, dt \\ &= V_{\text{rms}_1}^2 + V_{\text{rms}_2}^2 + \lim_{T\to\infty} \frac{1}{T} \int_{-T/2}^{T/2} v_1(t)v(t)_2 \, dt \end{aligned} \tag{7.58}$$

If the average of the product of the two signals is zero,

$$\lim_{T\to\infty} \frac{1}{T} \int_{-T/2}^{T/2} v_1(t)v_2(t) \, dt = 0 \tag{7.59}$$

then we say that voltage v_1 and v_2 are *orthogonal.* Equation (7.58) then simplifies to

$$V_{\text{rms}}^2 = V_{\text{rms}_1}^2 + V_{\text{rms}_2}^2 \tag{7.60}$$

Because the average power delivered to a resistance is proportional to the mean-square value of the resistance voltage (or current), Eq. (7.60) tells us that superposition of average power holds for two orthogonal signals.

In general, if a voltage $v(t)$ is the sum of N voltages,

$$v(t) = v_1(t) + v_2(t) + \cdots + v_N(t) \tag{7.61}$$

and the N voltages are orthogonal,

$$\lim_{T\to\infty} \frac{1}{T} \int_{-T/2}^{T/2} v_n(t)v_m(t) \, dt = 0\ddagger \qquad n \neq m \tag{7.62}$$

then the mean square value of $v(t)$ is given by

$$V_{\text{rms}}^2 = V_{\text{rms}_1}^2 + V_{\text{rms}_2}^2 + \cdots + V_{\text{rms}_N}^2 \tag{7.63}$$

This gives the following result:

RMS Value of the Sum of Orthogonal Signals

For *orthogonal signals* (voltages or currents), the squares of rms voltages add, and

$$V_{\text{rms}} = \sqrt{\sum_{n=1}^{N} V_{\text{rms}_n}^2} \tag{7.64}$$

† This section can be postponed until Chapter 15.
‡ If complex signals are considered, $v_m(t)$ must be replaced by its complex conjugate.

For N *orthogonal* signals,

$$P_{\mathrm{av}} = \frac{1}{R} V_{\mathrm{rms}}^2 = \frac{1}{R} \sum_{n=1}^{N} V_{\mathrm{rms}_n}^2 = \sum_{n=1}^{N} P_{\mathrm{av}_n} \tag{7.65}$$

where P_{av_n} is the average power absorbed from voltage component v_n. This proves that *for orthogonal signals, the average powers add. This superposition of power does not hold for signals that are not orthogonal.* Typically

$$P_{\mathrm{av}} \neq \sum_{n=1}^{N} P_{\mathrm{av}_n} \tag{7.66}$$

For sinusoids of different frequencies ($\omega_n \neq 0$) with amplitudes V_{m_n}, Eq. (7.64) becomes

$$V_{\mathrm{rms}} = \sqrt{\sum_{n=1}^{N} \frac{1}{2} V_{m_n}^2} \tag{7.67}$$

We can easily show that sinusoids satisfy Eq. (7.59) and are orthogonal for the following cases:

1. Over the infinite interval, sines and cosines of the same frequency are orthogonal to each other if they have the same phase angle.
2. Cosines are orthogonal to sines and cosines of different frequencies.
3. Sines are orthogonal to cosines and sines of different frequencies.

EXAMPLE 7.11 For the network shown in Fig. 7.15, find the rms value of v and the average power absorbed by the resistance.

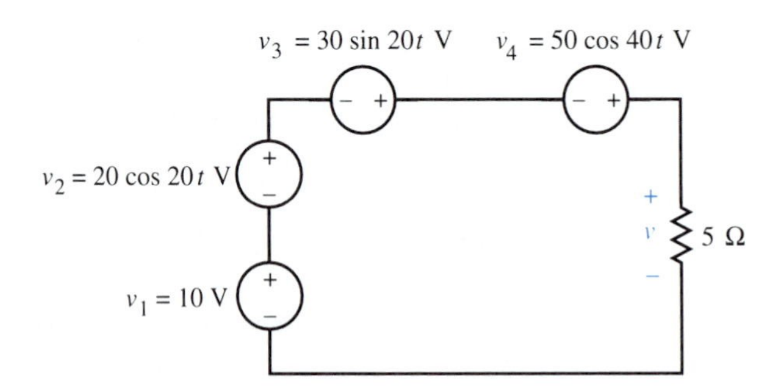

FIGURE 7.15 Voltage $v = \sum_{n=1}^{4} v_n$ is the sum of orthogonal voltages

Solution The four signals are orthogonal [see the statements below Eq. (7.67)]; therefore

$$V_{\mathrm{rms}} = \sqrt{10^2 + \left(\frac{20}{\sqrt{2}}\right)^2 + \left(\frac{30}{\sqrt{2}}\right)^2 + \left(\frac{50}{\sqrt{2}}\right)^2} = \sqrt{2000} = 44.72 \text{ V}$$

and

$$P = \frac{1}{R} V_{\mathrm{rms}}^2 = \frac{1}{5}(2000) = 400 \text{ W}$$

Note that the powers supplied by each source acting alone add to give

$$P = P_1 + P_2 + P_3 + P_4$$
$$= \frac{1}{5}(10)^2 + \frac{1}{5}\left(\frac{20}{\sqrt{2}}\right)^2 + \frac{1}{5}\left(\frac{30}{\sqrt{2}}\right)^2 + \frac{1}{5}\left(\frac{50}{\sqrt{2}}\right)^2$$
$$= 20 + 40 + 90 + 250 = 400 \text{ W}$$

so superposition of power holds, but only because the four voltages are orthogonal.

We will now work an example in which not all voltages are orthogonal.

EXAMPLE 7.12 Repeat Example 7.11 with v_4 changed to

$$v_4 = 50 \cos(20t - 53.13°) \text{ V}$$

Solution The rms values of v_1 through v_4 are the same, but v_4 is no longer orthogonal to v_2 and v_3. The problem could be worked by applying the definitions and integrating. A simpler way is to use the trigonometric identity

$$\cos(A + B) = \cos B \cos A - \sin B \sin A$$

This trigonometric identity lets us write voltage v_4 as a sum of a cosine and a sine component:

$$v_4 = 30 \cos 20t + 40 \sin 20t \text{ V}$$

Addition of v_1, v_2, v_3, and v_4 gives us

$$v = 10 + 50 \cos 20t + 70 \sin 20t \text{ V}$$

The three components are now orthogonal to each other, so

$$V_{\text{rms}} = \sqrt{10^2 + \left(\frac{50}{\sqrt{2}}\right)^2 + \left(\frac{70}{\sqrt{2}}\right)^2}$$
$$= \sqrt{3800} = 61.64 \text{ V}$$
$$P = \frac{1}{R} V_{\text{rms}}^2 = \frac{1}{5}(3800) = 760 \text{ W}$$

Note that the powers supplied by each source acting alone add to give

$$P_1 + P_2 + P_3 + P_4 = 400 \text{ W} \neq P$$

so superposition of power did not hold.

Remember Over the infinite interval, sines and cosines of different frequencies are orthogonal. Sines and cosines of the same nonzero frequency are orthogonal to each other if they have the same phase shift. Cosines are orthogonal to cosines of different frequencies, and sines are orthogonal to sines of different frequencies. Superposition of mean square values applies only to orthogonal signals. Superposition of power applies only to orthogonal signals.

EXERCISES

12. Use trigonometric identities and integration to show that $v_1(t) = 2 \sin 2t$ V and $v_2(t) = 4 \sin 4t$ V are orthogonal over the infinite interval.

13. Determine the rms value of the periodic voltages or currents given. (The components in part (e) are not orthogonal.)
    (a) $v_a = 4 \cos \omega t + 3 \cos \omega t$ V — *answer:* 4.95 V
    (b) $v_b = 4 \cos \omega t + 3 \sin \omega t$ V — *answer:* 3.54 V
    (c) $i_c = 4 \cos 5t + 4 \cos 10t$ A — *answer:* 4 A
    (d) $i_d = 4 + 3 \cos 2t$ A — *answer:* 4.53 A
    (e) $i_e = 2 \cos 2t + 2 \sin (2t + 45°)$ A — *answer:* 2.61 A

7.7 Summary

This chapter introduced the modeling of signals (voltages and currents) by mathematical functions. We first introduced the unit step function. The unit step function is 0 for negative time and takes on the value 1 for positive time. This provides a convenient notation to describe signals that are defined differently in separate time intervals. In an optional section we introduced the delta function, which is the idealized model for a very short duration pulse of unit area. (Delta functions are used in Chapter 14.) We then showed how the real exponential signal, which naturally occurs in linear systems, can be generalized to the complex exponential. This generalization lets us write trigonometric sines and cosines as functions of complex exponentials. The complex exponential representation for sinusoids is fundamental to the analysis contained in the remainder of this text.

We followed the sections on signal modeling by a discussion of average and rms values.

We concluded the chapter by showing that superposition of power applies to orthogonal signals but not to signals in general.

KEY FACTS

	Concept	Equation	Section	Page
❑	A *dc signal* is constant for all time.	(7.1)	7.1	200
❑	A *unit step* $u(t)$ has a value of 0 when the argument is negative and a value of 1 when the argument is positive.	(7.2)	7.1	200
❑	We can visualize the *delta function* $\delta(t)$ as a pulse of very short duration and unit area.	(7.5)	7.2	206
❑	The *real exponential* $e^{-t/\tau}$, where τ is the time constant, is usually considered negligible compared to its original value after five time constants.	(7.12)	7.3	208
❑	*Euler's identity* is $e^{j\omega t} = \cos \omega t + j \sin \omega t$.	(7.31)	7.4	211
❑	The *complex exponential* e^{st}, where $s = \sigma + j\omega$, is equal to $e^{j\omega t}(\cos \omega t + j \sin \omega t)$.	(7.38)	7.4	212
❑	The *dc component* of a signal is the average value. For periodic signals the average calcu-	(7.41)	7.5	215

KEY FACTS	Concept	Equation	Section	Page
	lated over one period is the same as the average over the infinite interval $-\infty < t < +\infty$.			
❑	The average power absorbed by a resistance is given by $P = (1/R)\, V_{\text{rms}}^2 = RI_{\text{rms}}^2$.	(7.50) (7.52)	7.5	216
❑	The *rms* value of a signal is the *effective* heating value. For periodic signals the rms value calculated over one period is the same as the rms value calculated over the infinite interval $-\infty < t < +\infty$.	(7.53)	7.5	217
❑	We can obtain the *rms value of a sinusoid* by dividing the magnitude by the square root of 2. This relation is *not* valid for most other signals (Examples 7.9 and 7.10).		7.5	217 218
❑	Sinusoids of different frequencies are orthogonal.		7.6	220
❑	For orthogonal signals, superposition applies to mean-square values and to power. The rms value of the sum of orthogonal signals is the square root of the sum of the squares of the rms values. For sinusoidal signals the rms value is the square root of one-half of the sum of the magnitudes squared.	(7.64) (7.65) (7.67)	7.6	219 220
❑	Superposition of mean-square values and superposition of power do *not* apply to signals that are not orthogonal.	(7.66)	7.6	220

PROBLEMS

Section 7.1

1. Sketch the following functions for $0 < t < 8$ s, and write them in terms of the unit step.

(a) $v(t) = \begin{cases} 0 & t < 0 \\ 10e^{-t/4}\ \text{V} & t > 0 \end{cases}$

(b) $v(t) = \begin{cases} 0 & t < 2 \\ 10e^{-t/4}\ \text{V} & t > 2 \end{cases}$

(c) $v(t) = \begin{cases} 0 & t < 2 \\ 10e^{-(t-2)/4}\ \text{V} & t > 2 \end{cases}$

2. Sketch the following functions and write them in terms of the unit step.

(a) $i(t) = 10 \sin\left[\left(\frac{2\pi}{T}\right)t\right] \Pi\left[\frac{t - T/2}{T}\right]$ A

(b) $i(t) = \sum_{n=0}^{\infty} (n+1)\, \Pi\left(t - \frac{1}{2} - n\right)$

3. Write the following two signals in terms of the unit step and/or unit ramp.

(a)

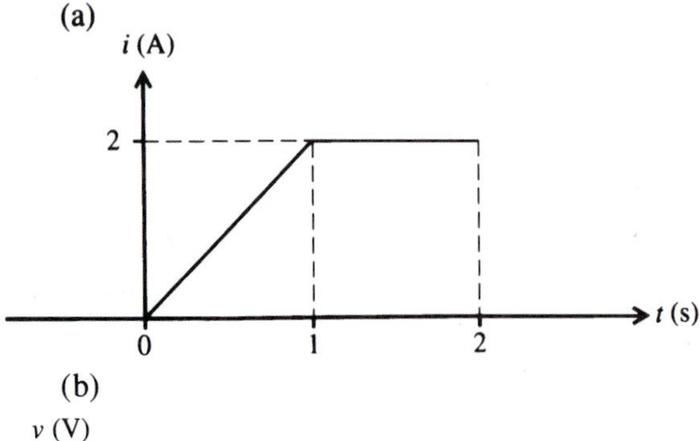

(b)

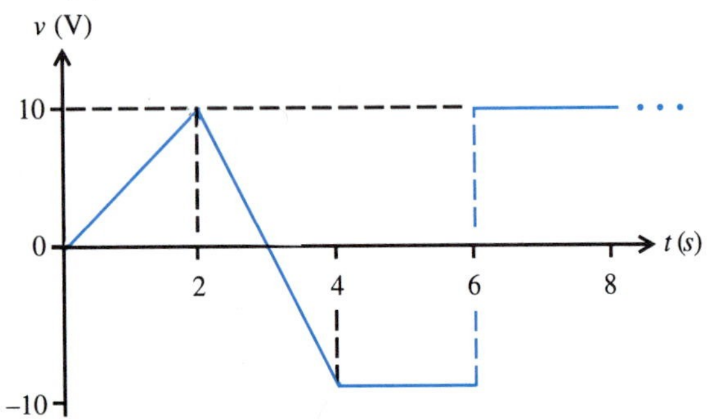

4. The current in Problem 3a passes through a 0.5-F capacitance. If $v(-\infty) = 0$, plot $v(t)$ for $-1 < t < 3$ s.

5. The following functions specify the voltage in volts applied to a 0.1-H inductance. Determine the inductance current as a function of time if $i(-\infty) = 0$. Sketch $i(t)$.

(a) $2u(t)$ (b) $2u(t-1)$
(c) $2e^{-2t}u(t)$ (d) $2e^{-2(t-1)}u(t-1)$
(e) $2[u(t) - u(t-1)]$ (f) $2e^{2t}u(-t)$

Section 7.2

6. Determine the integral over the interval from $-\infty$ to $+\infty$ of the following signals.

(a) $e^{-6t}\delta(t)$ (b) $(10 \cos 2t)\, \delta(t)$
(c) $(10 \sin \pi t)\, \delta(t)$ (d) $t^2\delta(t-3)$

7. Determine the voltage $v(t)$ across a 1-F capacitance if the current in amperes is as specified below.

(a) $\delta(t)$ (b) $\delta(t-1)$
(c) $\delta(t) + \delta(t-1)$ (d) $\sum_{n=0}^{\infty} \delta(t - nT)$

Section 7.3

8. Evaluate the following signals at $t = 0.2$ s.

(a) $4e^{-3t}$ (b) $6e^{4t}$

9. For what value of t do the following signals have a value of 6?

(a) $20e^{-2t}$ (b) $9e^{5t}$

10. The following voltages in volts appear across a 2 μF capacitance. Determine $i(0.4)$.

(a) $5e^{2t}$ (b) $10e^{-4t}$

11. What is the smallest integer number of time constants required for an exponential to decay to less than 10 percent of its initial value? What percentage of its initial value is it at this time?

Section 7.4

12. Determine the phase angle in degrees for each of the following signals, and evaluate the signal at $t = 0.2$ s.

(a) $25 \cos [2\pi 2(t - 0.1)]$
(b) $4 \cos (12t + \pi/6)$
(c) $25 \cos [2\pi 2(t + 0.1)]$
(d) $4 \cos (4t - 45°)$

13. Sketch the following signals and then write each as the real or imaginary part of a complex exponential.

(a) $10 \cos 2\pi 6t$
(b) $5 \sin (30t + 15°)$
(c) $100 \cos (50t + \pi/4)$
(d) $25e^{-3t} \cos (2\pi 5t + 30°)$

14. Write the following signals as the *sum* of complex exponentials.

(a) $20 \cos 3t$ (b) $10 \sin 5t$
(c) $10 \cos (2t + \pi/6)$ (d) $30 \sin (6t + \pi/4)$

15. Evaluate the following at $t = 1$.

(a) e^{3t} (b) e^{j3t}
(c) $e^{(3+j4)t}$ (d) $(d/dt)e^{(3+j4)t}$
(e) $e^{3t} \cos 4t$ (f) $(d/dt)e^{3t} \cos 4t$

16. Write the following signals in terms of the trigonometric sine and cosine.

(a) $10e^{j2t}$
(b) $20e^{(-2+j4)t}$
(c) $8e^{-2+j3t}$
(d) $30e^{-5t}(e^{j6t} + e^{-j6t})$

Section 7.5

17. The following current i is periodic. Determine the dc component (average value) of current i.

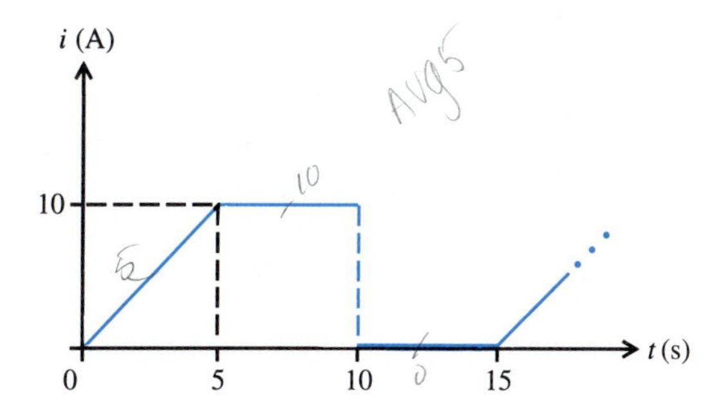

18. Current i is periodic as shown above. Determine the rms value (effective value) of current i. What average power will this current deliver to a 12-Ω resistance?

19. Determine the average value of the following periodic signals.

(a)

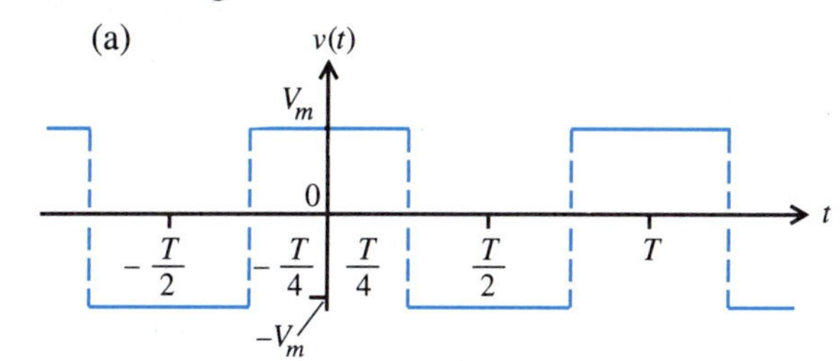

(b)

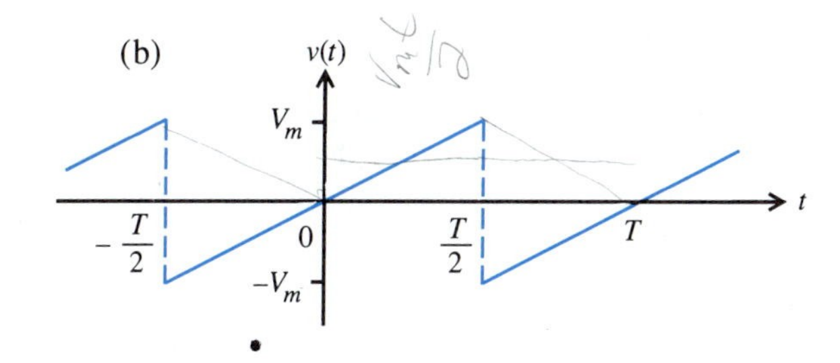

(c)

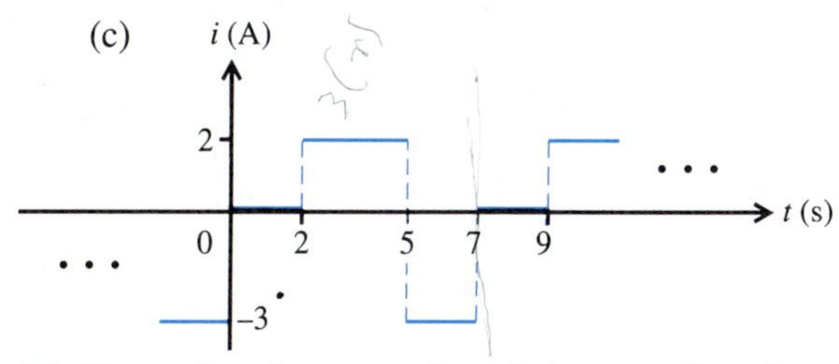

20. Determine the rms value of the preceding three periodic signals. What average power will the signals deliver to a 6-Ω resistance?

21. Determine the dc component of the following signals.

(a) $3 + 4 \cos 2t$ (b) $3 \cos t + 4 \cos 2t$

(c) $3 \cos t + 4 \cos t$ (d) $3 \cos 2t + 4 \sin 2t$

(e) $3 + 4 \sin 2t$ (f) $4 \cos^2 \omega t$

22. Determine the rms value (in A) of the preceding six signals. What average power will each current deliver to a 5-Ω resistance?

Section 7.6

23. Determine the rms value of the following signals.

(a) $3 + 4 \cos t$ (b) $3 \cos t + 4 \cos 2t$

(c) $3 \cos t + 4 \sin t$ (d) $3 \cos 3t + 4 \cos 3t$

(e) $10 \cos 3t + 4 \cos 6t + 8 \cos 9t$

(f) $2 \cos 4t - 10 \cos (4t - 36.87°)$

24. Show that

$$\lim_{T \to \infty} \frac{1}{T} \int_{-T/2}^{T/2} \cos \omega_1 t \cos \omega_2 t \, dt = \begin{cases} 0 & \omega_1 \neq \pm\omega_2 \\ 1 & \omega_1 = \omega_2 = 0 \\ \frac{1}{2} & \omega_1 = \pm\omega_2 \neq 0 \end{cases}$$

This establishes the fact that cosines of different frequencies are orthogonal over the infinite interval.

Supplementary Problems

25. The voltage across a parallel RLC circuit is given by

$$v = (2 + j3)e^{(-4+j5)t} + (2 - j3)e^{(-4-j5)t}$$

(a) Write this voltage as an exponential times a cosine plus a sine.

(b) Write voltage v as an exponential times a cosine with a phase angle.

26. Verify that $i = Ae^{-4t}$ satisfies the differential equation

$$\frac{d}{dt} i + 4i = 0$$

27. Verify that both $v_1 = A_1 e^{-t}$ and $v_2 = A_2 e^{-2t}$ satisfy the differential equation

$$\frac{d^2}{dt^2} v + 3 \frac{d}{dt} v + 2v = 0$$

28. Verify that both $v_1 = A_1 e^{j2t}$ and $v_2 = A_2 e^{-j2t}$ satisfy the differential equation

$$\frac{d^2}{dt^2} v + 4v = 0$$

29. For the following periodic current:
    (a) Sketch the signal.
    (b) Calculate the dc component of the current.
    (c) Calculate the rms current.

$$i(t) = A \sum_{n=-\infty}^{\infty} \Pi\left(\frac{t - nT}{\tau}\right) \text{ A} \qquad \tau \leq T$$

30. Determine the dc component and rms value of the following periodic current.

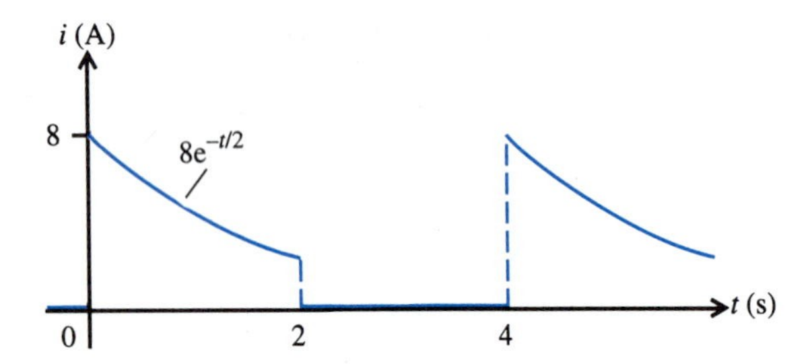

31. Voltmeters that read the average or dc value of a voltage are often adapted to read the rms value of a sinusoid. A *rectifier circuit* is added to take the absolute value of the voltage. The meter thus responds to V_{aav}, the average absolute value of v.
    (a) What constant must V_{aav} be multiplied by to indicate V_{rms} for a sinusoid?
    (b) The meter is made to read V_{rms} directly for a sinusoid by recalibration of the scale with the constant found in (a). If a square wave is applied to the meter and a value of 10 V (rms) is indicated, what is the true rms value of the square wave?

32. We often start induction motors over 50 hp at a reduced voltage in order to lessen the impact on the power system. The motor voltage is 0 for $t < 0$, $(V_m/2)\cos 2\pi 60t$ V for $0 < t < t_0$ and $V_m \cos 2\pi 60t$ for $t > t_0$.
    (a) Write an expression for $v(t)$ for $-\infty < t < +\infty$ by use of the unit step function.
    (b) If $t_0 \gg 1/60$ s, determine the rms value over the interval $0 \leq t \leq 2t_0$.

33. The current $i(t)$ through the horizontal deflection coil of the cathode-ray tube used in a television set is periodic with a frequency f_s ($f_s = 15{,}750$ sweeps per second, but we don't need this information). The period is $T = 1/f_s$. One period of the current is given by

$$s(t) = (I_m/0.42T)(t - 0.42T) \text{ A}$$

for $\quad 0 \leq t \leq 0.84T \quad$ and

$$s(t) = I_m \cos\left[(2\pi/0.32T)(t - 0.84T)\right] \text{ A}$$

for $\quad 0.84T < t < T.$

    (a) Sketch $s(t)$ and write $s(t)$ in terms of the unit step.
    (b) Determine the average value of $i(t)$.
    (c) Determine the rms value of $i(t)$.
    (d) The horizontal deflection coil is modeled as an inductance of value L henries. (i) Determine and sketch, with reasonable accuracy, the voltage (passive sign convention) across the inductance for $0 \leq t < T$. (ii) Determine the maximum value of inductance L and the minimum current I_m that can be used if the peak energy stored in the inductance is W_m joules, and the peak voltage cannot exceed a magnitude of V_m volts.

34. The 250-W resistive heating element of an electric coffeepot is designed to operate from a 60-Hz sinusoidal power source with an rms voltage of 120 V, but it is connected to an inverter that converts 24 V dc to a symmetrical square wave. The square wave has a zero volt dc value and a peak-to-peak value of 240 V. What power is absorbed by the coffeepot?

8

First-Order Circuits

In this chapter we solve the differential equation that describes a circuit with a single energy-storage element (a first-order circuit). We approach the problem in two steps.

For the first step, we consider the response of a circuit after the electric power is turned off. This *source-free response* is determined by the energy stored in the circuit capacitance or inductance. It is the *natural response* of the circuit. We first solve the differential equation that determines the response of a *source-free* circuit. We then show how to calculate the *time constant,* which determines the natural response, and use this to write the source-free response without actually writing a differential equation.

The second step is to determine the response of a circuit when the electric power is turned on or abruptly changes value. (The stored energy need not be zero when this occurs.) We first solve the differential equation that describes these circuits and find that the *complete response* is the sum of two components: the natural response and the *particular response.* The form of the natural response is determined by the source-free circuit, and the form of the particular response is determined by the input (the source). We find that we can obtain the particular response for some of our most important inputs by solving an algebraic equation.

8.1 The Source-Free *RC* Circuit

Most electronic circuits contain capacitors that store energy for some time after the power is turned off. We will now write and solve the equation that characterizes simple circuits of the type.

The voltage across the capacitance of Fig. 8.1 has a value of $v(0) = V_0$ volts at $t = 0$, so the energy stored in the capacitance is $w(0) = \frac{1}{2}CV_0^2$. The capacitance voltage immediately after the switch is opened at $t = 0$ is $v(0^+) = v(0) = V_0$ V, because the stored energy cannot change instantaneously. We wish to solve for v for all $t > 0$.

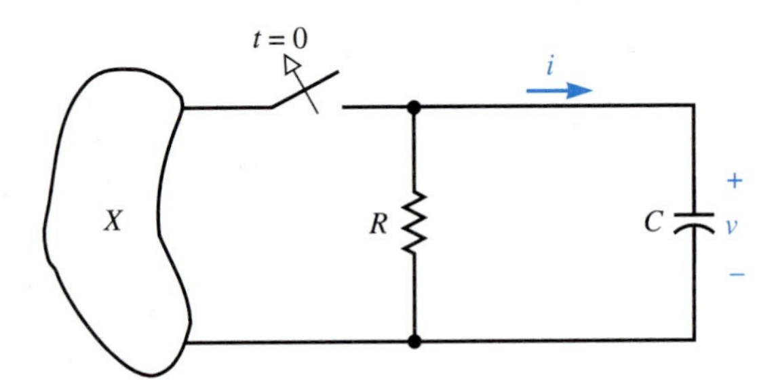

FIGURE 8.1
An *RC* circuit that is source-free for $t > 0$

Application of KCL to a surface enclosing the top node yields the differential equation

$$C\frac{d}{dt}v + \frac{1}{R}v = 0 \qquad t > 0 \tag{8.1}$$

which we can write as

$$\frac{d}{dt}v + \frac{1}{RC}v = 0 \qquad t > 0 \tag{8.2}$$

Several techniques are available to solve a differential equation like Eq. (8.2). We choose a method found in many calculus books and the more popular books on differential equations.† If we multiply Eq. (8.2) by the *integrating factor* $e^{(1/RC)t}$ and use the rule for the derivative of the product of two functions, we obtain

$$e^{(1/RC)t}\frac{d}{dt}v + \frac{1}{RC}e^{(1/RC)t}v = \frac{d}{dt}(e^{(1/RC)t}v) = 0 \tag{8.3}$$

Therefore the solution for v in Eq. (8.2) is the same as the solution for v in

$$\frac{d}{dt}(e^{(1/RC)t}v) = 0 \tag{8.4}$$

We integrate both sides of this equation with respect to time to obtain

$$\int \frac{d}{dt}(e^{(1/RC)t}v)dt = \int d(e^{(1/RC)t}v)$$
$$= e^{(1/RC)t}v - A = 0 \tag{8.5}$$

† Zill, Denis G., *A First Course in Differential Equations,* 5th ed. (Boston: PWS/Kent, 1993).

where A is a constant of integration. Multiplication of Eq. (8.5) by $e^{-(1/RC)t}$ gives us the solution to Eq. (8.2):

$$v = Ae^{-(1/RC)t} \qquad t > 0 \tag{8.6}$$

We now use the value of v at $t = 0^+$ (the *initial condition*†) and Eq. (8.6) evaluated at $t = 0^+$ to calculate the constant of integration:

$$v(0^+) = V_0 = Ae^0 = A \tag{8.7}$$

Thus the *source-free response* for the capacitance voltage is

$$v = V_0 e^{-t/RC} \qquad t > 0 \tag{8.8}$$

This is shown in Fig. 8.2.

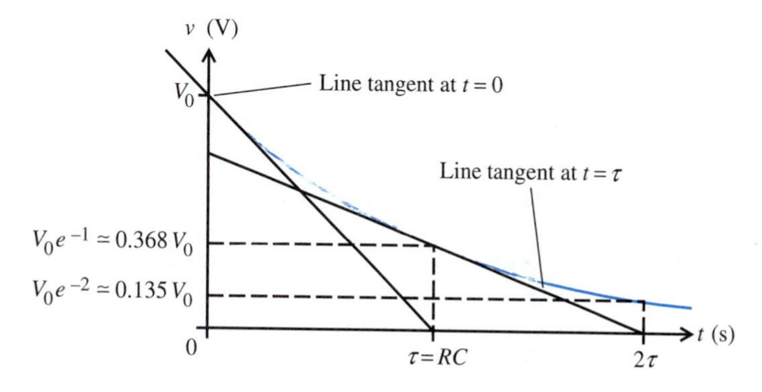

FIGURE 8.2 Source-free response of a simple *RC* circuit

With the *time constant* defined as

$$\tau = RC \tag{8.9}$$

the source-free response is

$$v = V_0 e^{-t/\tau} \qquad t > 0 \tag{8.10}$$

Note that after one time constant,

$$v(\tau) = V_0 e^{-1} = 0.368V_0 \tag{8.11}$$

after two time constants,

$$v(2\tau) = V_0 e^{-2} = 0.135V_0 \tag{8.12}$$

and after five time constants,

$$v(5\tau) = V_0 e^{-5} = 0.0067V_0 \tag{8.13}$$

For this reason we often say that in a source-free circuit the voltage across a capacitance is negligible, compared to its original value, after five time constants. We also see that if the voltage continued to change at its initial rate, it would reach its final value in one time constant, as indicated by the tangent lines in Fig. 8.2.

Now that we have determined voltage v, we can easily find the capacitance current from the terminal equation

$$i = C\frac{d}{dt}v = C\frac{d}{dt}V_0 e^{-t/RC} = -\frac{1}{R}V_0 e^{-t/RC} \qquad t > 0 \tag{8.14}$$

† We will learn to determine initial conditions in Section 8.3.

or by applying KCL and Ohm's law to the original circuit:

$$i = -\frac{1}{R}v = -\frac{1}{R}V_0 e^{-t/RC} \qquad t > 0 \tag{8.15}$$

We can see from Eqs. (8.8) and (8.14) that both the voltage and current asymptotically approach zero with time. Energy stored in the capacitance is gradually dissipated in the resistance. A loss of energy reduces the capacitance voltage, so the voltage continues to decrease, but at a reduced rate. If we calculate the energy dissipated in the resistance between $t = 0^+$ and infinity, we obtain

$$\begin{aligned} W &= \int_0^\infty \frac{1}{R} v^2\, dt = \int_0^\infty \frac{1}{R} V_0^2 e^{-2(1/RC)t}\, dt \\ &= \frac{1}{2} C V_0^2 \end{aligned} \tag{8.16}$$

The right-hand side of Eq. (8.16) is the energy stored in the capacitance at $t = 0^+$. Therefore all energy stored in the capacitance is eventually dissipated in the resistance.

First-Order Equations

We will now introduce some mathematical terminology relating to Eq. (8.2). This is a *first-order differential equation,* because it contains only the first derivative of the unknown time function v. Only a first derivative is present because the equation describes a *first-order circuit* (a circuit with only one energy-storage element). Equation (8.2) also describes a *source-free circuit,* and we say that this differential equation is *source-free* or *homogeneous,* because there is no forcing function present (the right side of this equation is zero). We call the solution to a source-free equation the *source-free response,* the *complementary response,* or the *natural response.*

We will now generalize the results of our solution for the natural response of source-free *RC* circuit in order to solve other source-free differential equations without resorting to integration. We can write Eq. (8.2) as

$$\frac{d}{dt}v - s_1 v = 0 \tag{8.17}$$

where

$$s_1 = -\frac{1}{RC} = -\frac{1}{\tau} \tag{8.18}$$

for the *RC* circuit of Fig. 8.1. We know that the solution to a first-order differential equation is an exponential of the form Ae^{st}. We substitute this into Eq. (8.17) and perform the differentiation to obtain

$$(s - s_1)Ae^{st} = 0 \tag{8.19}$$

We know that Ae^{st} is not zero except for the trivial case where $A = 0$. Therefore, for Ae^{st} to be a solution, s must satisfy the *characteristic equation*

$$\mathscr{A}(s) = s - s_1 = 0 \tag{8.20}$$

We see that the root s_1 of the characteristic equation determines the exponent in the natural response

$$v = Ae^{s_1 t} = Ae^{-t/\tau} = Ae^{-t/RC} \tag{8.21}$$

and thus determines the source-free response, except for the constant of integration A. Observe that to obtain the characteristic equation [Eq. (8.20)], we replace differentiation in Eq. (8.17) with multiplication by s and divide by v. We summarize the procedure as follows:

> To solve any first-order source-free differential equation, replace differentiation with multiplication by s and divide by the dependent variable (v) to obtain the characteristic equation. The natural response is $Ae^{s_1 t}$, where s_1 is the root of the characteristic equation and A is a constant of integration. Evaluate this natural response at the initial time and equate it to the initial value of the variable to solve for A. This gives the source-free response.

We now apply the procedure to obtain the source-free response of a simple circuit.

EXAMPLE 8.1 Determine the response for the *RC* circuit shown in Fig. 8.3 if $v(0^+) = 9$ V.

3 Ω
$\frac{1}{8}$ F
+
v
−
6 Ω

FIGURE 8.3
A first-order *RC* circuit

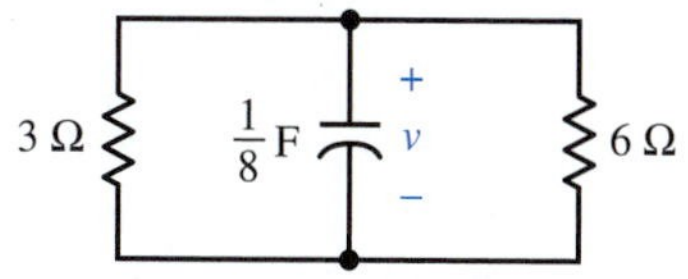

Solution Application of KCL to the top node gives us

$$\frac{1}{3}v + \frac{1}{8}\frac{d}{dt}v + \frac{1}{6}v = 0 \qquad t > 0$$

which reduces to

$$\frac{1}{8}\frac{d}{dt}v + \frac{1}{2}v = 0 \qquad t > 0$$

Multiplication of the preceding equation by 8 yields the characteristic equation

$$s + 4 = 0$$

with root

$$s_1 = -4$$

This gives the natural response

$$v = Ae^{-4t}$$

We now equate the value of v at $t = 0^+$ and the preceding equation evaluated at $t = 0^+$ to calculate the constant of integration.

$$v(0^+) = 9 = Ae^0 = A$$

This gives the response

$$v = 9e^{-4t}\text{ V} \qquad t > 0$$

The Time-Constant Method

If we had calculated the equivalent resistance of the two parallel resistances in Fig. 8.3, we would have found $R = 2\ \Omega$ and the time constant in seconds would be

$$\tau = RC = 2\left(\frac{1}{8}\right) = \frac{1}{4} = \frac{1}{s_1} \tag{8.22}$$

and we could immediately write

$$v = v(0^+)e^{-t/\tau} = 9e^{-4t}\text{ V} \qquad t > 0 \tag{8.23}$$

This method provides a quick way to write the source-free response of a first-order circuit:

The source-free response for any voltage or current in a first-order circuit is always given by

$$v = v(0^+)e^{-t/\tau} \qquad t > 0 \tag{8.24}$$

or

$$i = i(0^+)e^{-t/\tau} \qquad t > 0 \tag{8.25}$$

where $v(0^+)$ and $i(0^+)$ are initial conditions, and τ is the time constant calculated by use of the equivalent resistance seen by the energy-storage component.

Remember The source-free response (voltage or current) of an *RC* circuit is of the form $Ae^{-t/\tau}$, where the time constant is $\tau = RC$. The root of the characteristic equation determines the natural response except for the constant of integration.

EXERCISES

1. Determine the source-free response of an *RC* circuit described by each of the following differential equations.

   (a) $\frac{d}{dt}v + 6v = 0, \quad v(0^+) = 12\text{ V}$ answer: $12e^{-6t}$ V

   (b) $3\frac{d}{dt}v + 6v = 0, \quad v(0^+) = 12\text{ V}$ answer: $12e^{-2t}$ V

   (c) $i + 4\int_{-\infty}^{t} i d\lambda = 0,\ i(0^+) = 7\text{ A}$ *Hint:* First differentiate to eliminate the integral. answer: $7e^{-4t}$ A

2. Use KCL to obtain a differential equation in terms of voltage v, and solve for the response in the source-free circuit shown if $v(0^+) = 10$ V.

   answer: $10e^{-2t/3}$ V for $t > 0$

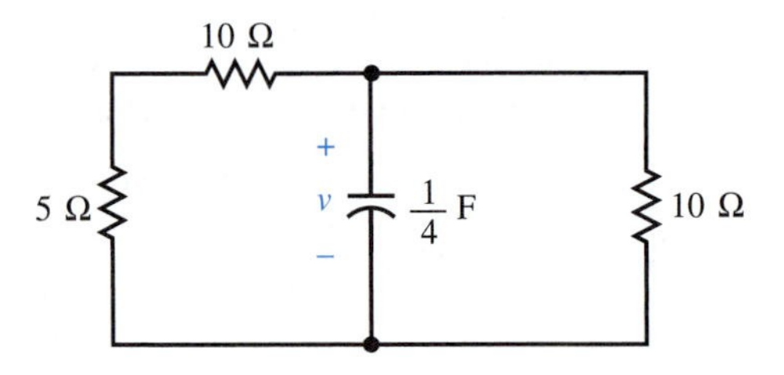

3. For the following circuit, use KVL to write a differential equation in terms of current i, and solve for the source-free current response if $v(0^+) = 40$ V. (Source-

free implies no independent sources.) *Hint:* Use $v(0^+)$ and evaluate the KVL equation at $t = 0^+$ to determine $i(0^+)$. Then differentiate the original KVL equation to eliminate the integral. Solve this equation for the source-free response i for $t > 0$.

answer: $2e^{-t/5}$ A for $t > 0$

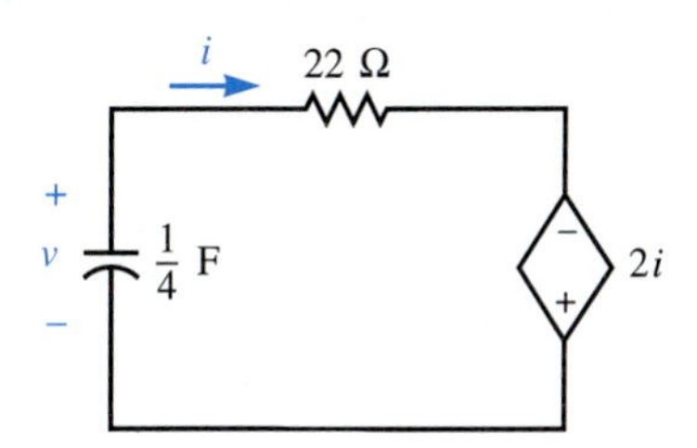

4. The following circuit is source-free for $t > 0$, but $v_C(0^+) = 100$ V. Calculate the equivalent resistance R seen by the capacitance. Use R and $v_C(0^+)$ to calculate $i_C(0^+)$. Use this current and the current-divider relation to calculate $i(0^+)$. Use R and the capacitance value to calculate the time constant τ. Now use the time-constant method to write $i(t)$ for $t > 0$, without solving for $v(t)$.

answer: $6e^{-2t}$ A for $t > 0$

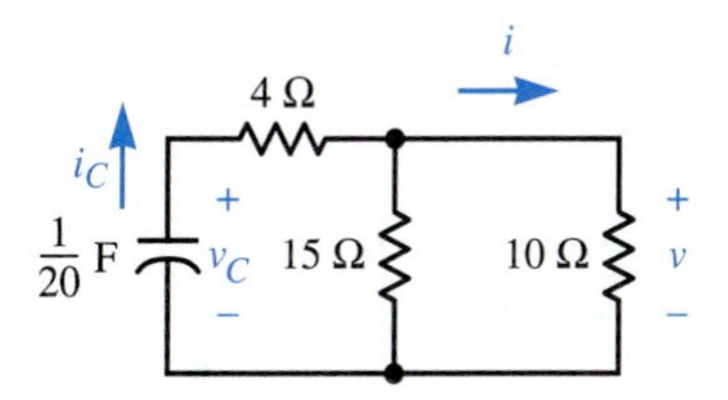

8.2 The Source-Free *RL* Circuit

Some systems make use of the energy stored in an inductor after the power source is disconnected. The classic breaker-point ignition system for an automobile is one example. We will now analyze simple circuits of this type.

The current through the inductance of Fig. 8.4 has a value $i(0) = I_0$ A at $t = 0$, so the energy stored in the inductance is $w(0) = \frac{1}{2}LI_0^2$. The inductance current immediately after the switch is opened at $t = 0$ is $i(0^+) = i(0) = I_0$ A, because the stored energy cannot change instantaneously. We wish to determine i for $t > 0$.

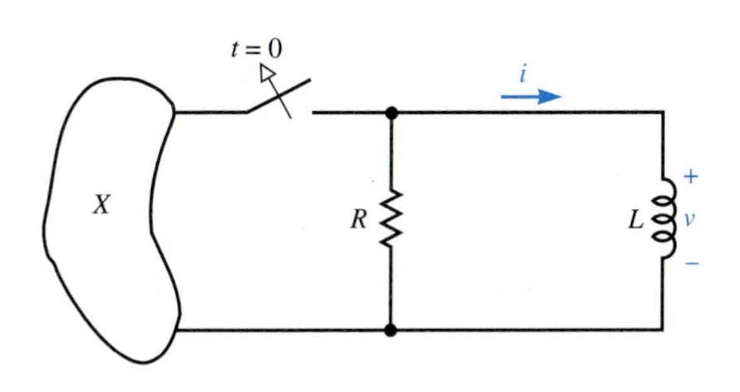

FIGURE 8.4
An *RL* circuit that is source-free for $t > 0$

Application of KVL around the loop gives us the differential equation

$$L\frac{d}{dt}i + Ri = 0 \qquad t > 0 \tag{8.26}$$

We can divide Eq. (8.26) by L to get

$$\frac{d}{dt}i + \frac{R}{L}i = 0 \qquad t > 0 \tag{8.27}$$

To obtain the characteristic equation we replace the derivative in Eq. (8.27) with s and divide by i. This gives us

$$s + \frac{R}{L} = 0 \tag{8.28}$$

This characteristic equation has the root

$$s_1 = -\frac{R}{L} \tag{8.29}$$

and we have the source-free response

$$i = Ae^{-(R/L)t} \qquad t > 0 \tag{8.30}$$

To determine the constant of integration, evaluate Eq. (8.30) at $t = 0^+$ and equate this to $i(0^+)$:

$$i(0^+) = I_0 = Ae^0 = A \tag{8.31}$$

Thus the inductance current is

$$i = I_0e^{-(R/L)t} \qquad t > 0 \tag{8.32}$$

This response is shown in Fig. 8.5. With the time constant given by

$$\tau = \frac{L}{R} \tag{8.33}$$

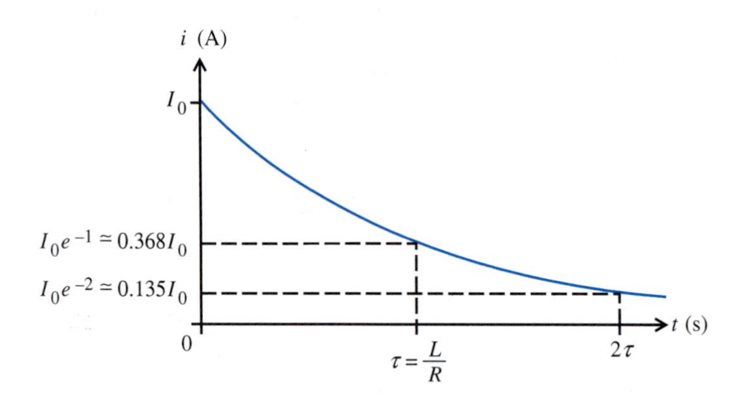

FIGURE 8.5
Source-free response of a simple *RL* circuit

the source-free response for the inductance current is

$$i = I_0e^{-t/\tau} \qquad t > 0 \tag{8.34}$$

We can calculate the inductance voltage v from the terminal equation:

$$v = L\frac{d}{dt}i = L\frac{d}{dt}I_0e^{-(R/L)t} = -RI_0e^{-(R/L)t} \qquad t > 0 \tag{8.35}$$

or by applying KVL and Ohm's law to the series *RL* circuit:

$$v = -Ri = -RI_0e^{-(R/L)t} \qquad t > 0 \tag{8.36}$$

As with the source-free *RC* circuit, all of the energy $w(0^+) = (\frac{1}{2})LI_0^2$ stored in the circuit at $t = 0$ is eventually dissipated in the resistance. (The results obtained for this *RL* circuit can also be obtained directly from those for the *RC* circuit by the use of duality.)

The Time-Constant Method

For more complicated first-order *RL* circuits, we often calculate the equivalent resistance R seen by the inductance and use this value of R to calculate the time constant:

$$\tau = \frac{L}{R} \tag{8.37}$$

The solution for any voltage or current in the circuit is then given by

$$i = i(0^+)e^{-t/\tau} \qquad t > 0 \tag{8.38}$$

or

$$v = v(0^+)e^{-t/\tau} \qquad t > 0 \tag{8.39}$$

The procedure is the same as the time-constant method for *RC* circuits. Only the time-constant calculation is different.

EXAMPLE 8.2 Use the time-constant method to write $i(t)$ for $t > 0$ in the circuit of Fig. 8.6 if $i_L(0^+) = 16$ A.

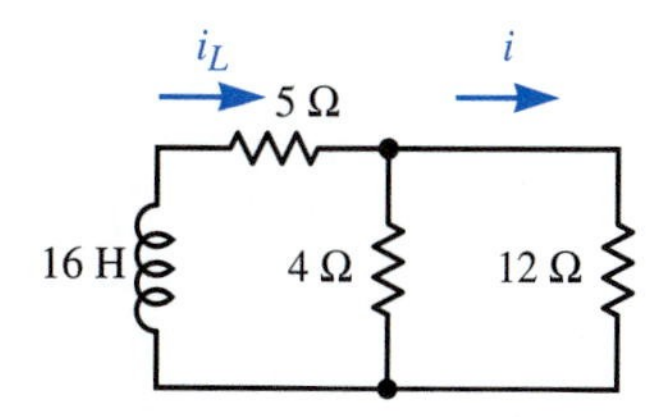

FIGURE 8.6
A first-order *RL* circuit

Solution The current-divider relation gives

$$i(0^+) = \frac{1/12}{1/12 + 1/4}i_L(0^+) = 4 \text{ A}$$

The equivalent resistance is

$$R = 5 + \frac{1}{1/12 + 1/4} = 8\ \Omega$$

The time constant in seconds is

$$\tau = L/R = 16/8 = 2$$

We can now write

$$i = i(0^+)e^{-t/\tau} = 4e^{-t/2} \text{ A} \qquad t > 0$$

Remember The source-free response (current or voltage) of an RL circuit is of the form $Ae^{-t/\tau}$. The time constant is $\tau = L/R$.

EXERCISES

5. Integrate the power, $p = Ri^2 = R(I_0e^{-tR/L})^2$, absorbed by the resistance in a source-free RL circuit from $t = 0^+$ to $t = \infty$ to show that the resistance eventually absorbs all of the energy stored in the inductance at $t = 0^+$.

6. Use KVL to obtain a differential equation in terms of current i, and solve for the response in the following source-free circuit if $i(0^+) = 10$ A.
answer: $10e^{-t/6}$ A for $t > 0$

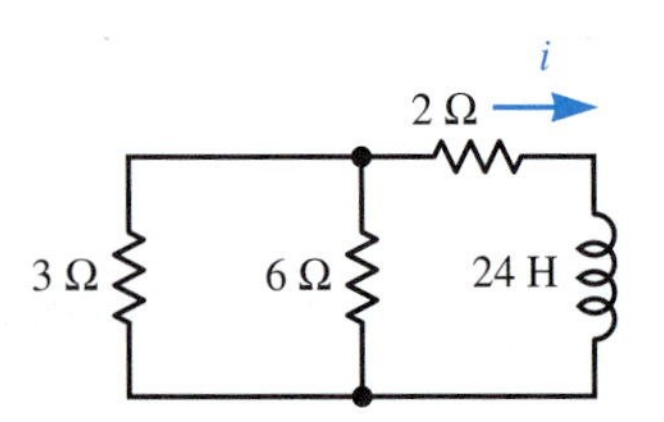

7. For the following circuit, use KCL to write a differential equation in terms of voltage v, and solve for the source-free response if $i(0^+) = 10$ A. (Source-free implies no independent sources.) *Hint:* Use $i(0^+)$ and evaluate the KCL equation at $t = 0^+$ to determine $v(0^+)$. Next differentiate the original KCL equation to eliminate the integral. Solve this differential equation for the source-free response v when $t > 0$.
answer: $-60e^{-72t}$ V for $t > 0$

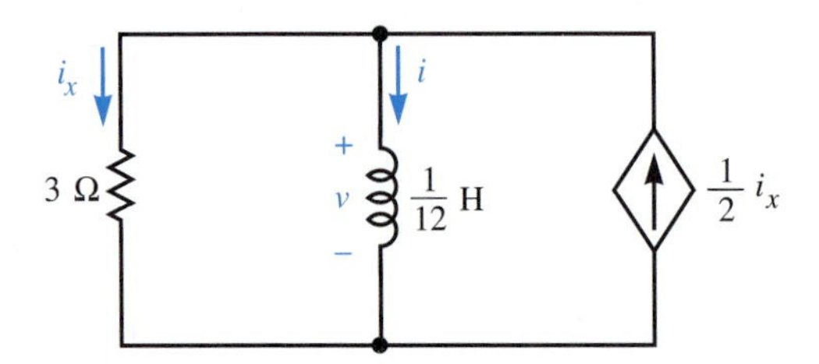

8. Determine the Thévenin resistance seen by the inductance in the preceding circuit. If $i(0^+) = 10$ A, use the time-constant method to write
   (a) the inductance current i for $t > 0$, and — answer: $10e^{-72t}$ A for $t > 0$
   (b) the voltage v for $t > 0$. — answer: $-60e^{-72t}$ V for $t > 0$

8.3 The Driven *RC* Circuit

When we first connect a voltage or current source to a circuit with an energy-storage element, the voltages and currents do not immediately reach their steady-state value. There is a transient time during which the natural response may be a significant, or even major, part of the complete response.

We begin with the analysis of a first-order *RC* circuit driven by an independent source (a *driven circuit*) and see that the complete response is the sum of the natural response component, with a form determined by the source-free circuit, and the particular response component, with a form determined by the source. We find that for a very important class of sources *(inputs)* we can determine the response without resorting to integration.

The following KCL equation describes the response of the first-order *RC* circuit shown in Fig. 8.7:

$$C\frac{d}{dt}v + \frac{1}{R}(v - v_s) = 0 \tag{8.40}$$

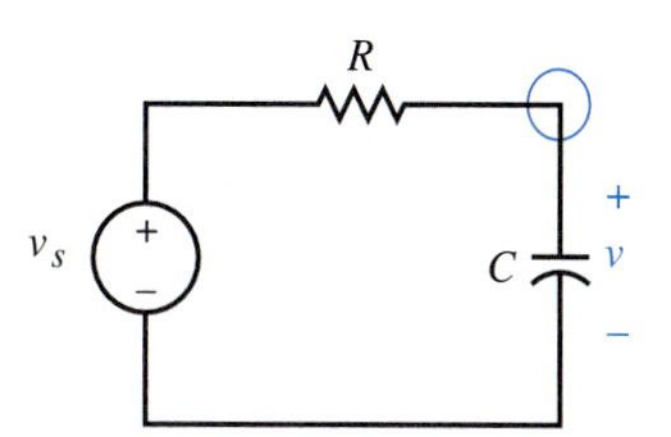

FIGURE 8.7
A first-order *RC* circuit

If we divide Eq. (8.40) by C we obtain

$$\frac{d}{dt}v + \frac{1}{RC}v = \frac{1}{RC}v_s \tag{8.41}$$

Equation (8.41) is a driven, or forced, first-order differential equation. We will next show how to solve equations of this type.

The Driven First-Order Equation

Inspection of Eq. (8.41) reveals that it is of the form

$$\frac{d}{dt}v - s_1 v = g(t) \tag{8.42}$$

where

$$s_1 = -\frac{1}{\tau} = -\frac{1}{RC} \tag{8.43}$$

We will solve Eq. (8.42) by the same method that we used for the source-free equation.† We multiply Eq. (8.42) by the integrating factor $e^{-s_1 t}$, where s_1 is the root of the characteristic equation

$$\mathcal{A}(s) = s - s_1 = 0 \tag{8.44}$$

to obtain

$$e^{-s_1 t}\frac{d}{dt}v - e^{-s_1 t}s_1\, v = e^{-s_1 t}g(t) \tag{8.45}$$

† Zill, *A First Course in Differential Equations.*

We know from the rule for the derivative of the product of two functions that Eq. (8.45) is equivalent to

$$\frac{d}{dt}(e^{-s_1 t}v) = e^{-s_1 t}g(t) \tag{8.46}$$

We integrate both sides of this equation with respect to time and multiply the result by $e^{s_1 t}$ to obtain

$$v = Ae^{s_1 t} + e^{s_1 t}\int e^{-s_1 t}g(t)dt \tag{8.47}$$

We see that the *complete response* for a first-order circuit is

$$v = v_n + v_p \tag{8.48}$$

where the *natural response* (the complementary response) v_n is as found in Section 8.1:

$$v_n = Ae^{s_1 t} \tag{8.49}$$

and the *particular response* v_p is determined by

The Particular Response

$$v_p = e^{s_1 t}\int e^{-s_1 t}g(t)dt \tag{8.50}$$

To determine the constant of integration A in Eq. (8.49), we *evaluate the complete response,* as given by Eq. (8.48), at $t = 0^+$ and equate it to the *known initial value* $v(0^+)$:

$$\begin{aligned} v(0^+) &= v_n(0^+) + v_p(0^+) \\ &= Ae^0 + v_p(0^+) \end{aligned} \tag{8.51}$$

We rearrange Eq. (8.51) to obtain

$$A = v_p(0^+) - v(0^+) \tag{8.52}$$

Therefore the complete response for a first-order circuit is given by

The Complete Response

$$v(t) = v_p(t) + [v(0^+) - v_p(0^+)]e^{s_1 t} \qquad t > 0 \tag{8.53}$$

Equation (8.53) is quite general. We can use Ohm's law and KVL to show that *any* voltage v in a first-order circuit is given by an equation of this form. The only requirement for a bounded forcing function $g(t)$ is that the particular response is defined and continuous for $t > 0$. Ohm's law and KCL will show that *any current i* in the circuit satisfies a similar equation.

We will now concentrate on the evaluation of Eq. (8.50) to determine the particular response for our most important inputs.

Exponential Inputs

The dc and sinusoidal signals are among the more important signals that occur in engineering practice. In Chapter 7 we saw that a dc signal was just a special case of an exponential signal, and Euler's identity lets us represent sinusoidal signals in terms of complex exponentials. Therefore, when we obtain a solution to Eq. (8.50) for the exponential *forcing function*

$$g(t) = Ke^{s_p t} \tag{8.54}$$

where s_p can be zero, real, imaginary, or complex, we will have found the solution for our most important inputs. Substitution of Eq. (8.54) into Eq. (8.50) and integration gives the solution to Eq. (8.50) as

$$v_p = \begin{cases} \dfrac{1}{s_p - s_1} Ke^{s_p t} & s_p \neq s_1 \quad (8.55a) \\ Kte^{s_p t} & s_p = s_1 \quad (8.55b) \end{cases}$$

Equation (8.55a) represents the usual situation. Equation (8.55b) is a rather unusual case and seldom encountered in practice.

We can rewrite Eq. (8.55a) using the characteristic polynomial $\mathscr{A}(s) = s - s_1$:

$$v_p = \frac{1}{\mathscr{A}(s_p)} Ke^{s_p t} \qquad \mathscr{A}(s_p) \neq 0 \tag{8.56}$$

If we rearrange Eq. (8.56) as

$$\mathscr{A}(s_p)v_p = Ke^{s_p t} \qquad \mathscr{A}(s_p) \neq 0 \tag{8.57}$$

we observe the following important result:

The Particular Response to an Exponential Input for $\mathscr{A}(s_p) \neq 0$

To obtain the particular response for an exponential forcing function $Ke^{s_p t}$, *replace differentiation with multiplication* by s_p.

This result is logical, because differentiation of $e^{s_p t}$ with respect to time is equivalent to multiplication by s_p. (This insight is the foundation of steady-state ac circuit analysis introduced in Chapter 10.)

EXAMPLE 8.3 Determine the complete response v of the *RC* circuit in Fig. 8.7, if $R = 16\ \Omega$, $C = \frac{1}{128}$ F, $v(0^+) = 4$ V, and $v_s(t) = 12e^{-4t}$ V for $t > 0$.

Solution The *RC* circuit of Fig. 8.7 is described by the KCL equation

$$\frac{1}{128}\frac{d}{dt}v + \frac{1}{16}v = \frac{1}{16}v_s$$

Multiplication of the preceding equation by 128 simplifies the algebra:

$$\frac{d}{dt}v + 8v = 8v_s$$

The characteristic equation

$$s + 8 = 0$$

has the root $s_1 = -8$, so the natural response is

$$v_n = Ae^{-8t} \qquad t > 0$$

We substitute the input

$$v_s = 12e^{-4t} \qquad t > 0$$

into the differential equation and replace d/dt with $s_p = -4$ to obtain:

$$-4v_p + 8v_p = 8(12e^{-4t}) \qquad t > 0$$

This gives the particular response

$$v_p = 24e^{-4t} \text{ V} \qquad t > 0$$

The complete response is the sum of the natural response and the particular response [Eq. (8.48)]:

$$v = Ae^{-8t} + 24e^{-4t} \qquad t > 0$$

We can evaluate this equation at $t = 0^+$ and equate it to $v(0^+)$ to determine the constant of integration A:

$$v(0^+) = 4 = Ae^0 + 24e^0$$

which yields

$$A = -20$$

and gives the complete response

$$v = -20e^{-8t} + 24e^{-4t} \text{ V} \qquad t > 0$$

This result is also obtained by substitution of the appropriate voltages into Eq. (8.53). The response is shown in Fig. 8.8.

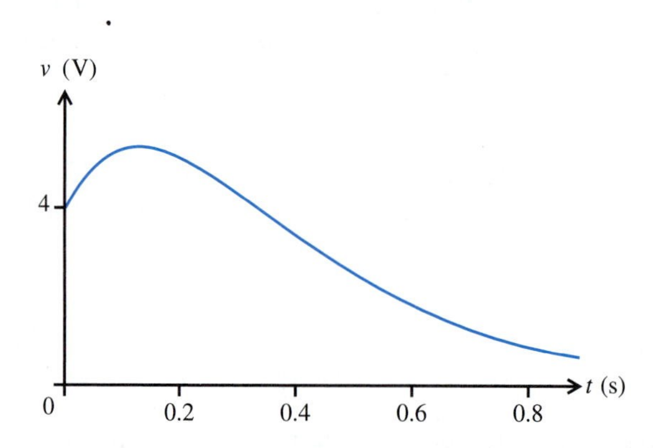

FIGURE 8.8 The complete response of an *RC* circuit when the capacitance has an initial voltage $v(0^+) =$ 4 V and the source is an exponential $v_s =$ $12e^{-4t}$ V for $t > 0$

We will now make two observations that we use to determine initial conditions when a source abruptly changes value. We have stated that an instantaneous change in capacitance voltage requires an instantaneous change in stored energy, which is a physical impossibility. This is also demonstrated by the terminal equation

$$i_C = C\frac{d}{dt}v_C \tag{8.58}$$

If the voltage $v_C(t)$ changed instantaneously, the capacitance current would be infinite, which is physically impossible. Mathematically we say that

Capacitance Voltage Is Continuous

$$v_C(t_0^+) = v_C(t_0) = v_C(t_0^-) \tag{8.59}$$

We can also reason that

Inductance Current Is Continuous

$$i_L(t_0^+) = i_L(t_0) = i_L(t_0^-) \tag{8.60}$$

for finite voltage.

Constant Inputs

A constant K is just a special, but important, case of an exponential: $K = Ke^{s_p t}$, with $s_p = 0$. Therefore:

The Particular Response for a Constant Input

To obtain the particular response for a constant input, replace differentiation with multiplication by zero.

The preceding result is logical, because a constant source gives a constant particular response, and the derivative of a constant is zero.

EXAMPLE 8.4 Determine the response v of the *RC* circuit in Fig. 8.7 to the step input $v_s = V_s u(t)$ V.

Solution The source is $v_s(t) = 0$ for all $t < 0$, so there is no energy stored in the capacitance at $t = 0^-$. This gives $v(0^-) = 0$, and continuity of capacitance voltage assures us that the initial condition for capacitance voltage is

$$v(0^+) = v(0^-) = 0$$

The circuit is described by Eq. (8.41), which we repeat:

$$\frac{d}{dt}v + \frac{1}{RC}v = \frac{1}{RC}v_s$$

and the natural response is

$$v_n = Ae^{-(1/RC)t} \qquad t > 0$$

We substitute the input

$$v_s = V_s e^0 = V_s \qquad t > 0$$

into the differential equation and replace d/dt with $s_p = 0$ to obtain:

$$0(v_p) + \frac{1}{RC} v_p = \frac{1}{RC} V_s$$

This gives the particular response

$$v_p = V_s \qquad t > 0$$

The complete response is the sum of the natural response and the particular response [Eq. (8.48)]:

$$v = Ae^{-(1/RC)t} + V_s \qquad t > 0$$

We can evaluate this equation at $t = 0^+$ and equate it to $v(0^+)$ to determine the constant of integration A:

$$v(0^+) = 0 = Ae^0 + V_s$$

which yields

$$A = -V_s$$

The result,

$$v = V_s - V_s e^{-(1/RC)t} \qquad t > 0$$

is depicted in Fig. 8.9. This result is also obtained by substituting the appropriate voltages into Eq. (8.53).

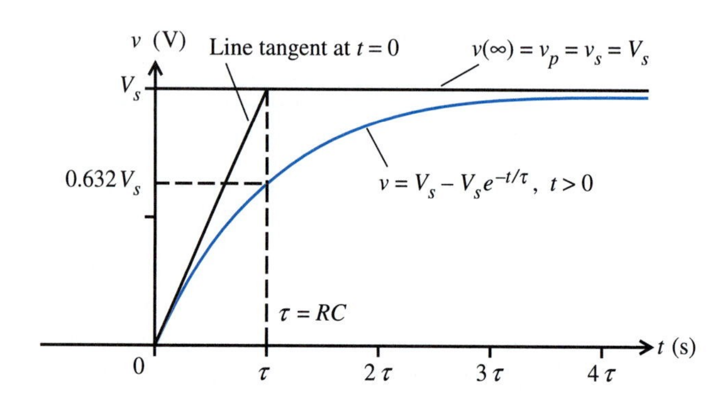

FIGURE 8.9 The complete response of an RC circuit to the step input $v_s = V_s u(t)$ V

We observed that the natural response of the circuits we have analyzed tends to zero with time. (This is clearly the case in Fig. 8.9.) We conclude that:

For Constant Sources

If a constant source has been applied to a stable circuit for *a very long time* only the constant particular response remains. That is,

$$v(\infty) = v_p \tag{8.61}$$

and

$$i(\infty) = i_p \tag{8.62}$$

are constant for any voltage v or current i in the circuit.

We found that the particular response is constant for a *constant source.* This result implies that the particular response component of capacitance current is zero:

$$i_{Cp} = C\frac{d}{dt}v_{Cp} = 0 \tag{8.63}$$

and the particular response component of inductance voltage is zero:

$$v_{Lp} = L\frac{d}{dt}i_{Lp} = 0 \tag{8.64}$$

for a *constant source.* Equations (8.63) and (8.64) provide the means to determine the particular response for constant sources *directly from the network.*

The Particular Response for a Constant Source

The *particular response* for a stable network with *constant sources* is found by replacing all capacitances by open circuits ($i_{Cp} = 0$) and all inductances by short circuits ($v_{Lp} = 0$).

(Although we have yet to determine the response of second-order circuits, Eqs. (8.61) through (8.64) and the preceding conclusions still apply.)

EXAMPLE 8.5 The voltage source in the circuit of Fig. 8.10 has the value

$$v_s = \begin{cases} 50\text{ V} & t < 0 \\ 20\text{ V} & t \geq 0 \end{cases}$$

(a) Determine $v(0^-)$, and (b) $v(t)$ for $t > 0$.

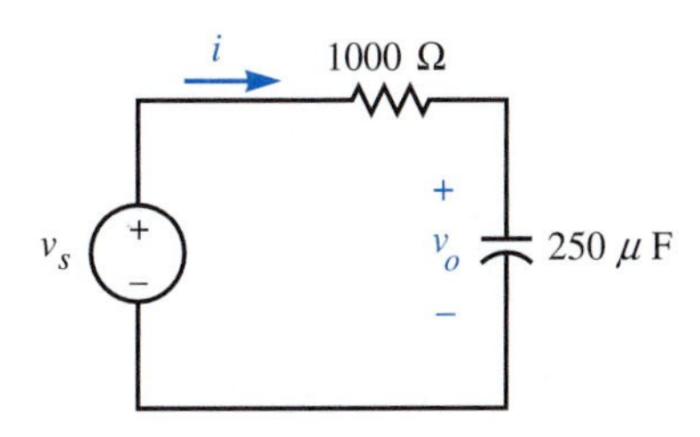

FIGURE 8.10 The *RC* circuit for Example 8.5

Solution (a) The voltage has been constant for a very long time, so only the particular response remains at $t = 0^-$. We can obtain the particular response directly from the circuit for a constant source. Replace the capacitance by an open circuit as shown in Fig. 8.11. From KVL

$$-50 + 1000(0) + v_p = 0 \qquad t < 0$$

so

$$v_p(0^-) = 50\text{ V}$$

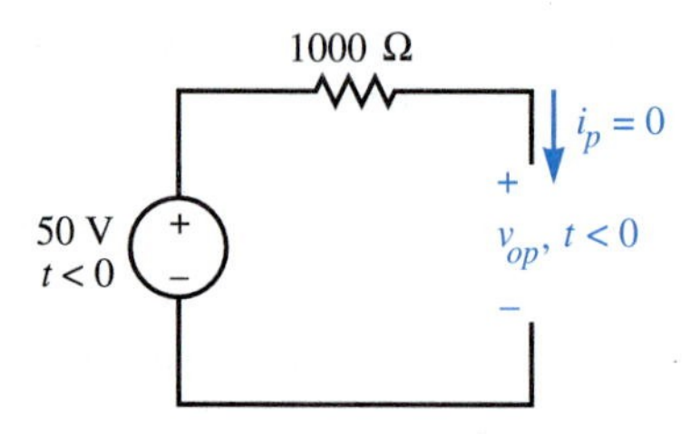

FIGURE 8.11 Equivalent circuit to determine the particular response when $t < 0$

(b) To obtain the particular response for $t > 0$, we replace the capacitance with an open circuit. The source now has a value of 20 V as shown in Fig. 8.12. Application of KVL obviously yields

$$v_p = 20 \text{ V} \qquad t > 0$$

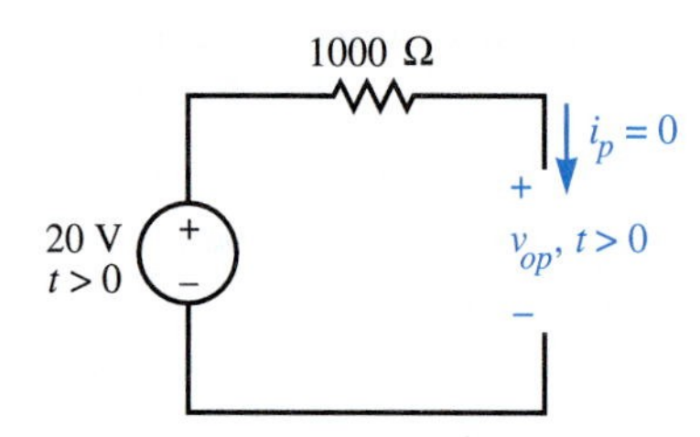

FIGURE 8.12 Equivalent circuit to determine the particular response for $t > 0$

Application of KCL to a surface enclosing the top right-hand node of Fig. 8.10 yields, after multiplication by 4000,

$$\frac{d}{dt}v + 4v = 4(20) \qquad t > 0$$

(We could easily find the particular response for $t > 0$ from the preceding differential equation. Just replace differentiation with multiplication by zero.)

We recognize that the characteristic equation is

$$s + 4 = 0$$

with root

$$s_1 = -\frac{1}{\tau} = -4$$

This gives the natural response

$$v_n = Ae^{-4t} \qquad t > 0$$

and the complete response is

$$v = v_p + v_n = 20 + Ae^{-4t} \qquad t > 0$$

Because capacitance voltage is continuous,

$$v(0^+) = v(0^-) = 50 \text{ V}$$

We evaluate $v(t)$ at $t = 0^+$ and equate it to the initial value $v(0^+)$:

$$v(0^+) = 50 = 20 + Ae^0$$

so

$$A = 30$$

and

$$v = 20 + 30e^{-4t} \qquad t > 0$$

The response is sketched in Fig. 8.13.

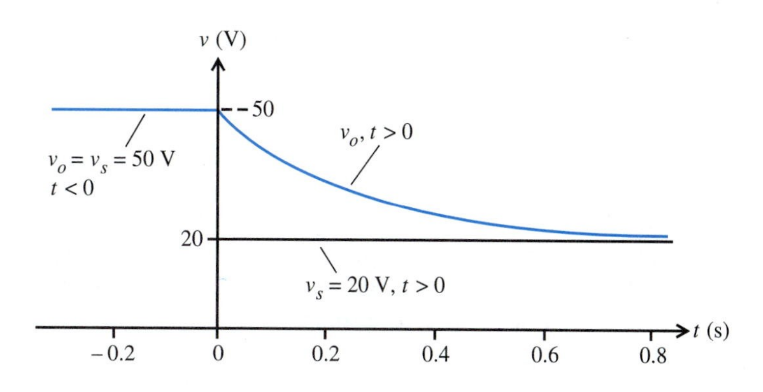

FIGURE 8.13 The input voltage v_s and the response v for the network of Fig. 8.10

We now see how we can solve problems of this type without writing a differential equation.

The Time-Constant Method

Substitution of Eqs. (8.61) and (8.62) into Eq. (8.53) gives us the useful result:

The Complete Response for a Constant Source

Any voltage v or current i in a first-order circuit driven by a source that is *constant* for $t > 0$ is given by

$$v(t) = v(\infty) + [v(0^+) - v(\infty)]e^{-t/\tau} \qquad t > 0 \qquad (8.65)$$

or

$$i(t) = i(\infty) + [i(0^+) - i(\infty)]e^{-t/\tau} \qquad t > 0 \qquad (8.66)$$

where $v(\infty) = v_p$ and $i(\infty) = i_p$ are the constant particular responses.

To apply Eq. (8.65) or (8.66), we must first calculate the initial capacitance voltage $v_C(0^+)$, if it is not given. We then use this voltage *and* the source value at $t = 0^+$ to calculate the initial value of the voltage, $v(0^+)$ [or current, $i(0^+)$], that we wish to determine. We calculate the Thévenin resistance R_{Th} seen by the source and use this to determine the time constant $\tau = R_{Th}C$. We determine $v(\infty) = v_p$ [or $i(\infty) = i_p$] by replacing the capacitance by an open circuit. To complete the solution, substitute $v(0^+)$, $v(\infty)$, and τ into Eq. (8.65) [or (8.66)]. We demonstrate the procedure with an example.

EXAMPLE 8.6 Determine the resistance current i in the circuit of Fig. 8.14 for $t > 0$, if

$$v_p = \begin{cases} 50 \text{ V} & t < 0 \\ 100 \text{ V} & t \geq 0 \end{cases}$$

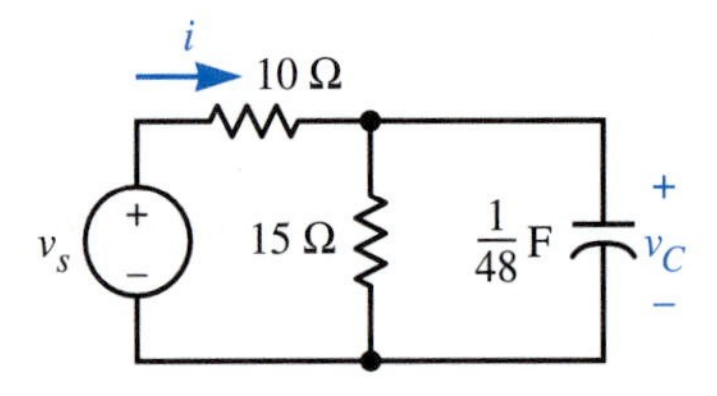

FIGURE 8.14
The circuit for Example 8.6

Solution Replace the capacitance with an open circuit to obtain the particular response:

$$i_p = \frac{1}{25} 50 = 2 \text{ A} \qquad t < 0$$

This will be the current i for $-\infty < t < 0$ because the source has been constant since $t = -\infty$. The particular response component of the capacitance voltage is given by the voltage-divider relation:

$$v_{Cp} = \frac{15}{10 + 15} 50 = 30 \text{ V} \qquad t < 0$$

The source has been constant for a very long time and capacitance voltage is continuous, so

$$v_C(0^+) = v_C(0^-) = v_{Cp}(0^-) = 30 \text{ V}$$

We use KVL, $v_s(0^+)$, and $v_C(0^+)$ to determine $i(0^+)$:

$$i(0^+) = \frac{1}{10}[v_s(0^+) - v_C(0^+)] = 7 \text{ A}$$

[Note that $i(0^+) = 7 \text{ A} \neq i(0^-) = i_p(0^-) = 2$ A. Resistance current need not be continuous. Resistances do not store electric energy.]

To obtain the particular response for $t > 0$, replace the capacitance with an open circuit. The source voltage is now 100 V, so

$$i(\infty) = i_p = \frac{1}{25} 100 = 4 \text{ A} \qquad t > 0$$

The Thévenin resistance seen by the source is

$$R_{Th} = \frac{1}{1/10 + 1/15} = 6 \ \Omega$$

and the time constant in seconds is

$$\tau = R_{Th}C = 6\frac{1}{48} = \frac{1}{8}$$

Substitution of these values of $i(\infty)$, $i(0^+)$ and τ into Eq. (8.66) gives

$$i = 4 + (7 - 4)e^{-8t} \text{ A} \qquad t > 0$$

This response is sketched in Fig. 8.15.

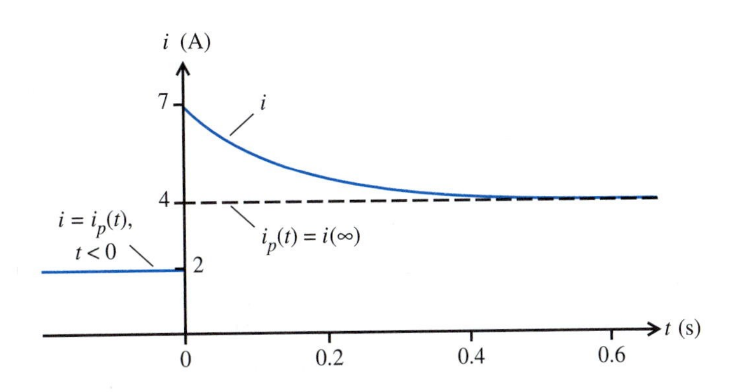

FIGURE 8.15 The response of the circuit in Fig. 8.14 to a change in source value

Remember

We can obtain the particular response of a circuit to an exponential source $V_s e^{s_p t}$ directly from the differential equation, if $\mathscr{A}(s_p) \neq 0$. We simply replace differentiation with multiplication by s_p in the differential equation, and solve this algebraic equation for the particular response.

For a *constant source* ($s_p = 0$) we can obtain the particular response directly from the circuit by replacing all capacitances by open circuits and all inductances by short circuits. If a constant source has been applied for a very long time, only the particular response remains.

We use the fact that capacitance voltage and inductance current are continuous to solve for initial conditions. Always use the complete response to solve for constants of integration.

EXERCISES

9. Write a KCL equation and solve this differential equation for the capacitance voltage in the following *RC* circuit, if the current source i_s is as given below:
   (a) $10\,u(t)$ A — *answer:* $10(1 - e^{-4t})u(t)$ V
   (b) $10e^{-2t}u(t)$ A — *answer:* $20(e^{-2t} - e^{-4t})u(t)$ V
   (c) $10e^{-4t}u(t)$ A [Use Eq. (8.55b) to determine v_p] — *answer:* $40te^{-4t}u(t)$ V

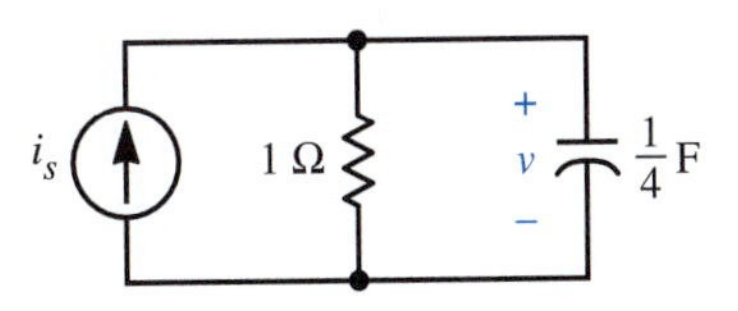

10. For the following circuit, on page 248, v_s is 50 V for $t < 0$ and 150 V for $t > 0$.
    (a) Replace the capacitance by an open circuit, and solve for $v(0^-) = v_p(0^-)$ and $i(0^-) = i_p(0^-)$ directly from the circuit. — *answer:* 50 V, 0 A
    (b) Use continuity of capacitance voltage to determine $v(0^+)$. — *answer:* 50 V
    (c) Write a KVL equation around the loop in the original circuit. Substitute $v(0^+)$ and $v_s(0^+)$ into the KVL equation to determine $i(0^+)$ (capacitance current can be discontinuous). — *answer:* 5 A
    (d) Write a KVL equation around the loop in terms of current i. Differentiate this equation to eliminate the integral, and solve this differential equation for $i(t)$ for $t > 0$. Sketch this response. — *answer:* $5e^{-2t}$ A $\quad t > 0$

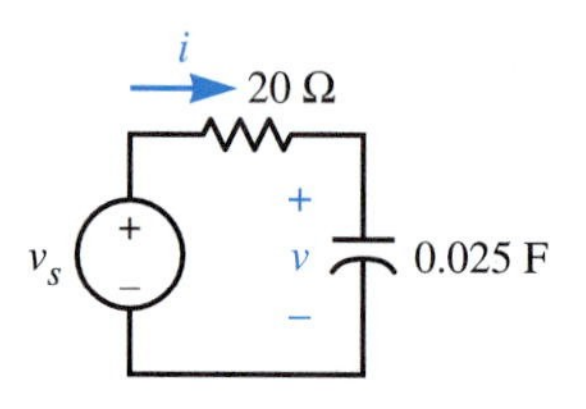

11. Use the time-constant method to solve for current i in the preceding RC circuit, if v_s is 50 V for $t < 0$ and 150 V for $t > 0$. answer: $5e^{-2t}$ A $t > 0$
12. The switch in the following circuit has been open for a very long time and is closed at $t = 0$. Use the time-constant method to calculate $v(t)$ for $t > 0$. (Be sure to calculate the time constant with the switch closed.)
answer: $20(1 - e^{-25t})$ V $t > 0$

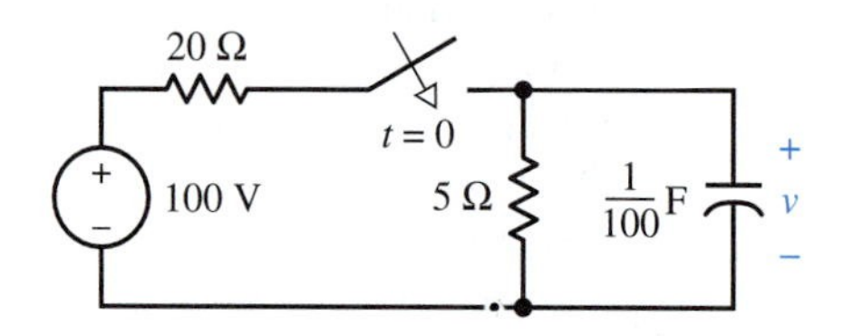

8.4 The Driven *RL* Circuit

We will now write the KVL equation that describes the driven RL circuit of Fig. 8.16:

$$-v_s + Ri + L\frac{d}{dt}i = 0 \tag{8.67}$$

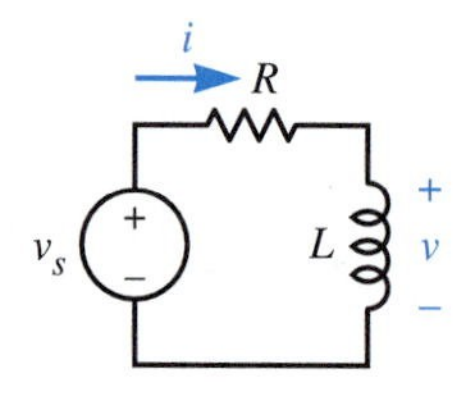

FIGURE 8.16
A first-order RL circuit

We can rearrange this equation to obtain

$$\frac{d}{dt}i + \frac{R}{L}i = \frac{1}{L}v_s \tag{8.68}$$

We recognize that this equation is in the form of Eq. (8.42):

$$\frac{d}{dt}i - s_1 i = g(t) \tag{8.69}$$

The characteristic equation obviously has the root

$$s_1 = -\frac{1}{\tau} = -\frac{R}{L} \tag{8.70}$$

where

$$\tau = \frac{L}{R} \tag{8.71}$$

is the *time constant.* The complete response is in the form of Eq. (8.53), with v replaced by i:

$$i(t) = i_p(t) - [i(0^+) - i_p(0^+)]e^{s_1 t} \qquad t > 0 \tag{8.72}$$

For a constant input, this reduces to the form of Eq. (8.66):

$$i(t) = i(\infty) - [i(0^+) - i(\infty)]e^{-t/\tau} \qquad t > 0 \tag{8.73}$$

The particular response i_p is found by any of the procedures developed in Section 8.3. We see that the procedure to analyze a first-order *RL* circuit is essentially the same as that for a first-order *RC* circuit. We only need to keep in mind the differences already noted:

1. Inductance current is continuous (capacitance voltage is continuous).
2. To obtain the particular response directly from the circuit for a constant source, replace the inductance with a short circuit (replace a capacitance with an open circuit).
3. The time constant is $\tau = L/R$ (for an *RC* circuit the time constant is $\tau = RC$).

EXAMPLE 8.7 Determine the response of the *RL* circuit of Fig. 8.16 to the step input $v_s = V_s u(t)$ V.

Solution We will use the time-constant method. The source has been zero for all $t < 0$, so the energy stored in the inductance at $t = 0^-$ must be zero. This fact and the continuity of inductance current gives us

$$i(0^+) = i(0^-) = 0$$

We replace the inductance by a short circuit to find the particular response for $t > 0$. Ohm's law and KVL applied around the loop in Fig. 8.16 gives

$$i_p = \frac{1}{R} V_s \qquad t > 0$$

Equation (8.73) gives the step response as

$$i(t) = \frac{1}{R} V_s - \frac{1}{R} V_s e^{-t/\tau} \qquad t > 0$$

where the time constant is given by Eq. (8.71):

$$\tau = \frac{L}{R}$$

This current is sketched in Fig. 8.17.

We can determine the inductance voltage from the terminal equation, by the time-constant method, or by using $i(t)$, Ohm's law, and KVL. By any method the result is

$$v = V_s e^{-t/\tau} \qquad t > 0$$

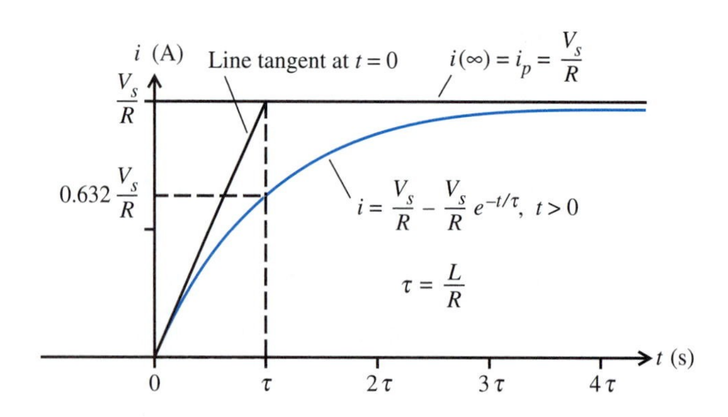

FIGURE 8.17 The response of an RL circuit to the step input $v_s = V_s u(t)$ V

EXAMPLE 8.8 The switch shown in Fig. 8.18 has been closed for a very long time and is opened at $t = 0$. Solve for $i(t)$ for $t > 0$. Sketch this current for $-0.1 < t < 0.3$ s.

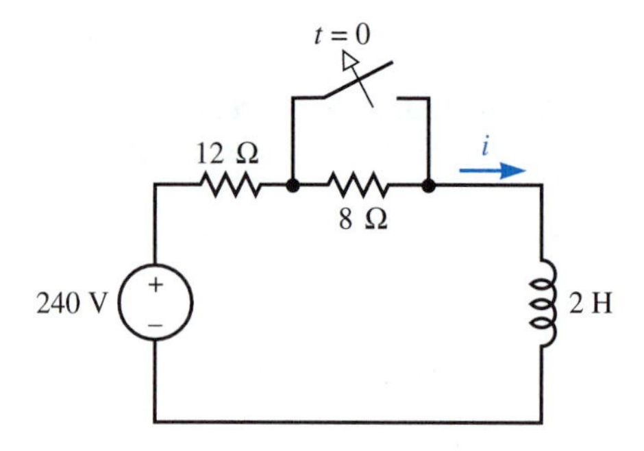

FIGURE 8.18 A switched RL circuit

Solution The source is constant for $t < 0$, so we can obtain the particular response by replacing the inductance with a short circuit. Because the source has been constant for a very long time, only the particular response remains. This fact and the continuity of inductance current gives us the initial condition

$$i(0^+) = i(0^-) = i_p(0^-) = \frac{1}{12} 240 = 20 \text{ A}$$

The source is constant after the switch is opened, so we can obtain the particular response by replacing the inductance with a short circuit. Ohm's law and KVL gives us

$$i_p = \frac{1}{12 + 8} 240 = 12 \text{ A} \qquad t > 0$$

The time constant in seconds *with the switch open* is

$$\tau = \frac{L}{R} = \frac{2}{12 + 8} = \frac{1}{10} \qquad t > 0$$

We can immediately write the complete response by the use of Eq. (8.66) or (8.73):

$$i = i_p + i_n = 12 + [20 - 12]e^{-10t} \text{ A} \qquad t > 0$$

This response is sketched in Fig. 8.19.

The circuits in Examples 8.4 through 8.8 were operating in the *steady state* at $t = 0^-$. That is, the source had been applied for such a long time that the natural

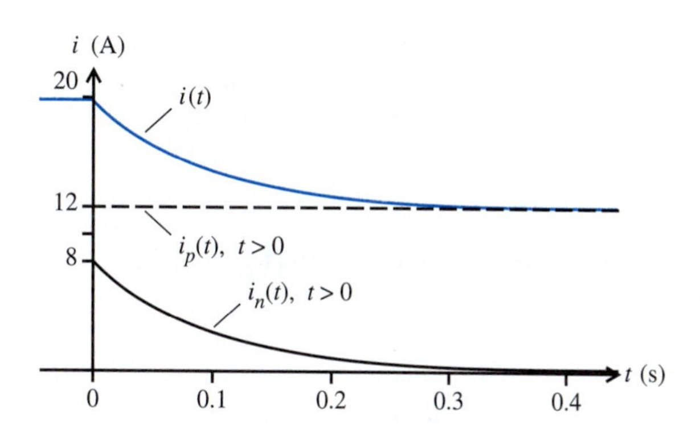

FIGURE 8.19 The response of a switched *RL* circuit

response had decayed to zero and only the particular response remained at $t = 0^-$. The values of $v(0^-)$ and $i(0^-)$ were determined solely by the particular responses $v_p(0^-)$ and $i_p(0^-)$. In the next section we consider an example where we must use the complete response to determine initial conditions.

Remember

We can obtain the particular response of an *RL* circuit to a constant source by replacing the inductance by a short circuit. If the source has been constant for a very long time, only the particular response remains. We use this and the fact that inductance current is continuous to solve for initial conditions. For a switched circuit, be sure to use the time constant that corresponds to the switch position under consideration.

EXERCISES

13. Use the shortcut method for analysis by mesh currents to obtain a differential equation and solve for the inductance current $i(t)$ for $t > 0$, if $i_s(t)$ in the following circuit has the specified value:
    (a) $2u(t)$ A — *answer:* $2(1 - e^{-2t})$ A $\quad t > 0$
    (b) $[2 + 2u(t)]$ A — *answer:* $(4 - 2e^{-2t})$ A $\quad t > 0$
    (c) 2 A for $t < 0$ and $2e^{-6t}$ A for $t > 0$ — *answer:* $3e^{-2t} - e^{-6t}$ A $\quad t > 0$

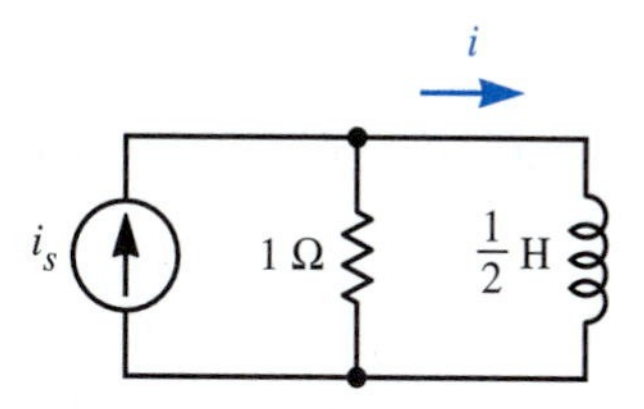

14. The switch in the following circuit has been open for a very long time and is closed at $t = 0$. Use the time-constant method to determine current $i(t)$ for $t > 0$, if $R_1 = 12\ \Omega$, $R_2 = 8\ \Omega$, and $L = 2$ H. — *answer:* $20 - 8e^{-6t}$ A $\quad t > 0$

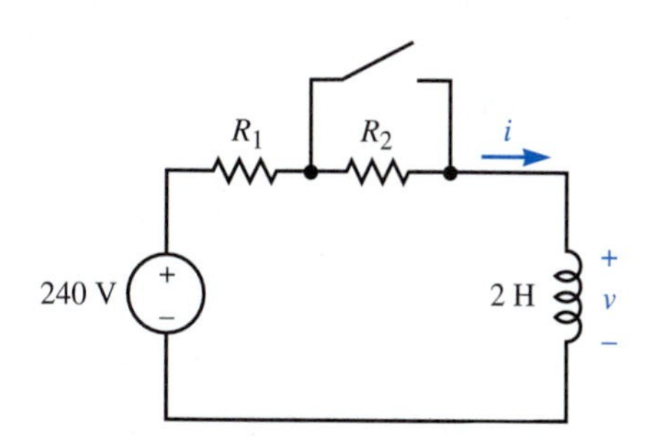

15. Use the time-constant method to determine voltage v for $t > 0$ in the preceding circuit, if the switch has been closed for a very long time and is opened at $t = 0$. The element values are $R_1 = 4\ \Omega$, $R_2 = 396\ \Omega$, and $L = 2$ H.

answer: $-23{,}760e^{-200t}$ V $\quad t > 0$

8.5 Circuits Not in the Steady State

We have analyzed circuits where the source abruptly changed value or a switch position was changed. In all cases the natural response had decayed to zero prior to the change. The circuit was operating in the *steady state,* and the particular response was all that remained. We will now consider the rather common case when a change occurs before a circuit reaches the steady state. For this situation we must use the complete response to determine the voltages and currents just before the change.

EXAMPLE 8.9 The circuit of Example 8.8 is repeated in Fig. 8.20. The switch has been closed for a very long time, is opened at $t = 0$, and then reclosed at $t = t_0 = 0.1$ s. Solve for $i(t)$.

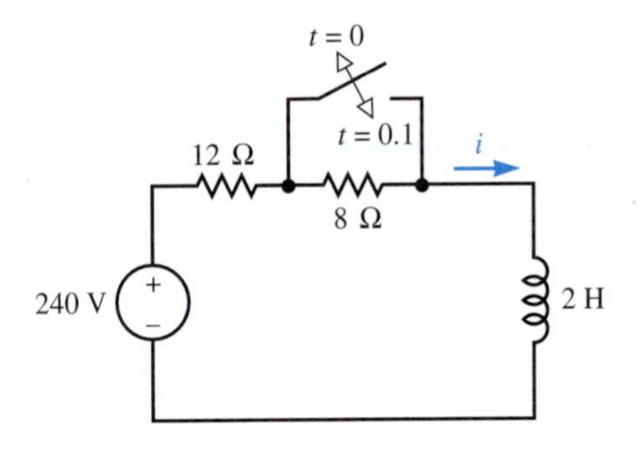

FIGURE 8.20 A circuit switched prior to the steady state

See Problem 8.3 in the PSpice manual.

Solution The circuit is operating in the steady state for $t < 0$ so the particular response i_{p0} is the complete response i for $t < 0$. We easily find this by replacing the inductance by a short circuit and applying KVL and Ohm's law:

$$i(t) = i_{p0}(t) = \frac{240}{12} = 20 \text{ A} \qquad t < 0$$

From Example 8.8 we know that the response after the switch is *opened* at $t = 0$ is

$$\begin{aligned} i(t) &= i_{p1}(t) + [i(0^+) - i_{p1}(0^+)]e^{-t/\tau_1} && 0 < t < t_0 \\ &= 12 + [20 - 12]e^{-10t} \text{ A} && 0 < t < t_0 \end{aligned}$$

where i_{p1} and τ_1 are with the particular response and time constant with the switch *open.*

The circuit is not operating in the steady state for $t = t_0^-$, so we must use the *complete response* to determine $i(t_0^-)$. Inductance current is continuous, so the initial condition at $t = t_0^+$ is

$$i(t_0^+) = i(t_0^-) = 12 + 8e^{-t_0/\tau_1} \text{ A}$$

The KVL equation with the switch closed is

$$-240 + 12i + 2\frac{d}{dt}i = 0 \qquad t > t_0$$

We determine the particular response by replacing differentiation with multiplication by zero in the preceding equation or by replacing the inductance with a short circuit. In either case the particular response with the switch *closed* is the same as for $t < 0$:

$$i_{p2}(t) = i_{p0}(t) = 20 \text{ A} \qquad t > t_0$$

From the KVL equation we recognize that the characteristic equation with the switch closed is

$$s + 6 = 0 \qquad t > t_0$$

with root

$$s_2 = -\frac{1}{\tau_2} = -6 \qquad t > t_0$$

Alternatively we can calculate the time constant τ_2, with the switch *closed,* without writing a differential equation. Use the resistance seen by the inductance when the switch closed. The time constant in seconds is

$$\tau_2 = \frac{L}{R} = \frac{2}{12} = \frac{1}{6}$$

Either way, the natural response with the switch closed is

$$i_{n2} = A_2 e^{-t/\tau_2} \qquad t > t_0$$

and the complete response is

$$\begin{aligned} i(t) &= i_{p2}(t) + i_{n2}(t) \\ &= i_{p2}(t) + A_2 e^{-t/\tau_2} \qquad t > t_0 \end{aligned}$$

To determine A_2 we evaluate this equation at $t = t_0^{+}$ and equate it to the initial value of $i(t_0^{+})$ obtained in our third equation:

$$i(t_0^{+}) = i_{p2}(t_0^{+}) + A_2 e^{-t_0/\tau_2}$$

This gives the constant of integration

$$A_2 = [i(t_0^{+}) - i_{p2}(t_0^{+})]e^{t_0/\tau_2} \qquad t > t_0$$

Substitution of this into the equation for current gives the complete response for $t > t_0$:

$$i(t) = i_{p2}(t) + [i(t_0^{+}) - i_{p2}(t_0^{+})]e^{-(t-t_0)/\tau_2} \qquad t > t_0$$

We see that $i(t)$ is obtained by a simple time shift in our second equation and the proper choice of time constant and particular response. This result is quite general and applies to any current in a first-order *RL* or *RC* circuit. If we replace i with v, the result applies to any voltage in the circuit. The source is *constant* for $t > t_0$ so

$$i_{p2}(t) = i(\infty)$$

In this case, we can rewrite the response in the *standard form*

$$i(t) = i(\infty) + [i(t_0^{+}) - i(\infty)]e^{-(t-t_0)/\tau}$$

where τ is the time constant for $t > t_0$.

We evaluate our solution at $t_0 = 0.1$ and obtain

$$i = \begin{cases} 20 \text{ A} & t < 0 \\ 12 + 8e^{-10t} \text{ A} & 0 < t < 0.1 \text{ s} \\ 20 - 5.057e^{-6(t-0.1)} \text{ A} & t > 0.1 \end{cases}$$

This response is sketched in Fig. 8.21.

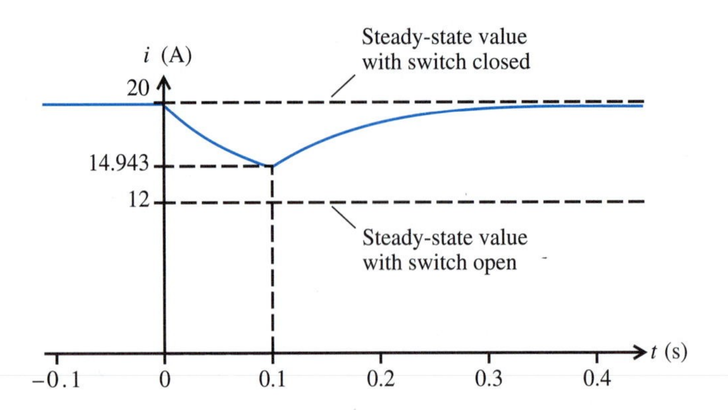

FIGURE 8.21 Response of a circuit switched prior to the steady state

Remember When a circuit has not reached steady state, the natural response is not zero. We must use the complete response to determine the voltage and current values prior to a change in a source or a switch position. For switched circuits, be sure to use the circuit model that corresponds to the switch position for the time interval considered.

EXERCISE

16. Determine and plot the complete response $v(t)$ for $0 < t < 4$ s, if the switch in the following circuit has been open for a very long time, is closed at $t = 0$, and reopened at $t = 1$ s.

answer: $10 - 10e^{-t}$ V, $0 < t < 1$ s, and $6.32e^{-(t-1)/2}$ V, $t > 1$ s

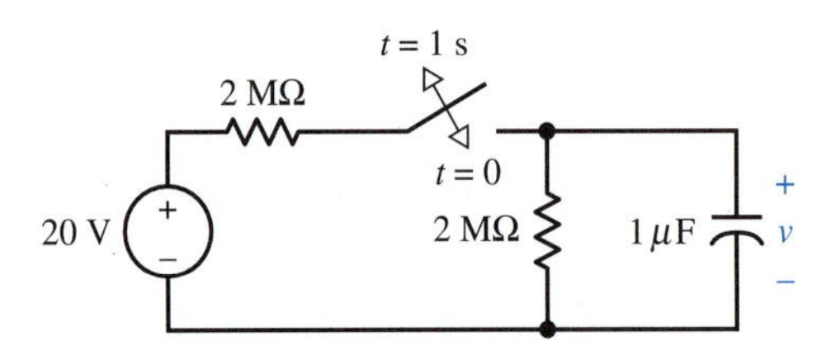

8.6 Superposition

In electronics we often have voltages or currents that are easily represented as a sum of two components. We can analyze these circuits by the use of *superposition.* Equation (8.50) gives the particular response as the integral of an exponential times the forcing function. We know that the integral of a sum of two functions is the sum of the integral of each function. Therefore, *superposition applies to the particular response* component of the solution.

EXAMPLE 8.10 The circuit in Fig. 8.22 represents the small-signal equivalent circuit of a single-stage amplifier with capacitive loading on the input and output. Determine $v_1(t)$ and $v_2(t)$, if the input signal is the step function $V_s u(t)$ V.

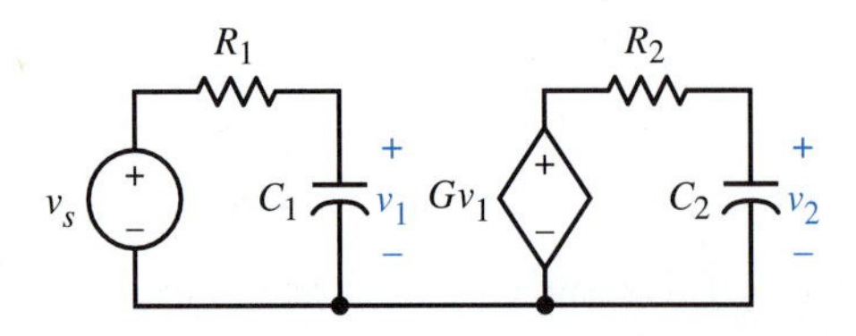

FIGURE 8.22 An amplifier with capacitive loading on the input and output

Solution We first define the time constants

$$\tau_1 = R_1 C_1$$

and

$$\tau_2 = R_2 C_2$$

We recognize that the input stage is a simple *RC* circuit with $v_1(0^+) = 0$ and $v_s = V_s$ is a constant for $t > 0$. We have previously analyzed this circuit, so we can write

$$v_1(t) = \begin{cases} 0 & t < 0 \\ V_s - V_s e^{-t/\tau_1} & t > 0 \end{cases}$$

Voltage $v_1(t)$ gives us the input to the right-hand part of the circuit as a sum of two voltages, GV_s and $GV_s e^{-t/\tau_1}$. The KVL equation that describes the right-hand loop reduces to

$$\frac{d}{dt} v_2 + \frac{1}{\tau_2} v_2 = \frac{G}{\tau_2} V_s - \frac{G}{\tau_2} V_s e^{-t/\tau_1} \qquad t > 0$$

We will use superposition to calculate the particular response component of $v_2(t)$. For the constant input V_s, we replace differentiation with multiplication by zero. The differential equation gives us

$$0 + \frac{1}{\tau_2} v'_{2p} = \frac{G}{\tau_2} V_s$$

which yields the first component of the particular response:

$$v'_{2p} = GV_s \qquad t > 0$$

For the exponential input we replace differentiation with multiplication by $-1/\tau_1$ in the differential equation (we assume that $\tau_1 \neq \tau_2$) to obtain:

$$-\frac{1}{\tau_1} v''_{2p} + \frac{1}{\tau_2} v''_{2p} = -\frac{G}{\tau_2} V_s e^{-t/\tau_1}$$

and we have

$$v''_{2p} = -\frac{\tau_1 G}{\tau_1 - \tau_2} V_s e^{-t/\tau_1} \qquad t > 0$$

This gives the particular response

$$v_{2p} = v'_{2p} + v''_{2p} = G\left(1 - \frac{\tau_1}{\tau_1 - \tau_2} e^{-t/\tau_1}\right) V_s \qquad t > 0$$

From the differential equation we see that the natural response is

$$v_{2n} = A_2 e^{-t/\tau_2} \qquad t > 0$$

We know that $v_2(0^+) = 0$, because the input is zero for $t < 0$ and capacitance voltage is continuous. We evaluate the complete response at $t = 0^+$ and equate it to $v_2(0^+)$ to solve for A_2:

$$v_2(0^+) = G\left(1 - \frac{\tau_1}{\tau_1 - \tau_2}\right) V_s + A_2 = 0$$

with the result that

$$A_2 = \frac{\tau_2}{\tau_1 - \tau_2} GV_s$$

The complete response $v_2(t)$ for $t > 0$ is the sum of v_{2p} and v_{2n}, with the constant of integration A_2 given by the preceding equation.

In the next section we see that superposition allows us to analyze circuits with sinusoidal sources.

Remember

We can use superposition to obtain the particular response for a sum of forcing functions. We use the complete response to determine the constants of integration.

EXERCISE

17. Determine voltage $v(t)$ for $t > 0$ in the following circuit if $v(0^+) = 6$ V.

answer: $4 + 12e^{-t} - 10e^{-3t}$ V for $t > 0$

See Problem 8.4 in the PSpice manual.

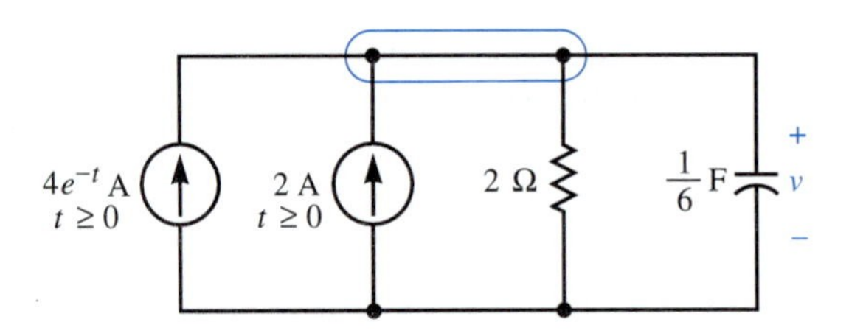

8.7 Sinusoidal Inputs

Virtually all electric power is generated and transmitted with voltages in the form of a sinusoid:

$$v(t) = V_m \cos(\omega_p t + \phi) \tag{8.74}$$

For this reason, and other reasons that are discussed in later chapters, we are very interested in calculating the response of a circuit to sinusoidal inputs.

Euler's identity gives the complex exponential as

$$V_m e^{j(\omega_p t + \phi)} = V_m \cos(\omega_p t + \phi) + jV_m \sin(\omega_p t + \phi) \tag{8.75}$$

We observe that

$$V_m e^{j(\omega_p t + \phi)} = V_m e^{j\phi} e^{j\omega_p t} \tag{8.76}$$

This is in the form of

$$\mathbf{v}(t) = \mathbf{V} e^{s_p t} \tag{8.77}$$

where

$$\mathbf{V} = V_m e^{j\phi} \tag{8.78}$$

and

$$s_p = j\omega_p \tag{8.79}$$

To solve for the particular response to a sinusoidal input, we first solve for the particular response to a complex exponential input. By superposition, the real part of the response for the complex exponential input is the response for the cosine input. This gives us the following procedure to solve for the particular response for a cosine input. [We assume that $\mathcal{A}(j\omega_p) \neq 0$.]

1. Replace the input function $V_m \cos(\omega_p t + \phi)$ in the differential equation (For the most general case, the forcing function can include derivatives of the input.)

$$\left(a_1 \frac{d}{dt} + a_0\right) v(t) = \left(b_1 \frac{d}{dt} + b_0\right) V_m \cos(\omega_p t + \phi) \tag{8.80}$$

with $V_m e^{j(\omega_p t + \phi)}$ to obtain (we write the equation in operator form)

$$\left(a_1 \frac{d}{dt} + a_0\right) \mathbf{v}(t) = \left(b_1 \frac{d}{dt} + b_0\right) V_m e^{j(\omega_p t + \phi)} \tag{8.81}$$

2. Replace differentiation with multiplication by $j\omega_p$ [see Eq. (8.57)] to obtain the algebraic equation.

$$(a_1 j\omega_p + a_0)\mathbf{v}_p(t) = (b_1 j\omega + b_0) V_m e^{j(\omega_p t + \phi)} \tag{8.82}$$

which is of the form

$$\mathcal{A}(j\omega)\mathbf{v}_p(t) = \mathcal{B}(j\omega) V_m e^{j(\omega_p t + \phi)} \tag{8.83}$$

Solve this equation for the complex response $\mathbf{v}_p(t)$.

3. The particular response $v_p(t)$ for the cosine input is the real part of $\mathbf{v}_p(t)$:

$$v_p(t) = \mathcal{Re}\{\mathbf{v}_p(t)\} = \mathcal{Re}\left\{\frac{\mathcal{B}(j\omega_p)}{\mathcal{A}(j\omega_p)} e^{j(\omega_p t + \phi)}\right\} \tag{8.84}$$

The following example demonstrates that the procedure is actually quite simple.

EXAMPLE 8.11 Solve for $v(t)$ in the circuit of Fig. 8.23 if $v_s(t) = 0$ for $t < 0$, and $v_s(t) = 15/\sqrt{2}\cos 3t$ V for $t > 0$.

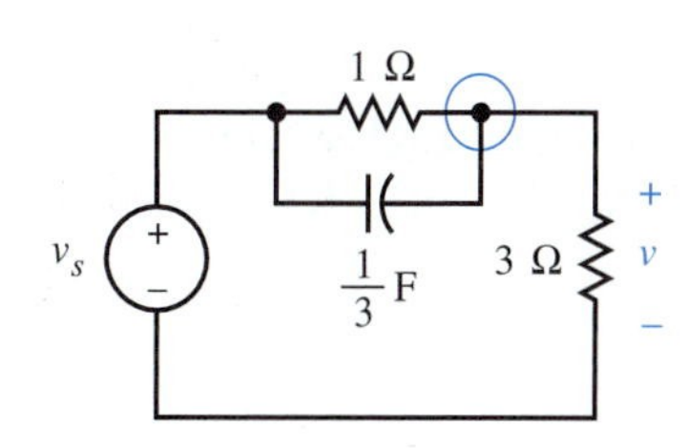

FIGURE 8.23
An *RC* circuit driven by a sinusoid

Solution A KCL equation for the indicated node is

$$\frac{1}{3}\frac{d}{dt}(v - v_s) + (v - v_s) + \frac{1}{3}v = 0$$

See Problem 8.5 in the PSpice manual.

which we can write in operator form as

$$\left(\frac{d}{dt} + 4\right)v = \left(\frac{d}{dt} + 3\right)v_s$$

$$= \left(\frac{d}{dt} + 3\right)\frac{15}{\sqrt{2}}\cos 3t \qquad t > 0$$

Replace cos $3t$ with e^{j3t}, and replace differentiation with multiplication by $j3$ in this equation to obtain

$$(j3 + 4)\mathbf{v}_p(t) = (j3 + 3)\frac{15}{\sqrt{2}}e^{j3t}$$

This gives the particular response for the complex exponential input:

$$\mathbf{v}_p(t) = \frac{(j3 + 3)}{(j3 + 4)}\frac{15}{\sqrt{2}}e^{j3t}$$

We see that this equation is of the form

$$\mathbf{v}_p(t) = \frac{\mathcal{B}(j3)}{\mathcal{A}(j3)}\frac{15}{\sqrt{2}}e^{j3t}$$

Now take the real part of $\mathbf{v}_p(t)$ to obtain the particular response for the cosine input:

$$\begin{aligned}
v_p(t) &= \mathcal{R}e\left\{\frac{\mathcal{B}(j3)}{\mathcal{A}(j3)}\frac{15}{\sqrt{2}}e^{j3t}\right\} \\
&= \mathcal{R}e\left\{\frac{3 + j3}{4 + j3}\frac{15}{\sqrt{2}}e^{j3t}\right\} \\
&= \mathcal{R}e\left\{\frac{\sqrt{3^2 + 3^2}e^{j\arctan(3/3)}}{\sqrt{4^2 + 3^2}e^{j\arctan(3/4)}}\cdot\frac{15}{\sqrt{2}}e^{j3t}\right\} \\
&= \mathcal{R}e\left\{\frac{45}{5}e^{j[3t + \arctan(3/3) - \arctan(3/4)]}\right\} \\
&= \mathcal{R}e\left\{\frac{45}{5}e^{j[3t + (\pi/4) - 0.644]}\right\} \\
&= 9\cos\left(3t + \frac{\pi}{4} - 0.644\right)\ \text{V} \\
&= 9\cos(3t + 8.13^\circ)\ \text{V}
\end{aligned}$$

From our second equation we see that the natural response is

$$v_n = Ae^{-4t} \qquad t > 0$$

The source has been zero for all time less than zero, so the initial *capacitance voltage* is zero. We write a KVL equation around the inner loop at $t = 0^+$

$$-\frac{15}{\sqrt{2}}\cos 0^+ + 0 + v(0^+) = 0$$

to solve for

$$v(0^+) = \frac{15}{\sqrt{2}}$$

We evaluate the complete response at $t = 0^+$ to solve for the constant of integration:

$$v(0^+) = A + 9\cos(0 + 8.13^\circ) = \frac{15}{\sqrt{2}}$$

This yields

$$A = 1.697$$

and

$$v(t) = 1.697e^{-4t} + 9\cos(3t + 8.13°)\ \text{V} \qquad t > 0$$

This response is plotted in Fig. 8.24.

Inspection of Fig. 8.24 reveals that after a few time constants the natural response has become negligible, and only the particular response remains. We say that the circuit is then operating in the *sinusoidal steady state.* The transient can be small and of short duration, as in the preceding example, or it can be comparable in amplitude to the steady-state response and last several cycles. In many applications, we are primarily interested in the steady-state response. Beginning with Chapter 10, most of the remaining chapters are devoted to steady-state sinusoidal analysis, popularly called *ac circuit analysis.*

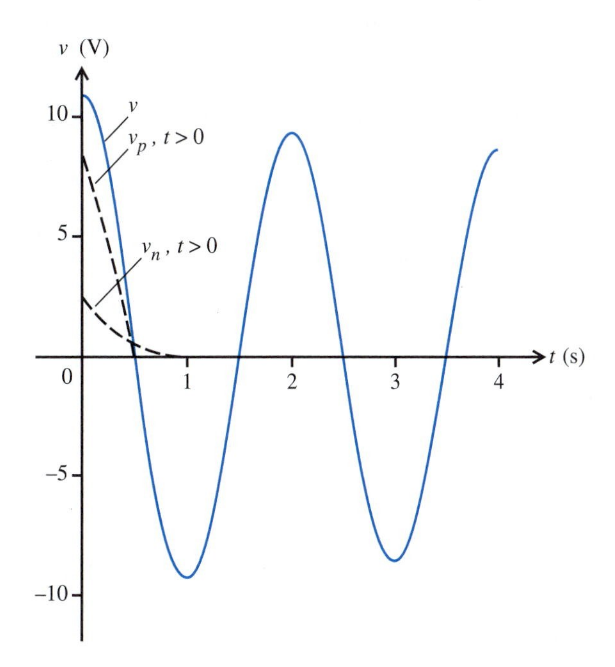

FIGURE 8.24
The response of an *RC* circuit to a sinusoidal input

Remember

To determine the particular response for a cosine source, we solve for the particular response for a complex exponential source and then take the real part of this response.

EXERCISES

18. Determine voltage $v(t)$ for $t > 0$ in the following circuit, if $v_s(t) = 20$ V for $t < 0$, and $v_s(t) = 50\cos 3t$ V for $t > 0$. Plot the response for $t > 0$.
answer: $40\cos(3t - 36.87°) - 12e^{-4t}$ V for $t > 0$

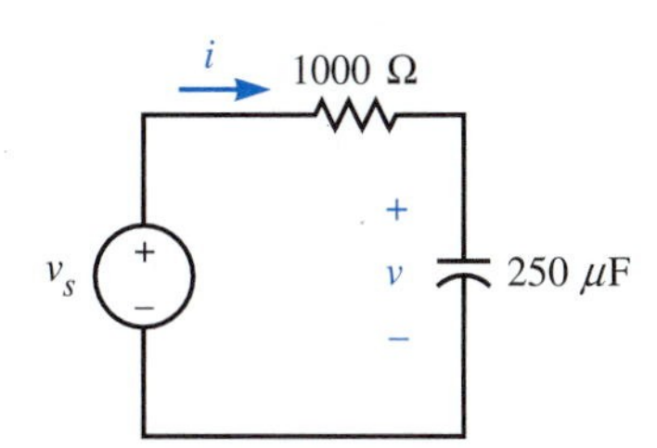

19. Determine the response $v(t)$ in the preceding circuit if $v_s(t) = 50 \cos 3t$ V for $t < 0$, and $v_s(t) = 20$ V for $t > 0$.

   answer: $v = 40 \cos(3t - 36.87°)$ V for $t < 0$, and $v(t) = 20 + 12e^{-4t}$ V for $t > 0$

8.8 Stability

In all of the examples we have considered, the natural response asymptotically approaches zero as time becomes infinite. Physically, resistance absorbs energy stored in the capacitance or inductance and causes the natural response to decay. Mathematically, the natural response decays because the root of the characteristic equation is negative. This gives a natural response that is an exponential with a negative exponent. We now analyze a network that includes a dependent source that can supply power and cause the circuit to be unstable.

EXAMPLE 8.12 The value of current i in the circuit of Fig. 8.25 is 10 A at time zero. Determine $i(t)$ for $t \geq 0$.

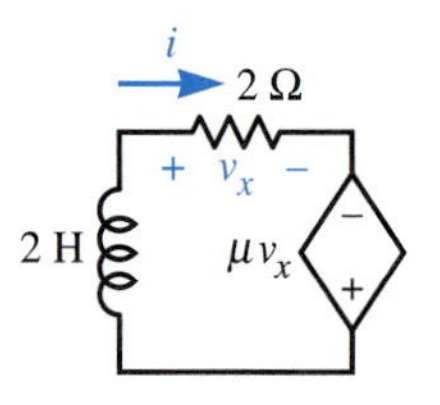

FIGURE 8.25 A first-order circuit that can be unstable

Solution We apply KVL around the loop to obtain

$$2\frac{d}{dt}i + 2i - \mu(2i) = 0$$

This first-order equation is easily solved to give

$$i = 10e^{(\mu-1)t} \text{ A} \qquad t \geq 0$$

The response found in the preceding example is sketched in Fig. 8.26 for three values of μ. We see that for $\mu = 0.9$, the negative root of the characteristic equation ($s_1 = -0.1$) causes the natural response to asymptotically decay to zero. This is a *stable* circuit.

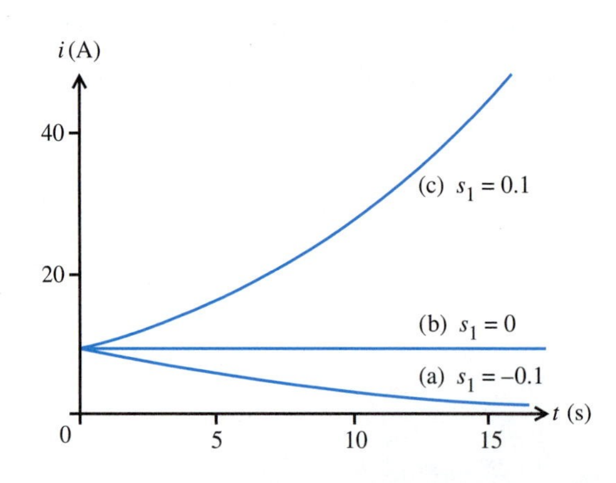

FIGURE 8.26 The natural response of a first-order circuit is determined by the root s_1 of the characteristic equation: (a) stable, $s_1 < 0$; (b) marginally stable, $s_1 = 0$; and (c) unstable, $s_1 > 0$

With $\mu = 1.1$, the positive root ($s_1 = 0.1$) causes the response to increase without bound. (Physically this is impossible. Some device will become nonlinear or be destroyed.) This is an *unstable* circuit.

With $\mu = 1$, the root of the characteristic equation is $s_1 = 0$. The response has the constant value $i(t) = 10$ A for $t \geq 0$ and does not decay to zero as time goes to infinity. However, the response does not increase without bound as it did for the unstable circuit. This is a *marginally stable* circuit, which is often considered to be a special case of an unstable circuit.

The sign on the root of the characteristic equation determines whether the natural response decays with time or grows without bound. Therefore the stability of a circuit is determined by the characteristic equation. The input and the particular response have no bearing on the stability of a linear circuit.

Remember

A first-order circuit is stable if the root of the characteristic equation is less than zero. The input to a linear system has no effect on the stability.

EXERCISE

20. Determine and sketch the voltage $v(t)$ for $t \geq 0$ in the following source-free circuit, if $v(0) = 10$ V (source-free implies no independent sources). Is the circuit stable?

answer: $10e^{t}$ V for $t \geq 0$, No

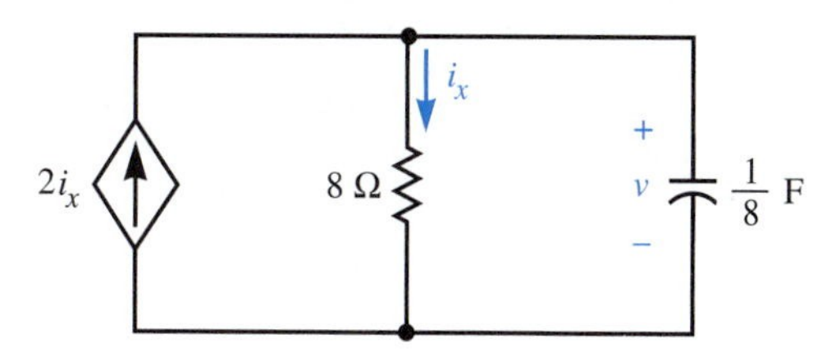

8.9 Summary

We began by solving the first-order linear differential equation that describes a source-free circuit with a single energy-storage element (capacitance or inductance). This gave the source-free response of the circuit. This response is an exponential $Ae^{s_1 t}$, and is the natural response of the circuit. The differential equation determined s_1, and the stored energy determined the constant of integration A. We then wrote the natural response as $Ae^{-t/\tau}$ where $\tau = -1/s_1$ is the time constant. For a first-order RC circuit $\tau = RC$ and for a first-order RL circuit $\tau = L/R$, where R is the equivalent resistance seen by the capacitance or inductance. This let us write the response of a first-order circuit without solving a differential equation.

We used an integrating factor to solve the first-order differential equation that describes a circuit with a source and one energy-storage element. We observed that the complete response has two parts. One part is the natural response that includes the constant of integration. The second part is the particular response due to the input (source). We then showed that for the usual case we can obtain the particular response to an exponential input $Ke^{s_p t}$ by simply replacing differentiation with multiplication by s_p. (For a constant input $s_p = 0$, so this is equivalent to replacing capacitance with an open circuit and inductance with a short circuit.) We extended the procedure to sinusoidal inputs by use of the complex exponential $Ke^{j\omega_p t}$.

We examined the stability of a first-order circuit, and saw that this was determined by the characteristic equation and did not depend on the particular input.

KEY FACTS

Concept	Equation	Section	Page
❑ A first-order differential equation describes a *source-free RC* circuit with a single capacitance.	(8.2) (8.17)	8.1	228 230
❑ The solution to this source-free equation is the *source-free response* and is determined by the *natural response* of the circuit.	(8.21)	8.1	230
❑ The time constant of an *RC* circuit is $\tau = RC$, where R is the equivalent resistance seen by the source.	(8.9) (8.22)	8.1	229 232
❑ Any voltage in a source-free *RC* circuit is given by $v(t) = v(0^+)e^{-t/\tau} \quad t > 0,$ and any current is given by $i(t) = i(0^+)e^{-t/\tau} \quad t > 0.$	 (8.24) (8.25)	8.1	232
❑ The time constant for an *RL* circuit is given by $\tau = L/R$, where R is the equivalent resistance seen by the inductance.	(8.33) (8.37)	8.2	234 235
❑ Any current in a source-free *RL* circuit is given by $i(t) = i(0^+)e^{-t/\tau} \quad t > 0,$ and any voltage is given by $v(t) = v(0^+)e^{-t/\tau} \quad t > 0.$	 (8.38) (8.39)	8.2	235
❑ The complete response for a network with a source is the sum of two parts: $v = v_n + v_p$, where the natural response v_n is of the same form as in the source-free circuit, and the v_p is obtained by integration.	 (8.48) (8.49) (8.50) (8.53)	8.3	238
❑ The particular response to an exponential input $Ke^{s_p t}$, is obtained by replacing differentiation with multiplication by s_p for the usual case.	(8.57)	8.3	239
❑ A constant input is a special case of an exponential input, where $s_p = 0$. We can obtain the particular response for constant inputs to a stable circuit directly from the network diagram by replacing each capacitance by an open circuit and each inductance by a short circuit.		8.3	241
❑ Capacitance voltage is continuous.	(8.59)	8.3	241
❑ Inductance current is continuous.	(8.60)	8.3	241
❑ If a dc source has been applied to a stable circuit for a very long time, only the particular response will be of significant amplitude.	(8.61) (8.62)	8.3	242

KEY FACTS Concept	Equation	Section	Page
❑ Any voltage in a first-order circuit driven by a source that is constant for $t > 0$ is given by $v(t) = v(\infty) + [v(0^+) - v(\infty)]e^{-t/\tau}$, and any current is given by $i(t) = i(\infty) + [i(0^+) - i(\infty)]e^{-t/\tau}$, where $v(\infty) = v_p$ and $i(\infty) = i_p$ are the constant particular responses.	(8.65) (8.66) (8.61)	8.3	245 245 245 242
❑ The particular response for the sum of two inputs is the sum of the particular response for each input (superposition). (Example 8.10)		8.6	254
❑ For a cosine input, the particular response is the real part of the particular response for a complex exponential input. For a stable circuit this particular response is called the *steady-state sinusoidal response* or the *ac response*. (Example 8.11)	(8.84)	8.7	257

PROBLEMS

Section 8.1

1. For the following circuit, $v_1(0^+) = 4$ V and $v_2(0^+) = -2$ V. Write a KVL equation for $t = 0^+$ and solve for $i(0^+)$. Write a KVL equation in terms of $i(t)$. Differentiate this once to eliminate the integral, and solve this differential equation for $i(t)$ when $t > 0$. Also solve for $v_1(t)$ and $v_2(t)$ when $t > 0$.

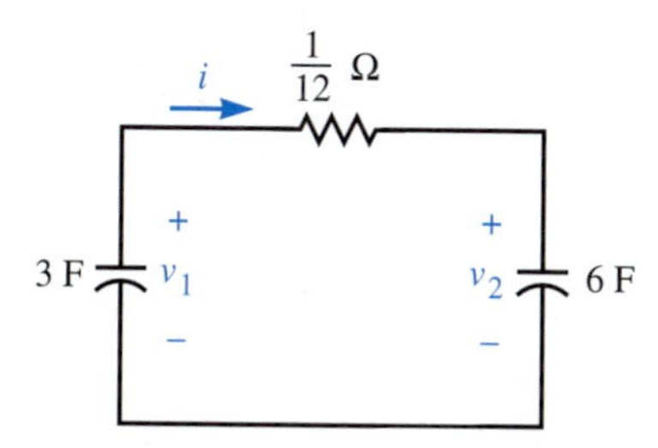

2. For the preceding circuit, $v_1(0^+) = 0$ V and $v_2(0^+) = 6$ V. Solve for $i(t)$, $v_1(t)$, and $v_2(t)$ for $t > 0$. [*Hint:* $v_1(\infty) = v_2(\infty) \neq 0$]

3. Use equivalent circuits to reduce the following networks to simple parallel *RC* circuits, then calculate v_{12} for $t > 0$, if $v_{12}(0^+) = 10$ V.

(a)

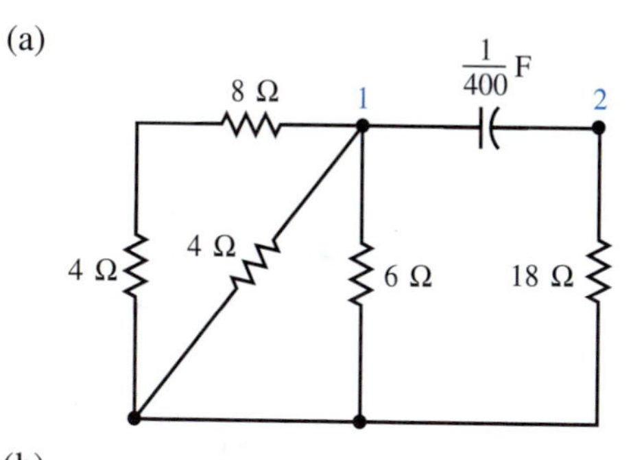

(b)

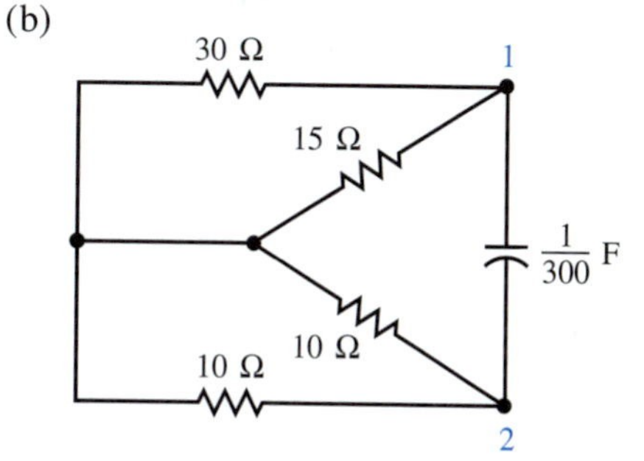

4. In the following source-free circuit, write a differential equation, and determine the parameter β that will give a time constant of 0.02 s. Use this

value of β and calculate $v(t)$, $t > 0$, if $v(0^+) =$ 40 V. 4.1

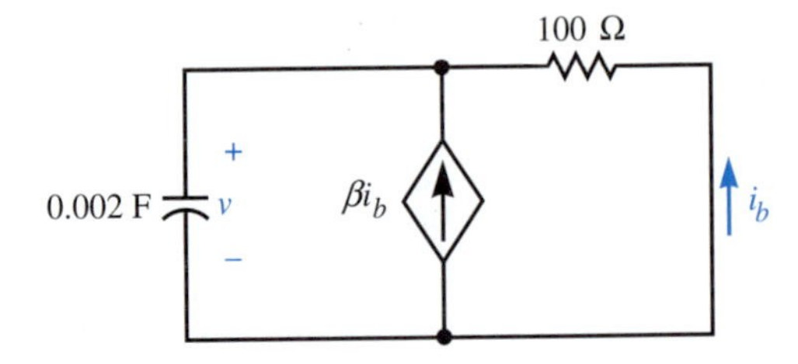

5. For an RC circuit, explain why
   (a) increasing the capacitance increases the time constant, and
   (b) increasing the resistance increases the time constant.

Section 8.2

6. For the following circuit, $i(0^+) = 8$ A. Write a KVL equation to solve for $i = i(t)$ for $t > 0$.

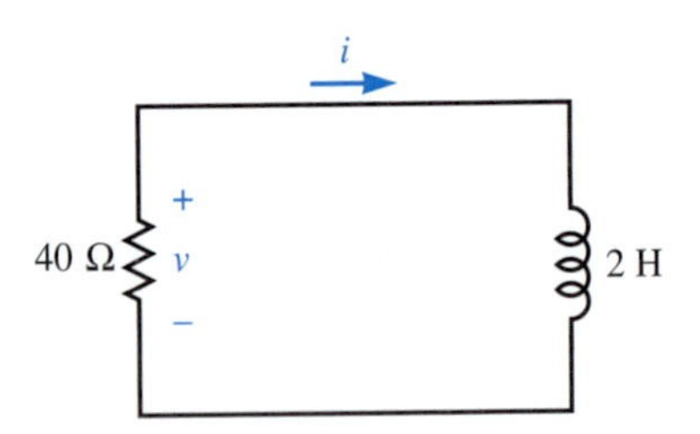

7. For the preceding circuit, $i(0^+) = 8$ A. Use this and Ohm's law to calculate $v(0^+)$. Write a KCL equation (in terms of voltage v) for the top node. Differentiate this integral equation once to obtain a first-order differential equation. Solve this equation for $v = v(t)$, $t > 0$, and use the value of $v(0^+)$ to calculate the constant of integration.
8. Write a differential equation, and determine the time constant of the following source-free network. Find $i(t)$ for $t > 0$, if $i(0^+) = 80$ A. 8.2

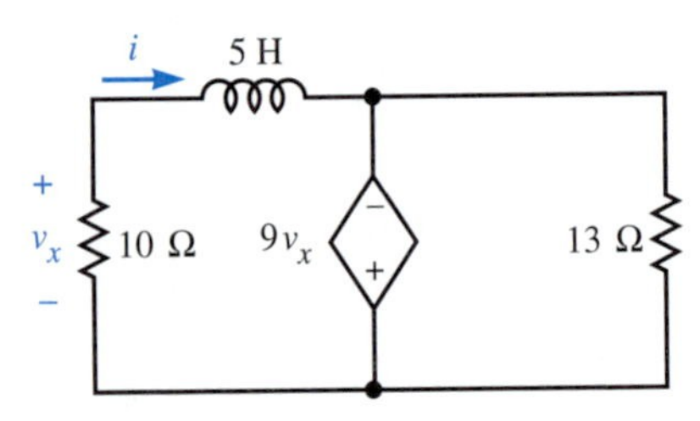

9. For the following circuit, the initial current is $i(0^+) = 24$ A. Calculate $i(t)$ for $t > 0$. What is the time constant of the circuit?

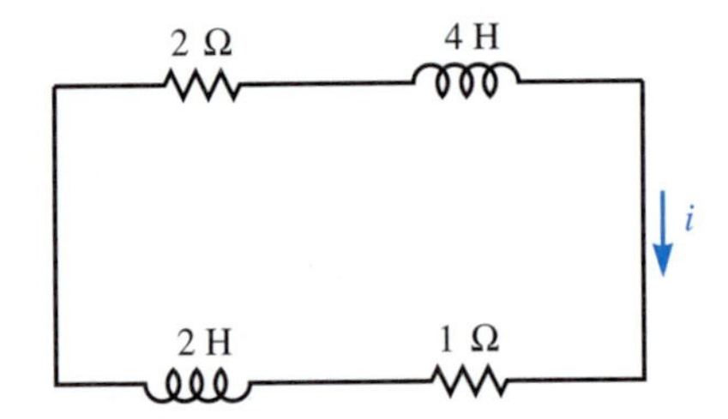

10. For the amplifier circuit that follows, $v_s = 0$ for $t > -1$ s and $i(0^+) = 10$ A. Use Thévenin's theorem to simplify the circuit and calculate the time constant. Determine and sketch current i for $0 \leq t \leq 1$ s.

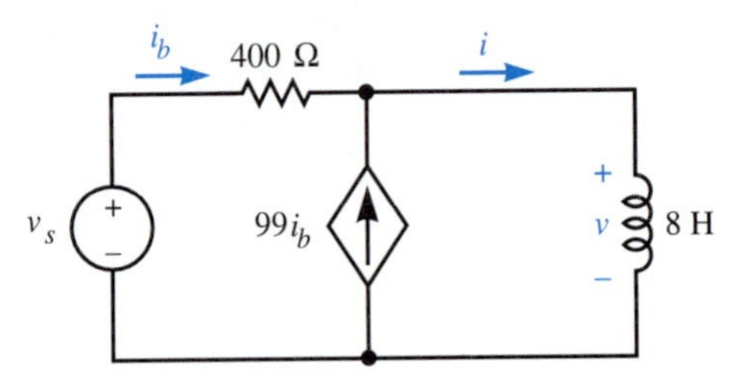

11. For the preceding amplifier circuit, use the short-cut method for analysis by mesh currents to obtain one KVL equation in terms of current i. Solve for $i(t)$ for $t > 0$, if $v_s = 0$ for $t > -1$ s, and $i(0^+) = 10$ A.
12. For an RL circuit, explain why
    (a) increasing the inductance increases the time constant, and
    (b) increasing the resistance decreases the time constant.

Section 8.3

13. In the following circuit, determine voltage v for the specified source values.
    (a) $v_s = 24$ V for $t < 0$ and $v_s = 48$ V for $t > 0$
    (b) $v_s = 48$ V for $t < 0$ and $v_s = 24e^{-8t}$ V for $t > 0$

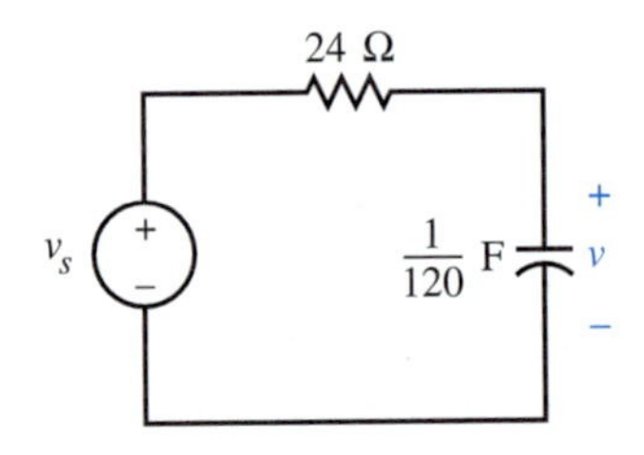

14. Determine voltage v in the following circuit for the specified source values.

(a) $i_s = 10u(t)$ A
(b) $i_s = 10e^{-2t}u(t)$ A
(c) $i_s = 10e^{-10t}u(t)$ A [use Eq. (8.55b)]

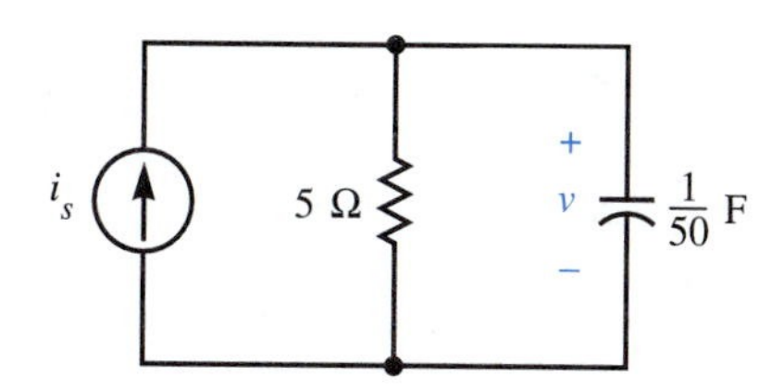

15. Determine voltage v in the following circuit if $v_s = 50[2 + u(t)]$ V.

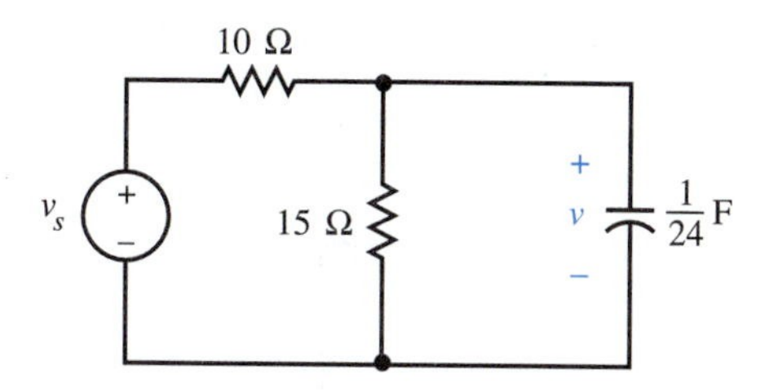

16. Solve for voltage v and current i in the following circuit.

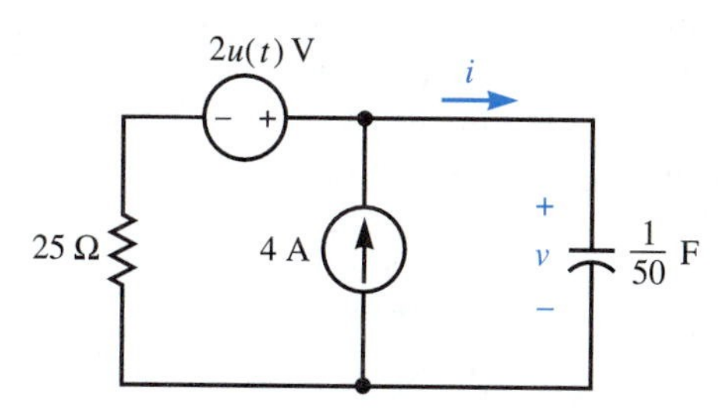

17. The switch in the following network has been closed for a very long time and is opened at $t = 0$.
   (a) Make a Norton equivalent circuit for the network to the left of terminals a and b, and use this to solve for voltage v for $t > 0$.
   (b) Use the shortcut method for analysis by node voltages to obtain a single KCL equation and solve it for v when t is greater than zero.

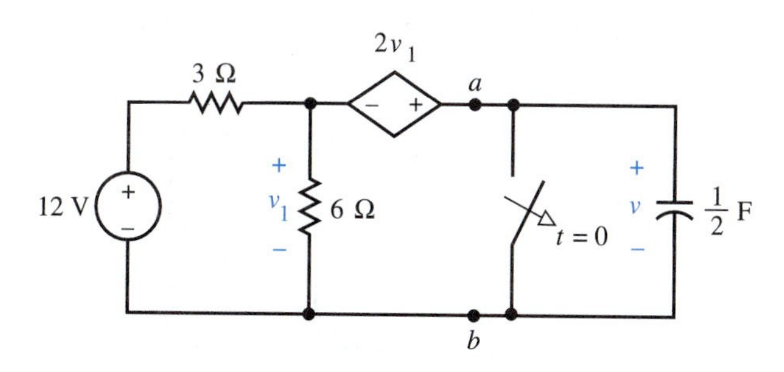

18. Determine voltage v for $t > 0$ in the following circuit with the specified switch positions.
   (a) Both switches have been open for a very long time, and switch 1 is closed at $t = 0$.
   (b) Switch 2 has been open for a very long time, switch 1 has been closed for a very long time, and switch 1 is opened at $t = 0$.
   (c) Both switches have been open for a very long time, and switch 2 is closed at $t = 0$.
   (d) Switch 1 has been open for a very long time, switch 2 has been closed for a very long time, and switch 2 is opened at $t = 0$.
   (e) What would happen if both switches 1 and 2 were closed at the same time?

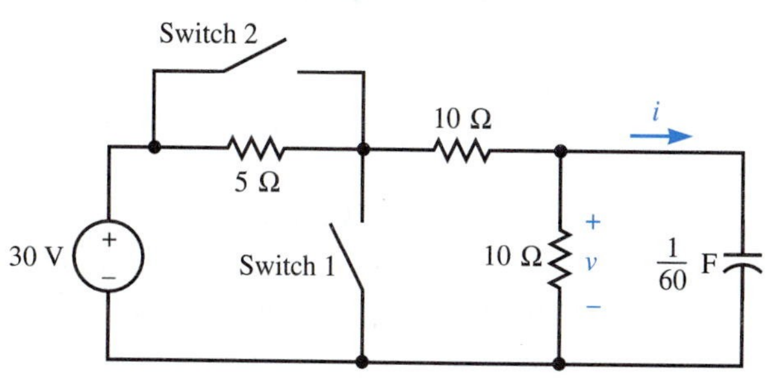

Section 8.4

19. Determine current i in the following circuit for the specified source values.
   (a) $24u(t)$ V
   (b) $24e^{-3t}u(t)$ V
   (c) $24e^{-12t}u(t)$ V [use Eq. (8.55b)]
   (d) $24tu(t)$ V [use Eq. (8.50)]

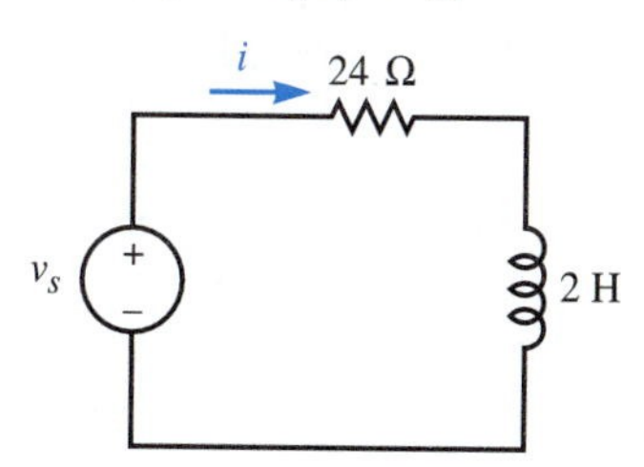

20. Determine current i for $t > 0$ in the following circuit if $i_s = 30[2 - u(t)]$ A.

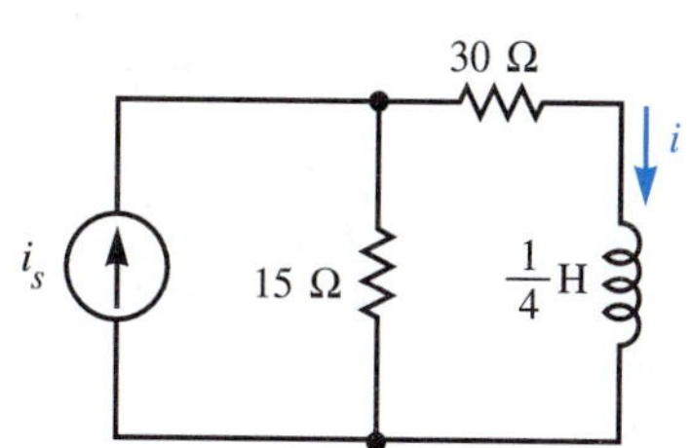

21. Solve for current i in the following circuit.

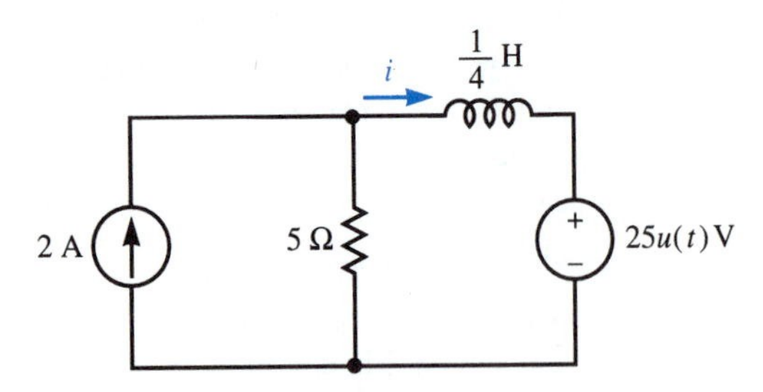

22. The switch in the following circuit has been closed for a very long time and is opened at $t = 0$. Find current i for $t > 0$.

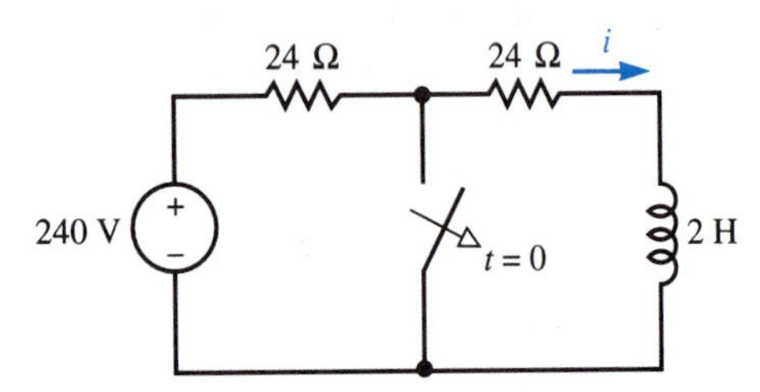

23. Determine current i for $t > 0$ in the following circuit with the specified switch positions.
    (a) Both switches have been open for a very long time. Switch 1 is closed at $t = 0$.
    (b) Switch 2 has been open for a very long time, switch 1 has been closed for a very long time, and switch 1 is opened at $t = 0$.
    (c) Both switches have been open for a very long time, and switch 2 is closed at $t = 0$.
    (d) Switch 1 has been open for a very long time, switch 2 has been closed for a very long time, and switch 2 is opened at $t = 0$.

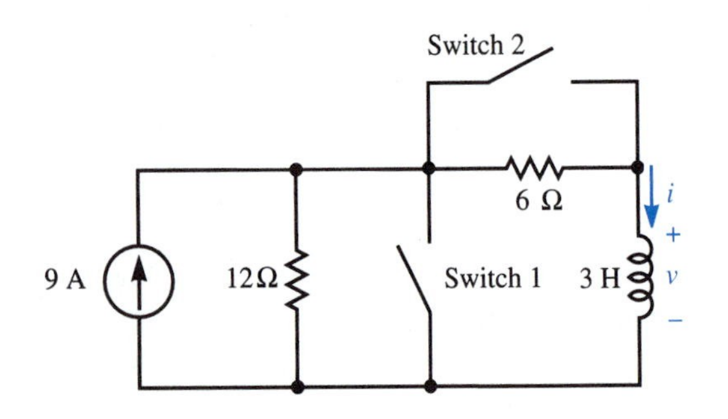

Section 8.5

24. Determine $i(t)$ for $t > 0$ in the preceding circuit if both switches have been open for a very long time, switch 2 is closed at $t = 0$, and reopened at $t = 0.2$ s.

25. Determine $v(t)$ for $t > 0$ in the following circuit if the switch has been open for a very long time, is closed at $t = 0$, and reopened at $t = 0.1$ s.

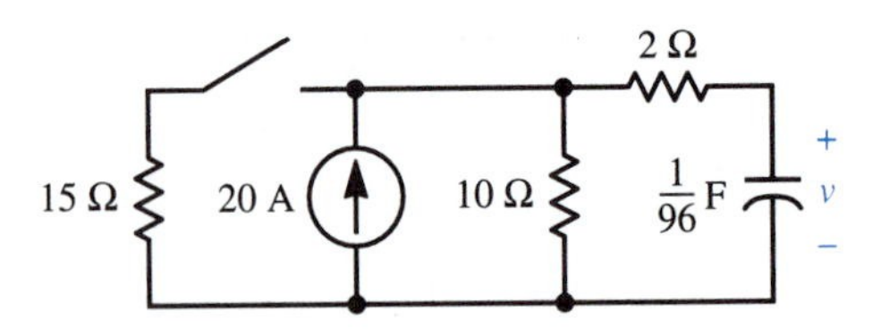

26. The input to the following circuit is $v_s = 10[u(t) - u(t - 0.001)]$ V. Determine $v(t)$ for the following capacitance values: (a) 0.1 μF, (b) 1 μF, and (c) 10 μF. Plot the results on one graph for $0 < t < 3$ ms.

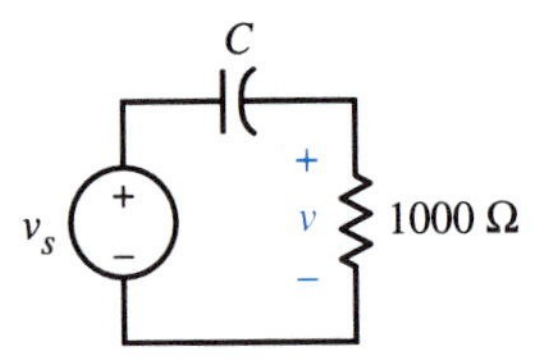

27. The input to the following circuit is $v_s = 10[u(t) - u(t - 0.001)]$ V. Determine $v(t)$ for the following capacitance values: (a) 0.1 μF, (b) 1 μF, and (c) 10 μF. Plot the results on one graph for $0 < t < 3$ ms.

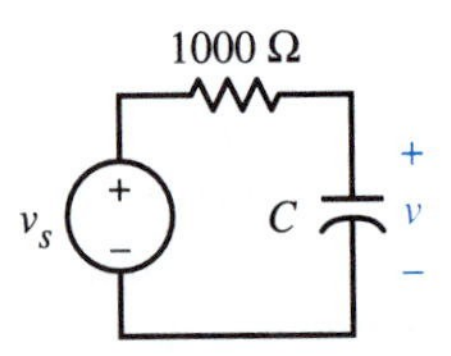

28. Determine $v(t)$ in the preceding circuit and accurately sketch $v(t)$ if $C = 1$ μF, and $v_s = 10[u(t) - 2u(t - 0.001) + u(t - 0.002)]$ V.

Section 8.6

29. Use superposition to determine voltage v for $t > 0$ in the following circuit if $v(0^+) = 40$ V.

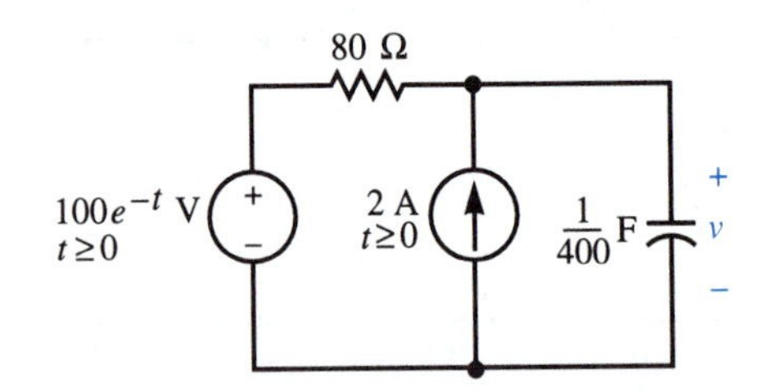

30. Use superposition to determine voltage v for $t > 0$ in the following circuit if $v(0^+) = 0$.

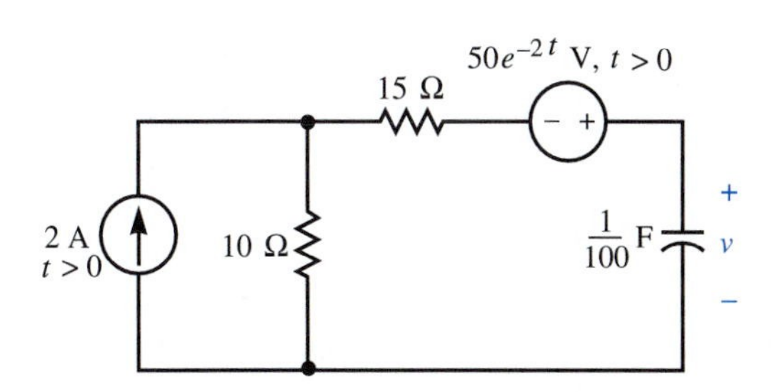

31. Use superposition to determine voltage v for $t > 0$ in the following circuit if $v_s = 36e^{-2t}u(t)$ V and $i_s = 2$ A for all time.

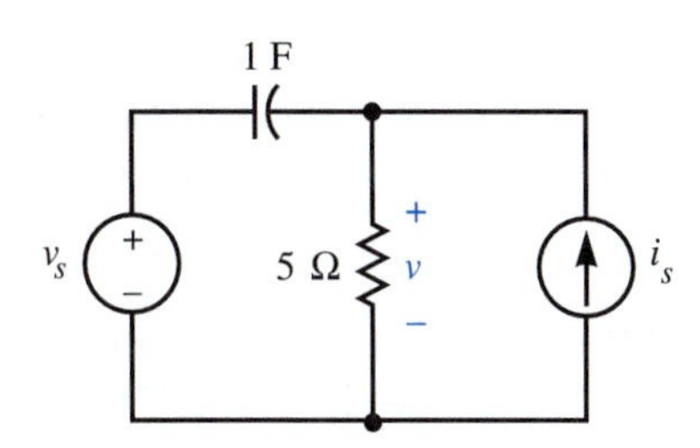

32. Use superposition to determine current i in the following circuit.

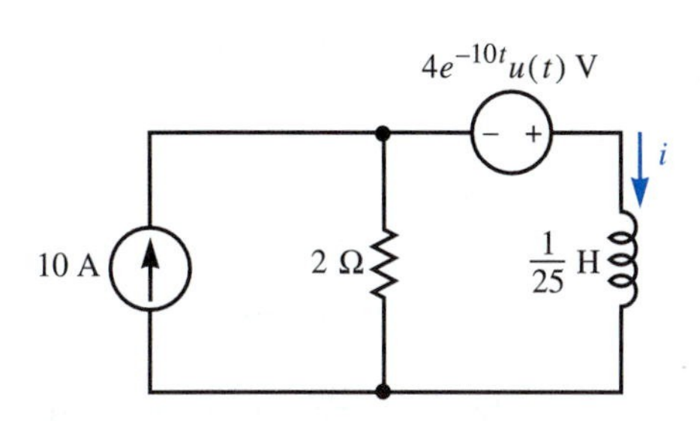

Section 8.7

33. Use superposition to determine voltage v for $t > 0$ in the following circuit if $i_s = 5 + (20 \cos 3t)u(t)$ A.

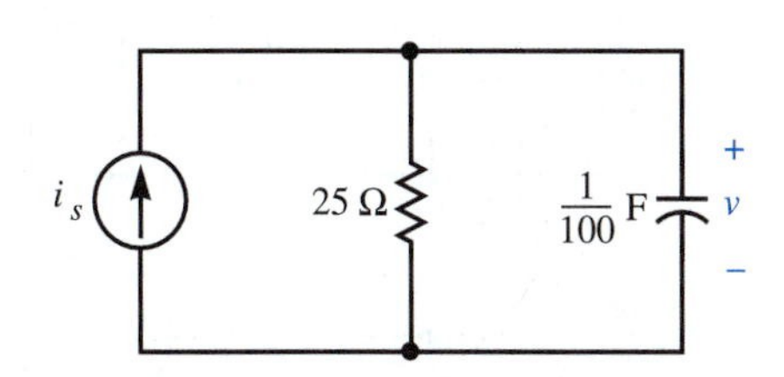

34. After having been open for a very long time, the switch in the following circuit is closed at $t = 0$. Determine current i for $t > 0$ if $v_a = 16 \cos 2t$ V and $v_b = 32$ V for all time.

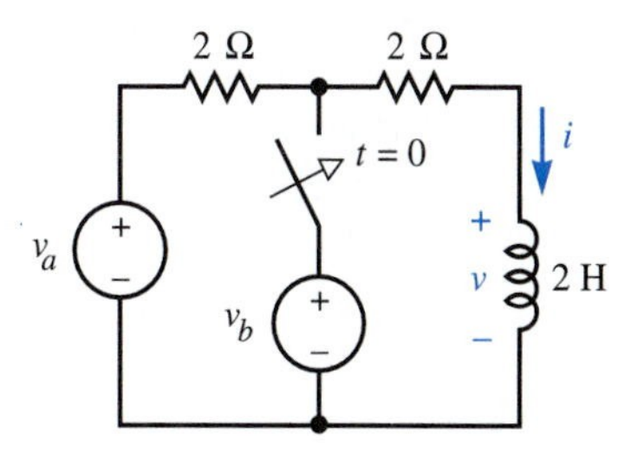

Section 8.8

35. For what values of β is the circuit of Problem 4 stable?

36. The dependent source in Problem 8 has a voltage gain of 9. What values of voltage gain yield a stable circuit?

Supplementary Problems

37. Calculate voltage v in the following circuit for the specified source voltage v_s.

    (a) $4e^{6t}u(-t) + 2u(t)$ V

    (b) $8e^{-3|t|}$ V

    (c) $(20 \cos 4t)[1 - u(t)]$ V

    (d) $(20 \cos 4t)u(t)$ V

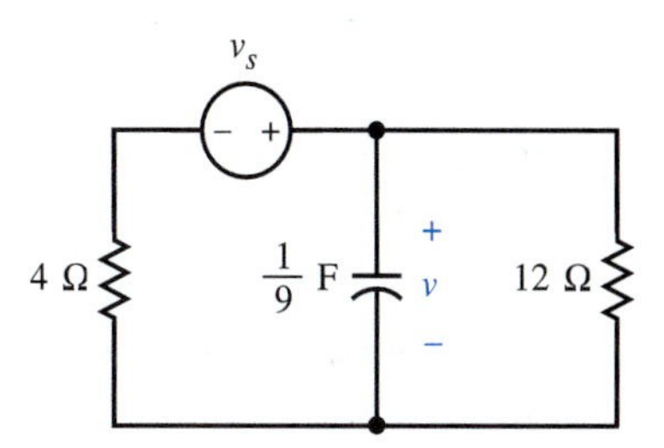

38. The power supply for the large-screen monitor used on a computer provides 20 kV dc. When the unit is unplugged, the capacitor in the power supply retains a voltage of 20 kV for several minutes. This voltage must decay to a safe value of 40 V in the one minute required to remove the case. Select the value of resistance that must be connected in parallel with the capacitance of 0.06 μF so that this safety requirement is satisfied. The resistance must consume as little power as posssible when the monitor is in operation. What power is consumed in the resistance that you selected?

39. The safety brake on a hoist operates when the current through a solenoid drops below 2 A. The current through the solenoid is 10 A when the

hoist is in normal operation. The solenoid model is an inductance of 10 H in series with a resistance of 2.4 Ω. For safety, the solenoid current must be reduced as rapidly as possible when the safety switch is opened. However, the voltage across the solenoid must not exceed 240 V, or the switch will arc and eventually fail.

(a) What value of resistance must be placed in parallel with the solenoid to satisfy these requirements?

(b) How much time is required for the solenoid current to drop below 2 A?

(c) What power does the resistance consume while the solenoid is energized?

(d) What is the peak power absorbed by the resistance when the switch is opened?

40. Determine the output voltage v_o for $t > 0$ in the following op-amp circuit if $v_s(t) = 10u(t)$ V.

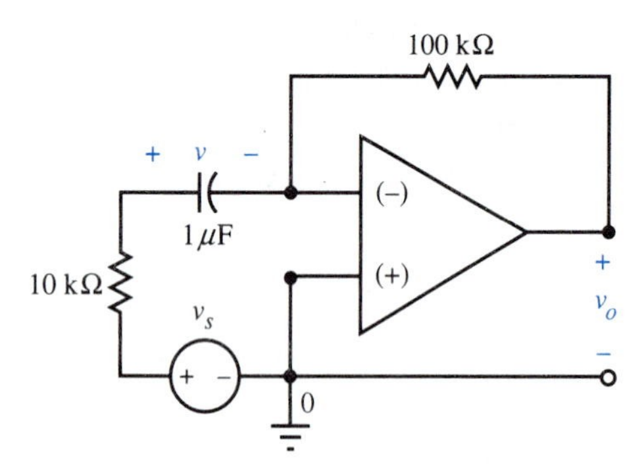

41. Determine v_o for $t > 0$ in the following op-amp circuit if $v_s(t) = 10u(t)$ V. What is the time constant of this circuit?

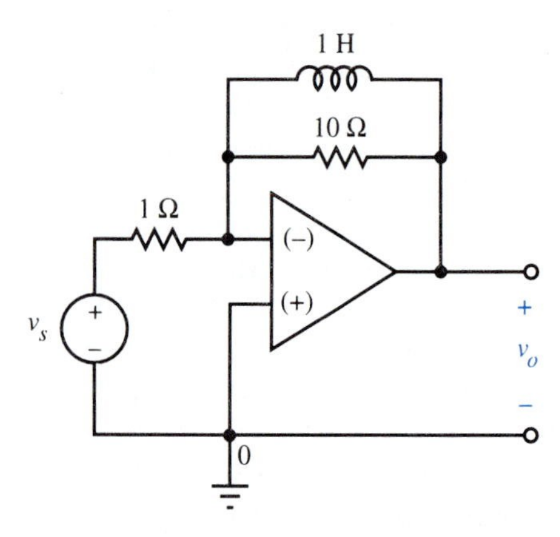

42. For the following circuit, determine the time constant as a function of dependent voltage source gain A. What is the effective capacitance seen by the 10-kΩ resistance? (This is a function of the gain A.)

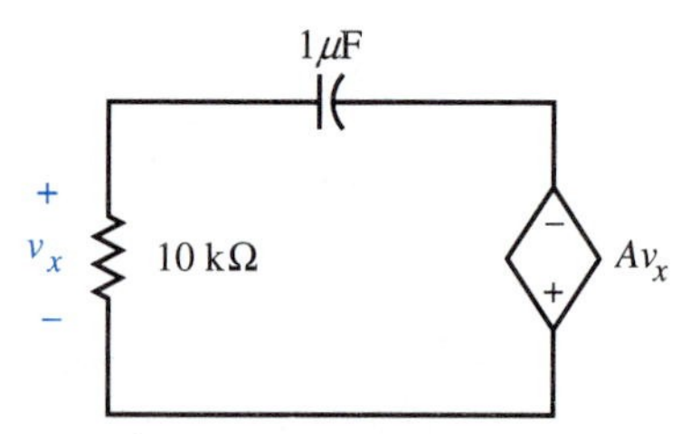

43. The following problem occurs in digital logic circuits. When circuits are connected together by a twisted pair (two wires twisted together), the capacitive coupling between wires can falsely trigger a logic gate. The capacitance per foot of a 30 AWG wire size twisted pair is approximately 15 pF. If the voltage v introduced on the logic gate exceeds 1.4 V in peak value, a false trigger may result. The equivalent circuit is shown below, where

$$v_s = 1 \times 10^9[r(t) - r(t - 3 \times 10^{-9})] \text{ V}$$

with $r(t)$ the unit ramp $[r(t) = tu(t)]$. Will false triggering occur if the twisted pair has a length of 2 ft? What minimum length of twisted pair will cause v to exceed 1.4 V in peak value?

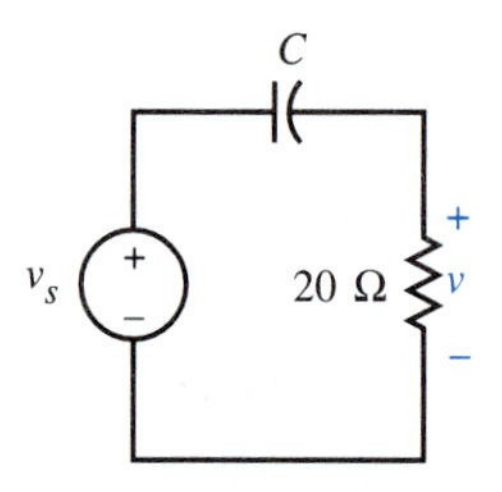

44. In each of the following three networks, find current i and voltage v_x for $t > 0$.

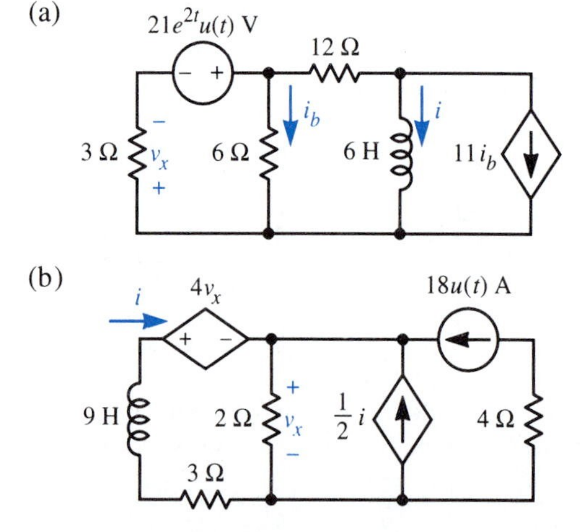

(c)

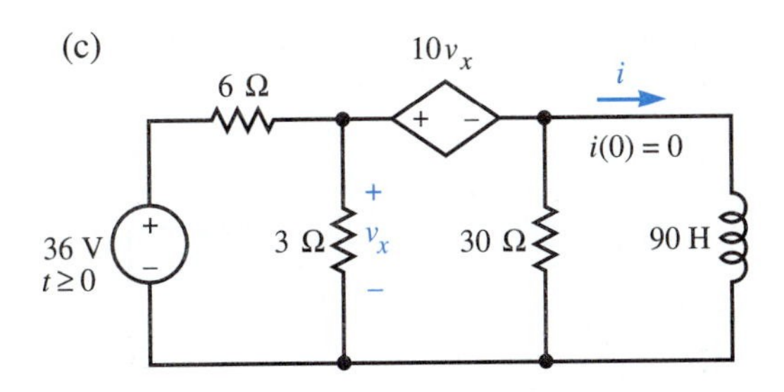

45. For the following network, find $v(t)$ if $C = \frac{1}{24}$ F, $v_{s1} = 5e^{-4t}u(t)$ V, and $v_{s2} = 10$ V for $t > -\infty$.

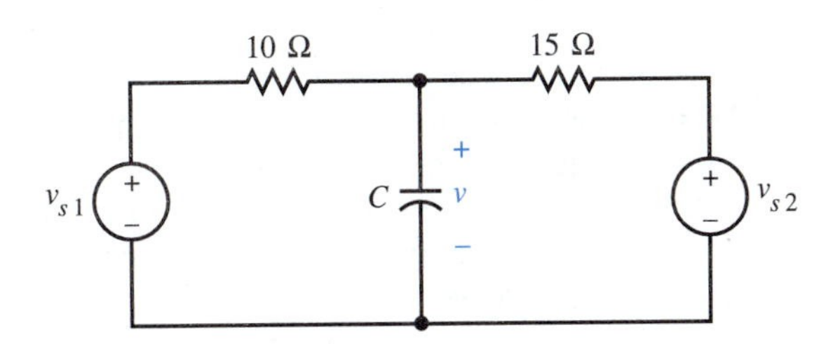

46. For the preceding circuit, $v_{s2} = 0$ V, $v_{s1} = 15\Pi(t - 0.5)$ V [$\Pi(t)$ is the unit pulse function], and (a) $C = 1/60$ F; (b) $C = 1/12$ F; (c) $C = 1/6$ F; (d) $C = 1/3$ F; (e) $C = 5/3$ F. On the same graph, plot $v(t)$ for all values of capacitance in (a) through (e) for $0 \leq t \leq 4$ s.

47. For the following networks, find voltage v_x for $t > 0$. What is the equivalent capacitance seen by the network to the left of terminals a and b?

(a)

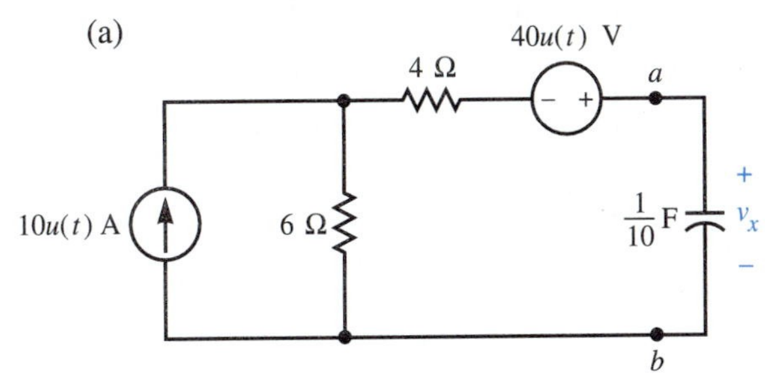

(b)

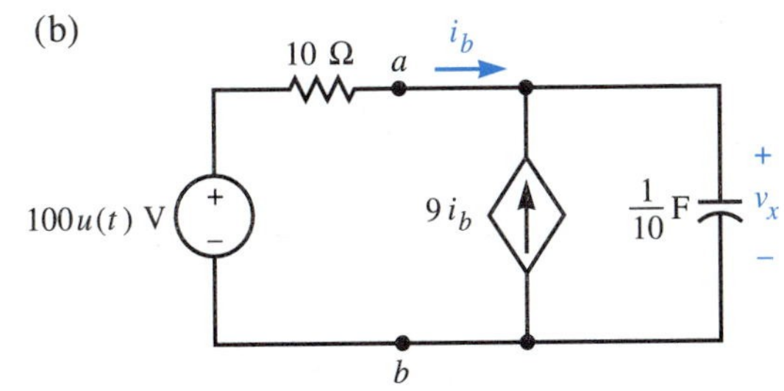

(c)

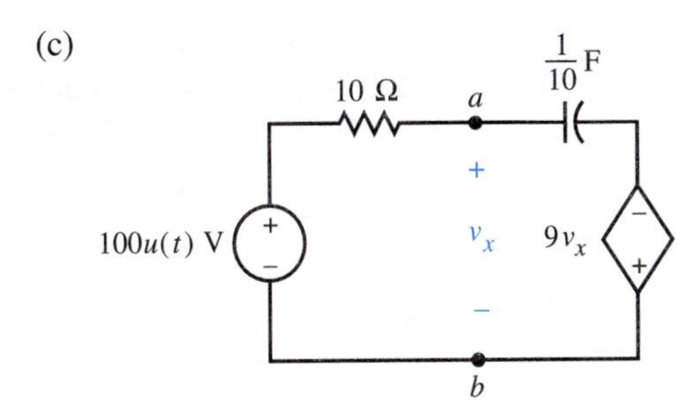

48. The switch in the following circuit has been closed for a very long time and is opened at $t = 0$. Find voltage v for $t > 0$.

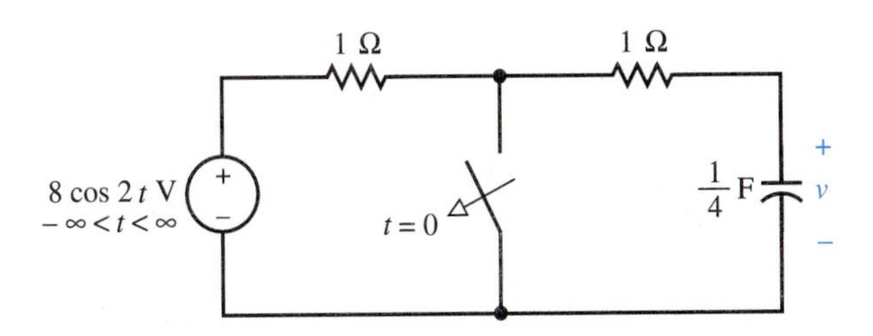

49. The switch in the preceding circuit has been open for a very long time and is closed at $t = 0$. Find voltage v for $t > 0$.

50. The 5-V source and 180-Ω resistance in the following circuit model the supply line to a transistor in a digital circuit. The switch and 20-Ω resistance model the transistor. What is the minimum capacitance C for a bypass capacitor so that voltage v does not drop below 4.5 V when the switch is closed for 10 ns with an initial capacitance voltage of 4.9 V? For this value of C, how long after the switch opens will it take the voltage to recover to 4.9 V?

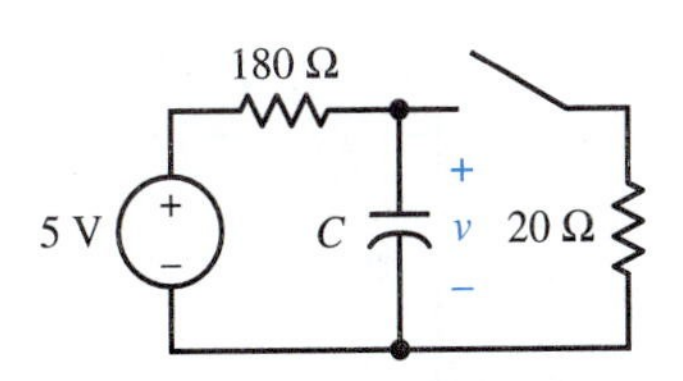

51. The input signal to a logic gate is modeled as a voltage source $v_s = 5u(t)$ V in series with a resistance R_s. The input terminal pair for the logic gate is modeled as a capacitance of 10 pF. The logic gate switches when the capacitance voltage exceeds 3 V. What is the maximum value for R_s so that the delay introduced by charging the capacitance does not exceed 5 ns?

52. The following is an equivalent circuit for an amplifier with capacitive coupling on the input and output. Determine v_1 and v_2 if $v_s = V_s u(t)$ V.

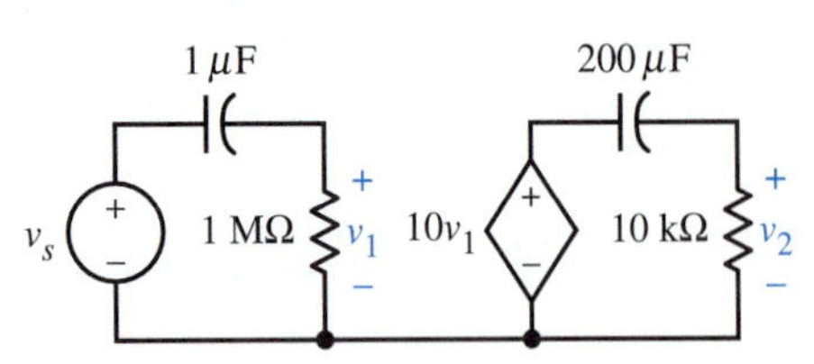

53. Determine the output voltage of the following op-amp circuit when:
 (a) $v_s = 10u(t)$ V
 (b) $v_s = (10 \cos 100t)u(t)$ V

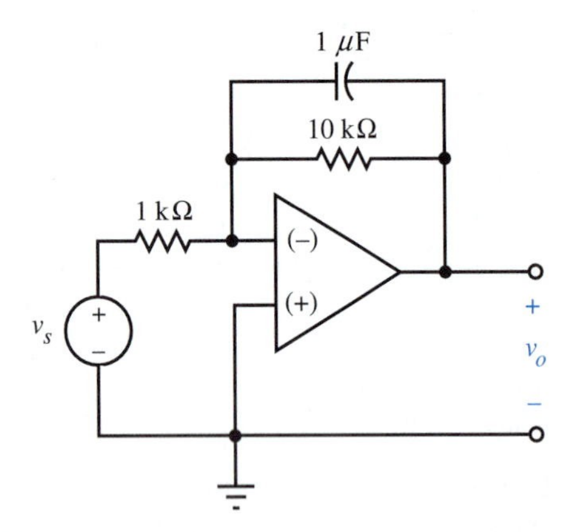

54. The output of the preceding circuit drives a second op-amp circuit so that the input to the second circuit is v_o rather than v_s. The second circuit is the same as that shown above, except that the capacitance is 2 μF. Determine the output v_2 of the second op-amp if $v_s = 0.1u(t)$ V.

55. Determine the steady-state output v_{2p} of the following op-amp circuit if $v_s = \cos \omega_p t$ V.

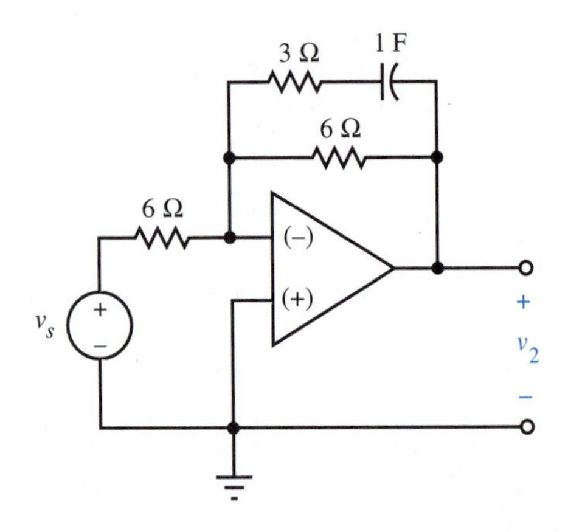

56. Determine voltage v in the following circuit if $v_s = 10u(t)$ V.

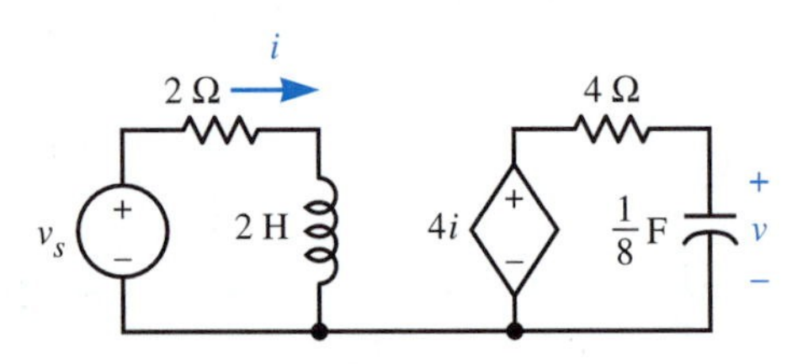

57. Replace the 4-Ω resistance in the preceding circuit with 8 Ω. Determine voltage v if $v_s = 10u(t)$ V.

58. The following network is a model for an oscillator using an electromechanical relay. Switch S_2 is normally closed, and switch S1 is manually closed at $t = 0$. When current $i \geq 6(1 - e^{-1})$ A, the relay, modeled by 2 H in series with 4 Ω, causes switch S2 to be opened. When $i < 6e^{-1}(1 - e^{-1})$ A, the relay causes switch S2 to be closed. The time required for the mechanical switching, as well as the change in inductance as the switching occurs, is ignored for simplicity. Calculate i, and draw a graph showing i as a function of time from $t = 0$ until the third opening of S2. Is i periodic for $t > t_0$? If so, for what value of t_0, and what is the period T?

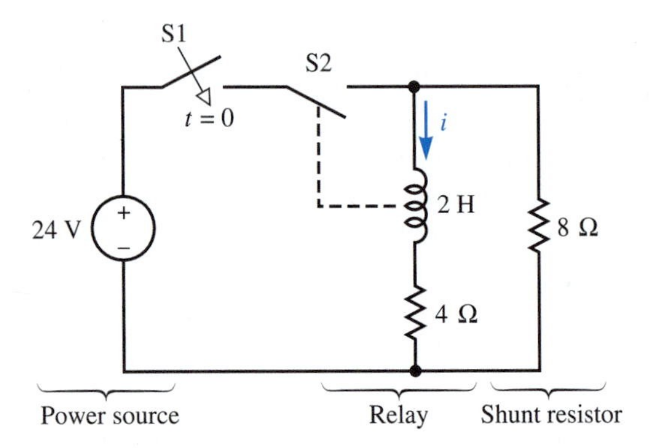

9

Second-Order Circuits

Many practical circuits contain two or more energy-storage elements. Common examples are a tuner that selects one radio station from the numerous ones that are broadcast, and an automobile ignition system that generates a high-voltage spark from 12 V dc. In this chapter we solve the second-order equation that describes a circuit with two energy-storage elements (a second-order circuit). The procedure is similar to that for a first-order equation. We find that the *complete response* is the sum of two components: the *natural response* of the circuit and the *particular response.* The form of the natural response is determined by the source-free circuit and can be a sum of two exponentials or the product of an exponential and a sinusoid. This means that the natural response can be oscillatory. An oscillatory natural response arises from an interchange of stored energy between inductance and capacitance. The particular response is determined by the input and may be calculated by integration. We shall see, however, that we can more simply obtain the particular response for some of our most important inputs by solving an algebraic equation.

9.1 Source-Free *RLC* Circuits

We begin by examining the *RLC* circuit of Fig. 9.1. Energy has been stored in the circuit by a previously connected source. The resulting *initial conditions,* $i(0^+) = I_0$ and $v(0^+) = V_0$, determine the stored energy. We wish to calculate the voltage v for $t > 0$.

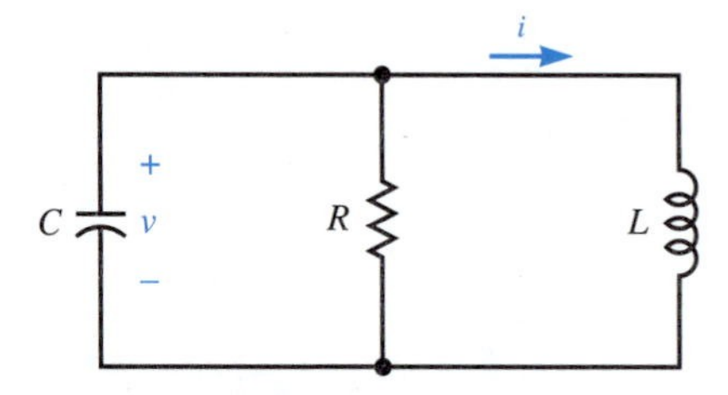

FIGURE 9.1
A source-free parallel *RLC* circuit

The KCL equation for a surface enclosing the top node is

$$C\frac{d}{dt}v + \frac{1}{R}v + \frac{1}{L}\int_{-\infty}^{t} v\,d\lambda = 0 \tag{9.1}$$

After dividing by C, we can differentiate this integro-differential equation to obtain the second-order differential equation

$$\frac{d^2}{dt^2}v + \frac{1}{RC}\frac{d}{dt}v + \frac{1}{LC}v = 0 \tag{9.2}$$

We now consider the solution of equations of this type.

The Second-Order Equation

We see that Eq. (9.2) is of the form

$$\frac{d^2}{dt^2}v + a_1\frac{d}{dt}v + a_0 v = 0 \tag{9.3}$$

Equation (9.3) is a *second-order* differential equation, because two is the highest-order derivative of the unknown variable v. In Appendix C we solve this source-free equation by the use of integrating factors, but here we will use a more intuitive approach. From our work with first-order equations, we might expect the natural response to consist of exponential terms. If we substitute an exponential $v = Ae^{st}$ into Eq. (9.3) and perform the differentiation, we obtain

$$(s^2 + a_1 s + a_0)Ae^{st} = 0 \tag{9.4}$$

We have $Ae^{st} \neq 0$ except for the trivial case where $A = 0$. This tells us that for Ae^{st} to be a solution to Eqs. (9.3) and (9.4), s must be a root of the *characteristic equation*

$$\mathcal{A}(s) = s^2 + a_1 s + a_0 = 0 \tag{9.5}$$

Since Eq. (9.5) has two roots, s_1 and s_2, both $A_1e^{s_1t}$ and $A_2e^{s_2t}$ are solutions to Eq. (9.3). Because each exponential is a solution, the sum of the two exponentials is also a solution, and the natural response of a second-order circuit is

$$v_n = A_1e^{s_1t} + A_2e^{s_2t} \qquad s_1 \neq s_2 \tag{9.6}$$

where s_1 and s_2 are the roots of the characteristic equation, and A_1 and A_2 are the constants of integration. This completely solves the problem, except for the *unusual* case where $s_1 = s_2$. In Appendix C we use integration to show that the natural response is $A_1 e^{s_1 t} + A_2 t e^{s_1 t}$ for $s_1 = s_2$. Therefore the solution to the second-order source-free equation is

The Natural Response of a Second-Order Circuit

$$v_n = \begin{cases} A_1 e^{s_1 t} + A_2 e^{s_2 t} & s_1 \neq s_2 \quad (9.7a) \\ (A_1 + A_2 t) e^{s_1 t} & s_1 = s_2 \quad (9.7b) \end{cases}$$

where s_1 and s_2 are the roots of the characteristic equation, and A_1 and A_2 are constants of integration.

In order to compare the response of different circuits, we often write a second-order differential equation in the *standard form*

$$\frac{d^2}{dt^2} v + 2\zeta\omega_0 \frac{d}{dt} v + \omega_0^2 v = 0 \tag{9.8}$$

where ω_0 is the *undamped natural frequency,* and ζ is the *damping ratio* (we will see the reason for these names as we analyze some circuits). The characteristic equation becomes

$$\mathscr{A}(s) = s^2 + 2\zeta\omega_0 s + \omega_0^2 = 0 \tag{9.9}$$

with roots

$$s_1 = -\zeta\omega_0 + \omega_0\sqrt{\zeta^2 - 1} \tag{9.10a}$$

$$= -\alpha + \omega_0\sqrt{\zeta^2 - 1} \tag{9.10b}$$

and

$$s_2 = -\zeta\omega_0 - \omega_0\sqrt{\zeta^2 - 1} \tag{9.11a}$$

$$= -\alpha - \omega_0\sqrt{\zeta^2 - 1} \tag{9.11b}$$

where the *damping coefficient* α is given by

$$\alpha = \zeta\omega_0 \tag{9.12}$$

There are three possible cases:

1. Overdamped: $\zeta > 1$. The two roots are real, negative, and distinct [the response is given by Eq. (9.7a)].
2. Underdamped: $\zeta < 1$. The two roots are imaginary or complex with a negative real part [the response is given by Eq. (9.7a), but later we will write the response in a more convenient form with the aid of Euler's identity].
3. Critically damped: $\zeta = 1$. The two roots are real and negative, but identical [the response is given by Eq. (9.7b)].

In the following examples we will analyze circuits for each of these three cases and learn how to use initial conditions to evaluate the constants of integration.

Remember

The roots of the characteristic equation determine the natural response of a circuit. We use initial conditions to evaluate the constants of integration.

EXERCISE

1. Write the characteristic equation for each of the following second-order differential equations. Determine the roots of the characteristic equation, and write the natural response by the use of Eq. (9.7). (There is insufficient information given to evaluate the constants of integration.)

(a) $\frac{d^2}{dt^2}v + 6\frac{d}{dt}v + 8v = 0$ answer: $A_1e^{-2t} + A_2e^{-4t}$

(b) $\frac{d^2}{dt^2}v + 4\frac{d}{dt}v + 4v = 0$ answer: $(A_1 + A_2t)e^{-2t}$

(c) $\frac{d^2}{dt^2}v + 6\frac{d}{dt}v + 25v = 0$ answer: $A_1e^{(-3+j4)t} + A_2e^{(-3-j4)t}$

(d) $\frac{d^2}{dt^2}v + 4v = 0$ answer: $A_1e^{j2t} + A_2e^{-j2t}$

9.2 The Parallel *RLC* Circuit

We now consider the differential equation that describes the parallel *RLC* circuit of Fig. 9.1. We write Eq. (9.2) in operational form,

$$\left(\frac{d^2}{dt^2} + \frac{1}{RC}\frac{d}{dt} + \frac{1}{LC}\right)v = 0 \tag{9.13}$$

and replace d/dt with s in the operator and divide by v to obtain the characteristic equation:

$$\mathcal{A}(s) = s^2 + \frac{1}{RC}s + \frac{1}{LC} = 0 \tag{9.14}$$

This characteristic equation has the roots

$$s_1 = -\frac{1}{2RC} + \sqrt{\left(\frac{1}{2RC}\right)^2 - \frac{1}{LC}} \tag{9.15}$$

and

$$s_2 = -\frac{1}{2RC} - \sqrt{\left(\frac{1}{2RC}\right)^2 - \frac{1}{LC}} \tag{9.16}$$

If we equate the coefficients in Eqs. (9.8) and (9.13), we see that for a *parallel RLC* circuit

$$\omega_0 = \frac{1}{\sqrt{LC}} \tag{9.17}$$

$$\zeta = \frac{1}{2R}\sqrt{\frac{L}{C}} \tag{9.18}$$

and

$$\alpha = \frac{1}{2RC} \tag{9.19}$$

We will first consider the case of a parallel *RLC* circuit where the roots of the characteristic equation are real and distinct.

The Overdamped Circuit

We will now calculate the source-free response for a parallel *RLC* circuit that is overdamped ($\zeta = \alpha/\omega_0 > 1$). We will see that the natural response of an overdamped circuit is not oscillatory (a mechanical analogy is the ride of an automobile with very stiff shock absorbers). The response may pass through zero once, depending on the initial conditions, but will then asymptotically approach zero. Because the circuit is described by a second-order equation, we need to evaluate two constants of integration. The following example shows how this is done.

EXAMPLE 9.1 For the parallel *RLC* circuit shown in Fig. 9.1, let $R = \frac{8}{25}\ \Omega$, $L = \frac{1}{100}$ H, and $C = \frac{1}{64}$ F. Determine the source-free response if $i(0^+) = 16$ A and $v(0^+) = 8$ V.

Solution The KCL equation for a surface enclosing the upper node is

$$\frac{1}{64}\frac{d}{dt}v + \frac{25}{8}v + 100\int_{-\infty}^{t} v\, d\lambda = 0 \qquad t > 0$$

We differentiate this equation to eliminate the integral and multiply it by 64:

$$\frac{d^2}{dt^2}v + 200\frac{d}{dt}v + 6400v = 0 \qquad t > 0$$

The characteristic equation

$$\mathscr{A}(s) = s^2 + 200s + 6400 = 0$$

has roots

$$s_1 = -40 \qquad \text{and} \qquad s_2 = -160$$

The roots are real, negative, and distinct, so the circuit is overdamped and the source-free response is given by Eq. (9.7a).

$$v = A_1 e^{-40t} + A_2 e^{-160t} \qquad t > 0$$

We determine the constants of integration by use of the initial conditions. If we evaluate the response at $t = 0^+$ and equate it to $v(0^+)$, we obtain

$$A_1 e^{-40(0)} + A_2 e^{-160(0)} = v(0^+) = 8$$

This gives the first equation needed to evaluate the constants of integration:

$$A_1 + A_2 = 8$$

Knowledge of the first derivative of v is needed to obtain a second equation containing A_1 and A_2. This derivative is easily found from the original KCL equation evaluated at $t = 0^+$. The KCL equation [Eq. (9.1)]

$$C\frac{d}{dt}v + \frac{1}{R}v + \frac{1}{L}\int_{-\infty}^{t} v\, d\lambda = 0$$

becomes

$$C\frac{d}{dt}v\bigg|_{t=0^+} + \frac{1}{R}v(0^+) + i(0^+) = 0$$

The required derivative is

$$\frac{d}{dt}v\bigg|_{t=0^+} = (-40A_1e^{-40t} - 160A_2e^{-160t})\bigg|_{t=0^+} = -40A_1 - 160A_2$$

If we substitute this result and the numerical values for R, C, $v(0^+)$, and $i(0^+)$ into the preceding equation, we obtain

$$\frac{5}{2}A_1 + \frac{5}{2}A_2 = 41$$

We solve this and the earlier equation simultaneously for $A_1 = -\frac{56}{5}$ and $A_2 = \frac{96}{5}$. This gives the response

$$v = -\frac{56}{5}e^{-40t} + \frac{96}{5}e^{-160t}\text{ V} \qquad t > 0$$

We can equate the coefficients in Eq. (9.8) and our second-order differential equation to show that $\zeta = 5/4 > 1$. The response of this overdamped circuit is plotted in Fig. 9.2.

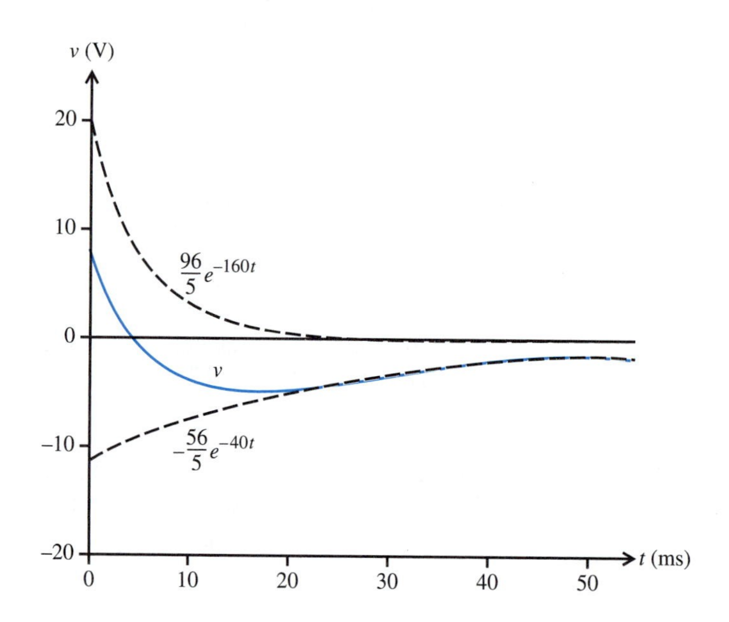

FIGURE 9.2 Source-free response of a second-order circuit with an overdamped response ($\zeta = 5/4$)

Remember

The characteristic equation for an overdamped circuit ($\zeta > 1$) has distinct real roots, s_1 and s_2. The natural response is of the form $A_1e^{s_1t} + A_2e^{s_2t}$. We need two equations to evaluate the constants of integration A_1 and A_2. To obtain the first equation, we evaluate our solution at $t = 0^+$ and equate it to the initial value of the voltage $v(0^+)$. To obtain the second equation, we substitute our solution into the original KCL equation, evaluate the derivative at $t = 0^+$, and substitute in the initial values, $v(0^+)$ and $i(0^+)$. We solve these two equations for the constants of integration.

EXERCISE

2. For a parallel *RLC* circuit, $C = 1$ F, $R = \frac{1}{3}\ \Omega$, and $L = \frac{1}{2}$ H. Determine the circuit voltage v, if $v(0^+) = 2$ V, and the initial inductance current is $i(0^+) = 4$ A (passive sign convention). Also calculate the damping ratio.

answer: $-6e^{-t} + 8e^{-2t}$ V for $t > 0$, $\frac{3}{4}\sqrt{2}$

The Underdamped Circuit

If the resistance in a parallel *RLC* circuit is increased sufficiently, the power (v^2/R) absorbed by the resistance decreases to the point that the circuit becomes oscillatory (a mechanical analogy is an automobile with very weak shock absorbers). We say that such a circuit is underdamped ($\zeta < 1$). The two roots of the characteristic equation, as given by Eqs. (9.10a) and (9.11a),

$$s = -\alpha \pm \omega_0\sqrt{\zeta^2 - 1} \tag{9.20}$$

become complex, so we rewrite equation (9.20) as

$$s = -\alpha \pm j\omega_0\sqrt{1 - \zeta^2} \tag{9.21}$$

We usually define the *damped natural frequency*

$$\omega_d = \omega_0\sqrt{1 - \zeta^2} \tag{9.22}$$

The roots of the characteristic equation are then

$$s = -\alpha \pm j\omega_d \tag{9.23}$$

Equation (9.7a) gives the natural response as

$$v = A_1e^{(-\alpha+j\omega_d)t} + A_2e^{(-\alpha-j\omega_d)t} \tag{9.24}$$

Euler's identity lets us rewrite this equation as

$$\begin{aligned} v &= A_1e^{-\alpha t}(\cos\omega_d t + j\sin\omega_d t) + A_2e^{-\alpha t}(\cos\omega_d t - j\sin\omega_d t) \\ &= e^{-\alpha t}[(A_1 + A_2)\cos\omega_d t + j(A_1 - A_2)\sin\omega_d t] \end{aligned} \tag{9.25}$$

If we redefine the constants of integration by

$$A = (A_1 + A_2) \tag{9.26}$$

and

$$B = j(A_1 - A_2) \tag{9.27}$$

we can write the natural response for an underdamped circuit as:

The Natural Response for an Underdamped Circuit

$$v = e^{-\alpha t}(A\cos\omega_d t + B\sin\omega_d t) \tag{9.28}$$

where the roots of the characteristic equation are $s = -\alpha \pm j\omega_d$, and A and B are the constants of integration.

We see that the damped natural frequency ω_d determines the angular frequency of the oscillatory terms, and the undamped natural frequency ω_0 is the angular frequency when the damping ratio ζ, and thus the damping factor α, are zero.

EXAMPLE 9.2 For the parallel RLC circuit shown in Fig. 9.1, let $R = 2\ \Omega$, $L = \frac{1}{100}$ H, and $C = \frac{1}{64}$ F. Determine the source-free response if $i(0^+) = 16$ A and $v(0^+) = 8$ V.

Solution The KCL equation for a surface enclosing the upper node is

$$\frac{1}{64}\frac{d}{dt}v + \frac{3}{2}v + 100\int_{-\infty}^{t} v\,d\lambda = 0 \qquad t > 0$$

or, after differentiation and multiplication by 64,

$$\frac{d^2}{dt^2}v + 96\frac{d}{dt}v + 6400v = 0 \qquad t > 0$$

The characteristic equation

$$\mathcal{A}(s) = s^2 + 96s + 6400 = 0$$

has roots

$$s = -48 \pm j64$$

The roots are complex, so the network is underdamped, and the complementary (natural) response is

$$v = e^{-48t}(A\cos 64t + B\sin 64t) \qquad t > 0$$

The constants of integration are easily determined from the initial conditions. Evaluate this last equation at $t = 0^+$ and equate this to the value of v at $t = 0^+$:

$$e^{-48(0^+)}\{A\cos[64(0^+)] + B\sin[64(0^+)]\} = v(0^+) = 8$$

This yields

$$A = 8$$

Knowledge of the derivative of v is needed to obtain a second equation containing A and B. This is easily found from the original KCL equation evaluated at $t = 0^+$. The KCL equation when evaluated at $t = 0^+$, becomes

$$C\frac{d}{dt}v\bigg|_{t=0^+} + \frac{1}{R}v(0^+) + \frac{1}{L}\int_{-\infty}^{0^+} v\,d\lambda = 0$$

The derivative of v is

$$\begin{aligned}\frac{d}{dt}v\bigg|_{t=0^+} &\\ &= e^{-48t}(-64A\sin 64t + 64B\cos 64t) - 48e^{-48t}(A\cos 64t + B\sin 64t)\bigg|_{t=0^+}\\ &= 64B - 48A = 64B - 384\end{aligned}$$

so

$$C(64B - 384) + \frac{1}{R}v(0^+) + i(0^+) = 0$$

Substitution of numerical values for R, C, $i(0^+)$, and $v(0^+)$ gives

$$\frac{1}{64}(64B - 384) + \frac{3}{2}(8) + 16 = 0$$

and

$$B = -22$$

The resulting solution is

$$v = e^{-48t}(8\cos 64t - 22\sin 64t)\ \text{V} \qquad t > 0$$

It follows from the element values that the damping ratio

$$\zeta = \frac{3}{5} < 1$$

is less than one as expected for an underdamped circuit.

The response for the source-free parallel *RLC* circuit of Example 9.2 is shown in Fig. 9.3. Although the response is oscillatory, the damping factor of $\zeta = 0.6$ causes the oscillations to die out quickly. If the resistance is doubled to $R = \frac{4}{3}\ \Omega$ for a damping ratio of 0.3, the oscillatory nature of the response is more apparent, as shown in Fig. 9.3. We will next analyze the unusual case of a critically damped circuit.

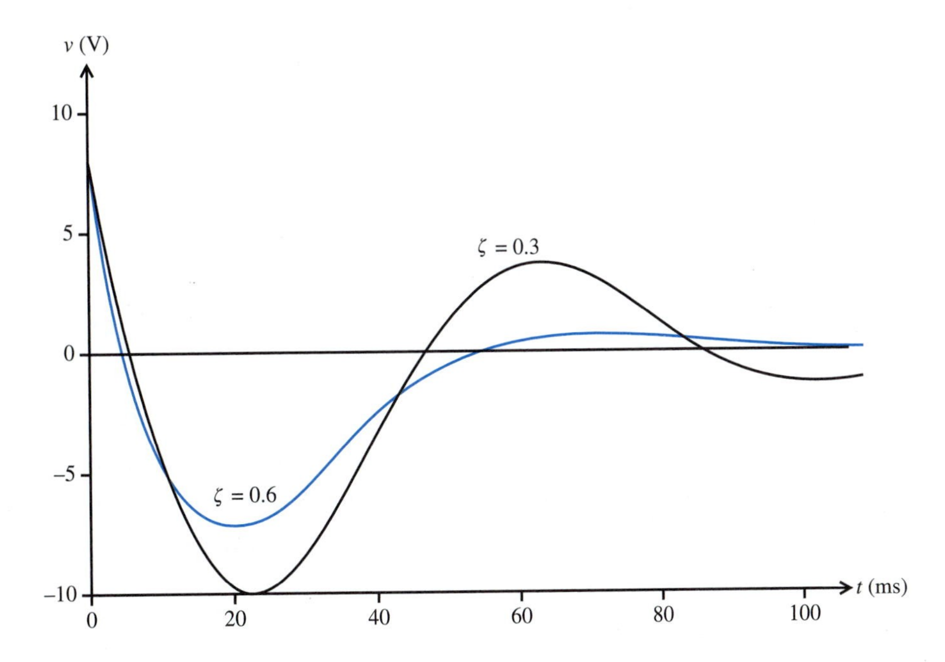

FIGURE 9.3 The natural response of an underdamped *RLC* circuit

Remember

Increasing the resistance in a parallel *RLC* circuit decreases the resistive losses and therefore decreases the damping. If the resistive losses become sufficiently small, the response becomes oscillatory ($\zeta < 1$). The roots of the characteristic equation become complex, $s = -\alpha \pm j\omega_d$, and the natural response is of the form $e^{-\alpha t}(A\cos\omega_d t + B\sin\omega_d t)$. We use the initial conditions $v(0^+)$ and $i(0^+)$ to evaluate the constants of integration A and B. The procedure is the same as for an overdamped circuit.

EXERCISE 3. For a parallel *RLC* circuit $C = 2$ F, $R = \frac{1}{4}\ \Omega$, and $L = \frac{1}{10}$ H. Determine the voltage v, if $v(0^+) = 12$ V, and the initial inductance current is $i(0^+) = -40$ A (passive sign convention).
answer: $e^{-t}(12 \cos 2t + 4 \sin 2t)$ V

The Critically Damped Circuit

A parallel *RLC* circuit is critically damped if the damping ratio ζ is *exactly* one. We might argue that critically damped circuits are impossible to build, because of the finite precision of our components. Nevertheless, the critically damped case is of interest, because it represents the dividing line between overdamped and underdamped circuits. Critical damping implies that the two roots, s_1 and s_2, of the characteristic equation are identical. In this situation the natural response is of the form given by Eq. (9.7b):

$$v = (A_1 + A_2 t)e^{-s_1 t} \tag{9.29}$$

Other than this one change, the analysis of a critically damped *RLC* circuit is identical to that of an underdamped or overdamped circuit.

EXAMPLE 9.3 Change the resistance to $R = \frac{2}{5}\ \Omega$ in the parallel *RLC* circuit used in Examples 9.1 and 9.2. Leave the inductance, capacitance, and initial conditions the same. Determine the source-free response.

Solution The KCL equation for a surface enclosing the top node is

$$\frac{1}{64}\frac{d}{dt}v + \frac{5}{2}v + 100\int_{-\infty}^{t} v\, d\lambda = 0 \qquad t > 0$$

or, after differentiation and multiplication by 64,

$$\frac{d^2}{dt^2}v + 160\frac{d}{dt}v + 6400v = 0 \qquad t > 0$$

The characteristic equation

$$\mathcal{A}(s) = s^2 + 160s + 6400 = 0$$

has roots

$$s_1 = s_2 = -80$$

We see that the characteristic equation has a second-order root and the circuit is critically damped. The source-free response is of the form

$$v = (A_1 + A_2 t)e^{-80t}$$

The constants of integration are evaluated as in the previous example:

$$[A_1 + A_2(0^+)]e^{-80(0^+)} = v(0^+) = 8$$

which gives

$$A_1 = 8$$

The derivative of v evaluated at $t = 0^+$ is

$$\left.\frac{d}{dt}v\right|_{t=0^+} = -80A_1 + A_2 = -640 + A_2$$

If we substitute this result, $v(0^+) = 8$ V, and $i(0^+) = 16$ A into our original KCL equation we obtain

$$A_2 = -1664$$

with the result

$$v = (8 - 1664t)e^{-80t} \text{ V} \qquad t > 0$$

This source-free response is shown in Fig. 9.4.

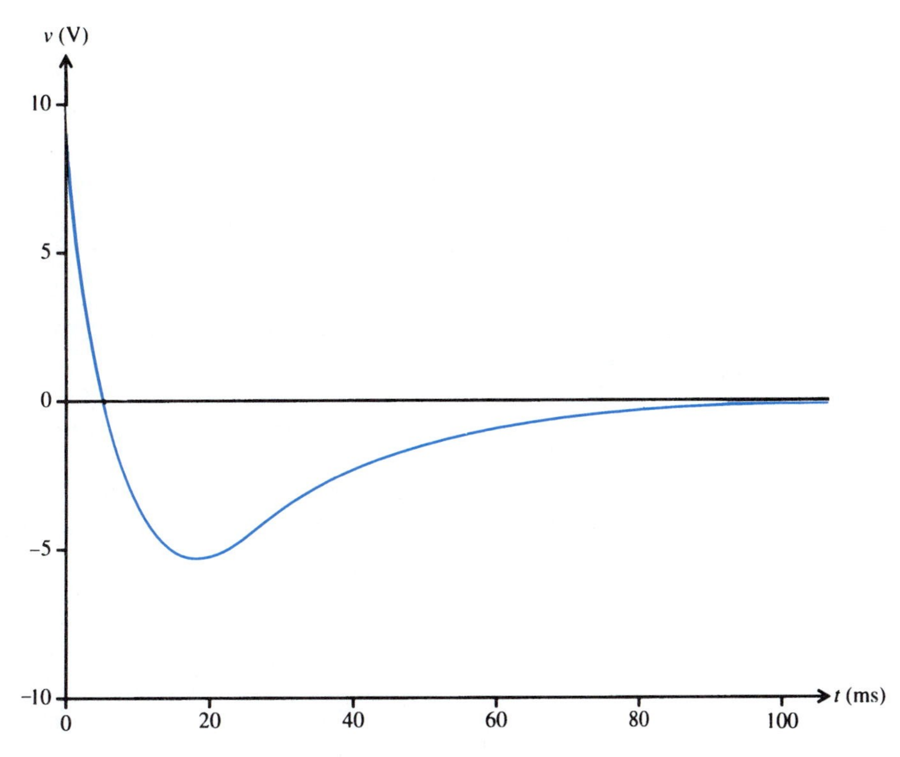

FIGURE 9.4 The source-free response of a critically damped parallel *RLC* circuit

Remember

If the roots of the characteristic equation are identical, $s_1 = s_2$, the natural response is of the form $(A_1 + A_2t)e^{s_1t}$, where the constants of integration, A_1 and A_2, are evaluated in the same manner as for overdamped or underdamped circuits.

EXERCISE

4. A parallel *RLC* circuit is source-free for $t > 0$ and has $C = 5$ F, $L = \frac{1}{125}$ H, and $R = \frac{1}{50}\ \Omega$. The initial voltage is $v(0^+) = 4$ V and the inductance current is $i(0^+) = 200$ A (passive sign convention). Determine v for $t > 0$.

answer: $(4 - 60t)e^{-5t}$ V for $t > 0$

9.3 The Series *RLC* Circuit

The procedure to analyze a series *RLC* circuit is essentially the same as for a parallel *RLC* circuit. We write a network equation (in the case of a series circuit this will be a KVL equation rather than a KCL equation) and identify the characteristic equation. We solve for the roots of the characteristic equation and write the natural response. To complete the solution we use the initial conditions to evaluate the constants of integration.

Write a KVL equation around the loop of the series *RLC* circuit shown in Fig. 9.5. We obtain

$$L\frac{d}{dt}i + Ri + \frac{1}{C}\int_{-\infty}^{t} i\, d\lambda = 0 \tag{9.30}$$

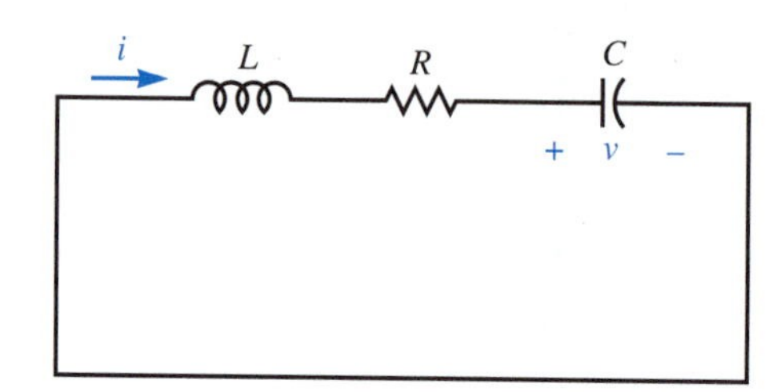

FIGURE 9.5
A source-free series *RLC* circuit

After dividing by L we differentiate this equation once and write it in operator notation as

$$\left(\frac{d^2}{dt^2} + \frac{R}{L}\frac{d}{dt} + \frac{1}{LC}\right) i = 0 \tag{9.31}$$

We can equate the coefficients in Eqs. (9.8) and (9.31) to show that the undamped natural frequency ω_0, the damping ratio ζ, and the damping coefficient α are given by

$$\omega_0 = \frac{1}{\sqrt{LC}} \tag{9.32}$$

$$\zeta = \frac{R}{2}\sqrt{\frac{C}{L}} \tag{9.33}$$

and

$$\alpha = \frac{R}{2L} \tag{9.34}$$

We will now analyze a series *RLC* circuit that is overdamped.

The Overdamped Circuit

If the resistance is sufficiently large in a series *RLC* circuit, the resistive losses (Ri^2) are large enough that the circuit is overdamped ($\zeta > 1$). The current will be the sum of two exponentials, and we must determine two constants of integration. We demonstrate the procedure in the following example.

EXAMPLE 9.4 The series *RLC* circuit shown in Fig. 9.5 has $R = \frac{25}{8}\ \Omega$, $L = \frac{1}{64}$ H, and $C = \frac{1}{100}$ F. Determine the source-free response if the initial conditions are $i(0^+) = 8$ A and $v(0^+) = 16$ V.

Solution The KVL equation around the loop in Fig. 9.5 is

$$\frac{1}{64}\frac{d}{dt}i + \frac{25}{8}i + 100\int_{-\infty}^{t} i\,d\lambda = 0 \qquad t > 0$$

We differentiate this equation to eliminate the integral and multiply it by 64:

$$\frac{d^2}{dt^2}i + 200\frac{d}{dt}i + 6400i = 0 \qquad t > 0$$

The characteristic equation

$$\mathcal{A}(s) = s^2 + 200s + 6400 = 0$$

has roots

$$s_1 = -40 \text{ and } s_2 = -160$$

The roots are real, negative, and distinct, so the circuit is overdamped and the source-free response is given by Eq. (9.7a).

$$i = A_1e^{-40t} + A_2e^{-160t} \qquad t > 0$$

We will now evaluate the constants of integration by use of the initial conditions. Evaluate the solution for i at $t = 0^+$ and equate it to $i(0^+)$. We obtain

$$A_1e^{-40(0)} + A_2e^{-160(0)} = i(0^+) = 8$$

This gives the first equation needed to evaluate the constants of integration:

$$A_1 + A_2 = 8$$

Knowledge of the first derivative of i is needed to obtain a second equation containing A_1 and A_2. This derivative is easily found from the original KVL equation evaluated at $t = 0^+$. The KVL equation

$$L\frac{d}{dt}i + Ri + \frac{1}{C}\int_{-\infty}^{t} i\,d\lambda = 0$$

becomes

$$L\frac{d}{dt}i\bigg|_{t=0^+} + Ri(0^+) + v(0^+) = 0$$

The required derivative is

$$\begin{aligned}\frac{d}{dt}i\bigg|_{t=0^+} &= (-40A_1e^{-40t} - 160A_2e^{-160t})\bigg|_{t=0^+} \\ &= -40A_1 - 160A_2\end{aligned}$$

Substitution of this result and the numerical values for R, L, $i(0^+)$, and $v(0^+)$ into the KVL equation gives us

$$\frac{5}{8}A_1 + \frac{5}{2}A_2 = 41$$

We solve for $A_1 = -\frac{56}{5}$ and $A_2 = \frac{96}{5}$ to obtain

$$i = -\frac{56}{5}e^{-40t} + \frac{96}{5}e^{-160t}\text{ V} \qquad t > 0$$

We can equate the coefficients in Eqs. (9.8) and our differential equation to show that $\zeta = 5/4$ as it was in Example 9.1. The response of this overdamped circuit is given by Fig. 9.2, if the ordinate label is changed from v(V) to i(A).

The natural response of an overdamped circuit is never oscillatory. The response may pass through zero once, depending on the initial conditions, but will then asymptotically approach zero.

Underdamped and Critically Damped Circuits

If the resistance in a series *RLC* circuit is decreased sufficiently, the power (Ri^2) absorbed by the resistance becomes so small that the natural response becomes oscillatory, and the circuit is underdamped ($\zeta < 1$). The procedure to analyze an underdamped series RLC circuit is essentially the same as that for an overdamped series *RLC* circuit. Only the form of the natural response is changed. For example, if the resistance in the series *RLC* circuit of Example 9.3 is decreased to $R = \frac{3}{2}\ \Omega$, and the capacitance, inductance, and initial conditions are unchanged, the KVL equation around the loop becomes

$$\frac{1}{64}\frac{d}{dt}i + \frac{3}{2}i + 100\int_{-\infty}^{t} i d\lambda = 0 \qquad t > 0 \tag{9.35}$$

and the source-free response is

$$i = e^{-48t}(8\cos 64t - 22\sin 64t)\text{ A} \qquad t > 0 \tag{9.36}$$

This response is given by Fig. 9.3 if the label on the ordinate is changed from v(V) to i(A).

With the proper choice of resistance, any *RLC* circuit is critically damped ($\zeta = 1$), and the two roots, s_1 and s_2, of the characteristic equation are identical. In this situation the natural response for a series *RLC* circuit is of the form given by Eq. (9.7b). Other than this single change, the analysis of a critically damped *RLC* circuit is identical to that of an underdamped or overdamped circuit. For example, if we change the resistance to $R = \frac{5}{2}\ \Omega$ in the series *RLC* circuit used in Example 9.4, leave the inductance, capacitance, and initial conditions the same, the KVL equation for the loop becomes

$$\frac{1}{64}\frac{d}{dt}i + \frac{5}{2}i + 100\int_{-\infty}^{t} i\, d\lambda = 0 \qquad t > 0 \tag{9.37}$$

and the source-free response is

$$i = (8 - 1664t)e^{-80t}\text{ A} \qquad t > 0 \tag{9.38}$$

This source-free response is given by Fig. 9.4, if the label on the ordinate is changed from v(V) to i(A).

Remember

The characteristic equation for an overdamped circuit ($\zeta > 1$) has distinct real roots, s_1 and s_2. The natural response is of the form $A_1e^{s_1t} + A_2e^{s_2t}$. Decreasing the resistance in a series *RLC* circuit decreases the resistive losses and therefore decreases damping. If the resistive losses become sufficiently small, the response becomes oscillatory ($\zeta < 1$). The roots of the characteristic equation become complex, $s = -\alpha \pm j\omega_d$, and the natural response is of the form $e^{-\alpha t}(A\cos\omega_d t + B\sin\omega_d t)$. If the roots of the characteristic equation are identical, $s_1 = s_2$, the natural response is of the form $(A_1 + A_2t)e^{s_1t}$. In any case, we need two equations to evaluate the constants of integration A_1 and A_2. To obtain the first equation for a series circuit, we evaluate our solution at $t = 0^+$ and equate it to the initial value of the current $i(0^+)$. To obtain the second equation, we substitute our solution into the original KVL equation, evaluate the derivative at $t = 0^+$, and substitute in the initial values, $i(0^+)$ and $v(0^+)$. We solve these two equations for the constants of integration.

EXERCISES

5. An inductance $L = 2$ H, a resistance $R = 10\ \Omega$, and a capacitance $C = \frac{1}{8}$ F are connected in series. Determine the current if $i(0^+) = 12$ A, and the initial capacitance voltage is $v(0^+) = 36$ V (passive sign convention).
   answer: $-10e^{-t} + 22e^{-4t}$ A for $t > 0$
6. An inductance $L = 5$ H, a resistance $R = 30\ \Omega$, and a capacitance $C = \frac{1}{125}$ F are connected in series. Determine the current if $i(0^+) = 4$ A and the initial capacitance voltage is $v(0^+) = 200$ V (passive sign convention).
   answer: $e^{-3t}(4 \cos 4t - 13 \sin 4t)$ A for $t > 0$
7. A series *RLC* circuit is source-free for $t > 0$ and has $L = 5$ H, $C = \frac{1}{125}$ F, and $R = 50\ \Omega$. The initial current is $i(0^+) = 4$ A and the capacitance voltage is $v(0^+) = 200$ V (passive sign convention). Determine i for $t > 0$.
   answer: $(4 - 60t)e^{-5t}$ A for $t > 0$

9.4 Second-Order Circuits with Sources

Energy is stored in a circuit by capacitance and inductance, and we have seen that stored energy causes a response in a circuit that is source-free. We now see how a source establishes capacitance voltage and inductance current. We begin by examining the *RLC* circuit of Fig. 9.6. To analyze this circuit we can use mesh currents and write two KVL equations, or we can use the shortcut method for node voltages and write one KCL equation. The choice is obvious.

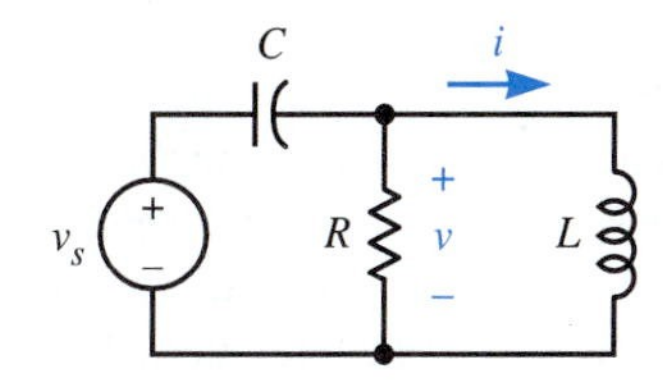

FIGURE 9.6
An *RLC* circuit

The KCL equation for a surface enclosing the top right-hand node is

$$C\frac{d}{dt}v + \frac{1}{R}v + \frac{1}{L}\int_{-\infty}^{t} v\,d\lambda = C\frac{d}{dt}v_s \tag{9.39}$$

After dividing by C, we differentiate this integro-differential equation to obtain the second-order differential equation

$$\frac{d^2}{dt^2}v + \frac{1}{RC}\frac{d}{dt}v + \frac{1}{LC}v = \frac{d^2}{dt^2}v_s \tag{9.40}$$

We will now see how to solve differential equations of this type.

The Forced Second-Order Equation

We see that Eq. (9.40) includes a second derivative of the source. For the most general case, the differential equation that describes a second-order circuit has the form

$$\begin{aligned}\frac{d^2}{dt^2}v + a_1\frac{d}{dt}v + a_0v &= b_2\frac{d^2}{dt^2}i + b_1\frac{d}{dt}i + b_0i \\ &= g(t)\end{aligned} \tag{9.41}$$

We can write Eq. (9.41) in operator form as

$$\begin{aligned}\left(\frac{d^2}{dt^2} + a_1\frac{d}{dt} + a_0\right)v &= \left(b_2\frac{d^2}{dt^2} + b_1\frac{d}{dt} + b_0\right)i \\ &= g(t)\end{aligned} \tag{9.42}$$

Although we leave the details to Appendix C, we can use the integrating factors e^{-s_1t} and e^{-s_2t}, where s_1 and s_2 are the roots of the characteristic equation, and two integrations to solve Eq. (9.41) or (9.42). The *complete response* v is the sum of two parts,

$$v = v_n + v_p \tag{9.43}$$

just as it was for first-order circuits.

The two integrations show that the *natural response* v_n is given by Eq. (9.7a) or (9.7b) in Section 9.1, where A_1 and A_2 are constants of integration determined from the initial conditions and the complete response. The integrations also yield the *particular response* v_p, but we need not actually perform the integrations for exponential inputs. (Exponential inputs include the widely used constant and sinusoidal inputs as special cases.)

Exponential Sources

If we evaluate $g(t)$ as given by Eq. (9.42) for the input $i = Ke^{s_pt}$, we have

$$\begin{aligned}g(t) &= \left(b_2\frac{d^2}{dt^2} + b_1\frac{d}{dt} + b_0\right)Ke^{s_pt} \\ &= (b_2s_p^2 + b_1s_p + b_0)Ke^{s_pt} = \mathcal{B}(s_p)Ke^{s_pt}\end{aligned} \tag{9.44}$$

which is simply the exponential input Ke^{s_p} multiplied by the constant $\mathcal{B}(s_p)$. (We use a subscript p on s to emphasize that we are looking for the particular response for a particular value of s.) Therefore, the particular response v_p must be an exponential function $K_2e^{s_pt}$ so that a weighted sum of the particular response v_p and its first two derivatives [the left side of Eq. (9.42)] is equal to the exponential $\mathcal{B}(s_p)Ke^{s_pt}$. We know that differentiation of e^{s_pt} is equivalent to multiplication by s_p, so Eq. (9.42) becomes

$$(s_p^2 + a_1s_p + a_0)v_p = (b_2s_p^2 + b_1s_p + b_0)Ke^{s_pt} \qquad \mathcal{A}(s_p) \neq 0 \tag{9.45}$$

or in more compact form

$$\mathcal{A}(s_p)v_p = \mathcal{B}(s_p)Ke^{s_pt} \qquad \mathcal{A}(s_p) \neq 0 \tag{9.46}$$

where

$$\mathcal{A}(s_p) = s_p^2 + a_1s_p + a_0 \tag{9.47}$$

is the characteristic polynomial for Eq. (9.42) evaluated at $s = s_p$. Use of the integrating factors and two integrations (see Appendix C) yields the same result.

If we compare Eqs. (9.42) and (9.45), we can conclude the following:

The Particular Response to an Exponential Input with $\mathcal{A}(s_p) \neq 0$

To obtain the particular response to an exponential input $Ke^{s_p t}$, replace differentiation with multiplication by s_p.

Although we will not go through the details, the particular response to an exponential input for a circuit of *any* order is found in this manner if $\mathcal{A}(s_p) \neq 0$.

EXAMPLE 9.5 Determine the particular response for the circuit of Fig. 9.6 on page 285, if $R = \frac{1}{6}\ \Omega$, $L = \frac{1}{4}$ H, $C = 2$ F, and $v_s = 12e^{-4t}$ V for $t > 0$.

Solution Substitution of the values for R, L, C, and v_s into Eq. (9.40) gives

$$\frac{d^2}{dt^2} v + 3 \frac{d}{dt} v + 2v = \frac{d^2}{dt^2} 12e^{-4t} \qquad t > 0$$

which we can write in operator notation as

$$\left(\frac{d^2}{dt^2} + 3 \frac{d}{dt} + 2\right)v = \frac{d^2}{dt^2} 12e^{-4t} \qquad t > 0$$

We replace differentiation with multiplication by $s_p = -4$ in this equation to get

$$[(-4)^2 + 3(-4) + 2]v_p = (-4)^2 12e^{-4t}$$

with the result that

$$v_p = 32e^{-4t} \text{ V} \qquad t > 0$$

In Chapter 8 we observed that a constant input was just a special case of an exponential input with $s_p = 0$. Again this implies that for constant sources we replace differentiation with multiplication by zero. Therefore we can obtain the particular response directly from the network in the same way that we did for first-order circuits.

The Particular Response for a Constant Source

To determine the particular response for a stable network with a constant source, replace all capacitances with open circuits ($i_{Cp} = 0$) and all inductances with short circuits ($v_{Lp} = 0$).

For the two unusual cases where $\mathcal{A}(s_p) = 0$, we show in Appendix C that

$$v_p = \begin{cases} \dfrac{\mathcal{B}(s_p)}{s_p - s_1} tKe^{s_p t} & s_1 \neq s_p = s_2 \qquad (9.48a) \\[2ex] \dfrac{1}{2} \mathcal{B}(s_p) t^2 Ke^{s_p t} & s_p = s_1 = s_2 \qquad (9.48b) \end{cases}$$

where

$$\mathcal{A}(s_p) = s_p^2 + a_1 s_p + a_0 = (s_p - s_1)(s_p - s_2) \qquad (9.49)$$

is the characteristic polynomial evaluated at $s = s_p$. These situations are seldom encountered.

Remember

To obtain the particular response to an input $Ke^{s_p t}$ from the differential equation, replace differentiation with multiplication by s_p, if $\mathscr{A}(s_p) \neq 0$. For a constant input, $s_p = 0$. In this case we can obtain the particular response directly from the circuit. Simply replace capacitance with an open circuit and inductance with a short circuit.

EXERCISE

8. Determine the particular response for each of the following differential equations.

(a) $\frac{d^2}{dt^2}v + 6\frac{d}{dt}v + 8v = 12e^{-t}$ — *answer:* $4e^{-t}$ V

(b) $\frac{d^2}{dt^2}v + 6\frac{d}{dt}v + 25v = 68e^{-4t}$ — *answer:* $4e^{-4t}$ V

(c) $\frac{d^2}{dt^2}v + 4\frac{d}{dt}v + 4v = 24$ [*Hint:* $s_p = 0$] — *answer:* 6 V

(d) $\frac{d^2}{dt^2}v + 6\frac{d}{dt}v + 8v = 12e^{-2t}$ [*Hint:* Use Eq. (9.48a)] — *answer:* $6te^{-2t}$ V

Parallel *RLC* Circuits

For constant (dc) sources $s_p = 0$, so we can determine the particular response by replacing capacitance with an open circuit and inductance with a short circuit, just as we did for first-order circuits. The complete response is the sum of the natural response and this particular response. The natural response includes the constants of integration, but we must use the complete response to determine these constants. We demonstrate this in the following example. (Although the circuit of Fig. 9.7 is not a parallel circuit, we see that when $v_s = 0$, the resulting source-free circuit, which determines the natural response, is parallel.)

EXAMPLE 9.6 Determine voltage $v(t)$ for $t > 0$ in the circuit of Fig. 9.7, if $v_s(t) = 100u(t)$ V.

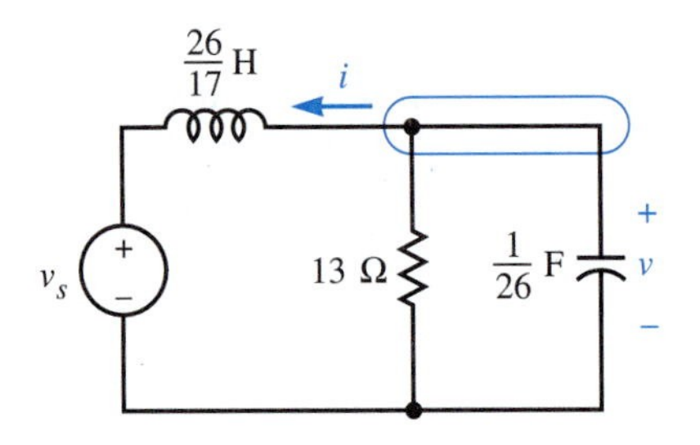

FIGURE 9.7 The second-order circuit for Example 9.6

Solution The source is zero for all $t < 0$, so the initial energy stored must be zero. This fact and the continuity of capacitance voltage and inductance current assures us that

$$v(0^+) = v(0^-) = 0$$

and

$$i(0^+) = i(0^-) = 0$$

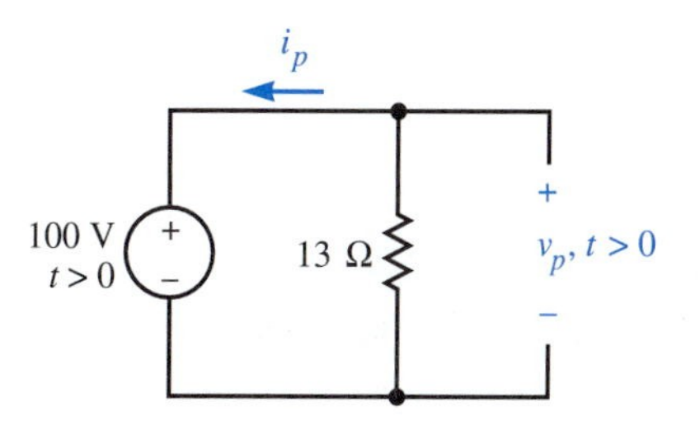

FIGURE 9.8
The equivalent circuit used to determine the particular response in the circuit of Fig. 9.7 for $t > 0$

The source is constant ($s_p = 0$) for the $t > 0$, so we can determine the particular response for $t > 0$ by replacing the capacitance with an open circuit and the inductance with a short circuit. These steps give the equivalent circuit for $t > 0$, shown in Fig. 9.8. This equivalent circuit gives the particular response component of the capacitance voltage:

$$v_p = 100 \text{ V} \qquad t > 0$$

The KCL equation in terms of v for the top node in Fig. 9.7 is

$$\frac{1}{26}\frac{d}{dt}v + \frac{1}{13}v + \frac{17}{26}\int_{-\infty}^{t}(v - v_s)\,d\lambda = 0$$

Multiplication of this equation by 26 and differentiation yields

$$\left(\frac{d^2}{dt^2} + 2\frac{d}{dt} + 17\right)v = 17v_s$$

(We could have found the particular response v_p for our *constant* source from this equation by replacing differentiation with multiplication by $s_p = 0$.)

The characteristic equation

$$\mathcal{A}(s) = s^2 + 2s + 17 = 0$$

has roots

$$s = -1 \pm j4$$

so the complete response is

$$\begin{aligned} v &= v_n + v_p \\ &= 100 + e^{-t}(A \cos 4t + B \sin 4t) \end{aligned}$$

To determine the constants of integration, we first evaluate $v(t)$ at $t = 0^+$ and equate it to the known value of $v(0^+)$:

$$100 + e^0(A \cos 0^+ + B \sin 0^+) = v(0^+) = 0$$

which gives

$$A = -100$$

We then calculate dv/dt, substitute this into the original KCL equation and evaluate this equation at $t = 0^+$:

$$\frac{1}{26}\frac{d}{dt}[100 + e^{-t}(A \cos 4t + B \sin 4t)]\bigg|_{t=0^+} + \frac{1}{13}v(0^+) + i(0^+) = 0$$

Completing the differentiation and substitution of the numerical values for A, $v(0^+)$, and $i(0^+)$ gives us

$$\frac{1}{26}[e^0(-100 \sin 0^+ + 4B \cos 0^+) - e^0(-100 \cos 0^+ + B \sin 0^+)] = 0$$

which yields

$$B = -25$$

The voltage response for the network is then

$$v = 100 + e^{-t}(-100 \cos 4t - 25 \sin 4t) \text{ V} \qquad t > 0$$

The input voltage v_s and response v are sketched in Fig. 9.9.

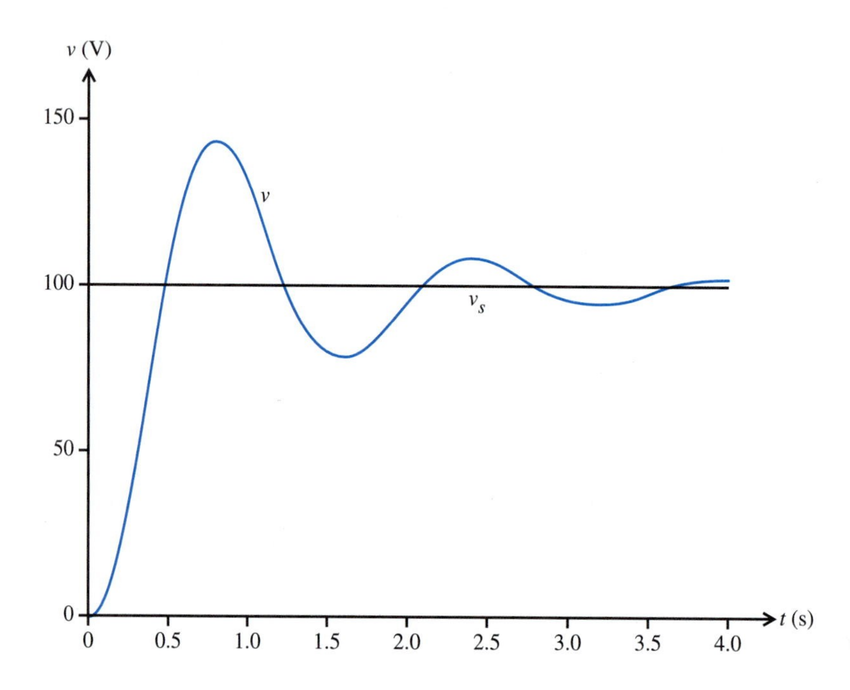

FIGURE 9.9 Input v_s and response v for the underdamped network of Fig. 9.7

We will now work an example where the source is not constant for $t > 0$.

EXAMPLE 9.7 Determine the particular response for $t > 0$ in the circuit of Fig. 9.10, if $i_s = 40e^{-3t}u(t)$ A.

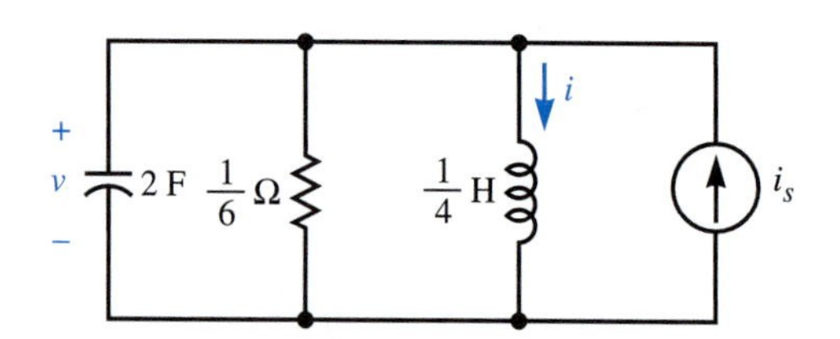

FIGURE 9.10 Network for Example 9.7

Solution The KCL equation for a surface enclosing the top node is

$$2\frac{d}{dt}v + 6v + 4\int_{-\infty}^{t} v\, d\lambda - i_s = 0$$

See Problem 9.3 in the PSpice manual.

We differentiate this KCL equation to eliminate the integral and divide by 2:

$$\left(\frac{d^2}{dt^2} + 3\frac{d}{dt} + 2\right)v = \frac{1}{2}\frac{d}{dt}40e^{-3t} \qquad t > 0$$

Replace differentiation with multiplication by $s_p = -3$ in this second-order equation:

$$\left[(-3)^2 + 3(-3) + 2\right]v_p = \frac{1}{2}(-3)40e^{-3t}$$

This gives

$$v_p = -30e^{-3t}\text{ V} \qquad t > 0$$

Inspection of our second-order differential equation gives us the characteristic equation:

$$s^2 + 3s + 2 = 0$$

with roots $s_1 = -1$ and $s_2 = -2$. Therefore the complete response is

$$v(t) = v_p(t) + v_n(t) = -30e^{-3t} + A_1e^{-t} + A_2e^{-2t} \qquad t > 0$$

The initial conditions will be $v(0^+) = v(0^-) = 0$ and $i(0^+) = i(0^-) = 0$, because the source is zero for all $t < 0$. To evaluate the constants of integration, we first evaluate $v(t)$ at $t = 0^+$ and equate it to the known value of $v(0^+)$:

$$-30 + A_1 + A_2 = v(0^+) = 0$$

We will need to determine

$$\frac{d}{dt}v(t)\bigg|_{t=0^+} = \frac{d}{dt}(-30e^{-3t} + A_2e^{-t} + A_1e^{-2t})\bigg|_{t=0^+} = -90 - A_1 - 2A_2$$

We substitute this, the initial conditions, $v(0^+)$ and $i(0^+)$, and the value of $i_s(0^+)$ into our original KCL equation to obtain the second equation that we need to solve for the constants of integration:

$$2(-90 - A_1 - 2A_2) + 6(0) + 0 - 40 = 0$$

We solve for $A_1 = 130$ and $A_2 = -100$. The complete response is then

$$v(t) = -30e^{-3t} + 130e^{-t} - 100e^{-2t}\text{ V} \qquad t > 0$$

If the source is not zero for all $t < 0$, we determine the initial conditions in the same way that we did for first-order circuits. If the source has been constant for a very long time prior to $t = 0$, only the particular response remains, so $v(0^-) = v_p(0^-)$ and $i(0^-) = i_p(0^-)$. We then use continuity of capacitance voltage and inductance current to determine the initial conditions.

Remember

We determine the particular response to Ke^{s_pt} by replacing differentiation with multiplication by s_p for the usual case where $\mathcal{A}(s_p) \neq 0$. For constant sources (dc), $s_p = 0$, so we can obtain the particular response directly from the circuit by replacing capacitance with an open circuit and inductance with a short circuit. To evaluate the constants of integration, we must use the *complete response.*

EXERCISES

9. Determine voltage v for $t > 0$ in the following circuit (on page 292), if $v_s = 300$ V for $t < 0$, and $v_s = 150$ V for $t > 0$. (*Hint:* $v(0^+) \neq 0$, and $i(0^+) \neq 0$)
answer: $150 + e^{-t}(150\cos 3t + 50\sin 3t)$ V for $t > 0$

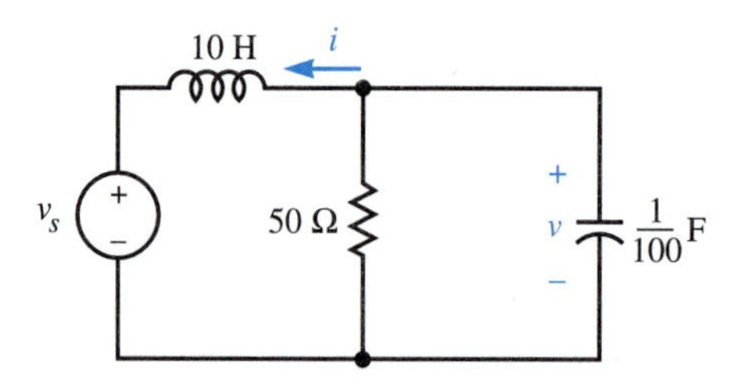

10. Solve for voltage v in the circuit of Fig. 9.6 if $C = 1$ F, $R = \frac{1}{3}\,\Omega$, $L = \frac{1}{2}$ H, and $v_s = (3e^{-4t})u(t)$ V. [*Hint:* $v(0^+) \neq 0$] answer: $(e^{-t} - 6e^{-2t} + 8e^{-4t})u(t)$ V

Step Response of a Series *RLC* Circuit

We will now determine the step response of the series *RLC* circuit of Fig. 9.11, where the response of interest is the capacitance voltage v. The most straightforward way to analyze a series circuit is to solve for the loop current. After the current is known, we can use the terminal equations to calculate the component voltages. We do not need integration to calculate the capacitance voltage. Just calculate the resistance voltage and the inductance voltage, then use KVL to determine the capacitance voltage. There is an alternative. We recognize that the natural response has the same form for each voltage and current in a circuit, because the voltages and currents are related to each other by the component equations, KCL, and KVL. Therefore, we can directly determine the capacitance voltage v. We first determine the particular response component v_p of v. We then obtain the characteristic equation from a KVL equation in terms of i, and use this to determine the natural response v_n. We will demonstrate the procedure.

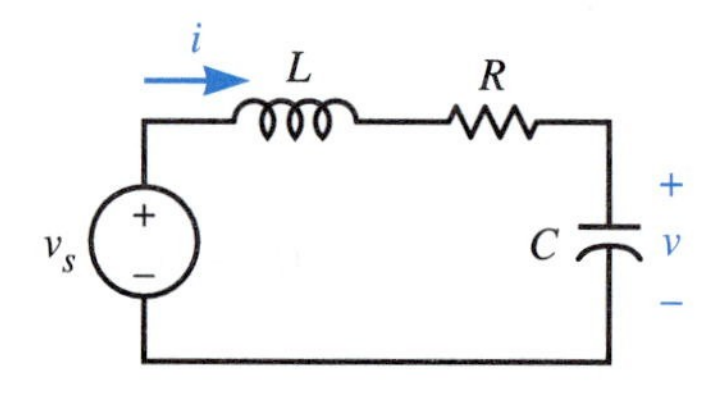

FIGURE 9.11
A series *RLC* circuit with step input $v_s = V_s u(t)$ V

Because $v_s = V_s u(t)$, the source is zero for all $t < 0$. This and continuity of capacitance voltage and inductance current gives us

$$i(0^+) = i(0^-) = 0 \tag{9.50}$$

and

$$v(0^+) = v(0^-) = 0 \tag{9.51}$$

The source is constant for all $t > 0$, so we can determine the particular response by replacing inductance with a short circuit and capacitance with an open circuit. This obviously gives

$$i_p = 0 \qquad t > 0 \tag{9.52}$$

and

$$v_p = V_s \quad t > 0 \tag{9.53}$$

We apply KVL around the loop to get

$$L\frac{d}{dt}i + Ri + \frac{1}{C}\int_{-\infty}^{t} i\,d\lambda = v_s \tag{9.54}$$

From Eq. (9.54) we recognize that the characteristic equation is

$$s^2 + \frac{R}{L}s + \frac{1}{LC} = 0 \tag{9.55}$$

which is in the standard form

$$s^2 + 2\zeta\omega_0 s + \omega_0^2 = 0 \tag{9.56}$$

with roots

$$s_1 = -\zeta\omega_0 + \omega_0\sqrt{\zeta^2 - 1} \tag{9.57}$$

and

$$s_2 = -\zeta\omega_0 - \omega_0\sqrt{\zeta^2 - 1} \tag{9.58}$$

where

$$\omega_0 = \frac{1}{\sqrt{LC}} \tag{9.59}$$

and

$$\zeta = \frac{R}{2}\sqrt{\frac{C}{L}} \tag{9.60}$$

For the overdamped case ($\zeta > 1$) the response will be

$$v(t) = v_p(t) + v_n(t) = V_s + A_1e^{s_1t} + A_2e^{s_2t} \tag{9.61}$$

To evaluate the constants of integration, we evaluate Eq. (9.61) at $t = 0^+$:

$$v(0^+) = V_s + A_1 + A_2 = 0 \tag{9.62}$$

We also know that

$$i(0^+) = C\frac{d}{dt}v\bigg|_{t=0^+} = C(s_1A_1 + s_2A_2) = 0 \tag{9.63}$$

We solve Eqs. (9.62) and (9.63) to obtain

$$v(t) = V_s + \frac{s_2}{s_1 - s_2}V_se^{s_1t} - \frac{s_1}{s_1 - s_2}V_se^{s_2t}\text{ V} \qquad t > 0 \tag{9.64}$$

We could derive similar general results for the critically damped and underdamped case. The responses for several values of ζ with ω_0 fixed are shown in Fig. 9.12.

We observe that the rise time decreases as the damping ratio decreases, but when the damping ratio is less than one, the voltage overshoots the final value and is oscillatory. The critically damped case ($\zeta = 1$) gives the fastest response possible without overshoot or oscillation.

Remember

The same characteristic equation describes all voltages and currents in a series or parallel *RLC* circuit.

FIGURE 9.12
The capacitance voltage in a series *RLC* circuit for several values of ζ with ω_0 fixed and input $v_s = V_s u(t)$ V

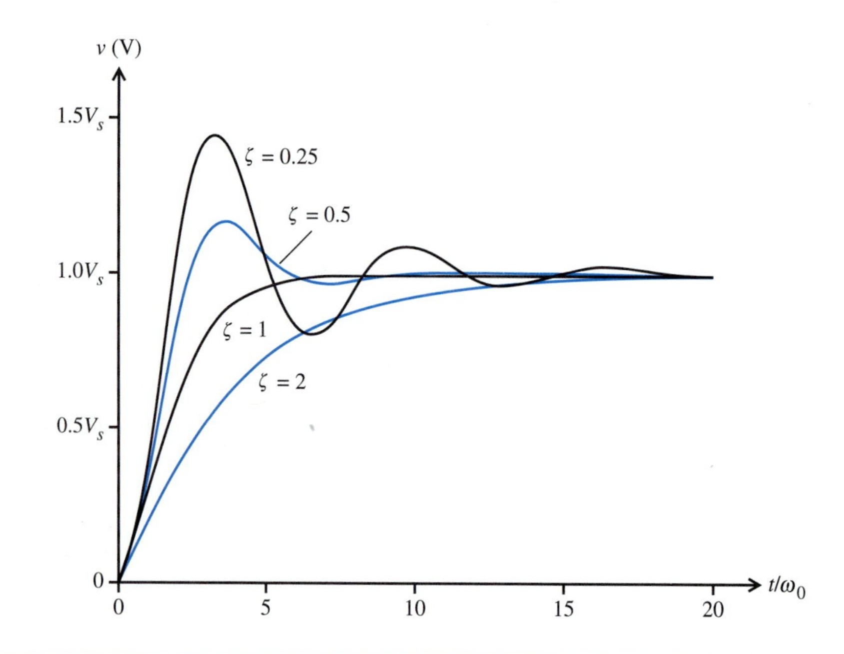

EXERCISE 11. Determine the capacitance voltage v in the series *RLC* circuit of Fig. 9.11 with $v_s = V_s u(t)$ V, if $R = 2\ \Omega$, $C = \frac{1}{10}$ F, and $L = 1$ H. (The circuit is underdamped.)

answer: $V_s[1 - e^{-t}(\cos 3t + \frac{1}{3}\sin 3t)]u(t)$ V

9.5 Sinusoidal Inputs

We determine the particular response to exponential inputs the same way for circuits of any order, if $\mathcal{A}(s_p) \neq 0$. This implies that we can determine the response to sinusoidal inputs for second-order circuits in the same way that we did for first-order circuits (Section 8.7). To determine the particular response for an input $K \cos(\omega_p t + \phi)$, we first determine the response for $Ke^{j(\omega_p t + \phi)}$. The particular response for the cosine is the real part of the response for the complex exponential input. (We generalize this method to LTI circuits of any order in Chapter 10.)

EXAMPLE 9.8 Determine the complete response i for $t > 0$ in the circuit of Fig. 9.13, if $v_s(t) = [300 \cos 4t]u(t)$ V.

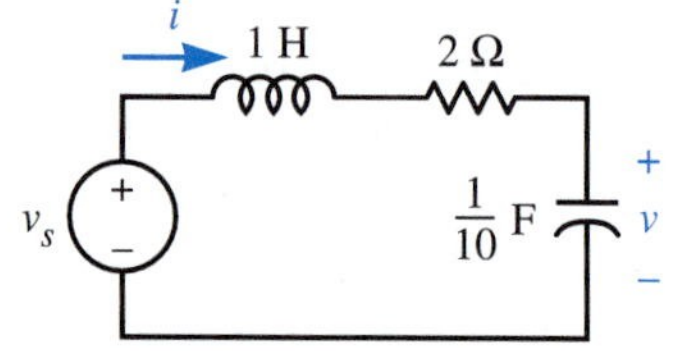

FIGURE 9.13
An *RLC* circuit excited by a sinusoidal source for $t > 0$

Solution First write a KVL equation for the loop:

$$-300 \cos 4t + \frac{d}{dt} i + 2i + 10 \int_{-\infty}^{t} i\, d\lambda = 0 \qquad t > 0$$

Differentiate this equation once to eliminate the integral and write the equation in operator form:

$$\left(\frac{d^2}{dt^2} + 2\frac{d}{dt} + 10\right)i = \frac{d}{dt}300\cos 4t \qquad t > 0$$

Replace $\cos 4t$ with e^{j4t} and replace differentiation with multiplication by $j4$:

$$[(j4)^2 + 2(j4) + 10]\mathbf{i}_p(t) = (j4)300e^{j4t} \qquad t > 0$$

this gives

$$\mathbf{i}_p(t) = \frac{j4}{-6 + j8}300e^{j4t} = 120e^{j(4t - 36.87°)} \qquad t > 0$$

and

$$i_p(t) = \mathcal{Re}\{\mathbf{i}_p\} = 120\cos(4t - 36.87°)\ \text{A} \qquad t > 0$$

We call this particular response the *steady-state sinusoidal response* or the *ac response.*

We see from our second-order differential equation that the characteristic equation has the roots

$$s = -1 \pm j3$$

so the complete response is

$$i = i_p + i_n = 120\cos(4t - 36.87°) + e^{-t}[A\cos 3t + B\sin 3t]\ \text{A} \qquad t > 0$$

Because the source is zero for $t < 0$, we have zero initial conditions: $i(0^+) = 0$, and $v(0^+) = 0$. We evaluate $i(t)$ at $t = 0^+$ and equate it to the known value of $i(0^+)$:

$$120\cos(-36.87°) + A = i(0^+) = 0$$

to find

$$A = -96$$

Next evaluate the derivative of i at $t = 0^+$:

$$\left.\frac{d}{dt}i\right|_{t=0^+} = -480\sin(-36.87°) - A + 3B = 384 + 3B$$

and substitute this, $v(0^+)$, $i(0^+)$, and $v_s(0^+)$ into our original KCL equation to get

$$B = -16$$

Substitution of the values of A and B completes the solution:

$$i = 120\cos(4t - 36.87°) - e^{-t}[96\cos 3t + 16\sin 3t]\ \text{A} \qquad t > 0$$

The sketch in Fig. 9.14 reveals that the complete response i is approximately equal to the particular response i_p after a few cycles of the input sinusoid. The circuit is then operating in the sinusoidal steady state.

Remember

The particular response to a cosine source is the real part of the response for a complex exponential source. The particular response is the steady-state response for a stable circuit.

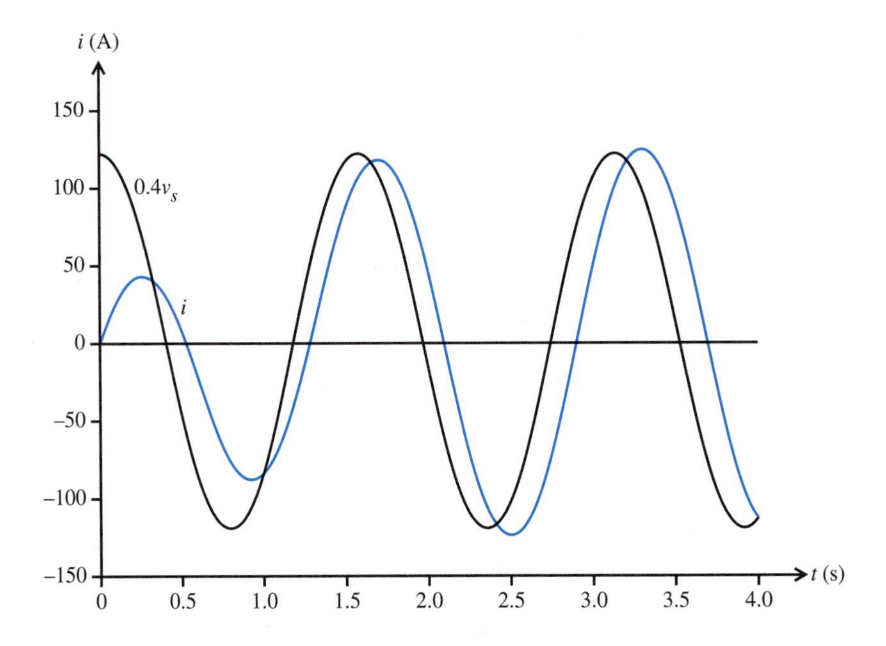

FIGURE 9.14 The response of an underdamped series *RLC* circuit to a sinusoid applied at $t = 0$ ($0.4v_s$ is shown for reference)

EXERCISE

12. Determine the steady-state voltage response (the particular response v_p) for a parallel *RLC* circuit driven by a current source of value $i_s = 30 \cos 3t$ A, if $R = \frac{1}{2}\ \Omega$, $L = \frac{1}{10}$ H, and $C = 2$ F [i_s into (+) mark]. *answer:* $9 \cos (3t - 53.13°)$ V

9.6 Operator Notation

In the following chapters we will be solving for the sinusoidal-steady-state response of more general circuits. The notation becomes cumbersome because these circuits are described by higher-order differential equations. For an *n*th-order linear-time-invariant (LTI) circuit, the differential equation has the form

$$a_n \frac{d^n}{dt^n} v + a_{n-1} \frac{d^{n-1}}{dt^{n-1}} v + \cdots + a_1 \frac{d}{dt} v + a_0 v$$
$$= b_m \frac{d^m}{dt^m} i + b_{m-1} \frac{d^{m-1}}{dt^{m-1}} i + \cdots + b_1 \frac{d}{dt} i + b_0 i \quad (9.65)$$

where i is the input, v is output, and the coefficients, $a_n, \ldots, a_0, b_m, \ldots, b_0$, depend on the circuit elements. We also write Eq. (9.65) in the operator form

$$\left(a_n \frac{d^n}{dt^n} + a_{n-1} \frac{d^{n-1}}{dt^{n-1}} + \cdots + a_1 \frac{d}{dt} + a_0\right) v$$
$$= \left(b_m \frac{d^m}{dt^m} + b_{m-1} \frac{d^{m-1}}{dt^{m-1}} + \cdots + b_1 \frac{d}{dt} + b_0\right) i \quad (9.66)$$

To simplify our work, we will use the D operator notation commonly found in calculus books:

$$Dv = \frac{d}{dt} v \quad (9.67)$$

Because any-order derivative is simply the derivative of the next-lower-order derivative, we also define

$$D^n v = \frac{d^n}{dt^n} v \tag{9.68}$$

The D operator lets us write Eq. (9.66) as

$$(a_n D^n + a_{n-1} D^{n-1} + \cdots + a_1 D + a_0)v = (b_m D^m + b_{m-1} D^{m-1} + \cdots + b_1 D + b_0)i \tag{9.69}$$

which is written compactly as

$$\mathcal{A}(D)v = \mathcal{B}(D)i \tag{9.70}$$

where

$$\mathcal{A}(D) = (a_n D^n + a_{n-1} D^{n-1} + \cdots + a_1 D + a_0) \tag{9.71}$$

and

$$\mathcal{B}(D) = (b_m D^m + b_{m-1} D^{m-1} + \cdots + b_1 D + b_0) \tag{9.72}$$

In our future work, we will refer to equations such as Eqs. (9.65) and (9.70) as LTI equations, because they describe linear time-invariant circuits.

The D operator notation is particularly convenient when we determine the particular response for the exponential $i = Ke^{s_p t}$. For this case, we simply substitute s_p for D, so the particular response v_p for Eq. (9.70) is given by

$$\mathcal{A}(s_p)v_p = \mathcal{B}(s_p)i \tag{9.73}$$

We will regularly use this notation for steady-state sinusoidal analysis.

We can also use a shorthand notation for integration. For a 1-F capacitance

$$v = \frac{1}{D} i = \int_{-\infty}^{t} i \, d\lambda \tag{9.74}$$

If we differentiate Eq. (9.74), we get

$$Dv = D\left(\frac{1}{D} i\right) = i \tag{9.75}$$

Therefore, multiplication by D (differentiation) is the inverse operation of multiplication by $1/D$ (integration), and we can write

$$D\frac{1}{D} = 1 \tag{9.76}$$

[Interchanging the order of the operations to $(1/D)D$ introduces a constant of integration. This is of no concern, because we always differentiate to eliminate integrals.]

Remember

We use the operator $D^n = d^n/dt^n$ to simplify our notation. To obtain the particular response for an exponential input $Ke^{s_p t}$, simply replace D with s_p.

EXERCISE

13. Determine the particular response component v_p for the differential equation $(D^3 + 2D^2 + 3D + 1)v = 50e^{-2t}$.
answer: $-10e^{-2t}$

9.7 More General Circuits†

Not all second-order *RLC* circuits are simple parallel or series circuits. For example, a practical inductor always has some resistance. If the resistance of an inductor is not negligible, we need to model a parallel connection of a capacitor, resistor, and inductor as shown in Fig. 9.15.

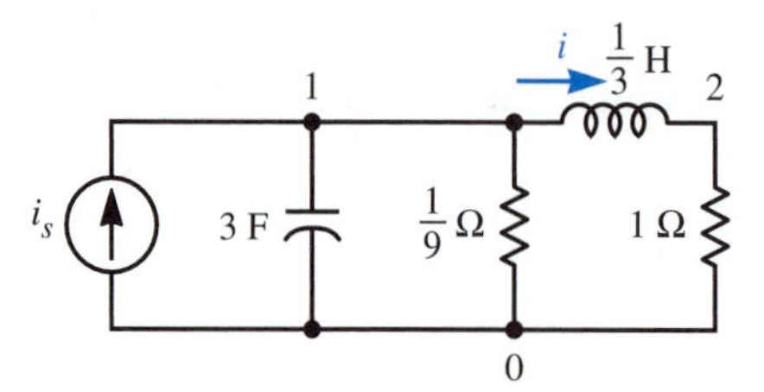

FIGURE 9.15
A more general *RLC* circuit

In Chapters 4 and 5, we developed several methods to write differential equations that describe a network. We now show how to reduce a set of two node-voltage equations to two differential equations, each of which contains only one node voltage.

EXAMPLE 9.9 The current source in the circuit of Fig. 9.15 has the value $i_s = 10u(t)$ A. Determine the node voltages for $t > 0$.

Solution Because the current source is zero for all $t < 0$, the initial capacitance voltage is $v_1(0^+) = v_1(0^-) = 0$, and the initial inductance current is $i(0^+) = i(0^-) = 0$. We use Ohm's law to obtain the initial value of $v_2(0^+)$:

$$v_2(0^+) = 1i(0^+) = 0$$

Application of KCL to a surface enclosing node 1 yields

$$3\frac{d}{dt}v_1 + 9v_1 + 3\int_{-\infty}^{t}(v_1 - v_2)\,d\lambda = i_s$$

Similarly, KCL applied to a surface enclosing node 2 gives us

$$v_2 + 3\int_{-\infty}^{t}(v_2 - v_1)\,d\lambda = 0$$

We use operator notation to write these two KCL equations as

$$\begin{bmatrix} 3D + 9 + 3\dfrac{1}{D} & -3\dfrac{1}{D} \\ -3\dfrac{1}{D} & 1 + 3\dfrac{1}{D} \end{bmatrix} \begin{bmatrix} v_1 \\ v_2 \end{bmatrix} = \begin{bmatrix} i_s \\ 0 \end{bmatrix}$$

Although we can use substitution to obtain differential equations that include only one node voltage, we choose Cramer's rule as the more convenient method:

† This section can be omitted without loss of continuity.

$$\begin{vmatrix} 3D + 9 + 3\dfrac{1}{D} & -3\dfrac{1}{D} \\ -3\dfrac{1}{D} & 1 + 3\dfrac{1}{D} \end{vmatrix} v_1 = \begin{vmatrix} i_s & -3\dfrac{1}{D} \\ 0 & 1 + 3\dfrac{1}{D} \end{vmatrix}$$

Evaluation of the determinants yields

$$\left(3D + 18 + 30\frac{1}{D}\right)v_1 = \left(1 + 3\frac{1}{D}\right)i_s$$

Differentiation (multiplication by D) and division by 3 reduces this to

$$(D^2 + 6D + 10)v_1 = \left(\frac{1}{3}D + 1\right)i_s$$

Cramer's rule also yields an equation in terms of v_2:

$$(D^2 + 6D + 10)v_2 = i_s$$

The source $i_s = 10u(t)$ V is constant for $t > 0$, so replace differentiation with multiplication by zero (or replace the capacitance with an open circuit and the inductance with a short circuit) to determine the particular response components of the node voltages:

$$v_{1p} = v_{2p} = 1 \text{ V}$$

The characteristic equation in each case has roots $s = -3 \pm j1$, so the complete responses are

$$v_1 = 1 + e^{-3t}(A_1 \cos t + B_1 \sin t) \qquad t > 0$$

and

$$v_2 = 1 + e^{-3t}(A_2 \cos t + B_2 \sin t) \qquad t > 0$$

To determine the constants of integration, we first evaluate v_1 at $t = 0^+$ and equate it to the initial value of $v_1(0^+)$:

$$1 + e^0(A_1 \cos 0^+ + B_1 \sin 0^+) = v_1(0^+) = 0$$

This gives $A_1 = -1$. To obtain a second equation, we evaluate the first node voltage equation at $t = 0^+$:

$$3\frac{d}{dt}[1 + e^{-3t}(A_1 \cos t + B_1 \sin t)]\bigg|_{t=0} + 9v_1(0^+) + i(0^+) = 10$$

This gives us $B_1 = -\frac{1}{3}$. The resulting solution for v_1 is

$$v_1 = 1 - e^{-3t}\left(\cos t - \frac{1}{3}\sin t\right) \text{ V} \qquad t > 0$$

We leave the determination of the constants of integration for v_2 to an exercise.

Remember

We use any convenient technique to write our network equations. We then use Cramer's rule or substitution to reduce simultaneous equations to differential equations in terms of only one variable.

EXERCISE 14. Evaluate the constants of integration in the equation for v_2 [*Hint:* Because the second node-voltage equation does not contain a derivative of v_2, we must differentiate this equation and evaluate it at $t = 0^+$ in order to obtain the second equation that we need.] *answer:* $A_2 = -1$, $B_2 = -3$

9.8 Summary

The response of a source-free second-order circuit is the natural response of the circuit. We determined the natural response from the characteristic equation, and the constants of integration by use of the initial conditions. We showed that the natural response is the sum of two exponentials if the roots of the characteristic equation are real and distinct. This is the underdamped case. If the two roots of the characteristic equation are identical, we must multiply one of the exponentials by t. This is the critically damped case. If the roots are complex, we can conveniently write the sum of the exponentials as a sum of damped sinusoids. This is the underdamped case.

We found that the response of a circuit with sources is the sum of two components: the natural response as determined by the source-free circuit, and the particular response. We used this complete response to calculate the constants of integration. We showed that for the usual case we can find the particular response to an exponential source $Ke^{s_p t}$ by replacing differentiation with multiplication by s_p. We recognized that for a constant source $s_p = 0$, so we replace capacitance with an open circuit and inductance with a short circuit to calculate the particular response. We concluded the chapter by showing how to use the complex exponential input to calculate the particular response for sinusoidal inputs, and how to reduce two simultaneous differential equations to differential equations in terms of only one unknown voltage or current.

KEY FACTS

Concept	*Equation*	*Section*	*Page*
❑ A second-order differential equation describes a circuit with two energy-storage elements.	(9.2) (9.8)	9.1	272 273
❑ The roots of the characteristic equation give the natural response, except for the constants of integration.	(9.7)	9.1	273
❑ If the roots s_1 and s_2 of the characteristic equation are real, and distinct, the natural response is of the form $A_1e^{s_1t} + A_2e^{s_2t}$. The circuit is *overdamped.* (Examples 9.1 and 9.4)	(9.7a)	9.1 9.2	273 275 282
❑ If the roots of the characteristic equation are complex, $s = -\alpha \pm j\omega_d$, we can use Euler's identity to write the natural response as $e^{-\alpha t}(A_1 \cos \omega_d t + A_2 \sin \omega_d t)$. The circuit is *underdamped.* (Example 9.2)	(9.28)	9.2	277 278
❑ If the two roots of the characteristic equation are identical, $s_1 = s_2$, the natural response is $(A_1 + A_2t)e^{s_1t}$. The circuit is *critically damped.* (Example 9.3)	(9.7b)	9.1 9.2 9.3	273 274 280

KEY FACTS *Concept*	*Equation*	*Section*	*Page*
❑ We determine the constants of integration from the initial conditions. (Examples 9.1–9.4)		9.2 9.3	274 281
❑ The complete response for a network with sources is the sum of two parts: the natural response and the particular response.	(9.43)	9.4	286
❑ The natural response contains the constants of integration.	(9.43)	9.4	286
❑ The complete response must be used to evaluate the constants of integration. (Examples 9.6–9.9)		9.4	286 288
❑ For an exponential input $Ke^{s_p t}$, the particular response is obtained by replacing differentiation with multiplication by s_p, if $\mathcal{A}(s_p) \neq 0$.	(9.45)	9.4	286
❑ A constant input is a special case of an exponential input, where $s_p = 0$.		9.4	287
❑ We can obtain the particular response for constant inputs to a stable circuit directly from the network diagram by replacing each capacitance by an open circuit and each inductance by a short circuit. (Example 9.6)		9.4	287 288
❑ If a source has been applied for a very long time, only the particular response remains. This and continuity of capacitance voltage and inductance current determine the initial conditions. (Example 9.6)		9.4	287 288
❑ For a second-order *RLC* circuit, the natural-response component of every voltage and current has the same form.		9.4	292
❑ For a cosine input, the particular response is the real part of the particular response for a complex exponential input. For a stable circuit this particular response is called the *steady-state sinusoidal response* or the *ac response*. (Example 9.8)		9.5	294
❑ The derivative operator $D^n = d^n/dt^n$ simplifies our notation.	(9.68)	9.6	297
❑ We can convert two simultaneous differential equations in terms of two node voltages or mesh currents to two differential equations, each of which includes only one node voltage or mesh current. (Example 9.9)		9.7	298

PROBLEMS

Section 9.2

1. For the following overdamped circuit, $C = 2$ F, $R = \frac{1}{6}\ \Omega$, and $L = \frac{1}{4}$ H. Determine the response $i = i(t)$ for $t > 0$ when $v(0^+) = 6$ V, and $i(0^+) = 16$ A. 9.1

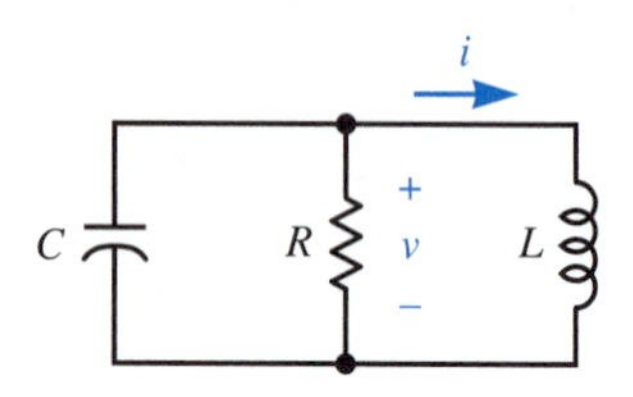

2. Determine the response, $v(t)$, of the following overdamped circuit when $t > 0$, if $v(0^+) = 20$ V and $i(0^+) = 2$ A. (Note the reference direction for i.)

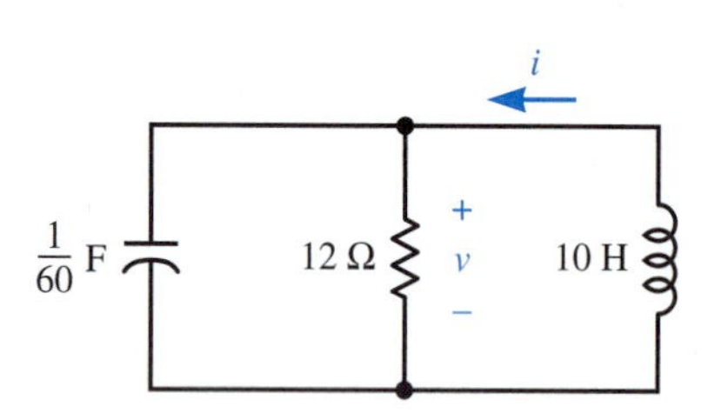

3. For the following underdamped circuit, determine $v(t)$ for $t > 0$, if $v(0^+) = 11$ V and $i(0^+) = -4$ A.

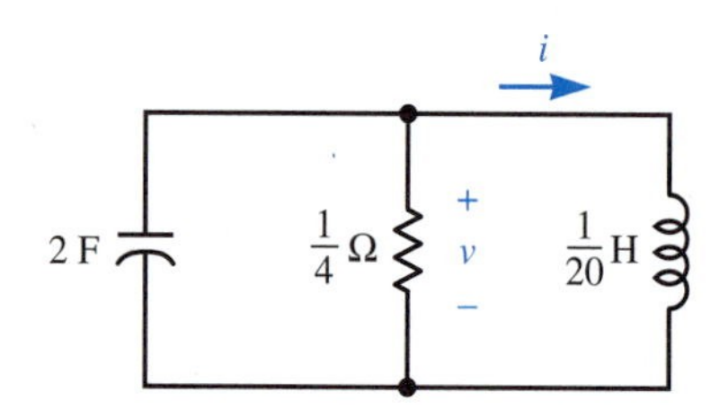

4. Given the following underdamped circuit, solve for $v(t)$, $t > 0$, if $v(0^+) = 20$ V and $i(0^+) = 1$ A. (Note the reference direction for i.)

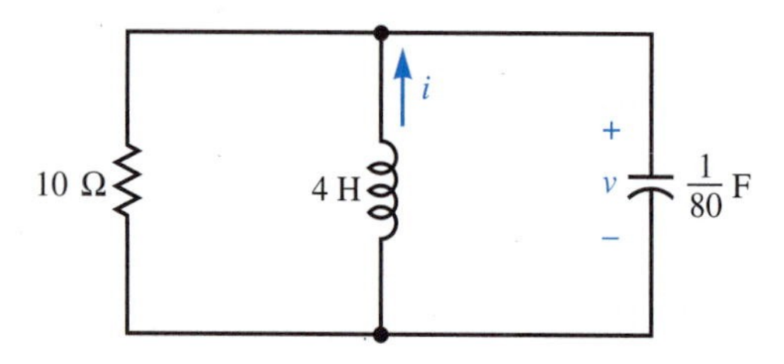

5. A parallel *RLC* circuit is critically damped. The initial value of the capacitance voltage is $v(0^+) = 2$ V and the initial inductance current is $i(0^+) = 4$ A (passive sign convention). Determine $i(t)$ for $t > 0$, if $C = 2$ F, $R = \frac{1}{8}\ \Omega$, and $L = \frac{1}{8}$ H.

6. In the following circuit, voltage v_s is zero for $t \geq 0$, $v(0^+) = 10$ V, $i(0^+) = 20$ A, and $\beta = 49$. Determine $v(t)$ for $t > 0$.

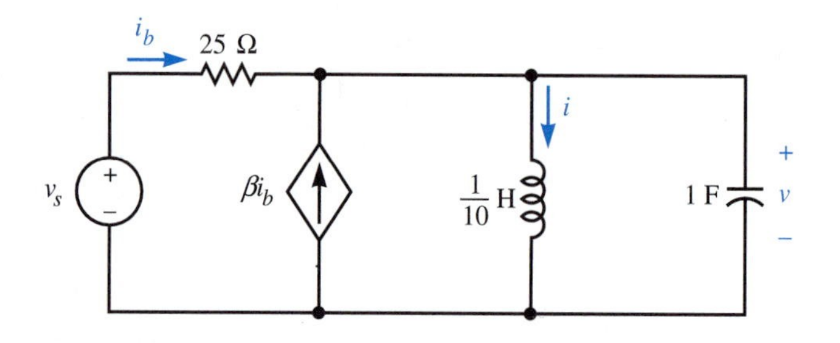

Section 9.3

7. In the following overdamped circuit, $L = 2$ H, $R = 6\ \Omega$, and $C = \frac{1}{4}$ F. If $v_s = 0$ for $t > 0$, determine the response $i = i(t)$ for $t > 0$ when $i(0^+) = 6$ A and $v(0^+) = 16$ V. 9.2

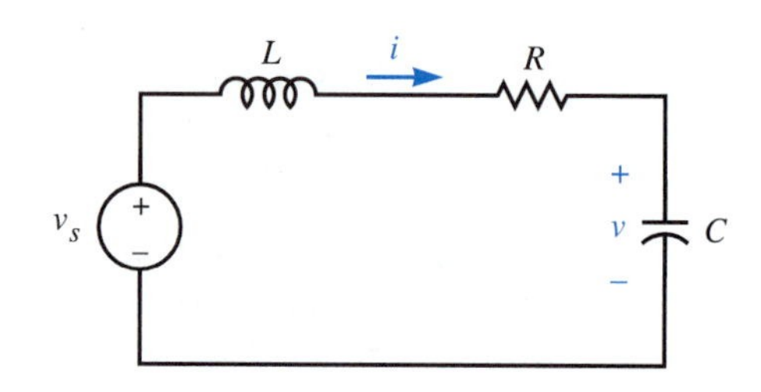

8. For the following overdamped circuit, determine $i(t)$ for $t > 0$, if $i(0^+) = 1$ A and $v(0^+) = 10$ V. (Note the reference marks for v.)

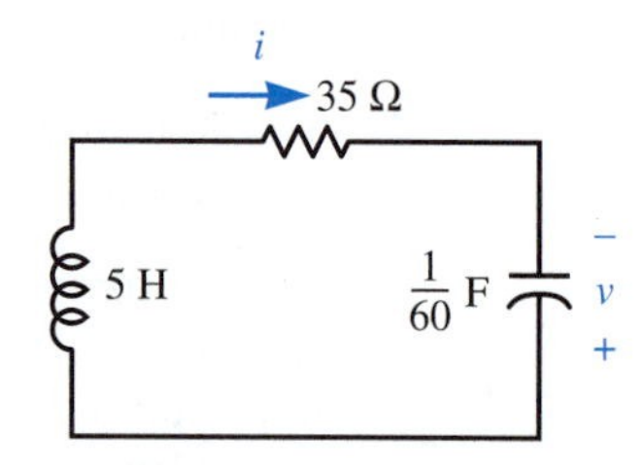

9. For the following underdamped circuit, $L = 2$ H,

$R = 8\ \Omega$, and $C = \frac{1}{40}$ F. Solve for $i(t)$ when $t > 0$, if $i(0^+) = 8$ A and $v(0^+) = 48$ V.

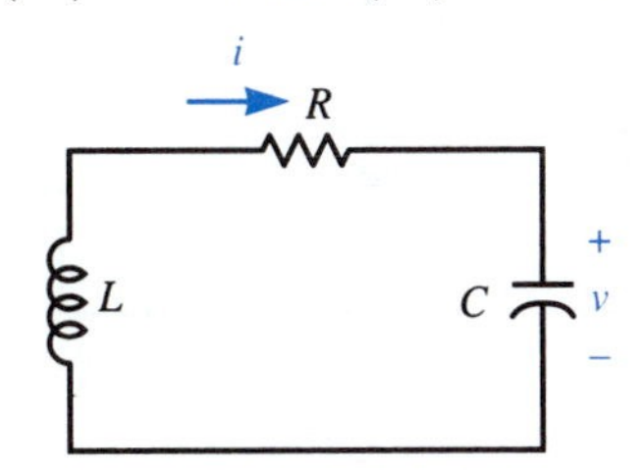

10. For the preceding underdamped circuit, $L = 5$ H, $R = 40\ \Omega$, and $C = \frac{1}{100}$ F. Solve for $i(t)$ when $t > 0$, if $i(0^+) = 5$ A, and $v(0^+) = 10$ V.
11. A series *RLC* circuit is critically damped with $L = 2$ H, $R = 12\ \Omega$, and $C = \frac{1}{18}$ F. Solve for $i(t)$ when $t > 0$, if $i(0^+) = 2$ A, and $v(0^+) = -38$ V (passive sign convention).
12. Determine the source-free response, $i(t)$, for $t > 0$ in the following circuit if $v_1(0^+) = 20$ V and $i(0^+) = 2$ A.

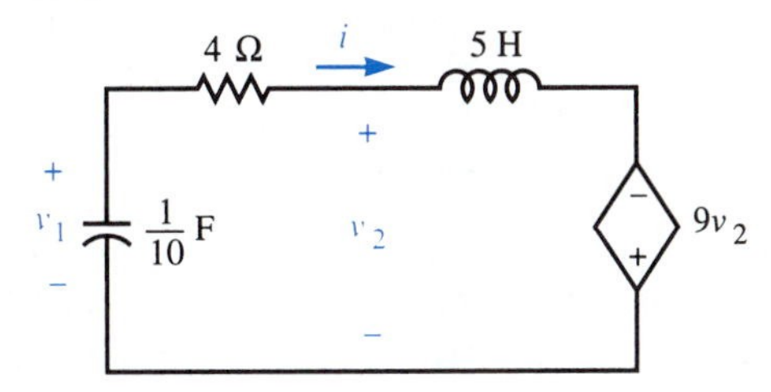

Section 9.4

13. For the following network, find voltage v for $t > 0$, when $v(0^+) = 100$ V and $i(0^+) = 2$ A are given, and $v_s = 50$ V for $t > 0$.

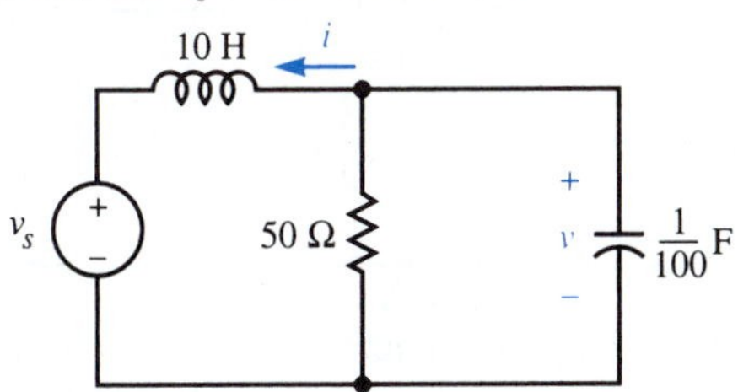

14. Determine voltage v in the following circuit for the source that is specified.

    (a) $v_s = 36u(t)$ V (b) $v_s = 36e^{-3t}u(t)$ V

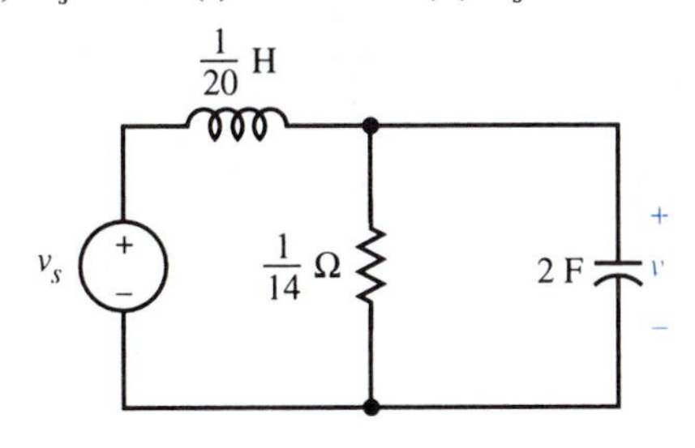

15. Determine voltage v for $t > 0$ in the following circuit with the specified values of R, L, and C.

    (a) $5\ \Omega$, 1 H, and $\frac{1}{10}$ F (underdamped)

    (b) $400\ \Omega$, $\frac{20}{3}$ H, and $10\ \mu$F (overdamped)

    (c) $25\ \Omega$, 10 H, and $4000\ \mu$F (critically damped)

    (d) $25\ \Omega$, 125 mH, and $1250\ \mu$F (underdamped)

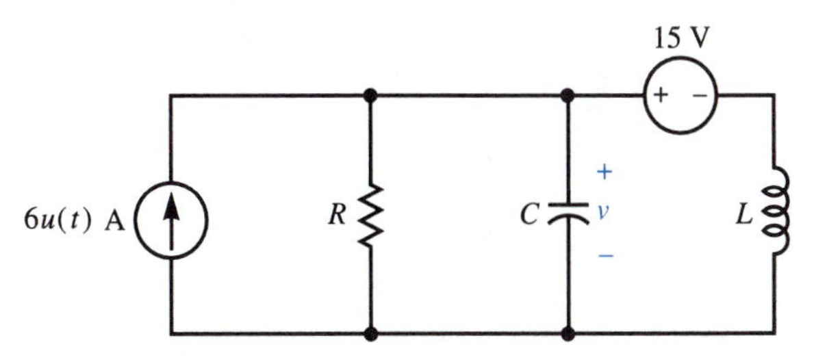

16. For the following network, the switch has been open for a very long time and is closed at $t = 0$. Determine current i for $t > 0$.

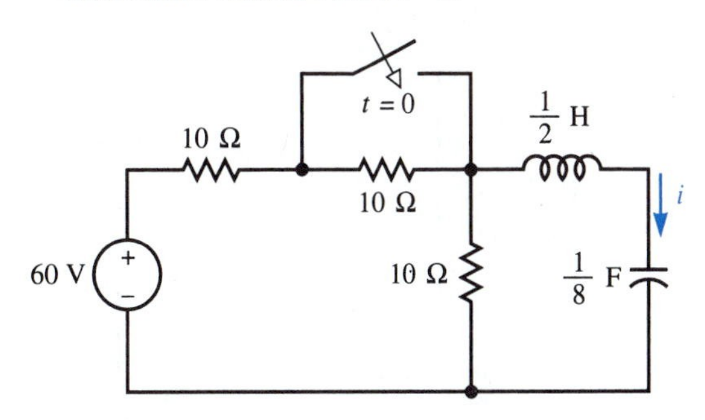

17. Find current i for $t > 0$ in the following circuit, if $v_s = 200[1 + u(t)]$ V.

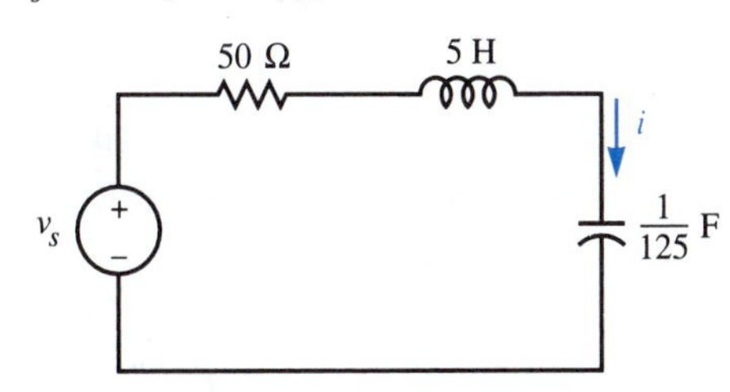

18. Determine current i in the following network for the specified source current.

    (a) $36u(t)$ A (b) $36e^{-2t}u(t)$ A

    (c) $36e^{-5t}u(t)$ A

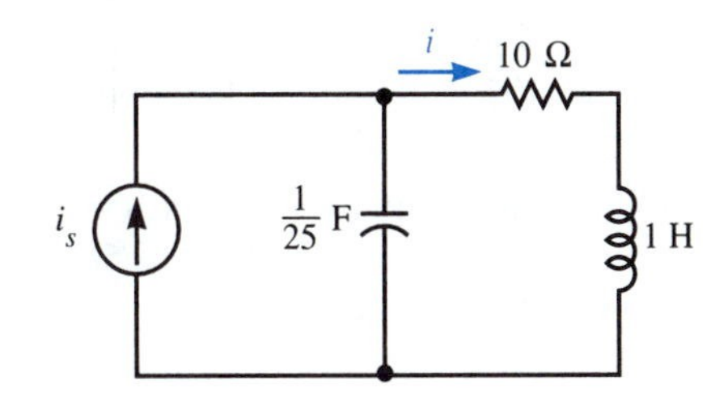

19. For the following network, find current i for $t > 0$ if $v(0^+) = 500$ V, $i(0^+) = 15$ A, and $i_s = 12e^{-3t}$ A for $t > 0$.

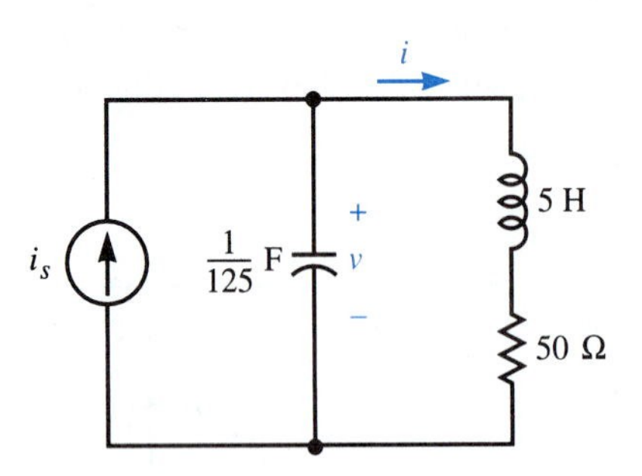

20. Determine current i for $t > 0$ in the following circuit if the switch has been closed for a very long time and is opened at $t = 0$. 9.4

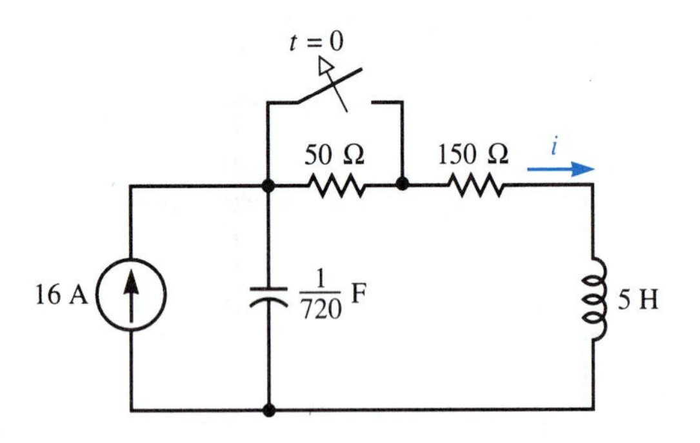

21. The switch in the following network has been open for a very long time and is closed at $t = 0$. Determine voltage v for $t > 0$.

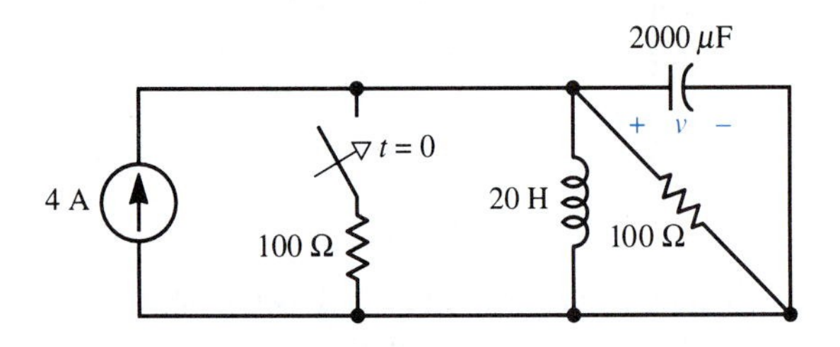

Section 9.5

22. A critically damped series *RLC* circuit is being driven by a voltage source of value $v_s = (200 \cos 2t)u(t)$ V. Determine the loop current if $L = 1$ H, $R = 8\ \Omega$, and $C = \frac{1}{16}$ F.

23. Determine voltage v in the following circuit if the source voltage is $v_s = 100[(\cos 5t)u(-t) + (\cos 3t)u(t)]$ V.

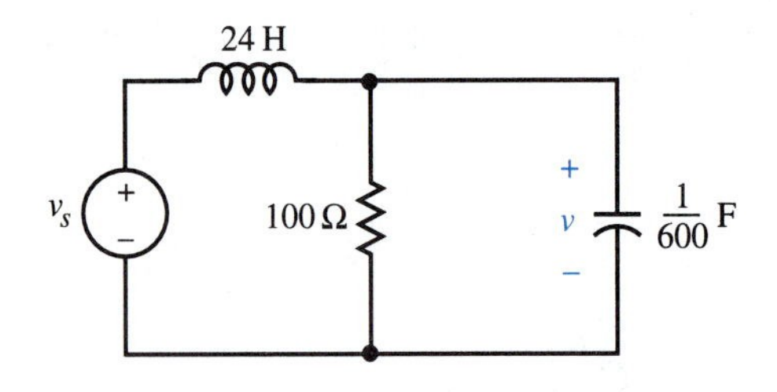

Section 9.7

24. For the following network, $v(0^+) = 3$ V and $i(0^+) = 1$ A. Use the shortcut method for analysis by node voltages to obtain two KCL equations. Reduce these two equations to a single differential equation in terms of voltage v. Determine voltage v for $t > 0$, if $v_s = 12$ V for $t > 0$.

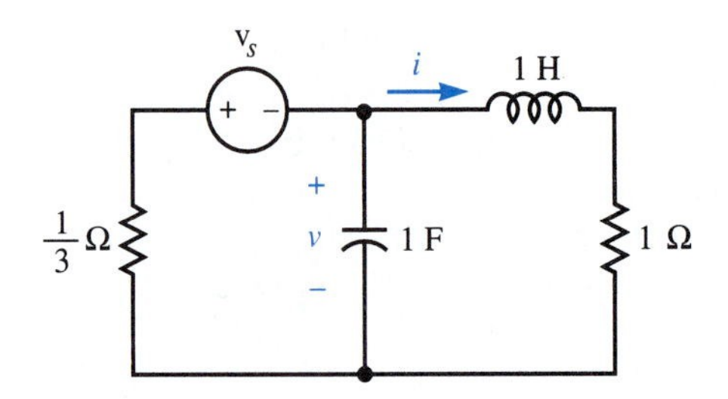

25. Use mesh current equations to determine current i in the preceding circuit if $v_s = 12$ V for $t > 0$, and $i(0^+) = 1$ A and $v(0^+) = 3$ V.

26. Determine the mesh currents in the following circuit if $v_s = 12u(t)$ V.

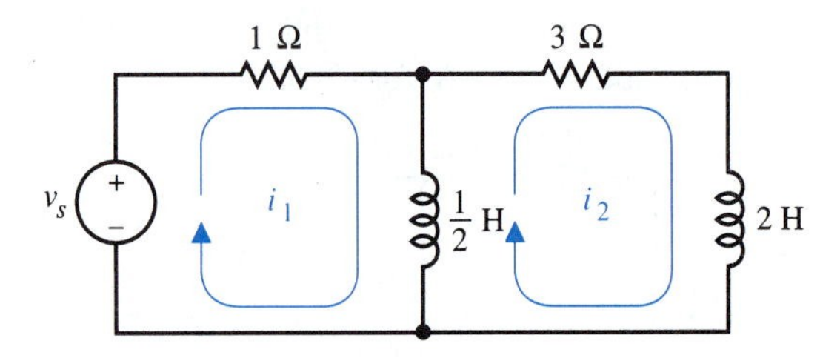

27. Solve for the mesh currents for $t > 0$ in the following network if $i_1(0^+) = 1$ A, $i_2(0^+) = 2$ A, and $v_s = 120e^{-3t}$ V for $t \geq 0$.

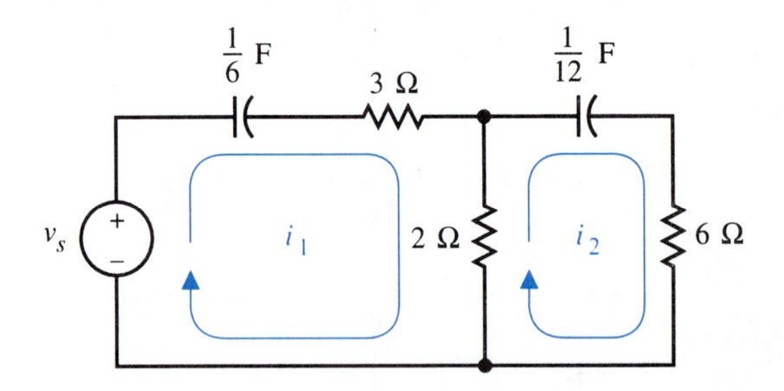

28. The initial conditions for the following network are $v_1(0^+) = 6$ V and $v_2(0^+) = 5$ V. The current source is $i_s = 20$ A, $t \geq 0$. Determine voltages v_1 and v_2 for $t > 0$.

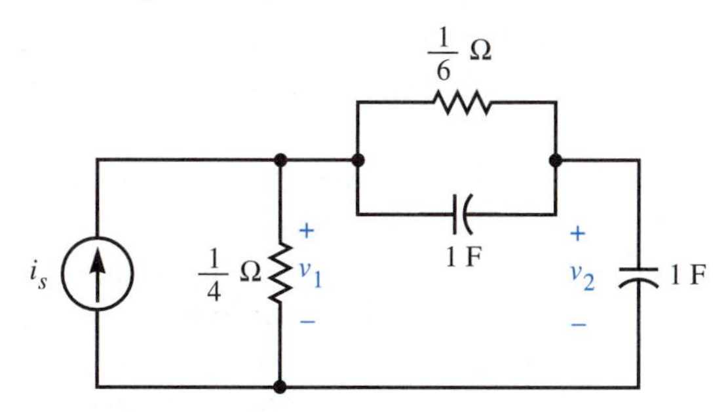

29. Solve for node voltages v_1 and v_2 in the following network, if $i_s = 3 + 6u(t)$ A.

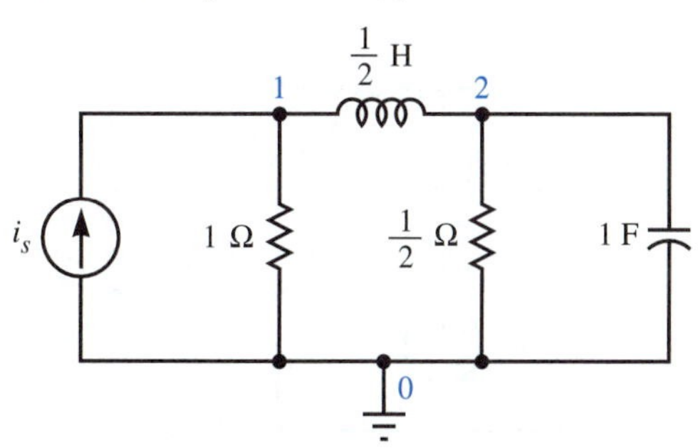

Supplementary Problems

30. Write a KCL equation for a parallel *RLC* circuit with $C = \frac{1}{80}$ F and $L = 4$ H. Determine
    (a) The range of resistance values for which the circuit is overdamped.
    (b) The resistance value for which the circuit is critically damped.
    (c) The range of resistance values for which the circuit is underdamped.

31. Write a KVL equation for a series *RLC* circuit with $L = 2$ H and $C = 4$ F. Determine
    (a) The range of resistance values for which the circuit is overdamped.
    (b) The resistance value for which the circuit is critically damped.
    (c) The range of resistance values for which the circuit is underdamped.

32. The following *LC* network has the initial conditions $v(0^+) = 40$ V and $i(0^+) = 10$ A. Determine $v(t)$ and $i(t)$ for $t > 0$. Calculate the energy stored in the capacitance and the inductance as a function of time. Then calculate the total energy stored in the circuit for $t > 0$. This is an example of a second-order circuit that is marginally stable. The energy stored in the circuit does not dissipate with time, because there are no resistive losses. A marginally stable circuit must have at least one root of the characteristic equation with zero real part, and no roots with a positive real part.

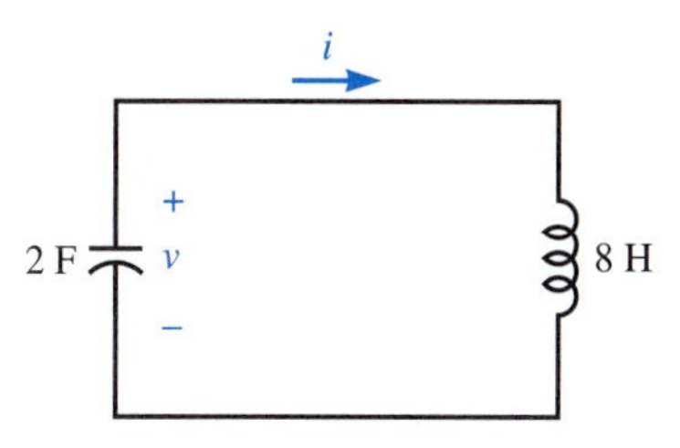

33. For a stable circuit the natural response decays to zero with time, and all energy initially stored in the source-free circuit is dissipated. The *real part* of the roots of the characteristic equation must be negative for a stable circuit. Determine the values of β for which the following circuit is stable. (Let $v_s = 0$. Stability depends only on the natural response.)

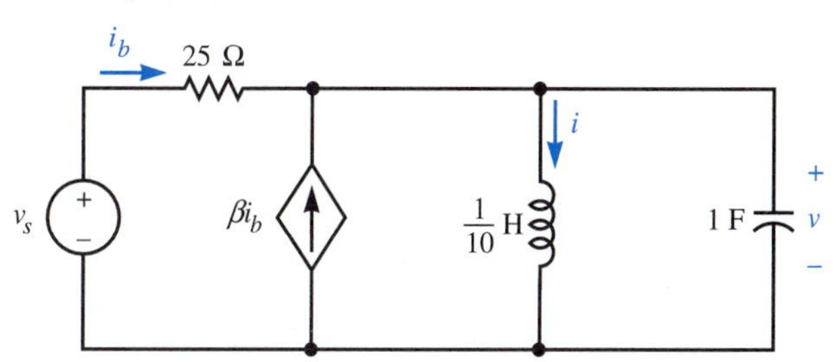

34. For the following circuit, calculate $i(t)$ for $t > 0$ if $v_1(0^+) = 60$ V and $i(0^+) = 6$ A. Is this circuit stable?

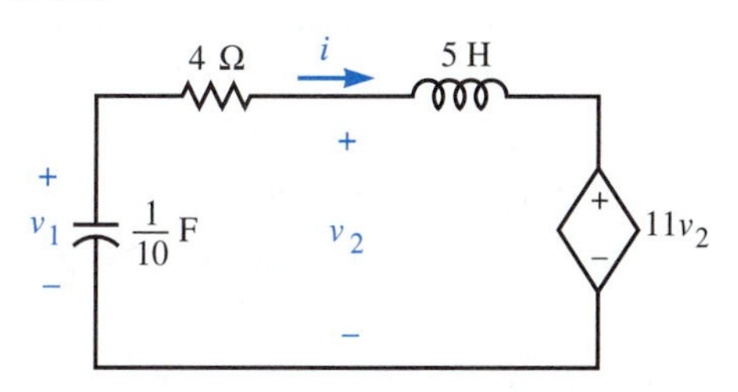

35. A simplified equivalent circuit of a tuned-drain oscillator that uses a field-effect transistor is shown in the following figure. The oscillator will oscillate at an angular frequency ω if the two roots of the characteristic equation are $s = \pm j\omega$. Show that this occurs if

$$\omega = \frac{1}{\sqrt{LC}}\sqrt{1+\frac{R}{r}}$$

and

$$M = \frac{CrR + L}{gr}$$

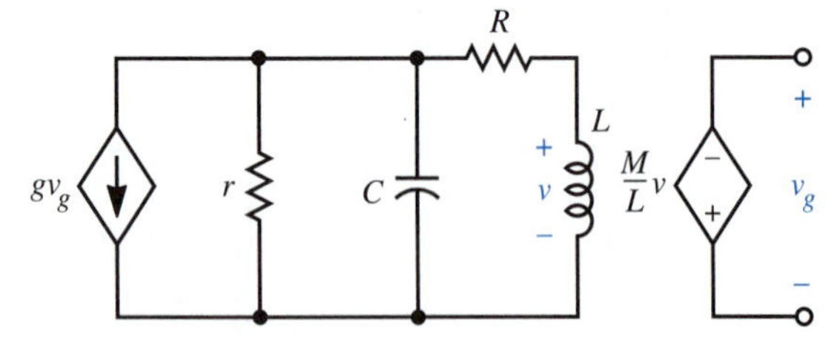

36. The switch in the following network has been closed for a very long time and is opened at $t = 0$. Find:
    (a) $v_C(0^-)$, $v_L(0^-)$, $i_C(0^-)$, and $i_L(0^-)$
    (b) $v_C(0^+)$, $v_L(0^+)$, $i_C(0^+)$, and $i_L(0^+)$
    (c) $v_C(\infty)$, $v_L(\infty)$, $i_C(\infty)$, and $i_L(\infty)$

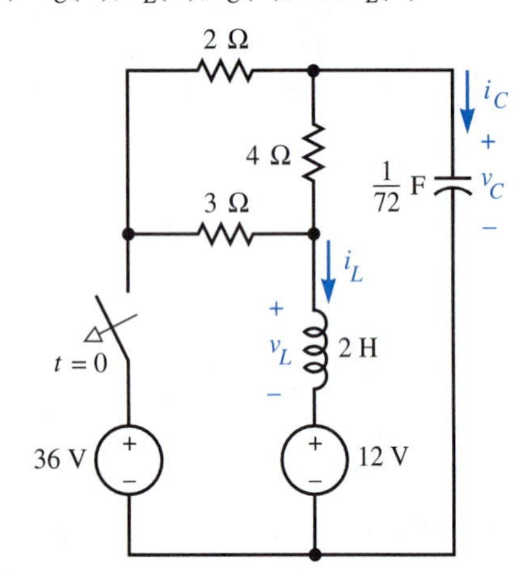

37. The switch in the preceding network has been open for a very long time and is closed at $t = 0$. Find:
    (a) $v_C(0^-)$, $v_L(0^-)$, $i_C(0^-)$, and $i_L(0^-)$
    (b) $v_C(0^+)$, $v_L(0^+)$, $i_C(0^+)$, and $i_L(0^+)$

38. In the following circuit determine voltage v for $t > 0$ if the switch has been open for a very long time and is closed at $t = 0$.

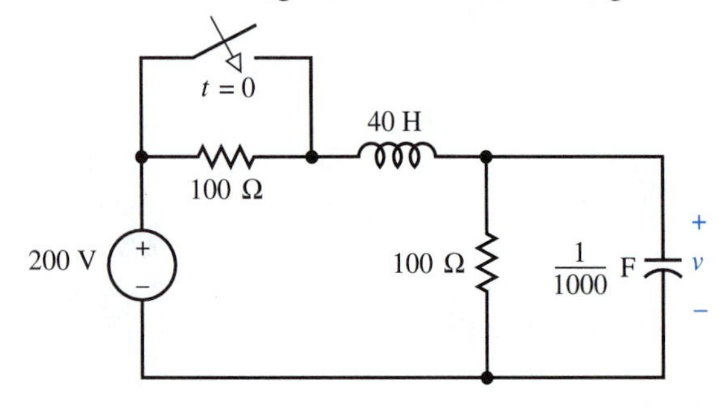

39. Determine voltage v_1 for $t > 0$ in the following network if the switch positions are as specified.
    (a) Switch 1 has been open for a very long time, switch 2 has been closed for a very long time, and switch 1 is closed at $t = 0$.
    (b) Switches 1 and 2 have been closed for a very long time, and switch 1 is opened at $t = 0$.
    (c) Switches 1 and 2 have been open for a very long time, and switch 2 is closed at $t = 0$.
    (d) Switch 1 has been open for a very long time, switch 2 has been closed for a very long time, and switch 2 is opened at $t = 0$. (This part requires the solution of simultaneous equations.)

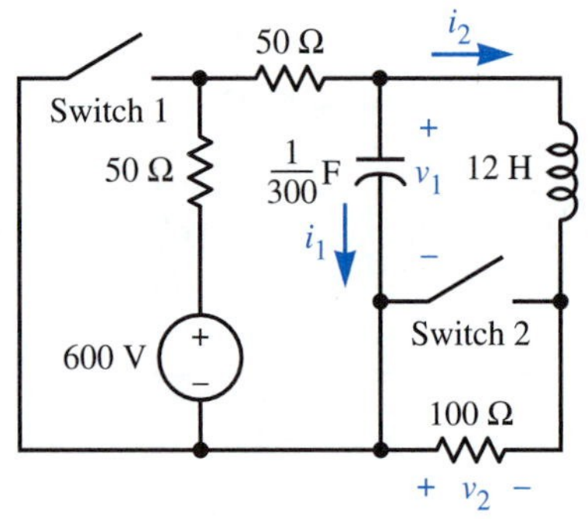

40. The following is an equivalent circuit for part of an automobile ignition system.
    (a) What is the value of current i if the switch has been closed for a very long time?
    (b) How long must the switch be closed so that $i \cong 4(1 - e^{-1})$ A? How fast can an engine turn (in revolutions per minute) if the current must reach this value for proper ignition, and the switch is closed for 22.5° of a crankshaft revolution (assume that $i = 0$ and $v = 0$) when the switch is first closed?
    (c) Neglect the losses in the 1.2 Ω resistance and determine the maximum $|v|$ if the inductance current is 4 A when the switch is opened. (Use stored energy.)
    (d) Determine the actual maximum $|v|$ if the circuit is in the steady state when the switch is

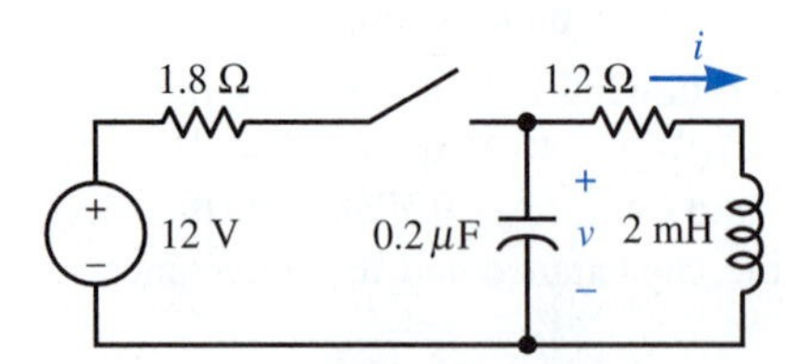

opened. How does this compare with the estimate obtained in part (c)?

41. Solve for voltages v_1 and v_2 in the following network for $t > 0$ if $i_s = 30 + 60u(t)$ A. (*Hint:* You can write a KCL equation in terms of v_2 that does not include v_1.)

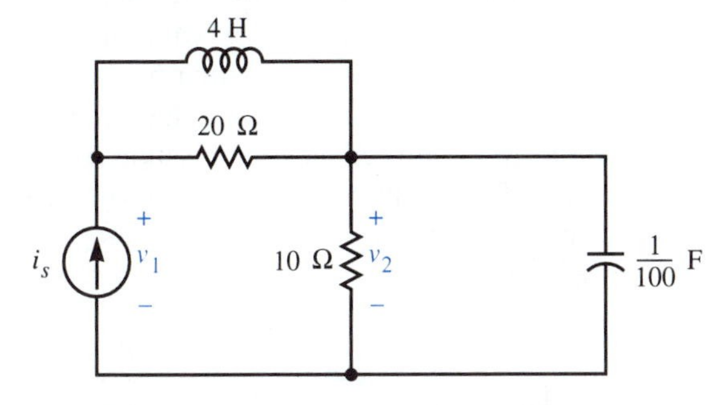

42. For each of the following circuits, determine voltage v if $v_s(t) = V_s u(t)$ V. (These are very peculiar circuits.)

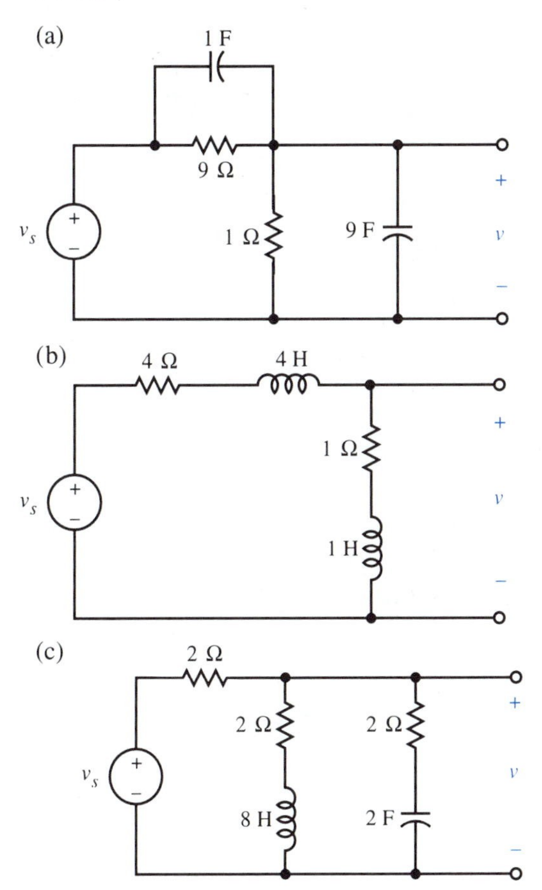

43. Determine the particular response component v_p of voltage v in the following circuit. (*Hint:* Use superposition.)

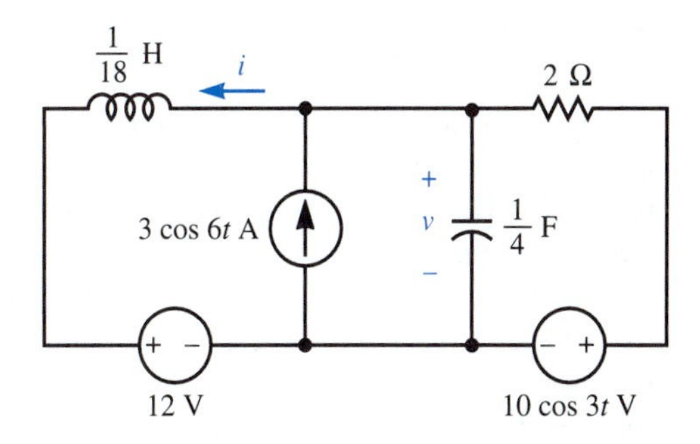

44. Solve for currents i_1, i_2, and i_3 in the following circuit if $v_s = 12e^{-t}u(t)$ V.

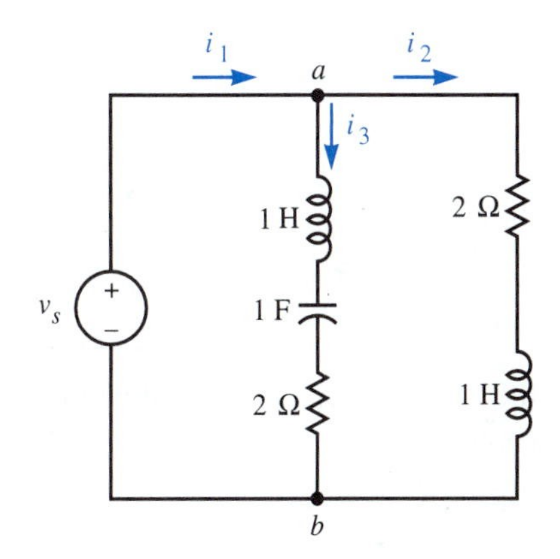

45. Solve for voltage v_{ab} in the following network if $i_s = 10u(t)$ A.

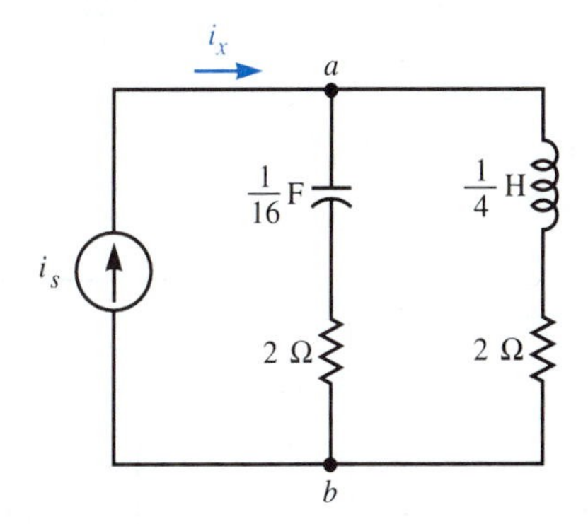

46. Replace the current source in the preceding network with a voltage source of value $20u(t)$ V [(+) terminal at the top] and solve for current i_x.

47. We can connect a resistor in parallel with a solenoid (inductor) to limit the voltage across the solenoid when the current is interrupted by a switch. The resistor consumes some power when the solenoid is energized. One way to avoid this is to connect a resistor and series capacitor in parallel with the solenoid. The model for this follows on the next page. The initial capacitance voltage is

$v_c(0^+) = 24$ V, and the initial inductance current is $i(0^+) = 10$ A. If possible, select R and C so that the absolute value of $v(t)$ does not exceed 240 V, and the absolute value of $i(t)$ does not exceed 2 A after 0.6 s. (The design may not be unique.) What is the maximum absolute value of v_C for your design? The capacitor must be rated to withstand this voltage. If you have access to the computer program PSpice, use this program to calculate and plot $i(t)$, $v(t)$, and $v_C(t)$ for $0 \leq t \leq 2$ s.

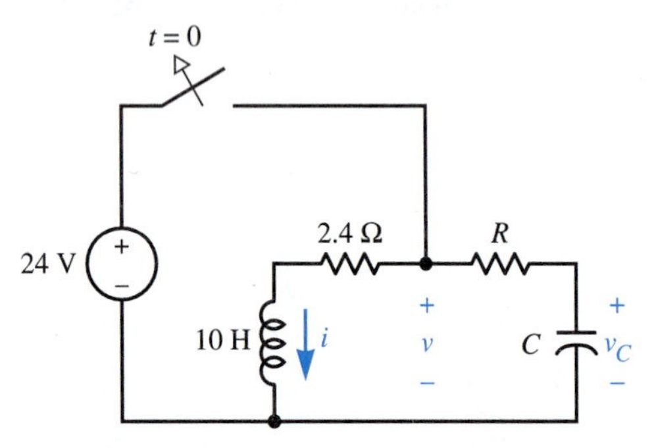

48. For the following op-amp circuit, the initial conditions are $v_a(0^+) = 0$ and $v_b(0^+) = 10$ V. Determine $v_o(t)$ for $t > 0$.

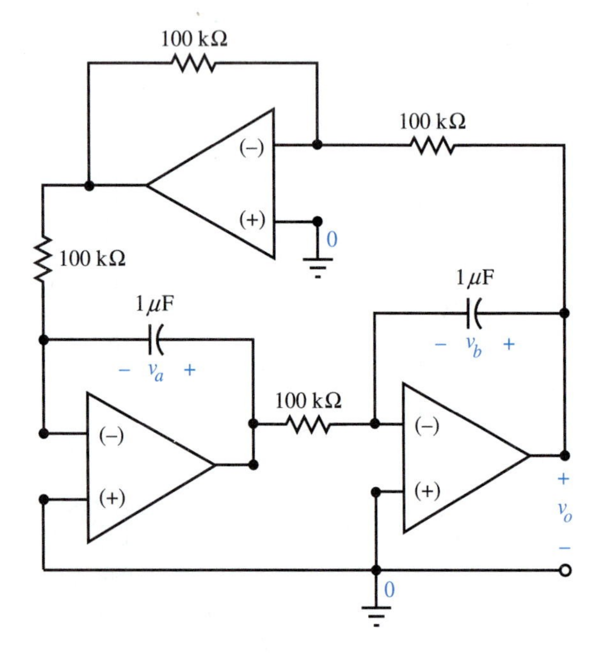

49. Determine the output voltage v_{ab} of the following circuit if $v_s = V_s u(t)$ V, and $R = 1\ \Omega$ and $C = 1$ F. This is a third-order circuit. The natural response will be the sum of three exponentials with the exponents determined by the roots of the characteristic equation. The particular response is found in the same way as it is for a second-order circuit (replace each capacitance with an open circuit). This network is used with a field-effect transistor operating in the grounded drain configuration to form an oscillator.) Evaluation of the constants of integration is tedious, so you can omit this step and use the computer program PSpice to plot the output voltage for $0 < t < 4$ s.

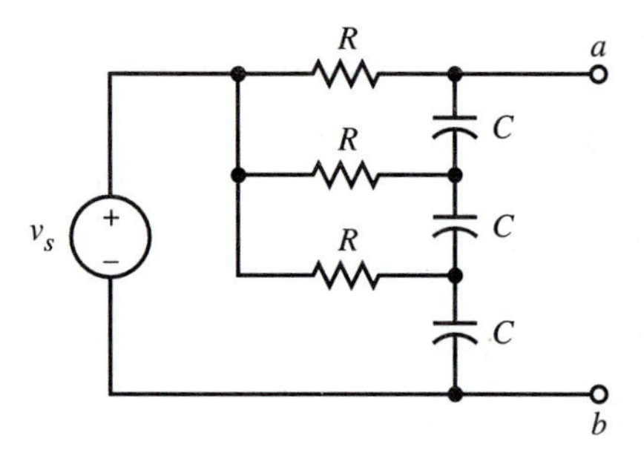

10

The Sinusoidal Steady State

In Chapters 8 and 9 we saw that when a sinusoidal input is applied to a stable LTI circuit, the steady-state response is always sinusoidal. This is one of the most important phenomena in electrical engineering. Of all possible periodic inputs, the sinusoidal signal is the *only* real one that reproduces itself throughout an arbitrary stable LTI circuit without changing its functional form or "shape."

We shall see in Chapter 15 that physical signals can be represented as sums or integrals of sinusoids. By studying the sinusoidal steady state, therefore, we prepare a foundation for a new and deep understanding of how circuits respond to all physical signals.

Our objective in this chapter is to describe the sinusoidal steady state with the use of the new concepts of *phasor, impedance, admittance,* and *transfer function.* These concepts are central to much of electrical engineering and to all that follows in this book. They are described using complex arithmetic. We encourage you, therefore, to master Appendix B before proceeding.

10.1 Pertinent Results from Chapters 8 and 9

We learned in Chapters 8 and 9 that the complete response, $v(t)$, of an LTI circuit is given by the sum of the particular response, $v_p(t)$, and the natural response, $v_n(t)$:

$$v(t) = v_p(t) + v_n(t) \tag{10.1}$$

We also learned that the steady-state response of a stable LTI circuit to a sinusoidal input is another sinusoid with the same frequency as the input. The steady-state response is sinusoidal because: (1) the particular response, $v_p(t)$, is a sinusoid with the same frequency as the input, and (2) the natural response, $v_n(t)$, approaches zero as t increases.

Figure 10.1 illustrates what happens when a sinusoid (Fig. 10.1b) is suddenly applied to an arbitrary stable LTI circuit. Even though the particular response, $v_p(t)$, is sinusoidal, the complete response, $v(t)$, is not sinusoidal initially because of the natural response, $v_n(t)$. As time increases, however, the natural response approaches zero asymptotically. Therefore, $v_n(t)$ eventually becomes negligibly small compared to $v_p(t)$ and the output is essentially sinusoidal. When this point is reached, we say that the circuit is operating in the *sinusoidal steady state.* The frequency of the sinusoidal steady-state response always equals the frequency of the input.

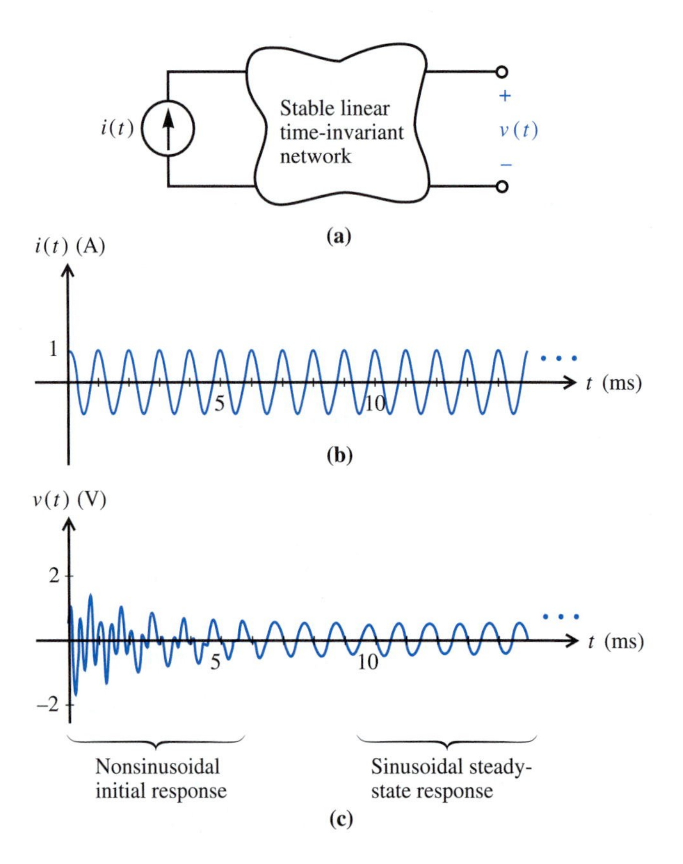

FIGURE 10.1 Typical response of a stable, linear, time-invariant network to a sinusoidal input applied at $t = 0$. (*a*) Network with input $i(t)$ and output $v(t)$. (*b*) Input $i(t)$. The input equals zero for $t < 0$ and is sinusoidal for $t > 0$. (*c*) Output $v(t)$. The particular response is sinusoidal with the same frequency as the input. Initially the complete response is not sinusoidal because of the natural response. As t increases, the natural response approaches zero asymptotically. Eventually, therefore, the complete response is sinusoidal with the same frequency as the input.

In the following section we shall introduce the new concepts of phasors and impedance to focus on the relation between a sinusoidal input and the sinusoidal steady-state response.

Remember The steady-state response of a stable LTI network to a sinusoidal input is a sinusoid with the same frequency as the input.

10.2 Phasors and Impedance

When a circuit has reached the sinusoidal steady state, all the current and voltage waveforms in the circuit are sinusoidal with the same frequency as the source. *Phasors* are complex constants used to represent the amplitudes and phases of these sinusoidal waveforms. *Impedances* are other complex constants used to relate voltage phasors to current phasors. In Chapter 11 we shall see that by using phasors and impedance, we can find the amplitudes and phases of sinusoidal waveforms by using only algebra. Our present goal is to explain in detail the concepts of phasor and impedance and to show how they are related.

Phasors

Let us begin with the definition of a phasor.

DEFINITION Phasor

The **phasor** corresponding to a sinusoid

$$i(t) = I_m \cos(\omega t + \phi_{\mathbf{I}}) \tag{10.2}$$

is defined as

$$\mathbf{I} = I_m \underline{/\phi_{\mathbf{I}}} \tag{10.3}$$

The phasor **I** represents the amplitude and phase of $i(t)$, as illustrated in Fig. 10.2. The subscript m in I_m signifies that I_m is the *m*aximum value or amplitude of $i(t)$ and that I_m is the *m*agnitude of the phasor **I**. The subscript **I** in $\phi_{\mathbf{I}}$ denotes that $\phi_{\mathbf{I}}$ is the phase angle of $i(t)$ and the angle of **I**. As an example, if $i(t) = 3\cos(100t + 25°)$, then $\mathbf{I} = 3\underline{/25°}$. The units of a phasor are always the same as those of the associated sinusoidal function. In Eq. (10.2) we have assumed that the sinusoidal function is a current, $i(t)$. Therefore, like $i(t)$, the phasor **I** has the unit ampere.

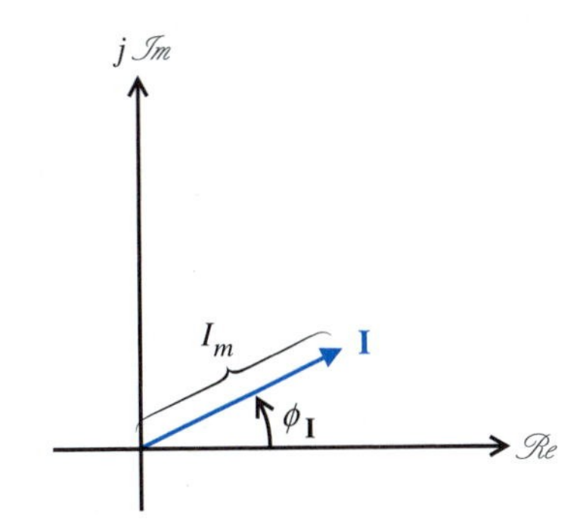

FIGURE 10.2 The phasor $\mathbf{I} = I_m \underline{/\phi_{\mathbf{I}}}$

Recall that in Section 9.5 we used complex exponentials to determine the particular response of LTI circuits when the input was a sinusoid.† We replaced $i(t)$ with a complex exponential

$$\mathbf{i}(t) = \mathbf{I}e^{j\omega t} = I_m e^{j(\omega t + \phi_\mathbf{I})} \tag{10.4}$$

because we knew from Euler's identity that

$$\begin{aligned} i(t) &= I_m \cos(\omega t + \phi_\mathbf{I}) \\ &= \mathcal{R}e\{I_m e^{j(\omega t + \phi_\mathbf{I})}\} \\ &= \mathcal{R}e\{\mathbf{I}e^{j\omega t}\} \end{aligned} \tag{10.5}$$

As shown in Fig. 10.3a, we can visualize the complex exponential $\mathbf{i}(t)$ as a *rotating phasor.* The real input, $i(t)$, is the projection of the rotating phasor $\mathbf{i}(t)$ onto the real axis. A plot of the real input $i(t)$ versus t is shown in Fig. 10.3b.

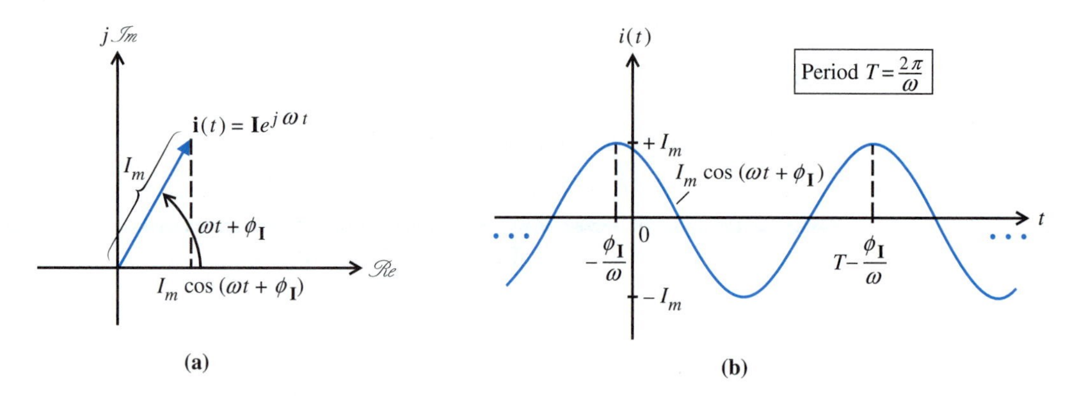

FIGURE 10.3
(*a*) Complex exponential (rotating phasor) $\mathbf{i}(t) = \mathbf{I}e^{j\omega t}$. The factor $e^{j\omega t}$ causes the phasor $\mathbf{I}$ to rotate counterclockwise with angular velocity ω. The real current is obtained by taking the real part of $\mathbf{i}(t)$. (*b*) Corresponding real current $i(t) = I_m \cos(\omega t + \phi_\mathbf{I})$.

Impedance

If the input to an LTI equation is the complex exponential $\mathbf{i}(t)$ of Eq. (10.4), then the particular response $\mathbf{v}_p(t)$ is also a complex exponential. This result follows directly from Eq. (9.73) with $s_p = j\omega$. Thus, *we can write the particular response to the rotating phasor input* $\mathbf{i}(t)$ *as another rotating phasor:*

$$\mathbf{v}_p(t) = \mathbf{V}e^{j\omega t} \tag{10.6}$$

Substitution of Eqs. (10.4) and (10.6) into the LTI equation

$$\mathcal{A}(D)v = \mathcal{B}(D)i \tag{10.7}$$

yields

$$\mathcal{A}(D)\mathbf{V}e^{j\omega t} = \mathcal{B}(D)\mathbf{I}e^{j\omega t} \tag{10.8}$$

† In Chapters 8 and 9, we used subscripts p on s and ω to emphasize that we were dealing with the particular response. In this and later chapters, we drop the subscript p for notational simplicity.

After differentiations, this becomes

$$\mathcal{A}(j\omega)\mathbf{V}e^{j\omega t} = \mathcal{B}(j\omega)\mathbf{I}e^{j\omega t} \tag{10.9}$$

and we see that we have simply replaced the derivative operator D in Eq. (10.8) by $j\omega$. If both sides are multiplied by $e^{-j\omega t}$, this becomes

$$\mathcal{A}(j\omega)\mathbf{V} = \mathcal{B}(j\omega)\mathbf{I} \tag{10.10}$$

which is a complex algebraic equation that does not depend on time. We can solve this equation for the voltage phasor:

$$\mathbf{V} = \mathbf{Z}\mathbf{I} \tag{10.11}$$

where we define the *impedance,* $\mathbf{Z}$, as follows.

DEFINITION Impedance

The **impedance** of a stable circuit whose LTI equation is

$$\mathcal{A}(D)v = \mathcal{B}(D)i \tag{10.12}$$

defined as the ratio of phasor output voltage $\mathbf{V}$ to phasor input current $\mathbf{I}$:

$$\mathbf{Z} = \frac{\mathbf{V}}{\mathbf{I}} \tag{10.13}$$

This impedance is given by

$$\mathbf{Z} = \frac{\mathcal{B}(j\omega)}{\mathcal{A}(j\omega)} \tag{10.14}$$

Like resistance, impedance has the unit ohm. Notice that the impedance depends on the angular frequency ω. Impedance, $\mathbf{Z}(j\omega)$, is not defined if $\mathcal{A}(j\omega) = 0$.

What is the point of Eq. (10.11)? The answer is that $\mathbf{V}$ is a phasor. Just as the phasor $\mathbf{I}$ tells us the amplitude and the phase of a sinusoidal input function [see Eq. (10.2)], the phasor $\mathbf{V}$ tells us the amplitude and the phase of the steady-state sinusoidal response. Equation (10.11) is sometimes called the *ac version of Ohm's law.*

It follows from Eq. (10.11) that the impedance $\mathbf{Z}$ determines the magnitude, $V_m = |\mathbf{V}|$, and the angle, $\phi_\mathbf{V} = \angle\mathbf{V}$, of the phasor $\mathbf{V}$:

$$V_m = |\mathbf{Z}|I_m \tag{10.15}$$

$$\phi_\mathbf{v} = \angle\mathbf{Z} + \phi_\mathbf{I} \tag{10.16}$$

where $|\mathbf{Z}|$ and $\angle\mathbf{Z}$ are, respectively, the magnitude and the angle of $\mathbf{Z}$. We see from Eq. (10.15) that the magnitude of the voltage phasor equals the *product* of the magnitude of the impedance and the magnitude of the current phasor. We see from Eq. (10.16) that the angle of the voltage phasor equals the *sum* of the angle of the impedance and the angle of the current phasor.

We can represent the relation between $\mathbf{V}$ and $\mathbf{I}$ in circuit form, as illustrated in Fig. 10.4. The circuit is characterized by its impedance $\mathbf{Z}$. Equation (10.10) is called the

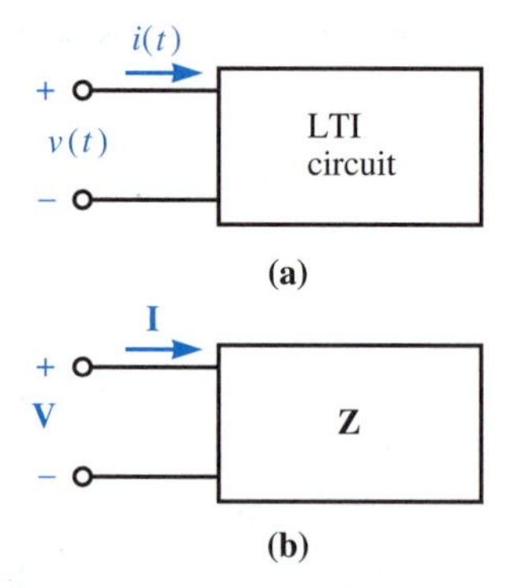

FIGURE 10.4 Representations of LTI circuit: (*a*) time domain, $\mathcal{A}(D)v(t) = \mathcal{B}(D)i(t)$; (*b*) frequency domain, $\mathbf{V} = \mathbf{ZI}$

frequency-domain version of the time-domain LTI Eq. (10.7). Figure 10.4b is called the *frequency-domain* representation of the *time-domain* LTI circuit of Fig. 10.4a. In much of our future work, we shall be concerned with frequency-domain equations and representations, because they describe sinusoidal steady-state voltages and currents by algebra instead of calculus.

Remember

If the input to a stable LTI circuit is a rotating phasor $\mathbf{i}(t) = \mathbf{I}e^{j\omega t}$ then the particular response is the rotating phasor $\mathbf{v}_p(t) = \mathbf{V}e^{j\omega t}$ where $\mathbf{V} = \mathbf{ZI}$. The impedance is given by $\mathbf{Z} = \mathcal{B}(j\omega)/\mathcal{A}(j\omega)$ where $\mathcal{B}(D)$ and $\mathcal{A}(D)$ are the operators found in the circuit's LTI equation $\mathcal{A}(D)v(t) = \mathcal{B}(D)i(t)$.

EXERCISES

1. Find the phasors corresponding to the following sinusoids.
   (a) $6 \cos (200t + 50°)$ mA
   (b) $6 \cos (400t + 50°)$ mA
   (c) $10 \sin (100t + 30°)$ mV *Hint:* Use the identity $\sin \phi = \cos (\phi - 90°)$.
   answer: (a) $6\underline{/50°}$ mA, (b) $6\underline{/50°}$ mA, (c) $10\underline{/-60°}$ mV
2. Find V_m and ϕ_v when $\mathbf{V} = \mathbf{ZI}$, where:
   (a) $\mathbf{Z} = 0.5\underline{/25°}$ kΩ and $\mathbf{I} = 2.77\underline{/50°}$ mA
   (b) $\mathbf{Z} = 1 + j$ kΩ and $\mathbf{I} = 100\underline{/-15°}$ μA
   (c) $\mathbf{Z} = 9e^{j30°}$ Ω and $\mathbf{I} = 3 + j4$ A.
   answer: (a) $1.38\underline{/75°}$ V, (b) $141\underline{/30°}$ mV, (c) $45\underline{/83.1°}$ V
3. Define in your own words the following terms (a) phasor, (b) rotating phasor, and (c) impedance.
4. Discuss the equation $\mathbf{V} = \mathbf{ZI}$. Specifically, explain (a) how this equation was obtained, and (b) how it relates the magnitude and angle of **V** to the magnitude and angle of **I**.

10.3 Impedances of *R*, *L*, and *C*

Until now the network components *R*, *L*, and *C* have been defined by differential equations relating terminal current $i(t)$ and voltage $v(t)$. In the sinusoidal steady state, these terminal equations can, in effect, be replaced by algebraic equations of the form $\mathbf{V} = \mathbf{ZI}$, where **Z** is the component's impedance, and phasors **I** and **V** represent terminal current and voltage. As a prerequisite step in the development of the totally algebraic network analysis of Chapter 11, we now derive the impedances of *R*, *L*, and *C*.

The terminal equation of a resistance is

$$v(t) = Ri(t) \tag{10.17}$$

By substituting Eqs. (10.4) and (10.6) into this, we obtain

$$\mathbf{V}e^{j\omega t} = R\mathbf{I}e^{j\omega t} \tag{10.18}$$

or

$$\mathbf{V} = R\mathbf{I} \tag{10.19}$$

Therefore the impedance of a resistance is

$$\mathbf{Z} = R = R\underline{/0^\circ} \tag{10.20}$$

The same result can be obtained directly from Eq. (10.14) by use of $\mathcal{A}(D) = 1$ and $\mathcal{B}(D) = R$. (The two methods are equivalent.)

For an inductance, we have

$$v(t) = L\frac{d}{dt}i \tag{10.21}$$

which by substitution of Eqs. (10.4) and (10.6) becomes

$$\mathbf{V}e^{j\omega t} = j\omega L\mathbf{I}e^{j\omega t} \tag{10.22}$$

or

$$\mathbf{V} = j\omega L\mathbf{I} \tag{10.23}$$

Therefore, for an inductance,

$$\mathbf{Z} = j\omega L = \omega L\underline{/90^\circ} \tag{10.24}$$

Again, this result could have been obtained from Eq. (10.14) by the use of $\mathcal{A}(D) = 1$ and $\mathcal{B}(D) = LD$.

Finally, for a capacitance,

$$C\frac{d}{dt}v = i \tag{10.25}$$

Here we will use Eq. (10.14) for variety. Since $\mathcal{A}(D) = CD$ and $\mathcal{B}(D) = 1$ in Eq. (10.25), Eq. (10.14) yields

$$\mathbf{Z} = \frac{1}{j\omega C} = \frac{1}{\omega C}\underline{/-90^\circ} \tag{10.26}$$

The frequency-domain circuits for R, L, and C are shown in Fig. 10.5. The phasor diagrams adjacent to each circuit illustrate the relationships between the voltage and current phasors. The equations for V_m and ϕ_v shown in the phasor diagrams were obtained with the use of Eqs. (10.15) and (10.16) and the impedances for each element. The quantities V_m and $\phi_\mathbf{V}$ tell us the amplitude and the phase of the sinusoidal terminal voltage $v(t) = V_m \cos(\omega t + \phi_\mathbf{V})$ caused by the sinusoidal input current $i(t) = I_m \cos(\omega t + \phi_\mathbf{I})$.

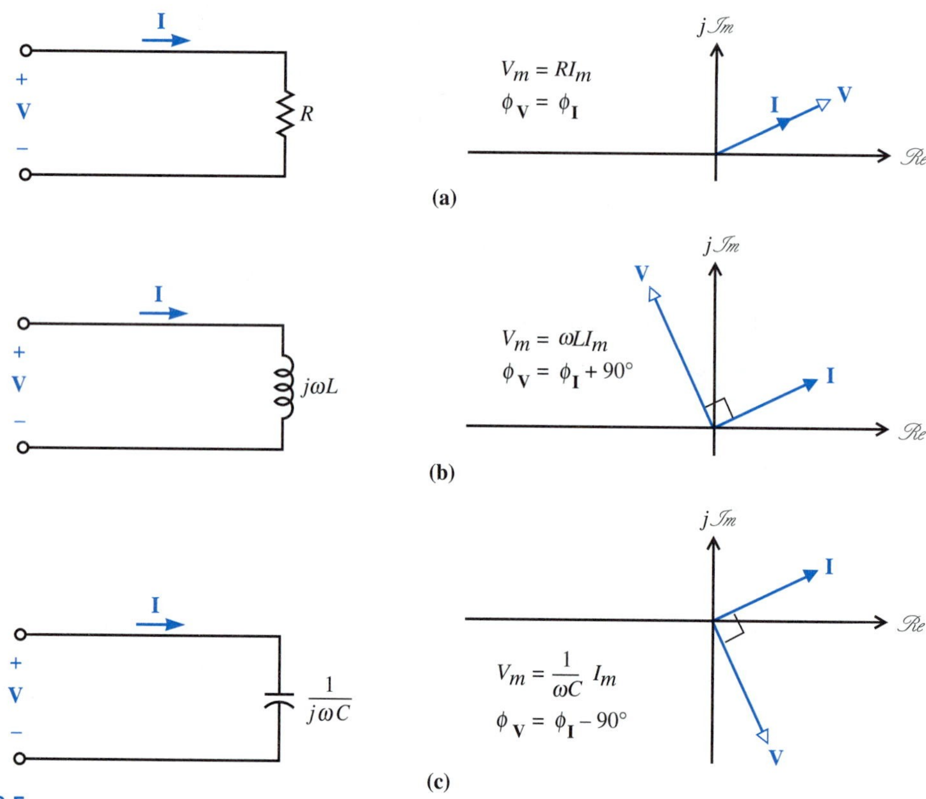

FIGURE 10.5
Frequency-domain representation of (*a*) resistance, $\mathbf{V} = R\mathbf{I}$, (*b*) inductance, $\mathbf{V} = j\omega L\mathbf{I}$, and (*c*) capacitance, $\mathbf{V} = (1/j\omega C)\mathbf{I}$

Remember The impedances of the elements R, L, and C are R, $j\omega L$, and $1/j\omega C$ ohms, respectively.

EXERCISES

5. Assume that $R = 100\ \Omega$, $C = 1000\ \mu F$, $L = 2$ mH, and $\mathbf{i}(t) = (3\underline{/45°})e^{j1000t}$ mA. Redraw Fig. 10.5 with numerical values for the magnitudes and angles of all the phasors indicated.

   answer: $\mathbf{I} = 3\underline{/45°}$ mA; resistance $\mathbf{V} = 300\underline{/45°}$ mV; inductance $\mathbf{V} = 6\underline{/135°}$ mV; capacitance $\mathbf{V} = 3\underline{/-45°}$ mV

6. Derive the impedance of a resistance R using Eq. (10.14).
7. Derive the impedance of an inductance L using Eq. (10.14).
8. Derive the impedance of a capacitance C from the LTI terminal equation, $\mathbf{i}(t) = \mathbf{I}e^{j\omega t}$, and $\mathbf{v}_p(t) = \mathbf{V}e^{j\omega t}$.

In Exercises 9 and 10, assume that $R = 1\ k\Omega$, $C = 1\ \mu F$, $L = 1$ H, and that $i(t) = 1 \cos(1000t - 45°)$ mA.

9. Redraw the frequency-domain components of Fig. 10.5, and indicate the appropriate numerical values for the phasors and impedances.

   answer: $\mathbf{I} = 1\underline{/-45°}$ mA; resistance $\mathbf{V} = 1\underline{/-45°}$ V; inductance $\mathbf{V} = 1\underline{/45°}$ V; capacitance $\mathbf{V} = 1\underline{/-135°}$ V

10. Draw the numerical-valued *rotating* current and voltage phasors, and use trigonometry to find the formulas for their projections on the real axis.

10.4 Relation of Phasors and Impedance to Real Input and Output

In this section we describe the relation of phasors and impedance to real inputs and outputs, and present several illuminating examples. Recall from Sec. 10.2 that the particular response to a rotating phasor input $\mathbf{i}(t) = \mathbf{I}e^{j\omega t}$ is a rotating phasor $\mathbf{v}_p(t) = \mathbf{V}e^{j\omega t}$, where $\mathbf{V} = \mathbf{ZI}$. The rotating phasors $\mathbf{i}(t)$ and $\mathbf{v}_p(t)$ can be broken down into their real and imaginary parts. The particular response $v_p(t)$ to the real input $i(t) = \mathcal{Re}\{\mathbf{I}e^{j\omega t}\} = I_m \cos(\omega t + \phi_\mathbf{I})$ is the real part of $\mathbf{v}_p(t)$:

$$v_p(t) = \mathcal{Re}\{\mathbf{V}e^{j\omega t}\} = V_m \cos(\omega t + \phi_\mathbf{V}) \tag{10.27}$$

where V_m and $\phi_\mathbf{V}$ are given by Eqs. (10.15) and (10.16), respectively. The relationship between the phasors $\mathbf{V}$ and $\mathbf{I}$ is illustrated in Fig. 10.6. The rotating phasors $\mathbf{i}(t) = \mathbf{I}e^{j\omega t}$ and $\mathbf{v}_p(t) = \mathbf{V}e^{j\omega t}$ are illustrated in Fig. 10.7a.

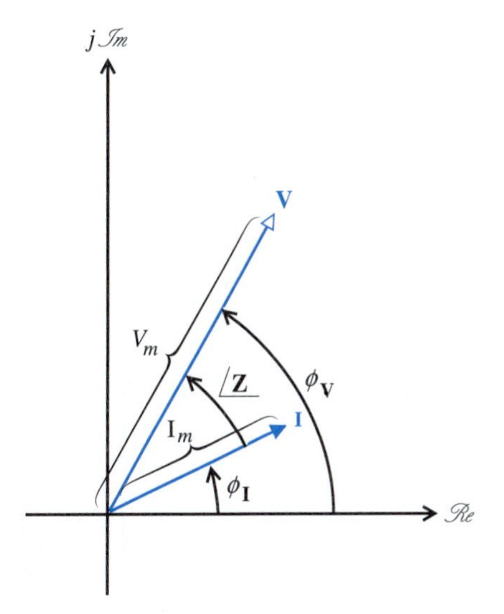

FIGURE 10.6 The phasors $\mathbf{V}$ and $\mathbf{I}$, where $\mathbf{V} = \mathbf{ZI}$. In the figure, $V_m = |\mathbf{Z}|I_m$ and $\phi_\mathbf{V} = \underline{/\mathbf{Z}} + \phi_\mathbf{I}$

The fundamental result is shown in Fig. 10.7. In this figure, the real input waveform,

$$i(t) = I_m \cos(\omega t + \phi_\mathbf{I}) \tag{10.28}$$

is the projection of the rotating current phasor $\mathbf{i}(t)$ onto the real axis, and the associated real steady-state output waveform

$$\begin{aligned} v_p(t) &= \mathcal{Re}\{\mathbf{V}e^{j\omega t}\} \\ &= V_m \cos(\omega t + \phi_\mathbf{V}) \end{aligned} \tag{10.29}$$

is the projection of the rotating voltage phasor $\mathbf{v}_p(t)$ onto the real axis.

Thus we see that the phasor $\mathbf{V}$ tells us the amplitude and the phase of the sinusoidal particular response, $v_p(t)$, just as the phasor $\mathbf{I}$ tells us the amplitude and phase of the sinusoidal input, $i(t)$. Plots of $i(t)$ and $v_p(t)$ versus t are given in Fig. 10.7b.

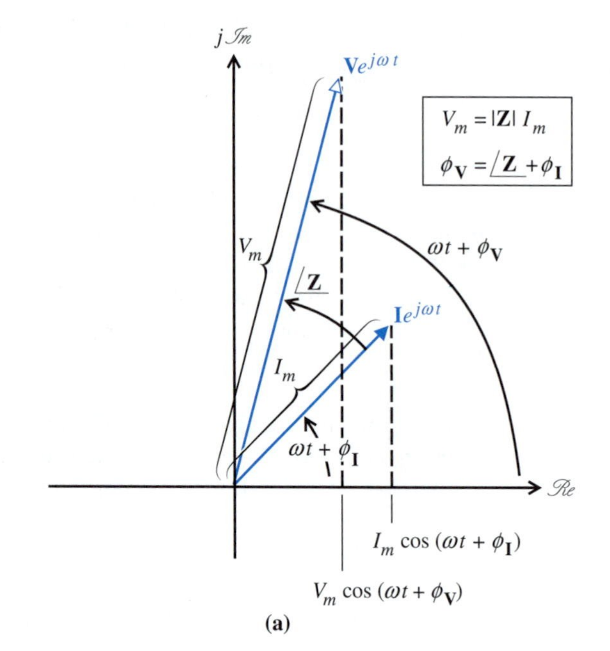

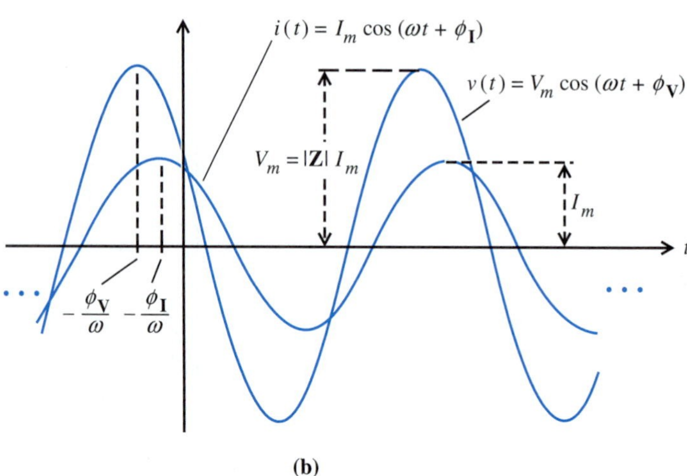

FIGURE 10.7 (*a*) Rotating current and voltage phasors. Both phasors rotate counterclockwise with angular velocity ω. (*b*) The sinusoidal current and voltage are the real parts of the rotating phasors. Both waveforms are sinusoidal with angular frequency ω

EXAMPLE 10.1 The input to a certain circuit is $i(t) = 7\cos(100t + 25°)$ A. The impedance is $\mathbf{Z}(j100) = 15\underline{/20°}\ \Omega$. Find (a) the output voltage phasor, $\mathbf{V}$, and (b) the particular response to $i(t)$.

Solution (a) The output phasor is given by $\mathbf{V} = \mathbf{ZI}$ where $\mathbf{Z} = 15\underline{/20°}\ \Omega$ and $\mathbf{I} = 7\underline{/25°}$ A. Therefore, $\mathbf{V} = (15\underline{/20°})\,(7\underline{/25°}) = 105\underline{/45°}$ V.

(b) The particular response to the input $i(t) = 7\cos(100t + 25°)$ A is therefore $v_p(t) = 105\cos(100t + 45°)$ V.

EXAMPLE 10.2 The relation between input current i and output voltage v of a certain circuit is given by the LTI equation

$$\frac{d}{dt}v + 2v = 4\frac{d}{dt}i + 4i$$

Assume that $i(t) = 5\cos(\omega t + 45°)$ A, where $\omega = 3$ rad/s. Find (a) the phasor **I**, (b) the impedance **Z**, (c) the phasor **V**, and (d) the particular response $v_p(t)$. Interpret your results graphically.

Solution (a) It follows from Eqs. (10.2) and (10.3) that

$$\mathbf{I} = I_m \underline{/\phi_{\mathbf{I}}} = 5\underline{/45°} \text{ A}$$

(b) The given LTI equation has the form $\mathcal{A}(D)v = \mathcal{B}(D)i$, where $\mathcal{A}(D) = D + 2$ and $\mathcal{B}(D) = 4D + 4$. The impedance follows from this and Eq. (10.14):

$$\mathbf{Z} = \frac{\mathcal{B}(j\omega)}{\mathcal{A}(j\omega)} = \frac{4j\omega + 4}{j\omega + 2}$$

Since the input angular frequency is $\omega = 3$ rad/s, this becomes

$$\mathbf{Z} = \frac{4 + j12}{2 + j3} = \frac{12.65\underline{/71.6°}}{3.61\underline{/56.3°}} = 3.5\underline{/15.3°}\ \Omega$$

(c) The voltage phasor is given by Eq. (10.13):

$$\mathbf{V} = \mathbf{ZI} = (3.5\underline{/15.3°})(5\underline{/45°}) = 17.5\underline{/60.3°} \text{ V}$$

(d) The result of (c) tells us that the amplitude of the voltage is 17.5 V and the phase is 60.3°. Thus

$$v_p(t) = 17.5\cos(\omega t + 60.3°) \text{ V}$$

where the angular frequency ω is the same as that of the input, $\omega = 3$ rad/s. The phasors **I** and **V** are illustrated in Fig. 10.8. Figure 10.9a depicts the rotating current and voltage phasors and their projections onto the real axis. These projections are, respectively, the sinusoidal input current $i(t)$ and the sinusoidal steady-state output voltage $v(t)$. Plots of $i(t)$ and $v(t)$ versus t are given in Fig. 10.9b.

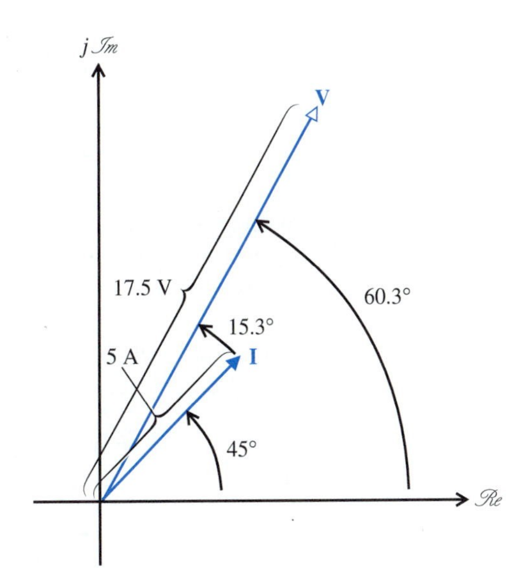

FIGURE 10.8 Phasors for Example 10.2. The impedance $\mathbf{Z} = 3.5\underline{/15.3°}\ \Omega$

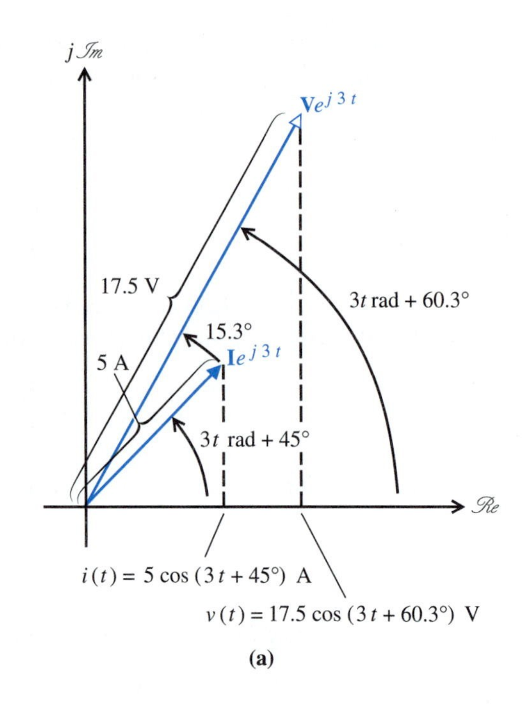

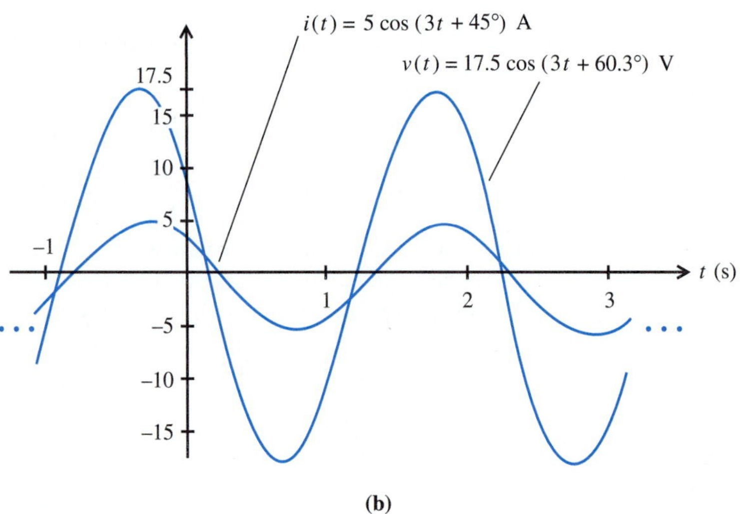

FIGURE 10.9
(*a*) Rotating current and voltage phasors for Example 10.2. The current and voltage phasors both rotate counterclockwise with angular velocity 3 rad/s. (*b*) Corresponding real waveforms. The sinusoidal current and voltage waveforms have the same angular frequency, 3 rad/s

EXAMPLE 10.3 Use phasors to find the steady-state sinusoidal voltage $v(t)$ across (a) a resistance R, (b) an inductance L, and (c) a capacitance C, if $i(t) = I_m \cos(\omega t + \phi_{\mathbf{I}})$. Check your result with the use of the time-domain terminal equations.

Solution In each part we use the equation $\mathbf{V} = \mathbf{ZI}$ with $\mathbf{I} = I_m\underline{/\phi_{\mathbf{I}}}$.

(a) For a resistance, $\mathbf{Z} = R$ so

$$\mathbf{V} = R\mathbf{I} = (R)(I_m\underline{/\phi_{\mathbf{I}}}) = RI_m\underline{/\phi_{\mathbf{I}}}$$

Therefore,

$$v(t) = RI_m \cos(\omega t + \phi_{\mathbf{I}})$$

The check on this result is trivial. Direct substitution into the time-domain terminal equation for a resistance yields

$$v(t) = Ri(t) = RI_m \cos(\omega t + \phi_{\mathbf{I}})$$

(b) For an inductance, $\mathbf{Z} = j\omega L$ so

$$\mathbf{V} = j\omega L\mathbf{I} = (\omega L\underline{/90^\circ})(I_m\underline{/\phi_{\mathbf{I}}}) = \omega LI_m\underline{/90^\circ + \phi_{\mathbf{I}}}$$

Therefore,

$$v(t) = \omega LI_m \cos(\omega t + \phi_{\mathbf{I}} + 90^\circ)$$

To check this result, we use the time-domain terminal equation

$$v(t) = L\frac{di}{dt} = L\frac{d}{dt}I_m \cos(\omega t + \phi_{\mathbf{I}}) = -\omega LI_m \sin(\omega t + \phi_{\mathbf{I}})$$

This result agrees with the original answer, because $\cos(\omega t + \phi_{\mathbf{I}} + 90^\circ) = -\sin(\omega t + \phi_{\mathbf{I}})$.

(c) For a capacitance, $\mathbf{Z} = 1/(j\omega C)$ so

$$\mathbf{V} = \frac{1}{j\omega C}\mathbf{I} = \left(\frac{1}{\omega C}\underline{/-90^\circ}\right)(I_m\underline{/\phi_{\mathbf{I}}}) = \frac{I_m}{\omega C}\underline{/\phi_{\mathbf{I}} - 90^\circ}$$

Therefore,

$$v(t) = \frac{I_m}{\omega C}\cos(\omega t + \phi_{\mathbf{I}} - 90^\circ)$$

To check this result, we use the time-domain terminal equation

$$v(t) = \frac{1}{C}\int i(t)dt = \frac{1}{C}\int I_m \cos(\omega t + \phi_{\mathbf{I}})\, dt = \frac{I_m}{\omega C}\cos(\omega t + \phi_{\mathbf{I}} - 90^\circ)$$

Since we are evaluating the particular response, the constant of integration equals zero.

EXAMPLE 10.4 (a) Write the differential equation relating $v(t)$ to $i(t)$ for the series *RLC* circuit of Fig. 10.10.

(b) Use your result of part (a) to find the impedance of the circuit.

(c) Assume that $i(t) = 3\cos(10^6 t + 15^\circ)$ mA, $R = 1$ kΩ, $L = 2$ mH, and $C = 0.001$ μF. Find the steady-state voltage $v(t)$.

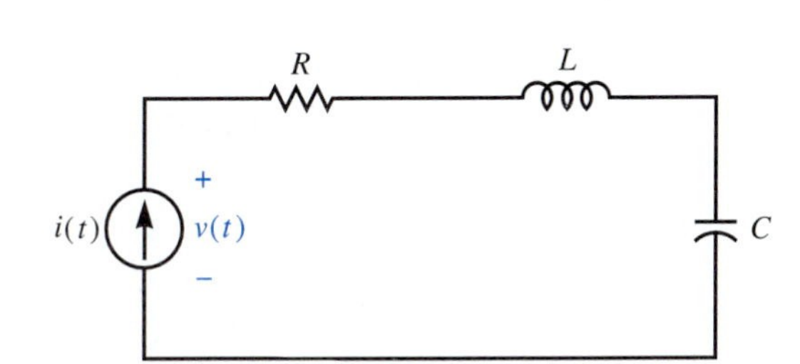

FIGURE 10.10
Series *RLC* circuit

Solution (a) The terminal equation of the series RLC circuit is, by KVL,

See Problems 10.1 through 10.3 in the PSpice manual.

$$v(t) = L\frac{d}{dt}i + Ri + \frac{1}{C}\int_{-\infty}^{t} i(\lambda)d\lambda$$

Differentiated with respect to t, this equation becomes

$$\frac{d}{dt}v = L\frac{d^2}{dt^2}i + R\frac{d}{dt}i + \frac{1}{C}i$$

(b) It follows from the above that $\mathcal{A}(D) = D$ and $\mathcal{B}(D) = LD^2 + RD + 1/C$. Therefore, by Eq. (10.14)

$$\mathbf{Z} = \frac{L(j\omega)^2 + Rj\omega + 1/C}{j\omega} = j\omega L + R + \frac{1}{j\omega C}$$

Notice that the impedance of a series RLC circuit equals the sum of the impedances of R, L, and C.

(c) For the given numerical values, we have

$$\mathbf{I} = 3\underline{/15^\circ} \text{ mA}$$

and

$$\mathbf{Z} = j10^6 \times 2 \times 10^{-3} + 10^3 + \frac{1}{j10^6 \times 10^{-9}} = 10^3 + j10^3\ \Omega$$
$$= \sqrt{2}\underline{/45^\circ}\ \text{k}\Omega$$

Therefore

$$\mathbf{V} = \mathbf{ZI} = 3\sqrt{2}\underline{/60^\circ}\ \text{V}$$

and the steady-state response is

$$v(t) = 3\sqrt{2}\cos(10^6 t + 60^\circ)\ \text{V}$$

Remember

The phasors $\mathbf{I} = I_m\underline{/\phi_\mathbf{I}}$ and $\mathbf{V} = V_m\underline{/\phi_\mathbf{V}}$ contain the amplitude and phase information of the sinusoidal waveforms $i(t) = I_m\cos(\omega t + \phi_\mathbf{I})$ and $v_p(t) = V_m\cos(\omega t + \phi_\mathbf{V})$, respectively.

EXERCISES

11. Explain in your own words the *practical* significance of the equation $\mathbf{V} = \mathbf{ZI}$.
12. A certain LTI circuit has input $i = 10\cos(100t + 20^\circ)$ mA and steady-state output $v = 50\cos(100t + 65^\circ)$ V.
    (a) Write down the phasors $\mathbf{I}$ and $\mathbf{V}$.
    (b) Determine $\mathbf{Z}$. Express your answer in polar form.
    answer: (a) $\mathbf{I} = 10\underline{/20^\circ}$ mA, $\mathbf{V} = 50\underline{/65^\circ}$ V, (b) $\mathbf{Z} = 5\underline{/45^\circ}$ kΩ
13. The impedance in Example 10.2 was derived with the use of Eq. (10.14). Show that the same result is obtained by substituting $\mathbf{i}(t) = \mathbf{I}e^{j\omega t}$ and $\mathbf{v}_p(t) = \mathbf{V}e^{j\omega t}$ into the given differential equation and solving for $\mathbf{V}/\mathbf{I}$.
14. Show that the magnitude and angle of $\mathbf{Z} = j\omega L + R + 1/j\omega C$ in Example 10.4 equal

$$|\mathbf{Z}| = \sqrt{R^2 + \left(\omega L - \frac{1}{\omega C}\right)^2} \quad \text{and} \quad \underline{/\mathbf{Z}} = \arctan\left(\frac{\omega L - 1/\omega C}{R}\right)$$

respectively. Use this result to show that the steady-state voltage is

$$v_p(t) = \sqrt{R^2 + \left(\omega L - \frac{1}{\omega C}\right)^2}\, I_m \cos\left\{\omega t + \phi_{\mathrm{I}} + \arctan\left[\frac{\omega L - 1/\omega C}{R}\right]\right\}$$

15. Show that the phasor voltage **V** corresponding to the voltage $v(t)$ in Fig. 10.10 may be written as $\mathbf{V} = \mathbf{V}_L + \mathbf{V}_R + \mathbf{V}_C$ where $\mathbf{V}_L$, $\mathbf{V}_R$, and $\mathbf{V}_C$ are the phasor voltage drops across the L, R, and C respectively. *Hint:* This result follows directly from the ac version of Ohm's law and the expression for **Z** found in Example 10.4.

10.5 Concepts Related to Impedance

The concepts of phasor and impedance give rise to several important related concepts. For easy reference, we place their definitions together in this section.

Driving-Point and Transfer Impedance

When we move beyond the two-terminal elements R, L, and C to networks, two situations can arise, as illustrated in Fig. 10.11. If the voltage phasor response **V**, appears across the terminals of the current phasor source **I**, as shown in Fig. 10.11a, then **Z** is called a *driving-point impedance.* If the voltage response appears across *another pair* of terminals, as shown in Fig. 10.11b, then **Z** is called a *transfer impedance.*

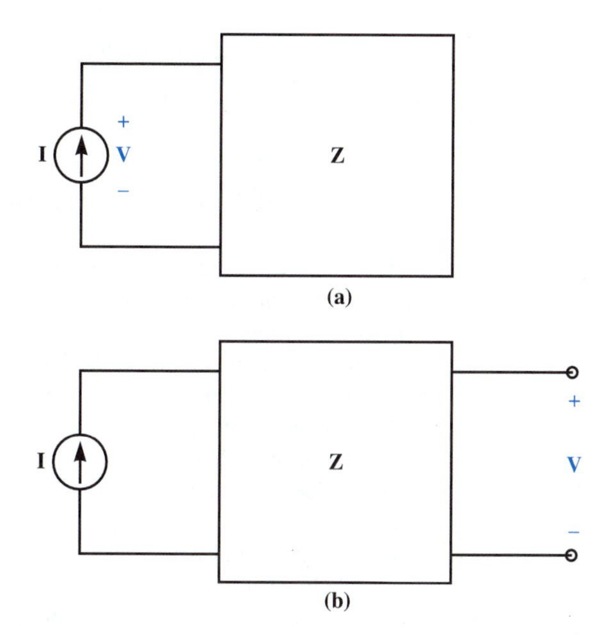

FIGURE 10.11 Driving-point and transfer impedance are distinct quantities. However, both are denoted by the same symbol **Z**. (*a*) Driving-point impedance and (*b*) transfer impedance

Driving-Point and Transfer Admittance

Assume that a sinusoidal voltage $v(t)$ is applied to a stable LTI circuit. We know that the steady-state current response $i(t)$ is sinusoidal with the same frequency. Thus we can represent $v(t)$ and $i(t)$ with the use of phasors **V** and **I**. The ratio of current phasor response to voltage phasor input $\mathbf{Y} = \mathbf{I}/\mathbf{V}$ is called an *admittance* and has the unit of siemens. If the current response is in the same branch as the voltage source as shown in

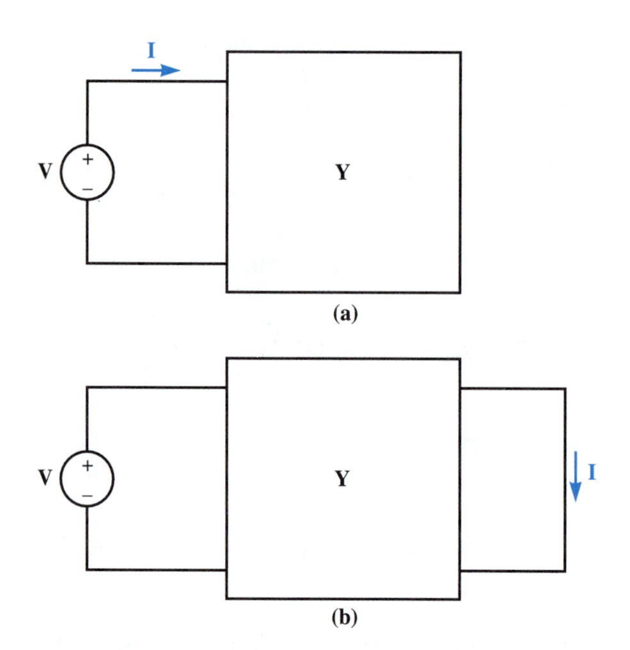

FIGURE 10.12 Driving-point and transfer admittance are distinct quantities denoted by the same symbol, **Y**. (*a*) Driving-point admittance and (*b*) transfer admittance

Fig. 10.12a, then **Y** is called a *driving-point admittance.* The driving-point admittance of any LTI circuit equals the reciprocal of the corresponding driving-point impedance

$$\mathbf{Y} = \mathbf{Z}^{-1} \tag{10.30}$$

If the current response is in a branch other than that of the voltage source as shown in Fig. 10.12b, then **Y** is called a *transfer admittance.* Transfer admittances and impedances do not satisfy Eq. (10.30).

Components of Impedance and Admittance

Every impedance can be expressed in the rectangular form $\mathbf{Z}(j\omega) = R(\omega) + jX(\omega)$. $R(\omega)$ and $X(\omega)$ are the *resistive* and the *reactive components* of the impedance, respectively, or more simply, the *resistance* and the *reactance.* Every admittance can be expressed in the rectangular form $\mathbf{Y}(j\omega) = G(\omega) + jB(\omega)$. $G(\omega)$ and $B(\omega)$ are the *conductive* and the *susceptive components* of the admittance, respectively, or more simply, the *conductance* and the *susceptance.* The units of $\mathbf{Z}(\omega)$, $R(\omega)$, and $X(\omega)$ are ohms, whereas those of $\mathbf{Y}(\omega)$, $G(\omega)$, and $B(\omega)$ are siemens.

To determine the relationships between driving-point resistance, reactance, conductance, and susceptance, we simply substitute $\mathbf{Y} = G(\omega) + jB(\omega)$ and $\mathbf{Z} = R(\omega) + jX(\omega)$ into $\mathbf{Y} = \mathbf{Z}^{-1}$, and equate the real and the imaginary parts of both sides. The results are tabulated in Table 10.1. Notice from the table that conductance is *not* the reciprocal of resistance, but depends on both resistance and reactance. Table 10.1 does not apply to transfer impedance or transfer admittance.

Inductive, Resistive, and Capacitive Networks

We know from Eq. (10.16) that the phase $\phi_{\mathbf{V}}$ of a sinusoidal voltage waveform equals the sum of the angle of the impedance, $\angle\mathbf{Z}$, and the phase of the current, $\phi_{\mathbf{I}}$. If the impedance angle $\angle\mathbf{Z}$ is positive, then the voltage is said to *lead* the current. An induc-

TABLE 10.1 The relationship between resistance $R(\omega)$, reactance $X(\omega)$, conductance $G(\omega)$, and susceptance $B(\omega)$. $\mathbf{Z}(j\omega)$ is any driving-point impedance, and $\mathbf{Y}(j\omega)$ is the corresponding driving-point admittance: $\mathbf{Z}(j\omega) = \mathbf{Y}^{-1}(j\omega)$. This table does not apply to transfer impedance or transfer admittance.

$\mathbf{Y}(j\omega) = G(\omega) + jB(\omega)$	$\mathbf{Z}(j\omega) = R(\omega) + jX(\omega)$
$G(\omega) = \dfrac{R(\omega)}{R^2(\omega) + X^2(\omega)}$	$R(\omega) = \dfrac{G(\omega)}{G^2(\omega) + B^2(\omega)}$
$B(\omega) = \dfrac{-X(\omega)}{R^2(\omega) + X^2(\omega)}$	$X(\omega) = \dfrac{-B(\omega)}{G^2(\omega) + B^2(\omega)}$

tance is the most elementary example of an impedance that has a positive angle ($\underline{/j\omega L} = 90°$). Consequently, any network that has a positive driving-point impedance angle is called *inductive.* If an impedance angle is negative, then we say that the voltage *lags* the current. A capacitance is the most elementary example of an impedance that has a negative angle ($\underline{/1/j\omega C} = -90°$). Consequently, any network that has a negative driving-point impedance angle is called *capacitive.* If an impedance angle is zero, then the current and voltage are said to be *in phase.* A resistance is the most elementary example of an impedance whose angle is zero. Any network whose driving-point impedance angle is zero is called *resistive.*

EXAMPLE 10.5 The driving-point impedance of a certain circuit operating at 100 rad/s is $\mathbf{Z}(j100) = 5\underline{/36.87°}\ \Omega$.

(a) Find the driving-point resistance and reactance.
(b) Find the driving-point admittance, conductance, and susceptance.
(c) Is the network inductive, resistive, or capacitive?

Solution (a) Writing $\mathbf{Z}$ in rectangular form, we find that $\mathbf{Z}(j100) = 4 + j3\ \Omega$. The resistance is therefore $R(100) = 4\ \Omega$, and the reactance is $X(100) = 3\ \Omega$

(b) The driving-point admittance is

$$\mathbf{Y}(j100) = \frac{1}{\mathbf{Z}(j100)} = \frac{1}{5}\underline{/-36.87°}\text{ S}$$

which in rectangular form equals $\mathbf{Y}(j100) = 0.16 - j0.12$ S. The conductance is therefore $G(100) = 0.16$ S, and the susceptance is $B(100) = -0.12$ S.

(c) Since the impedance angle is 36.87°, the voltage leads the current by 36.87°. Therefore the circuit is inductive.

EXAMPLE 10.6 (a) Determine the driving-point admittance of the parallel *RLC* circuit of Fig. 10.13. Comment on your result.

(b) Assume that $v(t) = 25\cos(10^4 t + 70°)$ mV, $R = 10\ \Omega$, $L = 1$ mH, and $C = 30\ \mu$F. Find the steady-state current $i(t)$.

FIGURE 10.13
Parallel *RLC* circuit

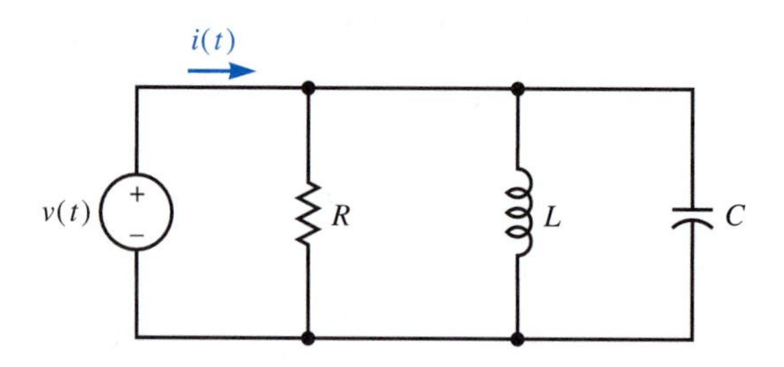

Solution (a) The driving-point admittance can be found from $\mathbf{Y} = 1/\mathbf{Z}$, where $\mathbf{Z}$ is the driving-point impedance. Since $\mathbf{Z}$ is unknown, however, it is easier to obtain $\mathbf{Y}$ from $\mathbf{Y} = \mathbf{I}/\mathbf{V}$. To obtain the ratio $\mathbf{I}/\mathbf{V}$, we can start with the circuit's terminal equation, which, by KCL, is

$$i(t) = C\frac{d}{dt}v + \frac{1}{R}v + \frac{1}{L}\int_{-\infty}^{t} v(\lambda)\, d\lambda$$

where we have taken care to write the input variable v on the right-hand side. When differentiated, this becomes

$$\frac{d}{dt}i = C\frac{d^2}{dt^2}v + \frac{1}{R}\frac{d}{dt}v + \frac{1}{L}v$$

If we substitute $\mathbf{I}e^{j\omega t}$ and $\mathbf{V}e^{j\omega t}$ into this and solve for $\mathbf{I}/\mathbf{V}$ we get the answer:

$$\mathbf{Y} = \frac{\mathbf{I}}{\mathbf{V}} = \frac{C(j\omega)^2 + (1/R)j\omega + (1/L)}{j\omega} = j\omega C + \frac{1}{R} + \frac{1}{j\omega L}$$

Notice that the admittance of the parallel *RLC* network is given by the sum of the admittances of *R*, *L*, and *C*.

(b) For the given numerical values, we have

$$\mathbf{V} = 25\underline{/70^\circ} \text{ mV}$$

and

$$\mathbf{Y} = j10^4 \times 30 \times 10^{-6} + \frac{1}{10} + \frac{1}{j10^4 \times 10^{-3}} = 0.1 + j0.2 \text{ S}$$
$$= 0.224\underline{/63.4^\circ} \text{ S}$$

Therefore

$$\mathbf{I} = \mathbf{VY} = 5.6\underline{/133.4^\circ} \text{ mA}$$

and

$$i(t) = \mathcal{R}e\{(5.6\underline{/133.4^\circ})e^{j10^4 t}\}$$
$$= 5.6 \cos(10^4 t + 133.4^\circ) \text{ mA}$$

Remember

Only driving-point impedance and admittance satisfy $\mathbf{Z} = \mathbf{Y}^{-1}$. A network is called inductive, resistive, or capacitive if the angle of its driving-point impedance is positive, zero, or negative, respectively.

EXERCISES

16. Define the following terms: (a) driving-point impedance, (b) transfer impedance, (c) driving-point admittance, and (d) transfer admittance.

17. Define the following terms: (a) resistance, (b) reactance, (c) conductance, and (d) susceptance.
18. The driving-point admittance of a certain circuit is $\mathbf{Y} = 3 + j4$ S. Find the conductance, susceptance, resistance, and reactance. State whether the circuit is inductive, resistive, or capacitive.
answer: $G = 3$ S, $B = 4$ S, $R = 0.12\ \Omega$, $X = -0.16\ \Omega$. The circuit is capacitive.
19. Find the expressions for the driving-point resistance, reactance, conductance, susceptance, and admittance of a series *RLC* circuit.
answer: $R(\omega) = R$, $X(\omega) = \omega L - 1/\omega C$,
$G(\omega) = \omega^2 RC^2/[(1 - \omega^2 LC)^2 + (\omega RC)^2]$,
$B(\omega) = \omega C(1 - \omega^2 LC)/[(1 - \omega^2 LC)^2 + (\omega RC)^2]$.
20. Devise a circuit composed of two resistances R_1 and R_2 to illustrate that Eq. (10.30) applies for driving-point admittance and does *not* apply for transfer admittance.

10.6 Transfer Functions

Impedance, $\mathbf{Z}(j\omega)$, is an example of what is more generally referred to as a *transfer function.* Consider any stable LTI circuit that is driven by exactly one sinusoidal input. The steady-state response of this circuit is sinusoidal with the same frequency as the input. Therefore we can represent both the input and the output with the use of phasors. The transfer function gives us the ratio of the output phasor to the input phasor. The transfer function itself is given by a ratio of polynomials in $j\omega$ of the form $\mathcal{B}(j\omega)/\mathcal{A}(j\omega)$, where $\mathcal{B}(D)$ and $\mathcal{A}(D)$ are, respectively, the differential operators operating on the network's time-domain input and output. An arbitrary transfer function is usually denoted by $\mathbf{H}(j\omega)$, or more simply $\mathbf{H}$. Thus,

$$\mathbf{H} = \frac{\mathbf{B}}{\mathbf{A}} = \frac{\mathcal{B}(j\omega)}{\mathcal{A}(j\omega)} \tag{10.31}$$

where $\mathbf{B}$ denotes the output phasor ($\mathbf{V}$ or $\mathbf{I}$), and $\mathbf{A}$ denotes the input phasor ($\mathbf{V}$ or $\mathbf{I}$). The term *transfer function* can be applied to both impedance and admittance, but it is most often used when the input and the output have the same dimensions. Any transfer function can be determined either directly from Eq. (10.31) or, equivalently, by substitution of the appropriate complex exponentials into the circuit's LTI equation.

In Section 10.4 we saw that the driving-point impedance of a series *RLC* circuit is given by the sum of the impedances of *R*, *L*, and *C*. We also saw that the driving-point admittance of a parallel *RLC* circuit is given by the sum of the admittances of *R*, *L*, and *C*. These results are reminiscent of the rules for combining series and parallel resistances, and suggest that circuits described by impedances can be analyzed with the use of only algebra—just as circuits composed of resistances can be analyzed with the use of only algebra. Indeed, we shall see in Chapter 11 that this is the case: *any transfer function can be obtained directly from the frequency-domain circuit with the use of only algebra.* This method bypasses calculus completely, and is one of the great techniques of circuit theory.

Remember

A circuit's *transfer function* equals the ratio of the output phasor to the input phasor.

EXERCISES

21. (a) Show that for the circuit shown, $\mathcal{A}(D)v_o(t) = \mathcal{B}(D)v_s(t)$ where $\mathcal{A}(D) = D + R/L$ and $\mathcal{B}(D) = D$.
    (b) Use Eq. (10.31) and your answer to part (a) to find the voltage transfer function, $\mathbf{H}(j\omega) = \mathbf{V}_o/\mathbf{V}_s$.
    answer: $\mathbf{H}(j\omega) = \mathcal{B}(j\omega)/\mathcal{A}(j\omega) = j\omega/(j\omega + R/L)$
    (c) Find the steady-state voltage output, $v_o(t)$, for $R = 2\ \Omega$, $L = 0.1$ H, and $v_s(t) = 10 \cos 8t$ V.
    answer: $v_o(t) = 3.714 \cos (8t + 68.2°)$ V

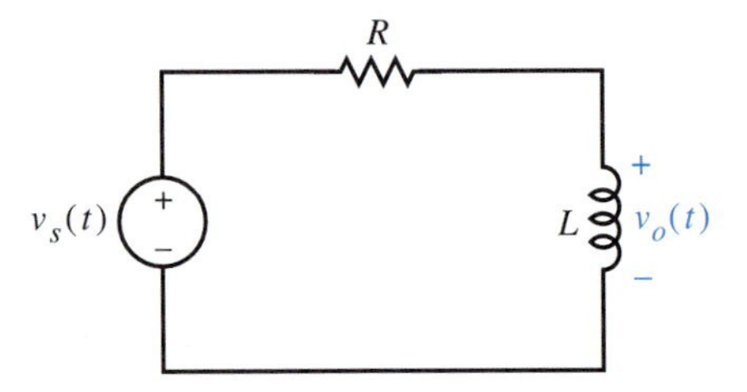

10.7 Summary

For *every* stable LTI circuit, a sinusoidal input produces a sinusoidal particular response. Since the natural response of a stable circuit always dies out as time increases, the complete response to a sinusoidal input will eventually be a sinusoid with the same frequency as the input. When this happens, the circuit has reached the sinusoidal steady state. The major topic of this chapter involved the use of phasors and transfer functions to determine the amplitude and the phase of the sinusoidal particular response. A phasor (for example, $\mathbf{I}$) is the complex amplitude of a complex exponential, as in $\mathbf{I}e^{j\omega t}$. If the complex exponential (or rotating phasor) $\mathbf{I}e^{j\omega t}$ is mathematically considered to be the input to a stable LTI circuit with input-output equation $\mathcal{A}(D)v(t) = \mathcal{B}(D)i(t)$, then the particular response is the rotating phasor $\mathbf{V}e^{j\omega t}$, where $\mathbf{V} = \mathbf{ZI}$ and $\mathbf{Z} = |\mathbf{Z}|\underline{/\theta} = \mathcal{B}(j\omega)/\mathcal{A}(j\omega)$ is the impedance. In practice, we are interested in a circuit's response to a real sinusoidal input, $I_m \cos (\omega t + \phi_\mathbf{I})$. By regarding this real input as the real part of $\mathbf{I}e^{j\omega t}$, we can obtain the real particular response by taking the real part of $\mathbf{V}e^{j\omega t}$, where $\mathbf{V} = \mathbf{ZI}$; that is, $\mathcal{R}e\{\mathbf{V}e^{j\omega t}\} = |\mathbf{Z}|I_m \cos (\omega t + \phi_\mathbf{I} + \theta)$. Thus, the sinusoidal input and output are projections of the rotating phasors onto the real axis. Figure 10.14 summarizes these key ideas. If you understand this figure and can apply the concepts involved to simple examples, then you have learned a great deal.

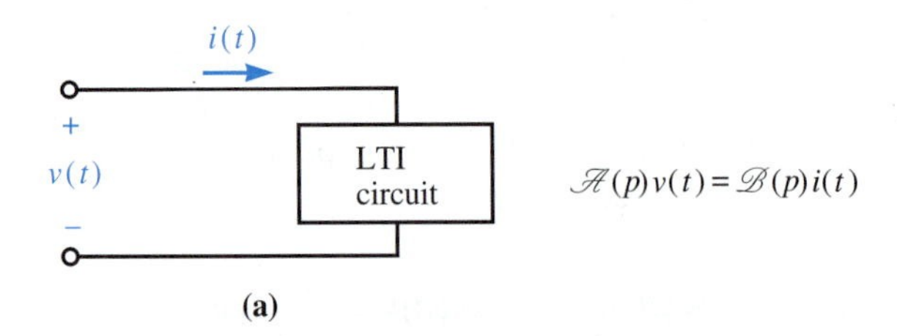

FIGURE 10.14 Evolution of the impedance concept. (a) Time-domain: The network is described by a linear differential equation with constant coefficients.

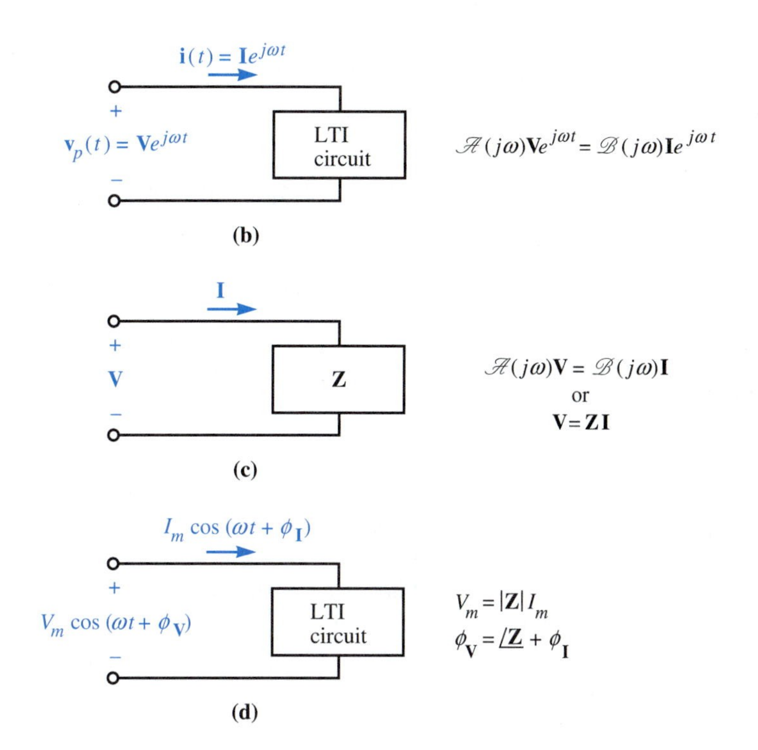

FIGURE 10.14 Continued (*b*) Particular response to input $\mathbf{I}e^{j\omega t}$: If $\mathbf{i}(t) = \mathbf{I}e^{j\omega t}$ then $\mathbf{v}_p(t) = \mathbf{V}e^{j\omega t}$ where $\mathbf{V} = \mathbf{ZI}$ and $\mathbf{Z} = \mathscr{B}(j\omega)/\mathscr{A}(j\omega)$. This solution requires that $\mathscr{A}(j\omega) \neq 0$, which is the case for every stable network. $\mathbf{Z}$ is the impedance of the network. (*c*) Frequency-domain: The relationship of $\mathbf{V}$ to $\mathbf{I}$ is indicated without $e^{j\omega t}$. (*d*) Particular response to sinusoidal inputs: The real part operator $\mathscr{R}e\{\ \}$ is applied to $\mathbf{I}e^{j\omega t}$ and $\mathbf{V}e^{j\omega t}$.

KEY FACTS

Concept	Equation	Section	Page
❑ The steady-state response of a stable LTI circuit to a sinusoidal input is a sinusoid with the same frequency as the input.		10.1	310
❑ A phasor, for example, $\mathbf{A}$, represents both the amplitude and the phase of a sinusoidal voltage or current. The magnitude of $\mathbf{A}$, $\lvert\mathbf{A}\rvert = A_m$, is the amplitude, and the angle of $\mathbf{A}$, $\angle\mathbf{A} = \phi_{\mathbf{A}}$, is the phase.	(10.2) (10.3)	10.2	311
❑ A cosine waveform is the projection of the rotating phasor $\mathbf{A}e^{j\omega t}$ onto the real axis of the complex plane: $A_m \cos(\omega t + \phi_{\mathbf{A}}) = \mathscr{R}e\{\mathbf{A}e^{j\omega t}\}$.		10.2	312
❑ A circuit's transfer function is given by $\mathbf{H} = \mathscr{B}(j\omega)/\mathscr{A}(j\omega)$, where $\mathscr{A}(D)$ and $\mathscr{B}(D)$ are the differential operators of the LTI equation relating excitation to response.	(10.31)	10.6	327
❑ The output phasor, for example, $\mathbf{B} = B_m\angle\phi_{\mathbf{B}}$, is obtained by multiplying the input phasor $\mathbf{A}$ by the transfer function: $\mathbf{B} = \mathbf{HA}$. Thus $B_m = \lvert\mathbf{H}\rvert A_m$ and $\phi_{\mathbf{B}} = \angle\mathbf{H} + \phi_{\mathbf{A}}$.	(10.31)	10.6	327

KEY FACTS	*Concept*	*Equation*	*Section*	*Page*
❑	When the input is $A_m \cos(\omega t + \phi_{\mathbf{A}}) = \mathcal{Re}\{\mathbf{A}e^{j\omega t}\}$, the steady-state output of a stable LTI circuit is $B_m \cos(\omega t + \phi_{\mathbf{B}}) = \mathcal{Re}\{\mathbf{B}e^{j\omega t}\}$, where $B_m = \lvert\mathbf{H}\rvert A_m$ and $\phi_{\mathbf{B}} = \angle\mathbf{H} + \phi_{\mathbf{A}}$.		10.4	317
❑	Impedance is the transfer function: $\mathbf{Z} = \mathbf{V}/\mathbf{I}$		10.5	323
❑	Admittance is the transfer function: $\mathbf{Y} = \mathbf{I}/\mathbf{V}$		10.5	323
❑	Impedance $\mathbf{Z} = R(\omega) + jX(\omega)$, where $R(\omega)$ is the resistance and $X(\omega)$ is the reactance.		10.5	323
❑	Admittance $\mathbf{Y} = G(\omega) + jB(\omega)$, where $G(\omega)$ is the conductance and $B(\omega)$ is the susceptance.		10.5	323
❑	Driving-point admittance $\mathbf{Y} = G(\omega) + jB(\omega)$ is related to driving-point impedance $\mathbf{Z} = R(\omega) + jX(\omega)$ by $\mathbf{Y} = \mathbf{Z}^{-1}$. The relation $\mathbf{Y} = \mathbf{Z}^{-1}$ does not hold for transfer admittances and impedances.	(10.30)	10.5	324

PROBLEMS

Section 10.1

1. Explain in your own words the meaning of the term *sinusoidal steady state.*

The natural response of a circuit is the solution to the homogeneous equation $\mathcal{A}(D)v_c(t) = 0$, for which the input is set to zero. Reasoning from physical grounds, we might expect that *every* physical circuit made up exclusively of *R*s, *L*s, and *C*s would be stable. [With the input set to zero, any energy initially present in the circuit should gradually be dissipated by the resistances; thus $v_n(t) \to 0$ as $t \to \infty$.] For the circuits shown in Problems 2 through 4, find $v_c(t)$ and discuss this idea. In Problem 4, consider three possibilities: $\beta < 1$, $\beta = 1$, and $\beta > 1$.

2.

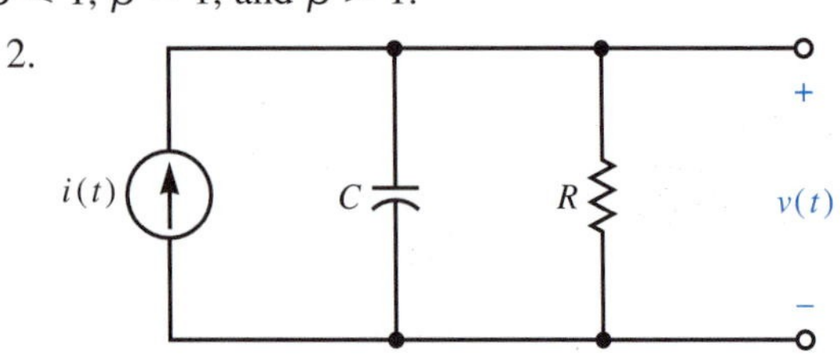

3.

4.

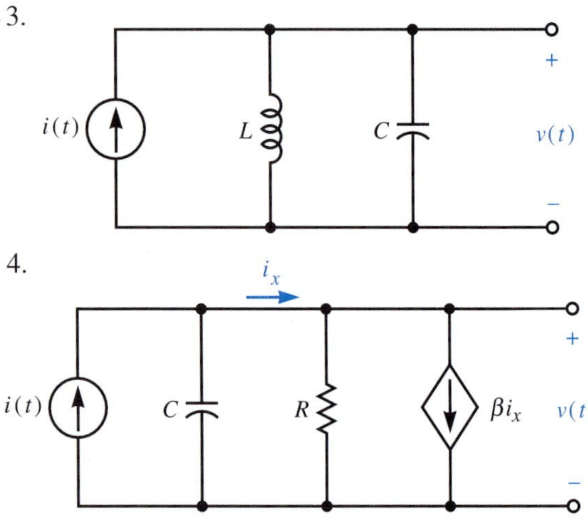

5. We know that for every LTI circuit, the output is related to the input by an equation of the form $\mathcal{A}(D)v = \mathcal{B}(D)i$. Each side of this equation is a weighted sum of derivatives.
   (a) Show that if you differentiate a sinusoid, $A\cos(\omega t + \phi)$, the result is another sinusoid,

$A' \cos(\omega t + \phi')$, having the same frequency, ω.

(b) Use trigonometric identities to show that the sum of any two sinusoids having the same frequency, ω, is another sinusoid that has frequency ω.

(c) Show that parts (a) and (b) imply that the particular response of an LTI circuit to a sinusoidal input is a sinusoid having the same frequency as the input. Are there any restrictions to this result?

Section 10.2

6. Let $\mathcal{A}(D) = a_n D^n + a_{n-1} D^{n-1} + \cdots + a_1 D + a_0$, where D denotes d/dt. An engineer claims that "since $\mathcal{A}(D)e^{j\omega t} = \mathcal{A}(j\omega)e^{j\omega t}$, by canceling the $e^{j\omega t}$'s it follows that $\mathcal{A}(D) = \mathcal{A}(j\omega)$." The engineer cannot possibly be correct, a differential operator is not a complex constant. What is wrong with the reasoning?

7. Give a detailed derivation of Eq. (10.9).

8. Think about the trigonometric function $i(t) = \sqrt{3}\cos(\omega t + 30°)$.

   (a) Use Euler's identity to prove that $i(t) = \mathcal{Re}\{\mathbf{I}e^{j\omega t}\}$ where $\mathbf{I} = \sqrt{3}\underline{/30°}$.

   (b) What is the rectangular form for the complex number $\sqrt{3}\underline{/30°}$?

   (c) Substitute your answer from part (b) into $\mathcal{Re}\{\mathbf{I}e^{j\omega t}\}$ and use Euler's identity to show that $\sqrt{3}\cos(\omega t + 30°) = \frac{3}{2}\cos\omega t - \frac{\sqrt{3}}{2}\sin\omega t$.

9. This problem generalizes the results of Problem 8 to $i(t) = I_m \cos(\omega t + \phi_\mathrm{I})$.

   (a) Use Euler's identity to prove that $i(t) = \mathcal{Re}\{\mathbf{I}e^{j\omega t}\}$ where $\mathbf{I} = I_m\underline{/\phi_\mathrm{I}}$.

   (b) What is the rectangular form for the complex number $I_m\underline{/\phi_\mathrm{I}}$?

   (c) Substitute your answer from part (b) into $\mathcal{Re}\{\mathbf{I}e^{j\omega t}\}$ and use Euler's identity to prove that $I_m \cos(\omega t + \phi_\mathrm{I}) = I_m \cos\phi_\mathrm{I} \cos\omega t - I_m \sin\phi_\mathrm{I} \sin\omega t$.

10. (a) Find the differential equation relating input $i(t)$ and output $v(t)$ for the circuit shown.

   (b) Use your answer to part (a) to write down the expressions for $\mathcal{A}(D)$ and $\mathcal{B}(D)$.

   (c) Use rotating phasors to derive the frequency-domain version of the differential equation of part (a). Your result should have the form $\mathcal{A}(j\omega)\mathbf{V} = \mathcal{B}(j\omega)\mathbf{I}$. *Hint:* See Eqs. (10.8) to (10.10).

   (d) Use your answer to part (c) to obtain the impedance seen by the source.

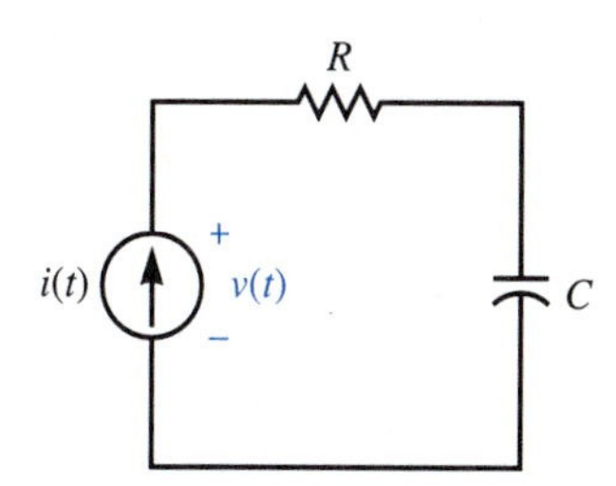

11. (a) Find the expression for the impedance of a parallel RL circuit. Put your result in rectangular form.

   (b) Assume that $R = 1\ \mathrm{k\Omega}$ and $L = 1$ mH. At what value of ω does $\underline{/\mathbf{Z}} = 45°$?

Section 10.3

12. A current $i(t) = 3\cos(1000t + 15°)$ mA is applied to a 2 mH inductance.

   (a) Find the current phasor, $\mathbf{I}$.

   (b) Find the impedance, $\mathbf{Z}$, of the inductance.

   (c) Find the voltage phasor $\mathbf{V} = \mathbf{ZI}$. Put your answer in polar form.

   (d) A student observes that since $\mathbf{V}$ is a phasor, its magnitude and angle are the amplitude and the phase, respectively, of the steady-state, sinusoidal voltage waveform, $v(t)$, across the inductance. Use this observation to write down the expression for $v(t)$.

   (e) Check your answer to part (d) by direct evaluation of the time-domain differential equation of the inductance: $v(t) = L\,di/dt$.

13. Repeat Problem 12 for a 1-kΩ resistance.

14. Repeat Problem 12 for a 0.001-μF capacitance.

Section 10.4

15. Phasors $\mathbf{V}$ and $\mathbf{I}$ are related by $\mathbf{V} = \mathbf{ZI}$. Find the corresponding real sinusoidal voltage $v(t)$ and current $i(t)$ if

   (a) $\mathbf{I} = 17.5\underline{/22°}$ mA, $\mathbf{Z} = 200\underline{/-60°}\ \Omega$ and $\omega = 100$ rad/s.

   (b) $\mathbf{I} = 10 + j10$ mA, $\mathbf{Z} = 100 + j100\ \mathrm{k\Omega}$ and $\omega = 150$ krad/s.

16. A certain device has input $i(t) = 3\cos(10t + 12.5°)$ A and output $v(t) = 6\cos(20t + 37°)$ V.

Find the impedance, if possible. If the impedance does not exist, give the reason.

17. (a) Show that $\mathcal{Re}\{\mathcal{A}(D)\mathbf{v}(t)\} = \mathcal{A}(D)\mathcal{Re}\{\mathbf{v}(t)\}$ if $\mathcal{A}(D) = 3D + 2$, $D = d/dt$, and $\mathbf{v}(t)$ is *any* complex function. [*Hint:* Write $v(t)$ in the rectangular form: $\mathbf{v}(t) = v_R(t) + jv_I(t)$.]
    (b) Generalize your analysis in (a) by assuming that $\mathcal{A}(D) = a_nD^n + a_{n-1}D^{n-1} + \cdots + a_0$, where the coefficients are arbitrary *real* numbers.
    (c) Show that $\mathcal{Re}\{\mathcal{B}(D)\mathbf{i}(t)\} = \mathcal{B}(D)\mathcal{Re}\{\mathbf{i}(t)\}$ if $\mathbf{i}(t)$ is *any* complex function, $\mathcal{B}(D) = b_mD^m + b_{m-1}D^{m-1} + \cdots + b_0$ and the coefficients are arbitrary *real* numbers.
    (d) What do the results of (b) and (c) have to do with Eq. (10.29)?
18. A linear time-invariant network has a current input $3 \cos(\omega t + 50°)$ A and a voltage output $9 \cos(\omega t + 85°)$ V. Find the associated impedance.
19. If the voltage output of the network in Problem 18 is $-3 \cos \omega t$ V, what is the input? Put your answer in the form $A \cos(\omega t + \phi)$.

Section 10.5

Refer to Table 10.1 for Problems 20 through 23.

20. Use the equation $\mathbf{Y} = \mathbf{Z}^{-1}$ to derive the expressions shown for $G(\omega)$ and $B(\omega)$.
21. Use the equation $\mathbf{Z} = \mathbf{Y}^{-1}$ to derive the expressions shown for $R(\omega)$ and $X(\omega)$.
22. Show that $R(\omega)G(\omega) = 1$ if and only if $X(\omega) = 0$.
23. Show that $B(\omega) = 0$ if and only if $X(\omega) = 0$.
24. Plot the impedance angle of a series RLC circuit versus ω, for $R = 1\ \Omega$, $L = 1$ mH, and $C = 1000\ \mu$F. Indicate the frequencies for which the circuit is (a) capacitive, (b) resistive, and (c) inductive.
25. Complete the table of driving-point quantities. The units are ohms and siemens. (Express **Z** and **Y** in polar form.)

	R	X	G	B	**Z**	**Y**
(a)	1	0				
(b)			1	0		
(c)	1	1				
(d)			1	1		

Section 10.6

26. (a) Derive the voltage transfer function $\mathbf{V}_o/\mathbf{V}_s$ of the following circuit from the appropriate differential equation. Does your result have the form of a voltage divider relation with resistances replaced by impedances?

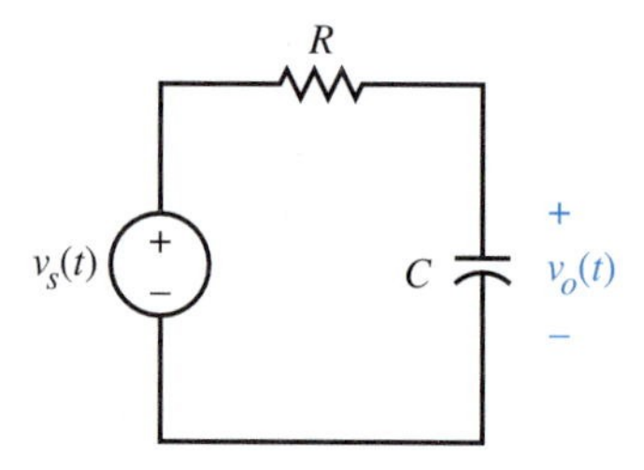

    (b) Use your results from part (a) to determine the steady-state response $v_o(t)$ in the preceding circuit for $v_s(t) = V_m \cos(\omega t + \phi_v)$.
27. (a) Find the current transfer function $\mathbf{I}_o/\mathbf{I}_s$ of the circuit shown below from the appropriate differential equation. Comment on any interesting feature of your result.

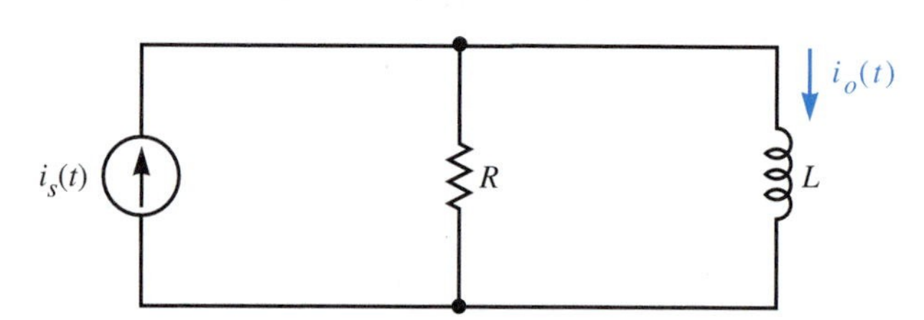

    (b) Use your answer to part (a) to determine the steady-state response $i_o(t)$ in the preceding circuit for $i_s(t) = I_m \cos(\omega t + \phi_I)$.
28. (a) What is the differential equation that relates the output $v_o(t)$ to the input $v_s(t)$ for the circuit shown below?

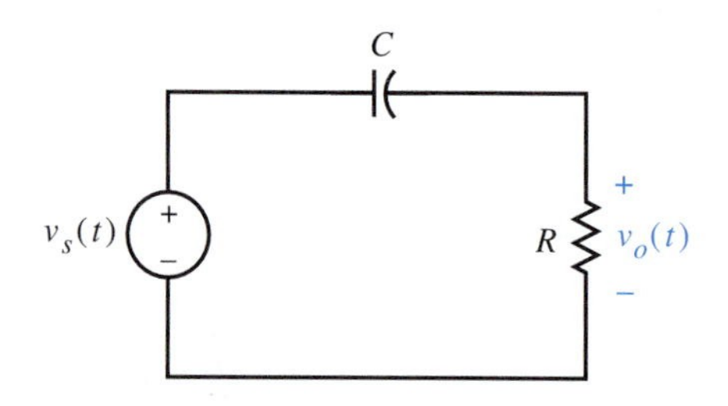

    (b) Use your answer to (a) to obtain the transfer function

$$\mathbf{H}(j\omega) = \frac{\mathbf{V}_o}{\mathbf{V}_s}$$

    (c) Give the magnitude and the angle of $\mathbf{H}(j\omega)$.

(d) If $\mathbf{v}_s(t) = \mathbf{V}_s e^{j\omega t}$, what is $\mathbf{v}_o(t)$?

(e) If $v_s(t) = V_{sm} \cos(\omega t + \phi_s)$, what is $v_o(t)$?

29. (a) Find the differential equation relating input $v_s(t)$ and output $v_o(t)$ for the circuit shown below.

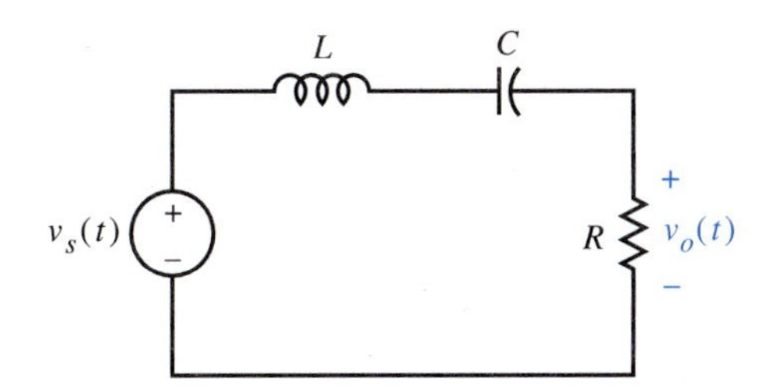

(b) Find the voltage transfer function.

(c) Find $v_{op}(t)$ if $L = 2$ H, $C = 1$ F, $R = 1\ \Omega$, and $v_s(t) = 100 \cos(t + 15°)$ mV.

Supplementary Problems

30. Complete Table 10.2.

31. Assuming that Table 10.2 refers to *driving-point* quantities, which of the networks for Problem 30 (rows a through h) is (*i*) capacitive, (*ii*) inductive, and (*iii*) resistive?

32. The switch closes at $t_0 = 0$ in the circuit shown below. A circuit designer wants the sinusoidal steady state to begin the instant the switch closes. Show how this is possible if the capacitance is given the right initial voltage $v_C(0^-)$, and state the value of $v_C(0^-)$.

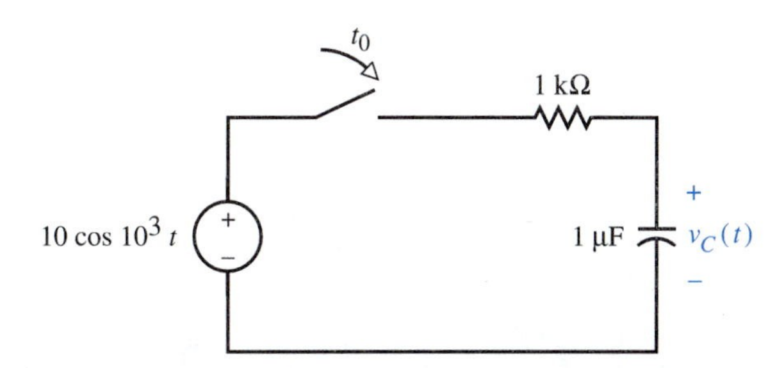

33. Determine the driving-point impedance $\mathbf{Z}_{ab}$ of the following circuit if $\omega = 2$ rad/s.

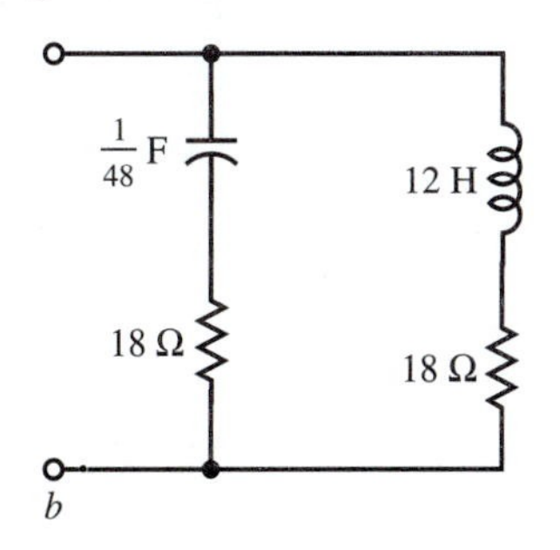

TABLE 10.2

	I ma	**Z** kΩ	*i*(*t*) ma	*v*(*t*) V
(a)	$10\angle 30°$	$5\angle 15°$		
(b)	$3\angle -25$	$1 + j$		
(c)	$3 + j4$	$1 + j$		
(d)			$2 \cos(10t + 20°)$	$6 \cos(10t + 65°)$
(e)		$10\angle 60°$	$2 \cos(50t + 20°)$	
(f)		$10\angle 60°$	$2 \sin(25t + 20°)$	
(g)		$\dfrac{100}{100 + j\omega}$	$2 \cos(100t + 20°)$	
(h)		$\dfrac{100}{100 + j\omega}$	$2 \cos(\omega t + 20°)$	

11

AC Circuit Analysis

When a stable, linear, time-invariant circuit is driven by a sinusoidal source, the particular response of every voltage and every current in the circuit is sinusoidal with the same frequency as the source. This fact implies that there is no need to use differential equations or calculus to find the voltage and current waveforms. Since all the waveforms are known to be sinusoidal with the same frequency, all that remains is to determine the amplitudes and phases. The amplitudes and phases are constants that can be calculated with the use of algebra.

The purpose of this chapter is to describe the algebraic techniques used to find the amplitudes and phases of sinusoidal steady-state response waveforms. We shall see that the algebraic circuit analysis techniques are simply reformulations of methods encountered earlier.

11.1 Foundations of AC Circuit Analysis

The term *ac circuit* is used to describe a linear, time-invariant circuit that is operating in the sinusoidal steady state. One of the most ingenious and useful ideas in circuit theory is to determine the amplitudes and phases of the sinusoidal waveforms in an ac circuit with the use of algebra—not calculus. The development of this algebra for ac circuit analysis involves the steps described in this section.

Frequency-Domain Independent Sources

The first step in the development of an algebraic system for ac circuit analysis was taken in Chapter 10, where we represented independent time-domain sources by corresponding phasors. The phasor sources are called *frequency-domain* sources and contain information equivalent to that contained in the corresponding time-domain sources. For example, the frequency-domain voltage source phasor $\mathbf{V}_s = V_{sm}\underline{/\phi_\mathbf{V}}$ represents the time-domain source

$$\begin{aligned} v_s(t) &= \mathcal{Re}\{\mathbf{V}_s e^{j\omega t}\} \\ &= V_{sm}\cos(\omega t + \phi_\mathbf{V}) \end{aligned} \tag{11.1}$$

As illustrated in Example 11.1, time-domain sources represented by a sine function must be converted to the cosine function with the use of the trigonometric identity $\sin x = \cos(x - 90°)$.

EXAMPLE 11.1 Figure 11.1a illustrates a linear network driven by three sinusoidal sources of the same frequency. Find and draw the frequency-domain equivalent.

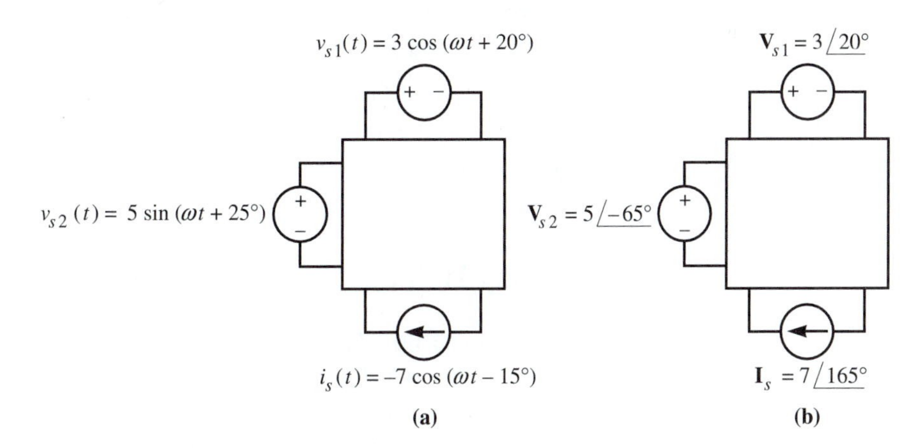

FIGURE 11.1 Time domain–frequency domain source transformations: (*a*) time domain; (*b*) frequency domain

Solution The time-domain voltage source $v_{s1}(t)$ of Fig. 11.1a is transformed into the frequency domain by noting that

$$\begin{aligned} v_{s1}(t) &= 3\cos(\omega t + 20°) \\ &= \mathcal{Re}\{3e^{j20°}e^{j\omega t}\} \end{aligned}$$

[as in Eq. (11.1)]. Therefore

$$\mathbf{V}_{s1} = 3\underline{/20°}$$

Similarly, for

$$i_s(t) = -7\cos(\omega t - 15^\circ)$$

we obtain

$$\mathbf{I}_s = -(7\angle -15^\circ) = 7\angle 165^\circ$$

The phasor representation of the third source is obtained by first converting the sine function into the cosine function by means of the identity

$$\sin(\omega t + \phi) = \cos(\omega t + \phi - 90^\circ)$$

Therefore

$$v_{s2}(t) = 5\sin(\omega t + 25^\circ)$$
$$= 5\cos(\omega t - 65^\circ)$$

and accordingly,

$$\mathbf{V}_{s2} = 5\angle -65^\circ$$

The frequency-domain representation of Fig. 11.1a is shown in Fig. 11.1b.

Notice that in Fig. 11.1a, the three independent sinusoidal sources all have the same frequency. If this were not the case, then Fig. 11.1b would have no meaning.

Remember

Use the identity $\sin x = \cos(x - 90^\circ)$ to convert all sine functions to cosine functions. In a frequency-domain representation like that of Fig. 11.1b, it is always necessary that the phasors refer to sinusoidal functions of time that have the same frequency.

EXERCISES

Give the phasor representation of the following sources.

1. $i(t) = 6\cos(1000t + 30^\circ)$ *answer:* $6\angle 30^\circ$
2. $v(t) = -5\cos(1000t + 30^\circ)$ *answer:* $5\angle -150^\circ$
3. $i(t) = 4\sin(1000t + 30^\circ)$ *answer:* $4\angle -60^\circ$
4. $v(t) = -3\sin(1000t + 30^\circ)$ *answer:* $3\angle 120^\circ$

Frequency-Domain Dependent Sources

Let us now extend the phasor source representation to include dependent sources. Consider Fig. 11.2a, which depicts a current-controlled voltage source.

In the sinusoidal steady state, $i(t)$ and $v(t)$ will be sinusoidal functions of t. To obtain the phasor representation of Fig. 11.2a, we replace $i(t)$ and $v(t)$ by rotating phasors:

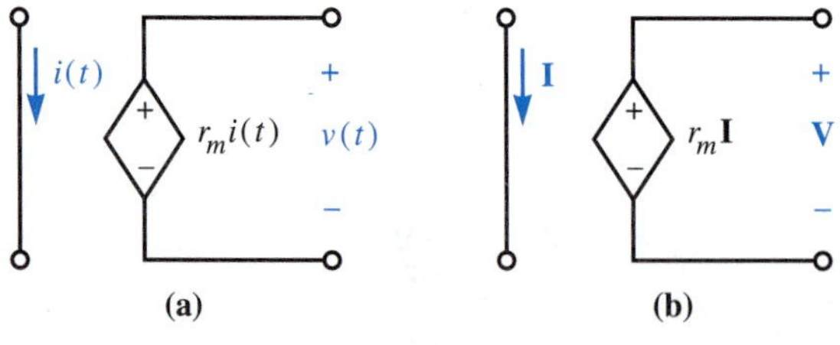

FIGURE 11.2
(*a*) Time-domain and (*b*) frequency-domain forms of a current-controlled voltage source

$$\mathbf{i}(t) = \mathbf{I}e^{j\omega t} \tag{11.2}$$

and

$$\mathbf{v}(t) = r\mathbf{i}(t)$$
$$= r\mathbf{I}e^{j\omega t} \tag{11.3}$$

Now Eq. (11.3) can be rewritten as

$$\mathbf{v}(t) = \mathbf{V}e^{j\omega t} \tag{11.4}$$

where

$$\mathbf{V} = r\mathbf{I} \tag{11.5}$$

We obtain the phasor or *frequency-domain* dependent source from this equation. The corresponding circuit symbol is shown in Fig. 11.2b. Results for the remaining types of dependent sources are obtained in the same way. Figure 11.3 summarizes the results. Notice that dependent sources work the same way in the frequency domain as they do in the time domain.

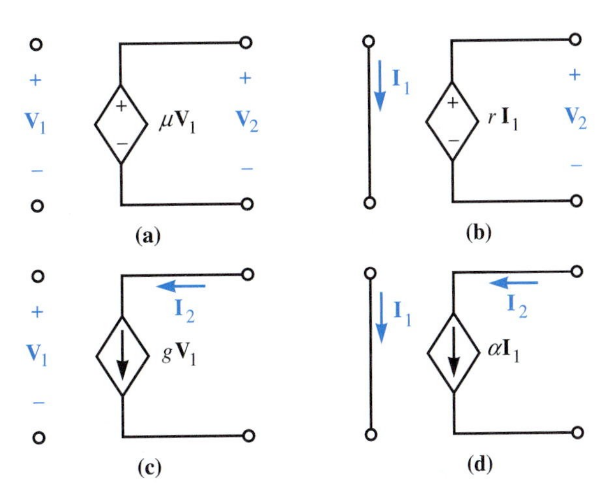

FIGURE 11.3 Frequency-domain versions of dependent sources: (a) voltage-controlled voltage source; (*b*) current-controlled voltage source: (*c*) voltage-controlled current source; (*d*) current-controlled current source

Remember Dependent sources work the same way in the frequency domain as they do in the time domain.

EXERCISE 5. Derive the entries in Fig. 11.3.

Frequency-Domain Terminal Equations for *R*, *L*, and *C*

The next step in our development is to use the rotating phasors $\mathbf{I}e^{j\omega t}$ and $\mathbf{V}e^{j\omega t}$ to transform the time-domain equations of the LTI network elements into frequency-domain equations. This step was taken in Chapter 10, where we obtained the frequency-domain equations for *R*, *L*, and *C* shown in Fig. 11.4. Each of the frequency-domain equations is an expression of the ac version of Ohm's law:

$$\mathbf{V} = \mathbf{Z}(j\omega)\mathbf{I} \tag{11.6}$$

or

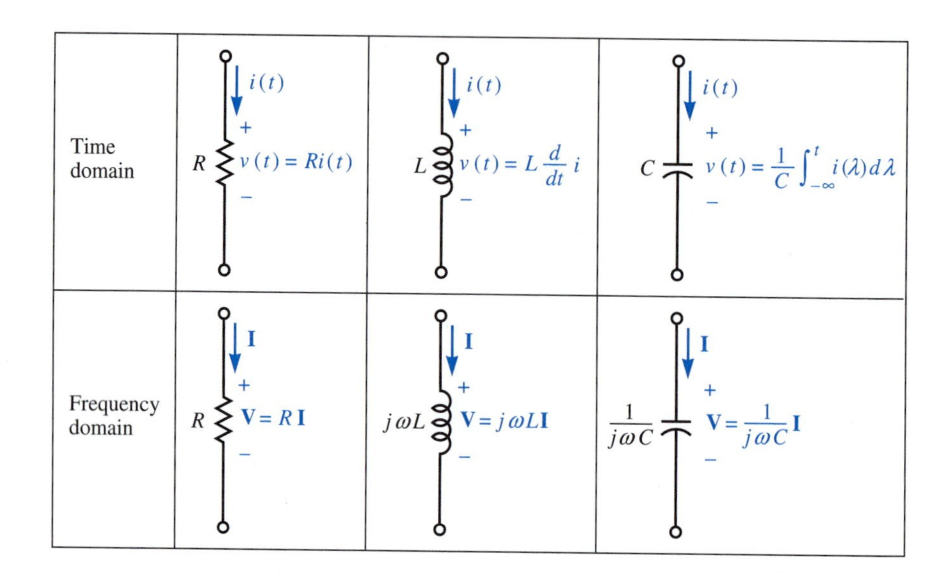

FIGURE 11.4 Time-domain and frequency-domain terminal equations for R, L, and C

$$\mathbf{I} = \mathbf{Y}(j\omega)\mathbf{V} \tag{11.7}$$

$\mathbf{Z}(j\omega)$ and $\mathbf{Y}(j\omega)$ are, respectively, the impedance and the admittance of the element, and $\mathbf{V} = V_m\underline{/\phi_\mathbf{V}}$ and $\mathbf{I} = I_m\underline{/\phi_\mathbf{I}}$ are voltage and current phasors. The real waveforms associated with Eq. (11.6) are the projections of the rotating phasors $\mathbf{I}e^{j\omega t}$ and $\mathbf{V}e^{j\omega t}$ onto the real axis:

$$i(t) = \mathcal{Re}\{\mathbf{I}e^{j\omega t}\} = I_m \cos(\omega t + \phi_\mathbf{I}) \tag{11.8}$$

and

$$\begin{aligned} v(t) &= \mathcal{Re}\{\mathbf{V}e^{j\omega t}\} \\ &= \mathcal{Re}\{\mathbf{Z}(j\omega)\mathbf{I}e^{j\omega t}\} \\ &= \underbrace{|\mathbf{Z}(j\omega)|I_m}_{V_m} \cos(\omega t + \underbrace{\underline{/\mathbf{Z}(j\omega)} + \phi_\mathbf{I}}_{\phi_\mathbf{V}}) \end{aligned} \tag{11.9}$$

Similar results apply for Eq. (11.7).

The Frequency-Domain Forms of KCL and KVL

We are now ready to take the final step in the development of the algebraic system for ac circuit analysis. *This step is to express Kirchhoff's current and voltage laws in terms of current and voltage phasors.* The results of this step are called *frequency-domain* or *phasor* versions of KCL and KVL. The reexpressed laws have exactly the same form as KCL and KVL for time-domain currents and voltages, with the exception that the currents and voltages are specified by phasors.

Recall that Kirchhoff's current law states that the algebraic sum of all the currents leaving any closed surface in a network is zero. That is,

$$\sum_{n=1}^{N} i_n(t) = 0 \tag{11.10}$$

where $i_n(t)$ is the nth current of the N currents leaving the closed surface at time t. In the sinusoidal steady state, we know that each of the currents $i_n(t)$, $1 \le n \le N$, is sinusoidal and has the same frequency. Therefore

$$i_n(t) = I_{mn} \cos(\omega t + \phi_n) \tag{11.11}$$

for $n = 1, 2, \ldots, N$, where the I_{mn}'s and ϕ_n's are the amplitudes and phases of the currents. Therefore, in the sinusoidal steady state, KCL, Eq. (11.10), becomes

$$\sum_{n=1}^{N} I_{mn} \cos(\omega t + \phi_n) = 0 \tag{11.12}$$

If we write each sinusoidal current in the sum (11.12) as the real part of a rotating phasor we obtain

$$\sum_{n=1}^{N} \mathcal{R}e\{\mathbf{I}_n e^{j\omega t}\} = 0 \tag{11.13}$$

We can interchange the summation and the real part operation in Eq. (11.13) to obtain

$$\mathcal{R}e\left\{\sum_{n=1}^{N} \mathbf{I}_n e^{j\omega t}\right\} = \mathcal{R}e\left\{\left(\sum_{n=1}^{N} \mathbf{I}_n\right) e^{j\omega t}\right\} = 0 \tag{11.14}$$

where the phasor $\mathbf{I}_n$ has magnitude I_{mn} and angle ϕ_n. The sum $\Sigma_{n=1}^{N}\mathbf{I}_n$ appearing in Eq. (11.14) equals some constant $\mathbf{K}$.

$$\sum_{n=1}^{N} \mathbf{I}_n = \mathbf{K} \tag{11.15}$$

Therefore, we can write Eq. (11.14) as

$$\mathcal{R}e\{\mathbf{K}e^{j\omega t}\} = 0 \tag{11.16}$$

Our problem now is to find the constant $\mathbf{K}$ that satisfies Eq. (11.16). The solution becomes obvious if we mentally picture $\mathbf{K}e^{j\omega t}$ as a rotating phasor: The real part of a rotating phasor equals zero for all t if and only if the phasor itself equals zero. Therefore, $\mathbf{K} = 0$ and we have the following result:

Frequency-Domain Form of Kirchhoff's Current Law

$$\sum_{n=1}^{N} \mathbf{I}_n = 0 \tag{11.17}$$

where $\mathbf{I}_n$ is the nth current phasor of the N current phasors leaving the closed surface.

The frequency-domain form of KCL states that the sum of all the phasor currents exiting any closed surface is zero. Equation (11.17) is also referred to as the *phasor form of KCL,* or simply as KCL.

EXAMPLE 11.2 Figure 11.5a illustrates an $N = 3$-branch node with known currents exiting via branches 1 and 2. The known currents are

$$i_1(t) = 5 \cos(\omega t + 30°)$$

and

$$i_2(t) = 5 \cos(\omega t + 150°)$$

Use the phasor form of KCL to determine $i_3(t)$.

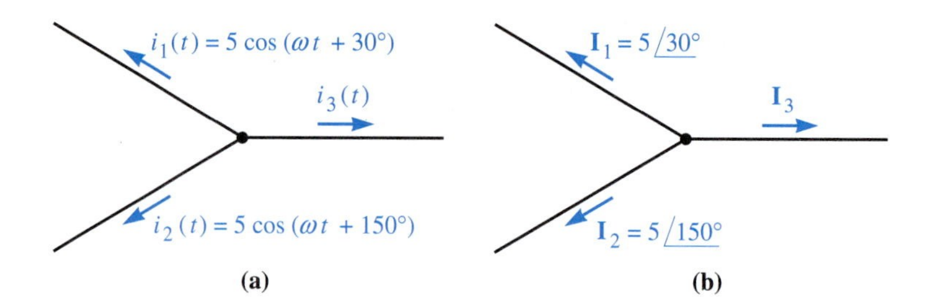

FIGURE 11.5
(*a*) Time-domain and (*b*) frequency-domain currents

Solution The currents $i_1(t)$, $i_2(t)$, and $i_3(t)$ are represented by phasors $\mathbf{I}_1$, $\mathbf{I}_2$, and $\mathbf{I}_3$, respectively, where $\mathbf{I}_1 = 5\angle 30°$ and $\mathbf{I}_2 = 5\angle 150°$. According to the phasor form of KCL, Eq. (11.17),

$$5\angle 30° + 5\angle 150° + \mathbf{I}_3 = 0$$

Therefore

$$\begin{aligned}\mathbf{I}_3 &= -5\angle 30° - 5\angle 150° \\ &= -(4.33 + j2.5) - (-4.33 + j2.5) \\ &= -j5 \\ &= 5\angle -90°\end{aligned}$$

The phasors are shown in Fig. 11.6. The time-domain current represented by $\mathbf{I}_3$ is

$$\begin{aligned}i_3(t) &= \mathcal{Re}\{\mathbf{I}_3 e^{j\omega t}\} \\ &= 5\cos(\omega t - 90°)\end{aligned}$$

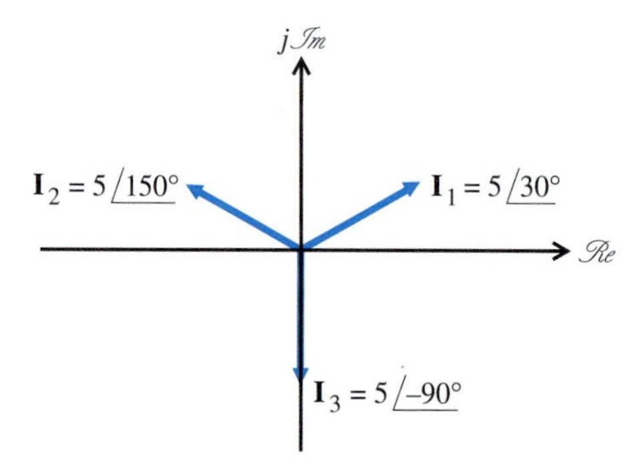

FIGURE 11.6
The current phasors sum to zero (KCL)

We emphasize that KCL indicates that the sum of all current *phasors* leaving any closed surface is zero. It does *not* follow that the peak or rms values of current sum to zero. For instance, in the above example, phasors $5\angle 30°$, $5\angle 150°$, and $5\angle -90°$ sum to zero, but the peak currents, 5, 5, and 5 do not.

Let us now consider Kirchhoff's voltage law. Kirchhoff's voltage law states that the algebraic sum of all voltage drops taken around any closed path in a network is zero. That is,

$$\sum_{n=1}^{N} v_n(t) = 0 \tag{11.18}$$

where $v_n(t)$ is the voltage drop, taken in the direction of the path along the nth segment of the N segments in the closed path. In the sinusoidal steady state, each of the voltages $v_n(t)$, $1 \le n \le N$, will be sinusoidal. At this point we recognize an obvious similarity between Eqs. (11.18) and (11.10). There is no need to repeat the derivation. If we follow reasoning similar to that which led to Eq. (11.17), we will obtain the following result:

Frequency-Domain Form of Kirchhoff's Voltage Law

$$\sum_{n=1}^{N} \mathbf{V}_n = 0 \tag{11.19}$$

where $\mathbf{V}_n$ is the phasor voltage drop, taken in the direction of the path, along the nth segment of the N segments in the closed path.

The frequency-domain form of KVL states that the sum of all the phasor voltage drops around a closed path is zero. Equation (11.19) is also referred to as the *phasor form of KVL,* or simply as KVL.

EXAMPLE 11.3 Figure 11.7 illustrates an $N = 4$-branch mesh with known voltage drops across three of its four components. These known voltages are given by the time-domain expressions

$$v_1(t) = \cos \omega t$$
$$v_2(t) = -2 \cos (\omega t + 45°)$$
$$v_3(t) = -3 \sin \omega t$$

Use the phasor form of KVL to determine $v_4(t)$.

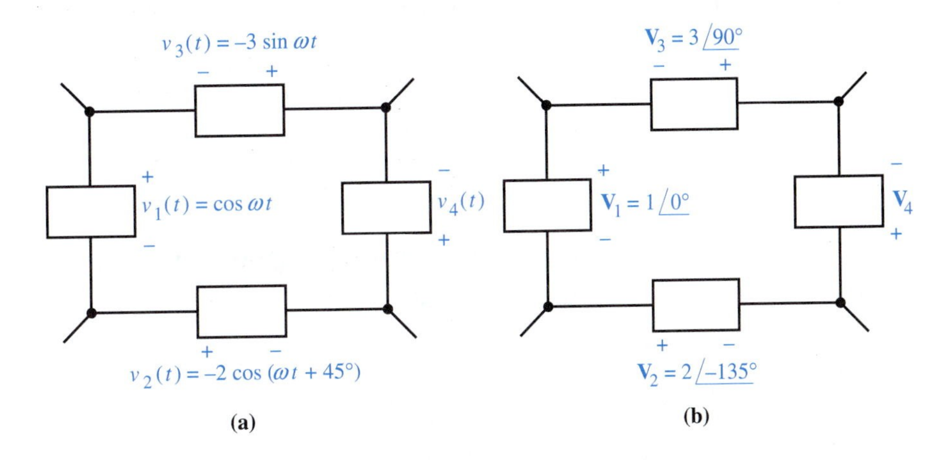

FIGURE 11.7 (*a*) Time-domain and (*b*) frequency-domain voltage drops

Solution We represent the voltages $v_1(t)$, $v_2(t)$, $v_3(t)$, and $v_4(t)$, respectively, as the real parts of the complex exponentials $\mathbf{V}_1 e^{j\omega t}$, $\mathbf{V}_2 e^{j\omega t}$, $\mathbf{V}_3 e^{j\omega t}$, and $\mathbf{V}_4 e^{j\omega t}$, where $\mathbf{V}_1 = 1\underline{/0°}$, $\mathbf{V}_2 = 2\underline{/-135°}$, and $\mathbf{V}_3 = 3\underline{/+90°}$. The phasor form of KVL, Eq. (11.19), tells us that

$$1\underline{/0°} + 2\underline{/-135°} + 3\underline{/+90°} + \mathbf{V}_4 = 0$$

and therefore

$$\begin{aligned}\mathbf{V}_4 &= -(1 + j0) - (-\sqrt{2} - j\sqrt{2}) - (0 + j3) \\ &= 0.4142 - j1.5858 \\ &= 1.6389\underline{/-75.36°}\end{aligned}$$

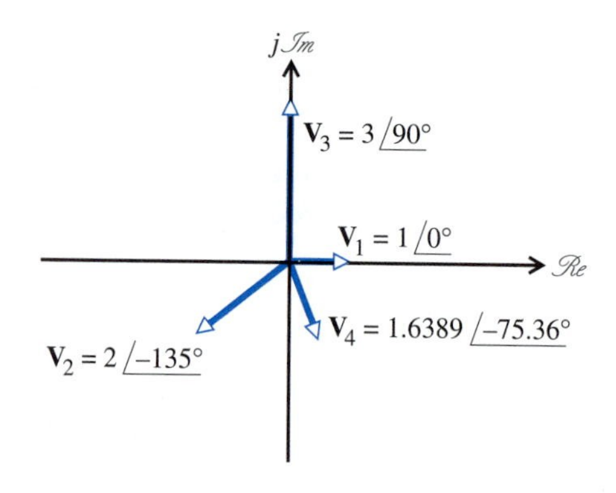

FIGURE 11.8
The voltage phasors sum to zero (KVL)

The phasors are shown in Fig. 11.8. The time-domain current represented by $\mathbf{V}_4$ is

$$\begin{aligned} v_4(t) &= \mathcal{Re}\{\mathbf{V}_4 e^{j\omega t}\} \\ &= 1.6389 \cos(\omega t - 75.36^\circ) \end{aligned}$$

Notice that KVL indicates that the sum of the *phasor* voltage drops around a closed path is zero. It does *not* indicate that the peak or rms values of voltage sum to zero. After all, peak and rms values are always nonnegative, so the *only* way they *could* sum to zero would be for each and every peak and rms value to be zero.

Remember

KCL and KVL work the same way in the frequency domain as they do in the time domain.

EXERCISE

6. Three phasor currents satisfy $\mathbf{I}_1 = \mathbf{I}_2 + \mathbf{I}_3$. Express I_{m1} in terms of I_{m_2}, I_{m3}, $\underline{/\mathbf{I}_2}$, and $\underline{/\mathbf{I}_3}$. (*Hint:* Draw the phasors $\mathbf{I}_1$, $\mathbf{I}_2$, and $\mathbf{I}_3$ as vectors in the complex plane and use trigonometry.)

answer: $I_{m1} = \sqrt{I_{m2}^2 + I_{m3}^2 - 2I_{m2}I_{m3}\cos(\phi_3 - \phi_2)}$, where $\phi_3 = \underline{/\mathbf{I}_3}$ and $\phi_2 = \underline{/\mathbf{I}_2}$

Relationship Between Time- and Frequency-Domain Circuits

We have noticed that the ac version of Ohm's law, $\mathbf{V} = \mathbf{ZI}$, has the same form as Ohm's law, $V = RI$. We also know that the frequency-domain versions of KCL and KVL have the same form as the time-domain versions of KCL and KVL. The profound implication of these facts is that we can use current and voltage phasors to analyze ac networks in the same way that we use dc current and voltage variables to analyze networks composed of resistances and dc sources. *This equivalence of ac and dc circuit analysis applies to the sinusoidal steady-state response of all LTI networks.*

A summary of the concepts underlying frequency-domain circuit analysis is given in Fig. 11.9. Figure 11.9a illustrates the time-domain representation of a simple circuit that has two sinusoidal sources with the same frequency ω that were connected a long time ago (mathematically, at $t = -\infty$). Because this is a stable LTI network, the complementary responses have decayed to zero, and all the voltages and currents are now sinusoidal with the same frequency as the sources. Thus, the circuit is operating in the sinusoidal steady state with frequency ω. The sinusoidal functions are indicated in Fig.

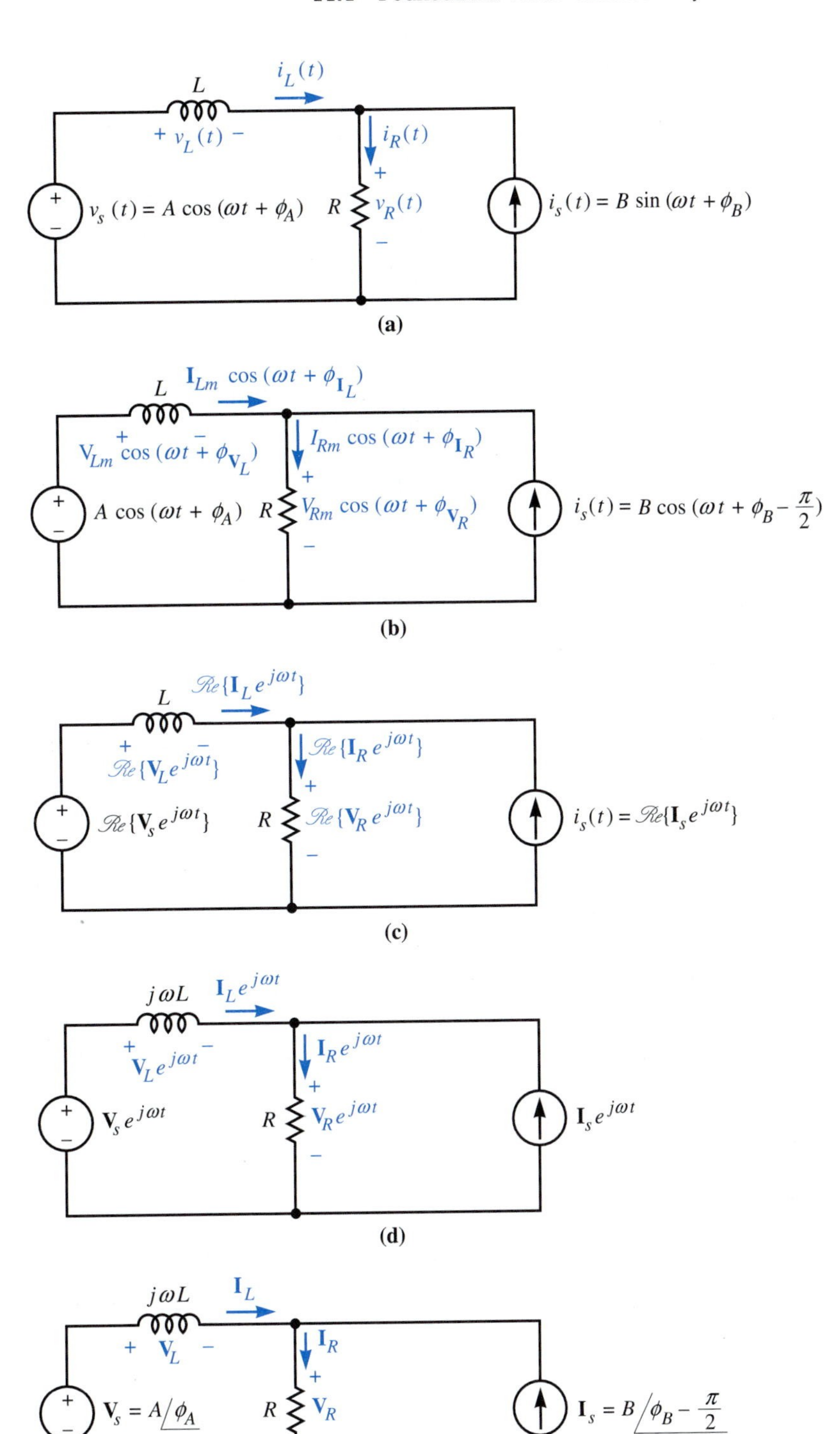

FIGURE 11.9 Illustration of time-frequency transformations in an ac circuit: (*a*) Circuit with two sinusoidal sources with same frequency, ω; (*b*) the circuit is operating in the sinusoidal steady state and all waveforms are expressed as cosine functions; (*c*) the cosine functions are expressed as the real parts of rotating phasors. The voltage-source and current-source phasors are $\mathbf{V}_s = A/\phi_A$ and $\mathbf{I}_s = B/\phi_B - 90°$, respectively. The response phasors are constants to be determined; (*d*) all voltages and currents are represented by rotating phasors; (*e*) all voltages and currents are represented by phasors.

11.9b as cosines, in which the amplitudes and phases are constants to be determined. Figure 11.9c shows that the cosine functions are the real parts of rotating phasors. Figure 11.9d shows the same circuit with the voltages and currents described by the rotating phasors. There is no physical meaning to these complex voltages and currents. We can progress from Fig. 11.9c to 11.9d because of *mathematical* properties of the operators $\mathcal{R}e\{\cdot\}$, $\mathcal{A}(D)$, and $\mathcal{B}(D)$ described earlier. Figure 11.9e is identical to Fig. 11.9d except that the $e^{j\omega t}$'s have been omitted. This corresponds to the cancellation of the $e^{j\omega t}$'s in all KCL, KVL, and element terminal equations in the network. Figure 11.9e is the frequency-domain representation of Fig. 11.9a.

When analyzing a circuit like that in Fig. 11.9a, you should understand the concepts underlying Fig. 11.9b through d, but draw only the result in Fig. 11.9e. The analysis can then be continued by application of the frequency-domain versions of KCL, KVL, and the component terminal equations.

EXAMPLE 11.4 Use the frequency-domain versions of KCL, KVL, and the element terminal equations to determine $i_L(t)$ in the circuit of Fig. 11.9a.

Solution Application of KCL at the top node in the network of Fig. 11.9e yields

$$\mathbf{I}_R = \mathbf{I}_L + \mathbf{I}_s$$

See Problem 11.1 in the PSpice manual.

Application of KVL yields

$$\mathbf{V}_R + \mathbf{V}_L = \mathbf{V}_s$$

The element terminal equations yield two more equations,

$$\mathbf{V}_L = j\omega L\mathbf{I}_L \qquad \text{and} \qquad \mathbf{V}_R = \mathbf{I}_R R$$

for a total of four equations in the four unknowns $\mathbf{I}_R$, $\mathbf{I}_L$, $\mathbf{V}_R$, and $\mathbf{V}_L$. These equations may be solved for $\mathbf{I}_L$. The result is

$$\mathbf{I}_L = \frac{\mathbf{V}_s}{R + j\omega L} - \frac{R\mathbf{I}_s}{R + j\omega L}$$

The corresponding time-domain solution is

$$i_L(t) = \mathcal{R}e\{\mathbf{I}_L e^{j\omega t}\} = I_{Lm}\cos(\omega t + \phi_{IL})$$

As a numerical example, assume that $\omega = 10^3$ rad/s, $R = 500\ \Omega$, $L = 0.5$ H, $A = 30$ V, $\phi_A = -25°$, $B = 20$ mA, and $\phi_B = 15°$. Then

$$\begin{aligned}\mathbf{I}_L &= \frac{30\angle -25°}{500 + j500} - \frac{10\angle 15° - 90°}{500 + j500} \\ &= \frac{24.601 - j3.019}{500 + j500} \\ &= 0.035\angle -52°\ \text{A}\end{aligned}$$

and

$$i_L(t) = 35\cos(1000t - 52°)\ \text{mA}$$

Example 11.4 was solved by direct application of KCL, KVL, and the element terminal equations. As in the time domain, this method is fundamental but often unnecessarily long. We will describe much faster methods later.

Before concluding the present topic, we notice that the transformation between time and frequency domains works both ways. Notice from Eq. (10.7) and (10.10) that the differential operator $D = d/dt$ enters a time-domain equation in exactly the same way that $j\omega$ enters a frequency-domain equation. In other words, differentiation in the time domain *always* corresponds to multiplication by $j\omega$ in the frequency domain. Therefore, if we start with a frequency-domain equation that has the form of Eq. (10.10), the corresponding time-domain equation can be derived simply by replacing $j\omega$ with $D = d/dt$ and phasors $\mathbf{V}$ or $\mathbf{I}$ with their time-domain counterparts $v(t)$ and $i(t)$. This process often provides an easy way to derive an LTI equation.

EXAMPLE 11.5 Find the differential equation relating $i_L(t)$ to the independent sources, $v_s(t)$ and $i_s(t)$, in the circuit of Fig. 11.9a.

Solution As we saw in Example 11.4, phasor $\mathbf{I}_L$ is related to phasors $\mathbf{V}_s$ and $\mathbf{I}_s$ by

$$\mathbf{I}_L = \frac{\mathbf{V}_s}{R + j\omega L} - \frac{R\mathbf{I}_s}{R + j\omega L}$$

If we substitute d/dt for $j\omega$, we obtain a mathematical jumble. However, by first multiplying both sides by $R + j\omega L$, we obtain

$$(R + j\omega L)\mathbf{I}_L = \mathbf{V}_s - R\mathbf{I}_s$$

The time-domain equation corresponding to this frequency-domain equation is

$$(R + LD)i_L(t) = v_s(t) - Ri_s(t)$$

or equivalently,

$$L\frac{d}{dt}i_L + Ri_L = v_s - Ri_s$$

The validity of this equation can be checked if we substitute in the complex exponentials $\mathbf{i}_L = \mathbf{I}_L e^{j\omega t}$, $\mathbf{v}_s = \mathbf{V}_s e^{j\omega t}$, and $\mathbf{i}_s = \mathbf{I}_s e^{j\omega t}$ and simplify. The result will be the original frequency-domain equation relating the phasors $\mathbf{I}_L$, $\mathbf{V}_s$, and $\mathbf{I}_s$. Of course, the validity can also be checked if we use the time-domain methods described in earlier chapters.

The axioms of network theory, KCL and KVL, have been shown to hold for phasors. The introduction of impedance has extended Ohm's law, $v(t) = Ri(t)$, to the phasor form $\mathbf{V} = \mathbf{ZI}$. It follows that any relationship in the time domain that is justified by KCL, KVL, and Ohm's law has an equivalent relationship in the frequency domain. *Therefore all the time-domain LTI network analysis techniques developed in earlier chapters can be reformulated in the frequency domain.*

In the following sections, we will accomplish this reformulation. This should serve as a review of many of the circuit analysis techniques we developed earlier.

Remember

Every frequency-domain concept and procedure we will encounter in this chapter is associated with a corresponding time-domain concept or procedure encountered earlier in this book.

EXERCISES

7. Explain in your own words the relationship between Fig. 11.9e and 11.9a.
8. The voltage transfer function of a certain circuit is given by

$$\mathbf{H}(j\omega) = \frac{3j\omega}{1 - \omega^2 + j\omega}$$

What is the differential equation relating the output voltage $v_o(t)$ to the input voltage $v_s(t)$? *Hint:* $-\omega^2 = (j\omega)(j\omega)$. answer: $(D^2 + D + 1)v_o = 3Dv_s$

11.2 Simple Circuits

Let us now apply the method of ac circuit analysis to some elementary circuits. In this section we consider only circuits that do not contain dependent sources. Ac circuits that do not contain dependent sources can often be analyzed by inspection. An ability to do this depends on knowledge of the simple circuits shown in Figs. 11.10 through 11.13. These circuits have already been drawn in the frequency domain, directing attention to the sinusoidal steady-state response of the corresponding time-domain circuits. The **Z**'s and **Y**'s represent arbitrary impedances and admittances.

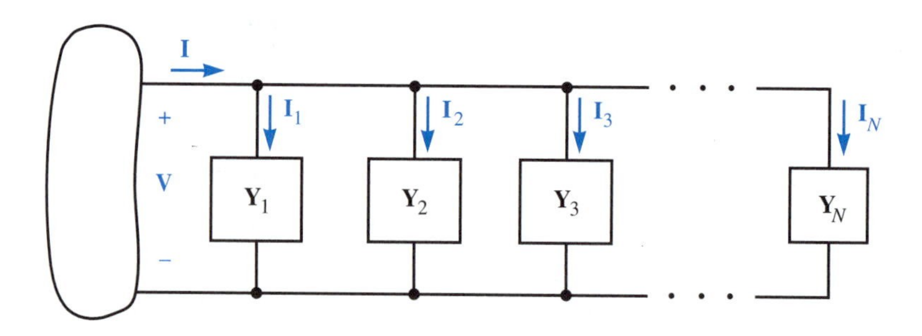

FIGURE 11.10 Admittances in parallel

Parallel Circuit

Consider the parallel circuit of Fig. 11.10. The voltage drop from the top to the bottom of each element of Fig. 11.10 is $+\mathbf{V}$ (by the phasor form of KVL). The current phasors $\mathbf{I}_n$ are related to $\mathbf{V}$ by $\mathbf{I}_n = \mathbf{Y}_n\mathbf{V}$, where $\mathbf{Y}_n$ is the admittance of the nth element. Application of the phasor form of KCL to the top node gives

$$\mathbf{I} = \mathbf{Y}_1\mathbf{V} + \mathbf{Y}_2\mathbf{V} + \cdots + \mathbf{Y}_N\mathbf{V} \tag{11.20}$$

or

$$\mathbf{I} = \mathbf{Y}_p\mathbf{V} \tag{11.21}$$

where $\mathbf{Y}_p$ is the equivalent admittance:

$$\mathbf{Y}_p = \mathbf{Y}_1 + \mathbf{Y}_2 + \cdots + \mathbf{Y}_N \tag{11.22}$$

This shows that *admittances that are in parallel add.* Expressed in terms of impedances $\mathbf{Z}_p = 1/\mathbf{Y}_p$ and $\mathbf{Z}_n = 1/\mathbf{Y}_n$, where $n = 1, 2, \ldots, N$, Eq. (11.22) becomes

$$\frac{1}{\mathbf{Z}_p} = \frac{1}{\mathbf{Z}_1} + \frac{1}{\mathbf{Z}_2} + \cdots + \frac{1}{\mathbf{Z}_N} \tag{11.23}$$

or

$$\mathbf{Z}_p = \frac{1}{\dfrac{1}{\mathbf{Z}_1} + \dfrac{1}{\mathbf{Z}_2} + \cdots + \dfrac{1}{\mathbf{Z}_N}} \tag{11.24}$$

For the special case $N = 2$, Eq. (11.24) becomes

$$\mathbf{Z}_p = \frac{1}{\dfrac{1}{\mathbf{Z}_1} + \dfrac{1}{\mathbf{Z}_2}} = \mathbf{Z}_1 \parallel \mathbf{Z}_2 \tag{11.25}$$

where the symbol "$\parallel$" should be read "in parallel with." This symbol provides a convenient notation for the mathematical operation given by Eq. (11.25). It is easy to verify that $\parallel$ is both commutative

$$\mathbf{Z}_1 \parallel \mathbf{Z}_2 = \mathbf{Z}_2 \parallel \mathbf{Z}_1 \tag{11.26}$$

and associative,

$$(\mathbf{Z}_1 \parallel \mathbf{Z}_2) \parallel \mathbf{Z}_3 = \mathbf{Z}_1 \parallel (\mathbf{Z}_2 \parallel \mathbf{Z}_3) \tag{11.27}$$

Since $\parallel$ is associative there is no need to use parentheses for the parallel combination of three or more impedances. For instance, the parallel combination in Eq. (11.27) can be written without ambiguity as $\mathbf{Z}_1 \parallel \mathbf{Z}_2 \parallel \mathbf{Z}_3$.

Series Circuit

Figure 11.11 shows a series circuit that is the dual of the network of Fig. 11.10. The current through each impedance is $\mathbf{I}$ (by the phasor form of KCL). The voltage phasors $\mathbf{V}_n$ are related to $\mathbf{I}$ by $\mathbf{V}_n = \mathbf{Z}_n\mathbf{I}$, where $\mathbf{Z}_n$ is the impedance of the nth element. Application of KVL around the path gives

$$\mathbf{V} = \mathbf{Z}_1\mathbf{I} + \mathbf{Z}_2\mathbf{I} + \cdots + \mathbf{Z}_N\mathbf{I} \tag{11.28}$$

or

$$\mathbf{V} = \mathbf{Z}_s\mathbf{I} \tag{11.29}$$

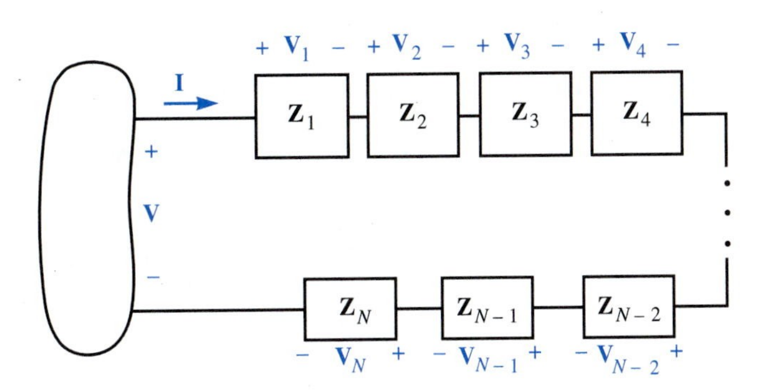

FIGURE 11.11
Impedances in series

where $\mathbf{Z}_s$ is the equivalent impedance:

$$\mathbf{Z}_s = \mathbf{Z}_1 + \mathbf{Z}_2 + \cdots + \mathbf{Z}_N \tag{11.30}$$

Therefore *impedances that are in series add.*

Current and Voltage Dividers

The network of Fig. 11.12 is an ac current divider. The network of Fig. 11.13 is an ac voltage divider. It is left as an exercise (Exercise 10) for you to show that for the current divider of Fig. 11.12,

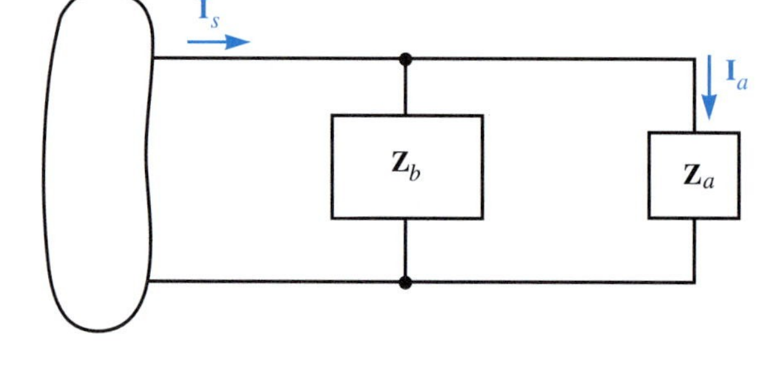

FIGURE 11.12 Current divider

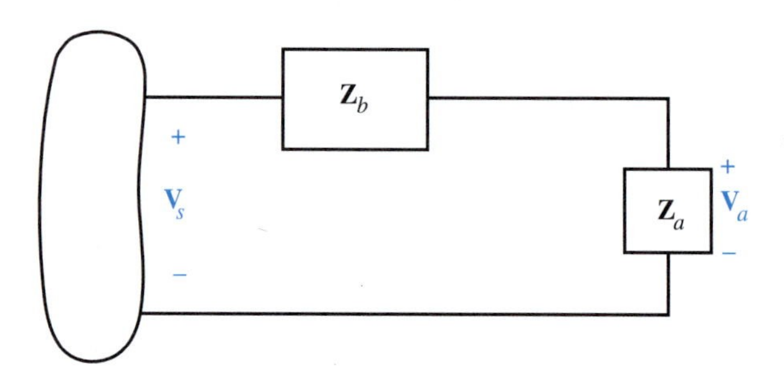

FIGURE 11.13 Voltage divider

$$\mathbf{I}_a = \frac{\mathbf{Z}_b}{\mathbf{Z}_a + \mathbf{Z}_b}\mathbf{I}_s = \frac{\mathbf{Y}_a}{\mathbf{Y}_a + \mathbf{Y}_b}\mathbf{I}_s \tag{11.31}$$

where $\mathbf{Y}_a = 1/\mathbf{Z}_a$ and $\mathbf{Y}_b = 1/\mathbf{Z}_b$. Similarly, for the voltage divider of Fig. 11.13,

$$\mathbf{V}_a = \frac{\mathbf{Z}_a}{\mathbf{Z}_a + \mathbf{Z}_b}\mathbf{V}_s = \frac{\mathbf{Y}_b}{\mathbf{Y}_a + \mathbf{Y}_b}\mathbf{V}_s \tag{11.32}$$

Circuit Analysis by Inspection

The preceding results for simple circuits can often be combined to provide immediate answers to seemingly complicated network problems. The method is called *circuit analysis by inspection.*

EXAMPLE 11.6 Find **V** by inspection of the network of Fig. 11.14.

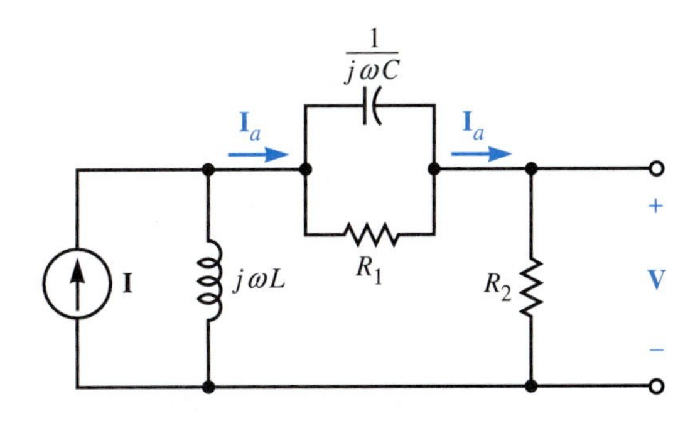

FIGURE 11.14
Circuit of Example 11.6

Solution One way to do this is to first determine $\mathbf{I}_a$ using the current-divider relation, Eq. (11.31), and then set $\mathbf{V} = \mathbf{I}_a R_2$. The result can be written by inspection with the use of Eq. (11.31) with $\mathbf{Z}_b = j\omega L$ and $\mathbf{Z}_a = R_1 \parallel (1/j\omega C) + R_2$:

$$\mathbf{V} = \frac{j\omega L}{j\omega L + R_1 \parallel \dfrac{1}{j\omega C} + R_2} R_2 \mathbf{I}$$

which becomes, after some routine algebraic manipulations,

$$\mathbf{V} = \frac{-\omega^2 + j\dfrac{\omega}{R_1 C}}{-\omega^2 + \left(\dfrac{R_1 + R_2}{L}\right)\left(\dfrac{1}{R_1 C}\right) + j\omega\left(\dfrac{1}{R_1 C} + \dfrac{R_2}{L}\right)} R_2 \mathbf{I}$$

We did not use numerical values in Example 11.6. If we had, we could have used numbers throughout the analysis to obtain a numerical value for **V**. By not including numerical values from the start, we obtained a formula for **V** that applies for *all* values of ω, R_1, R_2, C, and L—clearly a more general result. In our future work we will often work with variables rather than numbers in order to obtain similarly valuable results.

Many ac circuits cannot be readily analyzed by inspection. The systematic circuit analysis techniques described in the following two sections provide a means to analyze them.

Remember Admittances in parallel add as in Eq. (11.22). Impedances in series add as in Eq. (11.30). Currents and voltages divide as in Eqs. (11.31) and (11.32), respectively.

EXERCISES

9. Verify Eqs. (11.26) and (11.27).
10. Derive Eqs. (11.31) and (11.32).
11. Inspect the illustration on page 350 to find the quantities requested: (a) $\mathbf{V}/\mathbf{I}$, (b) $\mathbf{V}_o/\mathbf{V}$, (c) $\mathbf{I}_1/\mathbf{V}$, (d) $\mathbf{I}_2/\mathbf{V}$, (e) $\mathbf{I}_1/\mathbf{I}$, and (f) $\mathbf{I}_2/\mathbf{I}$.
    answer: (a) $R + j\omega RL/(R + j\omega L)$, (b) $j\omega L/(R + j\omega 2L)$, (c) $1/(R + j\omega 2L)$, (d) $j\omega L/(R^2 + j\omega 2LR)$, (e) $R/(R + j\omega L)$, (f) $j\omega L/(R + j\omega L)$

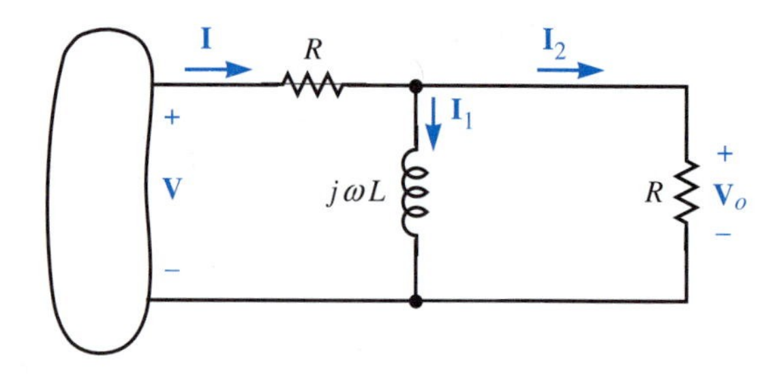

12. Evaluate the transfer function **V/I** of Example 11.6 in the limit $\omega \to \infty$. Explain why your result is correct by inspection of the circuit.
answer: $\mathbf{V/I} = R_2$ for $\omega \to \infty$ because the inductance and capacitance are equivalent to open and short circuits, respectively.

11.3 AC Node-Voltage Analysis

The technique of node-voltage analysis provides a way to systematically analyze any ac circuit.

Setting Up the Node-Voltage Equations

The procedure to set up the node-voltage equations in the frequency domain is basically the same as that in the time domain. Time-domain node-voltage analysis was described in Section 4.2. It is a good idea to review this material—particularly the shortcut method on page 102—before continuing with the following frequency-domain examples.

EXAMPLE 11.7 Set up and solve the node-voltage equations for the circuit of Fig. 11.15.

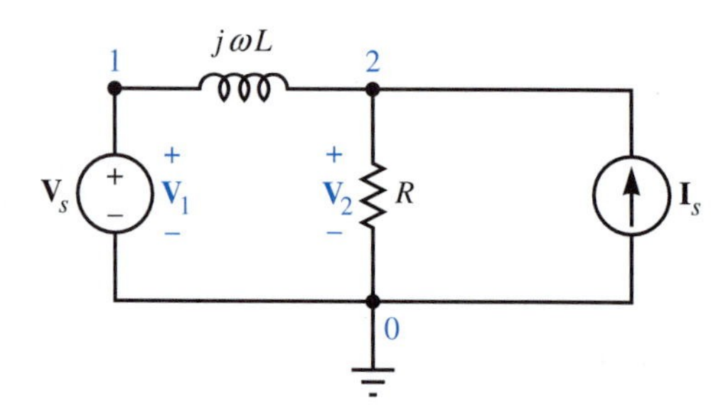

FIGURE 11.15
Circuit of Example 11.7

Solution Node voltage $\mathbf{V}_1$ is identical to $\mathbf{V}_s$. Application of KCL to a surface enclosing node 2 gives

$$\frac{1}{j\omega L}(\mathbf{V}_2 - \mathbf{V}_s) + \frac{1}{R}\mathbf{V}_2 - \mathbf{I}_s = 0$$

The solution of the preceding equation is

$$\mathbf{V}_2 = \frac{R}{R + j\omega L}\mathbf{V}_s + \frac{Rj\omega L}{R + j\omega L}\mathbf{I}_s$$

EXAMPLE 11.8 Write a set of node-voltage equations for the network shown in Fig. 11.16.

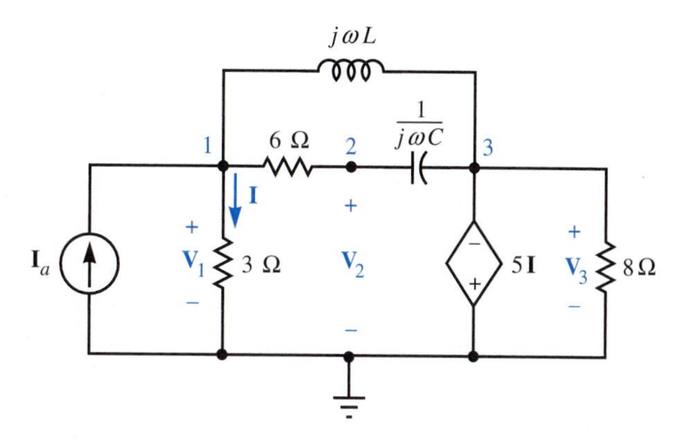

FIGURE 11.16
Circuit of Example 11.8

Solution For a surface enclosing node 1, KCL gives

$$-\mathbf{I}_a + \frac{1}{3}\mathbf{V}_1 + \frac{1}{6}(\mathbf{V}_1 - \mathbf{V}_2) + \frac{1}{j\omega L}(\mathbf{V}_1 - \mathbf{V}_3) = 0$$

or

$$\left(\frac{1}{2} + \frac{1}{j\omega L}\right)\mathbf{V}_1 - \frac{1}{6}\mathbf{V}_2 - \frac{1}{j\omega L}\mathbf{V}_3 = \mathbf{I}_a$$

For a surface enclosing node 2, KCL yields

$$\frac{1}{6}(\mathbf{V}_2 - \mathbf{V}_1) + j\omega C(\mathbf{V}_2 - \mathbf{V}_3) = 0$$

or

$$-\frac{1}{6}\mathbf{V}_1 + \left(j\omega C + \frac{1}{6}\right)\mathbf{V}_2 - j\omega C\mathbf{V}_3 = 0$$

KVL gives the third node-voltage equation:

$$\mathbf{V}_3 = -5\mathbf{I} = -5\left(\frac{1}{3}\mathbf{V}_1\right) = -\frac{5}{3}\mathbf{V}_1$$

The three node-voltage equations can be written in the matrix form

$$\begin{bmatrix} \frac{1}{2} + \frac{1}{j\omega L} & -\frac{1}{6} & -\frac{1}{j\omega L} \\ -\frac{1}{6} & j\omega C + \frac{1}{6} & -j\omega C \\ \frac{5}{3} & 0 & 1 \end{bmatrix} \begin{bmatrix} \mathbf{V}_1 \\ \mathbf{V}_2 \\ \mathbf{V}_3 \end{bmatrix} = \begin{bmatrix} \mathbf{I}_a \\ 0 \\ 0 \end{bmatrix}$$

where the left-hand matrix is the *node transformation matrix*.

EXAMPLE 11.9 Write two node-voltage equations involving $\mathbf{V}_1$ and $\mathbf{V}_3$ directly from the circuit of Fig. 11.16 by treating the series resistance and capacitance as a single impedance.

Solution Let $\mathbf{Z} = 6 + 1/j\omega C$ denote the impedance of the series resistance and capacitance. For a surface enclosing node 1, KCL gives us

$$-\mathbf{I}_a + \frac{1}{3}\mathbf{V}_1 + \frac{1}{\mathbf{Z}}(\mathbf{V}_1 - \mathbf{V}_3) + \frac{1}{j\omega L}(\mathbf{V}_1 - \mathbf{V}_3) = 0$$

or

$$\left(\frac{1}{3} + \frac{1}{\mathbf{Z}} + \frac{1}{j\omega L}\right)\mathbf{V}_1 - \left(\frac{1}{\mathbf{Z}} + \frac{1}{j\omega L}\right) = \mathbf{I}_a$$

As in Example 11.8, KVL gives the following node-voltage equation for $\mathbf{V}_3$:

$$\mathbf{V}_3 = -\frac{5}{3}\mathbf{V}_1$$

The two node-voltage equations can be written in the matrix form

$$\begin{bmatrix} \frac{1}{3} + \frac{1}{\mathbf{Z}} + \frac{1}{j\omega L} & -\frac{1}{\mathbf{Z}} - \frac{1}{j\omega L} \\ \frac{5}{3} & 1 \end{bmatrix} \begin{bmatrix} \mathbf{V}_1 \\ \mathbf{V}_3 \end{bmatrix} = \begin{bmatrix} \mathbf{I}_a \\ 0 \end{bmatrix}$$

EXAMPLE 11.10 Write the node-voltage equations for the network shown in Fig. 11.17.

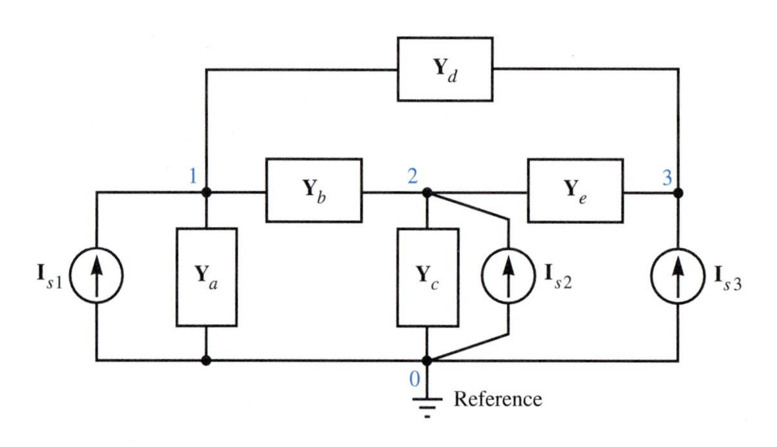

FIGURE 11.17 All node voltages are measured with respect to the reference

Solution Because this network contains only current sources, all the node equations will be KCL equations. These equations are

Node 1 $\mathbf{Y}_a\mathbf{V}_1 + \mathbf{Y}_b(\mathbf{V}_1 - \mathbf{V}_2) + \mathbf{Y}_d(\mathbf{V}_1 - \mathbf{V}_3) - \mathbf{I}_{s1} = 0$

Node 2 $\mathbf{Y}_b(\mathbf{V}_2 - \mathbf{V}_1) + \mathbf{Y}_c\mathbf{V}_2 + \mathbf{Y}_e(\mathbf{V}_2 - \mathbf{V}_3) - \mathbf{I}_{s2} = 0$

Node 3 $\mathbf{Y}_e(\mathbf{V}_3 - \mathbf{V}_2) + \mathbf{Y}_d(\mathbf{V}_3 - \mathbf{V}_1) - \mathbf{I}_{s3} = 0$

These can be written in matrix form as

$$\begin{bmatrix} \mathbf{Y}_a + \mathbf{Y}_b + \mathbf{Y}_d & -\mathbf{Y}_b & -\mathbf{Y}_d \\ -\mathbf{Y}_b & \mathbf{Y}_b + \mathbf{Y}_c + \mathbf{Y}_e & -\mathbf{Y}_e \\ -\mathbf{Y}_d & -\mathbf{Y}_e & \mathbf{Y}_e + \mathbf{Y}_d \end{bmatrix} \begin{bmatrix} \mathbf{V}_1 \\ \mathbf{V}_2 \\ \mathbf{V}_3 \end{bmatrix} = \begin{bmatrix} \mathbf{I}_{s1} \\ \mathbf{I}_{s2} \\ \mathbf{I}_{s3} \end{bmatrix}$$

Observe that the node transformation matrix on the left-hand side is symmetric. The node transformation matrix of a network containing no voltage sources and dependent sources can always be arranged to be symmetric.

Now we will describe a convenient interpretation of the KCL equations for networks composed of only admittances and independent current sources, as illustrated by Example 11.10. Consider the KCL equation for node 1. Notice that the coefficient of $\mathbf{V}_1$ is the sum of the admittances directly connecting node 1 to the other nodes in the network, the coefficient of $\mathbf{V}_2$ is the negative of the admittance directly connecting node 1 to node 2, and the coefficient of $\mathbf{V}_3$ is the negative of the admittance directly connecting node 1 to node 3. Now look at the KCL equation for node 2. The coefficient of $\mathbf{V}_1$ is the negative of the admittance directly connecting node 2 to node 1, the coefficient of $\mathbf{V}_2$ is the sum of the admittances directly connecting node 2 to other nodes in the network, and the coefficient of $\mathbf{V}_3$ is the negative of the admittance directly connecting node 3 to node 1. Finally, consider the KCL equation for node 3. The coefficients of $\mathbf{V}_1$, $\mathbf{V}_2$, and $\mathbf{V}_3$ in this equation have a similar interpretation. (What is it?) Notice also that the right-hand sides of the equations are simply the independent phasor current source inputs to the nodes for which KCL is written.

This interpretation can be generalized to an arbitrary $(N + 1)$-node network composed of only admittances and independent current sources. For this purpose we make the following definitions.

DEFINITION Self-Admittance

The **self-admittance** at node k, $\mathbf{Y}_{kk}$, is the sum of the admittances of the branches that connect node k directly to the other nodes of the network.

DEFINITION Mutual Admittance

The **mutual admittance** between nodes k and l, $\mathbf{Y}_{kl}$, is the negative of the sum of the admittances of the branches that connect node k directly to node l.

The node-voltage equations for a network composed of admittances and independent sources are all KCL equations and have the matrix form

$$\begin{bmatrix} \mathbf{Y}_{11} & \mathbf{Y}_{12} & \cdots & \mathbf{Y}_{1N} \\ \mathbf{Y}_{21} & \mathbf{Y}_{22} & \cdots & \mathbf{Y}_{2N} \\ \vdots & & & \\ \mathbf{Y}_{N1} & \mathbf{Y}_{N2} & \cdots & \mathbf{Y}_{NN} \end{bmatrix} \begin{bmatrix} \mathbf{V}_1 \\ \mathbf{V}_2 \\ \vdots \\ \mathbf{V}_N \end{bmatrix} = \begin{bmatrix} \mathbf{I}_{s1} \\ \mathbf{I}_{s2} \\ \vdots \\ \mathbf{I}_{sN} \end{bmatrix} \tag{11.33}$$

where $\mathbf{I}_{sk}$ is the sum of all source currents entering node k. The square matrix on the left-hand side is the network's *admittance matrix*. By definition, $\mathbf{Y}_{kl} = \mathbf{Y}_{lk}$, so that an admittance matrix is symmetric. An example of an admittance matrix was given in Example 11.10.

Let us now return to node-voltage analysis of more general circuits.

EXAMPLE 11.11 Set up the node-voltage equations for the ideal op-amp circuit of Fig. 11.18, where $\mathbf{Z}_1$, $\mathbf{Z}_2$, $\mathbf{Z}_a$, and $\mathbf{Z}_b$ are arbitrary impedances. Solve for $\mathbf{V}_o$.

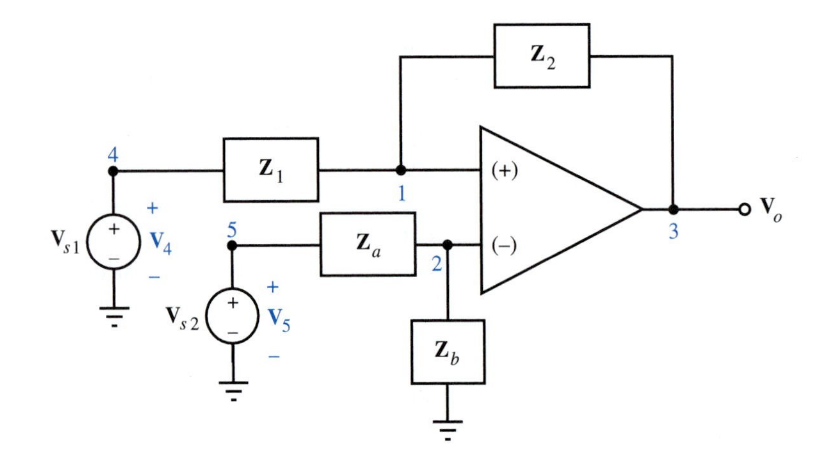

FIGURE 11.18 Op-amp circuit. All node voltages are measured with respect to the reference

Solution By inspection, $\mathbf{V}_4 = \mathbf{V}_{s1}$ and $\mathbf{V}_5 = \mathbf{V}_{s2}$. Using these results, we can apply KCL to a surface enclosing node 1:

$$\frac{1}{\mathbf{Z}_1}(\mathbf{V}_1 - \mathbf{V}_{s1}) + \frac{1}{\mathbf{Z}_2}(\mathbf{V}_1 - \mathbf{V}_3) = 0$$

Application of KCL to a surface enclosing node 2 yields

$$\frac{1}{\mathbf{Z}_a}(\mathbf{V}_1 - \mathbf{V}_{s2}) + \frac{1}{\mathbf{Z}_b}\mathbf{V}_1 = 0$$

where we have set $\mathbf{V}_2 = \mathbf{V}_1$ because the op amp is ideal. Solving for $\mathbf{V}_3 = \mathbf{V}_o$ yields

$$\mathbf{V}_o = -\frac{\mathbf{Z}_2}{\mathbf{Z}_1}\mathbf{V}_{s1} + \left(\frac{\mathbf{Z}_b}{\mathbf{Z}_a + \mathbf{Z}_b}\right)\left(\frac{\mathbf{Z}_1 + \mathbf{Z}_2}{\mathbf{Z}_1}\right)\mathbf{V}_{s2}$$

This example should be compared with Example 6.4.

We have shown how to set up the node-voltage equations. The next step is to solve them. There are several techniques to obtain numerical solutions, one of which is Cramer's rule. Other techniques (used in computers) are computationally more efficient. In the next section, we will use Cramer's rule to obtain the *theoretical* solution. This derivation will provide the start for our later investigation of Thévenin's theorem.

Remember Node-voltage analysis works the same way in the frequency domain as it does in the time domain.

EXERCISES

In Exercises 13 and 14, all elements are specified in terms of their admittances. The units are amperes, volts, and siemens.

13. Set up and solve the node-voltage equations for the circuit shown.

answer: $0.483\underline{/173^\circ}$ V, $0.117\underline{/-111^\circ}$ V

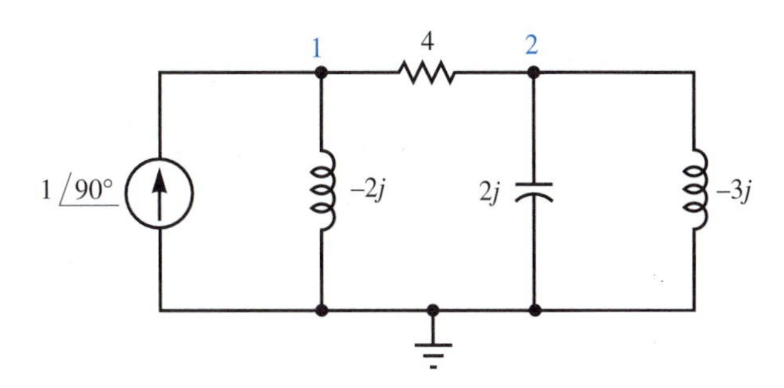

14. Set up and solve the node-voltage equations for the following circuit.

answer: $0.728\underline{/-74.0^\circ}$ V, $2.06\underline{/-119^\circ}$ V

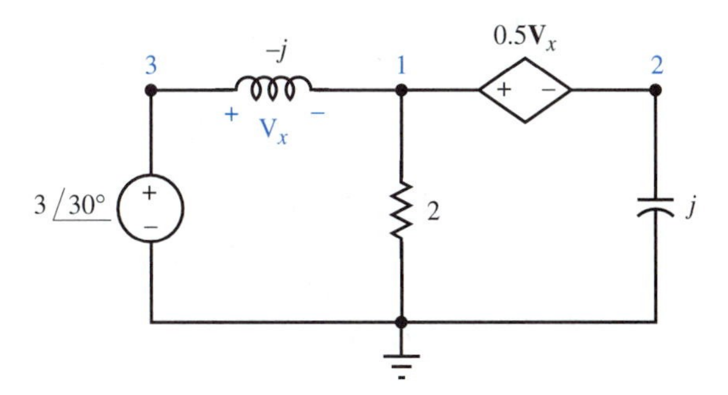

General Form of the Node-Voltage Equations and Solution by Cramer's Rule

Node-voltage analysis of an $(N + 1)$-node network with use of the procedure on page 101 yields a system of N equations:

$$\begin{matrix}\text{KCL} \\ \text{equations}\end{matrix}\left\{\begin{matrix}\\ \\ \\ \end{matrix}\right. \begin{matrix}\\ \\ \\ \text{KVL} \\ \text{equations}\end{matrix} \begin{bmatrix} \mathbf{T}_{11} & \mathbf{T}_{12} & \cdots & \mathbf{T}_{1N} \\ \vdots & \vdots & & \vdots \\ \mathbf{T}_{M1} & \mathbf{T}_{M2} & & \mathbf{T}_{MN} \\ \vdots & \vdots & & \vdots \\ \mathbf{T}_{N1} & \mathbf{T}_{N2} & \cdots & \mathbf{T}_{NN} \end{bmatrix} \begin{bmatrix} \mathbf{V}_1 \\ \vdots \\ \mathbf{V}_M \\ \vdots \\ \mathbf{V}_N \end{bmatrix} = \begin{bmatrix} \mathbf{I}_{s1} \\ \vdots \\ \mathbf{V}_{sM} \\ \vdots \\ \mathbf{V}_{sN} \end{bmatrix} \begin{matrix}\text{Due to current sources} \\ \\ \text{Due to voltage sources}\end{matrix} \tag{11.34}$$

The square matrix is the node transformation matrix. The letters $\mathbf{T}_{ij}$ are used to denote its elements, which, of course, depend on the circuit.

The form of the node-voltage equations is much simpler when the circuit is made up of passive elements and independent current sources only. Such a circuit was shown in Example 11.10, where it was noticed that we could set up the node-voltage equations by inspection with the use of the concepts of self- and mutual admittance.

The analytical expression for the phasor node voltages is given by Cramer's rule:

$$\mathbf{V}_n = \frac{\Delta_{1n}}{\Delta}\mathbf{I}_{s1} + \frac{\Delta_{2n}}{\Delta}\mathbf{I}_{s2} + \cdots + \frac{\Delta_{(M-1)n}}{\Delta}\mathbf{I}_{s(M-1)}$$
$$+ \frac{\Delta_{Mn}}{\Delta}\mathbf{V}_{sM} + \cdots + \frac{\Delta_{Nn}}{\Delta}\mathbf{V}_{sN} \tag{11.35}$$

for $n = 1, 2, \ldots, N$, where Δ is the determinant of the node transformation matrix, and Δ_{ij} is the cofactor for row i and column j (see Appendix A).

We can see from Eq. (11.35) that the superposition principle of Chapter 5 applies to frequency-domain circuit analysis. Equation (11.35) indicates that the nth node-voltage phasor is given by a weighted sum or superposition of voltages caused by the independent sources. A simple interpretation of the ratios Δ_{mn}/Δ can be obtained by setting all the independent source phasors at zero except for the term $m = k$. Equation (11.37) then becomes

$$\mathbf{V}_n = \begin{cases} \dfrac{\Delta_{kn}}{\Delta}\mathbf{I}_{sk} & \text{for } 1 \le k \le M-1 \\ \dfrac{\Delta_{kn}}{\Delta}\mathbf{V}_{sk} & \text{for } M \le k \le N \end{cases} \tag{11.36}$$

Therefore Δ_{kn}/Δ is simply the transfer function from source k to the voltage at node n. We see from Eq. (11.36) that the transfer function Δ_{kn}/Δ is

1. A driving-point impedance for $n = k = 1, 2, \ldots, M-1$.
2. A transfer impedance for $n \neq k$, $k = 1, 2, \ldots, M-1$.
3. A voltage transfer function for $k = M, M+1, \ldots, N$.

11.4 AC Mesh-Current Analysis

As we stated in Section 4.3, mesh-current circuit analysis applies only to planar circuits. The procedure summarized on page 119 applies to ac mesh-current analysis in the frequency domain as well as in the time domain. The following examples illustrate the method.

EXAMPLE 11.12 Set up and solve the mesh-current equations for the circuit of Fig. 11.19.

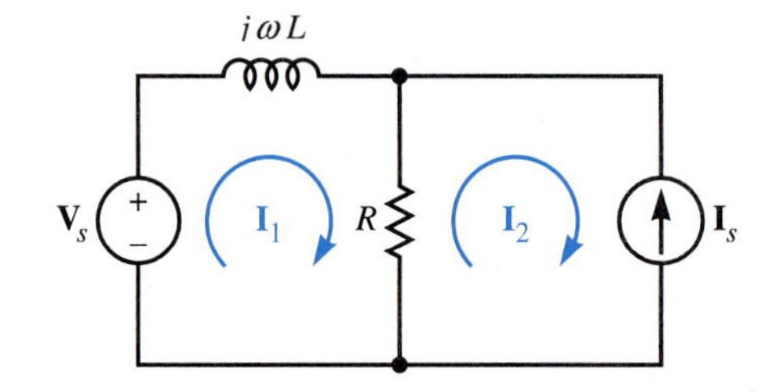

FIGURE 11.19 Circuit for Example 11.12

Solution We note that $\mathbf{I}_2$ is identical to $-\mathbf{I}_s$. The application of KVL around mesh 1 yields

$$-\mathbf{V}_s + j\omega L\mathbf{I}_1 + R(\mathbf{I}_1 - \mathbf{I}_s) = 0$$

The solution to the above equation is

$$\mathbf{I}_s = \frac{\mathbf{V}_s}{R + j\omega L} - \frac{R\mathbf{I}_s}{R + j\omega L}$$

EXAMPLE 11.13 Write a set of mesh-current equations for the network shown in Fig. 11.20.

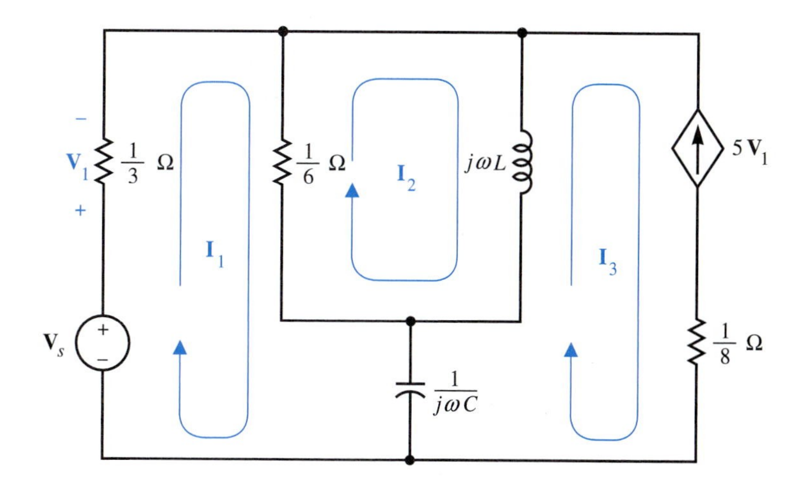

FIGURE 11.20
Circuit for Example 11.13

Solution Use the mesh currents assigned in Fig. 11.20. KVL for a closed path around mesh 1 gives the first mesh-current equation:

$$-\mathbf{V}_s + \frac{1}{3}\mathbf{I}_1 + \frac{1}{6}(\mathbf{I}_1 - \mathbf{I}_2) + \frac{1}{j\omega C}(\mathbf{I}_1 - \mathbf{I}_3) = 0$$

The second mesh-current equation is obtained by application of KVL to a closed path around mesh 2:

$$\frac{1}{6}(\mathbf{I}_2 - \mathbf{I}_1) + j\omega L(\mathbf{I}_2 - \mathbf{I}_3) = 0$$

KCL gives the third mesh-current equation:

$$\mathbf{I}_3 = -5\mathbf{V}_1 = -5\left(\frac{1}{3}\mathbf{I}_1\right) = -\frac{5}{3}\mathbf{I}_1$$

The three mesh-current equations can be written as

$$\begin{bmatrix} \frac{1}{2} + \frac{1}{j\omega C} & -\frac{1}{6} & -\frac{1}{j\omega C} \\ -\frac{1}{6} & j\omega L + \frac{1}{6} & -j\omega L \\ \frac{5}{3} & 0 & 1 \end{bmatrix} \begin{bmatrix} \mathbf{I}_1 \\ \mathbf{I}_2 \\ \mathbf{I}_3 \end{bmatrix} = \begin{bmatrix} \mathbf{V}_s \\ 0 \\ 0 \end{bmatrix}$$

where the left-hand matrix is the mesh transformation matrix.

EXAMPLE 11.14 Write the mesh-current equations for the network of Fig. 11.21.

Solution A KVL equation can be written for each mesh in this circuit. The KVL equations are

Mesh 1 $\mathbf{Z}_a\mathbf{I}_1 + \mathbf{Z}_b(\mathbf{I}_1 - \mathbf{I}_2) + \mathbf{Z}_d(\mathbf{I}_1 - \mathbf{I}_3) - \mathbf{V}_{s1} = 0$
Mesh 2 $\mathbf{Z}_b(\mathbf{I}_2 - \mathbf{I}_1) + \mathbf{Z}_c\mathbf{I}_2 + \mathbf{Z}_e(\mathbf{I}_2 - \mathbf{I}_3) - \mathbf{V}_{s2} = 0$
Mesh 3 $\mathbf{Z}_d(\mathbf{I}_3 - \mathbf{I}_1) + \mathbf{Z}_e(\mathbf{I}_3 - \mathbf{I}_2) + \mathbf{V}_{s3} = 0$

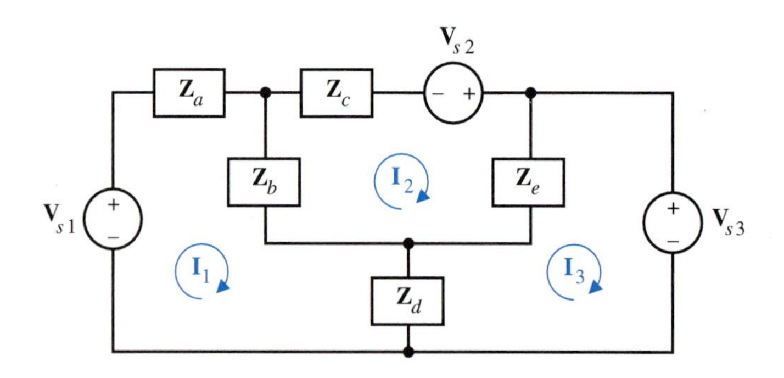

FIGURE 11.21
Circuit for Example 11.14

The matrix form is

$$\begin{bmatrix} \mathbf{Z}_a + \mathbf{Z}_b + \mathbf{Z}_d & -\mathbf{Z}_b & -\mathbf{Z}_d \\ -\mathbf{Z}_b & \mathbf{Z}_b + \mathbf{Z}_c + \mathbf{Z}_e & -\mathbf{Z}_e \\ -\mathbf{Z}_d & -\mathbf{Z}_e & \mathbf{Z}_d + \mathbf{Z}_e \end{bmatrix} \begin{bmatrix} \mathbf{I}_1 \\ \mathbf{I}_2 \\ \mathbf{I}_3 \end{bmatrix} = \begin{bmatrix} \mathbf{V}_{s1} \\ \mathbf{V}_{s2} \\ -\mathbf{V}_{s3} \end{bmatrix}$$

Notice that the mesh transformation matrix on the left-hand side is symmetric. The mesh transformation matrix of a network containing no current sources and dependent sources can always be arranged to be symmetric.

The mesh-current equations of networks composed of only impedances and independent voltage sources, such as that in Example 11.14, have an interpretation that is the dual of that occurring for Example 11.10. Consider the KVL equation for mesh 1. The coefficient of $\mathbf{I}_1$ is the sum of the impedances making up mesh 1, the coefficient of $\mathbf{I}_2$ is the negative of the impedance that is common to meshes 1 and 2, and the coefficient of $\mathbf{I}_3$ is the negative of the impedance that is common to meshes 1 and 3. Similarly, for the KVL equation of mesh 2, the coefficient of $\mathbf{I}_1$ is the negative of the impedance that is common to meshes 2 and 1, the coefficient of $\mathbf{I}_2$ is the sum of the impedances making up mesh 2, and the coefficient of $\mathbf{I}_3$ is the negative of the impedance that is common to meshes 3 and 1. The coefficients of $\mathbf{I}_1$, $\mathbf{I}_2$, and $\mathbf{I}_3$ for the third mesh equation have a similar interpretation. Notice further that the right-hand sides of the equations are simply the independent source phasor voltages in the meshes for which KVL is written. The above interpretations can be generalized to an arbitrary N-mesh network composed of only impedances and independent voltage sources. For this purpose we make the following definitions.

DEFINITION Self-Impedance

The **self-impedance** of mesh k, $\mathbf{Z}_{kk}$, is the sum of the impedances composing mesh k.

DEFINITION Mutual Impedance

The **mutual impedance** of meshes k and l, $\mathbf{Z}_{kl}$, is the negative of the sum of all impedances in common with meshes k and l.

The mesh equations for such a network are all KVL equations and have the matrix form:

$$\begin{bmatrix} \mathbf{Z}_{11} & \mathbf{Z}_{12} & \cdots & \mathbf{Z}_{1N} \\ \mathbf{Z}_{21} & \mathbf{Z}_{22} & \cdots & \mathbf{Z}_{2N} \\ \vdots & & & \\ \mathbf{Z}_{N1} & \mathbf{Z}_{N2} & & \mathbf{Z}_{NN} \end{bmatrix} \begin{bmatrix} \mathbf{I}_1 \\ \mathbf{I}_2 \\ \vdots \\ \mathbf{I}_N \end{bmatrix} = \begin{bmatrix} \mathbf{V}_{s1} \\ \mathbf{V}_{s2} \\ \vdots \\ \mathbf{V}_{sN} \end{bmatrix} \tag{11.37}$$

where $\mathbf{V}_{sk}$ is the sum of all independent phasor voltage source drops taken counterclockwise in mesh k (if we assume clockwise mesh currents). The square matrix on the left-hand side is the network's *impedance matrix.* By definition, $\mathbf{Z}_{kl} = \mathbf{Z}_{lk}$. Therefore the impedance matrix is symmetric. An example of an impedance matrix was given in Example 11.14.

Remember

Mesh-current analysis works the same way in the frequency domain as it does in the time domain.

EXERCISES

15. Set up and solve the mesh-current equations for the following circuit.

answer: $4.47\angle{-26.6^\circ}$ A, $4.47\angle{26.6^\circ}$ A

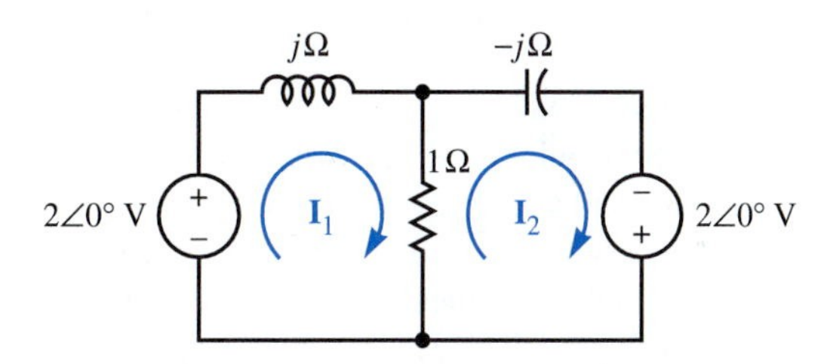

16. Find the mesh-currents for the following circuit.

answer: $485\angle{14^\circ}$ mA, $568\angle{47.2^\circ}$ mA, $175\angle{33.5^\circ}$ mA

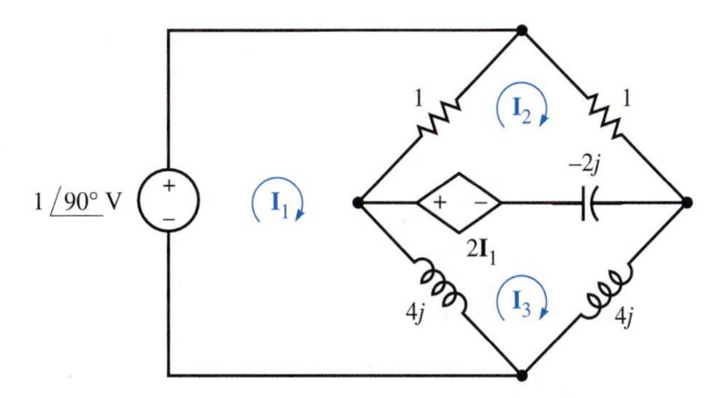

11.5 Superposition

The principle of superposition is fundamental to the theory of linear circuits. In Section 5.1 we said that the response of a linear network to a number of independent sources is equal to the sum of the responses to each independent source acting on the network alone. In Section 8.6 we learned that this *superposition* property also applies to the particular response. In this section, we use superposition to determine a network's steady-state response to two or more sinusoidal sources whose frequencies are different.

EXAMPLE 11.15 Two sinusoidal sources that have *different* frequencies have been driving the *RL* network of Fig. 11.22 for a long time (that is, since $t = -\infty$). Determine $v(t)$.

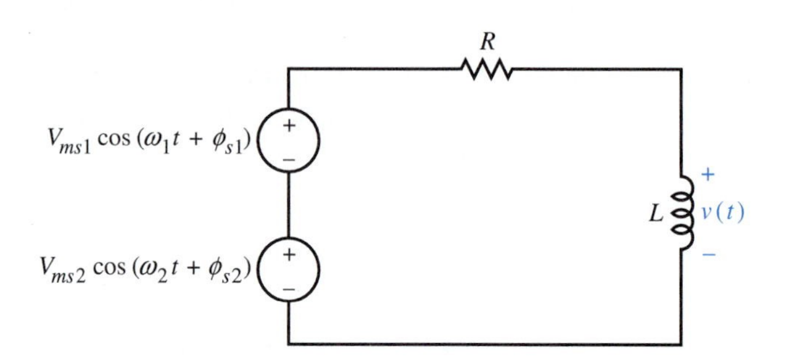

FIGURE 11.22 Circuit for Example 11.15

Solution The natural response $v_n(t)$ is zero by now (that is, for all finite t), because the network is stable and the sources were applied at $t = -\infty$. Therefore the complete response $v(t)$ is just the particular response. Because there are two sinusoidal inputs, the particular response of this network contains *two* sinusoidal terms (called *components*); one component has frequency ω_1, and the other has frequency ω_2. Frequency-domain concepts apply only to the *individual* components. The total response $v(t)$ is found by adding the individual sinusoidal responses caused by each source acting alone. The time- and frequency-domain networks for each source acting alone are shown in Fig. 11.23. The complex exponential factor $e^{j\omega_1 t}$ is implicit in Fig. 11.23b, and the factor $e^{j\omega_2 t}$ is implicit in Fig. 11.23d. The output-voltage phasors in Fig. 11.23b, d can be determined with the use of the voltage divider relation [Eq. (11.32)].

$$\mathbf{V}_1 = \frac{j\omega_1 L}{R + j\omega_1 L}\mathbf{V}_{s1} = \mathbf{H}(j\omega_1)\mathbf{V}_{s1}$$

$$\mathbf{V}_2 = \frac{j\omega_2 L}{R + j\omega_2 L}\mathbf{V}_{s2} = \mathbf{H}(j\omega_2)\mathbf{V}_{s2}$$

where we have defined

$$\mathbf{H}(j\omega) = \frac{j\omega L}{R + j\omega L}$$

The corresponding rotating phasors are

$$\mathbf{V}_1 e^{j\omega_1 t} = \mathbf{H}(j\omega_1)\mathbf{V}_{s1} e^{j\omega_1 t}$$

and

$$\mathbf{V}_2 e^{j\omega_2 t} = \mathbf{H}(j\omega_2)\mathbf{V}_{s2} e^{j\omega_2 t}$$

The real parts give the particular response in Fig. 11.23a, c:

$$v_1(t) = |\mathbf{H}(j\omega_1)| V_{ms1} \cos[\omega_1 t + \phi_{s1} + \underline{/\mathbf{H}(j\omega_1)}]$$

and

$$v_2(t) = |\mathbf{H}(j\omega_2)| V_{ms2} \cos[\omega_2 t + \phi_{s2} + \underline{/\mathbf{H}(j\omega_2)}]$$

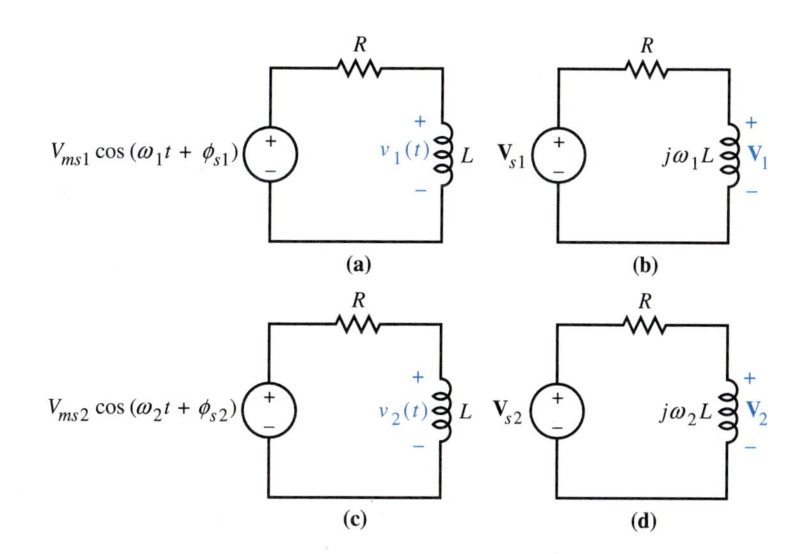

FIGURE 11.23 We consider one source at a time. (*a*) Time-domain circuit with source 2 set to zero. (*b*) Frequency-domain circuit corresponding to (*a*). (*c*) Time-domain circuit with source 1 set to zero. (*d*) Frequency-domain circuit corresponding to (*c*).

These are the individual components of the total particular response of the circuit of Fig. 11.22. The magnitude and angle of the transfer function are

$$|\mathbf{H}(j\omega)| = \frac{\omega L}{\sqrt{R^2 + (\omega L)^2}}$$

and

$$\angle\mathbf{H}(j\omega) = \frac{\pi}{2} - \arctan\left(\frac{\omega L}{R}\right)$$

By superposition, the total response is

$$\begin{aligned} v(t) &= v_1(t) + v_2(t) \\ &= \frac{\omega_1 L}{\sqrt{R^2 + \omega_1^2 L^2}} V_{ms1} \cos\left[\omega_1 t + \phi_{s1} + \frac{\pi}{2} - \tan^{-1}\left(\frac{\omega_1 L}{R}\right)\right] \\ &\quad + \frac{\omega_2 L}{\sqrt{R^2 + \omega_2^2 L^2}} V_{ms2} \cos\left[\omega_2 t + \phi_{s2} + \frac{\pi}{2} - \tan^{-1}\left(\frac{\omega_2 L}{R}\right)\right] \end{aligned}$$

EXAMPLE 11.16

In the circuit of Fig. 11.24, a sinusoidal current source with frequency ω_3 has been added to the *RL* network of the previous example. Determine $v(t)$.

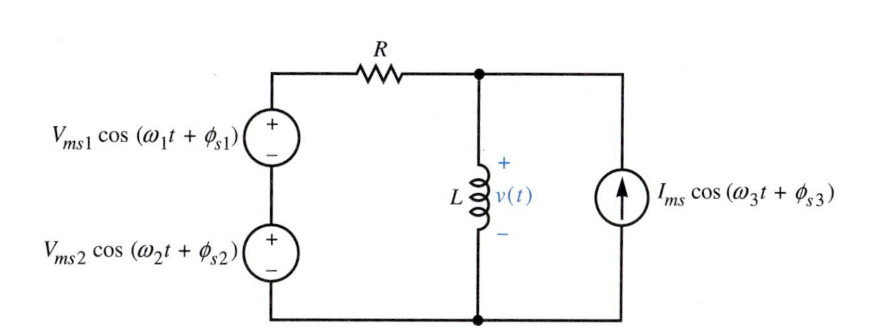

FIGURE 11.24 Independent sources with different frequencies must be considered one at a time

Solution Consider the three independent sources individually. The sinusoidal output components resulting from the voltage sources have already been determined for this network. (When the current source is set to zero, it is in effect an open circuit, and the network becomes that of Fig. 11.22.) The effect of the current source is found by setting the voltage sources to zero, as illustrated in Fig. 11.25.

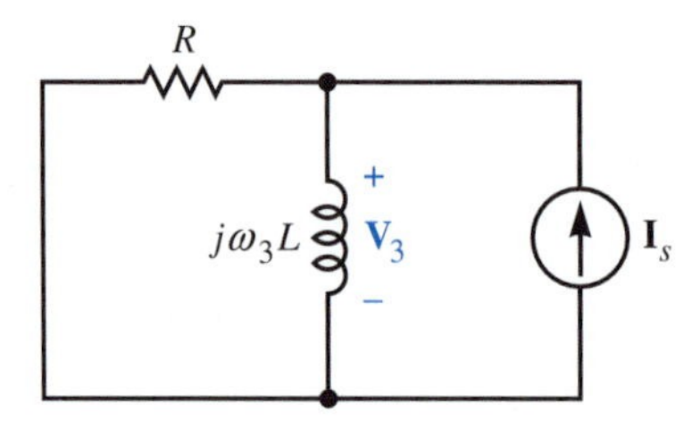

FIGURE 11.25 Voltage sources set to zero

The output-voltage phasor produced by the current source is, by inspection of Fig. 11.25,

$$\mathbf{V}_3 = (j\omega_3 L \parallel R)\mathbf{I}_s = \frac{j\omega_3 LR}{R + j\omega_3 L}\mathbf{I}_s$$

The complex exponential response is consequently

$$\mathbf{V}_3 e^{j\omega_3 t} = \frac{j\omega_3 LR}{R + j\omega_3 L}\mathbf{I}_s e^{j\omega_3 t}$$

The real part gives the sinusoidal steady-state response due to the current source:

$$v_3(t) = \frac{\omega_3 LR}{\sqrt{R^2 + \omega_3^2 L^2}} I_{ms} \cos\left[\omega_3 t + \phi_{s3} + \frac{\pi}{2} - \tan^{-1}\left(\frac{\omega_3 L}{R}\right)\right]$$

By superposition, the total response $v(t)$ in Fig. 11.24 is

$$\begin{aligned} v(t) &= v_1(t) + v_2(t) + v_3(t) \\ &= \frac{\omega_1 L}{\sqrt{R^2 + \omega_1^2 L^2}} V_{ms1} \cos\left[\omega_1 t + \phi_{s1} + \frac{\pi}{2} - \tan^{-1}\left(\frac{\omega_1 L}{R}\right)\right] \\ &\quad + \frac{\omega_2 L}{\sqrt{R^2 + \omega_2^2 L^2}} V_{ms2} \cos\left[\omega_2 t + \phi_{s2} + \frac{\pi}{2} - \tan^{-1}\left(\frac{\omega_2 L}{R}\right)\right] \\ &\quad + \frac{\omega_3 LR}{\sqrt{R^2 + \omega_3^2 L^2}} I_{ms} \cos\left[\omega_3 t + \phi_{s3} + \frac{\pi}{2} - \tan^{-1}\left(\frac{\omega_3 L}{R}\right)\right] \end{aligned}$$

Up to this point we have shown how to use superposition to find a network's response to sinusoidal sources with different frequencies. Superposition can also be used to advantage when the sources have the same frequency. Consider once again the two-source network in Fig. 11.9e. With the use of superposition, the current $\mathbf{I}_L$ can be written by *inspection* as

$$\mathbf{I}_L = \frac{\mathbf{V}_s}{j\omega L + R} - \frac{\dfrac{1}{j\omega L}}{\dfrac{1}{j\omega L} + \dfrac{1}{R}}\mathbf{I}_s \tag{11.38}$$

We obtain the terms in Eq. (11.38) by assuming that each source acts alone. The first term is the input (phasor) voltage divided by the driving-point impedance seen by the voltage source. The second term follows from the current divider relation [Eq. (11.31)]. For the numerical values used in Example 11.4 ($\omega = 10^3$ rad/s, $R = 500\ \Omega$, $L = 0.5$ H, $\mathbf{V}_s = 30\underline{/-25^\circ}$ V, and $\mathbf{I}_s = 20\underline{/-75^\circ}$ mA), the driving-point impedance is

$$R + j\omega L = 0.5\sqrt{2}\underline{/45^\circ}\ \text{k}\Omega \tag{11.39}$$

and the current transfer function is

$$-\frac{\dfrac{1}{j\omega L}}{\dfrac{1}{j\omega L} + \dfrac{1}{R}} = -(0.5\sqrt{2}\underline{/-45^\circ}) \tag{11.40}$$

Therefore

$$\begin{aligned}\mathbf{I}_L &= \frac{30\underline{/-25^\circ}}{0.5\sqrt{2}\underline{/45^\circ}} - (0.5\sqrt{2}\underline{/-45^\circ})(20\underline{/-75^\circ})\ \text{mA} \\ &= 30\sqrt{2}\underline{/-70^\circ} - (10\sqrt{2}\underline{/-120^\circ})\ \text{mA}\end{aligned} \tag{11.41}$$

and

$$\begin{aligned}i_L(t) &= \mathcal{R}e\{\mathbf{I}_L e^{j1000t}\} \\ &= 30\sqrt{2}\cos(1000t - 70^\circ) - 10\sqrt{2}\cos(1000t - 120^\circ)\ \text{mA}\end{aligned} \tag{11.42}$$

This result agrees with that given in Example 11.4, but it displays the contributions from the two sources individually.

Remember

When sources have different frequencies, superposition is applied to the *time-domain waveforms* arising from the different sources. *It is never correct to add nonrotating phasors that represent sinusoids with different frequencies.* For instance, in Example 11.15, it would be a *fundamental mistake* to define a total response phasor

$$\mathbf{V}_1 + \mathbf{V}_2 = \frac{j\omega_1 L}{R + j\omega_1 L}\mathbf{V}_{s1} + \frac{j\omega_2 L}{R + j\omega_2 L}\mathbf{V}_{s2} \tag{11.43}$$

Equation (11.35), however, shows us that we *can* add phasors that represent sinusoids having the *same* frequencies. An example of this is given by Eq. (11.38).

EXERCISES

17. Find $i(t)$ for the circuit on page 364. Express your answer in the form

$$i(t) = A\cos(\omega_1 t + \phi_A) + B\sin(\omega_2 t + \phi_B).$$

answer:
$$i(t) = \frac{V_m}{\sqrt{R^2 + (\omega_1 L)^2}}\cos[\omega_1 t + \phi_{\mathbf{V}} - \arctan(\omega_1 L/R)] + \frac{\omega_2 L I_m}{\sqrt{R^2 + (\omega_2 L)^2}}\cos[\omega_2 t + \phi_{\mathbf{I}} + 90^\circ - \arctan(\omega_2 L/R)]$$

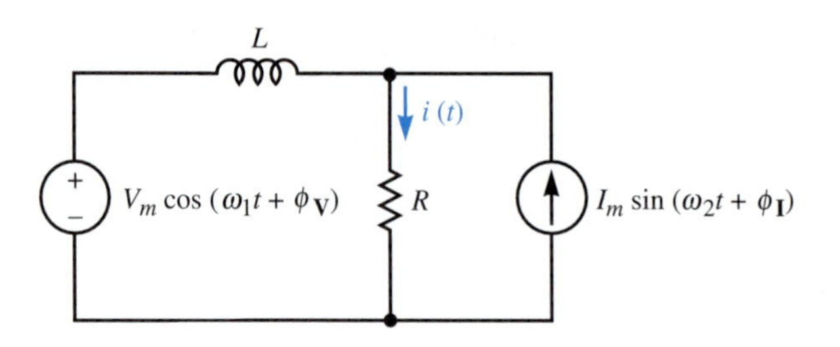

18. Find $v_o(t)$ in the following circuit. Express your answer in the form

$$v_o(t) = B_1 \cos(\omega_1 t + \phi_1) + B_2 \cos(\omega_2 t + \phi_2).$$

answer: $B_1 = \dfrac{\omega_1 L A_1}{\sqrt{R + (\omega_1 L)^2}}$, $\phi_1 = 90° - \arctan(\omega_1 L/R)$, $B_2 = \dfrac{RA_2}{\sqrt{R + (\omega_2 L)^2}}$, $\phi_2 = -\arctan(\omega_2 L/R)$

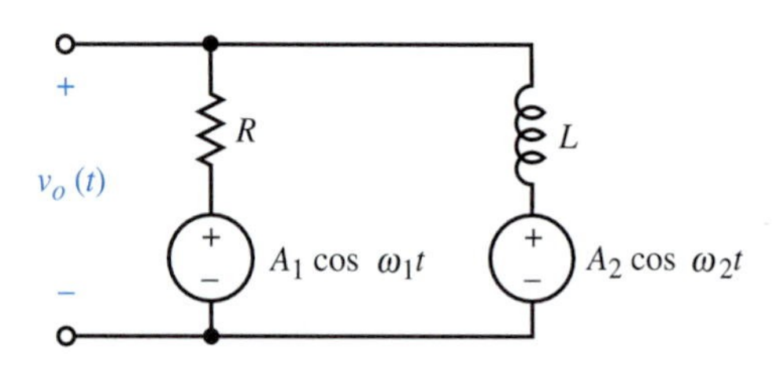

11.6 Thévenin and Norton Equivalent Circuits

Recall from Section 5.3 that the terminal equation of any network composed of resistances and sources has the form

$$v(t) = v_{oc}(t) - i(t)R_{Eq} \tag{11.44}$$

This result is illustrated in Fig. 11.26, where network A is the network composed of resistances and sources, $v_{oc}(t)$ is the voltage appearing across terminals a and b when the terminals are open-circuited, and R_{Eq} is the Thévenin equivalent resistance of network A. The two circuits enclosed by blue in Fig. 11.26b, c have the same terminal equation as network A of Fig. 11.26a. These circuits are called, respectively, the Thévenin equivalent circuit and the Norton equivalent circuit of network A. Thévenin and Norton equivalent circuits are important because we can use them in place of network A whenever we want to determine the effect of network A on another circuit.

In this section we will derive the Thévenin and Norton equivalents of an arbitrary ac network, which we call network A in Fig. 11.27. Network A may contain both dependent and independent sources. Network B (the load) is an arbitrary ac network that may contain both dependent and independent sources. No control voltage or current for a dependent source in one network (A or B), however, is found in the other. All independent sources generate sinusoidal waveforms with the same frequency ω.

We apply node-voltage analysis to network A to derive its Thévenin and Norton equivalents. The node-labeling conventions are shown in Fig. 11.27. We select terminal b as the reference node for the node-voltage analysis. Terminal a connects to some node

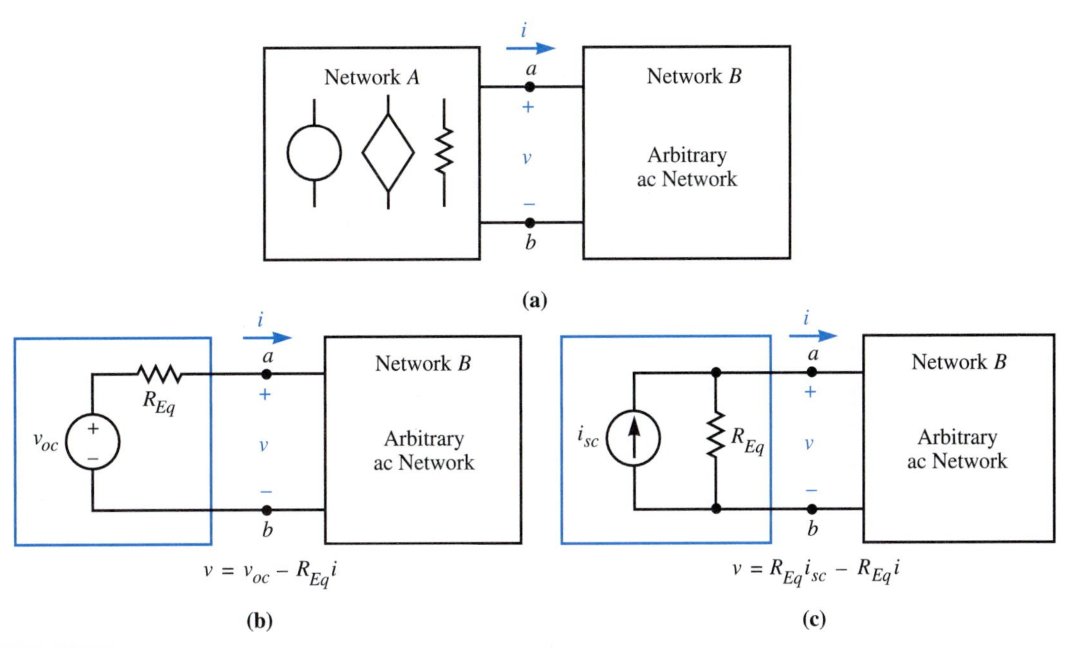

FIGURE 11.26
Replacing a network of resistances and independent and dependent sources by an equivalent network. (*a*) Networks coupled at only two terminals; (*b*) network *A* replaced by Thévenin equivalent (enclosed by blue); (*c*) network *A* replaced by Norton equivalent (enclosed by blue)

in network A. Wherever that node is, we label it node 1 of network A. We then number the remaining nodes in network A 2, 3, . . . , N in any convenient manner. We do not number the nodes in network B because all we are interested in is how network A relates terminal (load) current $\mathbf{I}$ to terminal voltage $\mathbf{V}_{ab}$.

The node-voltage equations for network A have a form similar to Eq. (11.34).

$$\begin{array}{l} \text{KCL equations} \\ \\ \text{KVL equations} \end{array} \left\{ \begin{bmatrix} \mathbf{T}_{11} & \mathbf{T}_{12} & \cdots & \mathbf{T}_{1N} \\ \vdots & \vdots & & \vdots \\ \mathbf{T}_{M1} & \mathbf{T}_{M2} & \cdots & \mathbf{T}_{MN} \\ \vdots & \vdots & & \vdots \\ \mathbf{T}_{N1} & \mathbf{T}_{N2} & \cdots & \mathbf{T}_{NN} \end{bmatrix} \right. \begin{bmatrix} \mathbf{V}_1 \\ \vdots \\ \mathbf{V}_M \\ \vdots \\ \mathbf{V}_N \end{bmatrix} = \begin{bmatrix} \mathbf{I}_{s1} - \mathbf{I} \\ \mathbf{I}_{s2} \\ \vdots \\ \mathbf{V}_{sM} \\ \vdots \\ \mathbf{V}_{sN} \end{bmatrix} \begin{array}{l} \text{Due to current sources} \\ \\ \text{Due to voltage sources} \end{array} \tag{11.45}$$

Load current (points to $\mathbf{I}_{s1} - \mathbf{I}$)

Node transformation matrix of network A

In Eq. (11.45), the phasor $\mathbf{I}_{s1}$ represents the net current entering node 1 from the independent current sources in network A. Similarly, $\mathbf{I}_{s2}$ represents the net current entering node 2 in network A from the independent current sources in network A, and so on. The phasor $\mathbf{V}_{sk}$ $(M \le k \le N)$ is the sum of the independent voltage-source phasors of network A in the loop for which KVL is written. The elements $\mathbf{T}_{mn}$ $(1 \le m \le N,$ $1 \le n \le N)$ are functions of $j\omega$, where ω is the common frequency of the sinusoidal waveforms present throughout the network. The difference between Eq. (11.45) and Eq.

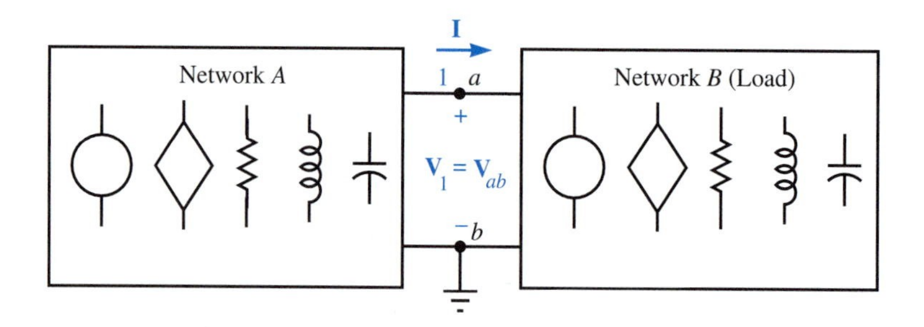

FIGURE 11.27 AC network coupled at only two terminals. Terminal a connects to node 1 of network A.

(11.34) is the extra term, $\mathbf{I}$, in Eq. (11.45). This term simply represents the current drawn out of node 1 of network A by the load.

By Cramer's rule,

$$\mathbf{V}_{ab} = \mathbf{V}_1$$
$$= \frac{\Delta_{11}}{\Delta}\mathbf{I}_{s1} + \frac{\Delta_{21}}{\Delta}\mathbf{I}_{s2} + \cdots + \frac{\Delta_{(M-1)1}}{\Delta}\mathbf{I}_{s(M-1)} + \frac{\Delta_{M1}}{\Delta}\mathbf{V}_{sM} + \cdots$$
$$+ \frac{\Delta_{N1}}{\Delta}\mathbf{V}_{sN} - \frac{\Delta_{11}}{\Delta}\mathbf{I} \tag{11.46}$$

The quantities Δ_{nm} are the nmth cofactors of the node transformation matrix, and Δ is its determinant. Consequently, the Δ_{nm}'s and Δ are functions of $j\omega$. We notice that if network B is an open circuit, then $\mathbf{I} = 0$ and $\mathbf{V}_{ab}$ is, by definition, the open-circuit voltage phasor, $\mathbf{V}_{oc}$:

$$\mathbf{V}_{oc} = \mathbf{V}_{ab}\bigg|_{\mathbf{I}=0}$$
$$= \frac{\Delta_{11}}{\Delta}\mathbf{I}_{s1} + \frac{\Delta_{21}}{\Delta}\mathbf{I}_{s2} + \cdots + \frac{\Delta_{(M-1)1}}{\Delta}\mathbf{I}_{s(M-1)}$$
$$+ \frac{\Delta_{M1}}{\Delta}\mathbf{V}_{sM} + \cdots + \frac{\Delta_{N1}}{\Delta}\mathbf{V}_{sN} \tag{11.47}$$

Substitution of Eq. (11.47) into Eq. (11.46) yields

$$\mathbf{V}_{ab} = \mathbf{V}_{oc} - \mathbf{Z}_{ab}(j\omega)\mathbf{I} \tag{11.48}$$

where we have defined

$$\mathbf{Z}_{ab}(j\omega) = \frac{\Delta_{11}(j\omega)}{\Delta(j\omega)} \tag{11.49}$$

$\mathbf{Z}_{ab}(j\omega)$ is called the *equivalent impedance* of network A looking into terminals a and b. We notice next that if network B is a short circuit, then $\mathbf{V}_{ab} = 0$ and $\mathbf{I}$ is, by definition, the *short-circuit phasor current* $\mathbf{I}_{sc}$. We can obtain an expression for $\mathbf{I}_{sc}$ by setting $\mathbf{V}_{ab} = 0$ in Eq. (11.48) and solving for $\mathbf{I}$. The result is

$$\mathbf{I}_{sc} = \frac{\mathbf{V}_{oc}}{\mathbf{Z}_{ab}(j\omega)} \tag{11.50}$$

Equation (11.50) can also be used to determine $\mathbf{Z}_{ab}(j\omega)$ if $\mathbf{I}_{sc} \neq 0$.

$$\mathbf{Z}_{ab}(j\omega) = \frac{\mathbf{V}_{oc}}{\mathbf{I}_{sc}} \tag{11.51}$$

Equations (11.48) and (11.51) are the fundamental definitions of the Thévenin and Norton equivalents of an ac circuit. They indicate that the terminal equation relating the phasors $\mathbf{V}_{ab}$ and $\mathbf{I}$ in the network of Fig. 11.27 is identical to the terminal equations in the networks of Fig. 11.28a, b. The networks of Fig. 11.28a, b are, respectively, the Thévenin and Norton equivalents of the network of Fig. 11.27. The equivalency of Figs. 11.27 and 11.28a, b means that the sinusoidal steady-state terminal voltage $v_{ab}(t)$ and current $i(t)$ will be identical for all three circuits. This is a very important result that applies to *any* ac network *A*.

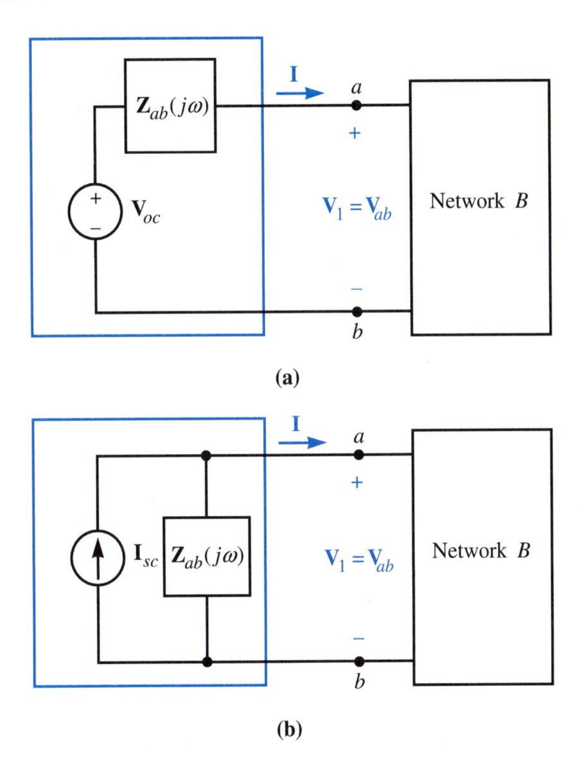

FIGURE 11.28 AC network *A* replaced by (*a*) Thévenin equivalent (enclosed by blue); (*b*) Norton equivalent (enclosed by blue)

As a matter of notation, $\mathbf{Z}_{ab}$ is sometimes denoted $\mathbf{Z}_{Th}$ for Thévenin equivalent impedance (Fig. 11.28a), $\mathbf{Z}_N$ for Norton equivalent impedance (Fig. 11.28b), or $\mathbf{Z}_{Eq}$ for either.

Procedure to Determine Z_{ab}

Let us next derive a procedure to determine $\mathbf{Z}_{ab}$. Refer to Eq. (11.47) to observe that the phasor $\mathbf{V}_{oc}$ is a weighted summation of all the independent source phasors. If all the independent sources are set to zero, then $\mathbf{V}_{oc}$ becomes zero and Eq. (11.48) becomes $\mathbf{V}_{ab} = -\mathbf{Z}_{ab}(j\omega)\mathbf{I}$. This suggests the following procedure to determine $\mathbf{Z}_{ab}(j\omega)$.

Disconnect the load from terminals *a* and *b* and set all independent sources of network *A* equal to zero, as shown in Fig. 11.29a. If no dependent sources are present, series and parallel combinations of impedance can often be used to find the equivalent impedance $\mathbf{Z}_{ab}(j\omega)$ looking into terminals *a* and *b*. If dependent sources are present, connect an independent source between terminals *a* and *b*, and define $\mathbf{I}_{ab}$ as the current entering network *A* at terminal *a*, as shown in Fig. 11.29b. If a current source is used, calculate $\mathbf{V}_{ab}$. If a voltage source is used, calculate $\mathbf{I}_{ab}$. The equivalent impedance of network *A* is $\mathbf{Z}_{ab}(j\omega) = \mathbf{V}_{ab}/\mathbf{I}_{ab}$.

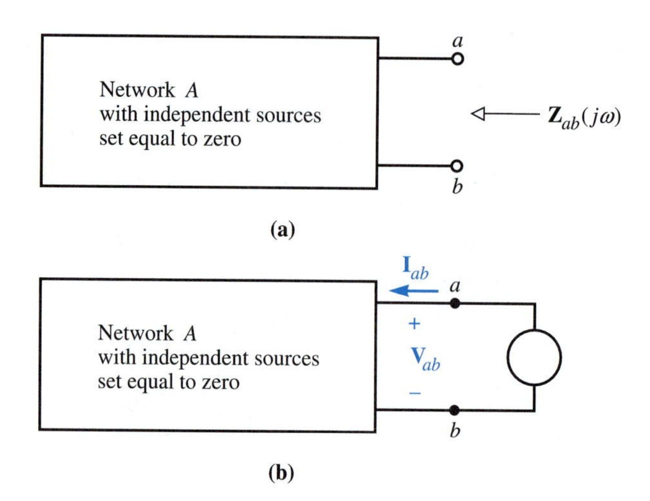

FIGURE 11.29 Network A with independent sources set to zero. (a) Meaning of $\mathbf{Z}_{ab}$; (b) calculation of $\mathbf{Z}_{ab}$ as ratio of driving-point voltage phasor to current phasor: $\mathbf{Z}_{ab} = \mathbf{V}_{ab}/\mathbf{I}_{ab}$. The external source can be either a voltage source or a current source

The following examples illustrate the procedure.

EXAMPLE 11.17 Find the Thévenin and Norton equivalents of the circuit of Fig. 11.30 (looking into terminals a and b).

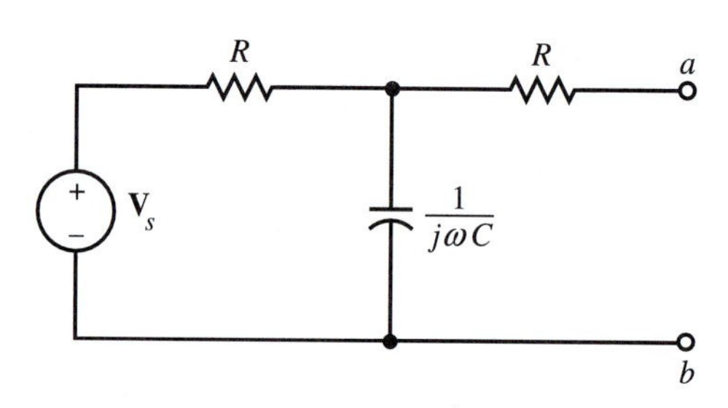

FIGURE 11.30 Circuit for Example 11.17

Solution We will first determine the open-circuit voltage $\mathbf{V}_{oc}$ by the voltage-divider equation:

$$\mathbf{V}_{oc} = \frac{\dfrac{1}{j\omega C}}{\dfrac{1}{j\omega C} + R}\mathbf{V}_s$$

$$= \frac{1}{1 + j\omega RC}\mathbf{V}_s$$

Next, we find the equivalent impedance $\mathbf{Z}_{ab}$ by setting $\mathbf{V}_s = 0$. After replacing the voltage source with a short circuit, we find by inspection that

$$\mathbf{Z}_{ab}(j\omega) = R + R \,\Big\|\, \frac{1}{j\omega C}$$

$$= \frac{R(2 + j\omega RC)}{1 + j\omega RC}$$

We use Eq. (11.50) to obtain the short-circuit current:

$$\mathbf{I}_{sc} = \frac{\mathbf{V}_{oc}}{\mathbf{Z}_{ab}(j\omega)} = \frac{\mathbf{V}_s}{R(2 + j\omega RC)}$$

The resulting Thévenin and Norton equivalents of the network of Fig. 11.30 are shown in Fig. 11.31.

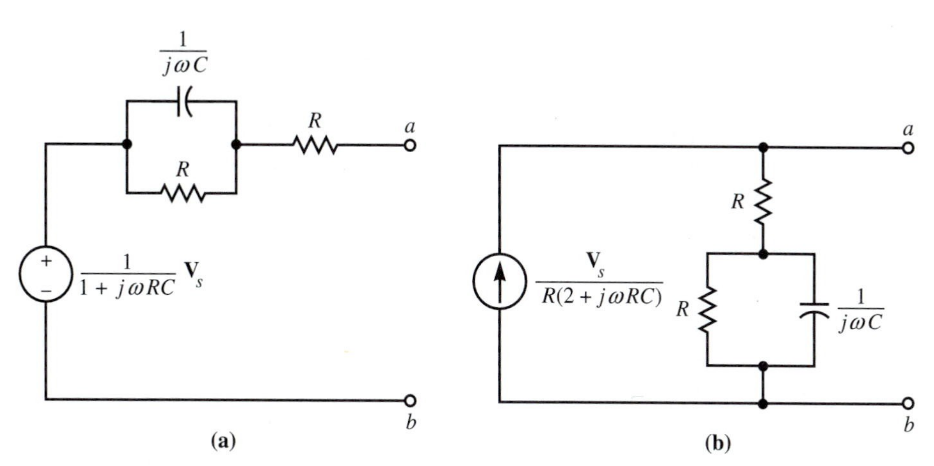

FIGURE 11.31
(*a*) Thévenin and (*b*) Norton equivalents for Example 11.17

EXAMPLE 11.18 Find the Thévenin and Norton equivalents of the network of Fig. 11.32.

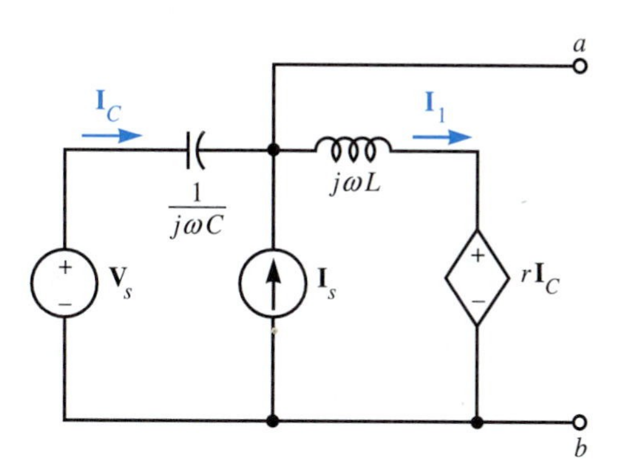

FIGURE 11.32
Circuit for Example 11.18

Solution First find $\mathbf{V}_{oc}$ (open terminals a and b). The inductance current phasor is $\mathbf{I}_1 = \mathbf{I}_s + \mathbf{I}_C$. KVL then yields

$$-\mathbf{V}_s + \frac{1}{j\omega C}\mathbf{I}_C + j\omega L(\mathbf{I}_C + \mathbf{I}_s) + r\mathbf{I}_C = 0$$

so that

$$\mathbf{I}_C = \frac{\mathbf{V}_s - j\omega L\mathbf{I}_s}{\dfrac{1}{j\omega C} + j\omega L + r}$$

The open-circuit voltage phasor is

$$\begin{aligned}\mathbf{V}_{oc} &= \mathbf{V}_s - \frac{1}{j\omega C}\mathbf{I}_C \\ &= \mathbf{V}_s - \left(\frac{\mathbf{V}_s - j\omega L\mathbf{I}_s}{1 - \omega^2 LC + j\omega rC}\right) \\ &= \frac{(-\omega^2 LC + j\omega rC)\mathbf{V}_s + j\omega L\mathbf{I}_s}{1 - \omega^2 LC + j\omega rC}\end{aligned}$$

We will find the Thévenin equivalent impedance $\mathbf{Z}_{ab}(j\omega)$ by setting the independent sources in the network to zero and applying a source to terminals a and b, as shown in Fig. 11.33. The driving-point impedance $\mathbf{Z}_{ab}(j\omega)$ is the ratio of the voltage phasor $\mathbf{V}_{ab}$ to the current phasor $\mathbf{I}_{ab}$. We can write a KCL equation at node a:

$$j\omega C\mathbf{V}_{ab} + (\mathbf{V}_{ab} - r\mathbf{I}_C)\frac{1}{j\omega L} = \mathbf{I}_{ab}$$

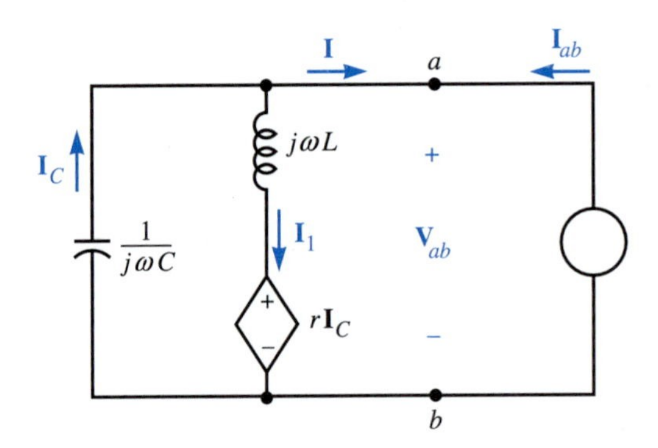

FIGURE 11.33 Method for finding $\mathbf{Z}_{ab}(j\omega)$

where

$$\mathbf{I}_C = -j\omega C\mathbf{V}_{ab}$$

By eliminating $\mathbf{I}_C$ from these equations, we find that

$$\mathbf{V}_{ab} = \frac{j\omega L}{1 - \omega^2 LC + j\omega rC}\mathbf{I}_{ab}$$

Therefore

$$\mathbf{Z}_{ab}(j\omega) = \frac{j\omega L}{1 - \omega^2 LC + j\omega rC}$$

Finally, to obtain $\mathbf{I}_{sc}$, we use Eq. (11.50), with $\mathbf{V}_{oc}$ and $\mathbf{Z}_{ab}(j\omega)$ as given above. The result is

$$\mathbf{I}_{sc} = \mathbf{I}_s + \left(j\omega + \frac{r}{L}\right)C\mathbf{V}_s$$

The Thévenin and Norton equivalents of the network of Fig. 11.32 are shown in Fig. 11.34.

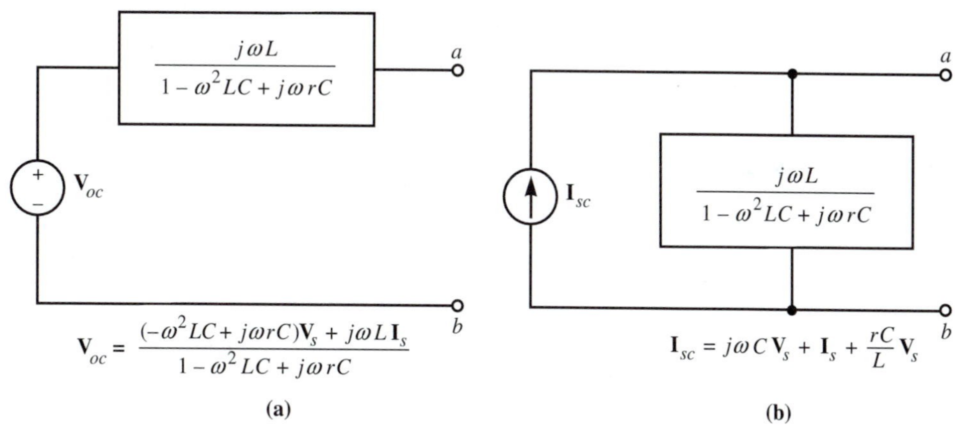

FIGURE 11.34
(*a*) Thévenin and (*b*) Norton equivalents for Example 11.18

We can see from the two preceding examples that the Thévenin and Norton equivalent impedance $\mathbf{Z}_{ab}(j\omega)$ of a network can be a complicated function of ω. However, for a fixed value of ω, denoted by $\omega = \omega_0$, the function $\mathbf{Z}_{ab}(j\omega)$ becomes simply a complex number, and in this case the Thévenin and Norton equivalent circuits can be simplified. To show this, we denote

$$\mathbf{Z}_{ab}(j\omega_0) = \mathbf{Z}_0 \tag{11.52}$$

We then write $\mathbf{Z}_0$ in rectangular form as

$$\mathbf{Z}_0 = R_0 + jX_0 \tag{11.53}$$

where

$$R_0 = \mathcal{R}e\{\mathbf{Z}_0\} \tag{11.54}$$

and

$$X_0 = \mathcal{I}m\{\mathbf{Z}_0\} \tag{11.55}$$

If X_0 is positive, then we can write Eq. (11.53) as

$$\mathbf{Z}_0 = R_0 + j\omega_0 L_0 \tag{11.56}$$

where L_0 is defined by

$$L_0 = \frac{X_0}{\omega_0} \tag{11.57}$$

and has the unit of henry. If X_0 is negative, then we can write Eq. (11.53) as

$$\mathbf{Z}_0 = R_0 + \frac{1}{j\omega_0 C_0} \tag{11.58}$$

where C_0 is defined by

$$C_0 = \frac{-1}{\omega_0 X_0} \tag{11.59}$$

and has the unit of farad. The implication of Eqs. (11.56) and (11.58) is that, *for a fixed frequency* $\omega = \omega_0$, the Thévenin and Norton equivalent impedance $\mathbf{Z}_{ab}(j\omega_0)$ can always be represented very simply as either a resistance R_0 in series with an inductance L_0, or a resistance R_0 in series with a capacitance C_0. (R_0 can be negative for a network containing dependent sources.)

EXAMPLE 11.19 Suppose that in Fig. 11.32, $C = 2\ \mu\text{F}$, $L = 1$ mH, $r = 30\ \Omega$, $\mathbf{V}_s = 10\underline{/0°}$ V, $\mathbf{I}_s = 5\underline{/90°}$ A, and $\omega = 10^4$ rad/s. Find the Thévenin and Norton equivalent circuits.

Solution For the given numerical values, the equivalent impedance (Fig. 11.34)

$$\mathbf{Z}_{ab}(j\omega) = \frac{j\omega L}{1 - \omega^2 LC + j\omega rC}$$

See Problem 11.2 in the PSpice manual.

becomes the complex number

$$\mathbf{Z}_{ab}(j10^4) = 6 + j8\ \Omega$$

Since the reactance is positive, Eqs. (11.56) and (11.57) apply, where

$$R_0 = 6\ \Omega$$

and

$$L_0 = \frac{8}{10^4} = 0.8 \text{ mH}$$

Similarly, from Fig. 11.34, we have

$$\mathbf{V}_{oc} = \frac{(-\omega^2 LC + j\omega rC)\mathbf{V}_s + j\omega L\mathbf{I}_s}{1 - \omega^2 LC + j\omega rC}$$

$$\mathbf{I}_{sc} = j\omega C\mathbf{V}_s + \mathbf{I}_s + \frac{rC}{L}\mathbf{V}_s$$

which become

$$\mathbf{V}_{oc} = 52.34\underline{/136.5°} \text{ V}$$

and

$$\mathbf{I}_{sc} = 5.234\underline{/83.4°} \text{ A}$$

The resulting Thévenin and Norton equivalents are shown in Fig. 11.35.

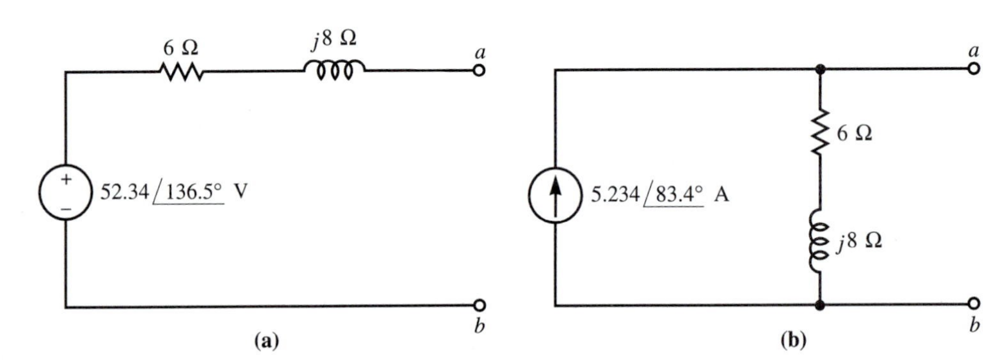

FIGURE 11.35
(*a*) Thévenin and (*b*) Norton equivalents for Example 11.19. The angular frequency is $\omega = 10^4$ rad/s.

Remember The Thévenin and Norton equivalent impedance of a network is, in general, a complicated function of frequency ω. For a *fixed* frequency, this equivalent impedance can always be represented either as a resistance in series with an inductance or as a resistance in series with a capacitance.

EXERCISES Find the Thévenin and Norton ac equivalent circuits of each circuit shown.

19.

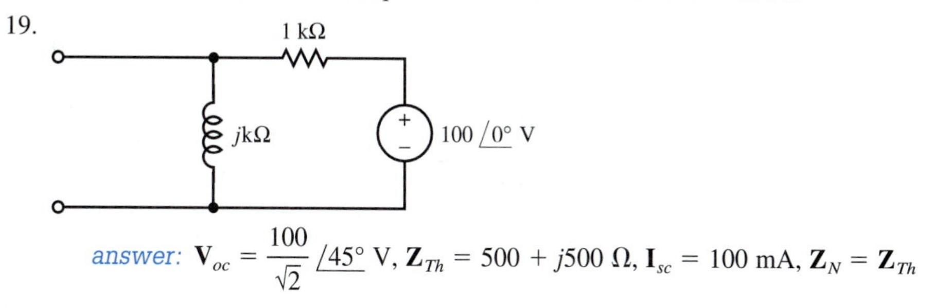

answer: $\mathbf{V}_{oc} = \frac{100}{\sqrt{2}} \underline{/45°}$ V, $\mathbf{Z}_{Th} = 500 + j500\ \Omega$, $\mathbf{I}_{sc} = 100$ mA, $\mathbf{Z}_N = \mathbf{Z}_{Th}$

20.

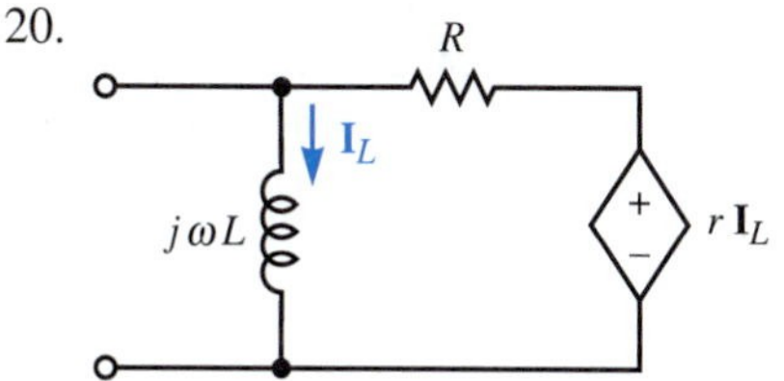

answer: $\mathbf{V}_{oc} = 0$, $\mathbf{Z}_{Th} = \frac{j\omega RL}{R - r + j\omega L}$, $\mathbf{I}_{sc} = 0$, $\mathbf{Z}_N = \mathbf{Z}_{Th}$

21. Determine the Thévenin and Norton equivalent circuits of the network shown.

answer: $\mathbf{V}_{oc} = 0$, $\mathbf{Z}_{Th} = \mathbf{Z}_N = \frac{10\sqrt{2}}{99} \underline{/135°}\ \Omega$

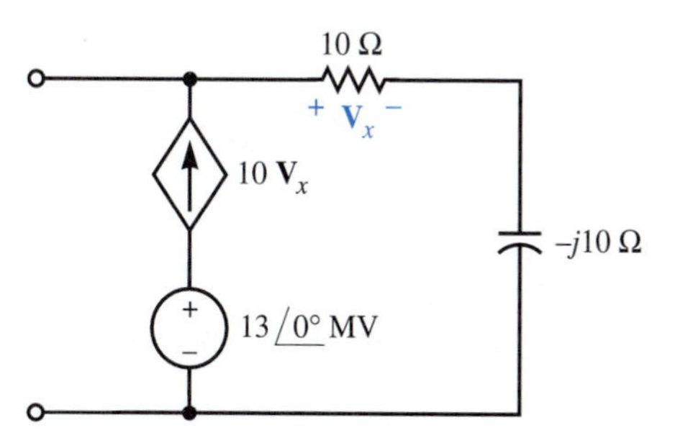

11.7 Summary

In this chapter, we described the methods of ac circuit analysis. These methods apply to circuits operating in the sinusoidal steady state, where every voltage and current waveform is sinusoidal with the same frequency ω. In ac circuit analysis, the amplitude and the phase of a sinusoid, for example, $v(t) = V_m \cos(\omega t + \phi)$, are compactly represented by a phasor, $\mathbf{V} = V_m \underline{/\phi}$. The ac version of Ohm's law, $\mathbf{V} = \mathbf{ZI}$, is used to relate the terminal voltage phasor $\mathbf{V}$ and current phasor $\mathbf{I}$ for R, L, and C. Similarly, algebraic terminal equations involving phasor voltage and current are used for controlled sources. The ac version of KCL,

$$\sum_{\substack{\text{Closed}\\\text{surface}}} \mathbf{I}_n = 0 \tag{11.60}$$

and KVL,

$$\sum_{\substack{\text{Closed}\\\text{path}}} \mathbf{V}_n = 0 \tag{11.61}$$

apply. The implication of all this is that ac circuits can be analyzed with the use of methods similar to those used for circuits composed of resistances and dc sources only. We showed, for example, that impedances in series add and admittances in parallel add. We developed voltage and current divider relations for impedances similar to those of Chapter 3. We formulated the systematic circuit analysis techniques of node-voltage analysis and mesh-current analysis of Chapter 4 for ac circuits. We then applied the principle of superposition to circuits containing more than one sinusoidal source. At this point we showed how to apply ac circuit analysis techniques to circuits containing sinusoidal sources with possibly *different* frequencies by analyzing the ac circuit for each frequency separately. We then extended the concept of Thévenin and Norton equivalent circuits to ac circuits.

KEY FACTS *Concept*	*Equation*	*Section*	*Page*
❑ Analysis of ac circuits relies on the ac version of Ohm's law, $\mathbf{V} = \mathbf{ZI}$, and the phasor versions of KCL,		11.1	335
$\sum_{\text{Closed surface}} \mathbf{I}_n = 0$	(11.17)		339
and KVL, $\sum_{\text{Closed path}} \mathbf{V}_n = 0$	(11.19)		341
❑ Impedances in series add.	(11.30)	11.2	348
❑ Admittances in parallel add.	(11.22)	11.2	346
❑ Algebraic voltage and current divider relations apply for phasor voltages and currents.	(11.31) (11.32)	11.2	348
❑ Algebraic node-voltage and mesh-current circuit analysis applies for phasor voltages and currents.		11.3 11.4	350 356
❑ When we apply superposition to ac circuit analysis, independent sources with different frequencies *must* be considered separately. It is *never* correct to add nonrotating phasors if they represent sinusoids that have different frequencies.		11.5	359

KEY FACTS	Concept	Equation	Section	Page
❑	The Thévenin equivalent of an ac circuit consists of the open-circuit voltage phasor, $\mathbf{V}_{oc}$, in series with the Thévenin equivalent impedance.		11.6	367
❑	The Norton equivalent of an ac circuit consists of the short-circuit current phasor, $\mathbf{I}_{sc}$, in parallel with the Norton equivalent impedance.		11.6	367
❑	The Thévenin and Norton equivalent impedances are equal and are given by $\mathbf{Z}_{Eq} = \mathbf{V}_{oc}/\mathbf{I}_{sc}$.	(11.51)	11.6	366

PROBLEMS

Section 11.1

Give the formula for the time-domain waveform associated with each of the following phasors. The units are milliamperes and volts. The angular frequency is ω.

1. $\mathbf{I} = 6\angle 30^\circ$
2. $\mathbf{I} = 1 + j$
3. $\mathbf{V} = 7e^{-j45^\circ}$
4. $\mathbf{V} = 10j$
5. (a) Find the numerical values of the phasor currents $\mathbf{I}_1$, $\mathbf{I}_2$, and $\mathbf{I}_3$ corresponding to the time-domain currents shown in Fig. 11.36. Put your answers in polar form.
   (b) Use your answer from part (a) to find $i_1(t)$. Express your answer in the form $i_1(t) = A\cos(100t + \phi)$.
   (c) Carefully plot $i_1(t)$, $i_2(t)$, and $i_3(t)$. Check your results by noting from the plots that $i_1(t) = i_2(t) + i_3(t)$.
   (d) Refer to a phasor diagram of the equation $\mathbf{I}_1 = \mathbf{I}_2 + \mathbf{I}_3$ to explain why the amplitude of $i_1(t)$ is *not* 50.
   (e) Explain why the amplitude of $i_1(t)$ is *not* 50 by referring to your plots from (c).
6. Assume that $v(t) = 3\cos 100t - 3\sin 100t$.
   (a) Find the phasor representation, $\mathbf{V}$, of $v(t)$.
   (b) Sketch $\mathbf{V}$ in the complex plane.
   (c) Use your answer from (a) and the formula $v(t) = \mathcal{R}e\{\mathbf{V}e^{j100t}\}$ to put $v(t)$ in the form $v(t) = A\cos(100t + \phi)$.
   (d) Check your answer to (c) using trigonometric identities.
7. For the circuit shown in Fig. 11.37, find $\mathbf{I}_1$ if $\mathbf{I}_4 = 4\angle 90^\circ$, $\mathbf{I}_5 = 5\angle 45^\circ$, and $\mathbf{I}_6 = 6\angle 135^\circ$ A.
8. Find the differential equation relating $v(t)$ to $i(t)$ for a circuit with transfer impedance

$$\mathbf{Z} = \frac{1}{1 + j\omega 3}\ \Omega$$

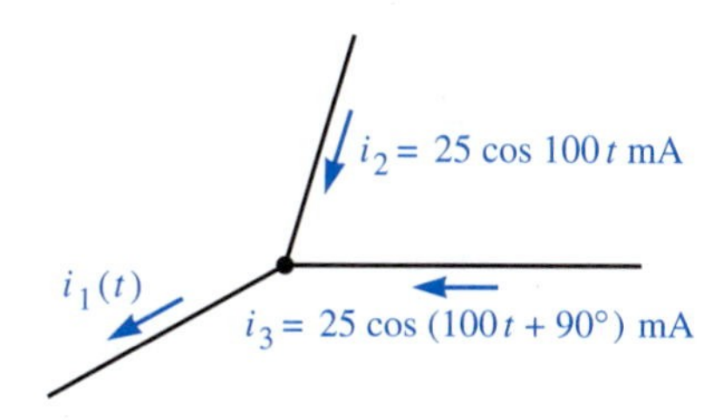

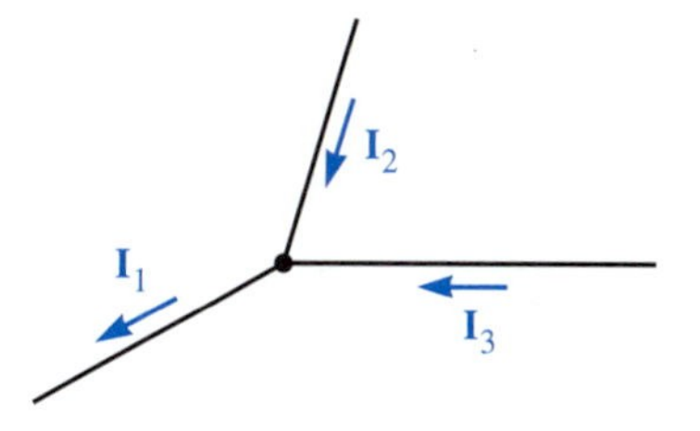

FIGURE 11.36

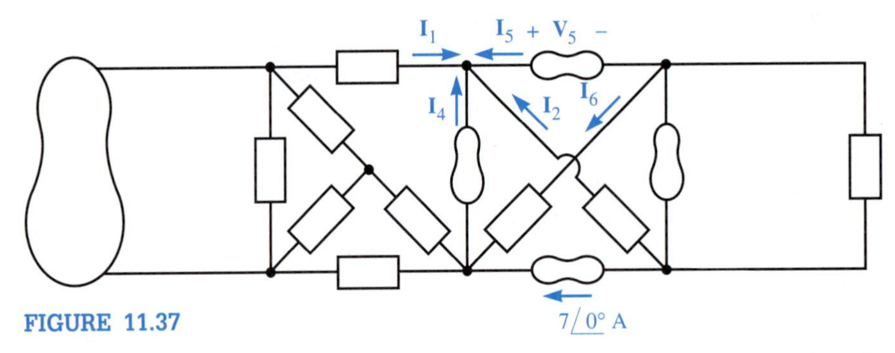

FIGURE 11.37

9. Find the differential equation relating $v(t)$ to $i(t)$ for a circuit with transfer admittance

$$\mathbf{Y} = \frac{j\omega}{2 + 4j\omega} \text{ S}$$

10. Find the differential equation relating $v(t)$ to $i(t)$ if

$$\frac{\mathbf{I}}{\mathbf{V}} = j\omega C + \frac{1}{R} + \frac{1}{j\omega L}$$

11. An ac circuit operates at $\omega = 100$ rad/s. We are given the system of equations

$$\begin{bmatrix} 3 & 2j \\ 1 & -2 \end{bmatrix} \begin{bmatrix} \mathbf{I}_1 \\ \mathbf{I}_2 \end{bmatrix} = \begin{bmatrix} 12\underline{/60^\circ} \\ 4 \end{bmatrix}$$

Find $i_1(t)$ and $i_2(t)$.

Section 11.2

12. Find the driving-point impedance of the following circuit by inspection.

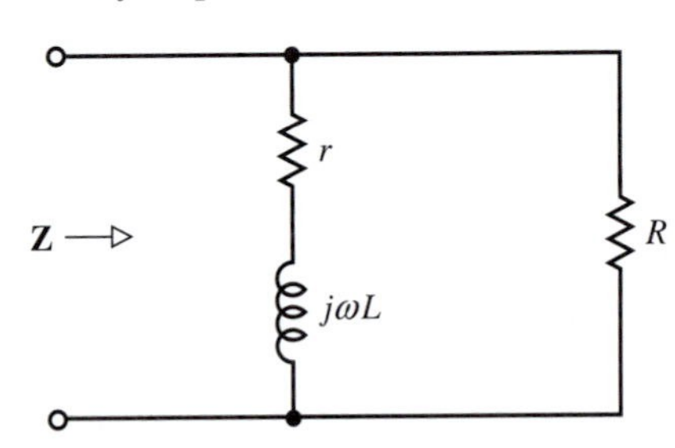

13. Find the driving-point impedance of the following circuit by inspection.

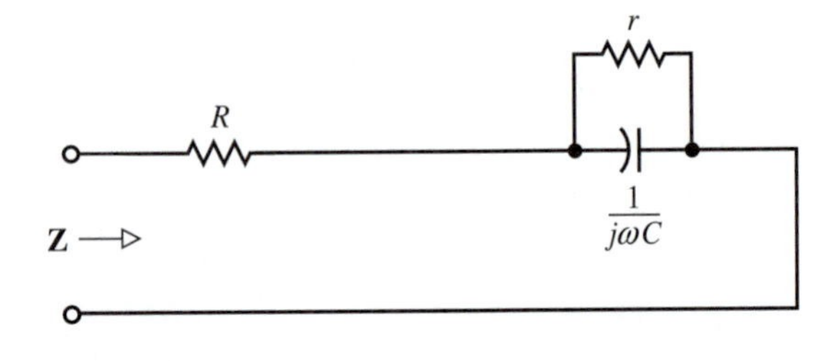

14. For the following circuit, find $\mathbf{V}_a$ and $\mathbf{I}_b$ by inspection.

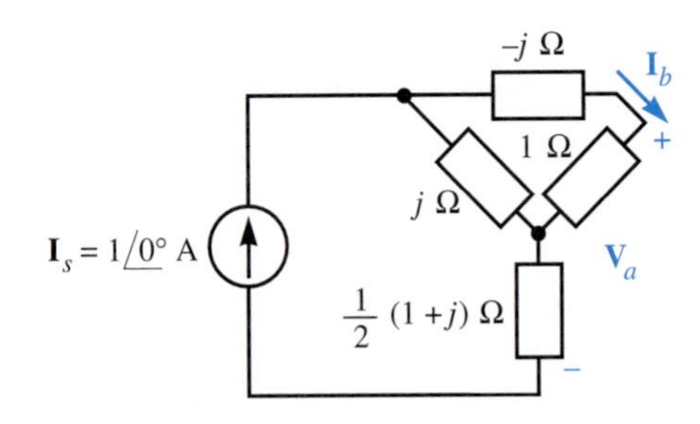

15. (a) Use the voltage-divider equation to determine $\mathbf{V}_R$, $\mathbf{V}_L$, and $\mathbf{V}_C$ in the series RLC circuit shown below.

(b) Write down the time-domain expressions for the voltages $v_L(t)$, $v_C(t)$, $v_R(t)$ and the source voltage $v_s(t)$. Assume that $\omega = 2000\pi$ rad/s.

(c) Your answer to part (b) should indicate that the voltage across the resistance, $v_R(t)$, is equal to the voltage across the voltage source, $v_s(t)$, even though the voltages $v_L(t)$ and $v_C(t)$ are not zero. Plot $v_L(t)$ and $v_C(t)$ and their sum, $v_L(t) + v_C(t)$.

(d) Since $v_L(t) + v_C(t) = 0$, one might assume that the series inductance and capacitance can be replaced with a short circuit without affecting the voltage across R. Is this assumption correct?

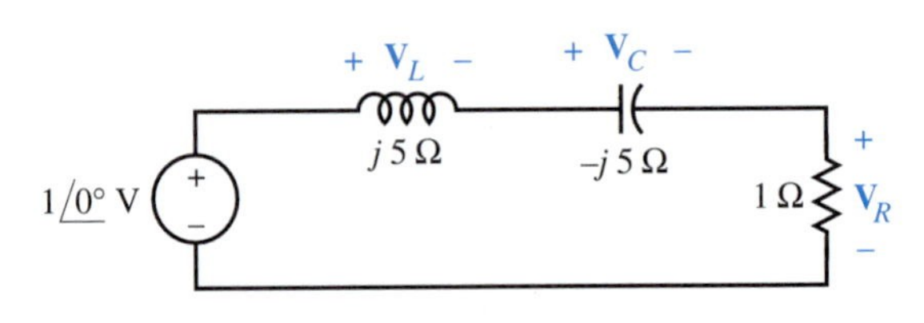

16. (a) For what value(s) of $\mathbf{Z}$ does the driving-point impedance $\mathbf{Z}_{\text{in}}$ of the circuit shown next equal $\mathbf{Z}$?

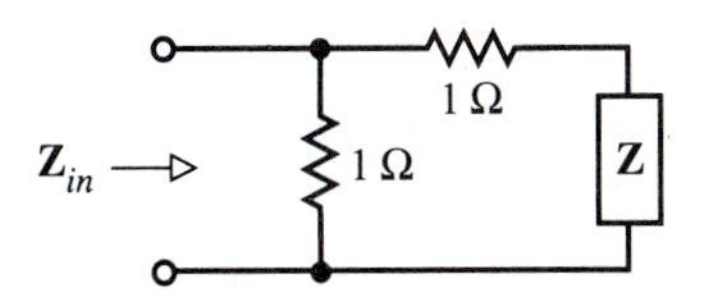

(b) Let $\mathbf{Z}_0$ denote the answer to (a). Show that the driving-point impedance of the circuit shown below also equals $\mathbf{Z}_0$.

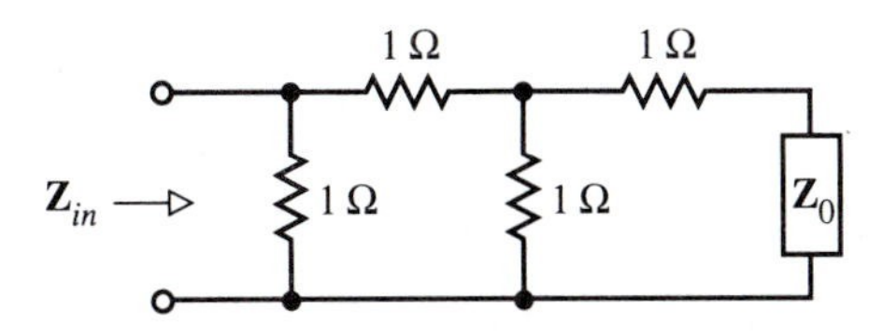

(c) Show that the driving-point impedance of the infinite ladder network shown below also equals $\mathbf{Z}_0$. *Hint*: Notice that we can make the network of part (b) one stage longer by replacing the rectangular element having impedance $\mathbf{Z}_0$ with the circuit of part (a) since both have impedance $\mathbf{Z}_0$.

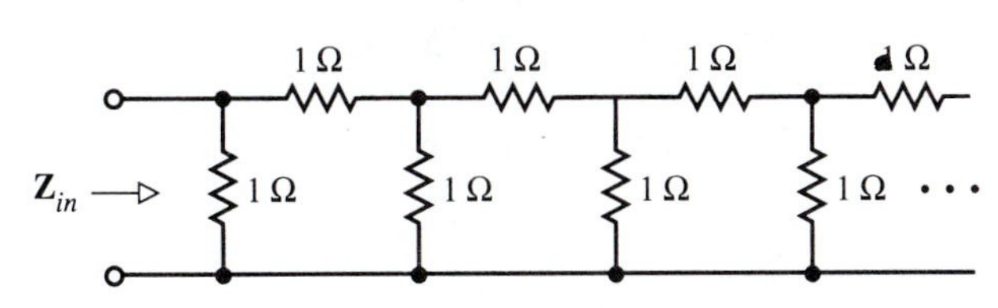

17. Compare the driving-point impedances of the circuits shown below.

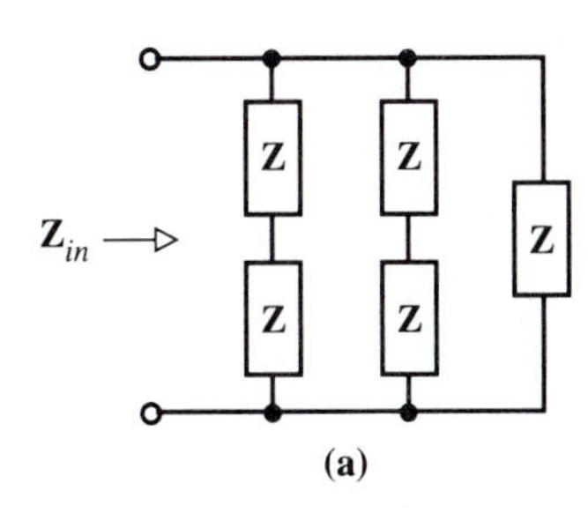

(a)

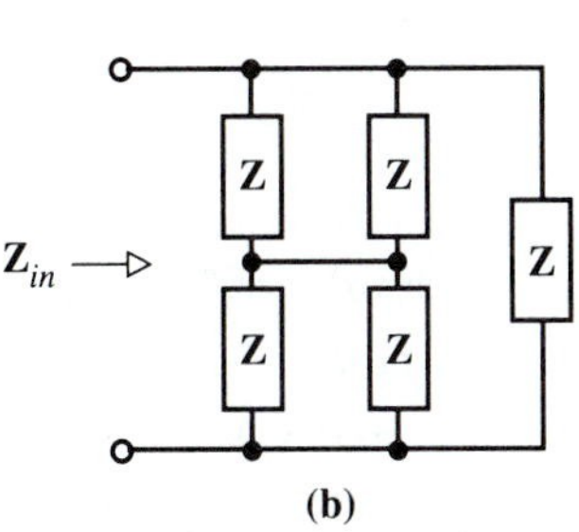

(b)

18. (a) Find the voltage transfer function $\mathbf{V}_2/\mathbf{V}_1 = \mathbf{H}_{21}(j\omega)$ by inspection of the circuit shown below.

(b) Now find $\mathbf{V}_3/\mathbf{V}_2 = \mathbf{H}_{32}(j\omega)$ by inspection.

(c) Multiply your results from (a) and (b) to determine $\mathbf{V}_3/\mathbf{V}_1 = \mathbf{H}_{31}(j\omega)$.

(d) Use your result from part (c) to find the differential equation relating $v_3(t)$ to $v_1(t)$.

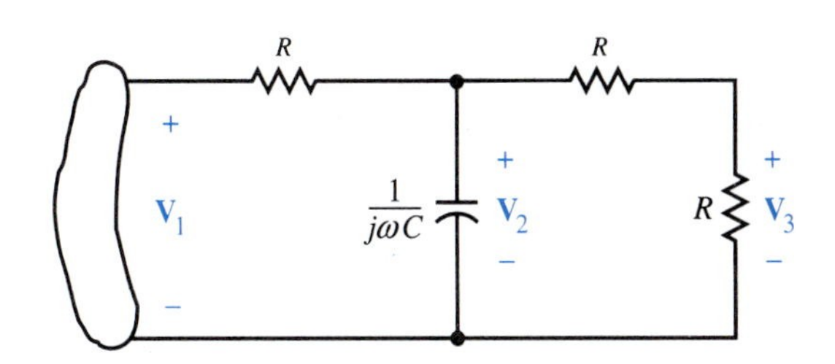

19. By inspection, find the formulas for $\mathbf{I}_2$, $\mathbf{I}_3$, $\mathbf{V}_1$, and $\mathbf{V}_4$ in the circuit below.

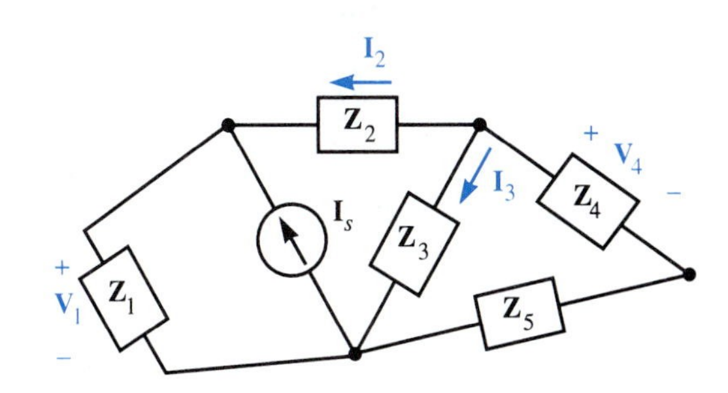

Section 11.3

20. Set up and solve the node-voltage equations for the circuit shown below.

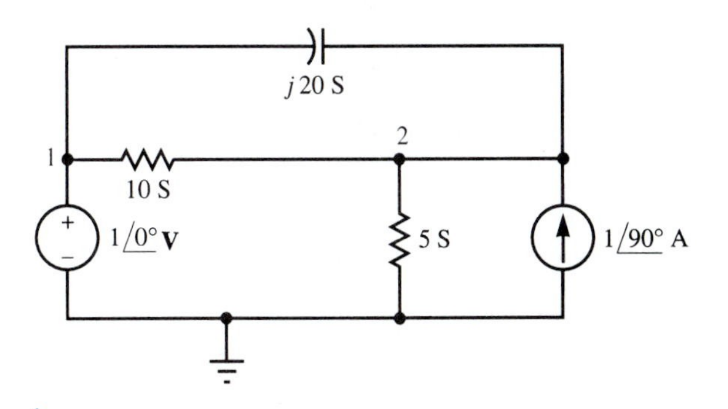

21. Transform the solutions to the node-voltage equations obtained in Problem 20 to the time domain. Assume that the capacitance has a value of 2 F.

22. Set up the node-voltage equations for the following circuit.

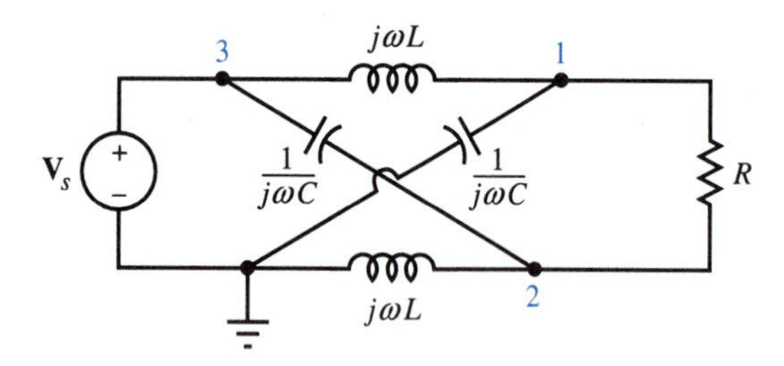

23. Set up and solve the node-voltage equations for the following circuit.

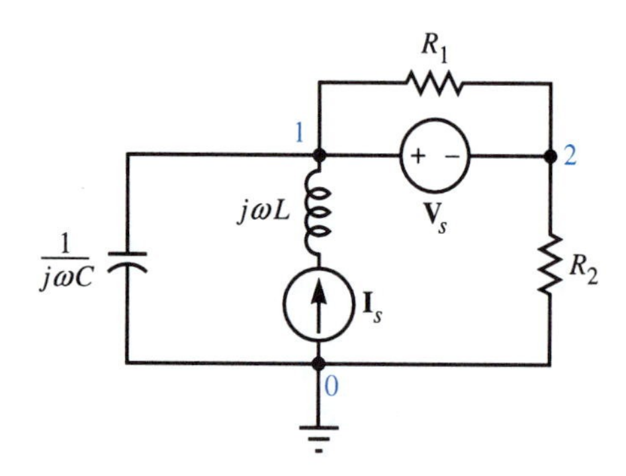

24. Set up and solve the node-voltage equations for the following circuit.

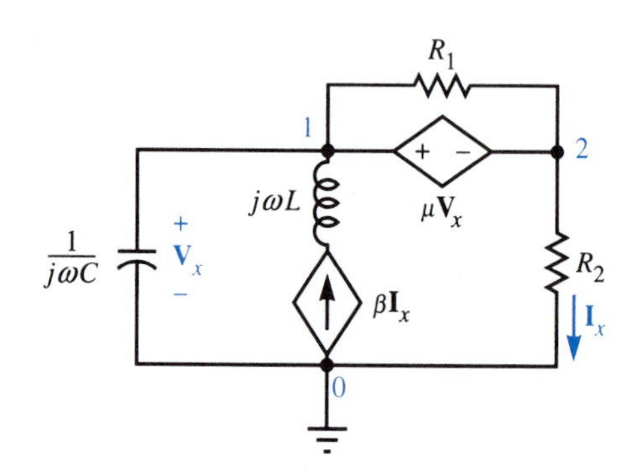

25. Set up the node-voltage equations for the following circuit.

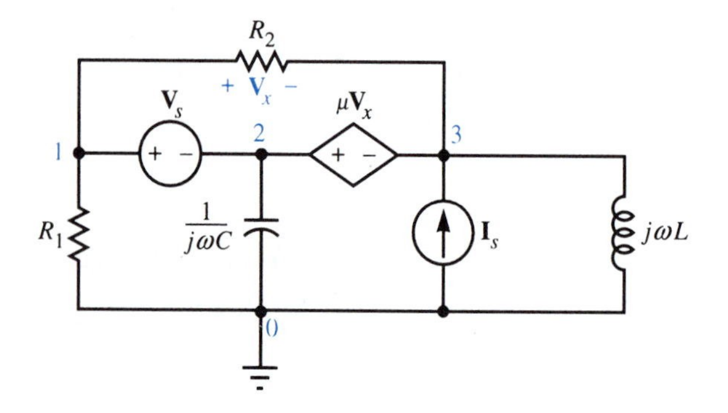

26. Set up and solve the node-voltage equations for the following circuit. 11.3

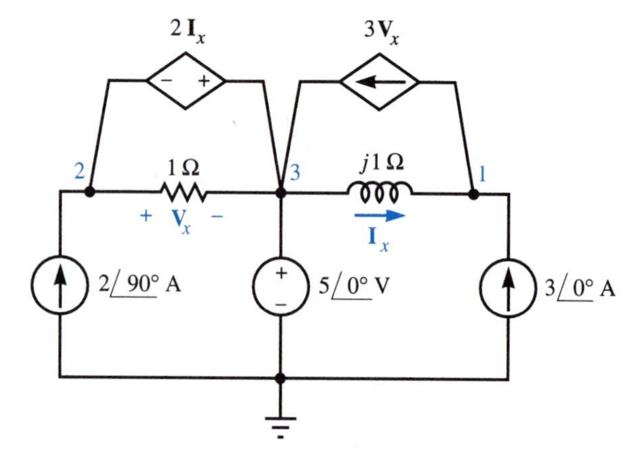

27. (a) Set up, in matrix form, the node-voltage equations for the Δ (top) and Y (bottom) circuits driven by current sources shown below.
    (b) Give the admittance matrix for each circuit.
    (c) Find the relationship between the Δ-circuit admittances and the Y-circuit admittances that would make the Δ and Y circuits electrically equivalent. Specifically, solve for $\mathbf{Y}_a$, $\mathbf{Y}_b$, and $\mathbf{Y}_c$ in terms of $\mathbf{Y}_1$, $\mathbf{Y}_2$, and $\mathbf{Y}_{12}$.

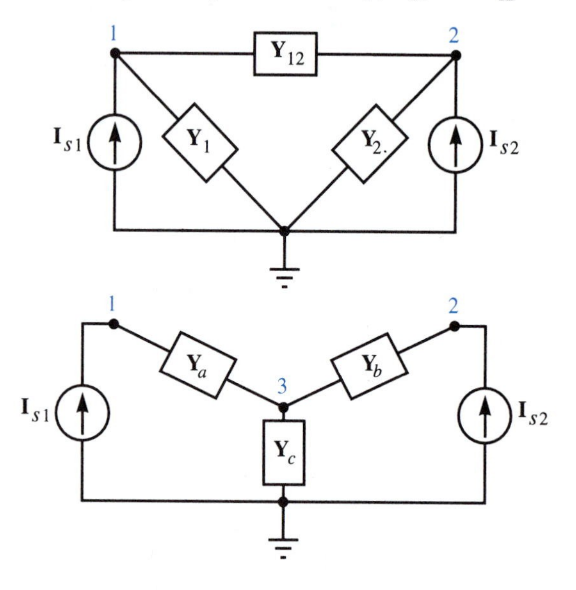

28. (a) Set up, in matrix form, the node-voltage equations for the circuit in Fig. 11.38.
    (b) What is the network's admittance matrix?

29. (a) Set up the node-voltage equations for the finite-gain op-amp circuit in Fig. 11.39.
    (b) Solve for $\mathbf{V}_o$.

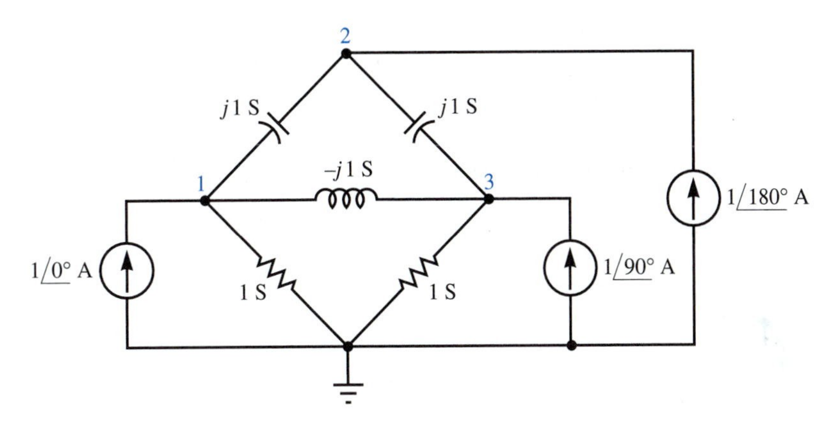

FIGURE 11.38

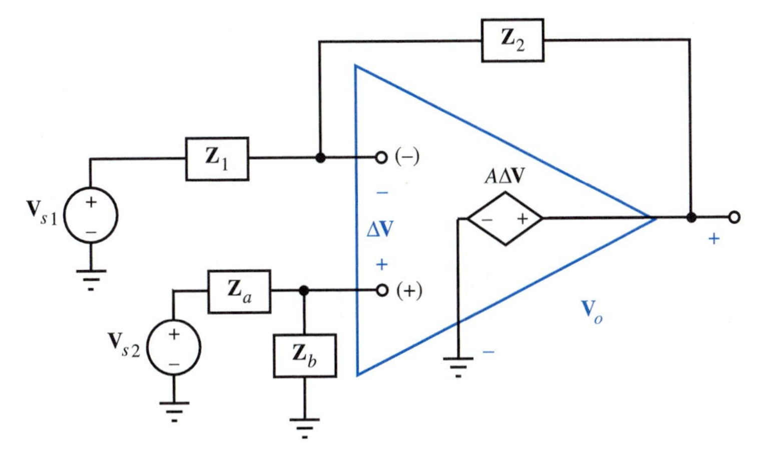

FIGURE 11.39

(c) Let $A \to \infty$ in your answer to (b). Does your result agree with that in Example 11.11?

30. The illustration below depicts a basic four-arm ac bridge used to measure impedance. The bridge is said to be balanced when $\mathbf{V}_{ac} = 0$. Show that the bridge is balanced if $\mathbf{Z}_1\mathbf{Z}_4 = \mathbf{Z}_2\mathbf{Z}_3$.

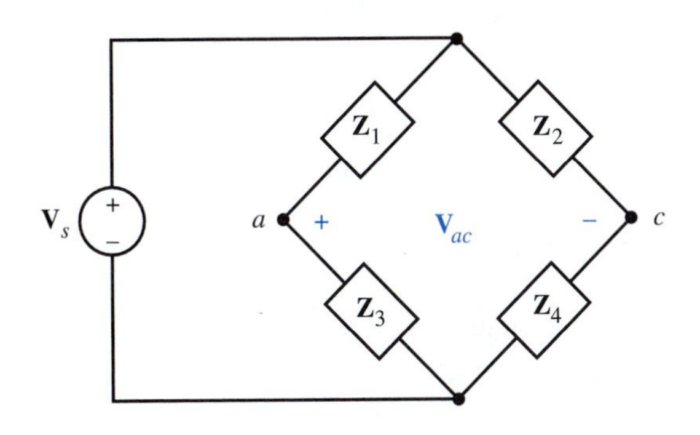

31. The ac bridge of Problem 30 can be used to measure a capacitor modeled by a capacitance C_x in series with a resistance R_x by setting $\mathbf{Z}_1 = R_1$, $\mathbf{Z}_2 = R_2$, $\mathbf{Z}_3 = R_3 + 1/j\omega C_3$, and $\mathbf{Z}_4 = \mathbf{Z}_x = R_x + 1/j\omega C_x$.

    (a) Show that when the bridge is balanced, C_x and R_x are given by $C_x = C_3R_1/R_2$ and $R_x = R_2R_3/R_1$, independent of ω.

    (b) How many of the bridge elements must be made adjustable to obtain a balance?

32. The ac bridge of Problem 31 can be used to measure an inductor (modeled by an inductance L_x in series with a resistance R_x) by setting $\mathbf{Z}_1 = R_1 \parallel (1/j\omega C_1)$, $\mathbf{Z}_2 = R_2$, $\mathbf{Z}_3 = R_3$, and $\mathbf{Z}_4 = \mathbf{Z}_x$. This is called a Maxwell bridge.

    (a) Show that when the bridge is balanced, L_x and R_x are given by $L_x = R_2R_3C_1$ and $R_x = R_2R_3/R_1$.

    (b) How many bridge elements must be made adjustable to obtain a balance?

33. (a) Set up, in matrix form, the node-voltage equations for $\mathbf{V}_1$ and $\mathbf{V}_2$ for the circuit shown.
    (b) Transform the equations of (a) into the time domain.
    (c) Solve the equations of (a) for $\mathbf{V}_2$. (Eliminate $\mathbf{V}_1$.)

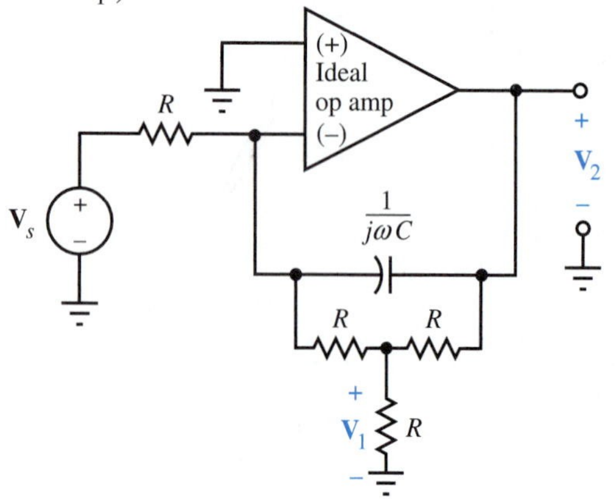

Section 11.4

34. Assign mesh currents to the circuit of Problem 23. Then set up and solve the mesh-current equations.
35. Set up the mesh-current equations for the following circuit.

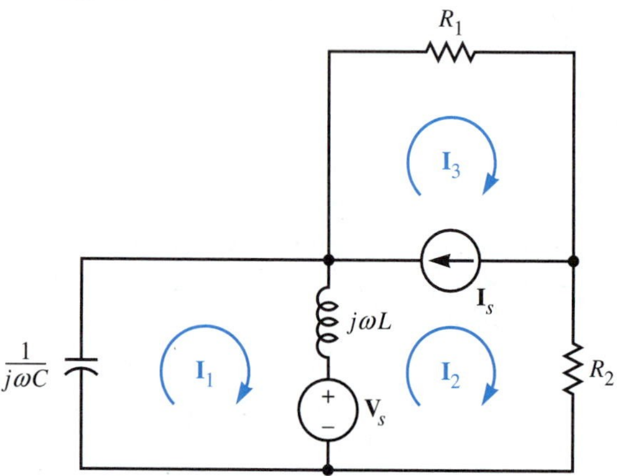

36. Write a set of mesh-current equations for the circuit shown.

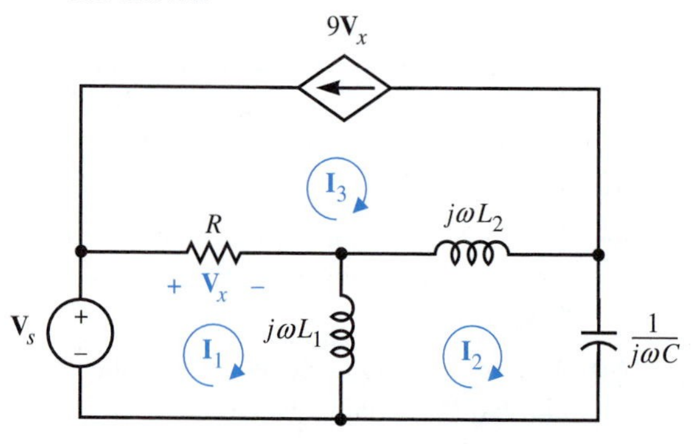

37. Set up the mesh-current equations for the following circuit.

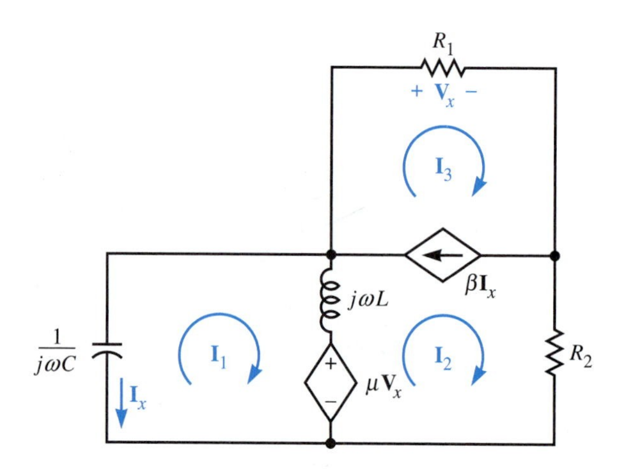

38. Set up and solve the mesh-current equations for the circuit shown.

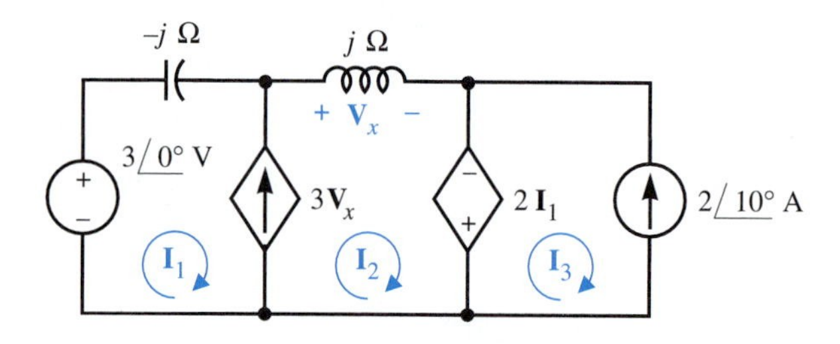

39. Set up the mesh-current equations for the following circuit.

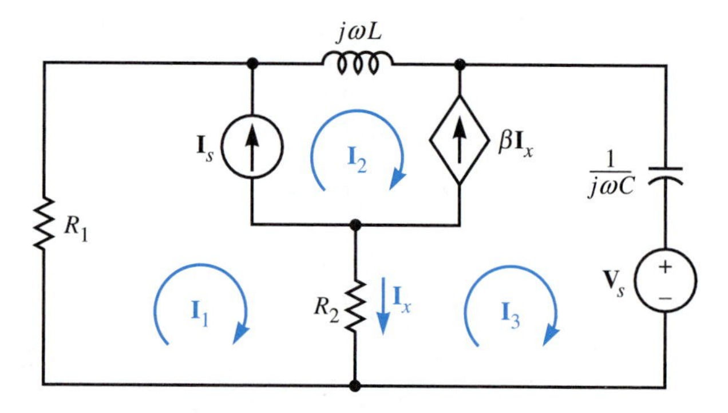

40. Draw a simple circuit that cannot be solved using mesh-current analysis.
41. (a) Set up, in matrix form, the mesh-current equations for the following circuit.
    (b) What is the circuit's impedance matrix?

(c) Solve for the mesh currents.

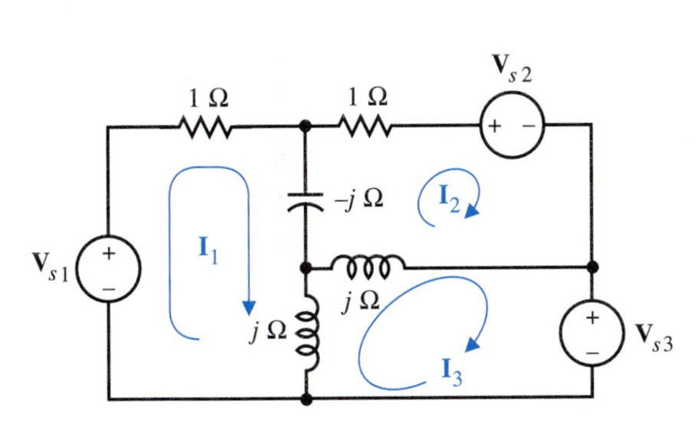

Section 11.5

42. Find $i_o(t)$ in the circuit shown below.

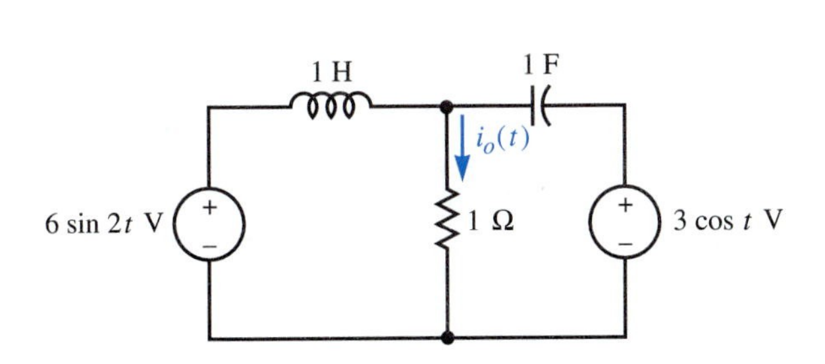

43. Find $v_o(t)$ in the following circuit.

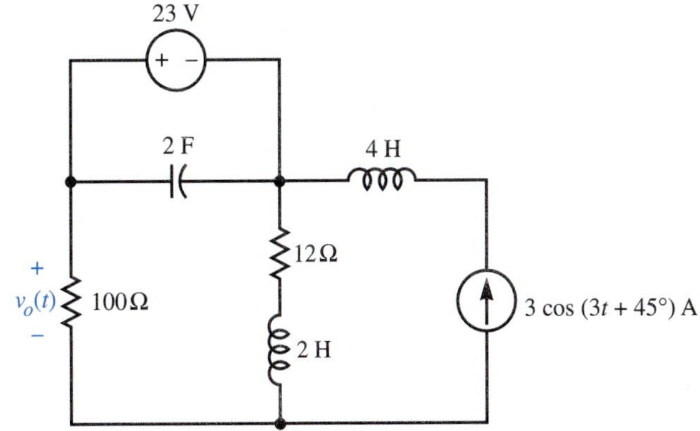

44. Find $i(t)$ in the circuit in Fig. 11.40.

45. Find $v(t)$ in the circuit in Fig. 11.41.

46. Find $v_o(t)$ in the circuit shown below.

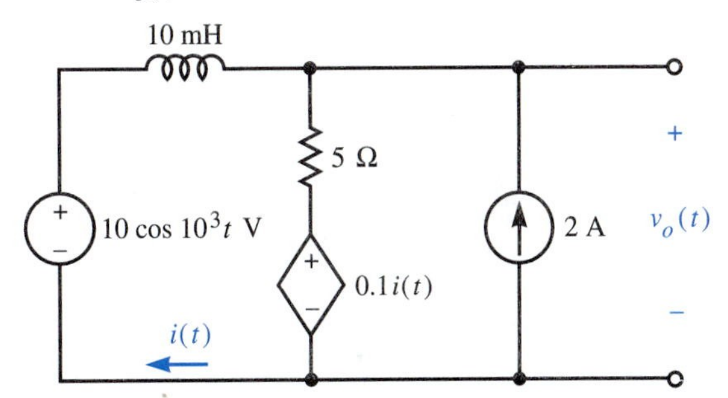

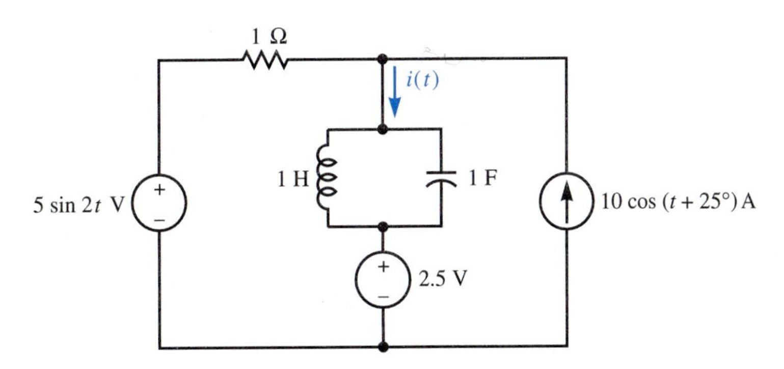

FIGURE 11.40

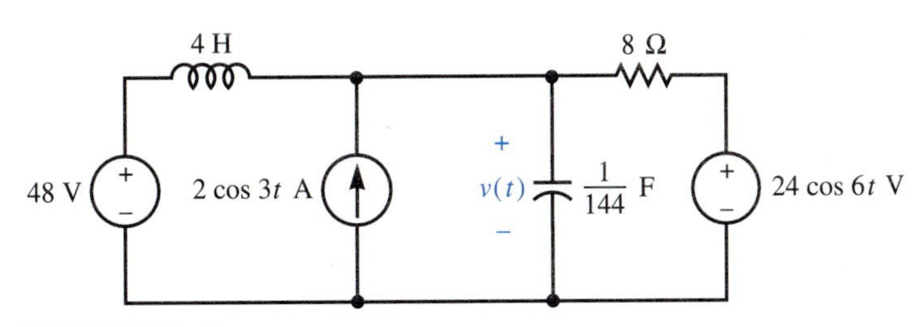

FIGURE 11.41

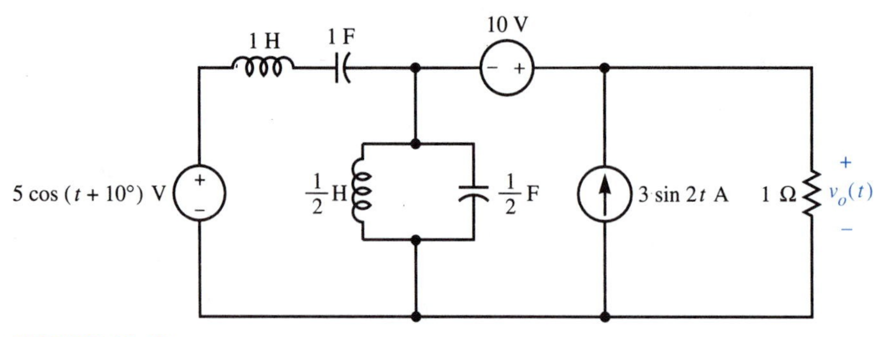

FIGURE 11.42

47. Find $v_o(t)$ in the circuit in Fig. 11.42.

48. Find $v_o(t)$ for the ideal op-amp circuit shown.

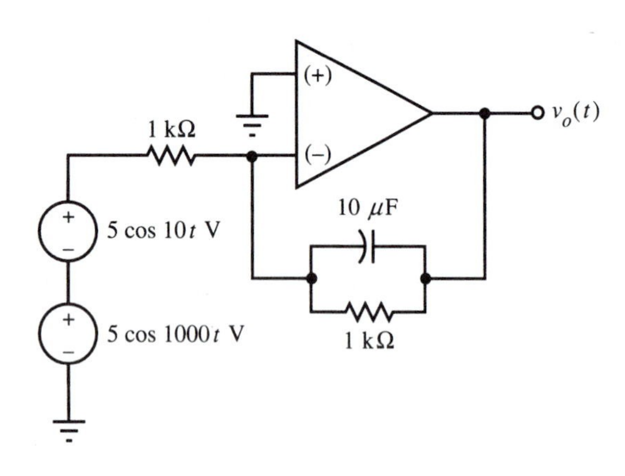

Section 11.6

Find the Thévenin and Norton equivalents of each of the following circuits.

49.

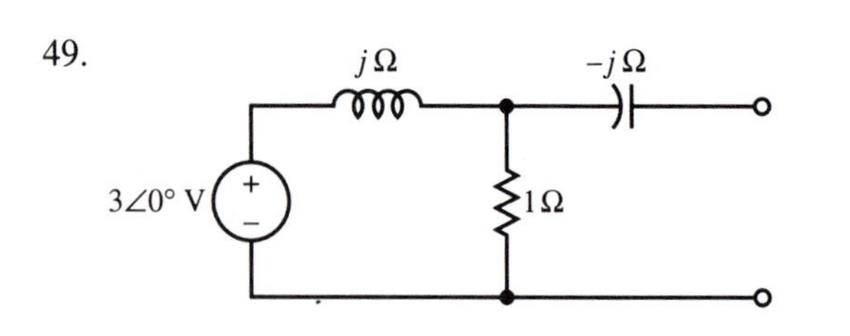

50.

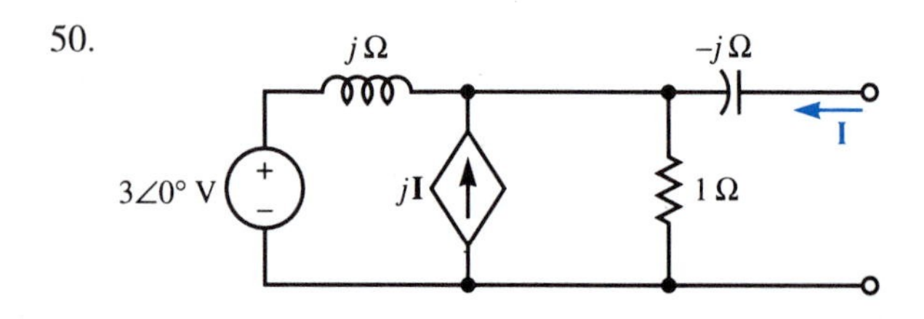

51.

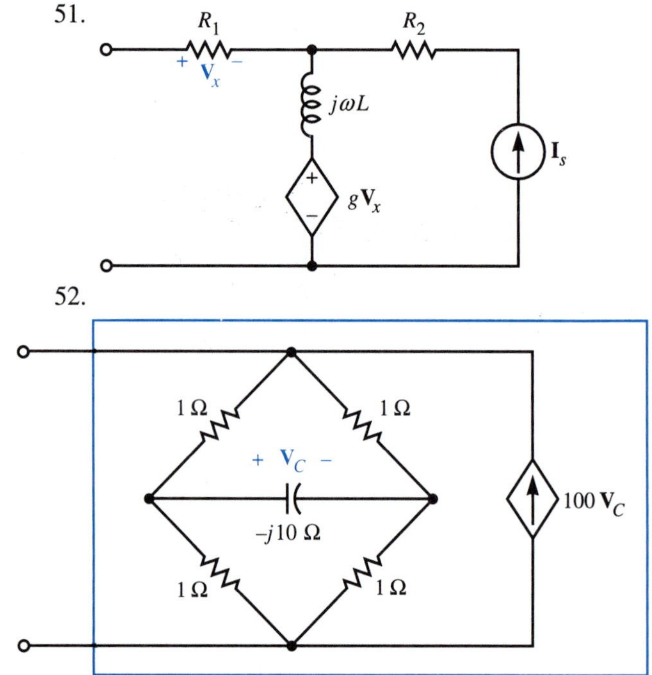

52.

53. (a) Find $\mathbf{V}_{oc}$ for the following circuit.
(b) Find $\mathbf{Z}_{Eq}$.
(c) Show that $\mathbf{I}_{sc}$ cannot be determined.

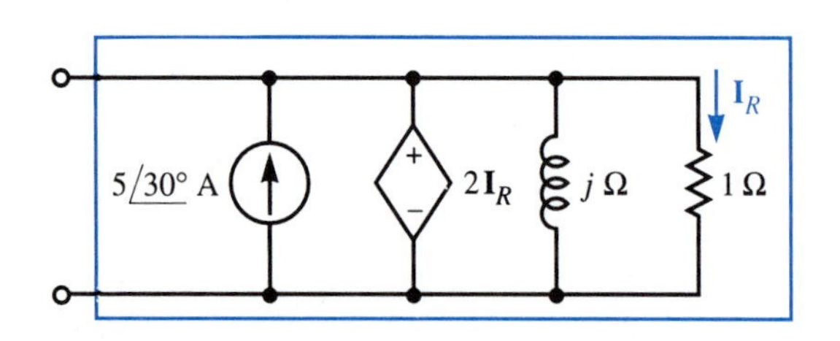

54. We are given the Thévenin equivalents of each of two separate circuits. The two circuits are then

connected in parallel. Find the Thévenin equivalent of the parallel circuit.

55. Find the Thévenin and Norton equivalents of the following circuit.

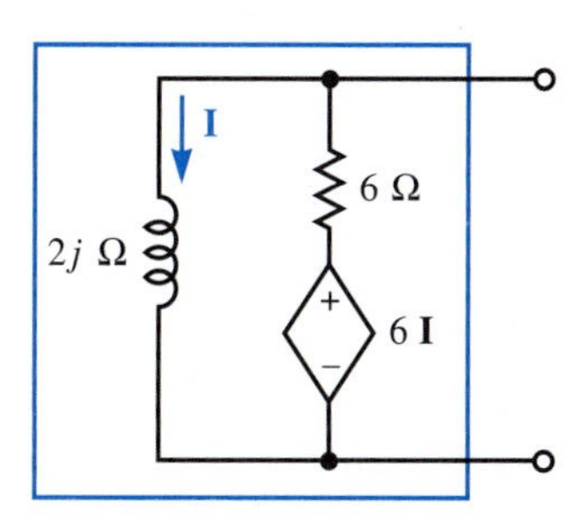

56. The following circuit does not have a Thévenin equivalent. Why not?

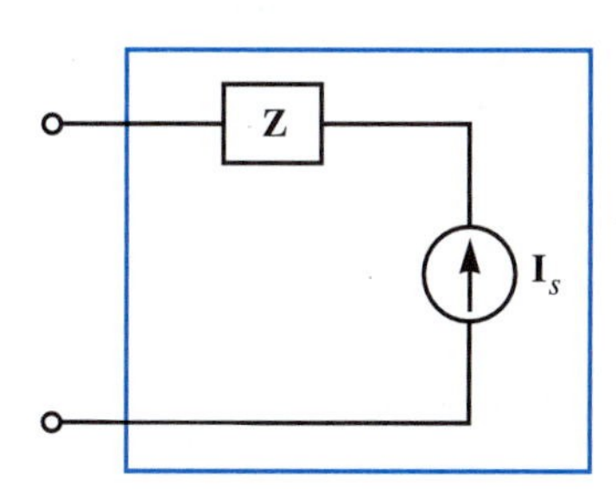

57. The following circuit does not have a Norton equivalent. Why not?

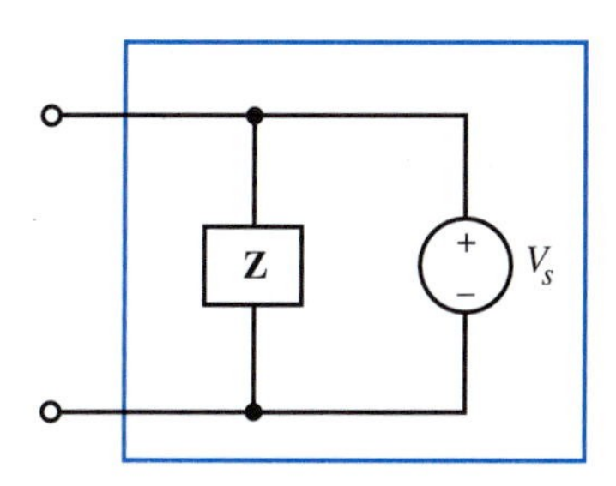

58. A linear time-invariant electrical network, network A, is terminated (separately) with a 1-Ω resistance and a 1000-μF capacitance as shown in the next figure. The respective currents (steady-state responses) are $i_1(t) = 5 \cos(1000t - 45°)$ A and $i_2(t) = 10 \cos(1000t - 45°)$ A.

(a) Find the Thévenin and Norton equivalents of network A.

(b) Assume that the capacitance or resistance is replaced with a 1-mH inductance. What current $i_3(t)$ will flow through the inductance?

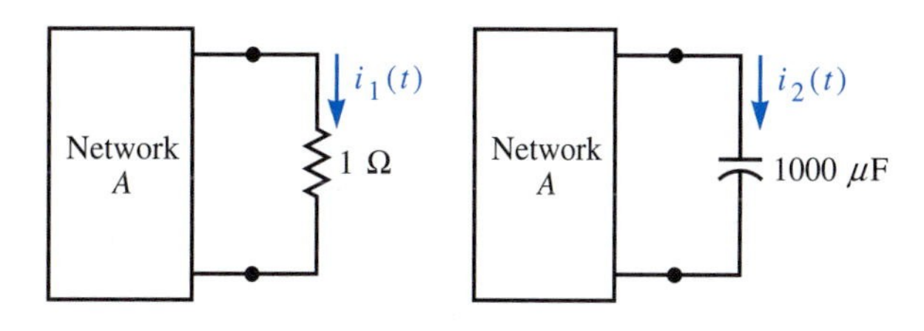

59. An unknown ac network with terminal pair a-b is terminated with a series RC network, where $R = 1\ \Omega$ and $C = 1\ \mu$F. The resulting steady-state current is $i_{ab}(t) = 5\ \cos(10^6 t + 45°)$ A. The series RC network is disconnected and the ac network is terminated with a parallel RLC network, where $R = 1\ \Omega$, $L = 1\ \mu$H, and $C = 1\ \mu$F. The resulting steady-state voltage is $v_{ab}(t) = (5/\sqrt{2}) \cos(10^6 t)$ V. The parallel RLC network is removed and the ac network is terminated with a 1-Ω resistance in series with a 1-μH inductance. Find the steady-state current, $i_{ab}(t)$.

60. When two ac networks, A and B, are connected together as shown in the first illustration below, $\mathbf{I}_{ab} = 1\angle 0°$ A and $\mathbf{V}_{aa'} = \sqrt{2}\angle -45°$ V. When the same networks are connected as shown in the second illustration, $\mathbf{I}_{ab'} = 3\angle 0°$ A and $\mathbf{V}_{aa'} = \sqrt{2}\angle 45°$ V. Find the Thévenin and Norton equivalents of each ac network A and B.

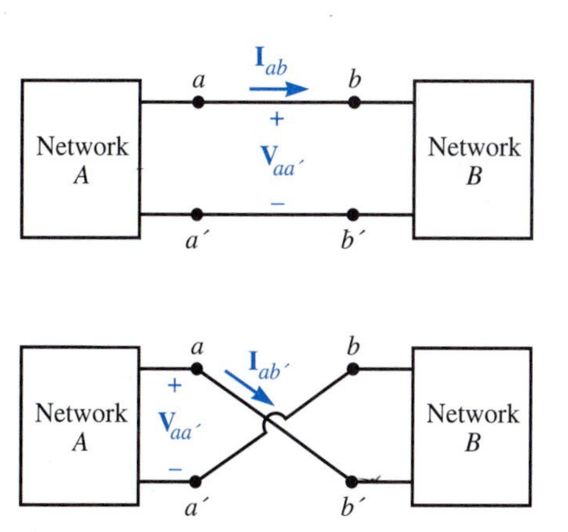

61. (a) Find the Thévenin equivalent circuit looking back into the output of the finite-gain op-amp circuit in Fig. 11.43 on the next page.

(b) What happens to your answer from (a) in the limit $A \rightarrow \infty$?

Supplementary Problems

62. Use any method to find the driving-point impedance of the network below.

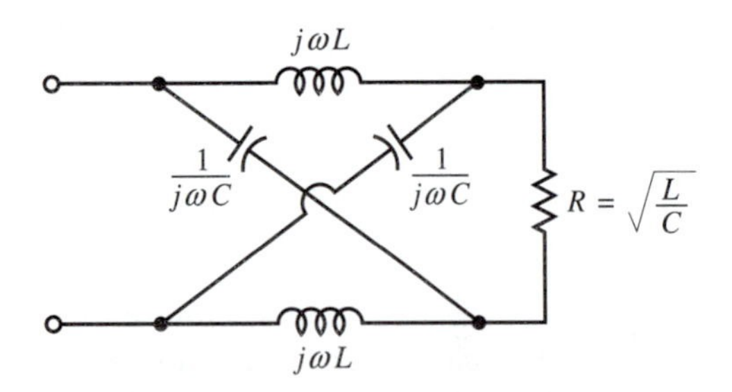

63. (a) The feedback ratio of the voltage amplifier circuit model shown in Fig. 11.44 is $\mathbf{V}_{Fb}/\mathbf{V}_o = \beta = \mathbf{Z}_a/(\mathbf{Z}_a + \mathbf{Z}_b)$. Show that the closed-loop gain $G = \mathbf{V}_o/\mathbf{V}_s$ is given by

$$G = \frac{A\beta}{A\beta + 1}\frac{1}{\beta}$$

where $A = A(j\omega)$ is the open-loop gain of the op amp.

(b) Under what condition does $G \simeq 1/\beta$?

64. Consider the circuit shown in Fig. 11.45.

(a) Show that

$$G = \frac{\mathbf{V}_o}{\mathbf{V}_s} = -\frac{\mathbf{Z}_b}{\mathbf{Z}_a}$$

for $|A\beta| \gg 1$, where β is the feedback factor, $\beta = \mathbf{Z}_a/(\mathbf{Z}_a + \mathbf{Z}_b)$.

(b) Assume that $\mathbf{Z}_a = 1/j\omega C$ and $\mathbf{Z}_b = R$. Use the approximation in (a) to obtain a differential equation relating $v_o(t)$ to $v_s(t)$.

(c) Repeat (b) for $\mathbf{Z}_a = R$ and $\mathbf{Z}_b = 1/j\omega C$.

65. The electrical energy required to keep an automobile battery charged is supplied by an alternating current generator, called an *alternator*, that is

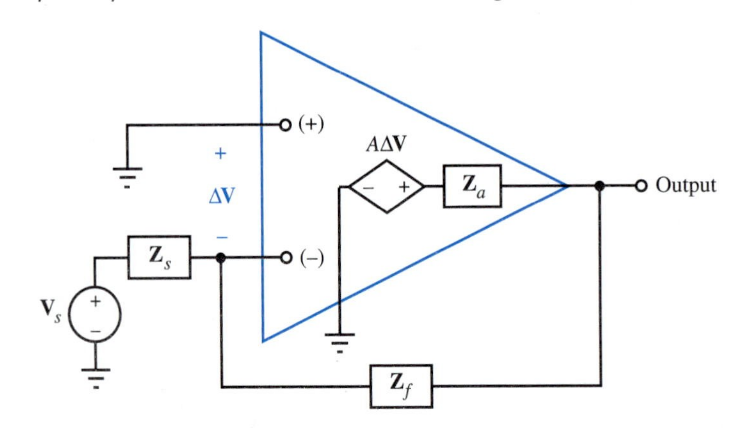

FIGURE 11.43

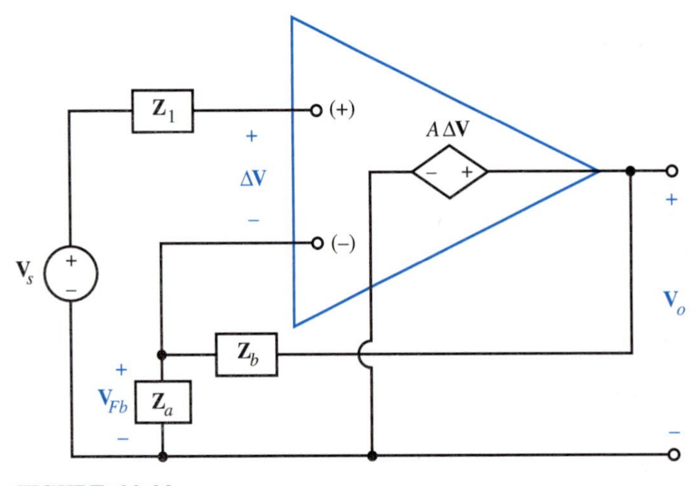

FIGURE 11.44

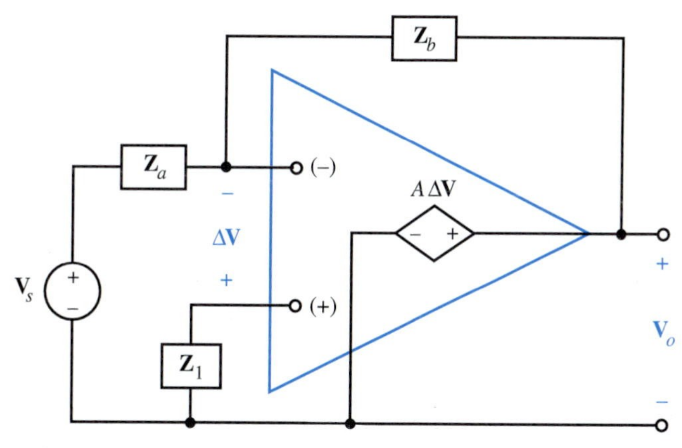

FIGURE 11.45

driven by the engine. The ac voltage from the alternator is converted to dc for the battery. For our purposes, we can model the alternator as a voltage source $v_s(t) = \alpha n \cos nt$ V in series with an inductance of value $L = 0.4$ mH, and the load as a variable resistance of value R_L as shown in the following circuit. The variable n is the engine speed, which has a minimum value of $n = n_1 = 1000$ rpm and a maximum value of $n = n_2 = 10{,}000$ rpm. The parameter α is adjusted by a *voltage regulator* that attempts to maintain a constant amplitude for voltage v_L. However, α has a maximum value of α_m. For $n = n_1$ this limits the maximum amplitude of the load current i_L to 40 A, if the amplitude of v_L is maintained at 12 V.

(a) What is the maximum amplitude αn of the alternator *internal voltage* $v_s(t)$ for $n = n_1$?

(b) What is the amplitude of the short-circuit current ($R_L = 0$) for $n = n_1$?

(c) The engine speed is increased to $n = n_2$. What is the amplitude of the internal voltage $v_s(t)$?

(d) What is the amplitude of the short-circuit current ($R_L = 0$) for $n = n_2$?

(e) What is the maximum amplitude that current i_L can have if the amplitude of the load voltage $v_L(t)$ is maintained at 12 V?

(f) Explain why an automobile alternator is not equipped with a *current regulator* to limit the maximum available current.

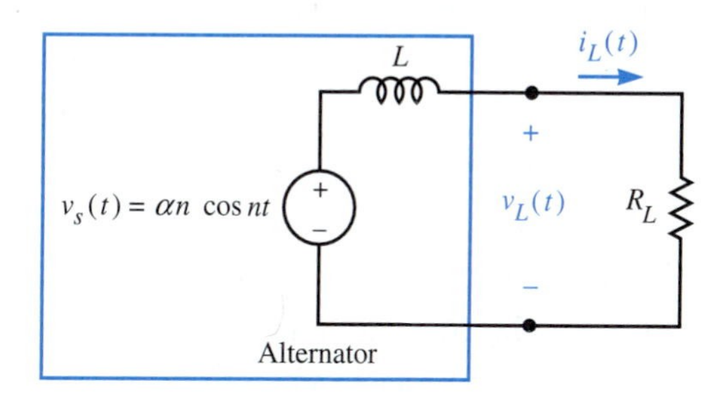

12

Power in AC Circuits

In this chapter, we investigate the flow of energy in ac circuits. In any ac circuit, the instantaneous power $p(t)$ delivered to any network part is given by the product of a sinusoidal voltage and a sinusoidal current. By plotting $p(t)$ and by using various trigonometric identities, we discover properties of energy flow in ac circuits that are both fascinating and practically useful. We find, for example, that ac networks not only consume energy at an average rate but also borrow and return energy to their sources. These and other related facts are important in the design of virtually every ac circuit and source, whether on the microelectronic scale or on the scale of large equipment used in factories.

We begin with a basic description of energy flow between an ac network and a source. This leads to the basic definitions of average power P, reactive power Q, and complex power $\mathbf{S}$. We describe the complex power balance theorem, which is akin to a conservation of energy law for an ac network. Finally, we derive the conditions under which maximum average power is exchanged between an ac source and an electrical load.

12.1 Instantaneous Power and Average Power

In this section we investigate the flow of energy between a sinusoidal source and a passive LTI network. The source can be a sinusoidal voltage source such as the ac outlet in your home, a sinusoidal current source, or, more generally, a network containing sinusoidal sources operating at a common frequency ω. The voltage and the current at the terminals of the network are shown in Fig. 12.1, where

$$v(t) = V_m \cos(\omega t + \phi_{\mathbf{V}}) \tag{12.1}$$

and

$$i(t) = I_m \cos(\omega t + \phi_{\mathbf{I}}) \tag{12.2}$$

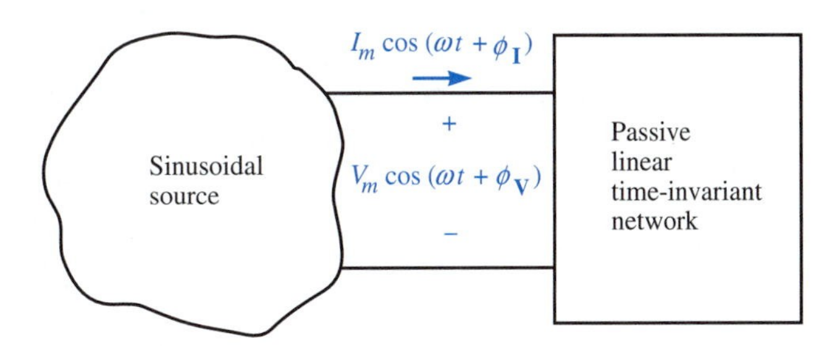

FIGURE 12.1 Sinusoidal source and load network

The instantaneous power entering the network is the product of voltage $v(t)$ and current $i(t)$:

$$p(t) = v(t)i(t) = V_m \cos(\omega t + \phi_{\mathbf{V}}) I_m \cos(\omega t + \phi_{\mathbf{I}}) \tag{12.3}$$

The unit of $p(t)$ is the watt. It is informative to write Eq. (12.3) in another form. Using the trigonometric identity

$$\cos A \cos B = \frac{1}{2}\cos(A - B) + \frac{1}{2}\cos(A + B) \tag{12.4}$$

we find that

$$p(t) = \frac{1}{2} V_m I_m \cos(\phi_{\mathbf{V}} - \phi_{\mathbf{I}}) + \frac{1}{2} V_m I_m \cos(2\omega t + \phi_{\mathbf{V}} + \phi_{\mathbf{I}}) \tag{12.5}$$

or

$$p(t) = \frac{1}{2} V_m I_m \cos\theta + \frac{1}{2} V_m I_m \cos(2\omega t + \phi_{\mathbf{V}} + \phi_{\mathbf{I}}) \tag{12.6}$$

where

$$\theta = \phi_{\mathbf{V}} - \phi_{\mathbf{I}} \tag{12.7}$$

is the angle of the network's driving-point impedance. The factor $\cos\theta$ in Eq. (12.6) is called the *power factor* of the network and is denoted by PF:

Power Factor

$$\text{PF} = \cos\theta \tag{12.8}$$

The angle θ is called the *power-factor angle*. If $\theta > 0$, then the current lags the voltage, and if $\theta < 0$, then the current leads the voltage. Because $\cos(-\theta) = \cos(+\theta)$, it is not possible to determine whether the current is leading or lagging the voltage from the value of PF alone. To convey this information, we refer to a *lagging power factor* for $\theta > 0$ (the current lags the voltage) and a *leading power factor* for $\theta < 0$ (the current leads the voltage). Substitution of Eq. (12.8) into Eq. (12.6) yields

$$p(t) = \frac{1}{2} V_m I_m \cdot \text{PF} + \frac{1}{2} V_m I_m \cos(2\omega t + \phi_V + \phi_I) \tag{12.9}$$

Equation (12.9) is plotted in Fig. 12.2. Notice that the instantaneous power entering the network is periodic with angular frequency 2ω. There are two components in $p(t)$: a constant component $\frac{1}{2}V_m I_m \cdot \text{PF}$ and a sinusoidal component with amplitude $\frac{1}{2}V_m I_m$ and angular frequency 2ω. According to Eq. (12.8), the largest possible value of a power factor is unity; $\text{PF} \leq 1$. Therefore $p(t)$ will generally assume both positive and negative values, as is illustrated in Fig. 12.2. When $p(t)$ is positive, energy flows into the network from the source. When $p(t)$ is negative, energy flows back out of the network into the source. Thus the source is not only delivering energy that is being consumed by the network, it is also lending energy that is returned by the network. The *average power* delivered to the network is the constant (first) term in Eq. (12.9), because the average value of a sinusoid is zero. The average power delivered to the network is also called the *real power* or the *active power*, and is denoted by P:

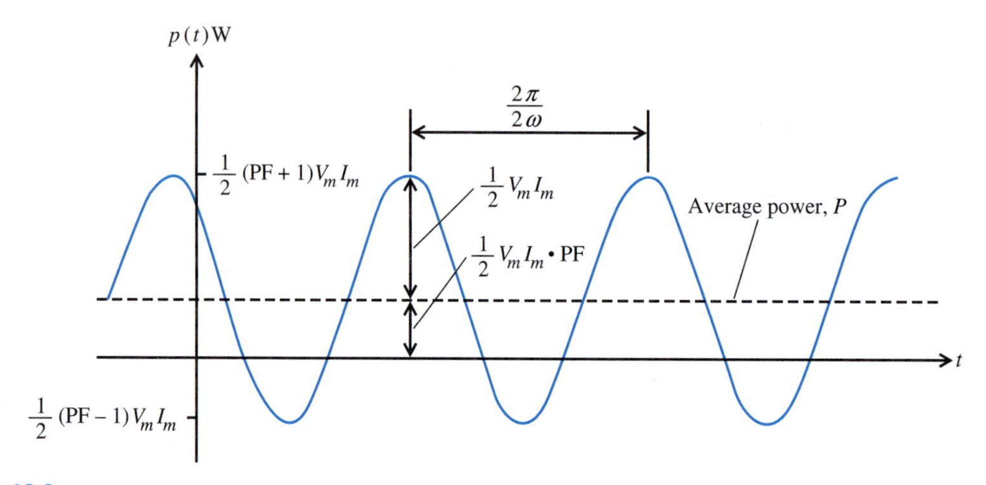

FIGURE 12.2
The instantaneous power $p(t)$ entering a passive LTI network [Plot of Eq. (12.9)]. $p(t)$ supplies average power $P = \frac{1}{2}V_m I_m \cdot \text{PF}$, but oscillates between positive and negative values with angular frequency 2ω.

$$P = \frac{1}{2} V_m I_m \cdot \text{PF} = \frac{1}{2} V_m I_m \cos\theta \tag{12.10}$$

The unit of average power is, of course, the watt.

Equation (12.10) has an important theoretical implication. Remember that P must be nonnegative for a passive load. Therefore $\cos\theta$ must be nonnegative, and so θ must be in the range $-\pi/2 \leq \theta \leq +\pi/2$. This means that *the resistive component of the driving-point impedance of every passive network is nonnegative*;

$$R(\omega) = \mathcal{R}e\{\mathbf{Z}(j\omega)\} \geq 0$$

EXAMPLE 12.1 A factory draws 120-A rms current from a 12,470-V rms 60-Hz source at a power factor of 0.75 lagging. What is the average power consumed?

Solution From Example 7.8, the peak voltage and current V_m and I_m in Eq. (12.10) are related to the rms (or effective) voltage V_{rms} and current I_{rms} by

$$V_m = \sqrt{2}\, V_{\mathrm{rms}}$$

and

$$I_m = \sqrt{2}\, I_{\mathrm{rms}}$$

Accordingly, Eq. (12.10) becomes

$$P = V_{\mathrm{rms}} I_{\mathrm{rms}} \cdot \mathrm{PF}$$

which is, in this example,

$$P = 12{,}470 \times 120 \times 0.75 = 1.12 \text{ MW}$$

EXERCISES

See Problem 12.1 in the PSpice manual.

1. Define and give the units of (a) instantaneous power, (b) average power, and (c) the power factor.
2. A sinusoidal current with amplitude 10 A is applied to an impedance $\mathbf{Z} = 10\underline{/-45^\circ}\ \Omega$.
   (a) Determine the power factor. (Be sure to indicate whether it is leading or lagging.) *answer:* 0.707 leading
   (b) Determine the average power supplied. *answer:* 353.6 W
3. Explain why driving-point resistance must be nonnegative for every passive network.

12.2 Apparent Power

Notice that the formula [Eq. (12.10)] for the average power supplied by a source depends not only on the amplitudes of voltage and current but also on the power factor. The product

$$\frac{1}{2} V_m I_m = \frac{P}{\mathrm{PF}} \tag{12.11}$$

is called the *apparent power*. The unit of apparent power is called the *volt-ampere* (VA) to emphasize that apparent power is simply one-half the product of the voltage and current amplitudes. This product is clearly *not* the average power P. As we see from Eq. (12.11), apparent power is computed by *dividing* the average power by the power factor.

Apparent power has practical significance for an electric utility company, because a utility company must supply both average and apparent power to a customer. To illustrate, suppose that a factory requires an average power P at a given voltage amplitude V_m. We can see from Eq. (12.11) that the *apparent* power will be larger than P if the power factor is less than 1. Thus, the current amplitude I_m that must be supplied will be larger for PF < 1 than it would be for PF $= 1$, even though the average power P supplied is the same in either case. A larger current amplitude cannot be supplied without additional cost to the utility company. The additional cost arises because greater current in the transmission wires connecting the factory to the utility company's generators will be accompanied by more power dissipation (through heat) in the transmission lines. For this reason, a utility company will often raise its rates for industrial customers who operate at low power factors.

EXAMPLE 12.2 A fully loaded 50-hp induction motor operates at 75 percent efficiency and a power factor of 0.8 lagging. Find the apparent power.

Solution Efficiency, η, is the ratio of output power to input power. The conversion between watts and horsepower is 1 hp = 745.7 W. Consequently, the input power P is

$$P = \frac{P_{\text{out}}}{\eta} = \frac{50 \times 745.7}{0.75}\ \text{W}$$
$$= 49.7\ \text{kW}$$

Therefore the apparent power is, from Eq. (12.11),

$$\frac{1}{2} V_m I_m = \frac{49.7}{0.8}\ \text{kVA} = 62.1\ \text{kVA}$$

EXERCISES

4. Define and give the unit of apparent power.
5. A sinusoidal current with amplitude 10 A is delivered to an impedance $\mathbf{Z} = 10\underline{/-45^\circ}\ \Omega$. Determine the apparent power supplied. *answer:* 500 VA
6. Explain the practical significance of apparent power.

12.3 Reactive Power

In this section we shall see that the instantaneous power supplied to any passive network can be divided into two parts. The first part describes energy that always flows into the network and is dissipated in it. This is the part that is responsible for the average power delivered to the network. The second part describes energy that is only borrowed and returned by the network.

The derivation is based on trigonometric identities. The first step is to use the identity $\cos(A + B) = \cos A \cos B - \sin A \sin B$ to rewrite $v(t)$ as

$$\begin{aligned} v(t) &= V_m \cos(\omega t + \phi_{\mathbf{V}}) \\ &= V_m \cos(\omega t + \phi_{\mathbf{I}} + \theta) \\ &= \underbrace{V_m \cos\theta \cos(\omega t + \phi_{\mathbf{I}})}_{\text{In phase with } i(t)} - \underbrace{V_m \sin\theta \sin(\omega t + \phi_{\mathbf{I}})}_{\text{In phase quadrature with } i(t)} \end{aligned} \tag{12.12}$$

We say that the first term on the right-hand side of Eq. (12.12) is *in phase* with $i(t) = I_m \cos(\omega t + \phi_I)$ because it has the same phase angle ϕ_I as $i(t)$. We say that the second term is *in phase quadrature* with $i(t)$ because it is 90° out of phase with $i(t)$ [$\cos(\omega t + \phi_I)$ leads $\sin(\omega t + \phi_I)$ by 90°]. By using Eq. (12.12), we can express the instantaneous power as

$$\begin{aligned} p(t) &= v(t)i(t) \\ &= V_m I_m \cos\theta \cos^2(\omega t + \phi_I) - V_m I_m \sin\theta \cos(\omega t + \phi_I)\sin(\omega t + \phi_I) \end{aligned} \tag{12.13}$$

We use $\cos^2 A = \frac{1}{2}(1 + \cos 2A)$ and $\cos A \sin A = \frac{1}{2}\sin 2A = -\frac{1}{2}\cos[2A + (\pi/2)]$ to obtain

$$p(t) = \frac{1}{2} V_m I_m \cos\theta[1 + \cos(2\omega t + 2\phi_I)] + \frac{1}{2} V_m I_m \sin\theta \cos\left(2\omega t + 2\phi_I + \frac{\pi}{2}\right) \tag{12.14}$$

which gives us the result

$$p(t) = \underbrace{P[1 + \cos(2\omega t + 2\phi_I)]}_{\text{Energy flow into the network}} + \underbrace{Q\cos\left(2\omega t + 2\phi_I + \frac{\pi}{2}\right)}_{\text{Energy borrowed and returned by the network}} \tag{12.15}$$

where

$$P = \frac{1}{2} V_m I_m \cos\theta \tag{12.16}$$

and

$$Q = \frac{1}{2} V_m I_m \sin\theta \tag{12.17}$$

The letter Q serves as a reminder that the second term in Eq. (12.15) comes from that part of $v(t)$ that is *in phase quadrature* with $i(t)$. The terms in Eq. (12.15) are plotted in Fig. 12.3 for $\theta = 45°$. The in-phase and quadrature-phase components of $v(t)$ are responsible for the first and second terms, respectively, on the right-hand side of Eq. (12.15). The first term is nonnegative if $\cos\theta \geq 0$, which must be the case for a passive network. Therefore the in-phase component of $v(t)$ produces a periodically fluctuating but continuous flow of energy into the network. The average value of the first term is P, as defined in Eq. (12.16). Therefore *the in-phase component of voltage is responsible for the delivery of the power consumed.* The second term in Eq. (12.15) is sinusoidal and therefore has an average value of zero. Therefore *the quadrature-phase component of voltage is responsible for the periodic two-way exchange of energy between the source and the network.* As far as the second term of Eq. (12.15) is concerned, the network is only borrowing energy and returning it to the source. The amount of energy borrowed depends on the amplitude, Q, of the second term in Eq. (12.15), as defined by Eq. (12.17). Q is called the *reactive* power and has the unit of *volt-ampere reactive* (VAR). As indicated earlier, the active power P is the average rate at which energy is consumed by the network. Equation (12.15) shows us that the reactive power Q is the maximum rate at which energy is borrowed by the network. Notice from Eqs. (12.16) and (12.17)

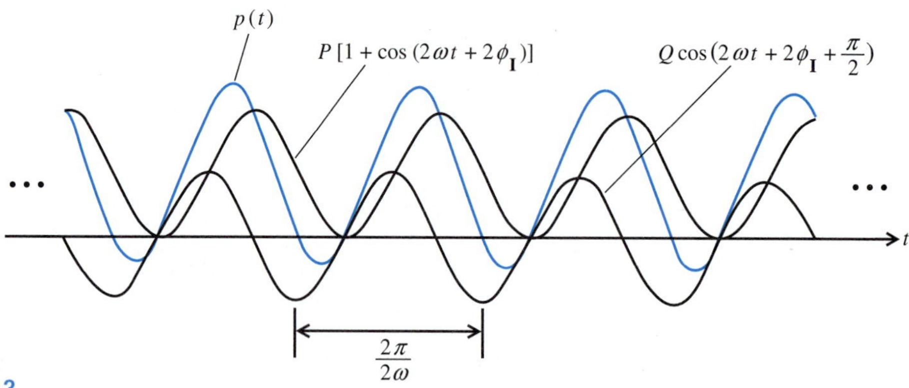

FIGURE 12.3
The terms in Eq. (12.15). Instantaneous power $p(t)$ can be represented as the sum of two components: The first component, $P[1 + \cos(2\omega t + 2\phi_{\mathbf{I}})]$, is oscillatory but never negative. This is the component that delivers the average power P to the network. The second component, $Q\cos(2\omega t + 2\phi_{\mathbf{I}} + \pi/2)$, has zero average value. It is the component that swaps energy to and from the source and network.

and $\theta = \underline{/\mathbf{Z}}$ that $Q = 0$ if $\mathbf{Z}$ is purely real, whereas $P = 0$ if $\mathbf{Z}$ is purely imaginary. Thus a resistance does not borrow and return energy, and an inductance and a capacitance do not consume energy.

Reactive Power and Stored Energy for *L* and *C*

There is an important relationship between the reactive power entering an inductance and the average energy that is stored in it. From the relation $\mathbf{V} = j\omega L\mathbf{I}$, it follows that $V_m = \omega L I_m$ and $\theta = 90°$. Use of these results in Eq. (12.17) leads to the following expression for the reactive power entering an inductance:

$$
\begin{aligned}
Q &= \frac{1}{2} V_m I_m \sin\theta \\
&= \frac{1}{2} I_m^2 \omega L
\end{aligned}
\tag{12.18}
$$

The *instantaneous energy* stored in the inductance is, according to Eq. (2.45),

$$
\begin{aligned}
w_L(t) &= \frac{1}{2} L i^2(t) \\
&= \frac{1}{2} L I_m^2 \cos^2(\omega t + \phi_{\mathbf{I}}) \\
&= \frac{1}{4} L I_m^2 + \frac{1}{4} L I_m^2 \cos(2\omega t + 2\phi_{\mathbf{I}})
\end{aligned}
\tag{12.19}
$$

The *average energy* stored in the inductance is the first term on the right-hand side of Eq. (12.19),

$$
W_{L,\text{ave}} = \frac{1}{4} L I_m^2 \tag{12.20}
$$

By comparing Eqs. (12.20) and (12.18), we can see that the reactive power entering the inductance can be expressed in terms of the average stored energy as

$$Q = 2\omega W_{L,\text{ave}} \tag{12.21}$$

That is, the maximum rate of energy exchange between a source and an inductance is equal to twice the product of the average energy stored in the inductance and the angular frequency of the source.

A similar result applies for capacitance. From Eq. (12.17) and the relation $\mathbf{I} = j\omega C\mathbf{V}$, it follows that the reactive power entering a capacitance is

$$Q = -\frac{1}{2}V_m^2\omega C \tag{12.22}$$

The instantaneous energy stored in a capacitance is, by Eq. (2.31),

$$\begin{aligned} w_C(t) &= \frac{1}{2}Cv^2(t) \\ &= \frac{1}{2}CV_m^2\cos^2(\omega t + \phi_{\mathbf{V}}) \\ &= \frac{1}{4}CV_m^2 + \frac{1}{4}CV_m^2\cos(2\omega t + 2\phi_{\mathbf{V}}) \end{aligned} \tag{12.23}$$

The average energy stored is

$$W_{C,\text{ave}} = \frac{1}{4}CV_m^2 \tag{12.24}$$

Therefore the reactive power entering a capacitance is related to the average stored energy by

$$Q = -2\omega W_{C,\text{ave}} \tag{12.25}$$

Notice that the reactive power entering a capacitance is negative, and that the reactive power entering an inductance is positive. For this reason, capacitors are said to be ''sources of reactive power,''† whereas inductors are said to ''consume reactive power.'' More generally, a passive network with a positive reactance will have a positive impedance angle, and will therefore be associated with a positive reactive power input Q. Such a network is inductive, and is said to ''consume reactive power.'' A passive network with a negative reactance is associated with a negative reactive power input Q. Such a network is capacitive, and is said to be a ''source of reactive power.'' If a network is inductive, some or all of its reactive power requirements can be supplied from a capacitive network rather than from the source. This procedure is illustrated in Example 12.4.

† Some people do not like this terminology because, after all, the difference in sign merely indicates that the borrowing and returning of energy occur at different times for the inductance and capacitance. Although this objection makes a valid point, the terminology is widely accepted.

EXERCISES

7. Define and give the unit of reactive power.
8. Define phase quadrature, and explain why the symbol Q is appropriate for reactive power.
9. A sinusoidal current with amplitude 10 A is input to an impedance $\mathbf{Z} = 10\underline{/-45^\circ}\ \Omega$. Determine the reactive power supplied. *answer:* -353.6 VAR

12.4 Complex Power

Convenient expressions for the average and the reactive power entering a network can be derived directly from the frequency-domain representation of Fig. 12.4, in which $\mathbf{V} = V_m e^{j\phi_V}$ and $\mathbf{I} = I_m e^{j\phi_I}$ are the phasors associated with $v(t)$ and $i(t)$, respectively. Here

$$\mathbf{V} = \mathbf{ZI} \quad \text{and} \quad \mathbf{I} = \mathbf{YV} \tag{12.26}$$

$$V_m = |\mathbf{Z}|I_m \quad \text{and} \quad I_m = |\mathbf{Y}|V_m \tag{12.27}$$

and

$$\phi_V = \underline{/\mathbf{Z}} + \phi_I \quad \text{and} \quad \phi_I = \underline{/\mathbf{Y}} + \phi_V \tag{12.28}$$

where $\mathbf{Z}$ and $\mathbf{Y}$ are the driving-point impedance and admittance of the network, respectively.

In connection with Fig. 12.4, we define the *complex power* entering the network as follows.

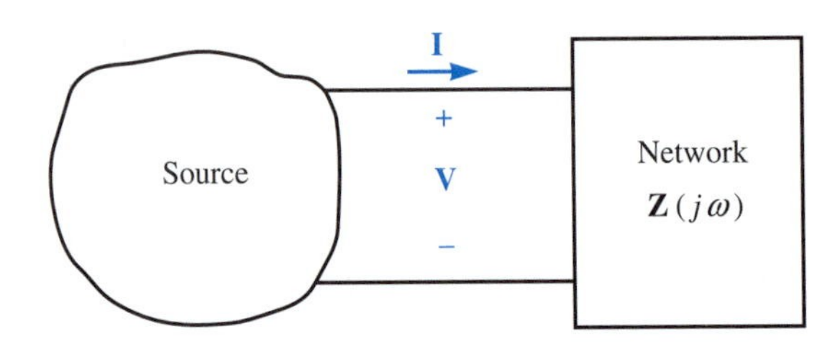

FIGURE 12.4 Frequency-domain representation of Fig. 12.1

DEFINITION Complex Power

The **complex power** entering the network of Fig. 12.4 is defined as

$$\mathbf{S} = \frac{1}{2}\mathbf{VI}^* \tag{12.29}$$

The unit of $\mathbf{S}$ is the volt-ampere (VA). Complex power is a purely mathematical concept. Like other complex quantities used in electrical engineering, complex power has no physical significance in itself, but provides a convenient way to derive physically meaningful quantities. For example, because $\mathbf{V} = V_m e^{j\phi_V}$ and $\mathbf{I} = I_m e^{j\phi_I}$, then

$$\mathbf{S} = \frac{1}{2} V_m I_m e^{j(\phi_V - \phi_I)} \tag{12.30}$$

or

$$\mathbf{S} = \frac{1}{2} V_m I_m e^{j\theta} = \frac{1}{2} V_m I_m \underline{/\theta} \tag{12.31}$$

where $\theta = \phi_\mathbf{V} - \phi_\mathbf{I}$. Therefore the magnitude of $\mathbf{S}$, $|\mathbf{S}|$, is the apparent power $\frac{1}{2}V_m I_m$, and the angle of $\mathbf{S}$, θ, is the angle of the network's impedance $\mathbf{Z}$. Furthermore, Euler's identity enables Eq. (12.31) to be rewritten as

$$\begin{aligned} \mathbf{S} &= \frac{1}{2} V_m I_m (\cos\theta + j\sin\theta) \\ &= \frac{1}{2} V_m I_m \cos\theta + j\frac{1}{2} V_m I_m \sin\theta \end{aligned} \tag{12.32}$$

so that from the definitions in Eqs. (12.16) and (12.17),

$$\mathbf{S} = P + jQ \tag{12.33}$$

Therefore the real part of $\mathbf{S}$ is the average power P, and the imaginary part of $\mathbf{S}$ is the reactive power Q. Equations (12.31) and (12.33) are illustrated in Fig. 12.5. Figure 12.6, which is derived directly from Fig. 12.5, is called the *power triangle*. Some interesting results can be derived by an inspection of the power triangle; for example, by the Pythagorean theorem,

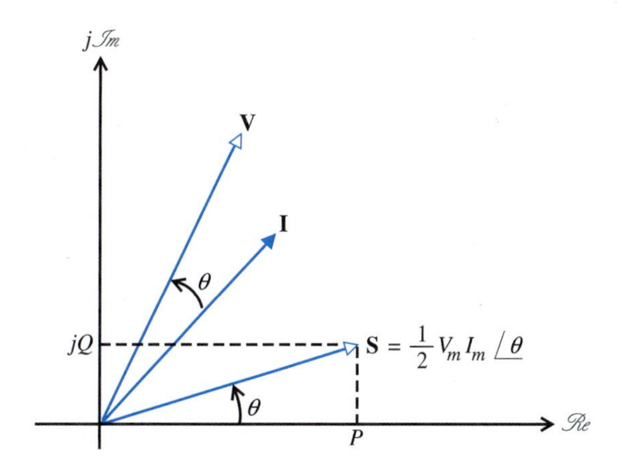

FIGURE 12.5 Complex power $\mathbf{S}$. Complex power is defined as $\mathbf{S} = \frac{1}{2}\mathbf{VI}^*$. The real part of $\mathbf{S}$ is the average power P. The imaginary part of $\mathbf{S}$ is the reactive power Q. The magnitude of $\mathbf{S}$ is the apparent power.

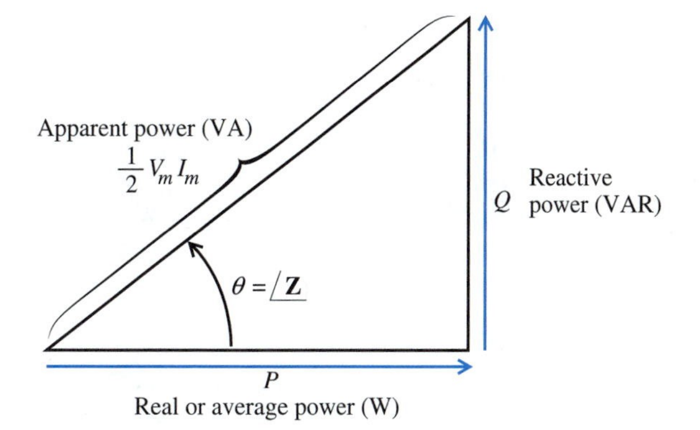

FIGURE 12.6 The power triangle

$$\frac{1}{2} V_m I_m = \sqrt{P^2 + Q^2} \tag{12.34}$$

and by the definition of $\cos\theta$,

$$\text{PF} = \frac{P}{\sqrt{P^2 + Q^2}} \tag{12.35}$$

Equations (12.34) and (12.35) are useful because they express apparent power and power factor as functions of average power and reactive power.

Important new results follow from Eq. (12.29) and the relation $\mathbf{V} = \mathbf{Z}(j\omega)\mathbf{I} = [R(\omega) + jX(\omega)]\mathbf{I}$:

$$\begin{aligned}\mathbf{S} &= \frac{1}{2}\mathbf{V}\mathbf{I}^* \\ &= \frac{1}{2}\mathbf{Z}I_m^2 \\ &= \frac{1}{2}I_m^2 R(\omega) + j\frac{1}{2}I_m^2 X(\omega)\end{aligned} \tag{12.36}$$

Therefore, by comparison with Eq. (12.33), we have

$$P = \frac{1}{2}I_m^2 R(\omega) \tag{12.37}$$

and

$$Q = \frac{1}{2}I_m^2 X(\omega) \tag{12.38}$$

According to Eqs. (12.37) and (12.38), for a given current amplitude I_m, the average power P entering a network depends on the network's resistance $R(\omega) = \mathcal{R}e\{\mathbf{Z}(\omega)\}$ and the reactive power Q depends on the network's reactance $X(\omega) = \mathcal{I}m\{\mathbf{Z}(j\omega)\}$. Similar results can be obtained with the use of Eq. (12.29) and the formula $\mathbf{I} = \mathbf{Y}(j\omega)\mathbf{V} = [G(\omega) + jB(\omega)]\mathbf{V}$:

$$\begin{aligned}\mathbf{S} &= \frac{1}{2}\mathbf{V}\mathbf{I}^* = \frac{1}{2}\mathbf{Y}^* V_m^2 \\ &= \frac{1}{2}V_m^2 G(\omega) - j\frac{1}{2}V_m^2 B(\omega)\end{aligned} \tag{12.39}$$

from which

$$P = \frac{1}{2}V_m^2 G(\omega) \tag{12.40}$$

and

$$Q = -\frac{1}{2}V_m^2 B(\omega) \tag{12.41}$$

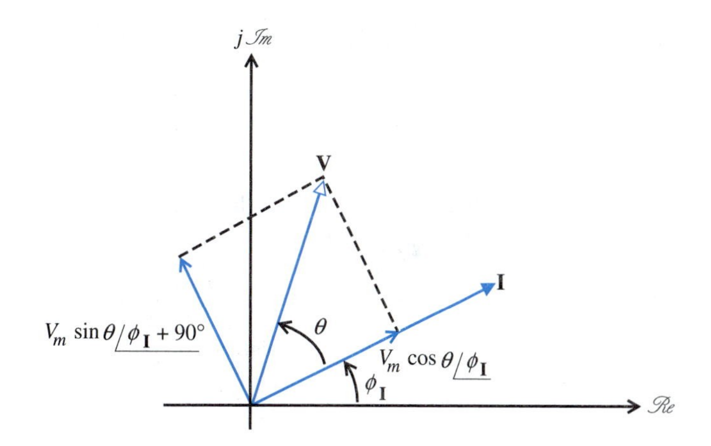

FIGURE 12.7 Decomposition of the phasor **V**. The component of **V** parallel to **I** is responsible for the average power supplied to the network. The component of **V** perpendicular to **I** is responsible for the reactive power supplied to the network.

A final insight is contained in Fig. 12.7, which shows the driving-point voltage and current phasors **V** and **I** as vectors in the complex plane. The vector **V** is viewed as having two components, one parallel to the vector **I** and the other perpendicular to **I**. These two components are $V_m \cos\theta \,/\phi_\mathbf{I}$ and $V_m \sin\theta \,/\phi_\mathbf{I} + 90°$, respectively, and they correspond to the in-phase and phase-quadrature components of $v(t)$ shown in Eq. (12.12). The substitution of

$$\mathbf{V} = \underbrace{V_m \cos\theta\, e^{j\phi_\mathbf{I}}}_{\text{Parallel to } \mathbf{I}} + \underbrace{V_m \sin\theta\, e^{j[\phi_\mathbf{I} + (\pi/2)]}}_{\text{Orthogonal to } \mathbf{I}} \tag{12.42}$$

into $\mathbf{S} = \frac{1}{2}\mathbf{VI}^*$ yields

$$\begin{aligned} \mathbf{S} &= \frac{1}{2}[V_m \cos\theta\, e^{j\phi_\mathbf{I}} + V_m \sin\theta\, e^{j[\phi_\mathbf{I} + (\pi/2)]}]\mathbf{I}^* \\ &= \frac{1}{2}(V_m \cos\theta\, e^{j\phi_\mathbf{I}} + jV_m \sin\theta\, e^{j\phi_\mathbf{I}})I_m e^{-j\phi_\mathbf{I}} \\ &= \frac{1}{2}V_m I_m \cos\theta + j\frac{1}{2}V_m I_m \sin\theta \\ &= P + jQ \end{aligned} \tag{12.43}$$

Therefore the component of **V** parallel to **I** is responsible for the average power $P = \mathcal{R}e\{\mathbf{S}\}$ supplied to the network. The component of **V** orthogonal to **I** is responsible for the reactive power $Q = \mathcal{I}m\{\mathbf{S}\}$ supplied to the network.† The component of **V** parallel to **I** corresponds to the component of $v(t)$ that is in phase with $i(t)$. The component of **V** that is orthogonal to **I** corresponds to the component of $v(t)$ that is in phase quadrature with $i(t)$.

EXAMPLE 12.3 A coil draws 1-A peak current at a 0.6 lagging power factor from a 120-V rms 60-Hz source. Assume that the coil is modeled by a series RL circuit (where R represents the winding resistance of the coil), and find (a) the complex power entering the coil and (b) the values of R and L.

† The term *orthogonal* is used here in the sense of *perpendicular*. It is worth noting that Eq. (12.42) is the frequency-domain form of Eq. (12.12). In Eq. (12.12), the component of $v(t)$ that is in phase quadrature with $i(t)$ is orthogonal to $i(t)$ in the sense discussed in Section 7.6.

Solution (a) The complex power entering the coil is $\mathbf{S} = P + jQ$, where from Eq. (12.16),

$$P = \frac{1}{2}(1)(\sqrt{2} \cdot 120)(0.6)$$
$$= 50.9 \text{ W}$$

Because PF $= \cos\theta = 0.6$ lagging, $\sin\theta = \sin(\cos^{-1} 0.6) = 0.8$. The reactive power is, from Eq. (12.17),

$$Q = \frac{1}{2}(1)(\sqrt{2} \cdot 120)(0.8)$$
$$= 67.9 \text{ VAR}$$

Therefore

$$\mathbf{S} = 50.9 + j67.8 \text{ VA}$$

The power triangle is shown in Fig. 12.8.

(b) The driving-point impedance of a series RL circuit is

$$\mathbf{Z}(j\omega) = R + j\omega L$$

Therefore, from Eq. (12.37) with $R(\omega) = R$,

$$R = \frac{2P}{I_m^2}$$
$$= \frac{(2)(50.9)}{1^2}$$
$$= 101.8\ \Omega$$

and from Eq. (12.38) with $X(\omega) = \omega L$,

$$L = \frac{2Q}{\omega I_m^2}$$
$$= \frac{(2)(67.8)}{(2\pi 60)(1^2)}$$
$$= 0.36 \text{ H}$$

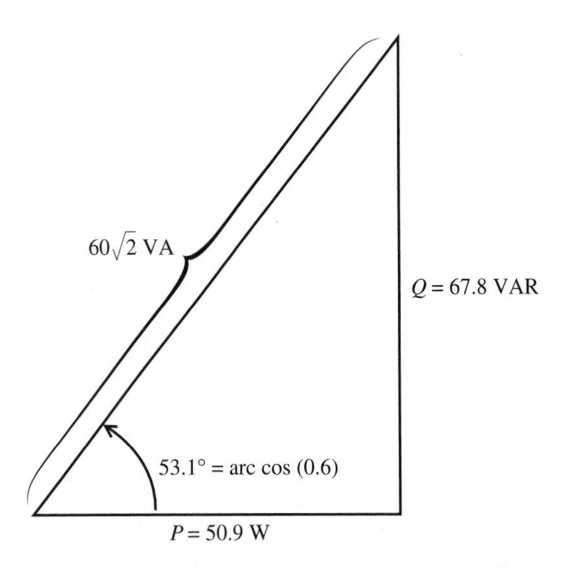

FIGURE 12.8 Power triangle for Example 12.3

Remember Average power, reactive power, apparent power, and power factor can be obtained from complex power **S**.

EXERCISES

10. Define and give the unit of complex power.
11. Sketch the power triangle from memory.
12. A sinusoidal current with an amplitude of 10 A is applied to an impedance $\mathbf{Z} = 10\underline{/-45^\circ}\ \Omega$.
    (a) Determine the complex power supplied. *answer:* $353.6 - j353.6$ VA
    (b) Sketch the power triangle. *answer:* 353.6 W, −353.6 VAR, 500 VA
    (c) Give the average power, the reactive power, and the apparent power.
13. A sinusoidal voltage with an effective amplitude of 120 V rms and frequency of 60 Hz is applied to a series *RL* circuit with impedance $\mathbf{Z} = 100 + j377\ \Omega$. Sketch the power triangle, and indicate the numerical values and engineering significance of its sides and angles. *answer:* 9.47 W, 35.7 VAR, 37 VA

12.5 The Complex Power Balance Theorem

This section describes an important property of energy flow in ac circuits. This property is an extension of the power balance equation that was described in Section 1.4.

We introduce the basic idea by considering two load impedances $\mathbf{Z}_a$ and $\mathbf{Z}_b$ connected in parallel across a sinusoidal voltage source (or ''ac line''), as illustrated in Fig. 12.9. The line current is

$$\mathbf{I}_s = \mathbf{I}_a + \mathbf{I}_b \tag{12.44}$$

where $\mathbf{I}_a$ and $\mathbf{I}_b$ are the phasor currents in the individual loads. The complex power *supplied* by the source is

$$\begin{aligned}\mathbf{S}_s &= \frac{1}{2}\mathbf{V}_s\mathbf{I}_s^* \\ &= \frac{1}{2}\mathbf{V}_s\mathbf{I}_a^* + \frac{1}{2}\mathbf{V}_s\mathbf{I}_b^* \\ &= \mathbf{S}_a + \mathbf{S}_b\end{aligned} \tag{12.45}$$

which is the *sum* of the complex powers *delivered* to the loads. This neat result can also be written as

$$-\mathbf{S}_s + \mathbf{S}_a + \mathbf{S}_b = 0 \tag{12.46a}$$

or

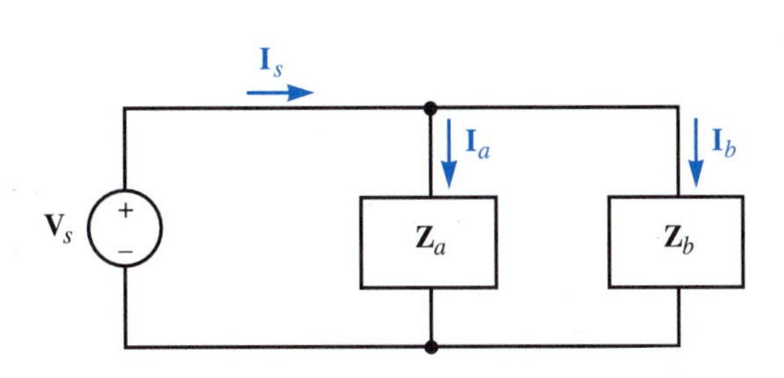

FIGURE 12.9 Parallel loads on an ac line

$$\sum \mathbf{S}_i = 0 \tag{12.46b}$$

where the sum is over the complex powers entering all components in the network, including the source.

Equations (12.45) and (12.46) are examples of the following theorem.

THEOREM Complex Power Balance Theorem

In an arbitrary ac network, the sum of the complex powers entering all network components (including the sources) equals zero.

This theorem is expressed mathematically as

$$\sum_{\substack{\text{All} \\ \text{components}}} \frac{1}{2}\mathbf{VI}^* = 0 \tag{12.47a}$$

or

$$\sum_{\substack{\text{All} \\ \text{components}}} \mathbf{S}_i = 0 \tag{12.47b}$$

In Eqs. (12.47a) and (12.47b), the passive sign convention is used for each component. $\mathbf{S}_i$ is the complex power delivered to the ith component, and the summation is over all components, including the sources. If the active sign convention is used for the sources, then we have

$$\sum_{\substack{\text{Supplied} \\ \text{by sources}}} \mathbf{S}_i = \sum_{\substack{\text{Delivered to} \\ \text{passive} \\ \text{elements}}} \mathbf{S}_i \tag{12.47c}$$

An example of this statement is given in Eq. (12.45), where the voltage source of Fig. 12.9 was the supplier of complex power.

The basic ingredients of a proof of the complex power balance equation can be understood if we consider the four-node network of Fig. 12.10. The components repre-

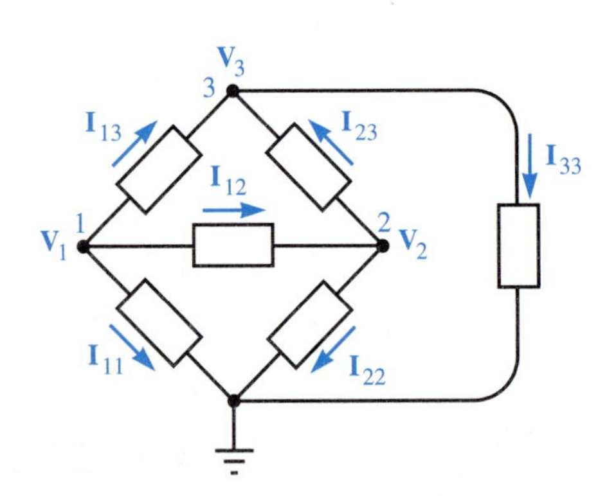

FIGURE 12.10 The node voltages are measured with respect to the reference.

sented by rectangles are arbitrary impedances (including open and short circuits), independent sources, and dependent sources. Subscripts pq have been used to identify the currents. The order of subscripts in $\mathbf{I}_{pq}$ signifies that the arrow points from node p to node q for $p \neq q$. For $p = q$, the arrow points from node p to the reference node (see Fig. 12.10). If we apply KCL at the (nonreference) nodes, we obtain the following three equations:

Node 1 $$\mathbf{I}_{11} + \mathbf{I}_{12} + \mathbf{I}_{13} = 0 \tag{12.48}$$

Node 2 $$-\mathbf{I}_{12} + \mathbf{I}_{22} + \mathbf{I}_{23} = 0 \tag{12.49}$$

Node 3 $$-\mathbf{I}_{13} - \mathbf{I}_{23} + \mathbf{I}_{33} = 0 \tag{12.50}$$

If we multiply the conjugates of Eqs. (12.48), (12.49), and (12.50) by $\frac{1}{2}\mathbf{V}_1$, $\frac{1}{2}\mathbf{V}_2$, and $\frac{1}{2}\mathbf{V}_3$, respectively, we obtain

$$\frac{1}{2}\mathbf{V}_1\mathbf{I}_{11}^* + \frac{1}{2}\mathbf{V}_1\mathbf{I}_{12}^* + \frac{1}{2}\mathbf{V}_1\mathbf{I}_{13}^* = 0 \tag{12.51}$$

$$-\frac{1}{2}\mathbf{V}_2\mathbf{I}_{12}^* + \frac{1}{2}\mathbf{V}_2\mathbf{I}_{22}^* + \frac{1}{2}\mathbf{V}_2\mathbf{I}_{23}^* = 0 \tag{12.52}$$

$$-\frac{1}{2}\mathbf{V}_3\mathbf{I}_{13}^* - \frac{1}{2}\mathbf{V}_3\mathbf{I}_{23}^* + \frac{1}{2}\mathbf{V}_3\mathbf{I}_{33}^* = 0 \tag{12.53}$$

Finally, by adding Eqs. (12.51), (12.52), and (12.53), we obtain

$$\frac{1}{2}\mathbf{V}_1\mathbf{I}_{11}^* + \frac{1}{2}(\mathbf{V}_1 - \mathbf{V}_2)\mathbf{I}_{12}^* + \frac{1}{2}(\mathbf{V}_1 - \mathbf{V}_3)\mathbf{I}_{13}^* + \frac{1}{2}\mathbf{I}_{22}^*\mathbf{V}_2$$
$$+ \frac{1}{2}(\mathbf{V}_2 - \mathbf{V}_3)\mathbf{I}_{23}^* + \frac{1}{2}\mathbf{I}_{22}^*\mathbf{V}_3 = 0 \tag{12.54}$$

The terms on the left-hand side of Eq. (12.54) represent the complex powers entering the individual components of the network of Fig. 12.10. [For example, $\frac{1}{2}(\mathbf{V}_1 - \mathbf{V}_2)\mathbf{I}_{12}^*$ is the complex power entering the component between nodes 1 and 2.] Equation (12.54) proves the complex power balance theorem for the network of Fig. 12.10. Notice that the proof depended only on KCL and KVL. The same approach can be taken to prove the complex power balance theorem for an arbitrary ac network.

EXAMPLE 12.4

See Problem 12.2 in the PSpice manual.

An inductive load draws 1 kW at PF = 0.9 lagging from a 120-V rms source. In an effort to raise the power factor seen by the source, a capacitive load is placed in parallel with the inductive load, as shown in Fig. 12.11. The capacitive load draws 10 W at PF = 0.02 leading.

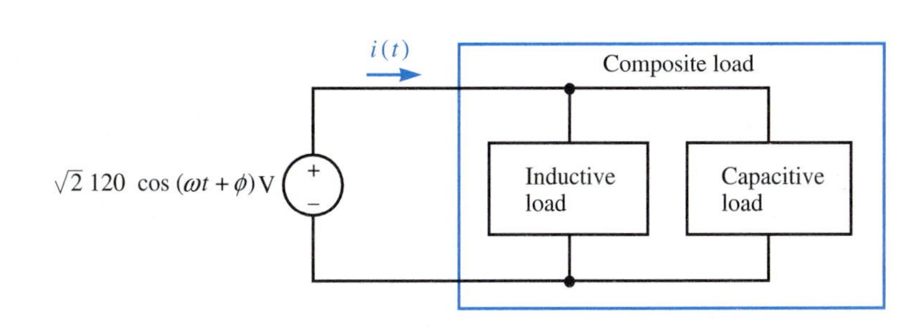

FIGURE 12.11
Circuit for Example 12.4

(a) Find the complex power supplied by the source.
(b) Find the rms (effective) current supplied by the source.
(c) Find the rms current supplied to the inductive load.
(d) Find the impedance of the composite load (enclosed in blue).
(e) Find the power factor of the composite load.
(f) Explain why the power factor of the composite load is greater than that of the inductive load.

Solution (a) By the complex power balance theorem, the complex power supplied by the source, $\mathbf{S}_s$, equals the sum of the complex powers entering the loads. Thus

$$\mathbf{S}_s = \mathbf{S}_1 + \mathbf{S}_2$$

where $\mathbf{S}_1$ and $\mathbf{S}_2$ denote the complex powers supplied to the inductive and capacitive loads, respectively. The complex power supplied to the inductive load is

$$\begin{aligned}\mathbf{S}_1 &= 1 \text{ kW} + j1 \tan(\cos^{-1} 0.9) \text{ kVAR} \\ &= 1000 + j484.3 \text{ VA}\end{aligned}$$

The complex power supplied to the capacitive load is

$$\begin{aligned}\mathbf{S}_2 &= 10 \text{ W} - j10 \tan(\cos^{-1} 0.02) \text{ VAR} \\ &= 10 - j499.9 \text{ VA}\end{aligned}$$

Therefore the source supplies

$$\mathbf{S}_s = 1010 - j15.6 \text{ VA}$$

The power triangles associated with $\mathbf{S}_1$, $\mathbf{S}_2$, and $\mathbf{S}_s$ are shown in Fig. 12.12.

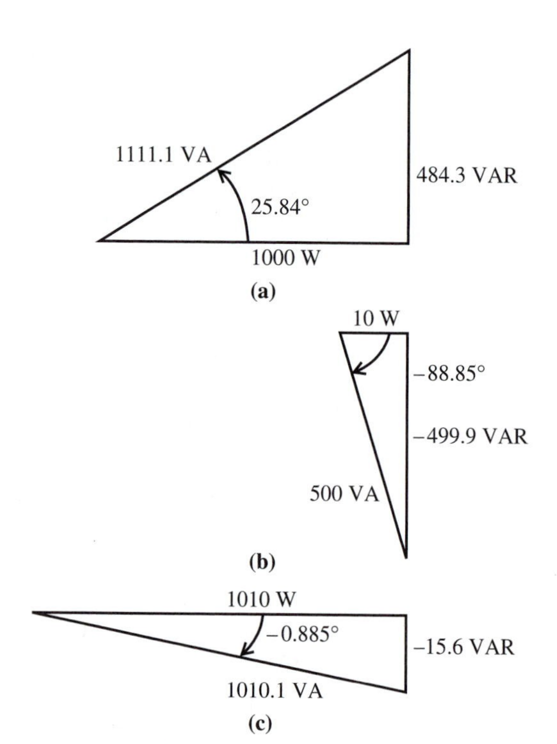

FIGURE 12.12 Power triangles for Example 12.4: (*a*) inductive load; (*b*) capacitive load; (*c*) composite load. (The power triangles are not drawn to scale.)

(b) The effective value of $i(t)$ can be determined from

$$|\mathbf{S}_s| = \frac{1}{2} V_m I_m$$

Recall that $I_m = \sqrt{2} I_{\text{rms}}$ and $V_m = \sqrt{2} V_{\text{rms}}$. Therefore

$$|\mathbf{S}_s| = V_{\text{rms}} I_{\text{rms}}$$

Solving for I_{rms}, we have

$$\begin{aligned} I_{\text{rms}} &= \frac{|\mathbf{S}_s|}{V_{\text{rms}}} \\ &= \frac{|1010 - j15.6|}{120} \text{ A rms} \\ &= \frac{1010.1}{120} = 8.417 \text{ A rms} \end{aligned}$$

(c) The rms current supplied to the inductive load follows from

$$I_{1\,\text{rms}} = \frac{|\mathbf{S}_1|}{V_{\text{rms}}} = \frac{1111.1}{120} = 9.259 \text{ A rms}$$

where we obtained $|\mathbf{S}_1|$ from the power triangle (Fig. 12.13a).

(d) Because $\mathbf{S}_s = \frac{1}{2}\mathbf{V}\mathbf{I}^*$ and $\mathbf{Z} = \mathbf{V}/\mathbf{I}$,

$$\mathbf{Z} = \frac{2\mathbf{S}_s}{|\mathbf{I}|^2} = \frac{2\mathbf{S}_s}{I_m^2} = \frac{\mathbf{S}_s}{I_{\text{rms}}^2} = 14.25 - j0.22 \ \Omega$$

(e)

$$\text{PF} = \cos(\angle\mathbf{Z}) = \cos\left(\tan^{-1}\frac{-0.22}{14.25}\right) = 0.9999 \text{ (leading)} \simeq 1$$

(f) The inductive load has a power factor of only 0.9, because it requires 484.3 VAR reactive power as well as 1 kW average power. When the capacitive load is placed in the circuit, its reactive power requirement (−499.9 VAR) almost cancels that of the inductive load (484.3 VAR). Thus the reactive power required by the composite load is only −15.6 VAR. In this sense, the capacitive load *supplies* the reactive power needed by the inductive load. The result is an increase in the power factor seen by the source from 0.9 (inductive load alone) to almost unity (composite load). The use of one load to supply reactive power needed by another load in an effort to raise power factor is called *power factor correction or improvement.*

EXAMPLE 12.5 A medium-size industrial facility, which consumes 2 MVA at PF = 0.65 lagging, is informed by the electric utility company that its rates will increase substantially unless it increases its power factor to at least 0.85 lagging. (a) Find the average power and the reactive power delivered to the facility. (b) Assuming that the average power is fixed, by what amount should the reactive power be decreased in order to achieve a power factor of at least 0.85? (c) What practical action can the facility take to raise its power factor to 0.85? (d) Is there any advantage, from the facility's point of view, to raising the power factor higher than 0.85?

Solution (a) The average power delivered is, by Eq. (12.16), $P = 2 \times 0.65$ MW $= 1.3$ MW. The reactive power delivered is, by Eq. (12.17), $Q = 2 \sin\theta$ MVAR, where $\theta = \arccos(0.65)$ and $\theta > 0$. This calculates to $Q = 1.5$ MVAR.

(b) We know from Eq. (12.35) that $\text{PF} = P/\sqrt{P^2 + Q^2}$. Therefore, for a fixed average power P power factor increases as the absolute value of the reactive power Q decreases. Let us denote the minimum permitted power factor by PF_{min}. Then to attain a power factor of at least PF_{min}, we need

$$\frac{P}{\sqrt{P^2 + Q^2}} > \text{PF}_{\text{min}}$$

By rearranging the above, we find that

$$|Q| < P\sqrt{1 - (\text{PF}_{\text{min}})^2}$$

We found in part (a) that $P = 1.3$ MW. To achieve a power factor of at least 0.85, therefore,

$$|Q| < 1.3\sqrt{1 - (0.85)^2} = 0.685 \text{ MVAR}$$

We also found in part (a) that the facility consumes 1.5 MVAR. Therefore, to attain a power factor of at least 0.85, the facility must decrease the reactive power it consumes by at least 1.5 MVAR − 0.685 MVAR = 815 kVAR.

(c) According to the complex power balance theorem, the complex power supplied by the utility company is equal to the sum of the complex powers delivered to the electrical components in the facility. It follows that the reactive power supplied equals the sum of the reactive powers delivered to all the electrical components in the facility. Therefore, the facility can increase its power factor to 0.85 by inserting capacitive reactance that consumes −815 kVAR (that is, *supplies* 815 kVAR) anywhere in the facility's electrical network. In practice, this may be done by installing a large bank of capacitors at the power substation for the facility. Alternatively, individual capacitors may be installed in parallel with the inductive components throughout the facility.

(d) Additional capacitance is required to raise the power factor further. Usually, there is no advantage from the facility's point of view to raising the power factor higher than the minimum required to achieve the lower electrical rate.

EXAMPLE 12.6 Analyze the distribution of complex power in the parallel *RLC* network of Fig. 12.13.

Solution The complex powers entering the resistance, inductance, and capacitance are, respectively,

$$\begin{aligned}\mathbf{S}_R &= P_R + jQ_R \\ &= \frac{1}{2}\frac{V_m^2}{R} + j0\end{aligned}$$

$$\begin{aligned}\mathbf{S}_L &= P_L + jQ_L \\ &= 0 + j\left(\frac{1}{2}V_m^2\frac{1}{\omega L}\right)\end{aligned}$$

and

$$\begin{aligned}\mathbf{S}_C &= P_C + jQ_C \\ &= 0 + j\left(-\frac{1}{2}V_m^2\omega C\right)\end{aligned}$$

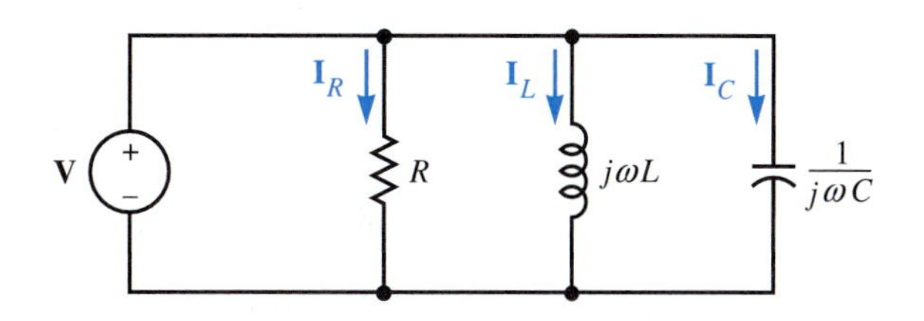

FIGURE 12.13
Parallel *RLC* circuit

By the complex power balance theorem, the sum of these complex powers equals the complex power supplied by the source $\mathbf{S}_s$:

$$\begin{aligned}\mathbf{S}_s &= \mathbf{S}_R + \mathbf{S}_L + \mathbf{S}_C \\ &= P_R + j(Q_L + Q_C) \\ &= \frac{1}{2}\frac{V_m^2}{R} + j\left(\frac{1}{2}V_m^2\frac{1}{\omega L} - \frac{1}{2}V_m^2\,\omega C\right)\end{aligned}$$

The power triangle is shown in Fig. 12.14. The real power supplied by the source is the term $\frac{1}{2}V_m^2/R$, which is the power dissipated by the resistance. The reactive power supplied by the source is the sum of the reactive powers consumed by the inductance and the capacitance. In view of the negative sign of Q_C, the capacitance can be thought of as a supplier of reactive power. By this interpretation, the reactive power supplied by the source is the difference between the reactive power consumed by the inductance and that supplied by the capacitance. Recall from Eqs. (12.21) and (12.25) that Q_L and Q_C can be expressed in terms of the average stored energies $W_{L,\text{ave}}$ and $W_{C,\text{ave}}$ in the inductance and the capacitance, respectively. This leads to

$$\mathbf{S}_s = \frac{1}{2}\frac{V_m^2}{R} + j2\omega(W_{L,\text{ave}} - W_{C,\text{ave}})$$

The reactive power supplied by the source is equal to 2ω times the *excess* of the average stored magnetic-field energy compared with the average stored electric-field energy.

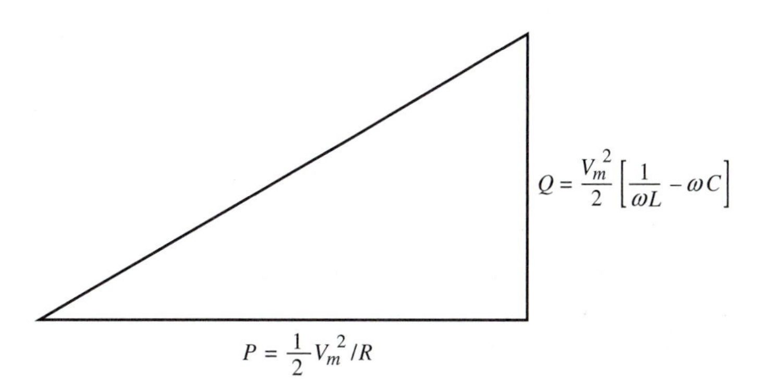

FIGURE 12.14
Power triangle for parallel *RLC* circuit

The analysis contained in Example 12.5 can be generalized. Consider an arbitrary network consisting of resistances, inductances, capacitances, and independent sources operating in the sinusoidal steady state at frequency ω. Because the passive elements are *R*s, *L*s, and *C*s, Eq. (12.47c) can be written as

$$\sum_{\substack{\text{All}\\\text{sources}}} \mathbf{S}_s = \sum_{\substack{\text{All}\\\text{resistances}}} \mathbf{S}_R + \sum_{\substack{\text{All}\\\text{inductances}}} \mathbf{S}_L + \sum_{\substack{\text{All}\\\text{capacitances}}} \mathbf{S}_C$$

$$= \sum_{\substack{\text{All}\\\text{resistances}}} \left(\frac{1}{2}\right)\frac{V_m^2}{R} + \sum_{\substack{\text{All}\\\text{inductances}}} j2\omega W_{L,\text{ave}} - \sum_{\substack{\text{All}\\\text{capacitances}}} j2\omega W_{C,\text{ave}}$$

$$= \sum_{\substack{\text{All}\\\text{resistances}}} \left(\frac{1}{2}\right)\frac{V_m^2}{R} + j2\omega\left(\sum_{\substack{\text{All}\\\text{inductances}}} W_{L,\text{ave}} - \sum_{\substack{\text{All}\\\text{capacitances}}} W_{C,\text{ave}}\right) \qquad (12.55a)$$

The first term on the right-hand side of Eq. (12.55a) is the total average power dissipated in the resistances. The second term is $j2\omega$ times the difference between the total average energy stored in the inductances and the total average energy stored in the capacitances. With the obvious definitions, Eq. (12.55a) can be rewritten as

$$\mathbf{S} = P + j2\omega\{W_{\text{tot ave magnetic}} - W_{\text{tot ave electric}}\} \qquad (12.55b)$$

where $\mathbf{S}$ is the total complex power supplied by the sources, P is the total average power dissipated in the resistances, $W_{\text{tot ave magnetic}}$ is the total average energy stored in the inductances, and $W_{\text{tot ave electric}}$ is the total average energy stored in the capacitances.

The power triangle is shown in Fig. 12.15. Equation (12.55b) represents a theoretical result that has remarkable generality, since it applies to *any* network made up of *R*s, *L*s, and *C*s. We can see from it that any such network will appear *resistive* to a source at a particular angular frequency ω if and only if the total average energy stored in the network's magnetic fields equals the total average energy stored in the network's electric fields.

Remember The sum of the complex powers entering all network components equals zero.

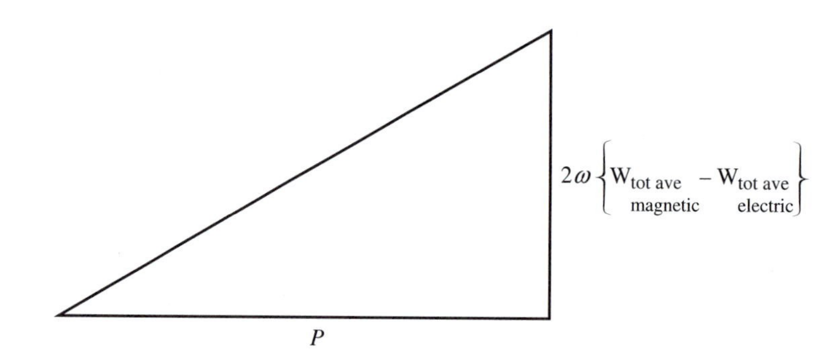

FIGURE 12.15 Power triangle for any circuit composed of *R*s, *L*s, and *C*s

EXERCISES

14. According to Eq. (12.47c), the complex power supplied by the sources in an ac network equals the complex power delivered to the passive components in the network.
    (a) Take the real parts of both sides of Eq. (12.47c) to derive a similar result concerning the average powers supplied and delivered. State the result in words.

    answer: $\sum_{\substack{\text{All}\\\text{Components}}} P = 0$

(b) Take the imaginary parts of both sides of Eq. (12.47c) to derive a similar result concerning the reactive powers supplied and delivered. State the result in words.

answer: $\sum_{\substack{\text{All}\\ \text{Components}}} Q = 0$

15. A fully loaded 5-hp induction motor operates at 70 percent efficiency and a power factor of 0.8 lagging.
   (a) Find the average power and the reactive power delivered to the motor.
   answer: 5.33 kW, 4 kVAR
   (b) By what amount should the reactive power be decreased in order to achieve a power factor of at least 0.95? *answer:* 2.34 kVAR
   (c) It is decided to raise the power factor to 0.95 lagging by placing capacitance in parallel with the power line. Find the value of C required if the line voltage equals 480 V rms and $f = 60$ Hz. *answer:* 27 μF

12.6 Maximum Power Transfer

In Section 5.4 we solved the problem of maximizing the power delivered to a load resistance R_L from a source with internal resistance R_s. In this section, we extend the problem to ac circuits. We consider the circuit of Fig. 12.16. In this figure the source is represented by its Thévenin equivalent circuit inside the blue rectangle. $\mathbf{V}_s$ is the open-circuit source (phasor) voltage, and $\mathbf{Z}_s$ is the equivalent source impedance. $\mathbf{Z}_L$ is an arbitrary load impedance. The load current and voltage are

$$\mathbf{I}_L = \frac{\mathbf{V}_s}{\mathbf{Z}_s + \mathbf{Z}_L} \tag{12.56}$$

and

$$\mathbf{V}_L = \frac{\mathbf{Z}_L}{\mathbf{Z}_s + \mathbf{Z}_L}\mathbf{V}_s \tag{12.57}$$

The complex power delivered to the load is

$$\begin{aligned}\mathbf{S}_L &= \frac{1}{2}\mathbf{V}_L\mathbf{I}_L^* \\ &= \frac{1}{2}\frac{|\mathbf{V}_s|^2\mathbf{Z}_L}{|\mathbf{Z}_s + \mathbf{Z}_L|^2}\end{aligned} \tag{12.58}$$

Notice that the complex power delivered depends on the amplitude of the voltage source $V_{sm} = |\mathbf{V}_s|$ but does not depend on its phase $\phi_s = \underline{/\mathbf{V}_s}$. The impedances $\mathbf{Z}_L$ and $\mathbf{Z}_s$ in the right-hand side of Eq. (12.58) are both functions of the source frequency ω. Therefore, if the source frequency is varied, $\mathbf{S}_L = P_L + jQ_L$ will vary. One way to maximize the average power delivered to the load is to leave the circuit as it is but choose the source frequency to maximize P_L. This approach is related to the topic of *frequency response,* described in Chapter 13. In the analysis that follows, we will assume that the frequency of the source is fixed.

For a fixed source frequency ω and open-circuit voltage amplitude V_{sm}, there are two cases of practical interest.

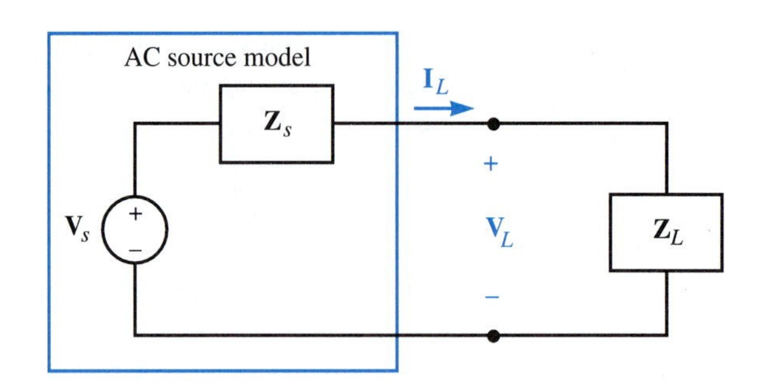

FIGURE 12.16
AC source and load impedance

Case 1 Choose the source impedance $\mathbf{Z}_s$ to maximize $P_L = \mathcal{Re}\{\mathbf{S}_L\}$ for a fixed load impedance $\mathbf{Z}_L$.

Case 2 Choose the load impedance $\mathbf{Z}_L$ to maximize P_L for a fixed source impedance $\mathbf{Z}_s$.

An analysis of both problems is facilitated by expressing $\mathbf{Z}_L$ and $\mathbf{Z}_s$ in terms of their resistive and reactive parts:

$$\mathbf{Z}_L = R_L + jX_L \tag{12.59}$$

$$\mathbf{Z}_s = R_s + jX_s \tag{12.60}$$

By substituting Eqs. (12.59) and (12.60) into (12.58), we find that

$$P_L = \frac{1}{2} \frac{V_{sm}^2 R_L}{(R_s + R_L)^2 + (X_s + X_L)^2} \tag{12.61}$$

and

$$Q_L = \frac{1}{2} \frac{V_{sm}^2 X_L}{(R_s + R_L)^2 + (X_s + X_L)^2} \tag{12.62}$$

Consider case 1, in which $\mathbf{Z}_L$ is fixed and $\mathbf{Z}_s$ is to be chosen for maximum power transfer. Because $\mathbf{Z}_L$ is fixed, it suffices to choose R_s and X_s to minimize the denominator in the preceding expression for P_L, Eq. (12.61). We can do this by choosing R_s to minimize $(R_s + R_L)^2$ and X_s to minimize $(X_s + X_L)^2$. Unless we wish to construct a negative resistance, the quantity $(R_s + R_L)^2$ is minimized by choosing

$$R_s = 0 \tag{12.63}$$

Negative reactance is easy to obtain. Therefore we minimize the quantity $(X_s + X_L)^2$ by choosing

$$X_s = -X_L \tag{12.64}$$

The resulting maximum average power delivered is obtained by substituting Eqs. (12.63) and (12.64) into Eq. (12.61). The result is

$$\underset{\mathbf{Z}_s}{\max} \{P_L\} = \frac{1}{2} \frac{V_{sm}^2}{R_L} \text{ W} \quad \text{(Case 1)} \tag{12.65}$$

which depends only on the open-circuit voltage amplitude and the load resistance $R_L = \mathcal{Re}\{\mathbf{Z}_L\}$. Representative plots of Eq. (12.61) are given in Fig. 12.17. (The figure assumes that $-X_L$ and R_L are fixed positive numbers.) Notice in Fig. 12.17b that re-

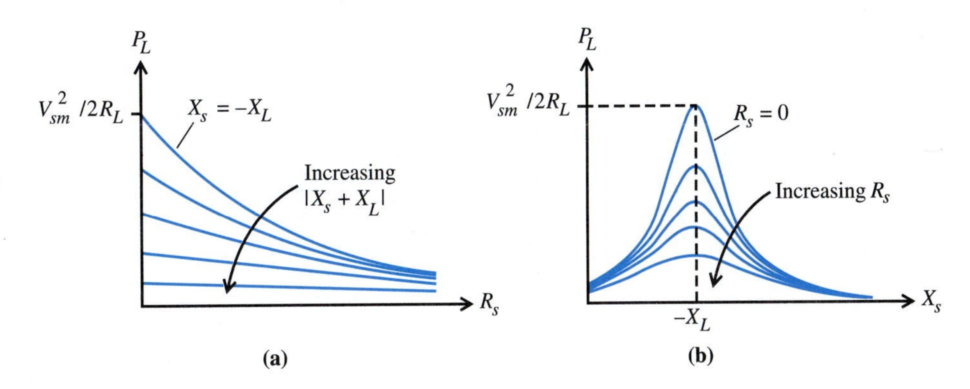

FIGURE 12.17 P_L versus R_s and X_s for a fixed load impedance. The average power delivered to the load is maximum when $R_s = 0$ and $X_s = -X_L$.

gardless of the value of R_s, a peak occurs when $X_s = -X_L$. Notice in Fig. 12.17a that regardless of the value of X_s, a peak occurs at $R_s = 0$. Therefore P_L *is always increased by a decrease in source resistance or by a decrease in the absolute value of the sum of source and load reactance.*

Refer now to case 2, in which $\mathbf{Z}_s$ is fixed and $\mathbf{Z}_L$ is to be chosen for maximum power transfer. It is easy to see that if we choose

$$X_L = -X_s \tag{12.66}$$

we minimize the term $(X_s + X_L)^2$ in Eq. (12.61). This leaves

$$P_L = \frac{1}{2}\frac{V_{sm}^2 R_L}{(R_s + R_L)^2} \tag{12.67}$$

which (by elementary calculus) has a maximum when

$$R_L = R_s \tag{12.68}$$

The maximum delivered power is

$$\max_{\mathbf{Z}_L} \{P_L\} = \frac{V_{sm}^2}{8R_s} \quad \text{(Case 2)} \tag{12.69}$$

The above value of P_L is called the *available power* of the source. It depends solely on the open-circuit voltage amplitude and the source resistance $R_s = \mathcal{Re}\{\mathbf{Z}_s\}$. The conditions for maximum power transfer stated in Eqs. (12.66) and (12.68) can be written more compactly as

$$\mathbf{Z}_L = R_s - jX_s \tag{12.70a}$$

or equivalently,

$$\mathbf{Z}_L = \mathbf{Z}_s^* \tag{12.70b}$$

A load impedance satisfying Eq. (12.70) is said to be *matched* to the source. Representative plots of Eq. (12.61) are given in Fig. 12.18. (The figure assumes that X_s is a negative number.) Notice in Fig. 12.18b that regardless of the value of R_L, a peak occurs when $X_L = -X_s$. Notice in Fig. 12.18a that a peak occurs at $R_L = R_s$ *only if* $X_L = -X_s$.

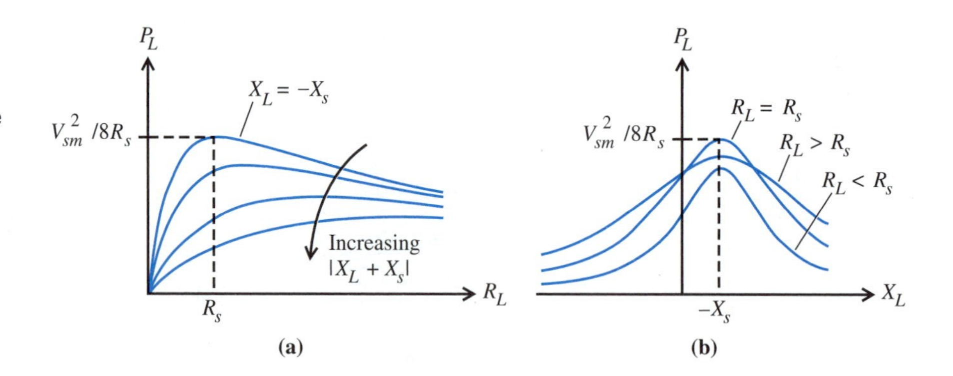

FIGURE 12.18 P_L versus R_L and X_L for a fixed source impedance. The average power delivered to the load is maximum when $R_L = R_s$ and $X_L = -X_s$.

See Problem 12.3 in the PSpice manual.

EXAMPLE 12.7 (a) Find the load impedance $\mathbf{Z}_L$ that maximizes the average power drawn from the network shown in Fig. 12.19, where $\omega = 10$ Mrad/s, $R = 2\ \Omega$, $C = 0.1\ \mu$F, $L = 0.2\ \mu$H, and $\mathbf{V}_s = 10\ \underline{/37.6°}$ mV.

(b) Find the average power delivered to the load impedance from (a).

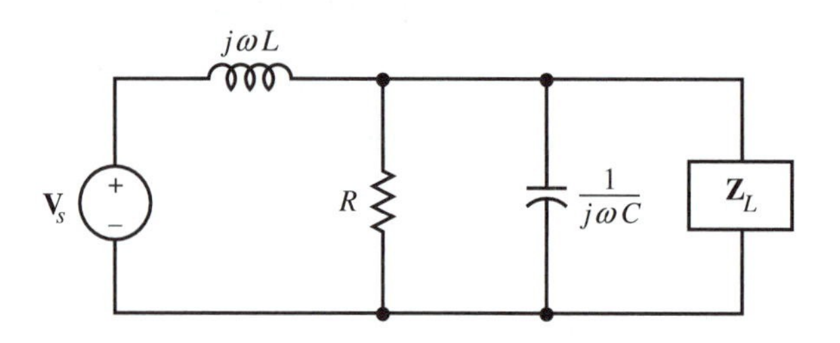

FIGURE 12.19 Circuit for Example 12.7

Solution (a) Maximum power transfer occurs when $\mathbf{Z}_L = \mathbf{Z}_s^*$, where $\mathbf{Z}_s$ is the equivalent impedance of the source. The equivalent impedance $\mathbf{Z}_s$ can be found by inspection if we set the independent source in the source network to zero. Thus

$$\mathbf{Z}_S = j\omega L \parallel R \parallel \frac{1}{j\omega C}$$
$$= \frac{G}{G^2 + (\omega C - 1/\omega L)^2} + j\frac{-(\omega C - 1/\omega L)}{G^2 + (\omega C - 1/\omega L)^2}$$

where $G = 1/R$. For the given numerical values, this simplifies to

$$\mathbf{Z}_s = 1 - j\ \Omega$$

The matched load impedance is $\mathbf{Z}_L = \mathbf{Z}_s^*$; that is,

$$\mathbf{Z}_L = 1 + j\ \Omega$$

Note that $\mathbf{Z}_L$ can be synthesized by a 1-Ω resistance in series with a 0.1-μH inductance.

(b) The power delivered under matched conditions can be found from Eq. (12.69) with the use of $R_s = 1$ and $V_{sm} = |\mathbf{V}_{oc}|$, where $\mathbf{V}_{oc}$ is the Thévenin equivalent voltage of the source network. Disconnecting $\mathbf{Z}_L$, we find

$$\mathbf{V}_{oc} = \frac{R \parallel (1/j\omega C)}{R \parallel (1/j\omega C) + j\omega L}\mathbf{V}_s$$

For the given numerical values, this simplifies to

$$\mathbf{V}_{oc} = \left[\frac{-2j/(2-j)}{-2j/(2-j)+2j}\right] 10 \underline{/37.6^\circ}$$
$$= \left(\frac{-2j}{2+2j}\right) 10 \underline{/37.6^\circ}$$
$$= 5\sqrt{2} \underline{/-97.4^\circ} \text{ mV}$$

Therefore

$$V_{sm} = 5\sqrt{2} \text{ mV}$$

and

$$P_L = \frac{(5\sqrt{2})^2}{8} \mu W$$
$$= 6.25 \mu\text{W}$$

Remember

We maximize the average power dissipated in a fixed load impedance $\mathbf{Z}_L = R_L + jX_L$ by choosing the source impedance as $\mathbf{Z}_s = 0 - jX_L$. We maximize the average power supplied by a source having fixed source impedance $\mathbf{Z}_s = R_s + jX_s$ by choosing the load impedance as $\mathbf{Z}_L = R_s - jX_s$.

EXERCISES

16. The short-circuit current $\mathbf{I}_{sc}$ from an ac network is $2 \underline{/20^\circ}$ A, and the open-circuit voltage $\mathbf{V}_{oc}$ is $10 \underline{/-65^\circ}$ V. What is the available power? *answer:* 28.68 W

For Exercises 17 through 20, use the circuit shown below.

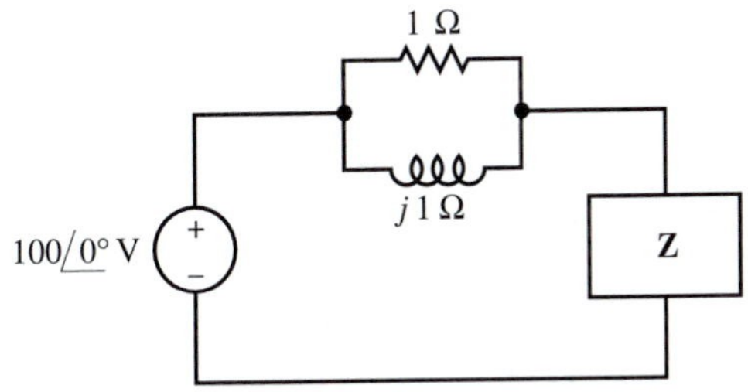

17. What is the value of the maximum average power that can be delivered to impedance **Z**? *answer:* 2.5 kW
18. What value of impedance **Z** will consume the maximum average power? *answer:* $0.5 - j0.5\ \Omega$
19. What is the value of the maximum average power that can be delivered to the 1-Ω resistance? *answer:* 10 kW
20. What value of impedance **Z** is required to maximize the average power delivered to the 1-Ω resistance? *answer:* $-j0.5\ \Omega$

12.7 Summary

The topic of this chapter was energy flow in ac circuits. We began by describing instantaneous power, $p(t) = v(t)i(t)$, supplied to a circuit operating in the sinusoidal steady state. This led to the conclusion that the average power entering the circuit is $P = \frac{1}{2}V_m I_m \cos\theta$. We defined and discussed the significance of apparent power, $\frac{1}{2}V_m I_m$, and reactive power, $Q = \frac{1}{2}V_m I_m \sin\theta$. We defined complex power as $\mathbf{S} = \frac{1}{2}\mathbf{V}\mathbf{I}^*$. Complex power has no physical meaning in itself. However, since $\mathbf{S} = P + jQ = \frac{1}{2}V_m I_m\underline{/\theta}$, complex power does provide a convenient analytical tool to calculate average, reactive, and apparent power. We then described the complex power balance theorem. According to this theorem, the total complex power delivered to all elements in an ac circuit (including sources) equals zero; that is, the total complex power supplied by the sources equals the total complex power delivered to the remaining elements of the circuit. Using this theorem, we showed that the complex power supplied to any network composed of *R*s, *L*s, and *C*s is given by

$$\mathbf{S} = P + j2\omega\{W_{\text{tot ave magnetic}} - W_{\text{tot ave electric}}\}$$

Finally, we extended the maximum power transfer problem treated in Chapter 5 to ac circuits. We considered two cases. In the first case, the load impedance was fixed, but the source impedance could be chosen. Here, maximum average power was delivered to the fixed load impedance $\mathbf{Z}_L$ when the source impedance $\mathbf{Z}_s$ was chosen so that $\mathcal{Re}\{\mathbf{Z}_s\} = 0$, $\mathcal{Im}\{\mathbf{Z}_s\} = -\mathcal{Im}\{\mathbf{Z}_L\}$. In the second case, the source impedance was fixed, but the load impedance could be chosen. Here, the maximum average power was supplied by the fixed source when $\mathbf{Z}_L = \mathbf{Z}_s^*$.

KEY FACTS

Concept	Equation	Section	Page
❑ Instantaneous power: $p(t) = v(t)i(t)$ W	(12.3)	12.1	387
❑ Average power: $P = \frac{1}{2}V_m I_m \cdot \text{PF W}$ where PF is the power factor PF $= \cos\theta$ with $\theta = \underline{/\mathbf{Z}}$.	(12.10)	12.1	388
❑ P is the average rate at which energy is *dissipated* by the network.		12.1	388
❑ Apparent power is $\frac{1}{2}V_m I_m$ VA, and so is simply one-half the product of the voltage and current amplitudes.	(12.11)	12.2	389
❑ Reactive power: $Q = \frac{1}{2}V_m I_m \sin\theta$ VAR	(12.17)	12.3	391
❑ Q is the peak rate at which energy is *borrowed* and *returned* by the network.		12.3	391
❑ Complex power: $\mathbf{S} = \frac{1}{2}\mathbf{V}\mathbf{I}^*$ VA	(12.29)	12.4	394

KEY FACTS Concept	Equation	Section	Page
Complex power: $\mathbf{S} = P + jQ = \frac{1}{2} V_m I_m \underline{/\theta}$ VA	(12.31) (12.33)	12.4	395
Complex power balance theorem: $\underset{\text{Supplied by sources}}{\sum \mathbf{S}_i} = \underset{\text{Delivered to passive elements}}{\sum \mathbf{S}_i}$	(12.47)	12.5	400
Complex power: $\mathbf{S} = P + j2\omega\{W_{\text{tot ave magnetic}} - W_{\text{tot ave electric}}\}$	(12.55b)	12.5	406
Maximum average power is delivered to a fixed impedance $\mathbf{Z}_L = R_L + jX_L$ when the source impedance $\mathbf{Z}_S$ is chosen as $\mathbf{Z}_S = -jX_L$.	(12.63) (12.64)	12.6	408
Maximum average power is supplied by a source with internal impedance $\mathbf{Z}_s = R_s + jX_s$ when the load impedance $\mathbf{Z}_L$ is chosen as $\mathbf{Z}_L = R_s - jX_s = \mathbf{Z}_s^*$. When $\mathbf{Z}_L = \mathbf{Z}_s^*$, we say that $\mathbf{Z}_L$ is *matched* to $\mathbf{Z}_s$.	(12.70)	12.6	409

PROBLEMS

Sections 12.1 to 12.3

1. A current $i(t) = I_m \cos(\omega t)$ is applied to a resistance, R.
   (a) Plot $i(t)$.
   (b) Plot the terminal voltage, $v(t)$.
   (c) Find and plot the instantaneous power supplied, $p(t)$.
   (d) Find the apparent power supplied.
   (e) Find the average power supplied.
   (f) Find the reactive power supplied.
2. Repeat Problem 1 for an inductance, L.
3. Repeat Problem 1 for a capacitance, C. Assume that the dc voltage across the capacitance is zero.
4. An ac current, $i(t) = 10 \cos(120\pi t)$ mA, is applied to a 1.5-V battery. The battery can be modeled by a 1.5-V dc voltage source in series with a 50-Ω resistance.
   (a) Find and plot the instantaneous power supplied to the battery.
   (b) Find the average power supplied to the battery.
   (c) Find the reactive power supplied to the battery. Be careful . . . this one is tricky!
5. A 1-kHz source delivers 10 mW at PF = 0.5 leading.
   (a) Find the peak instantaneous total power supplied.
   (b) Find the peak instantaneous power *dissipated.*
   (c) Find the peak instantaneous power *loaned.*
6. A 60-Hz source delivers 480 VA at PF = 0.707 lagging.
   (a) Assume that the current phase angle $\phi_{\mathbf{I}}$ equals zero, and plot the instantaneous power delivered $p(t)$.
   (b) What is the peak instantaneous power delivered by the source?

For Problems 7 through 11, the instantaneous power supplied by a voltage source is given by $p(t) = 50 + 10\cos(200t - 20°) + 5\sin(200t - 20°)$ mW.

7. Find P.
8. Find Q.
9. Find the apparent power.
10. Find the peak instantaneous power.
11. Find the frequency of the voltage source in hertz.
12. A series RLC circuit is driven by a sinusoidal source with angular frequency ω. Find the relation between R, L, C, and ω such that the reactive power delivered equals zero.
13. Repeat Problem 12 for a parallel RLC circuit.
14. A sinusoidal source delivers 14.14 VA at $f =$ 10 Hz to a parallel RC circuit. Find the values of R and C if PF = 0.707 leading and the source voltage is $4\sqrt{2}$ V rms.
15. A voltage source is supplying 2 VAR to a series RL network at a frequency of 60 Hz. The circuit element values are $R = 3.77\ \Omega$ and $L = 10$ mH. Find
    (a) The power factor of the network.
    (b) The average power supplied.
    (c) The apparent power supplied.
    (d) The effective voltage of the source.
    (e) The effective current.

Section 12.4

16. A 120-V rms 60-Hz sinusoidal voltage is applied to a load impedance **Z**. Measurements reveal that the apparent power entering the load is 60 VA and that the power factor is PF = 0.866 lagging.
    (a) What is the peak current supplied?
    (b) State the value of **Z** in polar form.
    (c) State the complex power **S** in polar form.
    (d) Sketch the power triangle, indicating the values of $|\mathbf{S}|$, $\angle\mathbf{S}$, P, and Q.
    (e) Assume that the load is a series RL circuit, so that $\mathbf{Z} = R + j\omega L$. Determine the values of R and L.
17. A load draws 10,000 VA from a 60-Hz sinusoidal source at a power factor of 0.707 leading.
    (a) What is the average power delivered to the load?
    (b) What is the reactive power delivered to the load?
    (c) Find the peak current if the effective source voltage is 120 V rms.
    (d) The load consists of a resistance R in parallel with a capacitance C. What is the expression for the complex power $\mathbf{S} = \frac{1}{2}\mathbf{VI}^*$ entering the load? State your answer in terms of V_{eff}, R, and $j\omega C$.
    (e) Find R and C if $V_{\text{eff}} = 120$ V rms.
18. A certain capacitor is modeled by a capacitance C in parallel with a resistance R.
    (a) Find the admittance of the parallel RC model.
    (b) The capacitor consumes $\mathbf{S} = 0.05 - j5$ VA when driven by a 7.07-V rms 60-Hz sinusoidal source. Determine the values of R and C.
    (c) What is the average stored energy?
19. Derive the expression for the complex power entering a series RLC circuit as a function of R, L, C, ω, and the rms terminal current I_{rms}. State the resulting expressions for P and Q.
20. Derive the expression for the complex power entering a series RLC circuit as a function of R, L, C, ω, and the rms terminal voltage V_{rms}. State the resulting expressions for P and Q.
21. Repeat Problem 19 for a parallel RLC circuit.
22. Repeat Problem 20 for a parallel RLC circuit.
23. A current $i(t) = 30\cos(5t + 20°)$ A is applied to a circuit. The resulting driving-point terminal voltage is $v(t) = 10\cos(5t + 65°)$ V.
    (a) What is the driving-point impedance of the circuit? Put your answer in polar form.
    (b) What is the complex power entering the circuit? Put your answer in polar form.
    (c) What is the apparent power entering the circuit?
24. A sinusoidal current $i(t) = I_m\cos(\omega t + \phi_I)$ is applied to a series RLC circuit.
    (a) Determine the complex power entering each element.
    (b) Sketch the power triangle of the series circuit and indicate its parameters.
    (c) At what frequency does the source supply only real power?
    (d) Under what condition does the source supply only reactive power?
25. A 1-V rms sinusoidal source with angular frequency ω is connected to a series RLC circuit in

which $R = 0.1\ \Omega$, $L = 1\ \mu H$, and $C = 1\ \mu F$. Make smooth plots of the following quantities versus ω for $0 \le \omega \le 2$ Mrad/s:

(a) power factor,

(b) average power delivered,

(c) reactive power delivered, and

(d) apparent power. Comment on your results. (*Hint:* The curves vary rapidly in the vicinity of $\omega = 1$ Mrad/s.)

Section 12.5

26. A 120-V rms sinusoidal voltage source provides power to two loads connected in series. Load 1 consumes 10 W at PF = 0.8 lagging. Load 2 consumes 14 W at PF = 0.9 leading. Find

    (a) The peak current.

    (b) The rms voltage across *each* load.

    (c) The power factor of the series combination of loads 1 and 2.

    (d) The complex power supplied by the source.

27. Two loads are connected in parallel across a sinusoidal source. The effective current supplied is 10 A rms. Load 1 consumes 12 kW at PF = 0.707 leading. Load 2 consumes 16 kW at PF = 0.866 lagging. Find

    (a) The rms current through each load.

    (b) The peak voltage.

    (c) The power factor of the parallel combination.

    (d) The complex power supplied by the source.

28. A 120-V rms 60-Hz sinusoidal voltage source drives a 50-kW induction motor that operates with a lagging power factor of 0.8. What is the peak current supplied by the source?

29. A 120-V rms 60-Hz sinusoidal voltage source drives a load composed of a motor and a capacitance connected in parallel. The motor consumes 50 kW at a lagging power factor of 0.8, and the capacitance has value C.

    (a) What value of C is required to make the power factor of the parallel load equal unity?

    (b) Assuming that the capacitance is as computed in (a), what is the peak current supplied by the source? Compare your answer with the answer to Problem 28.

30. Two loads are in parallel across a 70.7-V rms line. Load 1 draws 100 VA at a lagging power factor of 0.342. Load 2 draws −33.55 VAR at a power factor of 0.819.

    (a) What is the peak line voltage?

    (b) What is the complex power entering load 1?

    (c) What is the complex power entering load 2?

    (d) What is the rms current supplied by the source?

    (e) A third load is placed in parallel across the 70.7-V rms line. What should the reactance of the third load be in order to make the power factor of the combined loads equal unity? Assume that the third load is purely reactive.

31. A muffler repair shop uses either one, two, or three arc welders as required as customers bring their cars in for repair. When in use, each welder draws 70 A rms at a power factor of 0.7 lagging from the shop's 240-V rms 60-Hz power line.

    (a) Find the effective current that must be supplied when one, two, and three welders are in use.

    (b) In order to reduce the effective current drawn from the power line, it is decided to increase the power factors of each arc welder to 0.9 lagging by placing a capacitance C across the input voltage terminals of each welder. Find the value of C required.

    (c) Find the effective current that must be supplied when one, two, and three increased-power-factor welders are in use.

    (d) Since the power factors of the individual welders have been improved to 0.9 lagging, one might assume that the power factor of the combined welder load will be 0.9 lagging, regardless of how many welders are in use. Is this assumption true? Prove it.

32. In this problem we analyze the relation between power factor improvement and the reduction in the power wasted in the line that supplies power to the load. Assume that the power factor is increased by the insertion of pure capacitive or inductive reactance in parallel with the load and that the voltage supplied to the load is unchanged.

    (a) Let PF and PF′ denote the original and improved power factors, and I_{eff} and I'_{eff} denote the original and reduced effective currents. The percent reduction in effective current is defined as

$$\eta = \frac{I_{eff} - I'_{eff}}{I_{eff}} 100\%$$

Show that

$$\eta = \left(1 - \frac{\text{PF}}{\text{PF}'}\right) 100\%$$

(b) Show that $I'_{eff} = (\text{PF}/\text{PF}')I_{eff}$.

(c) Before power factor improvement, the average power dissipated in the line supplying power to the load is $(I_{eff})^2 r$ where r is the line resistance. Show that power factor improvement results in a reduction of this wasted power by $[1 - (\text{PF}/\text{PF}')^2]100\%$.

33. (a) Analyze the distribution of complex power in an ac series RLC circuit driven by a current source $\mathbf{I}_s$.

(b) Draw the associated power triangle.

Section 12.6

34. A series RC circuit is driven with an ac source. The open-circuit source voltage is $\mathbf{V}_S = 10\underline{/0°}$ V.

(a) Find the expression for the source impedance that maximizes the average power dissipated in the series RC load.

(b) Let $R = 10\ \Omega$, $C = 1\ \mu$F, and $\omega = 10^6$ rad/s. Represent the source impedance you determined in part (a) with the use of a series RL circuit, and specify numerical values for source resistance and inductance.

(c) Find the average power and the reactive power delivered to the load for the numerical values of part (b).

35. An ac source has open-circuit voltage $\mathbf{V}_s = 10\underline{/0°}$ V and source impedance $\mathbf{Z}_s = R + (1/j\omega C)$.

(a) Find the expression for the load impedance that maximizes the average power delivered by the source.

(b) Let $R = 10\ \Omega$, $C = 1\ \mu$F, and $\omega = 10^6$ rad/s. Represent the load impedance you determined in part (a) with the use of a series RL circuit, and specify numerical values for load resistance and inductance.

(c) Find the average power and the reactive power delivered to the load for the numerical values of part (b).

36. A practical ac source has open-circuit voltage $\mathbf{V}_s = 1\underline{/0°}$ V and Thévenin equivalent source impedance $\mathbf{Z}_s$. The available power from the source is 0.125 W. If an impedance $\mathbf{Z}_L = j1\ \Omega$ is attached to the terminals of the source, the voltage across that impedance is $\mathbf{V}_L = 1\underline{/90°}$ V. Find $\mathbf{Z}_s$.

37. When a LTI network is terminated in a 1-Ω resistance, the steady-state current in the resistance is $i_R(t) = 5\cos(1000t - 45°)$ A. When the same network is terminated in a 1000-μF capacitance, the steady-state current is $i_C(t) = 10\cos(1000t - 45°)$ A. Suppose that the network is now terminated in an arbitrary load impedance $\mathbf{Z}_L$.

(a) Give the expression for the average power supplied to $\mathbf{Z}_L$. Use numerical values wherever possible.

(b) Find the value of $\mathbf{Z}_L$ that consumes the maximum average power.

38. When two ac networks A and B are joined together as in connection 1 shown below, $\mathbf{I}_{ab} = 1\underline{/0°}$ A and $\mathbf{V}_{aa'} = \sqrt{2}\underline{/-45°}$ V. When the same networks are joined as in connection 2, $\mathbf{I}_{ab'} = 3\underline{/0°}$ A and $\mathbf{V}_{aa'} = \sqrt{2}\underline{/45°}$ V.

(a) Find the maximum average power that can be dissipated in a load impedance that is connected from terminal a to terminal a' in connection 1 (with networks A and B in place).

(b) Find the load impedance of (a) that consumes maximum average power.

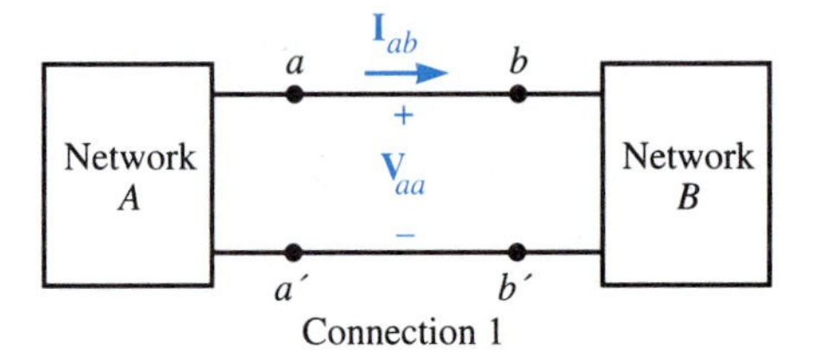

Connection 1

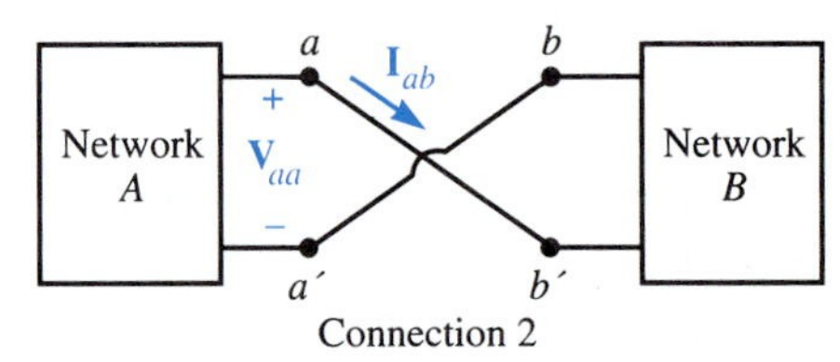

Connection 2

39. Repeat Problem 38 if the load impedance is connected from terminal a to terminal a' in connection 2 (with networks A and B in place).

40. Consider the active network on the next page.

(a) What load impedance should be connected to the network to consume the maximum average power?

(b) What is the maximum average power that can be obtained from the network?

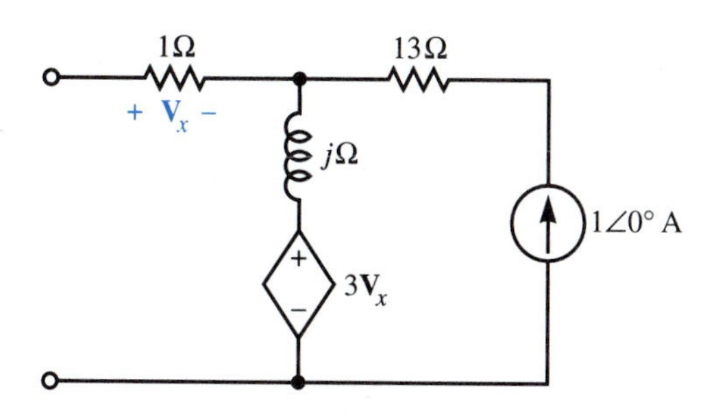

Supplementary Problems

41. You rented a low-budget, 90-year-old apartment. The apartment is wired with 15-A rms circuits. The voltage is 120 V rms, 60 Hz. Your television requires 3 A rms and is a resistive load. You find a bargain on a 120-V rms air conditioner, which requires 15 A rms at a 0.8 lagging power factor, and you wish to use both units at once. Neglect turn on transients.

    (a) What total complex power is required by the combined load?

    (b) What real power is required by the combined load?

    (c) What is the current required by the combined load?

    (d) You inadvertently install a 20-A rms fuse. What is the percentage increase in current and resistive heating in the apartment wires above the value at 15 A rms? Is this safe?

    (e) You find a bargain in capacitors at the surplus store. Can you use these to reduce the current to 15 A rms for the combined loads? If so, what value of C is required?

42. Two ac loads having identical power factor PF_0 are connected in series. Show that the power factor of the combined load equals PF_0.

43. Two inductive loads with power factors PF_1 and PF_2 are connected in series. Assume that $PF_1 \leq PF_2$. Show that the power factor PF_{comb} of the combined load satisfies the inequality $PF_1 \leq PF_{comb} \leq PF_2$. *Hint:* This problem can be solved using a simple graphical construction involving power triangles.

44. This problem is a generalization of Problem 43. Consider a network composed only of N inductive loads, but which is otherwise arbitrary. Denote the minimum and maximum power factors among the N loads by PF_{min} and PF_{max}, respectively. Show that the power factor PF_{net} of the network satisfies the inequality $PF_{min} \leq PF_{net} \leq PF_{max}$.

45. A sinusoidal source is applied to a network made up of Rs, Ls, and Cs. Use Eq. (12.55b) and the relations $\mathbf{S}_s = \frac{1}{2}\mathbf{V}_s\mathbf{I}_s^*$ and $\mathbf{Z}(j\omega) = \mathbf{V}_s/\mathbf{I}_s$ to show that the driving-point impedance of the network can be written in the form

$$\mathbf{Z}(j\omega) = \frac{P + j2\omega(W_{\text{tot ave magnetic}} - W_{\text{tot ave electric}})}{\frac{1}{2}I_{sm}^2}$$

13

Frequency Response

The topic of frequency response is important in the design or analysis of virtually every physical system or phenomenon that is linear and time-invariant.

We find a circuit's *frequency response* by plotting the circuit's transfer function $\mathbf{H}(j\omega)$ versus ω. This plot reveals that LTI circuits can be used to separate or *filter* sinusoidal signals that have different frequencies. Such filters are used in radio and television tuners to separate one broadcast channel from another.

We begin this chapter by giving some simple examples of frequency response. Then we embark on our major development, which is to discover the underlying reason any given circuit has the frequency response it has. During this development, we encounter several basically new concepts, including *resonance, complex frequency,* and *poles and zeros.* We conclude with a description of Bode plots, which are a practical and commonly used format to plot frequency response.

13.1 A New Viewpoint: Response Is a Function of Frequency

The frequency-domain description of circuits is much more than a labor-saving device for us to compute sinusoidal steady-state output. It also provides us with insight into how we might use a circuit. This statement is illustrated in the following discussion.

High-Pass Filter

See Problem 13.1 in the PSpice manual.

Figure 13.1 depicts an ordinary *RL* voltage divider. By examining the transfer function of this circuit, we will find that we can use the circuit to *separate* or *filter* sinusoidal waveforms.

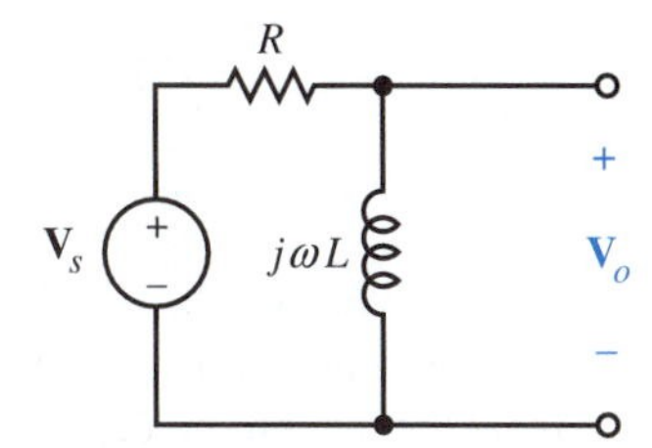

FIGURE 13.1
A high-pass filter

The circuit's voltage transfer function is given by the voltage-divider relation

$$\mathbf{H}(j\omega) = \frac{\mathbf{V}_o}{\mathbf{V}_s} = \frac{j\omega L}{R + j\omega L} \tag{13.1a}$$

which has magnitude

$$|\mathbf{H}(j\omega)| = \frac{\omega L}{\sqrt{R^2 + \omega^2 L^2}} = \frac{\dfrac{\omega L}{R}}{\sqrt{1 + \left(\dfrac{\omega L}{R}\right)^2}} \tag{13.1b}$$

and angle

$$\underline{/\mathbf{H}(j\omega)} = \frac{\pi}{2} - \text{arc tan}\,\frac{\omega L}{R} \tag{13.1c}$$

Let us plot the magnitude and angle of the voltage transfer function versus angular frequency ω. The results are shown in Fig. 13.2. When viewed as a function of frequency† this way, $|\mathbf{H}(j\omega)|$ and $\underline{/\mathbf{H}(j\omega)}$ are called the *amplitude* and *phase characteristics* of the circuit, respectively, and $\mathbf{H}(j\omega) = |\mathbf{H}(j\omega)|\underline{/\mathbf{H}(j\omega)}$ is called the *frequency response* of the circuit. The plots in Fig. 13.2 show us the relative amplitude and phase of the sinusoidal steady-state output:

$$v_o(t) = |\mathbf{H}(j\omega)|\,V_{sm}\cos\left(\omega t + \phi_{\mathbf{v}_s} + \underline{/\mathbf{H}(j\omega)}\right) \tag{13.2}$$

with respect to the circuit input

$$v_s(t) = V_{sm}\cos\left(\omega t + \phi_{\mathbf{v}_s}\right) \tag{13.3}$$

at any frequency ω.

† The term *frequency* is used in this book to denote either angular frequency ω (in radians per second) or "frequency" f (in hertz). Since $\omega = 2\pi f$, ω and f differ only in the choice of units: One hertz is equivalent to 2π radians per second.

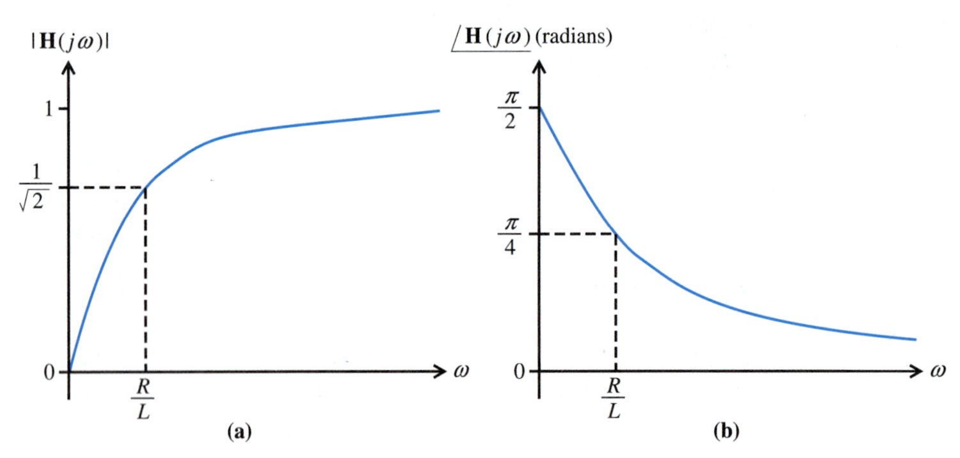

FIGURE 13.2
Frequency response characteristics of the circuit of Fig. 13.1. (*a*) Amplitude characteristic; (*b*) phase characteristic

The point R/L on the ω axis provides a convenient (but basically arbitrary) demarcation separating the frequency axis into two parts. We see by an inspection of Fig. 13.2a that the magnitude of the voltage transfer function $|\mathbf{H}(j\omega)|$ is small, $|\mathbf{H}(j\omega)| \ll 1$, for $\omega \ll R/L$. This tells us that the amplitude of the output sinusoid is small compared with the amplitude of the input sinusoid when the frequency of the input, ω, is small compared with R/L. We also see from Fig. 13.2 that $|\mathbf{H}(j\omega)|$ is nearly unity for $\omega \gg R/L$. This tells us that the amplitudes of the sinusoidal output and input are nearly the same when the frequency of the input is large compared with R/L.

It is easy to explain the shape of the amplitude characteristic of Fig. 13.2a. As ω approaches zero, the magnitude of the impedance of the inductance $|j\omega L|$ approaches zero. For $\omega = 0$, this magnitude is zero and the inductance is equivalent to a short circuit, which, of course, leads to zero voltage output. As ω approaches infinity, $|j\omega L|$ also approaches infinity. In the limit as $\omega \rightarrow \infty$, the inductance is equivalent to an open circuit, and this gives us unity voltage gain, because there is then no voltage drop across R.

Now let us consider how the behavior depicted by Fig. 13.2a can be used to separate one signal from another. Suppose that the input to the circuit is the sum of two sinusoidal waveforms as illustrated in Fig. 13.3. For the numerical circuit values given and excluding round-off error, the value of R/L equals 3.14×10^5 rad/s. The angular frequencies of the input sinusoids are $\omega = 3.14 \times 10^4$ rad/s and $\omega = 3.14 \times 10^6$ rad/s. Notice that the amplitudes of the input sinusoids each equal 3 V. It follows directly from Eqs. (13.1b) and (13.1c) that the circuit's transfer function is $\mathbf{H}(j3.14 \times 10^4) = 0.0995 \angle 84.3°$ for $\omega = 3.14 \times 10^4$ and $\mathbf{H}(j3.14 \times 10^6) = 0.995 \angle 5.7°$ for $\omega = 3.14 \times 10^6$. The steady-state output is, by superposition, the sum:

$$\begin{aligned} v_o(t) &= (0.0995)(3.0) \cos(3.14 \times 10^4 t + 10° + 84.3°) \\ &\quad + (0.995)(3.0) \cos(3.14 \times 10^6 t + 27° + 5.7°) \\ &= 0.299 \cos(3.14 \times 10^4 t + 94.3°) + 2.99 \cos(3.14 \times 10^6 t + 32.7°) \text{ V} \end{aligned} \quad (13.4)$$

The remarkable feature of this expression is that the amplitudes of the two output sinusoidal signals differ by a factor of 10. We see that the circuit has passed the higher-frequency input component with relatively little change in amplitude and phase, but has attenuated or *rejected* the lower-frequency input component by a factor of approximately 10. In effect, the circuit does indeed *filter* its input signals, *passing* the

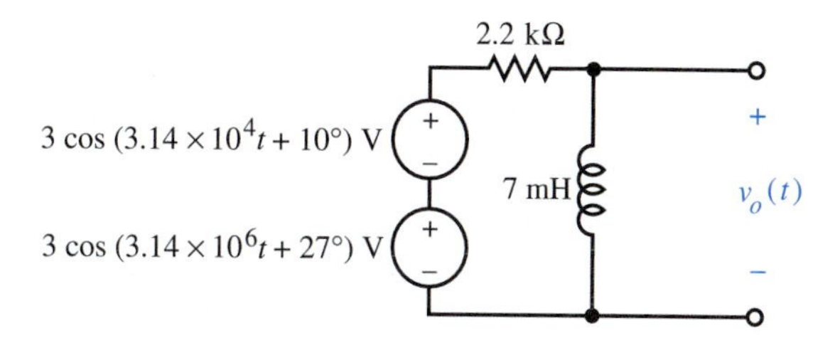

FIGURE 13.3
High-pass filter with $R/L = 3.14 \times 10^5$ rad/s

higher-frequency component and (approximately) removing or *rejecting* the lower-frequency component. For this reason we call the circuit a *high-pass filter.*

Notice that the amplitude characteristic of Fig. 13.2 is down by a factor of $1/\sqrt{2}$ at $\omega = R/L$ compared with its maximum of unity. A factor of $1/\sqrt{2}$ in amplitude corresponds to a factor of $(1/\sqrt{2})^2 = 1/2$ in power. For this reason we call R/L the *half-power* frequency of the filter. We call the frequency interval $0 \leq \omega \leq R/L$ the *stopband* of the filter: Signals whose frequencies are in this range are (approximately) stopped or rejected by the filter. The frequency interval $R/L < \omega < \infty$ is called the *passband* of the filter: Signals whose frequencies are in this range are (approximately) passed without change by the filter.

EXERCISES

1. Use Eq. (13.1a) to show that $\mathbf{H}(j\omega) = (1/\sqrt{2})\underline{/45^\circ}$ for $\omega = R/L$.
2. Find $v_o(t)$ in Fig. 13.3 if the angular frequencies of the inputs to the circuit are 3.14×10^3 rad/s and 3.14×10^7 rad/s rather than 3.14×10^4 rad/s and 3.14×10^6 rad/s. answer: $0.0299 \cos(10^3\pi t + 99.4^\circ) + 2.999 \cos(10^7\pi t + 27.57^\circ)$ V
3. A voltage $A_1 \sin(\omega_1 t + \phi_1)$ is applied across the terminals of a resistance of R. The average power that is delivered is 1 W. The voltage is then changed to $A_2 \cos(\omega_2 t + \phi_2)$, where $A_2 = A_1(1/\sqrt{2})$, $\omega_2 = 16\omega_1$, and $\phi_2 = \phi_1 + 27^\circ$. What is the average power delivered? answer: 0.5 W
4. For the circuit shown below:
   (a) Determine and sketch the frequency response characteristics.
   (b) Find the half-power frequency.
   (c) Identify the stopband and the passband.
   answer: (b) $1/RC$, (c) Stopband: $\omega < 1/RC$. Passband: $\omega > 1/RC$.

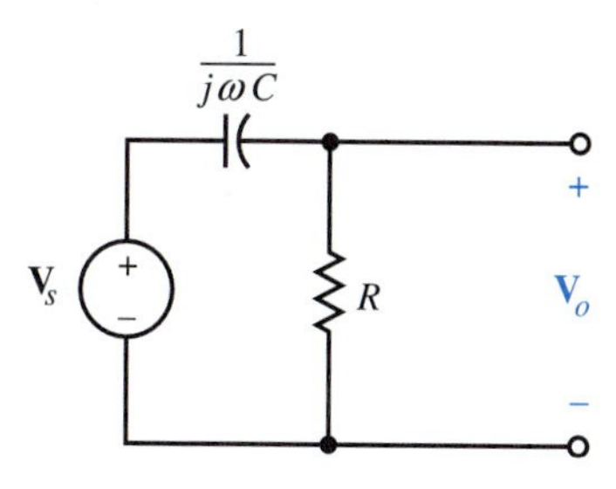

See Problem 13.2 in the PSpice manual.

Low-Pass Filter

A second example of frequency response is provided by the RC voltage divider of Fig. 13.4. Its frequency response

$$\mathbf{H}(j\omega) = \frac{1}{1 + j\omega RC} = \frac{1}{\sqrt{1 + (\omega RC)^2}}\underline{/-\arctan(\omega RC)} \tag{13.5}$$

FIGURE 13.4
A low-pass filter

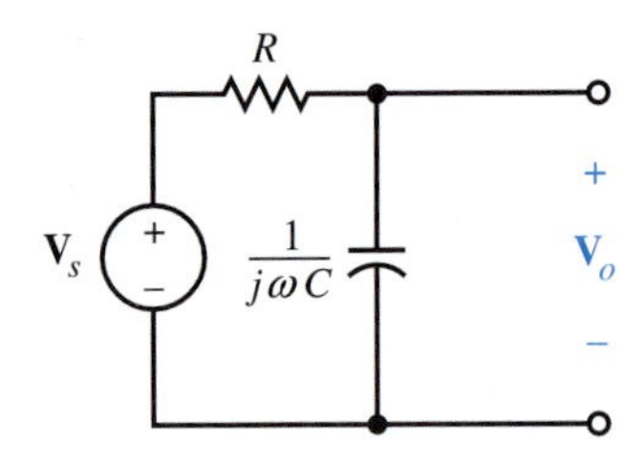

is plotted in Fig. 13.5. We can see from the plots that the circuit passes low-frequency sinusoidal inputs with little change in their amplitude and phase, but rejects high-frequency inputs. The demarcation between "low frequency" and "high frequency" occurs in the vicinity of the half-power frequency $\omega = 1/RC$. This circuit is called a *low-pass filter.* The frequency interval $0 \leq \omega \leq 1/RC$ is the *passband* of this filter, and $1/RC$ is its *half-power bandwidth* in radians per second. The *stopband* is the interval $1/RC < \omega < \infty$.

The *RL* and *RC* circuits of Figs. 13.1 and 13.4 are elementary examples of high-pass and low-pass filters. Additional examples of elementary high-pass and low-pass filters appear in Exercises 4 and 5. Much better filters can be obtained with the use of sophisticated circuit design techniques.

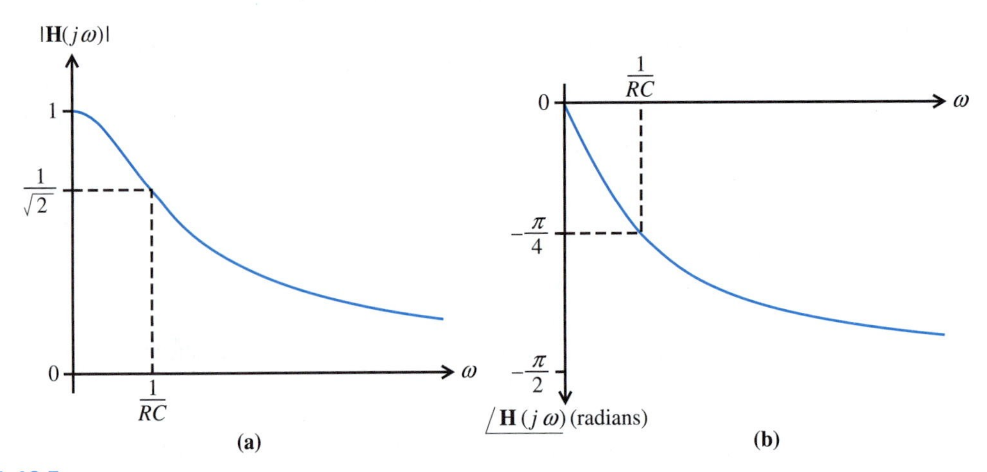

FIGURE 13.5
Frequency response characteristics of the circuit of Fig. 13.4. (*a*) Amplitude characteristic; (*b*) phase characteristic

EXERCISES 5. Repeat Exercise 4 for the circuit shown below.

answer: (b) R/L, (c) Stopband: $\omega > R/L$. Passband: $\omega < R/L$.

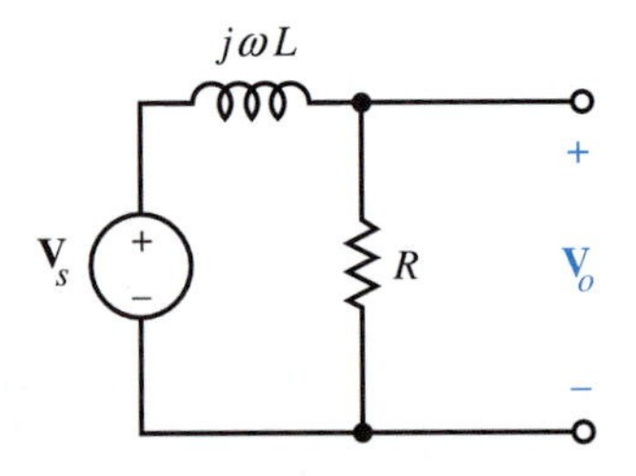

6. Assume that the input to the circuit shown below is $v_s(t) = 10 + 10 \cos 0.1t + 10 \cos t + 10 \cos 10t$. Find $v_o(t)$.

answer: $10 + 9.96 \cos (0.1t - 5.71°) + 5\sqrt{2} \cos (t - 45°) + 0.997 \cos (10t - 84.29°)$ V

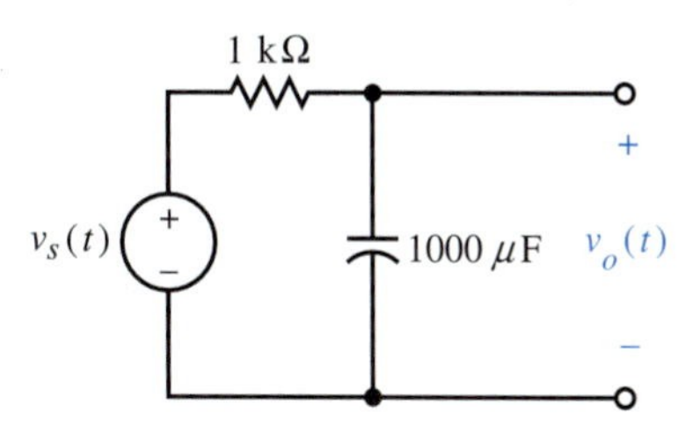

13.2 Resonance

See Problem 13.3 in the PSpice manual.

In the preceding section we showed that circuits can be used to filter sinusoidal signals that have different frequencies. We saw that the *kind* of filtering done by a circuit depends on the shape of the circuit's amplitude response characteristic. This observation leads us to ask *why* a given circuit has an amplitude characteristic with a particular shape. This is an important question because, if we know the reasons for the shape of an amplitude characteristic, we will be able to design better filters. The answer is intimately connected to the phenomenon of resonance, which is the topic we consider now.

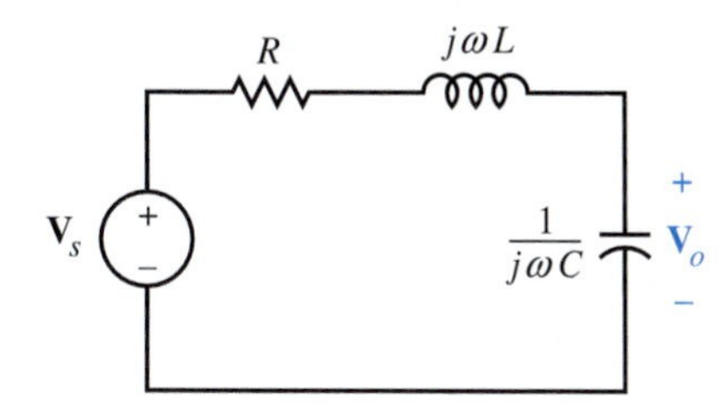

FIGURE 13.6
RLC circuit where $R = 0.1\ \Omega$, $L = 1\ \mu$H, and $C = 1\ \mu$F

Consider the series *RLC* circuit of Fig. 13.6. The voltage transfer function of this circuit is (by the voltage divider relation)

$$\mathbf{H}(j\omega) = \frac{\mathbf{V}_o}{\mathbf{V}_s} = \frac{1/j\omega C}{R + j\omega L + 1/j\omega C} = \frac{1}{1 - \omega^2 LC + j\omega RC} \tag{13.6}$$

for which

$$|\mathbf{H}(j\omega)| = \frac{1}{\sqrt{(1 - \omega^2 LC)^2 + (\omega RC)^2}} \tag{13.7}$$

and

$$\underline{/\mathbf{H}(j\omega)} = -\arctan \frac{\omega RC}{1 - \omega^2 LC} \tag{13.8}$$

Figure 13.7 shows the amplitude and phase characteristics for $R = 0.1\ \Omega$, $L = 1\ \mu$H, and $C = 1\ \mu$F.

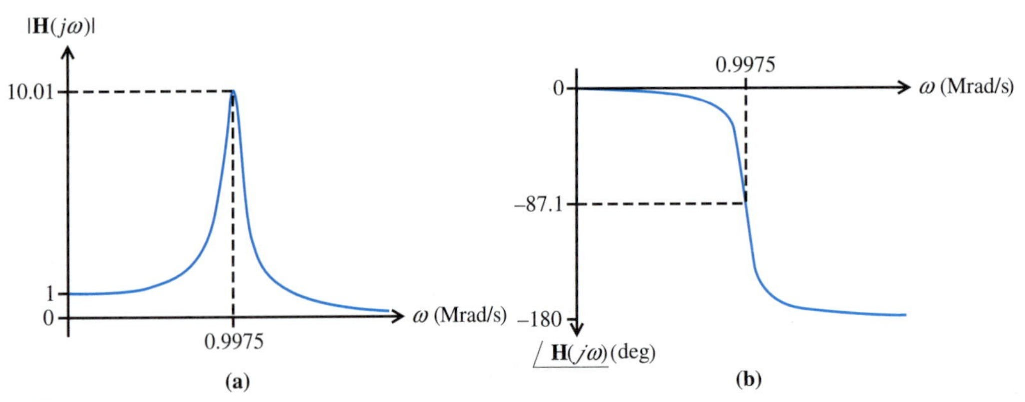

FIGURE 13.7
Frequency response characteristics of the circuit of Fig. 13.6 for $R = 0.1\ \Omega$, $L = 1\ \mu\text{H}$, and $C = 1\ \mu\text{F}$. (*a*) Amplitude characteristic; (*b*) phase characteristic

The most prominent feature of the frequency response is the sharp peak that occurs in the amplitude characteristic. We call this peak a *resonance* peak. We call the value of ω at which the peak occurs the *maximum-response resonance frequency,* ω_{mr}. A formula for ω_{mr} can be found if we set the derivative of $|\mathbf{H}(j\omega)|$ with respect to ω to zero and solve for ω. The result is

$$\omega_{mr} = \sqrt{\frac{1}{LC} - 2\left(\frac{R}{2L}\right)^2}$$
$$= 997.5 \text{ krad/s} = 0.9975 \text{ Mrad/s} \tag{13.9}$$

at which $|\mathbf{H}| = 10.01$ and $\angle\mathbf{H} = -87.1°$.

Since the amplitude characteristic for frequencies near 0.9975 Mrad/s is relatively large compared with that for other frequencies, a practical use of the circuit would be to increase the amplitudes of sinusoidal voltages whose frequencies are near 0.9975 Mrad/s relative to those with other frequencies. We will examine this application in detail in Section 13.5. In the next subsection, we will examine the more basic issue of *why* the peak occurs.

Why Resonance Occurs

Our objective in this subsection is to understand why there is a peak in the series *RLC* circuit's amplitude characteristic (Fig. 13.7a). How is it possible for the output amplitude to be a factor of 10.01 *greater* than the input amplitude? Let us first try to discover an underlying mathematical explanation. We can do this by writing the denominator of $\mathbf{H}(j\omega)$ [Eq. (13.6)],

$$\mathcal{A}(j\omega) = 1 - \omega^2 LC + j\omega RC \tag{13.10}$$

in factored form. To factor $\mathcal{A}(j\omega)$, we first find the roots to the characteristic equation

$$\mathcal{A}(s) = 1 + s^2 LC + sRC = 0 \tag{13.11}$$

For the circuit values given, the roots are complex. The roots are

$$s_1 = -\alpha + j\omega_d \tag{13.12}$$

and

$$s_2 = -\alpha - j\omega_d \tag{13.13}$$

where $\alpha = R/2L = 4 \times 10^4$ is the *damping coefficient* and $\omega_d = \sqrt{(1/LC) - (R/2L)^2} =$ 0.9987 Mrad/s is the *damped natural frequency.* Recall that the quantities s_1, s_2, α, and ω_d were encountered previously in connection with the natural response of underdamped *RLC* circuits (Sections 9.2 and 9.3). With s_1 and s_2 determined, we can now write $\mathscr{A}(s)$ in its factored form:†

$$\mathscr{A}(s) = LC(s - s_1)(s - s_2) \tag{13.14}$$

Therefore

$$\mathscr{A}(j\omega) = LC(j\omega - s_1)(j\omega - s_2) \tag{13.15}$$

and we can write the transfer function as follows:

$$\mathbf{H}(j\omega) = \frac{1}{1 - \omega^2 LC + j\omega RC} = \frac{1}{LC(j\omega - s_1)(j\omega - s_2)} \tag{13.16}$$

The advantage of writing the transfer function in factored form is that we can draw the factors $j\omega - s_1$ and $j\omega - s_2$ as vectors in the complex plane, as shown in Fig. 13.8. The factor

$$j\omega - s_1 = j\omega - (-\alpha + j\omega_d) = \alpha + j(\omega - \omega_d) \tag{13.17}$$

in Eq. (13.16) is represented as a vector starting at the root s_1 and ending at the point $j\omega$. Similarly, the factor

$$j\omega - s_2 = j\omega - (-\alpha - j\omega_d) = \alpha + j(\omega + \omega_d) \tag{13.18}$$

is represented as a vector starting at the root s_2 and ending at $j\omega$. The amplitude characteristic is determined by the inverse of the product of the lengths of these two vectors; that is,

$$|\mathbf{H}(j\omega)| = \frac{1/LC}{|j\omega - s_1|\,|j\omega - s_2|} \tag{13.19}$$

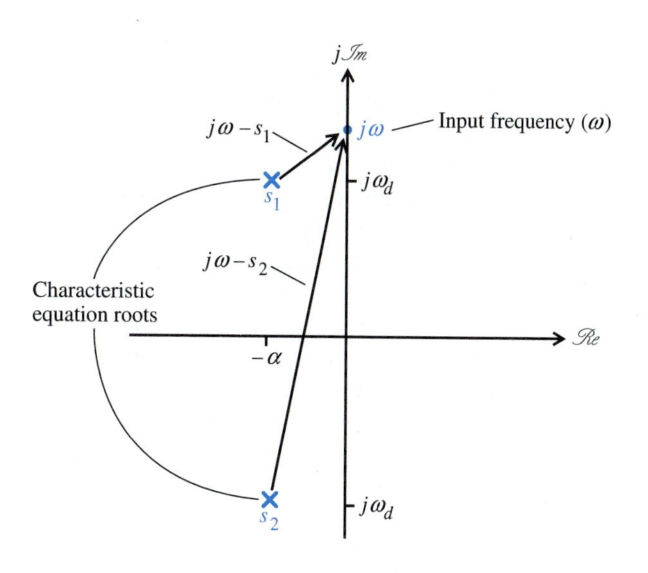

FIGURE 13.8 Characteristic equation roots and input frequency as points in the complex plane

† The equality $s^2LC + sRC + 1 = LC(s - s_1)(s - s_2)$ is established by the fact that both sides are polynomials with the same degree, 2, have the same roots, s_1 and s_2, and have the same coefficient, LC, multiplying the highest-degree variable, s^2.

The phase characteristic is determined by the negative of the sum of the angles of the two vectors

$$\underline{/\mathbf{H}(j\omega)} = -(\underline{/j\omega - s_1} + \underline{/j\omega - s_2}) \tag{13.20}$$

Figure 13.8 provides a *geometric* interpretation of the combined influence of the input frequency ω and the characteristic equation roots s_1 and s_2 in the determination of $\mathbf{H}(j\omega)$ of Eq. (13.16). The mathematical reason for the peak in $|\mathbf{H}(j\omega)|$ and for the abrupt phase change in $\underline{/\mathbf{H}(j\omega)}$ in Fig. 13.7 is related to the behavior of the vector $j\omega - s_1$.

We can see from Fig. 13.8 that if ω is varied in the range $\omega_d - \alpha$ to $\omega_d + \alpha$, the vector $j\omega - s_1$ undergoes a relatively pronounced change in both its length $|j\omega - s_1|$ and its angle $\underline{/j\omega - s_1}$. The magnitude $|j\omega - s_1|$ is minimum when $\omega = \omega_d$, and therefore the maximum of $|\mathbf{H}(j\omega)|$ occurs at approximately this frequency. The conclusion is as follows.

It is the ''closeness'' of the input frequency (represented as the point $j\omega$ in Fig. 13.8) to the circuit's characteristic equation root s_1 (shown as the point s_1) that causes the resonance peak in the amplitude characteristic.

This is the *mathematical* explanation for the peak in the amplitude characteristic of Fig. 13.7a.

Now that we have found the mathematical explanation for the resonance peak of Fig. 13.7a, we can also understand the physical reason for this peak. Recall from Section 9.1 that the *natural* response of a circuit is determined from the roots of the circuit's characteristic equation. For the series *RLC* circuit of Fig. 13.6, the roots are complex and the natural response has the form

$$v_{oc}(t) = e^{-\alpha t}(A \cos \omega_d t + B \sin \omega_d t) \tag{13.21}$$

where the constants A and B depend on initial conditions. What is important here is the fact that the natural response of the series *RLC* circuit of Fig. 13.6 is a *damped sinusoidal oscillation* with angular frequency ω_d. We have seen that the peak in $|\mathbf{H}(j\omega)|$ occurs when the circuit is driven by a sinusoidal source whose frequency ω is approximately equal to ω_d. We conclude that the resonance peak of Fig. 13.7a is caused by the periodic *reinforcement* that occurs when the circuit is driven at a frequency that is approximately equal to the circuit's damped natural frequency.

Examples of similar interactions between an input function and a system's natural response abound in nature. Consider a child on a swing. To go higher and higher, the child pumps the swing at the times that best reinforce the swing's natural oscillations. A series of small but correctly timed periodic inputs by the child results in a large oscillatory response. Another example of resonance occurs when a soprano sings a certain note that makes a glass goblet shatter. Another occurs when an earthquake tremor is amplified in a poorly designed structure. These examples serve to illustrate that resonance can have harmful as well as beneficial consequences.

In this section we have gained insight into the frequency response of a series *RLC* circuit by plotting the input frequency and the characteristic equation roots in the complex plane as in Fig. 13.8. In the next section, we will obtain even greater insight by combining the graphical approach of Fig. 13.8 with a new concept called *complex frequency*.

Remember

A characteristic equation root close to the imaginary axis of the complex plane causes a resonance peak in the amplitude characteristic.

EXERCISES

7. Assume that the input to the circuit of Fig. 13.6 is $v_s(t) = 10 + 7 \cos 10^6 t + 3 \sin (5 \times 10^6 t)$ V. Use the amplitude and phase characteristics of Fig. 13.7 to write down the (approximate) steady-state response, $v_o(t)$.

   answer: $10 + 70 \cos (10^6 t - 90°)$ V

8. This exercise will give you additional insight into the circuit of Fig. 13.6.
   (a) Find the driving-point impedance, $\mathbf{Z}(j\omega)$, of the circuit.
   (b) Show that the amplitude of the current in the circuit of Fig. 13.6 is maximum when the impedances of the inductance and capacitance cancel one another, that is when $j\omega L + 1/j\omega C = 0$.
   (c) Show that the frequency at which the maximum current amplitude occurs is given by $\omega_{mr} = 1/\sqrt{LC}$.
   (d) Evaluate ω_{mr} and plot the driving-point admittance magnitude $|\mathbf{Y}(j\omega)| = 1/|\mathbf{Z}(j\omega)|$ versus ω for the numerical values of Fig 13.7.
   (e) Compare your plot with Fig. 13.7a and comment.

   answer: (a) $R + j\omega L + 1/j\omega C$

9. Show the root of the characteristic equation of the following circuit as a point in the complex plane. Use a ruler and protractor to help plot the amplitude and phase characteristics. Assume that $R = 1$ kΩ and $C = 1000$ μF. (*Hint:* To use a ruler and protractor, you need to represent $\mathbf{H}(j\omega)$ by a drawing analogous to Fig. 13.8.)

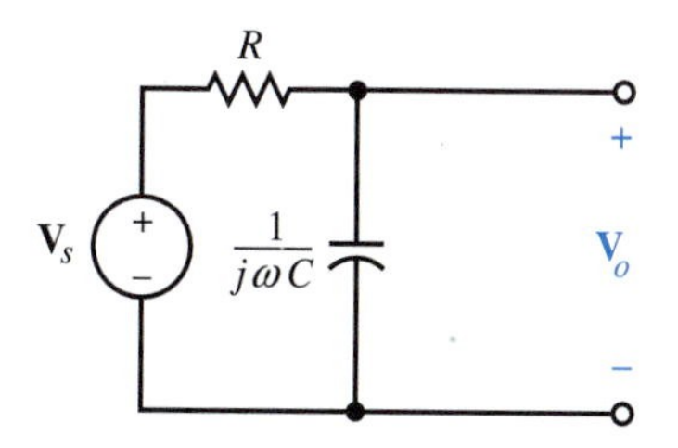

13.3 Complex Frequency

In the previous section, we interpreted the shape of an *RLC* circuit's amplitude and phase characteristics geometrically by plotting the input frequency and the roots of the characteristic equation as points in the complex plane. The geometric interpretation was somewhat limited because the input frequency was plotted as the point $j\omega$ and was therefore constrained to lie along the imaginary axis. In this section we will show that additional insight into the shape of the amplitude characteristic can be obtained by generalizing the notion of frequency to include any point in the complex plane. This is the basic idea behind complex frequency.

Complex frequency is defined as $s = \sigma + j\omega$, where $\sigma = \mathcal{R}e\{s\}$ is the *neper*† frequency and $\omega = \mathcal{I}m\{s\}$ is the *angular* frequency. The domain of all possible values of s is called the *complex-frequency plane* or the *s plane.* A particular value of s is a point in the s plane, as depicted in Fig. 13.9. In the illustration, the neper frequency is −0.02 Mnepers per second (MNp/s) and the angular frequency is 1.1 Mrad/s.

† The term *neper* is derived from the Naperian (natural) logarithm ln.

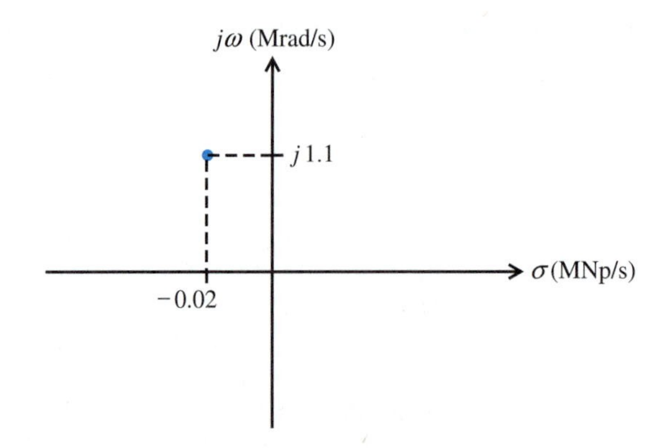

FIGURE 13.9
Depiction of a particular value of s as a point in the s plane. Here $s = -0.02$ MNp/s + $j1.1$ Mrad/s

Let us investigate the results of replacing the argument, $j\omega$, in the transfer function of Eq. (13.6) by $s = \sigma + j\omega$. The substitution of s for $j\omega$ in Eq. (13.6) yields

$$\mathbf{H}(s) = \frac{\mathbf{V}_o}{\mathbf{V}_s} = \frac{1/sC}{R + sL + (1/sC)} = \frac{1}{s^2LC + sRC + 1} = \frac{1/LC}{(s - s_1)(s - s_2)} \tag{13.22}$$

where s_1 and s_2 are the roots of the characteristic Eq. (13.11) given by Eqs. (13.12) and (13.13). Since we have called s the complex frequency, we will now call s_1 and s_2 *complex natural frequencies* of the circuit. It follows from Eq. (13.22) that

$$|\mathbf{H}(s)| = \frac{1/LC}{|s - s_1|\,|s - s_2|} \tag{13.23}$$

and

$$\underline{/\mathbf{H}(s)} = -\underline{/s - s_1} - \underline{/s - s_2} \tag{13.24}$$

Equations (13.22) through (13.24) can be interpreted geometrically in terms of the vectors $s - s_1$ and $s - s_2$ shown in Fig. 13.10. The factors $|s - s_1|$ and $|s - s_2|$ in Eq. (13.23) are the lengths of the vectors $s - s_1$ and $s - s_2$, respectively. We see, therefore, that $|\mathbf{H}(s)|$ becomes arbitrarily large as $s \to s_1$ or $s \to s_2$. The function $|\mathbf{H}(s)|$ is plotted as a function of s in Fig. 13.11. Notice that $|\mathbf{H}(s)|$ looks like a thin rubber sheet that is supported over the s plane by poles located on the complex natural frequencies s_1 and s_2. Observe that the intersection of $|\mathbf{H}(s)|$ with the plane $\sigma = 0$ is the amplitude characteristic $|\mathbf{H}(j\omega)|$ previously shown in Fig. 13.7a for $\omega > 0$.

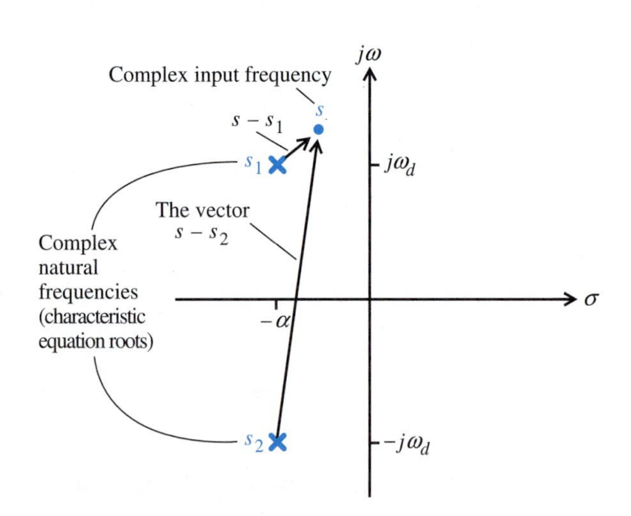

FIGURE 13.10
Characteristic equation roots and complex input frequency as points in the complex plane

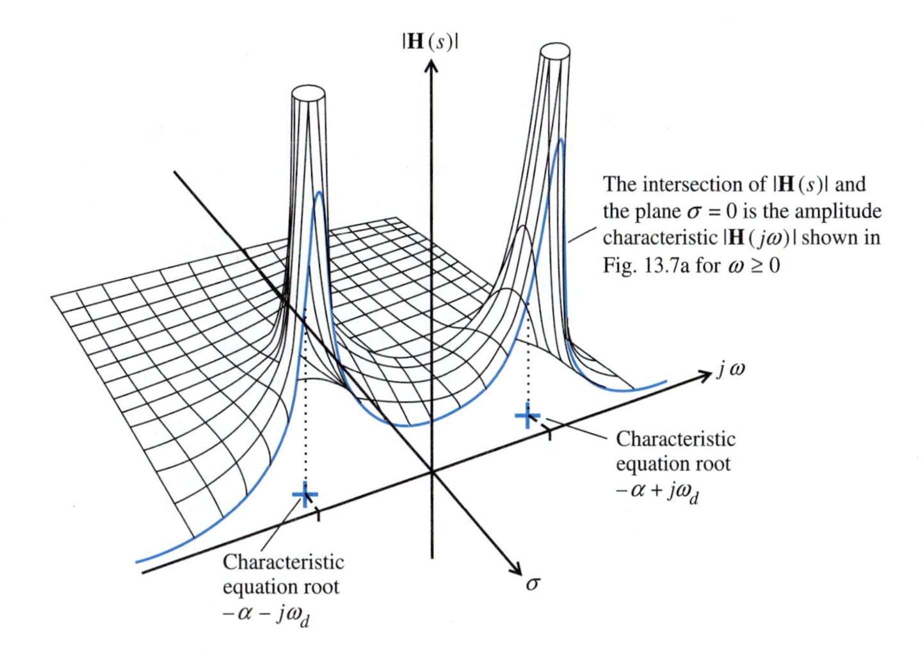

FIGURE 13.11 Frequency response magnitude as a function of complex frequency

Figure 13.11 illustrates one advantage of using the complex frequency s in place of $j\omega$ in a transfer function $\mathbf{H}(j\omega)$. The substitution of s for $j\omega$ has provided us with the *rubber sheet analogy.* The rubber sheet analogy provides the means to make a quick, approximate sketch of a circuit's amplitude characteristic by inspection of the factored algebraic expression for $\mathbf{H}(s)$. The amplitude characteristic $|\mathbf{H}(j\omega)|$ is always given by the intersection of $|\mathbf{H}(s)|$ (represented by the rubber sheet) and the plane defined by $\sigma = 0$. We will describe the rubber sheet analogy in greater detail in Section 13.4.

In this subsection we have defined the concept of complex frequency and have illustrated one way in which it can be used. In the following subsection we show that the complex frequency variable s describes spiraling phasors and damped sinusoids.

EXERCISE

10. For the *RC* low-pass filter shown:
    (a) Make a rough three-dimensional sketch (analogous to that in Fig. 13.11) of the voltage transfer function magnitude, $|\mathbf{V}_o/\mathbf{V}_s| = |\mathbf{H}(s)|$.
    (b) Explain in your own words how your sketch is related to the amplitude characteristic $|\mathbf{H}(j\omega)|$ shown in Fig. 13.5a.

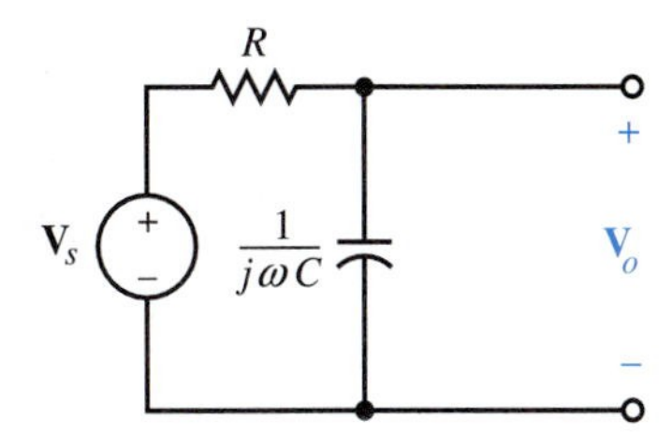

Spiraling Phasors and Damped Sinusoids

We know from our work in Chapter 10 that the complex exponential $\mathbf{V}e^{st}$ can be interpreted as a *rotating phasor* when $s = j\omega$. We now ask, ''How can we interpret $\mathbf{V}e^{st}$ when $s = \sigma + j\omega$ is an *arbitrary* complex number? To answer this question, we write

$$\mathbf{V}e^{st} = (V_m e^{j\phi_\mathbf{V}})e^{(\sigma + j\omega)t} = V_m e^{\sigma t} e^{j(\omega t + \phi_\mathbf{V})} = V_m e^{\sigma t} \underline{/\omega t + \phi_\mathbf{V}} \tag{13.25}$$

According to Eq. (13.25), multiplication of phasor $\mathbf{V}$ by e^{st} causes the phasor to rotate at an angular rate ω and vary in magnitude at an exponential rate σ, as illustrated in Fig. 13.12. Accordingly, $\mathbf{V}e^{st}$ may be interpreted as a *spiraling phasor*. The tip of the phasor spirals outward for $\sigma > 0$ and inward for $\sigma < 0$. Rotation without spiraling occurs for $\sigma = 0$. The projection of the spiraling phasor onto the real axis is the damped sinusoid.

$$\mathcal{R}e\{\mathbf{V}e^{st}\} = V_m e^{\sigma t} \cos(\omega t + \phi_\mathbf{V}) \tag{13.26}$$

The neper frequency σ determines the rate of exponential growth ($\sigma > 0$) or decay ($\sigma < 0$) of the damped sinusoid. For $\sigma = 0$, the amplitude of the sinusoid is constant, because the phasor rotates without spiraling. The $\sigma = 0$ case is the one we studied in Chapters 10 through 12.

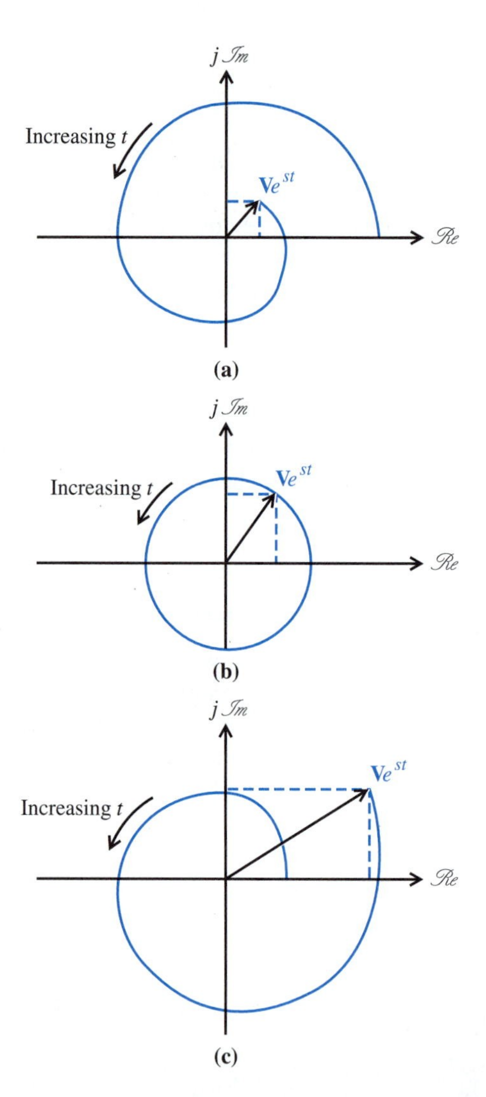

FIGURE 13.12 The spiraling phasor $\mathbf{V}e^{st} = V_m e^{\sigma t} \underline{/\omega t + \phi_V}$. The projection on the $\mathcal{R}e$ axis is $V_m e^{\sigma t} \cos(\omega t + \phi_\mathbf{V})$. (a) $\sigma < 0$; (b) $\sigma = 0$; (c) $\sigma > 0$

Relation of Spiraling Phasors to Real Input and Output†

Recall that in Chapter 10 we used rotating phasors to find the particular response of an LTI circuit to a sinusoidal input. We can similarly use spiraling phasors to find the particular response of a circuit to a damped sinusoidal input. The method is based on two facts.

The first fact is that the particular response of an LTI equation

$$\mathscr{A}(D)v_{op}(t) = \mathscr{B}(D)v_s(t) \tag{13.27}$$

to a spiraling phasor input‡

$$\mathbf{v}_s(t) = \mathbf{V}_s e^{st} \tag{13.28}$$

is also a spiraling phasor:

$$\mathbf{v}_{op}(t) = \mathbf{V}_o e^{st} \tag{13.29}$$

The input and output phasors are related by

$$\mathbf{V}_o = \mathbf{H}(s)\mathbf{V}_s \tag{13.30}$$

$\mathbf{H}(s)$ is the transfer function

$$\mathbf{H}(s) = \frac{\mathbf{V}_o}{\mathbf{V}_s} = \frac{\mathscr{B}(s)}{\mathscr{A}(s)} \tag{13.31}$$

We can establish Eq. (13.30) by simply substituting Eqs. (13.28) and (13.29) into Eq. (13.27), performing the differentiations, and solving for $\mathbf{V}_o$.

The second fact is that the particular response of the LTI equation, Eq. (13.27), to a damped sinusoidal input

$$v_s(t) = \mathscr{R}e\{\mathbf{V}_s e^{st}\} = V_{sm}e^{\sigma t}\cos(\omega t + \phi_{\mathbf{V}_s}) \tag{13.32}$$

is the damped sinusoid

$$v_{op}(t) = \mathscr{R}e\{\mathbf{V}_o e^{st}\} = V_{om}e^{\sigma t}\cos(\omega t + \phi_{\mathbf{V}_o}) \tag{13.33}$$

The relationship between the input and output damped sinusoids is illustrated in Fig. 13.13. The damped sinusoidal input, Eq. (13.32), is the projection of the spiraling input phasor, Eq. (13.28), onto the real axis. The corresponding damped sinusoidal particular response, Eq. (13.33), is the projection of the spiraling output phasor, Eq. (13.29), onto the real axis. Notice from Eqs. (13.32) and (13.33) that the output damped sinusoid has the same neper frequency σ and angular frequency ω as the input.

Equation (13.33) can be proved by reasoning similar to that which led to Eq. (8.84). The essential difference is that $j\omega$ is replaced by s and sinusoids are replaced by damped sinusoids. The practical result is that phasors can be used to calculate LTI circuit response to a damped sinusoidal input.

In the preceding discussion we assumed that the input and output were both voltages. Of course, similar considerations apply for currents.

† This section may be omitted without loss in continuity.

‡ Do not let the notation confuse you. Subscript s still denotes *source* as it did in previous chapters. The s in the exponent is the complex frequency $s = \sigma + j\omega$.

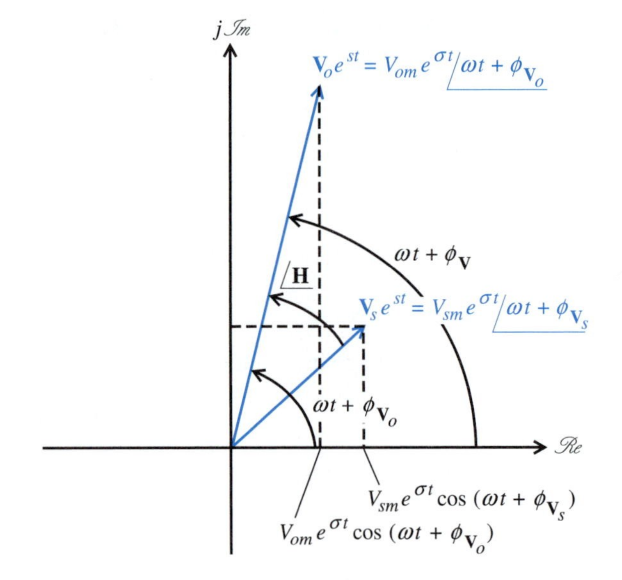

FIGURE 13.13 Input and output spiraling phasors and their projections. The real input and output are the projections of the spiraling phasors onto the real axis. $V_{om} = |\mathbf{H}|\, V_{sm}$; $\phi_{\mathbf{V}_o} = \underline{/\mathbf{H}} + \phi_{\mathbf{V}_s}$

EXAMPLE 13.1 (a) Start with the appropriate differential equation and determine the impedance $\mathbf{Z}(s)$ of a series *RLC* circuit.

(b) Find the particular response voltage at the input terminals to a damped sinusoidal input $i(t) = 3e^{\sigma t}\cos(\omega t + 10°)$ A, where $\sigma = -0.02$ MNp/s and $\omega = 1.1$ Mrad/s. Assume that $R = 0.1\ \Omega$, $L = 1\ \mu$H, and $C = 1\ \mu$F.

Solution (a) The differential equation relating driving-point current to voltage is

$$v = Ri + L\frac{d}{dt}i + \frac{1}{C}\int_{-\infty}^{t} i\, d\lambda$$

or

$$\frac{d}{dt}v = L\frac{d^2}{dt^2}i + R\frac{d}{dt}i + \frac{1}{C}i$$

which is

$$\mathcal{A}(D)v = \mathcal{B}(D)i$$

where $\mathcal{A}(D) = D$ and $\mathcal{B}(D) = LD^2 + RD + (1/C)$. The impedance is given by

$$\mathbf{Z}(s) = \frac{\mathbf{V}}{\mathbf{I}} = \frac{\mathcal{B}(s)}{\mathcal{A}(s)} = sL + R + \frac{1}{sC}$$

We can obtain the same result by substituting $\mathbf{i}(t) = \mathbf{I}e^{st}$ and $\mathbf{v}_p(t) = \mathbf{V}e^{st}$ into the differential equation and solving for $\mathbf{V}/\mathbf{I}$. $\mathbf{Z}(s)$ exists provided that $s \neq 0$.

(b) The input is a damped sinusoid: $i(t) = 3e^{\sigma t}\cos(\omega t + 10°)$ A, where $\sigma = -0.02$ MNp/s and $\omega = 1.1$ Mrad/s. The corresponding phasor current is

$$\mathbf{I} = 3\ \underline{/10°}\ \text{A}$$

and the associated complex frequency is $s = -0.02$ MNp/s $+\ j1.1$ Mrad/s. To find the particular response voltage, we evaluate the impedance $\mathbf{Z}(s)$ of (a) for $s = -2 \times 10^4 + j1.1 \times 10^6$. This yields

$$\mathbf{Z}(-2 \times 10^4 + j1.1 \times 10^6) = \underbrace{-0.02 + j1.1}_{sL} + \underbrace{0.1}_{R} + \underbrace{\frac{1}{-0.02 + j1.1}}_{\frac{1}{sC}}$$

$$= 0.202 \angle 71.6^\circ \ \Omega$$

The output phasor is

$$\mathbf{V} = \mathbf{ZI} = (0.202 \angle 71.6^\circ)(3 \angle 10^\circ) = 0.606 \angle 81.6^\circ \ \text{V}$$

Therefore the particular response to the damped sinusoidal input $i(t)$ is the damped sinusoid

$$v_p(t) = 0.606e^{\sigma t} \cos (\omega t + 81.6^\circ) \ \text{V}$$

The values of σ and ω are always the same as those of the input: $\sigma = -0.02$ MNp/s and $\omega = 1.1$ Mrad/s.

Remember

The particular response of an LTI equation to a spiraling phasor input $\mathbf{V}_s e^{st}$ is a spiraling phasor $\mathbf{H}(s)\mathbf{V}_s e^{st}$ where $\mathbf{H}(s)$ is the transfer function. The projections of the spiraling phasors onto the real axis of the complex plane are the corresponding damped sinusoidal input and output.

EXERCISES

11. A 1-Ω resistance is connected in parallel with a 1-mH inductance.
    (a) Starting from the appropriate differential equation, show that the driving-point impedance $\mathbf{Z}(s)$ is given by

$$\mathbf{Z}(s) = \frac{sLR}{sL + R}$$

    (b) A current $i(t) = 10e^{-3000t} \cos (2000t + 20^\circ)$ mA is applied to the parallel circuit. Determine the particular response voltage $v_p(t)$ appearing across the input terminals. *answer:* (b) $12.7e^{-3000t} \cos (2000t + 31.31^\circ)$ mV
12. Repeat Exercise 11 for $i(t) = 10 \cos (1000t + 20^\circ)$ mA. *answer:* $7.07 \cos (1000t + 65^\circ)$ mV
13. Repeat Exercise 11 for $i(t) = 10e^{-3000t}$ mA. *answer:* $15e^{-3000t}$ mV
14. Derive Eq. (13.30) by substituting Eqs. (13.28) and (13.29) into Eq. (13.27).

13.4 *s*-Domain Circuit Analysis

In Chapter 10 we used rotating phasors $\mathbf{i}(t) = \mathbf{I}e^{j\omega t}$ and $\mathbf{v}(t) = \mathbf{V}e^{j\omega t}$ to transform the differential time-domain terminal equations of R, L, and C into algebraic frequency-domain equations. We can similarly use spiraling phasors $\mathbf{i}(t) = \mathbf{I}e^{st}$ and $\mathbf{v}(t) = \mathbf{V}e^{st}$ to transform the differential terminal equations of R, L, and C into algebraic equations of the form

$$\mathbf{V} = \mathbf{Z}(s)\mathbf{I} \tag{13.34}$$

TABLE 13.1
s-Domain circuit elements

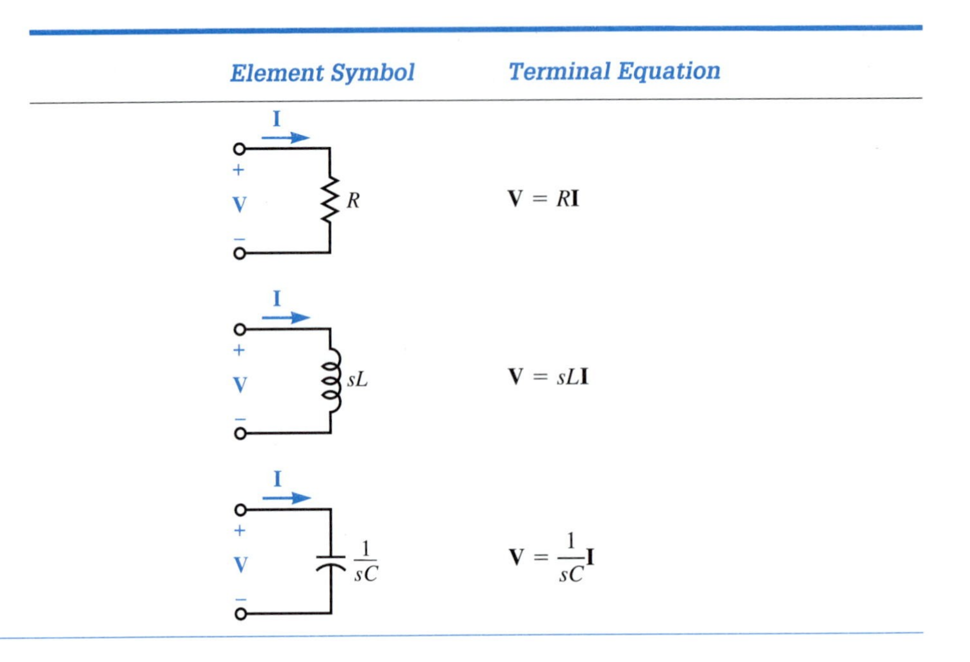

where $\mathbf{Z}(s)$ is the impedance of the element. The derivations are identical to those on pages 314 to 316 with the exception that $j\omega$ is replaced by s. The results are summarized in Table 13.1.

Spiraling phasors also transform Kirchhoff's laws and the terminal equations of dependent sources into algebraic equations involving phasors. The results are identical to Eqs. (11.17) and (11.19) and Fig. 11.3, with the understanding that $j\omega$ is replaced by s. The practical result is that the s-domain transfer function of any LTI circuit can be obtained by the phasor circuit analysis methods discussed in Chapter 11.

A simple example is provided by our problem of finding the driving-point impedance of a series *RLC* circuit. The complex-frequency (or s-domain) circuit is shown in Fig. 13.14. By using the fact that impedances in series add, we can write $\mathbf{Z}(s) = R + sL + (1/sC)$ by inspection. Notice that this result agrees with that obtained in Example 13.1(a) as it must.

Remember

We can use phasor circuit analysis to find the transfer function $\mathbf{H}(s) = \mathcal{B}(s)/\mathcal{A}(s)$ of any LTI circuit.

FIGURE 13.14
s-Domain circuit

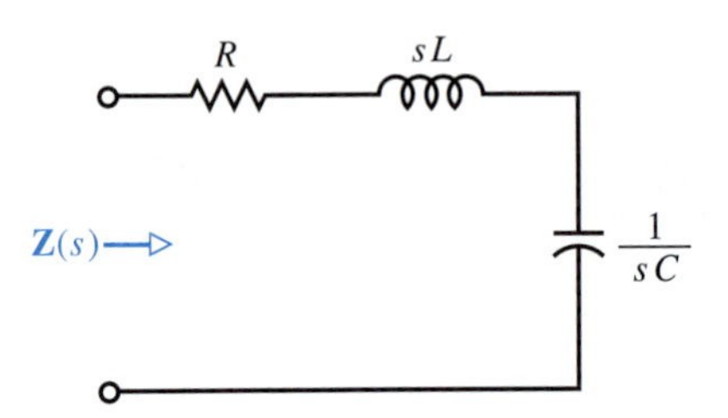

EXERCISES

15. Derive the entries in Table 13.1.
16. A current $i(t)$ is applied to a parallel *RL* circuit, where $R = 1\ \text{k}\Omega$ and $L = 1$ mH.
    (a) Draw the time-domain circuit.

(b) Draw the s-domain circuit.
(c) Write the driving-point impedance $\mathbf{Z}(s)$ by inspection.
(d) Find the particular response voltage for $i(t) = 5 \cos 10^6 t$ mA. *Hint:* To obtain $\mathbf{Z}(j\omega)$, just set $s = j\omega$ in your answer to (c).

answer: (c) $\dfrac{sRL}{R + sL}$; (d) $3535.5 \cos (10^6 t + 45°)$ V

17. Show that when Kirchhoff's current law is transformed into the complex-frequency domain, the result is Eq. (11.17). *Hint:* Start with Eq. (11.10) and specialize to $i_n(t) = I_{mn} e^{\sigma t} \cos (\omega t + \phi_n) = \mathcal{Re}\{\mathbf{I}_n e^{st}\}$ where $\mathbf{I}_n = I_{mn} \underline{/\phi_n}$ and $s = \sigma + j\omega$.

Poles and Zeros

In Section 13.2 we saw that the shape of an *RLC* circuit's frequency response characteristic is critically influenced by the roots of the circuit's characteristic equation. We obtained this insight by factoring the denominator of the circuit's transfer function $\mathbf{H}(s)$. This simple step resulted in a graphical interpretation for $\mathbf{H}(s)$. In this section we generalize this procedure to arrive at a completely graphical description of any transfer function.

In general, both the numerator and the denominator of a transfer function are polynomials in s:

$$\mathbf{H}(s) = \frac{\mathcal{B}(s)}{\mathcal{A}(s)} \tag{13.35}$$

Let us now write both the numerator and the denominator of Eq. (13.35) in factored form. To factor the denominator, we first find the roots to the characteristic equation

$$\mathcal{A}(s) = a_n s^n + a_{n-1} s^{n-1} + \cdots + a_0 = 0 \tag{13.36}$$

From now on, we will denote the distinct roots of the characteristic equation by s_{p1}, $s_{p2}, \ldots, s_{pl}$. We let m_i denote the order (multiplicity) of root s_{pi}, $i = 1, 2, \ldots, l$. The factored form of $\mathcal{A}(s)$ is given by

$$\mathcal{A}(s) = a_n (s - s_{p1})^{m_1} (s - s_{p2})^{m_2} \cdots (s - s_{pl})^{m_l} \tag{13.37}$$

Similarly, to factor the numerator, we first find the roots of the equation

$$\mathcal{B}(s) = b_m s^m + b_{m-1} s^{m-1} + \cdots + b_0 = 0 \tag{13.38}$$

We let $s_{z1}, s_{z2}, \ldots, s_{zr}$ denote the distinct roots and let q_i denote the order of root s_{zi}, $i = 1, 2, \ldots, r$. The factored form of $\mathcal{B}(s)$ is

$$\mathcal{B}(s) = b_m (s - s_{z1})^{q_1} (s - s_{z2})^{q_2} \cdots (s - s_{zr})^{q_r} \tag{13.39}$$

By substituting Eqs. (13.37) and (13.39) into Eq. (13.35), we obtain

$$\mathbf{H}(s) = K \frac{(s - s_{z1})^{q_1} (s - s_{z2})^{q_2} \cdots (s - s_{zr})^{q_r}}{(s - s_{p1})^{m_1} (s - s_{p2})^{m_2} \cdots (s - s_{pl})^{m_l}} \tag{13.40}$$

where $K = b_m / a_n$. The following definitions apply to Eq. (13.40).

1. The transfer function is said to have *poles* at the complex natural frequencies $s = s_{p1}, s_{p2}, \ldots, s_{pl}$. These frequencies are the roots of the characteristic equation $\mathcal{A}(s) = 0$. Since the poles are the roots of the characteristic equation, they determine the form of the circuit's natural response.
2. m_i is the *order* of the pole s_{pi}. The pole is called *simple* if $m_i = 1$.

3. The transfer function is said to have *zeros* at the complex frequencies $s_{z1}, s_{z2}, \ldots, s_{zr}$. These frequencies are the roots of the equation $\mathcal{B}(s) = 0$.
4. q_i is called the *order* of the zero s_{zi}. The zero is called *simple* if $q_i = 1$.
5. The poles and the zeros are called the *critical frequencies* of the circuit.
6. A plot of the critical frequencies of $\mathbf{H}(s)$ as points in the plane is called a *pole-zero plot* or *pole-zero constellation.* The poles are depicted by $\times$s, and the zeros are depicted by Os. All poles and zeros are assumed to be simple unless indicated otherwise, and the value of the constant K is shown in a box. A typical pole-zero plot is shown in Fig. 13.15.

We can see from Eq. (13.40) that the transfer function of a circuit is completely determined by the pole-zero plot and the constant K. Indeed, the differential equation relating the output and input of a circuit can be derived from the pole-zero plot. Therefore, the pole-zero plot provides a *complete* description of the circuit's input-output properties.

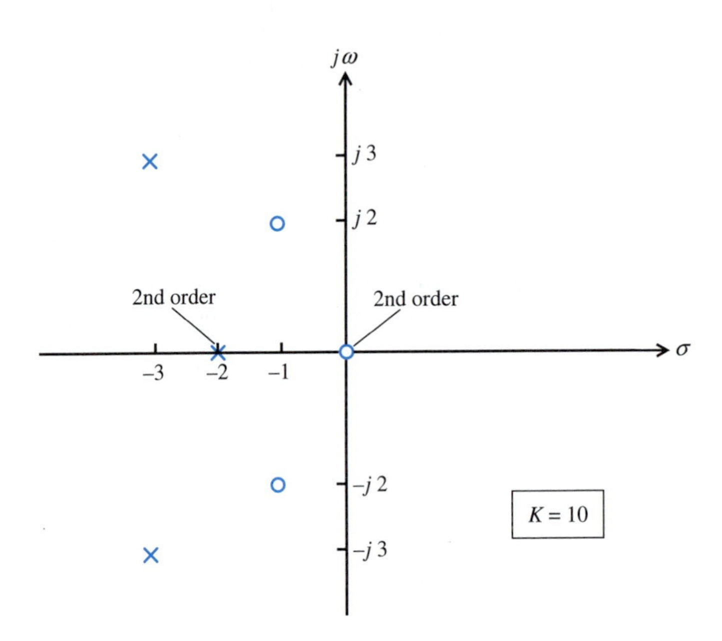

FIGURE 13.15
A typical pole-zero plot

EXAMPLE 13.2 Assume that the pole-zero plot of Fig. 13.15 refers to an impedance, $\mathbf{Z}(s)$. Find
(a) The critical frequencies.
(b) $\mathbf{Z}(s)$.
(c) The input/output equation in the time domain.
(d) The complex natural frequencies.
(e) The particular response for an input $i(t) = 3\cos(4t + 17°)$ A.

Solution (a) The critical frequencies are the poles $s_{p1} = -2$, $s_{p2} = -3 + j3$, and $s_{p3} = -3 - j3$ and the zeros $s_{z1} = 0$, $s_{z2} = -1 + j2$, and $s_{z3} = -1 - j2$.

(b) From Eq. (13.40) and the answer to (a),

$$\mathbf{Z}(s) = 10\frac{(s - s_{z1})^2(s - s_{z2})(s - s_{z3})}{(s - s_{p1})^2(s - s_{p2})(s - s_{p3})} = 10\frac{s^2(s + 1 - j2)(s + 1 + j2)}{(s + 2)^2(s + 3 - j3)(s + 3 + j3)}$$

$$= 10\frac{s^4 + 2s^3 + 5s^2}{s^4 + 10s^3 + 46s^2 + 96s + 72}$$

(c) Cross-multiply

$$\frac{\mathbf{V}}{\mathbf{I}} = 10\,\frac{s^4 + 2s^3 + 5s^2}{s^4 + 10s^3 + 46s^2 + 96s + 72}$$

to obtain

$$s^4\mathbf{V} + 10s^3\mathbf{V} + 46s^2\mathbf{V} + 96s\mathbf{V} + 72\mathbf{V} = 10s^4\mathbf{I} + 20s^3\mathbf{I} + 50s^2\mathbf{I}$$

The corresponding time-domain equation is

$$\frac{d^4}{dt^4}v + 10\frac{d^3}{dt^3}v + 46\frac{d^2}{dt^2}v + 96\frac{d}{dt}v + 72v = 10\frac{d^4}{dt^4}i + 20\frac{d^3}{dt^3}i + 50\frac{d^2}{dt^2}i$$

(d) The complex natural frequencies are the poles $s_{p1} = -2$, $s_{p2} = -3 + j3$, and $s_{p3} = -3 - j3$. Notice that s_{p1} is a second-order pole.

(e) Note that $3\cos(4t + 17°) = \mathcal{Re}\{\mathbf{I}e^{st}\}$, where $\mathbf{I} = 3\,\underline{/17°}$ and $s = j4$. The particular response, therefore, is given by $v_p(t) = \mathcal{Re}\{\mathbf{V}e^{st}\}$ where $\mathbf{V} = \mathbf{Z}(s)\mathbf{I}$ and $s = j4$. The impedance function $\mathbf{Z}(s)$ was determined in (b). By evaluating the second expression for $\mathbf{Z}(s)$ from (b) with $s = j4$, we obtain

$$\mathbf{Z}(j4) = 10\,\frac{(j4)^2(1 + j2)(1 + j6)}{(2 + j4)^2(3 + j)(3 + j7)} = 4.52\,\underline{/111.8°}\ \Omega$$

The output voltage phasor is, accordingly,

$$\mathbf{V} = \mathbf{ZI} = (4.52\,\underline{/111.8°})(3\,\underline{/17°}) = 13.56\,\underline{/128.8°}\ \mathrm{V}$$

Therefore the particular response to

$$i(t) = 3\cos(4t + 17°)\ \mathrm{A}$$

is

$$v_p(t) = 13.56\cos(4t + 128.8°)\ \mathrm{V}$$

Notice that the pole-zero plot of Fig. 13.15 has mirror symmetry about the σ axis. We can show that pole-zero plots describing real circuits *always* have mirror symmetry about the σ axis. To establish this property, we use the fact that the coefficients of the polynomial $\mathcal{A}(s) = a_n s^n + a_{n-1}s^{n-1} + \cdots + a_0$ are real. Since $a_n, a_{n-1}, \ldots, a_o$ are real,

$$\begin{aligned}\mathcal{A}^*(s) &= (a_n s^n)^* + (a_{n-1}s^{n-1})^* + \cdots + a_0^* \\ &= a_n(s^*)^n + a_{n-1}(s^*)^{n-1} + \cdots + a_0 = \mathcal{A}(s^*)\end{aligned} \qquad (13.41)$$

where the asterisk denotes complex conjugation. The ith pole of $\mathbf{H}(s)$ satisfies

$$\mathcal{A}(s_{pi}) = 0 \qquad (13.42)$$

If we conjugate Eq. (13.42), we obtain

$$\mathcal{A}^*(s_{pi}) = 0 \qquad (13.43)$$

which, with the aid of Eq. (13.41), becomes

$$\mathcal{A}(s_{pi}^*) = 0 \qquad (13.44)$$

We conclude that if s_{pi} is a pole, then so is s_{pi}^*. Therefore complex poles always occur in conjugate pairs. The same arguments apply to the zeros of $\mathbf{H}(s)$. *The result is that all pole-zero plots describing real LTI systems have mirror symmetry about the σ axis, as illustrated by Fig. 13.15.*

A final noteworthy feature of pole-zero plots follows from the Cramer's rule solution for the node voltages of an arbitrary circuit. The key result was given in Eq. (11.35), which we repeat below.

$$\mathbf{V}_i = \frac{\Delta_{1i}}{\Delta}\mathbf{I}_{s1} + \frac{\Delta_{2i}}{\Delta}\mathbf{I}_{s2} + \cdots + \frac{\Delta_{(M-1)i}}{\Delta}\mathbf{I}_{s(M-1)} + \frac{\Delta_{Mi}}{\Delta}\mathbf{V}_{sM} + \cdots + \frac{\Delta_{Ni}}{\Delta}\mathbf{V}_{sN} \tag{13.45}$$

In the above equation, the phasors $\mathbf{I}_{s1}, \mathbf{I}_{s2}, \ldots, \mathbf{V}_{sN}$ are independent source phasors, and $\mathbf{V}_i$ is the output-voltage phasor for node i. (The general Cramer's rule solution for $\mathbf{V}_i$ included the possibility that there is more than one independent source.) The factors $\Delta_{1i}/\Delta, \Delta_{2i}/\Delta, \ldots, \Delta_{Ni}/\Delta$ represent transfer impedances and voltage transfer functions between the independent sources and the output $\mathbf{V}_i$. Observe that regardless of the location of the output node (that is, regardless of the value of i), the denominator of the transfer function is always the same function $\Delta = \Delta(s)$. Since the denominators of the transfer functions are the same, the poles of the transfer functions are also the same. Thus, if we were to draw pole-zero plots of all the possible voltage transfer functions and transfer impedances of a circuit, the poles in the various plots would all be the same!†

As should now be evident, pole-zero plots are very useful tools to analyze both the natural response and the sinusoidal steady-state response of LTI circuits. In the following section we describe how the pole-zero plot provides the basis for the rubber sheet analogy.

Remember

A transfer function $\mathbf{H}(s) = \mathcal{B}(s)/\mathcal{A}(s)$ can be represented by a pole-zero plot. The poles and zeros are the roots of $\mathcal{A}(s)$ and $\mathcal{B}(s)$, respectively. The poles are the complex natural frequencies that appear in the circuit's complementary response.

EXERCISES

18. In the circuit shown, $R = 1\ \text{M}\Omega$ and $C = 1\ \mu\text{F}$.
    (a) Use any method to find the transfer function
    $$\mathbf{H}(s) = \frac{\mathbf{V}_o}{\mathbf{V}_s}$$
    (b) Draw the pole-zero plot.
    (c) Find the complex natural frequencies.
    (d) What is the form of the natural response?
    (e) Find $v_{op}(t)$ if $v_s(t) = 10\cos(t + 20°)$ V.

    answer: (a) Voltage divider: $\frac{1}{1+s}$, (c) -1, (d) Ae^{-t}, (e) $7.07\cos(t - 25°)$ V

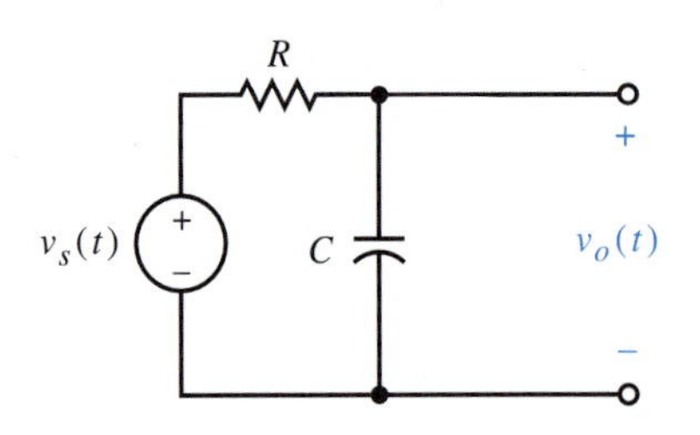

† Exceptions arise where the numerator Δ_{ki} contains a zero that cancels a particular pole in the denominator Δ.

19. Repeat Exercise 18 for the following circuit, where $R = 1\ \text{M}\Omega$ and $C = 1\ \mu\text{F}$.

answer: (a) Voltage divider: $\dfrac{s}{1+s}$, (c) -1, (d) Ae^{-t}, (e) $7.07 \cos(t + 65°)$ V

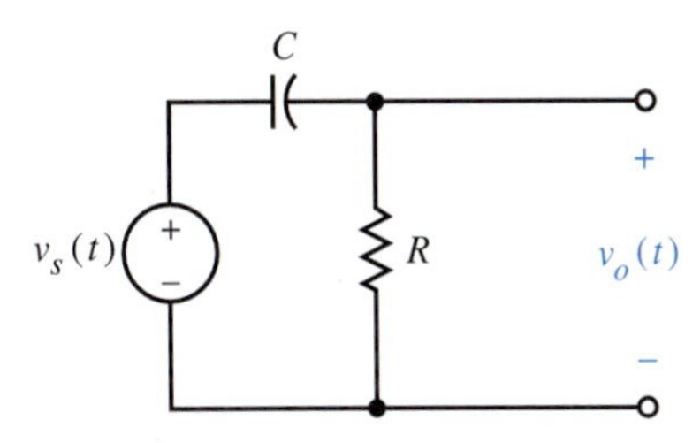

20. The differential equation relating the input $i(t)$ to the output $v(t)$ of a circuit is

$$\frac{d^2}{dt^2}v + 15\frac{d}{dt}v + 50v = \frac{d^2}{dt^2}i + 7\frac{d}{dt}i$$

Assume that the input is $\mathbf{i}(t) = \mathbf{I}e^{st}$ where $\mathbf{I} = I_m\underline{/\phi_{\mathbf{I}}}$ and $s = \sigma + j\omega$.
(a) Show that the particular solution has the form $\mathbf{v}_p(t) = \mathbf{V}e^{st}$.
(b) Using your analysis of (a), write $\mathbf{V}$ as $\mathbf{V} = \mathbf{Z}(s)\mathbf{I}$. Give the expression for $\mathbf{Z}(s)$ as the ratio of two factored polynomials in s.
(c) Express $\mathbf{Z}(s)$ in terms of the circuit's critical frequencies and sketch the corresponding pole-zero plot.
(d) Evaluate $\mathbf{Z}(s)$ at $s = j5$.
(e) Evaluate $\mathbf{V}$ for $s = j5$ and $I = 10\underline{/0°}$.
(f) Find $v_p(t)$ for $i(t) = 10 \cos 5t$.

answer: (b) $\dfrac{s(s+7)}{(s+10)(s+5)}$, (d) $0.544\underline{/54°}$, (e) $5.44\underline{/54°}$, (f) $5.44 \cos(5t + 54°)$

21. Repeat Exercise 20 for the equation

$$\frac{d^2}{dt^2}v + 15\frac{d}{dt}v + 50v = \frac{d^2}{dt^2}i + 6\frac{d}{dt}i + 34i$$

answer: (b) $\dfrac{(s+3+j5)(s+3-j5)}{(s+10)(s+5)}$, (d) $0.396\underline{/1.73°}$, (e) $3.96\underline{/1.73°}$, (f) $3.96 \cos(5t + 1.73°)$

The Rubber Sheet Analogy

We saw that the amplitude characteristic $|\mathbf{H}(s)|$ of a resonant *RLC* circuit resembles a rubber sheet above the complex-frequency plane (Fig. 13.11). Additional examples of the rubber sheet analogy are shown in Figs. 13.16 and 13.17.

Figures 13.16 and 13.17 depict three-dimensional plots of the amplitude characteristics $|\mathbf{H}(s)|$ of the high-pass *RL* filter and the low-pass *RC* filter described at the beginning of this chapter. We see again that the amplitude characteristics have the appearance of thin rubber sheets stretched above the s plane. The sheets appear to be *supported by poles* at the complex natural frequencies (or poles) and *held down by nails* at the zeros. The amplitude characteristic $|\mathbf{H}(j\omega)|$ can be visualized as the intersection of the rubber sheet with the $\sigma = 0$ plane.

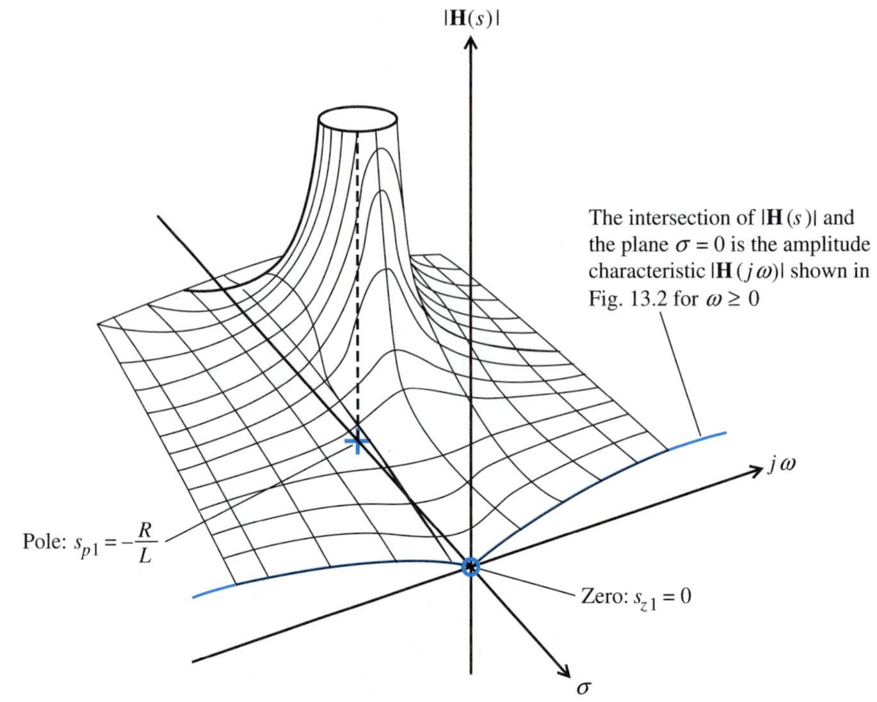

FIGURE 13.16 Magnitude of the transfer function of the high-pass *RL* circuit of Fig. 13.1

A significant feature of the rubber sheet analogy is that it illustrates the fundamental importance of the poles and zeros in determining the shape of $|\mathbf{H}(j\omega)|$. We see clearly from the analogy that the poles are responsible for the peaks in $|\mathbf{H}(j\omega)|$, whereas the zeros are responsible for the dips. Thus, we have found the answer to the question we asked at the start of Section 13.2.

In general, the behavior of the amplitude characteristic (or rubber sheet) as $|s| \rightarrow \infty$ is determined by the degrees of $\mathcal{A}(s)$ and $\mathcal{B}(s)$. As $|s| \rightarrow \infty$, the expression

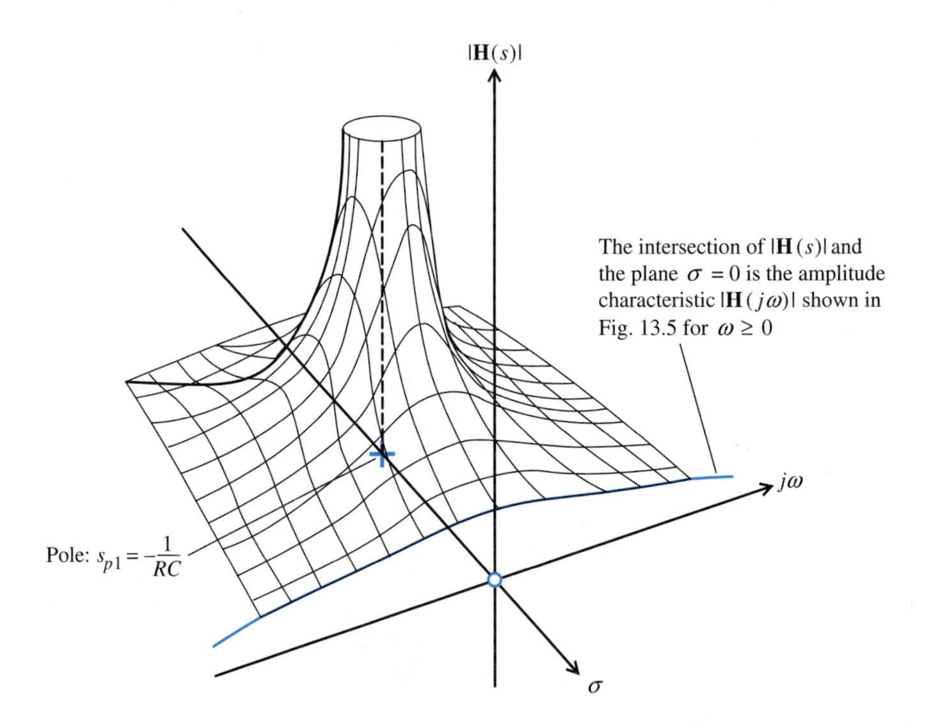

FIGURE 13.17 Magnitude of the transfer function of the low-pass *RC* circuit of Fig. 13.4

$$\mathbf{H}(s) = \frac{\mathcal{B}(s)}{\mathcal{A}(s)} = \frac{b_m s^m + \cdots + b_0}{a_n s^n + \cdots + a_0} \tag{13.46}$$

becomes

$$\mathbf{H}(s) = \frac{b_m}{a_n} s^{m-n} = K s^{m-n} \tag{13.47}$$

Therefore, if there are more poles than zeros ($n > m$), the amplitude characteristic eventually tends toward zero as $|s|$ increases. This is illustrated by Figs. 13.11 and 13.17. If the number of poles equals the number of zeros ($n = m$), as illustrated by Fig. 13.16, then the amplitude characteristic approaches the constant K as $|s| \to \infty$.

Remember

The rubber sheet analogy shows that the poles cause the peaks in $|\mathbf{H}(j\omega)|$, and the zeros cause the dips.

EXERCISES

22. Use the rubber sheet analogy to sketch the approximate amplitude characteristic $|\mathbf{H}(j\omega)|$ corresponding to the following pole-zero plot.

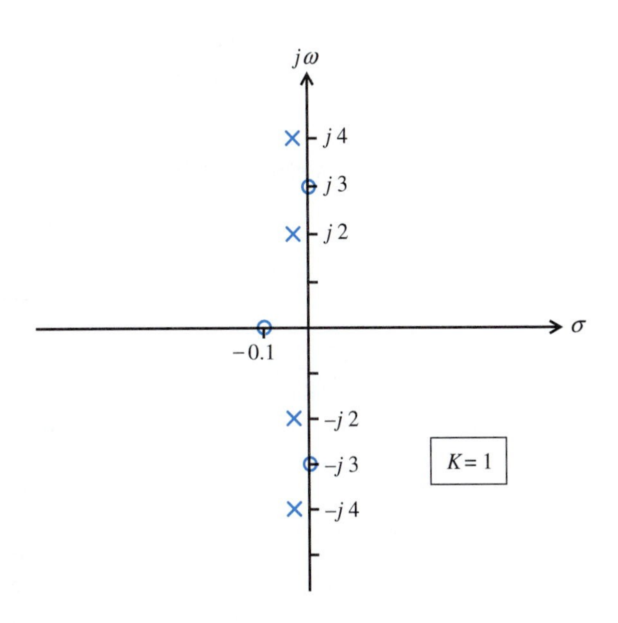

13.5 Band-Pass Filter

Band-pass filters are very important in radio and television systems, where they are used to separate one broadcast channel from another. The most important (and the simplest) examples of band-pass filters are provided by the series and parallel *RLC* circuits. We consider the series *RLC* circuit first.

Series *RLC* Circuit

To use a series *RLC* circuit as a band-pass filter, we drive it with a voltage source and take the output voltage across the resistance, as illustrated in Fig. 13.18.

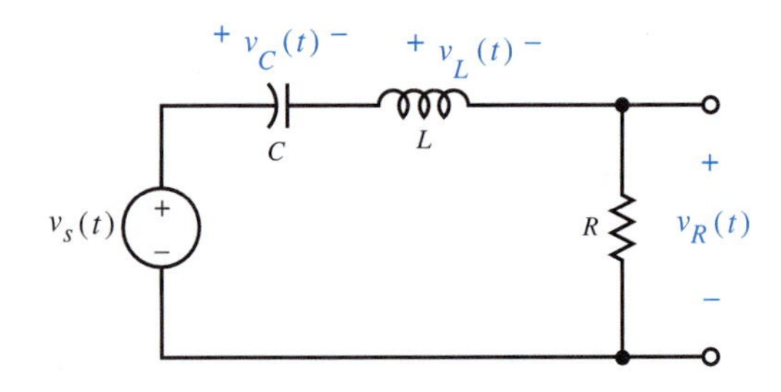

FIGURE 13.18
Series *RLC* band-pass filter

The voltage transfer function is given by

$$\mathbf{H}_R(s) = \frac{\mathbf{V}_R}{\mathbf{V}_s} = \frac{R}{R + sL + 1/sC} = \frac{sR/L}{s^2 + (R/L)s + 1/LC} = \frac{sR/L}{(s - s_{p1})(s - s_{p2})} \tag{13.48}$$

(We use a subscript R on $\mathbf{H}_R(s)$ to emphasize that the output voltage is taken across the resistance.) The pole-zero plot is shown in Fig. 13.19, where

$$\alpha = \frac{R}{2L} \tag{13.49}$$

is the damping coefficient,

$$\omega_0 = \frac{1}{\sqrt{LC}} \tag{13.50}$$

is the undamped natural frequency,

$$\omega_d = \sqrt{\omega_0^2 - \alpha^2} \tag{13.51}$$

is the damped natural frequency†, and

$$\beta = \sqrt{\alpha^2 - \omega_0^2} \tag{13.52}$$

We see from Fig. 13.19 that there are three distinct cases, depending on the values of α and ω_0. As the damping coefficient α is varied from 0 to ω_0 with ω_0 held constant, the poles move toward each other along the circular path defined by Eq. (13.51) and shown in Fig. 13.19a. The poles lie on the $j\omega$ axis at $\pm j\omega_0$ when $\alpha = 0$ and meet on the σ axis at $\sigma = -\omega_0$ when $\alpha = \omega_0$. When the poles are on the circular path, the damping ratio $\zeta = \alpha/\omega_0$ lies in the range $0 \le \zeta < 1$, which means that the circuit is *underdamped.* When the poles meet on the σ axis, the damping ratio is unity, and the circuit is *critically damped* (Fig. 13.19b). As α increases beyond ω_0, the natural frequencies separate with s_{p1} approaching zero and s_{p2} approaching minus infinity (Fig. 13.19c). The damping ratio exceeds unity, $\zeta > 1$, and the circuit is *overdamped* for $\alpha > \omega_0$. We use the *underdamped* series *RLC* circuit for the band-pass filter application.

Plots of $\mathbf{H}_R(j\omega)$ are shown in Fig. 13.20 on page 444 for $0 \le \zeta \le 1$. Notice that the peak in the amplitude characteristic, found by solving

$$\frac{d|\mathbf{H}_R(j\omega)|}{d\omega} = 0 \tag{13.53}$$

occurs at $\omega = \omega_0$. It is instructive to compare the frequency-domain characteristics with the circuit's natural response. The natural response that is shown in Fig. 13.21 results

† Don't let the terms *undamped natural frequency* and *damped natural frequency* confuse you: they are simply the universally accepted names for the quantities shown.

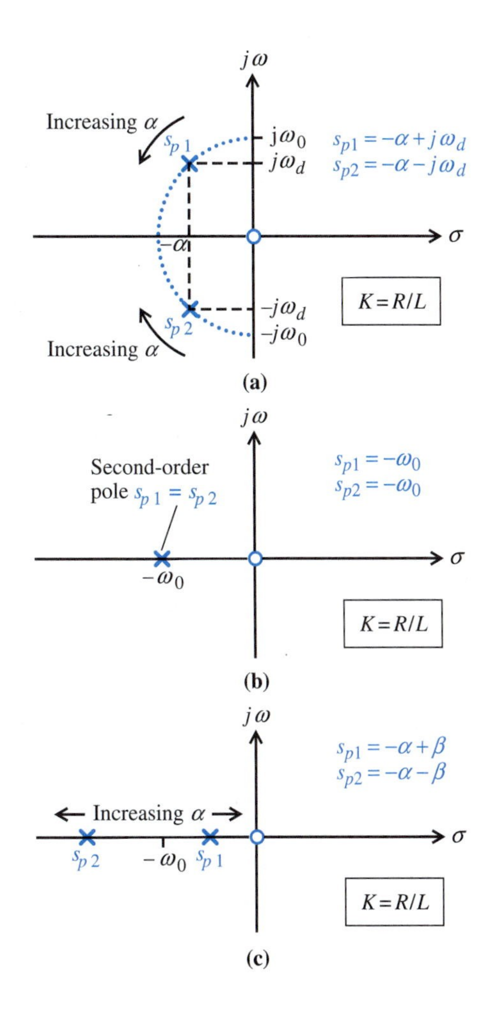

FIGURE 13.19
Pole-zero plot of $\mathbf{H}_R(s) = (sR/L)/(s - s_{p1})(s - s_{p2})$. (*a*) Underdamped ($\alpha < \omega_0$); (*b*) critically damped ($\alpha = \omega_0$); (c) overdamped ($\alpha > \omega_0$)

from a nonzero initial current $i_L(0^-)$ in the inductance. The natural response is a damped sinusoid with damping coefficient α and damped natural frequency $\omega_d = \omega_0\sqrt{1 - \zeta^2}$. When the damping ratio is small, the poles lie close to the $j\omega$ axis, the amplitude characteristic is sharply peaked, and the natural response is highly oscillatory. As the damping ratio increases, the poles move away from the $j\omega$ axis, so the amplitude characteristic is less peaked and the natural response is less oscillatory. The natural response is not oscillatory for $\zeta \geq 1$.

In the band-pass filter application, the input voltage $v_s(t)$ contains two or more sinusoidal components with different frequencies. Only the sinusoidal components in $v_s(t)$ that have frequencies near ω_0 can pass to the output, $v_R(t)$, without significant attenuation. The *passband* is defined as the band of frequencies between ω_L and ω_U, $\omega_L < \omega < \omega_U$, where ω_L and ω_U are the lower and upper half-power frequencies of the filter shown in Fig. 13.22. The *stopband* includes all frequencies outside of the passband. The *half-power bandwidth* is given by

$$\text{BW} = \omega_U - \omega_L \tag{13.54}$$

We will now derive formulas for ω_L, ω_U, and BW. The frequencies ω_L and ω_U are the positive solutions to the equation

$$|\mathbf{H}_R(j\omega)| = \frac{1}{\sqrt{2}} \tag{13.55}$$

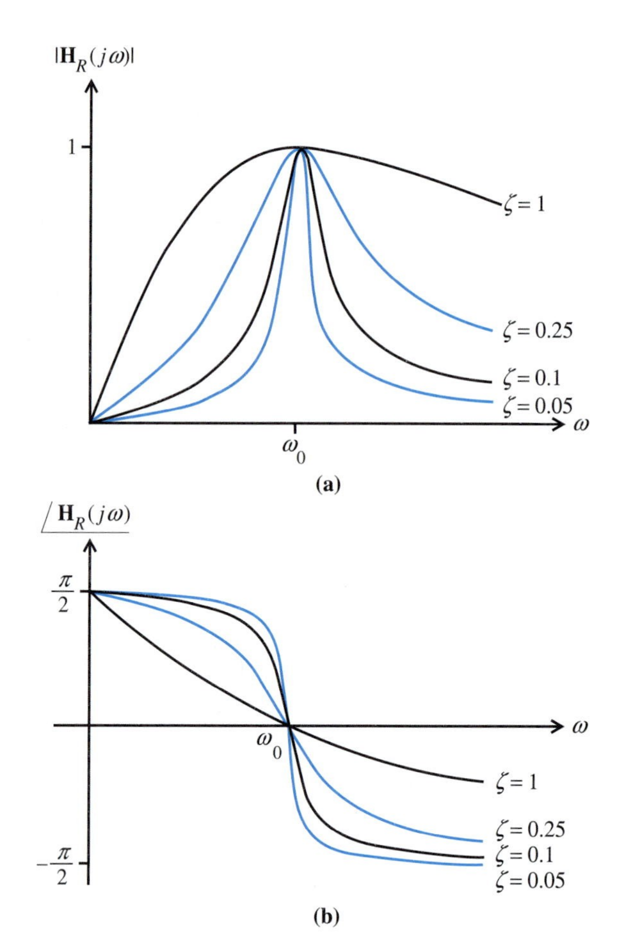

FIGURE 13.20
Frequency response of series *RLC* band-pass filter. Damping ratio $\zeta = \alpha/\omega_0$.
(*a*) Amplitude; (*b*) phase

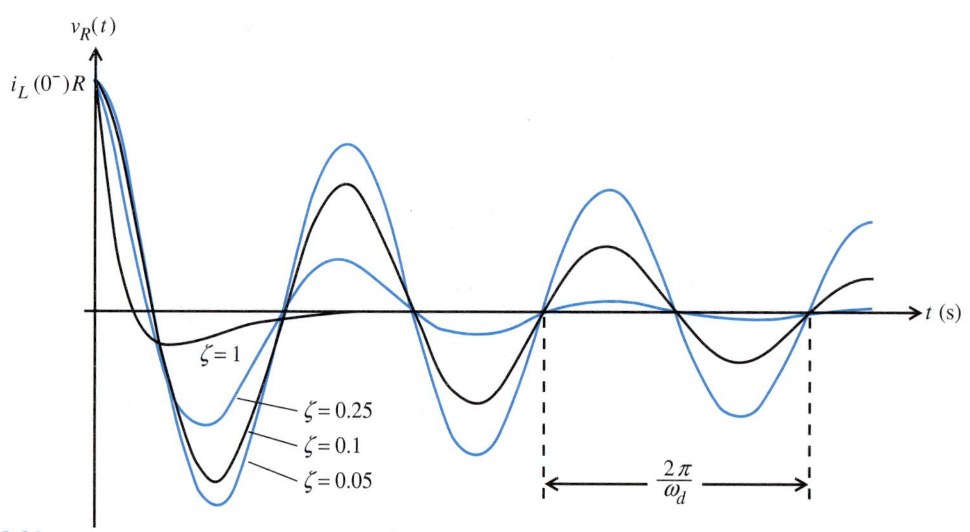

FIGURE 13.21
Natural response of series *RLC* band-pass filter. The waveform shown occurs when $v_s(t) = 0$ and $v_C(0^-) = 0$. $i_L(0^-)$ is a nonzero initial current in the inductance. The response is a damped sinusoid with frequency ω_d and damping coefficient α for $0 \le \zeta < 1$

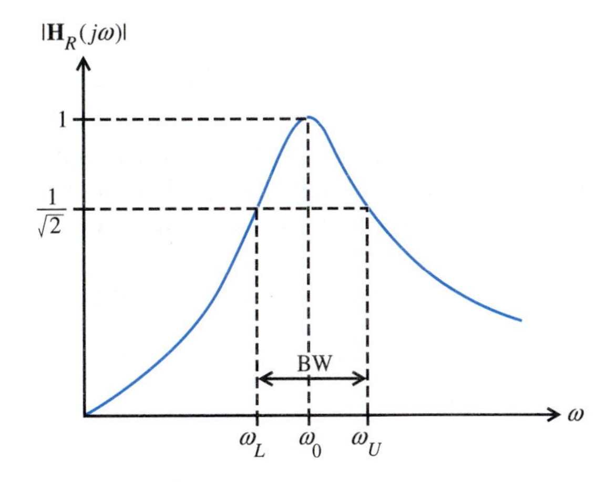

FIGURE 13.22 Illustration of half-power frequencies ω_L and ω_U

where $\mathbf{H}_R(j\omega)$ is given by Eq. (13.48) with $s = j\omega$. Thus Eq. (13.55) becomes

$$\left| \frac{R}{R + j\left(\omega L - \dfrac{1}{\omega C}\right)} \right| = \frac{1}{\sqrt{2}} \tag{13.56}$$

After we square both sides, Eq. (13.56) becomes

$$\frac{R^2}{R^2 + \left(\omega L - \dfrac{1}{\omega C}\right)^2} = \frac{1}{2} \tag{13.57}$$

and we see that

$$\left(\omega L - \frac{1}{\omega C}\right)^2 = R^2 \tag{13.58}$$

Therefore either

$$\omega L - \frac{1}{\omega C} = +R \tag{13.59}$$

or

$$\omega L - \frac{1}{\omega C} = -R \tag{13.60}$$

The solutions to Eq. (13.59) are

$$\omega_1 = \frac{R}{2L} + \sqrt{\frac{1}{LC} + \left(\frac{R}{2L}\right)^2} \tag{13.61}$$

and

$$\omega_2 = \frac{R}{2L} - \sqrt{\frac{1}{LC} + \left(\frac{R}{2L}\right)^2} \tag{13.62}$$

and the solutions to Eq. (13.60) are

$$\omega_3 = -\frac{R}{2L} + \sqrt{\frac{1}{LC} + \left(\frac{R}{2L}\right)^2} \tag{13.63}$$

and

$$\omega_4 = -\frac{R}{2L} - \sqrt{\frac{1}{LC} + \left(\frac{R}{2L}\right)^2} \tag{13.64}$$

We see from Fig. 13.22 that the quantities ω_L and ω_U we are solving for are nonnegative, and ω_U is greater than ω_L. Roots ω_2 and ω_4 are negative and can be discarded. The half-power frequencies ω_L and ω_U are given by ω_3 and ω_1, respectively:

$$\begin{aligned}\omega_L &= -\frac{R}{2L} + \sqrt{\frac{1}{LC} + \left(\frac{R}{2L}\right)^2} \\ &= -\alpha + \sqrt{\omega_0^2 + \alpha^2}\end{aligned} \tag{13.65}$$

and

$$\begin{aligned}\omega_U &= +\frac{R}{2L} + \sqrt{\frac{1}{LC} + \left(\frac{R}{2L}\right)^2} \\ &= +\alpha + \sqrt{\omega_0^2 + \alpha^2}\end{aligned} \tag{13.66}$$

By subtracting ω_L from ω_U, we see that the half-power bandwidth, Eq. (13.54), is given by the simple formula

$$\text{BW} = 2\alpha = \frac{R}{L} \tag{13.67}$$

We have seen that the series *RLC* circuit can be used as a band-pass filter. Band-pass filters without inductance can be constructed with the use of op amps. A typical op-amp band-pass filter is described in Problem 46.

Remember

The amplitude characteristic $|\mathbf{H}_R(j\omega)|$ of a resonant series *RLC* circuit has a peak at $\omega_0 = 1/\sqrt{LC}$ and a half-power bandwidth BW $= 2\alpha = R/L$.

EXERCISE

23. (a) Use the equivalent source concept to represent the circuit shown as a series *RLC* circuit.
(b) Use your answer to (a) to determine the transfer function $\mathbf{H}_o(s) = \mathbf{V}_o/\mathbf{V}_1$.

answer: $K\dfrac{s}{s^2 + s(R + R_s)/L + 1/LC}$, where $K = \dfrac{RR_2}{(R + R_2)L}$ and $R_s = R_1 \parallel R_2$

(c) Find an expression for the half-power bandwidth BW in terms of R_1, R_2, R, and L.

answer: $\dfrac{R + R_s}{L}$

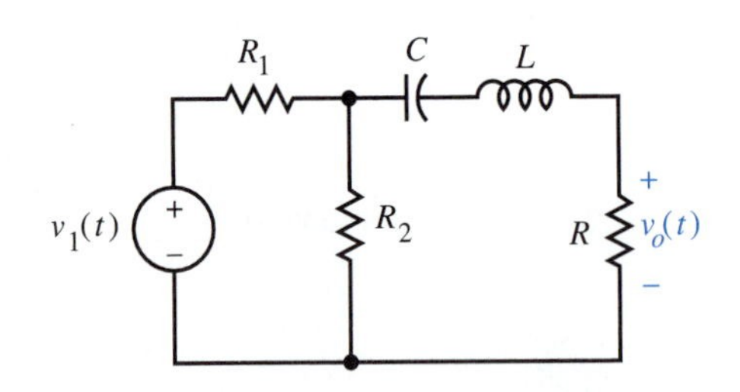

See Problem 13.4 in the PSpice manual.

Parallel *RLC* Circuit

To use a parallel *RLC* circuit as a band-pass filter, we drive it with a current and take the output as the voltage across the resistance, as shown in Fig. 13.23. Thus, the band-pass transfer function is simply the driving-point impedance. If instead we took the output as the current through the resistance, then the analysis of the parallel circuit would be the exact dual of the analysis of the series circuit of Fig. 13.18. Because of this close connection between the series and parallel *RLC* band-pass filters, we leave the development of the parallel filter to the following exercises.

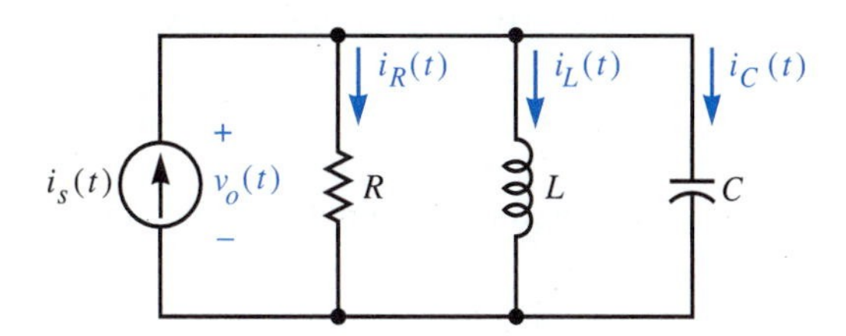

FIGURE 13.23 Parallel *RLC* band-pass filter

EXERCISES

24. For the circuit of Fig. 13.23:
    (a) Use the current divider relation to find $\mathbf{H}_G(s) = \mathbf{I}_R/\mathbf{I}_s$.
    (b) Use duality and Eq. (13.48) to check your result in (a).
    (c) Use duality to show that $|\mathbf{H}_G(j\omega)|$ has a peak when $\omega = \omega_0$, where $\omega_0 = 1/\sqrt{LC}$.
    (d) Use duality to show that the half-power bandwidth of $|\mathbf{H}_G(j\omega)|$ is given by $\mathrm{BW} = 1/RC$.

    answer: (a) $\dfrac{s/RC}{s^2 + s/RC + 1/LC}$

25. Assume that $C = 1\ \mu\mathrm{F}$ and $L = 2.5$ mH in the band-pass filter of Fig. 13.23. The output is the voltage $\mathbf{V}_o$.
    (a) What value of R is required for a half-power bandwidth of 1 krad/s?
    (b) Sketch the amplitude and phase characteristics for the value of R determined in (a).

    answer: (a) 1000 Ω

13.6 Unity-Power-Factor Resonance

There is a traditionally accepted way to define resonance that does not explicitly refer to the network's peak response. In this definition, resonance occurs when the power factor of the network, cos ($\angle\mathbf{Z}$), equals unity. This is *unity-power-factor resonance.* Unity-power-factor resonance occurs when the network appears purely resistive to the source. Thus no reactive power is supplied to a network operating at unity-power-factor resonance.

The unity-power-factor resonance frequency is determined by finding $\omega = \omega_{upf}$ for which

$$\mathcal{I}m\{\mathbf{Z}(j\omega)\} = 0 \tag{13.68}$$

or, equivalently,

$$\mathcal{I}m\{\mathbf{Y}(j\omega)\} = 0 \tag{13.69}$$

where **Z** and **Y** are, respectively, the network's driving-point impedance and admittance.

EXAMPLE 13.3 Refer to the series *RLC* circuit of Fig. 13.18.
(a) Determine the unity-power-factor resonance frequency.
(b) Determine $i(t)$, $v_L(t)$, and $v_C(t)$ when $\omega = \omega_{upf}$.
(c) Analyze the flow of stored energy in the circuit when $\omega = \omega_{upf}$.

Solution (a) The driving-point impedance $\mathbf{Z}(j\omega)$ is

$$\mathbf{Z}(j\omega) = R + j\left(\omega L - \frac{1}{\omega C}\right)$$

and therefore the unity-power-factor resonance frequency is found by solving

$$\omega L - \frac{1}{\omega C} = 0$$

This yields

$$\omega_{upf} = \sqrt{\frac{1}{LC}}$$
$$= \omega_0$$

which is recognized as the maximum-response resonance frequency associated with $\mathbf{H}_R(j\omega)$ (see Fig. 13.20).

(b) The driving-point impedance at unity-power-factor resonance is

$$\mathbf{Z}(j\omega_0) = R$$

This results in a driving-point current phasor

$$\mathbf{I} = \frac{\mathbf{V}_s}{\mathbf{Z}(j\omega_0)} = \frac{\mathbf{V}_s}{R}$$

a voltage phasor

$$\mathbf{V}_L = j\omega_0 L\mathbf{I} = j\omega_0 L\frac{\mathbf{V}_s}{R} = j\frac{1}{2\zeta}\mathbf{V}_s$$

across the inductance, and a voltage phasor

$$\mathbf{V}_C = \frac{1}{j\omega_0 C}\mathbf{I} = -j\omega_0 L\frac{\mathbf{V}_s}{R} = -j\frac{1}{2\zeta}\mathbf{V}_s$$

across the capacitance. These quantities are summarized in Fig. 13.24. Note that because $\mathbf{V}_L + \mathbf{V}_C = 0$, the voltage phasor $\mathbf{V}_R$ must equal the source phasor $\mathbf{V}_s$.

The waveforms associated with **I**, $\mathbf{V}_L$, and $\mathbf{V}_s$ are

$$i(t) = \frac{V_{sm}}{R}\cos(\omega_0 t + \phi_{\mathbf{V}})$$

$$v_L(t) = \frac{V_{sm}}{2\zeta}\cos(\omega_0 t + \phi_{\mathbf{V}} + 90°)$$

and

$$v_C(t) = \frac{V_{sm}}{2\zeta}\cos(\omega_0 t + \phi_{\mathbf{V}} - 90°)$$

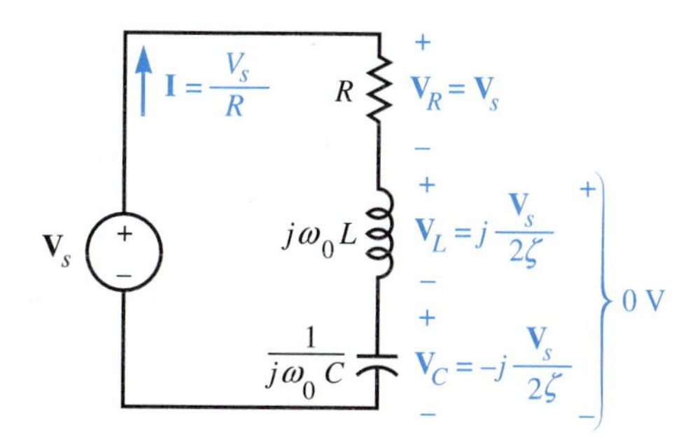

FIGURE 13.24 Series *RLC* at unity-power-factor resonance. The magnitudes of $\mathbf{V}_L$ and $\mathbf{V}_C$ greatly exceed the source voltage amplitude for $\zeta \ll 1$. However, the sum $\mathbf{V}_L + \mathbf{V}_C$ equals zero.

The amplitudes of $v_L(t)$ and $v_C(t)$ will be larger than the amplitude of the voltage source if the damping ratio is less than 0.5. However, because $v_L(t)$ and $v_C(t)$ have equal amplitudes and differ by 180° in phase, the combined voltage drop across the inductance and the capacitance is zero. It is this fact that makes the circuit appear resistive to the source for $\omega = \omega_0$.

(c) No reactive power enters the *RLC* network at unity-power-factor resonance. Nevertheless, the network does contain stored energy. With the use of Eqs. (2.31) and (2.45), we find that the instantaneous energies stored in the inductance and the capacitance are, respectively,

$$\begin{aligned} w_L(t) &= \frac{1}{2} L i^2(t) = \frac{1}{2} L \left(\frac{V_{sm}}{R}\right)^2 \cos^2(\omega_0 t + \phi_{\mathbf{V}}) \\ &= \frac{1}{4} L \left(\frac{V_{sm}}{R}\right)^2 + \frac{1}{4} L \left(\frac{V_{sm}}{R}\right)^2 \cos(2\omega_0 t + 2\phi_{\mathbf{V}}) \end{aligned}$$

and, similarly

$$\begin{aligned} w_C(t) &= \frac{1}{2} C v_C^2(t) \\ &= \frac{1}{4} C \left(\frac{V_{sm}}{2\zeta}\right)^2 + \frac{1}{4} C \left(\frac{V_{sm}}{2\zeta}\right)^2 \cos(2\omega_0 t + 2\phi_{\mathbf{V}} - 180^\circ) \\ &= \frac{1}{4} L \left(\frac{V_{sm}}{R}\right)^2 + \frac{1}{4} L \left(\frac{V_{sm}}{R}\right)^2 \cos(2\omega_0 t + 2\phi_{\mathbf{V}} - 180^\circ) \end{aligned}$$

where the last step follows with the use of $\zeta = \alpha/\omega_0$, where $\alpha = R/2L$ and $\omega_0 = 1/\sqrt{LC}$. The instantaneous energies are plotted in Fig. 13.25. We see from this figure and from the above expressions that $w_L(t)$ and $w_C(t)$ are equal except for a time delay, and that their sum equals a constant:

$$w_L(t) + w_C(t) = \frac{1}{2} L \left(\frac{V_{sm}}{R}\right)^2 \tag{13.70}$$

Equation (13.70) is a special case of an important general result. Recall that the instantaneous power delivered to an arbitrary circuit composed of *R*'s, *L*'s, and *C*'s is, by Eq. (12.15),

$$p(t) = \underbrace{P[1 + \cos(2\omega t + 2\phi_{\mathbf{I}})]}_{\text{Energy flow into the network}} + \underbrace{Q\cos\left(2\omega t + 2\phi_{\mathbf{I}} + \frac{\pi}{2}\right)}_{\text{Energy borrowed and returned by the network}} \tag{13.71}$$

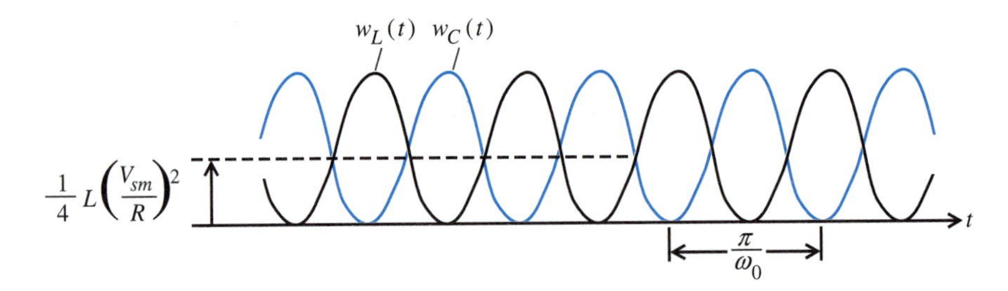

FIGURE 13.25 Instantaneous energies stored in inductance and capacitance of series *RLC* circuit when $\omega = \omega_{upf}$

where from Eq. (12.55b),

$$Q = 2\omega(W_{\text{tot ave magnetic}} - W_{\text{tot ave electric}}) \tag{13.72}$$

When unity-power-factor resonance occurs, the reactive power Q supplied to a network equals zero. By setting the right-hand side of Eq. (13.72) to zero, we obtain the following fundamental result.

> When unity-power-factor resonance occurs in any circuit made up of R's, L's, and C's, the total average energy stored in the magnetic fields equals the total average energy stored in the electric fields.

Remember At unity-power-factor resonance, a network appears purely resistive to the source.

EXERCISES

26. Show that the unity-power-factor resonance frequency of a parallel *RLC* circuit is $\omega_0 = 1/\sqrt{LC}$. Use duality to check your result.
27. Show that when the parallel *RLC* circuit shown is operated at its unity-power-factor resonance frequency, the phasor currents are $\mathbf{I}_R = \mathbf{I}_s$, $\mathbf{I}_L = -j\omega_0 RC\mathbf{I}_s$, and $\mathbf{I}_C = +j\omega_0 RC\mathbf{I}_s$. Explain why no current enters the part of the circuit enclosed by blue when $\omega = \omega_{upf}$.

   answer: The admittance of the enclosed part of the circuit equals zero when $\omega = \omega_{upf}$.

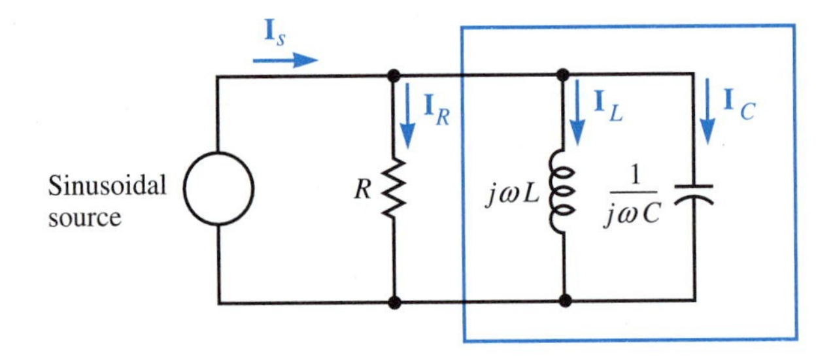

13.7 Quality Factor

The *quality factor Q* provides a figure of merit for ac circuits and circuit components.† Q is a dimensionless quantity that has been defined somewhat differently by various authors. We define Q as follows:

† Quality factor Q should not be confused with reactive power (also symbolized as Q). They are totally different quantities.

DEFINITION Quality Factor Q

$$Q = 2\pi \frac{\text{peak energy stored in the circuit}}{\text{energy dissipated by the circuit in one period}} \tag{13.73}$$

The value of Q depends on the frequency ω at which it is evaluated; hence

$$Q = Q(\omega) \tag{13.74}$$

Series and Parallel *RL* and *RC* Circuits

For illustration, consider the series *RL* circuit of Fig. 13.26a. To determine the Q of this circuit, we must assume that the circuit is operating in the sinusoidal steady state at some frequency ω. This means that a sinusoidal current

$$i(t) = I_m \cos(\omega t + \phi_I) \tag{13.75}$$

is flowing through the resistance and the inductance, as illustrated in Fig. 13.26b. We can evaluate the expression for Q, Eq. (13.73) by recognizing that the inductance L stores energy and that the resistance R dissipates energy. The instantaneous value of the energy stored is

$$\begin{aligned} w_L(t) &= \frac{1}{2} L i^2(t) \\ &= \frac{1}{2} L I_m^2 \cos^2(\omega t + \phi_I) \end{aligned} \tag{13.76}$$

Therefore

$$\begin{aligned} \text{Peak stored energy} &= w_L(t)\big|_{\text{peak}} \\ &= \frac{1}{2} L I_m^2 \end{aligned} \tag{13.77}$$

The average power dissipated by the resistance is

$$P = \frac{1}{2} I_m^2 R \tag{13.78}$$

The energy loss in one period is the product of the average power dissipated times the period $T = 2\pi/\omega$. Therefore

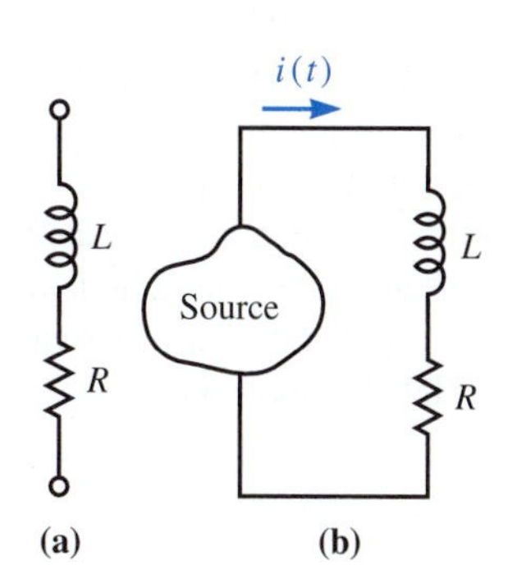

FIGURE 13.26 Series *RL* circuit: (*a*) alone; (*b*) with sinusoidal source

$$\text{Energy loss in one period} = \frac{1}{2} I_m^2 R \frac{2\pi}{\omega} \tag{13.79}$$

The substitution of Eqs. (13.77) and (13.79) into Eq. (13.73) yields the result

$$Q = 2\pi \frac{(1/2)LI_m^2}{(1/2)I_m^2 R(2\pi/\omega)} = \frac{\omega L}{R} \tag{13.80}$$

Observe that this result has the form

$$Q = \frac{|X|}{R} \tag{13.81}$$

where $X = \omega L$ is the reactance of the inductance.

The series RL circuit of Fig. 13.26 is sometimes used to model a physical inductor. In the model, R represents the internal or *winding resistance* of the inductor and L represents the inductance. According to this model, an inductor whose inductive reactance is large compared with its winding resistance has a high Q. Also according to this model, the Q of the inductor decreases with decreasing frequency ω. Because winding resistance is unavoidable, it is difficult to construct inductors that have high Qs at low frequencies.

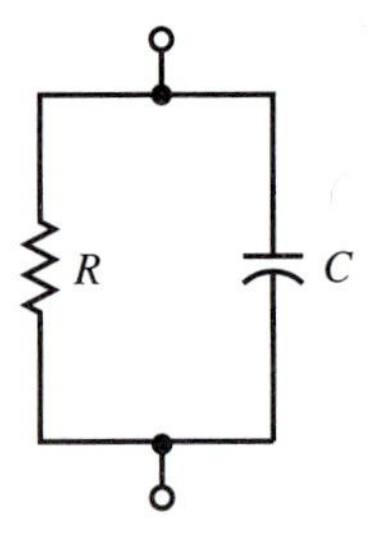

FIGURE 13.27
Parallel RC circuit

A second illustration is provided by the parallel RC circuit of Fig. 13.27. One way to determine the Q of this circuit is to apply the definition in Eq. (13.73). The Q of the parallel RC circuit can also be determined easily if we recognize that this circuit is the dual of the series RL circuit. Substituting C for L and $1/R$ for R in Eq. (13.80) leads immediately to

$$Q = \omega CR \tag{13.82}$$

or

$$Q = \frac{R}{|X|} \tag{13.83}$$

where $X = -1/\omega C$ is the reactance of the capacitance.

The parallel RC circuit is sometimes used to model a physical capacitor. In the model, R represents the capacitor's *leakage resistance* and C represents the capacitance. Leakage resistance arises from charge carriers in the dielectric material separating the plates of the capacitor. According to the model, capacitors whose capacitive reactances are small compared with their leakage resistances have high Qs.

Additional illustrations are provided by the parallel RL and series RC circuits of Figs. 13.28 and 13.29. Derivations similar to the preceding ones reveal that the Q of the parallel RL circuit is

$$Q = \frac{R}{\omega L} = \frac{R}{|X|} \tag{13.84}$$

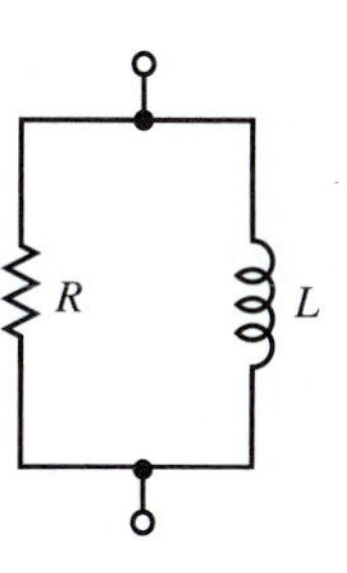

FIGURE 13.28
Parallel RL circuit

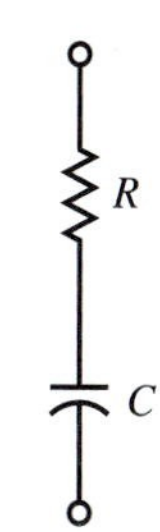

FIGURE 13.29
Series RC circuit

where $X = \omega L$, and that the Q of its dual series RC circuit is

$$Q = \frac{1}{\omega CR} = \frac{|X|}{R} \tag{13.85}$$

where $X = -1/\omega C$.

The results for the four circuits considered so far are conveniently summarized in Fig. 13.30. In this figure, the subscripts s and p refer to series and parallel circuits, respectively, and X refers to the reactance of either an inductance or a capacitance, as appropriate. The results indicated in the figure for the series circuit follow from Eqs. (13.81) and (13.85), which indicate that the expression for the Q of either the RC or the RL series circuit can be written as

$$Q_s = \frac{|X_s|}{R_s} \tag{13.86}$$

The results indicated in the figure for the parallel circuit follow from Eqs. (13.83) and (13.84), which indicate that the expression for the Q of either the RC or the RL parallel circuit can be written as

$$Q_p = \frac{R_p}{|X_p|} \tag{13.87}$$

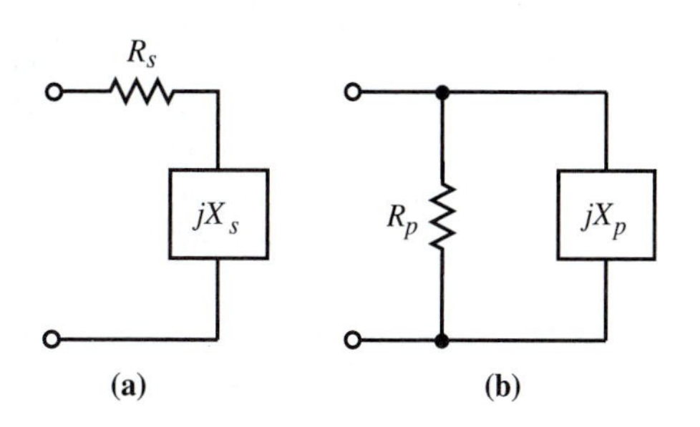

FIGURE 13.30
Series and parallel networks. The boxes contain *only* an inductance or a capacitance. (*a*) Series network $Q_s = |X_s|/R_s$; (*b*) parallel network $Q_p = R_p/|X_p|$

Remember

The quality factors of the series and parallel RC or RL networks of Fig. 13.30 are given by Eqs. (13.86) and (13.87), respectively.

Series *RLC* Circuit

A final and especially important illustration of Q is provided by the series RLC circuit. The Q of this circuit is ordinarily evaluated at the circuit's unity-power-factor resonance frequency $\omega_{upf} = \sqrt{1/LC} = \omega_0$, where it is denoted as Q_0. As was shown in Example 13.3, the total instantaneous energy stored in this circuit is constant at unity-power-factor resonance. The peak value of a constant is just that constant. Therefore, from Example 13.3,

$$\text{Peak stored energy} = \frac{1}{2} L \left(\frac{V_{sm}}{R} \right)^2 \tag{13.88}$$

The energy loss in one period is simply the average power dissipated $P = V_{sm}^2/2R$ times the period $T_0 = 2\pi/\omega_0$:

$$\text{Energy loss in one period} = \frac{V_{sm}^2}{2R} \frac{2\pi}{\omega_0} \tag{13.89}$$

Therefore, the Q of a series RLC circuit at the unity-power-factor resonance frequency, $\omega_0 = 1/\sqrt{LC}$, is

$$Q_0 = 2\pi \frac{\frac{1}{2} L \left(\frac{V_{sm}}{R} \right)^2}{\frac{V_{sm}^2}{2R} \frac{2\pi}{\omega_0}}$$

$$= \frac{\omega_0 L}{R} = \frac{\omega_0}{2\alpha} = \frac{1}{2\zeta} \tag{13.90}$$

Notice that $\frac{1}{2}\zeta$ is the factor appearing in $\mathbf{V}_L$ and $\mathbf{V}_C$ of Fig. 13.24 on page 449. At unity-power-factor resonance the voltage amplitudes across the inductance and capacitance are each Q_0 times the voltage amplitude of the source.

The quality factor Q_0 is also an indicator of the sharpness of the peak occurring in the amplitude characteristic of Fig. 13.20: The higher the value of Q_0, the sharper the resonance peak. Remember that the half-power bandwidth associated with $|\mathbf{H}_R(j\omega)|$ is exactly 2α [Eq. (13.68)]. If we combine Eqs. (13.67) and (13.90), we obtain

$$Q_0 = \frac{\omega_0}{\text{BW}} \tag{13.91}$$

or

$$\text{BW} = \frac{\omega_0}{Q_0} \tag{13.92}$$

We see from Eq. (13.92) that the half-power bandwidth is small compared with the "center frequency" ω_0 when Q_0 is large ($Q_0 \gg 1$).

Remember

The quality factor of a series RLC circuit at angular frequency $\omega_0 = 1/\sqrt{LC}$ is $Q_0 = \omega_0/\text{BW}$ where $\text{BW} = 2\alpha = R/L$.

EXERCISES

28. A series *RLC* circuit has a Q of 10 at its unity-power-factor resonance frequency of 100 kHz, and it dissipates 2 mW average power when driven by a 4-V rms 100-kHz sinusoidal voltage source.
    (a) What is the half-power bandwidth?
    (b) Determine the numerical values of R, L, and C.

    answer: (a) $20{,}000\pi$ rad/s, (b) 8000 Ω, 0.127 H, 20 pF

29. A parallel *RLC* circuit has a Q of 100 at its unity-power-factor resonance frequency of 1 kHz, and it dissipates 1 W average power when driven by a 1-A rms 1-kHz sinusoidal current source.
    (a) What is the half-power bandwidth?
    (b) Determine the numerical values of R, L, and C.

    answer: (a) 20π rad/s, (b) 1 Ω, 15.9 μH, 15.9 mF

13.8 The Universal Resonance Characteristic

In this section we will show that all resonance peaks caused by isolated simple poles located close to the $j\omega$ axis have approximately the same shape. The reason for this important fact can be understood if we consider a typical transfer function, for example,

$$\mathbf{H}(s) = \mathrm{K}\frac{(s - s_{z1})^2(s - s_{z2})(s - s_{z3})}{(s - s_{p1})(s - s_{p2})(s - s_{p3})(s - s_{p4})} \tag{13.93}$$

An exact calculation of the transfer function involves the lengths and the angles of the vectors from the poles and zeros of $\mathbf{H}(s)$ to the point s as illustrated in Fig. 13.31. In this illustration, s is assumed to be close to the isolated simple pole at s_{p1}.

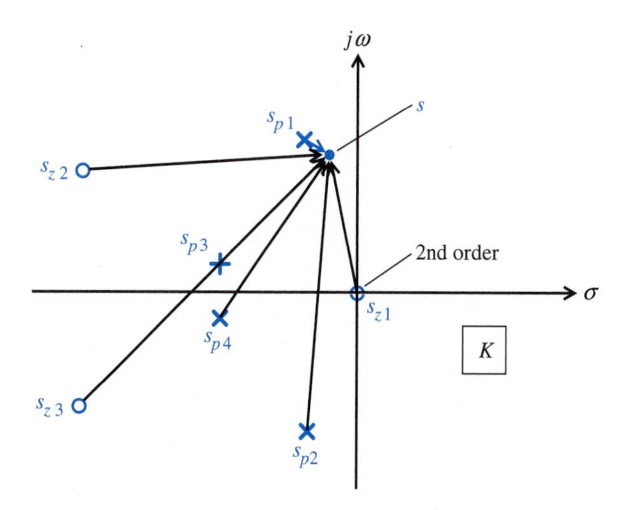

FIGURE 13.31 Graphical evaluation of $\mathbf{H}(s)$

If the point s is moved about in the vicinity of s_{p1}, the long vectors (in black) will remain nearly constant both in length and in angle, but the short vector $s - s_{p1}$ (in blue) will vary by a relatively large amount in length and angle. Thus, for s sufficiently close to s_{p1}, each of the long vectors can be approximated with reasonable accuracy as

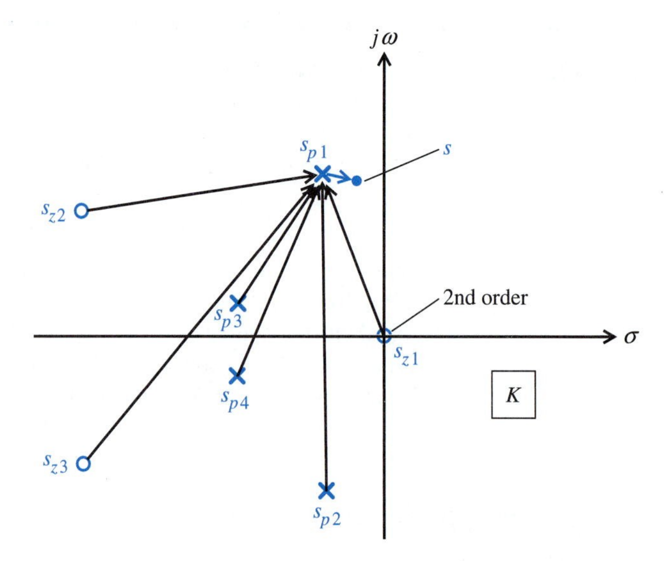

FIGURE 13.32 Approximations leading to Eq. (13.95) LTI circuit.

constants. A standard choice for the constant vectors is shown in Fig. 13.32, where each long vector has been approximated by a vector starting at the corresponding critical frequency and ending at s_{p1}. By means of this approximation, we see that

$$\mathbf{H}(s) \simeq K \frac{(s_{p1} - s_{z1})^2(s_{p1} - s_{z2})(s_{p1} - s_{z3})}{(s - s_{p1})(s_{p1} - s_{p2})(s_{p1} - s_{p3})(s_{p1} - s_{p4})} \tag{13.94}$$

for $s \simeq s_{p1}$.† Notice that only one factor in Eq. (13.94) depends on s. Therefore we can write

$$\mathbf{H}(s) \simeq \frac{k_{11}}{s - s_{p1}} \qquad \text{for } s \simeq s_{p1} \tag{13.95}$$

The constant

$$k_{11} = K \frac{(s_{p1} - s_{z1})^2(s_{p1} - s_{z2})(s_{p1} - s_{z3})}{(s_{p1} - s_{p2})(s_{p1} - s_{p3})(s_{p1} - s_{p4})} \tag{13.96}$$

is called the *residue* of $\mathbf{H}(s)$ at the simple pole s_{p1} and can be expressed as

$$k_{11} = (s - s_{p1})\mathbf{H}(s)\big|_{s=s_{p1}} \tag{13.97}$$

The technique leading to Eqs. (13.95) and (13.97) can be used to determine the approximate form of *any* transfer function in the vicinity of an isolated simple pole. The result will always have the form of Eq. (13.95), where the residue k_{11} is given by Eq. (13.97). It follows that if the isolated pole is close to the $j\omega$ axis, then the frequency response in the vicinity of the resonance peak is approximately

$$\mathbf{H}(j\omega) \simeq \frac{k_{11}}{j\omega - s_{p1}} \tag{13.98}$$

† A complex quantity z is approximately equal to another complex quantity z_1 if $|z - z_1| < \varepsilon$, where ε is a small positive number. Therefore the region of the s plane for which $s \simeq s_{p1}$ is a circle centered at s_{p1} with radius ε.

Another form for the approximation in Eq. (13.98) can be obtained if we set

$$s_{p1} = -\alpha + j\omega_d \tag{13.99}$$

where α and ω_d are the damping coefficient and the damped natural frequency associated with s_{p1}. The pole s_{p1} is close to the $j\omega$ axis if and only if $\alpha \ll \omega_d$.† Using Eq. (13.99), Eq. (13.98) becomes

$$\mathbf{H}(j\omega) \simeq \frac{k_{11}}{j(\omega - \omega_d) + \alpha} \tag{13.100}$$

This approximation to $\mathbf{H}(j\omega)$ can be simplified further by momentarily setting $\omega = \omega_d$, which shows that

$$\mathbf{H}(j\omega_d) \simeq \frac{k_{11}}{\alpha} \tag{13.101}$$

By substituting Eq. (13.101) into Eq. (13.100), we obtain the final result:

$$\mathbf{H}(j\omega) \simeq \frac{1}{1 + j\gamma}\mathbf{H}(j\omega_d) \tag{13.102}$$

where

$$\gamma = \frac{\omega - \omega_d}{\alpha} \tag{13.103}$$

The factor $1/(1 + j\gamma)$ is referred to as the *normalized universal resonance characteristic.* The normalized amplitude and phase characteristics are plotted in Fig. 13.33. These plots describe the approximate shapes of the amplitude and phase characteristics of *every* circuit in the vicinity of a resonance peak caused by an isolated simple pole near the $j\omega$ axis of the s plane. The peak in the universal amplitude characteristic occurs when $\gamma = 0$, or, equivalently, when $\omega = \omega_d$. The half-power frequencies occur when $\gamma = 1$, or, equivalently, when $\omega = \omega_d \pm \alpha$. Therefore the half-power bandwidth associated with the universal resonance peak is $\mathrm{BW} = 2\alpha$. The horizontal axis in Fig. 13.33 can be interpreted as the number of *half bandwidths,* $\alpha = \frac{1}{2}\mathrm{BW}$, by which ω exceeds the maximum-response resonance frequency. For example, if ω is three half bandwidths above ω_d, then $\omega = \omega_d + 3\alpha$ and $\gamma = 3$.

The normalized universal resonance characteristic is important because it describes the *approximate* shape of *all* resonance peaks caused by isolated simple poles close to the $j\omega$ axis. The universal resonance characteristic does not apply, however, for frequencies that are far removed from resonance. To show this, consider the following voltage transfer functions associated with the series *RLC* circuit of Fig. 13.18.

$$\mathbf{H}_R(s) = \frac{\mathbf{V}_R}{\mathbf{V}_s} = \frac{sR/L}{(s - s_{p1})(s - s_{p2})} \tag{13.104}$$

$$\mathbf{H}_C(s) = \frac{\mathbf{V}_C}{\mathbf{V}_s} = \frac{1/LC}{(s - s_{p1})(s - s_{p2})} \tag{13.105}$$

$$\mathbf{H}_L(s) = \frac{\mathbf{V}_L}{\mathbf{V}_s} = \frac{s^2}{(s - s_{p1})(s - s_{p2})} \tag{13.106}$$

† We assume that $\alpha > 0$ (stable circuit).

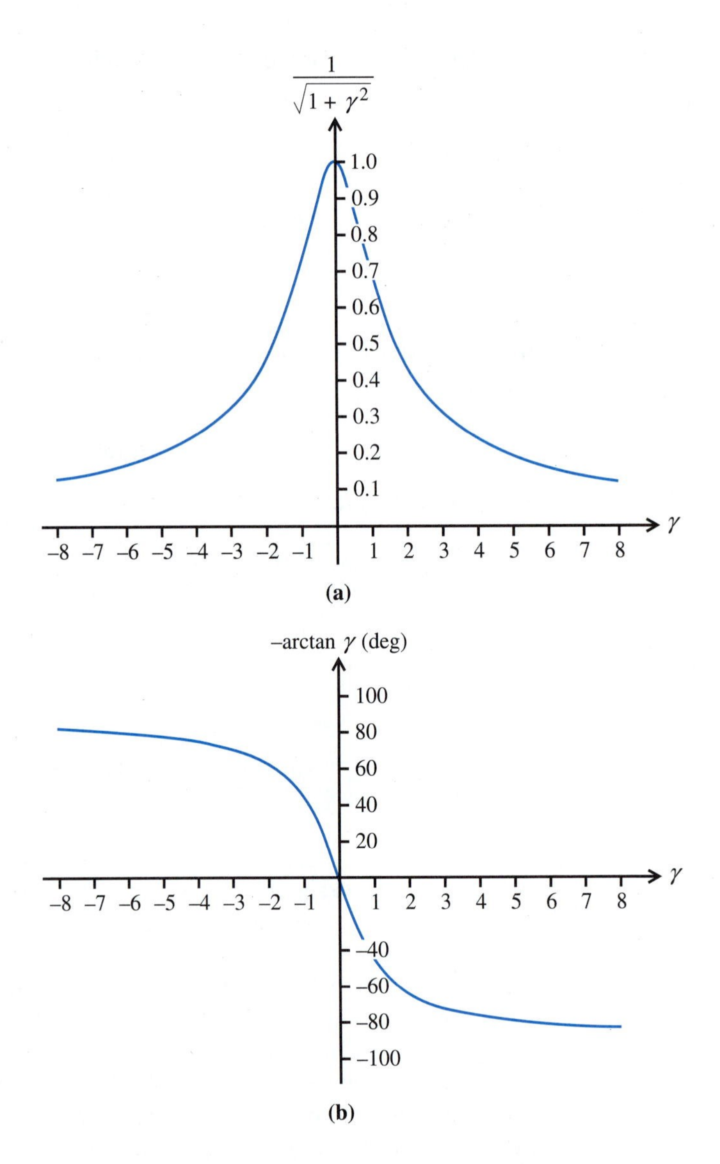

FIGURE 13.33 The normalized universal resonance characteristic. (*a*) Amplitude; (*b*) phase

The subscripts *R*, *L*, and *C* denote the circuit component across which the output voltage is measured. Recall that $\mathbf{H}_R(s)$ and $\mathbf{H}_C(s)$ were described previously in Sections 13.5 [Eq. (13.48)] and 13.3 [Eq. (13.22)], respectively. All three of these transfer functions can easily be derived by inspection of Fig. 13.18. Exact plots of the amplitude characteristics associated with the three transfer functions are shown in Fig. 13.34 for a damping ratio $\zeta = \alpha/\omega_0 = 0.05$. We can see from this figure that the locations, bandwidths, and shapes of the three resonance peaks are nearly equal. However, the amplitude characteristics are substantially different for frequencies far removed from resonance. As predicted by the universal resonance characteristic, all three maximum-response resonance frequencies approximately equal the damped natural frequency $\omega_d = \omega_0\sqrt{1 - \zeta^2}$ if the damping ratio is small.

Remember

All resonance peaks caused by isolated simple poles close to the $j\omega$ axis have approximately the same shape.

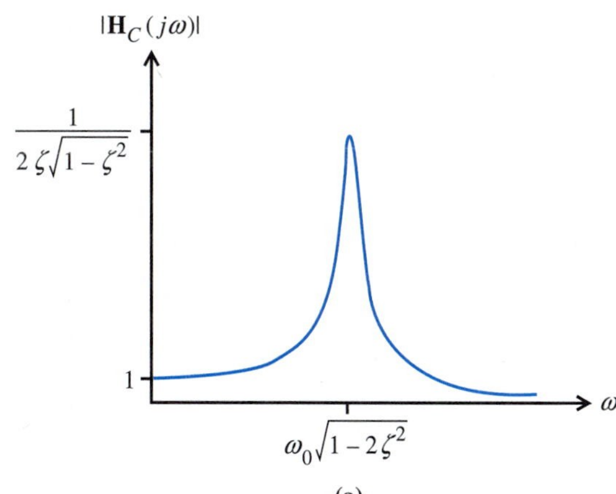

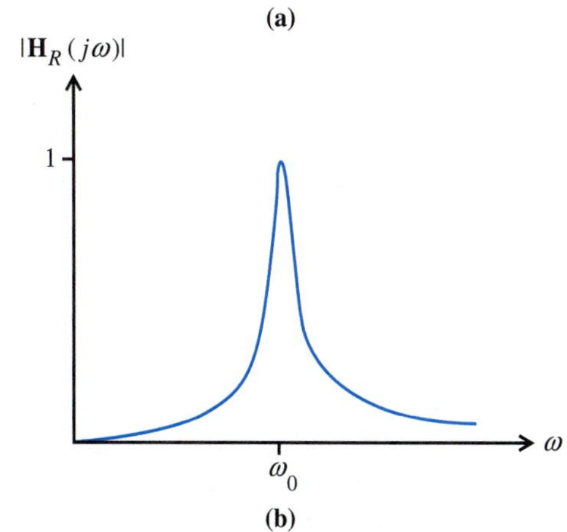

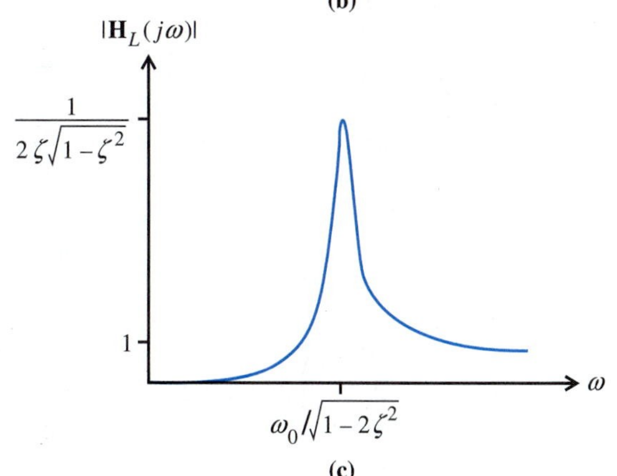

FIGURE 13.34 The amplitude characteristics (*a*) $|\mathbf{H}_C(j\omega)|$, (*b*) $|\mathbf{H}_R(j\omega)|$, and (*c*) $|\mathbf{H}_L(j\omega)|$ for $\zeta = 0.05$

EXERCISES

30. The amplitude characteristic of an unknown circuit has a peak of 20 at $\omega_r =$ 1 krad/s, with associated half-power bandwidth of 4 krad/s. Assuming that the resonance is due to an isolated simple pole, what is the value of the amplitude characteristic at $\omega = 6$ krad/s? *answer:* 7.43
31. Repeat Exercise 30 for $\omega = 8$ krad/s. *answer:* 5.49
32. Derive the expressions for the three maximum-response resonance frequencies shown in Fig. 13.34.

33. Refer again to Fig. 13.34. Explain in simple circuit terms why:
    (a) $\mathbf{H}_C(0) = 1$
    (b) $\mathbf{H}_L(\infty) = 1$
    (c) $\mathbf{H}_R(0) = \mathbf{H}_R(\infty) = 0$

13.9 Bode Plots

Pocket calculators and personal computers make it easy to accurately compute and plot a circuit's amplitude and phase characteristics $|\mathbf{H}(j\omega)|$ and $\underline{/\mathbf{H}(j\omega)}$. A standard way to present this information is by means of Bode plots. The primary advantages of our representing amplitude and phase information in the form of Bode plots are (1) Bode plots can be sketched rapidly with the use of only straight-line approximations, (2) the Bode amplitude and frequency scales make it possible to display a large range of numerical values compactly, and (3) Bode plots are an industrial standard. Another advantage, which is by no means insignificant, is that human sensory response to both audio and video stimuli seems naturally matched to the logarithmic measures used in Bode plots.

In the *Bode amplitude plot,*

$$|\mathbf{H}(j\omega)|_{\text{dB}} = 20 \log_{10} |\mathbf{H}(j\omega)| \tag{13.107}$$

is plotted versus frequency. The function $|\mathbf{H}(j\omega)|_{\text{dB}}$ has a *decibel (dB) amplitude* scale when $\mathbf{H}(j\omega)$ is dimensionless, a *decibel (dB) ohms* scale when $\mathbf{H}(j\omega)$ is an impedance in ohms, and a *decibel (dB) siemens* scale when $\mathbf{H}(j\omega)$ is an admittance in siemens. In the *Bode phase plot,* $\underline{/\mathbf{H}(j\omega)}$ is plotted (in degrees) versus frequency. Both amplitude and phase plots are drawn on semilog graph paper.

Bode plots are based on the fact that only seven different factors appear in any transfer function. these factors are a gain K, poles and zeros at the origin, poles and zeros on the real axis, and complex-conjugate poles and zeros. In making a Bode plot, we make straight-line plots of each of these factors. (A straight-line plot is just a plot of straight-line segments.) We then construct a straight-line plot of the entire transfer function by *adding* the straight-line magnitude plots and *adding* the straight-line phase plots. The result is a straight-line approximation to the actual amplitude and phase characteristics. The final step is to use a table of correction factors to obtain the actual characteristics.

We will explain how to draw Bode plots by starting with a few simple examples. It is important to study the following three examples carefully, because they provide the information you will need to progress easily to the general case.

EXAMPLE 13.4 Construct a Bode plot for the low-pass *RC* filter of Fig. 13.35.

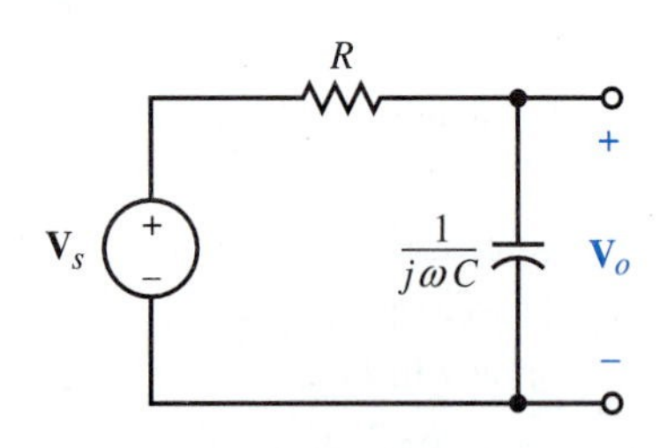

FIGURE 13.35
Low-pass filter

Solution The voltage transfer function is

$$\mathbf{H}_{LP}(s) = \frac{\dfrac{1}{sC}}{R + \dfrac{1}{sC}} = \frac{1}{s\tau + 1}$$

where $\tau = RC$ is the time constant. We see that $\mathbf{H}_{LP}(s)$ contains a real-axis pole at $s_{p1} = -1/\tau$. By setting $s = j\omega$, we obtain

$$\mathbf{H}_{LP}(j\omega) = \frac{1}{j\omega\tau + 1} = \frac{1}{|j\omega\tau + 1|} \underline{/-\text{arc tan } \omega\tau}$$

The dB amplitude characteristic, Eq. (13.107), is

$$|\mathbf{H}_{LP}(j\omega)|_{\text{dB}} = 20 \log |\mathbf{H}_{LP}(j\omega)| = -20 \log |j\omega\tau + 1|$$

In Bode's method, we start by approximating the dB amplitude characteristic by straight lines. The straight-line approximation to $|\mathbf{H}_{LP}(j\omega)|_{\text{dB}}$ is obtained from the fact that

$$|j\omega\tau + 1| = \sqrt{(\omega\tau)^2 + 1} \simeq \begin{cases} 1 & \text{for } \omega \ll \dfrac{1}{\tau} \\ \omega\tau & \text{for } \omega \gg \dfrac{1}{\tau} \end{cases}$$

Therefore

$$|\mathbf{H}_{LP}(j\omega)|_{\text{dB}} \simeq \begin{cases} 0 & \text{for } \omega \ll \dfrac{1}{\tau} \\ -20 \log \omega\tau & \text{for } \omega \gg \dfrac{1}{\tau} \end{cases}$$

The functions appearing on the right-hand side of this equation are the low-frequency and high-frequency asymptotes of $|\mathbf{H}_{LP}(j\omega)|_{\text{dB}}$. Both are straight lines when plotted versus $\log \omega$ and are shown as black lines in Fig. 13.36a.

The frequency at which the two asymptotes meet is called a *corner frequency*. The corner frequency can be found if we solve

$$-20 \log \omega\tau = 0$$

which yields

$$\omega = \frac{1}{\tau}$$

Thus the corner frequency is just the reciprocal of the time constant.

The slope of the asymptote $-20 \log \omega\tau$ can be obtained if we notice that if ω is increased by a factor of 2, then $-20 \log \omega\tau$ decreases by $20 \log 2 = 6.02$ dB:

$$-20 \log 2\omega\tau = -20 \log \omega\tau - 20 \log 2 = -20 \log \omega\tau - 6.02 \text{ dB}$$

A factor of 2 in frequency is called an *octave*. Therefore the asymptote $-20 \log \omega\tau$ has a slope of approximately -6 dB *per octave*. Similarly, if we increase ω by a factor of 10 (a *decade*), then $-20 \log \omega\tau$ decreases by $20 \log 10 = 20$ dB:

$$-20 \log 10\omega\tau = -20 \log \omega\tau - 20 \log 10 = -20 \log \omega\tau - 20 \text{ dB}$$

Therefore 6.02 dB per octave is equivalent to 20 dB per decade.

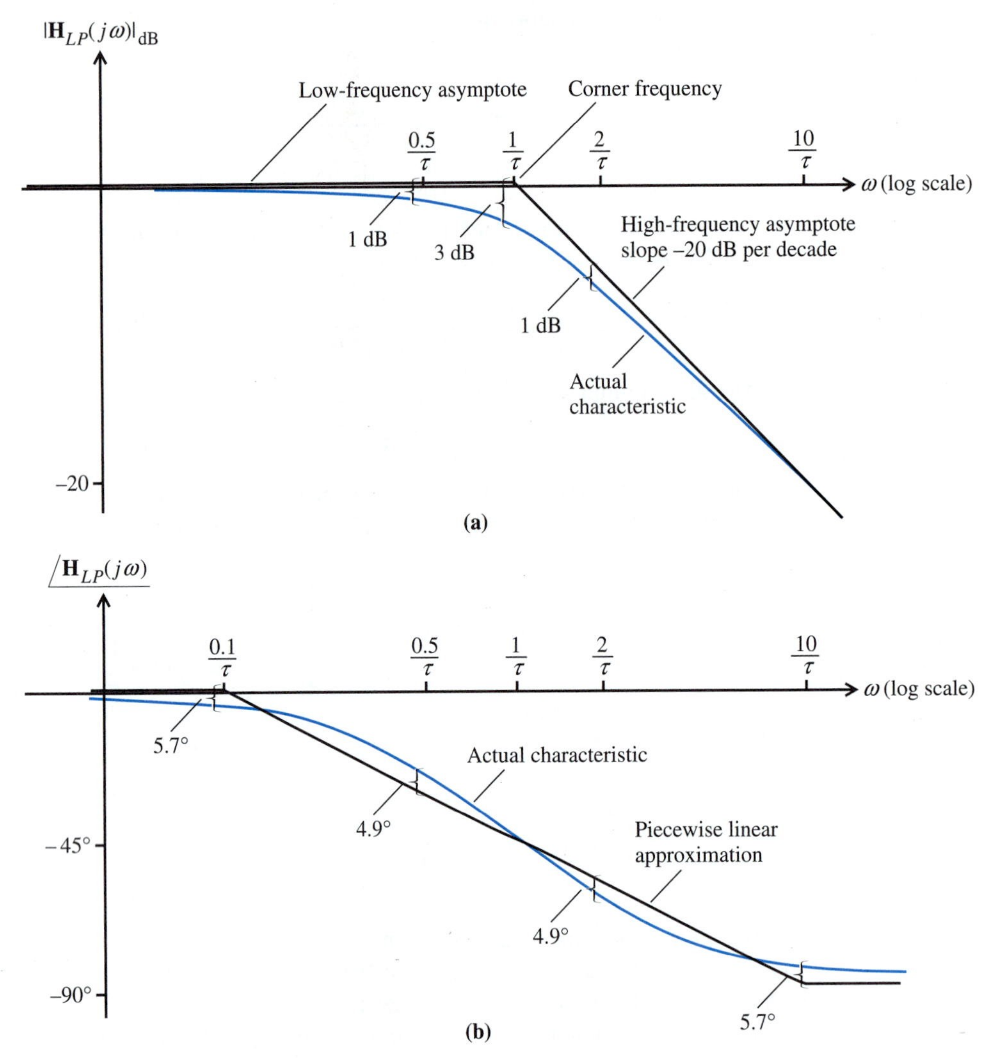

FIGURE 13.36
Bode plot of $\mathbf{H}_{LP}(j\omega)$. (*a*) Amplitude; (*b*) phase

To examine how good the straight-line approximation to $|\mathbf{H}_{LP}(j\omega)|_{dB}$ is, we have plotted the exact amplitude characteristic as the blue curve in Fig. 13.36a. We see from this figure that the actual characteristic lies below the approximate characteristic. The approximation error is *geometrically symmetrical* about the corner frequency $\omega = 1/\tau$. This means that the error at $\omega = \beta/\tau$ is the same as the error at $\omega = 1/\beta\tau$, where β is any positive number. The approximation error is 3 dB at $\omega = 1/\tau$, 2 dB at $\omega = 0.76/\tau$ and $\omega = 1.31/\tau$, and 1 dB at $\omega = 0.5/\tau$ and $2/\tau$. These numbers are recorded in Table 13.2. We will use this table to correct similar piecewise-linear characteristics in the future.

The phase characteristic of the low-pass circuit is given by

$$/\mathbf{H}_{LP}(j\omega) = -\arctan \omega\tau$$

Bode's straight-line phase plot is based on the fact that

TABLE 13.2 Magnitudes of corrections for corners due to first-order† real poles or zeros

$\omega\tau$	±*Amplitude Correction (dB)*	±*Phase Correction (Degrees)*
1	3	0
0.76 and 1.31	2	2.4
0.5 and 2	1	4.9
0.1 and 10	0.04	5.7

† Multiply the corrections by N for an Nth-order real pole or zero.

$$-\arctan \omega\tau \simeq \begin{cases} 0^\circ & \text{for } \omega < \dfrac{0.1}{\tau} \\ -90^\circ & \text{for } \omega > \dfrac{10}{\tau} \end{cases}$$

It follows from this that for $\omega < 0.1/\tau$ and for $\omega > 10/\tau$, we may approximate $\underline{/\mathbf{H}_{LP}(j\omega)}$ by the constants 0° and −90°, respectively. These constants are plotted as the horizontal black lines in Fig. 13.36b. These lines are the low- and high-frequency asymptotes of the phase characteristic. To complete the straight-line phase plot, we join the two horizontal asymptotes with a line that passes through −45° at $\omega = 1/\tau$ and intersects the asymptotes at $\omega = 0.1/\tau$ and $\omega = 10/\tau$. The slope of this line is −45° per decade. The exact phase characteristic is plotted as the blue curve in Fig. 13.36b.

By examining Fig. 13.36b, we see that the approximation error is geometrically symmetrical about $\omega = 1/\tau$. The magnitude of the error is 5.7° at the two *breakpoints*, $\omega = 0.1/\tau$ and $\omega = 10/\tau$, and 4.9° at $\omega = 0.5/\tau$ and $\omega = 2/\tau$. These numbers are included in Table 13.2.

EXAMPLE 13.5 Construct a Bode plot for the high-pass *RL* filter of Fig. 13.37.

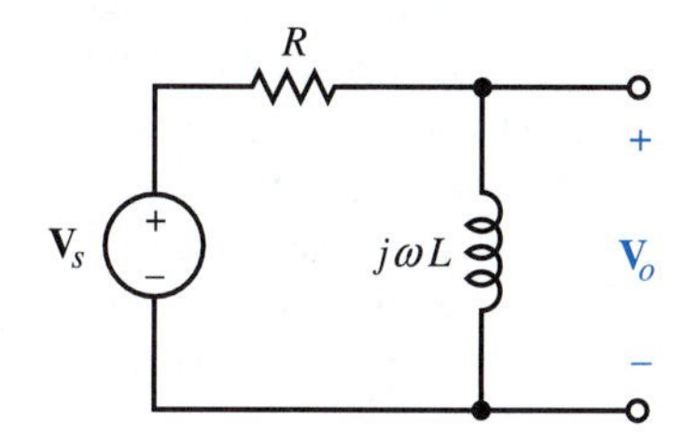

FIGURE 13.37 High-pass filter

Solution The voltage transfer function is

$$\mathbf{H}_{HP}(s) = \frac{sL}{R + sL} = \frac{s\tau}{s\tau + 1}$$

where $\tau = L/R$ is the time constant. Notice that the basic difference between $\mathbf{H}_{HP}(s)$ and $\mathbf{H}_{LP}(s)$ of the previous example is the numerator factor, $s\tau$, in $\mathbf{H}_{HP}(s)$. This factor means, of course, that $\mathbf{H}_{HP}(s)$ contains a zero at the origin. By setting $s = j\omega$, we have

$$\mathbf{H}_{HP}(j\omega) = \frac{j\omega\tau}{j\omega\tau + 1} = \frac{\omega\tau}{|j\omega\tau + 1|} \underline{/\frac{\pi}{2} - \arctan \omega\tau}$$

The dB amplitude characteristic, Eq. (13.107), is

$$|\mathbf{H}_{HP}(j\omega)|_{\mathrm{dB}} = 20 \log \frac{\omega\tau}{|j\omega\tau + 1|} = 20 \log \omega\tau - 20 \log |j\omega\tau + 1|$$

Since the logarithm of a *product* equals the *sum* of logarithms, the dB amplitude characteristic equals the sum of two functions. We have already plotted the function $-20 \log |j\omega\tau + 1|$ in Fig. 13.36a. The other function, $20 \log \omega\tau$, is a straight line with a slope of 6.02 dB per octave (20 dB per decade) when plotted versus $\log \omega$. The location of this line can be determined from the fact that $\log \omega\tau = 0$ at $\omega = 1/\tau$. The line $20 \log \omega\tau$ is plotted in Fig. 13.38 along with the straight-line approximation to $-20 \log |j\omega\tau + 1|$ obtained in Example 13.4.

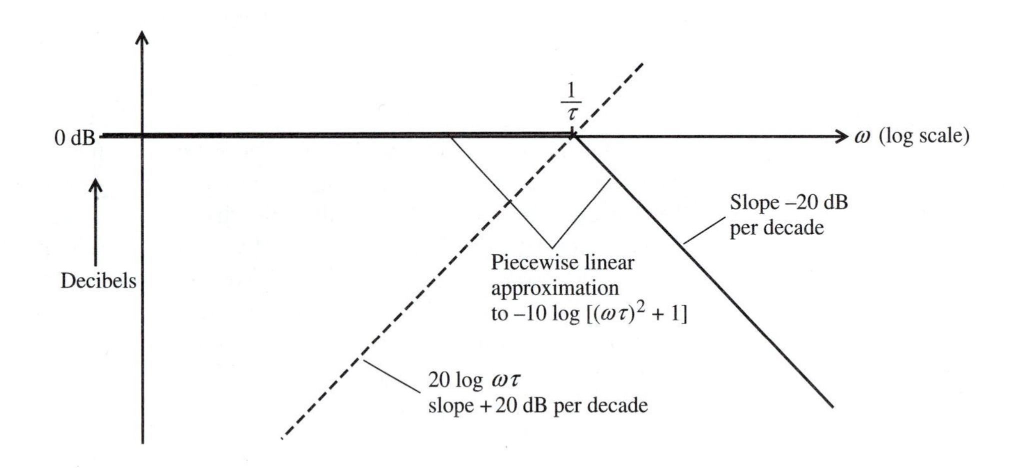

FIGURE 13.38 Pertinent to the construction of $|\mathbf{H}_{HP}(j\omega)|$

We can obtain a straight-line approximation to $|\mathbf{H}(j\omega)|_{\mathrm{dB}}$ by simply adding the two straight-line plots of Fig. 13.38. This sum is plotted as black lines on Fig. 13.39a. The exact amplitude characteristic is plotted as the blue curve.

We can see from Fig. 13.39 that the approximation error is 3 dB at the corner frequency $\omega = 1/\tau$ and geometrically decreases symmetrically on either side of $\omega = 1/\tau$ in agreement with the entries in Table 13.2.

The phase characteristic of the high-pass filter is

$$\underline{/\mathbf{H}_{HP}(j\omega)} = \frac{\pi}{2} - \arctan \omega\tau$$

which is the sum of the constant $\pi/2$ plus the function $-\arctan \omega\tau$ described in the preceding example. The straight-line approximation to $\underline{/\mathbf{H}_{HP}(j\omega)}$ is shown in Fig. 13.39b along with the exact phase curve. The difference between the two is in agreement with Table 13.2.

Now that we have considered the Bode plots of simple low-pass and high-pass filters, let us consider that of a resonant *RLC* circuit.

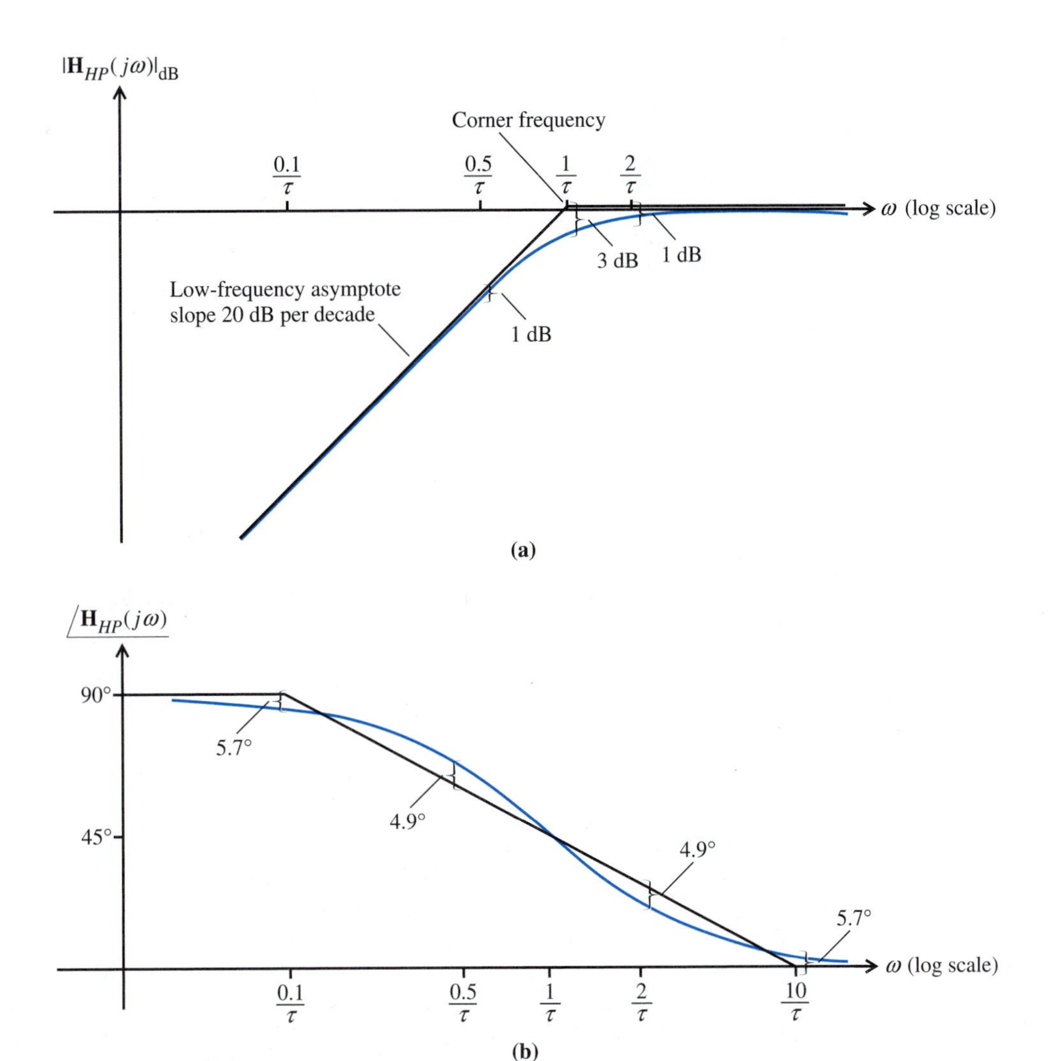

FIGURE 13.39
Bode plot of $\mathbf{H}_{HP}(j\omega)$. (*a*) Amplitude; (*b*) phase

EXAMPLE 13.6 Construct a Bode plot of the voltage transfer function for the *RLC* circuit of Fig. 13.40. Assume that the circuit is underdamped.

FIGURE 13.40
Circuit for Example 13.6

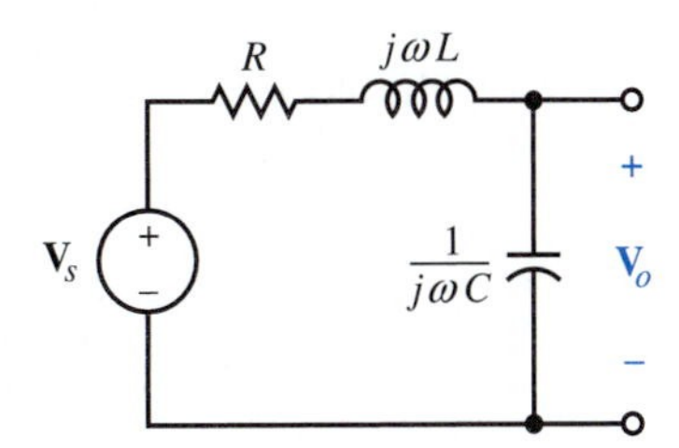

Solution The transfer function is given by

$$\mathbf{H}(s) = \frac{1/sC}{R + sL + (1/sC)} = \frac{1}{s^2LC + sRC + 1} = \frac{1/LC}{(s - s_{p1})(s - s_{p2})}$$

For an underdamped circuit, the poles are complex and are given by $-\alpha \pm j\omega_d$, where $\alpha = R/2L$ and $\omega_d = \sqrt{\omega_0^2 - \alpha^2}$ with $\omega_0 = 1/\sqrt{LC}$. If we set $s = j\omega$ and define $T = \sqrt{LC}$, the above becomes

$$\mathbf{H}(j\omega) = \frac{1}{1 - \omega^2T^2 + j\omega 2\zeta T}$$

where $\zeta = RC/2T$ is the damping ratio. The amplitude characteristic in decibels is

$$|\mathbf{H}(j\omega)|_{\text{dB}} = -20 \log |1 - \omega^2T^2 + j\omega 2\zeta T|$$

Because

$$|1 - \omega^2T^2 + j\omega 2\zeta T| = \sqrt{(1 - \omega^2T^2)^2 + (\omega 2\zeta T)^2} \simeq \begin{cases} 1 & \text{for } \omega \ll \dfrac{1}{T} \\ \omega^2T^2 & \text{for } \omega \gg \dfrac{1}{T} \end{cases}$$

then $|\mathbf{H}(j\omega)|_{\text{dB}}$ is an asymptotically linear function of $\log \omega$ for both low and high frequencies:

$$|\mathbf{H}(j\omega)|_{\text{dB}} \simeq \begin{cases} 0 & \text{for } \omega \ll \dfrac{1}{T} \\ -40 \log \omega T & \text{for } \omega \gg \dfrac{1}{T} \end{cases}$$

The high-frequency asymptote, $-40 \log \omega T$, has slope -12.04 dB per octave (or -40 dB per decade) and meets the low-frequency asymptote at the corner frequency $\omega = 1/T$, as illustrated in Fig. 13.41a (black lines).

Exact plots of the function $-20 \log |1 - \omega^2T^2 + j\omega 2\zeta T|$ are shown in Fig. 13.42a; these can be used to correct the straight-line approximation. We will illustrate this process by assuming that in the present example, $T = 0.01$ and $\zeta = 0.1$. We then find from Fig. 13.42a on page 468 that the actual characteristic is approximately 14 dB higher than the piecewise-linear approximation at $\omega = 100$, 9 dB higher at $\omega = 80$ and 125, and 2 dB higher at $\omega = 45$ and 222. The blue curve in Fig. 13.41a results when these corrections are added to the piecewise-linear approximation.

The Bode straight-line approximation to the phase characteristic

$$\underline{/\mathbf{H}(j\omega)} = -\arctan\left(\frac{\omega 2\zeta T}{1 - \omega^2T^2}\right)$$

is given by the black lines in Fig. 13.41b. The outer two black lines in the approximation are based on the fact that

$$-\arctan\left(\frac{\omega 2\zeta T}{1 - \omega^2T^2}\right) \simeq \begin{cases} 0^\circ & \text{for } \omega < \dfrac{0.1}{T} \\ -180^\circ & \text{for } \omega > \dfrac{10}{T} \end{cases}$$

The constants 0° and -180° are, respectively, the low-frequency and high-frequency asymptotes of $-\text{arc tan }[\omega 2\zeta T/(1 - \omega^2T^2)]$. The middle segment of the straight-line approximation intersects the outer horizontal asymptotes at the breakpoints $\omega = 0.1/T$ and $\omega = 10/T$. Thus, it has a slope of -90° per decade.

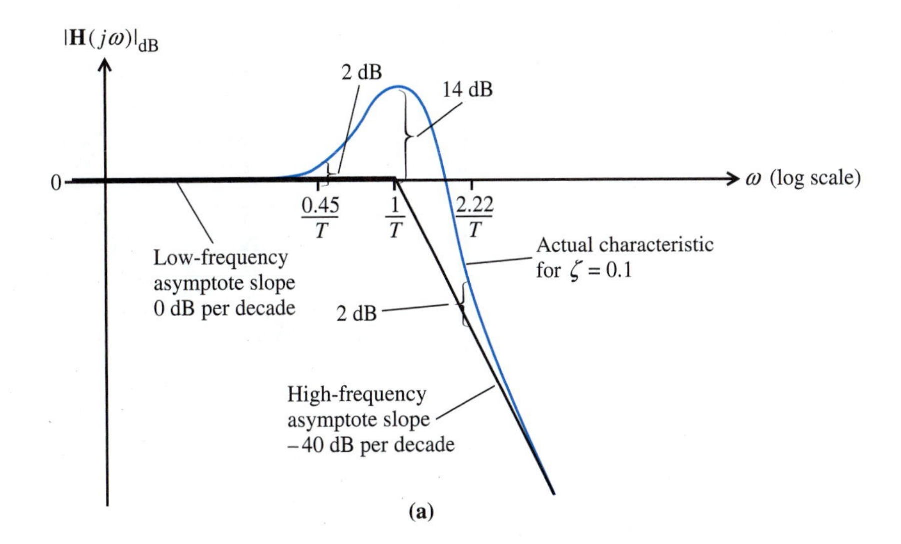

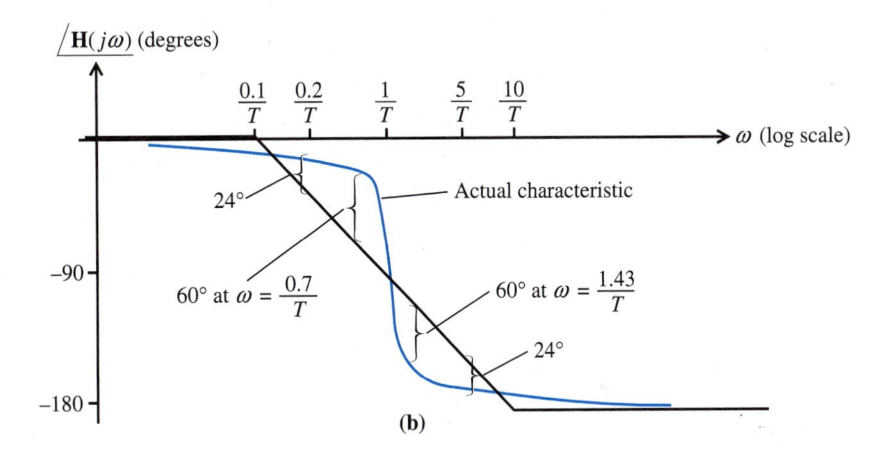

FIGURE 13.41 Bode plot of $\mathbf{H}(j\omega)$. (*a*) Amplitude; (*b*) phase. The black lines depict the piecewise-linear approximations. The blue curves are the exact characteristics for $\zeta = 0.1$

Exact plots of $/\mathbf{H}(j\omega)$ are shown in Fig. 13.42b. To illustrate the use of Fig. 13.42b, we again assume that $T = 0.01$ and $\zeta = 0.1$. We see from Fig. 13.42b that the magnitude of the phase correction is approximately 24° at $\omega = 20$ and 500, 40° at $\omega = 30$ and 333, and 60° at $\omega = 70$ and 143. The corrected phase characteristic is shown as the blue curve in Fig. 13.41b.

In the preceding three examples we showed how to construct Bode plots for transfer functions with a real-axis pole, a real-axis pole and a zero at the origin, and a pair of complex-conjugate poles. Equipped with this experience, we now progress to the general case.† The transfer function under consideration is

$$\mathbf{H}(s) = \frac{\mathcal{B}(s)}{\mathcal{A}(s)} \tag{13.108}$$

The amplitude characteristic is drawn in four steps.

† We limit our description to the most common class of transfer functions, called *minimum-phase* transfer functions. The zeros of a minimum-phase transfer function all lie in the left half of the s plane.

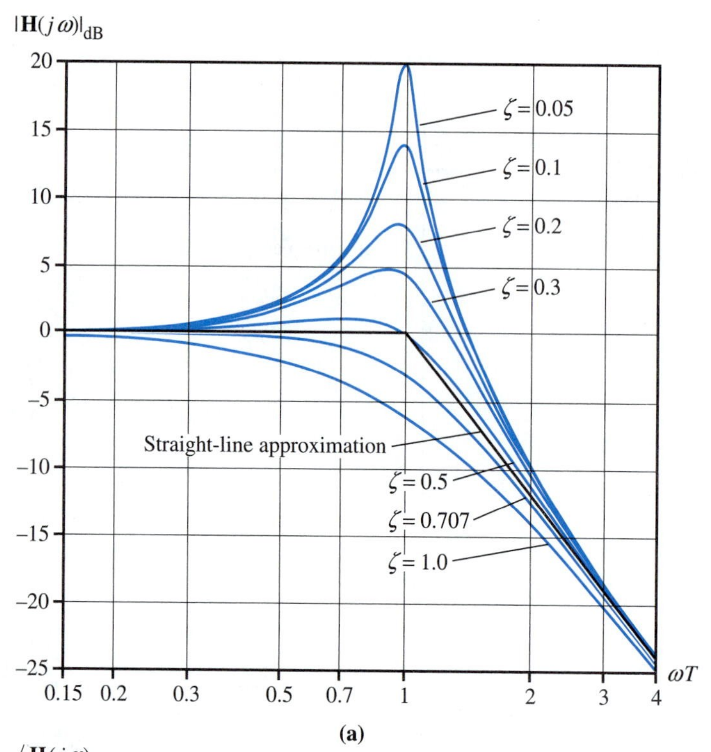

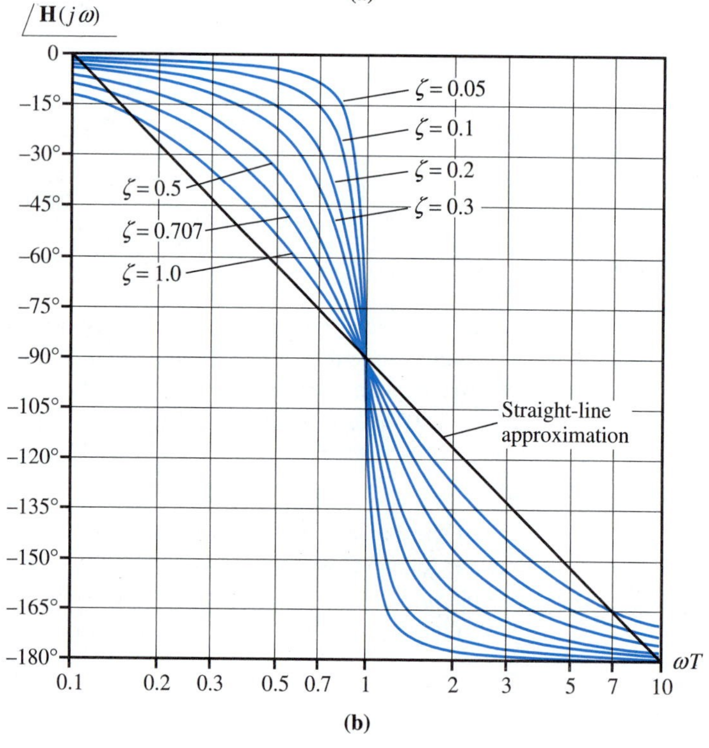

FIGURE 13.42 Bode plot of $\mathbf{H}(j\omega) = 1/(1 - \omega^2T^2 + j\omega2\zeta T)$. (*a*) Amplitude; (*b*) phase

Step 1 *Write* $\mathbf{H}(s)$ *as a product of factors of the forms*

(a) Constant factor K

(b) Poles or zeros at the origin $s^{\pm N}$

(c) Real poles or zeros not at the origin $(s\tau + 1)^{\pm N}$

(d) Complex-conjugate poles or zeros $(s^2T^2 + s2\zeta T + 1)^{\pm N}$

Step 2 *Recognize that the amplitude characteristic in decibels,* $|\mathbf{H}(j\omega)|_{\mathrm{dB}} = 20 \log |\mathbf{H}(j\omega)|$, *is the* sum *of terms corresponding to forms a through d above.*

(a) Constant: $20 \log |K|$

(b) Straight line: $\pm 20N \log |j\omega| = \pm 20N \log \omega$

(c) Curves: $\pm 20N \log |j\omega\tau + 1|$

(d) Curves: $\pm 20N \log |-\omega^2T^2 + 1 + j\omega 2\zeta T|$

In (b) through (d), the + is associated with zeros, whereas the − is associated with poles. The curves in (b) through (d) are the same as those described in the three preceding examples except for the factor $\pm N$ and are readily approximated by straight-line characteristics. Draw the straight-line approximations for the individual forms (a) through (d) in Step 1.

Step 3 *Draw the actual curves with the forms a through d in Step 1 by applying the corrections of Table 13.2 to the straight-line approximations or by using Table 13.3.*

Step 4 *Obtain the straight-line and the actual characteristic for* $|\mathbf{H}(j\omega)|_{\mathrm{dB}}$ *by adding the straight-line and the actual curves, respectively, from Step 3.*

A similar procedure is used for the construction of the phase characteristic. Table 13.3 summarizes the straight-line amplitude and phase approximations needed.

The following example demonstrates the procedure for an op-amp circuit whose transfer function contains real poles and zeros.

EXAMPLE 13.7 Draw the Bode plots for the voltage amplifier of Fig. 13.43. Assume that $R_1 = 400\ \Omega$, $R_F = 10\ \mathrm{k}\Omega$, $C_1 = 100\ \mu\mathrm{F}$, $C_F = 2\ \mu\mathrm{F}$, and $A = \infty$ (ideal op amp).

FIGURE 13.43
Circuit for Example 13.7

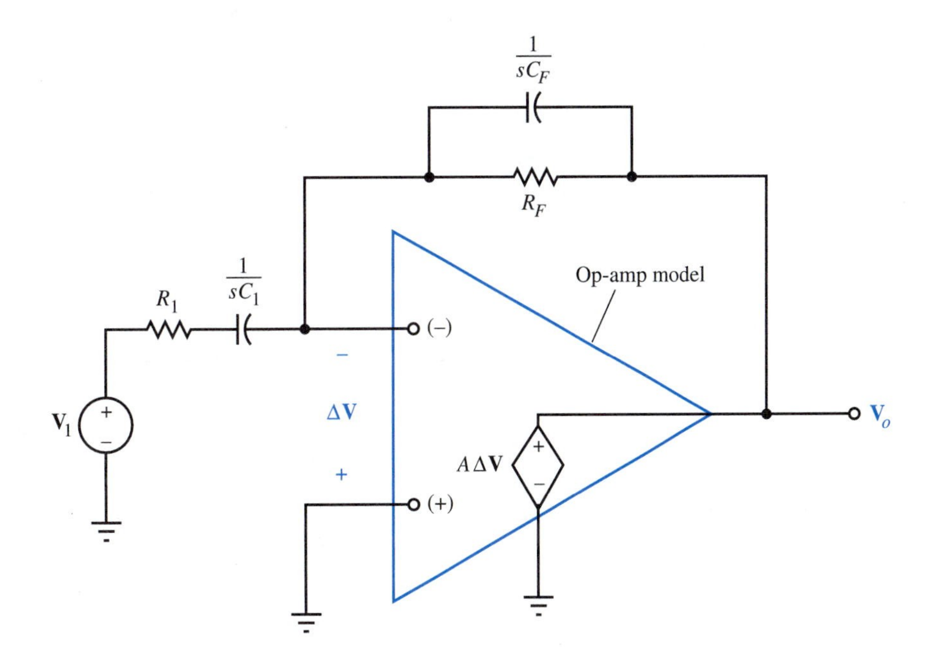

TABLE 13.3 Summary of Bode straight-line amplitude and phase approximations (20 dB/decade is equivalent to 6.02 dB/octave)

Factor	Pole-Zero Plot	dB Amplitude versus ω(log scale)	Phase versus ω(log scale)
s^N	$j\omega$; Nth order; σ	$20N$ dB/decade; 1; ω	$90N°$; ω
$\frac{1}{s^N}$	$j\omega$; Nth order; σ	$-20N$ dB/decade; 1; ω	ω; $-90N°$
$(\tau s + 1)^N$	$j\omega$; Nth order; σ	$20N$ dB/decade; ω; $\frac{1}{\tau}$	$90N°$; $0°$; ω; $\frac{0.1}{\tau}$; $\frac{1}{\tau}$; $\frac{10}{\tau}$
$\frac{1}{(\tau s + 1)^N}$	$j\omega$; Nth order; σ	$\frac{1}{\tau}$; ω; $-20N$ dB/decade	$\frac{0.1}{\tau}$; $\frac{1}{\tau}$; $\frac{10}{\tau}$; ω; $0°$; $-90N°$
$(s^2T^2 + s2\zeta T + 1)^N$	Nth order; $j\omega$; σ	$40N$ dB/decade; ω; $\frac{1}{T}$	$180N°$; $0°$; ω; $\frac{0.1}{T}$; $\frac{1}{T}$; $\frac{10}{T}$
$\frac{1}{(s^2T^2 + s2\zeta T + 1)^N}$	Nth order; $j\omega$; σ	$\frac{1}{T}$; ω; $-40N$ dB/decade	$\frac{0.1}{T}$; $\frac{1}{T}$; $\frac{10}{T}$; ω; $0°$; $-180N°$

Solution

Step 1 The first step is to obtain the transfer function and to write it in the form of Step 1 of the procedure. The KCL equation at the (−) op-amp terminal is

See Problem 13.5 in the PSpice manual.

$$\frac{-\Delta\mathbf{V} - \mathbf{V}_1}{R_1 + \dfrac{1}{sC_1}} + \frac{-\Delta\mathbf{V} - \mathbf{V}_o}{R_F \,\|\, \dfrac{1}{sC_F}} = 0$$

where $\Delta\mathbf{V} = \mathbf{V}_o/A = 0$ because $A \to \infty$ for an ideal op amp. If we set $\Delta\mathbf{V} = 0$, the above rearranges to

$$\frac{\mathbf{V}_o}{\mathbf{V}_1} = \mathbf{H}(s) = -\left(\frac{sC_1}{s\tau_1 + 1}\right)\left(\frac{R_F}{s\tau_F + 1}\right)$$

where $\tau_1 = R_1C_1$ and $\tau_F = R_FC_F$. For the numbers given, the transfer function is

$$\mathbf{H}(s) = -\frac{s}{(0.02s + 1)(0.04s + 1)}$$

Step 2 We now write the transfer function as

$$|\mathbf{H}(j\omega)|_{\text{dB}} = 20 \log \omega - 10 \log |0.02j\omega + 1| - 10 \log |0.04j\omega + 1|$$

and

$$\underline{/\mathbf{H}(j\omega)} = -90° - \underline{/0.02j\omega + 1} - \underline{/0.04j\omega + 1}$$

The straight-line approximations (black) and exact curves (blue) for each of the terms in $|\mathbf{H}(j\omega)|_{\text{dB}}$ and $\underline{/\mathbf{H}(j\omega)}$ are shown in Fig. 13.44. We obtain the straight-line approximations to $|\mathbf{H}(j\omega)|_{\text{dB}}$ and $\underline{/\mathbf{H}(j\omega)}$ by adding the straight-line approximations from Fig. 13.44. The results are given by the black line in Fig. 13.45. The actual amplitude and phase characteristics, shown in blue in Fig. 13.45, are obtained by adding the blue curves from Fig. 13.44.

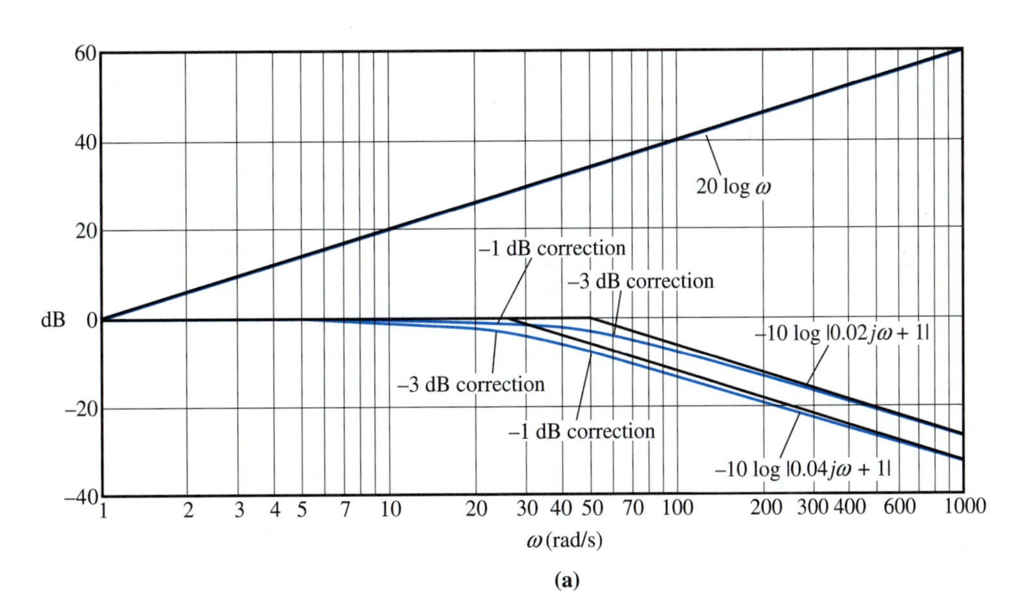

FIGURE 13.44
Bode plot components for Example 13.7. (*a*) Individual terms in $|\mathbf{H}(j\omega)|_{\text{dB}}$

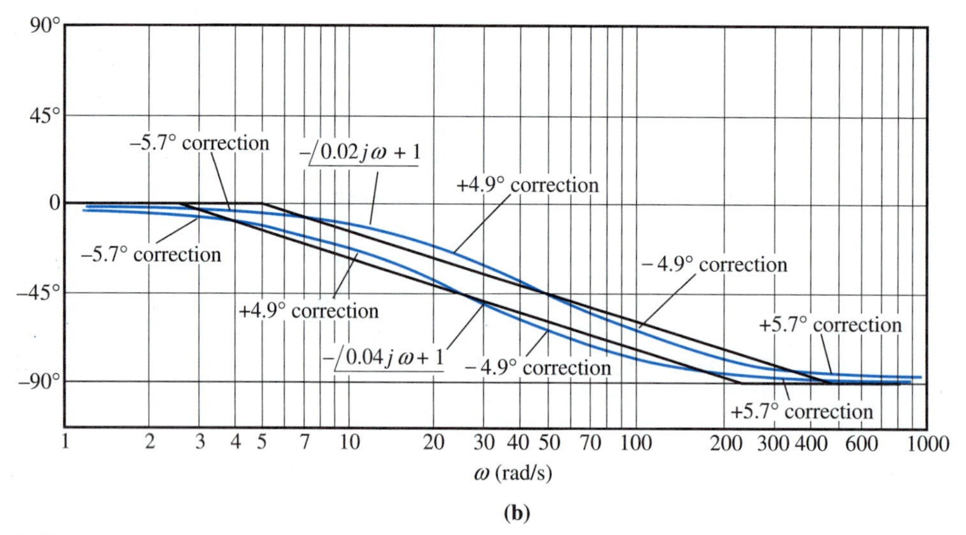

FIGURE 13.44
Continued (*b*) individual terms in $\underline{/\mathbf{H}(j\omega)}$

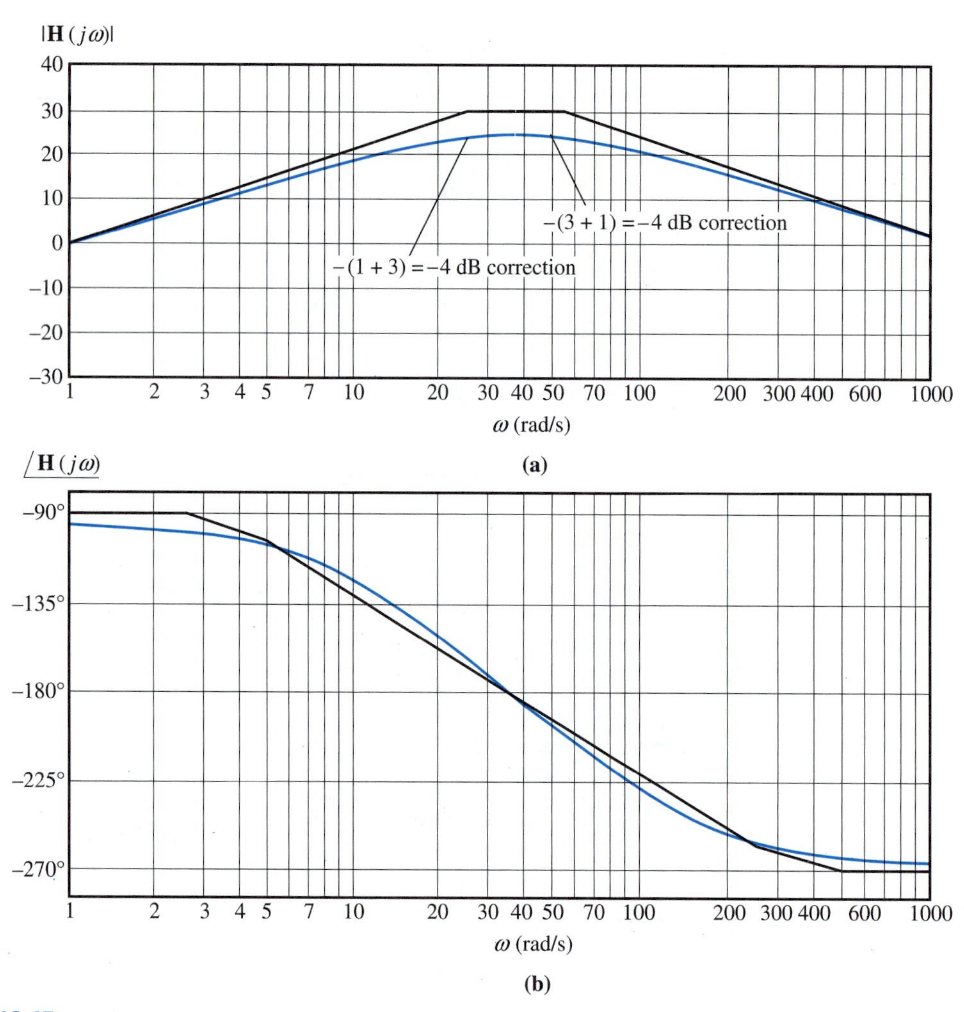

FIGURE 13.45
(*a*) Bode amplitude plot; (*b*) Bode phase plot

There is another procedure for drawing Bode amplitude plots that is often faster than the one we just discussed. The alternative procedure uses the simple fact that the *zeros* in a transfer function cause *upward* turns at corner frequencies, whereas the *poles* cause *downward* turns. The amounts of upward and downward turn depend on the orders of the zeros and poles, and whether they are real or complex. Nth-order real zeros cause upward turns of $20N$ dB per decade. Nth-order real poles cause downward turns of $20N$ dB per decade. Nth-order complex zeros and poles cause, respectively, upward and downward turns of $40N$ dB per decade. The details are best learned from an example.

EXAMPLE 13.8 Draw the Bode amplitude plot corresponding to

$$\mathbf{H}(s) = \frac{120s(s + 10^3)^2}{(s + 30)(s + 2 \times 10^4)^2[(3 \times 10^5)^{-2}s^2 + 0.2 \times (3 \times 10^5)^{-1}s + 1]}$$

Solution We draw the straight-line approximation first. The corner frequencies are determined by the poles and zeros of $\mathbf{H}(s)$. We make a *mental* note of the facts tabulated in Table 13.4. We begin the plot to the left of all the corner frequencies. Since the smallest corner frequency is 30 rad/s, we begin the plot with $\omega \ll 30$. For $|s| \ll 30$, we can neglect all the s's in $\mathbf{H}(s)$ except the one not in parentheses. That is, we approximate

$$\mathbf{H}(s) \simeq \frac{120s(10^3)^2}{(30)(2 \times 10^4)^2} = 0.01s \qquad \text{for } |s| \ll 30$$

and therefore

$$|\mathbf{H}(j\omega)|_{\text{dB}} = 20 \log |0.01j\omega| = -40 + 20 \log \omega \qquad \text{for } \omega \ll 30$$

We can conclude from this result that the zero at $s = 0$ causes a $+20$ dB per decade asymptote to the left of $\omega = 30$ rad/s as shown by the black line in Fig. 13.46. We can find the absolute level of this asymptote by substituting a convenient value of ω. Putting $\omega = 1$ gives

$$|\mathbf{H}(j1)|_{\text{dB}} = -40$$

To continue the straight-line plot, we simply increase ω, and change the slope of the line segments in accordance with the factors in $\mathbf{H}(s)$ as noted in Table 13.4. Thus, at

TABLE 13.4
Table for Example 13.8

Factor	Type	Corner Frequency (rad/s)	Slope Change (dB per decade)
s	1st-order real zero at $s = 0$	—	+20
$s + 30$	1st-order real pole at $s = -30$	30	−20
$(s + 10^3)^2$	2nd-order real zero at $s = -10^3$	10^3	+40
$(s + 2 \times 10^4)^2$	2nd-order real pole at $s = -2 \times 10^4$	2×10^4	−40
$(3 \times 10^5)^{-2}s^2 + 0.2 \times (3 \times 10^5)^{-1}s + 1$	1st-order complex pole	3×10^5	−40

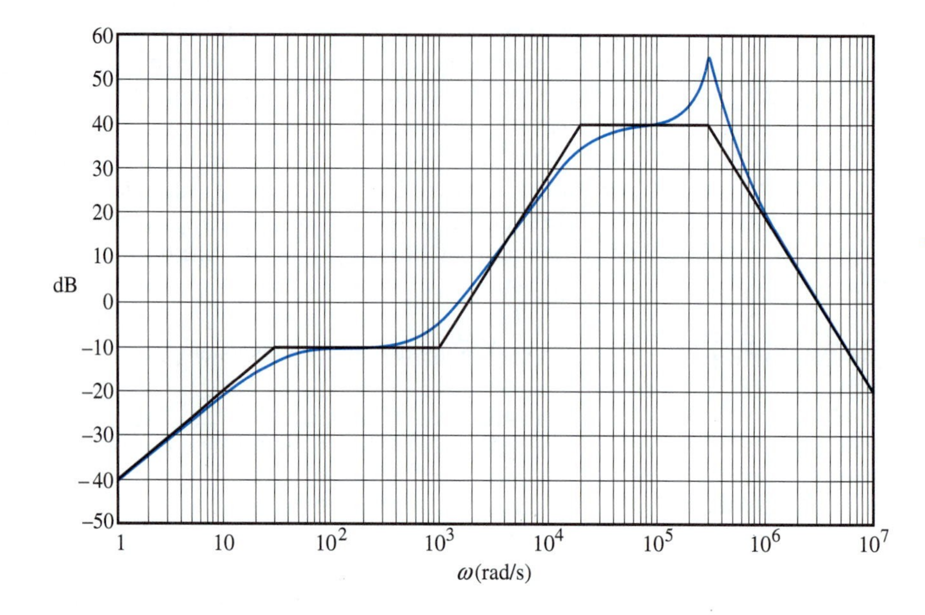

FIGURE 13.46
Bode amplitude plot for Example 13.8

$\omega = 30$ rad/s, there is a slope change of -20 dB per decade due to the first-order real pole at $s = -30$. This slope change causes a leveling off of the amplitude characteristic for $30 < \omega < 10^3$, as seen in Fig. 13.46. At $\omega = 10^3$ rad/s there is an upward turn of 40 dB per decade due to the second-order real zero at $s = -10^3$. Then, at $\omega = 2 \times 10^4$ there is a downward turn of 40 dB per decade due to the second-order real pole at $s = -2 \times 10^4$. A final slope change of -40 dB per decade occurs at $\omega = 3 \times 10^5$ rad/s due to the first-order complex pole. This completes the straight-line plot.

We next draw the actual amplitude characteristic by incorporating the corrections of Table 13.2 and Fig. 13.42a with $\zeta = 0.1$. Since the corner frequencies are all separated by more than a decade, the corrections are easy.

At $\omega = 30$, the actual characteristic is 3 dB below the straight-line approximation. At $\omega = \frac{30}{2} = 15$ and $\omega = 2 \times 30 = 60$, the actual characteristic is 1 dB below the straight-line approximation. We have a $2 \times 3 = 6$-dB correction at the corner frequency $\omega = 10^3$ because the associated zero is second-order. Also, the corrections are $2 \times 1 = 2$ dB at one octave to either side of the corner (at $\omega = 10^3/2 = 500$ and $\omega = 2 \times 10^3$). Similar corrections are made at the corner frequency $\omega = 2 \times 10^4$. We use Fig. 13.42a to help draw the actual characteristic in the vicinity of the $\omega = 3 \times 10^5$ rad/s corner frequency. We see from Fig. 13.42a with $\zeta = 0.1$ that the required correction is approximately 14 dB at the corner frequency and approximately 2 dB at $0.45 \times 3 \times 10^5 = 1.35 \times 10^5$ rad/s and $2.22 \times 3 \times 10^5 = 6.66 \times 10^5$ rad/s. The actual amplitude characteristic is given by the blue curve in Fig. 13.46.

The procedure we have just described is faster than the one on page 470 if we draw only the straight-line approximation to the Bode amplitude plot. It is also faster when we draw the actual amplitude characteristic *provided* the corner frequencies are all separated by more than a decade. When the corner frequencies are closer than a decade, however, as was the case in Example 13.7, it is usually faster to use the procedure on page 470 because we must then add together corrections from adjacent corners.

EXERCISES

34. Draw the Bode amplitude and phase plots associated with the circuit shown. Assume that $R_1 = R_2 = 10$ kΩ and $C = 1000$ μF.

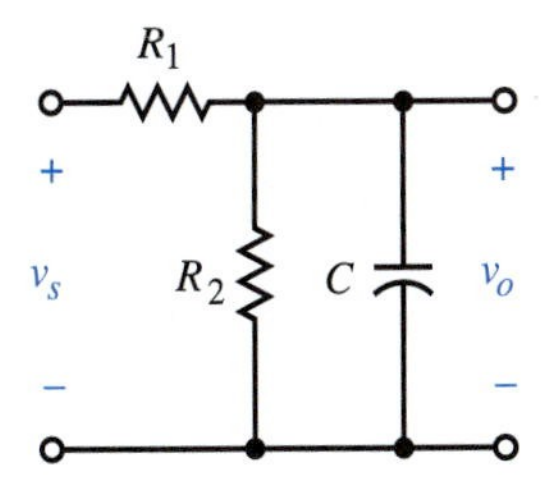

35. Repeat Exercise 34 for the circuit shown.

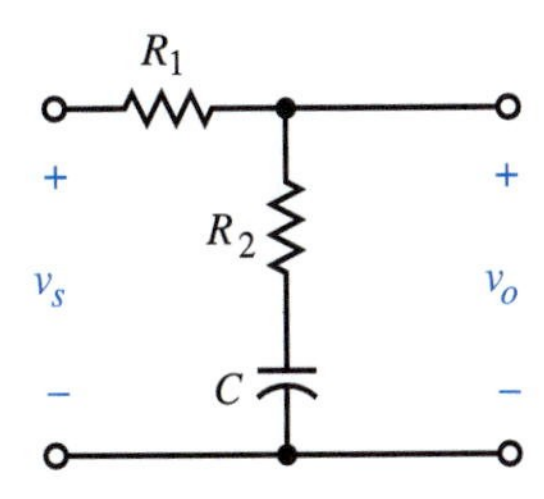

36. Repeat Exercise 34 for the circuit shown.

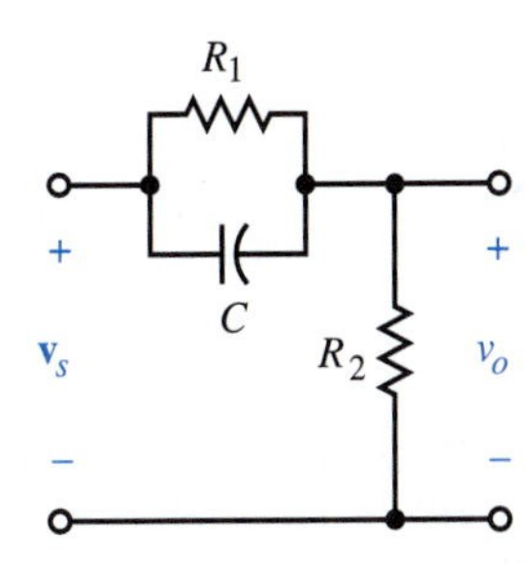

37. Repeat Exercise 34 for the circuit shown.

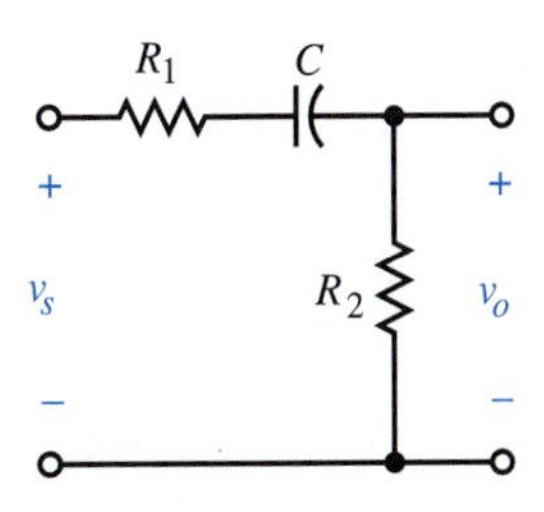

38. (a) Use an ideal op-amp model and determine the voltage transfer function $\mathbf{V}_o/\mathbf{V}_1$ of the following amplifier. Assume that $R = 1$ kΩ and $C = 1\mu$F.

(b) Sketch the Bode amplitude and phase plots.

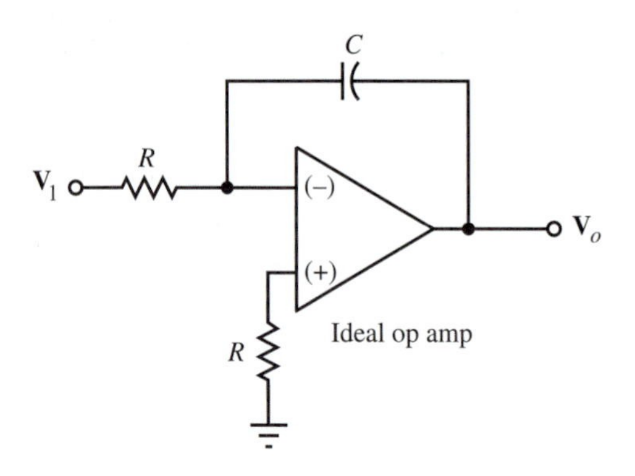

39. Draw the Bode amplitude and phase plots associated with the *RLC* circuit of Fig. 13.6. Assume that $R = 0.1\ \Omega$, $L = 1\ \mu$H, and $C = 1\ \mu$F.
40. Repeat Exercise 39 with $\mathbf{V}_o$ taken across the resistance instead of the capacitance.
41. Repeat Exercise 39 with $\mathbf{V}_o$ taken across the inductance instead of the capacitance.

13.10 Summary

In this chapter we saw that circuits can be used as filters. We described three types of filters: high-pass, low-pass, and band-pass.

To provide a better understanding of filters, we introduced the concept of complex frequency, $s = \sigma + j\omega$, and we examined the transfer function $\mathbf{H}(s) = \mathcal{B}(s)/\mathcal{A}(s)$. We showed that $\mathbf{H}(s)$ can be determined from the circuit's LTI equation and also from the circuit itself. We used spiraling phasors to show that a damped sinusoidal input $v_s(t) = V_{sm}e^{\sigma t}\cos(\omega t + \phi_s)$ produces a damped sinusoidal particular response $v_{op}(t) = V_{sm}|\mathbf{H}(s)|e^{\sigma t}\cos(\omega t + \phi_s + \angle\mathbf{H}(s))$.

We defined the poles and the zeros of $\mathbf{H}(s)$ to be the roots to $\mathcal{A}(s) = 0$ and $\mathcal{B}(s) = 0$, respectively, and we showed that $\mathbf{H}(s)$ is completely specified by its pole-zero plot. By using the rubber sheet analogy, we saw that the amplitude characteristic $|\mathbf{H}(j\omega)|$ is given by the intersection of the two-dimensional surface $|\mathbf{H}(s)|$ and the plane $\sigma = 0$. Poles near the $j\omega$ axis cause resonance peaks in the amplitude characteristic, whereas zeros cause dips. Since the poles are the roots to $\mathcal{A}(s) = 0$, the poles also determine the form of the circuit's natural response. Thus we saw that a filter's frequency response $\mathbf{H}(j\omega)$ is closely related to its natural response.

By referring to pole-zero diagrams, we showed that all resonance peaks caused by isolated simple poles near the $j\omega$ axis have approximately the same shape. An isolated simple pole, $s_{p1} = -\alpha + j\omega_d$, is near the $j\omega$ axis when $\alpha \ll \omega_d$. The maximum-response resonance frequency ω_{mr} associated with such a pole is always *approximately* ω_d, and the half-power bandwidth BW of the resonance peak is always *approximately* 2α. We showed that for a series *RLC* band-pass filter, ω_{mr} is *exactly* equal to $\omega_0 = 1/\sqrt{LC}$, and the half-power bandwidth is *exactly* equal to 2α. We also showed that for the series *RLC* circuit, $\text{BW} = \omega_0/Q_0$, where Q_0 is the quality factor of the circuit evaluated at ω_0.

We defined the unity-power-factor resonance frequency ω_{upf} as the frequency at which the circuit's driving-point impedance is purely real. In general, $\omega_{upf} \neq \omega_{mr}$.

When $\omega = \omega_{upf}$, the total average energy stored in the magnetic and the electric fields of a circuit composed of Rs, Ls, and Cs are equal.

We concluded this chapter with a description of Bode plots. The Bode plot format, in which the amplitude characteristic is plotted in decibels and the frequency axis is logarithmic, has become an industrial standard. (See Table 13.3 for a summary of Bode's straight-line amplitude and phase approximations.) Computer-generated frequency response plots often use the Bode plot format.

KEY FACTS

Concept	Equation	Section	Page
❑ Complex frequency s is defined by $s = \sigma + j\omega$, where $\sigma = \mathcal{R}e\{s\}$ is the neper frequency and $\omega = \mathcal{I}m\{s\}$ is the angular frequency.		13.3	427
❑ The poles of a transfer function $\mathbf{H}(s) = \frac{\mathcal{B}(s)}{\mathcal{A}(s)}$ are values of s for which $\mathcal{A}(s) = 0$. The poles are the roots of the characteristic equation.	(13.36)	13.4	435
❑ The zeros of $\mathbf{H}(s)$ are the values of s for which $\mathcal{B}(s) = 0$.	(13.38)	13.4	435
❑ When the complex exponential $\mathbf{V}_s e^{st}$ is applied to the LTI equation $\mathcal{A}(p)v_o(t) = \mathcal{B}(p)v_s(t)$, the particular response is $\mathbf{V}_o e^{st}$, where $\mathbf{V}_o = \mathbf{H}(s)\mathbf{V}_s$ provided that the complex input frequency s is not a pole.	(13.30)	13.3	431
❑ A complex exponential $\mathbf{V}e^{st}$, where $\mathbf{V} = V_m\angle\phi_{\mathbf{V}}$, may be interpreted as a spiraling phasor. The projection of the spiraling phasor onto the real axis is the damped sinusoid $V_m e^{\sigma t}\cos(\omega t + \phi_{\mathbf{V}})$.		13.3	430
❑ When the damped sinusoid $V_{sm}e^{\sigma t}\cos(\omega t + \phi_{\mathbf{V}_s})$ is applied to an LTI circuit, the particular response is $V_{om}e^{\sigma t}\cos(\omega t + \phi_{\mathbf{V}_o})$, where the amplitudes and phases are related by $\mathbf{V}_o = \mathbf{H}(s)\mathbf{V}_s$, with $s = \sigma + j\omega$.	(13.30) (13.33)	13.3	431
❑ The rubber sheet analogy provides a quick way to visualize the function $\lvert\mathbf{H}(s)\rvert$ as a surface above the s plane. A circuit's amplitude characteristic $\lvert\mathbf{H}(j\omega)\rvert$ is the intersection of the function $\lvert\mathbf{H}(s)\rvert$ with the $\sigma = 0$ plane.		13.4 13.3	439 429
❑ Every LTI circuit's transfer function in the vicinity of an isolated simple pole s_{p1} is given approximately by $\mathbf{H}(s) \simeq \frac{k_{11}}{s - s_{p1}}$ where	(13.95) (13.97)	13.8	456

KEY FACTS ***Concept***	***Equation***	***Section***	***Page***
$k_{11} = (s - s_{p1})\mathbf{H}(s)\Big\vert_{s=s_{p1}}$ is the residue of $\mathbf{H}(s)$ at the pole.			
❑ If the isolated pole $s_{p1} = -\alpha + j\omega_d$ is close to the $j\omega$ axis ($\alpha \ll \omega_d$), the circuit's frequency response characteristic is given approximately by $\mathbf{H}(j\omega) \simeq \dfrac{k_{11}}{j(\omega - \omega_d) + \alpha}$ for $\omega \simeq \omega_d$. We have called this characteristic the *universal resonance characteristic.*	(13.100)	13.8	457
❑ The half-power bandwidth of *every* resonance peak caused by an isolated pole $s_{p1} = -\alpha + j\omega_d$ near the $j\omega$ axis is given approximately by $\text{BW} \simeq 2\alpha$		13.8	457
❑ A circuit's maximum-response resonance frequency is the frequency at which the amplitude characteristic $\lvert\mathbf{H}(j\omega)\rvert$ is maximum.		13.2	424
❑ A circuit's unity-power-factor resonance frequency is the frequency at which the driving-point impedance is purely real.	(13.68)	13.6	447
❑ At unity-power-factor resonance, the total average energies stored in the magnetic and the electric fields of a circuit composed of Rs, Ls, and Cs are equal.		13.6	450
❑ Quality factor is defined by $Q(\omega) = 2\pi \dfrac{\text{peak energy stored in the circuit}}{\text{energy dissipated by the circuit in one period}}$	(13.73)	13.7	451
❑ For a band-pass series RLC circuit, the maximum-response and unity-power-factor resonance frequencies both occur at $\omega_0 = 1/\sqrt{LC}$ rad/s. The half-power bandwidth is given by $\text{BW} = \omega_0/Q_0$ rad/s, where $Q_0 = \omega_0 L/R$ is the quality factor evaluated at $\omega = \omega_0$.	(13.90) (13.92)	13.7	454

PROBLEMS

Section 13.1

For circuits of Problems 1 through 4 below: (a) Find the voltage transfer function, $\mathbf{V}_o/\mathbf{V}_s = \mathbf{H}(j\omega)$. (b) Use your answer to part (a) to find the expressions for $\mathbf{H}(j0)$ and $\mathbf{H}(j\infty)$. (c) Explain your answers to part (b) by inspection of the circuit with the capacitance or inductance replaced with open or short circuits. (d) Plot the amplitude and phase characteristics.

1.

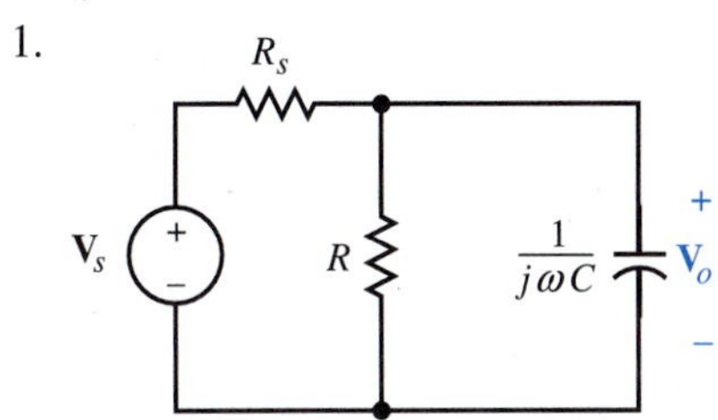

2.

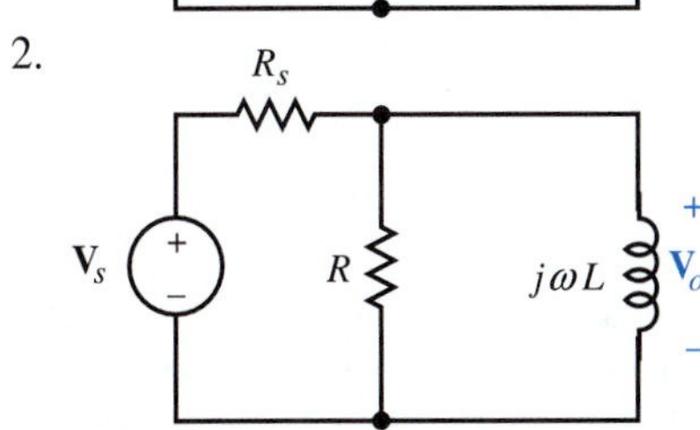

3.

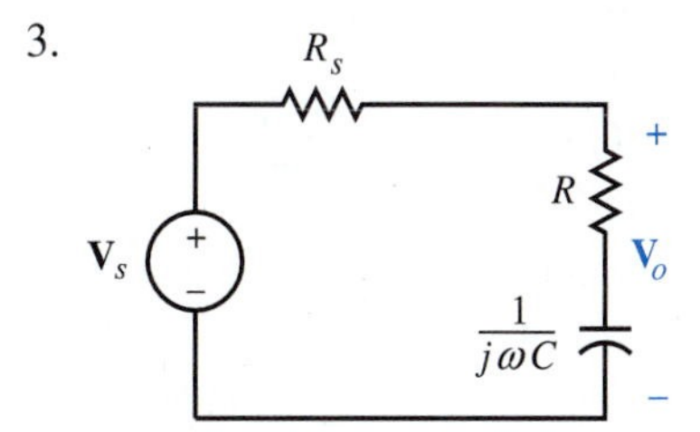

4.

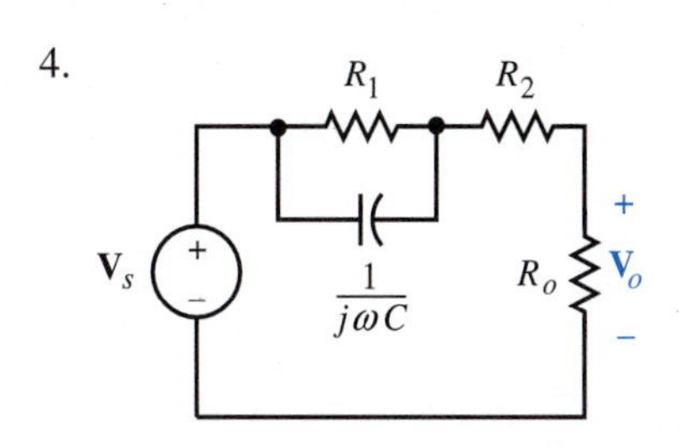

5. The hypothetical circuit has the frequency response characteristics shown in Fig. 13.47. Find the steady-state response to the input $v_s(t) = 10 + 3\cos 5t + 7\sin 10t + 4\cos(15t + 30°)$ V.

6. The following circuit is a *preemphasis filter*. A preemphasis filter is used in a commercial FM

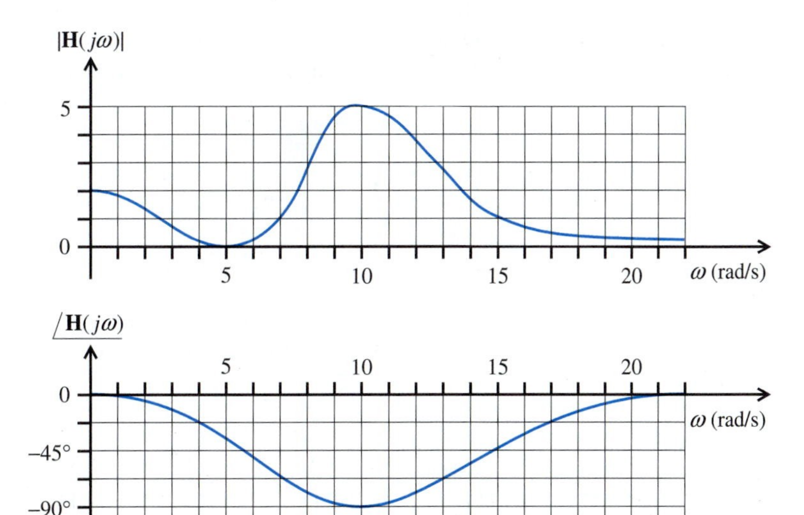

FIGURE 13.47

transmitter to boost high-frequency components in the audio signal before the signal is modulated and transmitted. The voltage transfer function can be written as follows:

$$H_p(j\omega) = K\frac{1 + j\omega\tau_1}{1 + j\omega\tau_2}$$

(a) Give the formulas for K, τ_1, and τ_2 as functions of R_1, R_2, and C.

(b) The time constants τ_1 and τ_2 are chosen to be 75 μs and 6.5 μs, respectively. Find R_1, R_2, and K if $C = 0.001$ μF.

(c) Plot the amplitude and the phase characteristics versus f, where $f = \omega/(2\pi)$.

(d) Find the filter output voltage signal, $v_{sig2}(t)$, for the audio signal input voltage

$$v_{sig1}(t) = \sum_{n=1}^{5} 10 \cos(2\pi n3 \times 10^3\, t) \text{ mV}$$

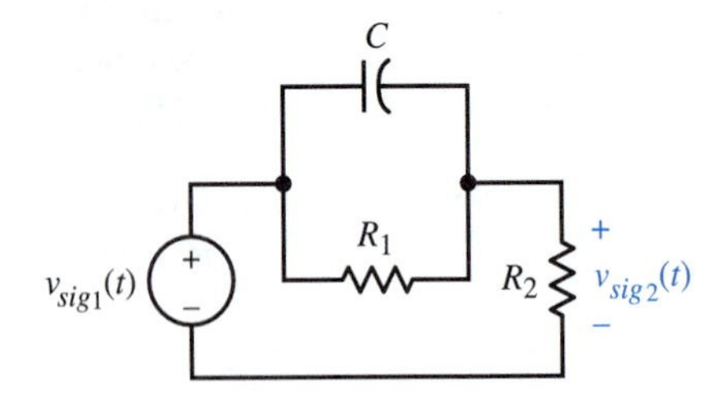

7. The circuit shown is a *deemphasis filter*. A deemphasis filter is used in an FM receiver to attenuate high-frequency noise that might be present with the demodulated audio signal. The voltage transfer function can be written in the following form

$$H_d(j\omega) = \frac{1}{1 + j\omega\tau_1}$$

where $\tau_1 = RC$. (a) Find R if $C = 0.001$ μF and $\tau_1 = 75$ μs. (b) Plot the amplitude and the phase characteristics versus f, where $f = \omega/(2\pi)$. (c) Find the filter output, $v_{noio}(t)$ if the input to the filter is noise, which we model as

$$v_{noi1}(t) = \sum_{n=1}^{5} 50 \cos(2\pi 4n \times 10^3 t + \phi_i) \quad \mu\text{V}$$

where the ϕ_i are arbitrary phase angles.

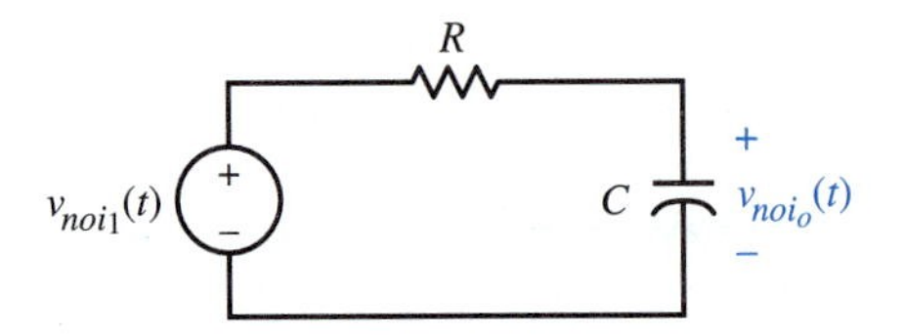

8. FM communication systems often use the preemphasis and deemphasis filters of Problems 6 and 7 as depicted in Fig. 13.48. We will use the voltages $v_{sig1}(t)$, $v_{sig2}(t)$, $v_{noi1}(t)$, and $v_{noio}(t)$ of Problems 6 and 7 to analyze the performance of the FM system. As in Problem 6, the audio signal $v_{sig1}(t)$ is applied to the preemphasis filter. The preemphasis filter output, $v_{sig2}(t)$, is then applied to the FM modulator and transmitter. The voltage $v_x(t) =$

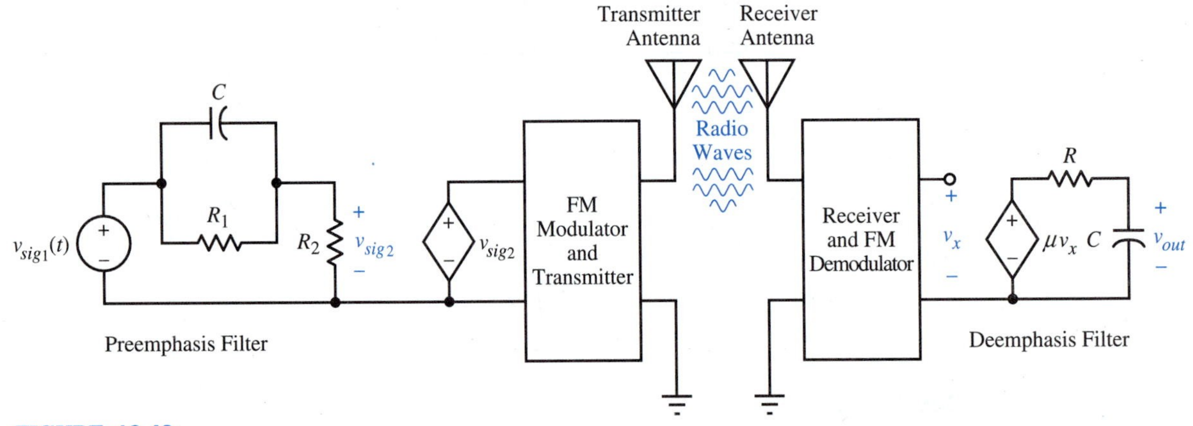

FIGURE 13.48

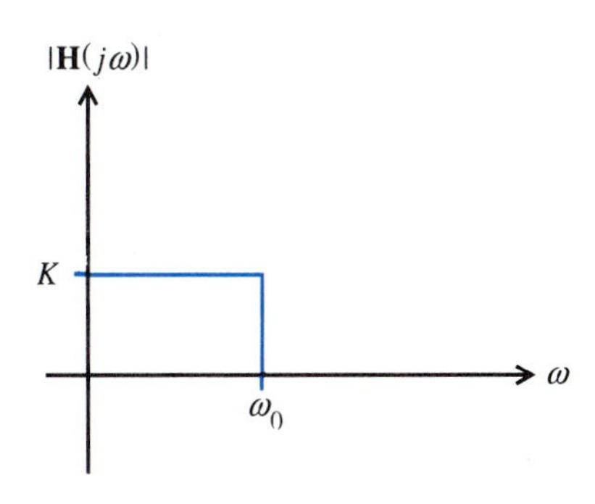

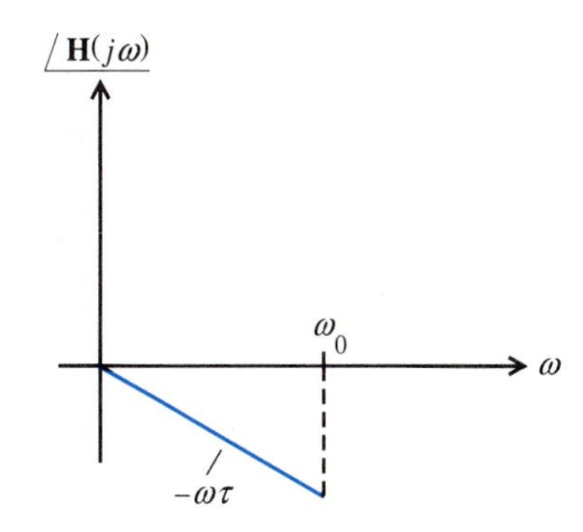

FIGURE 13.49

$v_{sig2}(t) + v_{noi1}(t)$ appears at the demodulator output of the receiver where $v_{noi1}(t)$ is noise that is unavoidably picked up in the communication process. The input to the deemphasis filter is $\mu v_x(t)$ where $\mu = 1/K$ and K is the constant appearing in $H_p(j\omega)$. The deemphasis filter output, $v_{out}(t)$, is applied to a speaker (not shown) which converts it to sound.

(a) Find the deemphasis filter output, $v_{out}(t)$. Write your answer as a sum of signal and noise: $v_{out}(t) = v_{sigo}(t) + v_{noio}(t)$. Give the formulas for $v_{sigo}(t)$ and $v_{noio}(t)$.

(b) Is the *signal* part of the output of the deemphasis filter, $v_{sigo}(t)$, approximately equal to the audio signal, $v_{sig1}(t)$, that is to be communicated?

(c) Find the rms value of the deemphasis filter input noise, $v_{noi1}(t)$.

(d) Find the rms value of the deemphasis filter output noise, $v_{noio}(t)$.

(e) What useful function is performed by the preemphasis/deemphasis filter combination?

9. The frequency response characteristics of an *ideal low-pass filter* are depicted in Fig 13.49 where K is the gain, ω_0 is the angular cutoff frequency or bandwidth, and τ is the *signal delay*. (The ideal low-pass filter is an analytically convenient but fictitious system whose transfer function can only be approximated using physical elements.)

(a) Find the steady-state response to an input $v_1(t) = V_{1m} \cos(\omega_1 t + \phi_{\mathbf{V}_1})$. Assume that $\omega_1 > \omega_0$.

(b) Show that if $\omega_1 < \omega_0$, the steady-state response to the input from (a) is $v_o(t) = KV_{1m} \cos[\omega_1(t - \tau) + \phi_{\mathbf{V}_1}]$.

(c) Assume that $K = 1$, $\omega_0 = 1000$ rad/s, and $\tau = 1$ ms. Find the steady-state response to the input $v_1(t) = 6 + 2\cos(300t + 20°) + 4\sin 800t + 17\cos(1100t + 45°) - 5\cos 2000t$.

10. Let $\mathbf{Z}_a = \mathbf{Z}_1 = 100\ \Omega$, $\mathbf{Z}_b = 1/j\omega C$, and $\mathbf{Z}_2 = R$ in the ideal op-amp circuit in Fig. 13.50.

(a) Choose R and C such that the amplitude characteristic approximates that of the ideal low-pass filter shown in Problem 9 with $K = 20$ and bandwidth 1000 rad/s. *Hint:* Set $\tau = RC = 1$ ms.

(b) Plot the amplitude and phase characteristics of your op-amp filter and compare them with those shown in Problem 9.

(c) Find the steady-state response of your filter to the input $v_1(t) = 6 + 2\cos(300t + 20°) + 4\sin 800t + 17\cos(1100t + 45°) - 5\cos 2000t$ and compare this with the ideal low-pass filter response for $\tau = 1$ ms.

11. The frequency response characteristics of an *ideal high-pass filter* are depicted in Fig. 13.51, where K is the gain, ω_0 is the angular cutoff frequency, and τ is the signal delay. (As with the ideal low-pass filter, the ideal high-pass filter is not physically realizable.)

(a) Find the steady-state response to an input $v_1(t) = V_{1m}\cos(\omega_1 t + \phi_{\mathbf{V}_1})$ for (i) $\omega_1 < \omega_0$ and (ii) $\omega_1 > \omega_0$.

(b) Assume that $K = 1$, $\omega_0 = 1000$ rad/s, and $\tau = 1$ ms. Find the steady-state response to the input $v_1(t) = 6 + 2\cos(300t + 20°) + 4\sin 800t + 17\cos(1100t + 45°) - 5\cos 2000t$.

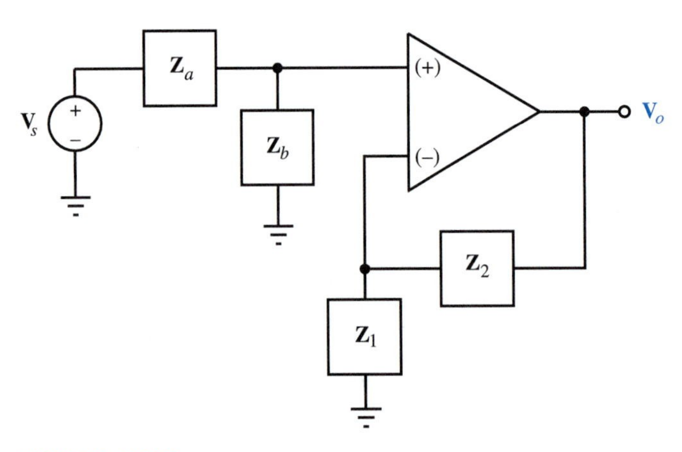

FIGURE 13.50

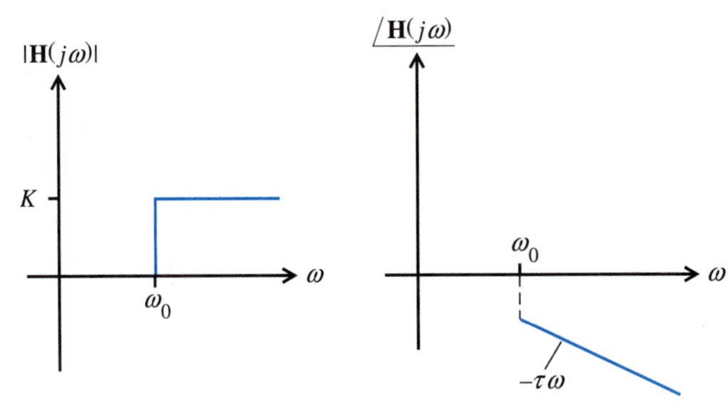

FIGURE 13.51

Section 13.2

12. Draw the dual of the series *RLC* circuit of Fig. 13.6 and specify the dual transfer function $\mathbf{H}(j\omega) = \mathbf{I}_o/\mathbf{I}_s$.

13. For the circuit shown, determine and plot the amplitude and phase characteristics of $\mathbf{V}_R/\mathbf{V}_s$ and $\mathbf{V}_{LC}/\mathbf{V}_s$.

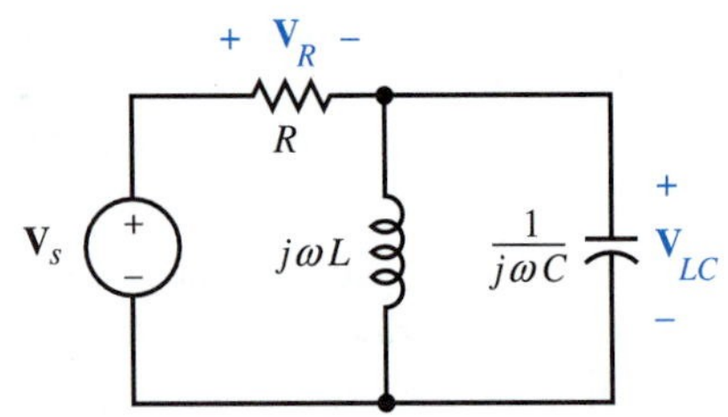

14. The circuit shown is a *band-rejection* filter. (a) Find the transfer function $\mathbf{H}(j\omega)$. (b) Use your answer to part (a) to find $\mathbf{H}(j0)$, $\mathbf{H}(j\infty)$, and $\mathbf{H}(j\omega_0)$ where $\omega_0 = \sqrt{1/LC}$. (c) Explain your answers to part (b) by replacing impedances with open or short circuits. *Hint*: What is the value of $j\omega L + 1/j\omega C$ for $\omega = \omega_0$? (d) Plot the amplitude and phase characteristics if $R = 0.1\ \Omega$, $L = 1\,\mu\text{H}$, and $C = 1\,\mu\text{F}$. (e) Explain why the name *band-rejection filter* is appropriate.

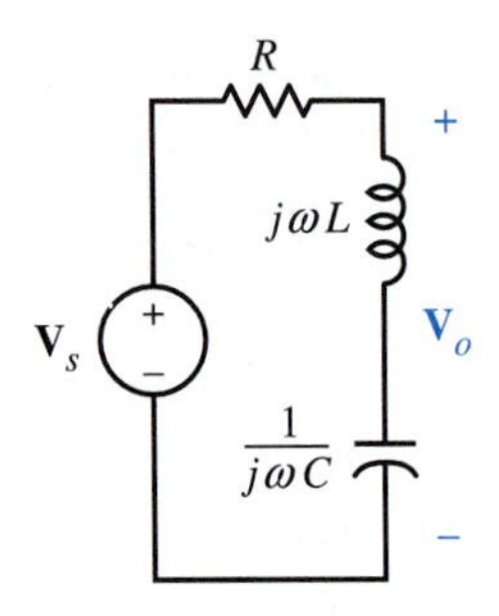

15. In the circuit shown, $L = 1\ \mu\text{H}$, $C = 1\ \mu\text{F}$, and $r = 1\ \text{k}\Omega$.

    (a) Show that the parallel *LC* combination can be

replaced with an open circuit when $\omega = 1/\sqrt{LC}$.

(b) Choose the value of R such that $\mathbf{V}_o/\mathbf{V}_s = 0.5$ for $\omega = 1/\sqrt{LC}$.

(c) Use the value of R determined in (a), and calculate and compare the currents $\mathbf{I}_s$, $\mathbf{I}_L$, $\mathbf{I}_C$, and $\mathbf{I}_R$ for $\omega = 1/\sqrt{LC}$ if $\mathbf{V}_s = 10\underline{/0^\circ}$ V. What is happening here?

(d) Repeat part (c) with $\omega = 0$.

(e) Repeat part (c) with $\omega = \infty$.

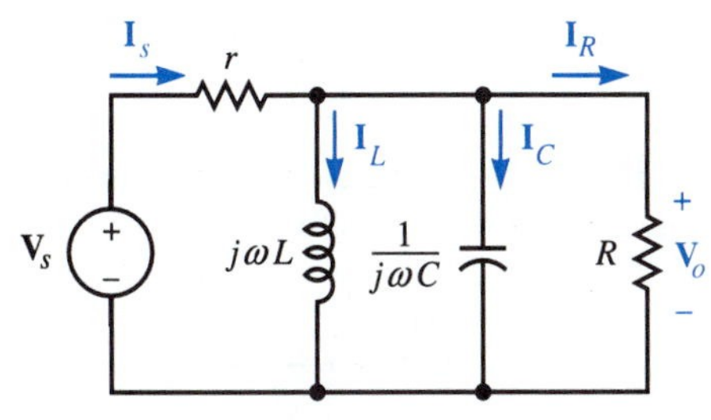

16. For the ideal op-amp circuit shown in Fig. 13.52,

(a) Write the node-voltage equations.

(b) Solve the node-voltage equations for $\mathbf{V}_o$ to show that the voltage transfer function is given by

$$\mathbf{H} = \frac{\mathbf{V}_o}{\mathbf{V}_s} = \frac{-\mathbf{Y}_1\mathbf{Y}_3}{\mathbf{Y}_5(\mathbf{Y}_1 + \mathbf{Y}_2 + \mathbf{Y}_3 + \mathbf{Y}_4) + \mathbf{Y}_3\mathbf{Y}_2}$$

(c) Let $\mathbf{Y}_1 = R_1^{-1}$, $\mathbf{Y}_2 = G_2$, $\mathbf{Y}_3 = R_3^{-1}$, $\mathbf{Y}_4 = j\omega C_4$, and $\mathbf{Y}_5 = j\omega C_5$. Show that $\mathbf{H}(j\omega) =$

$$\frac{-1}{(j\omega)^2 C_4 C_5 R_1 R_3 + j\omega C_5(R_1 + R_3 + R_1 G_2 R_3) + R_1 G_2}$$

(d) Let $R_1 G_2 = 1$, $C_4 C_5 R_1 R_3 = 10^{-12}$ s^2, and $C_5(R_1 + R_3 + R_1 G_2 R_3) = 10^{-7}$ s. Plot the amplitude and phase characteristics and compare them with Fig. 13.7.

Section 13.3

17. The complex functions $\mathbf{i}(t)$ and $\mathbf{v}(t)$ satisfy the equation

$$\frac{d^2}{dt^2}\mathbf{v} + 15\frac{d}{dt}\mathbf{v} + 50\mathbf{v} = \frac{d^2}{dt^2}\mathbf{i} + 7\frac{d}{dt}\mathbf{i}$$

(a) Without solving the equation, show that the real functions $i_R(t) = \mathcal{R}e\{\mathbf{i}(t)\}$ and $v_R(t) = \mathcal{R}e\{\mathbf{v}(t)\}$ must also satisfy the same equation.

(b) Without solving the equation, show that the real functions $i_I(t) = \mathcal{I}m\{\mathbf{i}(t)\}$ and $v_I(t) = \mathcal{I}m\{\mathbf{v}(t)\}$ must also satisfy the same equation.

18. Find the impedance $\mathbf{Z}(s)$ associated with the LTI equation of Problem 17.

19. The driving-point impedance of a network is

$$\mathbf{Z}(s) = \frac{s}{s+1}\ \Omega$$

(a) A current $i(t) = 2e^{-4t}\cos(3t + 15^\circ)$ A is applied to the network. Determine the particular response.

(b) Repeat part (a) if $i(t) = 2e^{-4t}$ A.

(c) Repeat part (a) if $i(t) = 2$ A.

20. The driving-point impedance of a network is

$$\mathbf{Z}(s) = 1 + \frac{2s}{1+2s}\ \Omega$$

Find the differential equation relating input $i(t)$ to output $v(t)$.

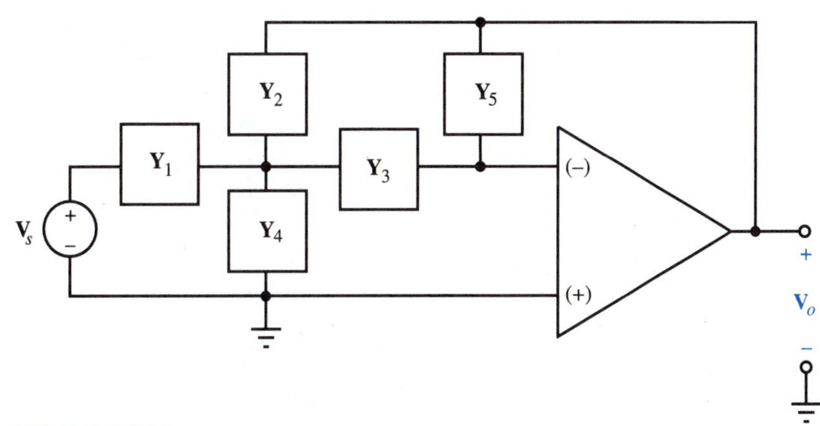

FIGURE 13.52

21. The following circuit depicts a simple model of a one-stage transistor circuit.
    (a) Draw the s-domain circuit.
    (b) Find $\mathbf{H}(s) = \mathbf{V}_o/\mathbf{V}_s$.
    (c) Let $v_s(t) = 7\cos(2000t + 30°)$ V. What is the value of complex frequency s for this input?
    (d) Find the particular response $v_{op}(t)$ in terms of r, R, C, and β for the input of (c) by first determining $\mathbf{V}_o$.
    (e) Use $\mathbf{H}(s)$ to determine the LTI equation relating $v_o(t)$ to $v_s(t)$.

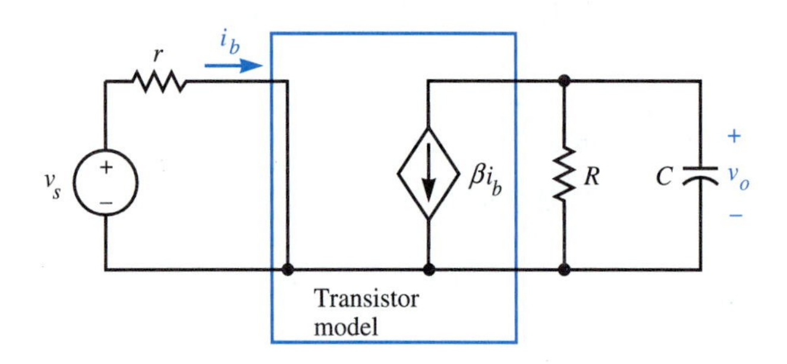

Section 13.4

22. Draw the pole-zero plot for the FM preemphasis filter of Problem 6.

23. Draw the pole-zero plot for the FM deemphasis filter of Problem 7.

For the pole-zero plots in Problems 24 and 25

(a) Express $\mathbf{H}(s)$ as the ratio of two polynomials.

(b) Find the value of the transfer function at a frequency $\omega = 2$ rad/s (neper frequency $\sigma = 0$).

24.

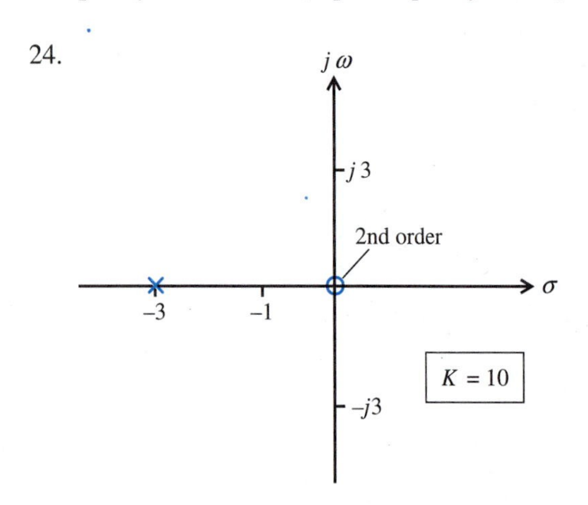

25.

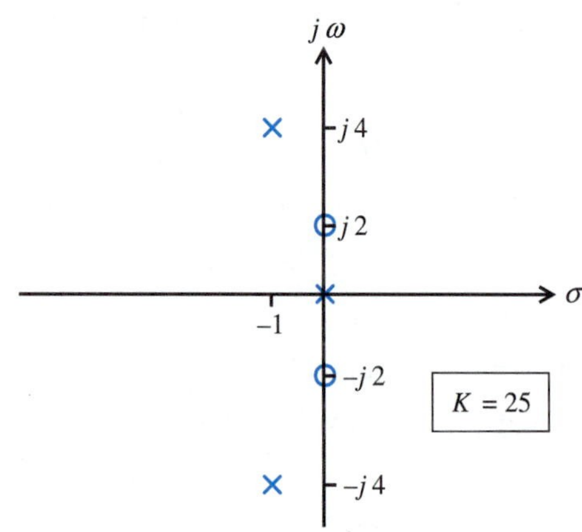

26. Find $\mathbf{H}(s) = \mathbf{V}_2/\mathbf{V}_1$ and draw the corresponding pole-zero plot for the following circuit.

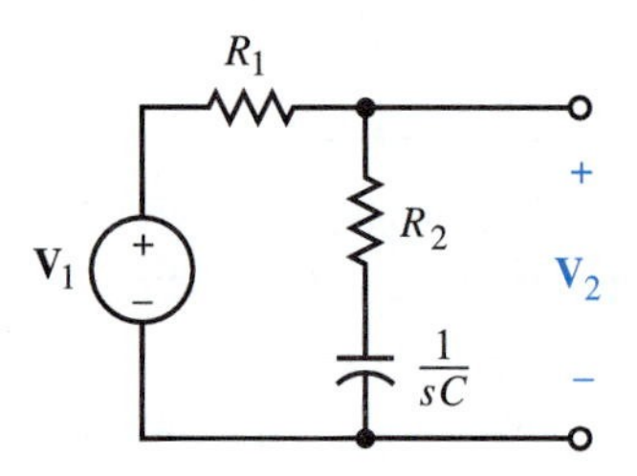

27. The source in the circuit shown below is a current or voltage source.
    (a) Determine the driving-point impedance $\mathbf{Z}(s)$, and draw the corresponding pole-zero plot.
    (b) Determine the driving-point admittance $\mathbf{Y}(s)$ and draw the corresponding pole-zero plot.
    (c) Determine the form of the natural response if the source is an ideal *current* source.
    (d) Determine the form of the natural response if the source is an ideal *voltage* source.

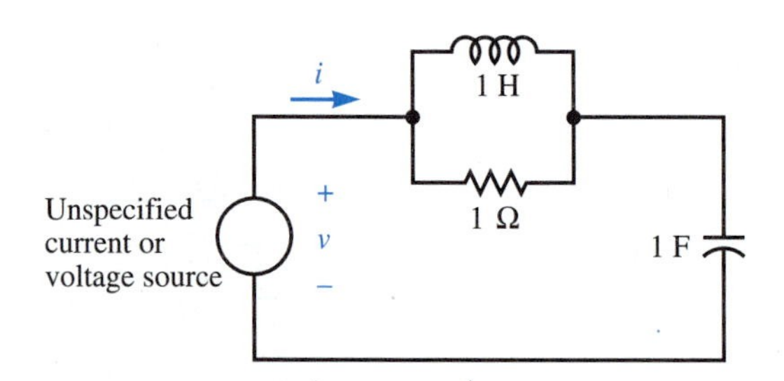

28. (a) Determine the s-domain transfer functions $\mathbf{H}_R(s) = \mathbf{V}_R/\mathbf{V}_s$ and $\mathbf{H}_{LC}(s) = \mathbf{V}_{LC}/\mathbf{V}_s$ for the circuit of Problem 13.
    (b) Draw the pole-zero plots, and explain how they differ.

29. Use the rubber sheet analogy to.make an approximate sketch of the amplitude response characteristic $|\mathbf{H}(j\omega)|$ associated with the pole-zero constellations of Problems 24 and 25.

30. Use the rubber sheet analogy to make an approximate sketch of the amplitude characteristic $|\mathbf{H}(j\omega)|$ for the circuit of Problem 26.

31. The input to the circuit shown below is $\mathbf{v}_s(t) = \mathbf{V}_s e^{st}$. For what value of s is the particular response, $v_{op}(t)$ zero for all t?

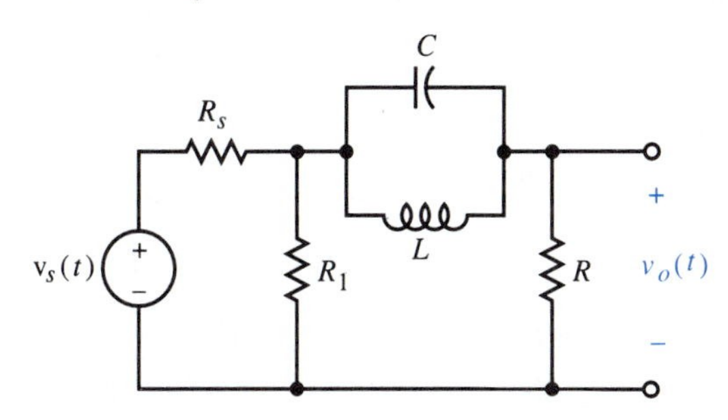

32. For what complex frequencies is the impedance of a series RC circuit (a) zero, (b) infinite?

33. Repeat Problem 32 for a series RL circuit.

34. Repeat Problem 32 for a series LC circuit.

35. Repeat Problem 32 for a series RLC circuit.

36. Repeat Problem 32 for a parallel RC circuit.

37. Repeat Problem 32 for a parallel RL circuit.

38. Repeat Problem 32 for a parallel LC circuit.

39. Repeat Problem 32 for a parallel RLC circuit.

40. For what values of r is the circuit shown unstable?

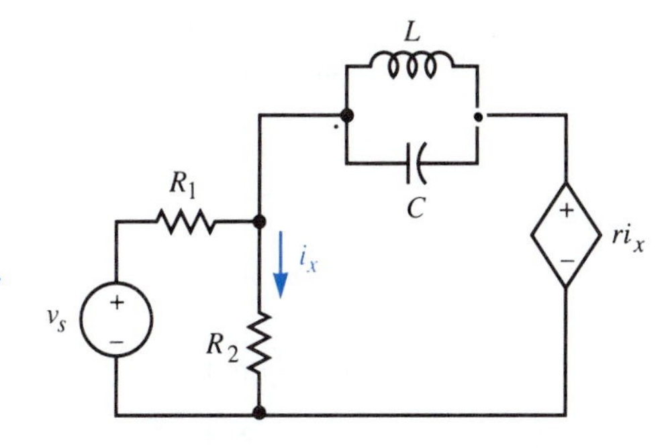

41. The circuit shown is called a *twin-T band-rejection* filter. The output of the filter is $\mathbf{V}_3$.

    (a) Set up the node-voltage equations.

    (b) Use Cramer's rule to solve for $\mathbf{V}_3$ and from this find the voltage transfer function $\mathbf{H}(s) = \mathbf{V}_3/\mathbf{V}_s$.

    (c) Write $\mathbf{H}(s)$ in factored form and draw the pole-zero plot.

    (d) Let $R = 1$ kΩ and $C = 1$ μF. Use the rubber sheet analogy to sketch $|\mathbf{H}(j\omega)|$. Explain why the circuit is called a band-rejection filter.

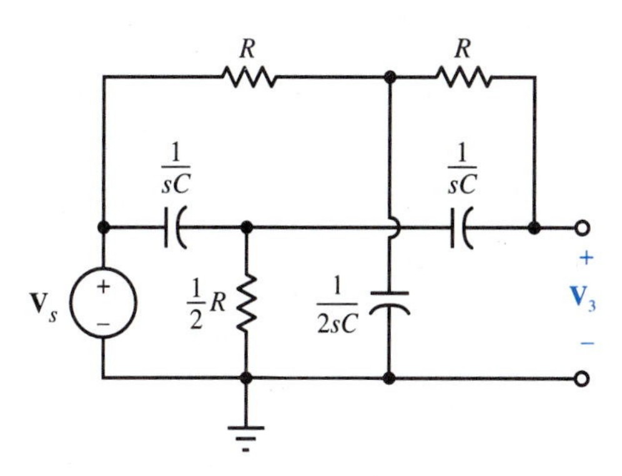

42. An *nth-order Butterworth filter* is a low-pass filter with the transfer function

$$\mathbf{H}(s) = \frac{k\omega_0^n}{(s - s_1)(s - s_2)\cdots(s - s_n)}$$

where k is a real constant and where the poles are symmetrically spaced about a semicircle of radius ω_0 in the left half of the s plane, as shown in the figure on page 486 for $n = 5$. The amplitude characteristic of an nth-order Butterworth filter is called *maximally flat,* meaning that

$$\left.\frac{d^i}{d\omega^i}|\mathbf{H}(j\omega)|\right|_{\omega=0} = 0$$

for $i = 1, 2, \ldots, n - 1$.

(a) Show that

$$|\mathbf{H}(j\omega)| = \frac{|k|\,\omega_0^n}{\sqrt{\omega^{2n} + \omega_0^{2n}}}$$

Hint:

$$\mathbf{H}(s)\mathbf{H}(-s) = \frac{k^2\omega_0^{2n}}{s^{2n} + \omega_0^{2n}}$$

because the poles of $\mathbf{H}(s)$ are the left-half s-plane roots to $s^{2n} + \omega_0^{2n} = 0$.

(b) Using your result from (a), show that ω_0 is the half-power bandwidth in radians per second. That is, show that $|\mathbf{H}(j\omega_0)| = (1/\sqrt{2})|\mathbf{H}(0)|$.

(c) Sketch $|\mathbf{H}(j\omega)|$ in the limit $n \to \infty$.

(d) Plot the amplitude characteristic for a fifth-order Butterworth filter.

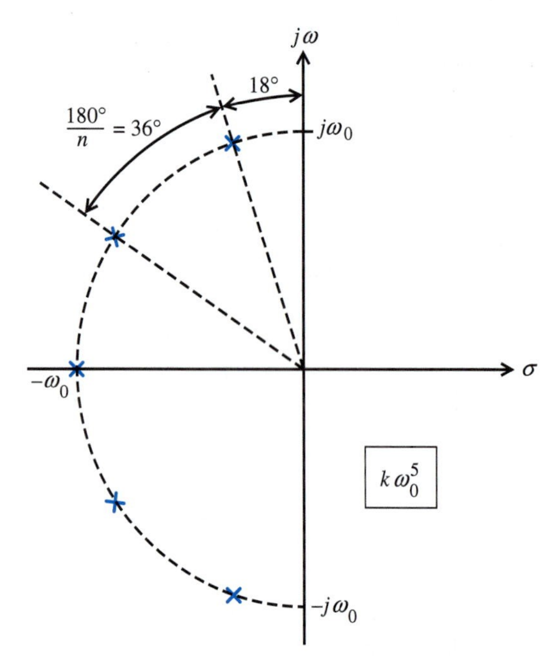

43. Butterworth filters can be synthesized by cascading op-amp circuits as shown in Fig. 13.53 for a fourth-order filter.

    (a) Assuming ideal op amps, show that the voltage transfer function $\mathbf{V}_2/\mathbf{V}_1$ is given by

$$\mathbf{H}(s) = \frac{-k_1}{a_1 s^2 + b_1 s + 1} \cdot \frac{-k_2}{a_2 s^2 + b_2 s + 1}$$

    For $i = 1, 2$, give the formulas for the constants k_i and for a_i and b_i in terms of the circuit elements.

    (b) Choose k_i, a_i, and b_i, $i = 1, 2$, such that $\mathbf{H}(0) = K$ and the poles of $\mathbf{H}(s)$ are located at the left-half s-plane roots of $s^{2n} + \omega_0^{2n} = 0$ where $n = 4$. Write your answers in terms of ω_0 and K.

    (c) Draw a circuit with three op amps that can be used to synthesize a sixth-order Butterworth filter.

    (d) Draw a circuit using three op amps that can be used to synthesize a fifth-order Butterworth filter.

Section 13.5

44. For the circuit shown,

    (a) Show that the transfer impedance is

$$\mathbf{H}(s) = \frac{\mathbf{V}_o}{\mathbf{I}_s} = \frac{\dfrac{sRR_s}{L}}{s^2 + s\left(\dfrac{R + R_s}{L}\right) + \dfrac{1}{LC}}$$

    (b) Find the form of the natural response $v_c(t)$.

    (c) Find the exact expression for the maximum-response resonance frequency ω_{mr} associated with $|\mathbf{H}(j\omega)|$.

    (d) Find the exact expression for the half-power bandwidth associated with $|\mathbf{H}(j\omega)|$.

    (e) Specify your answers to (a) through (d) using $R_s = 6\ \Omega$, $R = 20\ \Omega$, $L = 13$ mH, and $C = 2.2\ \mu$F.

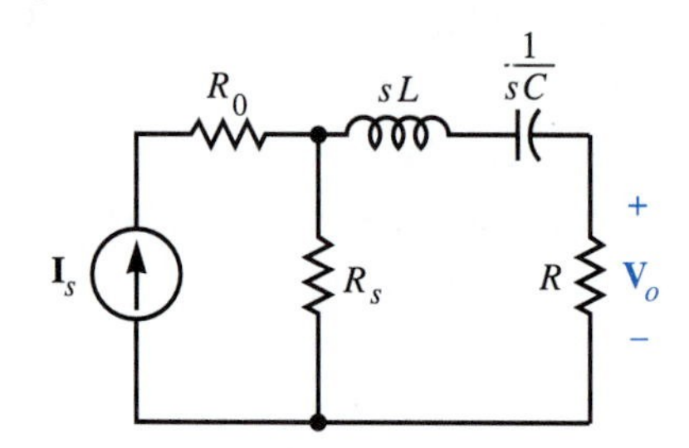

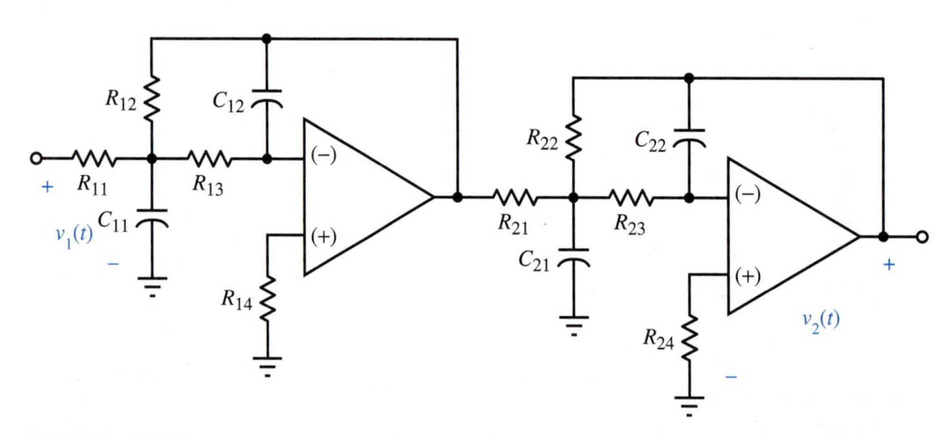

FIGURE 13.53

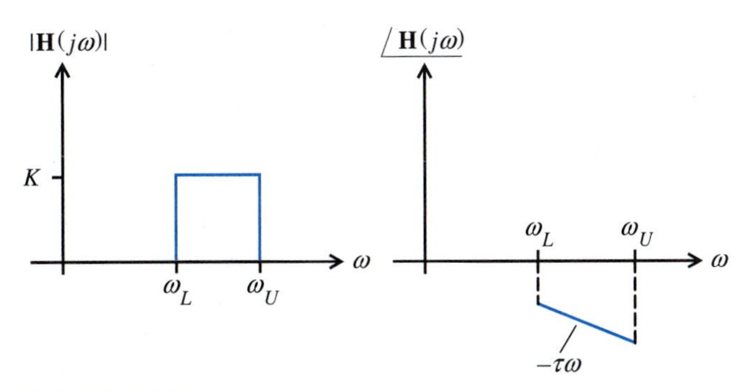

FIGURE 13.54

45. The frequency response characteristics of an *ideal band-pass filter* are depicted in Fig 13.54, where K is the gain, ω_L and ω_U are the lower and upper angular cutoff frequencies, respectively, and τ is the *signal delay.* (As with the ideal low-pass and high-pass filters, the ideal band-pass filter is an analytically convenient but physically unrealizable system.)
    (a) Find the steady-state response to the input $v_1(t) = V_{1m} \cos(\omega_1 t + \phi_{\mathbf{V}1})$ for (i) $\omega_1 < \omega_L$, (ii) $\omega_L < \omega_1 < \omega_U$, and (iii) $\omega_U < \omega_1$.
    (b) Assume that $K = 1$, $\omega_L = 500$ rad/s, $\omega_U = 1200$ rad/s, and $\tau = 1$ ms. Find the steady-state response to the input

$$v_1(t) = 6 + 2\cos(300t + 20°) + 4\sin 800t + 17\cos(1100t + 45°) - 5\cos 2000t$$

46. A band-pass filter employing an ideal op amp is shown below. Show that the voltage transfer function has the form

$$\mathbf{H}(s) = \frac{\mathbf{V}_o}{\mathbf{V}_s} = \frac{-\alpha A \omega_0 s}{s^2 + 2\alpha s + \omega_0^2}$$

Express α, ω_0, and A in terms of the circuit elements.

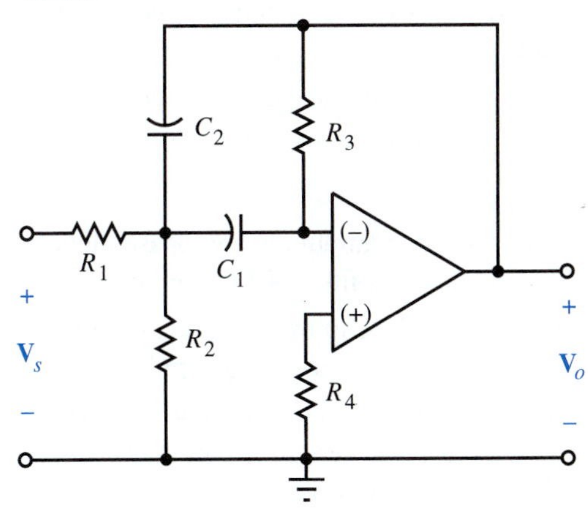

47. (a) Explain technically, but in your own words, why an RLC band-pass filter that has a small damping ratio $\zeta = \alpha/\omega_0$ must have a highly oscillatory natural response. Use mathematics and plots to support your explanation.
    (b) Find a formula that gives the half-power bandwidth of the band-pass filter in terms of the damping ratio and the pass-band "center frequency" ω_0.

Section 13.6

48. Compare the meanings of unity-power-factor resonance and maximum power transfer.

49. For the circuit shown,
    (a) Determine the unity-power-factor resonance frequency.
    (b) Determine the input impedance at unity-power-factor resonance.
    (c) Show that $W_{L,\text{ave}} = W_{C,\text{ave}}$ at the frequency determined in (a).

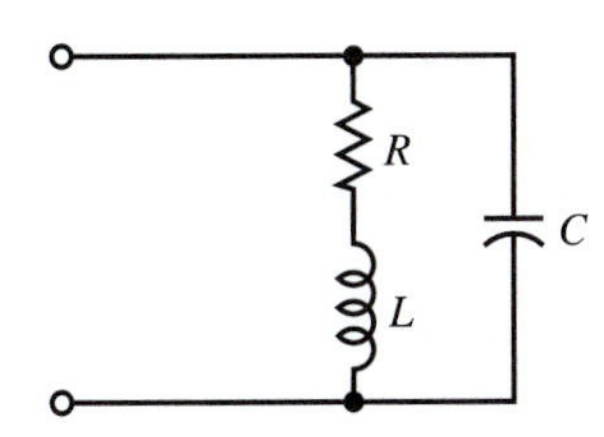

50. For the circuit shown on the next page,
    (a) Use the fact that $W_{L,\text{ave}} = W_{C,\text{ave}}$ to determine the unity-power-factor resonance frequency.
    (b) Verify your answer to (a) by showing that $\mathbf{V}/\mathbf{I}$ is purely real when $\omega = \omega_{upf}$.

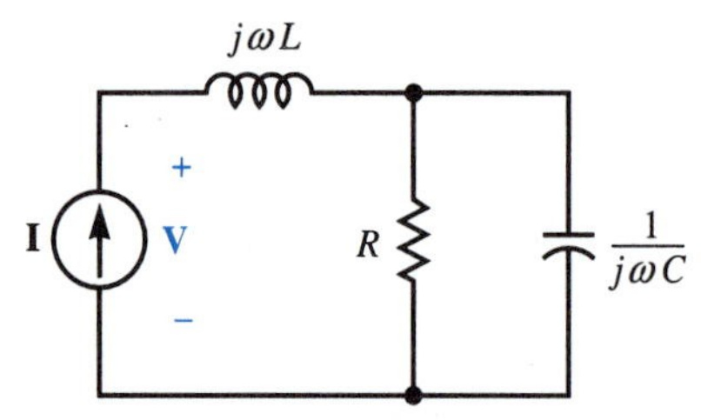

51. Find the unity-power-factor resonance frequency for the circuit shown, where $L = 1$ mH, $C = 1000\ \mu$F, $R = 100\ \Omega$, and $r = 0.5\ \Omega$.

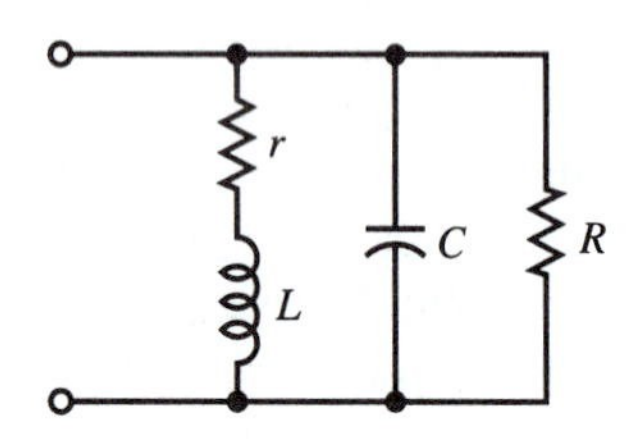

Section 13.7

52. The circuit shown below is a model of a capacitor. Determine its Q.

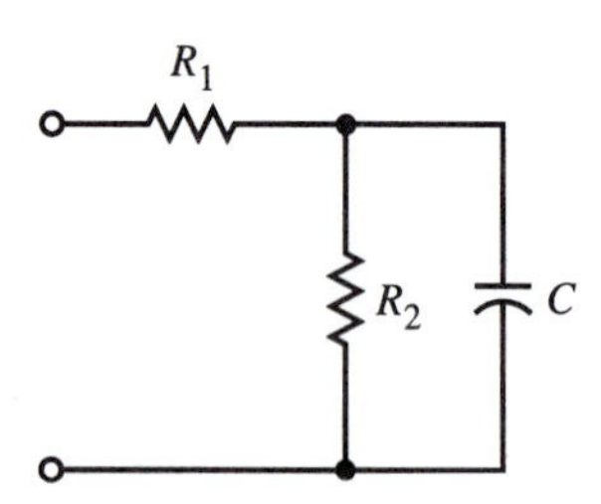

53. In the following two circuits, the reactances X_s and X_p are due to either an inductance or a capacitance.

    (a) Suppose we want to design the parallel circuit so that it has the same impedance as the series circuit at a fixed angular frequency ω. Show that $\mathbf{Z}_p$ will equal $\mathbf{Z}_s$ if we use the following design equations:

    $$R_p = (1 + Q_s^2)R_s \quad \text{and} \quad X_p = \frac{1 + Q_s^2}{Q_s^2} X_s$$

    where $Q_s = |X_s|/R_s$ is the quality factor of the series circuit.

    (b) Assume instead that we want to design the series circuit so that it has the same impedance as the parallel circuit at a fixed frequency ω. Show that $\mathbf{Z}_s$ will equal $\mathbf{Z}_p$ if we use the following design equations:

    $$R_s = \frac{1}{1 + Q_p^2} R_p \quad \text{and} \quad X_s = \frac{Q_p^2}{1 + Q_p^2} X_p$$

    where $Q_p = R_p/|X_p|$ is the quality factor of the parallel circuit.

    (c) Compare (a) with (b), to show that if $\mathbf{Z}_s$ and $\mathbf{Z}_p$ are equal, then $Q_s = Q_p$.

    (d) Use the results of (a) through (c) to find the series RC circuit that has the same impedance at $\omega = 1$ krad/s as a parallel RC circuit in which $R_p = 100$ kΩ and $C_p = 1\ \mu$F. Evaluate $\mathbf{Z}_p$, $\mathbf{Z}_s$, Q_p, and Q_s.

    (e) Find the parallel RL circuit that has the same impedance at $\omega = 20$ Mrad/s as a series RL circuit in which $R_s = 100\ \Omega$, $L = 100\ \mu$H. Evaluate $\mathbf{Z}_p$, $\mathbf{Z}_s$, Q_p, and Q_s.

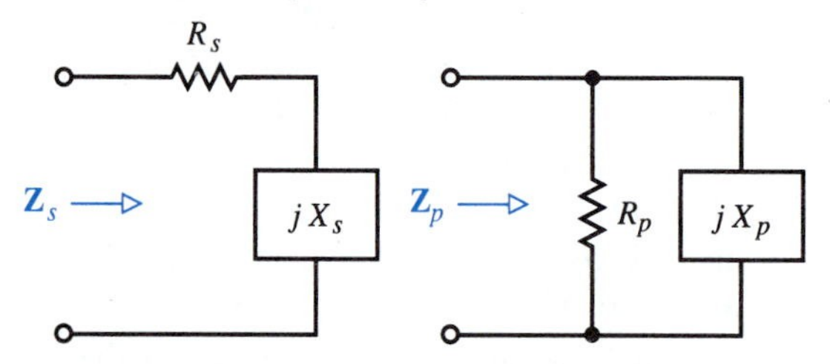

Section 13.8

54. Use approximations similar to those in Fig. 13.32 to explain why the maximum-response resonance frequencies and the half-power bandwidths associated with $\mathbf{H}_R(j\omega)$, $\mathbf{H}_C(j\omega)$, and $\mathbf{H}_L(j\omega)$ of Fig. 13.34 are nearly equal for $\zeta \ll 1$.

55. The amplitude response of an unknown circuit has a peak of 100 at 1 Mrad/s and equals 10 at 200 krad/s. Assuming that the peak is due to an isolated simple pole, sketch the (approximate) amplitude and phase characteristics.

56. Measurements of the amplitude response of a certain LTI system reveal that $|\mathbf{H}(j100)| = 316$, $|\mathbf{H}(j110)| = 447$, and $|\mathbf{H}(j120)| = 707$. Assuming that the response for these frequencies is caused primarily by a simple isolated pole, estimate:

    (a) The maximum response frequency

    (b) The peak value of $|\mathbf{H}|$

Section 13.9

57. Draw the Bode amplitude and phase plots for the FM preemphasis filter of Problem 6.

58. Draw the Bode amplitude and phase plots for the FM deemphasis filter of Problem 7.

59. The Bode straight-line amplitude plot asymptotes associated with a certain circuit are shown in Fig. 13.55 on the next page.

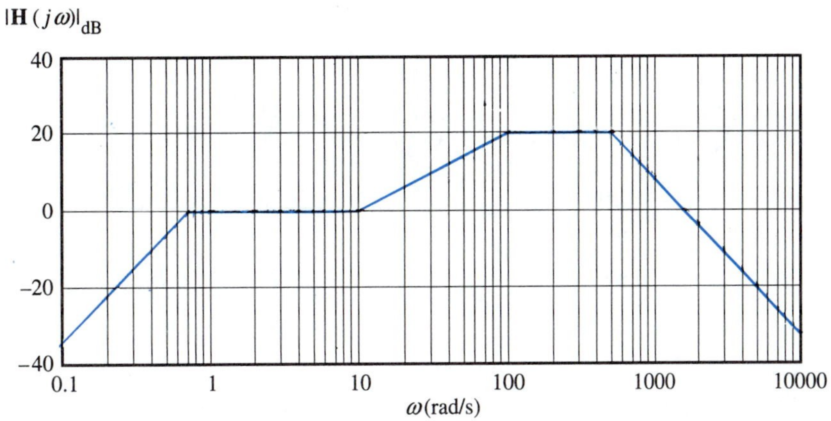

FIGURE 13.55

(a) Determine the poles of the circuit and their order.

(b) Determine the zeros of the circuit and their order.

(c) What is the transfer function of the circuit?

(d) Are the solutions to (a) through (c) necessarily unique? That is, is a circuit's transfer function completely specified by the associated straight-line amplitude plot? Explain.

60. The circuit in Fig. 13.56 represents three identical *RC* low-pass filters separated by buffer amplifiers.

(a) Show that the voltage transfer function of the circuit is

$$\frac{\mathbf{V}_o}{\mathbf{V}_s} = \mathbf{H}(s) = \mathbf{H}_1^3(s)$$

where

$$\mathbf{H}_1(s) = \frac{1}{1 + sRC}$$

(b) Draw the Bode amplitude and phase plots associated with $\mathbf{H}(s)$.

61. (a) Determine the transfer function $\mathbf{H}(s) = \mathbf{V}_o/\mathbf{V}_s$ for the circuit shown below.

(b) Draw the Bode amplitude and phase plots for $R = 1\text{ k}\Omega$ and $C = 1\mu\text{F}$.

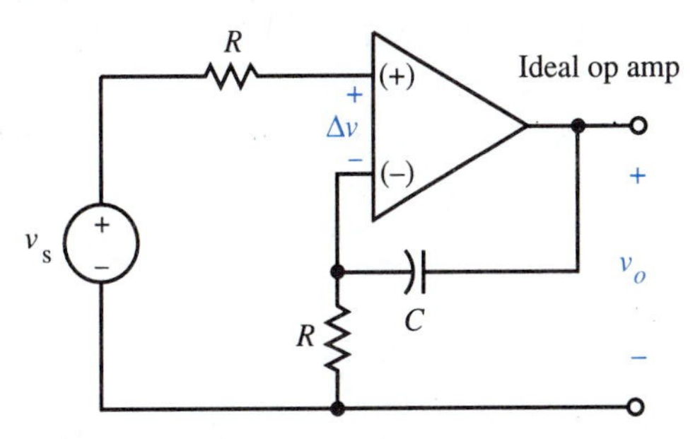

62. In Example 13.7, we assumed that the op-amp open-loop gain was infinite, $A = \infty$. A practical op-amp open-loop gain will be finite. Redraw the Bode amplitude and phase plots for the amplifier of Example 13.7 when $A = 10{,}000$.

Supplementary Problems

63. For the circuit configuration of Fig. 13.50.

(a) Use an ideal op amp, two 1-kΩ resistances, a resistance R_x, and a capacitance C_x to design

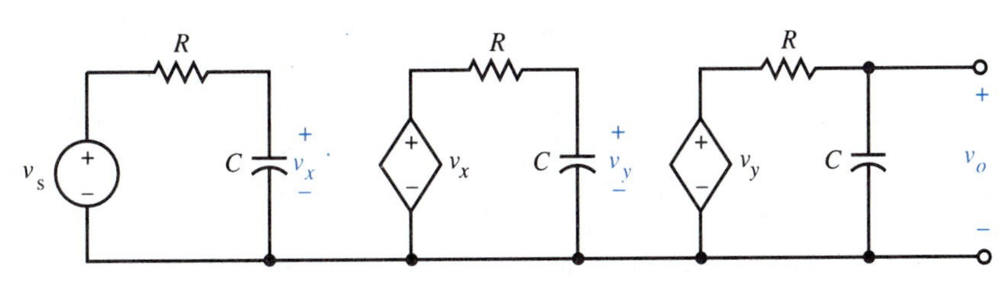

FIGURE 13.56

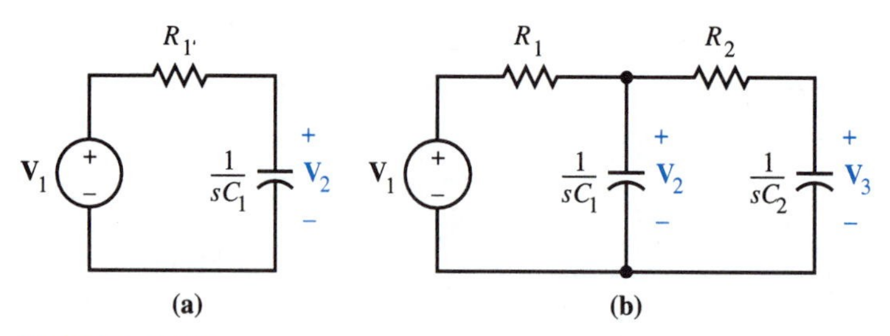

FIGURE 13.57

a high-pass filter. Choose R_x and C_x such that $\mathbf{H}(j\infty) = 20$ and $|\mathbf{H}(j1000)| = 20/\sqrt{2}$.

(b) Plot the amplitude and phase characteristics for the circuit you designed.

64. (a) Determine the transfer function $\mathbf{H}_1(s) = \mathbf{V}_2/\mathbf{V}_1$ for the circuit shown in Fig. 13.57a, if $R_1 = 1\ \text{k}\Omega$ and $C_1 = 1\ \mu\text{F}$.

(b) Determine the transfer function $\mathbf{H}_2(s) = \mathbf{V}_2/\mathbf{V}_1$ and $\mathbf{H}_3(s) = \mathbf{V}_3/\mathbf{V}_1$ for the circuit shown in Fig. 13.57b, for $R_1 = R_2 = 1\ \text{k}\Omega$ and $C_1 = C_2 = 1\ \mu\text{F}$.

(c) Does $\mathbf{H}_2(s) = \mathbf{H}_1(s)$? Explain.

(d) Does $\mathbf{H}_3(s) = \mathbf{H}_1^2(s)$? Explain.

65. A time-varying force $i(t)$ is applied to a sled having mass m by rocket thrusters causing it to slide with velocity $v(t)$. The equation of motion is $mDv + \mu v = i$ where μ is the coefficient of friction and D is the derivative operator. This equation neglects the change in mass due to fuel consumption.

(a) Find the frequency-domain version of the equation of motion.

(b) Use your result from part (a) to find the transfer function of the mechanical system, $\mathbf{H} = \mathbf{V}/\mathbf{I}$.

(c) Find the steady-state velocity, $v(t)$, of the mass to a sinusoidal force $i(t) = I_m \cos(\omega t + \phi_1)$.

(d) Does the velocity waveform lead or lag the force waveform?

(e) The system is called a *mechanical analogy* for a parallel *RC* circuit, for which output voltage, $v(t)$, is related to input current, $i(t)$, by $CDv + (1/R)v = i$. Analogous systems are described by differential equations having identical form. Describe any advantages that come to mind that pertain to the mechanical analogy concept.

66. (a) Determine the transfer function $\mathbf{H}(s) = \mathbf{V}_x/\mathbf{V}$ for the following circuit.

(b) Give the formula for $|\mathbf{H}(j\omega)|$.

(c) Give the formula for $\underline{/\mathbf{H}(j\omega)}$.

(d) What must be the relationship between the components, R, L, and C for $|\mathbf{H}(j\omega)|$ to be constant?

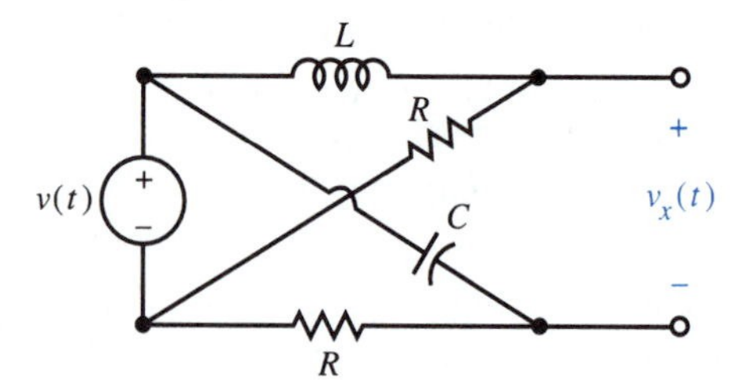

67. The circuit of Problem 66 can be used as a *cross-over network* for an audio system in which the resistance R connected to the inductance L is a model for the resistance of a low-range speaker (woofer), and the resistance R connected to the capacitance C models the resistance of the mid-range speaker. Assuming that $R = 4\Omega$, determine inductance and capacitance values so that the sum of the average powers delivered to the two speakers is constant, and the average power delivered to each speaker is the same at $\omega = 1000$ rad/s.

68. For the circuit shown on the next page, $R_1 = 10\ \Omega$, $R_2 = 20\ \Omega$, and $R_3 = 5\ \Omega$.

(a) Find the voltage transfer function, $\mathbf{H}(s) = \mathbf{V}_2/\mathbf{V}_1$.

(b) Use your answer to part (a) to find $\mathbf{H}(0)$. Check your result by replacing the capacitances with open circuits.

(c) Use your answer to part (a) to find $\mathbf{H}(j\infty)$. Check your result by replacing the capacitances with short circuits.

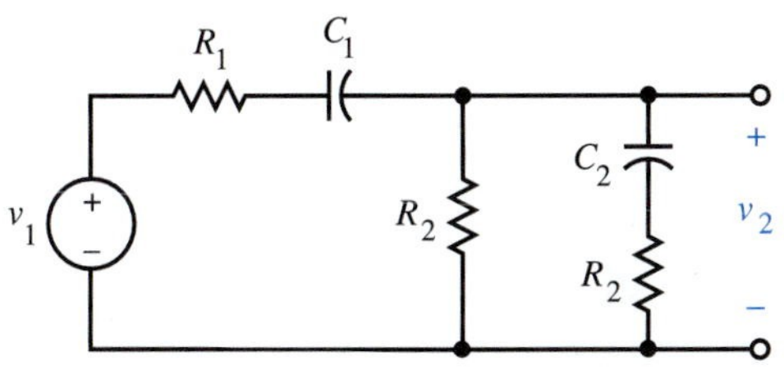

69. The following circuit illustrates a *scope probe attenuator* and oscilloscope input circuit model. $v_2(t)$ is the voltage that will be amplified and appear on the oscilloscope screen, and $v_1(t)$ is the voltage input to the probe. Capacitance C_2 models the input capacitance of the oscilloscope and coaxial cable, and R_2 is the input resistance of the oscilloscope.

   (a) Determine the transfer function $\mathbf{H}(s) = \mathbf{V}_2/\mathbf{V}_1$.

   (b) If R_1 is αR_2, what is the value of $\mathbf{H}(0)$? (Typically $\alpha = 9$.)

   (c) What value must C_1 have so that $\mathbf{H}(s)$ is a constant?

   (d) What equivalent capacitance and parallel resistance is seen by the source for the condition in (c)?

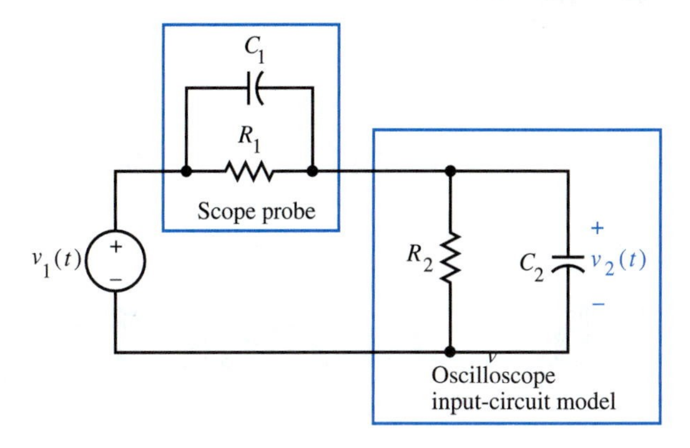

14

The Laplace Transform

One of the easiest ways to solve circuit problems involving initial conditions is by means of the Laplace transform. The principal property of the Laplace transform is that it transforms differential operators into algebraic operators. By exploiting this property, we can obtain the *complete* response of an LTI circuit to almost any input signal by using only algebra and table look-up. In addition to its value in the solution of circuit problems, the Laplace transform provides valuable theoretical insight into the interaction of signals and LTI systems.

14.1 Definitions

The Laplace transform† of a signal $v(t)$ is defined as follows:

DEFINITION Laplace Transform

$$V(s) = \int_{0^-}^{\infty} v(t)e^{-st}\,dt \qquad \mathcal{Re}\{s\} > \sigma_0 \tag{14.1}$$

where the notation 0^- indicates that the point $t = 0$ is included in the integration. The integral in Eq. (14.1) converges only for values of s that satisfy $\mathcal{Re}\{s\} > \sigma_0$, where σ_0 is a constant determined by $v(t)$. The region $\mathcal{Re}\{s\} > \sigma_0$ is the *region of convergence,* and σ_0 is the *abscissa of convergence.* We say that $V(s)$ *exists* in the region of convergence and is *undefined* outside of the region of convergence. When we study specific examples of the Laplace transform, we will see that the condition $\mathcal{Re}\{s\} > \sigma_0$ arises naturally as part of the development.

Associated with the *direct* Laplace transform, Eq. (14.1), is the *inverse* Laplace transform, which is given by:

Inverse Laplace Transform

$$v(t) = \int_{c-j\infty}^{c+j\infty} V(s)e^{st}\,\frac{ds}{2\pi j} \qquad c > \sigma_0 \tag{14.2}$$

In future work we will not have to evaluate the integral in Eq. (14.2) analytically. Instead, we will simply use a look-up table. We develop the look-up table in Section 14.3.

It is important to note that the direct Laplace transform $V(s)$, defined by Eq. (14.1), depends only on $v(t)$ for $t \geq 0$. The right-hand side of Eq. (14.2) converges to $v(t)$ for $t \geq 0$ but to 0 for $t < 0$. This result means that the transformation in Eq. (14.1) is one-to-one‡ for waveforms $v(t)$ that equal zero for $t < 0$. Waveforms of this type are called *causal* waveforms. Causal waveforms occur frequently in circuit analysis problems. Moreover, many noncausal waveforms can be made causal by the simple use of a time-shift operation, $t' = t - t_0$. Unless stated to the contrary, we will assume throughout this chapter that $v(t)$ is a causal waveform, and write Eqs. (14.1) and (14.2), respectively, as

† The Laplace transform comes in two varieties, the *unilateral* or *one-sided* Laplace transform and the *bilateral* or *two-sided* Laplace transform. Only the unilateral Laplace transform is considered in this book, where it is referred to as simply the Laplace transform.

‡ We will ignore the mathematical exceptions, because they do not correspond to physically measurable quantities.

$$V(s) = \mathcal{L}\{v(t)\} \tag{14.3a}$$

and

$$v(t) = \mathcal{L}^{-1}\{V(s)\} \tag{14.3b}$$

or

$$v(t) \leftrightarrow V(s) \tag{14.3c}$$

The functions $v(t)$ and $V(s)$ taken together are called a *Laplace transform pair.*

To develop a feeling for Eq. (14.1), we now derive the Laplace transforms of three simple but important waveforms. Study Example 14.1, and notice how the condition $\mathcal{Re}\{s\} > \sigma_0$ arises.

EXAMPLE 14.1 Determine the Laplace transform of $v(t) = e^{at}u(t)$, where a is an arbitrary real number.

Solution Substitution of $v(t) = e^{at}u(t)$ into Eq. (14.1) yields

$$\begin{aligned} V(s) &= \int_{0^-}^{\infty} e^{at}e^{-st}\,dt \\ &= \int_{0^-}^{\infty} e^{(a-s)t}\,dt = \frac{1}{a-s}e^{(a-s)t}\Big|_{0^-}^{\infty} \end{aligned}$$

which is

$$V(s) = \frac{1}{a-s}e^{(a-s)t}\Big|_{t=\infty} - \frac{1}{a-s}e^{(a-s)t}\Big|_{t=0^-}$$

With $\sigma + j\omega$ substituted for s in the exponent, this becomes

$$V(s) = \frac{1}{a-s}e^{(a-\sigma)t}e^{-j\omega t}\Big|_{t=\infty} - \frac{1}{a-s}e^{(a-\sigma)t}e^{-j\omega t}\Big|_{t=0^-}$$

The second term is simply

$$-\frac{1}{a-s}e^{(a-\sigma)0^-}e^{-j\omega 0^-} = -\frac{1}{a-s}$$

Evaluation of the first term requires care. The term $[1/(a-s)]e^{(a-\sigma)t}e^{-j\omega t}\big|_{t=\infty}$ should be interpreted as the limit $\lim_{t\to\infty}[1/(a-s)]e^{(a-\sigma)t}e^{-j\omega t}$. If $\sigma \le a$, then this limit is not a well-defined mathematical quantity. If $\sigma > a$, then the limit is well defined because the factor $e^{(a-\sigma)\infty}$ equals zero for $a - \sigma < 0$. Therefore

$$V(s) = \frac{1}{s-a}$$

provided that $\mathcal{Re}\{s\} > a$, and we have the result

$$e^{at}u(t) \leftrightarrow \frac{1}{s-a} \qquad \mathcal{Re}\{s\} > a$$

The abscissa of convergence is $\sigma_0 = a$.

The region of convergence is illustrated in Fig. 14.1a for $a < 0$ and in Fig. 14.1b for $a > 0$.

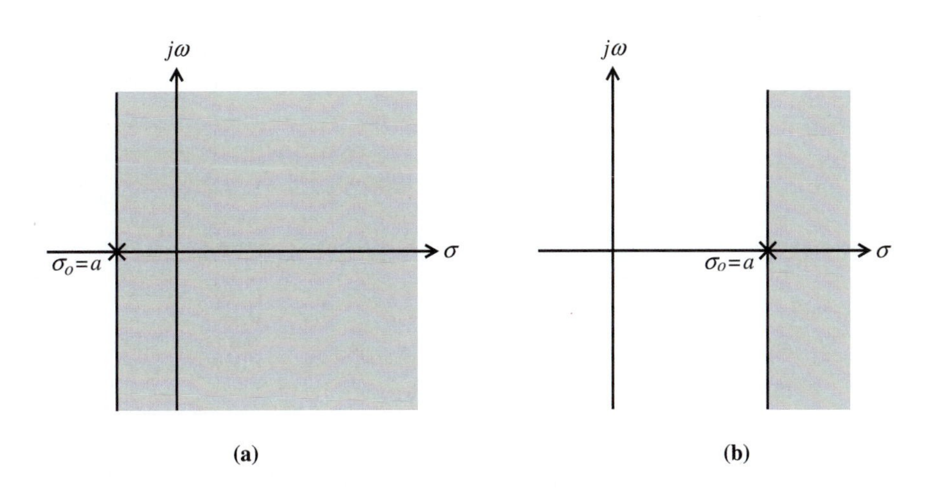

FIGURE 14.1 The region of convergence (shaded) of $V(s) = \mathcal{L}\{e^{at}u(t)\} = 1/(s-a)$. Note that $V(s)$ has a pole $s = a$. (*a*) $a < 0$; (*b*) $a > 0$

EXAMPLE 14.2 Determine the Laplace transform of a unit step function $v(t) = u(t)$.

Solution This function is a special case of Example 14.1 with $a = 0$. Consequently

$$u(t) \leftrightarrow \frac{1}{s} \qquad \sigma > 0$$

The region of convergence is illustrated in Fig. 14.2.

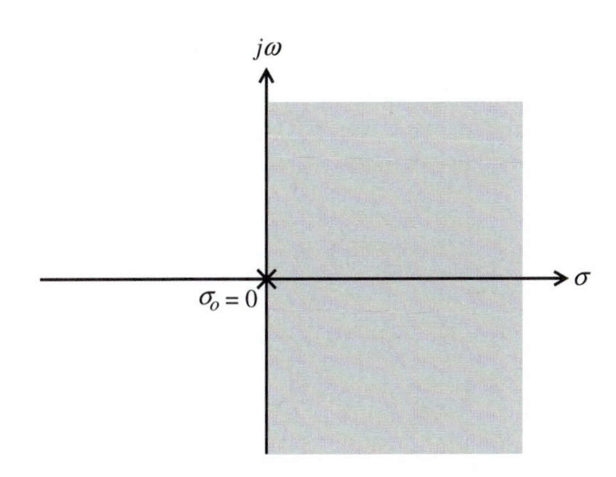

FIGURE 14.2 The region of convergence (shaded) of $V(s) = \mathcal{L}\{u(t)\} = 1/s$

EXAMPLE 14.3 Determine the Laplace transform of the unit impulse $v(t) = \delta(t)$.

Solution Substitution of $\delta(t)$ into Eq. (14.1) yields

$$V(s) = \int_{0-}^{\infty} \delta(t)e^{-st}\,dt = 1 \qquad \text{for all } s$$

Thus

$$\delta(t) \leftrightarrow 1$$

The abscissa of convergence is $\sigma_0 = -\infty$, because $V(s)$ exists for all s. The region of convergence is illustrated in Fig. 14.3.

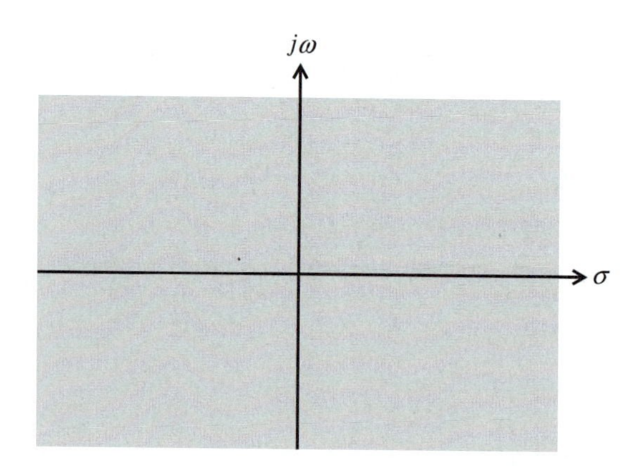

FIGURE 14.3 The region of convergence (shaded) of $V(s)$ for $v(t) = \delta(t)$

In the preceding examples, we have found the Laplace transforms of only three of an infinite number of possible waveforms. These three waveforms, however, are encountered repeatedly in circuit analysis and provide the building blocks for many more waveforms.

Remember

The Laplace transform of $e^{at}u(t)$, where a is any real number, is $1/(s - a)$, where $\mathcal{Re}\{s\} > a$.

EXERCISES Determine the Laplace transform of the signals in Exercises 1 through 5. Show the poles and zeros of $V(s)$ in the s plane, and shade the region of convergence of $V(s)$.

1. $v(t) = u(t - 1)$ answer: $\dfrac{e^{-s}}{s}, \mathcal{Re}\{s\} > 0$

2. $v(t) = u(t) - u(t - 1)$ answer: $\dfrac{1}{s} - \dfrac{e^{-s}}{s}, \mathcal{Re}\{s\} > 0$

3. $v(t) = e^{-3t}u(t) + e^{-5t}u(t)$ answer: $\dfrac{1}{s+3} + \dfrac{1}{s+5}, \mathcal{Re}\{s\} > -3$

4. $v(t) = e^{-3t}u(t) + e^{5t}u(t)$ answer: $\dfrac{1}{s+3} + \dfrac{1}{s-5}, \mathcal{Re}\{s\} > 5$

5. Find $v(t)$ if $V(s) = 1/(s + 2)$ where $\mathcal{Re}\{s\} > -2$. answer: $e^{-2t}u(t)$

6. Find $v(t)$ if

$$V(s) = 1 + \frac{1}{s} + \frac{1}{s+2} \qquad \mathcal{Re}\{s\} > 0$$

(*Hint*: Use the results of Examples 14.1 through 14.3.)
answer: $\delta(t) + u(t) + e^{-2t}u(t)$

14.2 Properties of the Laplace Transform

The Laplace transform provides a means to solve LTI circuit problems through the use of algebra and table look-up only. In this section we describe the Laplace transform properties that make this method of solution possible.

TABLE 14.1
Properties of the Laplace transform

Property	Signal	Laplace Transform
Linearity	$a_1v_1(t) + a_2v_2(t)$	$a_1V_1(s) + a_2V_2(s)$
Differentiation	$\dfrac{d^nv}{dt^n}$	$-v^{(n)}(0^-) - sv^{(n-1)}(0^-) - \cdots - s^{n-1}v(0^-) + s^nV(s)$
Special case, $n = 1$	$\dfrac{dv}{dt}$	$-v(0^-) + sV(s)$
Special case, $n = 2$	$\dfrac{d^2v}{dt^2}$	$-v^{(1)}(0^-) - sv(0^-) + s^2V(s)$
Integration	$\int_{-\infty}^{t} v(\lambda)\, d\lambda,\ t > 0$	$\dfrac{1}{s}V(s) + \dfrac{1}{s}\int_{-\infty}^{0-} v(\lambda)\, d\lambda$
Scaling	$v\left(\dfrac{t}{a}\right)$	$aV(as),\ a > 0$
Time delay	$v(t - t_d)u(t - t_d),\ t_d \geq 0$	$e^{-st_d}V(s)$
s shift	$v(t)e^{s_0t}$	$V(s - s_0)$
Convolution	$\int_0^t v_1(\lambda)v_2(t - \lambda)d\lambda$	$V_1(s)V_2(s)$
Initial value theorem	$v(0^+) = \lim_{\mathcal{R}e\{s\} \to \infty} sV(s)$	Provided the limit exists
Final value theorem	$\lim_{t \to \infty} v(t) = \lim_{s \to 0} sV(s)$	Provided the limit exists

Table 14.1 lists several important properties of the Laplace transform. The first column of the table simply lists the names of the properties. The second column lists various operations on a time-domain signal and the third column lists the Laplace transforms that result from the operations. For example, the *linearity* property tells us that if we scale and add two time-domain signals, then the Laplace transform of the result is simply the scaled and added Laplace transforms of the individual signals. Another example is given by the *differentiation* property, which gives us a way to find the Laplace transform of the derivative of a signal. Consider the first derivative where $n = 1$ in the table. The differentiation property tells us that the Laplace transform of dv/dt is $-v(0^-) + sV(s)$ where $v(0^-)$ is the value of the signal just before time $t = 0$ and $V(s)$ is the Laplace transform of $v(t)$. The point is that Table 14.1 provides us with a means to easily find the Laplace transforms of many signals without the need to evaluate Eq. (14.1).

Proofs of the properties most important to us are given later in this chapter. Before delving into the proofs, let us see from specific examples how we can use the properties to derive new Laplace transform pairs.

EXAMPLE 14.4 Use the linearity and s-shift properties to find the Laplace transform of $v(t) = \cos(\omega_0 t)u(t)$.

Solution By using Euler's identity we can write $v(t)$ as

$$v(t) = \frac{1}{2}e^{j\omega_0 t}u(t) + \frac{1}{2}e^{-j\omega_0 t}u(t)$$

It follows by the linearity property that

$$V(s) = \frac{1}{2}\mathcal{L}\{e^{j\omega_0 t}u(t)\} + \frac{1}{2}\mathcal{L}\{e^{-j\omega_0 t}u(t)\}$$

We know from Example 14.2 that

$$u(t) \leftrightarrow \frac{1}{s} \qquad \mathcal{Re}\{s\} > 0$$

Therefore, from the s-shift property,

$$\mathcal{L}\{e^{j\omega_0 t}u(t)\} = \frac{1}{s - j\omega_0} \qquad \mathcal{Re}\{s\} > 0$$

and

$$\mathcal{L}\{e^{-j\omega_0 t}u(t)\} = \frac{1}{s + j\omega_0} \qquad \mathcal{Re}\{s\} > 0$$

By combining the above results, we obtain

$$V(s) = \frac{1}{2}\frac{1}{s - j\omega_0} + \frac{1}{2}\frac{1}{s + j\omega_0} \qquad \mathcal{Re}\{s\} > 0$$

which simplifies to

$$V(s) = \frac{s}{s^2 + \omega_0^2} \qquad \mathcal{Re}\{s\} > 0$$

Therefore

$$\cos(\omega_0 t)u(t) \leftrightarrow \frac{s}{s^2 + \omega_0^2} \qquad \mathcal{Re}\{s\} > 0$$

In the preceding example, the region of convergence of each Laplace transform we encountered was $\mathcal{Re}\{s\} > 0$. In Example 14.5, we work with Laplace transforms that have different regions of convergence.

EXAMPLE 14.5 Assume that

$$v(t) = 2\delta(t) + 3e^{-2t}u(t) + \cos(2t)u(t) + 4e^{3t}u(t)$$

(a) Find $V(s)$.
(b) Plot the poles of $V(s)$ in the complex plane and show the region of convergence.

Solution (a) We can find $V(s)$ using the linearity property and the results of Examples 14.1, 14.3, and 14.4.

By the linearity property, we know that

$$V(s) = 2\mathcal{L}\{\delta(t)\} + 3\mathcal{L}\{e^{-2t}u(t)\} + \mathcal{L}\{\cos(2t)u(t)\} + 4\mathcal{L}\{e^{3t}u(t)\}$$

It follows from Examples 14.1, 14.3, and 14.4 that

$$\mathcal{L}\{\delta(t)\} = 1 \qquad \text{for all } s$$

$$\mathcal{L}\{e^{-2t}u(t)\} = \frac{1}{s + 2} \qquad \text{for } \mathcal{Re}\{s\} > -2$$

$$\mathcal{L}\{\cos(2t)u(t)\} = \frac{s}{s^2 + 4} \qquad \text{for } \mathcal{Re}\{s\} > 0$$

and

$$\mathcal{L}\{e^{3t}u(t)\} = \frac{1}{s-3} \qquad \text{for } \mathcal{R}e\{s\} > 3$$

Therefore

$$V(s) = 2 + \frac{3}{s+2} + \frac{s}{s^2+4} + \frac{4}{s-3}$$
$$= \frac{2s^4 + 6s^3 - 6s^2 + 14s - 52}{(s+2)(s^2+4)(s-3)}$$

(b) The poles are the values of s that cause the denominator of $V(s)$ to equal zero. Therefore the poles of $V(s)$ are given by $s = -2$, $s = j2$, $s = -j2$, and $s = 3$. These are plotted in Fig. 14.4. We can determine the region of convergence of $V(s)$ by writing $V(s)$ in its original form and noting the region of convergence of each term:

$$V(s) = \underbrace{2}_{\text{All } s} + \underbrace{\frac{3}{s+2}}_{\mathcal{R}e\{s\} > -2} + \underbrace{\frac{s}{s^2+4}}_{\mathcal{R}e\{s\} > 0} + \underbrace{\frac{4}{s-3}}_{\mathcal{R}e\{s\} > 3}$$

$V(s)$ exists if and only if *all four* terms exist. We see that all four terms exist if and only if $\mathcal{R}e\{s\} > 3$. The region of convergence of $V(s)$ is therefore given by $\mathcal{R}e\{s\} > 3$, which is the shaded region in Fig. 14.4. Note that the region of convergence of $V(s)$ lies just to the right of the poles of $V(s)$.

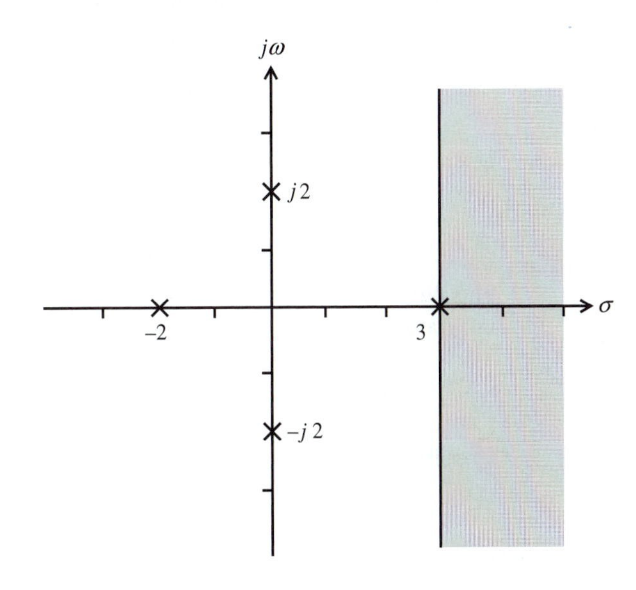

FIGURE 14.4 The region of convergence (shaded) for $V(s) = (2s^4 + 6s^3 - 6s^2 + 14s - 52)/[(s+2)(s^2+4)(s-3)]$. Notice that the region of convergence lies to the immediate right of the poles. The abscissa of convergence is $\sigma_0 = 3$

In Examples 14.1 through 14.5, the region of convergence was given by $\mathcal{R}e\{s\} > \sigma_0$, which was the region just to the right of the poles of $V(s)$. By using methods similar to those in Example 14.5b, we can show that the region of convergence of *every* Laplace transform $V(s)$ lies to the immediate right of the poles of $V(s)$. This result applies to all subsequent examples in this chapter. The formal proof is based on *partial fractions*, which are described in Section 14.4.

Now that we have developed some experience with the use of Table 14.1, let us prove some of the properties.

Linearity

The linearity property is used extensively in solving differential equations. The property states that the Laplace transform of the scaled sum of two signals equals the scaled sum of the Laplace transform of each signal. This property is very easy to prove from the definition of the Laplace transform, Eq. (14.1). The Laplace transform of $a_1v_1(t) + a_2v_2(t)$ is, by definition,

$$L\{a_1v_1(t) + a_2v_2(t)\} = \int_{0^-}^{+\infty} \{a_1v_1(t) + a_2v_2(t)\}\, e^{-st}\, dt \tag{14.4}$$

The integral of a sum equals the sum of integrals, therefore,

$$\begin{aligned} L\{a_1v_1(t) + a_2v_2(t)\} &= a_1 \int_{0^-}^{+\infty} v_1(t)\, e^{-st}\, dt + a_2 \int_{0^-}^{+\infty} v_2(t)\, e^{-st}\, dt \\ &= a_1\, L\{v_1(t)\} + a_2\, L\{v_2(t)\} \end{aligned} \tag{14.5}$$

In other words,

$$a_1v_1(t) + a_2v_2(t) \leftrightarrow a_1V_1(s) + a_2V_2(s) \tag{14.6}$$

which establishes the linearity property.

Differentiation

The differentiation property states that the Laplace transform of the nth derivative of $v(t)$ is given by

$$\mathcal{L}\left\{\frac{d^n v}{dt^n}\right\} = -v^{(n)}(0^-) - sv^{(n-1)}(0^-) - \cdots - s^{n-1}v(0^-) + s^nV(s)$$

where

$$v^{(k)}(0^-) = \left.\frac{d^k v}{dt^k}\right|_{t=0^-} \qquad \text{for } k = 1, 2, \ldots \tag{14.7}$$

This property is important in the solution of LTI equations. We can prove it for $n = 1$ by integrating

$$\mathcal{L}\left\{\frac{dv}{dt}\right\} = \int_{0^-}^{\infty} \frac{dv}{dt}\, e^{-st}\, dt \tag{14.8}$$

by parts. If we set $y = e^{-st}$ and $dx = (dv/dt)\, dt$, the identity

$$\int_{0^-}^{\infty} y\, dx = xy\Big|_{0^-}^{\infty} - \int_{0^-}^{\infty} x\, dy \tag{14.9}$$

yields

$$\begin{aligned} \mathcal{L}\left\{\frac{dv}{dt}\right\} &= v(t)e^{-st}\Big|_{0^-}^{\infty} + s\int_{0^-}^{\infty} v(t)e^{-st}\, dt \\ &= v(t)e^{-st}\Big|_{t=\infty} - v(t)e^{-st}\Big|_{t=0^-} + s\mathcal{L}\,\{v(t)\} \end{aligned} \tag{14.10}$$

The first term on the right-hand side equals zero for values if s is within the region of convergence. Therefore the above equation becomes

$$\mathcal{L}\left\{\frac{dv}{dt}\right\} = -v(0^-) + s\mathcal{L}\{v(t)\} \tag{14.11}$$

Equation (14.11) establishes the differentiation property for $n = 1$. To establish it for $n = 2$, we simply replace $v(t)$ with dv/dt in Eq. (14.11) to obtain

$$\mathcal{L}\left\{\frac{d^2v}{dt^2}\right\} = -\left.\frac{dv}{dt}\right|_{t=0^-} + s\mathcal{L}\left\{\frac{dv}{dt}\right\} \tag{14.12}$$

Substitution of Eq. (14.11) into Eq. (14.12) yields

$$\mathcal{L}\left\{\frac{d^2v}{dt^2}\right\} = -\left.\frac{dv}{dt}\right|_{t=0^-} - sv(0^-) + s^2\mathcal{L}\{v(t)\} \tag{14.13}$$

which establishes the differentiation property for $n = 2$. If we repeat the process we obtain the general result

$$\mathcal{L}\left\{\frac{d^nv}{dt^n}\right\} = -v^{(n)}(0^-) - sv^{(n-1)}(0^-) - \cdots - s^{n-1}v(0^-) + s^nV(s) \tag{14.14}$$

Integration

According to this property, the Laplace transform of the integral of $v(t)$ is given by

$$\mathcal{L}\left\{\int_{-\infty}^{t} v(\lambda)\,d\lambda\right\} = \frac{1}{s}V(s) + \frac{1}{s}\int_{-\infty}^{0^-} v(\lambda)\,d\lambda$$

where $t > 0$. This result is obtained from Eq. (14.11) by defining

$$v(t) = \int_{-\infty}^{t} g(\lambda)\,d\lambda \tag{14.15}$$

where $g(t)$ is not necessarily zero for $t < 0$. It follows that

$$\frac{dv}{dt} = g(t) \tag{14.16}$$

and

$$v(0^-) = \int_{-\infty}^{0^-} g(\lambda)\,d\lambda \tag{14.17}$$

so that when Eqs. (14.16) and (14.17) are substituted into Eq. (14.11), we obtain

$$\mathcal{L}\{g(t)\} = -\int_{-\infty}^{0^-} g(\lambda)\,d\lambda + s\mathcal{L}\left\{\int_{-\infty}^{t} g(\lambda)\,d\lambda\right\} \tag{14.18}$$

which can be solved to give

$$\mathcal{L}\left\{\int_{-\infty}^{t} g(\lambda)\,d\lambda\right\} = \frac{1}{s}\mathcal{L}\{g(t)\} + \frac{1}{s}\int_{-\infty}^{0^-} g(\lambda)\,d\lambda \tag{14.19}$$

A purely notational change then yields

$$\mathcal{L}\left\{\int_{-\infty}^{t} v(\lambda)\,d\lambda\right\} = \frac{1}{s}V(s) + \frac{1}{s}\int_{-\infty}^{0^-} v(\lambda)\,d\lambda \tag{14.20}$$

which is the integration property. Of course, the integral on the right-hand side of Eq. (14.20) will equal zero whenever $v(t)$ is causal.

The initial and final value theorems in Table 14.1 are useful because they enable us to compute initial and final values of $v(t)$ without computing the inverse transform of $V(s)$.

The Initial Value Theorem

The *initial value theorem* states that the value of $v(0^+)$ is given by

$$v(0^+) = \lim_{\mathcal{Re}\{s\}\to\infty} sV(s)$$

provided that the limit exists. To prove the initial value theorem, we begin with the differentiation specialized to $n = 1$:

$$\int_{0^-}^{\infty} \frac{dv}{dt} e^{-st}\, dt = sV(s) - v(0^-) \tag{14.21}$$

The left-hand side can be broken up into an integral from 0^- to 0^+ and an integral from 0^+ to ∞. This step yields

$$\int_{0^-}^{\infty} \frac{dv}{dt} e^{-st}\, dt = v(0^+) - v(0^-) + \int_{0^+}^{\infty} \frac{dv}{dt} e^{-st}\, dt \tag{14.22}$$

provided that $v(t)$ does not contain an impulse or derivatives of an impulse at $t = 0$. Substitution of Eq. (14.22) into Eq. (14.21) yields

$$v(0^+) + \int_{0^+}^{\infty} \frac{dv}{dt} e^{-st}\, dt = sV(s) \tag{14.23}$$

If we now let $\mathcal{Re}\{s\} \to \infty$, the factor e^{-st} in the integrand will force the integral in the above equation to equal zero. Therefore,

$$v(0^+) = \lim_{\mathcal{Re}\{s\}\to\infty} sV(s) \tag{14.24}$$

provided the limit exists. This proves the initial value theorem.

The Final Value Theorem

The *final value theorem* gives the following formula for the final value of $v(t)$:

$$\lim_{t\to\infty} v(t) = \lim_{s\to 0} sV(s)$$

provided $v(t)$ *has* a final value.†

To prove the final value theorem, we again begin with the differentiation property where $n = 1$, namely, Eq. (14.21). In the limit $s \to 0$, e^{-st} becomes unity. Thus, with $s \to 0$, Eq. (14.21) becomes

$$\int_{0^-}^{\infty} \frac{dv}{dt}\, dt = \lim_{s\to 0} sV(s) - v(0^-) \tag{14.25}$$

But

$$\int_{0^-}^{\infty} \frac{dv}{dt}\, dt = \lim_{t\to\infty} v(t) - v(0^-) \tag{14.26}$$

† For example, the final value theorem cannot be used to find the final value of $v(t) = \cos t$ because $\cos t$ oscillates between $+1$ and -1 forever. However, the initial value theorem *can* be used to find the value of $\cos 0^+$ because $\lim_{t\to 0} \cos t$ is well defined.

By substituting Eq. (14.26) into Eq. (14.25), we get

$$\lim_{t\to\infty} v(t) = \lim_{s\to 0} sV(s) \tag{14.27}$$

provided the limit exists. This proves the final value theorem.

We illustrate the initial and final value theorems in the following example.

EXAMPLE 14.6 Let $V_o(s) = (5s - 10{,}000)/(s^2 + 1000s)$ for $\mathcal{Re}\{s\} > 0$. Use the initial and final value theorems to find (a) $v(0^+)$ and (b) $\lim_{t\to\infty} v(t)$.

Solution (a) According to the initial value theorem,

$$v(0^+) = \lim_{\mathcal{Re}\{s\}\to\infty} sV(s) = \lim_{\mathcal{Re}\{s\}\to\infty} \left\{\frac{5s^2 - 10{,}000s}{s^2 + 1000s}\right\} = 5$$

(b) According to the final value theorem,

$$\lim_{t\to\infty} v(t) = \lim_{s\to 0} sV(s) = \lim_{s\to 0} \left\{\frac{5s^2 - 10{,}000s}{s^2 + 1000s}\right\} = -10$$

At this point we have proven five of the properties in Table 14.1. The proofs of the scaling and time-delay properties are straightforward and we relegate them to Problems 27 and 28. The proof of the convolution property is given in Appendix 14.1. In the following section we use Table 14.1 to develop a look-up table of important Laplace transform pairs. We will defer the exercises for Table 14.1 until we have developed Table 14.2.

14.3 Important Laplace Transform Pairs

To solve LTI circuit problems through the use of algebra and table look-up, we need a table of basic Laplace transform pairs in addition to a table of properties. Table 14.2 contains 12 Laplace transform pairs that are used pervasively in circuit analysis. In subsequent sections we will see how to use this table. This section is devoted totally to establishing Table 14.2. Transform pair LT1 was derived in Example 14.2. We can derive all the remaining entries in Table 14.2 by applying the properties from Table 14.1 to LT1. The essential steps are as follows.

To derive LT2 we begin with the fact that

$$\int_{-\infty}^{t} \delta(\lambda)\, d\lambda = u(t) \tag{14.28}$$

and

$$\int_{-\infty}^{0^-} \delta(\lambda)\, d\lambda = 0 \tag{14.29}$$

Therefore, if we set $v(t) = \delta(t)$ and $V(s) = 1$ and apply the integration property, we obtain

$$u(t) \leftrightarrow \frac{1}{s} \tag{14.30}$$

which is LT2.

TABLE 14.2
Laplace transform pairs†

	Signal v(t)	Laplace Transform V(s)
LT1	$\delta(t)$	1
LT2	$u(t)$	$\frac{1}{s}$
LT3	$tu(t)$	$\frac{1}{s^2}$
LT4	$\frac{1}{n!}t^n u(t)$	$\frac{1}{s^{n+1}}$
LT5	$e^{s_0 t}u(t)$	$\frac{1}{s - s_0}$
LT6	$te^{s_0 t}u(t)$	$\frac{1}{(s - s_0)^2}$
LT7	$\frac{1}{n!}t^n e^{s_0 t}u(t)$	$\frac{1}{(s - s_0)^{n+1}}$
LT8	$\lvert\mathbf{K}_0\rvert \cos(\omega_0 t + \angle\mathbf{K}_0)u(t)$	$\frac{(1/2)\mathbf{K}_0}{s - j\omega_0} + \frac{(1/2)\mathbf{K}_0^*}{s + j\omega_0}$
LT9	$\lvert\mathbf{K}_0\rvert e^{\sigma_0 t} \cos(\omega_0 t + \angle\mathbf{K}_0)u(t)$	$\frac{(1/2)\mathbf{K}_0}{s - s_0} + \frac{(1/2)\mathbf{K}_0^*}{s - s_0^*}$
LT10	$\frac{\lvert\mathbf{K}_0\rvert t^{n-1}}{(n-1)!} e^{\sigma_0 t} \cos(\omega_0 t + \angle\mathbf{K}_0)u(t)$	$\frac{(1/2)\mathbf{K}_0}{(s - s_0)^n} + \frac{(1/2)\mathbf{K}_0^*}{(s - s_0^*)^n}$
LT11	$e^{-\alpha t} \cos(\omega_d t)u(t)$	$\frac{s + \alpha}{(s + \alpha)^2 + \omega_d^2}$
LT12	$e^{-\alpha t} \sin(\omega_d t)u(t)$	$\frac{\omega_d}{(s + \alpha)^2 + \omega_d^2}$

† The region of convergence of each transform listed is the region to the immediate right of the poles of $V(s)$. For entries LT5 through LT10, $s_0 = \sigma_0 + j\omega_0$ denotes an arbitrary point in the complex plane. For entries LT8 through LT10, $\mathbf{K}_0$ is a complex constant.

We derive LT3 by repeating the above steps. We start with

$$\int_{-\infty}^{t} u(\lambda)\, d\lambda = tu(t) \tag{14.31}$$

and

$$\int_{-\infty}^{0-} u(\lambda)\, d\lambda = 0 \tag{14.32}$$

The integration property yields [with $v(t) = u(t)$ and $V(s) = 1/s$]

$$tu(t) \leftrightarrow \frac{1}{s^2} \tag{14.33}$$

Iterating n times yields LT4:

$$\frac{1}{n!}\, t^n u(n) \leftrightarrow \frac{1}{s^{n+1}} \tag{14.34}$$

LT5 follows by applying the s-shift property to LT2. This shows immediately that

$$e^{s_0 t}u(t) \leftrightarrow \frac{1}{s - s_0} \tag{14.35}$$

Notice that this result is a generalization of the special case in which s_0 is a real constant $s_0 = a$, derived in Example 14.1.

LT7 is obtained by application of the s-shift property to LT4. This yields

$$\frac{1}{n!}\,t^n e^{s_0 t}u(t) \leftrightarrow \frac{1}{(s - s_0)^{n+1}} \tag{14.36}$$

LT6 is, of course, a special case of LT7 where $n = 1$.

LT10 is obtained from LT7 and the linearity property. If we replace s_0 with s_0^* in Eq. (14.36) and apply the linearity property, we find that

$$\frac{\mathbf{K}_0}{(n-1)!}\,t^{n-1}e^{s_0 t}u(t) + \frac{\mathbf{K}_0^*}{(n-1)!}\,t^{n-1}e^{s_0^* t}u(t) \leftrightarrow \frac{\mathbf{K}_0}{(s - s_0)^n} + \frac{\mathbf{K}_0^*}{(s - s_0^*)^n} \tag{14.37}$$

where $\mathbf{K}_0$ is a constant. With $s_0 = \sigma_0 + j\omega_0$, the left-hand side of Eq. (14.37) becomes

$$\frac{\mathbf{K}_0}{(n-1)!}\,t^{n-1}e^{s_0 t}u(t) + \frac{\mathbf{K}_0^*}{(n-1)!}\,t^{n-1}e^{s_0^* t}u(t) = \frac{2|\mathbf{K}_0|}{(n-1)!}\,t^{n-1}e^{\sigma_0 t}\cos(\omega_0 t + \angle\mathbf{K}_0) \tag{14.38}$$

Therefore

$$\frac{2|\mathbf{K}_0|}{(n-1)!}\,t^{n-1}e^{\sigma_0 t}\cos(\omega_0 t + \angle\mathbf{K}_0) \leftrightarrow \frac{\mathbf{K}_0}{(s - s_0)^n} + \frac{\mathbf{K}_0^*}{(s - s_0^*)^n} \tag{14.39}$$

which is LT10. LT8 and LT9 are special cases of this result.

Transform pairs LT11 and LT12 are special cases of LT9. We leave the details to Exercises 18 and 19.

We have listed only the most basic Laplace transform pairs in Table 14.2. However, by using Tables 14.1 and 14.2, we can derive the Laplace transform of virtually every signal encountered in everyday circuit analysis. We shall encounter several pertinent examples in Sections 14.5 through 14.7.

EXERCISES

Use Tables 14.1 and 14.2 to obtain the Laplace transforms of the following signals.

7. $tu(t)$ — answer: $\dfrac{1}{s^2}$

8. $(t - t_0)u(t - t_0)$ where $t_0 > 0$ — answer: $\dfrac{e^{-st_0}}{s^2}$

9. $tu(t - t_0)$ where $t_0 > 0$ — answer: $(1 + st_0)\dfrac{e^{-st_0}}{s^2}$

10. $5te^{7t}u(t) + 2\cos(\omega_0 t + 45°)u(t)$ — answer: $\dfrac{5}{(s-7)^2} + \dfrac{1\angle 45°}{s + j\omega_0} + \dfrac{1\angle -45°}{s + j\omega_0}$

11. $\Pi\left(\dfrac{t - 0.5\tau}{\tau}\right)$ — answer: $\dfrac{1}{s} - \dfrac{e^{-s\tau}}{s}$

Use Tables 14.1 and 14.2 to determine the inverse Laplace transforms of the following functions.

12. $\dfrac{1}{s}e^{-s}$ answer: $u(t-1)$

13. $\dfrac{1}{s - j30 + 10}$ answer: $e^{(-10+j30)t}u(t)$

14. $\dfrac{s-2}{s^2 - 2s + 2}$ answer: $e^t[\cos(t) - \sin(t)]u(t)$

15. Use Tables 14.1 and 14.2 to determine the inverse Laplace transform of $V(s) = (s+1)/[(s+2)(s+3)]$. [*Hint*: First write $V(s)$ in partial fraction form, $V(s) = a/(s+2) + b/(s+3)$, where a and b are constants to be determined by you.] answer: $-e^{-2t}u(t) + 2e^{-3t}u(t)$

16. Use LT11 of Table 14.2 to show that

$$\cos(\omega_0 t)u(t) \leftrightarrow \frac{s}{s^2 + \omega_0^2}$$

17. Use LT12 of Table 14.2 to show that

$$\sin(\omega_0 t)u(t) \leftrightarrow \frac{\omega_0}{s^2 + \omega_0^2}$$

18. Use LT9 to show that

$$e^{-\alpha t}\cos(\omega_d t)u(t) \leftrightarrow \frac{s+\alpha}{(s+\alpha)^2 + \omega_d^2}$$

19. Use LT9 to show that

$$e^{-\alpha t}\sin(\omega_d t)u(t) \leftrightarrow \frac{\omega_d}{(s+\alpha)^2 + \omega_d^2}$$

Find the initial and final values of the signals with the following Laplace transforms.

20. $\dfrac{1}{s}e^{-3s}$ answer: $v(0^+) = 0,\ v(\infty) = 1$

21. $\dfrac{s}{(s+1)(s+2)}$ answer: $v(0^+) = 1,\ v(\infty) = 0$

22. $\dfrac{s+3}{s^2 + 6s + 18}e^{-s}$ answer: $v(0^+) = 0,\ v(\infty) = 0$

14.4 Partial Fraction Expansions

Partial fractions provide the key to the table look-up technique to determine inverse Laplace transforms. In most problems, inverse Laplace transforms are required for functions of the form

$$G(s) = \frac{N(s)}{D(s)} \tag{14.40}$$

where $N(s)$ and $D(s)$ are polynomials in s. The function $G(s)$ is called a *rational algebraic* function of s. A specific example of a rational algebraic function is

$$G(s) = \frac{s+1}{(s+2)^2(s-1)} \tag{14.41a}$$

for which the numerator polynomial is

$$N(s) = s + 1 \tag{14.41b}$$

and the denominator polynomial is

$$D(s) = (s+2)^2(s-1) = s^3 + 3s^2 - 4 \tag{14.41c}$$

Arbitrary rational algebraic functions like Eq. (14.41a) do not appear in Table 14.2. However, rational algebraic functions can be written as a sum of partial fractions, each of which does appear in Table 14.1. Therefore, by relying on the linearity of the Laplace transform, we can determine the inverse transform of a rational algebraic function using table look-up.

Partial Fraction Form

The first step to determine the partial fraction of $G(s)$ in Eq. (14.40) is to write down its *form*. The form of the partial fraction expansion of $G(s)$ depends on the degrees of the numerator and denominator polynomials in Eq. (14.40). In what follows, we denote the degrees of $N(s)$ and $D(s)$ by $deg\{N(s)\}$ and $deg\{D(s)\}$, respectively.

If the degree of the numerator polynomial is less than the degree of the denominator polynomial, $deg\{N(s)\} < deg\{D(s)\}$, then the partial fraction expansion of $G(s) = N(s)/D(s)$ has l terms, where l is the number of distinct poles of $G(s)$:

$$G(s) = L_1(s) + L_2(s) + \cdots + L_l(s) \tag{14.42a}$$

Each term, $L_i(s)$, for $1 \le i \le l$, is associated with a distinct pole s_i. If a pole, s_i, has order $m_i = 1$, then

$$L_i(s) = \frac{k_{i1}}{s - s_i} \tag{14.42b}$$

where k_{i1} is a constant. If any pole, s_i, has order m_i—not necessarily equal to one—then

$$L_i(s) = \frac{k_{i1}}{s - s_i} + \frac{k_{i2}}{(s - s_i)^2} + \cdots + \frac{k_{im_i}}{(s - s_i)^{m_i}} \tag{14.42c}$$

where $k_{i1}, k_{i2}, \ldots, k_{im_i}$ are constants. The constant k_{i1} has special significance and is called the *residue* of $G(s)$ at the pole s_i.

EXAMPLE 14.7 Determine the *form* of the partial fraction expansion of

$$G(s) = \frac{s}{(s+2)(s+4)(s+6)} = \frac{s}{s^3 + 12s^2 + 44s + 48}$$

Solution The degree of the numerator, $deg\{s\} = 1$, is less than the degree of the denominator, $deg\{s^3 + 12s^2 + 44s + 48\} = 3$. There are three distinct poles: $s_1 = -2$, $s_2 = -4$, $s_3 = -6$. Therefore Eq. (14.42a) applies with $l = 3$. Since each pole has order 1, then Eq. (14.42b) applies for $i = 1, 2, 3$. The partial fraction expansion of $G(s)$ has the form

$$G(s) = \frac{s}{(s+2)(s+4)(s+6)} = \frac{k_{11}}{s+2} + \frac{k_{21}}{s+4} + \frac{k_{31}}{s+6}$$

where k_{11}, k_{21}, and k_{31} are the residues of $G(s)$ at -2, -4, and -6, respectively, and are still to be determined.

EXAMPLE 14.8 Determine the *form* of the partial fraction expansion of

$$G(s) = \frac{s+1}{(s+2)^2(s-1)} = \frac{s+1}{s^3+3s^2-4}$$

Solution The degree of the numerator, $deg\{s+1\} = 1$, is less than the degree of the denominator, $deg\{s^3+3s^2-4\} = 3$. There are two distinct poles: $s_1 = -2$, and $s_2 = 1$. Therefore Eq. (14.42a) applies with $l = 2$. Equation (14.42c) applies for pole s_1 because s_1 has order $m_1 = 2$. Equation (14.42b) applies for pole s_2 because s_2 has order $m_2 = 1$. The partial fraction expansion of $G(s)$ has the form

$$G(s) = \frac{s+1}{(s+2)^2(s-1)} = \underbrace{\frac{k_{11}}{s+2} + \frac{k_{12}}{(s+2)^2}}_{L_1(s)} + \underbrace{\frac{k_{21}}{s-1}}_{L_2(s)}$$

where k_{11}, k_{12}, and k_{21} are constants to be determined. The residues of $G(s)$ at -2 and $+1$ are k_{11} and k_{21}, respectively.

If the degree of $N(s)$ equals or exceeds the degree of $D(s)$, $deg\{N(s)\} \geq deg\{D(s)\}$, then the partial fraction expansion of $G(s) = N(s)/D(s)$ will have the form

$$G(s) = p(s) + L_1(s) + L_2(s) + \cdots + L_l(s) \tag{14.43a}$$

where $p(s)$ is a polynomial in s with degree

$$deg\{p(s)\} = deg\{N(s)\} - deg\{D(s)\} \tag{14.43b}$$

and the form of the $L_i(s)$ is as shown in Eq. (14.42).

EXAMPLE 14.9 Determine the *form* of the partial fraction expansion of

$$G(s) = \frac{s^4+2s^3+s^2+1}{s(s+3)} = \frac{s^4+2s^3+s^2+1}{s^2+3s}$$

Solution The degree of the numerator exceeds the degree of the denominator by 2. Therefore, Eq. (14.43) applies, in which $p(s)$ has degree 2, and $G(s)$ has the form

$$G(s) = \frac{s^4+2s^3+s^2+1}{s(s+3)} = \underbrace{\alpha_2 s^2 + \alpha_1 s + \alpha_0}_{p(s)} + \underbrace{\frac{k_{11}}{s}}_{L_1(s)} + \underbrace{\frac{k_{21}}{s+3}}_{L_2(s)}$$

where α_0, α_1, α_2, k_{11}, and k_{21} are constants to be determined. The constants k_{11} and k_{21} are residues of $G(s)$ at 0 and -3, respectively.

Evaluation of Partial Fraction Constants

After the form of a partial fraction expansion has been determined, the next step is to evaluate the constants. A general formulation of the solution is straightforward, but notationally cumbersome. Therefore it is better to proceed with examples.

EXAMPLE 14.10 Determine the partial fraction expansion of

$$G(s) = \frac{s}{(s+2)(s+4)(s+6)} = \frac{k_{11}}{s+2} + \frac{k_{21}}{s+4} + \frac{k_{31}}{s+6}$$

Solution An easy way to determine k_{11} is to first multiply the above equation by $s + 2$ to obtain

$$(s+2)G(s) = \frac{s}{(s+4)(s+6)} = k_{11} + \frac{(s+2)k_{21}}{(s+4)} + \frac{(s+2)k_{31}}{s+6}$$

Now if we set $s = -2$, the above becomes

$$(s+2)G(s)\Big|_{s=-2} = \frac{-2}{(-2+4)(-2+6)} = -\frac{1}{4} = k_{11}$$

The above steps can be written compactly as

$$k_{11} = (s+2)G(s)\Big|_{s=-2}$$

Similarly, we have

$$k_{21} = (s+4)G(s)\Big|_{s=-4} = \frac{-4}{(-4+2)(-4+6)} = 1$$

and

$$k_{31} = (s+6)G(s)\Big|_{s=-6} = \frac{-6}{(-6+2)(-6+4)} = -\frac{3}{4}$$

The method of Example 14.10 worked because the poles in $G(s)$ had order 1. We will show later that if $G(s)$ is *any* rational algebraic function with a simple (that is, first-order) pole at $s = s_i$, then the residue of $G(s)$ at s_i is given by

$$k_{i1} = (s - s_i)G(s)\Big|_{s=s_i} \tag{14.44}$$

The following example will illustrate how to proceed when the poles of $G(s)$ are not simple.

EXAMPLE 14.11 Determine the partial fraction expansion of

$$G(s) = \frac{s+1}{(s+2)^2(s-1)}$$

Solution According to the result of Example 14.8, the form of the partial fraction expansion is

$$\frac{s+1}{(s+2)^2(s-1)} = \frac{k_{11}}{s+2} + \frac{k_{12}}{(s+2)^2} + \frac{k_{21}}{s-1} \tag{14.45}$$

We will describe three methods for finding the values of the constants. Methods 1 and 2 have the advantage of being straightforward and applicable in general. Method 3 has the advantage of being faster, but it yields values only for k_{12} and k_{21}.

Method 1

Multiply both sides of Eq. (14.45) by $(s + 2)^2(s - 1)$ to obtain

$$s + 1 = k_{11}(s + 2)(s - 1) + k_{12}(s - 1) + k_{21}(s + 2)^2$$

and write both sides as second-degree polynomials in s:

$$0s^2 + 1s + 1 = (k_{11} + k_{21})s^2 + (k_{11} + k_{12} + 4k_{21})s - 2k_{11} - k_{12} + 4k_{21}$$

If equality is to hold for all values of s, then the coefficients of like powers of s on either side of this equation must be equal. Thus

$$\begin{aligned} k_{11} + k_{21} &= 0 \\ k_{11} + k_{12} + 4k_{21} &= 1 \\ -2k_{11} - k_{12} + 4k_{21} &= 1 \end{aligned}$$

This system can be solved to yield $k_{11} = -\frac{2}{9}$, $k_{12} = \frac{1}{3}$, and $k_{21} = \frac{2}{9}$. The conclusion is

$$\frac{s + 1}{(s + 2)^2(s - 1)} = -\frac{2/9}{s + 2} + \frac{1/3}{(s + 2)^2} + \frac{2/9}{(s - 1)}$$

Method 2

Substitute three convenient values for s to obtain three equations in three unknowns. For example, with $s = 0$, Eq. (14.45) becomes

$$-\frac{1}{4} = \frac{1}{2}k_{11} + \frac{1}{4}k_{12} - k_{21}$$

and with $s = -1$, Eq. (14.45) becomes

$$0 = k_{11} + k_{12} - \frac{1}{2}k_{21}$$

Finally, with $s = 2$, Eq. (14.45) becomes

$$\frac{3}{16} = \frac{1}{4}k_{11} + \frac{1}{16}k_{12} + k_{21}$$

The solution to the above system of equations is $k_{11} = -\frac{2}{9}$, $k_{12} = \frac{1}{3}$, and $k_{21} = \frac{2}{9}$, which agrees with the previous result, as it must.

Method 3

Observe that if we multiply both sides of Eq. (14.45) by $s - 1$,

$$(s - 1)G(s) = \frac{s + 1}{(s + 2)^2} = (s - 1)\frac{k_{11}}{s + 2} + (s - 1)\frac{k_{12}}{(s + 2)^2} + k_{21}$$

and if we set $s = 1$, we obtain

$$(s - 1)G(s)\bigg|_{s=1} = \frac{2}{9} = k_{21}$$

Similarly, if we multiply both sides of Eq. (14.45) by $(s+2)^2$,

$$(s+2)^2G(s) = \frac{s+1}{s-1} = (s+2)k_{11} + k_{12} + (s+2)^2\frac{k_{21}}{s-1}$$

and if we set $s=-2$, we have

$$(s+2)^2G(s)\bigg|_{s=-2} = \frac{1}{3} = k_{12}$$

Therefore Eq. (14.45) becomes

$$\frac{s+1}{(s+2)^2(s-1)} = \frac{k_{11}}{s+2} + \frac{1/3}{(s+2)^2} + \frac{2/9}{s-1}$$

This method does not work for k_{11}. However, k_{11} can still be found by method 1 or 2. Perhaps the easiest approach is to set $s=-1$ in the above equation, to get

$$0 = k_{11} + \frac{1}{3} - \frac{1}{9}$$

which reveals that

$$k_{11} = -\frac{2}{9}$$

Example 14.12 will illustrate how to proceed when the degree of $N(s)$ exceeds that of $D(s)$.

EXAMPLE 14.12 Determine the partial fraction expansion of

$$G(s) = \frac{s^4+2s^3+s^2+1}{s(s+3)}$$

Solution According to Example 14.9, the form of the partial fraction expansion is

$$G(s) = \underbrace{\alpha_2s^2 + \alpha_1s + \alpha_0}_{p(s)} + \frac{k_{11}}{s} + \frac{k_{21}}{s+3}$$

Again, there are several ways to obtain the values of the constants. Methods 1 and 2 of Example 14.11 will work, but they should be avoided because they are too lengthy. (In using method 2, we would have to substitute five values for s and solve the resulting system of five equations for α_2, α_1, α_0, k_{11}, and k_{21}.) The recommended approach is to find α_2, α_1, and α_0 first by long division, stopping the division process just before a term with a negative exponent is obtained:

$$\begin{array}{r|l}
 & s^2 - s + 4 \\
s^2+3s & s^4 + 2s^3 + s^2 + 0s + 1 \\
 & \underline{s^4 + 3s^3} \\
 & \quad -s^3 + s^2 + 0s + 1 \\
 & \quad \underline{-s^3 - 3s^2} \\
 & \qquad 4s^2 + 0s + 1 \\
 & \qquad \underline{4s^2 + 12s} \\
 & \qquad\quad -12s + 1 \quad \text{Remainder} \\
 & \qquad\quad \text{—STOP—}
\end{array}$$

It follows that

$$G(s) = \frac{s^4 + 2s^3 + s^2 + 1}{s(s+3)} = \underbrace{s^2 - s + 4}_{p(s)} + \frac{-12s + 1}{s(s+3)}$$

which shows that $\alpha_2 = 1$, $\alpha_1 = -1$, and $\alpha_0 = 4$. The partial fraction expansion of the fractional term [call it $G_1(s)$] can now be taken

$$G_1(s) = \frac{-12s + 1}{s(s+3)} = \frac{k_{11}}{s} + \frac{k_{21}}{s+3}$$

Equation (14.44) works beautifully for determining k_{11} and k_{21}. To find k_{11}, multiply both sides of $G_1(s)$ by s and set $s = 0$. Thus

$$k_{11} = sG_1(s)\big|_{s=0} = \left.\frac{-12s + 1}{s + 3}\right|_{s=0} = \frac{1}{3}$$

Similarly,

$$k_{21} = (s+3)G_1(s)\big|_{s=-3} = \left.\frac{-12s + 1}{s}\right|_{s=-3} = -\frac{37}{3}$$

Therefore,

$$G_1(s) = \frac{-12s + 1}{s(s+3)} = \frac{1/3}{s} - \frac{37/3}{s+3}$$

and, combining the above with $p(s)$, we obtain the result

$$\frac{s^4 + 2s^3 + s^2 + 1}{s(s+3)} = s^2 - s + 4 + \frac{1/3}{s} - \frac{37/3}{s+3}$$

A partial (but reassuring) check can be made by setting $s = 1$. This yields

$$\frac{1 + 2 + 1 + 1}{4} = 1 - 1 + 4 + \frac{1}{3} - \frac{37}{12}$$

or

$$\frac{5}{4} = \frac{15}{12}$$

which is, of course, a true statement.

Examples 14.10 through 14.12 have illustrated methods that can be applied to any rational algebraic function $G(s) = N(s)/D(s)$. The methods apply for both real and complex poles.† Whenever $deg\{N(s)\} \geq deg\{D(s)\}$, we can use long division as in Example 14.12 to find the constants appearing in the polynomial $p(s)$ of Eq. (14.43). The remaining fractional term $G_1(s) = [\text{Remainder}] \div D(s)$ will always be a rational algebraic function whose numerator has a smaller degree than the denominator. Thus $G(s)$ is easily put in the form $G(s) = p(s) + G_1(s)$, where $deg\{p(s)\} = deg\{N(s)\} - deg\{D(s)\}$. The function $G_1(s)$ can then be put in partial fraction form, Eq. (14.42), and the constants k_{ij} may be evaluated by any of the three methods of Example 14.11. Whenever $deg\{N(s)\} < deg\{D(s)\}$, long division is no longer necessary, since $p(s) = 0$.

† A helpful shortcut for complex poles is described in Example 14.15.

Heaviside's Theorem

We have stated that whenever a pole s_i is simple (has order $m_i = 1$), then the residue k_{i1} at that simple pole can be found from Eq. (14.44), which we repeat below:

$$k_{i1} = (s - s_i)G(s)\Big|_{s=s_i} \tag{14.44}$$

This formula applies regardless of the relative degrees of $N(s)$ and $D(s)$ and can be established by writing

$$G(s) = \frac{N(s)}{c(s - s_1)^{m_1}(s - s_2)^{m_2} \cdots (s - s_i)^1 \cdots (s - s_l)^{m_l}}$$

Simple pole s_i

$$= \underbrace{p(s)}_{\substack{\text{Present if} \\ deg\{N(s)\} \geq deg\{D(s)\}}} + \underbrace{L_1(s) + \cdots + \frac{k_{i1}}{s - s_i} + \cdots + L_l(s)}_{G_1(s)} \tag{14.46}$$

By multiplying both sides by $(s - s_i)$, we obtain

$$(s - s_i)G(s) = \frac{\cancel{(s - s_i)}N(s)}{c(s - s_1)^{m_1} \cdots \cancel{(s - s_i)} \cdots (s - s_l)^{m_l}}$$

$$= (s - s_i)p(s) + (s - s_i)L_1(s) + \cdots + k_{i1} + \cdots + (s - s_i)L_l(s) \tag{14.47}$$

Equation (14.44) is a direct consequence of the fact that all terms on the right-hand side of Eq. (14.47) except k_{i1} equal zero for $s = s_i$. Equation (14.44) is sometimes referred to as *Heaviside's theorem.*

For completeness, we state the following general formula for obtaining any of the coefficients k_{ij} appearing in Eqs. (14.42).

$$k_{ij} = \frac{1}{(m_i - j)!} \frac{d^{m_i - j}}{ds^{m_i - j}} [(s - s_i)^{m_i} G(s)] \Big|_{s=s_i} \tag{14.48}$$

for $j = 1, 2, \ldots, m_i$, where m_i is the order of s_i. Equation (14.48) can be verified if we multiply both sides of Eq. (14.42a) by $(s - s_i)^{m_i}$ and differentiate as indicated. Although Eq. (14.48) gives a complete theoretical solution to the problem of determining the partial fraction expansion constants k_{ij}, it is often less convenient to apply than methods 1 through 3 of Example 14.11.

EXERCISES Determine the partial fraction expansion of the rational algebraic functions in Exercises 23 through 28.

23. $\dfrac{1}{s^2 + 1}$ *answer:* $\dfrac{j0.5}{s + j} - \dfrac{j0.5}{s - j}$

24. $\dfrac{s}{(s + 1)^2}$ *answer:* $\dfrac{1}{s + 1} - \dfrac{1}{(s + 1)^2}$

25. $\dfrac{s^2 + 2s + 2}{s^2 + 7s + 12}$ *answer:* $1 + \dfrac{5}{s+3} - \dfrac{10}{s+4}$

26. $\dfrac{s^3 + 3s^2 + 2s + 1}{s+2}$ *answer:* $s^2 + s + \dfrac{1}{s+2}$

27. $\dfrac{(s^3 + 3s^2 + 2s + 1)(s^2 + 2s + 2)}{(s+2)(s^2 + 7s + 12)}$

answer: $s^2 - 4s + 20 - \dfrac{239}{s+2} + \dfrac{923}{s+3} - \dfrac{259}{s+4}$

28. $\dfrac{s(s+3+j)(s+3-j)}{(s+1+3j)(s+1-3j)(s+5)}$ *answer:* $1 - \dfrac{1}{s+5} - \dfrac{8}{(s+1)^2 + 9}$

29. Suppose that $G_a(s) = k_a/(s - s_a)$ and $G_b(s) = k_b/(s - s_b)$. Determine the partial fraction expansion of $G_a(s)G_b(s)$ if $s_a \neq s_b$.

answer: $\dfrac{\alpha}{s - s_a} - \dfrac{\alpha}{s - s_b}$ where $\alpha = \dfrac{k_a k_b}{s_a - s_b}$

30. Repeat Exercise 29 for $s_a = s_b$. *answer:* $\dfrac{k_a k_b}{(s - s_a)^2}$

Determine the partial fraction expansions of the following rational algebraic functions.

31. $\dfrac{s}{s^2 + \omega_0^2}$ *answer:* $\dfrac{0.5}{s - j\omega_0} + \dfrac{0.5}{s + j\omega_0}$

32. $\dfrac{\omega_0}{s^2 + \omega_0^2}$ *answer:* $\dfrac{-j0.5}{s - j\omega_0} + \dfrac{j0.5}{s + j\omega_0}$

33. $\dfrac{s + \alpha}{(s+\alpha)^2 + \omega_d^2}$ *answer:* $\dfrac{0.5}{s + \alpha - j\omega_d} + \dfrac{0.5}{s + \alpha + j\omega_d}$

34. $\dfrac{\omega_d}{(s+\alpha)^2 + \omega_d^2}$ *answer:* $\dfrac{-j0.5}{s + \alpha - j\omega_d} + \dfrac{j0.5}{s + \alpha + j\omega_d}$

14.5 Solution of LTI Equations

One of the easiest ways to solve linear differential equations with constant coefficients and arbitrary initial conditions is to use the Laplace transform. Examples 14.13 through 14.16 show how it is done.

EXAMPLE 14.13 The input to the RC circuit of Fig. 14.5 is $v_s(t) = u(t)$ V. The circuit is at rest for $t < 0$. Find the output $v_o(t)$ for $t \geq 0$.

Solution The differential equation relating output to input is

$$2\frac{dv_o}{dt} + v_o(t) = v_s(t)$$

where $v_o(0^-) = 0$ because the circuit is at rest for $t < 0$. Obviously, both sides of this equation are functions of t. Because these functions of t are equal, so must be their Laplace transforms. Thus

$$\mathcal{L}\left\{2\frac{dv_o}{dt} + v_o(t)\right\} = \mathcal{L}\{v_s(t)\}$$

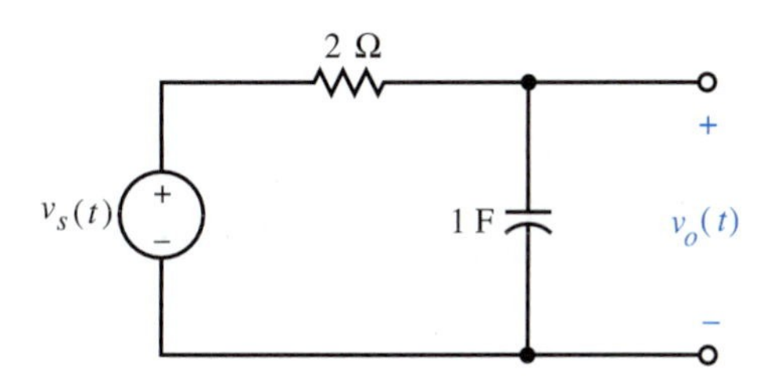

FIGURE 14.5
RC circuit

which, by the linearity property, becomes

$$2\mathscr{L}\left\{\frac{dv_o}{dt}\right\} + \mathscr{L}\{v_o(t)\} = \mathscr{L}\{v_s(t)\}$$

By using the differentiation property, we can put this in the form

$$2[sV_o(s) - v_o(0^-)] + V_o(s) = V_s(s)$$

Solving for $V_o(s)$ gives

$$V_o(s) = \frac{1}{2s+1}V_s(s) + \frac{2v_o(0^-)}{2s+1}$$
$$= \frac{0.5}{s+0.5}V_s(s)$$

where we have used the fact that $v_o(0^-) = 0$. It follows from LT2 of Table 14.2 that

$$V_s(s) = \mathscr{L}\{u(t)\} = \mathscr{L}\{u(t)\} = \frac{1}{s}$$

so

$$V_o(s) = \frac{0.5}{s+0.5}\cdot\frac{1}{s} = \frac{0.5}{s(s+0.5)}$$

The complete response for $t \geq 0$ is the inverse Laplace transform of $V_o(s)$. The easiest way to find the inverse transform of $V_o(s)$ is through the use of Tables 14.1 and 14.2. To do this, we need to first express $V_o(s)$ in partial fraction form. Since the degree of the numerator polynomial is less than that of the denominator polynomial, this form is given by Eq. (14.42):

$$V_o(s) = \frac{0.5}{s(s+0.5)} = \frac{k_{11}}{s} + \frac{k_{21}}{s+0.5}$$

Each pole is simple, so Eq. (14.44) applies to each residue k_{11} and k_{21}:

$$k_{11} = \left.\frac{0.5}{s+0.5}\right|_{s=0} = 1$$
$$k_{21} = \left.\frac{0.5}{s}\right|_{s=-0.5} = -1$$

Therefore

$$V_o(s) = \frac{1}{s} - \frac{1}{s+0.5}$$

It follows from the above and linearity that

$$v_o(t) = \mathcal{L}^{-1}\{V_o(s)\} = \mathcal{L}^{-1}\left\{\frac{1}{s}\right\} - \mathcal{L}^{-1}\left\{\frac{1}{s+0.5}\right\}$$

which, by LT2 and LT5, yields the final result:

$$\begin{aligned} v_o(t) &= u(t) - e^{-0.5t}u(t) \\ &= (1 - e^{-0.5t})u(t) \text{ V} \end{aligned}$$

The next example is slightly more challenging. The techniques are the same as in Example 14.13.

EXAMPLE 14.14 The input to the *RC* circuit of Fig. 14.5 is $v_s(t) = 5tu(t)$ V, and the capacitance voltage is -3 V at $t = 0^-$. Find the output $v_o(t)$ for $t \geq 0$.

Solution The differential equation relating output to input is, for $t > 0$,

$$2\frac{dv_o}{dt} + v_o(t) = v_s(t)$$

By taking the Laplace transform of both sides, we obtain (in a manner identical to that of Example 14.13)

$$2[sV_o(s) - v_o(0^-)] + V_o(s) = V_s(s)$$

which yields

$$\begin{aligned} V_o(s) &= \frac{1}{2s+1}V_s(s) + \frac{2v_o(0^-)}{2s+1} \\ &= \frac{0.5}{s+0.5}V_s(s) - \frac{3}{s+0.5} \end{aligned}$$

It follows from LT3 and the linearity property that $V_s(s) = 5/s^2$. Therefore

$$V_o(s) = \frac{2.5}{(s+0.5)s^2} - \frac{3}{s+0.5}$$

The complete response for $t \geq 0$ is given by the inverse Laplace transform of $V_o(s)$. Again, we will use partial fraction expansion and table look-up. The partial fraction expansion of the first term in the above equation is

$$\begin{aligned} G(s) &= \frac{2.5}{(s+0.5)s^2} \\ &= \frac{k_{11}}{s} + \frac{k_{12}}{s^2} + \frac{k_{13}}{s+0.5} \end{aligned}$$

where

$$k_{13} = (s+0.5)G(s)\Big|_{s=-0.5} = \frac{2.5}{s^2}\Big|_{s=-0.5} = 10$$

$$k_{12} = s^2G(s)\Big|_{s=0} = \frac{2.5}{s+0.5}\Big|_{s=0} = 5$$

By substituting these values into the expression for $G(s)$ and setting $s = 1$, we find that $k_{11} = 10$. Thus

$$G(s) = \frac{2.5}{(s+0.5)s^2} = \frac{10}{s} + \frac{5}{s^2} + \frac{10}{s+0.5}$$

which, when combined with the second term in $V_o(s)$, yields

$$V_o(s) = -\frac{10}{s} + \frac{5}{s^2} + \frac{7}{s+0.5}$$

We now have $V_o(s)$ in a form where we can use table look-up to obtain $v_o(t)$. It follows directly from LT1, LT2, and the linearity property that

$$\begin{aligned} v_o(t) &= \mathcal{L}^{-1}\left\{-\frac{10}{s} + \frac{5}{s^2} + \frac{7}{s+0.5}\right\} \\ &= -10\mathcal{L}^{-1}\left\{\frac{1}{s}\right\} + 5\mathcal{L}^{-1}\left\{\frac{1}{s^2}\right\} + 7\mathcal{L}^{-1}\left\{\frac{1}{s+0.5}\right\} \\ &= -10u(t) + 5tu(t) + 7e^{-0.5t}u(t) \text{ V} \end{aligned}$$

In the next example, the input to the RC circuit is a causal sinusoid. The method of solution is the same as that of Examples 14.13 and 14.14.

EXAMPLE 14.15 The input to the RC circuit in Fig. 14.5 is $v_s(t) = A\cos(\omega_0 t + \phi)u(t)$. The initial voltage on the capacitance is $v_o(0^-)$. Find $v_o(t)$ for $t \geq 0$ in terms of R, C, $v_o(0^-)$, A, ω_0, and ϕ.†

Solution The differential equation relating output to input is

$$RC\frac{dv_o}{dt} + v_o(t) = v_s(t)$$

Proceeding as in Examples 14.13 and 14.14, we take the Laplace transform of both sides to obtain

$$RC[sV_o(s) - v_o(0^-)] + V_o(s) = V_s(s)$$

Therefore

$$V_o(s) = \frac{\frac{1}{RC}}{s + \frac{1}{RC}}V_s(s) + \frac{v_o(0^-)}{s + \frac{1}{RC}}$$

which, by LT8, becomes

$$V_o(s) = \frac{\frac{A}{2RC}\angle\phi}{\left(s + \frac{1}{RC}\right)(s - j\omega_0)} + \frac{\frac{A}{2RC}\angle{-\phi}}{\left(s + \frac{1}{RC}\right)(s + j\omega_0)} + \frac{v_o(0^-)}{s + \frac{1}{RC}}$$

As in Examples 14.13 and 14.14, the next step is to write $V_o(s)$ in partial fraction expansion form. The partial fraction expansion of the first term is

† The use of symbols rather than numerical values can make a simple analysis appear complicated. The result, however, is often well worth the extra effort. The major advantage of expressing a solution in terms of symbols rather than numerical values arises in design problems where we must choose the values of one or more elements to produce a desired result.

$$\frac{\dfrac{A}{2RC}\angle\phi}{\left(s+\dfrac{1}{RC}\right)(s+j\omega_0)} = \frac{k_{11}}{s+\dfrac{1}{RC}} + \frac{k_{21}}{s-j\omega_0}$$

where

$$k_{11} = \left.\frac{\dfrac{A}{2RC}\angle\phi}{s-j\omega_0}\right|_{s=-1/RC} = -\frac{\dfrac{A}{2RC}\angle\phi}{\dfrac{1}{RC}+j\omega_0}$$

$$= \frac{-A/2}{\sqrt{1+(\omega_0RC)^2}}\angle\phi - \tan^{-1}\omega_0 RC$$

and

$$k_{21} = \left.\frac{\dfrac{A}{2RC}\angle\phi}{s+\dfrac{1}{RC}}\right|_{s=j\omega_0} = \frac{\dfrac{A}{2RC}\angle\phi}{\dfrac{1}{RC}+j\omega_0}$$

$$= \frac{A/2}{\sqrt{1+(\omega_0RC)^2}}\angle\phi - \tan^{-1}\omega_0 RC$$

We can obtain the partial fraction expansion of the second term in $V_o(s)$ in the same way. However, there is a shortcut that is often helpful when we deal with complex poles. Since the constants in the second term in $V_o(s)$ are the complex conjugates of those in the first term, we can immediately write the result

$$\frac{\dfrac{A}{2RC}\angle-\phi}{\left(s+\dfrac{1}{RC}\right)(s+j\omega_0)} = \frac{k_{11}^*}{s+\dfrac{1}{RC}} + \frac{k_{21}^*}{s+j\omega_0}$$

where k_{11} and k_{21} are given above. Substitution of the partial fraction expansions into the original expression for $V_o(s)$ yields

$$V_o(s) = \frac{k_{11}+k_{11}^*+v_o(0^-)}{s+\dfrac{1}{RC}} + \frac{k_{21}}{s-j\omega_0} + \frac{k_{21}^*}{s+j\omega_0}$$

It follows by LT5, LT8, and the linearity property that

$$v_o(t) = [v_o(0^-)+k_{11}+k_{11}^*]e^{-t/RC}u(t) + 2|k_{21}|\cos(\omega_0 t + \angle k_{21})u(t)$$

where k_{11} and k_{21} are given above. By substituting the expressions for k_{11} and k_{21}, we obtain the final result.

$$v_o(t) = \left[v_o(0^-) - \frac{A}{\sqrt{1+(\omega_0RC)^2}}\cos(\phi - \tan^{-1}\omega_0RC)\right]e^{-t/RC}u(t)$$
$$+ \frac{A}{\sqrt{1+(\omega_0RC)^2}}\cos(\omega_0 t + \phi - \tan^{-1}\omega_0RC)u(t)$$

In this section we have shown how the Laplace transform may be used to solve the differential equations that arise in LTI circuit problems. In summary, the method consists of the following four steps:

Step 1 *Take the Laplace transform of both sides of the time-domain equation.*

Step 2 *Solve for $V_o(s)$.*

Step 3 *Use partial fraction expansions to write $V_o(s)$ as a sum of terms of the forms included in Table 14.2.*

Step 4 *Use Tables 14.1 and 14.2 to inverse-transform $V_o(s)$ into the time domain.*

Section 14.7 contains a deeper theoretical look at each step. In Section 14.6, we will describe a way to use the Laplace transform that *bypasses* the differential equation of the circuit.

EXERCISES

35. Start from the appropriate differential equation, and use Laplace transforms to determine $v_R(t)$ for $v_s(t) = u(t)$ V in the circuit shown. The voltage $v_C(0^-)$ is zero.
answer: $e^{-t}u(t)$ V

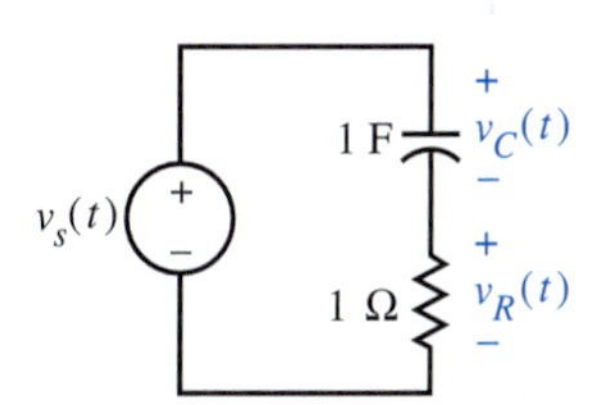

36. Repeat Exercise 35 for $v_s(t) = 10tu(t)$ V and $v_C(0^-) = 6$ V.
answer: $(10 - 16e^{-t})u(t)$ V

37. Start from the appropriate differential equation, and use Laplace transforms to determine $v_L(t)$ for $v_s(t) = u(t)$ in the following circuit. The voltage $v_R(0^-)$ is arbitrary.
answer: $[1 - v_R(0^-)]e^{-(R/L)t}u(t)$

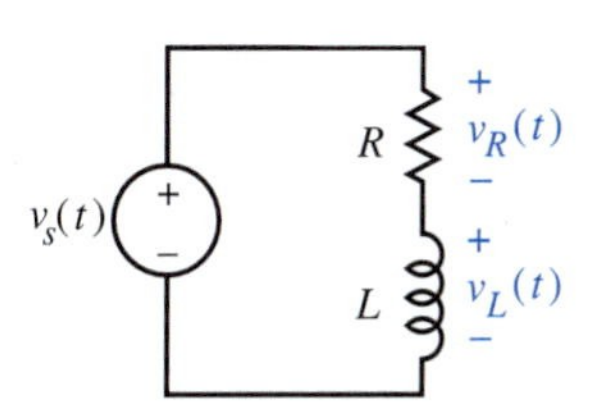

38. Repeat Exercise 37 for $v_s(t) = A\cos(\omega_0 t + \phi)u(t)$.

answer: $$\frac{A\omega_0}{\sqrt{(R/L)^2 + \omega_0^2}} \cos\left[\omega_0 t + \phi + 90° - \arctan\left(\frac{\omega_0 L}{R}\right)\right]u(t)$$
$$+ \left\{\frac{AR/L}{\sqrt{(R/L)^2 + \omega_0^2}} \cos\left[\phi + 180° - \arctan\left(\frac{\omega_0 L}{R}\right)\right] - v_R(0^-)\right\} e^{-(R/L)t}u(t)$$

39. Start from the appropriate differential equation, and determine $v_o(t)$ when $i_s(t) = \delta(t)$ in the circuit shown below, where $d = 1/2RC$ and $\omega_0 = 1/\sqrt{LC}$. Assume that the circuit is at rest for $t < 0$ and that the response is underdamped.

answer: $\frac{1}{C} e^{-\alpha t} \left[\cos \omega_0 t - \frac{\alpha}{\omega_0} \sin \omega_0 t \right] u(t)$

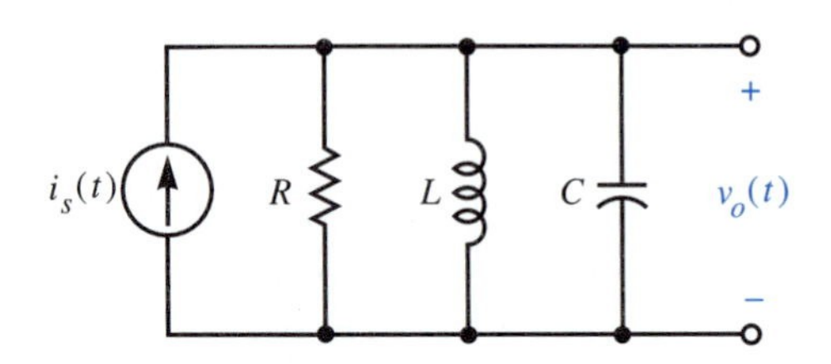

14.6 Laplace Transform Circuit Analysis

The Laplace transform can be used to transform any LTI circuit from the time domain to the frequency domain. This transformation includes the terminal equations of the circuit components R, L, and C, the terminal equations of the sources, and the underlying circuit laws KCL and KVL. The motivation is similar to that in Chapters 11 and 13: the transformation from time to frequency replaces the integro-differential representation of the circuit by an algebraic representation, which greatly simplifies the task of circuit analysis. As we will see, an advantage of the Laplace transform is that it not only transforms the circuit but also transforms the circuit's initial conditions. Once this is done, we can obtain the circuit's complete response by algebra and table look-up.

The s-domain representations of the circuit elements obtained from the Laplace transform are somewhat different from those obtained in Chapter 13. The reason is that the Laplace transform accounts for initial conditions. To illustrate, the terminal equation of inductance with initial condition $i(0^-) = I_o$ (Fig. 14.6a) is

$$v(t) = L\frac{di}{dt} \quad \text{given} \quad i(0^-) = I_o \tag{14.49}$$

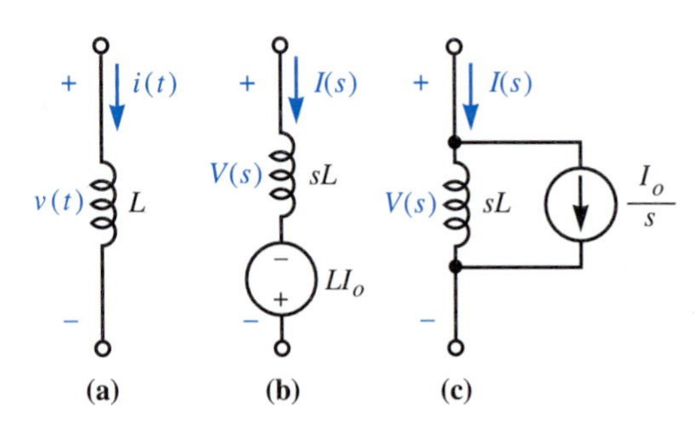

FIGURE 14.6
Time- and frequency-domain representations of inductance with initial conditions:
(*a*) time domain; (*b*), (*c*) frequency domain

By taking the Laplace transform, we have, according to the linearity and differentiation properties,

$$\begin{aligned} V(s) &= sLI(s) - Li(0^-) \\ &= sLI(s) - LI_o \end{aligned} \tag{14.50}$$

which is an algebraic equation relating frequency-domain terminal voltage $V(s)$ and current $I(s)$ to the inductance's impedance sL and initial current I_o. Equation (14.50) can be rearranged to read

$$I(s) = \frac{1}{sL} V(s) + \frac{I_o}{s} \tag{14.51}$$

Equations (14.50) and (14.51) are represented by Fig. 14.6b, c, which we can verify using KVL and KCL.

The frequency-domain representation of a capacitance can be obtained similarly. The result is shown in Fig. 14.7 along with the corresponding time-domain circuits. As in Chapter 13, the s-domain representation for a resistance R is simply that resistance R.

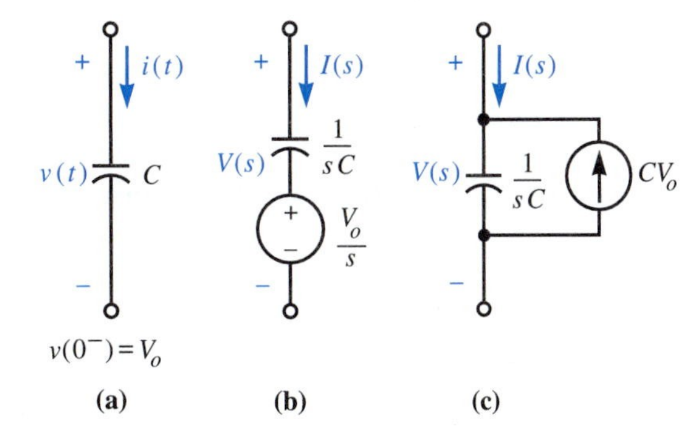

FIGURE 14.7 Time- and frequency-domain representations of capacitance with initial conditions: (*a*) time domain; (*b*) and (*c*) frequency domain

Usually the equivalent circuits in Figs. 14.6b and 14.7b are easier to use in circuit analysis involving mesh-current equations, whereas those in Figs. 14.6c and 14.7c are easier to use when writing node-voltage equations.

It is a straightforward exercise to transform the basic circuit theory laws KCL and KVL into the frequency domain. For example, by taking the Laplace transform of Kirchhoff's current law

$$\sum_{\substack{\text{Closed}\\\text{surface}}} i_n(t) = 0 \tag{14.52}$$

we obtain

$$\mathcal{L}\left\{\sum_{\substack{\text{Closed}\\\text{surface}}} i_n(t)\right\} = 0 \tag{14.53}$$

This becomes, by the linearity property,

$$\sum_{\substack{\text{Closed}\\\text{surface}}} \mathcal{L}\{i_n(t)\} = 0 \tag{14.54}$$

which is, by definition,

$$\sum_{\substack{\text{Closed}\\\text{surface}}} I_n(s) = 0 \tag{14.55}$$

Thus, the sum of the Laplace transforms of the currents leaving any closed surface in the network equals zero. A similar result applies to KVL, namely,

$$\sum_{\substack{\text{Closed} \\ \text{path}}} V_n(s) = 0 \tag{14.56}$$

Now that we have Laplace-transformed the circuit element terminal equations and the axioms KCL and KVL into the s domain, we are prepared to find the complete response of LTI circuits using only algebra and table look-up. The method consists of the following three steps.

Step 1 *Draw the Laplace-transformed circuit.*

Step 2 *Use any convenient frequency-domain circuit analysis method to relate the Laplace transform output [for example, $V_o(s)$] to the Laplace transform input(s).*

Step 3 *Find the time-domain output [for example, $v_o(t)$] by taking the inverse Laplace transform of $V_o(s)$.*

Examples 14.16 through 14.19 illustrate the method.

EXAMPLE 14.16 Solve the problem in Example 14.14 with the use of Laplace transform circuit analysis.

Solution The first step is to determine the frequency-domain (that is, Laplace-transformed) circuit, using Figs. 14.6 and 14.7 for help. Since the RC circuit of Fig. 14.5 is a series circuit, it is convenient to select the series frequency-domain representation shown in Fig. 14.7b for the capacitance, setting $V_o = -3$ V. The frequency-domain representation of the source voltage $v_s(t) = 5tu(t)$ follows from LT3 and the linearity property, as $V_s(s) = 5/s^2$ V. The resulting frequency-domain circuit is shown in Fig. 14.8. The next step is to determine the equation relating $V_o(s)$ to $V_s(s)$. We will leave it to you as an easy exercise in circuit analysis to derive the equation

$$V_o(s) = \left(\frac{1/s}{2 + 1/s}\right) V_s(s) + \left(\frac{-3/s}{2 + 1/s}\right) 2$$

The substitution $V_s(s) = 5/s^2$ leads to

$$V_o(s) = \frac{2.5}{(s + 0.5)s^2} - \frac{3}{s + 0.5}$$

Notice that this result is identical to that in Example 14.14, which was derived from the circuit's differential equation. We can now determine the complete response $V_o(t)$ by taking the inverse Laplace transform of $V_o(s)$. These steps in the analysis were presented in Example 14.14.

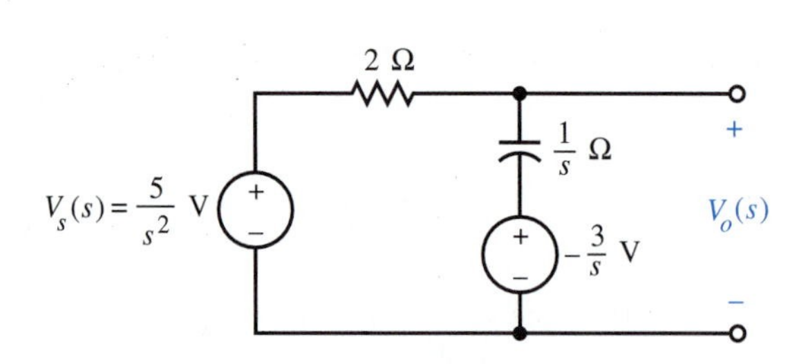

FIGURE 14.8
Laplace-transformed circuit

EXAMPLE 14.17 The input to the ideal op-amp circuit of Fig. 14.9 is $v_1(t) = 10u(t)$ V, and the capacitance has an initial voltage $v_C(0^-) = 5$ V. Find and sketch $v_o(t)$ for $t \geq 0$ if $R_1 = R_2 = 1$ kΩ and $C = 1$ μF.

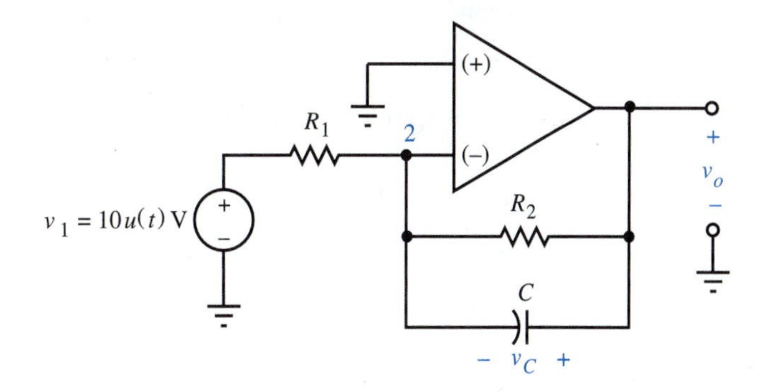

FIGURE 14.9 Op-amp circuit with step input and initial condition $v_C(0^-) = 5$ V

Solution The Laplace-transformed circuit is shown in Fig. 14.10.

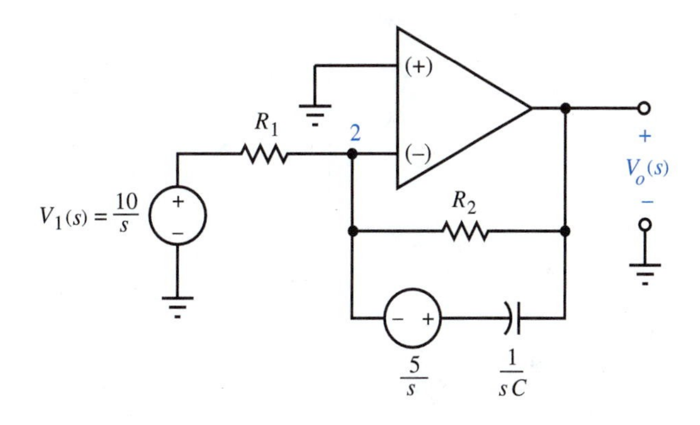

FIGURE 14.10 Laplace-transformed circuit

Application of KCL at node 2 yields

$$\frac{V_1(s)}{R_1} + sC\left[V_o(s) - \frac{5}{s}\right] + \frac{V_o(s)}{R_2} = 0$$

By setting $V_1(s) = 10/s$ and rearranging, we obtain

$$V_o(s) = \frac{5}{s + \dfrac{1}{R_2C}} - \frac{10}{sR_1C\left(s + \dfrac{1}{R_2C}\right)}$$

which becomes, for the given numerical values,

$$V_o(s) = \frac{5}{s + 1000} - \frac{10{,}000}{s(s + 1000)}$$

By expanding the second term into partial fractions and rearranging, we obtain

$$V_o(s) = \frac{15}{s + 1000} - \frac{10}{s}$$

Therefore, from Tables 14.1 and 14.2,

$$v_o(t) = 15e^{-1000t}u(t) - 10u(t) \text{ V}$$

This result assumes that the units of t are seconds. If we convert the units of t to milliseconds, we obtain

$$v_o(t) = 15e^{-t}u(t) - 10u(t) \text{ V}$$

A sketch of $v_o(t)$ is given in Fig. 14.11.

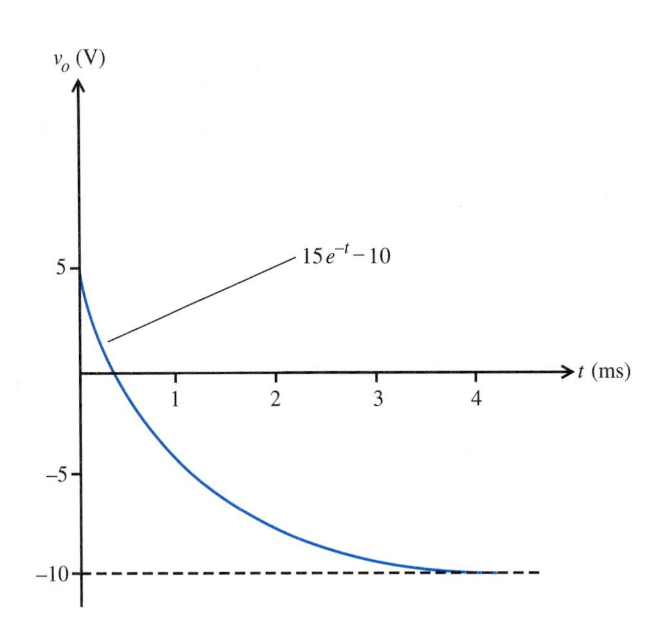

FIGURE 14.11 Ouptut of op-amp circuit for $t \geq 0$

Examples 14.18 and 14.19 are slightly more challenging.

EXAMPLE 14.18 Steady-state conditions exist in the circuit of Fig. 14.12 at $t = 0^-$, and switch S1 closes at $t = 0$. Determine $i_1(t)$ and $i_2(t)$ for $t \geq 0$.

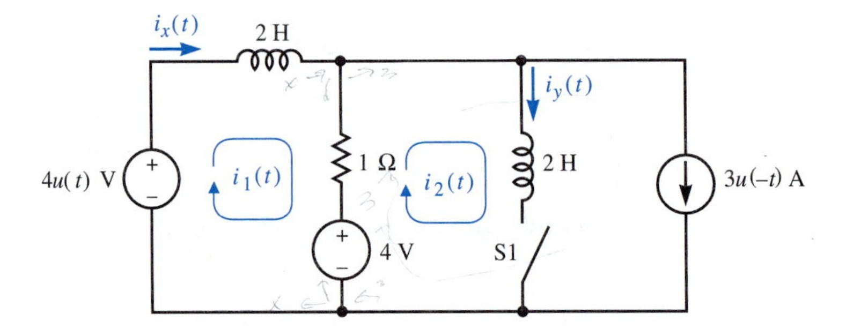

FIGURE 14.12 Circuit with switch S1

Solution By inspection, $i_y(0^-) = 0$ and $i_x(0^-) = -1$ A. The frequency-domain circuit is shown in Fig. 14.13.

Notice that the $3u(-t)$-A current source has been transformed to an open circuit and that the 4-V dc source has the same transform, $4/s$, as the source $4u(t)$ V. This is

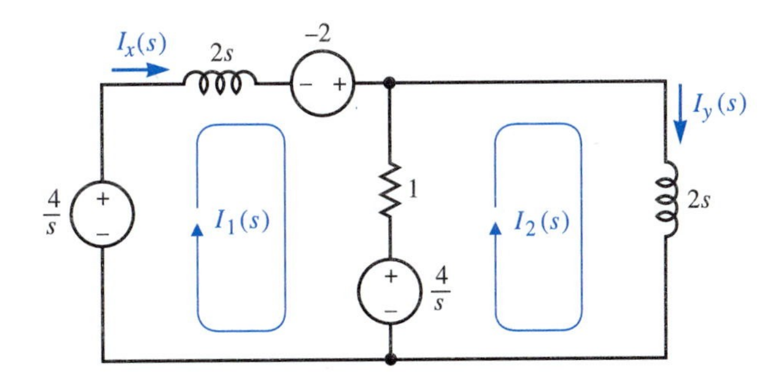

FIGURE 14.13 Frequency-domain circuit

because the (unilateral) Laplace transform of a function is determined only by the value of that function for $t \geq 0$.

The mesh-current equations are

$$\begin{bmatrix} 2s+1 & -1 \\ -1 & 2s+1 \end{bmatrix} \begin{bmatrix} I_1(s) \\ I_2(s) \end{bmatrix} = \begin{bmatrix} -2 \\ \frac{4}{s} \end{bmatrix}$$

Cramer's rule yields the solution

$$I_1(s) = \frac{\begin{vmatrix} -2 & -1 \\ \frac{4}{s} & 2s+1 \end{vmatrix}}{\begin{vmatrix} 2s+1 & -1 \\ -1 & 2s+1 \end{vmatrix}} = \frac{-4s - 2 + (4/s)}{4s^2 + 4s + 1 - 1}$$

$$I_2(s) = \frac{\begin{vmatrix} 2s+1 & -2 \\ -1 & \frac{4}{s} \end{vmatrix}}{\begin{vmatrix} 2s+1 & -1 \\ -1 & 2s+1 \end{vmatrix}} = \frac{6 + (4/s)}{4s^2 + 4s + 1 - 1}$$

By simplifying and taking partial fractions, we get

$$I_1(s) = \frac{-2s^2 - s + 2}{2s^2(s+1)} = \frac{1/2}{s+1} - \frac{3/2}{s} + \frac{1}{s^2}$$

$$I_2(s) = \frac{3s+2}{2s^2(s+1)} = \frac{-1/2}{s+1} + \frac{1/2}{s} + \frac{1}{s^2}$$

which yield for $t \geq 0$,

$$i_1(t) = \frac{1}{2}e^{-t} - \frac{3}{2} + t \text{ A}$$

$$i_2(t) = -\frac{1}{2}e^{-t} + \frac{1}{2} + t \text{ A}$$

EXAMPLE 14.19 The circuit of Fig. 14.14 is at rest at $t = 0^-$. Find $v_C(t)$ for $t \geq 0$.

Solution The transformed circuit is shown in Fig. 14.15.

By KCL,

$$\frac{V_C(s) - V_1(s)}{s} + sV_C(s) = I_1(s)$$

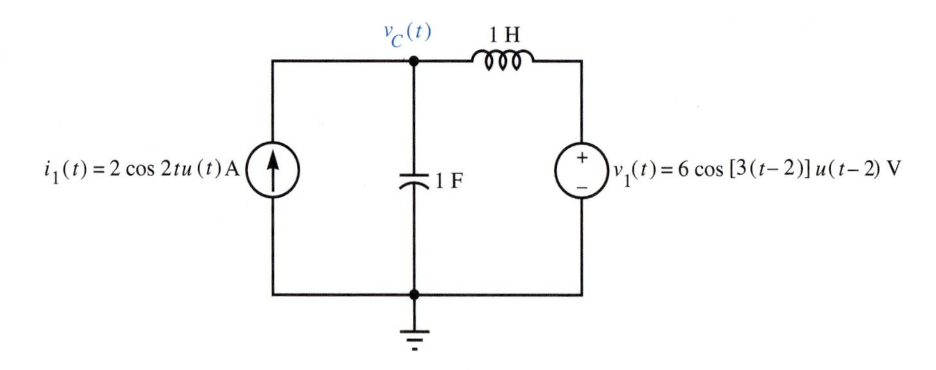

FIGURE 14.14
Circuit for Example 14.19

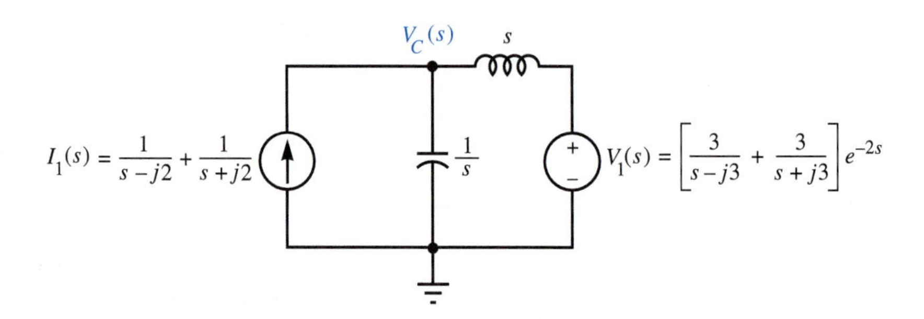

FIGURE 14.15
Laplace-transformed circuit

which rearranges to

$$V_C(s) = \frac{V_1(s)}{s^2+1} + \frac{s}{s^2+1}I_1(s)$$

$$= \frac{s}{(s+j)(s-j)(s-j2)} + \frac{s}{(s+j)(s-j)(s+j2)}$$

$$+ \left[\frac{3}{(s+j)(s-j)(s-j3)} + \frac{3}{(s+j)(s-j)(s+j3)}\right]e^{-2s}$$

By using partial fraction expansions and combining terms, we obtain

$$V_C(s) = \frac{j1/3}{s-j} - \frac{j1/3}{s+j} - \frac{j2/3}{s-j2} + \frac{j2/3}{s+j2}$$

$$+ \left[\frac{3/8}{s-j} + \frac{3/8}{s+j} - \frac{3/8}{s-j3} - \frac{3/8}{s+j3}\right]e^{-2s}$$

which, by the time-delay property of Table 14.1 and LT8 of Table 14.2, has inverse transform

$$v_C(t) = \left[\frac{2}{3}\cos(t+90°) + \frac{4}{3}\cos(2t-90°)\right]u(t)$$

$$+ \left\{\frac{3}{4}\cos(t-2) + \frac{3}{4}\cos[3(t-2)+180°]\right\}u(t-2) \text{ V}$$

Remember

In Laplace transform circuit analysis, the initial conditions are included in the frequency-domain circuit elements as shown in Figs. 14.6 and 14.7.

EXERCISES

40. In the circuit shown, $v_1(t) = 4u(t)$ V. Use Laplace transform circuit analysis to determine $i_R(t)$ for $t \geq 0$. answer: $8e^{-t}u(t)$ A

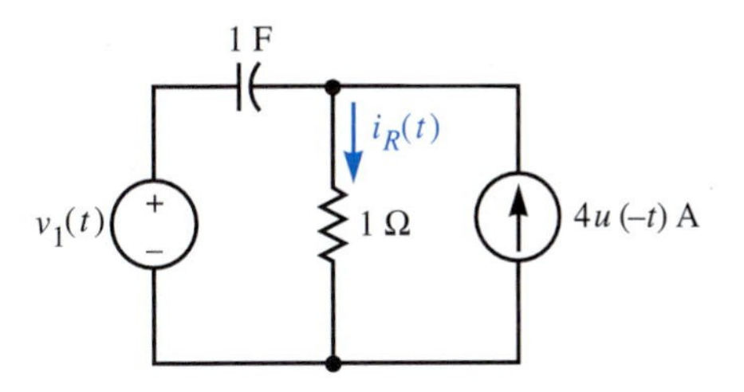

41. Repeat Exercise 40 for $v_1(t) = 4$ V. answer: $4e^{-t}u(t)$ A
42. Repeat Exercise 40 for $v_1(t) = 4u(t-1)$ V.
answer: $4e^{-(t-1)}u(t-1) + 4e^{-t}u(t)$ A
43. Repeat Exercise 40 for $v_1(t) = 8\cos(t + 30°)u(t)$.
answer: $[9.464e^{-t} + 1.464\cos(t) - 5.464\sin(t)]u(t)$
44. Switch S1 in the circuit shown opens at $t = 0$. Use Laplace transform circuit analysis to determine $v_1(t)$ and $v_2(t)$ for $t \geq 0$.
answer: $v_1(t) = [10 + 2.5e^{-0.25t}]u(t)$, $v_2(t) = -2.5e^{-0.25t}u(t)$

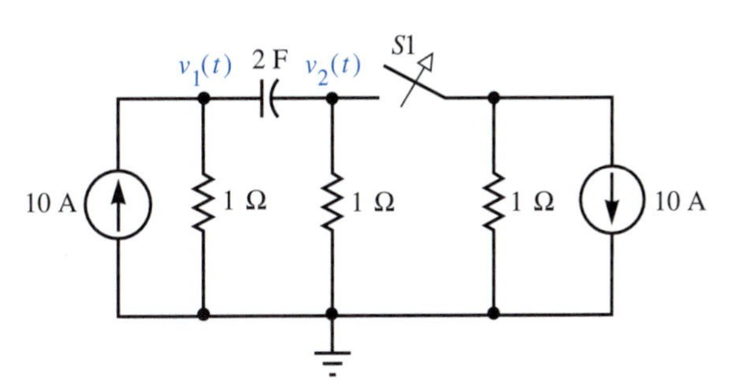

14.7 Impulse Response and Convolution

In this section we use the Laplace transform to obtain an important new way to calculate the response of an LTI circuit. The new result involves the concepts of *impulse response* and *convolution,* which are central topics in systems and signal processing. We begin with the LTI equation relating a circuit's input $v_s(t)$ and output $v_o(t)$

$$\mathcal{A}(D)v_o(t) = \mathcal{B}(D)v_s(t) \tag{14.57}$$

We are interested in obtaining the complete response for $t \geq 0$. We assume that the initial conditions $v_o(0^-)$, $v_o^{(1)}(0^-)$, . . . , $v_o^{(n-1)}(0^-)$ are known. Let us see what happens when we solve this problem in a general way using Laplace transforms.

The Laplace transform of the left-hand side of Eq. (14.57) is

$$\mathcal{L}\{\mathcal{A}(D)v_o(t)\} = \mathcal{L}\left\{a_n\frac{d^nv_o}{dt^n} + a_{n-1}\frac{d^{n-1}v_o}{dt^{n-1}} + \cdots + a_0v_o(t)\right\}$$

$$= a_n\mathcal{L}\left\{\frac{d^nv_o}{dt^n}\right\} + a_{n-1}\mathcal{L}\left\{\frac{d^{n-1}v_o}{dt^{n-1}}\right\} + \cdots + a_0\mathcal{L}\{v_o(t)\} \quad (14.58)$$

According to the differentiation property

$$\mathcal{L}\left\{\frac{d^nv_o}{dt^n}\right\} = s^nV_o(s) - \sum_{k=0}^{n-1} s^{n-1-k}v_o^{(k)}(0^-)$$

$$\mathcal{L}\left\{\frac{d^{n-1}v_o}{dt^{n-1}}\right\} = s^{n-1}V_o(s) - \sum_{k=0}^{n-2} s^{n-2-k}v_o^{(k)}(0^-)$$

$$\vdots \qquad\qquad \vdots$$

$$\mathcal{L}\left\{\frac{dv_o}{dt}\right\} = sV_o(s) - v_o(0^-)$$

$$\mathcal{L}\{v_o(t)\} = V_o(s) \quad (14.59)$$

Combining Eqs. (14.58) and (14.59), we obtain

$$\mathcal{L}\{\mathcal{A}(D)v_o(t)\} = \mathcal{A}(s)V_o(s) + E(s) \quad (14.60)$$

where

$$E(s) = -a_n\sum_{k=0}^{n-1} s^{n-1-k}v_o^{(k)}(0^-) - a_{n-1}\sum_{k=0}^{n-2} s^{n-2-k}v_o^{(k)}(0^-) - \cdots - a_1v_o(0^-) \quad (14.61)$$

is an $(n-1)$st-degree polynomial that depends on the initial conditions $v_o(0^-)$, $v_o^{(1)}(0^-)$, $v_o^{(n-1)}(0^-)$ and the coefficients $a_1, a_2, \ldots, a_n$. The Laplace transform of the right-hand side of Eq. (14.57) can be obtained similarly. The result is

$$\mathcal{L}\{\mathcal{B}(D)v_s(t)\} = \mathcal{B}(s)V_s(s) + F(s) \quad (14.62)$$

where

$$F(s) = -b_m\sum_{k=0}^{m-1} s^{m-1-k}v_s^{(k)}(0^-) - b_{m-1}\sum_{k=0}^{m-2} s^{m-2-k}v_s^{(k)}(0^-) - \cdots - b_1v_s(0^-) \quad (14.63)$$

is an $(m-1)$st-degree polynomial that depends on "initial" input values $v_s(0^-)$, $v_s^{(1)}(0^-), \ldots, v_s^{(m-1)}(0^-)$ and the coefficients $b_1, b_2, \ldots, b_m$. It follows from Eqs. (14.60) and (14.62) that the Laplace transform of LTI Eq. (14.57) is

$$\mathcal{A}(s)V_o(s) + E(s) = \mathcal{B}(s)V_s(s) + F(s) \quad (14.64)$$

The solution follows easily

$$V_o(s) = \frac{\mathcal{B}(s)}{\mathcal{A}(s)}V_s(s) + \frac{F(s) - E(s)}{\mathcal{A}(s)} \quad (14.65)$$

We know from Chapter 13 that the ratio $\mathcal{B}(s)/\mathcal{A}(s)$ is the circuit's transfer function $H(s)$:

$$H(s) = \frac{\mathcal{B}(s)}{\mathcal{A}(s)} \quad (14.66)$$

The second term on the right-hand side of Eq. (14.65) is a function of the initial conditions $v_o(0^-)$, $v_o^{(1)}(0^-)$, . . . , $v_o^{(n-1)}(0^-)$, and $v_s(0^-)$, $v_s^{(1)}(0^-)$, . . . , $v_s^{(m-1)}(0^-)$. This term is called the *stored-energy response,* because it represents that portion of the output $v_o(t)$, $t \geq 0$, caused by the energy that is stored in the circuit at $t = 0^-$. The first term on the right-hand side of Eq. 14.65 is called the *forced response,* because it represents that portion of the output caused by the input $v_s(t)$ for $t \geq 0$. Thus

$$V_o(s) = \underbrace{H(s)V_s(s)}_{\substack{\longleftrightarrow \\ \text{Forced} \\ \text{response}}} + \underbrace{\frac{F(s) - E(s)}{\mathcal{A}(s)}}_{\substack{\longleftrightarrow \\ \text{Stored-energy} \\ \text{response}}} \tag{14.67}$$

The complete response, $v_o(t)$, is given by the inverse Laplace transform of Eq. (14.67). Let us assume that the circuit is initially at rest. A circuit initially at rest, by definition, has zero initial conditions and contains no stored energy. For such a circuit the stored-energy response is zero and Eq. (14.67) simplifies to

$$V_o(s) = H(s)V_s(s) \tag{14.68}$$

If we apply the convolution property of Table 14.1 to Eq. (14.68), we obtain the following basic result:

The forced response of any LTI circuit is given by

$$v_o(t) = \int_{0^-}^{t} h(t - \lambda)v_s(\lambda)\, d\lambda \tag{14.69}$$

where

$$h(t) = \mathcal{L}^{-1}\{H(s)\} \tag{14.70}$$

The forced response is the complete response when the initial conditions are all zero. The integral appearing in Eq. (14.69) is called a *convolution integral.* We see that Eq. (14.69) provides us with a completely new way to compute the response of an LTI circuit: we can now evaluate a convolution integral instead of solving a differential equation, Eq. (14.57), or using Laplace transforms. Equation (14.69) also provides us with a completely new understanding of the way output is related to input. Let us now see why.

An obviously important quantity in Eq. (14.69) is the function $h(t)$. What is the significance of this function? To answer this question, assume that a unit impulse is applied to the circuit; that is, assume that $v_s(t) = \delta(t)$. According to Eq. (14.69), the circuit's response to the impulse is

$$v_o(t) = \int_{0^-}^{t} h(t - \lambda)\delta(\lambda)\, d\lambda = h(t) \tag{14.71}$$

where we used the sifting property (Section 7.2) to evaluate the integral. Equation (14.71) tells us the significance of the function $h(t)$: $h(t)$ is the *impulse response* of the circuit. The three essential facts are summarized below.

1. The impulse response, $h(t)$, of a circuit is the response of the circuit to a unit impulse.
2. A circuit's impulse response and the transfer function are Laplace transform pairs: $h(t) \leftrightarrow H(s)$.
3. The forced response of an LTI circuit is given by the convolution of the input and the impulse response (Eq. 14.69).

Calculation of a convolution integral is not always easy. The topic is included in most signals and system courses. We will present one example to illustrate the main ideas.

EXAMPLE 14.20 The RC circuit of Fig. 14.5, on page 515, is at rest for $t \leq 0$.
(a) Find the impulse response of the RC circuit.
(b) Assume that $v_s(t) = u(t)$ V. Find the output $v_o(t)$ for $t \geq 0$.

Solution (a) According to Eq. (14.70) we can find the impulse response by taking the inverse Laplace transform of the transfer function. We see from Fig. 14.5 that the circuit is a voltage divider. Thus the transfer function is obtained easily and is

$$H(s) = \frac{\frac{1}{sC}}{R + \frac{1}{sC}} = \frac{\frac{1}{s}}{2 + \frac{1}{s}} = \frac{0.5}{s + 0.5}$$

Therefore, by the linearity property and LT5,

$$h(t) = 0.5e^{-0.5t}u(t)$$

(b) The output can be found from Eq. (14.71) with $v_s(t) = u(t)$ and $h(t) = 0.5e^{-0.5t}u(t)$,

$$\begin{aligned} v_o(t) &= \int_{0^-}^{t} h(t - \lambda)v_s(\lambda)\, d\lambda \\ &= \int_{0^-}^{t} 0.5e^{-0.5(t-\lambda)}u(t - \lambda)u(\lambda)\, d\lambda \end{aligned}$$

The two functions, $h(t - \lambda) = 0.5e^{-0.5(t-\lambda)}u(t - \lambda)$, and $v_s(\lambda) = u(\lambda)$ V appearing in the integral are plotted in Fig. 14.16a for $t < 0$ and in Fig. 14.16b for $t \geq 0$. Notice that the functions have been plotted versus λ. This is because to evaluate the integral, we must multiply the two functions together and integrate along the λ axis from $\lambda = 0^-$ to $\lambda = t$. Also notice that the step function $u(t - \lambda)$ makes $h(t - \lambda)$ equal zero for $\lambda > t$. The product is shown in Fig. 14.16c for $t < 0$ and in Fig. 14.16d for $t \geq 0$. The output $v_o(t)$ is given by the area under the products. For $t < 0$ this area is, by inspection of Fig. 14.16c, zero. For $t \geq 0$ the area is given by

$$\begin{aligned} v_o(t) &= \int_{0^-}^{t} 0.5e^{-0.5(t-\lambda)}\, d\lambda \\ &= 0.5e^{-0.5t} \int_{0^-}^{t} e^{0.5\lambda}\, d\lambda = 1 - e^{-5t} \text{ V} \end{aligned}$$

Thus, as in example 14.13, we find that $v_o(t) = 1 - e^{-5t}$ V for $t \geq 0$.

Remember

For every LTI circuit that is initially at rest, the output is given by the convolution of the input and the impulse response of the circuit. A circuit's impulse response and its transfer function are Laplace transform pairs.

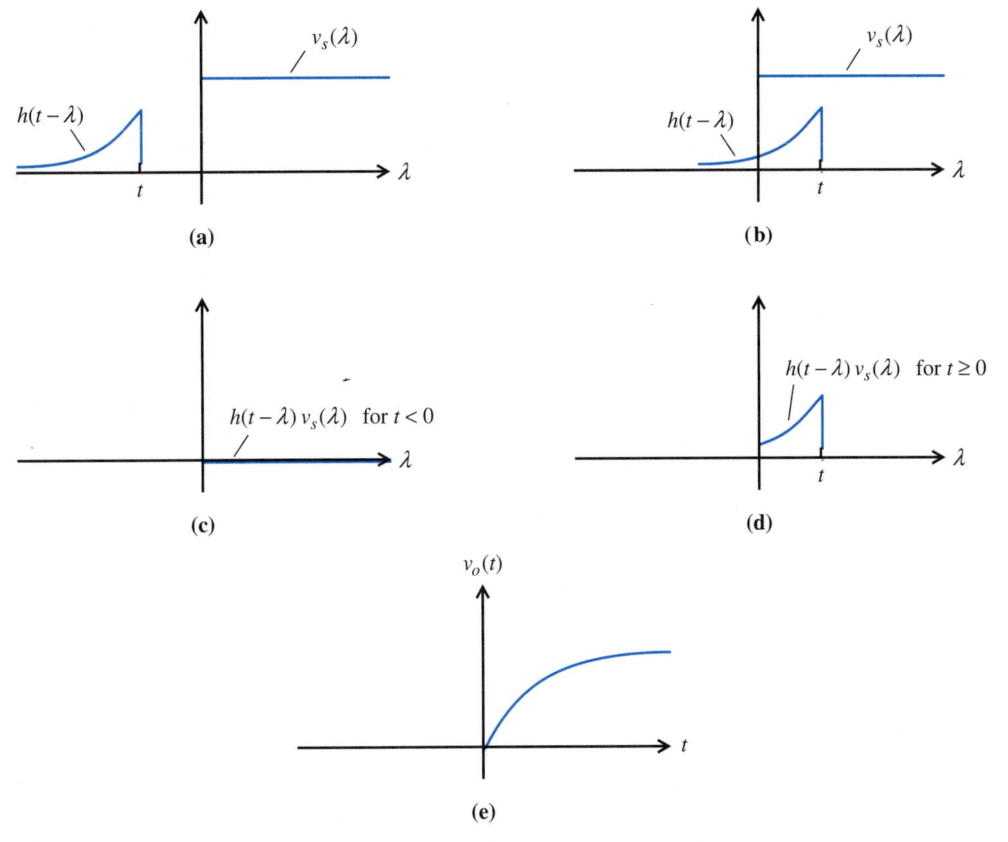

FIGURE 14.16
Graphical interpretation of convolution. As t increases, $h(t-\lambda)$ "slides" along the λ axis, causing the product $v_s(\lambda)h(t-\lambda)$ and its area v_o to vary. (*a*) $v_s(\lambda)$ and $h(t-\lambda)$ for $t<0$; (*b*) $v_s(\lambda)$ and $h(t-\lambda)$ for $t\geq 0$; (*c*) $v_s(\lambda)h(t-\lambda)$ for $t<0$; (*d*) $v_s(\lambda)h(t-\lambda)$ for $t\geq 0$; (*e*) $v_o(t)$.

EXERCISES

45. The impulse response of a hypothetical circuit is given by $h(t) = u(t)$.
    (a) Find the transfer function, $H(s)$. answer: $1/s$
    (b) Assume that the circuit is initially at rest. Use convolution to find the output, $v_o(t)$, for an input $v_s(t) = u(t)$. answer: $tu(t)$ (a ramp)
46. The impulse response of a hypothetical circuit is given by $h(t) = e^{-3t}u(t)$.
    (a) Find the transfer function, $H(s)$. answer: $1/(s+3)$
    (b) Use convolution to find the output for input $v_s(t) = u(t)$. Assume that the circuit is initially at rest. answer: $\frac{1}{3}(1-e^{-3t})u(t)$
47. Find the output to the circuit of Exercise 46 for an input $v_s(t) = \delta(t) + u(t)$. answer: $e^{-3t}u(t) + \frac{1}{3}(1-e^{-3t})u(t)$

14.8 Summary

This chapter described one of the most useful of all analytical aids to determine the response of an LTI circuit with arbitrary input and initial conditions. The Laplace transform $V(s)$ exists for virtually every signal that arises in circuit theory and engineering practice.

We saw that the Laplace transform can be applied to LTI circuit analysis with the use of one of two methods. In the first method, the Laplace transform is applied to the differential equation relating the input and output. This process yields an algebraic equation from which the Laplace transform of the output can be obtained. In the second method, the Laplace transform is applied to the circuit itself to obtain a frequency-domain circuit from which the Laplace transform of the response can be obtained. In both methods, all initial conditions are included in the analysis from the start, and only algebraic manipulations are required to obtain the Laplace transform of the response. The time-domain response is obtained by using partial fraction expansions and table look-up. In Section 14.7 we used the Laplace transform to uncover a relation between a circuit's response to any input and its response to an impulse. If the circuit is initially at rest, the output equals the convolution of the input and the circuit's impulse response. This is the forced response. If the circuit is not initially at rest, the output contains another term called the stored-energy response.

KEY FACTS

Concept	Equation	Section	Page
❑ The Laplace transform of a signal $v(t)$ is defined as $$V(s) = \int_{0-}^{\infty} v(t)e^{-st}\,dt \qquad \mathcal{Re}\{s\} > \sigma_0$$ where σ_0 is the abscissa of convergence.	(14.1)	14.1	493
❑ The Laplace transform is one-to-one if $v(t) = 0$ for $t < 0$.		14.1	493
❑ The inverse Laplace transform of $V(s)$ is given by $$v(t) = \int_{c-j\infty}^{c+j\infty} V(s)e^{st}\,\frac{ds}{2\pi j} \qquad c > \sigma_0$$	(14.2)	14.1	493
❑ The region of convergence, $\mathcal{Re}\{s\} > \sigma_0$, of every rational algebraic $V(s)$ lies to the immediate right of the poles of $V(s)$.		14.2	499
❑ Laplace transform properties and pairs are given in Tables 14.1 and 14.2, respectively.		14.2 14.3	497 504
❑ A rational algebraic function of s may be expanded into *partial fractions.* This step simplifies the task of finding the inverse Laplace transform by look-up table.		14.4	506
❑ There are two useful techniques to apply the Laplace transform to LTI circuit analysis. The first is to apply the Laplace transform to the LTI equation relating output to input. The second is to use the Laplace transform to transform the circuit itself into the frequency domain.		14.5 14.6	514 520

KEY FACTS *Concept*	*Equation*	*Section*	*Page*
❑ For a circuit initially at rest, the output is equal to the convolution of the input and the circuit's impulse response.	(14.69)	14.7	529
❑ A circuit's impulse response and transfer function are Laplace transform pairs.	(14.70)	14.7	529
❑ A circuit's forced response is solely due to the input. It is given by the convolution of the input and the circuit's impulse response.	(14.69)	14.7	529
❑ A circuit's stored-energy response is solely due to nonzero initial conditions.	(14.67)	14.7	529

Appendix 14.1 Proof of the Convolution Property

In this appendix we prove the convolution property given in Table 14.1. This property states that the Laplace transform of the convolution of two causal signals equals the product of the Laplace transforms of the signals. That is

$$\mathcal{L}\left\{\int_{0^-}^{t} v_1(t-\lambda)v_2(\lambda)\,d\lambda\right\} = V_1(s)V_2(s) \tag{14.72}$$

Since $v_1(t-\lambda) = 0$ for $\lambda > t$, we can change the upper limit of the integral in (14.72) from t to ∞ without changing the value of the integral. By making this change and writing out the definition of the Laplace transform, the left-hand side of (14.72) becomes

$$L\left\{\int_{0^-}^{t} v_1(t-\lambda)v_2(\lambda)\,d\lambda\right\} = \int_{0^-}^{\infty}\left\{\int_{0^-}^{\infty} v_1(t-\lambda)v_2(\lambda)\,d\lambda\right\} e^{-st}\,dt \tag{14.73}$$

We next interchange the order of integration in the right-hand side of Eq. (14.73) to obtain

$$L\left\{\int_{0^-}^{t} v_1(t-\lambda)v_2(\lambda)\,d\lambda\right\} = \int_{0^-}^{\infty}\left\{\int_{0^-}^{\infty} v_1(t-\lambda)e^{-st}\,dt\right\} v_2(\lambda)\,d\lambda \tag{14.74}$$

The integral inside the braces on the right-hand side of (14.74) is the Laplace transform of the time-delayed signal $v_1(t-\lambda)$. We know from the time-delay property that this Laplace transform is just $V_1(s)e^{-s\lambda}$. Therefore, (14.74) becomes

$$\begin{aligned} L\left\{\int_{0^-}^{t} v_1(t-\lambda)v_2(\lambda)\,d\lambda\right\} &= \int_{0^-}^{\infty} \{V_1(s)e^{-s\lambda}\}v_2(\lambda)\,d\lambda \\ &= V_1(s)\int_{0^-}^{\infty} e^{-s\lambda}v_2(\lambda)\,d\lambda \\ &= V_1(s)V_2(s) \end{aligned} \tag{14.75}$$

Thus, we have established the convolution property.

PROBLEMS

Section 14.1

For Problems 1 through 6, find the Laplace transforms of the given signals and specify the region of convergence.

1. $e^{-3t}u(t)$
2. $e^{t}u(t)$
3. $e^{3(t-1)}u(t-1)$
4. $e^{t}u(t+1)$
5. $u(t) + e^{3t}u(t) + e^{-3t}u(t)$
6. $\cos(4t)u(t)$

For Problems 7 and 8, explain why the unilateral Laplace transforms of the pairs of functions of t are identical.

7. $u(t)$ and 1
8. $u(-t)$ and 0

For Problems 9 through 11, find the signals that have the given Laplace transforms.

9. $\dfrac{1}{s+6}$ for $\mathcal{R}e\{s\} > -6$
10. $\dfrac{1}{s-6}$ for $\mathcal{R}e\{s\} > 6$
11. $\dfrac{s+7}{s+6}$ for $\mathcal{R}e\{s\} > -6$ *Hint:* $s + 7 = (s+6) + 1$
12. Determine the Laplace transform of $v(t) = \mathbf{V}e^{s_0 t}u(t)$ where $s_0 = \sigma_0 + j\omega_0$ is an arbitrary complex number. Show the poles and zeros of $V(s)$ in the s plane and shade the region of convergence.
13. Determine the Laplace transform of $v(t) = V_m e^{\sigma_0 t}\cos(\omega_0 t + \phi_0)u(t)$, where σ_0, ω_0, and ϕ_0 are arbitrary real numbers. Show the poles and zeros of $V(s)$ in the s plane, and shade the region of convergence. [*Hint:* Express $v(t)$ as the sum of two causal exponential signals.]

For Problems 14 through 17, use Tables 14.1 and 14.2 to find the Laplace transform of the given signal.

14. $e^{-3t}\cos 4t\, u(t)$
15. $12\cos(5t + 45°)u(t)$
16. $7\sin(10t + 30°)u(t)$
17. $12e^{-3t}\cos(5t + 45°)u(t)$

Sections 14.2 and 14.3

For Problems 18 through 21, sketch each signal and use Tables 14.1 and 14.2 to find their Laplace transforms.

18. $\Pi\left(\dfrac{t-t_0}{\tau}\right)$, where $t_0 > 0.5\tau$
19. $\cos(\omega_0 t)\Pi\left(\dfrac{t-t_0}{\tau}\right)$, where $t_0 > 0.5\tau$
20. $\cos[\omega_0(t-t_0)]\Pi\left(\dfrac{t-t_0}{\tau}\right)$, where $t_0 > 0.5\tau$
21. $tu(t) + (2-2t)u(t-1) + (t-2)u(t-2)$

For Problems 22 through 26, use Tables 14.1 and 14.2 to find the signals having the given Laplace transforms.

22. $\dfrac{e^{-s}}{s+2}$
23. $\dfrac{e^{-s}}{s}$
24. $\dfrac{s}{s^2+9}$
25. $\dfrac{s+2}{(s+2)^2+9}$
26. $\dfrac{s+4}{s^2+8s+41}$
27. According to the scaling property, $L\{v(t/a)\} = aV(as)$ for $a > 0$, where $L\{v(t)\} = V(s)$. Prove the scaling property by substituting $v(t/a)$ into Eq. (14.1) and changing the variable of integration.
28. $\mathcal{L}\{v(t-t_d)u(t-t_d)\} = e^{-st_d}V(s)$ for $t_d \geq 0$, according to the time-delay property. Prove this property by substituting $v(t-t_d)u(t-t_d)$ into Eq. (14.1) and changing the variable of integration.
29. Verify that the initial value theorem gives the correct result for $v(t) = Ae^{-\alpha t}u(t)$.
30. Verify that the final value theorem gives the correct result for $v(t) = A(1-e^{-\alpha t})u(t)$ where $\alpha > 0$.
31. Show that the final value theorem gives an incorrect result for $v(t) = A\cos(\omega_0 t)u(t)$. Discuss.

Section 14.4

32. Find $g(t) = \mathcal{L}^{-1}\{G(s)\}$, where

$$G(s) = \frac{2s^2 + 5s + 4}{(s + 1)(s^2 + 2s + 2)}$$

33. Find $g(t) = \mathcal{L}^{-1}\{G(s)\}$, where

$$G(s) = \frac{s^3 + 3s^2 + 2s + 2}{s^2 + 2s + 1}$$

34. The Laplace transform of a particular function $g(t)$ is $G(s) = N(s)/D(s)$, where $deg\{N(s)\} < deg\{D(s)\}$ and where all the roots $s_1, s_2, \ldots, s_n$ of the equation $D(s) = 0$ are distinct.

(a) Use Eq. (14.42) to show that the partial fraction expansion of $G(s)$ is given by

$$G(s) = \frac{k_{11}}{s - s_1} + \frac{k_{21}}{s - s_2} + \cdots + \frac{k_{n1}}{s - s_n} = \sum_{i=1}^{n} \frac{k_{i1}}{s - s_i}$$

where k_{i1} is the residue of $G(s)$ at s_i.

(b) Identify the entries in Tables 14.1 and 14.2 that enable you to conclude that the inverse Laplace transform of $G(s)$ is given by

$$g(t) = k_{11}e^{s_1 t}u(t) + k_{21}e^{s_2 t}u(t) + \cdots + k_{n1}e^{s_n t}u(t) = \sum_{i=1}^{n} k_{i1}e^{s_i t}u(t)$$

35. (a) Let $G(s) = N(s)/D(s)$ be a rational algebraic function with a simple pole at s_1. Assume that $deg\{N(s)\} < deg\{D(s)\}$. Show that the residue of $G(s)$ at s_1 is given by

$$k_{11} = \frac{N(s_1)}{D'(s_1)}$$

where $D'(s_1)$ denotes the derivative to $D(s)$ evaluated at s_1. [*Hint:* Write $D(s)$ as $D(s) = (s - s_1)\,Q(s)$ and note that by Eq. (14.44), $k_{11} = N(s_1)/Q(s_1)$. Then show that $D'(s_1) = Q(s_1)$.]

(b) Generalize the results from (a) to show that if all the poles $s_1, s_2, \ldots, s_n$ of the function $G(s) = N(s)/D(s)$ are simple, then the partial fraction expansion of $G(s)$ is

$$G(s) = \sum_{i=1}^{n} \frac{N(s_i)/D'(s_i)}{s - s_i}$$

This result is known as *Heaviside's expansion theorem.*

36. Use the Heaviside expansion theorem of Problem 35 to find the partial fraction expansion of

$$G(s) = \frac{2s^2 + 5s + 4}{(s + 3)(s^2 + 3s + 2)}$$

Then find $g(t) = \mathcal{L}^{-1}\{G(s)\}$.

37. Prove that the region of convergence of the Laplace transform lies to the immediate right of the poles of $V(s)$. {*Hint:* Write $V(s)$ in partial fraction form. If you are stuck, consider a simple example like $V(s) = 1/[(s + 1)(s - 3)]$, then generalize.}

Section 14.5

Use the Laplace transform to find the complete response of the following differential equations for $t > 0$.

38. $\dfrac{dv}{dt} + 4v = 100,\ v(0^-) = 20$

39. $\dfrac{dv}{dt} + 4v = 100t,\ v(0^-) = 20$

40. $\dfrac{dv}{dt} + 4v = 100 \cos t,\ v(0^-) = 20$

41. $\dfrac{d^2v}{dt^2} + 3\dfrac{dv}{dt} + 2v = 4$ where

$$v(0^-) = 0 \text{ and } \left.\frac{dv}{dt}\right|_{t=0^-} = 0$$

42. In the circuit shown, $R = 1\ \Omega$ and $C = 1$ F.

(a) Find $v_R(t)$ and $v_C(t)$ for $t \geq 0$ if $v_1(t) = 10u(t)$ V and $v_C(0^-) = -10$ V.

(b) Find $v_R(t)$ and $v_C(t)$ for $t \geq 0$ if $v_1(t) = Atu(t)$ V and $v_C(0^-) = 0$ V.

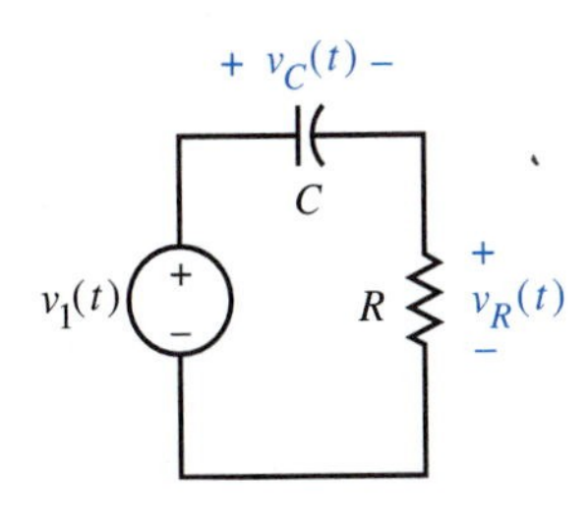

43. In the circuit shown on the next page, $R = 1\ \Omega$, $L = 1$ H, $v_1(t) = 10\,u(t)$ V and $v_R(0^-) = -10$ V. Find $v_R(t)$ and $v_L(t)$ for $t \geq 0$.

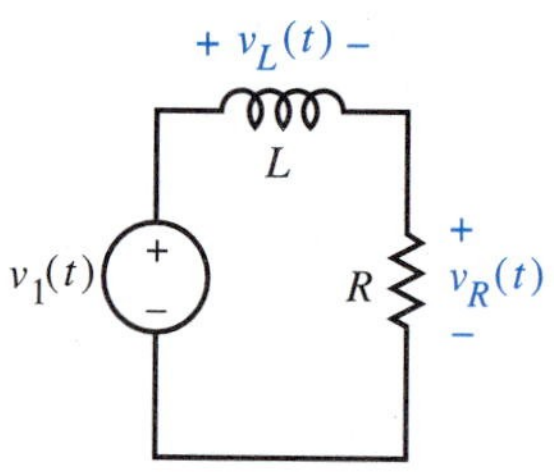

44. Use Laplace transforms to determine $v_R(t)$ and $v_L(t)$ for the circuit of Problem 43 if $v_1(t) = Atu(t)$ and $v_R(0^-) = 0$.

45. In the circuit shown, $L = 1$ H, $C = 1$ F, $v_1(t) = 10\ u(t)$ V, and there is no energy stored in the circuit for $t < 0$. Find $v_L(t)$ and $v_C(t)$ for $t \geq 0$.

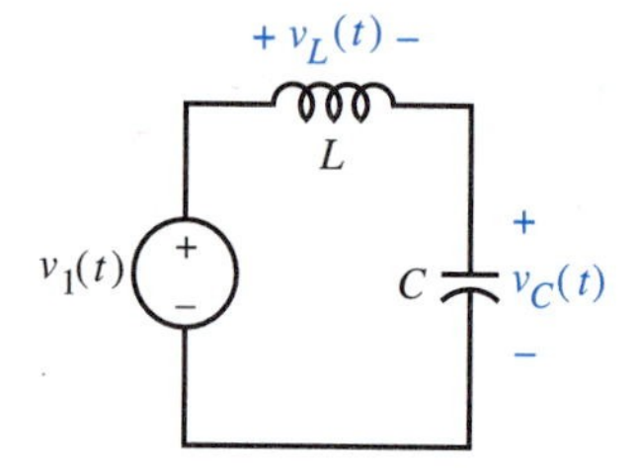

46. (a) Use Laplace transforms to determine $v_L(t)$ and $v_C(t)$ for the circuit of Problem 45 if $v_1(t) = Atu(t)$ and $i_L(0^-) = v_C(0^-) = 0$.
(b) Sketch $v_L(t)$ and $v_C(t)$.

Section 14.6

47. In the network shown, switch S1 has been closed for a very long time. At time $t = 0$, the switch is opened. Use Laplace transform circuit analysis to find v for $t > 0$.

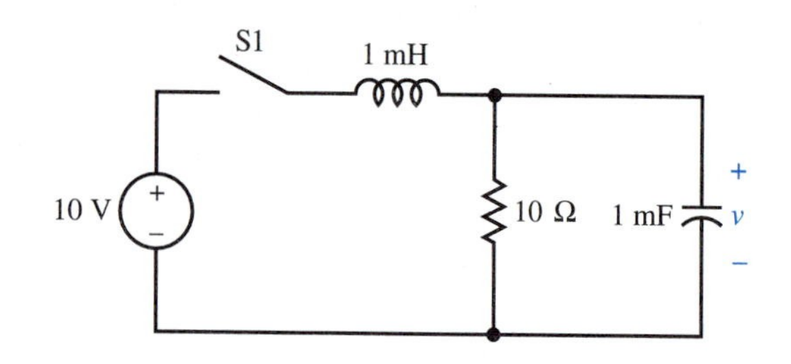

48. Repeat Problem 47 if switch S1 has been open for a very long time and closes at time $t = 0$.

49. In the circuit shown, switch S1 has been in position B for a long time. At $t = 0$ it switches to position A, where it remains until $t = 1$ ms, when it switches to position B. Determine and plot v for $t \geq 0$. Assume that the capacitances are uncharged for $t < 0$.

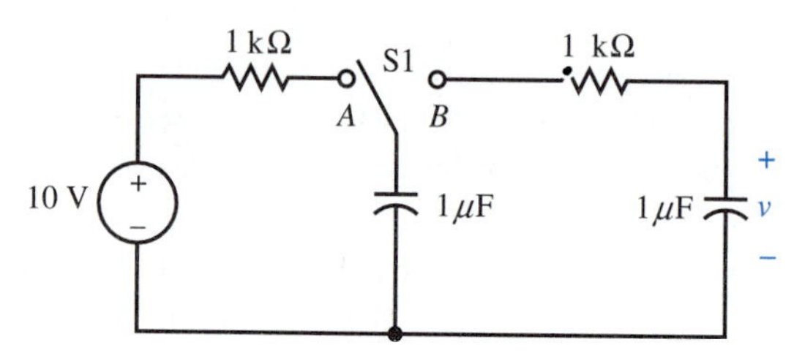

50. In the ideal op-amp circuit shown, we are given $v_C(0^-) = 5$ V and $v_1 = 10u(t)$ V. Find and sketch v_o for $t > 0$.

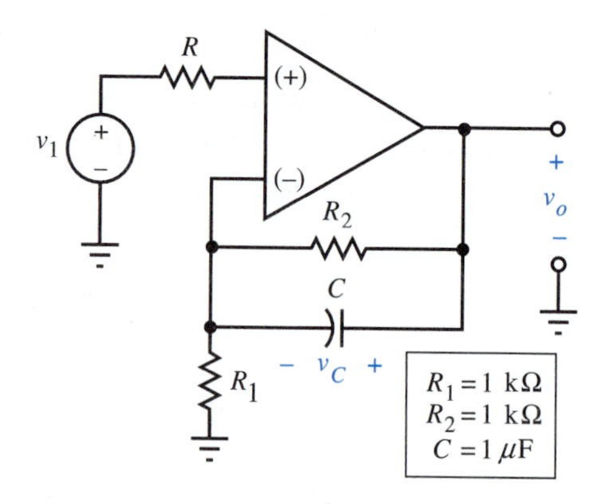

51. Repeat Problem 50 with $v_1 = tu(t)$ V.

52. Assume that the input to the ideal op-amp circuit of Problem 50 is $v_1 = A\cos(1000t)u(t)$. If $v_C(0^-) = 5$ V, what value of A causes the output v_o to be purely sinusoidal for $t > 0$?

53. Use Laplace transform circuit analysis to find the mesh currents for $t \geq 0$ in the circuit in Fig. 14.17.

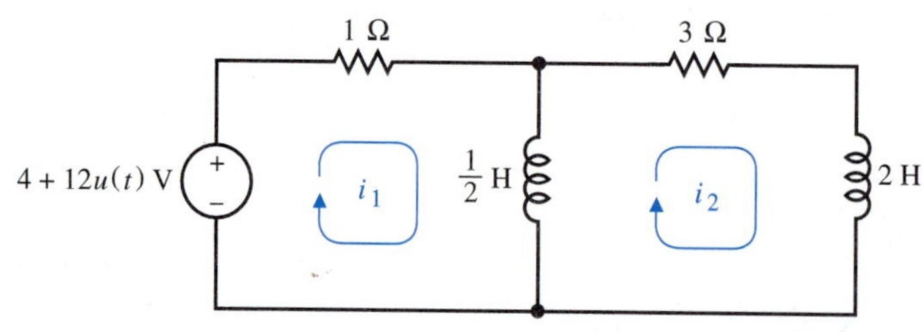

FIGURE 14.17

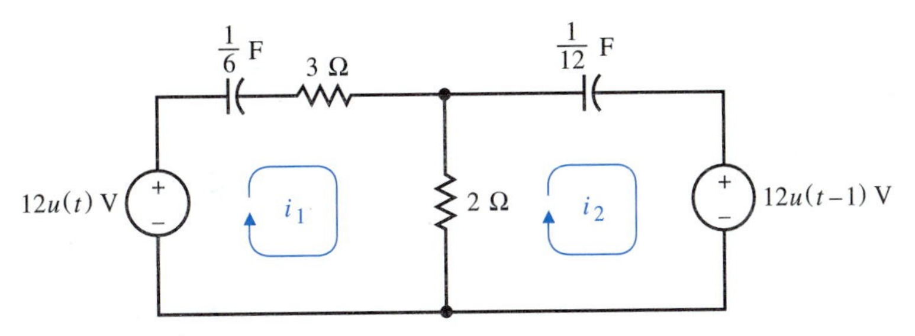

FIGURE 14.18

54. Solve for the mesh currents i_1 and i_2 for $t \geq 0$ in the circuit in Fig. 14.18. The circuit is at rest for $t < 0$.

55. In the circuit shown, the inductance has an initial current $i(0^-) = 3$ mA.

(a) Draw the Laplace-transformed circuit.

(b) Show that $I(s) = H(s)0.003/s$, where $0.003/s$ represents the initial current and $H(s)$ is a transfer function.

(c) Sketch the pole-zero plots of $H(s)$ for $\beta = 0$, -1, -2.

(d) Determine and sketch $i(t) = \mathcal{L}^{-1}\{I(s)\}$ for $\beta = 0, -1, -2$. Discuss your results.

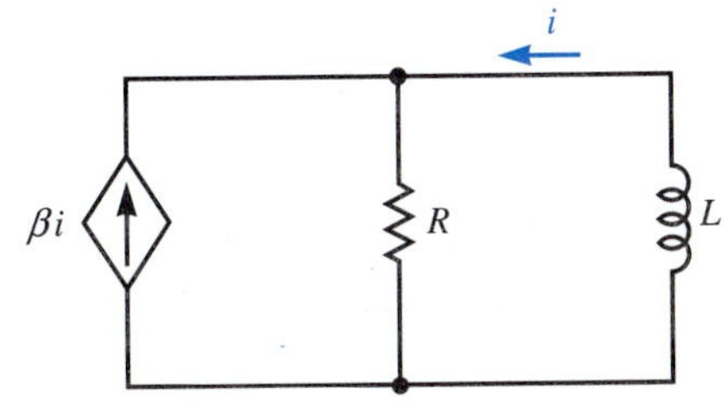

56. In the circuit shown, the capacitance has an initial voltage $v(0^-) = 2$ V.

(a) Draw the Laplace-transformed circuit.

(b) Show that $I(s) = Y(s)2/s$, where $2/s$ represents the initial voltage and $Y(s)$ is an admittance.

(c) Sketch the pole-zero plot of $Y(s)$ for $\beta = 0$, -1, -2.

(d) Determine and sketch $i(t) = \mathcal{L}^{-1}\{I(s)\}$ for $\beta = 0, -1, -2$. Discuss your results.

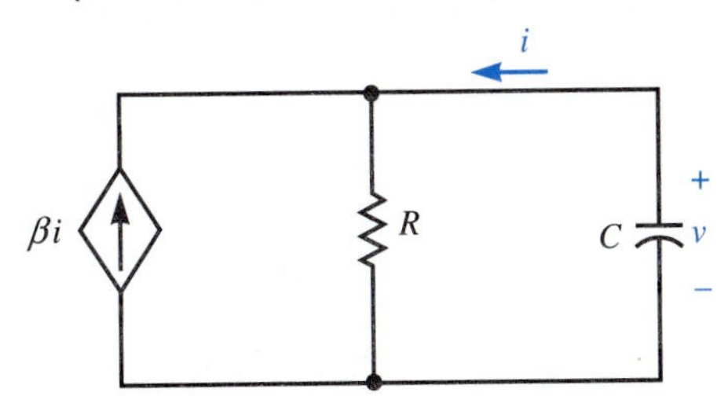

57. In the circuit shown, $v(0^-) = 2$ V. There is no current in the inductance at $t = 0^-$.

(a) Draw the Laplace-transformed circuit.

(b) Determine $I(s)$. Show that it can be written in the form $I(s) = Y(s)2/s$.

(c) Sketch the pole-zero plots of $Y(s)$ for $\beta = 0$, -1, -2.

(d) Determine and sketch $i(t) = \mathcal{L}^{-1}\{I(s)\}$ for $\beta = 0, -1, -2$. Discuss your results.

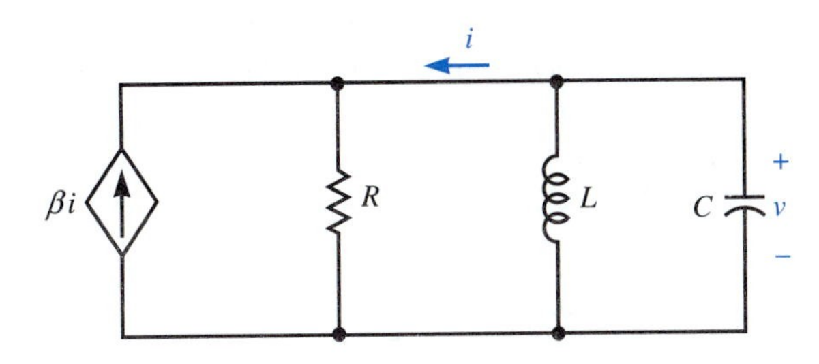

58. In the ideal op-amp circuit shown, $v_C(0^-) = 2$ V and $v_1 = 5u(t)$ V. Find and sketch v_o for $t \geq 0$.

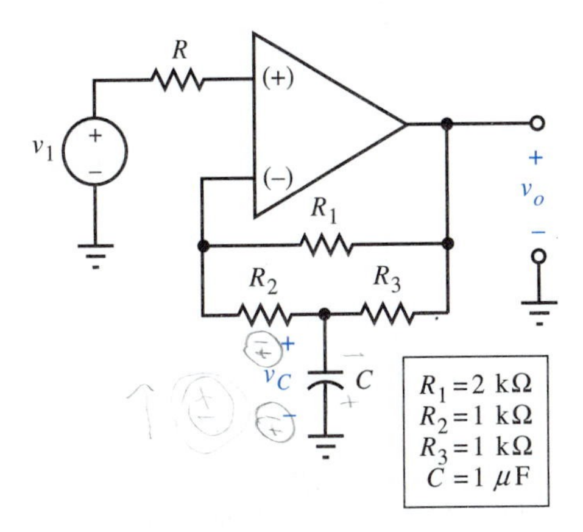

59. Assume that the input to the ideal op-amp circuit of Problem 58 is $v_1 = 5\cos(1000t)u(t)$ V. Find $v_C(0^-)$ such that the output v_o is purely sinusoidal for $t > 0$.

Section 14.7

60. A circuit has transfer function $H(s) = s/(s + 1)$. Find the impulse response, $h(t)$.

61. An input $v_s(t) = \Pi(t - 0.5)$ is applied to a circuit having impulse response $h(t) = e^{-t}u(t)$. Use convolution and plots similar to those in Fig. 14.16 to find the output. Assume that the circuit is initially at rest.

62. An input $v_s(t) = \Pi(t - 0.5)$ is applied to a certain LTI device having impulse response $h(t) = \Pi(t - 0.5)$. Use convolution and plots similar to those in Fig. 14.16 to find the forced response.

63. (a) An LTI network containing no independent sources is at rest at $t = 0^-$. An impulse is applied at $t = 0$ and the response is $e^{-2t}\cos(3t)u(t)$. Use Laplace transforms to determine the complete response of the same network to the input $\cos(2t)u(t)$.

    (b) What is the differential equation relating the input and output?

64. An ideal delay line is a device whose output, $v_o(t)$, is related to its output, $v_s(t)$, by the equation $v_o(t) = v_s(t - t_d)$ where t_d is the delay ($t_d \geq 0$). Find the impulse response and transfer function of an ideal delay line.

Supplementary Problems

65. Op amps are frequently configured to act as *buffers.* The purpose of a buffer is to prevent one circuit from drawing current from another. In part (a) of the following figure, the ideal op amp is the buffer.

    (a) Show that the voltage transfer function of the circuit in part (a) is $H(s) = H_1(s)H_2(s)$, where $H_1(s) = 1/(sR_1C_1 + 1)$ and $H_2(s) = 1/(sR_2C_2 + 1)$.

    (b) Assume that $R_1 = R_2 = 1\ \text{k}\Omega$, and $C_1 = C_2 = 1\ \mu\text{F}$, and $v_{C_1}(0^-) = v_{C_2}(0^-) = 0$. Determine the response of the circuit to $v_1(t) = \delta(t)$.

    (c) Find the voltage transfer function of the circuit of part (b) of this figure. How does your result compare to that of (a)?

    (d) Determine the response of the circuit in part (b) to $v_1(t) = \delta(t)$, with $R_1 = R_2 = 1\ \text{k}\Omega$, $C_1 = C_2 = 1\ \mu\text{F}$, and $v_{C_1}(0^-) = v_{C_2}(0^-) = 0$.

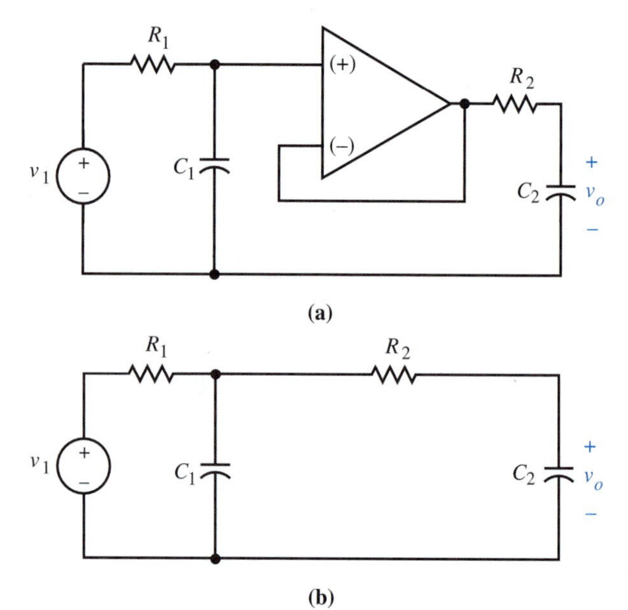

66. The circuit shown is an important part of multivibrator circuits discussed in electronics courses. The switch S1 has been open for all $t < 0$. Switch S2 is a controlled switch that remains closed if, and only if, the current through it remains positive: $i > 0$.

    (a) Show why switch S2 is closed for $t < 0$.

    (b) At $t = 0$, switch S1 closes. Determine and plot v for $t \geq 0$.

    (c) Find the value of t when $v(t)$ just reaches 0 V.

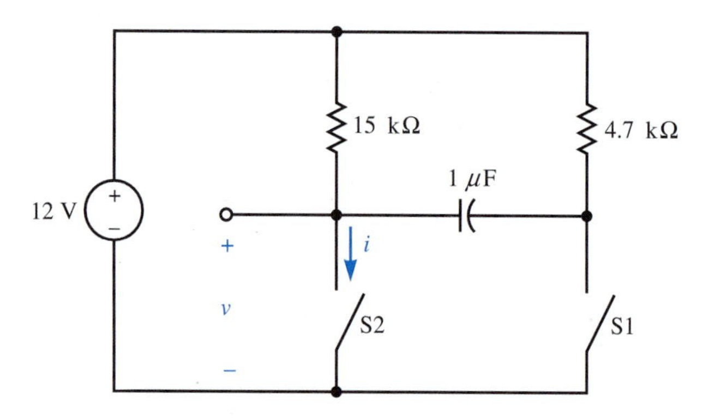

67. The transfer function

$$H_R(s) = \frac{sR/L}{(s - s_1)(s - s_2)}$$

of a band-pass *RLC* filter was examined in detail in Chapter 13. Use partial fraction expansions and Tables 14.1 and 14.2 to show that the impulse response $h(t)$ is as given on the next page.

(a) Overdamped circuit:

$$h(t) = \frac{R}{L \cosh \theta} \cosh (\beta t - \theta) e^{-\alpha t} u(t)$$

where $\theta = \text{arctanh}\,(\alpha/\beta)$.

(b) Critically damped circuit:

$$h(t) = \frac{R}{L}(1 - \alpha t) e^{-\alpha t} u(t)$$

(c) Underdamped circuit:

$$h(t) = \frac{R}{L \cos \phi} \cos (\omega_d t + \phi) e^{-\alpha t} u(t)$$

where $\phi = \arctan\,(\alpha/\omega_d)$.

15

Fourier Series

In this chapter we will describe the Fourier series and show how to use it in circuit analysis. In the Fourier series, a periodic function with period T is represented by a sum of sinusoidal functions whose frequencies are multiples of $1/T$. We use the Fourier series to help us find the response of a circuit to a periodic input waveform. The response of an LTI circuit to a periodic input waveform is found by adding the circuit's responses to the individual sinusoidal terms in the Fourier-series representation of the input.

The essential advantage of a Fourier series is that it represents a periodic input waveform by a sum of signals, *each of which is simpler to work with than the input periodic waveform.* The method is analogous to the technique used by physicists in which a force vector is decomposed into a vector sum of perpendicular force-vector components to simplify a problem in mechanics. In a Fourier series, the input waveform is decomposed into a sum of mutually orthogonal sinusoidal-waveform components. We shall see that the mutual orthogonality of the component waveforms simplifies the problem of finding the series representation of the input waveform.

15.1 Outline of the Fourier-Series Method

We will now outline the basic strategy used in determining LTI circuit response to a periodic input. For illustration, we will assume that the triangular wave of Fig. 15.1 is applied to an LTI circuit as shown in Fig. 15.2. The problem is to determine the complete response $v_o(t)$. The solution is easy with use of the Fourier series. The first step is to represent the input triangular wave by its Fourier series, which is a sum of sinusoidal terms. This is the *Fourier analysis* step. The Fourier series for the triangular wave shown in Fig. 15.1 is given in Fig. 15.2. Later sections will provide formulas to obtain this and other Fourier series. The second step is to determine the particular response of the LTI circuit to the *individual* sinusoidal terms shown in Fig. 15.2. This is an easy step because we can use ac circuit analysis. The third and final step is to apply superposition. Because the circuit is *linear,* the particular response to the triangular wave—a sum of sinusoids—is the sum of the particular responses to each sinusoid acting alone. Step three is the *Fourier synthesis* step, in which we construct the output waveform by adding together its sinusoidal terms.

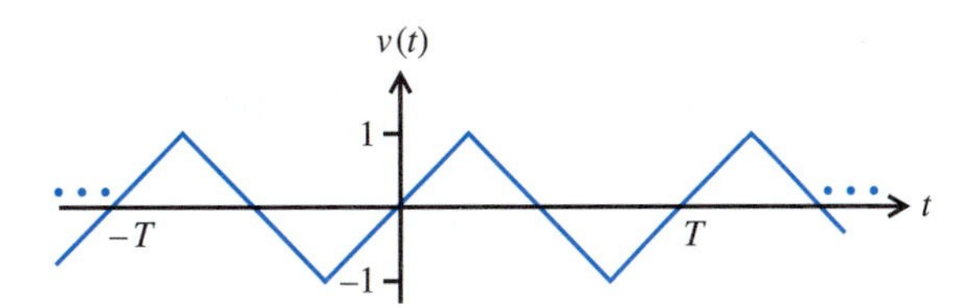

FIGURE 15.1
Triangular wave. According to Fourier, this, and every other periodic signal, can be represented by a sum of sinusoidal components

Notice that in this method the output is represented by a sum or series of sinusoidal terms. Often, this form for the output is all that we need. If we want a plot of the output, we can program a computer to sum the first hundred or so terms and plot the result. Examples of computer-generated plots are given later in this chapter.

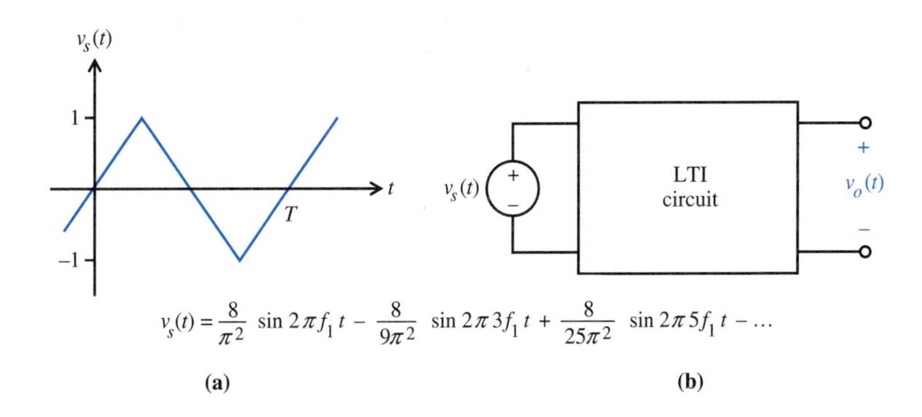

FIGURE 15.2
(*a*) Triangular wave input; (*b*) LTI circuit. Once it is recognized that $v_s(t)$ is equivalent to a sum of sinusoidal components, the output $v_o(t)$ can be determined with the use of ac circuit analysis and superposition

EXERCISES

1. The triangular voltage waveform of Fig. 15.1 can be represented as the infinite sum of sinusoids:

$$v(t) = \frac{8}{\pi^2}\sin 2\pi f_1 t - \frac{8}{9\pi^2}\sin 2\pi 3f_1 t + \frac{8}{25\pi^2}\sin 2\pi 5f_1 t - \cdots$$
$$= \sum_{n=1,3,5,\ldots}^{\infty} (-1)^{(n-1)/2}\frac{8}{(\pi n)^2}\sin 2\pi n f_1 t$$

(a) Calculate and plot the sum of the first three nonzero terms of the above formula,

$$v_5(t) = \frac{8}{\pi^2} \sin 2\pi f_1 t - \frac{8}{9\pi^2} \sin 2\pi 3 f_1 t + \frac{8}{25\pi^2} \sin 2\pi 5 f_1 t$$

where $f_1 = 1$ Hz. To make the plot, calculate and plot $v_5(t)$ for $t = 0, 0.05, 0.10, 0.15, \ldots, 0.95, 1.0$. Then connect your plotted points with a smooth curve. Compare your result with the triangular wave of Fig. 15.1. Notice that $v_5(t)$ is approximately equal to the triangular wave.

(b) Suppose that $v_5(t)$ is applied to a series RL circuit where $R = 1\ \Omega$, $L = 2$ H, and $f_1 = 1$ Hz. Determine the particular response voltage across the inductance by adding the particular responses due to the individual terms on the right-hand side of the above equation. Express the solution as a sum of three sinusoids.

answer: (b) $v_L(t) = 0.808 \sin(2\pi t + 4.55°) + 0.090 \sin(6\pi t + 1.52°) + 0.033 \sin(10\pi t + 0.91°)$ V

(c) Plot your answer to (b).

2. Use a calculator or a computer program to help you sum and plot the first seven nonzero terms of the summation in Exercise 1. Compare your result with Fig. 15.1.

15.2 The Sine-Cosine Form of the Fourier Series

The Fourier series can be written in any of three basic forms, all of which are used by engineers. In this section, we describe the sine-cosine form of the Fourier series. Let us now consider the problem of representing an arbitrary real periodic waveform $v(t)$ by a sum of sinusoidal terms or *components* as follows:

DEFINITION Sine-Cosine Form of Fourier Series

$$\begin{aligned} v(t) &= a_0 + a_1 \cos 2\pi f_1 t + \cdots + a_n \cos 2\pi n f_1 t + \cdots + b_1 \sin 2\pi f_1 t + \cdots \\ &\quad + b_n \sin 2\pi n f_1 t + \cdots \\ &= a_0 + \sum_{n=1}^{\infty} (a_n \cos 2\pi n f_1 t + b_n \sin 2\pi n f_1 t) \end{aligned} \tag{15.1}$$

where $f_1 = 1/T$ and T is the period of $v(t)$. We now ask, ''If we *can* express $v(t)$ as in Eq. (15.1), how *must* the coefficients a_0 and a_n, b_n, $n = 1, 2, 3, \ldots$, be related to the function $v(t)$?''

Part of the answer to this question follows from the observation that a_0 must be the average value of $v(t)$. This is because the average values of all the other terms on the right-hand side are zero. Therefore,

$$a_0 = \frac{1}{T} \int_{t_0}^{t_0+T} v(t)\, dt \tag{15.2}$$

where t_0 is arbitrary. [If you are not convinced, substitute the right-hand side of Eq. (15.1) into Eq. (15.2) and integrate. The result will be a_0.]

A formula for a_n $(n \geq 1)$ can be obtained if we multiply both sides of Eq. (15.1) by $\cos 2\pi n f_1 t$ and integrate over an interval of duration T:

$$\begin{aligned}\int_{t_0}^{t_0+T} v(t) \cos 2\pi n f_1 t \, dt &= \int_{t_0}^{t_0+T} a_0 \cos 2\pi n f_1 t \, dt \\ &+ \sum_{k=1}^{\infty} a_k \int_{t_0}^{t_0+T} \cos 2\pi k f_1 t \cos 2\pi n f_1 t \, dt \\ &+ \sum_{k=1}^{\infty} b_k \int_{t_0}^{t_0+T} \sin 2\pi k f_1 t \cos 2\pi n f_1 t \, dt \end{aligned} \tag{15.3}$$

Observe that the first integral on the right-hand side of Eq. (15.3) is zero because the area under an integer number of cycles of a cosine function is zero. Also observe that each of the remaining integrals on the right-hand side *except* the one involving the $\cos^2 2\pi n f_1 t$ (this is the term for $k = n$ in the first summation) is also zero. It is easy to see why. Because

$$\cos 2\pi k f_1 t \cos 2\pi n f_1 t = \frac{1}{2} \cos 2\pi (k + n) f_1 t + \frac{1}{2} \cos 2\pi (k - n) f_1 t$$

the integral

$$\int_{t_0}^{t_0+T} \cos 2\pi k f_1 t \cos 2\pi n f_1 t \, dt$$

is equal to one-half the area under $k + n$ cycles of a cosine function plus one-half the area under $k - n$ cycles of a cosine function. Both of these are zero if $k \neq n$.

Similar remarks apply to the integrals involving the products

$$\sin 2\pi k f_1 t \cos 2\pi n f_1 t$$

Because

$$\sin 2\pi k f_1 t \cos 2\pi n f_1 t = \frac{1}{2} \sin 2\pi (k + n) f_1 t + \frac{1}{2} \sin 2\pi (k - n) f_1 t$$

the integral

$$\int_{t_0}^{t_0+T} \sin 2\pi k f_1 t \cos 2\pi n f_1 t \, dt$$

equals zero for all integer values of k. Since all but one integral on the right-hand side of Eq. (15.3) equal zero, Eq. (15.3) becomes simply

$$\begin{aligned}\int_{t_0}^{t_0+T} v(t) \cos 2\pi n f_1 t \, dt &= a_n \int_{t_0}^{t_0+T} \cos^2 2\pi n f_1 t \, dt \\ &= a_n \frac{T}{2}\end{aligned} \tag{15.4}$$

and this yields

$$a_n = \frac{2}{T} \int_{t_0}^{t_0+T} v(t) \cos 2\pi n f_1 t \, dt \qquad \text{for } n = 1, 2, \ldots \tag{15.5}$$

The coefficient b_n can be obtained similarly if we multiply both sides of Eq. (15.1) by $\sin 2\pi n f_1 t$ and integrate over an interval T. All the integrals on the right-hand side except the one involving $\sin^2 2\pi n f_1 t$ will be zero. This leads to

$$b_n = \frac{2}{T}\int_{t_0}^{t_0+T} v(t) \sin 2\pi n f_1 t \, dt \qquad \text{for } n = 1, 2, \ldots \tag{15.6}$$

The a_n and b_n uniquely determined by Eqs. (15.2), (15.5), and (15.6) are the *real Fourier coefficients* of $v(t)$. Their units are the same as those of $v(t)$. With a_n and b_n so determined, the right-hand side of Eq. (15.1) is called the *sine-cosine* form of the *Fourier series* of $v(t)$.

Notice that the formulas for the Fourier coefficients given by Eq. (15.2), (15.5), and (15.6) are *extremely simple. Each* coefficient is computed directly from $v(t)$ *independently* of the other coefficients. This independence is a consequence of the fact that the terms on the right-hand side of Eq. (15.1) are *orthogonal* over an interval T:

$$\int_{t_0}^{t_0+T} x_1(t)x_2(t) \, dt = 0 \tag{15.7}$$

where $x_1(t)$ and $x_2(t)$ denote any two distinct terms (a_0, $a_k \cos 2\pi k f_1 t$, $b_k \sin 2\pi k f_1 t$) on the right-hand side of Eq. (15.1). As a consequence of Eq. (15.7), all but one integral on the right-hand side of Eq. (15.3) equaled zero—a significant simplification.

We use Eqs. (15.2), (15.5), and (15.6) to compute the values of the Fourier coefficients of $v(t)$. Once we have computed the values, we substitute them into Eq. (15.1) to obtain the Fourier-series representation of $v(t)$.

Parseval's Theorem for Real Periodic Waveforms

Parseval's theorem shows us that the mean square value of a periodic waveform can be expressed directly in terms of the real Fourier coefficients. We can derive Parseval's theorem by writing the mean square value of $v(t)$ as

$$\frac{1}{T}\int_{t_0}^{t_0+T} v^2(t) \, dt = \frac{1}{T}\int_{t_0}^{t_0+T} \left(a_0 + \sum_{n=1}^{\infty} a_n \cos 2\pi n f_1 t + \sum_{n=1}^{\infty} b_n \sin 2\pi n f_1 t \right)^2 dt \tag{15.8}$$

When we square the quantity in parentheses on the right-hand side of Eq. (15.8) as indicated, we obtain many cross products. All the cross products integrate to zero because the components in a Fourier series are orthogonal. The integrals of the non-cross-product terms are easily obtained. The result is

THEOREM Parseval's Theorem for Real Periodic Waveforms

$$\frac{1}{T}\int_{t_0}^{t_0+T} v^2(t) \, dt = a_0^2 + \sum_{n=1}^{\infty} \frac{1}{2} a_n^2 + \sum_{n=1}^{\infty} \frac{1}{2} b_n^2 \tag{15.9}$$

Parseval's theorem is often expressed in terms of normalized power. The *normalized power* of a signal is defined simply as the average power the signal dissipates when applied to a 1-Ω resistance.

It is easily shown that the quantity a_0^2 in Eq. (15.9) is the normalized power in the dc component a_0, $\frac{1}{2}a_n^2$ is the normalized power in the component $a_n \cos 2\pi n f_1 t$, and $\frac{1}{2}b_n^2$ is the normalized power in $b_n \sin 2\pi n f_1 t$. Therefore Parseval's theorem tells us that *the total normalized power in $v(t)$ equals the sum of the normalized powers in each of the Fourier-series components of $v(t)$*. This means that if $v(t)$ were applied across a resistance, then the total average power supplied would equal the sum of the average powers dissipated by the individual Fourier-series components.

Remember

A real periodic waveform can be represented by the sine-cosine form of the Fourier series given by Eq. (15.1). The real Fourier coefficients are given by Eqs. (15.2), (15.5), and (15.6).

EXERCISES

3. Find the period and the real Fourier-series coefficients of the periodic waveform $v(t) = 3 + 5 \cos 2\pi 10^3 t + 7 \sin 6\pi 10^3 t$.
   answer: $T = 1$ ms, $a_0 = 3$, $a_1 = 5$, $b_3 = 7$, and all other coefficients equal zero.
4. Use Parseval's theorem to find the mean square value of the waveform, $v(t)$, of exercise 3.
   answer: 46

15.3 Symmetry Properties

We now investigate the relationship between a waveform's symmetry and its Fourier series. The three symmetry properties described below are important because they tell us when certain Fourier coefficients equal zero *without the need to evaluate the integrals* in Eqs. (15.2), (15.5), and (15.6).

Symmetry Property 1

If $v(t)$ is an even function, that is, if $v(-t) = v(t)$, then the Fourier series for $v(t)$ contains only even component functions.

The even component functions are the dc term a_0 and the cosine terms $a_n \cos 2\pi n f_1 t$; $n = 1, 2, \ldots$. Therefore, if $v(t)$ is an even function, then $b_n = 0$ for $n = 1, 2, 3, \ldots$. An example of an even function is the trapezoidal pulse train of Fig. 15.3.

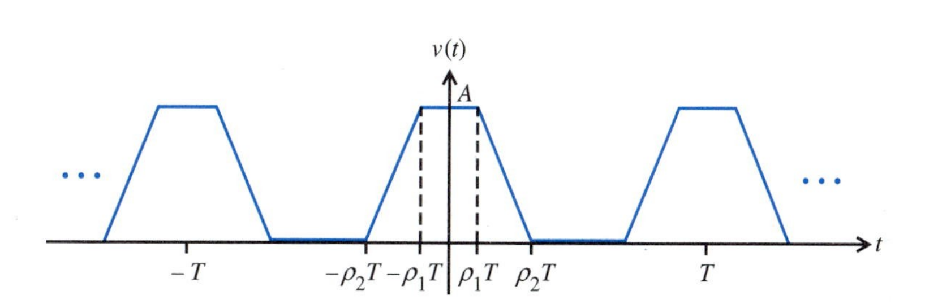

FIGURE 15.3
A function with even symmetry

PROOF Write Eq. (15.1) as

$$v(t) = v_a(t) + v_b(t) \tag{15.10}$$

where

$$v_a(t) = a_0 + \sum_{n=1}^{\infty} a_n \cos 2\pi n f_1 t \tag{15.11}$$

is an even function, and

$$v_b(t) = \sum_{n=1}^{\infty} b_n \sin 2\pi n f_1 t \tag{15.12}$$

is an odd function. We know that $v(t)$ is an even function. Therefore the right-hand side of Eq. (15.10) must be even:

$$v_a(-t) + v_b(-t) = v_a(t) + v_b(t) \tag{15.13a}$$

Since $v_a(t)$ is even and $v_b(t)$ is odd, Eq. (15.13a) can be written as

$$v_a(t) - v_b(t) = v_a(t) + v_b(t) \tag{15.13b}$$

from which follows the result

$$v_b(t) = 0 \tag{15.14}$$

It then follows from Eqs. (15.12) and (15.14) that

$$b_n = \frac{2}{T}\int_{t_0}^{t_0+T} v_b(t) \sin 2\pi n f_1 t \, dt = 0 \qquad n = 1, 2, \ldots \tag{15.15}$$

Symmetry Property 2

If $v(t)$ is an odd function, $v(-t) = -v(t)$, then its Fourier series contains only odd component functions.

The odd component functions are the sine terms $b_n \sin 2\pi n f_1 t$, $n = 1, 2, \ldots$. Therefore, if $v(t)$ is an odd function, then $a_n = 0$ for $n = 0, 1, 2, \ldots$. An example of an odd function is the sawtooth wave of Fig. 15.4.

Symmetry Property 2 can be proved in a manner similar to the proof of Symmetry Property 1. The details are in Problem 5.

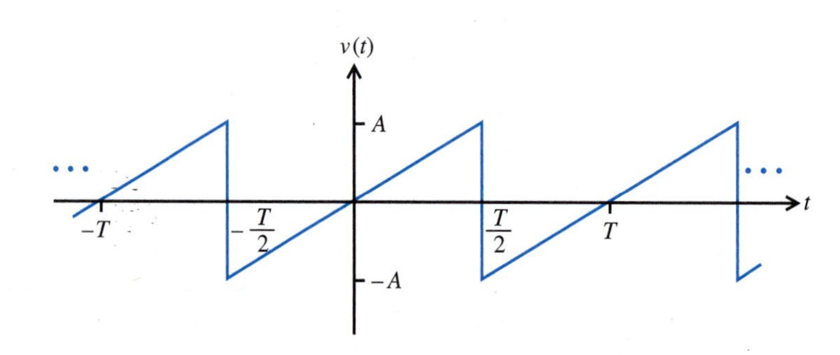

FIGURE 15.4
A function with odd symmetry

A periodic function satisfying $v(t) = -v(t - T/2)$ is said to be *half-wave symmetric*. An example of a half-wave-symmetric waveform is the double sawtooth wave shown in Fig. 15.5.

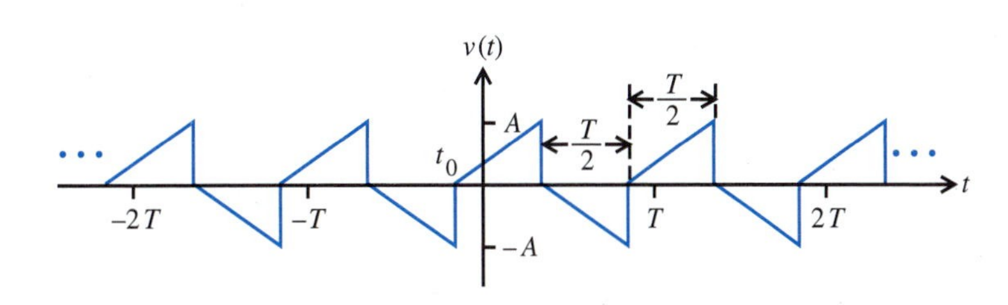

FIGURE 15.5 A function with half-wave symmetry

Symmetry Property 3

If $v(t)$ is half-wave symmetric, then its Fourier series contains only half-wave-symmetric component functions.

The half-wave-symmetric component functions are $a_n \cos 2\pi n f_1 t$ and $b_n \sin 2\pi n f_1 t$ for $n = 1, 3, 5, \ldots$. Only frequencies that are *odd* multiples of $f_1 = 1/T$ are present in a half-wave-symmetric function. Thus $a_0 = a_n = b_n = 0$ for $n = 2, 4, 6, \ldots$.

See Problem 6 for a proof of Symmetry Property 3. The following example illustrates how the symmetry properties can be used to simplify the problem of finding a Fourier series.

EXAMPLE 15.1 Find the Fourier series of the square wave illustrated in Fig. 15.6.

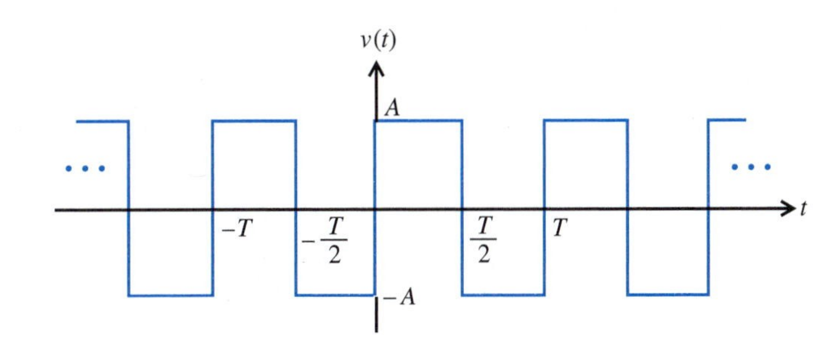

FIGURE 15.6 A square wave

Solution We can see by inspection of Fig. 15.6 that $v(t)$ is an odd function. By Symmetry Property 2, $a_n = 0$ for $n = 0, 1, \ldots$. We can also see from Fig. 15.6 that $v(t)$ is half-wave symmetric. By Symmetry Property 3, $a_n = b_n = 0$ for n even. Therefore, it suffices to evaluate Eq. (15.6) for $n = 1, 3, 5, \ldots$. It is convenient here to choose $t_0 = -T/2$. For the square wave of Fig. 15.6, Eq. (15.6) becomes

See Problem 15.1 in the PSpice manual.

$$b_n = \frac{2}{T}\int_{-T/2}^{T/2} v(t) \sin 2\pi n f_1 t \, dt$$

$$= -\frac{2}{T}\int_{-T/2}^{0} A \sin 2\pi n f_1 t \, dt + \frac{2}{T}\int_{0}^{T/2} A \sin 2\pi n f_1 t \, dt$$

The change of variable t to $-t$ in the first integral shows that the first integral is a disguised version of the second integral:

$$-\frac{2}{T}\int_{-T/2}^{0} A \sin 2\pi n f_1 t \, dt = \frac{2}{T}\int_{0}^{T/2} A \sin 2\pi n f_1 t \, dt$$

Therefore

$$\begin{aligned} b_n &= \frac{4}{T}\int_{0}^{T/2} A \sin 2\pi n f_1 t \, dt \\ &= \frac{4A}{2\pi n f_1 T}(1 - \cos \pi n f_1 T) \\ &= \frac{4A}{n\pi} \qquad n = 1, 3, 5, \ldots \end{aligned}$$

The Fourier series of the square wave is therefore

$$\begin{aligned} v(t) &= \frac{4A}{\pi}\sin 2\pi f_1 t + \frac{4A}{3\pi}\sin 2\pi 3 f_1 t + \frac{4A}{5\pi}\sin 2\pi 5 f_1 t + \cdots \\ &= \sum_{n=1,3,5,\ldots}^{\infty} \frac{4A}{n\pi}\sin 2\pi n f_1 t \end{aligned}$$

No computer can sum the infinite number of terms appearing in the Fourier-series representation of a waveform. To obtain a computer-generated plot, therefore, a computer is programmed to calculate the *partial* Fourier series of $v(t)$. The partial Fourier series $\hat{v}_N(t)$ is simply the sum of all the terms of the Fourier series up to and including those with frequencies nf_1 where $n = N$. Computer plots of the partial Fourier series of the square wave of Example 15.1 are shown in Fig. 15.23 on page 575. Notice that $\hat{v}_N(t)$ in Fig. 15.23 appears to converge to the square wave of Fig. 15.5 as N increases. This convergence is discussed in more detail in Section 15.9.

The next example illustrates Parseval's theorem.

EXAMPLE 15.2 Write Parseval's theorem for the special case that $v(t)$ is the square wave shown in Fig. 15.6.

Solution The mean square value of the square wave is, by Parseval's theorem and the results of Example 15.1,

$$\begin{aligned} \frac{1}{T}\int_{-T/2}^{T/2} v^2(t)\, dt &= \frac{1}{2}\left(\frac{4A}{\pi}\right)^2 + \frac{1}{2}\left(\frac{4A}{3\pi}\right)^2 + \frac{1}{2}\left(\frac{4A}{5\pi}\right)^2 + \cdots \\ &= \sum_{n=1,3,5,\ldots}^{\infty} \frac{1}{2}\left(\frac{4A}{n\pi}\right)^2 = A^2 \end{aligned}$$

Since $v^2(t) = A^2$ for all t (see Fig. 15.6), the evaluation of the integral tells us that the total normalized power in $v(t)$ is A^2. This is the term on the right-hand side of the preceding equation. The series expression shows the normalized powers contained in each of the Fourier components in $v(t)$. Parseval's theorem tells us that the sum of these normalized powers equals A^2.

Frequently, electronic circuits may add a dc voltage to a waveform that has either odd or half-wave symmetry. Strictly speaking, this will destroy the symmetry involved. However, the addition of a dc voltage to any periodic waveform changes only the dc coefficient, a_0 of the Fourier series. If you encounter a waveform that *would be* either odd symmetric or half-wave symmetric after the *subtraction* of a dc voltage, all the

Fourier coefficients of that waveform except a_0 will satisfy the conditions given in symmetry properties 2 and 3. Exercise 5 illustrates this simple point.

Remember

An even symmetric periodic waveform contains only dc and cosine components. An odd symmetric periodic waveform contains only sine components. A half-wave symmetric waveform contains only frequencies that are odd multiples of $1/T$.

EXERCISES

5. A dc voltage of A V is added to the odd, half-wave symmetric waveform $v(t)$ of Example 15.1, shown in Fig. 15.6, to produce a new waveform $v'(t) = v(t) + A$.
   (a) Sketch $v'(t)$.
   (b) Does $v'(t)$ have odd symmetry? *answer:* No
   (c) Does $v'(t)$ have half-wave symmetry? *answer:* No
   (d) Add A to the Fourier series for $v(t)$ derived in Example 15.1 to obtain the Fourier series for $v'(t)$.
   answer: $A + \sum_{n=1}^{\infty} (4A/n\pi) \sin(2\pi f_1 t)$
   (e) Discuss the difference between the Fourier-series coefficients of $v'(t)$ and $v(t)$.
   answer: Only the dc component has changed since only the average value has been changed.

6. State which of the waveforms shown in the following illustration contain (a) no sine components, (b) no cosine components, and (c) only frequencies that are odd multiples of $f_1 = 1/T$.
   answer: (a) $v_1(t)$ and $v_2(t)$, (b) $v_3(t)$ and $v_4(t)$, (c) $v_2(t)$ and $v_4(t)$.

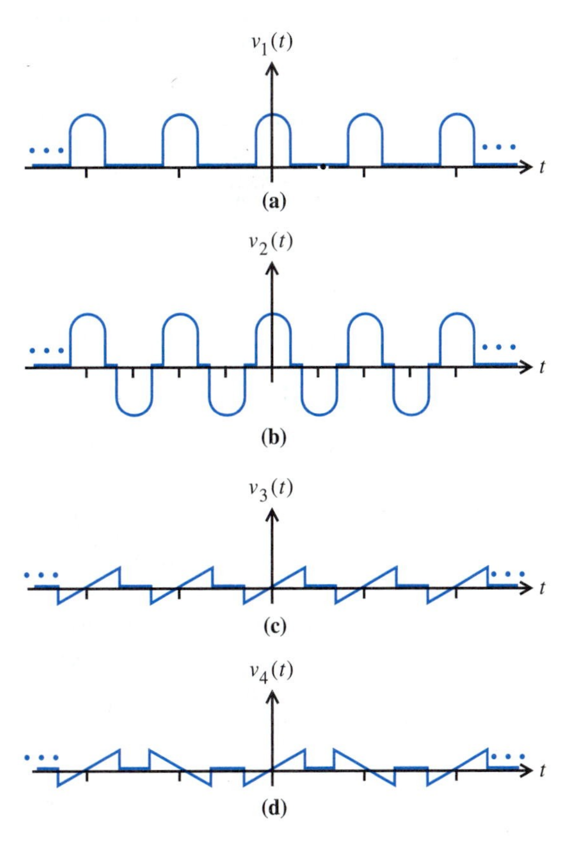

7. Derive the real Fourier series of the triangular wave of Fig. 15.1.

answer: $\sum_{n=1,3,5,\ldots}^{\infty} (-1)^{(n-1)/2}\, 8A/(\pi n)^2 \sin(2\pi n f_1 t)$

8. Derive the real Fourier series of the sawtooth wave $v(t)$ of Fig. 15.5.

answer: $\sum_{n=1}^{\infty} (-1)^{n+1} (2A/n\pi) \sin(2\pi n f_1 t)$

15.4 The Amplitude-Phase Form of the Fourier Series

The form of the Fourier series described in the previous sections is not the only form used. Another useful form of the Fourier series of a real periodic waveform can be obtained from the identity

$$A_n \cos(2\pi n f_1 t + \phi_n) = A_n \cos\phi_n \cos 2\pi n f_1 t - A_n \sin\phi_n \sin 2\pi n f_1 t \quad (15.16)$$

If we set

$$a_n = A_n \cos\phi_n \qquad n = 1, 2, \ldots \quad (15.17a)$$

$$b_n = -A_n \sin\phi_n \qquad n = 1, 2, \ldots \quad (15.17b)$$

in Eq. (15.1) and use Eq. (15.17), then Eq. (15.1) becomes

DEFINITION Amplitude-Phase Form of Fourier Series

$$v(t) = A_0 + A_1\cos(2\pi f_1 t + \phi_1) + \cdots + A_n \cos(2\pi n f_1 t + \phi_n) + \cdots$$
$$= A_0 + \sum_{n=1}^{\infty} A_n \cos(2\pi n f_1 t + \phi_n) \quad (15.18)$$

in which

$$A_0 = a_0 \quad (15.19a)$$

$$A_n = \sqrt{a_n^2 + b_n^2} \qquad n = 1, 2, \ldots \quad (15.19b)$$

and

$$\phi_n = \tan^{-1}\left(\frac{-b_n}{a_n}\right) \qquad n = 1, 2, \ldots \quad (15.19c)$$

We call Eq. (15.18) the *amplitude-phase* form of the Fourier series of $v(t)$. The A_n and ϕ_n are the *Fourier amplitudes* and *phases*. The sinusoidal components in Eq. (15.18) are called *tones* or *harmonics*: $A_1 \cos(2\pi f_1 t + \phi_1)$ is the *fundamental* or *first harmonic*, $A_2 \cos(2\pi 2 f_1 t + \phi_2)$ is the *second harmonic*, and so forth.† Doubling of the

† The dc term A_0 is occasionally called the zeroth harmonic.

frequency raises the harmonic by an *octave*. Thus the second harmonic is one octave above the first, and the fourth harmonic is two octaves above the first.

The terms *octave* and *harmonics* used by engineers are identical to those used by musicians. This common terminology is helpful in analyzing or designing an audio amplifier.

One-Sided Amplitude, Phase, and Power Spectra

A plot of the Fourier amplitudes versus frequency, as illustrated in Fig. 15.7a, is called the *one-sided amplitude spectrum* of $v(t)$. The corresponding plot of the Fourier phases in Fig. 15.7b is called the *one-sided phase spectrum* of $v(t)$. The plots indicate that at each frequency $f = nf_1, n = 0, 1, 2, \ldots$, there are sinusoidal components in $v(t)$. The amplitude spectrum tells us that the amplitude of the component with frequency nf_1 is A_n (designated by the height of the corresponding line). The phase spectrum tells us that the phase of the component with frequency nf_1 is ϕ_n (again designated by the height of the corresponding line). The amplitude and phase spectra are called *line* spectra because they consist of lines.

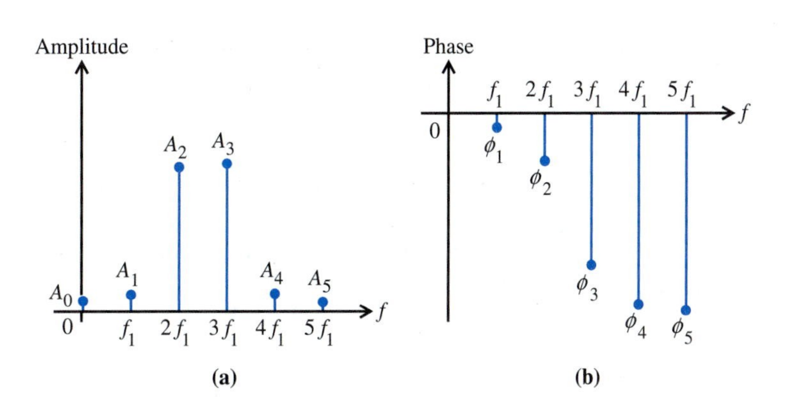

FIGURE 15.7 A typical one-sided spectrum: (*a*) amplitude; (*b*) phase

Restatement of Parseval's Theorem for Real Periodic Waveforms

The terms on the right-hand side of Eq. (15.18), like those in Eq. (15.1), are orthogonal over an interval T. Consequently, the normalized power in $v(t)$,

$$\frac{1}{T}\int_{t_0}^{t_0+T} v^2(t)\,dt = \frac{1}{T}\int_{t_0}^{t_0+T} [A_0 + \sum_{n=1}^{\infty} A_n \cos(2\pi n f_1 t + \phi_n)]^2\,dt \quad (15.20)$$

equals the sum of the normalized powers in the individual components.

THEOREM Parseval's Theorem for Real Periodic Waveforms

$$\frac{1}{T}\int_{t_0}^{t_0+T} v^2(t)\,dt = A_0^2 + \sum_{n=1}^{\infty} \frac{1}{2} A_n^2 \quad (15.21)$$

Equation (15.21) is a restatement of Parseval's theorem, Eq. (15.9). A plot of the normalized powers in the components in Eq. (15.18) versus frequency as illustrated in Fig. 15.8 is called the *one-sided power spectrum* of $v(t)$. The height of the line at $f = nf_1$ in the power spectrum is numerically equal to the average power that would be dissipated if the nth-harmonic voltage were applied across the terminals of a 1-Ω resistance. According to Parseval's theorem, the sum of the heights of all the lines in the power spectrum is numerically equal to the average power that would be dissipated if $v(t)$ were applied across the terminals of a 1-Ω resistance.

FIGURE 15.8 A typical one-sided power spectrum

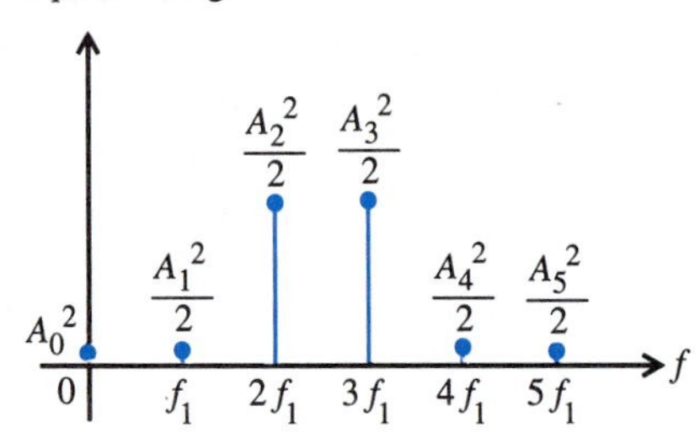

EXAMPLE 15.3

See Problem 15.2 in the PSpice manual.

For Fig. 15.9:

(a) Find the Fourier amplitudes and phases for the periodic voltage pulse train (where $A > 0$).

(b) Sketch the one-sided amplitude, phase, and power spectra for $A = 10$ V, $\tau = 0.2$ ms, and $T = 1$ ms.

(c) Assume that the pulse train is applied across the terminals of a 1-Ω resistance. Determine the percentage of power dissipation due to harmonics 0 through 5. Assume A, τ, and T as in (b).

FIGURE 15.9 Periodic pulse train

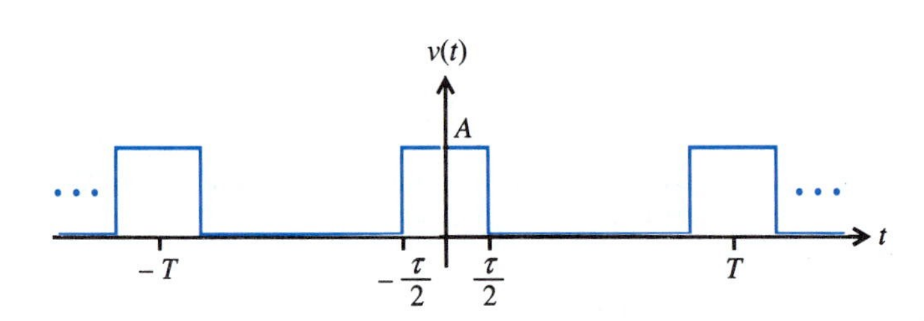

Solution (a) First determine the Fourier coefficients. Since $v(t)$ is even, $b_n = 0$ for $n = 1, 2, \ldots$. Choose $t_0 = -T/2$ for convenience. Then, from Eqs. (15.2) and (15.5),

$$a_0 = \frac{1}{T}\int_{-T/2}^{T/2} v(t)\,dt = \frac{1}{T}\int_{-\tau/2}^{\tau/2} A\,dt = A\,\frac{\tau}{T}$$

$$\begin{aligned} a_n &= \frac{2}{T}\int_{-T/2}^{T/2} v(t)\cos 2\pi n f_1 t\,dt \\ &= \frac{2}{T}\int_{-\tau/2}^{\tau/2} A\cos 2\pi n f_1 t\,dt \\ &= 2A\,\frac{\tau}{T}\,\frac{\sin \pi n f_1 \tau}{\pi n f_1 \tau} \qquad \text{for } n = 1, 2, \ldots \end{aligned}$$

b. (Phase / Power)?

where $f_1 = 1/T$. Then, by Eq. (15.19a),

$$A_0 = A\frac{\tau}{T}$$

and by Eq. (15.19b),

$$A_n = 2A\frac{\tau}{T}\left|\frac{\sin \pi n f_1 \tau}{\pi n f_1 \tau}\right| \qquad \text{for } n = 1, 2, \ldots$$

Because there are no sine components, the Fourier phases can equal either 0° or ±180° depending on the sign of the corresponding a_n: If $a_n > 0$, $\phi_n = 0°$, and if $a_n < 0$, $\phi_n = \pm 180°$. We arbitrarily choose the minus sign in front of the 180°. Then we write

$$\phi_n = \begin{cases} 0 & \dfrac{\sin \pi n f_1 \tau}{\pi n f_1 \tau} > 0 \\ -180° & \dfrac{\sin \pi n f_1 \tau}{\pi n f_1 \tau} < 0 \end{cases}$$

(b) The amplitude, phase, and power spectra are illustrated in Fig. 15.10 for $A = 10$ V, $\tau = 0.2$ ms, and $T = 1$ ms.

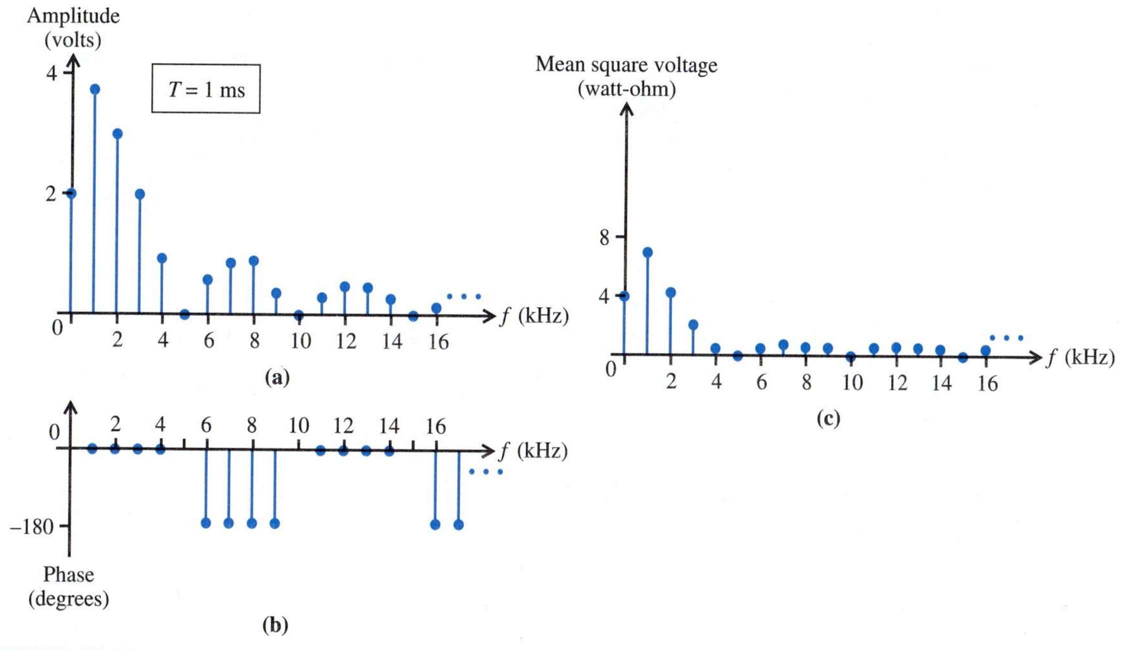

FIGURE 15.10
(*a*) One-sided amplitude, (*b*) phase, and (*c*) power spectra of a periodic pulse train

(c) The total average power dissipated in the 1-Ω resistance is

$$P = \frac{1}{T}\int_{-T/2}^{+T/2} v^2(t)\, dt = \frac{\tau}{T}A^2 = 20 \text{ W}$$

The average power dissipation due to the nth harmonic is

$$P_n = \begin{cases} \left(A\dfrac{\tau}{T}\right)^2 & \text{for } n = 0 \\ \dfrac{1}{2}\left(\dfrac{2A\tau}{T}\dfrac{\sin \pi n f_1 \tau}{\pi n f_1 \tau}\right)^2 & \text{for } n \neq 0 \end{cases}$$

which becomes, for the given values of A, T, and τ,

P_0:	4.00 W
P_1:	7.00
P_2:	4.58
P_3:	2.04
P_4:	0.44
P_5:	0.0
Total	18.06 W

Therefore harmonics 0 through 5 account for $100 \times (18.06/20)\% \simeq 90\%$ of the total average power dissipation.

Remember

A real periodic waveform can be represented by the amplitude-phase form of the Fourier series given by Eqs. (15.18) and (15.19). This form of the Fourier series is used to define the one-sided amplitude, phase, and power spectra of the waveform.

EXERCISE

9. (a) Sketch the one-sided amplitude, phase, and power spectra of the triangular waveform $v(t)$ of Fig. 15.1. Assume that $T = 1$ ms. [*Hint*: The real Fourier series of $v(t)$ is given in Exercise 1.]
   answer: $A_n = 8/n\pi$, $\theta_n = (-1)^{(n+1)/2}\, 90°$ for n odd. There are no even harmonics.

   (b) Assume that $v(t)$ is applied across the terminals of a 1-Ω resistance. What is the average power dissipated?
   answer: $P = \frac{1}{3}$ W

   (c) Determine the percentage of power dissipation due to harmonics 0 through 4.
   answer: 99.77%

15.5 The Exponential Form of the Fourier Series

The *exponential* or *complex form* of the Fourier series is a representation of a periodic waveform by a sum of complex exponentials. This form is sometimes more convenient to use than either of the real forms. In addition, it provides us with an analytical stepping stone to the Fourier transform.

The complex form of the Fourier series of a real periodic waveform can be obtained directly from the real form of the Fourier series. However, when we discuss the complex form of the Fourier series, it is better to generalize to include *complex* periodic waveforms. Therefore, we will make this generalization now. We consider an arbitrary real *or* complex periodic function $v(t)$, and write it in the exponential Fourier series:

DEFINITION Exponential Form of Fourier Series

$$v(t) = \sum_{n=-\infty}^{+\infty} V_n e^{j2\pi n f_1 t} \tag{15.22}$$

where $f_1 = 1/T$ and T is the period. The quantities V_n, $n = 0, \pm 1, \pm 2, \ldots$, are phasors.† They are called *complex Fourier coefficients*. Observe that the terms in the summation are rotating phasors. Those corresponding to $n > 0$ rotate counterclockwise, and those corresponding to $n < 0$ rotate clockwise. The sum of the rotating phasors is $v(t)$. A formula for the kth phasor V_k can be obtained if we multiply both sides of Eq. (15.22) by $e^{-j2\pi k f_1 t}$ and integrate from t_0 to $t_0 + T$, where t_0 is arbitrary. This yields

$$\begin{aligned}\int_{t_0}^{t_0+T} v(t)e^{-j2\pi k f_1 t}\,dt &= \int_{t_0}^{t_0+T} \sum_{n=-\infty}^{+\infty} V_n e^{j2\pi n f_1 t} e^{-j2\pi k f_1 t}\,dt \\ &= \sum_{n=-\infty}^{+\infty} V_n \int_{t_0}^{t_0+T} e^{j2\pi(n-k)f_1 t}\,dt \end{aligned} \tag{15.23}$$

By using Euler's identity, we can write

$$\int_{t_0}^{t_0+T} e^{j2\pi(n-k)f_1 t}\,dt = \int_{t_0}^{t_0+T} \cos 2\pi(n-k)f_1 t\,dt + j\int_{t_0}^{t_0+T} \sin 2\pi(n-k)f_1 t\,dt \tag{15.24}$$

Note that if $n \neq k$, then each of the two integrals on the right-hand side of Eq. (15.24) equals zero, because the area under an integer number of cycles of any sinusoid is zero. If $n = k$, the second integral on the right-hand side is still zero, because $\sin 0 = 0$. However, since $\cos 0 = 1$, the value of the first integral on the right-hand side is T for $n = k$:

$$\int_{t_0}^{t_0+T} e^{j2\pi(n-k)f_1 t}\,dt = \begin{cases} T & n = k \\ 0 & n \neq k \end{cases} \tag{15.25}$$

The substitution of Eq. (15.25) into Eq. (15.23) yields

$$\int_{t_0}^{t_0+T} v(t)e^{-j2\pi k f_1 t}\,dt = V_k T \tag{15.26}$$

where V_k is the kth complex Fourier coefficient. By substituting n for k in Eq. (15.26) and solving for V_n, we obtain

$$V_n = \frac{1}{T}\int_{t_0}^{t_0+T} v(t)e^{-j2\pi n f_1 t}\,dt \qquad n = 0, \pm 1, \pm 2, \ldots \tag{15.27}$$

Equation (15.27) is the formula for the nth complex Fourier series coefficient, V_n, of an arbitrary real or complex periodic waveform $v(t)$. Special results occur when $v(t)$ is real. These are explained in the following subsection.

† For notational simplicity we will drop the boldface convention for phasors when we write the Fourier series.

Relationship Between Real and Exponential Forms of the Fourier Series of a Real Signal

We will now examine the relationship between the real and the exponential forms of the Fourier series. With application of Euler's identity, Eq. (15.27) can be written as

$$V_n = \frac{1}{T}\int_{t_0}^{t_0+T} v(t) \cos 2\pi n f_1 t \, dt - j\frac{1}{T}\int_{t_0}^{t_0+T} v(t) \sin 2\pi n f_1 t \, dt \qquad (15.28)$$

for $n = 0, \pm 1, \pm 2, \ldots$. Comparing Eq. (15.28) with Eq. (15.5), (15.6), and (15.2), we see that for *real* $v(t)$,

$$V_n = \frac{1}{2} a_n - j\frac{1}{2} b_n \qquad \text{for } n = 1, 2, \ldots \qquad (15.29a)$$

and

$$V_0 = a_0 \qquad (15.29b)$$

It follows that the *real* Fourier coefficients a_0, a_n, and b_n, $n = 1, 2, \ldots$, of a *real* periodic waveform $v(t)$ can be obtained from the complex Fourier-series coefficients V_n of that waveform with the use of the following formulas:

$$\begin{aligned} a_n &= 2\mathcal{R}e\{V_n\} \qquad & n = 1, 2, \ldots \\ b_n &= -2\mathcal{I}m\{V_n\} \qquad & n = 1, 2, \ldots \\ a_0 &= V_0 \end{aligned} \qquad (15.30)$$

The preceding equations are very useful. A convenient feature of Eq. (15.27) is that it replaces three integrals, Eqs. (15.2), (15.5), and (15.6), with one. Thus, to compute the real Fourier-series coefficients of a real periodic waveform $v(t)$, we can now evaluate *one* integral, Eq. (15.27), and then apply Eq. (15.30).

We can also obtain the Fourier amplitudes A_n, $n = 0, 1, 2, \ldots$, and phases $\phi_n = 1, 2, 3, \ldots$, of a real periodic waveform $v(t)$ from the complex Fourier-series coefficients V_n of that waveform. To derive the appropriate formulas, we substitute Eqs. (15.17a), (15.17b), and (15.19a) into Eqs. (15.29a) and (15.29b), to obtain

$$\begin{aligned} V_n &= \frac{1}{2} A_n \cos \phi_n + j\frac{1}{2} A_n \sin \phi_n \\ &= \frac{1}{2} A_n e^{j\phi_n} \qquad n = 1, 2, 3, \ldots \end{aligned} \qquad (15.31)$$

and

$$V_0 = A_0 \qquad (15.32)$$

This yields the convenient formulas

$$\begin{aligned} A_0 &= V_0 \\ A_n &= 2|V_n| \qquad & n = 1, 2, 3, \ldots \\ \phi_n &= \angle V_n \qquad & n = 1, 2, 3, \ldots \end{aligned} \qquad (15.33)$$

A final observation regarding the complex Fourier coefficients of a real periodic waveform follows from Eq. (15.27) with n replaced by $-n$:

$$V_{-n} = \frac{1}{T} \int_{t_0}^{t_0+T} v(t) e^{+j2\pi n f_1 t} \, dt \tag{15.34}$$

Note that the substitution of $-n$ for n has changed the sign of the j in the exponent in the integral in Eq. (15.27). If $v(t)$ is real, then the change in the sign of the j in the exponent is equivalent to conjugation of the entire right-hand side. That is, for real $v(t)$,

$$V_{-n} = \left[\frac{1}{T} \int_{t_0}^{t_0+T} v(t) e^{-j2\pi n f_1 t} \, dt \right]^* \tag{15.35}$$

which, by comparison with Eq. (15.27), shows that

$$V_{-n} = V_n^* \tag{15.36}$$

Therefore, for $v(t)$ real, we can obtain the negatively indexed complex Fourier coefficients by conjugating the positively indexed coefficients as in Eq. (15.36).

Parseval's Theorem for Real or Complex Periodic Waveforms

The most general form of Parseval's theorem applies to both real and complex waveforms. We obtain the general form by conjugating both sides of Eq. (15.22), multiplying the result by $v(t)$, and integrating from t_0 to $t_0 + T$. This yields

$$\int_{t_0}^{t_0+T} v(t) v^*(t) \, dt = \int_{t_0}^{t_0+T} v(t) \sum_{n=-\infty}^{+\infty} V_n^* e^{-j2\pi n f_1 t} \, dt \tag{15.37}$$

By interchanging the order of summation and integration we obtain

$$\int_{t_0}^{t_0+T} v(t) v^*(t) \, dt = \sum_{n=-\infty}^{+\infty} V_n^* \int_{t_0}^{t_0+T} v(t) e^{-j2\pi n f_1 t} \, dt \tag{15.38}$$

which, with the aid of Eq. (15.27), becomes

$$\frac{1}{T} \int_{t_0}^{t_0+T} v(t) v^*(t) \, dt = \sum_{n=-\infty}^{+\infty} V_n V_n^* \tag{15.39}$$

or

THEOREM Parseval's Theorem for Real or Complex Periodic Waveforms

$$\frac{1}{T} \int_{t_0}^{t_0+T} |v(t)|^2 \, dt = \sum_{n=-\infty}^{+\infty} |V_n|^2 \tag{15.40}$$

This is Parseval's theorem for real or complex periodic waveforms. Of course, if $v(t)$ is real, then $|v(t)|^2 = v^2(t)$.

Two-Sided Amplitude, Phase, and Power Spectra

Plots of the magnitudes and phases of the complex Fourier coefficients versus frequency as illustrated in Fig. 15.11a, b are called the *two-sided amplitude spectrum* and the *two-sided phase spectrum* of $v(t)$, respectively. The *spectrum* of $v(t)$ is the pair of amplitude and phase spectra. Because $v(t)$ can be synthesized from its spectrum by means of Eq. (15.22), the spectrum provides a complete representation of $v(t)$.

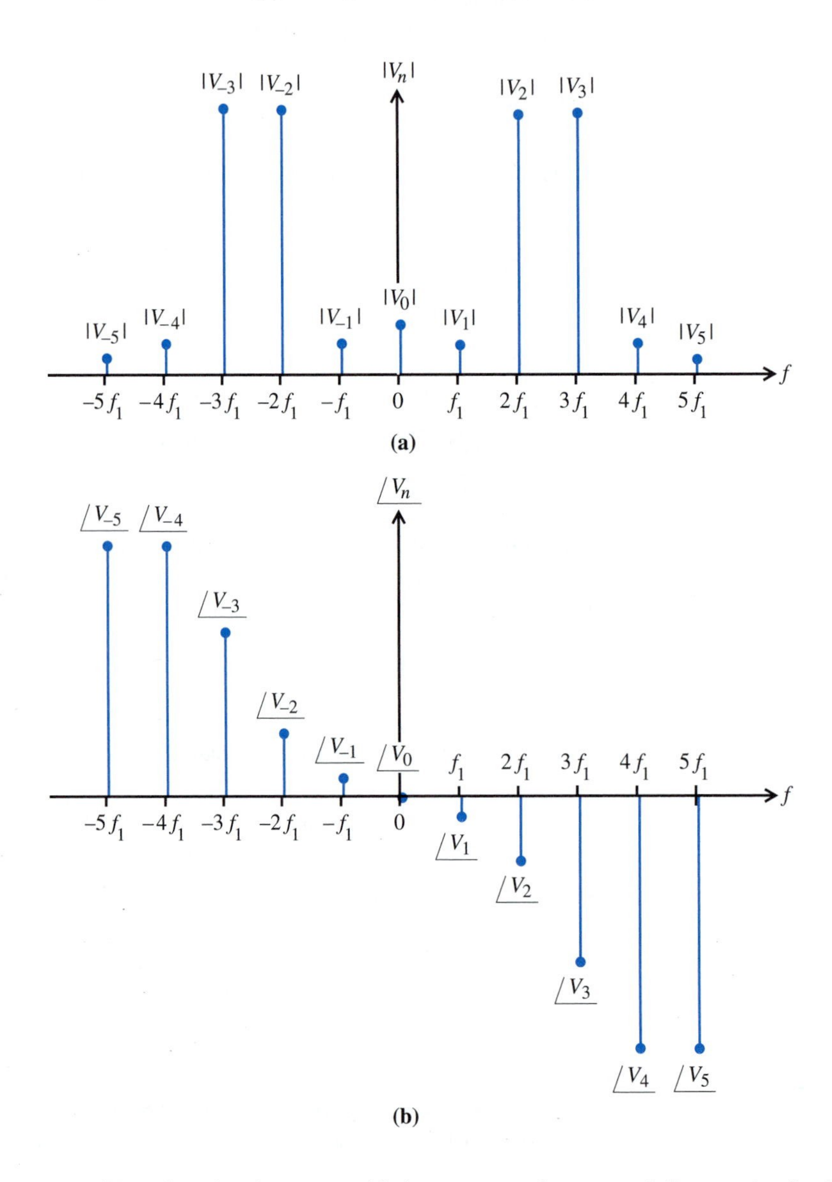

FIGURE 15.11 A typical two-sided spectrum: (*a*) amplitude; (*b*) phase

You will notice that in a two-sided spectrum, the spectral lines exist for both positive and negative frequencies. Negative frequencies do not physically exist. The concept of negative frequency arises because we represent sinusoids as rotating phasors. Consider the simplest example illustrated in Fig. 15.12. Since $\cos 2\pi ft = \mathcal{R}e\{e^{j2\pi ft}\} = \frac{1}{2}e^{j2\pi ft} + \frac{1}{2}e^{-j2\pi ft}$ the process of taking the real part of $e^{j2\pi ft}$ is equivalent to adding the counterclockwise rotating phasor $\frac{1}{2}e^{j2\pi ft}$ to the clockwise rotating phasor $\frac{1}{2}e^{-j2\pi ft}$. We see therefore that negative frequencies simply refer to phasors that rotate

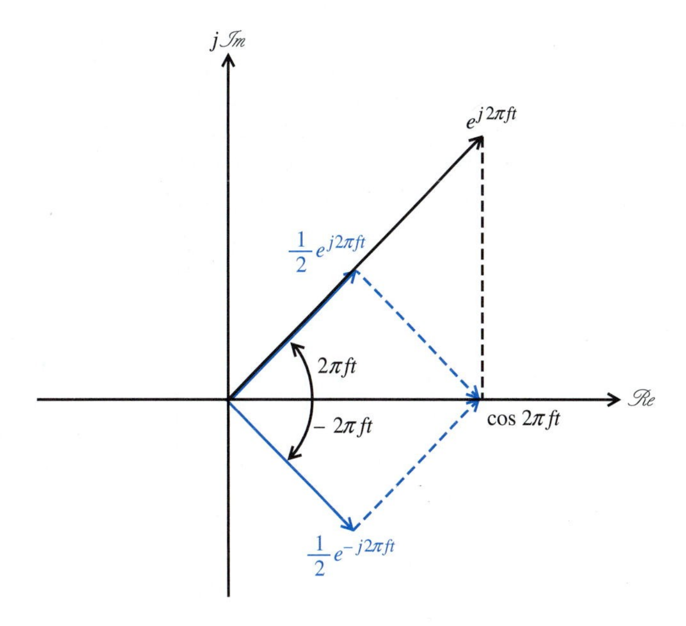

FIGURE 15.12 Rotating phasor representations of $\cos 2\pi ft$

clockwise. The lines plotted on the positive frequency axis in the two-sided spectrum of Fig. 15.11 refer to phasors in Eq. (15.22) that rotate counterclockwise ($n > 0$). The lines plotted on the *negative* frequency axis correspond to phasors that rotate clockwise ($n < 0$). Equation (15.36) applies for real $v(t)$. Therefore, for a real periodic waveform the amplitude spectrum has even symmetry, $|V_n| = |V_{-n}|$, and the phase spectrum has odd symmetry $\underline{/V_n} = -\underline{/V_{-n}}$.

A plot of $|V_n|^2$ versus frequency, as illustrated in Fig. 15.13, is called the *two-sided power spectrum* of $v(t)$. It follows from Eqs. (15.33) and (15.36) that for real $v(t)$,

$$|V_n|^2 + |V_{-n}|^2 = \frac{1}{2}A_n^2 \qquad n = 1, 2, 3, \ldots \tag{15.41}$$

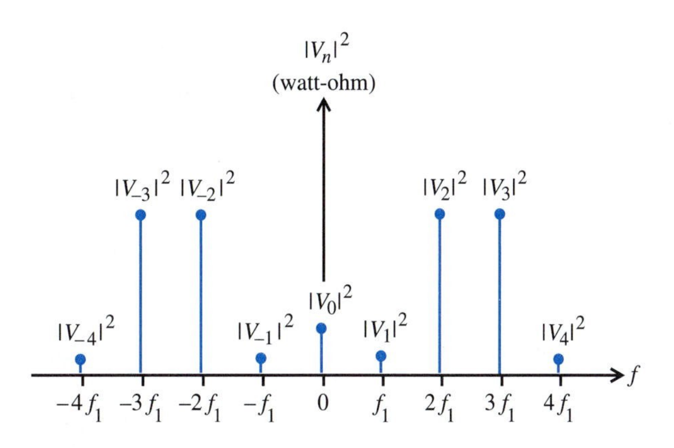

FIGURE 15.13 A typical two-sided power spectrum

The quantity on the right-hand side of Eq. (15.41) is the normalized power in the nth Fourier component $A_n \cos(2\pi nf_1t + \phi_n)$. Therefore, we find the normalized power carried by the nth harmonic of $v(t)$, $n \geq 1$, by adding $|V_n|^2$ and $|V_{-n}|^2$. The normalized power carried by the dc component is $|V_0|^2$. The total normalized power in $v(t)$ is, according to Parseval's theorem, Eq. (15.40), equal to the sum of all the line heights of the power spectrum.

EXAMPLE 15.4 The signal shown in Fig. 15.14 is commonly found in electronic clocks.

(a) Determine the complex Fourier coefficients of the waveform.

(b) Use your result from (a) to find the real Fourier coefficients a_n and b_n and the Fourier amplitudes and phases A_n and ϕ_n.

(c) Write the first several terms of each form of the Fourier series with the assumption that $T = 1$ ms, $\tau = 1/(2\pi)$ ms, and $A = 100$ V.

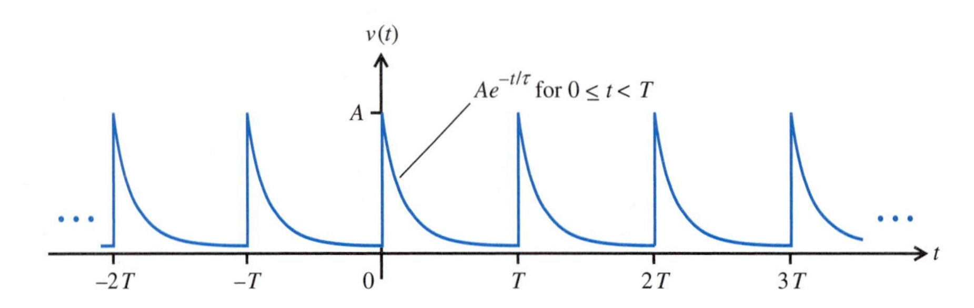

FIGURE 15.14 Periodically repeated exponential waveform

Solution (a) A convenient interval of integration for Eq. (15.27) is $0 \le t < T$:

$$\begin{aligned} V_n &= \frac{1}{T}\int_0^T v(t)e^{-j2\pi n f_1 t}\,dt \\ &= \frac{1}{T}\int_0^T Ae^{-t/\tau}e^{-j2\pi n f_1 t}\,dt \\ &= \frac{A}{T}\left.\frac{e^{[-(t/\tau)-j2\pi n f_1 t]}}{-(1/\tau) - j2\pi n f_1}\right|_0^T \end{aligned}$$

Because $f_1 T = 1$, this becomes

$$\begin{aligned} V_n &= A\frac{\tau}{T}\frac{1 - e^{-T/\tau}}{1 + j2\pi n f_1 \tau} \\ &= A\frac{\tau}{T}\frac{1 - e^{-T/\tau}}{\sqrt{1 + (2\pi n f_1 \tau)^2}}\underline{/-\tan^{-1}2\pi n f_1 \tau} \end{aligned}$$

(b) The following results follow easily from Eqs. (15.30) and (15.33),

$$a_0 = V_0 = 2A\frac{\tau}{T}(1 - e^{-T/\tau})$$

$$a_n = 2A\frac{\tau}{T}\frac{1 - e^{-T/\tau}}{1 + (2\pi n f_1 \tau)^2}$$

$$b_n = 2A\frac{\tau}{T}\frac{(1 - e^{-T/\tau})2\pi n f_1 \tau}{1 + (2\pi n f_1 \tau)^2} \qquad n = 1, 2, \ldots$$

$$A_0 = V_0 = A\frac{\tau}{T}(1 - e^{-T/\tau})$$

$$A_n = 2A\frac{\tau}{T}\frac{1 - e^{-T/\tau}}{\sqrt{1 + (2\pi n f_1 \tau)^2}} \qquad n = 1, 2, \ldots$$

$$\phi_n = -\tan^{-1} 2\pi n f_1 \tau \qquad n = 1, 2, \ldots$$

(c) Numerical evaluation of the first four complex Fourier coefficients in (b) with the use of $A = 100$ V, $T = 1$ ms, and $\tau = 1/(2\pi)$ ms $\simeq 0.159$ ms yields

$$V_0 \simeq 15.89 \text{ V}$$
$$V_1 \simeq 11.23\underline{/-45^\circ} \text{ V} = 7.94 - j7.94 \text{ V}$$
$$V_2 \simeq 7.10\underline{/-63.4^\circ} \text{ V} = 3.12 - j6.35 \text{ V}$$
$$V_3 \simeq 5.02\underline{/-71.6^\circ} \text{ V} = 1.58 - j4.76 \text{ V}$$

Then, with $V_{-n} = V_n^*$, the complex form of the Fourier series is

$$\begin{aligned} v(t) \simeq\ & 15.89 + (11.23\underline{/-45^\circ})e^{j2\pi f_1 t} + (11.23\underline{/+45^\circ})e^{-j2\pi f_1 t} \\ & + (7.10\underline{/-63.4^\circ})e^{j2\pi 2f_1 t} + (7.10\underline{/+63.4^\circ})e^{-j2\pi 2f_1 t} \\ & + (5.02\underline{/-71.6^\circ})e^{j2\pi 3f_1 t} + (5.02\underline{/+71.6^\circ})e^{-j2\pi 3f_1 t} + \cdots \text{ V} \end{aligned}$$

where $f_1 = 1000$ Hz. This leads to the real forms

$$\begin{aligned} v(t) \simeq\ & 15.89 + 22.46 \cos(2\pi 1000t - 45^\circ) \\ & + 14.20 \cos(2\pi 2000t - 63.4^\circ) \\ & + 10.04 \cos(2\pi 3000t - 71.6^\circ) + \cdots \text{ V} \end{aligned}$$

and

$$\begin{aligned} v(t) \simeq\ & 15.89 + 15.88 \cos 2\pi 1000t + 6.24 \cos 2\pi 2000t \\ & + 3.16 \cos 2\pi 3000t + \cdots \\ & + 15.88 \sin 2\pi 1000t + 12.7 \sin 2\pi 2000t \\ & + 9.52 \sin 2\pi 3000t + \cdots \text{ V} \end{aligned}$$

Remember

The complex form of the Fourier series applies to both complex and real periodic waveforms. When $v(t)$ is real, we can find its real Fourier series by first finding V_n by the use of Eq. (15.27) and then applying Eq. (15.30) or (15.33). The complex form of the Fourier series is used to define the two-sided amplitude, phase, and power spectra of the waveform. If $v(t)$ is real, then $V_{-n} = V_n^*$.

EXERCISES

10. (a) Use Eq. (15.27) to determine the complex Fourier-series coefficients of the periodic pulse train of Fig. 15.9.
    (b) Check your answer by referring to the results for a_n and b_n in Example 15.3 and applying Eq. (15.30).
    answer: $V_n = (A\tau/T)[\sin(n\pi f_1 \tau)/n\pi f_1 \tau]$ for $n \neq 0$, $V_0 = A\tau/T$
11. (a) Show that if $v(t)$ is a real, even function of t, then V_n is a real, even function of n.
    (b) Show that if $v(t)$ is a real, odd function of t, then V_n is an imaginary, odd function of n.
12. Use Eq. (15.27) to determine the complex Fourier-series coefficients of a periodic impulse train

$$v(t) = \sum_{i=-\infty}^{+\infty} \delta(t - iT)$$

    answer: $v_n = 1/T$ for $n = 0, \pm 1, \pm 2, \ldots$
13. Use the formula in Exercise 1 to determine the complex Fourier-series coefficients of the triangular waveform $v(t)$ of Fig. 15.1.
    answer: $V_n = j[4/(n\pi)^2](-1)^{(n+1)/2}$, for n odd. $V_n = 0$ for n even.

15.6 Fourier-Series Tables

Electrical engineers frequently use tables to save them from needlessly rederiving results that are already known. One such table is Table 15.1, which lists the complex Fourier-series coefficients for several periodic signals found in engineering. The corresponding results for the real series can be obtained easily from the table with the use of Eqs. (15.30) and (15.33).

TABLE 15.1 Some familiar periodic waveforms and their exponential Fourier-series coefficients

	Name	*Figure*	*Exponential Fourier Coefficient V_n*	
FS1	Square wave	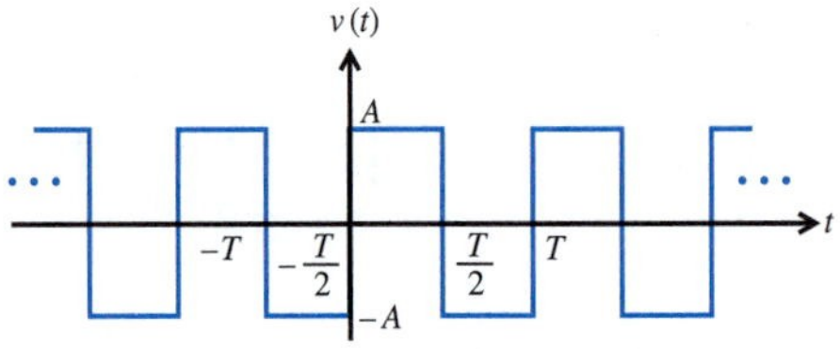	$-j\dfrac{2A}{n\pi}$	n odd
			0	n even
FS2	Periodic rectangular pulse (rectangular wave)	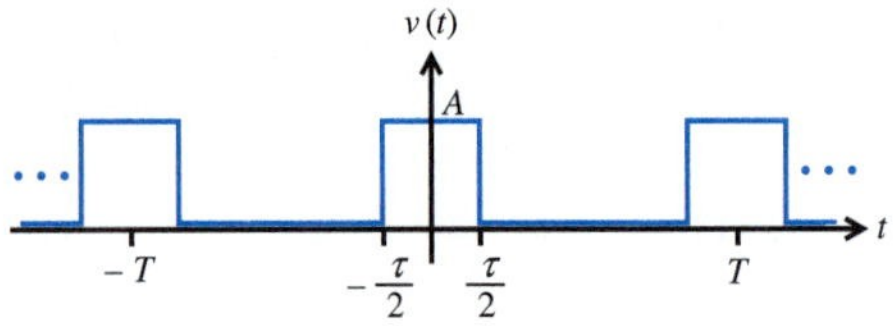	$\dfrac{A}{\pi n}\sin\dfrac{n\pi\tau}{T}$	$n \neq 0$
			$A\dfrac{\tau}{T}$	$n = 0$
FS3	Periodically repeated exponential	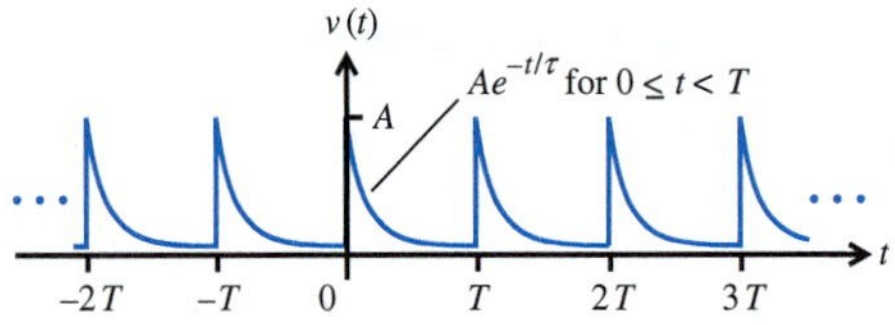	$A\dfrac{\tau}{T}\dfrac{1 - e^{-T/\tau}}{1 + j2\pi n\tau/T}$	
FS4	Sawtooth	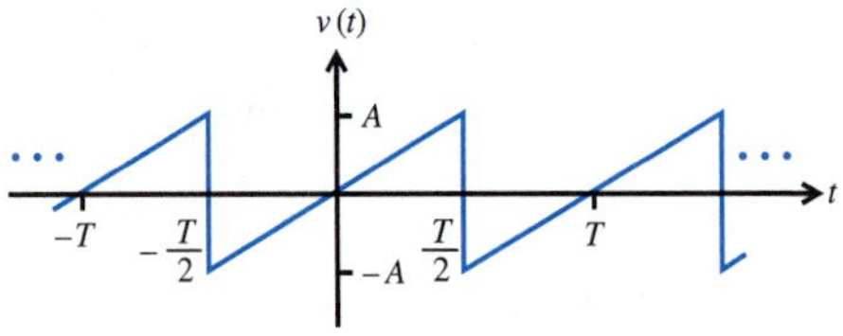	$j\dfrac{A}{n\pi}(-1)^n$	$n \neq 0$
			0	$n = 0$
FS5	Double sawtooth	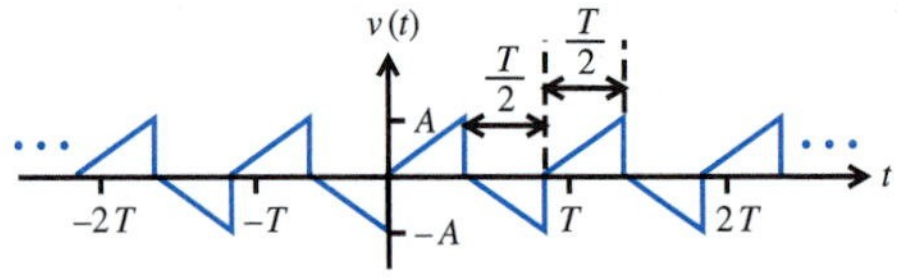	$-\dfrac{2A}{(\pi n)^2} - j\dfrac{A}{\pi n}$	n odd
			0	n even
FS6	Triangular wave	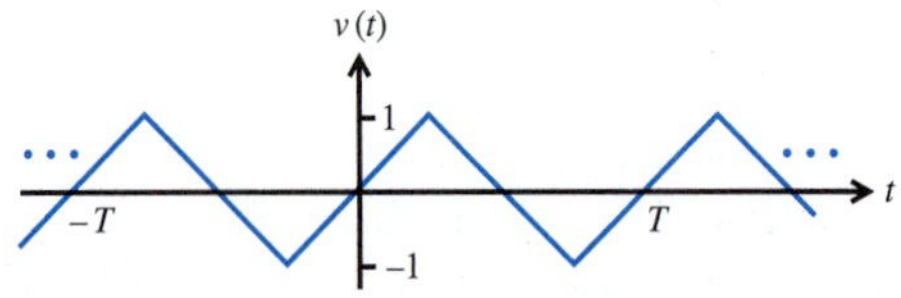	$j\dfrac{4}{\pi^2 n^2}(-1)^{(n+1)/2}$	n odd
			0	n even

TABLE 15.1 *Continued*

Name		Figure	Exponential Fourier Coefficient V_n
FS7	Trapezoidal wave	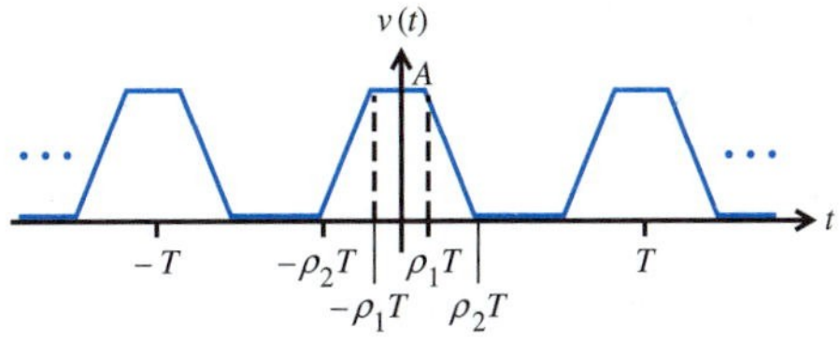 $v(t)$; A; $-T$, $-\rho_2 T$, $-\rho_1 T$, $\rho_1 T$, $\rho_2 T$, T; t	$\frac{A\tau}{T}\frac{\sin(\pi n t_0/T)}{\pi n}\frac{\sin(\pi n\tau/T)}{\pi n}$ where $t_0 = (\rho_1 + \rho_2)T$ and $\tau = (\rho_2 - \rho_1)T$
FS8	Half-rectified cosine	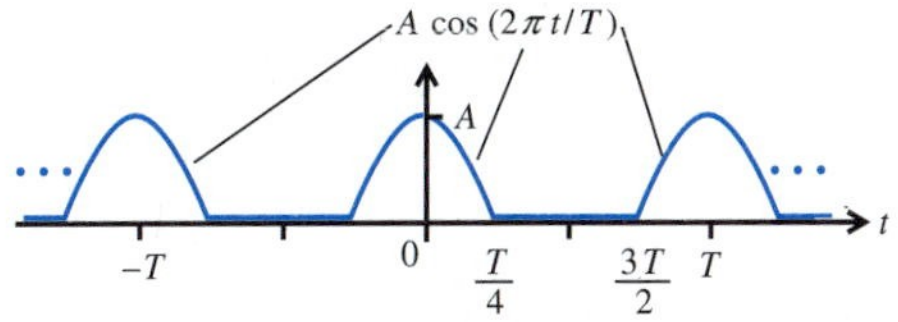 $A\cos(2\pi t/T)$; A; $-T$, 0, $\frac{T}{4}$, $\frac{3T}{2}$, T; t	$\frac{(A/\pi)\cos n\pi/2}{1 - n^2}$ $n \neq \pm 1$ $\frac{A}{4}$ $n = \pm 1$
FS9	Full-rectified cosine (where frequency n/T_o is associated with V_n)	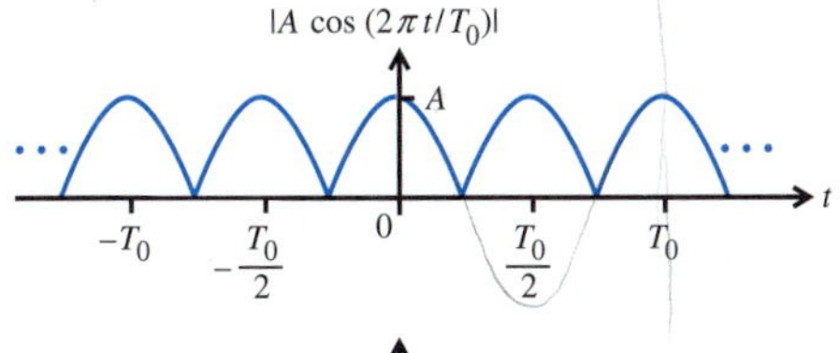 $\lvert A\cos(2\pi t/T_0)\rvert$; A; $-T_0$, $-\frac{T_0}{2}$, 0, $\frac{T_0}{2}$, T_0; t	$\frac{(2A/\pi)\cos n\pi/2}{1 - n^2}$ $n \neq \pm 1$ 0 $n = \pm 1$
FS10	Periodic impulse train	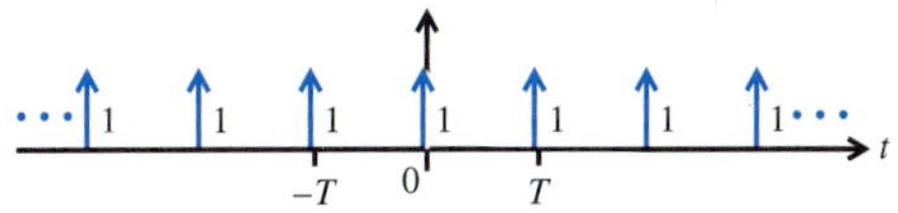 1, 1, 1, 1, 1, 1, 1; $-T$, 0, T; t	$\frac{1}{T}$

Table 15.2 can also save you time. It lists four basic properties of the complex Fourier series. These properties can be helpful to derive the Fourier coefficients of waveforms not shown in Table 15.1.

Linearity

The linearity property states that the nth complex Fourier coefficient of the weighted sum of two periodic waveforms $ax(t) + by(t)$, where $x(t)$ and $y(t)$ have period T, is simply $aX_n + bY_n$, where X_n and Y_n are the nth complex Fourier coefficients of $x(t)$ and $y(t)$, respectively. This property can be established by simply observing that

TABLE 15.2
Four properties of the exponential Fourier series

Property	Waveform	Complex Fourier Coefficient
Linearity	$ax(t) + by(t)$	$aX_n + bY_n$
Reversal	$v(-t)$	V_{-n}
Time shift	$v(t - t_d)$	$V_n e^{-j2\pi n f_1 t_d}$
Differentiation	$\frac{d}{dt}v(t)$	$j2\pi n f_1 V_n$

$$ax(t) + by(t) = \sum_{n=-\infty}^{+\infty} aX_n e^{j2\pi n f_1 t} + \sum_{n=-\infty}^{+\infty} bY_n e^{j2\pi n f_1 t}$$
$$= \sum_{n=-\infty}^{+\infty} (aX_n + bY_n) e^{j2\pi n f_1 t} \quad (15.42)$$

Therefore $aX_n + bY_n$ is the nth complex Fourier coefficient of $ax(t) + by(t)$.

Reversal

The reversal property states that the effect of reversing a periodic waveform in time, $v(t) \rightarrow v(-t)$, is to exchange V_n with V_{-n}. To establish this property, we compute the nth complex Fourier coefficient G_n of $g(t) = v(-t)$:

$$G_n = \frac{1}{T} \int_{t_0}^{t_0+T} v(-t) e^{-j2\pi n f_1 t}\, dt \quad (15.43)$$

By a change of variables, λ for $-t$, this becomes

$$G_n = \frac{1}{T} \int_{-t_0-T}^{-t_0} v(\lambda) e^{j2\pi n f \lambda}\, d\lambda = \frac{1}{T} \int_{t_0}^{t_0+T} v(\lambda) e^{j2\pi n f_1 \lambda}\, d\lambda \quad (15.44)$$

where the last step follows from the fact that $v(\lambda)$ has period T. But the nth complex Fourier coefficient of $v(t)$ is

$$V_n = \frac{1}{T} \int_{t_0}^{t_0+T} v(\lambda) e^{-j2\pi n f_1 \lambda}\, d\lambda \quad (15.45)$$

Therefore, by comparing Eqs. (15.44) and (15.45), we obtain

$$G_n = V_{-n} \quad (15.46)$$

Time Shift

According to the time-shift property, the nth complex Fourier coefficient of the time-shifted periodic waveform $v(t - t_d)$ equals V_n times $e^{-j2\pi n f_1 t_d}$. To establish this property, we observe that since

$$v(t) = \sum_{n=-\infty}^{+\infty} V_n e^{j2\pi n f_1 t} \quad (15.47)$$

then

$$v(t - t_d) = \sum_{n=-\infty}^{+\infty} V_n e^{j2\pi n f_1 (t - t_d)}$$
$$= \sum_{n=-\infty}^{+\infty} (V_n e^{-j2\pi n f_1 t_d}) e^{j2\pi n f_1 t} \quad (15.48)$$

Therefore $V_n e^{-j2\pi n f_1 t_d}$ is the nth complex Fourier coefficient of $v(t - t_d)$.

Differentiation

The differentiation property states that the nth complex Fourier coefficient of the derivative of $v(t)$ is the product $j2\pi n f_1 V_n$, where V_n is the nth complex coefficient of $v(t)$.

This property can be established simply by differentiating both sides of Eq. (15.22) with respect to t.

Another important property of the complex Fourier series is developed in Problem 29.

EXERCISES

14. Let $v(t)$ be the sum of the square wave and the half-rectified cosine wave, FS1 and FS8 in Table 15.1. Assume that their amplitudes A and their periods T are identical.
    (a) Sketch $v(t)$.
    (b) Determine the complex Fourier-series coefficients of $v(t)$. (*Hint:* Use the linearity property of Table 15.2.)
    (c) Sketch the two-sided amplitude and phase spectra.

    answer: (b) $-j\dfrac{2A}{n\pi}$ for n odd, $n \neq \pm 1$; $\dfrac{A}{4} - j\dfrac{2A}{n\pi}$ for $n = \pm 1$; $\dfrac{A(-1)^{(n-1)/2}}{\pi(1-n^2)}$ for n even.

15. Let $g(t)$ be the time-reversed version of a periodically repeated exponential wave: $g(t) = v(-t)$, where $v(t)$ is the waveform of Fig. 15.14.
    (a) Sketch $g(t)$.
    (b) Use Tables 15.1 and 15.2 to determine the complex Fourier-series coefficients of $g(t)$.
    (c) Sketch the two-sided amplitude and phase spectra of $v(t)$ and $g(t)$.

    answer: (b) $\dfrac{A\tau}{T}\dfrac{1-e^{-T/\tau}}{\sqrt{1+(4\pi^2n^2\tau^2/T^2)}}e^{j\arctan(2\pi n\tau/T)}$

16. Let $v_d(t) = v(t - t_d)$ denote the time-shifted version of the half-wave-rectified cosine wave, FS8 of Table 15.1. Assume that the time delay is $t_d = T/2$.
    (a) Sketch $v_d(t)$.
    (b) Use the time-shift property of Table 15.2 to determine the complex Fourier-series coefficients of $v_d(t)$.
    (c) Use your result from (b) and the linearity property to verify entry FS9 of Table 15.1.
    (d) Sketch the two-sided amplitude and phase spectra of entry FS9.

    answer: (b) $\dfrac{A\cos(n\pi/2)}{\pi(1-n^2)}$, $n \neq \pm 1$; $\dfrac{A}{4}e^{-j\pi n}$ for $n = \pm 1$

17. Differentiate the triangular wave of Fig. 15.1, and apply the differentiation property of Table 15.2 to FS6 of Table 15.1 to determine the complex Fourier series of a square wave. Explain why your answer differs from FS1 of Table 15.1.

    answer: $-\dfrac{8}{n\pi T}(-1)^{(n+1)/2}$ for n odd, 0 for n even. This disagrees with FS1 because the square wave in FS1 has odd symmetry and the square wave of this exercise has even symmetry.

15.7 Circuit Response to Periodic Input

Now that we know how to represent a periodic waveform by its Fourier series, we are prepared to determine the particular response of a stable LTI circuit to a periodic input.

Our method for obtaining the particular response uses the principle of superposition and ac circuit analysis. The Fourier series enables us to represent the input as a sum of sinusoidal components. Because the circuit is linear, the particular response to this sum is given by the sum of the particular responses to the individual components. Because the circuit is also time-invariant, the individual responses can be determined with the use of methods in Chapter 11. Because the circuit is also stable, the complete response to the periodic input will be equal to the particular response. Details are given in the following subsections.

Application of Real Fourier Series

We will show how to apply the real Fourier series to circuit analysis by considering a specific example. Assume that the particular response $v_o(t)$ of band-pass *RLC* circuit of Fig. 15.15 is to be determined for the periodically repeated exponential waveform of Fig. 15.14 in which $A = 100$, $T = 1$ ms, and $\tau = 1/(2\pi)$ ms.

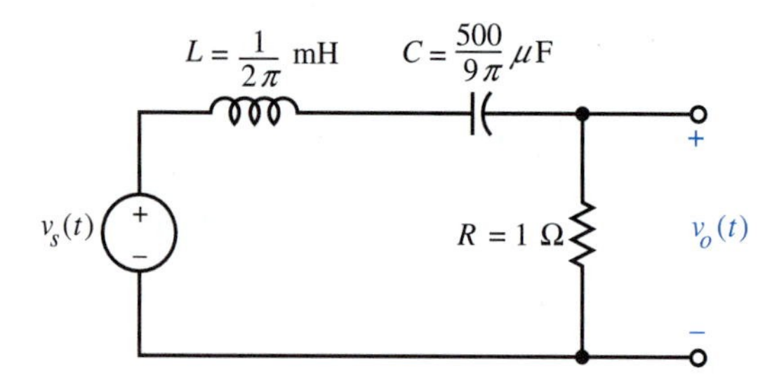

FIGURE 15.15
Band-pass *RLC* circuit

The first step in the solution is to express $v_s(t)$ in its Fourier-series form. Either the real or the complex form of the series can be used. Suppose that the real form of Eq. (15.22) is used. Then, according to Example 15.4,

$$\begin{aligned} v_s(t) = {} & 15.89 + 22.46 \cos(2\pi 1000t - 45°) + 14.20 \cos(2\pi 2000t - 63.4°) \\ & + 10.04 \cos(2\pi 3000t - 71.6°) + 7.70 \cos(2\pi 4000t - 76.6°) \\ & + 6.24 \cos(2\pi 5000t - 78.7°) + \cdots \end{aligned} \tag{15.49}$$

The second step is to determine the circuit's particular response to the *individual* components in Eq. (15.49). This step is easy because, by ac circuit analysis, the particular response to the individual harmonic $A_n \cos(2\pi n f_1 t + \phi_n)$ is simply $A_n |H(j2\pi n f_1)| \cos(2\pi n f_1 t + \phi_n + \underline{/H(j2\pi n f_1)})$, where $H(j\omega)$ is the transfer function of the circuit:†

$$\begin{aligned} H(j\omega) &= \frac{R}{R + j\left(\omega L - \dfrac{1}{\omega C}\right)} \\ &= \frac{1}{1 + j(0.001f - 9/0.001f)} \end{aligned} \tag{15.50}$$

† Again, for notational simplicity we drop the boldface notation, $\mathbf{H}(j\omega)$, that was used in Chapters 10 through 13.

The third step is to apply superposition. When the input is the sum in Eq. (15.49), the particular response is the sum

$$
\begin{aligned}
v_o(t) &= A_0H(0) + A_1|H(j2\pi f_1 t)|\cos(2\pi f_1 t + \phi_1 + \underline{/H(j\omega_1)}) \\
&\quad + A_2|H(j2\pi 2f_1)|\cos(2\pi 2f_1 t + \phi_2 + \underline{/H(j2\omega_1)}) + \cdots \\
&= (15.89)(0) + (22.46)(0.124)\cos(2\pi 1000t - 45^\circ + 82.87^\circ) \\
&\quad + (14.20)(0.371)\cos(2\pi 2000t - 63.4^\circ + 68.20^\circ) \\
&\quad + (10.04)(1)\cos(2\pi 3000t - 71.6^\circ + 0^\circ) + \cdots \\
&= 2.79\cos(2\pi 1000t + 37.87^\circ) \\
&\quad + 5.27\cos(2\pi 2000t + 4.8^\circ) \\
&\quad + 10.04\cos(2\pi 3000t - 71.6^\circ) + \cdots
\end{aligned}
\tag{15.51}
$$

This step is illustrated in Fig. 15.16.

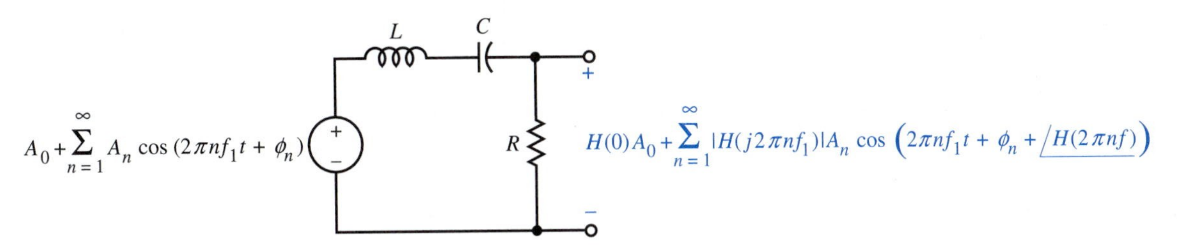

FIGURE 15.16
A sum of sinusoidal inputs yields a sum of sinusoidal outputs. Here $H(j\omega) = 1/\{R + j[\omega L - (1/\omega C)]\}$

A good way to view the relationship between the input and output is illustrated in Fig. 15.17. This figure contains plots of the one-sided spectrum of the input $v_s(t)$ and the transfer function of the circuit, all plotted versus f. Because the independent variable in these plots is f, it has been convenient to introduce the more economical notation

$$
\mathrm{H}(f) = H(j2\pi f) \tag{15.52}
$$

as indicated in Fig. 15.17. As is evident in Eq. (15.51), the amplitude of each sinusoidal component in $v_o(t)$ is given by the product $A_n|H(j2\pi n f_1)|$. Therefore, *the output waveform $v_o(t)$ has an amplitude spectrum that is the product of the amplitude spectrum of the input waveform and the amplitude characteristic of the filter.* It can also be seen from Eq. (15.51), that the phase of each sinusoidal component in $v_o(t)$ is given by the sum $\phi_n + \underline{/H(j2\pi n f_1)}$. Therefore, *the phase spectrum of $v_o(t)$ equals the sum of the phase spectrum of $v_s(t)$ and the phase characteristic of the filter at frequencies nf_1.*

Figure 15.18 contains plots of *one period* of the partial Fourier series $v_{oN}(t)$ of the output waveform of Eq. (15.51). This figure was generated by a computer, which summed the first N terms in Eq. (15.51) for $N = 1, 2, 3, 4, 5, 10$, and 15. (The dc term is zero and is not included in this count.) It can be seen from the figure that the convergence is rapid; there is very little change in the appearance of the plots for $N > 10$. A plot of two periods of the partial Fourier series of $v_o(t)$ with $N = 25$ is given in Fig. 15.19. Except for the small ripple (enclosed in the dashed circle), this plot can be considered to be the actual output $v_o(t)$. The ripple is related to the Gibbs phenomenon, which concerns the way in which a partial Fourier series converges. The Gibbs phenomenon is described in Section 15.8.

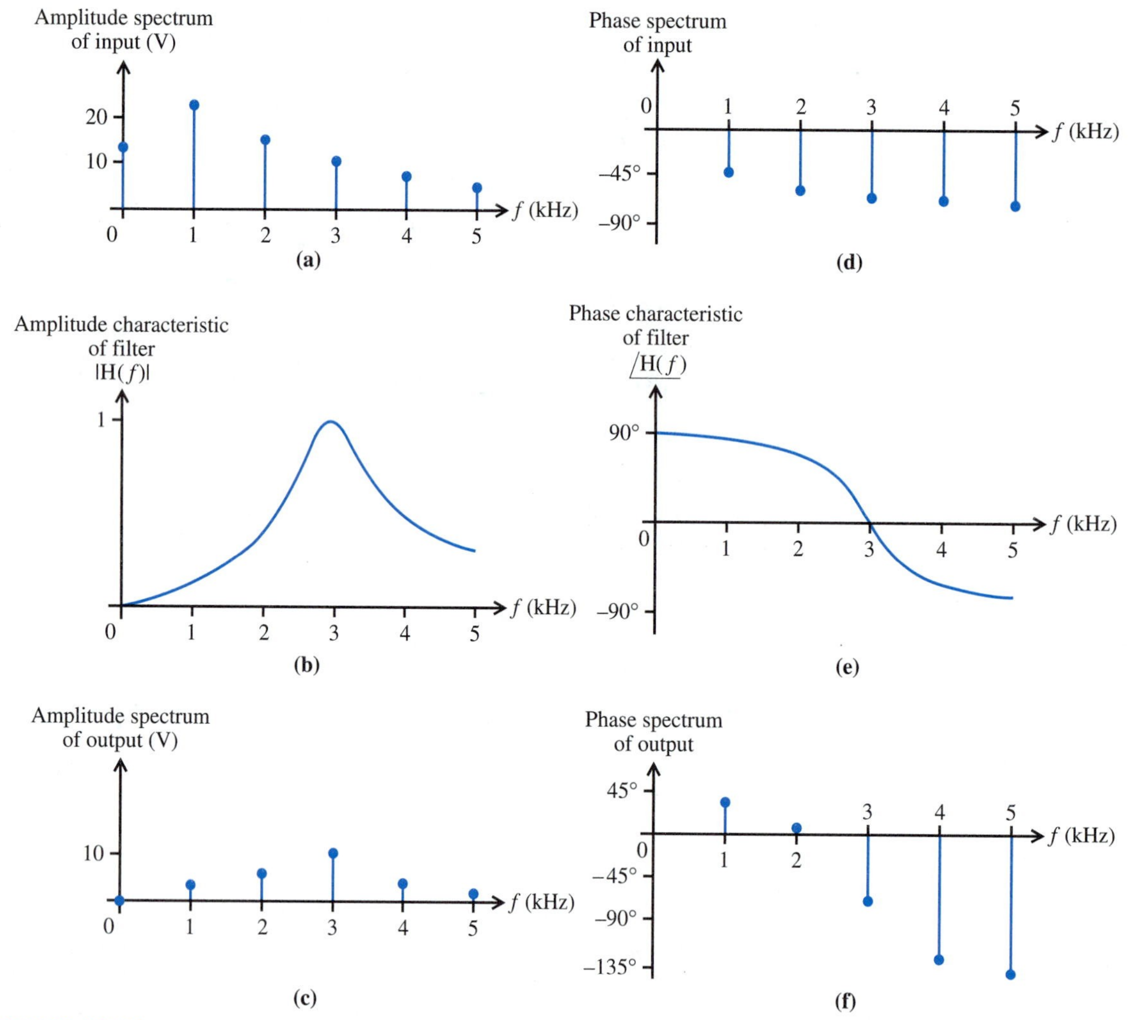

FIGURE 15.17
The product of (*a*) the input amplitude spectrum and (*b*) the filter's amplitude characteristic yields (*c*) the output amplitude spectrum. The sum of (*d*) the input phase spectrum and (*e*) the filter's phase characteristic at nf_1 yields (*f*) the output phase spectrum.

If we think about the output shown in Fig. 15.19 from either the frequency-domain viewpoint or the time-domain viewpoint, we realize that the output is what we would expect. Consider first the frequency-domain viewpoint. The *RLC* circuit is a band-pass filter (see Fig. 15.17b). The circuit passes the third ($f = 3$ kHz) harmonic of the input waveform while attenuating the other harmonics. We see that the output waveform of Fig. 15.19 exhibits three major cycles or swings each millisecond. These three cycles each millisecond are caused by the relatively large amplitude of the 3-kHz harmonic in the output. The output also contains harmonics at 1, 2, 4, 5, . . . kHz, but these are smaller than the 3-kHz harmonic. The presence of the smaller-amplitude harmonics at 1, 2, 4, 5, . . . kHz causes $v_o(t)$ to have the nonsinusoidal shape shown in the figure. If the values of the *RLC* circuit had been chosen so that the filter were more frequency-selective (that is, had a narrower pass band centered at 3 kHz), then $v_o(t)$ would look much more like a pure 3-kHz sinusoid.

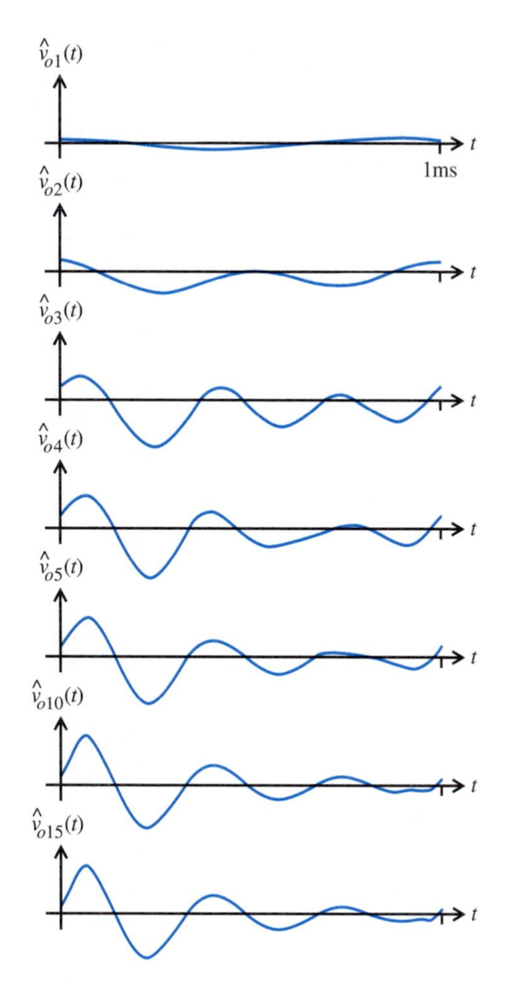

FIGURE 15.18
One period of the partial Fourier series $\hat{v}_{oN}(t)$ with $N =$ (*a*) 1, (*b*) 2, (*c*) 3, (*d*) 4, (*e*) 5, (*f*) 10, and (*g*) 15

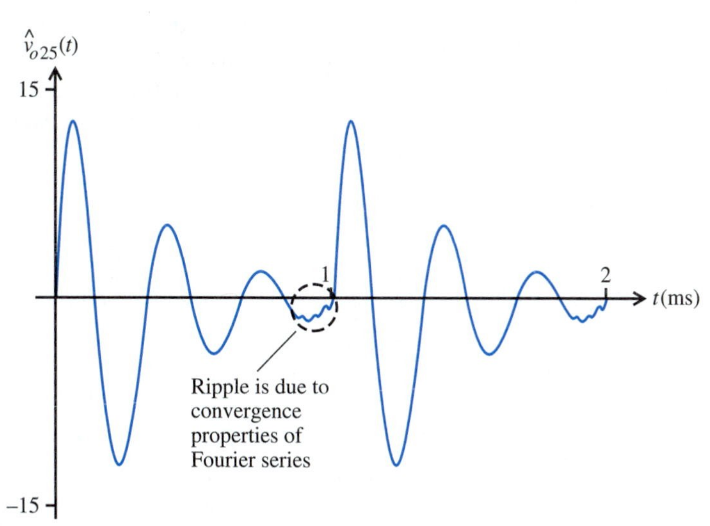

FIGURE 15.19
Two periods of the partial Fourier series $\hat{v}_{oN}(t)$ with $N = 25$

The second point of view for understanding the appearance of $v_o(t)$ is to consider the time domain. For the values given, the input waveform of Fig. 15.14, $v_s(t)$, resembles a sequence of narrow pulses that occur every $T = 1$ ms. We can show that the response of the band-pass *RLC* circuit to a single narrow pulse is approximately a damped sinusoid with frequency 3 kHz. The response to the sequence of pulses in $v_s(t)$ is, by superposition, approximately a sequence of damped 3-kHz sinusoids occurring every 1 ms, as seen in Fig. 15.19.

Application of Complex Fourier Series

We stated earlier that we can also find the response of an LTI circuit to a periodic input by using the complex form of the Fourier series. This method is illustrated in Fig. 15.20, in which the real periodic input waveform $v_s(t)$ is represented as the sum

$$v_s(t) = \sum_{n=-\infty}^{+\infty} V_{sn} e^{j2\pi n f_1 t} \tag{15.53}$$

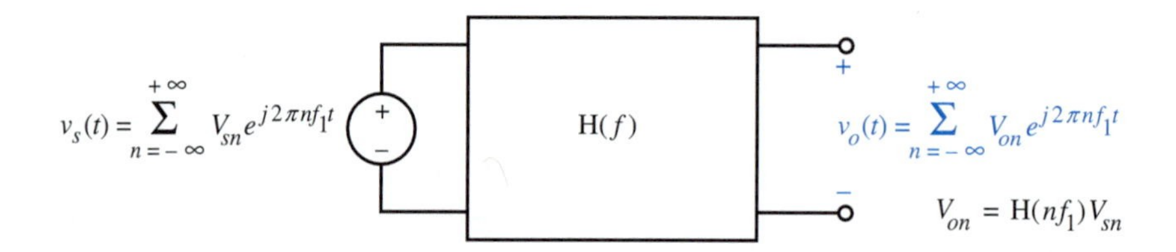

FIGURE 15.20 A sum of complex exponential inputs yields a sum of complex exponential outputs.

where V_{sn} is the complex Fourier coefficient

$$V_{sn} = \frac{1}{T}\int_{t_0}^{t_0+T} v_s(t) e^{-j2\pi n f_1 t}\, dt \tag{15.54}$$

As shown in Chapter 10, the individual complex exponential $V_{sn}e^{j2\pi n f_1 t}$ produces a complex exponential particular response $V_{on}e^{j2\pi n f_1 t}$, where $V_{on} = H(j2\pi n f_1)V_{sn}$, or in terms of the economical notation of Eq. (15.52), $V_{on} = \mathrm{H}(nf_1)V_{sn}$. From the principle of superposition, it follows that the response to the sum of complex exponentials $v_s(t)$ of Eq. (15.53) is simply the sum

$$v_o(t) = \sum_{n=-\infty}^{+\infty} V_{on} e^{jn\omega_1 t} \tag{15.55}$$

where

$$V_{on} = H(jn\omega_1)V_{sn} \tag{15.56}$$

or, equivalently, where

$$V_{on} = \mathrm{H}(nf_1)V_{sn} \tag{15.57}$$

The computation of the two-sided output spectrum is illustrated in Fig. 15.21. The height of each line in the output amplitude spectrum is the product of the height of the corresponding line of the input amplitude spectrum and the value of the amplitude characteristic $|\mathrm{H}(f)|$ at the associated frequency. The output phase spectrum is the sum of the input phase spectrum and the phase characteristic of the filter at frequencies nf_1.

It follows directly from Eq. (15.56) that the output power spectrum is equal to the product of the input power spectrum and the squared magnitude of the transfer function

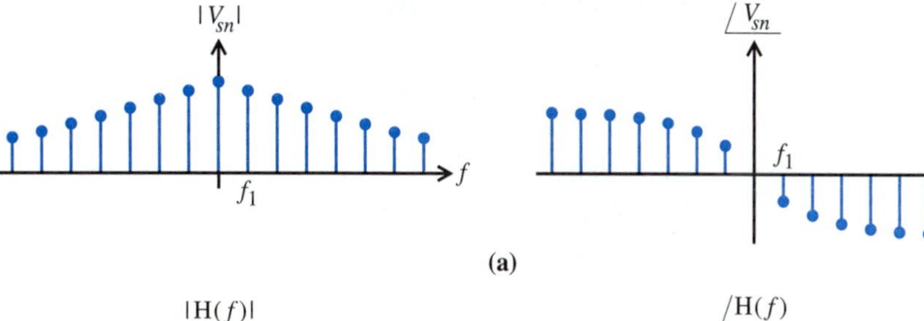

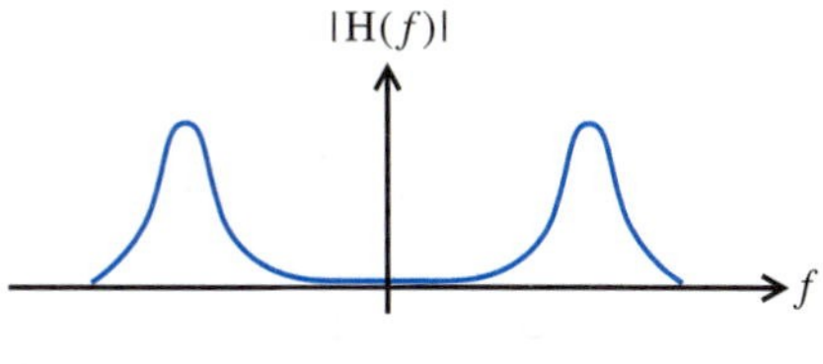

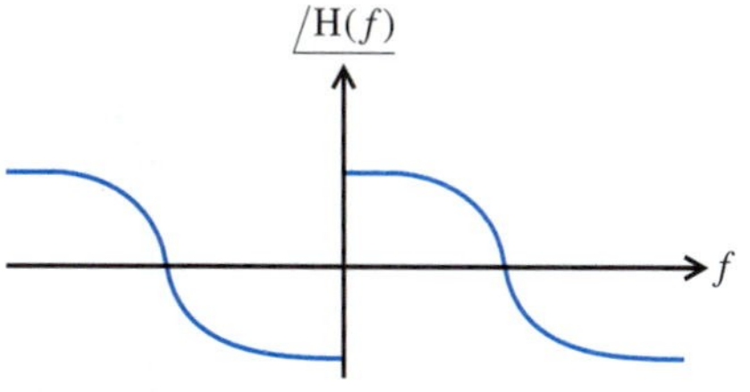

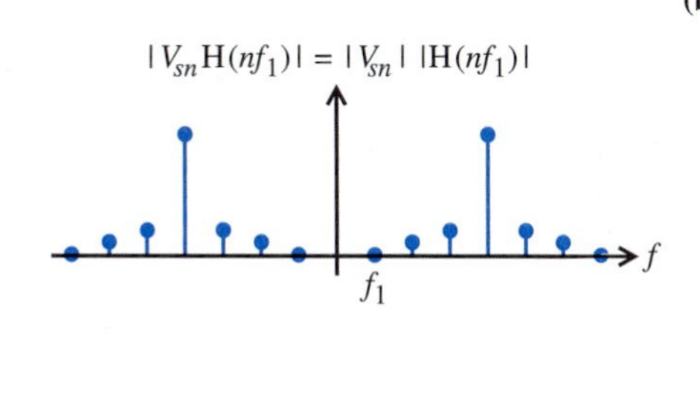

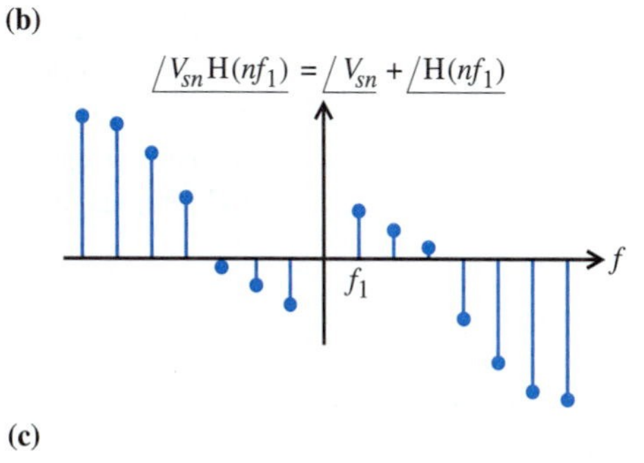

FIGURE 15.21 Computation of two-sided output amplitude and phase spectra: (*a*) input spectrum; (*b*) filter transfer function; (*c*) output spectrum

$$|V_{on}|^2 = |H(jn\omega_1)|^2 |V_{sn}|^2 \tag{15.58}$$

or

$$|V_{on}|^2 = |\mathrm{H}(nf_1)|^2 |V_{sn}|^2 \tag{15.59}$$

Remember

When the input to an LTI circuit is a periodic waveform, the output is also a periodic waveform. The amplitude spectrum of the output is given by the product of the amplitude spectrum of the input and the amplitude characteristic of the circuit. The output phase spectrum is given by the sum of the phase spectrum of the input and the phase characteristic of the circuit.

EXERCISES

18. The pulse train of Fig. 15.9 is applied to a low-pass *RC* circuit. The parameter values are $A = 10$ V, $T = 1$ ms, $\tau = 0.1$ ms, $R = 50\ \Omega$, and $C = 1\ \mu$F.
    (a) Plot the two-sided amplitude and phase characteristics of the circuit and the two-sided amplitude and phase spectra of the pulse train input.
    (b) Plot the two-sided amplitude and phase spectra of the output.
19. Repeat Exercise 18 with the low-pass *RC* circuit replaced by a high-pass *RC* circuit with the same values of *R* and *C*. (For a high-pass *RC* circuit, the output voltage is the voltage across *R*.)
20. Repeat Exercise 18 with the low-pass *RC* circuit replaced by the band-pass *RLC* circuit of Fig. 15.15.

15.8 Convergence and Existence of Fourier Series

An understanding of the way the Fourier series converges to $v(t)$ can be obtained from the problem illustrated in Fig. 15.22. In this figure, $v(t)$ is a real periodic waveform with period T, and $\hat{v}_N(t)$ is an approximation to $v(t)$ given by the *finite* trigonometric series

$$\hat{v}_N(t) = a_0 + \sum_{n=1}^{N} (a_n \cos 2\pi n f_1 t + b_n \sin 2\pi n f_1 t) \tag{15.60}$$

where $f_1 = 1/T$. The problem is to adjust the a_n's and b_n's so that the approximation error

$$e_N(t) = v(t) - \hat{v}_N(t) \tag{15.61}$$

is as small as possible. How do we choose the a_n and b_n to make the *waveform* $e_N(t)$ as small as possible?

The solution to this problem depends on what is meant by a small waveform. *One* way to define the size of a waveform is by the waveform's mean square value. In the problem at hand, both $v(t)$ and $\hat{v}_N(t)$ have period T. Therefore $e_N(t)$ also has period T, and its mean square value is given by

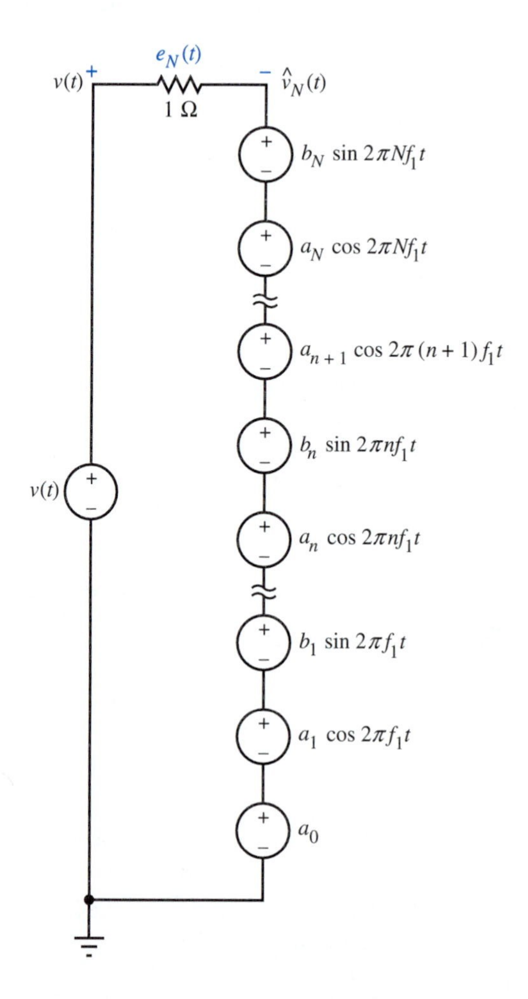

FIGURE 15.22 How should the dc voltage a_0 and the amplitudes a_n and b_n ($n = 1, 2, \ldots, N$) be chosen to make the waveform $e_N(t)$ as "small" as possible?

$$\varepsilon(N) = \frac{1}{T}\int_{t_0}^{t_0+T} e_N^2(t)\, dt \tag{15.62}$$

Recall that the mean square value of $e_N(t)$ is numerically equal to the average power that $e_N(t)$ supplies to the 1-Ω resistance of Fig. 15.22. In many engineering applications, it is reasonable to regard $\hat{v}_N(t)$ as the best possible approximation to $v(t)$ when the average power dissipated in the resistance is a minimum, that is, when $\varepsilon(N)$ is a minimum. This is the celebrated *minimum mean square error fidelity criterion.* To find the values of a_n that minimize $\varepsilon(N)$, we differentiate

$$\varepsilon(N) = \frac{1}{T}\int_{t_0}^{t_0+T} [v(t) - \hat{v}_N(t)]^2\, dt \tag{15.63}$$

with respect to a_n, $n = 1, 2, \ldots, N$, to obtain

$$\begin{aligned}\frac{d\varepsilon(N)}{da_n} &= \frac{-2}{T}\int_{t_0}^{t_0+T} [v(t) - \hat{v}_N(t)]\, \frac{d\hat{v}_N(t)}{da_n}\, dt \\ &= \frac{-2}{T}\int_{t_0}^{t_0+T} [v(t) - \hat{v}_N(t)] \cos 2\pi n f_1\, dt\end{aligned} \tag{15.64}$$

Then by setting the right-hand side to zero, we find

$$\frac{2}{T}\int_{t_0}^{t_0+T} \hat{v}_N(t) \cos 2\pi n f_1\, dt = \frac{2}{T}\int_{t_0}^{t_0+T} v(t) \cos 2\pi n f_1 t\, dt \tag{15.65a}$$

or

$$\begin{aligned}\frac{2}{T}\int_{t_0}^{t_0+T} \left(a_0 + \sum_{m=1}^{N} a_m \cos 2\pi m f_1 t + \sum_{m=1}^{N} b_m \sin 2\pi m f_1 t\right) \cos 2\pi n f_1 t\, dt \\ = \frac{2}{T}\int_{t_0}^{t_0+T} v(t) \cos 2\pi n f_1 t\, dt\end{aligned} \tag{15.65b}$$

All the terms in the left-hand integral integrate to zero except the one involving $a_m \cos 2\pi m f_1 t \cos 2\pi n f_1 t$ where $m = n$. By integration this term, Eq. (15.65b), becomes

$$a_n = \frac{2}{T}\int_{t_0}^{t_0+T} v(t) \cos 2\pi n f_1 t\, dt \tag{15.66}$$

which is recognized as the Fourier coefficient given by Eq. (15.5). To see whether the Fourier coefficient results in a minimum or a maximum value for $\varepsilon(N)$, we use Eq. (15.64) to compute

$$\begin{aligned}\frac{d^2\varepsilon(N)}{da_n^2} &= \frac{2}{T}\int_{t_0}^{t_0+T} \frac{d\hat{v}_N(t)}{da_n} \cos 2\pi n f_1 t\, dt \\ &= \frac{2}{T}\int_{t_0}^{t_0+T} \cos^2 2\pi n f_1 t\, dt \\ &= 1 \qquad n = 1, 2, \ldots, N\end{aligned} \tag{15.67}$$

Because the second derivative is positive, the mean square error $\varepsilon(N)$ obtained by the use of Eq. (15.66) is a minimum.

By differentiating $\varepsilon(N)$ with respect to a_0 and b_n, we similarly find that $\varepsilon(N)$ is minimized by

$$a_0 = \frac{1}{T}\int_{t_0}^{t_0+T} v(t)\, dt \tag{15.68}$$

which is the Fourier coefficient of Eq. (15.2), and

$$b_n = \frac{2}{T}\int_{t_0}^{t_0+T} v(t)\sin 2\pi n f_1 t\, dt \tag{15.69}$$

which is the Fourier coefficient of Eq. (15.6). Therefore

The Fourier coefficients make $\hat{v}_N(t)$ the best possible approximation to $v(t)$ in the sense that they cause the mean square approximation error $\varepsilon(N)$ to be a minimum.

In other words, the Fourier series converges in such a way that the power dissipated in the resistance of Fig. 15.22 is minimized. The convergence is as fast as possible in the sense that the mean square error is decreased as much as possible with the addition of each new term in the partial Fourier series.

Plancherel showed that if you keep increasing the number of terms in the partial Fourier series, the mean square error becomes vanishingly small:

$$\lim_{N\to\infty} \varepsilon(N) = 0 \tag{15.70}$$

The *only* restriction is that $v(t)$ have finite normalized power. The result is called *Plancherel's theorem.* A moment's thought will convince you of the importance of Plancherel's theorem: Every physical signal has finite power, and no physical device can detect an error that has no energy. This means that it is impossible physically to distinguish between a periodic signal and its (complete) Fourier series.

A mathematically stronger justification for the use of Fourier series was given by Dirichlet. Dirichlet showed that if the following three conditions are met, then the Fourier series representation of a periodic signal, $v(t)$, *equals* $v(t)$ except at isolated values of t where $v(t)$ is discontinuous.

Dirichlet Conditions

1. $v(t)$ is absolutely integrable over a period; that is,

$$\int_{-T/2}^{+T/2} |v(t)|\, dt < \infty$$

2. $v(t)$ has a finite number of maxima and minima within a period.
3. $v(t)$ has a finite number of discontinuities within a period. Each discontinuity must be finite.

The Dirichlet conditions are sufficient but not necessary conditions for $v(t)$ to have a Fourier series. That is, if $v(t)$ satisfies the three Dirichlet conditions, then it is guaranteed to have a Fourier-series representation that *equals* $v(t)$ except at the isolated discontinuities of $v(t)$. If $v(t)$ does not satisfy one or more of the Dirichlet conditions, then it *may or may not* have a Fourier-series representation. Virtually all periodic waveforms that can exist in a physical circuit satisfy the Dirichlet conditions.

The nature of the convergence of $\hat{v}_N(t)$ to $v(t)$ is illustrated in Fig. 15.23, which contains plots of the partial Fourier series

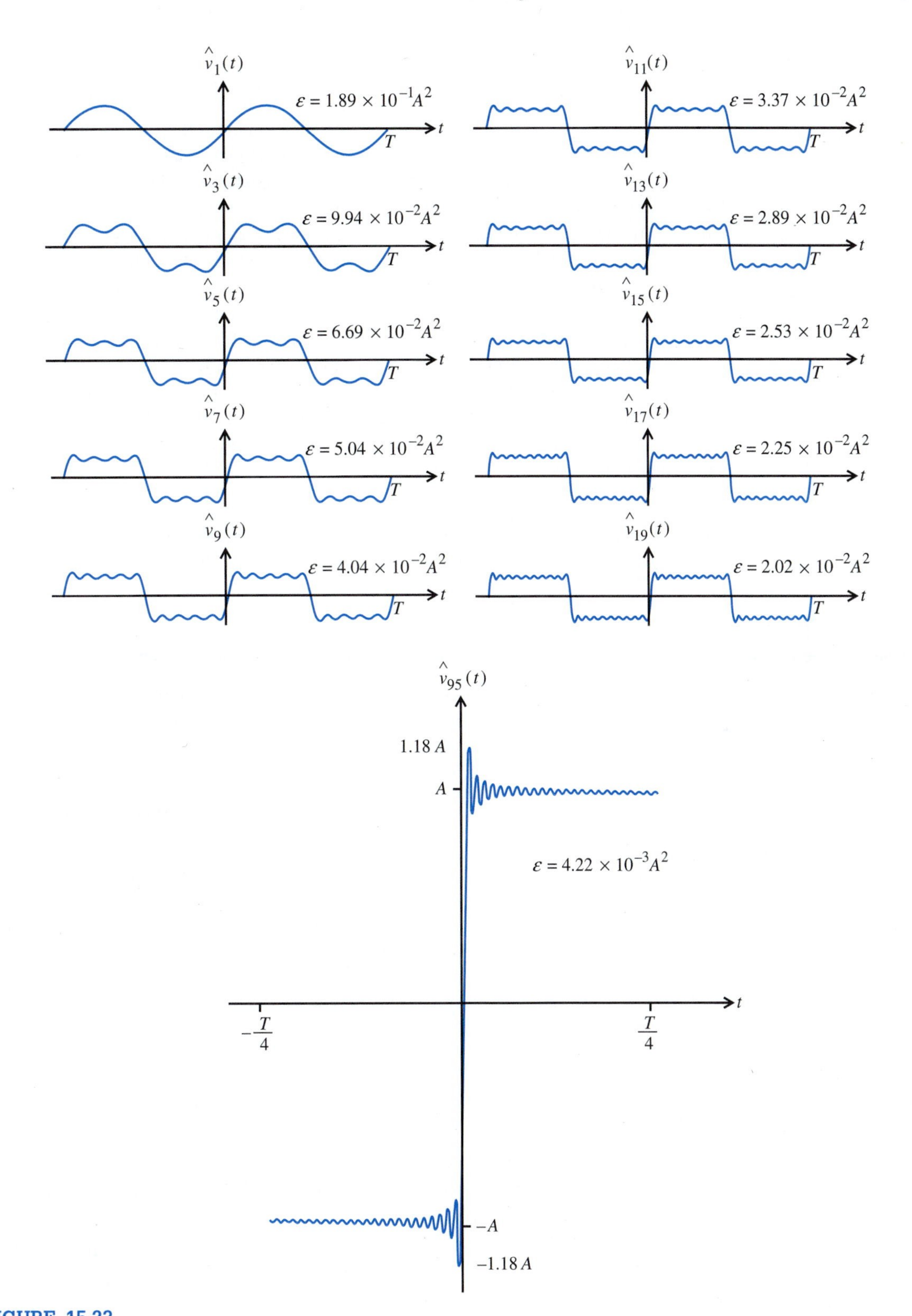

FIGURE 15.23
Partial Fourier series of the square wave of Fig. 15.6.

$$\hat{v}_N(t) = \sum_{n=1,3,5,\ldots}^{N} \frac{4A}{\pi n} \sin 2\pi n f_1 t \tag{15.71}$$

of the square wave of Fig. 15.6. Figure 15.23 also indicates the mean square approximation errors $\varepsilon(N)$ associated with each partial Fourier series. It can be seen from this figure that as N increases, $\hat{v}_N(t)$ approaches $v(t)$ in appearance and $\varepsilon(N)$ gets smaller. An interesting feature of the partial Fourier series occurs in the vicinity of the discontinuities of $v(t)$, where $\hat{v}_N(t)$ overshoots and oscillates. The amount of overshoot is approximately 9 percent of the amplitude of discontinuity of $v(t)$ for large but finite N *and does not decrease as N increases.* As N increases, the oscillations become more rapid and are grouped closer to the discontinuity. This oscillatory behavior of the partial Fourier series at the discontinuities of $v(t)$ can be seen in computer-generated plots of $\hat{v}_N(t)$, where N is necessarily finite. This feature of the partial Fourier series is called the *Gibbs phenomenon.* Because the square wave in this example satisfies the Dirichlet conditions, in the limit $N \rightarrow \infty$, $\hat{v}_N(t)$ converges to $v(t)$ for every value of t not corresponding to a discontinuity of $v(t)$. At a discontinuity, $\hat{v}_N(t)$ converges to a point halfway between the values on either side of the discontinuity.

Remember

The Fourier coefficients make $\hat{v}_N(t)$ the best possible approximation to $v(t)$ in the sense that they cause the mean square approximation error $\varepsilon(N)$ to be a minimum.

EXERCISES

21. (a) Determine the value of the constant α in the circuit below so that the average power dissipated in the resistance is minimized. The current $i_s(t)$ is an arbitrary periodic waveform with period T.
(b) What is the resulting minimum average power dissipation?

answer: (a) $\alpha = (2/T)\int_T i_s(t)\cos(2\pi f_1 t)\,dt = a_1$,

(b) $P = (R/T)\int_T i_s^2(t)\,dt - \alpha^2 R/2$

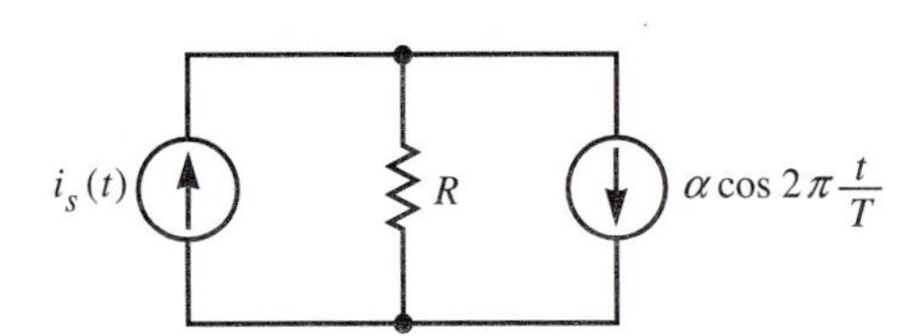

In Exercises 22 through 25, the functions are portions of signals that repeat every T seconds outside the given interval. Sketch two periods of each signal, and state which Dirichlet conditions, if any, are violated.

22. e^{-3t}, $0 \le t < T$ — answer: None

23. $1/t$, $0 \le t < T$ — answer: 1 and 3

24. $\cos(1/t)$, $0 \le t < T$ — answer: 2

25. $u(t) + u(t - \frac{1}{2}T) + u[t - \frac{3}{4}T] + u[t - \frac{7}{8}T] + \cdots$, $0 \le t < T$ — answer: 3

15.9 The Fourier Transform

We saw in Section 15.6 that a signal having period T can be expressed as a discrete sum of rotating phasors whose angular velocities, $2\pi n f_1$, are integer multiples of a fundamental angular velocity, $2\pi f_1$, where $f_1 = 1/T$. Fourier modified this exponential Fourier series to include *aperiodic* (not periodic) signals by the ingenious but simple step of taking the limit $T \rightarrow \infty$. This leads directly to the *inverse Fourier transform.* The inverse Fourier transform represents an aperiodic signal by an integral of rotating phasors whose angular frequencies vary continuously from minus to plus infinity.

The Fourier transform is one of the most widely used analytical tools in linear systems theory. It is a primary topic in higher-level courses in signals and systems. However, it is not widely used in circuit analysis; here the Laplace transform is generally the preferred technique. Our objective in this section is to introduce the Fourier transform in a way that will provide a helpful bridge between circuit analysis and higher-level courses in signals and systems.

Definitions and Examples

We begin by defining the Fourier transform:

DEFINITION Fourier Transform

The **direct Fourier transform** of a signal $v(t)$ is defined by

$$\mathrm{V}(f) = \int_{-\infty}^{+\infty} v(t) e^{-j2\pi ft}\, dt \tag{15.72}$$

The direct Fourier transform $\mathrm{V}(f)$ is a complete frequency-domain description of the time-domain signal $v(t)$. It is a *complete* description because we can obtain the original signal, $v(t)$, from $\mathrm{V}(f)$.

The formula used to recover $v(t)$ from $\mathrm{V}(f)$ is called the *inverse Fourier transform:*

Inverse Fourier Transform

$$v(t) = \int_{-\infty}^{+\infty} \mathrm{V}(f) e^{j2\pi ft}\, df \tag{15.73}$$

The *inverse* Fourier transform, Eq. (15.73), should be compared with Eq. (15.22), which is the corresponding representation for periodic signals and which is reproduced below.

$$v(t) = \sum_{n=-\infty}^{+\infty} V_n e^{j2\pi n f_1 t}$$

By comparing Eq. (15.73) with Eq. (15.22), we can acquire an intuitive understanding of Eqs. (15.72) and (15.73). The quantity $V(f)e^{j2\pi ft}\,df = [V(f)df]e^{j2\pi ft}$ appearing in Eq. (15.73) is a rotating phasor with angular velocity $2\pi f$ and infinitesimal complex amplitude $V(f)\,df$. It corresponds to the rotating phasor $V_n e^{j2\pi f_1 t}$ of Eq. (15.22) which has angular velocity $2\pi n f_1$ and finite complex amplitude V_n. Thus $V(f)$ of Eq. (15.72) tells us the essential information regarding the infinitesimal complex amplitudes in Eq. (15.73). If $v(t)$ has the unit volts, then $V(f)df$ also has the unit volts, and, in turn, the unit of $V(f)$ is volt-seconds or volts per hertz (V · s or V/Hz). $V(f)$ is often referred to as the *voltage density spectrum* of $v(t)$.

The Fourier transform $V(f)$ is generally a complex function of frequency f. The magnitude of $V(f)$, $|V(f)|$, is called the *amplitude density spectrum,* and the angle of $V(f)$, $\angle V(f)$, is called the *phase spectrum.* The transformation from $v(t)$ to $V(f)$ is called *Fourier analysis* or *spectrum analysis,* whereas that from $V(f)$ to $v(t)$ is called *Fourier synthesis.*

EXAMPLE 15.5 Determine and plot the Fourier transform of the one-sided decaying exponential

$$v(t) = e^{-t/\tau}u(t)$$

(where τ is positive), illustrated in Fig. 15.24.

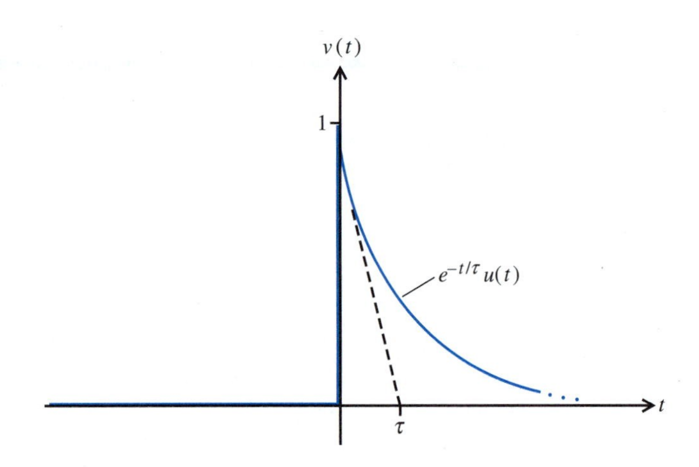

FIGURE 15.24 One-sided decaying exponential waveform

Solution For notational simplicity, we set $a = -1/\tau$. Note that a is negative because τ is positive. Substitution of $e^{at}u(t)$ into Eq. (15.72) yields

$$\begin{aligned} V(f) &= \int_{-\infty}^{+\infty} e^{at}u(t)e^{-j2\pi ft}\,dt \\ &= \int_{0}^{+\infty} e^{(a-j2\pi f)t}\,dt \\ &= \frac{1}{a-j2\pi f}e^{(a-j2\pi f)t}\Bigg|_0^{\infty} \\ &= \frac{1}{a-j2\pi f}e^{at}e^{-j2\pi ft}\Bigg|_{\infty} - \frac{1}{a-j2\pi f}e^{at}e^{-j2\pi ft}\Bigg|_0 \end{aligned}$$

Because a is negative, the factor e^{at} will equal zero for $t = \infty$. Therefore, the first term on the right-hand side of the above equation equals zero. This leaves

$$V(f) = -\frac{1}{a - j2\pi f} e^{at} e^{-j2\pi ft} \Big|_0 = \frac{1}{-a + j2\pi f}$$

or

$$V(f) = \frac{1}{(1/\tau) + j2\pi f}$$

Plots of $|V(f)|$ and $\underline{/V(f)}$ are given in Fig. 15.25.

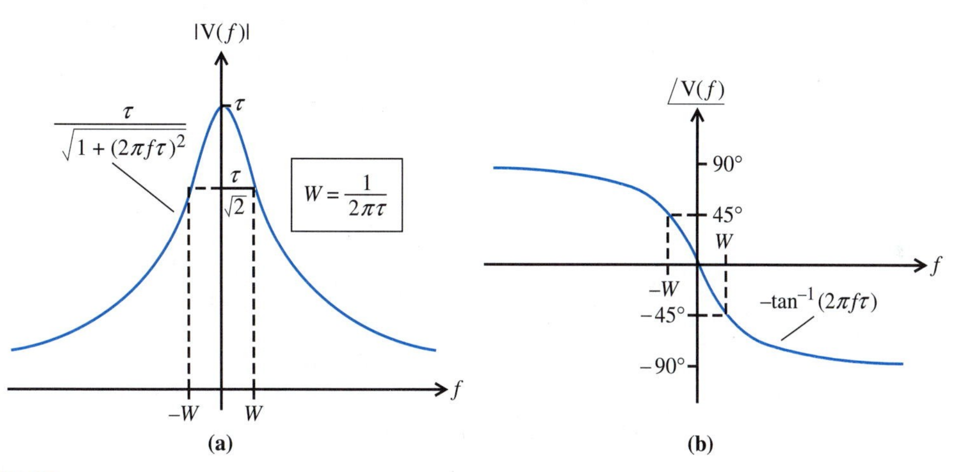

FIGURE 15.25
Fourier transform of the one-sided decaying exponential: (*a*) amplitude density; (*b*) phase

Analogous to Eq. (15.36) for Fourier coefficients, the Fourier transform of a real signal has the property that $V(-f) = V^*(f)$. It follows that the amplitude density spectrum of a real signal is an even function, $|V(-f)| = |V(f)|$, and the phase spectrum is an odd function, $\underline{/V(-f)} = -\underline{/V(f)}$. This property is proved in Problem 38.

EXAMPLE 15.6 Determine and plot the Fourier transform of the rectangular pulse

$$v(t) = \Pi\left(\frac{t}{\tau}\right)$$

illustrated in Fig. 15.26.

FIGURE 15.26
Rectangular pulse

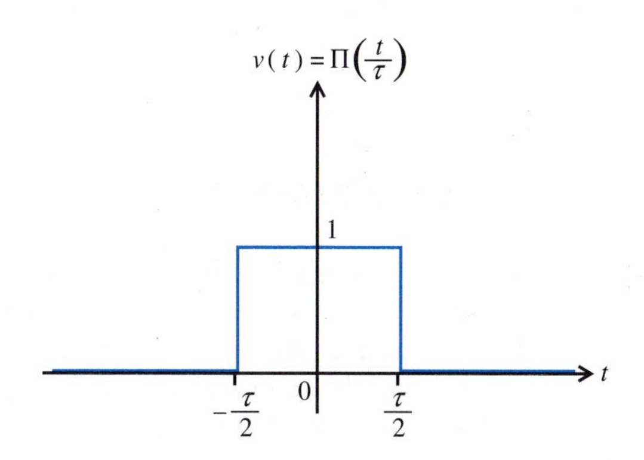

Solution Substitution of $\Pi(t/\tau)$ into Eq. (15.72) yields

$$\mathrm{V}(f) = \int_{-\infty}^{+\infty} \Pi\left(\frac{t}{\tau}\right) e^{-j2\pi ft}\,dt = \int_{-\tau/2}^{+\tau/2} e^{-j2\pi ft}\,dt$$

$$= \frac{1}{-j2\pi f} e^{-j2\pi ft}\Bigg|_{-\tau/2}^{+\tau/2} = \frac{1}{-j2\pi f}(e^{-j\pi f\tau} - e^{+j\pi f\tau})$$

$$= \tau \frac{\sin \pi f \tau}{\pi f \tau}$$

The Fourier transform is plotted in Fig. 15.27. Because $\mathrm{V}(f)$ is purely real, there is no need to plot the amplitude density and phase spectra separately.

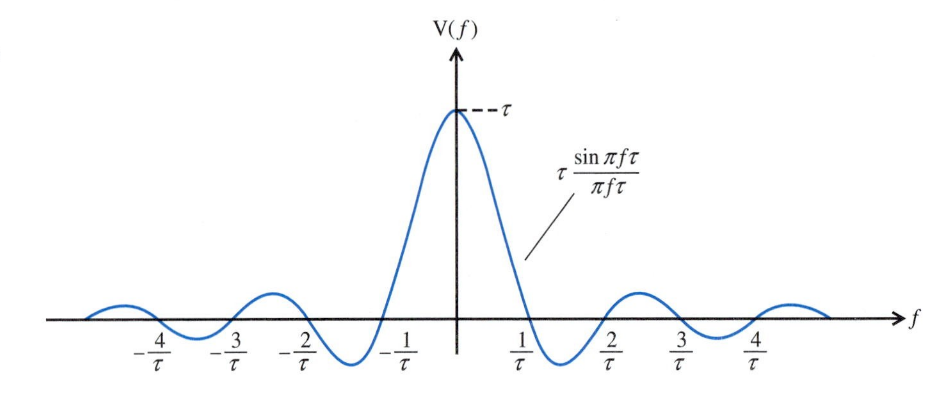

FIGURE 15.27
Fourier transform of the rectangular pulse

Remember

For a real signal $v(t)$, the amplitude density spectrum is an even function of f, and the phase spectrum is an odd function of f.

EXERCISES Use Eq. (15.72) to determine the Fourier transforms of the waveforms in Exercises 26 through 28. Sketch each waveform and its corresponding density spectrum.

26. $v(t) = \delta(t)$ answer: 1
27. $v(t) = e^{bt}u(-t),\ b > 0$ answer: $1/(b - j2\pi f)$
28. $v(t) = e^{at}u(t) + e^{bt}u(-t),\ a < 0,\ b > 0$ answer: $(a + b)/[(2\pi f)^2 + j2\pi f(b - a) + ab]$

Use Eq. (15.73) to determine the waveforms with the Fourier transforms given in Exercises 29 and 30. Sketch each waveform and its corresponding density spectrum.

29. $\mathrm{V}(f) = \delta(f)$ answer: 1
30. $\mathrm{V}(f) = \frac{1}{2}A\delta(f - f_0) + \frac{1}{2}A\delta(f + f_0)$ answer: $A \cos(2\pi f_0 t)$

Derivation of the Fourier Transform

We can derive the Fourier transform formula (Eq. 15.72) and its inverse Eq. (15.73) easily if we keep an example in mind. Consider the rectangular pulse, $v(t)$, shown in Fig. 15.28a. The first step in the derivation is to define the *periodic version*, $v_T(t)$, of the pulse $v(t)$ as shown in Fig. 15.28b. The periodic version $v_T(t)$ is simply the waveform $v(t)$ repeated every T seconds and we can write it as a Fourier series:

$$v_T(t) = \sum_{n=-\infty}^{+\infty} V_n e^{j2\pi n f_1 t} \tag{15.74}$$

where $f_1 = 1/T$ and

$$V_n = \frac{1}{T}\int_{-T/2}^{T/2} v_T(t) e^{-j2\pi n f_1 t}\, dt \tag{15.75}$$

Notice from Fig. 15.28 that

$$v(t) = \begin{cases} v_T(t) & \text{for } |t| \le \dfrac{T}{2} \\ 0 & \text{otherwise} \end{cases} \tag{15.76}$$

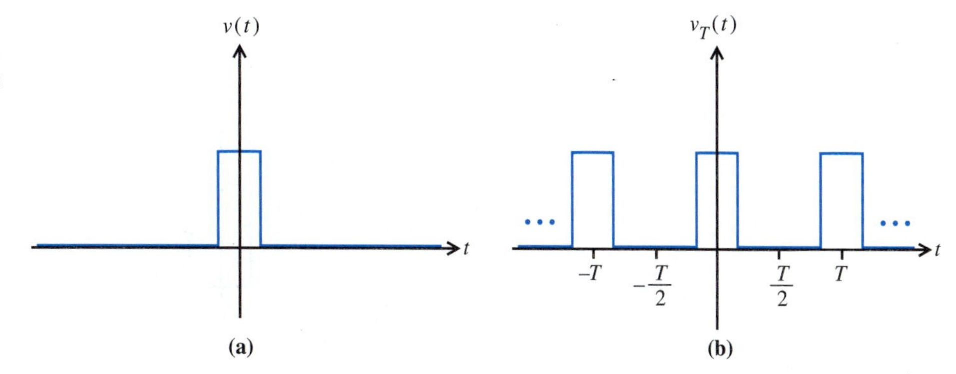

FIGURE 15.28 (*a*) An aperiodic signal $v(t)$ and (*b*) its periodic version $v_T(t)$

The next step is to use Eq. (15.76) to rewrite the nth complex Fourier coefficient V_n in Eq. (15.75) as

$$V_n = \frac{1}{T}\int_{-\infty}^{+\infty} v(t) e^{-j2\pi n f_1 t}\, dt \tag{15.77a}$$

or equivalently,

$$V_n = \frac{1}{T} \mathrm{V}(nf_1) \tag{15.77b}$$

where $V(f)$ is the Fourier transform of $v(t)$ defined in Eq. (15.72). The relationship between the Fourier transform of $v(t)$ and the Fourier coefficients of $v_T(t)$ expressed by Eq. (15.77b) is illustrated in Fig. 15.29. Observe that except for the scale factor $1/T$, the complex Fourier coefficients V_n of the periodic signal $v_T(t)$ can be regarded as ''samples'' (represented by the vertical lines) of the density spectrum $\mathrm{V}(f)$ of the nonperiodic signal $v(t)$ at the discrete frequencies nf_1, $n = 0, \pm 1, \pm 2, \ldots$. The substitution of Eq. (15.77b) into Eq. (15.75) yields

$$v_T(t) = \frac{1}{T} \sum_{n=-\infty}^{+\infty} \mathrm{V}(nf_1) e^{j2\pi n f_1 t} \tag{15.78}$$

The final step in the derivation is the key conceptual step. As we can see by referring to Fig. 15.28, the periodic signal $v_T(t)$ equals the aperiodic signal $v(t)$ if $T = \infty$. That is,

$$\begin{aligned} v(t) &= \lim_{T\to\infty} v_T(t) \\ &= \lim_{T\to\infty} \frac{1}{T} \sum_{n=-\infty}^{+\infty} \mathrm{V}(nf_1) e^{j2\pi n f_1 t} \end{aligned} \tag{15.79}$$

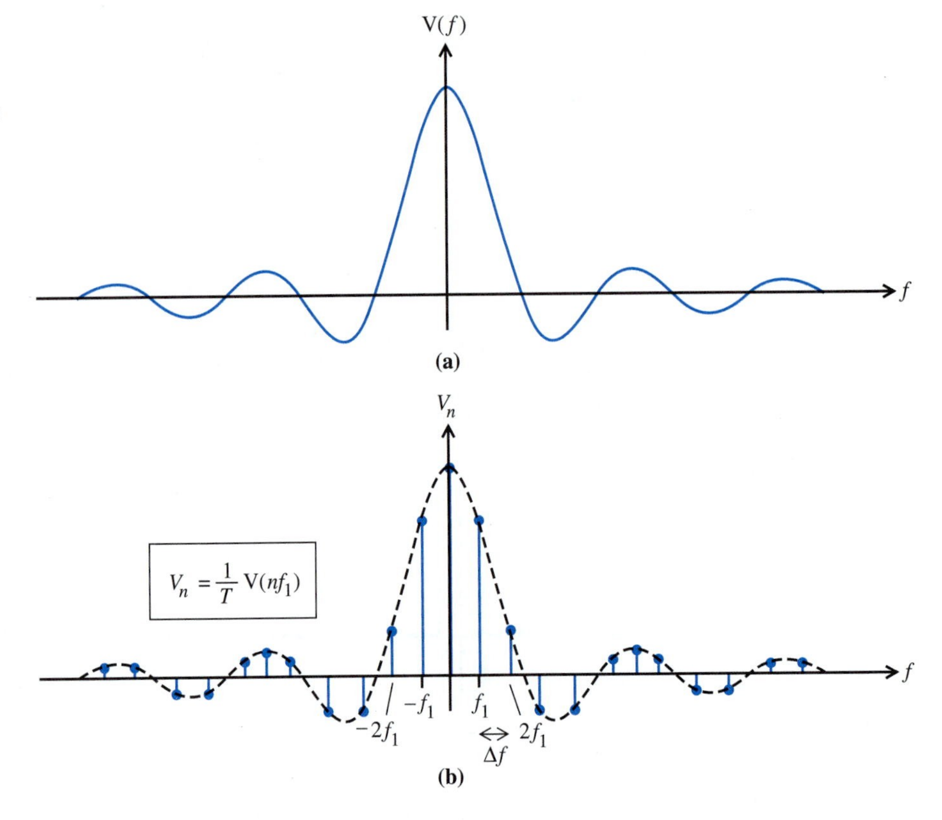

FIGURE 15.29 The relationship between the Fourier transform of $v(t)$ and the Fourier coefficients of $v_T(t)$. (*a*) Fourier transform of $v(t)$; (*b*) the Fourier coefficients of $v_T(t)$

In taking the limit in Eq. (15.79), it is useful to recognize that f_1 is not only the frequency of the fundamental component of the Fourier series, it is also the frequency interval

$$\Delta f = f_1 = \frac{1}{T} \tag{15.80}$$

separating the lines in the spectrum of $v_T(t)$ as shown in Fig. 15.29. Observe that as T increases to infinity, this frequency interval shrinks to zero. The substitution of Eq. (15.80) into Eq. (15.79) yields

$$\begin{aligned} v(t) &= \lim_{\Delta f \to 0} \sum_{n=-\infty}^{+\infty} V(n\Delta f) e^{j2\pi n \Delta f t} \, \Delta f \\ &= \lim_{\Delta f \to 0} \sum_{\substack{f = n\Delta f \\ n=-\infty}}^{+\infty} V(f) e^{j2\pi f t} \, \Delta f \end{aligned} \tag{15.81}$$

In the limit as $\Delta f \to 0$, Eq. (15.81) becomes, by the definition of an integral,

$$v(t) = \int_{-\infty}^{+\infty} V(f) e^{j2\pi f t} \, df \tag{15.82}$$

which is the inverse Fourier transform of $V(f)$ given by Eq. (15.73).

The preceding derivation has far-reaching implications. Because the inverse Fourier transform is a limiting form of the Fourier series, all the properties of the Fourier series can be extended to the inverse Fourier transform. At this point, we have accomplished our objective of introducing the Fourier transform. We will leave further developments to your future courses in signals and systems.

Remember

The Fourier transform and its inverse come directly from the Fourier series. The difference is that the Fourier series applies to periodic signals and the Fourier transform to aperiodic signals. A periodic signal contains *only* frequencies that are integer multiples of the fundamental frequency $f_1 = 1/T$ where T is the period of the signal. An aperiodic signal has no period: it contains *all* frequencies. Thus, Fourier showed that *all signals found in nature can be considered to be sums or integrals of sinusoidal components.*

15.10 Summary

Fourier claimed that every periodic waveform with period T can be represented as a sum of sinusoidal components with frequencies nf_1, where $n = 0, 1, 2, \ldots$ and $f_1 = 1/T$. This sum of sinusoids is the Fourier-series representation of the periodic waveform. The real form of the Fourier series consists of sines and cosines. The coefficients of the sines and cosines are called Fourier coefficients. We showed that even symmetric waveforms contain only dc and cosine Fourier components, odd symmetric waveforms contain only sine components, and half-wave symmetric waveforms contain only components whose frequencies are odd harmonics, $nf_1, n = 1, 3, 5, \ldots$. By means of standard trigonometric identities, we gave an alternative form of the Fourier series with terms of the form $A_n \cos(2\pi nf_1t + \phi_n)$, where A_n was called a Fourier amplitude and ϕ_n was called a Fourier phase. Finally, we described the exponential form of the Fourier series. The exponential form applies to both complex and real periodic waveforms. The real series Fourier coefficients, a_n and b_n, and the Fourier amplitudes and phases, A_n and ϕ_n, are related to the complex Fourier coefficients V_n by very simple formulas, Eqs. (15.30) and (15.33). All three forms of the Fourier series are used in practice. The Fourier series provides a convenient method to compute the particular response of an LTI circuit to a periodic input. An important topic of this chapter is that of the existence of the Fourier series. By referring to certain conditions established by Dirichlet, we were able to conclude that every periodic waveform that has engineering significance can be represented by a Fourier series. We then discussed the minimum mean square error convergence property of the Fourier series and the Gibbs phenomenon.

We concluded this chapter by introducing the Fourier transform. The Fourier transform provides the means to represent any physical signal $v(t)$ as an integral of complex exponential components. We showed that the inverse Fourier transform is the limiting form of the complex Fourier series as $T \to \infty$.

KEY FACTS	Concept	Equation	Section	Page
❑	Any real periodic waveform found in nature or engineering can be represented by a sum of sinusoidal components whose frequencies are multiples of $f_1 = 1/T$, where T is the period.		15.8	574
❑	There are two basic forms of the real Fourier series: A *sine-cosine* form $v(t) = a_0 + \sum_{n=1}^{\infty} a_n \cos 2\pi nf_1 t + \sum_{n=1}^{\infty} b_n \sin 2\pi nf_1 t$	(15.1)	15.2	542

KEY FACTS Concept	Equation	Section	Page
where the *Fourier coefficients* a_n and b_n are given by			
$a_0 = \frac{1}{T}\int_{t_0}^{t_0+T} v(t)\,dt$	(15.2)		
$a_n = \frac{2}{T}\int_{t_0}^{t_0+T} v(t)\cos 2\pi n f_1 t\,dt$ for $n = 1, 2, \ldots$	(15.5)		
$b_n = \frac{2}{T}\int_{t_0}^{t_0+T} v(t)\sin 2\pi n f_1 t\,dt$ for $n = 1, 2, \ldots$	(15.6)		
and an *amplitude-phase* form			
$v(t) = A_0 + \sum_{n=1}^{\infty} A_n \cos(2\pi n f_1 t + \phi_n)$	(15.18)	15.4	550
where the *Fourier amplitudes and phases*, A_n and ϕ_n, are given by			
$A_0 = a_0$; $A_n = \sqrt{a_n^2 + b_n^2}$ $n = 1, 2, \ldots$; $\phi_n = \tan^{-1}\left(\frac{-b_n}{a_n}\right)$ $n = 1, 2, \ldots$	(15.19)		
❑ If $v(t)$ is even, then $b_n = 0$ for $n = 1, 2, \ldots$.		15.3	545
❑ If $v(t)$ is odd, then $a_n = 0$ for $n = 0, 1, 2, \ldots$.		15.3	546
❑ If $v(t)$ is half-wave symmetric, then $a_0 = a_n = b_n = 0$ for $n = 2, 4, 6, \ldots$.		15.3	547
❑ The *one-sided amplitude spectrum* is a plot of the Fourier amplitudes A_n versus frequency, $f \geq 0$.		15.4	551
❑ The *one-sided phase spectrum* is a plot of the Fourier phases ϕ_n versus frequency, $f \geq 0$.		15.4	551
❑ The *one-sided power spectrum* of $v(t)$ is a plot of $A_0^2, \frac{1}{2}A_1^2, \frac{1}{2}A_2^2, \ldots$ versus frequency, $f \geq 0$.		15.4	552
❑ According to *Parseval's theorem*, for real periodic signals,		15.2, 15.4	544 551
$\frac{1}{T}\int_{t_0}^{t_0+T} v^2(t)\,dt = a_0^2 + \sum_{n=1}^{\infty}\frac{1}{2}a_n^2 + \sum_{n=1}^{\infty}\frac{1}{2}b_n^2$	(15.9)		
$= A_0^2 + \sum_{n=1}^{\infty}\frac{1}{2}A_n^2$	(15.21)		
❑ A real or complex periodic waveform $v(t)$ can be represented by the *exponential form* of the Fourier series,		15.5	555
$v(t) = \sum_{n=-\infty}^{+\infty} V_n e^{j2\pi n f_1 t}$	(15.22)		

KEY FACTS	Concept	Equation	Section	Page
	where the complex Fourier coefficient V_n is given by $$V_n = \frac{1}{T}\int_{t_0}^{t_0+T} v(t)e^{-j2\pi n f_1 t}\,dt \qquad n = 0, \pm 1, \pm 2, \ldots$$	(15.27)		
❑	If $v(t)$ is real, then $V_{-n} = V_n^*$.	(15.36)		557
❑	The *two-sided amplitude spectrum* is a plot of $\lvert V_n \rvert$ versus frequency, $-\infty < f < \infty$.		15.5	558
❑	The *two-sided phase spectrum* is a plot of $\angle V_n$ versus frequency, $-\infty < f < \infty$.		15.5	559
❑	According to *Parseval's theorem,* for real or complex periodic signals, $$\frac{1}{T}\int_{t_0}^{t_0+T} \lvert v(t)\rvert^2\,dt = \sum_{n=-\infty}^{+\infty} \lvert V_n \rvert^2$$	(15.40)	15.5	557
❑	By representing a periodic waveform by its Fourier series, the response of any LTI circuit to that periodic waveform may be obtained using ac circuit analysis and superposition.		15.7	565
❑	The amplitude spectrum of an LTI circuit's output is given by the *product* of the amplitude spectrum of the input and the amplitude characteristic of the circuit.		15.7	567
❑	The phase spectrum of the output is given by the *sum* of the phase spectrum of the input and the phase characteristic of the circuit.		15.7	567
❑	The *Dirichlet conditions* are sufficient conditions for the Fourier series of $v(t)$ to *equal* $v(t)$ except at isolated discontinuities.		15.8	574
❑	The Fourier coefficients make the partial Fourier series $\hat{v}_N(t)$ the best possible approximation to $v(t)$ in the sense of minimum mean square error.		15.8	574
❑	A 9 percent overshoot of a partial Fourier series occurs in the vicinity of a discontinuity of $v(t)$.		15.8	576
❑	The Fourier transform of a signal $v(t)$ is defined as $$\mathrm{V}(f) = \int_{-\infty}^{+\infty} v(t)e^{-j2\pi ft}\,dt$$	(15.72)	15.9	577
❑	The inverse Fourier transform of $\mathrm{V}(f)$ is given by $$v(t) = \int_{-\infty}^{+\infty} \mathrm{V}(f)e^{j2\pi ft}\,df$$	(15.73)	15.9	577

PROBLEMS

Section 15.2

1. A periodic waveform is given by $v(t) = 10 + 7 \cos 6\pi 10^3 t + 5 \cos(8\pi 10^3 t + 45°)$. Find the period T and the real Fourier-series coefficients.

2. Consider the time-shifted periodic impulse train

$$v(t) = \sum_{n=-\infty}^{+\infty} A\delta(t - t_d - nT)$$

where T is the period and t_d is an arbitrary time shift.

(a) Sketch $v(t)$ for $A = 5$, $T = 1$ ms, and $t_d = 0.25$ ms.

(b) Use Eqs. (15.2), (15.5), and (15.6) to determine the real Fourier coefficients of $v(t)$. (Assume that A, T, and t_d are arbitrary.)

(c) Write the real Fourier series of $v(t)$.

3. The periodic waveform

$$v(t) = x_0\sqrt{2} \cos 2\pi t + y_0 \sqrt{2} \sin 2\pi t + z_0 \sqrt{2} \cos 4\pi t$$

is obviously determined by the values of x_0, y_0, and z_0. The quantity $\mathbf{v} = (x_0, y_0, z_0)$ can be regarded as a vector in a three-dimensional cartesian coordinate system. Find a relationship between the mean square value of $v(t)$ and the length of the vector $\mathbf{v}$.

4. Let $\{\phi_i(t), i = 1, 2, \ldots\}$ be any set of real waveforms having the property that

$$\int_{t_0}^{t_0+T} \phi_i(t)\phi_k(t)\, dt = \begin{cases} c_i & i = k \\ 0 & i \neq k \end{cases}$$

where c_i is a constant that depends on i. Assume that

$$v(t) = \sum_{i=1}^{\infty} \alpha_i \phi_i(t) \qquad t_0 \le t \le t_0 + T \qquad (15.83)$$

(a) Show that if Eq. (15.83) is true, then

$$\alpha_i = \frac{1}{c_i} \int_{t_0}^{t_0+T} v(t)\phi_i(t)\, dt \qquad (15.84)$$

(b) Show that if Eq. (15.83) is true, then

$$\int_{t_0}^{t_0+T} v^2(t)\, dt = \sum_{i=1}^{\infty} c_i \alpha_i^2 \qquad (15.85)$$

(c) Show that Eqs. (15.1), (15.2), (15.5), (15.6), and (15.9) of the text are special cases of Eqs. (15.83), (15.84), and (15.85).

Section 15.3

5. To prove Symmetry Property 2, write the Fourier series of an arbitrary real periodic waveform $v(t)$ as $v(t) = v_a(t) + v_b(t)$, where

$$v_a(t) = a_0 + \sum_{n=1}^{\infty} a_n \cos 2\pi n f_1 t$$

and

$$v_b(t) = \sum_{n=1}^{\infty} b_n \sin 2\pi n f_1 t$$

(a) Show that $v_a(t)$ is an even function and $v_b(t)$ is an odd function.

(b) Use your result from (a) to show that $v(t)$ is an odd function if and only if $v_a(t) = 0$ for all t.

(c) Use your result from (b) to show that $v(t)$ is an odd function if and only if $a_n = 0$ for $n = 0, 1, 2, \ldots$.

6. To prove Symmetry Property 3, write the Fourier series of an arbitrary real periodic waveform as $v(t) = v_1(t) + v_2(t)$, where

$$v_1(t) = a_0 + \sum_{n=2,4,6,\ldots}^{\infty} (a_n \cos 2\pi n f_1 t + b_n \sin 2\pi n f_1 t)$$

and

$$v_2(t) = \sum_{n=1,3,5,\ldots}^{\infty} (a_n \cos 2\pi n f_1 t + b_n \sin 2\pi n f_1 t)$$

(a) Show that $v_1(t)$ has the property that $v_1(t) = v_1(t - T/2)$, and $v_2(t)$ has the property that $v_2(t) = -v_2(t - T/2)$.

(b) Use your result from (a) to show that $v(t)$ is half-wave symmetric if and only if $v_1(t) = 0$ for all t.

(c) Use your result from (b) to show that $v(t)$ is half-wave symmetric if and only if $a_0 = a_n = b_n = 0$ for $n = 2, 4, 6, \ldots$.

7. Suppose that A V dc is added to the odd symmetric waveform $v(t)$ of Fig. 15.4 to produce a new waveform $v'(t) = A + v(t)$.
   (a) Sketch $v'(t)$.
   (b) Is $v'(t)$ odd symmetric?
   (c) Show why $v'(t)$ contains no *cosine* functions.

8. Suppose that A V dc is added to the half-wave symmetric waveform $v(t)$ of Fig. 15.5 to produce a new waveform $v'(t) = A + v(t)$.
   (a) Sketch $v'(t)$.
   (b) Is $v'(t)$ half-wave symmetric?
   (c) Show why the Fourier coefficients of $v'(t)$ satisfy $a_0 \neq 0$, $a_n = b_n = 0$ for $n = 2, 4, 6, \ldots$.

9. A *full-wave rectified cosine* wave is defined by

$$v(t) = |A \cos 2\pi t/T_0|$$

   (a) Sketch $v(t)$ versus t.
   (b) What is the period?
   (c) Determine the real Fourier series.

Section 15.4

10. Plot the one-sided amplitude, phase, and power spectra of a delayed periodic impulse train

$$v(t) = \sum_{n=-\infty}^{+\infty} A\delta(t - t_d - nT)$$

where t_d is an arbitrary constant.

11. The periodic pulse train of Fig. 15.9 is applied across the terminals of a 3-kΩ resistance. Find the percentage of the total average power dissipation caused by the dc component. Write your answer in terms of τ and T.

12. The *duty factor* of the periodic pulse train $v(t)$ of Fig. 15.9 is defined by DF $= (\tau/T) \times 100\%$.
   (a) Sketch $v(t)$ for $A = 5$ and a duty factor of 50%.
   (b) Plot the one-sided amplitude, phase, and power spectra of $v(t)$ of (a).
   (c) Comment on any interesting features of your plots.

13. Sketch the periodic pulse train of Example 15.3 and its amplitude, phase, and power spectra for the following two special cases.
   (a) $\tau = T$
   (b) $\tau = \varepsilon$ and $A = \varepsilon^{-1}$, where ε is arbitrarily small ($\varepsilon \rightarrow 0$)

14. Plot the one-sided amplitude, phase, and power spectra of the periodic waveform

$$v(t) = 5 \cos 100\pi t + 17 - 3 \sin 300\pi t + 4 \sin 100\pi t$$

15. (a) Use Eq. (15.18) to show that the effect of replacing a periodic waveform $v(t)$ by $v'(t) = v(t - t_d)$ is to replace the Fourier phases ϕ_n, $n = 1, 2, \ldots$, by $\phi'_n = \phi_n - 2\pi n f_1 t_d$.
   (b) Illustrate your result from (a) by assuming that the periodic pulse train of Fig. 15.9 is delayed by $t_d = T/4$, where $A = 10$ V, $\tau = 0.2$ ms, and $T = 1$ ms. Plot the new waveform and its associated one-sided amplitude, phase, and power spectra.

Section 15.5

16. A periodic waveform $v(t)$ is obtained by repeating a truncated exponential pulse $p(t) = Ae^{at}[u(t) - u(t - T)]$ every T seconds, that is

$$v(t) = \sum_{n=-\infty}^{+\infty} p(t - nT)$$

   (a) Sketch $p(t)$ for $A = 5$, $a = 1$, and $T = 2$ s.
   (b) Sketch $v(t)$ for the parameters used in (a).
   (c) Find the complex Fourier series of $v(t)$. Express your answer in terms of A, a, T, and $f_1 = 1/T$.

17. (a) Derive the complex Fourier coefficients of the waveform shown.
   (b) Plot the two-sided amplitude, phase, and power spectra.

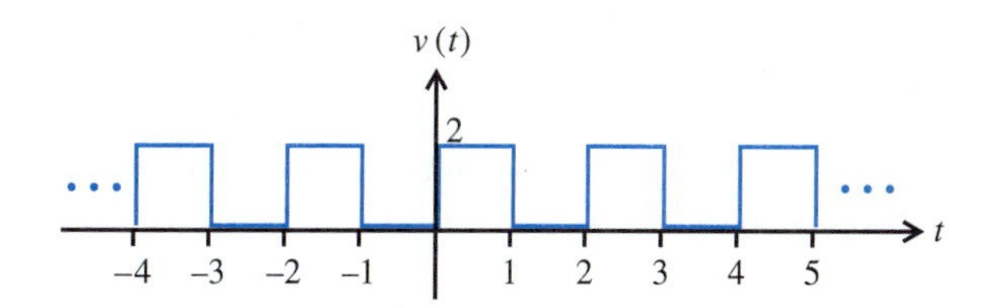

18. Assume that the voltage signal, $v(t)$, of Fig. 15.14 is applied across a resistance R.
   (a) Find the average power dissipated as a function of A, τ, and R.
   (b) Find the average power dissipated by the nth harmonic in $v(t)$.

19. (a) Derive the complex Fourier coefficients of a time-shifted periodic impulse train

$$v(t) = \sum_{n=-\infty}^{+\infty} \delta(t - nT - t_d)$$

where t_d is arbitrary.

(b) Plot the two-sided amplitude, phase, and power spectra.

(c) Use your answer from (a) to determine the Fourier amplitudes and phases A_n and ϕ_n.

(d) Plot the one-sided amplitude, phase, and power spectra.

(e) Use your answer to (a) to obtain the real Fourier coefficients a_n and b_n.

(f) Write the three forms of the Fourier series of $v(t)$.

Section 15.6

20. A waveform

$$v_a(t) = \sum_{n=-\infty}^{+\infty} V_{an} e^{j2\pi n f_1 t}$$

with period $T = 1/f_1$ is time-delayed $T/2$ s to produce another waveform $v_b(t) = v_a(t - T/2)$. The waveform $v_b(t)$ is then subtracted from $v_a(t)$ to produce a third waveform $v_c(t) = v_a(t) - v_b(t)$. An illustration of this procedure is given in Fig. 15.30 for the special case that $v_a(t)$ is a periodically repeated exponential signal.

(a) Use the time-shift property to show that, in general, the nth complex Fourier-series coefficient of $v_c(t)$ is given by

$$V_{cn} = \begin{cases} 2V_{an} & n \text{ odd} \\ 0 & n \text{ even} \end{cases}$$

(b) Does $v_c(t)$ (in general) have half-wave symmetry?

(c) Show that the nth complex Fourier-series coefficient of the alternating exponential waveform $v_c(t)$ shown in Fig. 15.30 is

$$V_{cn} = \begin{cases} \dfrac{2A\tau}{T} \dfrac{1 - e^{-T/\tau}}{1 + j2\pi n\tau/T} & n \text{ odd} \\ 0 & n \text{ even} \end{cases}$$

21. Use any method to verify or derive FS1 of Table 15.1.

22. Use any method to verify or derive FS2 of Table 15.1.

23. Use any method to verify or derive FS5 of Table 15.1.

24. Use any method to verify or derive FS6 of Table 15.1

25. Use any method to verify or derive FS7 of Table 15.1.

26. Use any method to verify or derive FS8 of Table 15.1.

27. Use any method to verify or derive FS9 of Table 15.1.

28. Let $x(t)$ and $y(t)$ be two periodic functions with period T. Show that

$$\frac{1}{T} \int_{t_0}^{t_0+T} x(t) y^*(t)\, dt = \sum_{n=-\infty}^{+\infty} X_n Y_n^*$$

where X_n and Y_n are the complex Fourier-series coefficients $x(t)$ and $y(t)$, respectively.

29. Let $x(t)$ and $y(t)$ be two periodic functions with period T. Define a new function

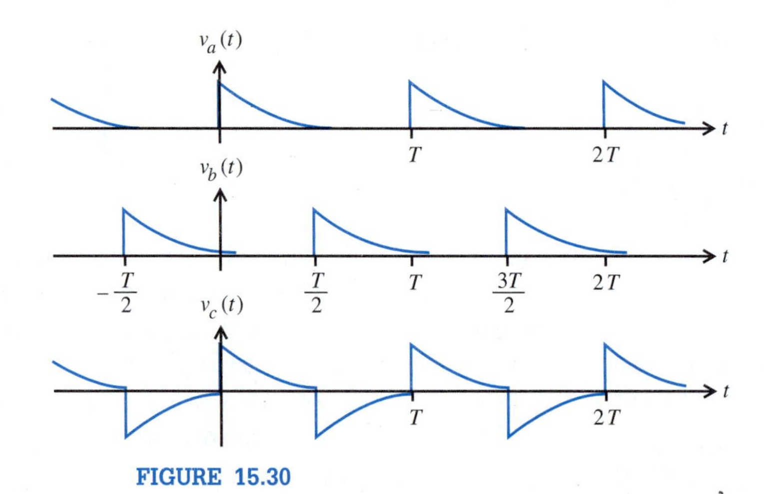

FIGURE 15.30

$$z(t) = \frac{1}{T}\int_{t_0}^{t_0+T} x(\lambda)y(t-\lambda)\,d\lambda$$

where t_0 is arbitrary. The function $z(t)$ is referred to as the *convolution* of the periodic functions $x(t)$ and $y(t)$.

(a) Show that $z(t)$ is periodic with period T.

(b) Show that the nth complex Fourier-series coefficient of $z(t)$, Z_n, is given by $Z_n = X_nY_n$, where X_n and Y_n are the nth complex Fourier-series coefficients of $x(t)$ and $y(t)$, respectively.

Section 15.7

30. A low-pass RC circuit has a periodic pulse train input, where $A = 1$ V, $\tau = 0.3$ ms, and $T = 1$ ms. The resistance is $R = 2$ kΩ. Choose C so that the fundamental component of the output waveform is 100 times smaller than the dc component of the output waveform.

31. A series band-pass RLC filter uses a resistance $R = 2.2$ kΩ.

(a) Choose the values of L and C to pass the fifth harmonic of a 10-kHz input triangular wave without attenuation and so that the seventh output harmonic is 25 times smaller in amplitude than the fifth.

(b) Make a rough sketch of the output to illustrate the predominant shape of the output waveform.

32. The full-wave-rectified cosine $v(t)$ of FS9 (Table 15.1) is applied to a low-pass RC filter.

(a) Use FS9 to write the real Fourier series of $v(t)$. Put your result in the form of Eq. (15.18). Assume that $A = 120\sqrt{2}$ and $T_0 = \frac{1}{60}$ s, and give the numerical values of the dc term and the amplitudes and phases of the first five harmonics.

(b) Plot the one-sided amplitude and phase spectra of $v(t)$ for $A = 120\sqrt{2}$ and $T_0 = \frac{1}{60}$ s.

(c) Find the voltage transfer function of the filter and plot the filter's amplitude and phase characteristics. Assume for your plots that $RC = 1/(120\pi)$ s.

(d) Plot the one-sided amplitude and phase spectra of the output $v_0(t)$ for the numerical values of (a) through (c).

(e) Write the series expression for the filter output, giving the numerical values of the dc term and the amplitudes and phases of the first five harmonics.

33. A hypothetical circuit called an *ideal low-pass filter* has the voltage transfer function illustrated in the figure. The input voltage is the half-rectified cosine of Table 15.1, where $A = 1$ V and $T = 1$ ms. Determine and sketch the output voltage for $W = 500$ Hz.

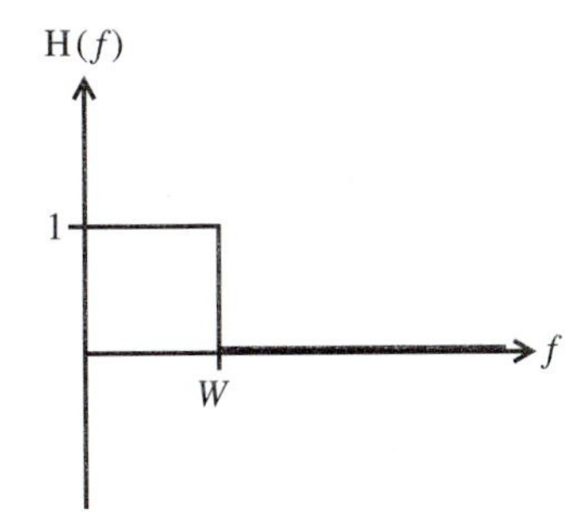

Section 15.9

34. (a) Determine the values of α and ϕ in the following circuit to minimize the average power dissipated in the resistance. The current $i_s(t)$ is an arbitrary periodic waveform with period T.

(b) What is the resulting minimum average power dissipation?

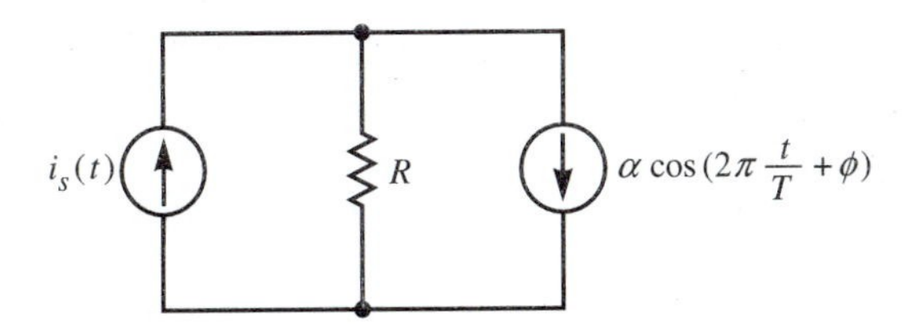

35. In what sense is

(a) Plancherel's theorem *more general* than the Dirichlet conditions?

(b) Plancherel's theorem weaker than the Dirichlet conditions?

36. Let $\{\phi_i(t),\ i = 1, 2, \ldots, I\}$ be any set of real functions satisfying

$$\int_{t_0}^{t_0+T} \phi_i(t)\phi_k(t)\,dt = \begin{cases} c_i & i = k \\ 0 & i \neq k \end{cases}$$

where c_i is a constant that depends on i. Let $v(t)$ be any waveform, and let $\hat{v}_I(t)$ be an approximation to $v(t)$:

$$\hat{v}_I(t) = \sum_{i=1}^{I} \alpha_i \phi_i(t)$$

(a) Show that the coefficients α_i minimizing the integral squared error

$$\varepsilon = \int_{t_0}^{t_0+T} [v(t) - \hat{v}_I(t)]^2 \, dt$$

are

$$\alpha_i = \frac{1}{c_i} \int_{t_0}^{t_0+T} v(t)\phi_i(t) \, dt; \; i = 1, 2, \ldots, I$$

(b) Show that

$$\varepsilon_{\min} = \int_{t_0}^{t_0+T} v^2(t) \, dt - \sum_{i=1}^{I} c_i \alpha_i^2$$

(c) Show that

$$\varepsilon_{\min} = \int_{t_0}^{t_0+T} v^2(t) \, dt - \int_{t_0}^{t_0+T} \hat{v}_I^2(t) \, dt$$

Section 15.10

37. (a) Find $V(f)$ if $v(t) = e^{-t/\tau}[u(t) - u(t - T)]$.
    (b) Sketch $v(t)$, $|V(f)|$, and $\underline{/V(f)}$ for $\tau = T = 1$ ms.

38. Show that if $v(t)$ is a real function, then $V(-f) = V^*(f)$. [*Hint:* This important result follows directly from Eq. (15.72).]

39. A rectangular pulse, $v(t) = \Pi(t/\tau)$, is repeated every T s, where $T = 5\tau$, to yield a periodic signal $v_T(t)$.
    (a) Write down the Fourier transform of $v(t)$ and the Fourier series of $v_T(t)$.
    (b) Sketch $v(t)$, $V(f)$, $v_T(t)$, and the spectrum of $v_T(t)$.
    (c) Discuss your results.

Supplementary Problems

40. Frequencies ω_1 and ω_2 are present in the waveform

$$v_s(t) = a \cos(\omega_1 t + \phi_1) + b \cos(\omega_2 t + \phi_2)$$

(a) Assume that $v_s(t)$ is the input to an LTI circuit. Show that the same frequencies ω_1 and ω_2 are present in the output $v_o(t)$.

(b) Assume that $v_s(t)$ is the input to a *square-law device* whose output is $v_o(t) = cv_s^2(t)$, where c is a real nonnegative constant. What frequencies are present in $v_o(t)$?

(c) Assume that $v_s(t)$ appears at the input to a modulator (multiplier), shown schematically below. The modulator output is $v_o(t) = v_s(t) \cos \omega_c t$. What frequencies are present in $v_o(t)$?

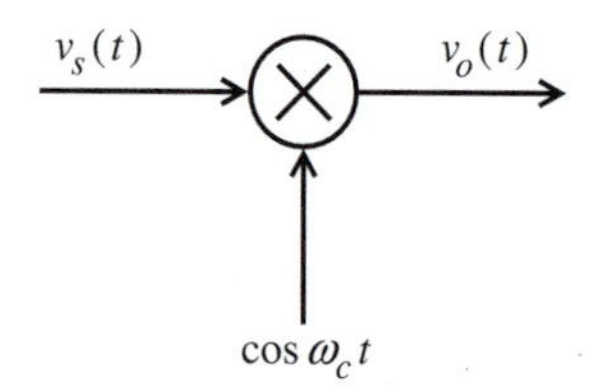

16

Equivalent Circuits for Three-Terminal Networks and Two-Port Networks

In Chapter 11 we showed that an arbitrary two-terminal network of sources and impedances can be represented by a Thévenin or Norton equivalent. In this chapter we will generalize the equivalent-circuit concept further to include arbitrary *three-terminal* networks composed of sources and impedances, and an important class of four-terminal networks called *two-port* networks.

The equivalent circuits developed here are used in electronics as linear models for physical devices. They are also used to simplify the analysis of circuits used in communication and power-transmission systems.

16.1 Equivalent Circuits for Three-Terminal Networks

Figure 16.1 depicts an arbitrary three-terminal LTI ac network or device containing impedances, dependent sources, and independent sources. Phasor currents $\mathbf{I}_1$ and $\mathbf{I}_2$ enter at terminals 1 and 2. Phasor voltage drops $\mathbf{V}_1$ and $\mathbf{V}_2$ appear from terminals 1 and 2, respectively, to terminal 0.

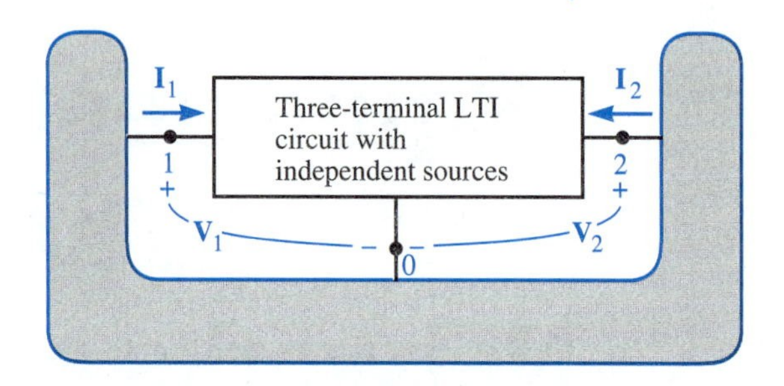

FIGURE 16.1
A general three-terminal network

We can develop equivalent circuits for the three-terminal network of Fig. 16.1 by regarding various pairs of the variables $\mathbf{I}_1$, $\mathbf{I}_2$, $\mathbf{V}_1$, and $\mathbf{V}_2$ as independent variables (inputs) and the remaining pairs as dependent variables (outputs). The possible combinations are listed in Table 16.1 along with the equivalent-circuit parameters with which they are associated.† The derivations of the first three equivalent circuits listed are in Sections 16.2, 16.3, and 16.6. The hybrid-**g** equivalent circuit, listed last, is developed in Problem 20.

TABLE 16.1
Viewpoints leading to three-terminal equivalent circuits

Variables Treated as Inputs	*Variables Treated as Outputs*	*Resulting Equivalent Circuit Parameters*
1. $\mathbf{I}_1, \mathbf{I}_2$	$\mathbf{V}_1, \mathbf{V}_2$	Impedance
2. $\mathbf{V}_1, \mathbf{V}_2$	$\mathbf{I}_1, \mathbf{I}_2$	Admittance
3. $\mathbf{I}_1, \mathbf{V}_2$	$\mathbf{I}_2, \mathbf{V}_1$	Hybrid-**h**
4. $\mathbf{I}_2, \mathbf{V}_1$	$\mathbf{I}_1, \mathbf{V}_2$	Hybrid-**g**

16.2 Three-Terminal Impedance-Parameter Equivalent Circuits

To derive the impedance-parameter equivalent network of the three-terminal LTI circuit of Fig. 16.1, we take the viewpoint that voltages $\mathbf{V}_1$ and $\mathbf{V}_2$ are dependent variables or

† No source exists that can simultaneously apply independent values of both driving-point voltage and driving-point current to a network. Therefore the pairs $(\mathbf{I}_1, \mathbf{V}_1)$ and $(\mathbf{I}_2, \mathbf{V}_2)$ are excluded from the first column of Table 16.1.

''outputs.'' The independent variables are taken to be the currents $\mathbf{I}_1$ and $\mathbf{I}_2$ (these are the ''inputs'') and the independent sources within the rectangle. Since the network is linear, we know from the superposition principle that outputs $\mathbf{V}_1$ and $\mathbf{V}_2$ can be obtained by linearly superimposing the voltage contributions caused by the independent sources acting one at a time. This means that we can write $\mathbf{V}_1$ and $\mathbf{V}_2$ as

$$\begin{bmatrix} \mathbf{V}_1 \\ \mathbf{V}_2 \end{bmatrix} = \begin{bmatrix} \mathbf{z}_{11} & \mathbf{z}_{12} \\ \mathbf{z}_{21} & \mathbf{z}_{22} \end{bmatrix} \begin{bmatrix} \mathbf{I}_1 \\ \mathbf{I}_2 \end{bmatrix} + \begin{bmatrix} \mathbf{V}_{1oc} \\ \mathbf{V}_{2oc} \end{bmatrix} \tag{16.1}$$

where the $\mathbf{z}_{ij}$ are proportionality constants and $\mathbf{V}_{1oc}$ and $\mathbf{V}_{2oc}$ represent the contributions from the independent sources inside the box. (The meaning of the subscripts will soon be clear.) We can derive formulas for the proportionality constants and $\mathbf{V}_{1oc}$ and $\mathbf{V}_{2oc}$ by recalling the following simple fact: The response caused by each independent source acting alone is found by setting the other independent sources to zero.

The components of $\mathbf{V}_1$ and $\mathbf{V}_2$ caused by the independent sources *within* the network can be found if we set $\mathbf{I}_1 = \mathbf{I}_2 = 0$ (open terminals 1 and 2), as illustrated in Fig. 16.2. The resulting values of $\mathbf{V}_1$ and $\mathbf{V}_2$ are called *open-circuit voltages:*

$$\mathbf{V}_{1oc} = \mathbf{V}_1 \Big|_{\substack{\mathbf{I}_1=0 \\ \mathbf{I}_2=0}} \tag{16.2}$$

$$\mathbf{V}_{2oc} = \mathbf{V}_2 \Big|_{\substack{\mathbf{I}_1=0 \\ \mathbf{I}_2=0}} \tag{16.3}$$

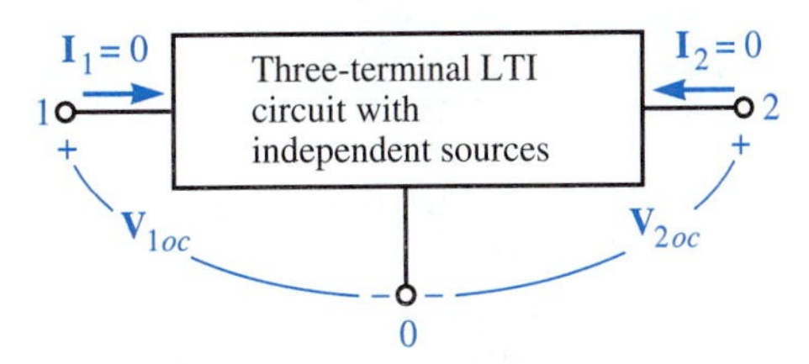

FIGURE 16.2 $\mathbf{V}_{1oc}$ and $\mathbf{V}_{2oc}$ are determined by setting $\mathbf{I}_1 = \mathbf{I}_2 = 0$.

The components of $\mathbf{V}_1$ and $\mathbf{V}_2$ due to the current $\mathbf{I}_1$ are determined by setting $\mathbf{I}_2 = 0$ (open terminal 2) and setting all independent sources inside the network to zero, as illustrated in Fig. 16.3. It follows from an inspection of this figure that the contribution to $\mathbf{V}_1$ due to $\mathbf{I}_1$ is

$$\mathbf{V}_1 \Big|_{\substack{\mathbf{I}_2=0 \\ \text{IIS}=0}} = \mathbf{z}_{11}\mathbf{I}_1 \tag{16.4}$$

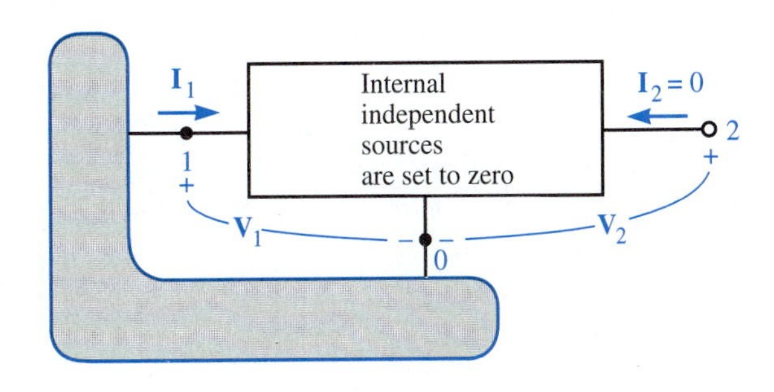

FIGURE 16.3 The open-circuit parameters $\mathbf{z}_{11}$ and $\mathbf{z}_{21}$ are determined by setting $\mathbf{I}_2$ and the internal independent sources to zero.

where IIS = 0 denotes that the independent internal sources are set to zero and where $\mathbf{z}_{11}$ is the network's driving-point impedance looking into terminals 1 and 0 when terminal 2 is open-circuited. We can now solve Eq. (16.4) for $\mathbf{z}_{11}$. We can calculate the remaining proportionality constants in a similar way to obtain:

$$\mathbf{z}_{11} = \left.\frac{\mathbf{V}_1}{\mathbf{I}_1}\right|_{\substack{\mathbf{I}_2=0\\ \text{IIS}=0}} \tag{16.5}$$

$$\mathbf{z}_{21} = \left.\frac{\mathbf{V}_2}{\mathbf{I}_1}\right|_{\substack{\mathbf{I}_2=0\\ \text{IIS}=0}} \tag{16.6}$$

$$\mathbf{z}_{12} = \left.\frac{\mathbf{V}_1}{\mathbf{I}_2}\right|_{\substack{\mathbf{I}_1=0\\ \text{IIS}=0}} \tag{16.7}$$

$$\mathbf{z}_{22} = \left.\frac{\mathbf{V}_2}{\mathbf{I}_2}\right|_{\substack{\mathbf{I}_1=0\\ \text{IIS}=0}} \tag{16.8}$$

We will show how to apply Eqs. (16.5) through (16.8) in Example 16.1.

The parameters $\mathbf{z}_{11}$, $\mathbf{z}_{21}$, $\mathbf{z}_{12}$, and $\mathbf{z}_{22}$ are called the *open-circuit impedance or* **z** *parameters* of the three-terminal network. The parameters are impedances since each parameter equals the ratio of phasor voltage to phasor current. They are *open-circuit* parameters because each parameter is found by setting either $\mathbf{I}_1$ or $\mathbf{I}_2$ equal to zero (open circuit). With the obvious definitions, Eq. (16.1) can be written as a matrix equation:

$$\mathbf{V} = \mathbf{ZI} + \mathbf{V}_{oc} \tag{16.9}$$

As far as external measurements are concerned, the three-terminal network of Fig. 16.1 is completely characterized by its open-circuit voltages $\mathbf{V}_{1oc}$ and $\mathbf{V}_{2oc}$ and its impedance parameters $\mathbf{z}_{11}$, $\mathbf{z}_{12}$, $\mathbf{z}_{21}$, and $\mathbf{z}_{22}$.

Consider now the circuit of Fig. 16.4. We can see by inspection that the terminal equations of this circuit are also given by Eq. (16.1). Consequently, the network of Fig. 16.1 can be replaced by the network of Fig. 16.4, called the *three-terminal* **z***-parameter equivalent.*

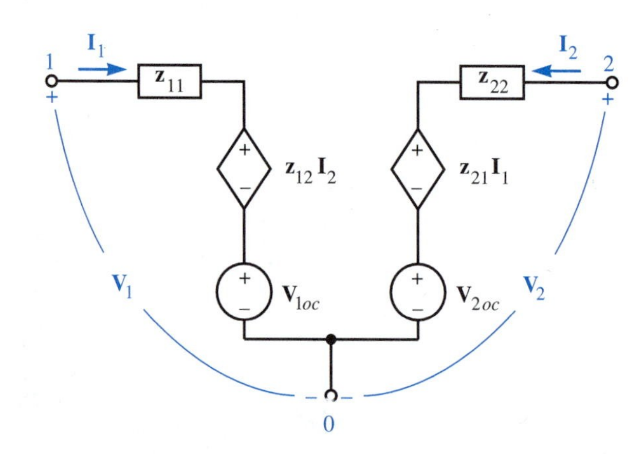

FIGURE 16.4 Three-terminal impedance-parameter equivalent network

An important special case arises when the network of Fig. 16.1 contains no independent sources. For this special case, $\mathbf{V}_{oc} = \mathbf{0}$, and the independent voltage sources in Fig. 16.1 are equivalent to short circuits.

For *reciprocal* three-terminal networks, $\mathbf{z}_{12} = \mathbf{z}_{21}$. All three-terminal networks composed of only *independent* sources, resistances, capacitances, inductances, and coupled inductances are reciprocal.

As illustrated by the following example, there is more than one way to determine an impedance-parameter equivalent circuit.

EXAMPLE 16.1

Find $\mathbf{V}_{1oc}$, $\mathbf{V}_{2oc}$, and the $\mathbf{z}$ parameters of the network in Fig. 16.5 by (a) direct application of Eqs. (16.2), (16.3), and (16.5) through (16.8) and (b) mesh-current analysis.

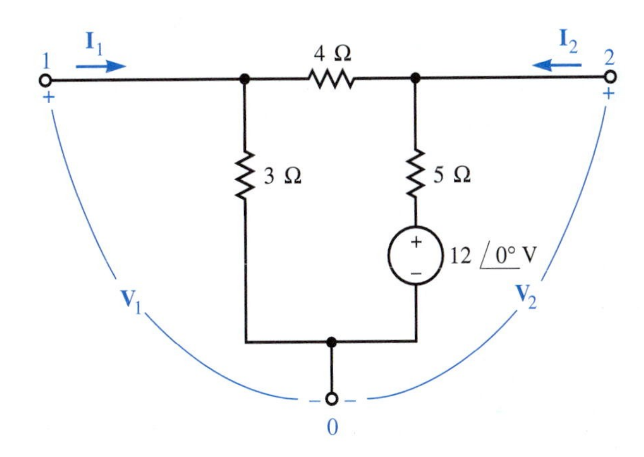

FIGURE 16.5 Three-terminal circuit

Solution (a) We will first find $\mathbf{V}_{1oc}$ and $\mathbf{V}_{2oc}$, which are the values of $\mathbf{V}_1$ and $\mathbf{V}_2$, respectively, when $\mathbf{I}_1 = \mathbf{I}_2 = 0$. We set $\mathbf{I}_1 = \mathbf{I}_2 = 0$ in Fig. 16.5 and use a voltage-divider equation to obtain

See Problem 16.1 in the PSpice manual.

$$\mathbf{V}_{1oc} = \frac{3}{3+4+5}\,12\underline{/0^\circ} = 3\underline{/0^\circ}\text{ V}$$

and

$$\mathbf{V}_{2oc} = \frac{3+4}{3+4+5}\,12\underline{/0^\circ} = 7\underline{/0^\circ}\text{ V}$$

The impedance parameters follow from Eqs. (16.5) through (16.8). The values of $\mathbf{z}_{11}$ and $\mathbf{z}_{21}$ can be found from Eqs. (16.5) and (16.6) and inspection of Fig. 16.6:

$$\mathbf{z}_{11} = \left.\frac{\mathbf{V}_1}{\mathbf{I}_1}\right|_{\substack{\mathbf{I}_2=0\\ \text{IIS}=0}} = 3 \parallel (4+5) = \frac{9}{4}\ \Omega$$

$$\mathbf{z}_{21} = \left.\frac{\mathbf{V}_2}{\mathbf{I}_1}\right|_{\substack{\mathbf{I}_2=0\\ \text{IIS}=0}} = \left.\frac{\mathbf{V}_2}{\mathbf{V}_1}\frac{\mathbf{V}_1}{\mathbf{I}_1}\right|_{\substack{\mathbf{I}_2=0\\ \text{IIS}=0}} = \left(\frac{5}{4+5}\right)\left(\frac{9}{4}\right) = \frac{5}{4}\ \Omega$$

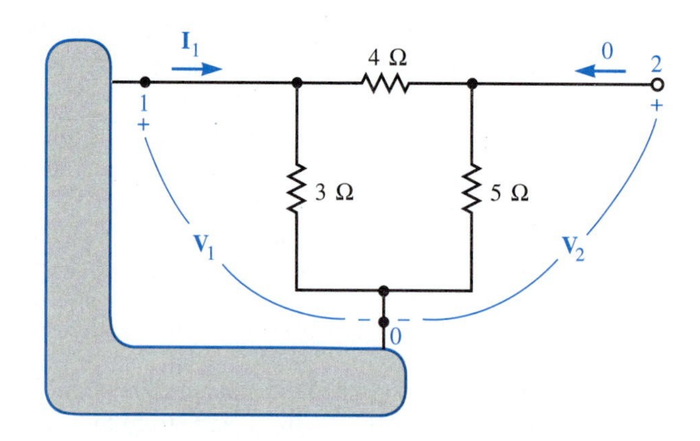

FIGURE 16.6
Circuit for finding $\mathbf{z}_{11}$ and $\mathbf{z}_{21}$

Similarly, from Fig. 16.7 and Eqs. (16.7) and (16.8),

$$\mathbf{z}_{22} = \left.\frac{\mathbf{V}_2}{\mathbf{I}_2}\right|_{\substack{\mathbf{I}_1=0 \\ \text{IIS}=0}} = (3 + 4) \parallel 5 = \frac{35}{12}\ \Omega$$

and

$$\mathbf{z}_{12} = \left.\frac{\mathbf{V}_1}{\mathbf{I}_2}\right|_{\substack{\mathbf{I}_1=0 \\ \text{IIS}=0}} = \left.\frac{\mathbf{V}_1}{\mathbf{V}_2}\frac{\mathbf{V}_2}{\mathbf{I}_2}\right|_{\substack{\mathbf{I}_1=0 \\ \text{IIS}=0}} = \left(\frac{3}{3+4}\right)\left(\frac{35}{12}\right) = \Omega$$

Notice that $\mathbf{z}_{12} = \mathbf{z}_{21}$.

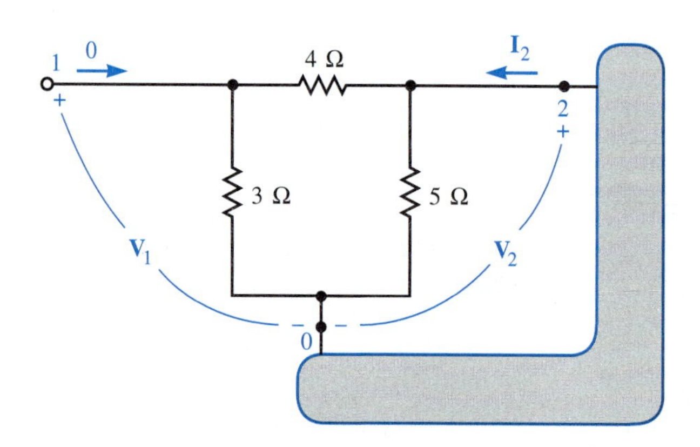

FIGURE 16.7
Circuit for finding $\mathbf{z}_{22}$ and $\mathbf{z}_{12}$

(b) We define mesh currents $\mathbf{I}_1$, $\mathbf{I}_2$, and $\mathbf{I}_3$ as shown in Fig. 16.8. The mesh-current equations are

$$\begin{aligned} 3\mathbf{I}_1 + 0\mathbf{I}_2 - 3\mathbf{I}_3 &= \mathbf{V}_1 \\ 0\mathbf{I}_1 + 5\mathbf{I}_2 + 5\mathbf{I}_3 &= \mathbf{V}_2 - 12\underline{/0^\circ} \\ -3\mathbf{I}_1 + 5\mathbf{I}_2 + 12\mathbf{I}_3 &= -12\underline{/0^\circ} \end{aligned}$$

By elimination of $\mathbf{I}_3$ these become

$$\mathbf{V}_1 = \frac{9}{4}\mathbf{I}_1 + \frac{5}{4}\mathbf{I}_2 + 3\underline{/0^\circ}$$

$$\mathbf{V}_2 = \frac{5}{4}\mathbf{I}_1 + \frac{35}{12}\mathbf{I}_2 + 7\underline{/0^\circ}$$

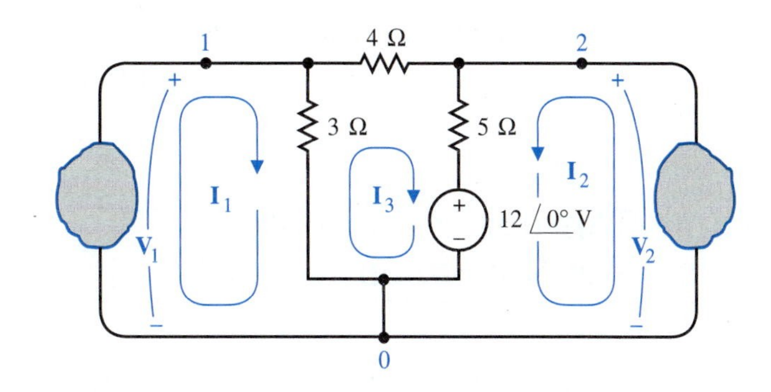

FIGURE 16.8
Mesh-current definitions

which is, in matrix form,

$$\begin{bmatrix} \mathbf{V}_1 \\ \mathbf{V}_2 \end{bmatrix} = \begin{bmatrix} \frac{9}{4} & \frac{5}{4} \\ \frac{5}{4} & \frac{35}{12} \end{bmatrix} \begin{bmatrix} \mathbf{I}_1 \\ \mathbf{I}_2 \end{bmatrix} + \begin{bmatrix} 3\angle 0^\circ \\ 7\angle 0^\circ \end{bmatrix}$$

Since this has the form of Eq. (16.9), we can read the impedance parameters and open-circuit voltages directly from it. We find as in (a) that $\mathbf{z}_{11} = \frac{9}{4}\ \Omega$, $\mathbf{z}_{12} = \mathbf{z}_{21} = \frac{5}{4}\ \Omega$, $\mathbf{z}_{22} = \frac{35}{12}\ \Omega$, $\mathbf{V}_{1oc} = 3\angle 0^\circ$ V, and $\mathbf{V}_{2oc} = 7\angle 0^\circ$ V.

Remember

We derive the impedance-parameter equivalent circuit by regarding $\mathbf{I}_1$ and $\mathbf{I}_2$ as independent variables.

EXERCISES

1. Repeat Example 16.1 (a) with the $+12\angle 0^\circ$-V independent source replaced with a current-controlled voltage source with terminal equation $\mathbf{V} = 2\mathbf{I}_1$.
   answer: $\mathbf{z}_{11} = \frac{11}{4}\ \Omega$, $\mathbf{z}_{12} = \frac{5}{4}\ \Omega$, $\mathbf{z}_{21} = \frac{29}{12}\ \Omega$, $\mathbf{z}_{22} = \frac{35}{12}\ \Omega$, $\mathbf{V}_{1oc} = \mathbf{V}_{2oc} = 0$ V
2. Repeat Example 16.1 (b) with the $+12\angle 0^\circ$-V independent source replaced with a current-controlled voltage source with terminal equation $\mathbf{V} = 2\mathbf{I}_1$.
   answer: $\mathbf{z}_{11} = \frac{11}{4}\ \Omega$, $\mathbf{z}_{12} = \frac{5}{4}\ \Omega$, $\mathbf{z}_{21} = \frac{29}{12}\ \Omega$, $\mathbf{z}_{22} = \frac{35}{12}\ \Omega$, $\mathbf{V}_{1oc} = \mathbf{V}_{2oc} = 0$ V
3. Determine the three-terminal **z**-parameter equivalent circuit of the network shown by direct application of Eqs. (16.2), (16.3), and (16.5) through (16.8).
   answer: $\mathbf{z}_{11} = 1\ \Omega$, $\mathbf{z}_{12} = 1\ \Omega$, $\mathbf{z}_{21} = 1\ \Omega$, $\mathbf{z}_{22} = 2 + j\ \Omega$, $\mathbf{V}_{1oc} = 5\angle 0^\circ$ V, $\mathbf{V}_{2oc} = 10\angle 0^\circ$ V

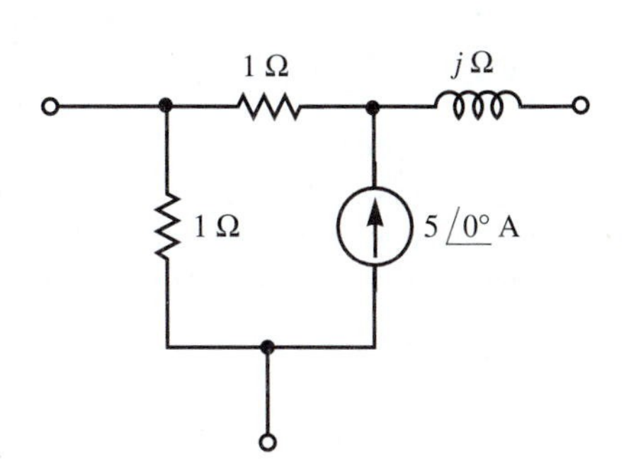

4. Repeat Exercise 3 using mesh-current analysis.

answer: $\mathbf{z}_{11} = 1\ \Omega$, $\mathbf{z}_{12} = 1\ \Omega$, $\mathbf{z}_{21} = 1\ \Omega$, $\mathbf{z}_{22} = 2 + j\ \Omega$, $\mathbf{V}_{1oc} = 5\underline{/0^\circ}$ V, $\mathbf{V}_{2oc} = 10\underline{/0^\circ}$ V

16.3 Three-Terminal Admittance-Parameter Equivalent Circuits

Another way to obtain terminal equations for the original three-terminal network of Fig. 16.1 is to regard the currents $\mathbf{I}_1$ and $\mathbf{I}_2$ as the outputs caused by two independent voltage sources $\mathbf{V}_1$ and $\mathbf{V}_2$ and the independent sources inside the network. Here we use the principle of superposition to write $\mathbf{I}_1$ and $\mathbf{I}_2$ as a linear combination of current components caused by the external voltage sources and the independent sources inside the network. That is, we write $\mathbf{I}_1$ and $\mathbf{I}_2$ as

$$\begin{bmatrix} \mathbf{I}_1 \\ \mathbf{I}_2 \end{bmatrix} = \begin{bmatrix} \mathbf{y}_{11} & \mathbf{y}_{12} \\ \mathbf{y}_{21} & \mathbf{y}_{22} \end{bmatrix} \begin{bmatrix} \mathbf{V}_1 \\ \mathbf{V}_2 \end{bmatrix} - \begin{bmatrix} \mathbf{I}_{1sc} \\ \mathbf{I}_{2sc} \end{bmatrix} \tag{16.10}$$

where the $\mathbf{y}_{ij}$ are proportionality constants and where $\mathbf{I}_{1sc}$ and $\mathbf{I}_{2sc}$ represent *outward* current contributions due to the independent sources inside the box.

The components arising from the independent sources within the network are obtained by setting voltages $\mathbf{V}_1 = \mathbf{V}_2 = 0$ (connect terminals 1 and 2 to terminal 0) as illustrated in Fig. 16.9. The resulting currents flowing from terminals 1 and 2 to terminal 0 are called *short-circuit currents* $\mathbf{I}_{1sc}$ and $\mathbf{I}_{2sc}$, respectively. Thus, comparing the sign convention of Fig. 16.1,

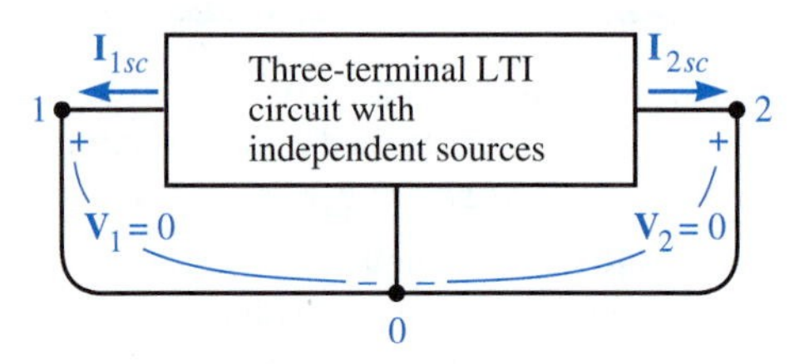

FIGURE 16.9 $\mathbf{I}_{1sc}$ and $\mathbf{I}_{2sc}$ are determined by setting $\mathbf{V}_1 = \mathbf{V}_2 = 0$.

$$\mathbf{I}_{1sc} = -\mathbf{I}_1 \Big|_{\substack{\mathbf{V}_1=0\\ \mathbf{V}_2=0}} \tag{16.11}$$

$$\mathbf{I}_{2sc} = -\mathbf{I}_2 \Big|_{\substack{\mathbf{V}_1=0\\ \mathbf{V}_2=0}} \tag{16.12}$$

The components in terminal currents $\mathbf{I}_1$ and $\mathbf{I}_2$ caused by the voltage source $\mathbf{V}_1$ can be determined if we set $\mathbf{V}_2 = 0$ (connect terminal 2 to terminal 0) and set the independent sources inside the network to zero (IIS = 0) as shown in Fig. 16.10. Then, by inspection of Fig. 16.10, the contribution to $\mathbf{I}_1$ from $\mathbf{V}_1$ is

$$\mathbf{I}_1 \Big|_{\substack{\mathbf{V}_2=0\\ \text{IIS}=0}} = \mathbf{y}_{11}\mathbf{V}_1 \tag{16.13}$$

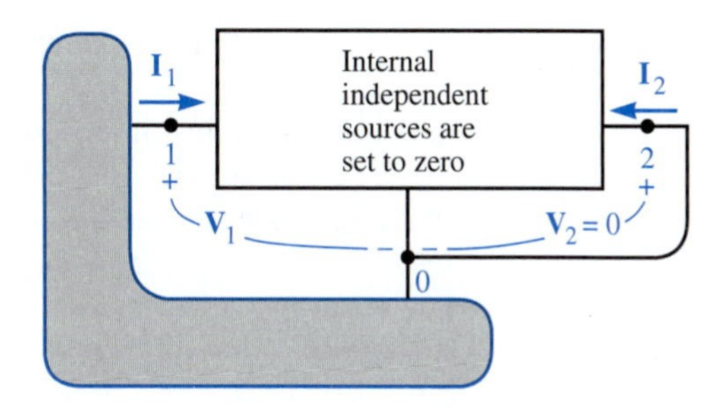

FIGURE 16.10 The short-circuit parameters. $\mathbf{y}_{11}$ and $\mathbf{y}_{21}$ are determined by setting $\mathbf{V}_2$ and the internal independent sources to zero.

where $\mathbf{y}_{11}$ is the network's driving-point admittance looking into terminals 1 and 0 when terminal 2 is shorted to terminal 0. Equation (16.13) can easily be solved for $\mathbf{y}_{11}$. We can calculate the remaining admittance parameters in a similar way to obtain

$$\mathbf{y}_{11} = \frac{\mathbf{I}_1}{\mathbf{V}_1}\bigg|_{\substack{\mathbf{V}_2=0\\ \text{IIS}=0}} \tag{16.14}$$

$$\mathbf{y}_{21} = \frac{\mathbf{I}_2}{\mathbf{V}_1}\bigg|_{\substack{\mathbf{V}_2=0\\ \text{IIS}=0}} \tag{16.15}$$

$$\mathbf{y}_{12} = \frac{\mathbf{I}_1}{\mathbf{V}_2}\bigg|_{\substack{\mathbf{V}_1=0\\ \text{IIS}=0}} \tag{16.16}$$

$$\mathbf{y}_{22} = \frac{\mathbf{I}_2}{\mathbf{V}_2}\bigg|_{\substack{\mathbf{V}_1=0\\ \text{IIS}=0}} \tag{16.17}$$

The parameters $\mathbf{y}_{11}$, $\mathbf{y}_{12}$, $\mathbf{y}_{21}$, and $\mathbf{y}_{22}$ are called the *short-circuit admittance parameters* of the three-terminal network of Fig. 16.1. For a reciprocal network, $\mathbf{y}_{12} = \mathbf{y}_{21}$. With the obvious definitions, Eq. (16.10) has the form

$$\mathbf{I} = \mathbf{YV} - \mathbf{I}_{sc} \tag{16.18}$$

As far as external measurements are concerned, the original network is completely characterized by its short-circuit currents $\mathbf{I}_{1sc}$ and $\mathbf{I}_{2sc}$ and its short-circuit admittance parameters.

We can see by inspection that the terminal equations of the circuit of Fig. 16.11 are also given by Eqs. (16.11) and (16.12). Consequently, the network of Fig. 16.1 can be

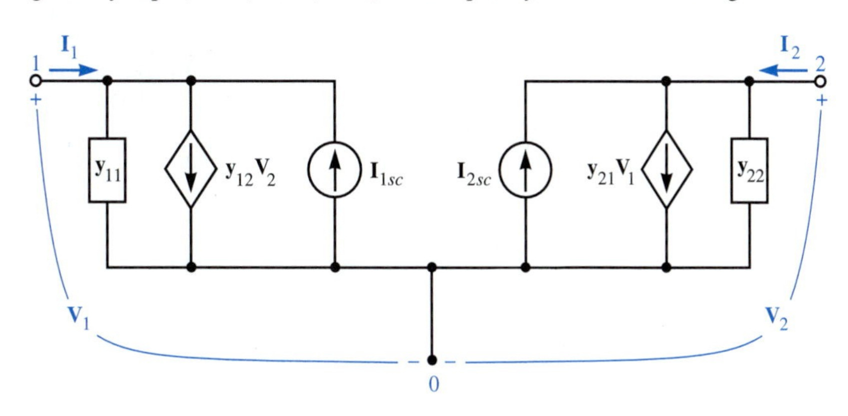

FIGURE 16.11 Three-terminal admittance-parameter equivalent network

replaced by the network of Fig. 16.11, called the *three-terminal* **y**-*parameter equivalent network.*

As with the three-terminal impedance-parameter equivalent, an important special case arises if the network of Fig. 16.1 contains no independent sources. For this special case, $\mathbf{I}_{sc} = \mathbf{0}$ and the independent current sources in Fig. 16.11 are equivalent to open circuits.

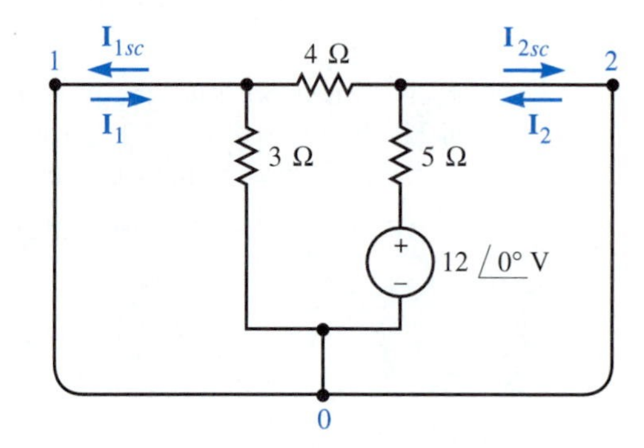

FIGURE 16.12 Circuit for finding $\mathbf{I}_{1sc}$ and $\mathbf{I}_{2sc}$

EXAMPLE 16.2 Find $\mathbf{I}_{1sc}$, $\mathbf{I}_{2sc}$, and the admittance parameters of the network of Example 16.1 by (a) direct application of Eqs. (16.11), (16.12), and (16.14) through (16.17) and (b) node-voltage analysis.

Solution (a) We will first find the short-circuit currents. It follows from Eqs. (16.11) and (16.12) and Fig. 16.12 that

$$\mathbf{I}_{1sc} = -\mathbf{I}_1 \Big|_{\substack{\mathbf{V}_1=0 \\ \mathbf{V}_2=0}} = 0 \text{ A}$$

and

$$\mathbf{I}_{2sc} = -\mathbf{I}_2 \Big|_{\substack{\mathbf{V}_1=0 \\ \mathbf{V}_2=0}} = \frac{12}{5} \underline{/0^\circ} \text{ A}$$

The values of $\mathbf{y}_{11}$ and $\mathbf{y}_{21}$ can be obtained by an inspection of Fig. 16.13 and Eqs. (16.14) and (16.15):

$$\mathbf{y}_{11} = \frac{\mathbf{I}_1}{\mathbf{V}_1} \Big|_{\substack{\mathbf{V}_2=0 \\ \text{IIS}=0}}$$

$$= \frac{1}{3} + \frac{1}{4} = \frac{7}{12} \text{ S}$$

and

$$\mathbf{y}_{21} = \frac{\mathbf{I}_2}{\mathbf{V}_1} \Big|_{\substack{\mathbf{V}_2=0 \\ \text{IIS}=0}}$$

$$= -\frac{1}{4} \text{ S}$$

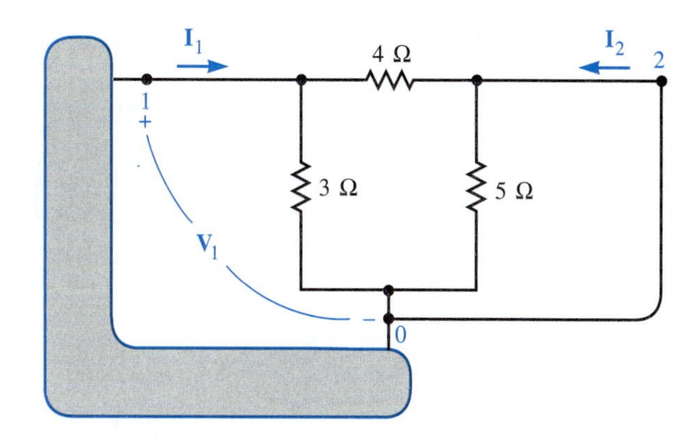

FIGURE 16.13 Circuit for finding $\mathbf{y}_{11}$ and $\mathbf{y}_{21}$

Similarly, we find from Fig. 16.14 and Eqs. (16.16) and (16.17):

$$\mathbf{y}_{12} = \frac{\mathbf{I}_1}{\mathbf{V}_2}\bigg|_{\substack{\mathbf{V}_1=0 \\ \text{IIS}=0}} = -\frac{1}{4}\ \text{S}$$

and

$$\mathbf{y}_{22} = \frac{\mathbf{I}_2}{\mathbf{V}_2}\bigg|_{\substack{\mathbf{V}_1=0 \\ \text{IIS}=0}} = \frac{1}{5} + \frac{1}{4} = \frac{9}{20}\ \text{S}$$

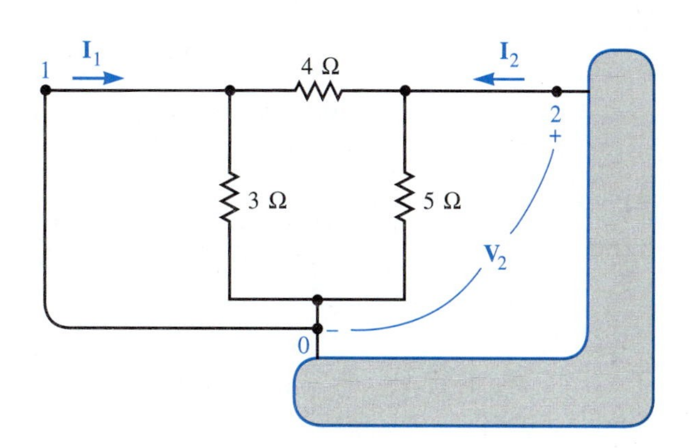

FIGURE 16.14 Circuit for finding $\mathbf{y}_{12}$ and $\mathbf{y}_{22}$

(b) The node-voltage definitions are shown in Fig. 16.15.

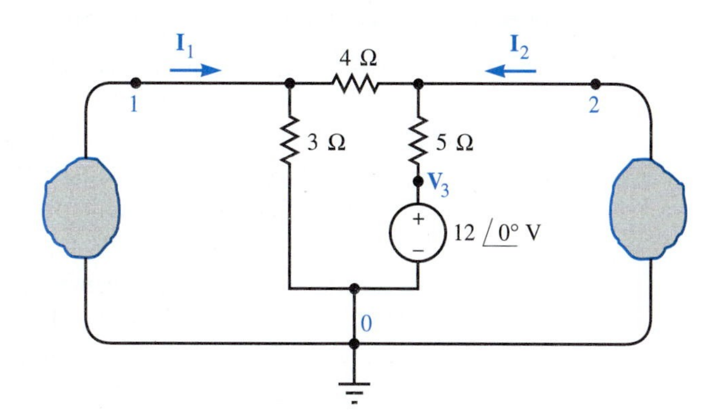

FIGURE 16.15 Node-voltage definitions

The node-voltage equations are

$$\left(\frac{1}{3}+\frac{1}{4}\right)\mathbf{V}_1 - \frac{1}{4}\mathbf{V}_2 + 0\mathbf{V}_3 = \mathbf{I}_1$$

$$-\frac{1}{4}\mathbf{V}_1 + \left(\frac{1}{4}+\frac{1}{5}\right)\mathbf{V}_2 - \frac{1}{5}\mathbf{V}_3 = \mathbf{I}_2$$

and

$$\mathbf{V}_3 = 12\underline{/0^\circ}$$

The elimination of $\mathbf{V}_3$ leads to

$$\frac{7}{12}\mathbf{V}_1 - \frac{1}{4}\mathbf{V}_2 = \mathbf{I}_1$$

and

$$-\frac{1}{4}\mathbf{V}_1 + \frac{9}{20}\mathbf{V}_2 - \frac{12}{5}\underline{/0^\circ} = \mathbf{I}_2$$

Therefore

$$\begin{bmatrix}\mathbf{I}_1\\ \mathbf{I}_2\end{bmatrix} = \begin{bmatrix}\frac{7}{12} & -\frac{1}{4}\\ -\frac{1}{4} & \frac{9}{20}\end{bmatrix}\begin{bmatrix}\mathbf{V}_1\\ \mathbf{V}_2\end{bmatrix} - \begin{bmatrix}0\\ \frac{12}{5}\underline{/0^\circ}\end{bmatrix}$$

which has the form of Eq. (16.18). Consequently, $\mathbf{y}_{11} = \frac{7}{12}$ S, $\mathbf{y}_{12} = \mathbf{y}_{21} = -\frac{1}{4}$ S, $\mathbf{y}_{22} = \frac{9}{20}$ S, $\mathbf{I}_{1sc} = 0$, and $\mathbf{I}_{2sc} = \frac{12}{5}\underline{/0^\circ}$ A as in (a).

Up to now we have obtained the impedance- and admittance-parameter equivalents of a three-terminal circuit. In the next section, we will show how these two equivalent circuits are related.

Remember

We derive the admittance-parameter equivalent circuit by regarding $\mathbf{V}_1$ and $\mathbf{V}_2$ as independent variables.

EXERCISES

5. Repeat Example 16.2(a) with the $+12\underline{/0^\circ}$-V independent source replaced with a current-controlled voltage source with terminal equation $\mathbf{V} = 2\mathbf{I}_1$.
   answer: $\mathbf{y}_{11} = \frac{7}{12}$ S, $\mathbf{y}_{12} = -\frac{1}{4}$ S, $\mathbf{y}_{21} = -\frac{29}{60}$ S, $\mathbf{y}_{22} = \frac{11}{20}$ S, $\mathbf{I}_{1sc} = \mathbf{I}_{2sc} = 0$ A
6. Repeat Example 16.2(b) with the $+12\underline{/0^\circ}$-V independent source replaced with a current-controlled voltage source with terminal equation $\mathbf{V} = 2\mathbf{I}_1$.
   answer: $\mathbf{y}_{11} = \frac{7}{12}$ S, $\mathbf{y}_{12} = -\frac{1}{4}$ S, $\mathbf{y}_{21} = -\frac{29}{60}$ S, $\mathbf{y}_{22} = \frac{11}{20}$ S, $\mathbf{I}_{1sc} = \mathbf{I}_{2sc} = 0$ A
7. Determine the three-terminal $\mathbf{y}$-parameter equivalent circuit of the network shown by direct application of Eqs. (16.11), (16.12), and (16.14) through (16.17).
   answer: $\mathbf{y}_{11} = 1.5 - j0.5$ S, $\mathbf{y}_{12} = -0.5 + j0.5$ S, $\mathbf{y}_{21} = -0.5 + j0.5$ S, $\mathbf{y}_{22} = 0.5 - j0.5$ S, $\mathbf{I}_{1sc} = 3.54\underline{/54^\circ}$ A, $\mathbf{I}_{2sc} = 3.53\underline{/-45^\circ}$ A

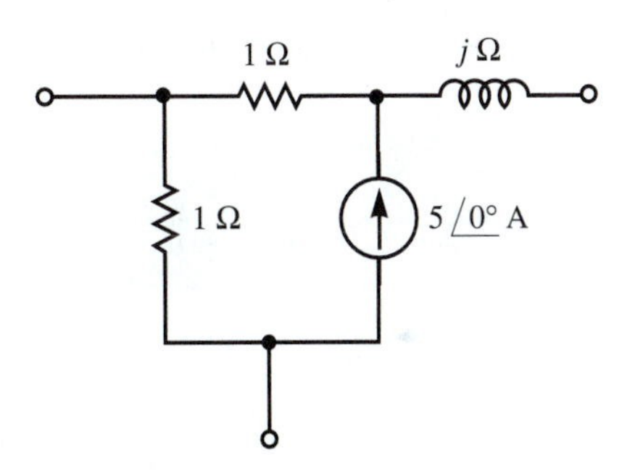

8. Repeat Exercise 7 with node-voltage analysis.

   answer: $\mathbf{y}_{11} = 1.5 - j0.5$ S, $\mathbf{y}_{12} = -0.5 + j0.5$ S, $\mathbf{y}_{21} = -0.5 + j0.5$ S, $\mathbf{y}_{22} = 0.5 - j0.5$ S, $\mathbf{I}_{1sc} = 3.54\angle 54°$ A, $\mathbf{I}_{2sc} = 3.53\angle -45°$ A

16.4 Relationship Between Impedance-Parameter and Admittance-Parameter Equivalent Circuits

Equations (16.9) and (16.18) and their corresponding circuits of Figs. 16.4 and 16.11 provide alternative equivalent circuits for a given three-terminal network. This equivalency implies that there is a relationship between $\mathbf{Z}$, $\mathbf{V}_{oc}$, $\mathbf{Y}$, and $\mathbf{I}_{sc}$. To obtain this relationship, solve Eq. (16.9), repeated below, for $\mathbf{I}_1$ and $\mathbf{I}_2$:

$$\begin{bmatrix} \mathbf{V}_1 \\ \mathbf{V}_2 \end{bmatrix} = \begin{bmatrix} \mathbf{z}_{11} & \mathbf{z}_{12} \\ \mathbf{z}_{21} & \mathbf{z}_{22} \end{bmatrix} \begin{bmatrix} \mathbf{I}_1 \\ \mathbf{I}_2 \end{bmatrix} + \begin{bmatrix} \mathbf{V}_{1oc} \\ \mathbf{V}_{2oc} \end{bmatrix} \tag{16.19}$$

Cramer's rule leads quickly to the solution

$$\begin{bmatrix} \mathbf{I}_1 \\ \mathbf{I}_2 \end{bmatrix} = \begin{bmatrix} \dfrac{\mathbf{z}_{22}}{|\mathbf{Z}|} & -\dfrac{\mathbf{z}_{12}}{|\mathbf{Z}|} \\ -\dfrac{\mathbf{z}_{21}}{|\mathbf{Z}|} & \dfrac{\mathbf{z}_{11}}{|\mathbf{Z}|} \end{bmatrix} \begin{bmatrix} \mathbf{V}_1 \\ \mathbf{V}_2 \end{bmatrix} - \begin{bmatrix} \dfrac{\mathbf{z}_{22}}{|\mathbf{Z}|} & -\dfrac{\mathbf{z}_{12}}{|\mathbf{Z}|} \\ -\dfrac{\mathbf{z}_{21}}{|\mathbf{Z}|} & \dfrac{\mathbf{z}_{11}}{|\mathbf{Z}|} \end{bmatrix} \begin{bmatrix} \mathbf{V}_{1oc} \\ \mathbf{V}_{2oc} \end{bmatrix} \tag{16.20}$$

where $|\mathbf{Z}|$ denotes the determinant† of $\mathbf{Z}$. We can write Eq. (16.20) as

$$\mathbf{I} = \mathbf{Z}^{-1}\mathbf{V} - \mathbf{Z}^{-1}\mathbf{V}_{oc} \tag{16.21}$$

where it has been natural to define the "$\mathbf{Z}$ inverse" matrix

$$\mathbf{Z}^{-1} = \begin{bmatrix} \dfrac{\mathbf{z}_{22}}{|\mathbf{Z}|} & -\dfrac{\mathbf{z}_{12}}{|\mathbf{Z}|} \\ -\dfrac{\mathbf{z}_{21}}{|\mathbf{Z}|} & \dfrac{\mathbf{z}_{11}}{|\mathbf{Z}|} \end{bmatrix} \tag{16.22}$$

Observe that $\mathbf{Z}^{-1}$ exists if and only if $|\mathbf{Z}| \neq 0$. We can see that Eq. (16.20) has the same form as Eq. (16.18), namely,

$$\mathbf{I} = \mathbf{Y}\mathbf{V} - \mathbf{I}_{sc} \tag{16.23}$$

† Determinants are described in Appendix A.

Therefore, by comparing Eqs. (16.21) and (16.23), we see that

$$\mathbf{Y} = \begin{bmatrix} \mathbf{y}_{11} & \mathbf{y}_{12} \\ \mathbf{y}_{21} & \mathbf{y}_{22} \end{bmatrix} = \begin{bmatrix} \dfrac{\mathbf{z}_{22}}{|\mathbf{Z}|} & -\dfrac{\mathbf{z}_{12}}{|\mathbf{Z}|} \\ -\dfrac{\mathbf{z}_{21}}{|\mathbf{Z}|} & \dfrac{\mathbf{z}_{11}}{|\mathbf{Z}|} \end{bmatrix} = \mathbf{Z}^{-1} \tag{16.24}$$

and

$$\mathbf{I}_{sc} = \begin{bmatrix} \mathbf{I}_{1sc} \\ \mathbf{I}_{2sc} \end{bmatrix} = \begin{bmatrix} \dfrac{\mathbf{z}_{22}}{|\mathbf{Z}|} & -\dfrac{\mathbf{z}_{12}}{|\mathbf{Z}|} \\ -\dfrac{\mathbf{z}_{21}}{|\mathbf{Z}|} & \dfrac{\mathbf{z}_{11}}{|\mathbf{Z}|} \end{bmatrix} \begin{bmatrix} \mathbf{V}_{1oc} \\ \mathbf{V}_{2oc} \end{bmatrix} = \mathbf{Z}^{-1}\mathbf{V}_{oc} \tag{16.25}$$

Equations (16.24) and (16.25) are formulas for computing the short-circuit admittance and current matrices $\mathbf{Y}$ and $\mathbf{I}_{sc}$ from the open-circuit impedance and voltage matrices $\mathbf{Z}$ and $\mathbf{V}_{oc}$.

The reverse calculation can be obtained if we solve Eq. (16.10) for $\mathbf{V}_1$ and $\mathbf{V}_2$. Cramer's rule leads quickly to the result

$$\mathbf{V} = \mathbf{Y}^{-1}\mathbf{I} + \mathbf{Y}^{-1}\mathbf{I}_{sc} \tag{16.26}$$

This has the form $\mathbf{V} = \mathbf{ZI} + \mathbf{V}_{oc}$ where

$$\mathbf{Z} = \mathbf{Y}^{-1} = \begin{bmatrix} \dfrac{\mathbf{y}_{22}}{|\mathbf{Y}|} & -\dfrac{\mathbf{y}_{12}}{|\mathbf{Y}|} \\ -\dfrac{\mathbf{y}_{21}}{|\mathbf{Y}|} & \dfrac{\mathbf{y}_{11}}{|\mathbf{Y}|} \end{bmatrix} \tag{16.27}$$

and

$$\mathbf{V}_{oc} = \mathbf{Y}^{-1}\mathbf{I}_{sc} = \mathbf{ZI}_{sc} \tag{16.28}$$

The matrix $\mathbf{Y}^{-1}$ ($\mathbf{Y}$ inverse) exists if and only if $|\mathbf{Y}| \neq 0$, where $|\mathbf{Y}|$ is the determinant of $\mathbf{Y}$. Equations (16.24), (16.25), (16.27), and (16.28) provide the means for converting a **z**-parameter equivalent circuit to a **y**-parameter equivalent circuit and vice versa. For example, if we wish to convert a **y**-parameter equivalent to a **z**-parameter equivalent, we simply use Eq. (16.27) to compute the **z** parameters and Eq. (16.28) to compute $\mathbf{V}_{oc}$.

The conversion from one equivalent to another is not possible, however, if $|\mathbf{Z}|$ or $|\mathbf{Y}|$ equals zero. We shall examine this issue in the next section.

EXERCISE

9. We found that the **z** parameters of the network in Example 16.1 are $\mathbf{z}_{11} = \frac{9}{4}$, $\mathbf{z}_{12} = \mathbf{z}_{21} = \frac{5}{4}$, and $\mathbf{z}_{22} = \frac{35}{12}$ ohms. The open-circuit voltages were found to be $\mathbf{V}_{1oc} = 3\underline{/0^\circ}$ V and $\mathbf{V}_{2oc} = 7\underline{/0^\circ}$ V. Use Eqs. (16.24) and (16.25) to find the **y** parameters and the short-circuit currents of the same network. Do your results agree with those determined in Example 16.2? *answer:* The answers agree.

16.5 Existence of Three-Terminal z- and y-Parameter Equivalent Circuits

It is possible for *theoretical* circuits to have impedance-parameter equivalent circuits but not admittance-parameter equivalent circuits and vice versa. The reason for this can be understood in terms of the simple example shown in Fig. 16.16.

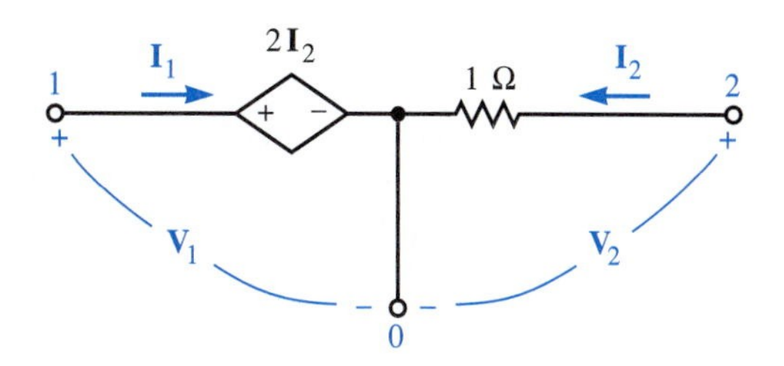

FIGURE 16.16 This circuit has a **z**-parameter equivalent circuit but no **y**-parameter equivalent circuit.

It follows by an inspection of the circuit shown that

$$\mathbf{V}_1 = 0\mathbf{I}_1 + 2\mathbf{I}_2 \tag{16.29}$$

and

$$\mathbf{V}_2 = 0\mathbf{I}_1 + 1\mathbf{I}_2 \tag{16.30}$$

It follows from the above equations that

$$\mathbf{Z} = \begin{bmatrix} 0 & 2 \\ 0 & 1 \end{bmatrix} \tag{16.31}$$

and

$$\mathbf{V}_{oc} = 0 \tag{16.32}$$

It follows from Eq. (16.31) and Appendix A that $|\mathbf{Z}| = 0$. Therefore $\mathbf{Y} = \mathbf{Z}^{-1}$ does not exist. As a consequence, the circuit cannot be characterized by a **y**-parameter equivalent circuit. The mathematical reason for this is that $\mathbf{V}_1$ and $\mathbf{V}_2$ cannot be chosen independently. The impossibility of a **y**-parameter equivalent for Fig. 16.16 can also be explained by examination of the circuit. Remember that circuit theory is based on KVL, KCL, and component definitions. The dependent source voltage in Fig. 16.16 is, by *definition,* $2\mathbf{I}_2$ V. If terminal 1 is connected to terminal 0, as depicted in Fig. 16.17, KVL is violated, which is impossible in the context of circuit theory. (The sum of voltage drops around the blue path in Fig. 16.17 is not zero for arbitrary values of $\mathbf{V}_2$.) In addition, terminal 2 *cannot,* within the context of circuit theory, be shorted to terminal 0 if an arbitrary voltage $\mathbf{V}_1$ is applied at terminal 1 as depicted in Fig. 16.18.

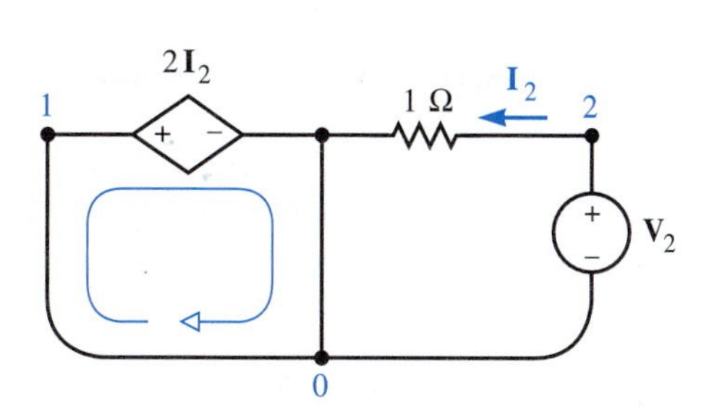

FIGURE 16.17 Shorting terminal 1 to terminal 0 results in an impossible circuit (KVL is violated).

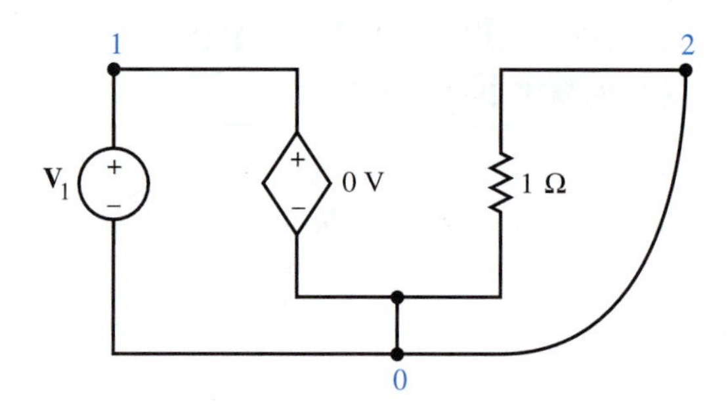

FIGURE 16.18 Shorting terminal 2 to terminal 0 results in an impossible circuit.

It is similarly possible for a circuit to have a **y**-parameter equivalent but not have a **z**-parameter equivalent. An example is shown in Fig. 16.19. Note that $\mathbf{I}_1$ and $\mathbf{I}_2$ cannot be chosen independently.

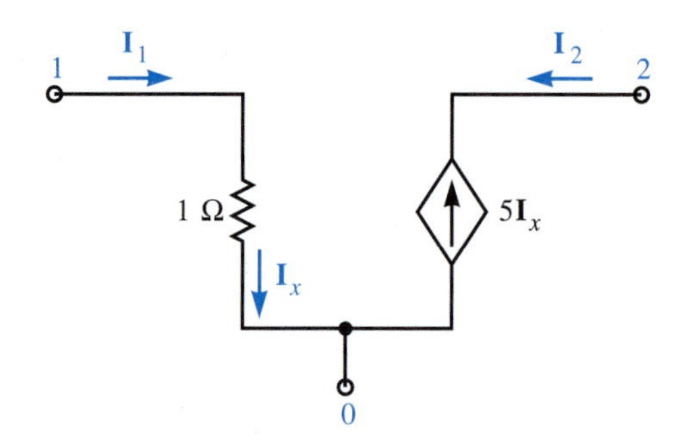

FIGURE 16.19 This circuit has a **y**-parameter equivalent circuit but no **z**-parameter equivalent circuit.

Finally, it is possible for a circuit to have neither **y**- nor **z**-parameter equivalent circuits. An example is shown in Fig. 16.20, where neither voltage $\mathbf{V}_1$ nor current $\mathbf{I}_2$ can be independent variables.

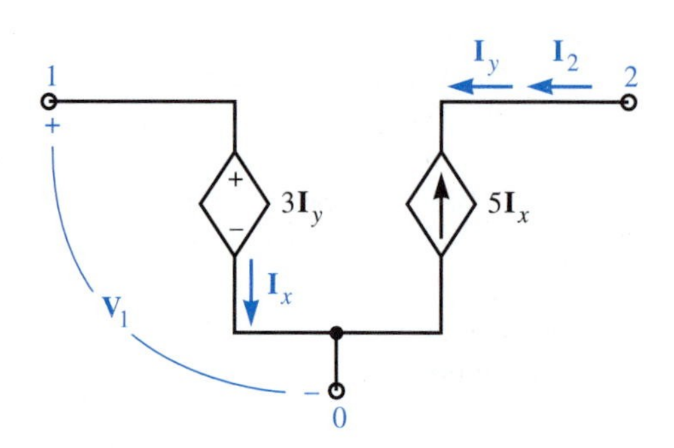

FIGURE 16.20 This circuit has neither a **z**- nor a **y**-parameter equivalent circuit.

EXERCISES

In Exercises 10 through 13, give an example of the circuits specified.

10. A circuit that has both **z**- and **y**-parameter equivalent circuits.
11. A circuit that has a **z**-parameter equivalent but not a **y**-parameter equivalent.
12. A circuit that has a **y**-parameter equivalent but not a **z**-parameter equivalent.
13. A circuit that has neither a **z**- nor a **y**-parameter equivalent.

16.6 Three-Terminal Hybrid-h-Parameter Equivalent Circuits

A third equivalent circuit for the network in Fig. 16.1 can be obtained if we assume that $\mathbf{I}_1$ and $\mathbf{V}_2$ are independent variables and $\mathbf{V}_1$ and $\mathbf{I}_2$ are dependent variables. Superposition indicates that $\mathbf{V}_1$ and $\mathbf{I}_2$ can then be expressed as linear combinations of $\mathbf{I}_1$, $\mathbf{V}_2$ and the independent sources in the network. Thus

$$\begin{bmatrix} \mathbf{V}_1 \\ \mathbf{I}_2 \end{bmatrix} = \begin{bmatrix} \mathbf{h}_{11} & \mathbf{h}_{12} \\ \mathbf{h}_{21} & \mathbf{h}_{22} \end{bmatrix} \begin{bmatrix} \mathbf{I}_1 \\ \mathbf{V}_2 \end{bmatrix} + \begin{bmatrix} \mathbf{V}_{1ocsc} \\ -\mathbf{I}_{2ocsc} \end{bmatrix} \tag{16.33}$$

The quantities $\mathbf{h}_{11}$, $\mathbf{h}_{12}$, $\mathbf{h}_{21}$, and $\mathbf{h}_{22}$ are the *hybrid-***h***-parameters* of the network of Fig. 16.1. The significance of the **h**-parameters can be determined directly from Eq. (16.33) if we observe that

$$\mathbf{h}_{11} = \left.\frac{\mathbf{V}_1}{\mathbf{I}_1}\right|_{\substack{\mathbf{V}_2=0 \\ \text{IIS}=0}} \tag{16.34}$$

$$\mathbf{h}_{21} = \left.\frac{\mathbf{I}_2}{\mathbf{I}_1}\right|_{\substack{\mathbf{V}_2=0 \\ \text{IIS}=0}} \tag{16.35}$$

$$\mathbf{h}_{12} = \left.\frac{\mathbf{V}_1}{\mathbf{V}_2}\right|_{\substack{\mathbf{I}_1=0 \\ \text{IIS}=0}} \tag{16.36}$$

$$\mathbf{h}_{22} = \left.\frac{\mathbf{I}_2}{\mathbf{V}_2}\right|_{\substack{\mathbf{I}_1=0 \\ \text{IIS}=0}} \tag{16.37}$$

As before, IIS $= 0$ denotes that the independent internal sources are set to zero. Observe that the units of $\mathbf{h}_{11}$ and $\mathbf{h}_{22}$ are ohms and siemens, respectively, whereas $\mathbf{h}_{12}$ and $\mathbf{h}_{21}$ are dimensionless. The fact that the four parameters do not all have the same units is the reason these parameters are referred to as *hybrid parameters.* It also follows from Eq. (16.33) that $\mathbf{V}_{1ocsc}$ and $\mathbf{I}_{2ocsc}$ are given by

$$\mathbf{V}_{1ocsc} = \left.\mathbf{V}_1\right|_{\substack{\mathbf{I}_1=0 \\ \mathbf{V}_2=0}} \tag{16.38}$$

$$\mathbf{I}_{2ocsc} = \left.-\mathbf{I}_2\right|_{\substack{\mathbf{I}_1=0 \\ \mathbf{V}_2=0}} \tag{16.39}$$

The equivalent circuit associated with Eq. (16.33) is shown in Fig. 16.21. For the special case in which there are no independent sources inside the original network, the independent voltage and current sources in Fig. 16.21 equal zero, and are equivalent to short and open circuits, respectively.

The elements in the **h**-parameter equivalent circuit can be related to those of the **z**- and **y**-parameter equivalent circuits by straightforward manipulations of the corresponding equations. A summary of the results is contained in Tables 16.2 through 16.5 of Section 16.8.

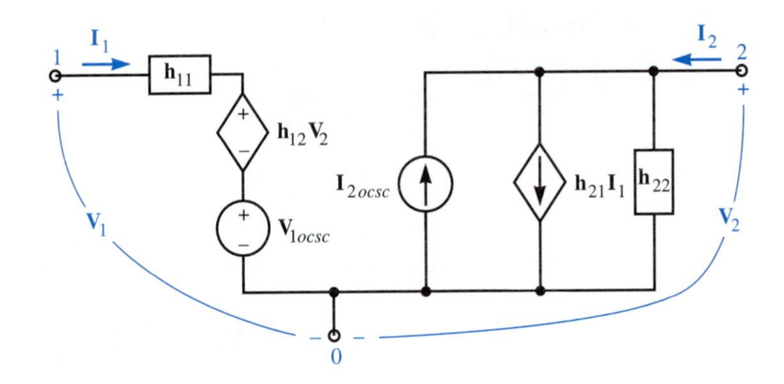

FIGURE 16.21 Three-terminal hybrid-**h**-parameter equivalent network

As with **z** and **y** parameters, there are several ways to determine the **h**-parameter equivalent circuit of a network. The following example illustrates one approach.

EXAMPLE 16.3 Find $\mathbf{V}_{1ocsc}$, $\mathbf{I}_{2ocsc}$, and the hybrid-**h** parameters for the circuit of Example 16.1.

Solution We find $\mathbf{V}_{1ocsc}$ and $\mathbf{I}_{2ocsc}$ by setting $\mathbf{I}_1 = 0$ and $\mathbf{V}_2 = 0$ as shown in Fig. 16.22. It follows from Eqs. (16.38) and (16.39) and an inspection of Fig 16.22 that

$$\mathbf{V}_{1ocsc} = \mathbf{V}_1 \bigg|_{\substack{\mathbf{I}_1 = 0 \\ \mathbf{V}_2 = 0}} = 0 \text{ V}$$

and

$$\mathbf{I}_{2ocsc} = -\mathbf{I}_2 \bigg|_{\substack{\mathbf{I}_1 = 0 \\ \mathbf{V}_2 = 0}} = \frac{12}{5} \underline{/0^\circ} \text{ A}$$

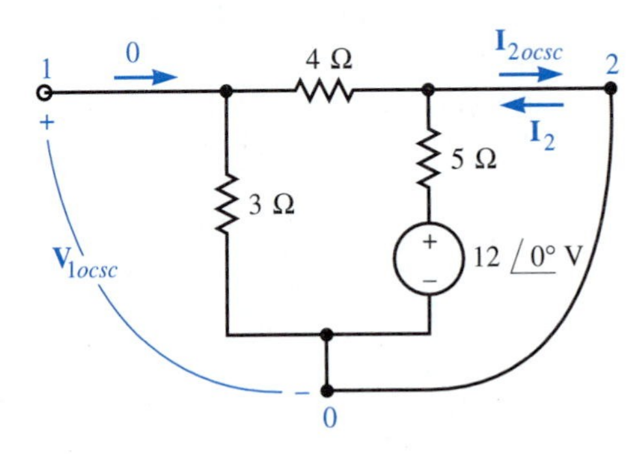

FIGURE 16.22 Circuit for finding $\mathbf{V}_{1ocsc}$ and $\mathbf{I}_{2ocsc}$

The parameters $\mathbf{h}_{11}$ and $\mathbf{h}_{21}$ can be obtained from Eqs. (16.34) and (16.35) and an inspection of Fig. 16.23:

$$\mathbf{h}_{11} = \frac{\mathbf{V}_1}{\mathbf{I}_1} \bigg|_{\substack{\mathbf{V}_2 = 0 \\ \text{IIS} = 0}} = 3 \parallel 4 = \frac{12}{7} \, \Omega$$

$$\mathbf{h}_{21} = \frac{\mathbf{I}_2}{\mathbf{I}_1} \bigg|_{\substack{\mathbf{V}_2 = 0 \\ \text{IIS} = 0}} = -\frac{3}{3+4} = -\frac{3}{7}$$

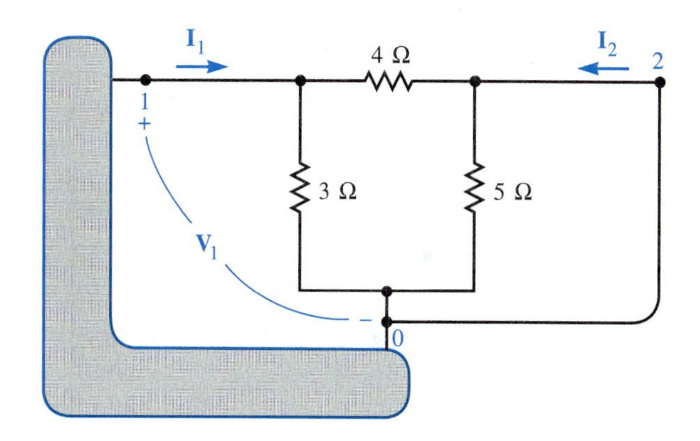

FIGURE 16.23 Circuit for finding $\mathbf{h}_{11}$ and $\mathbf{h}_{21}$

Similarly, from Eqs. (16.36) and (16.37) and Fig. 16.24 we obtain

$$\mathbf{h}_{12} = \left.\frac{\mathbf{V}_1}{\mathbf{V}_2}\right|_{\substack{\mathbf{I}_1=0\\ \text{IIS}=0}} = \frac{3}{3+4} = \frac{3}{7}$$

$$\mathbf{h}_{22} = \left.\frac{\mathbf{I}_2}{\mathbf{V}_2}\right|_{\substack{\mathbf{I}_1=0\\ \text{IIS}=0}} = \frac{1}{5} + \frac{1}{7} = \frac{12}{35}\text{ S}$$

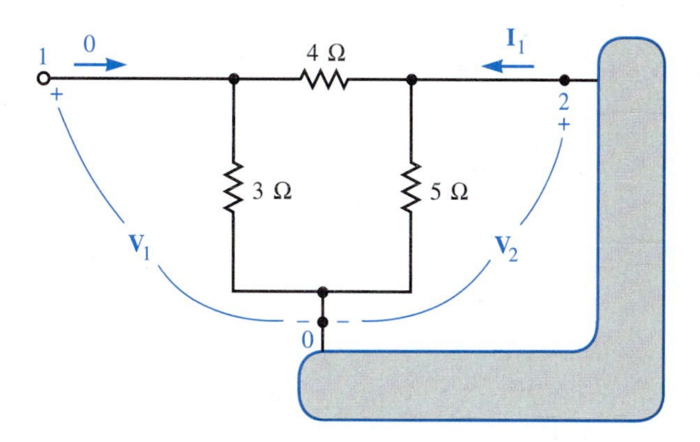

FIGURE 16.24 Circuit for finding $\mathbf{h}_{12}$ and $\mathbf{h}_{22}$

Hybrid-**h** equivalent circuits are frequently used to model bipolar junction transistors. In the next example we will see how a simple but remarkably useful hybrid-**h** transistor model is used to analyze a transistor amplifier.

EXAMPLE 16.4 Figure 16.25 depicts a single-stage transistor amplifier where B is a dc voltage used to supply power to the transistor. The capacitors are employed simply to prevent the dc voltages caused by B from reaching the source $v_s(t)$ and the load resistance R_L. Their values are large enough so that impedances, $1/j\omega C_1$ and $1/j\omega C_2$, are negligible. Use the simplified hybrid-**h** transistor model of Fig. 16.26, in which $\mathbf{h}_{21} = \beta$ is the only nonzero parameter, to determine the voltage transfer function $\mathbf{V}_o/\mathbf{V}_s$.

Solution The ac circuit model is shown in Fig. 16.27. Since we are interested only in the portion of the output due to the source v_s, we have set voltage B to zero and have replaced the capacitances by short circuits. This places R_1 and R_2 in parallel and R_C and R_L in parallel as indicated.

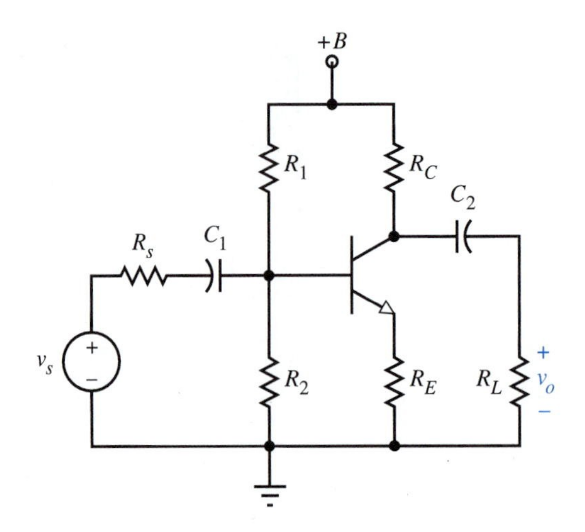

FIGURE 16.25
Transistor amplifier

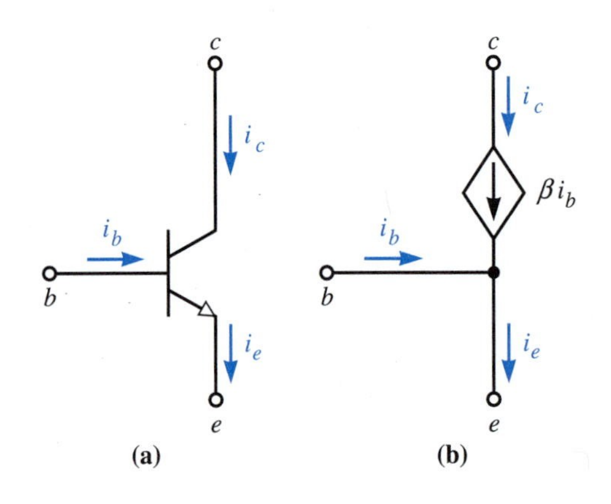

FIGURE 16.26
(*a*) Transistor and (*b*) simplified hybrid-**h** model. A transistor is composed of a base *b*, an emitter *e*, and a collector *c*.

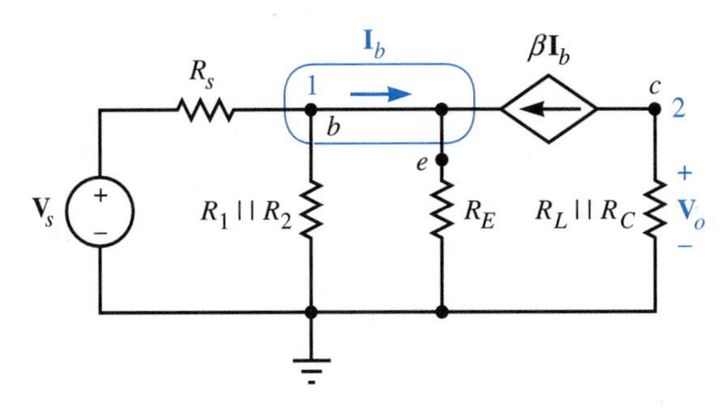

FIGURE 16.27
Circuit model

The application of KCL at the surface enclosing node 1 yields

$$\frac{\mathbf{V}_1 - \mathbf{V}_s}{R_s} + \frac{\mathbf{V}_1}{R_1 \| R_2} + \frac{\mathbf{V}_1}{R_E} = \beta \mathbf{I}_b$$

where

$$\mathbf{V}_1 = (\beta + 1)\mathbf{I}_b R_E$$

The KCL equation at node 2 is

$$\frac{\mathbf{V}_o}{R_L \| R_C} + \beta \mathbf{I}_b = 0$$

By combining the previous three equations, we obtain

$$\frac{\mathbf{V}_o}{\mathbf{V}_s} = -\left\{\left(\frac{\beta}{\beta+1}\right)\left[\frac{1}{1+\dfrac{R_s}{R_1 \parallel R_2}+\dfrac{R_s}{(\beta+1)R_E}}\right]\right\}\frac{R_L \parallel R_C}{R_E}$$

This expression simplifies considerably for typical circuit values for which $\beta \gg 1$, $R_s \ll R_1 \parallel R_2$, and $R_s \ll (\beta + 1)R_E$. For these values, the factor inside the braces is approximately 1. Thus

$$\frac{\mathbf{V}_o}{\mathbf{V}_s} \simeq -\frac{R_L \parallel R_C}{R_E}$$

which is a very convenient design formula.

We have now derived three of the four possible three-terminal equivalent circuits listed in Table 16.1. At this point you should be well prepared to derive the remaining (hybrid-**g**) equivalent. (See Tables 16.2 through 16.5 and Problem 20.) We turn to two-port networks next.

Remember

We derive the hybrid-**h**-parameter equivalent circuit by regarding $\mathbf{I}_1$ and $\mathbf{V}_2$ as independent variables.

EXERCISES

14. Show that the bipolar transistor model in Fig. 16.26 is an example of the three-terminal hybrid-**h**-parameter equivalent network of Fig. 16.21.
    answer: Set $\mathbf{h}_{11} = \mathbf{h}_{22} = 0$, $\mathbf{h}_{21} = \beta$, $\mathbf{V}_{1oc} = 0$, and $\mathbf{I}_{2sc} = 0$ to obtain Fig. 16.21.
15. Determine the three-terminal hybrid-**h**-parameter equivalent circuit of the network shown by direct application of Eqs. (16.34) through (16.39).
    answer: $\mathbf{h}_{11} = 0.6 + j0.2\ \Omega$, $\mathbf{h}_{12} = 0.4 - j0.2$, $\mathbf{h}_{21} = -0.4 + j0.2$, $\mathbf{h}_{22} = 0.4 - j0.2$ S, $\mathbf{V}_{1ocsc} = 2.24\ \underline{/63.4^\circ}$ V, $\mathbf{I}_{2ocsc} = 4.47\ \underline{/-26.6^\circ}$ A

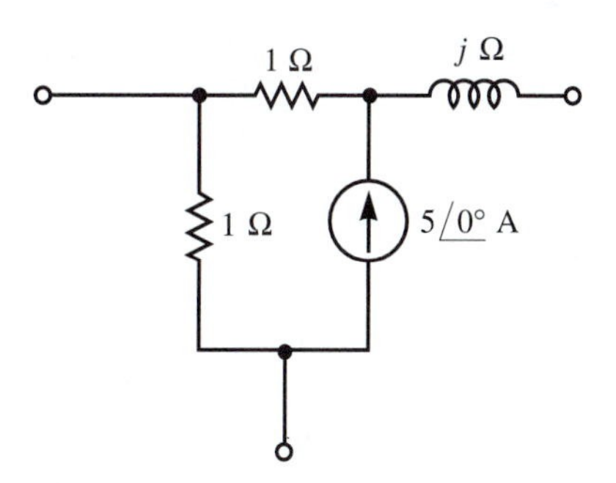

16.7 Two-Port Networks

The definition of a two-port network can be understood from Fig. 16.28, which depicts a four-terminal network with terminals a and a' and b and b' grouped into pairs. Notice that the four-terminal network shown has the special property that the *net current entering each terminal pair is zero.* In other words, the currents exiting at terminals a'

and b', $\mathbf{I}_1$ and $\mathbf{I}_2$, equal the corresponding currents entering at terminals a and b. This is called the *two-port* property. It is the key property that distinguishes a two-port network from an arbitrary four-terminal network.

The special constraint on the currents entering a two-port network is *sometimes* the result of the circuit inside the rectangle of Fig. 16.28. Frequently, however, the constraints on the currents are *not* a result of the circuit itself, but of the way it is connected to other circuits. This is illustrated in Fig. 16.29. By KCL, the net current entering each circuit, A and B, is zero. Therefore $\mathbf{I}_x = \mathbf{I}_1$, $\mathbf{I}_y = \mathbf{I}_2$, and the four-terminal network in the middle is a two-port. A more general configuration is shown in Fig. 16.30. Observe that each of the networks I, II, III, and IV is properly called a two-port network because the two-port property must be satisfied for each network. (Show why this is so.)

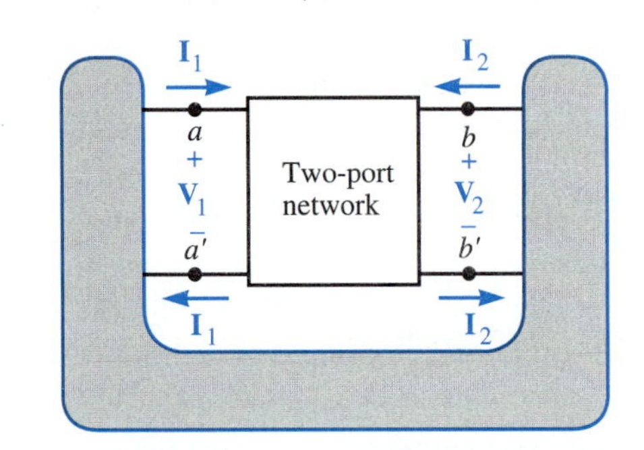

FIGURE 16.28 Two-port network

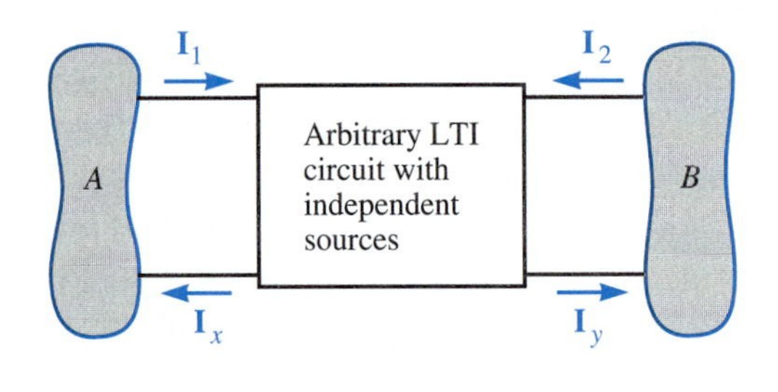

FIGURE 16.29 The arbitrary circuit enclosed in the rectangle is a two-port.

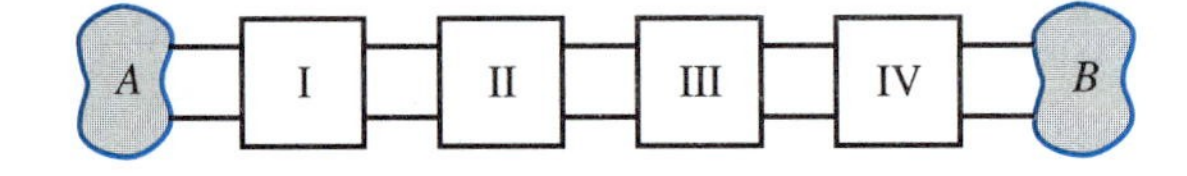

FIGURE 16.30 Networks I, II, III, and IV are two-port networks.

Notice that in the defining circuit of Fig. 16.28, *no statement is made* about voltage $\mathbf{V}_{a'b'}$ or $\mathbf{V}_{ab}$. The only variables of interest in defining the two-port network are the currents $\mathbf{I}_1$ and $\mathbf{I}_2$ and the voltages $\mathbf{V}_{aa'} = \mathbf{V}_1$ and $\mathbf{V}_{bb'} = \mathbf{V}_2$.

Three-terminal networks are frequently connected as four-terminal networks, as illustrated in Fig. 16.31. As a result, two-port networks and three-terminal networks are often confused. The distinction is as follows:

When a three-terminal network is connected as shown in Fig. 16.31, then, of course, the equations and equivalent circuits developed in Sections 16.2, 16.3, and 16.6 for the three-terminal network (inner rectangle) still apply. Therefore the relationships between the variables $\mathbf{I}_1$, $\mathbf{I}_2$, $\mathbf{V}_1$, and $\mathbf{V}_2$ of the four-terminal network (outer rectangle) are the same as those for the three-terminal network. It does *not* follow, however, that $\mathbf{I}_x = \mathbf{I}_1$ and $\mathbf{I}_y = \mathbf{I}_2$. The only constraint is that imposed by KCL, which states that $\mathbf{I}_1 + \mathbf{I}_2 = \mathbf{I}_x + \mathbf{I}_y$. The values of $\mathbf{I}_x$ and $\mathbf{I}_y$ are determined both by the sum $\mathbf{I}_1 + \mathbf{I}_2$ and

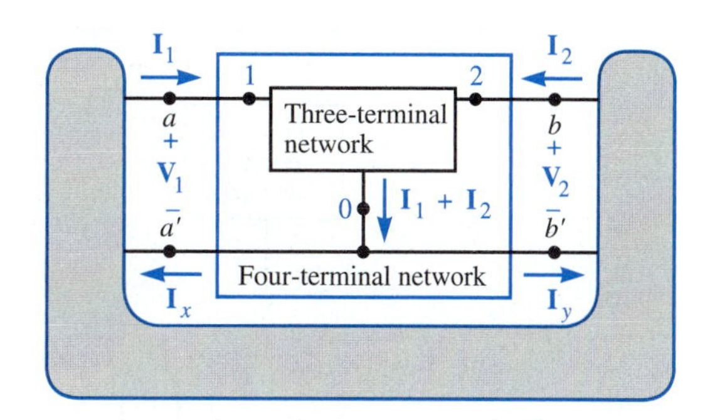

FIGURE 16.31 A three-terminal network connected as a four-terminal network

by the external circuit connected at terminals a, a', b, and b'. The external circuit determines whether or not the four-terminal network is a two-port.

Remember The net current entering each terminal pair of a two-port network equals zero.

EXERCISES

16. Prove that network III in Fig. 16.30 is a two-port network.
17. Network A shown below is an arbitrary four-terminal network. Is there a value of β that forces network A to satisfy the two-port property? If so, state the value(s) of β.

    answer: $\beta = -5$

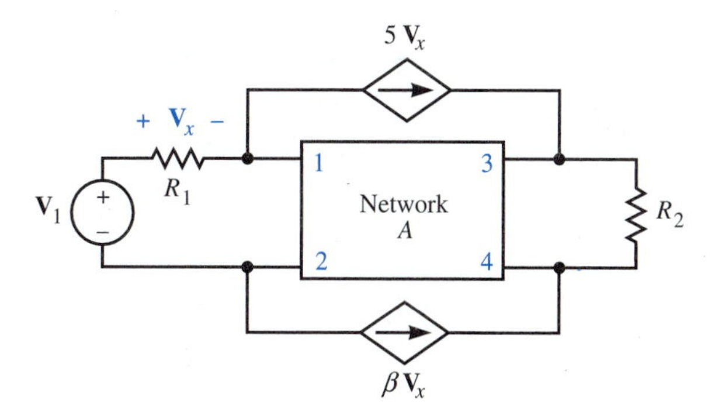

16.8 Equivalent Circuits for Two-Port Networks

Two-port networks are characterized by the equations relating $\mathbf{I}_1$, $\mathbf{I}_2$, $\mathbf{V}_1$, and $\mathbf{V}_2$. The derivations of the terminal equations of a two-port network are identical to those of three-terminal networks. The results are given in Tables 16.2 through 16.5. We must be cautious when defining equivalent circuits for two-port networks because the equations given in Tables 16.2 through 16.4 *make no statement* about the values of $\mathbf{V}_{ab}$ and $\mathbf{V}_{a'b'}$ in the two-port network of Fig. 16.28. The circuits of Fig. 16.32a–d on page 616 are frequently used as equivalent circuits for two-port networks even though they may not specify the correct values for $\mathbf{V}_{ab}$ and $\mathbf{V}_{a'b'}$. Example 16.5 illustrates this point.

TABLE 16.2
Summary of terminal equations for three-terminal and two-port networks

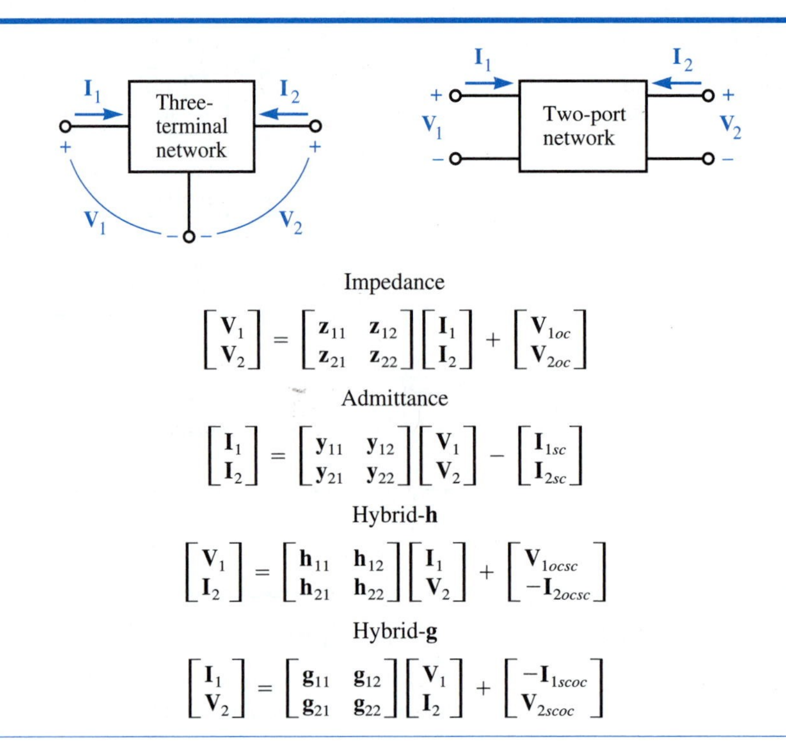

Impedance

$$\begin{bmatrix} \mathbf{V}_1 \\ \mathbf{V}_2 \end{bmatrix} = \begin{bmatrix} \mathbf{z}_{11} & \mathbf{z}_{12} \\ \mathbf{z}_{21} & \mathbf{z}_{22} \end{bmatrix} \begin{bmatrix} \mathbf{I}_1 \\ \mathbf{I}_2 \end{bmatrix} + \begin{bmatrix} \mathbf{V}_{1oc} \\ \mathbf{V}_{2oc} \end{bmatrix}$$

Admittance

$$\begin{bmatrix} \mathbf{I}_1 \\ \mathbf{I}_2 \end{bmatrix} = \begin{bmatrix} \mathbf{y}_{11} & \mathbf{y}_{12} \\ \mathbf{y}_{21} & \mathbf{y}_{22} \end{bmatrix} \begin{bmatrix} \mathbf{V}_1 \\ \mathbf{V}_2 \end{bmatrix} - \begin{bmatrix} \mathbf{I}_{1sc} \\ \mathbf{I}_{2sc} \end{bmatrix}$$

Hybrid-**h**

$$\begin{bmatrix} \mathbf{V}_1 \\ \mathbf{I}_2 \end{bmatrix} = \begin{bmatrix} \mathbf{h}_{11} & \mathbf{h}_{12} \\ \mathbf{h}_{21} & \mathbf{h}_{22} \end{bmatrix} \begin{bmatrix} \mathbf{I}_1 \\ \mathbf{V}_2 \end{bmatrix} + \begin{bmatrix} \mathbf{V}_{1ocsc} \\ -\mathbf{I}_{2ocsc} \end{bmatrix}$$

Hybrid-**g**

$$\begin{bmatrix} \mathbf{I}_1 \\ \mathbf{V}_2 \end{bmatrix} = \begin{bmatrix} \mathbf{g}_{11} & \mathbf{g}_{12} \\ \mathbf{g}_{21} & \mathbf{g}_{22} \end{bmatrix} \begin{bmatrix} \mathbf{V}_1 \\ \mathbf{I}_2 \end{bmatrix} + \begin{bmatrix} -\mathbf{I}_{1scoc} \\ \mathbf{V}_{2scoc} \end{bmatrix}$$

TABLE 16.3
Summary of formulas for three-terminal and two-port networks

Impedance

$$\mathbf{z}_{11} = \left.\frac{\mathbf{V}_1}{\mathbf{I}_1}\right|_{\substack{\mathbf{I}_2 = 0 \\ \text{IIS} = 0}} \qquad \mathbf{z}_{12} = \left.\frac{\mathbf{V}_1}{\mathbf{I}_2}\right|_{\substack{\mathbf{I}_1 = 0 \\ \text{IIS} = 0}}$$

$$\mathbf{z}_{21} = \left.\frac{\mathbf{V}_2}{\mathbf{I}_1}\right|_{\substack{\mathbf{I}_2 = 0 \\ \text{IIS} = 0}} \qquad \mathbf{z}_{22} = \left.\frac{\mathbf{V}_2}{\mathbf{I}_2}\right|_{\substack{\mathbf{I}_1 = 0 \\ \text{IIS} = 0}}$$

$$\mathbf{V}_{1oc} = \mathbf{V}_1\Big|_{\substack{\mathbf{I}_1 = 0 \\ \mathbf{I}_2 = 0}} \qquad \mathbf{V}_{2oc} = \mathbf{V}_2\Big|_{\substack{\mathbf{I}_1 = 0 \\ \mathbf{I}_2 = 0}}$$

Admittance

$$\mathbf{y}_{11} = \left.\frac{\mathbf{I}_1}{\mathbf{V}_1}\right|_{\substack{\mathbf{V}_2 = 0 \\ \text{IIS} = 0}} \qquad \mathbf{y}_{12} = \left.\frac{\mathbf{I}_1}{\mathbf{V}_2}\right|_{\substack{\mathbf{V}_1 = 0 \\ \text{IIS} = 0}}$$

$$\mathbf{y}_{21} = \left.\frac{\mathbf{I}_2}{\mathbf{V}_1}\right|_{\substack{\mathbf{V}_2 = 0 \\ \text{IIS} = 0}} \qquad \mathbf{y}_{22} = \left.\frac{\mathbf{I}_2}{\mathbf{V}_2}\right|_{\substack{\mathbf{V}_1 = 0 \\ \text{IIS} = 0}}$$

$$\mathbf{I}_{1sc} = -\mathbf{I}_1\Big|_{\substack{\mathbf{V}_1 = 0 \\ \mathbf{V}_2 = 0}} \qquad \mathbf{I}_{2sc} = -\mathbf{I}_2\Big|_{\substack{\mathbf{V}_1 = 0 \\ \mathbf{V}_2 = 0}}$$

Hybrid-**h**

$$\mathbf{h}_{11} = \left.\frac{\mathbf{V}_1}{\mathbf{I}_1}\right|_{\substack{\mathbf{V}_2 = 0 \\ \text{IIS} = 0}} \qquad \mathbf{h}_{12} = \left.\frac{\mathbf{V}_1}{\mathbf{V}_2}\right|_{\substack{\mathbf{I}_1 = 0 \\ \text{IIS} = 0}}$$

$$\mathbf{h}_{21} = \left.\frac{\mathbf{I}_2}{\mathbf{I}_1}\right|_{\substack{\mathbf{V}_2 = 0 \\ \text{IIS} = 0}} \qquad \mathbf{h}_{22} = \left.\frac{\mathbf{I}_2}{\mathbf{V}_2}\right|_{\substack{\mathbf{I}_1 = 0 \\ \text{IIS} = 0}}$$

$$\mathbf{V}_{1ocsc} = \mathbf{V}_1\Big|_{\substack{\mathbf{I}_1 = 0 \\ \mathbf{V}_2 = 0}} \qquad \mathbf{I}_{2ocsc} = -\mathbf{I}_2\Big|_{\substack{\mathbf{I}_1 = 0 \\ \mathbf{V}_2 = 0}}$$

Hybrid-**g**

$$\mathbf{g}_{11} = \left.\frac{\mathbf{I}_1}{\mathbf{V}_1}\right|_{\substack{\mathbf{I}_2 = 0 \\ \text{IIS} = 0}} \qquad \mathbf{g}_{12} = \left.\frac{\mathbf{I}_1}{\mathbf{I}_2}\right|_{\substack{\mathbf{V}_1 = 0 \\ \text{IIS} = 0}}$$

$$\mathbf{g}_{21} = \left.\frac{\mathbf{V}_2}{\mathbf{V}_1}\right|_{\substack{\mathbf{I}_2 = 0 \\ \text{IIS} = 0}} \qquad \mathbf{g}_{22} = \left.\frac{\mathbf{V}_2}{\mathbf{I}_2}\right|_{\substack{\mathbf{V}_1 = 0 \\ \text{IIS} = 0}}$$

$$\mathbf{I}_{1scoc} = -\mathbf{I}_1\Big|_{\substack{\mathbf{V}_1 = 0 \\ \mathbf{I}_2 = 0}} \qquad \mathbf{V}_{2scoc} = \mathbf{V}_2\Big|_{\substack{\mathbf{V}_1 = 0 \\ \mathbf{I}_2 = 0}}$$

TABLE 16.4 Summary of relationships among parameters for three-terminal and two-port networks

	Z	**Y**	**H**	**G**
Z	$\begin{matrix} \mathbf{z}_{11} & \mathbf{z}_{12} \\ \mathbf{z}_{21} & \mathbf{z}_{22} \end{matrix}$	$\begin{matrix} \frac{\mathbf{y}_{22}}{\lvert\mathbf{Y}\rvert} & -\frac{\mathbf{y}_{12}}{\lvert\mathbf{Y}\rvert} \\ -\frac{\mathbf{y}_{21}}{\lvert\mathbf{Y}\rvert} & \frac{\mathbf{y}_{11}}{\lvert\mathbf{Y}\rvert} \end{matrix}$	$\begin{matrix} \frac{\lvert\mathbf{H}\rvert}{\mathbf{h}_{22}} & \frac{\mathbf{h}_{12}}{\mathbf{h}_{22}} \\ -\frac{\mathbf{h}_{21}}{\mathbf{h}_{22}} & \frac{1}{\mathbf{h}_{22}} \end{matrix}$	$\begin{matrix} \frac{1}{\mathbf{g}_{11}} & -\frac{\mathbf{g}_{12}}{\mathbf{g}_{11}} \\ \frac{\mathbf{g}_{21}}{\mathbf{g}_{11}} & \frac{\lvert\mathbf{G}\rvert}{\mathbf{g}_{11}} \end{matrix}$
Y	$\begin{matrix} \frac{\mathbf{z}_{22}}{\lvert\mathbf{Z}\rvert} & -\frac{\mathbf{z}_{12}}{\lvert\mathbf{Z}\rvert} \\ -\frac{\mathbf{z}_{21}}{\lvert\mathbf{Z}\rvert} & \frac{\mathbf{z}_{11}}{\lvert\mathbf{Z}\rvert} \end{matrix}$	$\begin{matrix} \mathbf{y}_{11} & \mathbf{y}_{12} \\ \mathbf{y}_{21} & \mathbf{y}_{22} \end{matrix}$	$\begin{matrix} \frac{1}{\mathbf{h}_{11}} & -\frac{\mathbf{h}_{12}}{\mathbf{h}_{11}} \\ \frac{\mathbf{h}_{21}}{\mathbf{h}_{11}} & \frac{\lvert\mathbf{H}\rvert}{\mathbf{h}_{11}} \end{matrix}$	$\begin{matrix} \frac{\lvert\mathbf{G}\rvert}{\mathbf{g}_{22}} & \frac{\mathbf{g}_{12}}{\mathbf{g}_{22}} \\ -\frac{\mathbf{g}_{21}}{\mathbf{g}_{22}} & \frac{1}{\mathbf{g}_{22}} \end{matrix}$
H	$\begin{matrix} \frac{\lvert\mathbf{Z}\rvert}{\mathbf{z}_{22}} & \frac{\mathbf{z}_{12}}{\mathbf{z}_{22}} \\ -\frac{\mathbf{z}_{21}}{\mathbf{z}_{22}} & \frac{1}{\mathbf{z}_{22}} \end{matrix}$	$\begin{matrix} \frac{1}{\mathbf{y}_{11}} & -\frac{\mathbf{y}_{12}}{\mathbf{y}_{11}} \\ \frac{\mathbf{y}_{21}}{\mathbf{y}_{11}} & \frac{\lvert\mathbf{Y}\rvert}{\mathbf{y}_{11}} \end{matrix}$	$\begin{matrix} \mathbf{h}_{11} & \mathbf{h}_{12} \\ \mathbf{h}_{21} & \mathbf{h}_{22} \end{matrix}$	$\begin{matrix} \frac{\mathbf{g}_{22}}{\lvert\mathbf{G}\rvert} & \frac{-\mathbf{g}_{12}}{\lvert\mathbf{G}\rvert} \\ -\frac{\mathbf{g}_{21}}{\lvert\mathbf{G}\rvert} & \frac{\mathbf{g}_{11}}{\lvert\mathbf{G}\rvert} \end{matrix}$
G	$\begin{matrix} \frac{1}{\mathbf{z}_{11}} & -\frac{\mathbf{z}_{12}}{\mathbf{z}_{11}} \\ \frac{\mathbf{z}_{21}}{\mathbf{z}_{11}} & \frac{\lvert\mathbf{Z}\rvert}{\mathbf{z}_{11}} \end{matrix}$	$\begin{matrix} \frac{\lvert\mathbf{Y}\rvert}{\mathbf{y}_{22}} & \frac{\mathbf{y}_{12}}{\mathbf{y}_{22}} \\ -\frac{\mathbf{y}_{21}}{\mathbf{y}_{22}} & \frac{1}{\mathbf{y}_{22}} \end{matrix}$	$\begin{matrix} \frac{\mathbf{h}_{22}}{\lvert\mathbf{H}\rvert} & -\frac{\mathbf{h}_{12}}{\lvert\mathbf{H}\rvert} \\ -\frac{\mathbf{h}_{21}}{\lvert\mathbf{H}\rvert} & \frac{\mathbf{h}_{11}}{\lvert\mathbf{H}\rvert} \end{matrix}$	$\begin{matrix} \mathbf{g}_{11} & \mathbf{g}_{12} \\ \mathbf{g}_{21} & \mathbf{g}_{22} \end{matrix}$

TABLE 16.5 Summary of relationships among open-circuit voltages and short-circuit currents for three-terminal and two-port networks

$$\begin{bmatrix} \mathbf{V}_{1oc} \\ \mathbf{V}_{2oc} \end{bmatrix} = \begin{bmatrix} \mathbf{z}_{11} & \mathbf{z}_{12} \\ \mathbf{z}_{21} & \mathbf{z}_{22} \end{bmatrix} \begin{bmatrix} \mathbf{I}_{1sc} \\ \mathbf{I}_{2sc} \end{bmatrix} = \begin{bmatrix} 1 & -\frac{\mathbf{h}_{12}}{\mathbf{h}_{22}} \\ 0 & -\frac{1}{\mathbf{h}_{22}} \end{bmatrix} \begin{bmatrix} \mathbf{V}_{1ocsc} \\ -\mathbf{I}_{2ocsc} \end{bmatrix} = \begin{bmatrix} -\frac{1}{\mathbf{g}_{11}} & 0 \\ -\frac{\mathbf{g}_{21}}{\mathbf{g}_{11}} & 1 \end{bmatrix} \begin{bmatrix} -\mathbf{I}_{1scoc} \\ \mathbf{V}_{2scoc} \end{bmatrix}$$

$$\begin{bmatrix} \mathbf{I}_{1sc} \\ \mathbf{I}_{2sc} \end{bmatrix} = \begin{bmatrix} \mathbf{y}_{11} & \mathbf{y}_{12} \\ \mathbf{y}_{21} & \mathbf{y}_{22} \end{bmatrix} \begin{bmatrix} \mathbf{V}_{1oc} \\ \mathbf{V}_{2oc} \end{bmatrix} = \begin{bmatrix} \frac{1}{\mathbf{h}_{11}} & 0 \\ \frac{\mathbf{h}_{21}}{\mathbf{h}_{11}} & -1 \end{bmatrix} \begin{bmatrix} \mathbf{V}_{1ocsc} \\ -\mathbf{I}_{2ocsc} \end{bmatrix} = \begin{bmatrix} -1 & \frac{\mathbf{g}_{12}}{\mathbf{g}_{22}} \\ 0 & \frac{1}{\mathbf{g}_{22}} \end{bmatrix} \begin{bmatrix} -\mathbf{I}_{1scoc} \\ \mathbf{V}_{2scoc} \end{bmatrix}$$

$$\begin{bmatrix} \mathbf{V}_{1ocsc} \\ -\mathbf{I}_{2ocsc} \end{bmatrix} = \begin{bmatrix} 1 & -\frac{\mathbf{z}_{12}}{\mathbf{z}_{22}} \\ 0 & -\frac{1}{\mathbf{z}_{22}} \end{bmatrix} \begin{bmatrix} \mathbf{V}_{1oc} \\ \mathbf{V}_{2oc} \end{bmatrix} = \begin{bmatrix} \mathbf{y}_{11} & 0 \\ \frac{\mathbf{y}_{21}}{\mathbf{y}_{11}} & -1 \end{bmatrix} \begin{bmatrix} \mathbf{I}_{1sc} \\ \mathbf{I}_{2sc} \end{bmatrix} = -\begin{bmatrix} \mathbf{g}_{11} & \mathbf{g}_{12} \\ \mathbf{g}_{21} & \mathbf{g}_{22} \end{bmatrix} \begin{bmatrix} -\mathbf{I}_{1scoc} \\ \mathbf{V}_{2scoc} \end{bmatrix}$$

$$\begin{bmatrix} -\mathbf{I}_{1scoc} \\ \mathbf{V}_{2scoc} \end{bmatrix} = \begin{bmatrix} -\frac{1}{\mathbf{z}_{11}} & 0 \\ -\frac{\mathbf{z}_{21}}{\mathbf{z}_{11}} & 1 \end{bmatrix} \begin{bmatrix} \mathbf{V}_{1oc} \\ \mathbf{V}_{2oc} \end{bmatrix} = \begin{bmatrix} -1 & \frac{\mathbf{y}_{12}}{\mathbf{y}_{22}} \\ 0 & \frac{1}{\mathbf{y}_{22}} \end{bmatrix} \begin{bmatrix} \mathbf{I}_{1sc} \\ \mathbf{I}_{2sc} \end{bmatrix} = \begin{bmatrix} \mathbf{h}_{11} & \mathbf{h}_{12} \\ \mathbf{h}_{21} & \mathbf{h}_{22} \end{bmatrix} \begin{bmatrix} \mathbf{V}_{1ocsc} \\ -\mathbf{I}_{2ocsc} \end{bmatrix}$$

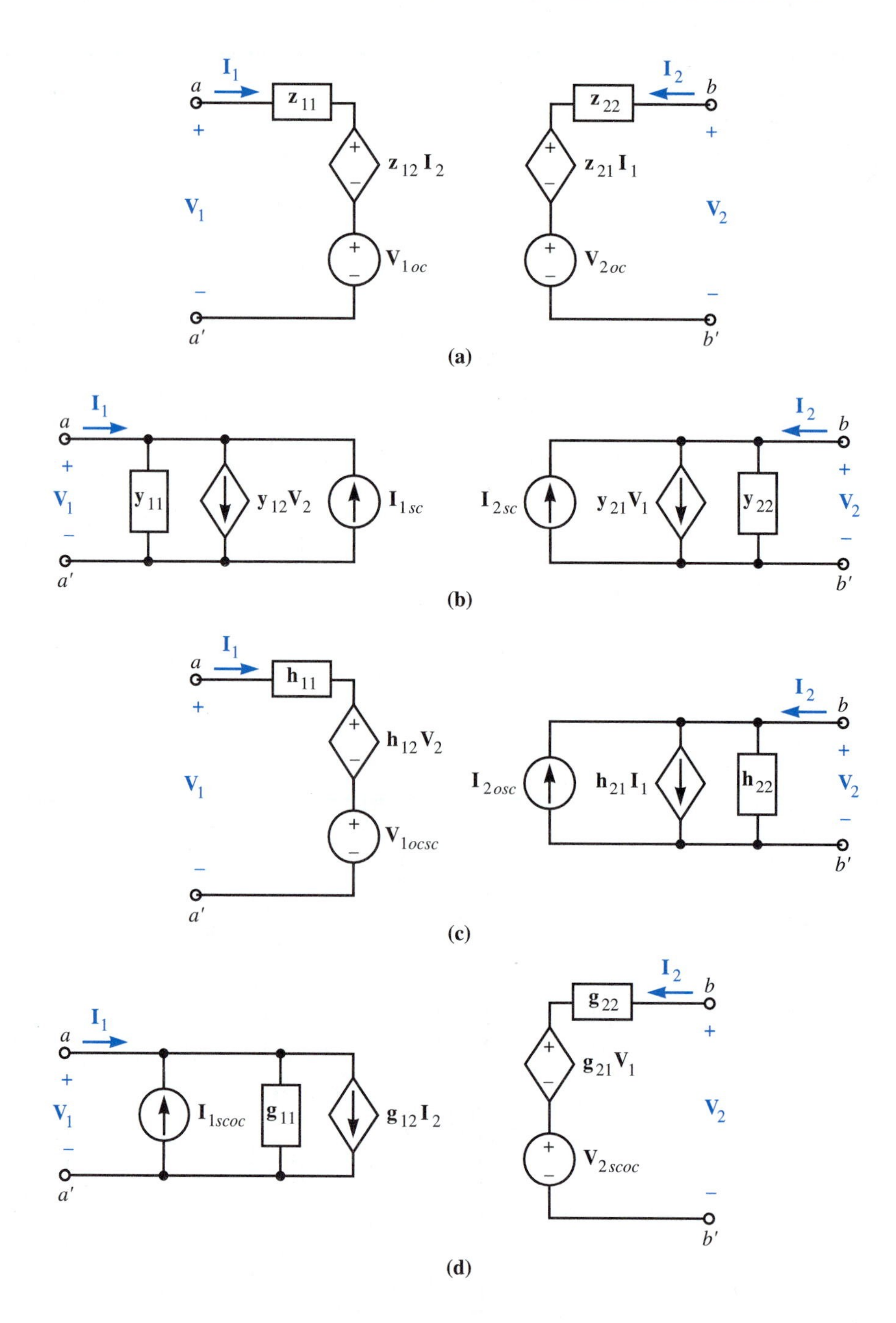

FIGURE 16.32 Equivalent circuits for two ports: (*a*) **z**-parameter; (*b*) **y**-parameter; (*c*) hybrid-**h**; (*d*) hybrid-**g**

EXAMPLE 16.5 Find a **z**-parameter equivalent circuit for the two-port network enclosed in the rectangle in Fig. 16.33.

Solution Observe first that the network in the rectangle does satisfy the two-port property because the net current entering network *A* or *B* must be zero. Therefore Table 16.2 applies. *Assuming* that we are not interested in the voltages $\mathbf{V}_{ab}$ and $\mathbf{V}_{a'b'}$, we can use the equivalent circuit of Fig. 16.32a. Let us now evaluate the parameters of the equivalent circuit.

See Problem 16.2 in the PSpice manual.

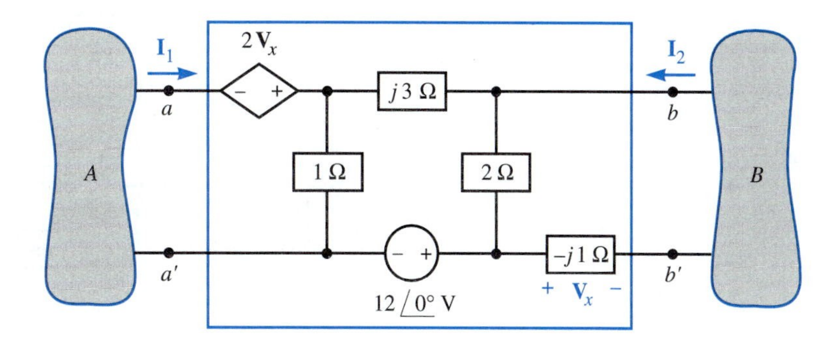

FIGURE 16.33 Two-port network

$\mathbf{V}_{oc}$ is determined by setting $\mathbf{I}_1 = \mathbf{I}_2 = 0$. The condition $\mathbf{I}_2 = 0$ leads immediately to $\mathbf{V}_x = 0$, and inspection yields

$$\mathbf{V}_{1oc} = \frac{1}{3 + j3} 12 \underline{/0°} = 2.83 \underline{/-45°} \text{ V}$$

$$\mathbf{V}_{2oc} = -\frac{2}{3 + j3} 12 \underline{/0°} = 5.66 \underline{/135°} \text{ V}$$

Parameter $\mathbf{z}_{11}$ is the ratio $\mathbf{V}_1/\mathbf{I}_1$ when $\mathbf{I}_2 = 0$ and the 12-V internal source is set to zero. With $\mathbf{I}_2 = 0$, and $\mathbf{V}_x = 0$, an inspection of Fig. 16.34 reveals that

$$\mathbf{z}_{11} = 1 \parallel (2 + j3) = 0.85 \underline{/11.3°} \ \Omega$$

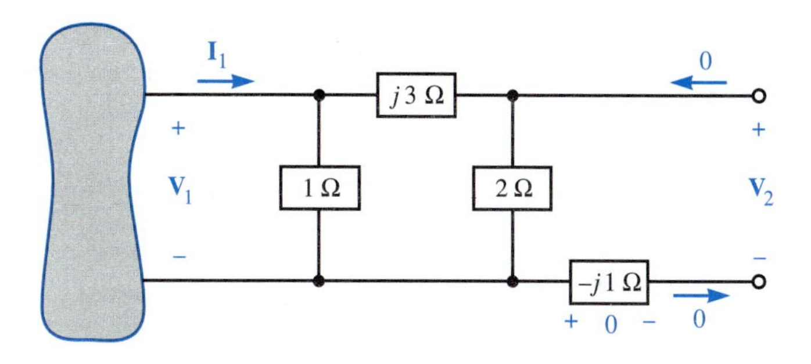

FIGURE 16.34 $\mathbf{I}_2$ and the internal independent source are set to zero.

Similarly,

$$\mathbf{z}_{21} \Bigg|_{\substack{\mathbf{I}_2 = 0 \\ \text{IIS} = 0}} = 2 \frac{1}{3 + j3} = 0.47 \underline{/-45°} \ \Omega$$

The configuration for computing $\mathbf{z}_{22}$ and $\mathbf{z}_{12}$ is shown in Fig. 16.35. Again, we find by inspection that

$$\mathbf{z}_{12} = \frac{\mathbf{V}_1}{\mathbf{I}_2} \Bigg|_{\substack{\mathbf{I}_1 = 0 \\ \text{IIS} = 0}} = 1 \frac{2}{3 + j3} + j2 = 1.7 \underline{/78.7°} \ \Omega$$

and

$$\mathbf{z}_{22} = \frac{\mathbf{V}_2}{\mathbf{I}_2} \Bigg|_{\substack{\mathbf{I}_1 = 0 \\ \text{IIS} = 0}} = 2 \parallel (1 + j3) - j1 = 1.37 \underline{/-14.0°} \ \Omega$$

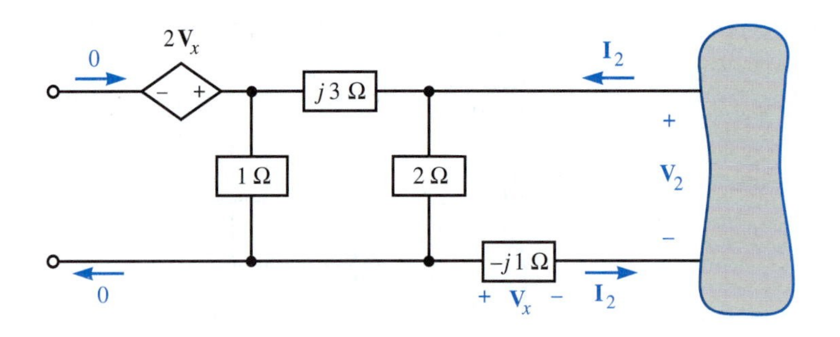

FIGURE 16.35
$\mathbf{I}_1$ and the internal independent source are set to zero.

Therefore, as in the top row of Table 16.2, we have

$$\begin{bmatrix}\mathbf{V}_1\\ \mathbf{V}_2\end{bmatrix} = \begin{bmatrix}0.85\,\underline{/11.3^\circ} & 1.7\,\underline{/78.7^\circ}\\ 0.47\,\underline{/-45^\circ} & 1.37\,\underline{/-14.0^\circ}\end{bmatrix}\begin{bmatrix}\mathbf{I}_1\\ \mathbf{I}_2\end{bmatrix} + \begin{bmatrix}2.83\,\underline{/-45^\circ}\\ 5.66\,\underline{/135^\circ}\end{bmatrix}$$

Remember

The circuits of Fig. 16.32 for the two-port network of Fig. 16.28 make no statement about $\mathbf{V}_{ab}$ and $\mathbf{V}_{a'b'}$.

EXERCISES

18. Give a possible **z, y,** and hybrid-**h**-parameter equivalent circuit for the two-port network shown. State why the term "equivalent circuit" should be used with caution.

answer: $\mathbf{z}_{11} = \frac{2}{3}\,\Omega$, $\mathbf{z}_{12} = \frac{1}{3}\,\Omega$, $\mathbf{z}_{21} = \frac{1}{3}\,\Omega$, $\mathbf{z}_{22} = \frac{2}{3}\,\Omega$, $\mathbf{V}_{1oc} = -4$ V, $\mathbf{V}_{2oc} = 4$ V
$\mathbf{y}_{11} = 2$ S, $\mathbf{y}_{12} = -1$ S, $\mathbf{y}_{21} = -1$ S, $\mathbf{y}_{22} = 2$ S, $\mathbf{I}_{1sc} = -12$ A, $\mathbf{I}_{2sc} = 12$ A
$\mathbf{h}_{11} = \frac{1}{2}\,\Omega$, $\mathbf{h}_{12} = \frac{1}{2}$, $\mathbf{h}_{21} = -\frac{1}{2}$, $\mathbf{h}_{22} = \frac{3}{2}$ S, $\mathbf{V}_{1ocsc} = -6$ V, $\mathbf{I}_{2ocsc} = 6$ A

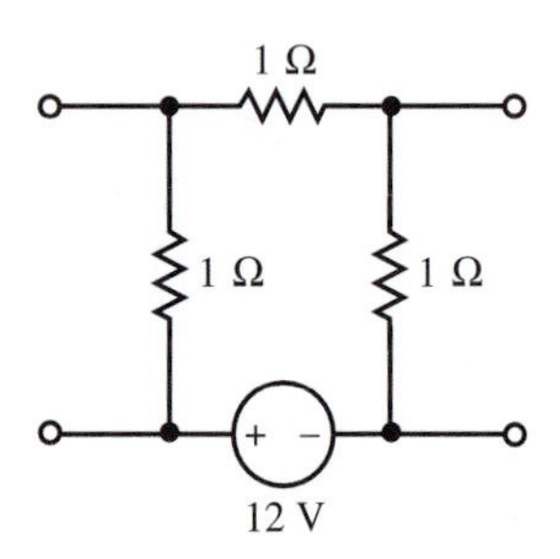

19. Repeat Exercise 18 for the circuit shown.

answer: $\mathbf{z}_{11} = R_1$, $\mathbf{z}_{12} = 0$, $\mathbf{z}_{21} = hR_o$, $\mathbf{z}_{22} = R_o$, $\mathbf{V}_{1oc} = \mathbf{V}_{2oc} = 0$
$\mathbf{y}_{11} = 1/R_1$, $\mathbf{y}_{12} = 0$, $\mathbf{y}_{21} = -h/R_1$, $\mathbf{y}_{22} = 1/R_o$, $\mathbf{I}_{1sc} = \mathbf{I}_{2sc} = 0$
$\mathbf{h}_{11} = R_1$, $\mathbf{h}_{12} = 0$, $\mathbf{h}_{21} = -h$, $\mathbf{h}_{22} = 1/R_o$, $\mathbf{V}_{1ocsc} = \mathbf{I}_{2ocsc} = 0$

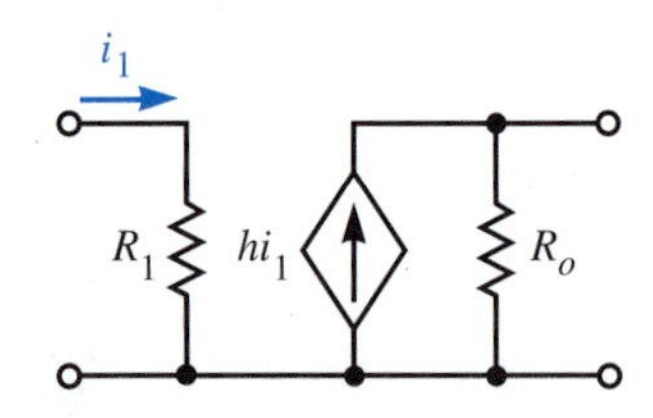

16.9 Delta-to-Wye and Wye-to-Delta Transformations

We conclude this chapter with a presentation of delta-wye and wye-delta transformations. We will see that these transformations occasionally simplify the analysis of a circuit. We begin by defining *wye* (Y) and *delta* (Δ) networks as shown in Fig. 16.36 (a) and (b), respectively. For obvious reasons, these wye and delta networks are also called *tee* (T) and *pi* (Π) networks. The Y and Δ networks shown are equivalent with respect to their effect on any other network if the impedances are related in the manner given by the equations of Fig. 16.36. A derivation of these Δ-to-Y and Y-to-Δ transformations follows Example 16.6.

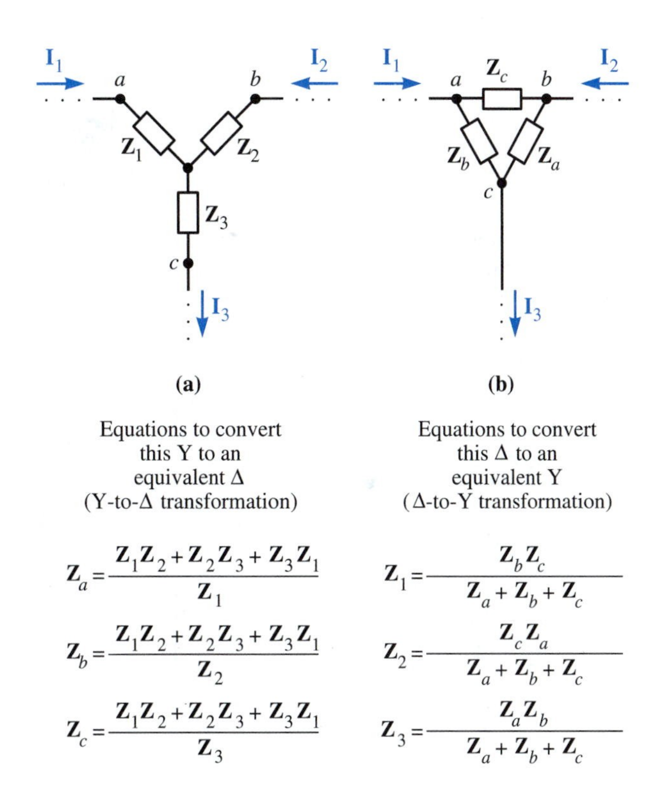

$$\mathbf{Z}_a = \frac{\mathbf{Z}_1\mathbf{Z}_2 + \mathbf{Z}_2\mathbf{Z}_3 + \mathbf{Z}_3\mathbf{Z}_1}{\mathbf{Z}_1} \qquad \mathbf{Z}_1 = \frac{\mathbf{Z}_b\mathbf{Z}_c}{\mathbf{Z}_a + \mathbf{Z}_b + \mathbf{Z}_c}$$

$$\mathbf{Z}_b = \frac{\mathbf{Z}_1\mathbf{Z}_2 + \mathbf{Z}_2\mathbf{Z}_3 + \mathbf{Z}_3\mathbf{Z}_1}{\mathbf{Z}_2} \qquad \mathbf{Z}_2 = \frac{\mathbf{Z}_c\mathbf{Z}_a}{\mathbf{Z}_a + \mathbf{Z}_b + \mathbf{Z}_c}$$

$$\mathbf{Z}_c = \frac{\mathbf{Z}_1\mathbf{Z}_2 + \mathbf{Z}_2\mathbf{Z}_3 + \mathbf{Z}_3\mathbf{Z}_1}{\mathbf{Z}_3} \qquad \mathbf{Z}_3 = \frac{\mathbf{Z}_a\mathbf{Z}_b}{\mathbf{Z}_a + \mathbf{Z}_b + \mathbf{Z}_c}$$

FIGURE 16.36 Equivalent Δ and Y networks: (*a*) Y-connected (T-connected); (*b*) Δ-connected (Π-connected)

The Δ and Y networks will be equivalent with respect to the effect on any other network if the impedances are related in the manner given by the equations of Fig. 16.36.

Example 16.6 illustrates how a Δ-to-Y transformation can simplify the analysis of a circuit.

EXAMPLE 16.6 Determine the Thévenin equivalent impedance for the all resistance network to the right of terminals a and b in Fig. 16.37. Use your result to find current **I**.

Solution The lower three resistances in Fig. 16.37 are connected in a Δ configuration and can be replaced with an equivalent Y. This will not alter the current through any other component.

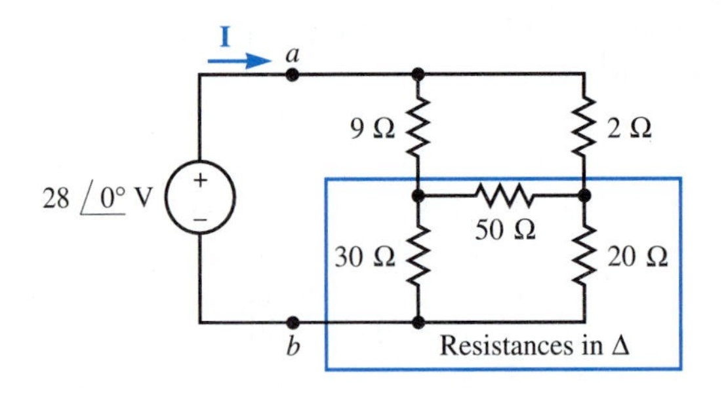

FIGURE 16.37 Network with resistances in Δ

We can obtain the impedance values from the equations given in Fig. 16.36:

$$\mathbf{Z}_1 = \frac{(30)(50)}{20 + 30 + 50} = \frac{1500}{100} = 15\ \Omega$$

$$\mathbf{Z}_2 = \frac{(50)(20)}{100} = 10\ \Omega$$

$$\mathbf{Z}_3 = \frac{(20)(30)}{100} = 6\ \Omega$$

Replacement of the Δ circuit of Fig. 16.37 with its equivalent Y gives the circuit shown in Fig. 16.38.

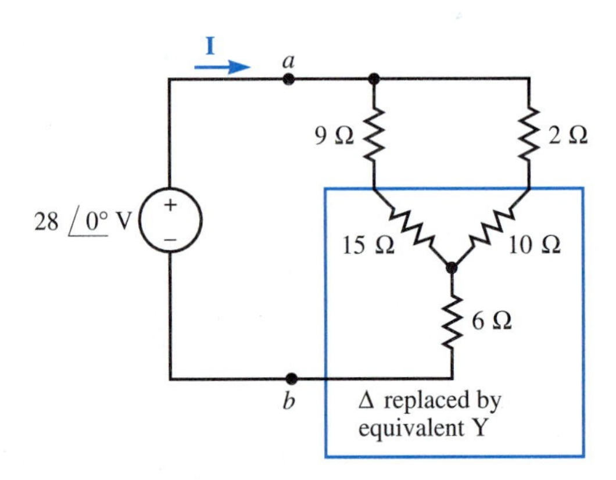

FIGURE 16.38 The first step in simplifying the network of Fig. 16.37

The 9-Ω and 15-Ω resistances are in series, for an equivalent resistance of 24 Ω, and the 10-Ω and 2-Ω resistances are in series, for an equivalent resistance of 12 Ω. The equivalent resistance of these 24-Ω and 12-Ω resistances in parallel is

$$R_p = \frac{1}{(1/24) + (1/12)} = 8\ \Omega$$

Thus the Thévenin equivalent impedance of the network is

$$\mathbf{Z} = 8 + 6 = 14\ \Omega$$

Therefore

$$\mathbf{I} = \frac{28/0^\circ}{\mathbf{Z}} = \frac{28/0^\circ}{14} = 2/0^\circ\ \text{A}$$

We used all resistances in the above example to keep the example simple. Of course, the same method applies for complex impedances. Additional examples of equivalent Δ-Y circuits for all resistance and complex impedance circuits are found in the problems.

Derivation of the Δ-to-Y and Y-to-Δ Transformations

Ohm's law and KVL applied to the network of Fig. 16.36a give

$$\begin{aligned}\mathbf{V}_{ac} &= \mathbf{Z}_1\mathbf{I}_1 + \mathbf{Z}_3\mathbf{I}_3 = \mathbf{Z}_1\mathbf{I}_1 + \mathbf{Z}_3(\mathbf{I}_1 + \mathbf{I}_2) \\ &= (\mathbf{Z}_1 + \mathbf{Z}_3)\mathbf{I}_1 + \mathbf{Z}_3\mathbf{I}_2\end{aligned} \tag{16.40}$$

where the second step required the use of KCL. In a similar manner,

$$\mathbf{V}_{bc} = \mathbf{Z}_3\mathbf{I}_1 + (\mathbf{Z}_2 + \mathbf{Z}_3)\mathbf{I}_2 \tag{16.41}$$

We can solve Eqs. (16.40) and (16.41) for currents $\mathbf{I}_1$ and $\mathbf{I}_2$ by the use of Cramer's rule. This gives

$$\mathbf{I}_1 = \frac{\mathbf{Z}_2 + \mathbf{Z}_3}{\mathbf{Z}_1\mathbf{Z}_2 + \mathbf{Z}_2\mathbf{Z}_3 + \mathbf{Z}_3\mathbf{Z}_1}\mathbf{V}_{ac} - \frac{\mathbf{Z}_3}{\mathbf{Z}_1\mathbf{Z}_2 + \mathbf{Z}_2\mathbf{Z}_3 + \mathbf{Z}_3\mathbf{Z}_1}\mathbf{V}_{bc} \tag{16.42}$$

$$\mathbf{I}_2 = -\frac{\mathbf{Z}_3}{\mathbf{Z}_1\mathbf{Z}_2 + \mathbf{Z}_2\mathbf{Z}_3 + \mathbf{Z}_3\mathbf{Z}_1}\mathbf{V}_{ac} + \frac{\mathbf{Z}_1 + \mathbf{Z}_3}{\mathbf{Z}_1\mathbf{Z}_2 + \mathbf{Z}_2\mathbf{Z}_3 + \mathbf{Z}_3\mathbf{Z}_1}\mathbf{V}_{bc} \tag{16.43}$$

We can easily analyze the network of Fig. 16.36b by the use of KCL and Ohm's law to give

$$\begin{aligned}\mathbf{I}_1 &= \frac{1}{\mathbf{Z}_b}\mathbf{V}_{ac} + \frac{1}{\mathbf{Z}_c}\mathbf{V}_{ab} = \frac{1}{\mathbf{Z}_b}\mathbf{V}_{ac} + \frac{1}{\mathbf{Z}_c}(\mathbf{V}_{ac} - \mathbf{V}_{bc}) \\ &= \left(\frac{1}{\mathbf{Z}_b} + \frac{1}{\mathbf{Z}_c}\right)\mathbf{V}_{ac} - \frac{1}{\mathbf{Z}_c}\mathbf{V}_{bc}\end{aligned} \tag{16.44}$$

where the second step required the use of KVL. In a similar manner,

$$\mathbf{I}_2 = -\frac{1}{\mathbf{Z}_c}\mathbf{V}_{ac} + \left(\frac{1}{\mathbf{Z}_a} + \frac{1}{\mathbf{Z}_c}\right)\mathbf{V}_{bc} \tag{16.45}$$

We can solve Eqs. (16.44) and (16.45) for voltages $\mathbf{V}_{ac}$ and $\mathbf{V}_{bc}$ by the use of Cramer's rule to give

$$\mathbf{V}_{ac} = \frac{\mathbf{Z}_a\mathbf{Z}_b + \mathbf{Z}_b\mathbf{Z}_c}{\mathbf{Z}_a + \mathbf{Z}_b + \mathbf{Z}_c}\mathbf{I}_1 + \frac{\mathbf{Z}_a\mathbf{Z}_b}{\mathbf{Z}_a + \mathbf{Z}_b + \mathbf{Z}_c}\mathbf{I}_2 \tag{16.46}$$

and

$$\mathbf{V}_{bc} = \frac{\mathbf{Z}_a\mathbf{Z}_b}{\mathbf{Z}_a + \mathbf{Z}_b + \mathbf{Z}_c}\mathbf{I}_1 + \frac{\mathbf{Z}_a\mathbf{Z}_b + \mathbf{Z}_c\mathbf{Z}_a}{\mathbf{Z}_a + \mathbf{Z}_b + \mathbf{Z}_c}\mathbf{I}_2 \tag{16.47}$$

The Δ-to-Y transformation of Fig. 16.36 can be established if we equate the coefficients for $\mathbf{I}_1$ and $\mathbf{I}_2$ in Eqs. (16.40) and (16.41) to the corresponding coefficients in Eqs. (16.46) and (16.47). If we equate the coefficients on $\mathbf{V}_{ac}$ and $\mathbf{V}_{bc}$ in Eqs. (16.44) and (16.45) to the corresponding coefficients in Eqs. (16.42) and (16.43), we establish the Y-to-Δ transformation.

EXERCISES

Transform each of the following Δ circuits to a Y circuit. Be sure to label the terminals for the equivalent circuit.

20.

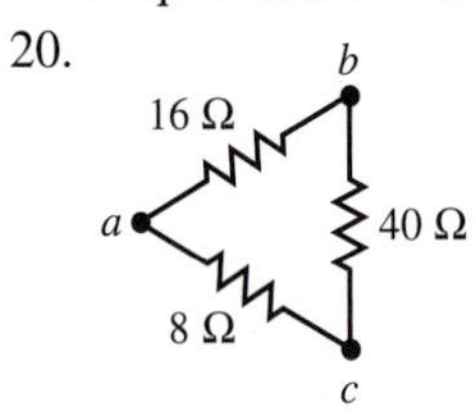

answer: $\mathbf{Z}_1 = 2\ \Omega$, $\mathbf{Z}_2 = 10\ \Omega$, $\mathbf{Z}_3 = 5\ \Omega$

21.

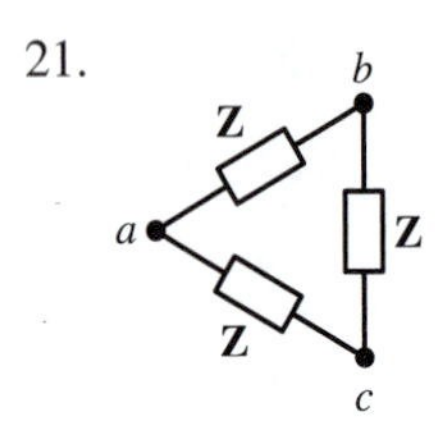

answer: $\mathbf{Z}_1 = \mathbf{Z}_2 = \mathbf{Z}_3 = \mathbf{Z}/3$

Transform each of the following Y circuits to a Δ circuit. Label the terminals for the equivalent circuit.

22.

answer: $\mathbf{Z}_a = 230\ \Omega$, $\mathbf{Z}_b = 184\ \Omega$, $\mathbf{Z}_c = 115\ \Omega$

23.

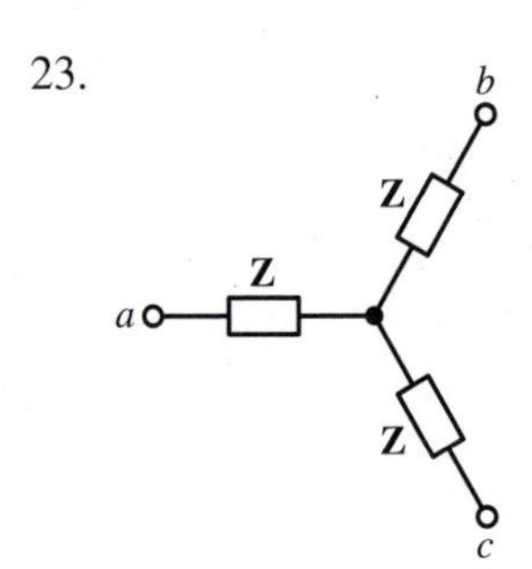

answer: $\mathbf{Z}_a = \mathbf{Z}_b = \mathbf{Z}_c = 3\mathbf{Z}$

24. In the circuit below, use a Y-Δ transformation to calculate currents $\mathbf{I}_1$ and $\mathbf{I}_2$.

answer: $\mathbf{I}_1 = \mathbf{I}_2 = 9$ A

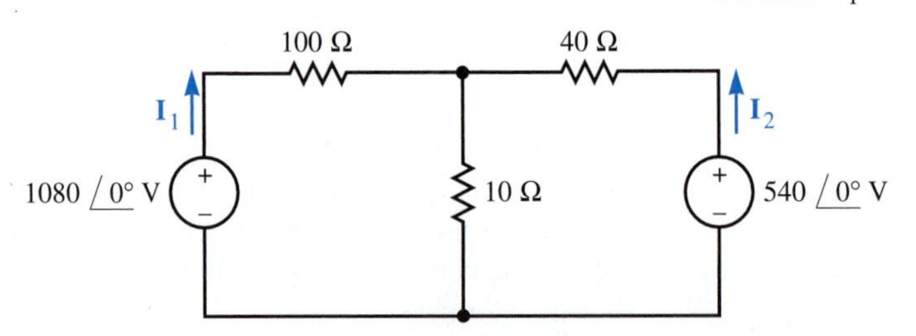

25. Find the equivalent resistance for the resistive network connected between terminals a and b in the circuit below. Use this to calculate current **I**.

answer: $\mathbf{I} = 4$ A

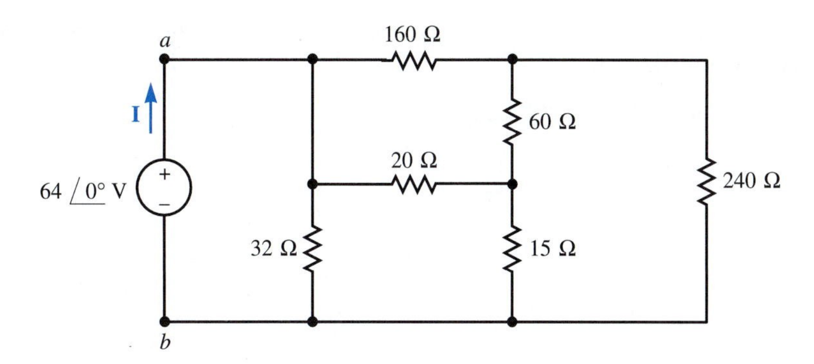

16.10 Summary

In this chapter we obtained **z**-, **y**-, **h**-, and **g**-parameter equivalent circuits for three-terminal and two-port networks. Tables 16.2 through 16.5 summarize the results. We also introduced Y-to-Δ and Δ-to-Y transformations that can be used for network simplification.

KEY FACTS	Concept	Equation	Section	Page
❑	For certain theoretical circuits, one or more of the equivalent circuits may not exist.		16.5	605
❑	Four equivalent circuits are available to model three-terminal and two-port networks. The four equivalent circuits are summarized in Tables 16.2 through 16.5.		16.7	611
❑	Two-port networks are four-terminal networks with the following special property: The net current entering the input terminal pair equals zero and the net current entering the output terminal pair equals zero.		16.7	612
❑	The equivalent circuits for two-port networks are limited because they do not describe the voltages V_{ab} and $V_{a'b'}$ (Fig. 16.28).		16.8	613
❑	A Y circuit can be transformed into an equivalent Δ circuit and vice versa. The transformations are given in Fig. 16.36.		16.9	619

PROBLEMS

Section 16.2

1. Find the impedance-parameter equivalent of the ladder network shown.

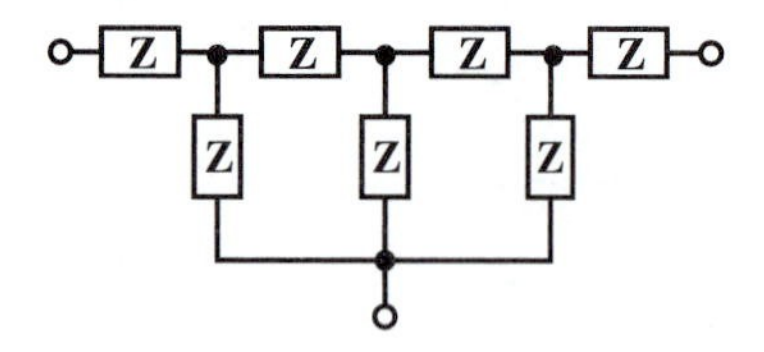

2. Find the impedance-parameter equivalent of the circuit shown.

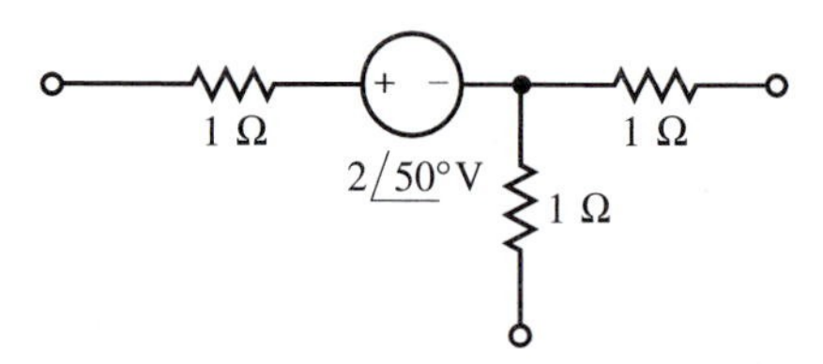

3. Find the impedance-parameter equivalent of the bridged-T circuit shown.

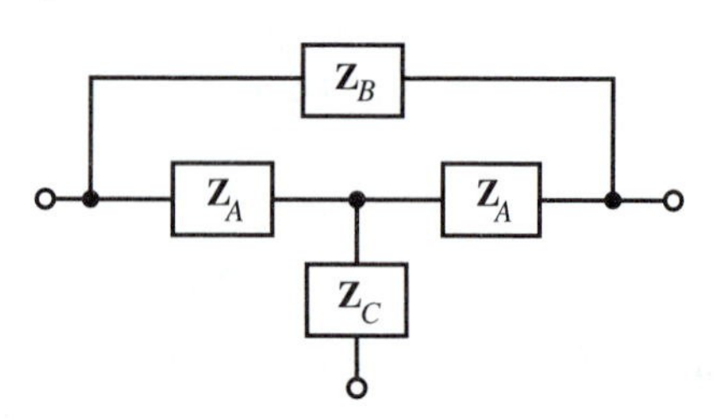

4. A three-terminal network is driven by a voltage source with source impedance $\mathbf{Z}_s$ as shown. Use the impedance-parameter equivalent network of Fig. 16.4, and determine the Thévenin equivalent looking into terminals a and b. (Find $\mathbf{V}_{Th}$ and $\mathbf{Z}_{Th}$.)

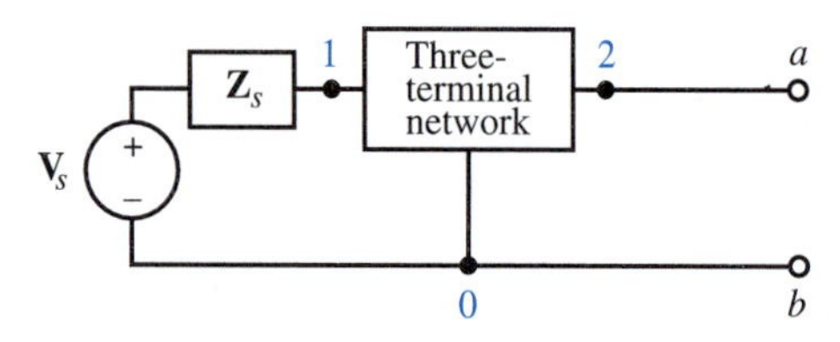

5. Assume that the three-terminal network shown in Problem 4 does not contain independent sources. Assume also that a load impedance $\mathbf{Z}_L$ is connected across terminals a and b. Determine the transfer admittance $\mathbf{I}_{ab}/\mathbf{V}_s$ in terms of $\mathbf{Z}_L$, $\mathbf{Z}_s$, and the impedance parameters.

6. Find the impedance-parameter equivalent circuit of the circuit shown.

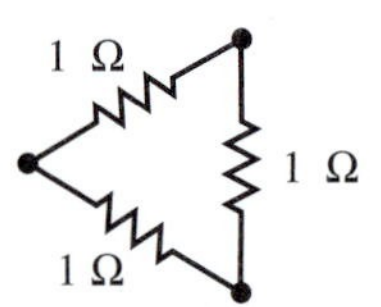

7. An engineer knows the open-circuit voltages $\mathbf{V}_{1oc}$ and $\mathbf{V}_{2oc}$ and the short-circuit currents $\mathbf{I}_{1sc}$ and $\mathbf{I}_{2sc}$ of a certain three-terminal network. By using only this information, can the engineer determine *any* of the impedance parameters? Explain.

Section 16.3

8. Find the admittance-parameter equivalent circuit of the network of Problem 1.

9. Find the admittance-parameter equivalent of the circuit of Problem 2.

10. Find the admittance-parameter equivalent of the parallel-T network shown.

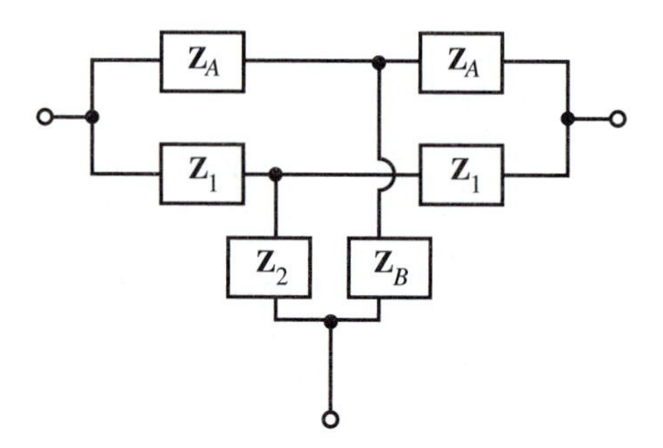

11. Assume that the three-terminal network shown does not contain independent sources. Determine the Norton equivalent of the circuit looking into terminals a and b in terms of the admittance parameters of the three-terminal network.

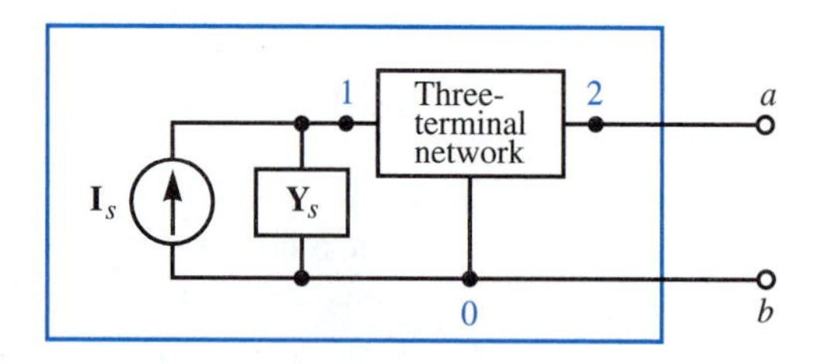

Section 16.4

12. A certain three-terminal network has admittance parameters $\mathbf{y}_{11} = 2\,\underline{/0^\circ}$ S, $\mathbf{y}_{22} = 1\,\underline{/0^\circ}$ S, $\mathbf{y}_{12} = 1\,\underline{/45^\circ}$ S and $\mathbf{y}_{21} = 1\,\underline{/-45^\circ}$ S. Find the impedance parameters.

13. The open-circuit voltages of the network of Problem 12 are given by

$$\mathbf{V}_{oc} = \begin{bmatrix} 3\,\underline{/60^\circ} \\ 1 + j \end{bmatrix} \text{V}$$

Find the short-circuit currents.

14. Discuss the similarities and the differences between Eqs. (16.27) and (16.28) and the corresponding results for Thévenin and Norton equivalents of two-terminal networks.

Section 16.5

15. We have shown that certain circuit models are not possible. On the other hand, at our own peril, we can interconnect physical components in any way we wish. Give some examples to show why it can be unwise to try to physically construct a circuit corresponding to an impossible circuit model.

Section 16.6

16. Find the hybrid-**h**-parameter equivalent of the ladder network of Problem 1.

17. Find the hybrid-**h**-parameter equivalent of the network of Problem 2.

18. Find the hybrid-**h**-parameter equivalent of the network shown below.

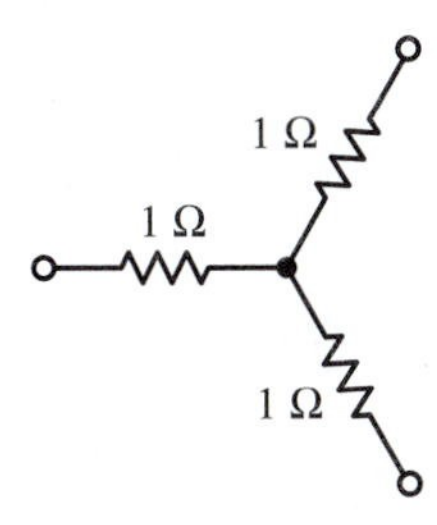

19. Use the simplified hybrid-**h** transistor model shown in Fig. 16.26 to determine the relationship between i_1 and i_2 in the circuit shown.

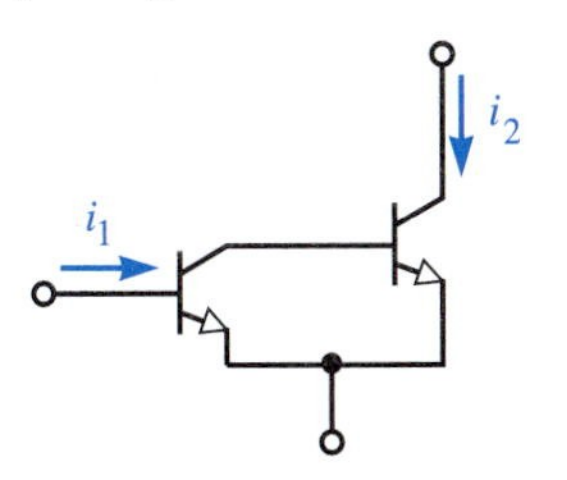

20. Hybrid-**g** equivalent circuits are frequently used to model field-effect transistors (FETs). The terminal equations for the hybrid-**g** representation of an ac network containing independent sources have the form

$$\mathbf{I}_1 = \mathbf{g}_{11}\mathbf{V}_1 + \mathbf{g}_{12}\mathbf{I}_2 - \mathbf{I}_{1scoc}$$
$$\mathbf{V}_2 = \mathbf{g}_{21}\mathbf{V}_1 + \mathbf{g}_{22}\mathbf{I}_2 + \mathbf{V}_{2scoc}$$

(a) Proceed in a manner similar to that of Sections 16.2, 16.3, and 16.6 to develop formulas for the constants appearing in the above equations.

(b) Draw the hybrid-**g** equivalent circuit.

(c) Show that

$$\begin{bmatrix} \mathbf{g}_{11} & \mathbf{g}_{12} \\ \mathbf{g}_{21} & \mathbf{g}_{22} \end{bmatrix} = \begin{bmatrix} \mathbf{h}_{11} & \mathbf{h}_{12} \\ \mathbf{h}_{21} & \mathbf{h}_{22} \end{bmatrix}^{-1}$$

where the $\mathbf{h}_{ij}$'s are the hybrid-**h** parameters of the circuit.

(d) Show that

$$\begin{pmatrix} -\mathbf{I}_{1scoc} \\ \mathbf{V}_{2scoc} \end{pmatrix} = -\begin{bmatrix} \mathbf{g}_{11} & \mathbf{g}_{12} \\ \mathbf{g}_{21} & \mathbf{g}_{22} \end{bmatrix} \begin{pmatrix} \mathbf{V}_{1ocsc} \\ -\mathbf{I}_{2ocsc} \end{pmatrix}$$

where $\mathbf{V}_{1ocsc}$ and $\mathbf{I}_{2ocsc}$ appear in the hybrid-**h** model of Fig. 16.21.

Section 16.7

21. In your own words, explain the difference between an ordinary four-terminal network and a two-port network.

22. In your own words, explain whether or not a three-terminal network can be treated as a two-port network.

Section 16.8

23. Give the **z**-, **y**-, and **h**-parameter equivalents for the circuit shown.

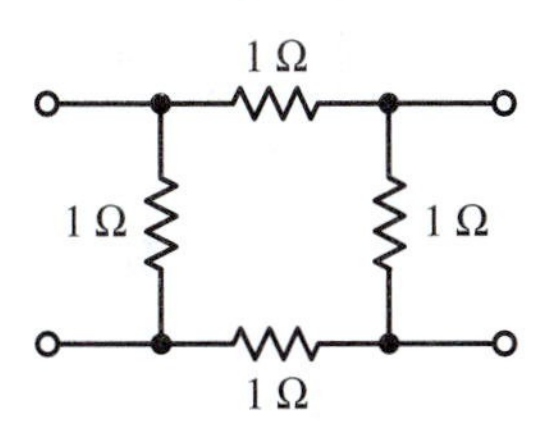

24. Repeat Problem 23 for the circuit shown.

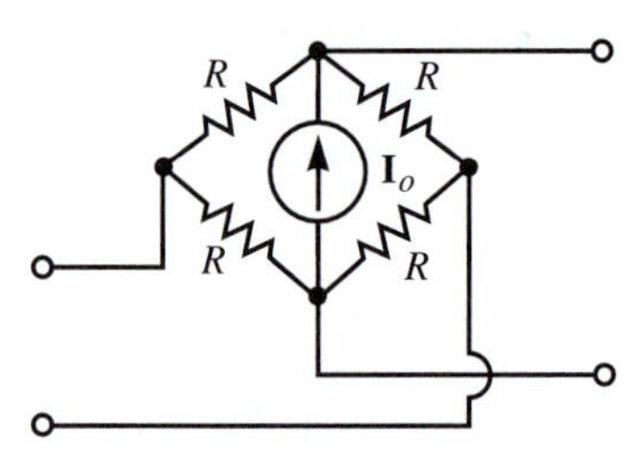

25. Repeat Problem 23 for the circuit shown.

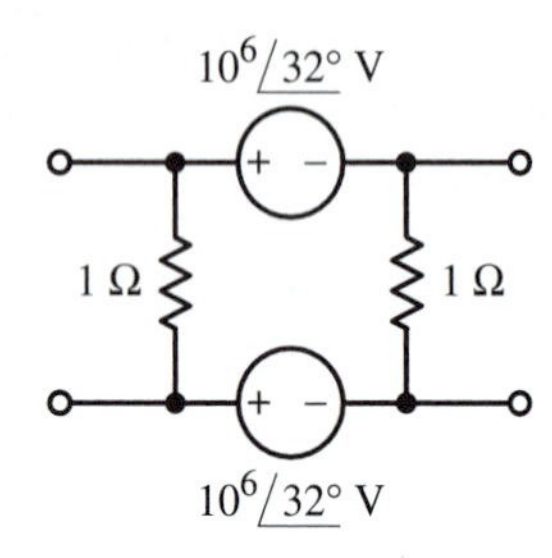

26. Repeat Problem 23 for the circuit shown.

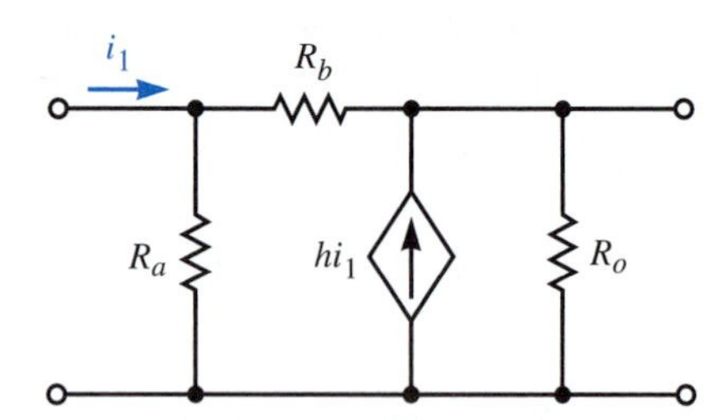

27. The term "equivalent circuit" should be used with special care for two-port networks. Explain why.

28. The figure shown is a *series* connection of two four-terminal networks *A* and *B*.
    (a) Show that if networks *A* and *B* each have the two-port property, then the composite network enclosed in blue automatically has the two-port property.
    (b) Show that if the two-port requirement is satisfied by the composite network (enclosed in blue), then the two-port requirement is *not necessarily* satisfied by either network *A* or network *B*.

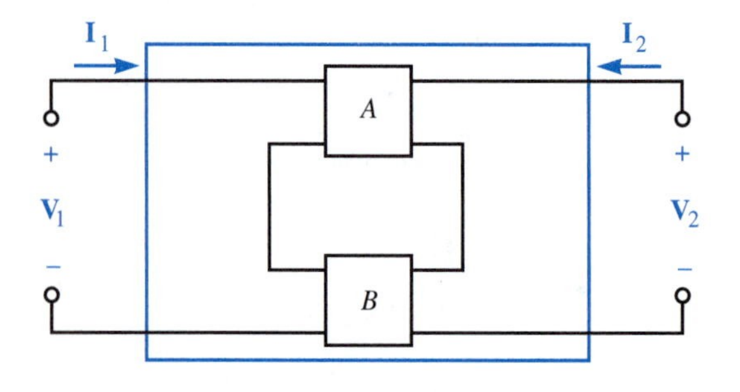

29. Show that if networks *A* and *B* of Problem 28 each satisfy the two-port requirement, then the **z**-parameter matrix of the composite network is given by the sum of the **z**-parameter matrices of networks *A* and *B*.

30. The network shown is a *parallel* connection of two four-terminal networks *A* and *B*.
    (a) Show that if networks *A* and *B* each have the two-port property, then the composite network (enclosed in blue) automatically has the two-port property.
    (b) Show that if the two-port requirement is satisfied by the composite network enclosed in blue, then the two-port requirement is *not necessarily* satisfied by either network *A* or network *B*.

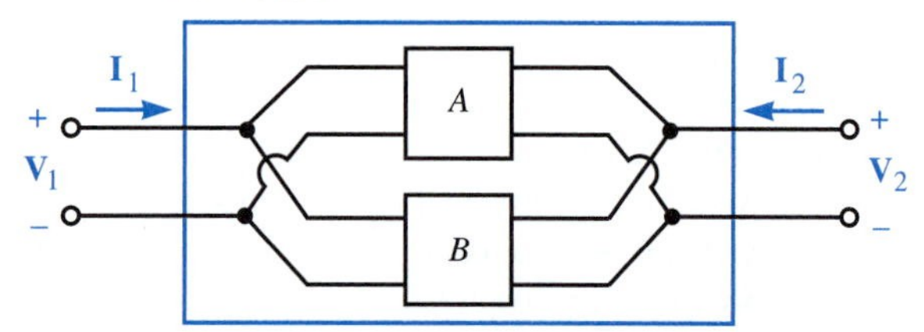

31. Show that if networks *A* and *B* of Problem 30 each satisfy the two-port requirement, then the **y**-parameter matrix of the composite network is given by the sum of the **y**-parameter matrices of networks *A* and *B*.

Section 16.9

Determine an equivalent Δ for each Y circuit and an equivalent Y for each Δ circuit. (Be sure to label the terminals on the equivalent circuit.)

32.

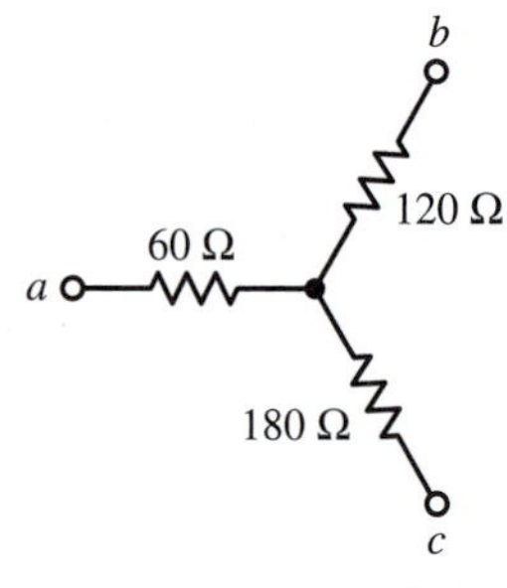

33.

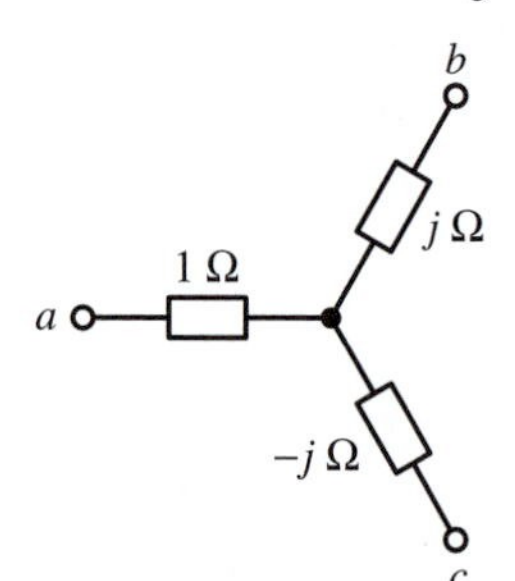

34.

35.

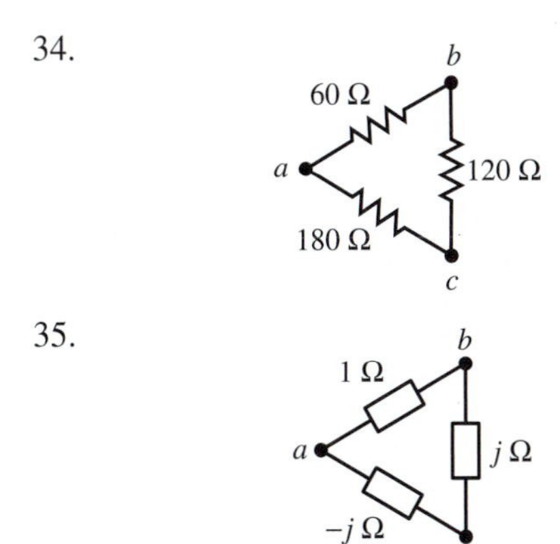

Find the current i or the voltage v for the networks in Problems 36 and 37.

36.

37.

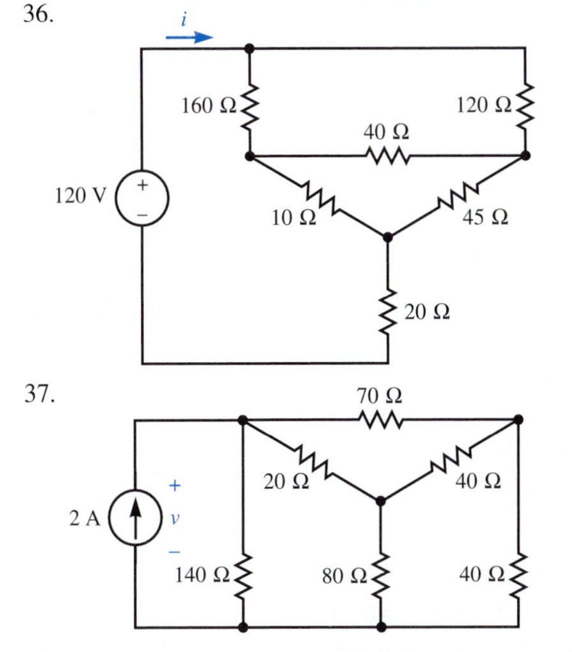

38. Determine $\mathbf{I}_{aA}$, $\mathbf{I}_{bB}$, and $\mathbf{I}_{cC}$ for the network shown below. Then make a Δ-Y transformation and calculate $\mathbf{V}_{AB}$, $\mathbf{V}_{BC}$, and $\mathbf{V}_{CA}$.

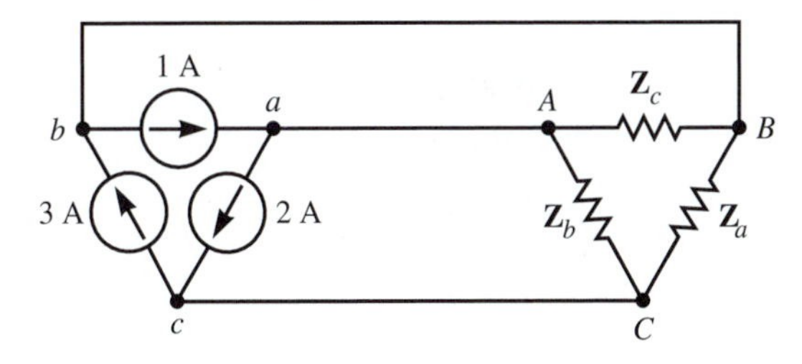

Supplementary Problems

39. The three-terminal network in the circuit shown contains no independent sources. Find an expression for the voltage transfer function $\mathbf{V}_o/\mathbf{V}_s$ in terms of the $\mathbf{z}$ parameters of the three-terminal network.

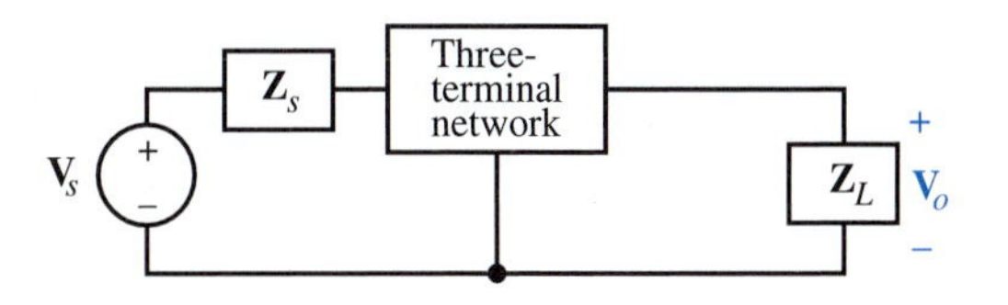

40. Repeat Problem 39 with the use of $\mathbf{y}$ parameters.

41. Repeat Problem 39 with the use of hybrid-$\mathbf{h}$ parameters.

42. Draw the Bode amplitude and phase plots of the voltage transfer function of the transistor amplifier of Fig. 16.25. Assume that $C_1 = C_2 = 1000\ \mu\text{F}$, $R_s = 100\ \Omega$, $R_L = 18\ \text{k}\Omega$, $R_1 = 48\ \text{k}\Omega$, $R_2 = 22\ \text{k}\Omega$, $R_E = 2.2\ \text{k}\Omega$, and $R_C = 4.1\ \text{k}\Omega$. Use the transistor model of Fig. 16.26 with $\beta = 100$. Compare your plots with the design formula given in the solution to Example 16.4.

43. Replace the network to the right of terminals *a*, *b*, and *c* in the following figure with a single Δ-connected network and then calculate i_1 and i_2.

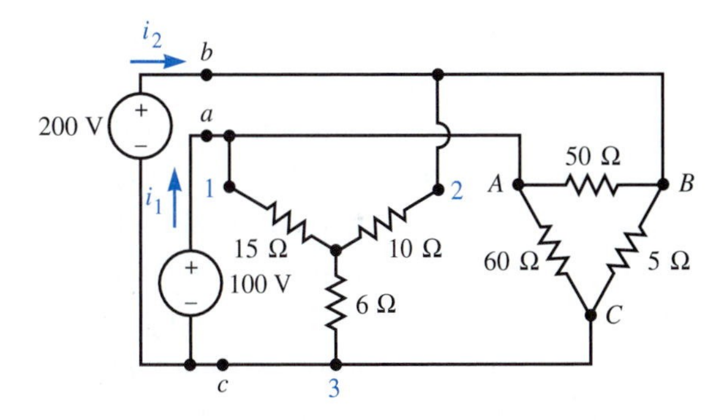

44. Determine $\mathbf{V}_{AB}$, $\mathbf{V}_{BC}$, and $\mathbf{V}_{CA}$ for the network shown below. Then make a Y-Δ transformation and calculate $\mathbf{I}_{aA}$, $\mathbf{I}_{bB}$, and $\mathbf{I}_{cC}$.

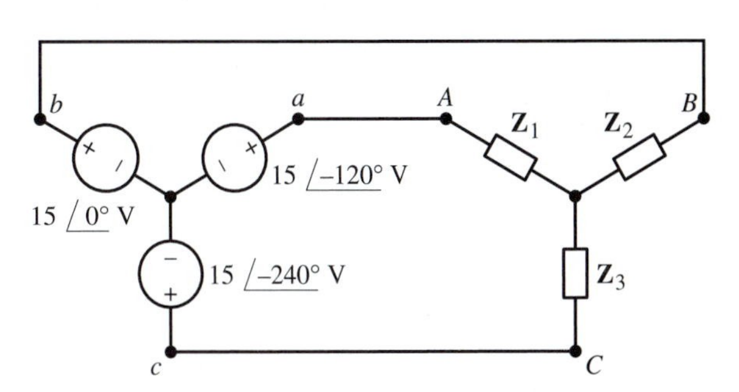

17

Mutual Inductance and Transformers

Tremendous excitement was created in the scientific community in 1831 when Michael Faraday experimentally established that the change in magnetic flux due to a change of current in one conductor induced a voltage in another conductor linked by that flux. Today we say that these conductors are *magnetically coupled.* This is an undesirable effect when power lines induce a voltage in adjacent telephone lines, but magnetically coupled inductors (coils) are beneficially used in electronics and power systems. For example, power systems use coupled inductors to form *transformers* that efficiently convert from the generation voltage to a higher voltage for power transmission. Other transformers convert this high voltage to a more convenient lower voltage for use by the customer.

In this chapter we analyze circuits with magnetic coupling and develop models for practical transformers. We also introduce our last network component, the *ideal transformer,* and analyze circuits that include these components.

17.1 Mutual Inductance in the Time Domain

Coils of wire placed in close proximity form coupled inductors. We usually place one coil inside the other, as shown in Fig. 17.1, so that a large fraction of the magnetic flux that links (is enclosed by) one coil will also link the other.

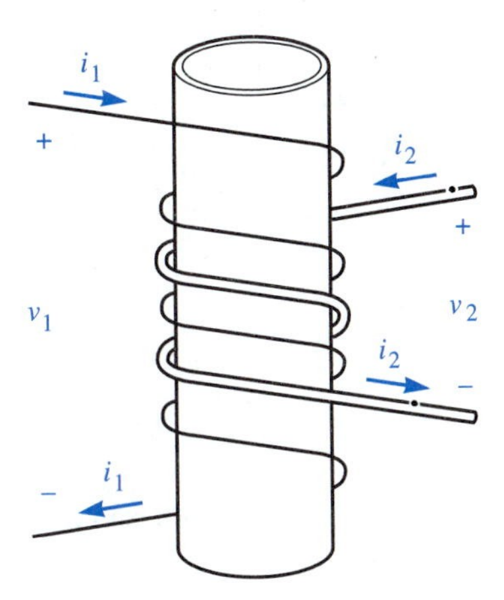

FIGURE 17.1 Two magnetically coupled coils wound on a plastic tube

In Fig. 17.2 we have drawn two coils side by side to permit a clearer picture. We have assigned a dot mark (•) to the end of the coils at terminals *a* and *c*. This is because current into the dot-marked end of each coil causes magnetic flux to pass through the coils in the same direction. Not all flux linking (passing through) one turn of a coil will link the others, so the fluxes shown are the average flux per turn. This simplified model yields the same result as one that considers the individual turns of the coils. With the wire resistance neglected, Faraday's law gives the voltage v_1 across the coil with N_1 turns:

$$v_1 = N_1 \frac{d}{dt} \phi_1 = N_1 \frac{d}{dt} \phi_{11} + N_1 \frac{d}{dt} \phi_{12} \tag{17.1}$$

where $\phi_1 = \phi_{11} + \phi_{12}$ is the total flux linking coil 1. For linear magnetic material, flux is proportional to current, so we can use Eq. (2.34) to write flux ϕ_{11} in terms of the current i_1 and the *self-inductance* L_1:

$$\phi_{11} = \frac{L_1 i_1}{N_1} \tag{17.2}$$

where L_1 is a constant. In a similar way, we can write flux ϕ_{12} in terms of current i_2 and the *mutual inductance* M_{12} between coil 1 and coil 2:

$$\phi_{12} = \frac{M_{12} i_2}{N_1} \tag{17.3}$$

where M_{12} is constant. We can now substitute Eqs. (17.2) and (17.3) into Eq. (17.1) to obtain voltage v_1 as a function of currents i_1 and i_2:

$$v_1 = L_1 \frac{d}{dt} i_1 + M_{12} \frac{d}{dt} i_2 \tag{17.4}$$

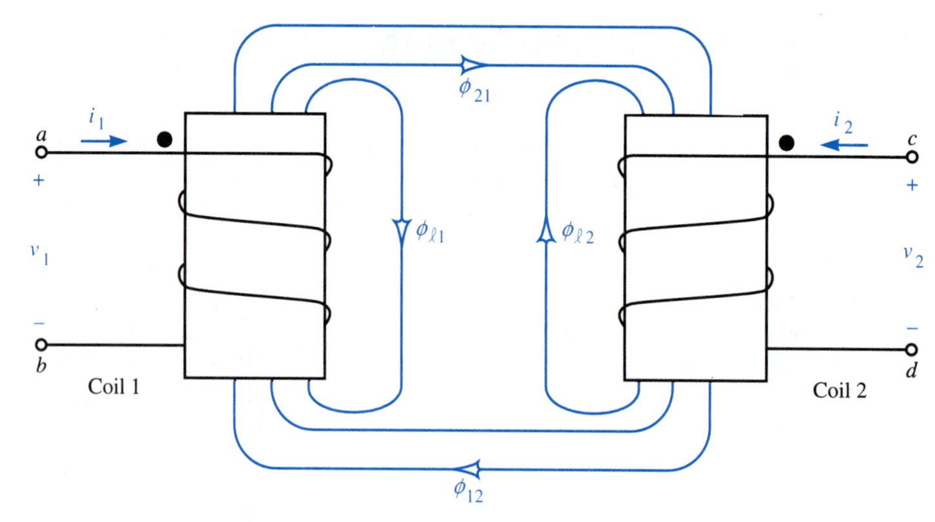

FIGURE 17.2
A simplified model for coupled coils. The dot marks (•) by terminals *a* and *c* of coils 1 and 2 indicate that currents into terminals *a* and *c* cause magnetic flux to pass through the two coils in the same direction.

$\phi_{\ell 1}$ The flux due to current i_1, which links only coil 1 (leakage flux)

$\phi_{\ell 2}$ The flux due to current i_2, which links only coil 2 (leakage flux)

ϕ_{21} The flux due to current i_1, which links coil 2

ϕ_{12} The flux due to current i_2, which links coil 1

$\phi_{11} = \phi_{\ell 1} + \phi_{21}$

The total flux due to current i_1, which links coil 1

$\phi_{22} = \phi_{\ell 2} + \phi_{12}$

The total flux due to current i_2, which links coil 2

We can also obtain voltage v_2 as a function of currents i_1 and i_2. We use the self-inductance L_2 of coil 2 to write

$$\phi_{22} = \frac{L_2 i_2}{N_2} \tag{17.5}$$

We define the mutual inductance M_{21} between coil 2 and coil 1 by

$$\phi_{21} = \frac{M_{21} i_1}{N_2} \tag{17.6}$$

This lets us write

$$v_2 = L_2 \frac{d}{dt} i_2 + M_{21} \frac{d}{dt} i_1 \tag{17.7}$$

As we establish later in this chapter,

$$M = M_{12} = M_{21} \tag{17.8}$$

so the subscripts will be omitted.

We modeled inductors with network components called inductances. We now extend this model to coupled coils by formally defining coupled inductances.

DEFINITION Coupled Inductances

A voltage is induced across an inductance that is proportional to the derivative of the current through the inductance. This constant of proportionality is the inductance or *self-inductance* L. For coupled inductances, an additional voltage is induced that is proportional to the derivative of the current through the second inductance. This constant of proportionality is the *mutual inductance* M. The terminal equations for coupled inductances are

$$v_1 = L_1 \frac{d}{dt} i_1 + M \frac{d}{dt} i_2 \tag{17.9}$$

$$v_2 = L_2 \frac{d}{dt} i_2 + M \frac{d}{dt} i_1 \tag{17.10}$$

and the component symbol is shown in Fig. 17.3.

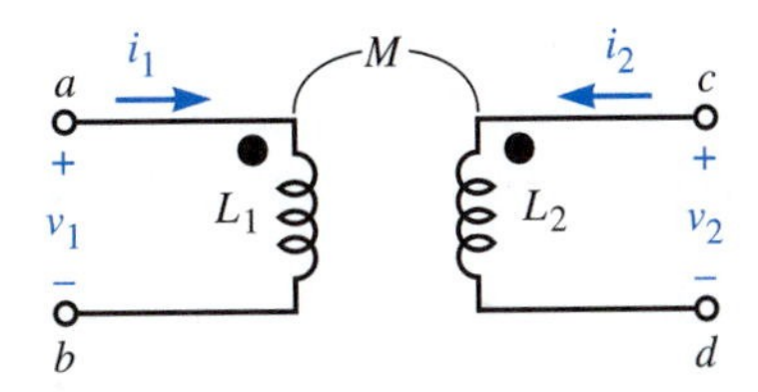

FIGURE 17.3
Symbol for coupled inductances

The unit of both self-inductance and mutual inductance is the henry (H). Equations (17.9) and (17.10) imply that currents i_1 and i_2 cause flux through the coils that is in the same direction (additive), as shown in Figs. 17.1 and 17.2. The relative directions in which the coils are wound are implied by the dot marks shown in Figs. 17.2 and 17.3. Current into the dot-marked end of each coil causes flux to pass through the coils in the same direction.

Remember that the sign on the self-induced voltage term is positive for the passive sign convention. The sign on the mutually induced voltage term is the same as the sign on the self-induced term, if the current reference arrows are oriented the same way relative to the corresponding dot marks. This is true because currents that are oriented the same way relative to the corresponding dot marks cause flux through the coils in the same direction.

EXAMPLE 17.1 Write a set of mesh-current equations for the network of Fig. 17.4.

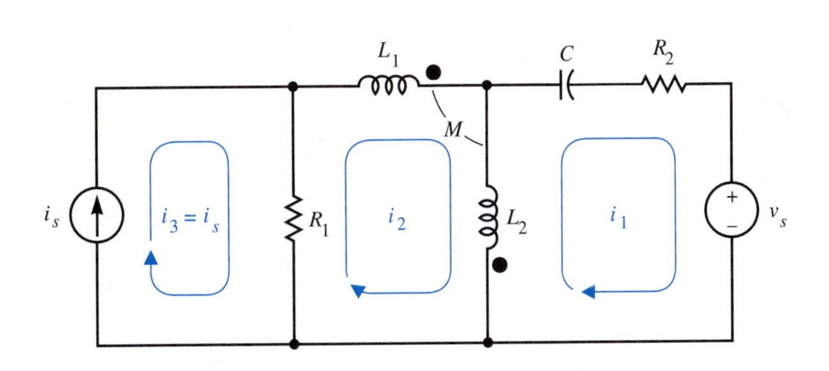

FIGURE 17.4
Mesh-current example with mutual inductance

Solution Assign the mesh currents as shown. Write a KVL equation around mesh 1 and around mesh 2:

$$L_2 \frac{d}{dt}(i_1 - i_2) + M\frac{d}{dt}(-i_2) + \frac{1}{C}\int_{-\infty}^{t} i_1\, d\lambda + R_2 i_1 + v_s = 0$$

$$R_1(i_2 - i_s) + L_1\frac{d}{dt}i_2 + M\frac{d}{dt}(i_2 - i_1) + L_2\frac{d}{dt}(i_2 - i_1) + M\frac{d}{dt}i_2 = 0$$

where the third mesh equation,

$$i_3 = i_s$$

was used to eliminate i_3 from the second mesh equation when it was written.

Because circuits with mutual inductance are most often analyzed in the frequency domain, this chapter is designed so that we will not lose continuity if we skip to Section 17.2 after working Exercises 1 and 2.

Remember

Mutual inductance is our network model that accounts for the transfer of electrical energy from one inductor to another by magnetic coupling. Mutual inductance accounts for a voltage induced across one coil as a result of a change in current through another coil. Current into like-marked ends of coils (dotted or undotted) will cause flux through the coils in the same direction.

EXERCISES

1. A network with two coupled inductances is shown below. Use derivatives to express v_{ab} and v_{cb} as functions of i_x and i_y.

   answer: $(L_1 + M)di_y/dt - Mdi_x/dt,\ -(L_2 + M)di_y/dt + L_2 di_x/dt$

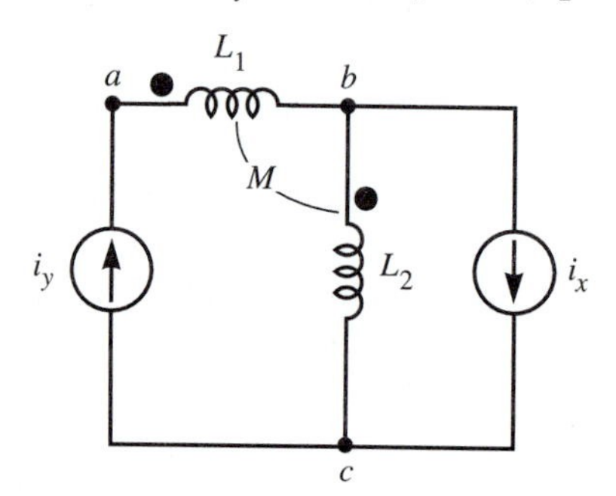

2. Write mesh-current equations for the network shown below.

   answer:
$$\begin{bmatrix} 4\frac{d}{dt} + 3 & -6\frac{d}{dt} \\ -6\frac{d}{dt} & 14\frac{d}{dt} + 5 + \frac{1}{3}\int_{-\infty}^{t} d\lambda \end{bmatrix} \begin{bmatrix} i_1 \\ i_2 \end{bmatrix} = \begin{bmatrix} v_s \\ 0 \end{bmatrix}$$

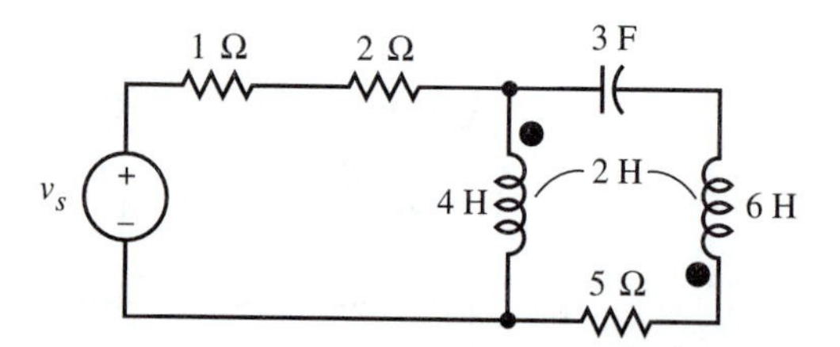

Other Considerations†

The *magnetic coefficient of coupling* k is defined by the equation

$$k = \frac{M}{\sqrt{L_1 L_2}} \tag{17.11}$$

With the magnetic flux variables defined as in Fig. 17.2, this can be written as

$$k = \frac{\sqrt{M_{21}}\,\sqrt{M_{12}}}{\sqrt{L_1 L_2}} = \frac{\sqrt{N_2 \dfrac{\phi_{21}}{i_1}}\sqrt{N_1 \dfrac{\phi_{12}}{i_2}}}{\sqrt{N_1 \dfrac{\phi_{11}}{i_1}}\sqrt{N_2 \dfrac{\phi_{22}}{i_2}}} = \sqrt{\frac{\phi_{21}}{\phi_{11}}}\sqrt{\frac{\phi_{12}}{\phi_{22}}} \tag{17.12}$$

Because ϕ_{21} is a fraction of ϕ_{11} and ϕ_{12} is a fraction of ϕ_{22}, the two ratios in the preceding equation lie between zero and one, and

$$0 \le k \le 1 \tag{17.13}$$

This bound can also be justified from energy considerations.

We know that magnetic fields store energy. For circuit analysis we need to relate the energy stored in coupled coils to voltage and current measurements. We begin by examining the input power to coupled inductances. The total power p absorbed by the coupled inductances of Fig. 17.3 is the sum of p_1, the power input to coil 1, and p_2, the power input to coil 2. Thus

$$\begin{aligned} p &= p_1 + p_2 \\ &= i_1 v_1 + i_2 v_2 \\ &= i_1\left(L_1 \frac{d}{dt} i_1 + M \frac{d}{dt} i_2\right) + i_2\left(L_2 \frac{d}{dt} i_2 + M \frac{d}{dt} i_1\right) \end{aligned} \tag{17.14}$$

Equation (17.14) can be written as

$$p = L_1 i_1 \frac{d}{dt} i_1 + M \frac{d}{dt}(i_1 i_2) + L_2 i_2 \frac{d}{dt} i_2 \tag{17.15}$$

The three terms on the right-hand side of the preceding equation are exact differentials. We can integrate these to show that the energy stored in coupled inductances is

$$w = \frac{1}{2} L_1 i_1^2 + M i_1 i_2 + \frac{1}{2} L_2 i_2^2 \tag{17.16}$$

for the current reference arrows oriented the same way with respect to the corresponding dot mark.

If the current reference arrows were oriented oppositely with respect to the dot marks, so that the currents would cause flux through the coils that opposed each other, the stored energy would be given by

† These topics can be omitted without loss of continuity.

$$w = \frac{1}{2}L_1 i_1^2 - M i_1 i_2 + \frac{1}{2}L_2 i_2^2 \qquad (17.17)$$

Remember

Coupled coils are passive. Energy is stored in a magnetic field as the magnetic flux increases and returned as the flux decreases.

EXERCISES

3. Determine the coupling coefficient for two inductances if $L_1 = 16$ H, $L_2 = 9$ H, and $M = 5$ H.
   answer: 5/12
4. Two coupled inductances have self-inductances of $L_1 = 25$ H and $L_2 = 9$ H. If the coupling coefficient is 0.4, determine the value of the mutual inductance M.
   answer: 6 H
5. Refer to the circuit used in Exercise 1. Determine the energy stored in the coupled inductances when $L_1 = 16$ H, $L_2 = 9$ H, and $M = 5$ H, if $i_x = 2e^{2t}$ A and $i_y = 4e^{2t}$ A.
   answer: $186e^{4t}$ J
6. The location of the dot is reversed on the 16 H inductance in the circuit used in Exercise 5. Determine the energy stored in the coupled inductances.
   answer: $106e^{4t}$ J

17.2 Mutual Inductance in the Frequency Domain

We remember that differentiation in the time domain transforms into multiplication by $j\omega$ in the frequency domain, so the terminal equations for coupled inductances [Eqs. (17.9) and (17.10)] give us the following:

Coupled Inductances in the Frequency Domain

In the frequency domain the terminal equations for coupled inductances are

$$\mathbf{V}_1 = j\omega L_1 \mathbf{I}_1 + j\omega M \mathbf{I}_2 \qquad (17.18)$$

$$\mathbf{V}_2 = j\omega L_2 \mathbf{I}_2 + j\omega M \mathbf{I}_1 \qquad (17.19)$$

and the frequency domain component symbol is shown in Fig. 17.5.

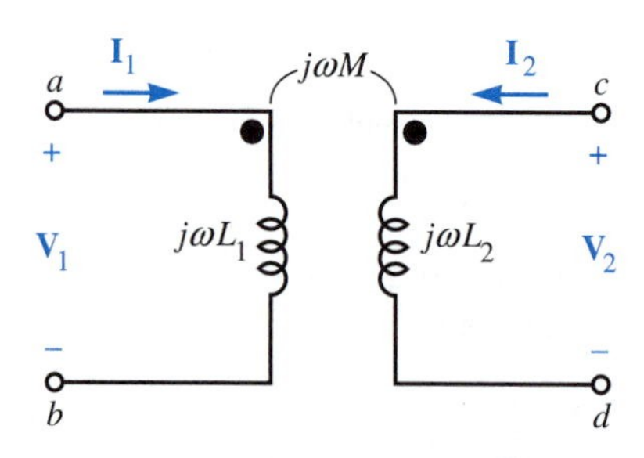

FIGURE 17.5 Frequency-domain symbol for coupled inductances

Remember

It is easy to assign the proper signs in Eqs. (17.18) and (17.19). Just remember that the sign on a self-induced voltage is positive for the passive sign convention. The sign on a mutually induced voltage is the same as the sign on the self-induced term in the same equation, if the currents are oriented the same way relative to the dot marks.

EXERCISES

7. The sources in Fig. 17.4 (Example 17.1), which is reproduced below, have angular frequency ω and phasor representations $\mathbf{I}_s$ and $\mathbf{V}_s$. Write the mesh-current equations.

answer:
$$\begin{bmatrix} j\omega L_2 + R_2 + \dfrac{1}{j\omega C} & -j\omega(L_2 + M) \\ -j\omega(L_2 + M) & j\omega(L_1 + L_2 + 2M) + R_1 \end{bmatrix} \begin{bmatrix} \mathbf{I}_1 \\ \mathbf{I}_2 \end{bmatrix} = \begin{bmatrix} -\mathbf{V}_s \\ R_1\mathbf{I}_s \end{bmatrix}$$

$$\mathbf{I}_3 = \mathbf{I}_s$$

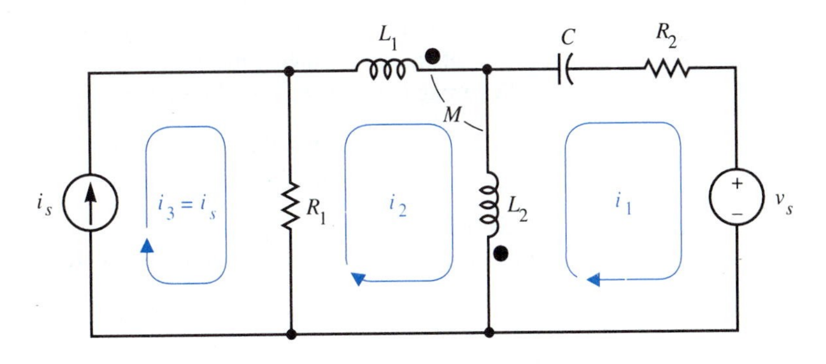

8. Use phasors to write the mesh-current equations for the network of Exercise 2, which is reproduced below, if $v_s = 24 \cos 2t$ V.

answer:
$$\begin{bmatrix} 3 + j8 & -j12 \\ -j12 & 5 + j\dfrac{167}{6} \end{bmatrix} \begin{bmatrix} \mathbf{I}_1 \\ \mathbf{I}_2 \end{bmatrix} = \begin{bmatrix} 24\underline{/0^\circ} \\ 0 \end{bmatrix}$$

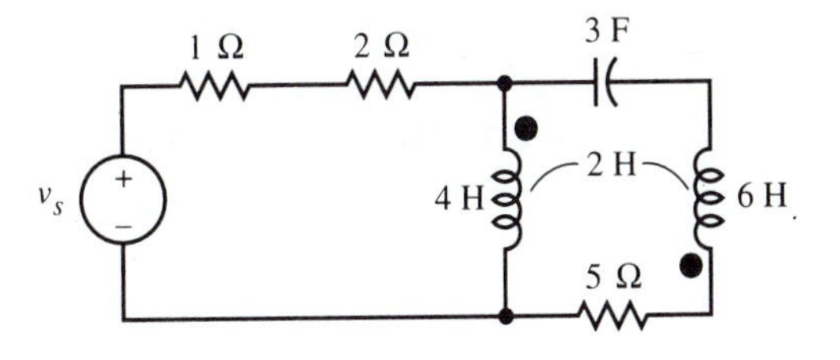

Linear Transformers

Electronic circuits, such as radio and television receivers, often use magnetically coupled coils, as shown in Fig. 17.6, to alter the effective input or output impedance of a portion of a circuit. These coupled coils *transform* an impedance, and we refer to them as *linear transformers* or *air-core transformers.* (These transformers are linear because they are wound on a magnetically linear material, such as a plastic tube.)

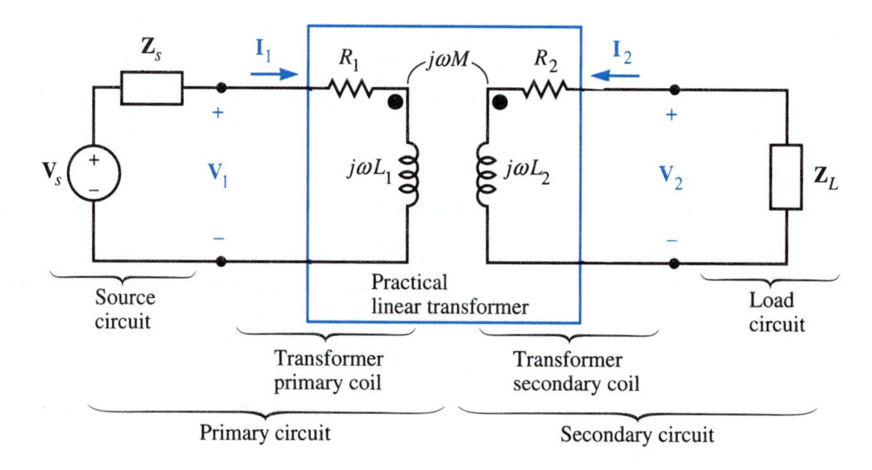

FIGURE 17.6 A circuit with a linear transformer

These circuits typically operate over a narrow range of frequencies and the coefficient of coupling $k = M/\sqrt{L_1 L_2}$ is usually less than 0.5. We approach the analysis of such circuits in the following manner.

Summing the voltages for closed paths on both sides of the transformer gives us

$$\begin{bmatrix} R_1 + j\omega L_1 & j\omega M \\ j\omega M & R_2 + j\omega L_2 + \mathbf{Z}_L \end{bmatrix} \begin{bmatrix} \mathbf{I}_1 \\ \mathbf{I}_2 \end{bmatrix} = \begin{bmatrix} \mathbf{V}_1 \\ 0 \end{bmatrix} \tag{17.20}$$

To simplify our notation, we define the following impedances. The *self-impedances* of the transformer primary and secondary sides† are

$$\mathbf{Z}_{11} = R_1 + j\omega L_1 \tag{17.21}$$

and

$$\mathbf{Z}_{22} = R_2 + j\omega L_2 \tag{17.22}$$

respectively, and the *mutual impedances* are

$$\mathbf{Z}_{12} = \mathbf{Z}_{21} = j\omega M \tag{17.23}$$

This lets us write Eq. (17.20) as

$$\begin{bmatrix} \mathbf{Z}_{11} & \mathbf{Z}_{12} \\ \mathbf{Z}_{21} & \mathbf{Z}_{22} + \mathbf{Z}_L \end{bmatrix} \begin{bmatrix} \mathbf{I}_1 \\ \mathbf{I}_2 \end{bmatrix} = \begin{bmatrix} \mathbf{V}_1 \\ 0 \end{bmatrix} \tag{17.24}$$

We can use Cramer's rule to solve for the ratio $\mathbf{V}_1/\mathbf{I}_1$. This gives the *input impedance*, $\mathbf{Z}_1$, seen by the source circuit:

$$\mathbf{Z}_1 = \frac{\mathbf{V}_1}{\mathbf{I}_1} = \frac{\mathbf{Z}_{11}(\mathbf{Z}_{22} + \mathbf{Z}_L) - \mathbf{Z}_{12}\mathbf{Z}_{21}}{\mathbf{Z}_{22} + \mathbf{Z}_L} \tag{17.25a}$$

$$= \mathbf{Z}_{11} - \mathbf{Z}_{12}\mathbf{Z}_{21}\frac{1}{\mathbf{Z}_{22} + \mathbf{Z}_L} \tag{17.25b}$$

We see from Eq. (17.25b) that the input impedance $\mathbf{Z}_1$ is equal to the impedance of the primary circuit $\mathbf{Z}_{11}$ in series with an equivalent impedance

$$\mathbf{Z}_{e1} = -\mathbf{Z}_{12}\mathbf{Z}_{21}\frac{1}{\mathbf{Z}_2} \tag{17.26}$$

† We can call either coil the primary and the other coil the secondary. If there is a source on only one side of the transformer, we usually call this the primary side.

The equivalent impedance $\mathbf{Z}_{e1}$ is the total impedance of the secondary circuit $\mathbf{Z}_2$:

$$\mathbf{Z}_2 = \mathbf{Z}_{22} + \mathbf{Z}_L \tag{17.27}$$

reflected to the primary circuit. The mutual impedances $\mathbf{Z}_{12}$ and $\mathbf{Z}_{21}$ effectively *transform* the impedance of the secondary circuit $\mathbf{Z}_2$ to a new value $\mathbf{Z}_{e1}$ and place it in series with the impedance of the primary coil $\mathbf{Z}_{11}$ to give the input impedance $\mathbf{Z}_1$ seen by the source network.

We now solve Eq. (17.20) for currents $\mathbf{I}_1$ and $\mathbf{I}_2$ and take the ratio to obtain the *forward current transfer function.*

$$\frac{\mathbf{I}_2}{\mathbf{I}_1} = -\frac{\mathbf{Z}_{21}}{\mathbf{Z}_{22} + \mathbf{Z}_L} = -\frac{\mathbf{Z}_{21}}{\mathbf{Z}_2} \tag{17.28}$$

We can now use $\mathbf{V}_2 = -\mathbf{Z}_L\mathbf{I}_2$ and a moderate amount of algebra to establish the *forward voltage transfer function.*

$$\frac{\mathbf{V}_2}{\mathbf{V}_1} = \frac{\mathbf{Z}_{21}\mathbf{Z}_L}{\mathbf{Z}_{11}(\mathbf{Z}_{22} + \mathbf{Z}_L) - \mathbf{Z}_{12}\mathbf{Z}_{21}} \tag{17.29}$$

Equations (17.28) and (17.29) provide an alternative interpretation of the transformer action of coupled inductances. The transformer lets us provide a voltage (or current) at the secondary terminals that is larger or smaller than the voltage (or current) at the primary terminals.

Linear transformers are frequently used in electronics to perform one or more of the following three functions.

1. The transformer action transforms the load impedance into a new value to match the impedance of the source for maximum power transfer.
2. The transformer provides dc voltage *isolation* between the source and load, because there is no dc path between the two coils.
3. When used in conjunction with capacitors, the inductances of the coils can provide a voltage transfer function that passes a sinusoid of one frequency and rejects a sinusoid of another frequency (the circuit is a *filter*).

An example of a circuit where a linear transformer performs all three functions is shown in Fig. 17.7.

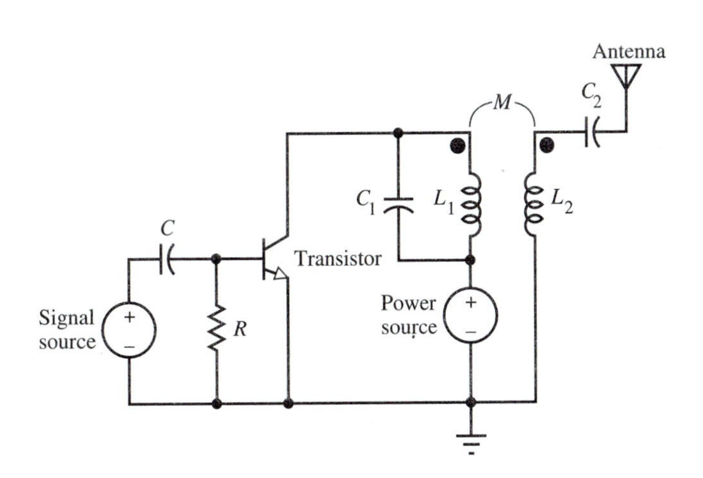

FIGURE 17.7 A radio-frequency amplifier with transformer-coupled output

In the circuit of Fig. 17.7, the transformer, consisting of coupled inductances L_1 and L_2, in conjunction with capacitances C_1 and C_2, matches the impedance of the antenna to the output impedance of the transistor. The combined effect of the inductances and capacitances provides an exact impedance match at only one frequency, thereby providing the filtering action. The coil L_2 is connected only to the reference (ground). This prevents the dc voltage of the power source from appearing on the antenna side of the transformer. Analysis of this application of transformers is outlined in Problems 32 and 34.

Remember

Coupled coils form a linear transformer that can alter the effective impedance of a device. Impedance matching is often used to maximize power transfer in electronic circuits.

EXERCISES

9. Assume that R_1 and R_2 are negligibly small in the circuit of Fig. 17.6.
   (a) Write Eq. (17.25b) in terms of $j\omega L_1$, $j\omega L_2$, $j\omega M$, and $\mathbf{Z}_L$.
   answer: $j\omega L_1 + \omega^2 M^2/(j\omega L_2 + \mathbf{Z}_L)$
   (b) Substitute $M = k\sqrt{L_1 L_2}$ into the result achieved in part (a).
   answer: $j\omega L_1 + \omega^2 k^2 L_1 L_2/(j\omega L_2 + \mathbf{Z}_L)$
   (c) Substitute $L_2 = (n_2/n_1)^2 L_1$ into the result obtained in part (b), where n_1/n_2 is the effective ratio of the number of turns on coil 1 to the number of turns on coil 2.
   answer: $j\omega L_1 + \omega^2 L_1^2 (n_2/n_1)^2 k^2/[(j\omega(n_2/n_1)^2 L_1 + \mathbf{Z}_L)]$
10. Use the expression obtained in part (a) of the preceding exercise with $\mathbf{Z}_L = R_L$.
   (a) Find the input impedance $\mathbf{Z}_{\text{in}}(j\omega) = \mathbf{V}_1/\mathbf{I}_1$.
   answer: $j\omega L_1 + \omega^2 M^2/(R_L + j\omega L_2)$
   (b) Determine the forward voltage transfer function $\mathbf{H}_\mathbf{V}(j\omega) = \mathbf{V}_2/\mathbf{V}_1$.
   answer: $j\omega M R_L/[j\omega L_1(R_L + j\omega L_2) + \omega^2 M^2]$
   (c) Find the forward-current transfer function $\mathbf{H}_\mathbf{I}(j\omega) = \mathbf{I}_2/\mathbf{I}_1$ circuit.
   answer: $-j\omega M/(R_L + j\omega L_2)$
   (d) Determine the complex-power gain $\mathbf{G}(j\omega) = -\mathbf{V}_2\mathbf{I}_2^*/\mathbf{V}_1\mathbf{I}_1^*$.
   answer: $\omega^2 M^2 R_L/\{[j\omega L_1(R_L + j\omega L_2) + \omega^2 M^2][R_L - j\omega L_2]\}$

Equivalent T and Π Networks†

With the exception of dc isolation, we can often duplicate a linear transformer by an equivalent inductive T or Π network. Assume that terminals *b* and *d* of two coupled inductances are directly connected together in Fig. 17.5.

The frequency-domain terminal equations that describe the coupled inductances are

$$\begin{bmatrix} \mathbf{V}_1 \\ \mathbf{V}_2 \end{bmatrix} = \begin{bmatrix} j\omega L_1 & j\omega M \\ j\omega M & j\omega L_2 \end{bmatrix} \begin{bmatrix} \mathbf{I}_1 \\ \mathbf{I}_2 \end{bmatrix} \tag{17.30}$$

We can write two KVL equations for the T network (Y network) of Fig. 17.8 to obtain the terminal equations:

$$\begin{bmatrix} \mathbf{V}_1 \\ \mathbf{V}_2 \end{bmatrix} = \begin{bmatrix} j\omega(L_a + L_c) & j\omega L_c \\ j\omega L_c & j\omega(L_b + L_c) \end{bmatrix} \begin{bmatrix} \mathbf{I}_1 \\ \mathbf{I}_2 \end{bmatrix} \tag{17.31}$$

† This topic can be omitted without loss of continuity. A T network is also called a Y network, and a Π network is also called a Δ network.

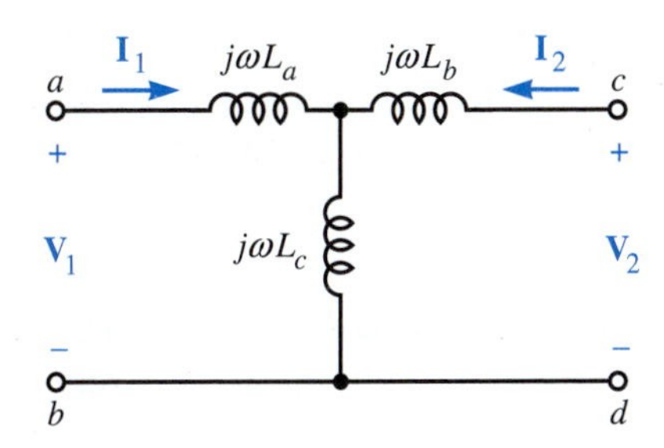

FIGURE 17.8
A T network (Y network)

To design a T network that is equivalent to the coupled inductances of Fig. 17.8 equate each element in the impedance matrix of Eq. (17.31) to the corresponding element in the impedance matrix of Eq. (17.30). Solve for L_a, L_b, and L_c in terms of L_1, L_2, and M, to obtain

Equivalent T Network

$$L_a = L_1 - M \tag{17.32}$$

$$L_b = L_2 - M \tag{17.33}$$

$$L_c = M \tag{17.34}$$

We can write two KCL equations for the coupled inductances of Fig. 17.5, or simply solve for the currents from Eq. 17.30, to obtain the terminal equations for coupled inductances in the following form:

$$\begin{bmatrix} \mathbf{I}_1 \\ \mathbf{I}_2 \end{bmatrix} = \begin{bmatrix} \dfrac{L_2}{j\omega(L_1L_2 - M^2)} & \dfrac{-M}{j\omega(L_1L_2 - M^2)} \\ \dfrac{-M}{j\omega(L_1L_2 - M^2)} & \dfrac{L_1}{j\omega(L_1L_2 - M^2)} \end{bmatrix} \begin{bmatrix} \mathbf{V}_1 \\ \mathbf{V}_2 \end{bmatrix} \tag{17.35}$$

We obtain the terminal equations for the Π network (Δ network) of Fig. 17.9 from two KCL equations:

$$\begin{bmatrix} \mathbf{I}_1 \\ \mathbf{I}_2 \end{bmatrix} = \begin{bmatrix} \dfrac{1}{j\omega L_A} + \dfrac{1}{j\omega L_C} & -\dfrac{1}{j\omega L_C} \\ -\dfrac{1}{j\omega L_C} & \dfrac{1}{j\omega L_B} + \dfrac{1}{j\omega L_C} \end{bmatrix} \begin{bmatrix} \mathbf{V}_1 \\ \mathbf{V}_2 \end{bmatrix} \tag{17.36}$$

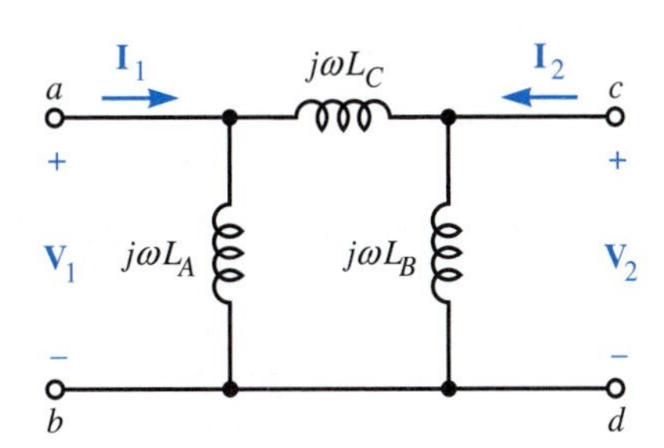

FIGURE 17.9
A Π network (Δ network)

To design a Π network that is equivalent to the coupled inductances of Fig. 17.5, equate each element in the admittance matrix of Eq. (17.36) to the corresponding element in the admittance matrix of Eq. (17.35) and solve for L_A, L_B, and L_C in terms of L_1, L_2, and M, to obtain

Equivalent Π Network

$$L_A = \frac{L_1L_2 - M^2}{L_2 - M} \tag{17.37}$$

$$L_B = \frac{L_1L_2 - M^2}{L_1 - M} \tag{17.38}$$

$$L_C = \frac{L_1L_2 - M^2}{M} \tag{17.39}$$

Although the mathematically equivalent T and Π networks always exist, they can be constructed only if the calculated inductances are nonnegative. Also, practical coils always have resistance, so the equivalent T and Π networks have a resistive component in the mutual impedance as well as the self-impedances. Magnetically coupled inductances have purely reactive mutual impedances.

If the circuit is operated with a sinusoidal signal of fixed angular frequency ω_0 and if either reactance in the T network,

$$\omega_0 L_a = \omega_0(L_1 - M) \tag{17.40}$$

or

$$\omega_0 L_b = \omega_0(L_2 - M) \tag{17.41}$$

is negative, we can still construct an equivalent T network by using a capacitance in place of the inductance to form the negative reactance. For instance, if the calculated reactance $\omega_0(L_1 - M)$ is negative, we replace the inductance L_a in the T network by a capacitance of value C_a that satisfies the equation

$$-\frac{1}{\omega_0 C_a} = \omega_0(L_1 - M) \tag{17.42}$$

Equation (17.42) gives a capacitance value of

$$C_a = \frac{1}{\omega_0^2(M - L_1)} \tag{17.43}$$

Similar results hold true for the Π network. The choice of whether to use coupled inductances, a T network, or a Π network for impedance matching is usually based on economics.

Remember

We can often replace coupled coils, if they have a common terminal, with an equivalent T network or Π network. If the calculated value of an inductance for the equivalent circuit is negative, we cannot construct the equivalent. However, if the circuit operates at a single fixed frequency, we can construct an equivalent that is valid for this frequency by using a capacitance in place of an inductance.

EXERCISES Replace the coupled inductance network in Fig. 17.5 with an equivalent T network and give numerical values for L_a, L_b, and L_c if $L_1 = 2$ H, $L_2 = 40$ H, and M has the value specified in Exercises 11 through 14.

11. 1 H — *answer:* 1 H, 39 H, 1 H
12. 2 H — *answer:* 0 H (a short circuit), 38 H, 2 H
13. 3 H — *answer:* −1 H, 37 H, 3 H
14. 6 H — *answer:* −4 H, 34 H, 6 H

Replace the coupled inductance network in Fig. 17.5 with an equivalent Π network and give numerical values for L_A, L_B, and L_C if $L_1 = 2$ H, $L_2 = 40$ H, and M has the value specified in Exercises 15 through 18.

15. 1 H — *answer:* $\frac{79}{39}$ H, 79 H, 79 H
16. 2 H — *answer:* 2 H, ∞ H (an open circuit), 38 H
17. 3 H — *answer:* $\frac{71}{37}$ H, −71 H, $\frac{71}{3}$ H
18. 6 H — *answer:* $\frac{22}{17}$ H, −22 H, $\frac{22}{3}$ H

Symmetry†

We now establish the symmetry of mutual inductance ($M_{12} = M_{21}$). If we show that M_{12} must equal M_{21} for some input currents, we will have established the equality for all input currents, because M_{12} and M_{21} are constants. Refer to the frequency domain model of Fig. 17.5. Assume input currents $\mathbf{I}_1 = I_{m1}\underline{/\phi_1}$ and $\mathbf{I}_2 = I_{m2}\underline{/\phi_2}$. The complex-power input to the coupled inductances is

$$\begin{aligned}\mathbf{S} &= \mathbf{S}_1 + \mathbf{S}_2 = \frac{1}{2}\mathbf{V}_1\mathbf{I}_1^* + \frac{1}{2}\mathbf{V}_2\mathbf{I}_2^* \\ &= \frac{1}{2}(j\omega L_1\mathbf{I}_1 + j\omega M_{12}\mathbf{I}_2)\mathbf{I}_1^* + \frac{1}{2}(j\omega L_2\mathbf{I}_2 + j\omega M_{21}\mathbf{I}_1)\mathbf{I}_2^* \\ &= \frac{1}{2}\omega(M_{12} - M_{21})I_{m1}I_{m2}\sin(\phi_2 - \phi_1) \\ &\quad + \frac{1}{2}j\omega[L_1 I_{m1}^2 + L_2 I_{m2}^2 + (M_{12} + M_{21})I_{m1}I_{m2}\cos(\phi_2 - \phi_1)] \\ &= P + jQ\end{aligned} \tag{17.44}$$

Because coupled inductances are passive,

$$P = \frac{1}{2}\omega(M_{12} - M_{21})I_{m1}I_{m2}\sin(\phi_2 - \phi_1) \geq 0 \tag{17.45}$$

This inequality is satisfied for all choices of $\mathbf{I}_1$ and $\mathbf{I}_2$ if and only if $M_{12} - M_{21} = 0$ because I_{m1}, I_{m2}, and ω are nonnegative; and the sine can be positive or negative.

Remember Because coupled inductances are passive, mutual inductance must be symmetric.

† This proof can be omitted without loss of continuity.

17.3 Ideal Transformers

Transformers in power systems enable power to be generated, transmitted, and utilized at different voltages. For example, an electrical utility will generate power at some convenient voltage, perhaps 13.8 kV, and then use transformers to increase this to 120 kV or more so that power can be transmitted at high voltage and low current over smaller wires. Transformers near the customer reduce this to a more convenient lower voltage (120 V for your lights). Equipment in a residence or business may contain additional transformers. For example, a transformer reduces the 120-V residential voltage to about 6 V for a model train.

All transformers used for 60-Hz power and many transformers used in electronics have ferromagnetic cores. The type of construction is often similar to that depicted in Fig. 17.10. This gives a very large inductance for a coil, compared with the value if the core is not ferromagnetic. This typically gives a large ratio of inductive reactance (ωL_1 or ωL_2) to resistance (R_1 or R_2) for the model of Fig. 17.6. The result is that, in many practical cases, we can neglect resistances R_1 and R_2. This leaves the coupled inductances of Fig. 17.5 as our transformer model. The coupling coefficient, $k = M/\sqrt{L_1 L_2}$ defined in Eq. (17.11), will also be close to one,† so we often assume that $M = \sqrt{L_1 L_2}$. Unfortunately, with this latter assumption the terminal equations [Eqs. (17.18) and (17.19)] for coupled inductances are linearly dependent. We will instead start with the linear-transformer model of Fig. 17.6 and matrix Eq. (17.20) that describes this circuit.

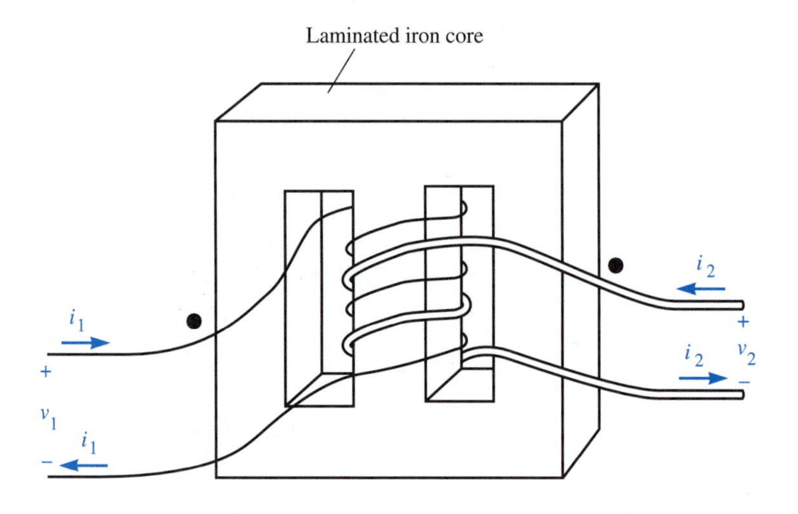

FIGURE 17.10 A transformer with a ferromagnetic core (a physical transformer may have hundreds of turns of wire in each coil)

We first assume that the resistances are negligibly small and use Eq. (17.28), the forward voltage transfer function, to write $\mathbf{V}_1$ as a function of $\mathbf{V}_2$.

$$\mathbf{V}_1 = \frac{\mathbf{Z}_{11}(\mathbf{Z}_{22} + \mathbf{Z}_L) - \mathbf{Z}_{12}\mathbf{Z}_{21}}{\mathbf{Z}_{21}\mathbf{Z}_L}\mathbf{V}_2$$

$$= \frac{j\omega L_1(j\omega L_2 + \mathbf{Z}_L) - (j\omega M)^2}{j\omega M \mathbf{Z}_L}\mathbf{V}_2 \qquad (17.46)$$

† Transformers with ferromagnetic cores can be constructed with magnetic coefficients of coupling greater than 0.995. Transformers with nonmagnetic cores rarely have a coupling coefficient greater than 0.5.

With the substitution of

$$M = k\sqrt{L_1 L_2} \tag{17.47}$$

Eq. (17.46) becomes

$$\mathbf{V}_1 = \frac{(k^2 - 1)\omega^2 L_1 L_2 + j\omega L_1 \mathbf{Z}_L}{j\omega k\sqrt{L_1 L_2}\,\mathbf{Z}_L}\mathbf{V}_2 \tag{17.48}$$

Unity coupling coefficient reduces Eq. (17.48) to

$$\mathbf{V}_1 = \sqrt{\frac{L_1}{L_2}}\mathbf{V}_2 \tag{17.49}$$

We define the effective turns ratio

$$\frac{n_1}{n_2} = \sqrt{\frac{L_1}{L_2}} \tag{17.50}$$

(We call this the effective turns ratio because for coils of a fixed geometry, the inductance is proportional to the square of the number of turns.) We now obtain our desired result:

$$\mathbf{V}_1 = \frac{n_1}{n_2}\mathbf{V}_2 \tag{17.51}$$

We now use the forward-current transfer ratio [Eq. (17.28)] to write Current $\mathbf{I}_2$ as a function of $\mathbf{I}_1$.

$$\mathbf{I}_2 = -\frac{\mathbf{Z}_{21}}{\mathbf{Z}_{22} + \mathbf{Z}_L}\mathbf{I}_1 \tag{17.52}$$

For $\omega L_2 \gg |\mathbf{Z}_L|$ the preceding equation becomes

$$\mathbf{I}_2 = -\frac{\mathbf{Z}_{21}}{\mathbf{Z}_{22}}\mathbf{I}_1 = -\frac{j\omega M}{j\omega L_2}\mathbf{I}_1 \tag{17.53}$$

Substitution of Eq. (17.47) into Eq. (17.53) gives us

$$\mathbf{I}_2 = -k\sqrt{\frac{L_1}{L_2}}\mathbf{I}_1 \tag{17.54}$$

With $k = 1$, this gives

$$\mathbf{I}_2 = -\frac{n_1}{n_2}\mathbf{I}_1 \tag{17.55}$$

Equations (17.51) and (17.55) determine the characteristics of coupled coils in the limiting case where the resistances of the coils are much less than their corresponding inductive reactances, the coupling coefficient is 1, the square root of the inductance ratio is n_1/n_2, and the inductive reactance ωL_2 is much greater than the magnitude of the load impedance $\mathbf{Z}_L$. This is the description of an ideal transformer. We can easily verify that Eqs. (17.51) and (17.55) imply that the complex power absorbed by an ideal transformer is zero (see Exercise 19).

DEFINITION Ideal Transformer

The complex power absorbed by an **ideal transformer** is zero, and the voltages are related by the turns ratio. This gives the terminal equations

$$\begin{bmatrix} \mathbf{V}_1 \\ \mathbf{I}_2 \end{bmatrix} = \begin{bmatrix} 0 & \dfrac{n_1}{n_2} \\ -\dfrac{n_1}{n_2} & 0 \end{bmatrix} \begin{bmatrix} \mathbf{I}_1 \\ \mathbf{V}_2 \end{bmatrix} \qquad (17.56)$$

and the component symbol is shown in Fig. 17.11.

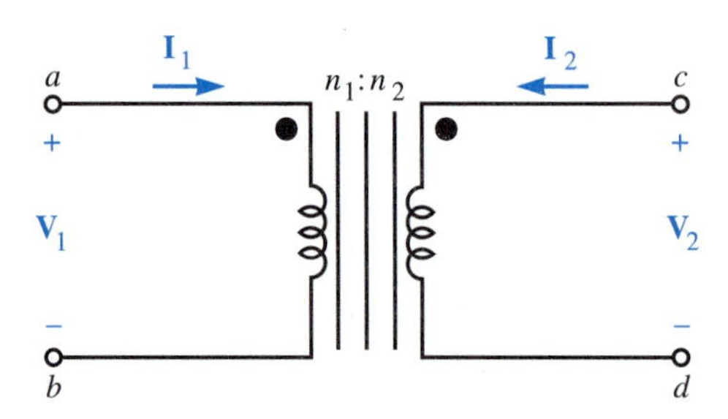

FIGURE 17.11 Symbol for an ideal transformer

Ideal transformers with more than two coils present no unusual problem. The voltages for any pair of coils are still related by the corresponding turns ratio, and the total complex power absorbed by any ideal transformer is zero.

Remember

The ideal transformer component absorbs zero complex power. The voltage ratio $\mathbf{V}_1/\mathbf{V}_2$ for an ideal transformer is equal to the effective turns ratio n_1/n_2. The current ratio $-\mathbf{I}_1/\mathbf{I}_2$ is *inversely* proportional to the turns ratio n_1/n_2.

EXERCISES

19. Use Eqs. (17.51) and (17.55) to show that the complex power,

$$\mathbf{S} = \mathbf{S}_1 + \mathbf{S}_2 = \mathbf{V}_1\mathbf{I}_1^* + \mathbf{V}_2\mathbf{I}_2^*$$

absorbed by an ideal transformer is zero. (Voltages and currents are in rms values.)

20. Assume that the transformer that supplies power to three machines is ideal. The primary voltage is 13.8 kV (rms), and the complex powers supplied to the machines are $\mathbf{S}_a = 2 + j1$ kVA, $\mathbf{S}_b = 3 + j1$ kVA, and $\mathbf{S}_c = 5 + j0$ kVA. (a) Determine the complex power supplied by the 13.8 kV system. (b) What is the primary current for the transformer? (c) If the secondary voltage is 240 V (rms), what is the secondary current for the transformer? (The system is single-phase. Three-phase systems are discussed in Chapter 18.) *answer:* $10 + j2$ kVA, 0.739 A, 42.5 A

Reflected Impedance

Connection of a load impedance $\mathbf{Z}_L$ to the secondary terminals (c and d) of an ideal transformer as shown in Fig. 17.12a gives

$$\mathbf{V}_2 = -\mathbf{I}_2\mathbf{Z}_L = -\left(-\frac{n_1}{n_2}\mathbf{I}_1\right)\mathbf{Z}_L \qquad (17.57)$$

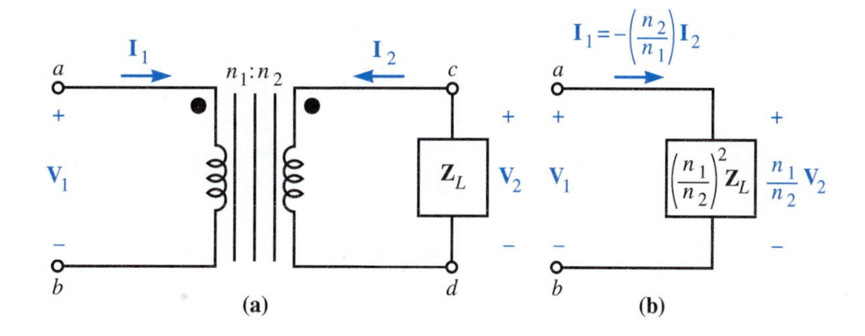

FIGURE 17.12 (*a*) An ideal transformer with load and (*b*) its equivalent circuit

Equations (17.51) and (17.57) yield

$$\mathbf{V}_1 = \frac{n_1}{n_2}\mathbf{V}_2 = \frac{n_1}{n_2}\frac{n_1}{n_2}\mathbf{I}_1\mathbf{Z}_L$$

$$= \left(\frac{n_1}{n_2}\right)^2 \mathbf{Z}_L\mathbf{I}_1 \tag{17.58}$$

Equation (17.58) is easily solved for the input impedance (Thévenin impedance) looking into terminals a and b:

$$\mathbf{Z}_1 = \frac{\mathbf{V}_1}{\mathbf{I}_1} = \left(\frac{n_1}{n_2}\right)^2 \mathbf{Z}_L \tag{17.59}$$

We say that impedance $\mathbf{Z}_1$ is the impedance of $\mathbf{Z}_L$ *reflected* to the primary side. Equations (17.51), (17.57), and (17.59) give the equivalent circuit shown in Fig. 17.12b. In the next section we will see how reflected impedances simplify the analysis of some circuits.

Networks with Ideal Transformers

Analysis of networks containing ideal transformers is no more difficult than the analysis of other linear networks. Example 17.2 illustrates how we can use reflected impedances to analyze some circuits with transformers.

EXAMPLE 17.2 Use the reflected impedances to find the impedance seen by the source in Fig. 17.13.

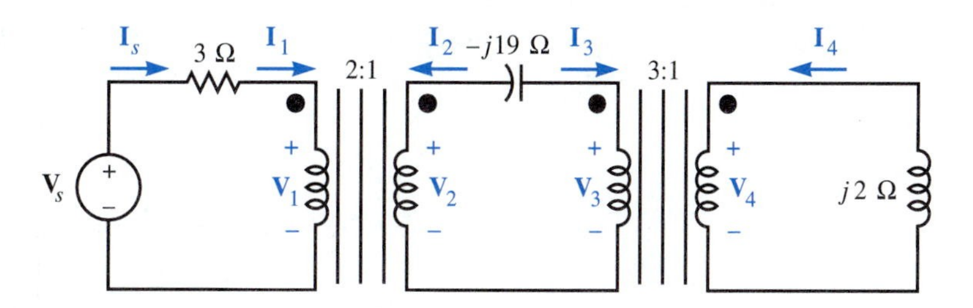

FIGURE 17.13 An ideal transformer example

Solution If we reflect the inductive load of $j2\ \Omega$ to the primary of the right-hand transformer we obtain the equivalent circuit in Fig. 17.14.

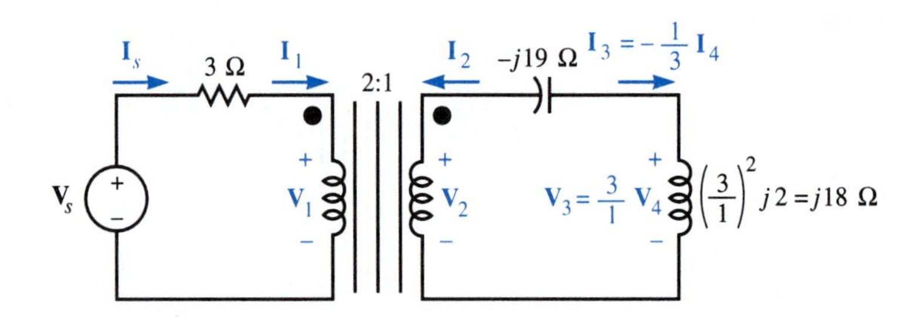

FIGURE 17.14 The first equivalent circuit

The effective load on the secondary side of the left-hand transformer is

$$\mathbf{Z}'_L = -j19 + j18$$
$$= -j1\ \Omega$$

Now reflect $\mathbf{Z}'_L$ to the primary side of the left-hand transformer to obtain the circuit of Fig. 17.15.

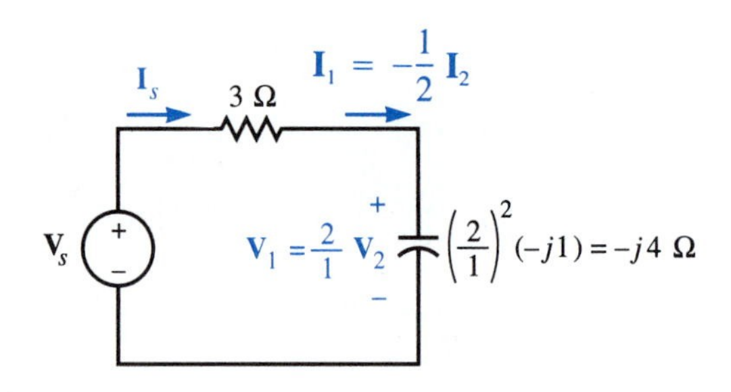

FIGURE 17.15 The second equivalent circuit

Source $\mathbf{V}_s$ sees impedance

$$\mathbf{Z} = 3 - j4\ \Omega$$
$$= 5\underline{/-53.13^\circ}\ \Omega$$

Remember An impedance $\mathbf{Z}_L$ is reflected from the secondary side to the primary side of a transformer by multiplying $\mathbf{Z}_L$ by the turns ratio squared $(n_1/n_2)^2$.

EXERCISES

21. For the following circuit, calculate: (a) $\mathbf{I}_s$, (b) $\mathbf{V}_1$, (c) $\mathbf{I}_1$, (d) $\mathbf{V}_2$, (e) $\mathbf{I}_2$, and (f) the impedance seen by the 360-V source. Use the method of reflected impedance.
answer: 12 A, 72 V, 4 A, 24 V, −12 A, 30 Ω

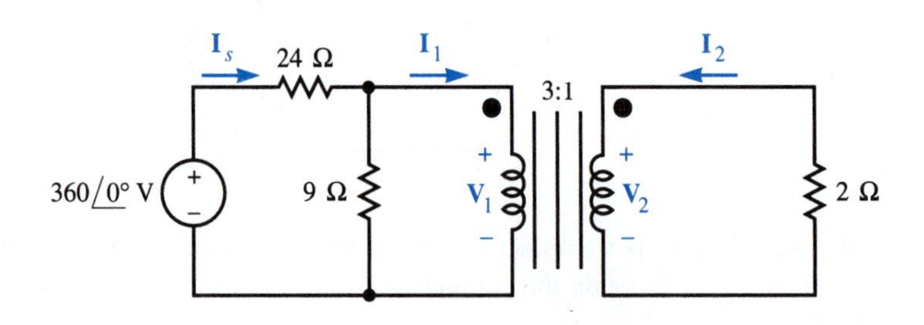

22. Refer to the preceding circuit. Use Thévenin's theorem, reflected impedance, and the terminal characteristics for the ideal transformer to replace all components to the left of the 2-Ω resistance with a voltage source and a series resistance.

answer: 32.7 V, 0.723 Ω, −12 A

17.4 Circuit Analysis with Ideal Transformers†

Some circuits contain transformers interconnected in a manner that makes analysis by reflected impedances impractical. An example of such a circuit is shown in Fig. 17.16. We will analyze this circuit by the method of mesh currents.

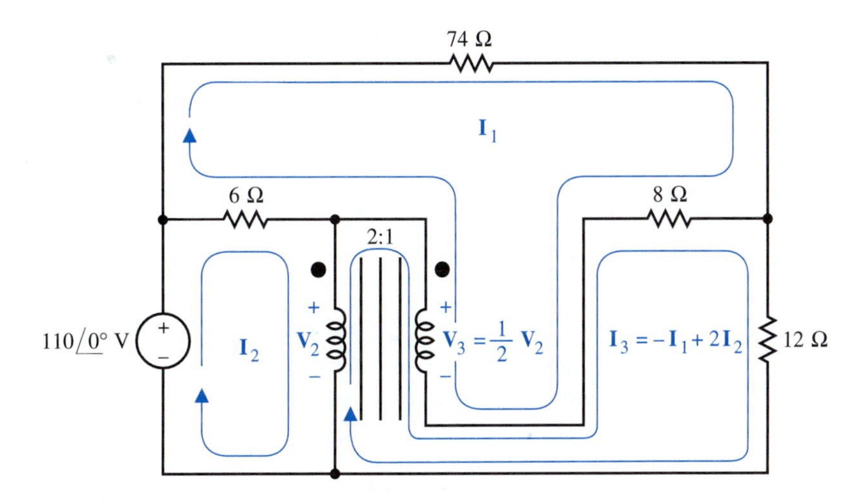

FIGURE 17.16 A mesh-current example with an ideal transformer

Mesh Currents

Analysis of networks containing ideal transformers by the method of mesh currents is straightforward. To do this, we assign a voltage variable to each coil of the transformer, and write the mesh-current equations as if the transformer coils were voltage sources. The transformer voltages are unknowns, so we obtain additional equations from the turns-ratio relations between primary and secondary voltages and primary and secondary currents. This technique is illustrated in the following example.

EXAMPLE 17.3 Solve for mesh currents $\mathbf{I}_1$, $\mathbf{I}_2$, and $\mathbf{I}_3$ in the circuit of Fig. 17.16.

Solution First assign voltages $\mathbf{V}_2$ and $\mathbf{V}_3$ to the transformer as shown on the figure.

The turns-ratio relation between primary and secondary voltage and primary and secondary current provides the two additional equations required by the introduction of $\mathbf{V}_2$ and $\mathbf{V}_3$. The voltage relationship is

$$\mathbf{V}_3 = \frac{1}{2}\mathbf{V}_2$$

See Problem 17.2 in the PSpice manual.

† This topic can be omitted without loss of continuity.

Write this on the network diagram. The current relationship is

$$\mathbf{I}_3 - \mathbf{I}_1 = -\frac{2}{1}(\mathbf{I}_2 - \mathbf{I}_3)$$

We solve this for one of the currents in terms of the other two.

$$\mathbf{I}_3 = -\mathbf{I}_1 + 2\mathbf{I}_2$$

Now write this on the circuit diagram. Application of KVL around mesh 1 gives

$$74\mathbf{I}_1 + 8[\mathbf{I}_1 - (-\mathbf{I}_1 + 2\mathbf{I}_2)] - (1/2)\mathbf{V}_2 + 6(\mathbf{I}_1 - \mathbf{I}_2) = 0$$

Summation of the voltages around mesh 2 gives

$$-110 + 6(\mathbf{I}_2 - \mathbf{I}_1) + \mathbf{V}_2 = 0$$

and KVL applied to mesh 3 gives

$$-\mathbf{V}_2 + \frac{1}{2}\mathbf{V}_2 + 8[(-\mathbf{I}_1 + 2\mathbf{I}_2) - \mathbf{I}_1] + 12(-\mathbf{I}_1 + 2\mathbf{I}_2) = 0$$

We can write these three equations as

$$\begin{bmatrix} 96 & -22 & -\frac{1}{2} \\ -6 & 6 & 1 \\ -28 & 40 & -\frac{1}{2} \end{bmatrix} \begin{bmatrix} \mathbf{I}_1 \\ \mathbf{I}_2 \\ \mathbf{V}_2 \end{bmatrix} = \begin{bmatrix} 0 \\ 110 \\ 0 \end{bmatrix}$$

Cramer's rule (or our calculator) gives the solution for $\mathbf{I}_1$, $\mathbf{I}_2$, and $\mathbf{V}_1$. We then use the turns ratio to obtain $\mathbf{I}_3$ and $\mathbf{V}_2$. The result is

$$\begin{bmatrix} \mathbf{I}_1 \\ \mathbf{I}_2 \\ \mathbf{I}_3 \end{bmatrix} = \begin{bmatrix} 1 \\ 2 \\ 3 \end{bmatrix} \text{A} \quad \text{and} \quad \begin{bmatrix} \mathbf{V}_2 \\ \mathbf{V}_3 \end{bmatrix} = \begin{bmatrix} 104 \\ 52 \end{bmatrix} \text{V}$$

Remember

We can adapt the method of analysis by mesh currents to any planar network that contains ideal transformers. We use the turns ratio of the transformer to write the voltage and current of one coil in terms of the voltage and current of the other.

EXERCISE

23. Use the ideal transformer model of Fig. 17.11 to solve for the mesh currents.

answer: 8 A, 4 A

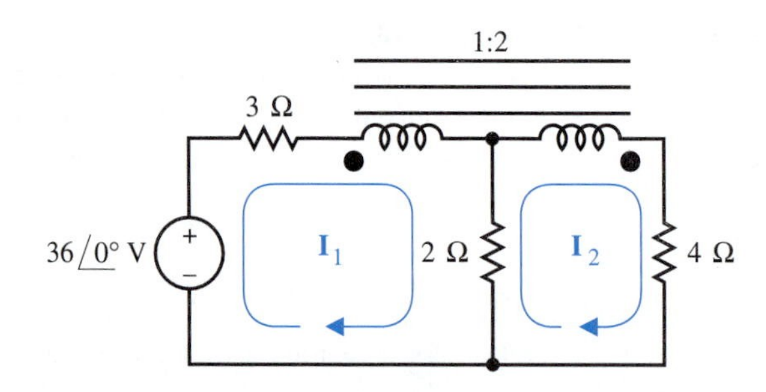

Node Voltages

We can easily adapt the method of analysis by node voltages to include ideal transformers. To do this we assign a current variable to each coil of the transformer and write the node-voltage equations as if the transformer coils were current sources. The transformer currents are unknowns, so we obtain additional equations from the turns-ratio relations between primary and secondary voltages and primary and secondary currents. The following example demonstrates the procedure.

EXAMPLE 17.4 Solve for the node voltages in the network of Fig. 17.17.

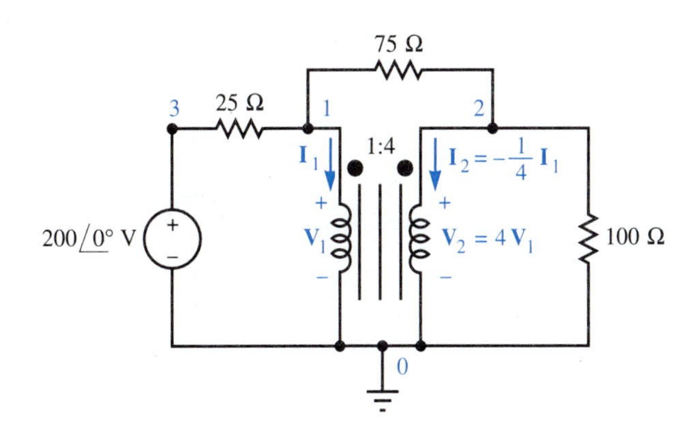

FIGURE 17.17 A node-voltage example with an ideal transformer

Solution We assign currents $\mathbf{I}_1$ and $\mathbf{I}_2$ to the two transformer coils as shown. This introduces two additional unknowns, but the transformer terminal equations provide the two extra equations that we need:

$$\mathbf{I}_2 = -\frac{1}{4}\mathbf{I}_1$$

$$\mathbf{V}_2 = 4\mathbf{V}_1$$

Write these relationships on the network diagram. The KCL equation for a surface enclosing node 1 is

$$\frac{1}{25}(\mathbf{V}_1 - 200) + \frac{1}{75}[\mathbf{V}_1 - (4\mathbf{V}_1)] + \mathbf{I}_1 = 0$$

and that for a surface enclosing node 2 is

$$\frac{1}{75}[(4\mathbf{V}_1) - \mathbf{V}_1] + \frac{1}{100}(4\mathbf{V}_1) + \left(-\frac{1}{4}\mathbf{I}_1\right) = 0$$

The preceding two equations are our node-voltage equations:

$$\begin{bmatrix} 0 & 1 \\ \frac{2}{25} & -\frac{1}{4} \end{bmatrix} \begin{bmatrix} \mathbf{V}_1 \\ \mathbf{I}_1 \end{bmatrix} = \begin{bmatrix} 8 \\ 0 \end{bmatrix}$$

We can solve the preceding matrix equation for $\mathbf{V}_1$ and $\mathbf{I}_1$. We obtain $\mathbf{V}_2$ and $\mathbf{I}_2$ from the turns ratio. The result is

$$\begin{bmatrix} \mathbf{V}_1 \\ \mathbf{V}_2 \end{bmatrix} = \begin{bmatrix} 25 \\ 100 \end{bmatrix} \text{V} \quad \text{and} \quad \begin{bmatrix} \mathbf{I}_1 \\ \mathbf{I}_2 \end{bmatrix} = \begin{bmatrix} 8 \\ -2 \end{bmatrix} \text{A}$$

Remember

We can use the method of analysis by node voltages for any network with ideal transformers. We use the turns ratio of the transformer to write the voltage and current of one coil in terms of the voltage and current of the other.

EXERCISE

24. Use the ideal transformer component of Fig. 17.11 to write a set of node-voltage equations for the following network and solve for the node voltages.

answer: 8 V, 4 V, 6 V

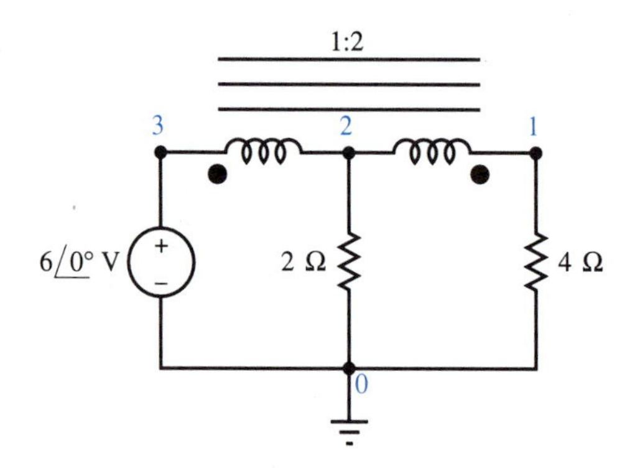

17.5 A Model for Practical Transformers†

In principle, we can analyze any linear transformer by the method used in Section 17.2. However, the coils of transformers used to provide ac electrical power are wound on laminated iron or steel cores and have very high impedances and coupling coefficients close to one. In this situation, the transformer parameters of Fig. 17.6 must be very precisely known or the analysis will be inaccurate. For this reason an alternative model is used.

To develop our new transformer model, we begin with the coupled inductance model of Fig. 17.6. We use

$$\mathbf{V}_2 = -\mathbf{Z}_L\mathbf{I}_2 \tag{17.60}$$

to write Eq. (17.20) as

$$\begin{bmatrix}\mathbf{V}_1\\ \mathbf{V}_2\end{bmatrix} = \begin{bmatrix}R_1 + j\omega L_1 & j\omega M\\ j\omega M & R_2 + j\omega L_2\end{bmatrix}\begin{bmatrix}\mathbf{I}_1\\ \mathbf{I}_2\end{bmatrix} \tag{17.61}$$

We first use the coupling coefficient k [originally defined in Eq. (17.11)]

$$k = \frac{M}{\sqrt{L_1 L_2}} \tag{17.62}$$

to express inductance L_1 in terms of two component inductances:

$$L_1 = (1 - k)L_1 + kL_1 = L_{\ell 1} + L_{M1} \tag{17.63}$$

† This section can be omitted without loss of continuity.

where we define the *primary leakage inductance* by

$$L_{\ell 1} = (1 - k)L_1 \tag{17.64}$$

and the *primary magnetizing inductance* by

$$L_{M1} = kL_1 \tag{17.65}$$

Physically the leakage inductance $L_{\ell 1}$ is due to the magnetic flux caused by current $\mathbf{I}_1$ that links coil 1, but not coil 2:

In a similar way we define

$$L_2 = (1 - k)L_2 + kL_2 = L_{\ell 2} + L_{M2} \tag{17.66}$$

where the secondary leakage inductance is

$$L_{\ell 2} = (1 - k)L_2 \tag{17.67}$$

and the secondary magnetizing inductance is

$$L_{M2} = kL_2 \tag{17.68}$$

We use the effective turns ratio defined in Eq. (17.50) to write

$$\sqrt{L_2} = \frac{n_2}{n_1}\sqrt{L_1} \tag{17.69}$$

We substitute Eq. (17.69) into Eq. (17.62) and solve for

$$M = k\frac{n_2}{n_1}L_1 = \frac{n_2}{n_1}L_{M1} \tag{17.70}$$

Substitution from Eqs. (17.63), (17.66), and (17.70) into Eq. (17.61) gives us

$$\begin{bmatrix} \mathbf{V}_1 \\ \mathbf{V}_2 \end{bmatrix} = \begin{bmatrix} R_1 + j\omega(L_{\ell 1} + L_{M1}) & j\omega L_{M1}\dfrac{n_2}{n_1} \\ j\omega L_{M1}\dfrac{n_2}{n_1} & R_2 + j\omega(L_{\ell 2} + L_{M2}) \end{bmatrix} \begin{bmatrix} \mathbf{I}_1 \\ \mathbf{I}_2 \end{bmatrix} \tag{17.71}$$

We can arrange the two equations contained in Eq. (17.71) in a manner that lets us develop a new equivalent circuit:

$$\mathbf{V}_1 = (R_1 + j\omega L_{\ell 1})\mathbf{I}_1 + j\omega L_{M1}\left(\mathbf{I}_1 + \frac{n_2}{n_1}\mathbf{I}_2\right) \tag{17.72}$$

$$\mathbf{V}_2 = (R_2 + j\omega L_{\ell 2})\mathbf{I}_2 + j\omega\frac{n_2}{n_1}L_{M1}\left(\mathbf{I}_1 + \frac{n_2}{n_1}\mathbf{I}_2\right) \tag{17.73}$$

Careful inspection of Eqs. (17.72) and (17.73) reveals that these KVL equations also describe the alternative transformer model shown in Fig. 17.18.

A similar analysis lets us construct an equivalent circuit that uses the magnetizing inductance of the secondary, as shown in Fig. 17.19.

The parameters in the transformer models of Figs. 17.18 and 17.19 are conveniently calculated from open-circuit and short-circuit tests. For example, in most power transformers $k > 0.95$, so the voltage across the magnetizing inductance of Fig. 17.18 is approximately equal to voltage $\mathbf{V}_1$ when the secondary is open-circuited, and we have

$$\frac{n_2}{n_1} \simeq \left.\frac{\mathbf{V}_2}{\mathbf{V}_1}\right|_{\mathbf{I}_2 = 0} \tag{17.74}$$

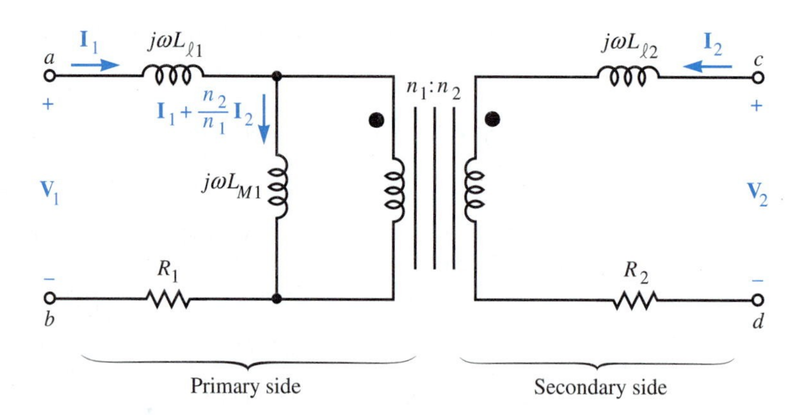

FIGURE 17.18
A transformer model that uses the primary magnetizing inductance

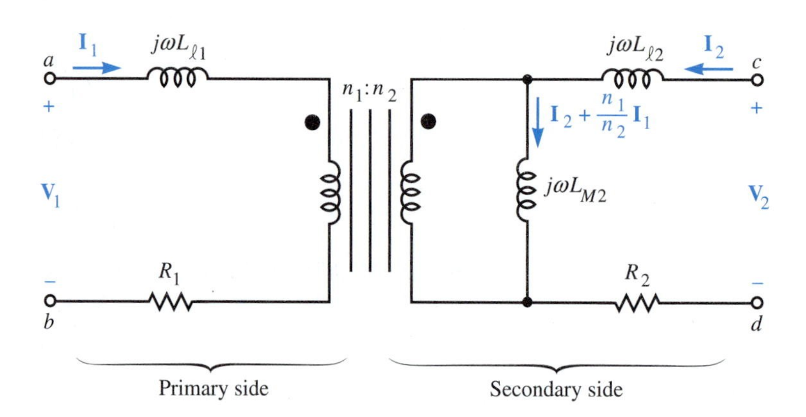

FIGURE 17.19
A transformer model that uses the secondary magnetizing inductance

(You should consult a text on electrical machinery if you need to accurately measure the parameters of a transformer.)

The voltage across the leakage inductance is much less than the corresponding terminal voltage for most transformers operated at design capacity. Therefore, we can reconnect the magnetizing inductance in the model of Fig. 17.18 to terminals *a* and *b* and simplify our analysis with little error. (For the model of Fig. 17.19, we would reconnect the magnetizing inductance to terminals *c* and *d*.)

Coupled inductances wound on ferromagnetic cores have heating losses in addition to the resistive Ri^2 losses in the wire. These losses, caused by the changing magnetic flux in the core, are due to two causes. The changing flux induces *eddy currents* (the current circulates in eddies) in the core. This produces resistive losses called eddy-current losses. Changing flux in the magnetic core also produces *hysteresis losses* in the core. We model these core losses with a resistance R_c placed in parallel with the magnetizing inductance. These core losses are small and are usually neglected in circuit analysis, as they were in all transformer models discussed in this chapter.

Because the magnetizing current is small relative to the full-load current in large power transformers, we often further simplify the model and assume that the magnetizing current is zero. In addition, leakage reactances, ωL_1 and ωL_2, are usually much larger than the corresponding winding resistances, R_1 and R_2, so these resistances are assumed to be zero. This gives a transformer model that includes only the leakage inductances and the ideal transformer. This simplified model is frequently used in the analysis of power systems.

Remember Leakage inductance is due to magnetic flux that links only one coil of a pair of magnetically coupled coils. Increasing the coupling coefficient k decreases the leakage inductance.

EXERCISE

25. For two magnetically coupled coils, $L_1 = 20$ H, $L_2 = 80$ H, and $M = 10$ H. The resistance values are $R_1 = 1\ \Omega$ and $R_2 = 2\ \Omega$, and the radian frequency is ω.
(a) Draw a transformer model like that shown in Fig. 17.18.
(b) Draw a transformer model like that shown in Fig. 17.19.
answer: $L_{\ell 1} = 15$ H, $L_{M1} = 5$ H, $L_{\ell 2} = 60$ H, $L_{M2} = 20$ H, $n_1/n_2 = 1/2$

17.6 Summary

We introduced mutual inductance to model the transfer of energy from one coil to another by magnetic coupling and applied time-domain analysis to circuits with coupled coils. We next established the frequency-domain model for coupled inductances and used phasor analysis to investigate the linear transformer and develop equivalent T and Π networks. The analysis of the linear transformer led us to the ideal-transformer model as a limiting case. We used the ideal transformer to analyze circuits by reflected impedances, and then demonstrated how to adapt node-voltage and mesh-current analysis to circuits with ideal transformers. We concluded by modeling a transformer in terms of leakage inductance and magnetizing inductance.

KEY FACTS

Concept	Equation	Section	Page
❑ A voltage is induced across an inductance that is proportional to the derivative of the current through the inductance. The constant of proportionality is the self-inductance L. For coupled inductances, an additional voltage is induced that is proportional to the derivative of the current through the second inductance. This constant of proportionality is the *mutual inductance M*.	(17.9) (17.10)	17.1	631
❑ The magnetic coefficient of coupling is $k = M/\sqrt{L_1L_2}$.	(17.11)	17.1	633
❑ We always have $0 \leq k \leq 1$	(17.13)	17.1	633
❑ Coupled inductances transform an impedance.	(17.26)	17.2	636
❑ Coupled inductances provide a useful current and voltage transfer function.	(17.28) (17.29)	17.2	637
❑ In some applications, a T network can replace coupled inductances.	(17.32) (17.34)	17.2	639
❑ In some applications, a Π network can replace coupled inductances.	(17.37) (17.39)	17.2	640

KEY FACTS	***Concept***	***Equation***	***Section***	***Page***
❑	An ideal transformer assumes:		17.3	642
	(a) Coil resistances are zero.			
	(b) The equivalent turns ratio is equal to the square root of the ratio of the primary self-inductance to the secondary self-inductance.	(17.50)		643
	(c) The reactance of the secondary self-inductance is infinite compared to the magnitude of the load impedance.	(17.53)		
	(d) The magnetic coefficient of coupling is one.	(17.55)		
❑	The total complex power input to an ideal transformer is zero.		17.3	644
❑	For an ideal transformer, the ratio of the primary voltage to the secondary voltage is equal to the equivalent turns ratio. The ratio of the secondary current to the primary current is equal to the negative of this turns ratio.	(17.56)	17.3	644
❑	Impedances are reflected as the turns ratio squared.	(17.59)	17.3	645
❑	An ideal transformer introduces two additional unknowns into our mesh-current or node-voltage equations. We obtain the necessary extra two equations from the turns-ratio relationship between primary and secondary voltages and currents.		17.4	647 648 649
❑	We improve the accuracy of the ideal transformer model if we include leakage inductance and magnetizing inductance.		17.5	652

PROBLEMS

Section 17.1

1. Two coupled coils are encapsulated in black plastic, so it is impossible to tell how they are wound. The name plate shows that coil 1 is connected to terminals a and b, and coil 2 is connected to terminals c and d. Both coils are listed as 0.1 H, and the mutual inductance is 0.08 H. The coil resistances are 1-Ω each. We need to determine the dot marks. We arbitrarily place a dot on terminal a. We then connect a 1-A ammeter between terminals c and d, so that we measure the current leaving terminal c of coil 2. We next connect the negative terminal of a 1.5-V flashlight battery to terminal b. When we touch the positive terminal of the battery to terminal a, the ammeter jumps up scale a bit. When we disconnect the battery, the meter jumps down scale. (a) Where should the dot mark be placed on coil 2? (b) What is the magnetic coefficient of coupling k?

2. For the following circuit, $i_s = 100 \cos 3t$ A. Calculate v_1 and v_2.

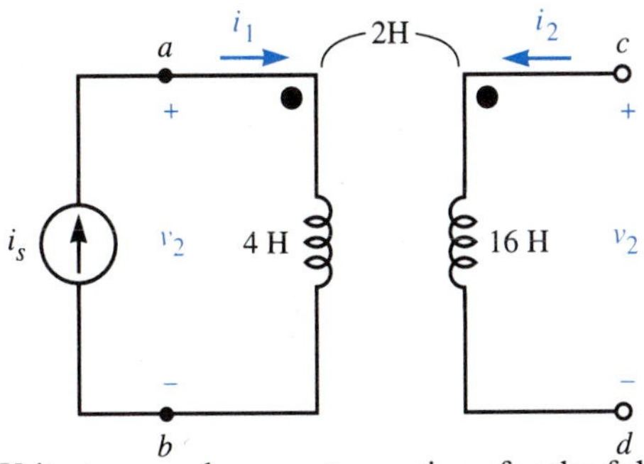

3. Write two mesh-current equations for the following circuit.

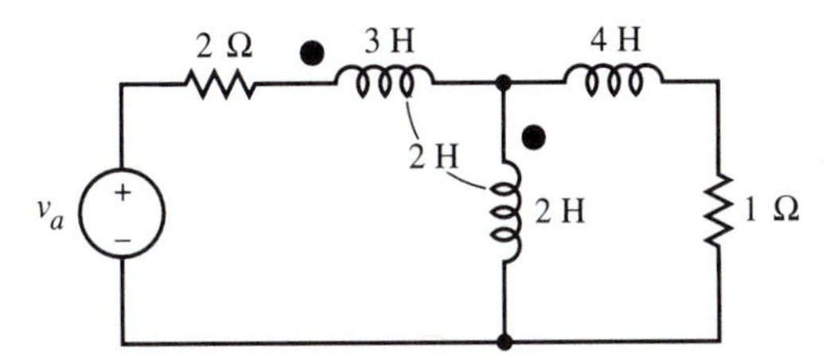

4. Use Eqs. (17.9), (17.10), and KVL to determine the equivalent inductance connected between terminals a and b in the following two illustrations.

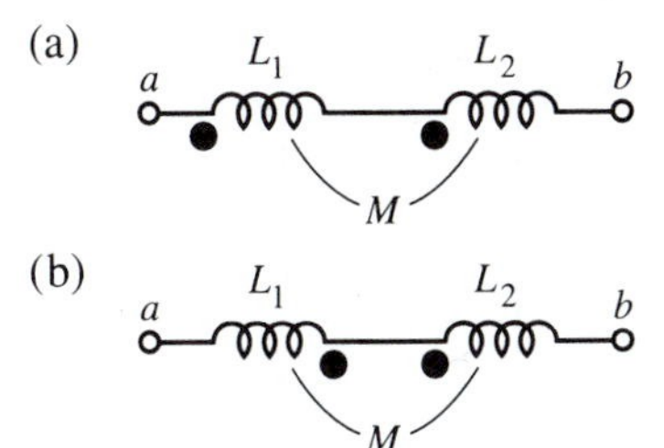

Section 17.2

5. The impedances of the coupled inductances shown below were measured at frequency $f_1 = 1$ kHz on an impedance bridge. (a) Give the time-domain circuit for the coupled inductances. (b) Give the frequency-domain equivalent circuit at frequency $f_2 = 500$ Hz.

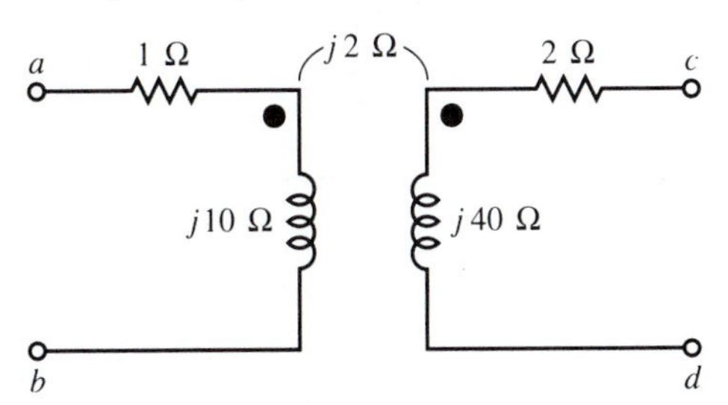

6. A voltage source $\mathbf{V}_s = 120\underline{/0^\circ}$ V [(+) reference at the top] is connected to terminals a and b in the following circuit. (a) Write the mesh current equations for the circuit. (b) Solve for the current input to terminal a. 17.1

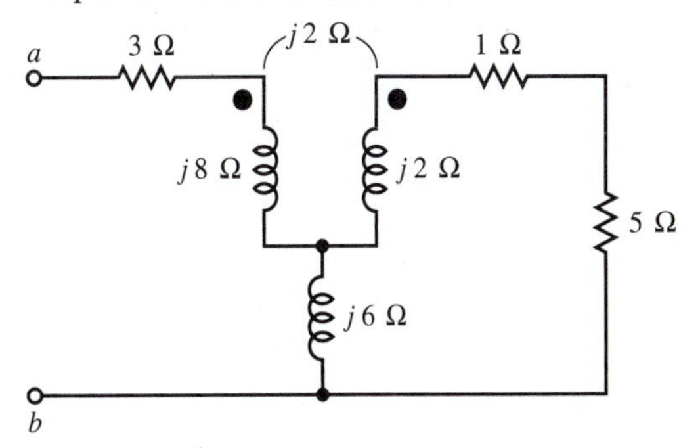

7. The primary impedance of a linear transformer is $\mathbf{Z}_{11} = R_1 + j\omega L_1 = 6 + j40\ \Omega$ and the secondary impedance is $\mathbf{Z}_{22} = 4 + j16\ \Omega$. The mutual impedance is $\mathbf{Z}_{12} = \mathbf{Z}_{21} = j\omega M = j20\ \Omega$. A load of $\mathbf{Z}_L = 8 + j0\ \Omega$ is connected between terminals c and d.
   (a) What is the impedance $\mathbf{Z}_1 = \mathbf{V}_{ab}/\mathbf{I}_{ab}$ looking into terminals a and b?
   (b) Calculate the magnetic coefficient of coupling $k = M/\sqrt{L_1 L_2}$ for the linear transformer.
   (c) Calculate $\mathbf{I}_{ab}$, $\mathbf{I}_{cd}$, $\mathbf{V}_{cd}$, and the complex power absorbed at terminals a and b ($\mathbf{S}_{ab} = \mathbf{V}_{ab}\mathbf{I}_{ab}^*$), if $\mathbf{V}_{ab} = 100\underline{/0^\circ}$ V rms.

8. Two coupled inductances, as shown in Fig. 17.5, have inductance values $L_1 = 18$ mH, $L_2 = 2$ mH, and $M = 3$ mH. Design and sketch a T network that is equivalent to these coupled inductances at an angular frequency of 10,000 rad/s. (Terminals b and d are connected for the coupled inductances.)

9. Two coupled inductances, as shown in Fig. 17.5, have inductance values $L_1 = 18$ mH, $L_2 = 2$ mH, and $M = 3$ mH. Design and sketch a Π network that is equivalent to these coupled inductances at an angular frequency of 10,000 rad/s. (Terminals b and d are connected for the coupled inductances.)

10. Replace the coupled inductances in the circuit of Problem 6 by an equivalent T network and use parallel and series combinations to calculate the impedance looking into terminals a and b.

Section 17.3

11. An ideal transformer supplies 2 A to a 6-V doorbell and is connected to the 120-V service in your

house. (a) What turns ratio must the transformer have? (b) What is the current on the 120-V side? (c) All voltages and currents are rms values. What is the input power to the transformer?

12. For the following circuit, resistance R_L is 3 Ω. Calculate currents $\mathbf{I}_1$ and $\mathbf{I}_2$ and voltages $\mathbf{V}_1$ and $\mathbf{V}_2$, if $\mathbf{V}_s = 110\angle 0°$ V rms.

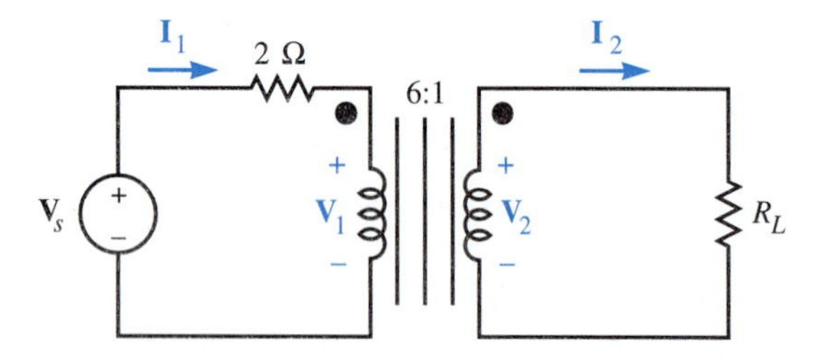

13. For the preceding circuit, calculate the value of load resistance R_L that will absorb the maximum power. What power is absorbed by this load resistance?

14. The load resistance R_L in Problem 12 has a value of 8 Ω. The transformer is replaced with a different one. What turns ratio must the transformer have to deliver the maximum power to the load?

15. For the following ideal transformer circuit, determine: (a) the source current **I**, (b) the complex power **S** supplied by the source, (c) the output voltage $\mathbf{V}_o$, and (d) the voltage on the primary winding $\mathbf{V}_1$, if $\mathbf{V}_s = 20\angle 0°$ V rms.

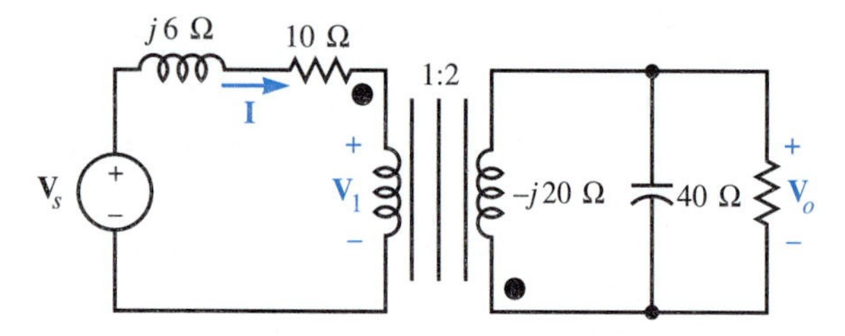

16. Find the phasor currents $\mathbf{I}_1$, $\mathbf{I}_2$, and $\mathbf{I}_3$, if $\mathbf{V}_s = 30\angle 30°$ V in the following circuit.

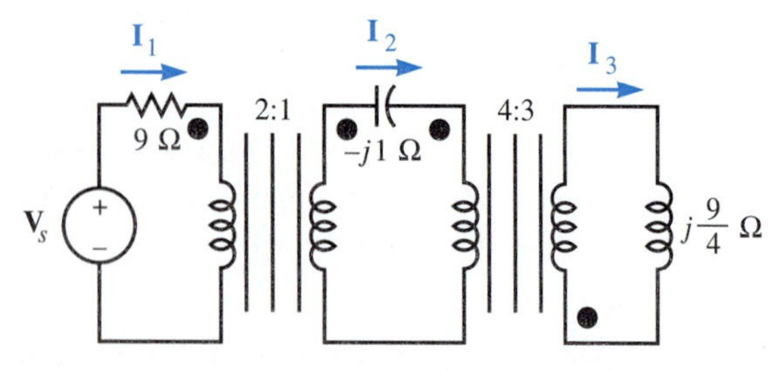

17. The connection of two transformer coils in series, as shown in the following circuit, is often called an *autotransformer*. Assume that the transformers are ideal. Calculate: (a) $\mathbf{V}_{cd}/\mathbf{V}_{ad}$, (b) $\mathbf{I}_2/\mathbf{I}_1$, and (c) $\mathbf{Z}_1 = \mathbf{V}_{ad}/\mathbf{I}_1$.

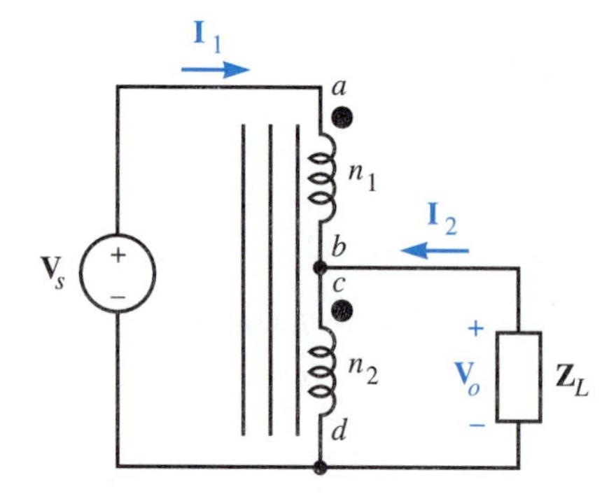

18. Repeat Problem 17 with the dot mark by terminal *c* moved to the end by terminal *d*.

Section 17.4

19. Use the method of Example 17.3 to solve for the mesh currents in the following circuit if $\mathbf{V}_s = 120\angle 0°$ V.

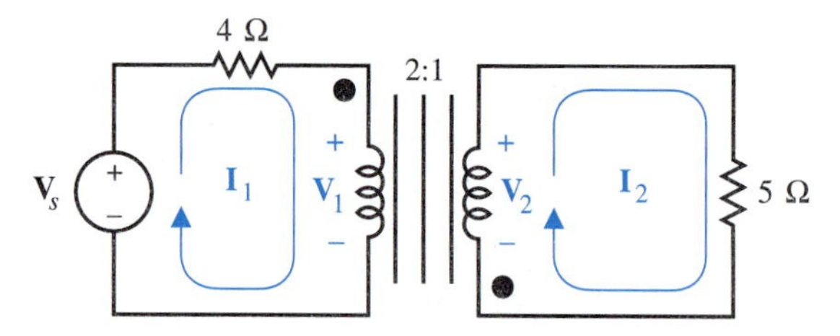

20. Solve for the mesh currents in the following circuit if $\mathbf{V}_s = 6\angle 0°$ V.

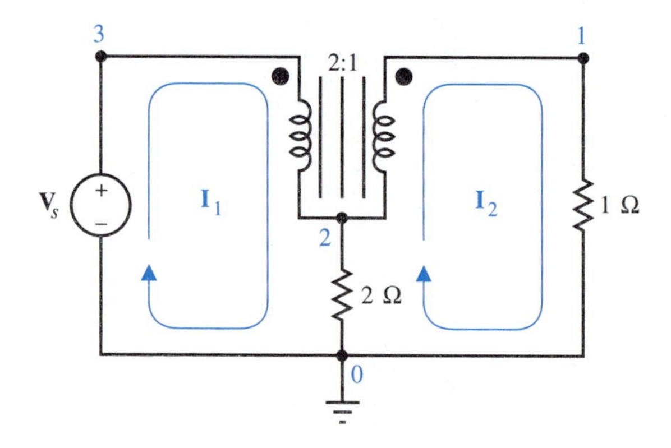

21. Use the method of node voltages described in Example 17.4 to solve for voltages $\mathbf{V}_1$ and $\mathbf{V}_2$ in the circuit of Problem 19 if $\mathbf{V}_s = 120\angle 0°$ V.

22. Solve for the node voltages in the following circuit if $\mathbf{V}_s = 24\angle 0°$ V.

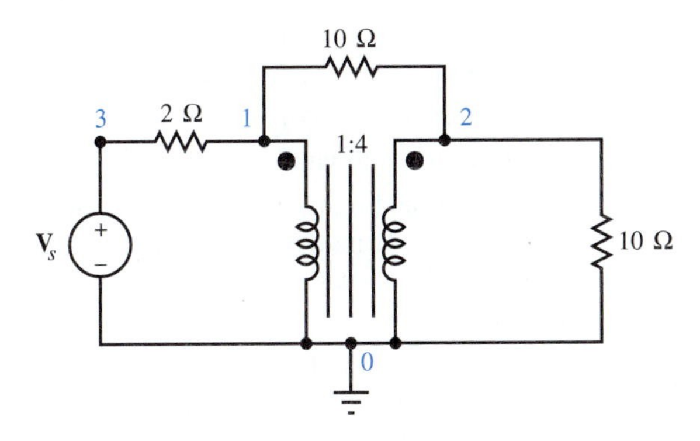

23. Solve for the mesh currents in the following circuit. Use these mesh currents and calculate voltages $\mathbf{V}_1$ and $\mathbf{V}_2$ if $\mathbf{V}_s = 120\underline{/0°}$ V.

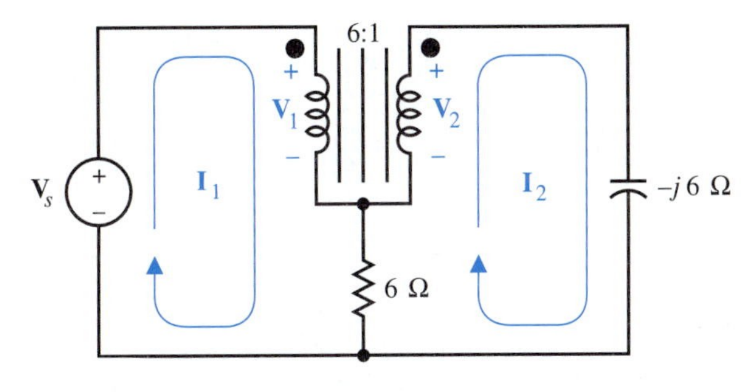

Section 17.5

24. Construct a model as shown in Fig. 17.18 for the coupled inductances in the circuit of Problem 5.

25. The following is an equivalent circuit for two coupled inductances. Compute L_1, L_2, M, and k.

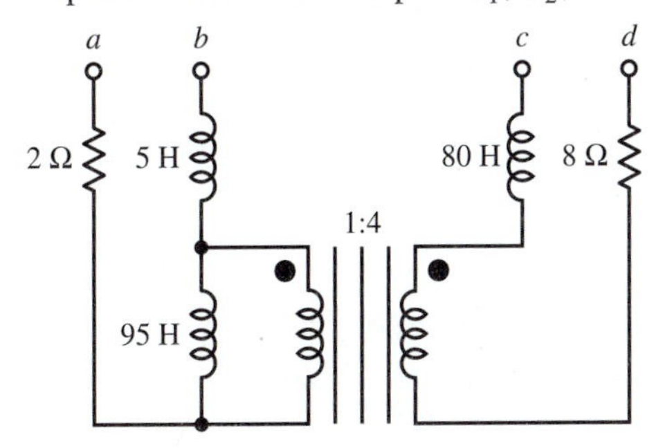

26. For the preceding circuit, reconnect the magnetizing inductance to terminals a and b to obtain a less exact model. Then reflect the inductance and resistance of the secondary to the primary to obtain the equivalent resistance and leakage inductance seen looking into the primary. (Neglect the magnetizing current for this step.)

27. Assume that the resistances of the transformer coils are negligibly small, and the coupling coefficient is expected to be greater than 0.9.

    (a) The secondary of the transformer is open-circuited and 120 V 60 Hz is applied to the primary. The primary current is 1.2 A, and the secondary voltage is 240 V. Estimate the magnetizing reactance X_{M1} and the effective turns ratio n_1/n_2.

    (b) The primary voltage is reduced to 12 V 60 Hz and the secondary is short-circuited. The short-circuit current in the primary is 24 A. Estimate the leakage reactance $X_{\ell p} = X_{\ell 1} + (n_1/n_2)^2 X_{\ell 2}$ seen looking into the primary.

    (c) Draw a model for this transformer that is similar to that in Fig. 17.18, but with all leakage reactance reflected to the primary (Show reactances, not inductances).

Supplementary Problems

28. For the following circuit,

    (a) A 60-Hz source is connected to terminals (a) and (b). Terminals c and d are open-circuited. We measure $V_{ab} = 85$ V rms, $V_{cd} = 40$ V rms, $I_{ab} = 2$ A rms, and the power absorbed by the transformer $P_{ab} = 25$ W. Determine R_1, L_1, and M.

    (b) A 60-Hz source is connected to terminals c and d. Terminals a and b are open-circuited. We measure $V_{cd} = 35$ V rms, $V_{ab} = 40$ V rms, $I_{cd} = 2$ A rms, and the power input to the transformer $P_{cd} = 15$ W. Determine R_2, L_2, and M.

    (c) Use the values of inductances calculated in (a) and (b) to calculate the magnetic coefficient of coupling, $k = M/\sqrt{L_1 L_2}$.

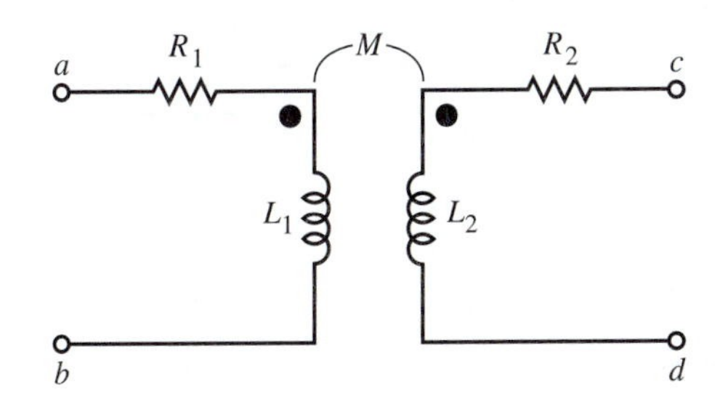

29. Use phasors and mesh currents to solve for currents i_a and i_b in the following circuit. (You need to use superposition, also.)

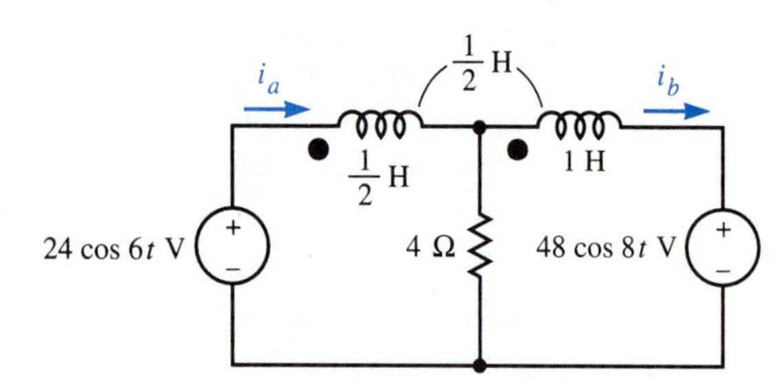

30. Use phasors, superposition, and node voltages to solve for the voltage across the 4-Ω resistance in the preceding circuit. [For the coupled inductances, use Eq. (17.35).]
31. The voltage sources in the circuit of Problem 29 are zero for $t < 0$, and $i_a(0^+) = 0$ and $i_b(0^+) = 0$. Solve for the voltage across the 4-Ω resistance. You can use mesh currents, or transform the terminal equations given in Eq. (17.35) to the time domain and use node voltages.
32. The following network is used as an impedance transformation device at $f_0 = 1000$ Hz.
    (a) Select capacitance C_2 so that the self-impedance of the secondary is real.
    (b) Select M so that the conductance looking into terminals a and b with C_1 disconnected is $\frac{1}{600}$ S.
    (c) Now select C_1 so that the driving-point impedance looking into terminals a and b is 600 Ω.
    (d) If a practical source, modeled as an ideal voltage source in series with a 600-Ω resistance, is connected to terminals a and b, what is the Thévenin impedance seen by the 100-Ω load?
    (e) Determine the transfer function $\mathbf{H}(j\omega_0) = \mathbf{V}_o/\mathbf{V}_{in}$, where $\omega_0 = 2\pi 1000$ rps.

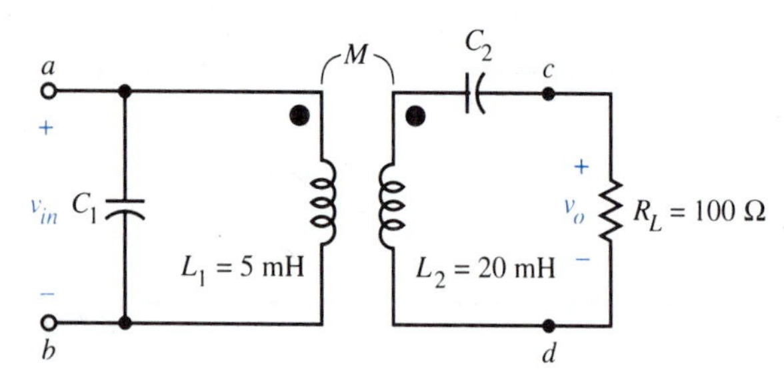

33. Use PSpice and the values of M, C_1, and C_2 calculated in Problem 32, to plot $|\mathbf{H}(j\omega)|$ for $\frac{1}{2}\omega_0 \leq \omega \leq 2\omega_0$.
34. The *doubly-tuned transformer* shown in the following circuit is used in radio receivers.
    (a) If $k = 0.04$, $R = 0.02$ Ω, $L = 1$ H, and $C = 1$ F, use PSpice to plot $|\mathbf{H}(j\omega)| = |\mathbf{V}_o/\mathbf{I}_s|$.
    (b) Magnitude- and frequency-scale the network so that the resonant frequency of the primary circuit and secondary circuit is 455 kHz and L is 1 mH.
    (c) After frequency scaling, what is the bandwidth for which the magnitude of the response is greater than or equal to 0.707 of the value at 455 kHz?

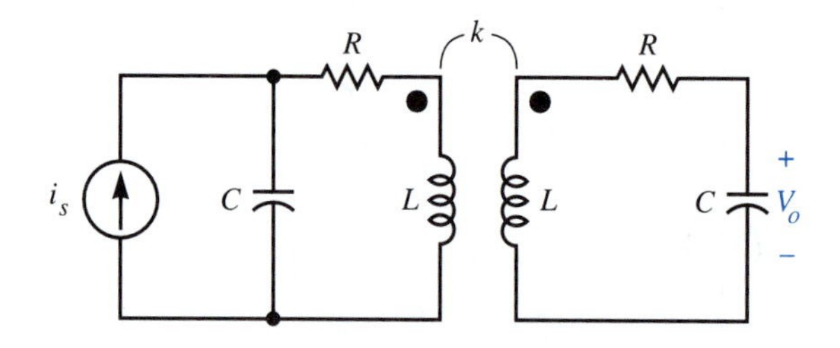

35. Calculate the value of a single inductance that is equivalent to the coupled inductances connected between terminals a and b. [Use Eq. (17.35).]
    (a)

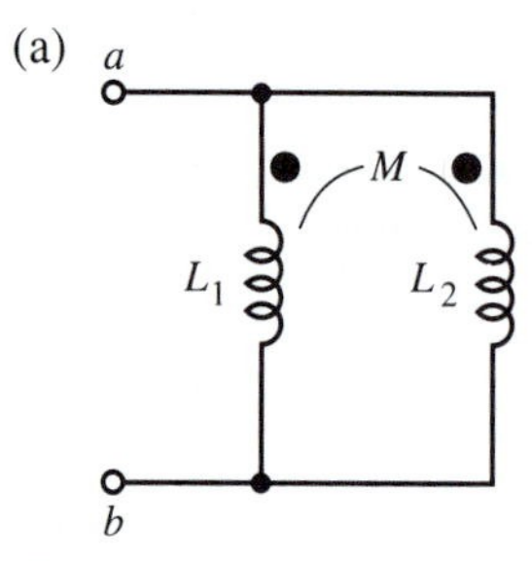

    (b)

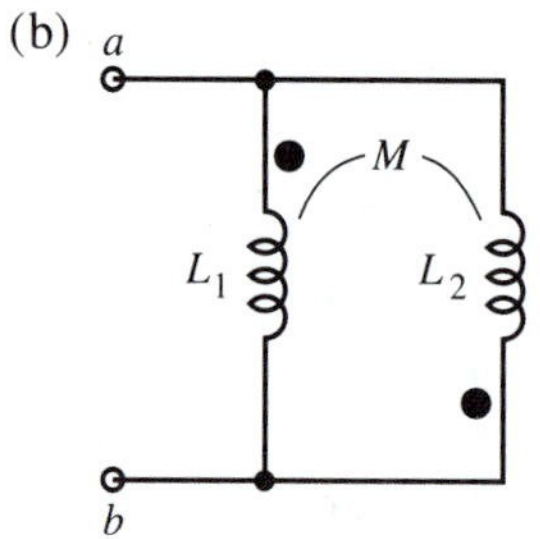

36. For the following circuit, determine the turns ratio n_1/n_2 and the reactance so that the load $\mathbf{Z}_L$ absorbs maximum power (n_1/n_2 is real, not complex). Calculate the power absorbed by this value of $\mathbf{Z}_L$, if $\mathbf{V}_s = 100\underline{/0°}$ V rms.

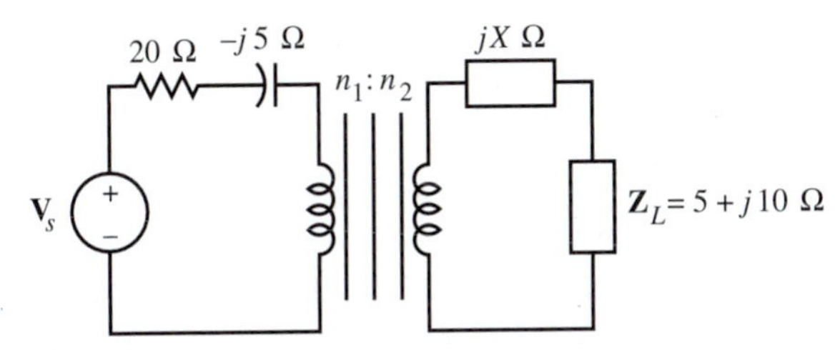

37. A voltage source of internal impedance $\mathbf{Z}_s = 300\underline{/0^\circ}\ \Omega$ is connected to a load of impedance $\mathbf{Z}_L = 50\underline{/0^\circ}\ \Omega$ through an ideal transformer with turns ratio n_1/n_2, where side 1 corresponds to the source.
    (a) What value of n_1/n_2 will result in the maximum power delivered to $\mathbf{Z}_L$?
    (b) Repeat part (a) if the transformer resistances $R_1 = 36\ \Omega$ and $R_2 = 1\ \Omega$ are included in the model.
38. Your fishing cabin is at the end of the rural distribution system. The line voltage is 109 V rms with or without the television set being turned on. The result is that your 12-inch television has an 11-inch picture. You buy a 120/12 V transformer (assume it is ideal) rated at 12 VA. Show how you would connect this as an auto-transformer to supply 120 V to the television set (show the dot marks). The television requires 100 VA. Will this overload the 12-VA transformer?
39. For the ideal transformer network below, solve for voltages $\mathbf{V}_1$ and V_2 if $\mathrm{V}_s = 72\underline{/0^\circ}$ V.

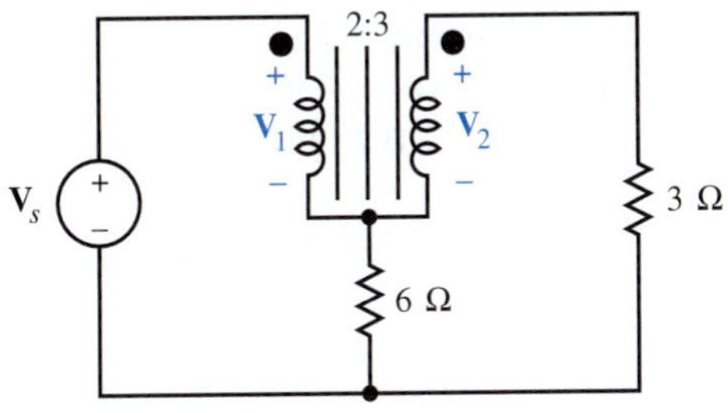

40. Calculate the transfer function $\mathbf{I}_2/\mathbf{I}_1$ and the impedance seen by the current source for the network shown in the following circuit.

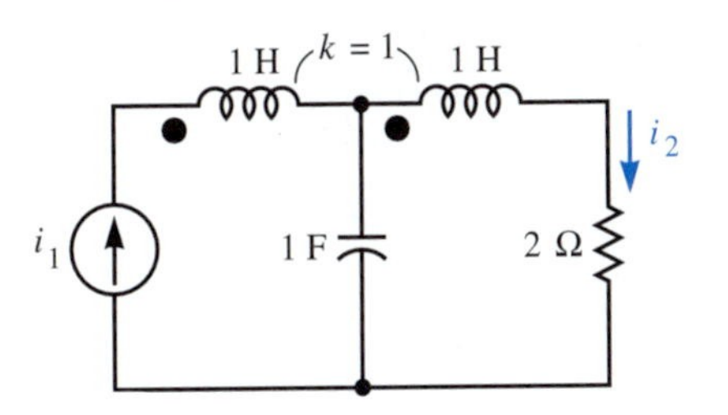

41. Construct a model that includes an ideal transformer for the preceding circuit.
42. The two coils of the ideal transformer in the following circuit are connected in series to form an autotransformer.
    (a) Find the Thévenin impedance with respect to terminals c and d.
    (b) Find the Thévenin impedance seen by the source when an impedance of $\mathbf{Z}_L$ is connected between terminals c and d.

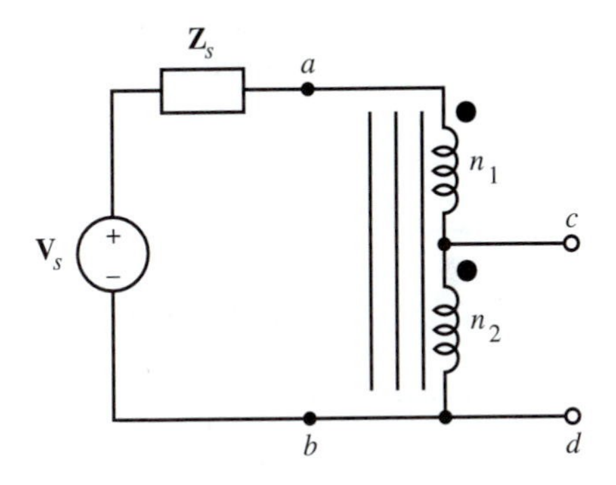

43. Repeat Problem 42 with the dot mark on the lower coil at the bottom.
44. The transformer shown in Problem 42 is not ideal. The coil with n_1 turns has inductance L_1, and the coil with n_2 turns has inductance L_2. The coefficient of coupling is 1.
    (a) What is the Thévenin impedance looking into terminals c and d?
    (b) What is the Thévenin impedance seen by the source when an impedance of $\mathbf{Z}_L$ is connected between terminals c and d?
45. Consider the ideal autotransformer connected as shown in the following circuit. (Remember that

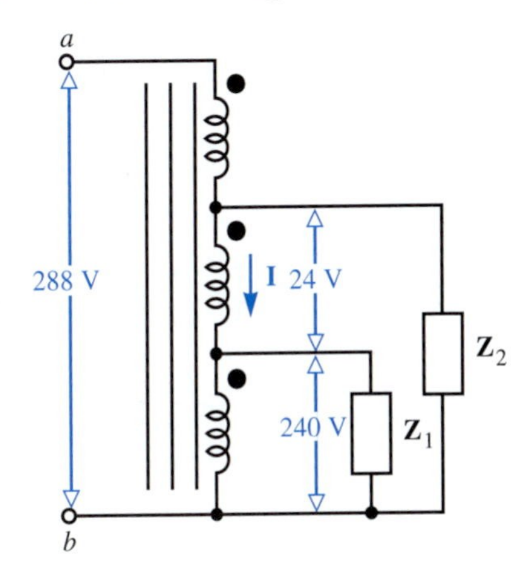

the complex power absorbed by an ideal transformer is zero.)

(a) What is the Thévenin impedance looking into terminals a and b?

(b) Find a relation between $\mathbf{Z}_1$ and $\mathbf{Z}_2$ so that current $\mathbf{I}$ is zero.

46. Find the Thévenin impedance looking into terminals a and b for the following network. If a resistance of value R_L is connected between terminals a and b, what resistance is seen between terminals c and d by the voltage sources? (Remember that the complex power absorbed by an ideal transformer is zero.)

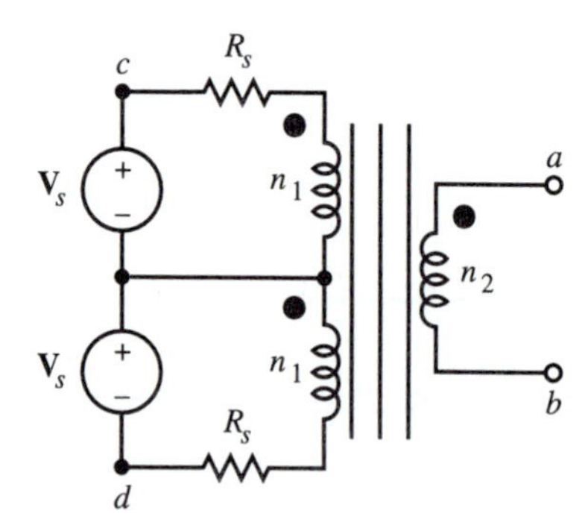

47. Calculate currents $\mathbf{I}_1$ and $\mathbf{I}_2$ in the network in Fig. 17.20.

48. For a linear transformer (Fig. 17.6), $R_1 = 1\ \Omega$, $R_2 = 2\ \Omega$, $L_1 = 10$ H and $L_2 = 40$ H. The load impedance is $\mathbf{Z}_L = R_L = 2\ \Omega$ and the source impedance is $\mathbf{Z}_s = 0$.

(a) Calculate the coefficient of coupling k for (i) $M = 1$ H, (ii) $M = 2$ H, (iii) $M = 3$ H, and (iv) $M = 6$ H. Use PSpice to solve parts (b) and (c).

(b) Use a linear scale to plot $|\mathbf{Z}_{in}(j\omega)|$ for $0 \le f \le 0.4$ Hz for the values of k found in part (a).

(c) Plot $|\mathbf{H_V}(j\omega)| = |\mathbf{V}_2/\mathbf{V}_1|$ for $0 \le f \le 0.05$ Hz for the values of k found in part (a).

49. Repeat Problem 48 when a 0.1-F capacitance is placed in series with R_1 and to the right of $\mathbf{V}_1$; and a 0.025-F capacitance is placed in series with R_2 and to the left of $\mathbf{V}_2$.

50. The load on the linear transformer shown in the following circuit is a capacitance of value C_2. The voltage $\mathbf{V}_1$ is supplied by a voltage source.

(a) Calculate the transfer function $\mathbf{V}_2/\mathbf{V}_1$.

(b) Define $\omega_0 = 1/\sqrt{L_2C_2} = 1/\sqrt{L_1C_1}$, $Q_1 = \omega_0 L_1/R_1$, $Q_2 = \omega_0 L_2/R_2$ and $k = M/\sqrt{L_1L_2}$. Write the transfer function in a form that does not contain M, L_1, L_2, or C_1.

(c) Define k_c to be the value of k for which $|\mathbf{V}_2/\mathbf{V}_1|$ is a maximum at $\omega = \omega_0$. Develop an expression for k_c in terms of Q_1 and Q_2.

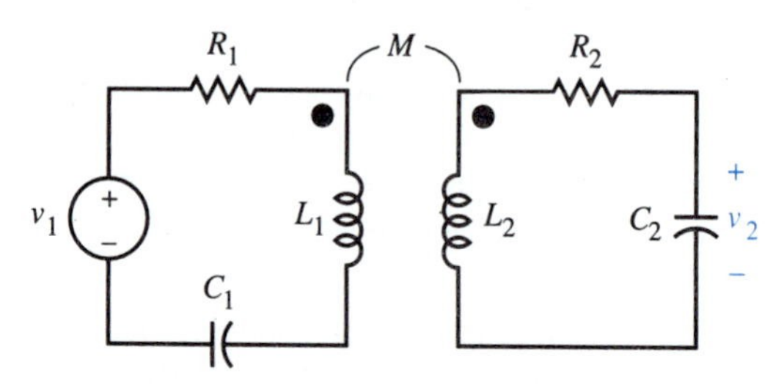

51. A small transformer has a turns ratio of 5/1 and is designed for a supply voltage of 120 V rms and a maximum output current of 2 A rms. Transformers of this size are typically far from ideal. The parameters of the model shown in Fig. 17.18a are $R_1 = 10\ \Omega$, $R_2 = 0.5\ \Omega$, $R_{c1} = 1.2\ \text{k}\Omega$, $X_{M1} = 5$

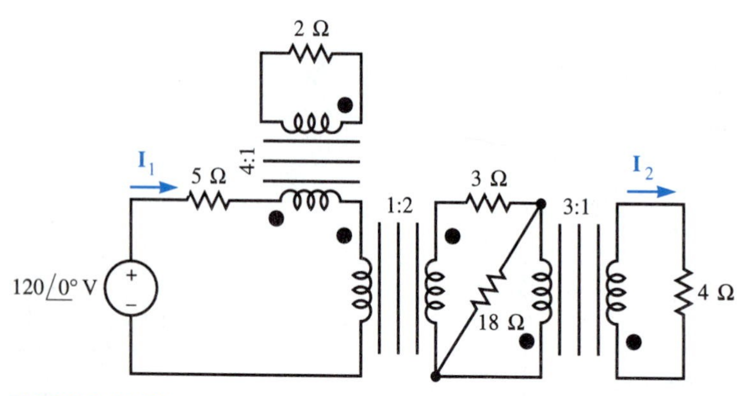

FIGURE 17.20

kΩ, $X_{\ell 1} = 40\ \Omega$, and $X_{\ell 2} = 2\ \Omega$. Calculate the load voltage when a 10-Ω resistive load is connected. What is the transformer input current?

52. A transformer has an effective turns ratio of 20, a primary resistance of $R_1 = 200\ \Omega$, and a secondary resistance of $R_2 = 0.5\ \Omega$. The core losses are negligible. The inductance of the primary with the secondary open-circuited is $L_1 = 5$ H. The inductance of the primary coil with the secondary coil short-circuited is $L_{1sc} = 0.1$ H. Construct a model for the transformer that incorporates an ideal transformer. You can assume that $\omega L_1 \gg R_1$.

53. Show that the terminal equations for each of the following two circuits are the same as for an ideal transformer. These circuits can be used to model an ideal transformer.

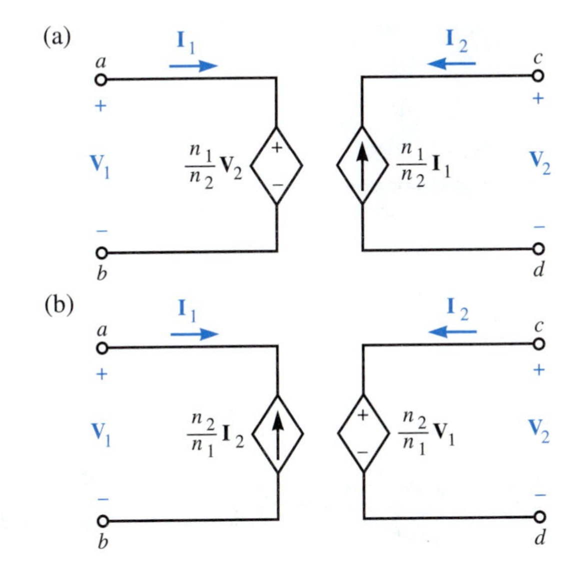

18

Single- and Three-Phase Power Circuits

The analyses of electrical generation, transmission, and distribution systems are specialized topics in electrical engineering that can fill several semester courses. Expertise in these subjects is typically required only of engineers employed by electrical utilities and consulting engineering firms and of plant engineers. Nevertheless, all engineers require some knowledge of how electric power is supplied.

We begin by describing three-wire single-phase systems, which supply power to nearly all residences in the United States. We then present a brief introduction to balanced three-phase circuits, which are used in industry and many commercial air conditioning systems. We see that an important property of a balanced three-phase system is that it delivers *constant* power. That is, the power delivered does not fluctuate with time as in a single-phase system.

18.1 Three-Wire Single-Phase Systems

In the United States nearly all residences and many small commercial facilities are supplied power by a 120/240-V 60-Hz three-wire connection to the power system. The power source and load can be modeled as in Fig. 18.1.

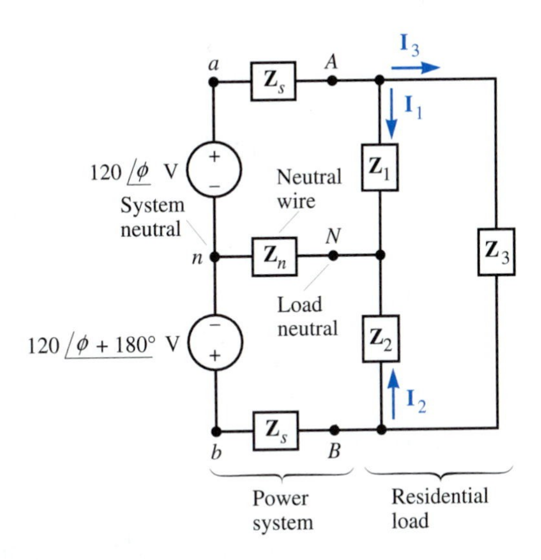

FIGURE 18.1 The three-wire single-phase system that supplies residential power

For electric power systems, the phasor voltages and currents are *universally specified in terms of the effective or rms value.*

In Fig. 18.1 the line-to-neutral voltages at the source are $\mathbf{V}_{an} = 120\underline{/\phi}$ V and $\mathbf{V}_{bn} = 120\underline{/\phi + 180°}$ V, where the system neutral (n) is typically connected to the earth (earth ground). Thus $\mathbf{V}_{ab} = \mathbf{V}_{an} + \mathbf{V}_{nb} = 240\underline{/\phi}$ V and the system supplies both 120- and 240-V service. The system impedances $\mathbf{Z}_s$ and $\mathbf{Z}_n$ are normally small enough that the load voltages $\mathbf{V}_{AN}$, $\mathbf{V}_{BN}$, and $\mathbf{V}_{AB}$ differ from the system voltages by only a few percent.

Low-power loads such as lighting and small appliances are typically designed for 120-V line voltages and are represented by impedances $\mathbf{Z}_1$ and $\mathbf{Z}_2$. Large loads such as cooking stoves and central air conditioners, which are usually connected to 240 V, are represented by impedance $\mathbf{Z}_3$. Equal 120-V loads $\mathbf{Z}_1$ and $\mathbf{Z}_2$ will result in *line currents* being related by $\mathbf{I}_{aA} = -\mathbf{I}_{bB}$ and *neutral current* $\mathbf{I}_{nN}$ equal to zero. Loads $\mathbf{Z}_1$ and $\mathbf{Z}_2$ are, of course, rarely balanced, but the neutral current is usually less than the line current, which results in reduced power losses. The connection shown in Fig. 18.1 is used for safety as well as efficiency. The system supplies 240-V service without any line having more than 120 V with respect to the system neutral, which is connected to earth ground.

For the unusual case where $\mathbf{Z}_1 = \mathbf{Z}_2$, the analysis can be simplified if we realize that because the neutral line, represented by impedance $\mathbf{Z}_n$, carries no current, we can replace it by an open circuit. Loads $\mathbf{Z}_1$ and $\mathbf{Z}_2$ are effectively in series, and this series combination is in parallel with $\mathbf{Z}_3$. Typically $\mathbf{Z}_1 \neq \mathbf{Z}_2$ and a technique such as mesh-current or node-voltage analysis is appropriate.

Although power is supplied to the load by three wires, this *is not* considered to be a three-phase system, because $\mathbf{V}_{an}$, $\mathbf{V}_{bn}$, and $\mathbf{V}_{ab}$ differ in phase by either 0° or 180°. The corresponding voltages in a balanced three-phase system would differ by 120°.

EXAMPLE 18.1 Refer to Fig. 18.1. The load represented by $\mathbf{Z}_1$ absorbs a complex power of $\mathbf{S}_1 = 1200\underline{/60^\circ}$ VA, $\mathbf{Z}_2$ absorbs $\mathbf{S}_2 = 600\underline{/0^\circ}$ VA, and $\mathbf{Z}_3$ absorbs $\mathbf{S}_3 = 2400\underline{/-45^\circ}$ VA. Neglect line impedances $\mathbf{Z}_s$ and neutral impedance $\mathbf{Z}_n$ when calculating the currents. Find $\mathbf{I}_1$, $\mathbf{I}_2$, $\mathbf{I}_3$, $\mathbf{I}_{aA}$, $\mathbf{I}_{bB}$, $\mathbf{I}_{nN}$, and the line losses.

Solution We use the relation†

$$\mathbf{S} = \mathbf{V}\mathbf{I}^*$$

to calculate the load currents

$$\mathbf{I}_1 = \left(\frac{\mathbf{S}_1}{\mathbf{V}_{AN}}\right)^* = \left(\frac{1200\underline{/60^\circ}}{120\underline{/\phi}}\right)^* = 10\underline{/(\phi - 60^\circ)}\text{ A}$$

$$\mathbf{I}_2 = \left(\frac{\mathbf{S}_2}{\mathbf{V}_{BN}}\right)^* = \left(\frac{600\underline{/0^\circ}}{120\underline{/\phi + 180^\circ}}\right)^* = 5\underline{/(\phi + 180^\circ)}\text{ A}$$

and

$$\mathbf{I}_3 = \left(\frac{\mathbf{S}_3}{\mathbf{V}_{AB}}\right)^* = \left(\frac{2400\underline{/-45^\circ}}{240\underline{/\phi}}\right)^* = 10\underline{/(\phi + 45^\circ)}\text{ A}$$

From KCL,

$$\mathbf{I}_{aA} = \mathbf{I}_1 + \mathbf{I}_3 = 12.18\underline{/\phi - 7.5^\circ}\text{ A}$$

$$\mathbf{I}_{bB} = \mathbf{I}_2 - \mathbf{I}_3 = 13.99\underline{/\phi - 149.64^\circ}\text{ A}$$

$$\mathbf{I}_{nN} - (\mathbf{I}_1 + \mathbf{I}_2) = 8.66\underline{/\phi + 90^\circ}\text{ A}$$

The power lost in the lines for the system shown in Fig. 18.1 is

$$P = |\mathbf{I}_{nN}|^2\mathcal{Re}\{\mathbf{Z}_n\} + |\mathbf{I}_{aA}|^2\mathcal{Re}\{\mathbf{Z}_s\} + |\mathbf{I}_{bB}|^2\mathcal{Re}\{\mathbf{Z}_s\}$$
$$= 75\mathcal{Re}\{\mathbf{Z}_n\} + 148.24\mathcal{Re}\{\mathbf{Z}_s\} + 195.71\mathcal{Re}\{\mathbf{Z}_s\}$$

If the resistances of the lines and neutral wire are the same,

$$R = \mathcal{Re}\{\mathbf{Z}_n\} = \mathcal{Re}\{\mathbf{Z}_s\}$$

then the power loss in the three wires is

$$P = (419.0)R\text{ W}$$

Remember In power systems, voltage and current are universally specified in terms of rms (effective) values. A three-wire single-phase system supplies power to most residences. Two lines are 120 V with respect to the neutral wire, and the neutral wire is connected to earth ground. The voltage between the two 120-V lines is 240 V, and voltages are supplied at a frequency of 60 Hz in the United States.

EXERCISE

1. Refer to Fig. 18.1. Assume that the angle ϕ is zero. The impedances are $\mathbf{Z}_s = j1\ \Omega$, $\mathbf{Z}_n = j2\ \Omega$, $\mathbf{Z}_1 = \mathbf{Z}_2 = 10\ \Omega$, and $\mathbf{Z}_3 = 10\ \Omega$. Find currents $\mathbf{I}_1$, $\mathbf{I}_2$, and $\mathbf{I}_3$ and

† The effective (rms) values for voltage and current are used in power systems, so a factor of $\frac{1}{2}$ is not included in the equation for complex power.

voltages $\mathbf{V}_{AN}$, $\mathbf{V}_{BN}$, and $\mathbf{V}_{Nn}$. Show these on a phasor diagram that also includes $\mathbf{V}_{an}$ and $\mathbf{V}_{bn}$.

answer: $11.49\angle -16.7°$ A, $11.49\angle 163.3°$ A, $22.99\angle -16.7°$ A, $114.9\angle -16.7°$ V, $114.9\angle 163.3°$ V, 0 V

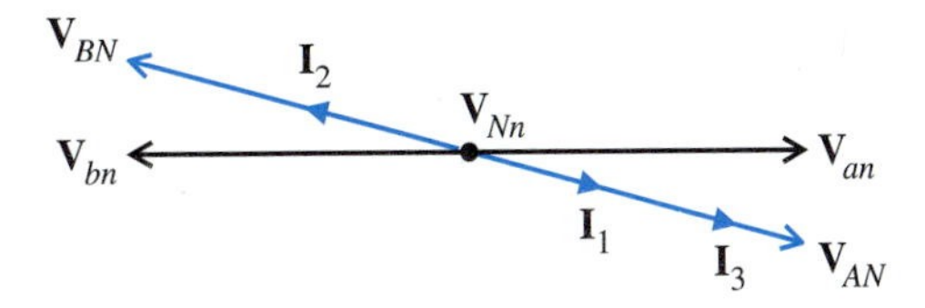

18.2 Three-Phase Sources

Although the direct conversion of heat or solar radiation to electrical energy is used in special applications, most electrical energy is supplied from electromechanical generators. More specifically, most electrical energy is generated by *synchronous machines,* called alternators, in which an electromagnet is rotated inside an internally slotted laminated-steel cylinder. The rotating magnetic field induces voltages in wires (conductors) placed in these slots. These conductors supply electrical energy to the connected load.

Actual alternator design and construction is rather involved, since the number and connection of the wires placed in each slot is chosen to make the resultant terminal voltage as nearly sinusoidal as practical. Nevertheless, the end view of a single-phase two-wire alternator can be represented pictorially as shown in Fig. 18.2a. Terminals a and a' represent the ends of a loop of wire going into and coming out of the page. As the electromagnet rotates, a voltage is generated. This is represented by voltage $v_{aa'}$ or phasor voltage $\mathbf{V}_{aa'}$.

Installation of three sets of windings displaced by 120 mechanical degrees, as depicted in Fig. 18.2b, gives three voltages $v_{aa'}$, $v_{bb'}$, and $v_{cc'}$, each displaced by 120 electrical degrees from each other. If all voltages are sinusoidal and of equal magnitude V_m, this is a *balanced* three-phase set of voltages.

For reasons that are detailed in power systems texts, generation, transmission, and utilization of three-phase power has some advantages over single-phase power. Even automobiles use three-phase alternators, in conjunction with rectifiers, to supply dc power.

Remember

The phase voltages for a balanced three-phase generator have equal magnitude and are displaced 120° with respect to each other. Phasor voltages and currents in power systems are expressed in effective (rms) values.

EXERCISE

2. The voltage of one phase of a balanced three-phase generator is known to be $v_{aa'} = 1600 \cos (2\pi 60t + 45°)$ V. What is the phasor value of $v_{aa'}$? Give possible values for the phasors $V_{bb'}$ and $V_{cc'}$.

answer: $1131\angle 45°$ V, $1131\angle -75°$ V, $1131\angle -195°$ V

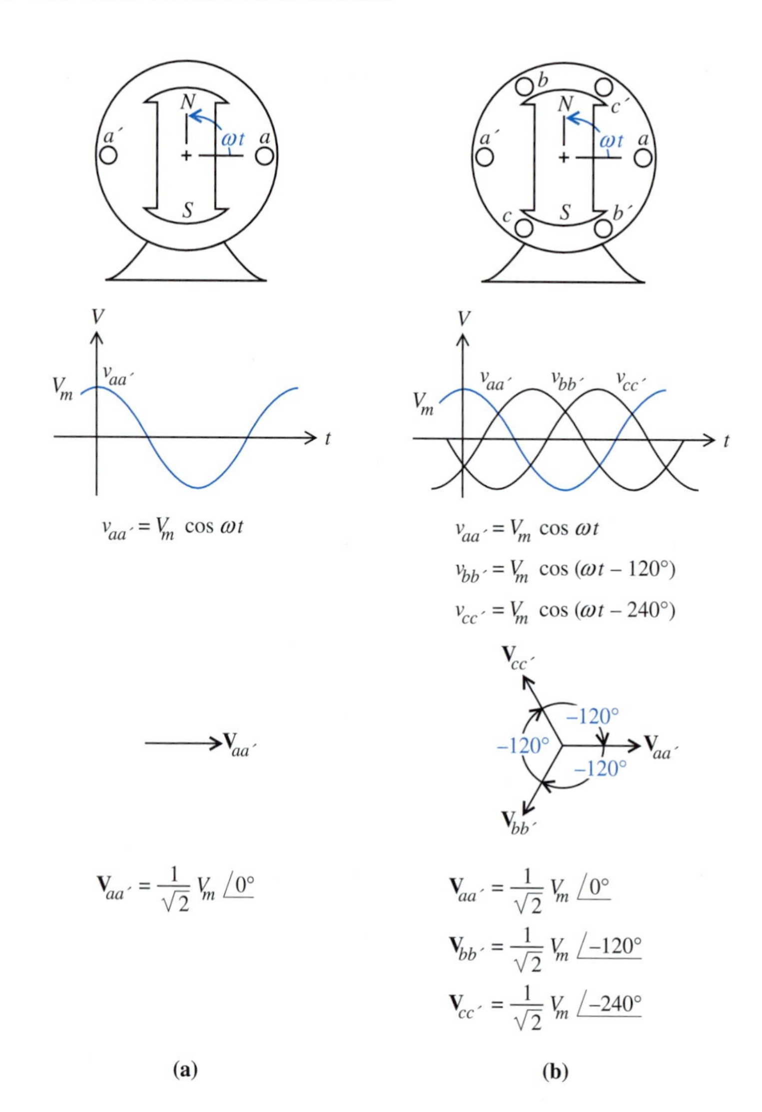

FIGURE 18.2 Simplified representations of alternators and the generated voltages: (*a*) single-phase; (*b*) three-phase

Phase Sequence

The generated voltage set $\mathbf{V}_{aa'} = V_{rms}\angle\phi$, $\mathbf{V}_{bb'} = V_{rms}\angle\phi - 120°$, and $\mathbf{V}_{cc'} = V_{rms}\angle\phi - 240°$, shown in Fig. 18.2b, has a *positive* or *abc phase sequence,* because $\mathbf{V}_{aa'}$ leads $\mathbf{V}_{bb'}$ by 120°, and $\mathbf{V}_{bb'}$ leads $\mathbf{V}_{cc'}$ by 120°.

If we interchange label *b* with *c* and *b′* with *c′* on the wires, we have the voltage set $\mathbf{V}_{aa'} = V_{rms}\angle\phi$, $\mathbf{V}_{cc'} = V_{rms}\angle\phi - 120°$, and $\mathbf{V}_{bb'} = V_{rms}\angle\phi - 240°$. These voltages have a negative or *acb* phase sequence, because $\mathbf{V}_{aa'}$ leads $\mathbf{V}_{cc'}$ by 120° and $\mathbf{V}_{cc'}$ leads $\mathbf{V}_{bb'}$ by 120°.

Remember

We change the phase sequence of a three-phase source when we interchange the labels on two phases.

EXERCISE

3. If $\mathbf{V}_{aa'} = 240\angle 30°$ V, write (a) the positive-phase-sequence set of voltages, $\mathbf{V}_{aa'}$, $\mathbf{V}_{bb'}$, and $\mathbf{V}_{cc'}$ and (b) write the negative-phase-sequence set of voltages $\mathbf{V}_{aa'}$, $\mathbf{V}_{bb'}$, and $\mathbf{V}_{cc'}$. Draw the phasor diagram in each case.

answer: (a) $240\underline{/30°}$ V, $240\underline{/-90°}$ V, $240\underline{/-210°}$ V;
(b) $240\underline{/30°}$ V, $240\underline{/150°}$ V, $240\underline{/270°}$ V

Y-Connected Source

Although three-phase power generation requires three windings that have a total of six terminals, these windings are interconnected, and only three or at most four wires are used to supply power. We will use three ideal voltage sources to represent the voltages generated in the alternator windings. A Y source connection is shown in Fig. 18.3. Terminals a', b', and c' are connected to a common point n called the *neutral*.

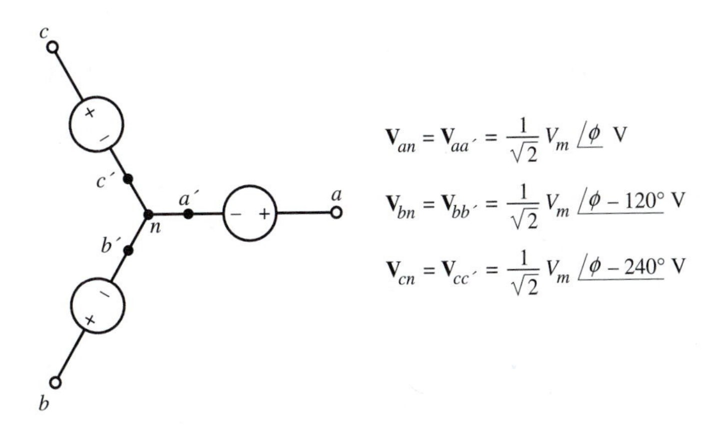

FIGURE 18.3 A balanced Y-connected source with positive (*abc*) phase sequence

As mentioned earlier, voltages in electric power systems are specified in rms or effective values unless otherwise noted. The rms value of the *line-to-neutral voltage* or *phase voltage* for the Y connection is

$$V_{LN} = \frac{1}{\sqrt{2}} V_m \tag{18.1}$$

where V_m is the maximum value of the sinusoidal line-to-neutral voltage.

Voltages $\mathbf{V}_{an}$, $\mathbf{V}_{bn}$, and $\mathbf{V}_{cn}$ are the phasor line-to-neutral voltages. For a Y connection, these are the same as the phase voltages $\mathbf{V}_{aa'}$, $\mathbf{V}_{bb'}$, and $\mathbf{V}_{cc'}$. *Line-to-line* or simply *line voltages* are easily calculated from the line-to-neutral voltages:

$$\begin{aligned}\mathbf{V}_{ab} &= \mathbf{V}_{an} - \mathbf{V}_{bn}\\ &= \mathbf{V}_{an} - \mathbf{V}_{an}(1\underline{/-120^\circ})\\ &= \mathbf{V}_{an}[1 - (\cos 120^\circ - j\sin 120^\circ)]\\ &= \mathbf{V}_{an}\left(\frac{3}{2} + j\frac{1}{2}\sqrt{3}\right)\\ &= \sqrt{3}\underline{/30^\circ}\,\mathbf{V}_{an}\end{aligned} \tag{18.2}$$

Thus the magnitude of any line voltage is the square root of three times the magnitude of the line-to-neutral voltage. For the positive (*abc*) phase sequence, the line-to-line voltages lead the corresponding line-to-neutral voltages by 30°. This relationship is more easily visualized with reference to the phasor diagram of Fig. 18.4a, where ϕ, the angle on $\mathbf{V}_{an}$, has been chosen as zero for convenience. If ϕ were unequal to zero, the entire phasor diagram would simply be rotated counterclockwise by angle ϕ.

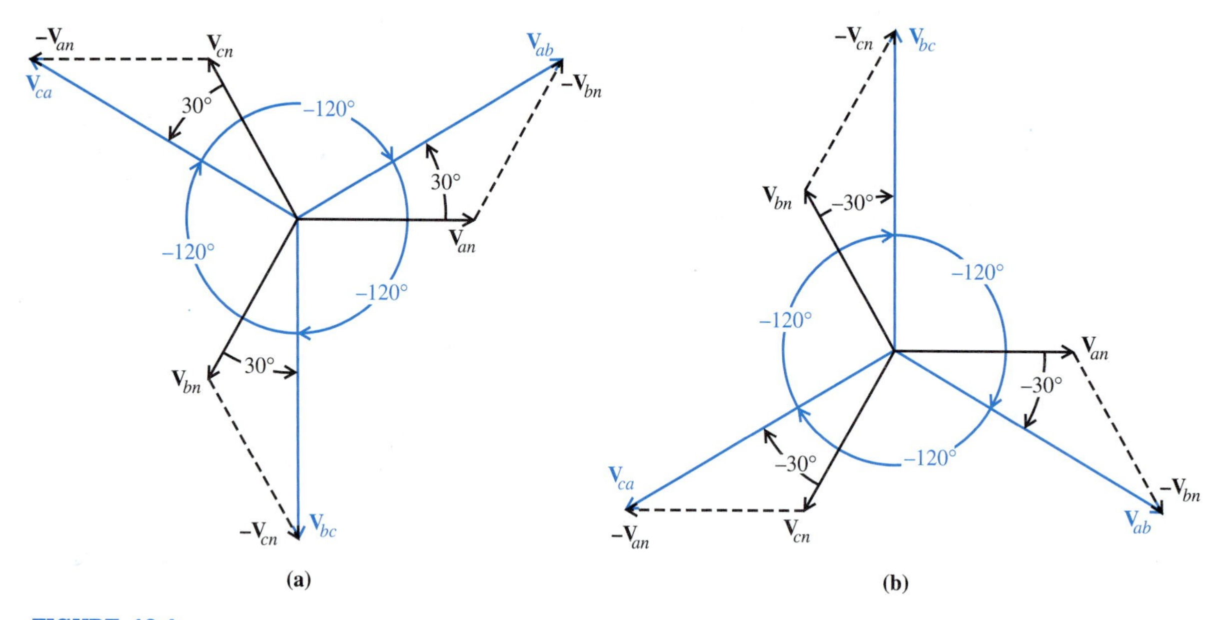

FIGURE 18.4
The relation between line-to-neutral voltages and line-to-line voltages for a balanced three-phase system. (a) positive (*abc*) phase sequence; (b) negative (*acb*) phase sequence

Interchanging the wire designations for *b* and *c* and for *b*′ and *c*′ would result in the phase voltages shown in Fig. 18.4b. The line voltages now have a *negative* or *acb phase sequence*†. Often the terminals *a*′, *b*′, and *c*′ are not accessible, so changing the phase sequence is accomplished by only interchanging the labeling of any two of the terminals *a*, *b*, or *c*. We should note that interchanging any two of the three line connections to a three-phase motor changes the phase sequence and therefore reverses the direction of rotation of the motor. The phase sequence of the line voltages is also important if two sources or transformers are to be connected in parallel.

† The negative phase sequence is also called a *cba* sequence.

Δ-Connected Source

Instead of a Y connection, the coils of a balanced three-phase generator can be connected in a Δ, as shown in Fig. 18.5. A quick check will reveal that

$$\mathbf{V}_{ab} + \mathbf{V}_{bc} + \mathbf{V}_{ca} = 0 \tag{18.3}$$

so KVL is not violated. (In practice, any slight imbalance is accommodated by the winding impedances, which have been neglected, but some current will circulate in the Δ mesh.) Line voltages are equal to the corresponding phase voltages for a Δ connection. Line voltages rather than line-to-neutral voltages would now be the same magnitude as the generated voltages and would be in phase with them.

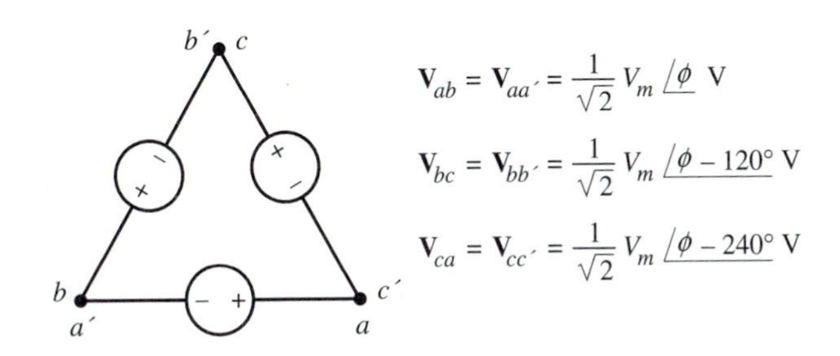

FIGURE 18.5
A Δ-connected source with positive (*abc*) phase sequence

Although Δ-connected sources have no neutral point, we still speak of line-to-neutral voltages, because the system usually has a grounding transformer or a load that connects the neutral point to earth ground. (Occasionally a small low-voltage three-phase system will have one corner of the Δ grounded.)

As a user of three-phase power, we usually are not interested in whether the supply is connected in delta or wye. We only need to know the line voltages. Unless otherwise noted, the voltage of a three-phase system is specified in terms of the rms value of the line voltage (line-to-line voltage). With few exceptions, the neutral of a three-phase system is connected to earth ground.

Remember

The line voltage (line-to-line voltage) is the square root of three times the line-to-neutral voltage. The line voltage is 30° out of phase with respect to the corresponding line-to-neutral voltage. Unless otherwise noted, a three-phase system is specified in terms of the rms value of the line voltage. Interchanging the labeling of any two lines in a three-phase system changes the phase sequence. Therefore, interchanging the connections to any two lines reverses the direction of rotation for a three-phase motor.

EXERCISE

4. The power service in a small factory is specified as 480 V, three phase. What is (a) the rms value of line (line-to-line) voltage, (b) the rms value of the line-to-neutral voltage, (c) the peak value of the line voltage, and (d) the phase difference between the line voltage and the corresponding line-to-neutral voltage?

   answer: 480 V, 277 V, 679 V, The line voltage leads the line-to-neutral voltage by 30° for a positive-phase-sequence system.

18.3 Three-Phase Loads

Like three-phase sources, three-phase loads can be connected in Δ or Y. In many instances, we may not know, or need to know, the internal connection of a three-phase machine, such as a motor. The nameplate simply specifies the machine rating in terms of its line voltage (line-to-line voltage) and the line current, or possibly in terms of the line voltage, kVA rating, and the power factor.

In some instances the internal connections of the windings can be changed between Δ and Y, so that the motor can be operated at different voltages. For this reason, we need to examine the relationship between line voltage and phase voltage and between line current and phase current for the two connections. Throughout the discussion we will assume that the loads are balanced (have identical phase impedances) and are supplied by a balanced positive-phase-sequence source.

Y-Connected Loads

Consider a Y-connected load as shown in Fig. 18.6. The neutral N may or may not be brought outside the machine and connected to the system neutral. It is comparatively easy to show that the current in any neutral wire will be zero in a balanced system. For a selected phase, the phase voltage (the voltage across $\mathbf{Z}_{LN}$) will be equal to the line-to-neutral voltage, the phase current will be equal to the line-to-neutral current and therefore equal to the line current. For example, for phase A:

$$\mathbf{V}_{AN} = V_{LN}\underline{/\phi} = \frac{1}{\sqrt{3}}\underline{/-30^\circ}\ \mathbf{V}_{ab} \tag{18.4}$$

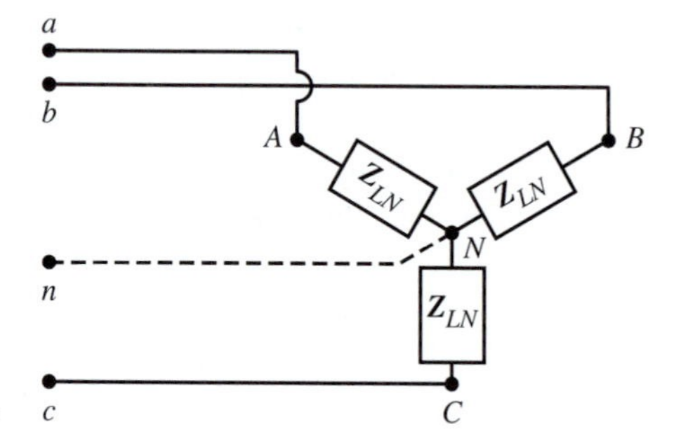

FIGURE 18.6
A Y-connected load

and

$$\mathbf{I}_{aA} = \mathbf{I}_{AN} = \frac{1}{\mathbf{Z}_{LN}}\mathbf{V}_{AN} = I_L\underline{/(\phi - \theta)} \tag{18.5}$$

where

$$I_L = |\mathbf{I}_{aA}| \tag{18.6}$$

and

$$\mathbf{Z}_{LN} = |\mathbf{Z}_{LN}|\underline{/\theta} \tag{18.7}$$

The line currents and phase currents for a Y-connected load are shown in Fig. 18.7 for $\phi = 0$ and $\theta > 0$. Voltages $\mathbf{V}_{AB}$ and $\mathbf{V}_{AN}$ are included for reference.

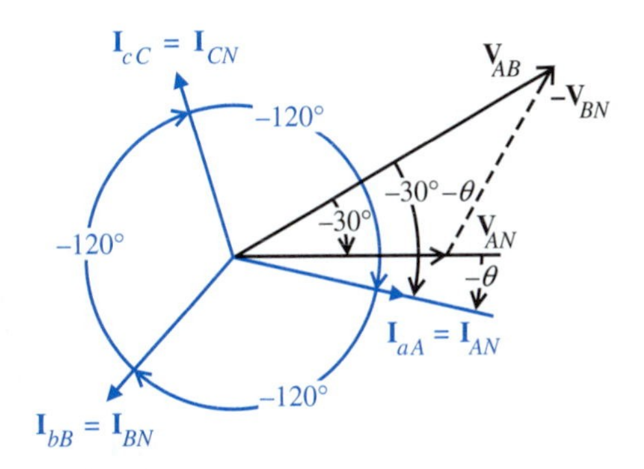

FIGURE 18.7 Phasor diagram for the line currents and phase currents in a balanced Y-connected three-phase load in an *abc* phase-sequence system

The complex power absorbed by a balanced Y-connected load is the sum of the power absorbed by each of the three phases. For a balanced load, the power absorbed by each phase is the same. Therefore, the total complex power absorbed by the load is

$$\mathbf{S} = \mathbf{S}_A + \mathbf{S}_B + \mathbf{S}_C = 3\mathbf{S}_A \tag{18.8}$$

which gives

$$\mathbf{S} = 3\mathbf{V}_{AN}\mathbf{I}_{AN}^{*} = 3V_{LN}I_{LN}\underline{/\theta}\ \text{VA} \tag{18.9}$$

Line voltage has a magnitude of

$$V_L = \sqrt{3}\ V_{LN}\ \text{V} \tag{18.10}$$

and line current has a magnitude of

$$I_L = I_{LN}\ \text{A} \tag{18.11}$$

Substitution of Eqs. (18.10) and (18.11) into Eq. (18.9) gives the following relationship for the total three-phase complex power absorbed by the load:

Complex Power Absorbed by Y-Connected Loads

$$\mathbf{S} = \sqrt{3}\ V_L I_L \underline{/\theta}\ \text{VA} \tag{18.12}$$

where V_L and I_L are in rms values. We can conveniently express the complex power absorbed by a Y-connected load as

$$\begin{aligned}\mathbf{S} &= 3\mathbf{V}_{AN}\mathbf{I}_{AN}^{*} = 3\mathbf{V}_{AN}\left(\frac{\mathbf{V}_{AN}}{\mathbf{Z}_{LN}}\right)^{*} \\ &= 3\frac{1}{\mathbf{Z}_{LN}^{*}}V_{LN}^{2} = 3\frac{1}{\mathbf{Z}_{LN}^{*}}\left(\frac{V_L}{\sqrt{3}}\right)^{2}\end{aligned} \tag{18.13}$$

which gives

Complex Power Absorbed by Y-Connected Loads

$$\mathbf{S} = \frac{1}{\mathbf{Z}_{LN}^{*}}V_L^{2} \tag{18.14}$$

EXAMPLE 18.2 Three impedances of $5\angle 36.87^\circ\ \Omega$ are Y-connected to a 440-V three-phase line. Find the complex power absorbed by the three-phase load and the magnitude of the rms line current.

Solution

$$\mathbf{S} = \frac{1}{\mathbf{Z}_{LN}^*} V_L^2 = \frac{440^2}{5\angle -36.87^\circ} = 38{,}720\angle 36.87^\circ \text{ VA}$$
$$= 30{,}976 + j23{,}232 \text{ VA}$$
$$I_L = \frac{|\mathbf{S}|}{\sqrt{3}\, V_L} = 50.81 \text{ A}$$

Remember For a balanced positive (abc) phase sequence three-phase system, the line voltage (line-to-line voltage) is $\sqrt{3}$ times the line-to-neutral voltage in magnitude and leads the line-to-neutral voltage by 30°. The phase current for a Y connection is the same as the corresponding line current.

EXERCISES A balanced positive (abc) phase sequence system with a balanced Y-connected load has a line-to-neutral impedance (phase impedance) of $\mathbf{Z}_{LN} = 20\angle 15^\circ\ \Omega$. The line voltage is $\mathbf{V}_{AB} = 208\angle 0^\circ$ V.

5. Calculate the line voltages $\mathbf{V}_{BC}$ and $\mathbf{V}_{CA}$. *answer:* $208\angle -120^\circ$ V, $208\angle -240^\circ$ V
6. Calculate the line-to-neutral voltages $\mathbf{V}_{AN}$, $\mathbf{V}_{BN}$, and $\mathbf{V}_{CN}$. *answer:* $120\angle -30^\circ$ V, $120\angle -150^\circ$ V, $120\angle -270^\circ$ V
7. Calculate the line-to-neutral currents $\mathbf{I}_{AN}$, $\mathbf{I}_{BN}$, and $\mathbf{I}_{CN}$. These are also the line currents. *answer:* $6\angle -45^\circ$ A, $6\angle -165^\circ$ A, $6\angle -285^\circ$ A
8. Find the complex power absorbed by the load. *answer:* $2160\angle 15^\circ$ VA
9. You wish to design a three-phase system so that 120-V lighting can be operated from the line-to-neutral connections of Y-connected transformers. What will be the three-phase line voltage (line-to-line voltage)? This system is widely used in industry. *answer:* 208 V

Δ-Connected Loads

Examine the Δ-connected load of Fig. 18.8. For a selected phase, the phase voltage (the voltage across a phase impedance $\mathbf{Z}_{LL}$) is the corresponding line voltage.

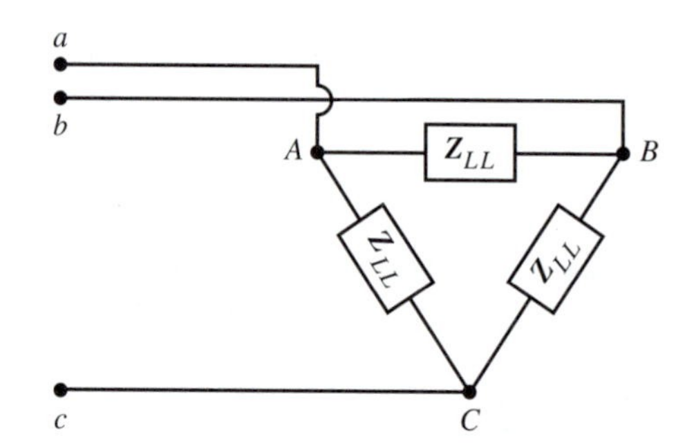

FIGURE 18.8
A Δ-connected load

Line-to-line currents or *phase currents* through impedances $\mathbf{Z}_{LL}$ for a Δ-connected load are easily found in terms of line voltages. Define the magnitude of the phase current by

$$I_{LL} = |\mathbf{I}_{AB}| \tag{18.15}$$

This lets us write

$$\mathbf{I}_{AB} = \frac{1}{\mathbf{Z}_{LL}} \mathbf{V}_{AB} = I_{LL} \underline{/\phi + 30^\circ - \theta} \tag{18.16}$$

$$\mathbf{I}_{BC} = \frac{1}{\mathbf{Z}_{LL}} \mathbf{V}_{BC} = \mathbf{I}_{AB}(1\underline{/-120^\circ}) = I_{LL}\underline{/\phi + 30^\circ - \theta - 120^\circ} \tag{18.17}$$

$$\mathbf{I}_{CA} = \frac{1}{\mathbf{Z}_{LL}} \mathbf{V}_{CA} = \mathbf{I}_{AB}(1\underline{/-240^\circ}) = I_{LL}\underline{/\phi + 30^\circ - \theta - 240^\circ} \tag{18.18}$$

Line currents are easily found in terms of the phase currents of the Δ. For example:

$$\begin{aligned} \mathbf{I}_{aA} &= \mathbf{I}_{AB} - \mathbf{I}_{CA} = \mathbf{I}_{AB}(1 - 1\underline{/-240^\circ}) \\ &= \mathbf{I}_{AB}\left(\frac{3}{2} - j\frac{1}{2}\sqrt{3}\right) = \sqrt{3}\underline{/-30^\circ}\,\mathbf{I}_{AB} \end{aligned} \tag{18.19}$$

Thus for a positive (*abc*) phase sequence, line currents lag the corresponding phase currents (line-to-line currents) for a Δ by 30° and are larger than the phase currents by a factor of $\sqrt{3}$. These relationships are shown in Fig. 18.9.

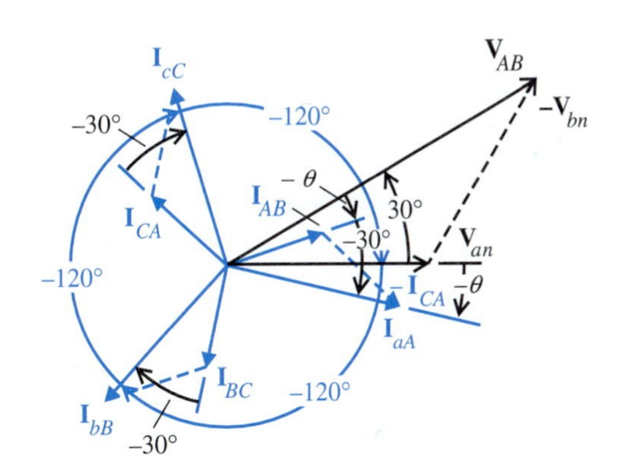

FIGURE 18.9 Phasor diagram for the line currents and phase currents in a balanced Δ-connected three-phase load in an *abc* phase-sequence system

The complex power absorbed by a balanced Δ-connected load is three times the power absorbed by one branch of the delta. Therefore

$$\begin{aligned} \mathbf{S} &= 3\mathbf{S}_a = 3\mathbf{V}_{AB}\mathbf{I}_{AB}^* \\ &= 3\mathbf{V}_{AB}\left(\frac{1}{\sqrt{3}}\underline{/30^\circ}\mathbf{I}_{aA}\right)^* \\ &= (3V_L\underline{/\phi + 30^\circ})\left(\frac{1}{\sqrt{3}} I_L\underline{/\phi + 30^\circ - \theta}\right)^* \end{aligned} \tag{18.20}$$

This gives

The Complex Power Absorbed by Δ-Connected Loads

$$\mathbf{S} = \sqrt{3}\, V_L I_L \underline{/\theta} \text{ VA} \tag{18.21}$$

This is *identical* to Eq. (18.12) for a Y-connected load. In terms of the line-to-line impedance $\mathbf{Z}_{LL}$ (phase impedance of a Δ-connected load),

$$\begin{aligned}\mathbf{S} &= 3\mathbf{V}_{AB}\mathbf{I}_{AB}^{*} \\ &= 3\mathbf{V}_{AB}\left(\frac{\mathbf{V}_{AB}}{\mathbf{Z}_{LL}}\right)^{*}\end{aligned} \tag{18.22}$$

which gives

The Complex Power Absorbed by a Δ-Connected Load

$$\mathbf{S} = 3\frac{1}{\mathbf{Z}_{LL}^{*}} V_L^2 \text{ VA} \tag{18.23}$$

We can obtain a relationship between the impedances of equivalent balanced Y and Δ loads by equating the power absorbed by a Δ-connected load and a Y-connected load:

$$\frac{1}{\mathbf{Z}_{LN}^{*}} V_L^2 = 3\frac{1}{\mathbf{Z}_{LL}^{*}} V_L^2 \tag{18.24}$$

If we take the complex conjugate of both sides of Eq. (18.24) we obtain

Equivalent Balanced Δ and Y Loads

$$\mathbf{Z}_{LN} = \frac{1}{3}\mathbf{Z}_{LL} \tag{18.25}$$

for equal complex power absorption from a power system.

EXAMPLE 18.3 Three impedances of $5\underline{/36.87^\circ}\ \Omega$ are connected in Δ to a 440-V three-phase line. Find the complex power absorbed by the three-phase load and the magnitude of the rms line current.

Solution

$$\begin{aligned}\mathbf{S} &= 3\frac{1}{\mathbf{Z}_{LL}^{*}} V_L^2 = 3\frac{440^2}{5\underline{/-36.87^\circ}} = 116{,}160\underline{/36.87^\circ} \text{ VA} \\ &= 92{,}928 + j69{,}969 \text{ VA}\end{aligned}$$

$$I_L = \frac{|\mathbf{S}|}{\sqrt{3}\, V_L} = 152.42 \text{ A}$$

Remember

For a Δ-connected load or source in a balanced system, the line voltage (line-to-line voltage) is the same as the corresponding phase voltage. The magnitude of the line current is $\sqrt{3}$ times the magnitude of the corresponding phase current (line-to-line current). The line current lags the corresponding phase current by 30° for a positive (*abc*) phase sequence.

EXERCISES

A balanced positive (*abc*) phase sequence system has a balanced Δ-connected load. The phase impedances of the Δ-connected load each have a value of $\mathbf{Z}_{LL} = 20\underline{/15^\circ}\ \Omega$. The line voltage is $\mathbf{V}_{AB} = 240\underline{/0^\circ}$ V.

10. Calculate the phase currents (line-to-line currents) $\mathbf{I}_{AB}$, $\mathbf{I}_{BC}$, and $\mathbf{I}_{CA}$ for the Δ-connected load. *answer:* $12\underline{/-15^\circ}$ A, $12\underline{/-135^\circ}$ A, $12\underline{/-255^\circ}$ A
11. Calculate the line currents $\mathbf{I}_{aA}$, $\mathbf{I}_{bB}$, and $\mathbf{I}_{cC}$. *answer:* $20.78\underline{/-45^\circ}$ A, $20.78\underline{/-165^\circ}$ A, $20.78\underline{/-285^\circ}$ A
12. Determine the complex power absorbed by the load. *answer:* $8640\underline{/15^\circ}$ VA

18.4 Three-Phase Systems

A power system may have both Δ- and Y-connected sources and usually includes both Δ- and Y-connected loads. Care is required when discussing phase voltages and currents, as the relation between line voltages and phase voltages and line currents and phase currents is different for a Δ and a Y.

EXAMPLE 18.4

For the balanced 2400-V three-phase system shown in Fig. 18.10, let $\mathbf{V}_{ab}$ be the reference; that is, let $\mathbf{V}_{ab} = 2400\underline{/0^\circ}$ V. Assume a positive (*abc*) phase sequence.
(a) Calculate currents $\mathbf{I}_{AA_1}$, $\mathbf{I}_{AA_2}$, $\mathbf{I}_{A_2B_2}$, and $\mathbf{I}_{aA}$.
(b) Determine the complex power absorbed by each load and the combined load.

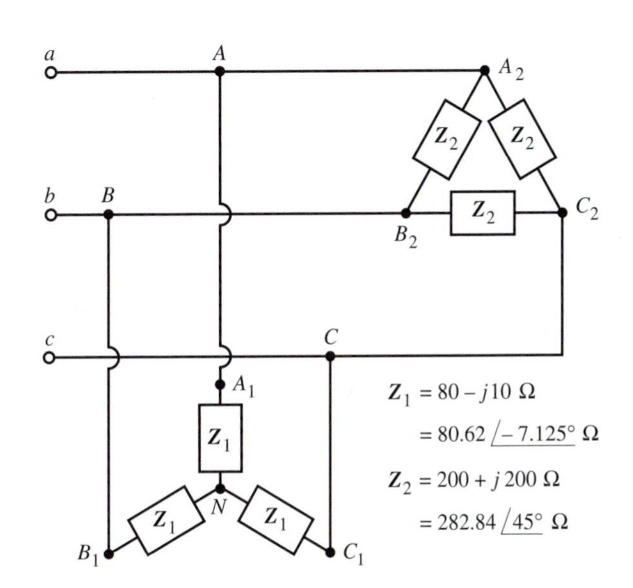

FIGURE 18.10 A system with combined Δ and Y loads

Solution (a) A straightforward way to calculate the requested line currents is

See Problem 18.1 in the PSpice manual.

$$\mathbf{I}_{AA_1} = \frac{1}{\mathbf{Z}_1}\frac{\mathbf{V}_{AB}}{\sqrt{3}}(1\underline{/-30^\circ}) = \frac{1}{80 - j10}\frac{2400\underline{/0^\circ}}{\sqrt{3}}(1\underline{/-30^\circ})$$
$$= 17.19\underline{/-22.88^\circ} = 15.84 - j6.68 \text{ A}$$

Similarly,

$$\mathbf{I}_{AA_2} = \frac{1}{\mathbf{Z}_2}\sqrt{3}\,\mathbf{V}_{AB}(1\underline{/-30^\circ}) = 14.70\underline{/-75^\circ} = 3.80 - j14.20 \text{ A}$$

$$\mathbf{I}_{A_2B_2} = \frac{1}{\mathbf{Z}_2}\mathbf{V}_{AB} = 8.49\underline{/-45^\circ} = 6 - j6 \text{ A}$$

and

$$\mathbf{I}_{aA} = \mathbf{I}_{AA_1} + \mathbf{I}_{AA_2} = 19.64 - j20.88 = 28.66\underline{/-46.75^\circ} \text{ A}$$

(b) The individual complex powers are most easily found in the following manner:

$$\mathbf{S}_1 = \frac{1}{\mathbf{Z}_1^*}V_L^2 = \frac{1}{80 + j10}2400^2$$
$$= 70{,}892 - j8{,}862 = 71{,}444\underline{/-7.125^\circ} \text{ VA}$$
$$\mathbf{S}_2 = 3\frac{1}{\mathbf{Z}_2^*}V_L^2 = 3\frac{1}{200 - j200}2400^2$$
$$= 43{,}200 + j43{,}200 = 61{,}094\underline{/45^\circ} \text{ VA}$$

and the total complex power is

$$\mathbf{S} = \mathbf{S}_1 + \mathbf{S}_2 = 114{,}092 + j34{,}338 = 119{,}148\underline{/16.76^\circ} \text{ VA}$$

The complex powers could have been found first and the currents calculated from them. The other line currents and line-to-line currents can easily be found from the 120° symmetry property.

When analyzing power systems, we usually have complex power specified and we are interested in the rms values of currents not the phasor values.

EXAMPLE 18.5 A balanced three-phase load draws 80 kW at a lagging power factor of 0.8 and a line voltage of 440 V. Find the complex power and the line current.

Solution We can write the complex power as

$$\mathbf{S} = P + jQ = |\mathbf{S}|\cos\theta + j|\mathbf{S}|\sin\theta$$

with the power-factor angle given by

$$\theta = \text{arc}\cos 0.8 = 36.87^\circ$$

Therefore, the magnitude of the complex power can be calculated from the real power by

$$|\mathbf{S}| = \frac{P}{\cos\theta} = \frac{80}{\cos 36.87^\circ} = 100 \text{ kVA}$$

and

$$\mathbf{S} = 100\underline{/36.87^\circ} \text{ kVA}$$

The line current is given by Eq. (18.12) or (18.21):

$$I_L = \frac{|\mathbf{S}|}{\sqrt{3}\,\mathbf{V}_L} = \frac{100 \times 10^3}{440\sqrt{3}} = 131.21 \text{ A}$$

We should note that if the complex power is known, we do not need to know whether the load is Δ-connected or Y-connected in order to calculate the line current.

We also rely on complex power to calculate line current for combined loads.

EXAMPLE 18.6 Two balanced three-phase loads, one drawing 80 kW at a power factor of 0.8 lagging and the second drawing 75 kVA at a power factor of 0.6 leading are connected to a 440-V three-phase line. What line current is required to supply the combined load?

Solution From the previous example the complex power for one load is

$$\mathbf{S}_1 = 100\underline{/36.87^\circ} \text{ kVA}$$

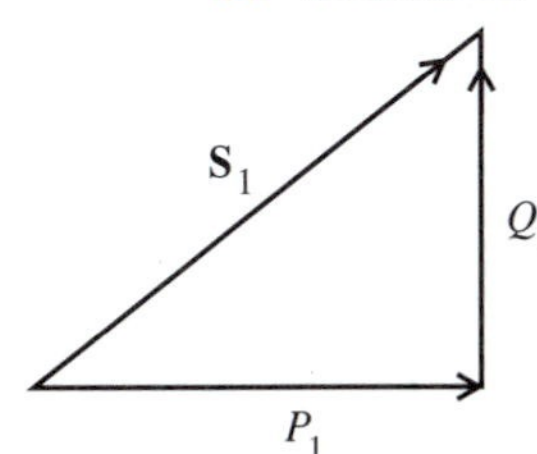

The power-factor angle for the second load is (the angle is negative because the power factor is leading)

$$\theta_2 = \text{arc cos}\,(0.6) = -53.13^\circ$$

The magnitude of the complex power is specified as 75 kVA, so

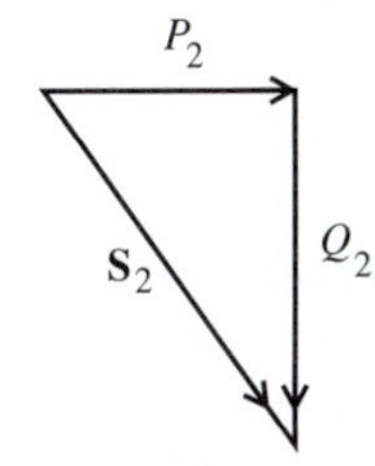

$$\mathbf{S}_2 = 75\underline{/-53.13^\circ} \text{ kVA}$$

Adding the two complex powers gives the total complex power absorbed by the loads:

$$\begin{aligned}\mathbf{S} = \mathbf{S}_1 + \mathbf{S}_2 &= 100\underline{/36.87^\circ} + 75\underline{/-53.13^\circ} \\ &= 80 + j60 + 45 - j60 = 125\underline{/0^\circ} \text{ kVA}\end{aligned}$$

This gives a line current for the combined loads of

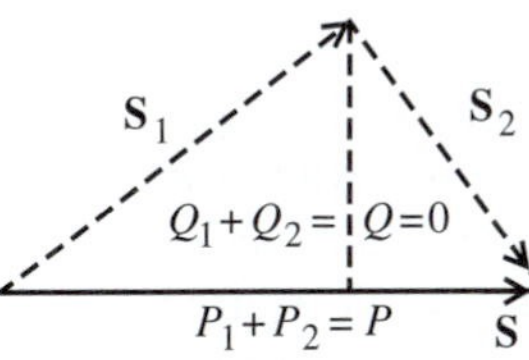

$$I_L = \frac{|\mathbf{S}|}{\sqrt{3}\,\mathbf{V}_L} = \frac{125{,}000}{440\sqrt{3}} = 164 \text{ A}$$

The AC analysis option of a circuit analysis program such as PSpice is very useful to analyze three-phase systems. The systems can be analyzed directly, or for a balanced system, all sources and loads can be converted to their equivalent Y. Then only one phase needs to be analyzed. The remaining voltages and currents can be obtained from the symmetry properties and by converting any equivalent Y connections back to the original Δ connection.

Remember

We can use the line voltage, line current, and power factor to calculate complex power. We do not need to know whether the load is connected in Δ or Y. Line current is easily calculated from the line voltage and the complex power.

EXERCISES

13. A line voltage in a balanced system is $\mathbf{V}_{AB} = 200\underline{/0^\circ}$ V and a line current is $\mathbf{I}_{aA} = 10\underline{/-90^\circ}$ A. Determine
    (a) The complex power absorbed by the load.
    (b) The real power absorbed by the load.

    answer: $3.464\underline{/60^\circ}$ kVA, 1.732 kW (positive-phase sequence assumed)

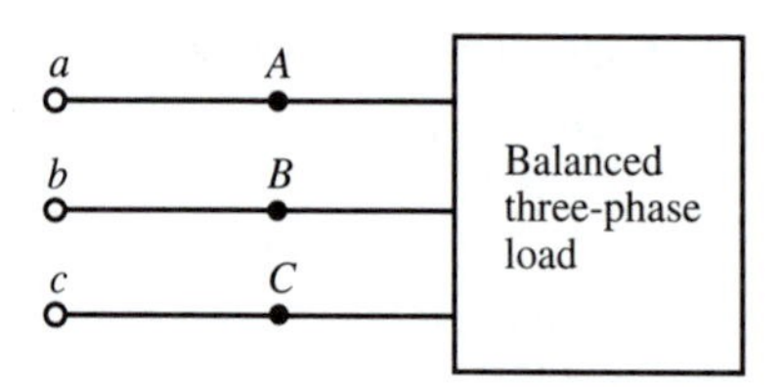

14. A balanced three-phase three-wire system contains a Δ-connected load with an impedance of $3 - j4\ \Omega$ per phase (line-to-line). A balanced Y-connected load with an impedance of $3 + j4\ \Omega$ per phase (line-to-neutral) is also connected in parallel with the Δ load. The line voltage is 480 V. Find:
    (a) The complex power absorbed by the Δ-connected load
    (b) The complex power absorbed by the Y-connected load
    (c) The complex power absorbed by the combined loads
    (d) The line current for the Δ load
    (e) The line current for the Y load
    (f) The line current for the combined loads

    answer: $138.24\underline{/-53.13^\circ}$ kVA, $46.08\underline{/53.13^\circ}$ kVA, $132.92\underline{/-33.69^\circ}$ kVA, 166.28 A, 55.43 A, 159.88 A

Power-Factor Correction

As we discussed in Example 12.5 industrial users of power are not only charged for the energy (kWh) that they use, but they are assessed a penalty for an excessively low power factor. Although reactive power Q does not require energy to generate, it does cost money to deliver to the customer. The electric utility must size its system to carry the increased current required by low power factors. For this reason, many industries do power-factor correction by connecting capacitors in parallel with their system load. (Few industries have a leading power factor.)

The real power P_2 consumed by a capacitor is usually negligible compared to its reactive power Q_2, but power-factor correction is often performed with synchronous motors that also perform useful work and therefore absorb significant real power P_2. The percent reduction in line current is small as the power factor is reduced from 0.85 to 1. For this reason the cost of correction to unity power factor is seldom economical. (An exception is the Pea Ridge Mine near Sullivan, Missouri. It is primarily powered by synchronous motors and normally corrects to unity power factor at no additional cost.)

We must correct our system to an overall power factor PF. The complex power to be corrected is

$$\mathbf{S}_1 = P_1 + jQ_1 \tag{18.26}$$

and the complex power absorbed by our power-factor-correction device is

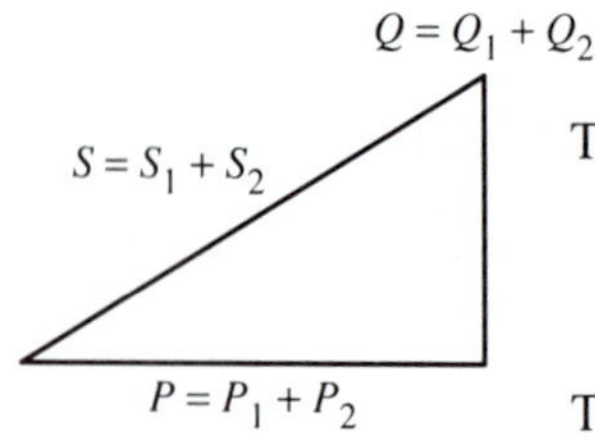

$$\mathbf{S}_2 = P_2 + jQ_2 \tag{18.27}$$

The total complex power will be

$$\begin{aligned}\mathbf{S} &= \mathbf{S}_1 + \mathbf{S}_2 \\ &= P_1 + P_2 + j(Q_1 + Q_2)\end{aligned} \tag{18.28}$$

The required power-factor angle for the combined load is

$$\theta = \text{arc cos PF} \tag{18.29}$$

where $\theta > 0$ for a lagging power factor. With a little trigonometry we can solve for the reactive power Q_2' needed to correct to a power-factor angle θ:

$$Q_2 = -Q_1 + (P_1 + P_2) \tan \theta \tag{18.30}$$

EXAMPLE 18.7 Load $\mathbf{S}_2$ in Example 18.6 is a synchronous motor used for power-factor correction. The real power P_2 will change by a negligible amount as the reactive power Q_2 changes. Determine the value of Q_2 necessary to correct to a lagging power factor of 0.95. What is the line current for PF = 0.95?

Solution Equation (18.29) gives the power-factor angle $\theta = \text{arc} \cos 0.95 = 18.19°$, Eq. (18.30) gives us

$$\begin{aligned} Q_2 &= -60 + (80 + 45) \tan 18.19° \\ &= -18.91 \text{ kVAR} \end{aligned}$$

The complex power for a power factor of 0.95 is

$$\begin{aligned} \mathbf{S} &= 80 + j60 + 45 - j18.91 \\ &= 125 + j41.09 \text{ kVA} \\ &= 131.6\underline{/18.19°} \text{ kVA} \end{aligned}$$

and the line current is

$$\begin{aligned} I_L &= \frac{131{,}600}{440\sqrt{3}} \\ &= 173 \text{ A} \end{aligned}$$

Remember Power-factor correction is used to reduce line current and avoid penalties assessed by electric utilities for low power factor.

EXERCISE

15. A three-phase 480-V (line-to-line) load has an input of 25 kVA with a power factor of 0.8 lagging. Find:
    (a) The complex power absorbed by the load
    (b) The line current
    (c) The kVAR rating of a lossless three-phase capacitor that must be connected in parallel with the motor to obtain a combined power factor of 0.9 lagging.
    (d) The kVAR rating of the capacitor needed to correct to a power factor of 0.9 leading.
    (e) Draw the complex-power triangle that shows the complex power of the load, the capacitor, and the combination.

    answer: $25\underline{/36.87°}$ kVA, 30.07 A, −5.3136 kVAR, −24.69 kVAR

18.5 Instantaneous Power

Any balanced polyphase system will have an instantaneous power that is constant. The following is a proof for a balanced positive-sequence three-phase load. The instantaneous power absorbed by such a load, written in terms of the rms line-to-neutral voltage V_{LN} and rms line current I_L, is

$$\begin{aligned}
p &= v_{AN}i_{AN} + v_{BN}i_{BN} + v_{CN}i_{CN} \\
&= 2V_{LN}\cos(\omega t + \phi)I_L\cos(\omega t + \phi - \theta) \\
&\quad + 2V_{LN}\cos(\omega t + \phi - 120°)I_L\cos(\omega t + \phi - 120° - \theta) \\
&\quad + 2V_{LN}\cos(\omega t + \phi - 240°)I_L\cos(\omega t + \phi - 240° - \theta) \\
&= V_{LN}I_L[3\cos\theta + \cos(2\omega t + 2\phi - \theta) + \cos(2\omega t + 2\phi - \theta - 240°) \\
&\quad + \cos(2\omega t + 2\phi - \theta - 480°)] \\
&= V_{LN}I_L[3\cos\theta + \cos(2\omega t + 2\phi - \theta) + \cos(2\omega t + 2\phi - \theta - 240°) \\
&\quad + \cos(2\omega t + 2\phi - \theta - 120°)] \qquad (18.31)
\end{aligned}$$

The three time-varying terms in Eq. (18.31) can be represented by three phasors that differ in phase by 120°. These terms thus sum to zero, leaving

$$\begin{aligned}
p &= 3V_{LN}I_L\cos\theta = 3\frac{V_L}{\sqrt{3}}I_L\cos\theta = \sqrt{3}\,V_LI_L\cos\theta \\
&= P \qquad (18.32)
\end{aligned}$$

The instantaneous power p is therefore equal to the average power P and is thus a constant. As the instantaneous power into a balanced three-phase load is constant, three-phase electric motors tend to produce less noise and vibration than single-phase motors of the same power.

Remember

The instantaneous power absorbed by a balanced three-phase load is equal to the average power.

EXERCISE

16. A two-phase power system has three lines a, b, and n, with $v_{an} = V_m\cos\omega t$ and $v_{bn} = V_m\cos(\omega t - 90°)$. Calculate the instantaneous power supplied to a balanced two-phase load with an impedance of $\mathbf{Z}_{an} = \mathbf{Z}$ from line a to line n, and an impedance of $\mathbf{Z}_{bn} = \mathbf{Z}$ from line b to line n. Sketch the power absorbed by phase a, the power absorbed by phase b, and the total power as a function of t. For the sketch, assume $\mathbf{Z} = |\mathbf{Z}|\underline{/45°}$. Is the instantaneous power equal to the average power?

answer: $p = (V_m^2/|\mathbf{Z}|)\cos\theta_{\mathbf{Z}} = P$

18.6 Power Measurements

Average power is usually measured by an electromechanical *wattmeter* represented schematically in Fig. 18.11. A wattmeter has two coils: a current coil, represented by the horizontal coil, which produces a magnetic field proportional to the current, and a potential or voltage coil, represented by the vertical coil, which draws a small amount of current proportional to, and in phase with, the voltage. The interaction of the magnetic field established by the current coil with the magnetic field produced by the voltage coil causes a force, and thus a torque, proportional to the product vi. This torque tends to rotate the voltage coil, and thus the indicating needle, against the force of a spring. The mass of the moving coil prevents it from following the 120-Hz torque pulses produced by the 60-Hz voltage and current, with the result that the needle deflection is determined

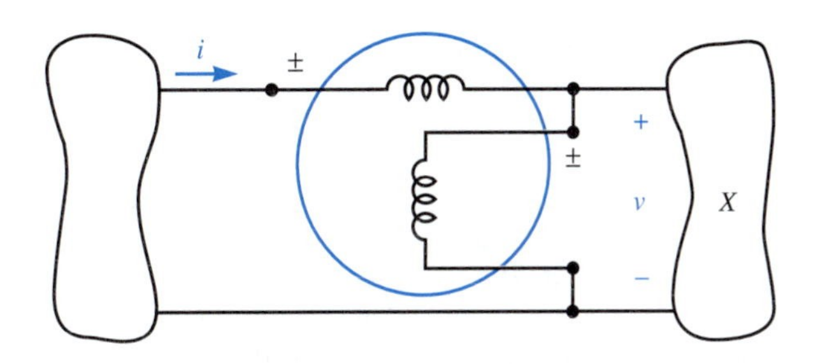

FIGURE 18.11 Measuring power absorbed with a wattmeter

by the average of the product vi. Proper scale calibration gives a meter that reads average power. A more detailed description can be found in other sources.

The (±) mark on the current coil in the wattmeter symbol shown in Fig. 18.11 indicates that the average power is calculated using the current into the (±) marked terminal. The (±) mark on the voltage coil indicates that the average power is calculated using the voltage with the (+) reference mark on this terminal and the (−) reference mark on the other voltage coil terminal. For an ideal wattmeter, the current coil is assumed to have zero impedance and the voltage coil is assumed to have infinite impedance. (For a very accurate analysis the actual impedance of the voltage and current coils must be used.) When dealing with ac circuits, we can write the power measured in the circuit of Fig. 18.11 as

$$P = \mathcal{R}e\{\mathbf{VI}^*\} \tag{18.33}$$

EXAMPLE 18.8 Assume an ideal wattmeter in Fig. 18.11. The voltage and current are

$$v = 120\sqrt{2}\cos(2\pi 60t + 15°) \text{ V}$$
$$i = 10\sqrt{2}\cos(2\pi 60t - 45°) \text{ A}$$

What is the wattmeter reading?

Solution

$$\mathbf{V} = 120\underline{/15°} \text{ V}$$
$$\mathbf{I} = 10\underline{/-45°} \text{ A}$$
$$\mathbf{S} = \mathbf{VI}^* = 1200\underline{/60°} \text{ VA}$$
$$= 600 + j600\sqrt{3} \text{ VA}$$
$$= P + jQ$$

The wattmeter reading is

$$P = 600 \text{ W}$$

Residential customers are charged for the energy they consume. This total energy is recorded on an instrument called a watthour meter.

$$1 \text{ Wh} = (1 \text{ W})(1 \text{ h}) = \left(1\,\frac{\text{J}}{\text{s}}\right)(3600 \text{ s})$$
$$= 3600 \text{ J} \tag{18.34}$$

The principle of operation of a watthour meter can be found in many texts on ac circuits, power systems, and electrical machinery.

Average power measurements are very straightforward for a three-phase load. If we have access to the neutral, we can connect a wattmeter current coil in each line and connect the corresponding voltage coil from the line to neutral. The power absorbed by the load is simply the sum of the three wattmeter readings:

$$\begin{aligned} P &= P_A + P_B + P_C \\ &= \mathcal{Re}\{\mathbf{V}_{AN}\mathbf{I}^*_{aA}\} + \mathcal{Re}\{\mathbf{V}_{BN}\mathbf{I}^*_{bB}\} + \mathcal{Re}\{\mathbf{V}_{CN}\mathbf{I}^*_{cC}\} \end{aligned} \tag{18.35}$$

In some instances the neutral is not accessible, but we can measure the power absorbed from a balanced three-phase source by *any* three-phase load with the use of only two wattmeters. For illustration, assume a Δ-connected load as in Fig. 18.8. Connect one wattmeter to measure current $\mathbf{I}_{aA}$ and voltage $\mathbf{V}_{AB}$. This gives

$$P_A = \mathcal{Re}\{\mathbf{V}_{AB}\mathbf{I}^*_{aA}\} = \mathcal{Re}\{\mathbf{V}_{AB}(\mathbf{I}_{AB} + \mathbf{I}_{AC})^*\} \tag{18.36}$$

Connect the second meter to read

$$P_C = \mathcal{Re}\{\mathbf{V}_{CN}\mathbf{I}^*_{cC}\} = \mathcal{Re}\{\mathbf{V}_{CB}(\mathbf{I}_{CB} + \mathbf{I}_{CA})^*\} \tag{18.37}$$

The two wattmeter readings add to give

$$\begin{aligned} P_A + P_C &= \mathcal{Re}\{\mathbf{V}_{AB}\mathbf{I}^*_{AB} + \mathbf{V}_{CB}\mathbf{I}^*_{CB} + (\mathbf{V}_{AB} - \mathbf{V}_{CB})\mathbf{I}^*_{AC}\} \\ &= \mathcal{Re}\{\mathbf{V}_{AB}\mathbf{I}^*_{AB} + \mathbf{V}_{CB}\mathbf{I}^*_{CB} + \mathbf{V}_{AC}\mathbf{I}^*_{AC}\} \\ &= P \end{aligned} \tag{18.38}$$

where the three terms in the last sum are the complex powers absorbed by the three branches of the delta. Because we can always perform a Δ-Y transformation, Eq. (18.38) gives the average power absorbed by *any* (balanced or unbalanced) three-phase load.

For a *balanced three-phase load* connected to a *balanced positive (abc) phase sequence three-phase source,* we can write the power read by the wattmeters as

$$\begin{aligned} P_A &= \mathcal{Re}\{\mathbf{V}_{AB}\mathbf{I}^*_{aA}\} \\ &= \mathcal{Re}\{\mathbf{V}_{AN}(\sqrt{3}\underline{/30^\circ})\mathbf{I}^*_{AN}\} \\ &= \mathcal{Re}\{\mathbf{V}_{AN}\mathbf{I}^*_{AN}(\sqrt{3}\underline{/30^\circ})\} \end{aligned} \tag{18.39}$$

$$\begin{aligned} P_C &= \mathcal{Re}\{\mathbf{V}_{CB}\mathbf{I}^*_{cC}\} \\ &= \mathcal{Re}\{-\mathbf{V}_{AN}(\sqrt{3}\underline{/30^\circ})(1\underline{/-120^\circ})[\mathbf{I}_{AN}(1\underline{/-240^\circ})]^*\} \\ &= \mathcal{Re}\{\mathbf{V}_{AN}\mathbf{I}^*_{AN}(\sqrt{3}\underline{/-30^\circ})\} \end{aligned} \tag{18.40}$$

We can use these wattmeter readings to calculate the power-factor angle:

$$\begin{aligned} \frac{P_A}{P_C} &= \frac{\mathcal{Re}\{\mathbf{V}_{AN}\mathbf{I}^*_{AN}\sqrt{3}\underline{/30^\circ}\}}{\mathcal{Re}\{\mathbf{V}_{AN}\mathbf{I}^*_{AN}\sqrt{3}\underline{/-30^\circ}\}} = \frac{V_{LN}I_L \cos(\theta + 30^\circ)}{V_{LN}I_L \cos(\theta - 30^\circ)} \\ &= \frac{\cos(\theta + 30^\circ)}{\cos(\theta - 30^\circ)} = \frac{\cos 30^\circ \cos\theta - \sin 30^\circ \sin\theta}{\cos 30^\circ \cos\theta + \sin 30^\circ \sin\theta} \\ &= \frac{\sqrt{3}\cos\theta - \sin\theta}{\sqrt{3}\cos\theta + \sin\theta} \end{aligned} \tag{18.41}$$

This is solved for

$$\begin{aligned} \tan\theta &= \frac{\sin\theta}{\cos\theta} \\ &= \sqrt{3}\,\frac{P_C - P_A}{P_C + P_A} \end{aligned} \tag{18.42}$$

The tangent of θ is positive for a lagging power factor and negative for a leading power factor. This gives

$$\theta = \arctan\left(\sqrt{3}\,\frac{P_C - P_A}{P_C + P_A}\right) \tag{18.43}$$

Thus, for a balanced positive-phase sequence three-wire three-phase system, the two wattmeter readings provide not only the power absorbed by the load, but the power factor as well. A similar relation for θ can be found for a negative-phase-sequence system.

A low-power-factor load can result in a negative wattmeter reading. Most practical wattmeters do not indicate negative quantities. As a result, the current coil must be reversed for negative power readings. For low-power-factor loads, P is calculated as the difference of two numbers of nearly the same magnitude. This can lead to large percentage errors, so three wattmeters are often used to measure low-power-factor loads.

EXAMPLE 18.9 Two wattmeters are connected as in Eqs. (18.36) and (18.37) to measure the power absorbed by the combined load of Example 18.4. Calculate

(a) The reading of each wattmeter.
(b) The power absorbed by the load.
(c) The power factor of the load.

Solution (a) The currents and voltages required are available from Example 18.4. The wattmeter in line aA indicates

$$\begin{aligned} P_A &= \mathcal{R}e\{\mathbf{V}_{AB}\mathbf{I}_{aA}^*\} \\ &= \mathcal{R}e\{(2400\underline{/0^\circ})(28.66\underline{/-46.75^\circ})^*\} \\ &= 47{,}130 \text{ W} \end{aligned}$$

The wattmeter in line cC indicates

$$\begin{aligned} P_C &= \mathcal{R}e\{\mathbf{V}_{CB}\mathbf{I}_{cC}^*\} = \mathcal{R}e\{(\mathbf{V}_{BC}\underline{/-180^\circ})\mathbf{I}_{cC}^*\} \\ &= \mathcal{R}e\{(2400\underline{/60^\circ})(28.66\underline{/-46.75^\circ - 240^\circ})^*\} \\ &= 66{,}960 \text{ W} \end{aligned}$$

(b) The power absorbed by the combined load is

$$\begin{aligned} P &= P_A + P_C = 47{,}130 + 66{,}960 \\ &= 114{,}090 \text{ W} \end{aligned}$$

(c) From Eq. (18.43)

$$\begin{aligned} \theta &= \arctan\left(\sqrt{3}\,\frac{(P_C - P_A)}{(P_A + P_C)}\right) \\ &= 16.75^\circ \end{aligned}$$

$$\text{PF} = \cos\theta = 0.96 \text{ lag}$$

The combined load presents a lagging power factor because $\theta > 0$.

Remember

Two wattmeters can be used to measure the average power for a three-phase load. If the voltages and load are balanced, the wattmeter readings can also be used to calculate the power factor.

EXERCISES

17. For the wattmeter connections given in Eqs. (18.36) and (18.37), the wattmeter readings are $P_A = 3.2$ kW and $P_C = 1.25$ kW.
    (a) Find the average power absorbed by the load.
    (b) If the power source is a positive-phase-sequence balanced three-phase source, and the three-phase load is balanced, find the load power factor.

    answer: 4.45 kW, 0.797 leading

18. Two wattmeters are connected to a Y-connected load with phase impedance of $10\underline{/36.87^\circ}\ \Omega$ and a 480-V line voltage. The coils are connected as in Eqs. (18.36) and (18.37). Calculate P_A, the power indicated by the wattmeter in line A, and P_C, the power indicated by the wattmeter in line C. Use P_A and P_C to calculate the total power and the power factor and indicate whether the power factor is leading or lagging.

    answer: 5.22 kW, 13.20 kW, 18.43 kW, 0.8 lagging

18.7 Summary

We began with a discussion of the single-phase three-wire system typically used for residential service in the United States. We then defined a balanced three-phase system and used the symmetry properties to simplify current and power calculations. (An advanced text on electric power systems should be consulted for the analysis of unbalanced three-phase systems.)

KEY FACTS

Concept	Equation	Section	Page
❑ Residences and small businesses are supplied with 120/240-V three-wire single-phase service.		18.1	663
❑ Most power is generated, transmitted, and supplied to industry by balanced three-phase systems.		18.2	665
❑ Voltages and currents are specified in rms (effective) values for power systems.	(18.1)	18.2	667
❑ A three-phase system is specified in terms of the rms value of its line voltage (line-to-line voltage) V_L.		18.2	667
❑ The phase voltage for a Y connection is the line-to-neutral voltage V_{LN}.	(18.4)	18.3	670
❑ For a positive (*abc*) phase sequence, the line-to-neutral voltage lags the corresponding line voltage by 30° and has an rms value V_{LN} equal to the rms line voltage V_L divided by the square root of 3.	(18.4)	18.3	670
❑ The phase current for a Y connection is the line current I_L.	(18.5)	18.3	670
❑ The phase voltage for a Δ connection is the line voltage V_L.		18.3	672

KEY FACTS	*Concept*	*Equation*	*Section*	*Page*
❑	The phase current for a Δ connection is the line-to-line current I_{LL}.	(18.15)	18.3	673
❑	For a positive (*abc*) phase sequence, the line-to-line current leads the corresponding line current by 30° and has an rms value I_{LL} equal to the rms line current I_L divided by the square root of 3.	(18.19)	18.3	673
❑	The complex power absorbed by *any* balanced three-phase load is $\mathbf{S} = \sqrt{3}\, V_L I_L \underline{/\theta}$ VA, where θ is the angle on the phase impedance.	(18.12) (18.21)	18.3	671 674
❑	The complex power absorbed by a Y-connected load with phase impedance $\mathbf{Z}_{LN}$ is $\mathbf{S} = V_L^2/\mathbf{Z}_{LN}^*$ VA.	(18.14)	18.3	671
❑	The complex power absorbed by a Δ-connected load with phase impedance $\mathbf{Z}_{LL}$ is $\mathbf{S} = 3V_L^2/\mathbf{Z}_{LL}^*$ VA.	(18.23)	18.3	674
❑	A balanced-Δ load absorbs the same complex power as a balanced-Y load if the phase impedances are related by $\mathbf{Z}_{LL} = 3\mathbf{Z}_{LN}$.	(18.25)	18.3	674
❑	We correct a lagging power factor with a capacitive load.	(18.30)	18.4	679
❑	The instantaneous power absorbed by a balanced three-phase load is equal to the average power.	(18.32)	18.5	680
❑	Two wattmeters can measure the power absorbed from any three-phase source by any load.	(18.38)	18.6	682
❑	Two wattmeters can measure the power-factor angle if both the source and load are balanced.	(18.43)	18.6	682

PROBLEMS

Section 18.1

1. Your desk lamp is rated at 100 W and 120 V. Its power is supplied by the 60-Hz power system. Write the lamp voltage $v(t)$. Assume that the lamp is a linear resistance, and write the lamp current $i(t)$.

2. Refer to Fig. 18.1. The line impedances are $\mathbf{Z}_s = j0.2\ \Omega$, and the neutral impedance is $\mathbf{Z}_n = j0.4\ \Omega$. The load represented by $\mathbf{Z}_1$ absorbs a complex power of $\mathbf{S}_1 = 240 + j0$ VA, and the load represented by $\mathbf{Z}_2$ absorbs a complex power of $\mathbf{S}_2 = 120 + j0$ VA when the load voltages are 120 V rms.

   (a) $\mathbf{Z}_3$ is an open circuit. What is the complex power absorbed by each load when connected as shown?

   (b) If $\mathbf{Z}_3$ absorbs $\mathbf{S}_3 = 5760\underline{/36.87^\circ}$ VA at 240 V rms, calculate the complex power absorbed by each load. What percent change in the power absorbed by the loads occurs as a result of the presence of the line and neutral impedances?

Section 18.2

3. A three-phase system has a positive phase sequence and phasor line-to-neutral voltage $\mathbf{V}_{an} = 480\underline{/0^\circ}$ V. Determine:
   (a) The three phasor line-to-neutral voltages.
   (b) The three phasor line voltages (line-to-line voltages).
   (c) Draw a phasor diagram that includes all line-to-neutral voltages and all line voltages.
4. A three-phase system has a positive phase sequence and a phasor line voltage $\mathbf{V}_{ab} = 480\underline{/0^\circ}$ V. Determine:
   (a) The three phasor line voltages (line-to-line voltages).
   (b) The three phasor line-to-neutral voltages.
   (c) Draw a phasor diagram that includes all line voltages and all line-to-neutral voltages.
5. Line-to-neutral voltage is measured and found to be $v_{an} = 169.7 \cos(2\pi 60t - 30^\circ)$ V. The three-phase source is Y-connected and has a positive phase sequence. Determine:
   (a) The three phasor line-to-neutral voltages.
   (b) The three phasor line voltages (line-to-line voltages).
   (c) Draw a phasor diagram that includes all line-to-neutral and all line voltages. Label all phasors.
6. A three-phase source is connected in a delta with $v_{ab} = 169.7 \cos(2\pi 60t - 30^\circ)$ V. If the phase sequence is positive, determine:
   (a) The three phasor line voltages (line-to-line voltages).
   (b) The three phasor line-to-neutral voltages.
   (c) Draw a phasor diagram that includes all line voltages and all line-to-neutral voltages. Label all phasors.
7. Repeat Problem 3 for a negative-phase-sequence system.
8. Repeat Problem 4 for a negative-phase-sequence system.

Section 18.3

9. A balanced Y-connected load with a phase resistance of 24 Ω and an inductive reactance of 32 Ω is connected to a balanced three-phase source with a line voltage of 480 V. Find the line current and complex power absorbed by the load.
10. A balanced Y-connected three-phase resistance heater of total complex power $\mathbf{S} = P$ is supplied power by a balanced three-wire three-phase system with an rms line voltage of V_L.
    (a) Find the line current.
    (b) If one wire is accidentally disconnected, is the load supplied single-phase, two-phase, or three-phase power?
    (c) Find the line current and the heater power with the line disconnected (assume that the load is a linear resistance).
11. A balanced Δ-connected load with a phase resistance of 24 Ω and an inductive reactance of 32 Ω is connected to a balanced three-phase source with a line voltage of 480 V. Find the line current and complex power absorbed by the load.
12. A balanced Δ-connected three-phase resistance heater of total complex power $\mathbf{S} = P$ is supplied power by a balanced three-wire three-phase system with an rms line voltage of V_L.
    (a) Find the line current.
    (b) If one wire is accidentally disconnected, is the load supplied single-phase, two-phase, or three-phase power?
    (c) Find the line current and the heater power with the line disconnected (assume that the load is a linear resistance).
13. A balanced Y-connected three-phase load absorbs a complex power of 100 kVA with a leading power factor of 0.8 when the rms line voltage is 2400 V.
    (a) What are the line-to-neutral impedances of the Y load?
    (b) What is the rms value of the line current?
    (c) If $\mathbf{V}_{AB} = 2400\underline{/0^\circ}$ V and the phase sequence is positive, what is the phasor value of the line-to-neutral current?
    (d) What would be the line-to-line impedances for a Δ-connected load that would absorb the same complex power?
    (e) For the equivalent Δ load, what is the value of the line-to-line current $\mathbf{I}_{AB}$? Does this correspond to any line current?

(f) If the load is a synchronous motor operating at an overall efficiency of 85 percent, what is the output horsepower of the motor (1 hp = 745.7 W)?

14. A balanced Δ-connected three-phase load absorbs a real power of 80 kW with a lagging power factor of 0.8 when the rms line voltage is 2400 V.

(a) What are the line-to-line impedances of the Δ load?

(b) What is the magnitude of the line current?

(c) If $\mathbf{V}_{AB} = 2400\underline{/0^\circ}$ V and the phase sequence is positive, what is the phasor value of the line-to-line current $\mathbf{I}_{AB}$?

(d) What would be the equivalent line-to-neutral impedances required for a Y-connected load that would absorb the same complex power?

(e) For the equivalent Y load, what is the value of the line-to-neutral current $\mathbf{I}_{AN}$? Does this correspond to any line current?

(f) If the load is an induction motor operating at an overall efficiency of 85 percent, what is the output horsepower of the motor (1 hp = 745.7 W)?

(g) What is the VAR rating of a three-phase capacitor bank connected in parallel with the load that will correct the power factor to 0.9 lagging? To unity power factor?

(h) What is the magnitude of the line current when the power factor is corrected to 0.9? What is the ratio of the line current after power-factor correction to that before?

(i) Repeat (h) for unity-power-factor correction. Can you say why industry might not correct their load to unity power factor? (Compare the decrease in line current per volt-ampere rating of the capacitance in each case.)

(j) Make a single phasor diagram of the line currents $\mathbf{I}_{aA}$ without power-factor correction and with the power factor corrected to 0.9 and to 1.

15. A balanced three-phase load is connected to a balanced three-phase power system. The line voltage is 480 V, and the line current is 10 A. The angle on the phase impedance of the load is 30°. Find the complex power **S**, the real power P, and the reactive power Q absorbed by the load. Is the load inductive or capacitive?

16. A balanced three-phase load is connected to a transformer by three lines, each of which has a resistance of 1 Ω and an inductive reactance of 4 Ω. Measurements at the transformer end of the line indicate that the line voltage is 240 V, the line current is 10 A, and the power factor is 0.6 leading.

(a) Calculate the complex power delivered by the transformer.

(b) Calculate the complex power absorbed by the load.

(c) Calculate the line voltage at the load.

Section 18.4

17. Two balanced three-phase loads are connected to a balanced three-phase system with a line voltage of 480 V at the loads. The line impedances are $\mathbf{Z}_L = 1 + j4\ \Omega$, the phase impedances for the Y-connected load are $\mathbf{Z}_{LN} = 2 + j2\ \Omega$, and the phase impedances for the Δ-connected load are $\mathbf{Z}_{LL} = 14 - j14\ \Omega$. Determine:

(a) The complex power absorbed by the Y load

(b) The complex power absorbed by the Δ load

(c) The complex power absorbed by the combined loads

(d) The line current required for the combined load

(e) The voltage at the source end of the line

18. Repeat Problem 17 when the line-to-line voltage at the source end is known to be 480 V. [For (e), find the line-to-line voltage at the load.]

19. A motor has an output horsepower of 20, a lagging power factor of 0.8, and operates at an efficiency of 90 percent. This motor is connected to a 2400-V three-phase system. The second load is an ozone generator that consumes 6 kW at a leading power factor of 0.6. Determine the complex power, power factor, and line current for the combined load (1 hp = 745.7 W).

20. The power factor for a 100-kVA three-phase inductive load is 0.8, and the line voltage is 480 V.

(a) Determine the line current, and draw a diagram (a *power triangle*) that represents the complex power as a vector.

(b) Determine the reactive power rating of the lossless three-phase capacitive load required to correct to a lagging power factor of 0.9.

Determine the line current for the combined load, and draw the power triangle.

(c) Determine the reactive power rating of the lossless three-phase capacitive load required to correct to unity power factor. What is the line current for the combined load?

(d) With $\mathbf{V}_{AN}$ as the reference, make a single phasor diagram that shows $\mathbf{I}_{AN}$ for a power factor of 0.8, 0.9, and 1. Do you see why we might not correct to unity power factor?

21. A 2400-V three-phase load has a lagging power factor of 0.707 and absorbs a real power of 100 kW.

(a) Determine the line current.

(b) Determine the kVA rating of a lossless three-phase capacitive load required to correct the power factor to 0.8.

(c) The capacitors used for power-factor correction are connected in a Y. If the frequency is 60 Hz, what is the capacitance of each of the three capacitors? What must be the ac voltage rating of the capacitors? What is the peak voltage across a capacitor in normal operation?

Section 18.6

22. Two wattmeters are connected to a balanced three-phase load as indicated by Eqs. (18.36) and (18.37). The load is supplied by a balanced positive-phase-sequence source. If the readings are $P_A = 10$ kW and $P_C = 2$ kW, calculate the real power absorbed by the load, the power factor, and the complex power absorbed by the load.

23. Two wattmeters are connected to a balanced three-phase load as indicated by Eqs. (18.36) and (18.37). The load is supplied by a balanced positive-phase-sequence source. If the wattmeter readings are $P_A = 2$ kW and $P_C = 10$ kW, calculate the real power absorbed by the load, the power factor, and the complex power absorbed by the load.

24. Two wattmeters are connected, as indicated by Eqs. (18.36) and (18.37), to measure the combined load of Example 18.4. What are the two wattmeter readings?

25. Two wattmeters are connected to a balanced three-phase load as indicated by Eqs. (18.36) and (18.37). The load is supplied by a balanced positive-phase-sequence source, and the load absorbs a complex power of $\mathbf{S} = 3 + j4$ kVA. Determine the two wattmeter readings.

26. Two wattmeters are connected to a balanced three-phase load as indicated by Eqs. (18.36) and (18.37). The load is supplied by a balanced positive-phase-sequence source, and the load absorbs a complex power of $\mathbf{S} = 3 - j4$ kVA. Determine the two wattmeter readings.

Supplementary Problems

27. A three-phase motor absorbs 10 kW of real power at a power factor of 0.707 lagging and a line voltage of 480 V.

(a) Find the complex power and line current.

(b) A three-phase capacitor bank, which can be assumed lossless, is connected in parallel with the motor and draws 2.5 kVA. For the combined load, find the complex power absorbed, the power factor, and the resulting line current.

28. A three-phase motor absorbs 10 kVA at a lagging power factor of 0.6 and a line voltage of 208 V.

(a) What is the line current?

(b) What size capacitor bank, in kilovoltamperes, must be connected in parallel with the load to reduce the line current for the combined load to 20.82 A?

29. A 2400-V three-phase motor operated at rated voltage has a rated output of 100 hp, a full load efficiency of 90 percent, and a lagging power factor of 0.8 at full load. When started, the motor will initially require four times the rated current, and the power factor can be approximated by zero. The motor is connected in parallel with a 100-kW balanced three-phase resistance heater. What is the line current for the combined load when the motor is started?

30. Replace the line-to-line impedances of the Δ-connected load of Example 18.4 by equivalent line-to-neutral impedances in a Y connection. Then make a single-phase circuit for the line-*a*-to-neutral phase. Solve this for $\mathbf{I}_{AA_1}$ $\mathbf{I}_{AA_2}$, and $\mathbf{I}_{aA}$. Calculate the complex power absorbed by each single-phase load and the combined single-phase complex power. Next, find the three-phase complex powers by multiplying these quantities by 3 and compare your results with those obtained in Example 18.4.

31. Our plant load is divided into three categories: the first consists of 10 kW of incandescent lighting; the second is a 100-hp load of induction motors operating at an efficiency of 90 percent and a lagging power factor of 0.8; and the third, an ozone generator for wastewater treatment, consumes 50 kVA at a leading power factor of 0.6. Determine the complex power absorbed by the factory, the plant power factor, and the line current for a line voltage of 2400 V.

32. The largest load in a plant is an electric-arc furnace. This gives a total plant load of 1 MVA and lagging power factor of 0.5. We choose to install a synchronous motor to drive a new blower in an attempt to improve the plant power factor. We calculate that the motor will require 200 kW of real power. What must be the kVA rating of the motor, if we are to correct the plant to a lagging power factor of 0.9?

33. The following circuit can be used to check for the phase sequence of a balanced set of three-phase voltages. When the phase sequence is *abc,* as shown in Fig. 18.4a, lamp *A* will be brighter than lamp *B*. When the phase sequence is *acb,* as shown in Fig. 18.4b, lamp *B* will be brighter than *A*. Calculate $\mathbf{V}_{AN}$ and $\mathbf{V}_{CN}$ for both phase sequences to verify that this is true. Assume that $\mathbf{V}_{ab} = 120\underline{/0^\circ}$ V.

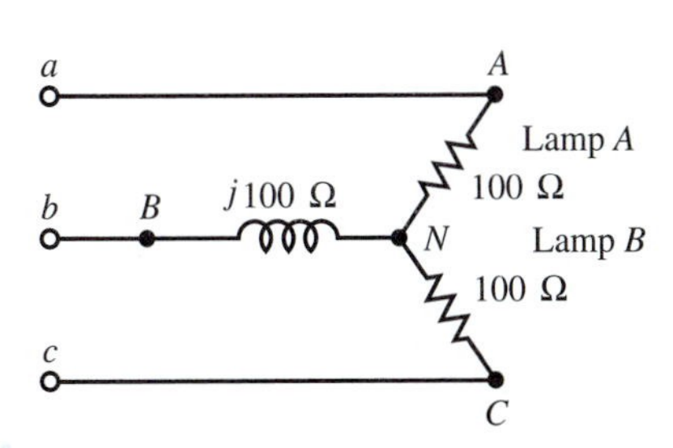

34. The following circuit can be used to check for the phase sequence of a balanced three-phase voltage. Assume that the voltmeter is ideal (draws no current) and indicates rms values and that the line-to-line voltage is 208 V. Find the voltmeter reading for an *abc* phase sequence and an *acb* phase sequence. In practice, the voltmeter is often replaced by a small neon lamp and series resistance. The brightness of the lamp is used to indicate the phase sequence.

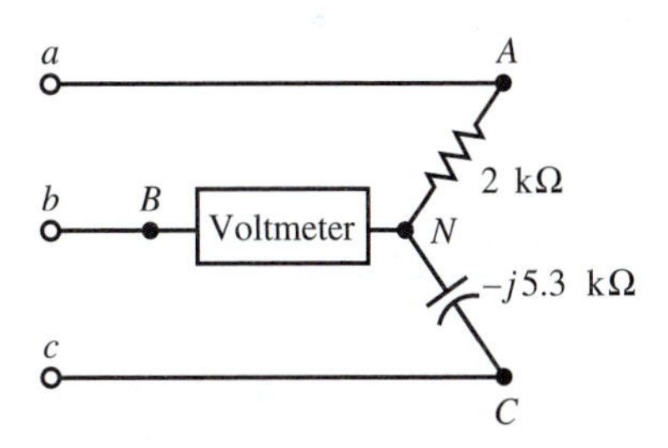

35. Two identical systems are connected together. The equivalent circuit looking into system 1 from the connection is $\mathbf{V}_1 = 1\underline{/0^\circ}$ V and $\mathbf{Z}_1 = j1\ \Omega$, and that looking into system 2 is $\mathbf{V}_2 = 1\ \underline{/0^\circ}$ V and $\mathbf{Z}_2 = j1\ \Omega$. The voltage at the interconnection is $\mathbf{V}$. This is depicted in the model shown in the following circuit. Under these conditions, $\mathbf{V} = 1\underline{/0^\circ}$ V and no power is delivered to system 2. The operator of system 1 wants to deliver real power to system 2.

(a) The operator of system 1 raises voltage $\mathbf{V}_1$ to $1.2\underline{/0^\circ}$ V. Find $\mathbf{V}$, $\mathbf{I}$, and the complex power $\mathbf{S} = \mathbf{VI}^* = P + jQ$ delivered to system 2 at the interconnection.

(b) The operator changes voltage $\mathbf{V}_1$ to $1\underline{/30^\circ}$ V. Find $\mathbf{V}$, $\mathbf{I}$, and the complex power $\mathbf{S} = P + jQ$ delivered to system 2.

(c) Do the results surprise you? Why? This is one reason *phase-shifting transformers* are used.

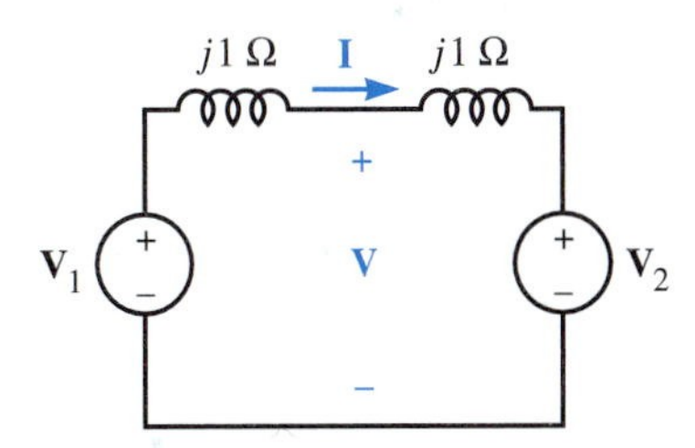

36. Determine the quantities specified for the power system shown on page 690.

(a) If $\mathbf{V}_1 = 1\underline{/0^\circ}$ V and $\mathbf{Z}_2 = 1.2 - j1.6\ \Omega$, find $\mathbf{I}$, $\mathbf{V}_2$, $\mathbf{S}_1$, and $\mathbf{S}_2$.

(b) If $\mathbf{V}_2 = 1\underline{/0^\circ}$ V and $\mathbf{Z}_2 = 1.2 - j1.6\ \Omega$, find $\mathbf{I}$, $\mathbf{V}_1$, $\mathbf{S}_1$, and $\mathbf{S}_2$.

(c) If $\mathbf{V}_1 = 1\underline{/0^\circ}$ V and $\mathbf{Z}_2 = 1.2 + j1.6\ \Omega$, find $\mathbf{I}$, $\mathbf{V}_2$, $\mathbf{S}_1$, and $\mathbf{S}_2$.

(d) If $\mathbf{V}_1 = 1\underline{/0^\circ}$ V and $\mathbf{S}_1 = 0.3 - j0.4$ VA, find $\mathbf{I}$, $\mathbf{V}_2$, $\mathbf{Z}_2$, and $\mathbf{S}_2$.

(e) If $\mathbf{V}_1 = 1\underline{/0^\circ}$ V and $\mathbf{S}_2 = 0.3 - j0.4$ VA, find $\mathbf{I}$, $\mathbf{V}_2$, $\mathbf{Z}_2$, and $\mathbf{S}_1$.

(f) If $P_2 = 0.3$ W, Q_2 is unknown, $\mathbf{V}_1 = 1\underline{/0^\circ}$ V and $|\mathbf{V}_2| = 1$ V, find $\mathbf{V}_2$, $\mathbf{I}$, $\mathbf{Q}_2$, and $\mathbf{Z}_2$.

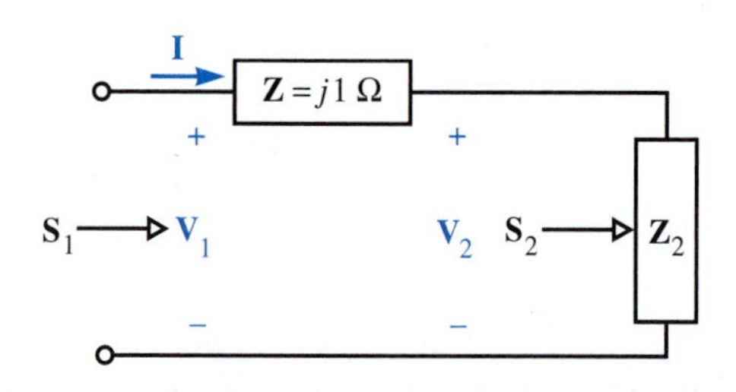

37. Three voltmeters, each with arbitrary impedance, are connected in a Y to the lines of a balanced positive-phase-sequence three-wire system. The meter connected to line a reads value A V, the one connected to line b reads B V, and the one connected to line c reads C V. Show that the line-to-line voltage can be found from the expression

$$V_L^2 = \left\{\frac{1}{2}[A^2 + B^2 + C^2] + \frac{1}{2}[6(A^2B^2 + B^2C^2 + C^2A^2) - 3(A^4 + B^4 + C^4)]^{1/2}\right\}$$

$$= \left\{\frac{1}{2}[A^2 + B^2 + C^2] + \frac{1}{2}[3(A^2 + B^2 + C^2)^2 - 6(A^4 + B^4 + C^4)]^{1/2}\right\}$$

Linear Equations and Determinants

If the number of simultaneous linear equations that we must solve is large, we usually use matrix notation as an aid in organizing and solving the equations.

A.1 Simultaneous Equations

We will now show how to write a set of simultaneous linear equations as a single matrix equation. Consider the set of equations

$$a_{11}v_1 + a_{12}v_2 + a_{13}v_3 + a_{14}v_4 = s_1 \tag{A.1}$$

$$a_{21}v_1 + a_{22}v_2 + a_{23}v_3 + a_{24}v_4 = s_2 \tag{A.2}$$

$$a_{31}v_1 + a_{32}v_2 + a_{33}v_3 + a_{34}v_4 = s_3 \tag{A.3}$$

$$a_{41}v_1 + a_{42}v_2 + a_{43}v_3 + a_{44}v_4 = s_4 \tag{A.4}$$

We define the *matrices*

$$\mathbf{v} = [v_j] = \begin{bmatrix} v_1 \\ v_2 \\ v_3 \\ v_4 \end{bmatrix} \tag{A.5}$$

and

$$\mathbf{s} = [s_i] = \begin{bmatrix} s_1 \\ s_2 \\ s_3 \\ s_4 \end{bmatrix} \tag{A.6}$$

These are four rows by one column or 4 × 1 matrices *(column vectors)*. We also define the four-row by four-column, or 4 × 4, matrix

$$\mathbf{a} = [a_{ij}] = \begin{bmatrix} a_{11} & a_{12} & a_{13} & a_{14} \\ a_{21} & a_{22} & a_{23} & a_{24} \\ a_{31} & a_{32} & a_{33} & a_{34} \\ a_{41} & a_{42} & a_{43} & a_{44} \end{bmatrix} \begin{matrix} \leftarrow \text{Row 1} \\ \\ \\ \leftarrow \text{Row 4} \end{matrix} \tag{A.7}$$

Column 1 Column 4

We write Eqs. (A.1)–(A.4) in matrix form as

$$\mathbf{av} = \mathbf{s} \tag{A.8}$$

Equation (A.1) contributes the first row of matrix **a** and **s**,

$$\begin{bmatrix} a_{11} & a_{12} & a_{13} & a_{14} \\ & & & \\ & & & \\ & & & \end{bmatrix} \begin{bmatrix} v_1 \\ v_2 \\ v_3 \\ v_4 \end{bmatrix} = \begin{bmatrix} s_1 \\ \\ \\ \\ \end{bmatrix} \tag{A.9}$$

Equation (A.2) contributes the second row,

$$\begin{bmatrix} & & & \\ a_{21} & a_{22} & a_{23} & a_{24} \\ & & & \\ & & & \end{bmatrix} \begin{bmatrix} v_1 \\ v_2 \\ v_3 \\ v_4 \end{bmatrix} = \begin{bmatrix} \\ s_2 \\ \\ \\ \end{bmatrix} \tag{A.10}$$

and so on. The pattern seems obvious. We can write equation n as

$$s_i = \sum_{j=1}^{4} a_{ij} v_j \qquad i = 1, 2, 3, 4 \tag{A.11}$$

The complete set of equations written as a single matrix equation is

$$\begin{bmatrix} a_{11} & a_{12} & a_{13} & a_{14} \\ a_{21} & a_{22} & a_{23} & a_{24} \\ a_{31} & a_{32} & a_{33} & a_{34} \\ a_{41} & a_{42} & a_{43} & a_{44} \end{bmatrix} \begin{bmatrix} v_1 \\ v_2 \\ v_3 \\ v_4 \end{bmatrix} = \begin{bmatrix} s_1 \\ s_2 \\ s_3 \\ s_4 \end{bmatrix} \tag{A.12}$$

EXAMPLE A.1 Write the following set of equations as a single matrix equation.

$$v_1 + 3v_2 + 2v_3 = 130$$
$$6v_1 + 5v_2 + 4v_3 = 280$$
$$9v_1 + 8v_2 + 7v_3 = 460$$

Solution

$$\begin{bmatrix} 1 & 3 & 2 \\ 6 & 5 & 4 \\ 9 & 8 & 7 \end{bmatrix} \begin{bmatrix} v_1 \\ v_2 \\ v_3 \end{bmatrix} = \begin{bmatrix} 130 \\ 280 \\ 460 \end{bmatrix}$$

EXAMPLE A.2 Write the following set of equations as a single matrix equation.

$$5i_1 + 6\frac{d}{dt}i_1 - 6\int i_3\,dt = -100\cos 6t$$

$$-24\frac{d}{dt}i_1 + 8\frac{d}{dt}i_2 + 4i_2 + 16\frac{d}{dt}i_3 = 0$$

$$24\frac{d}{dt}i_1 - 6\int i_1\,dt - 8\frac{d}{dt}i_2 - 16\frac{d}{dt}i_3 + 6\int i_3\,dt = 25$$

Solution

$$\begin{bmatrix} 5 + 6\dfrac{d}{dt} & 0 & -6\displaystyle\int dt \\ -24\dfrac{d}{dt} & 8\dfrac{d}{dt} + 4 & 16\dfrac{d}{dt} \\ 24\dfrac{d}{dt} - 6\displaystyle\int dt & -8\dfrac{d}{dt} & -16\dfrac{d}{dt} + 6\displaystyle\int dt \end{bmatrix} \begin{bmatrix} i_1 \\ i_2 \\ i_3 \end{bmatrix} = \begin{bmatrix} -100\cos 6t \\ 0 \\ 25 \end{bmatrix}$$

Solution of this matrix equation requires some knowledge of differential equations and will not be attempted in this appendix. When only resistive elements are present, the equations are algebraic. We consider the solution of simultaneous linear algebraic equations in the next section.

A.2 Determinants and Cramer's Rule

There are several ways to solve systems of linear algebraic equations. These methods include simple substitution, Cramer's rule, and numerical algorithms, such as gaussian elimination, that are well suited for the digital computer. Although Cramer's rule is not efficient for the solution of a large number of simultaneous linear equations, the formulation is valuable for the derivation of certain network properties and we present it here.

The *determinant* of the $n \times n$ square matrix $\mathbf{a} = [a_{ij}]$, denoted by

$$\begin{aligned} \Delta &= |\mathbf{a}| \\ &= |a_{ij}| \end{aligned} \tag{A.13}$$

is defined for any n in terms of sums of products of elements of $\mathbf{a}$.

The definition for arbitrary n, while compact, is not particularly convenient and will not be presented. For $n = 2$, the definition yields

$$\Delta = \begin{vmatrix} a_{11} & a_{12} \\ a_{21} & a_{22} \end{vmatrix} = a_{11}a_{22} - a_{12}a_{21} \tag{A.14}$$

and for $n = 3$ the definition yields

$$\Delta = \begin{vmatrix} a_{11} & a_{12} & a_{13} \\ a_{21} & a_{22} & a_{23} \\ a_{31} & a_{32} & a_{33} \end{vmatrix} = \begin{aligned} & a_{11}a_{22}a_{33} + a_{12}a_{23}a_{31} + a_{13}a_{21}a_{32} \\ & -a_{13}a_{22}a_{31} - a_{12}a_{21}a_{33} - a_{11}a_{23}a_{32} \end{aligned} \tag{A.15}$$

These relationships can be remembered if we recognize that the determinant for $n = 2$ is obtained from the product of the two terms on the diagonal line with negative slope minus the product of the two terms on the diagonal line with positive slope as shown below.

$$\begin{vmatrix} a_{11} & a_{12} \\ a_{21} & a_{22} \end{vmatrix} \qquad \Delta = -a_{12}a_{21} \quad +a_{11}a_{22} \tag{A.16}$$

Similarly, for $n = 3$ the six terms are as shown:

$$\Delta = \begin{matrix} -a_{11}a_{23}a_{32} \\ -a_{12}a_{21}a_{33} \\ -a_{13}a_{22}a_{31} \end{matrix} \quad \begin{vmatrix} a_{11} & a_{12} & a_{13} \\ a_{21} & a_{22} & a_{23} \\ a_{31} & a_{32} & a_{33} \end{vmatrix} \quad \begin{matrix} +a_{13}a_{21}a_{32} \\ +a_{12}a_{23}a_{31} \\ +a_{11}a_{22}a_{33} \end{matrix} \tag{A.17}$$

We will now show an alternative method for evaluating the determinant of a matrix of order greater than two. If the matrix $\mathbf{a}_{ij}$ is the matrix formed by removing row i and column j from matrix $\mathbf{a}$, the scalar

$$\Delta_{ij} = (-1)^{i+j}|\mathbf{a}_{ij}| \tag{A.18}$$

is called the *cofactor* of element a_{ij}.

EXAMPLE A.3 Determine the cofactors Δ_{12}, Δ_{22}, and Δ_{32} of the matrix

$$\mathbf{a} = \begin{bmatrix} 1 & 3 & 2 \\ 6 & 5 & 4 \\ 9 & 8 & 7 \end{bmatrix}$$

Solution

$$\Delta_{12} = (-1)^{1+2} \begin{vmatrix} 6 & 4 \\ 9 & 7 \end{vmatrix} = -\begin{vmatrix} 6 & 4 \\ 9 & 7 \end{vmatrix}$$

$$= -[(6)(7) - (4)(9)] = -6$$

$$\Delta_{22} = (-1)^{2+2} \begin{vmatrix} 1 & 2 \\ 9 & 7 \end{vmatrix} = \begin{vmatrix} 1 & 2 \\ 9 & 7 \end{vmatrix}$$

$$= [(1)(7) - (2)(9)] = -11$$

$$\Delta_{32} = (-1)^{3+2} \begin{vmatrix} 1 & 2 \\ 6 & 4 \end{vmatrix} = -\begin{vmatrix} 1 & 2 \\ 6 & 4 \end{vmatrix}$$

$$= -[(1)(4) - (2)(6)] = 8$$

The determinant Δ of the matrix **a** can be calculated from

Laplace's Expansion

$$\Delta = |\mathbf{a}| = \sum_{j=1}^{n} a_{ij}\Delta_{ij} \tag{A.19}$$

$$= \sum_{i=1}^{n} a_{ij}\Delta_{ij} \tag{A.20}$$

The sum in Eq. (A.19) is called the *Laplace expansion* of Δ about its ith row, and the sum in Eq. (A.20) is called the Laplace expansion of Δ about its jth column. We can expand the determinant about any row or column of matrix **a**.

EXAMPLE A.4 Evaluate

$$\Delta = \begin{vmatrix} 1 & 3 & 2 \\ 6 & 5 & 4 \\ 9 & 8 & 7 \end{vmatrix}$$

Solution Expand the determinant about the first column, as in Eq. (A.20):

$$\Delta = a_{11}\Delta_{11} + a_{21}\Delta_{21} + a_{31}\Delta_{31}$$

$$\Delta_{11} = (-1)^{1+1} \begin{vmatrix} 5 & 4 \\ 8 & 7 \end{vmatrix} = (35 - 32) = 3$$

$$\Delta_{21} = (-1)^{2+1} \begin{vmatrix} 3 & 2 \\ 8 & 7 \end{vmatrix} = -(21 - 16) = -5$$

$$\Delta_{31} = (-1)^{3+1} \begin{vmatrix} 3 & 2 \\ 5 & 4 \end{vmatrix} = (12 - 10) = 2$$

Equation (A.20) gives the determinant:

$$\begin{aligned} \Delta &= 1(3) + 6(-5) + 9(2) \\ &= -9 \end{aligned}$$

We will now see how to use determinants to solve simultaneous linear equations. We can write the set of linear equations

$$\begin{aligned} a_{11}x_1 + a_{12}x_2 + \cdots + a_{1n}x_n &= b_1 \\ a_{21}x_1 + a_{22}x_2 + \cdots + a_{2n}x_n &= b_2 \\ \vdots \qquad\qquad & \quad \vdots \\ a_{n1}x_1 + a_{n2}x_2 + \cdots + a_{nn}x_n &= b_n \end{aligned} \tag{A.21}$$

in matrix notation as

$$\begin{bmatrix} a_{11} & a_{12} & \cdots & a_{1n} \\ a_{21} & a_{22} & \cdots & a_{2n} \\ \vdots & & & \\ a_{n1} & a_{n2} & \cdots & a_{nn} \end{bmatrix} \begin{bmatrix} x_1 \\ x_2 \\ \vdots \\ x_n \end{bmatrix} = \begin{bmatrix} b_1 \\ b_2 \\ \vdots \\ b_n \end{bmatrix} \tag{A.22}$$

or more abstractly as

$$\mathbf{ax} = \mathbf{b} \tag{A.23}$$

If $\Delta^{(j)}$ is defined as the determinant of the matrix obtained by replacing the jth column of **a** by vector **b**, the value of the variable x_j is given by

Cramer's Rule

$$x_j = \frac{1}{\Delta} \Delta^{(j)} \qquad j = 1, 2, \ldots, n \tag{A.24}$$

Alternatively, if $\Delta^{(j)}$ is expanded about column j, we can write Cramer's rule as

$$x_j = \frac{1}{\Delta} \sum_{i=1}^{n} b_i \Delta_{ij} \qquad j = 1, 2, \ldots, n \tag{A.25}$$

If $\Delta = 0$, the equations are linearly dependent, and a unique solution does not exist. The determinant $\Delta^{(j)}$ can, of course, be expanded about any convenient row or column, as can Δ. An $n \times n$ square matrix with n linearly independent rows (rank n), and thus with $\Delta \neq 0$, is said to be *nonsingular*.

EXAMPLE A.5 Solve the set of equations

$$x_1 + 3x_2 + 2x_3 = 13$$
$$6x_1 + 5x_2 + 4x_3 = 28$$
$$9x_1 + 8x_2 + 7x_3 = 46$$

by the use of Cramer's rule.

Solution First write the equations in matrix notation:

$$\begin{bmatrix} 1 & 3 & 2 \\ 6 & 5 & 4 \\ 9 & 8 & 7 \end{bmatrix} \begin{bmatrix} x_1 \\ x_2 \\ x_3 \end{bmatrix} = \begin{bmatrix} 13 \\ 28 \\ 46 \end{bmatrix}$$

From Example A.4,

$$\Delta = -9$$
$$\Delta_{11}^{(1)} = \Delta_{11} = 3$$
$$\Delta_{21}^{(1)} = \Delta_{21} = -5$$
$$\Delta_{31}^{(1)} = \Delta_{31} = 2$$

Then

$$x_1 = \frac{1}{\Delta} \Delta^{(1)} = \frac{1}{-9} \begin{vmatrix} 13 & 3 & 2 \\ 28 & 5 & 4 \\ 46 & 8 & 7 \end{vmatrix}$$
$$= \frac{1}{-9} [13\Delta_{11}^{(1)} + 28\Delta_{21}^{(1)} + 46\Delta_{31}^{(1)}]$$
$$= \frac{1}{-9} [13(3) + 28(-5) + 46(2)]$$
$$= \frac{1}{-9} (-9) = 1$$

$$x_2 = \frac{1}{\Delta} \Delta^{(2)} = \frac{1}{-9} \begin{vmatrix} 1 & 13 & 2 \\ 6 & 28 & 4 \\ 9 & 46 & 7 \end{vmatrix}$$
$$= \frac{1}{-9} [13\Delta_{12}^{(2)} + 28\Delta_{22}^{(2)} + 46\Delta_{32}^{(2)}]$$

$$\Delta_{12}^{(2)} = \Delta_{12} = (-1)^{1+2} \begin{vmatrix} 6 & 4 \\ 9 & 7 \end{vmatrix} = -(42 - 36) = -6$$

$$\Delta_{22}^{(2)} = \Delta_{22} = (-1)^{2+2} \begin{vmatrix} 1 & 2 \\ 9 & 7 \end{vmatrix} = (7 - 18) = -11$$

$$\Delta_{32}^{(2)} = \Delta_{32} = (-1)^{3+2} \begin{vmatrix} 1 & 2 \\ 6 & 4 \end{vmatrix} = -(4 - 12) = 8$$

$$x_2 = \frac{1}{-9}[13(-6) + 28(-11) + 46(8)]$$

$$= \frac{1}{-9}(-18) = 2$$

$$x_3 = \frac{1}{\Delta}\Delta^{(3)} = \frac{1}{-9}\begin{vmatrix} 1 & 3 & 13 \\ 6 & 5 & 28 \\ 9 & 8 & 46 \end{vmatrix}$$

$$= \frac{1}{-9}[13\Delta_{13}^{(3)} + 28\Delta_{23}^{(3)} + 46\Delta_{33}^{(3)}]$$

$$\Delta_{13}^{(3)} = \Delta_{13} = (-1)^{1+3}\begin{vmatrix} 6 & 5 \\ 9 & 8 \end{vmatrix} = (48 - 45) = 3$$

$$\Delta_{23}^{(3)} = \Delta_{23} = (-1)^{2+3}\begin{vmatrix} 1 & 3 \\ 9 & 8 \end{vmatrix} = -(8 - 27) = 19$$

$$\Delta_{33}^{(3)} = \Delta_{33} = (-1)^{3+3}\begin{vmatrix} 1 & 3 \\ 6 & 5 \end{vmatrix} = (5 - 18) = -13$$

$$x_3 = \frac{1}{-9}[13(3) + 28(19) + 46(-13)]$$

$$= \frac{1}{-9}(-27) = 3$$

Determinants of higher order than three can be evaluated by repeated expansion with cofactors.

Cramer's rule is seldom used to solve numerical problems involving more than three simultaneous equations. Computer programs based on gaussian elimination, or a related technique, are used.

PROBLEMS

Section A.1

1. Write the equations

$$3v_1 + v_2 + 3v_3 = 14$$
$$2v_1 - 2v_2 - v_3 = -5$$
$$v_1 + v_2 + 2v_3 = 9$$

as a matrix equation of the form

$$\mathbf{av} = \mathbf{i}$$

2. Write the equations

$$\frac{d^2}{dt^2}v_1 - 2v_2 = 5\cos t$$

$$-3\frac{d}{dt}v_1 + 4\int v_2\,dt = e^{-4t}$$

as a single matrix equation.

Section A.2

3. Evaluate the determinant of matrix **a** given

$$\mathbf{a} = \begin{bmatrix} 3 & 1 & 3 \\ 2 & -2 & -1 \\ 1 & 1 & 2 \end{bmatrix}$$

(a) By multiplying the elements along the diagonals as shown in Eq. (A.17)

(b) By Laplace's expansion about the first column

(c) By Laplace's expansion about the second column

(d) By Laplace's expansion about the third column

(e) By Laplace's expansion about the second row

4. Use Cramer's rule to solve for the voltages of Problem 1.

5. Use Cramer's rule to solve for the currents:

$$\begin{bmatrix} 6 & -2 & -3 \\ -2 & 7 & -4 \\ -3 & -4 & 9 \end{bmatrix}\begin{bmatrix} i_1 \\ i_2 \\ i_3 \end{bmatrix} = \begin{bmatrix} -7 \\ 0 \\ 16 \end{bmatrix}$$

B

Complex Arithmetic

There are two standard analytical forms for a complex number, the *rectangular* and the *exponential* forms. In addition to these analytical forms, there is a standard notation for complex numbers called the *polar* form.

B.1 Rectangular Form

The *rectangular form* of a complex number z is

$$z = x + jy \tag{B.1}$$

in which x and y are real numbers, and the *imaginary unit* j is defined by $j^2 = -1$. In Eq. (B.1), the real number x is called the *real part* (or *real component*) of z, and can be identified by the *real part operator* $\mathcal{Re}\{\cdot\}$ as follows:

$$x = \mathcal{Re}\{z\} \tag{B.2}$$

Similarly, the real number y is called the *imaginary part* (or *imaginary component*) of z and can be identified by the *imaginary part operator* $\mathcal{Im}\{\cdot\}$ as

$$y = \mathcal{Im}\{z\} \tag{B.3}$$

If the real part, x, of z is nonzero and the imaginary part, y, is zero, then z is said to be *purely real.* If the imaginary part, y, of z is nonzero, and the real part, x, is zero, then z is said to be *purely imaginary.* For example, 3.1415 is purely real, and $-j7$ is purely imaginary. If both the real and the imaginary parts of z are zero, then z is said to be zero, and is denoted by $z = 0$.

Since it takes two real numbers to specify a complex number, it is convenient to display a complex number as a point in a plane, called the complex plane, as shown in Fig. B.1a. Each point in the complex plane corresponds to exactly one complex number. The abscissa in this representation is the coordinate of possible values of the real part, x. The ordinate† is the coordinate of j times possible values of the imaginary part, y. An alternative representation of the same complex number is shown in Fig. B.1b. In this figure, a directed line segment is drawn from the origin to the value of z, forming a vector in the complex plane. With some abuse of terminology‡, it will occasionally be convenient to refer to the complex number z as "the vector z." The projection of this vector onto the abscissa gives the value of the real part of z, whereas the projection onto the ordinate gives j times the value of the imaginary part of z.

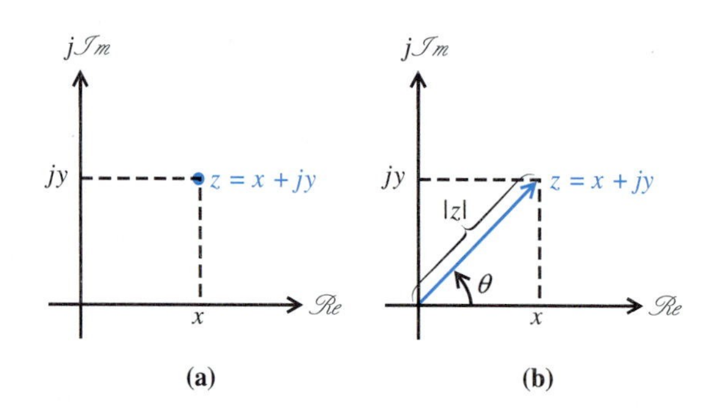

FIGURE B.1 Graphical representation of a complex number z: (*a*) as a point; (*b*) as a vector. The length of the vector, $|z|$, equals the magnitude of z. The angle of the vector, $\theta = \mathcal{arg}\{z\}$, equals the argument (angle) of z.

† Mathematics texts correctly use y as the ordinate rather than jy. The convention to label the ordinate jy is, unfortunately, prevalent in electrical engineering. Despite the notational problem, we will use the prevalent convention.

‡ Complex numbers are actually not vectors in the mathematical sense. They can be multiplied and divided by other complex numbers. Vectors cannot be multiplied or divided by other vectors.

B.2 Exponential Form

The *exponential form* of a complex number z involves its *magnitude* (or *modulus*), denoted by $|z|$, and its *argument* (or *angle in radians*), denoted by $\theta = \mathscr{arg}\{z\}$. These terms are defined in Fig. B.1b and can be seen to correspond to the length and the angle, respectively, of the vector z. It can readily be appreciated from Fig. B.1b that

$$x = |z| \cos \theta \tag{B.4}$$

and

$$y = |z| \sin \theta \tag{B.5}$$

so that

$$\begin{aligned} z &= x + jy \\ &= |z| \cos \theta + j|z| \sin \theta \\ &= |z|(\cos \theta + j \sin \theta) \\ &= |z|e^{j\theta} \end{aligned} \tag{B.6}$$

Equation (B.6) follows from Euler's identity, which was proved in Section 7.4:

$$e^{j\theta} = \cos \theta + j \sin \theta \tag{B.7}$$

The series of equalities in Eqs. (B.4) through (B.6) shows the relationship between the rectangular form and the exponential form:

$$z = |z|e^{j\theta} \tag{B.8}$$

From Eqs. (B.4) through (B.6) (or directly from Fig. B.1b), we find further that

$$|z| = \sqrt{x^2 + y^2} \tag{B.9}$$

and

$$\theta = \tan^{-1} \frac{y}{x} \tag{B.10}$$

Equation (B.10) contains a hidden trap. Consider the complex numbers $z_1 = -1 + j1$ and $z_2 = 1 - j1$. By sketching z_1 and z_2 as vectors in the complex plane, we find that the angles are, respectively, $\theta_1 = \mathscr{arg}\{-1 + j1\} = 135°$ and $\theta_2 = \mathscr{arg}\{1 - j1\} = -45°$. On the other hand, Eq. (B.10) yields $\theta_1 = \tan^{-1} \frac{1}{-1}$ and $\theta_2 = \tan^{-1} \frac{-1}{1}$. The trap lies in the fact that, although *algebraically* $\frac{1}{-1}$ does equal $\frac{-1}{1}$, θ_1 certainly does not equal θ_2. Therefore, when evaluating Eq. (B.10), remember to correctly account for the signs of y and x.

The angle θ is undefined for $x = y = 0$. Length is never complex or negative. Therefore, regardless of the value of z, its magnitude is a nonnegative real number,

$$|z| \geq 0 \tag{B.11}$$

The factor $e^{j\theta}$ in Eq. (B.8) is a *unit* vector with the angle θ in the complex plane. This observation makes it obvious that, for example, $j = e^{j\pi/2}$, $e^{j\pi} = -1$, and $e^{j2\pi} = 1$, and so forth. [The student should confirm this using Eq. (B.7).]

The exponential representation contains an inherent ambiguity, since the addition of 2π radians to the angle of z does not change the value of z. (Rotation of the vector z of Fig. B.1b by 2π radians does not change the vector. Equivalently, $e^{j(\theta+2\pi)} =$

$e^{j\theta}e^{j2\pi} = e^{j\theta}$.) This ambiguity is sometimes removed by *defining* the angle of a complex number to lie in some 2π range, such as $0 \leq \theta < 2\pi$ or $-\pi < \theta \leq \pi$. If the argument of z, θ, is written as $\theta = \theta_0 + 2\pi k$, where $-\pi < \theta_0 \leq +\pi$ and $k = 0, \pm 1, \ldots$, then θ_0 is called the *principal argument* of z and is denoted by $\mathcal{A}rg\{z\}$.

B.3 Polar Form

The *polar* form of a complex number z is simply the pair of its polar coordinates $|z|$ and θ of Fig. B.1b. There are two conventional notations used for the polar form,

$$z = |z|\underline{/\theta} \tag{B.12}$$

and

$$z = |z|\underline{/z} \tag{B.13}$$

Thus both $\underline{/\theta}$ and $\underline{/z}$ are sometimes used to denote the angle of z. Since (B.12) and (B.13) are definitions of notation rather than analytical formulas, the exponential or the rectangular forms are used in the analytical derivation of this book.

B.4 Arithmetic Operations

The rule of addition (or subtraction) of complex numbers is identical to the rule of addition (or subtraction) of vectors. Figure B.2 illustrates the addition of two complex numbers z_1 and z_2 as vector addition in the complex plane. The vector z_2 can be translated from the origin to the tip of vector z_1 when forming the sum. Translation of a vector in the complex plane does not change its value, since by definition a vector is a directed line segment. (The vector z_2 is completely specified by its length and angle or by the extent of its real and imaginary components. These quantities are invariant to a translation of the vector.)

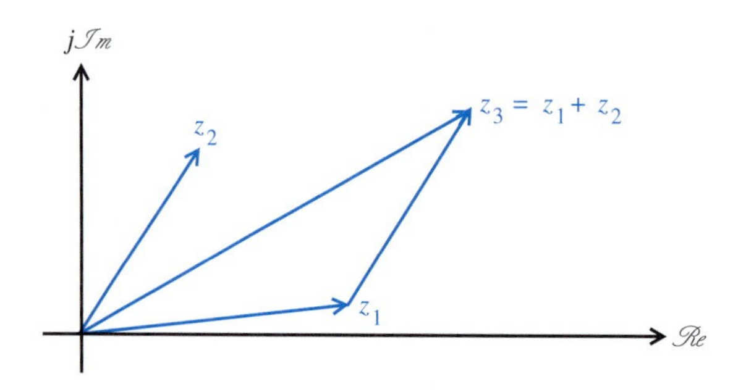

FIGURE B.2 Addition of complex numbers

It is more convenient to use the rectangular form than the polar form when we add complex numbers. The sum of two complex numbers $z_1 = x_1 + jy_1$ and $z_2 = x_2 + jy_2$ is then clearly seen to be a complex number $z_3 = x_3 + jy_3$ whose real component is the sum of the real components of z_1 and z_2 and whose imaginary component is the sum of the imaginary components of z_1 and z_2:

$$\begin{aligned} z_1 + z_2 &= x_1 + jy_1 + x_2 + jy_2 \\ &= x_1 + x_2 + j(y_1 + y_2) \\ &= x_3 + jy_3 \\ &= z_3 \end{aligned} \tag{B.14}$$

where $x_3 = x_1 + x_2$ and $y_3 = y_1 + y_2$.

The product of two complex numbers

$$\begin{aligned} z_1 z_2 &= |z_1| e^{j\theta_1} |z_2| e^{j\theta_2} \\ &= |z_1|\,|z_2| e^{j(\theta_1 + \theta_2)} \end{aligned} \tag{B.15}$$

has a magnitude that is the product of the magnitudes of z_1 and z_2 and an angle that is the sum of the angles of z_1 and z_2. Thus, for any two complex numbers z_1 and z_2,

$$|z_1 z_2| = |z_1|\,|z_2| \tag{B.16}$$

and

$$arg\{z_1 z_2\} = arg\{z_1\} + arg\{z_2\} \tag{B.17}$$

Alternatively, using the rectangular form,

$$\begin{aligned} z_1 z_2 &= (x_1 + jy_1)(x_2 + jy_2) \\ &= (x_1 x_2 - y_1 y_2) + j(x_1 y_2 + x_2 y_1) \end{aligned} \tag{B.18}$$

Similarly, the quotient of two complex numbers is (for $z_2 \neq 0$)

$$\begin{aligned} \frac{z_1}{z_2} &= \frac{|z_1| e^{j\theta_1}}{|z_2| e^{j\theta_2}} \\ &= \frac{|z_1|}{|z_2|} e^{j(\theta_1 - \theta_2)} \end{aligned} \tag{B.19}$$

Therefore

$$\left| \frac{z_1}{z_2} \right| = \frac{|z_1|}{|z_2|} \tag{B.20}$$

and

$$arg\left\{\frac{z_1}{z_2}\right\} = arg\{z_1\} - arg\{z_2\} \tag{B.21}$$

Alternatively, the division can be performed using the rectangular form as follows:

$$\begin{aligned} \frac{z_1}{z_2} &= \frac{x_1 + jy_1}{x_2 + jy_2} \\ &= \frac{x_1 + jy_1}{x_2 + jy_2} \cdot \frac{x_2 - jy_2}{x_2 - jy_2} \\ &= \frac{x_1 x_2 + y_1 y_2 + j(y_1 x_2 - y_2 x_1)}{x_2^2 + y_2^2} \\ &= \frac{x_1 x_2 + y_1 y_2}{x_2^2 + y_2^2} + j\,\frac{y_1 x_2 - y_2 x_1}{x_2^2 + y_2^2} \end{aligned} \tag{B.22}$$

The procedure used to remove the imaginary factor in the denominator as shown is called *rationalizing the denominator.*

Observe that the exponential form yields more insight into the nature of complex multiplication and division than does the rectangular form. The polar form is also easier to apply in analytical expressions involving several factors of complex quantities. However, the rectangular form is sometimes more convenient to use when we evaluate complex expressions.

B.5 The Complex Conjugate

The *complex conjugate* of a complex number z, denoted by z^*, is obtained by replacing the unit j by $-j$, which is the second root to $j^2 = -1$. For example, if $z = 3 + j2$, then $z^* = 3 - j2$. Replacement of the number j by $-j$ causes a reflection of z about the real axis, as shown in Fig. B.3. The rectangular form of the complex conjugate of $z = x + jy$ is, in general,

$$z^* = x - jy \tag{B.23}$$

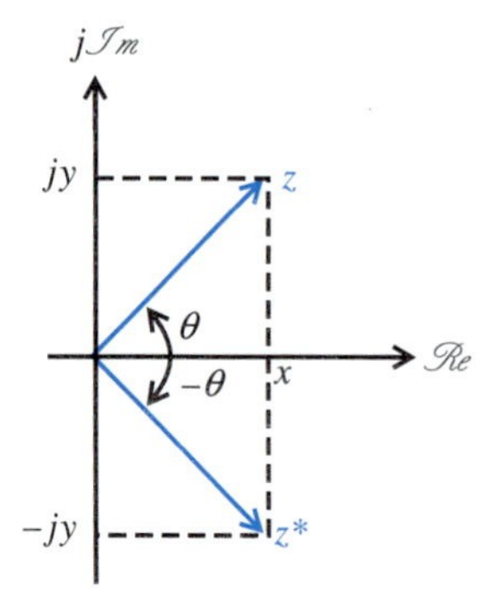

FIGURE B.3
The number z and its complex conjugate z^*

The polar form of the complex conjugate of $z = |z|e^{j\theta}$ is, in general:

$$z^* = |z|e^{-j\theta} \tag{B.24}$$

Therefore, complex conjugation reverses the sign of the imaginary part of z while leaving the real part of z unchanged. Equivalently, complex conjugation reverses the sign of the angle of z while leaving the magnitude of z unchanged.

Another important observation is that for any complex number z,

$$z + z^* = 2\mathscr{Re}\{z\} \tag{B.25}$$

and

$$z - z^* = 2j\mathscr{Im}\{z\} \tag{B.26}$$

These relationships are illustrated in Fig. B.4, and are easily verified with the use of the rectangular form. It is also useful to recognize that

$$zz^* = |z|^2 \tag{B.27}$$

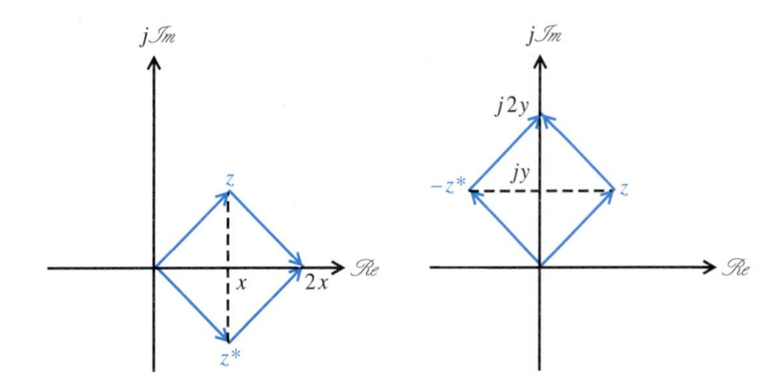

FIGURE B.4
Illustration of Eqs. (B.25) and (B.26)

which is easily verified with the use of either the rectangular or the exponential form. An application of Eq. (B.27) arises in rationalizing a denominator, as illustrated in Eq. (B.22). The numerator and the denominator were both multiplied by z_2^*, rendering the denominator a positive real number.

An important property of the operation of conjugation is that it commutes with respect to addition, multiplication, and division. The student should verify that

$$(z_1 + z_2)^* = z_1^* + z_2^* \tag{B.28}$$

$$(z_1 z_2)^* = z_1^* z_2^* \tag{B.29}$$

and

$$\left(\frac{z_1}{z_2}\right)^* = \frac{z_1^*}{z_2^*} \tag{B.30}$$

These relations make it easy to conjugate complex algebraic expressions. Simply replace j by $-j$ wherever it appears. For example,

$$\left[\frac{(3 + j)e^{j2}}{(4 - 5j)(7j)} + 2j\right]^* = \frac{(3 - j)e^{-j2}}{(4 + 5j)(-7j)} - 2j \tag{B.31}$$

B.6 Boldface Notation

Boldface notation is often used in electrical engineering to denote certain (but not all) complex quantities. The quantities that most often appear in boldface are current and voltage phasors **I** and **V**, respectively; impedance and admittance **Z** and **Y**, respectively; transfer function **H**; and complex power **S**. We use a subscript m to denote the magnitude of a phasor: $V_m = |\mathbf{V}|$ and $I_m = |\mathbf{I}_m|$. We often use boldface notation to denote complex voltage and current waveforms, $\mathbf{v}(t)$ and $\mathbf{i}(t)$ respectively.

B.7 Summary

The principal definitions and properties developed in this section are summarized in Table B.1.

TABLE B.1 Summary of complex arithmetic

Name	Figure	Definition	Properties
Rectangular form	B.1	$z = x + jy$	$x = \|z\| \cos \theta$, $y = \|z\| \sin \theta$
Exponential form	B.1	$z = \|z\| e^{j\theta}$	$\|z\| = \sqrt{x^2 + y^2}$, $\theta = \mathscr{arg}\{z\} = \tan^{-1}(y/x)$
Polar form		$z = \|z\| \underline{/\theta}$ or $z = \|z\| \underline{/z}$	
Addition	B.2	$z_1 + z_2 = (x_1 + x_2) + j(y_1 + y_2)$	—
Subtraction	—	$z_1 - z_2 = (x_1 - x_2) + j(y_1 - y_2)$	—
Multiplication	—	$z_1 z_2 = \|z_1\| \|z_2\| e^{j(\theta_1 + \theta_2)}$	$\|z_1 z_2\| = \|z_1\| \|z_2\|$ $\mathscr{arg}\{z_1 z_2\} = \mathscr{arg}\{z_1\} + \mathscr{arg}\{z_2\}$
Division	—	$z_1/z_2 = [\|z_1\|/\|z_2\|] e^{j(\theta_1 - \theta_2)}$	$\|z_1/z_2\| = \|z_1\|/\|z_2\|$ $\mathscr{arg}\{z_1/z_2\} = \mathscr{arg}\{z_1\} - \mathscr{arg}\{z_2\}$
Real part operator	B.1	$\mathscr{Re}\{z\} = x$	$\mathscr{Re}\{z_1 \pm z_2\} = \mathscr{Re}\{z_1\} \pm \mathscr{Re}\{z_2\}$
Imaginary part operator	B.1	$\mathscr{Im}\{z\} = y$	$\mathscr{Im}\{z_1 \pm z_2\} = \mathscr{Im}\{z_1\} \pm \mathscr{Im}\{z_2\}$
Conjugate operator	B.3	$z^* = x - jy = \|z\| e^{-j\theta}$	$(z_1 \pm z_2)^* = z_1^* \pm z_2^*$ $(z_1 z_2)^* = z_1^* z_2^*$, $(z_1/z_2)^* = z_1^*/z_2^*$
—	B.4	—	$z + z^* = 2\mathscr{Re}\{z\}$, $z - z^* = 2j\, \mathscr{Im}\{z\}$

PROBLEMS

1. Plot the following eight numbers as vectors in the complex plane: $z_1 = 3$, $z_2 = 1\underline{/\pi}$, $z_3 = 1\underline{/\pi/2}$, $z_4 = -2j$, $z_5 = 3e^{j\pi/6}$, $z_6 = 1 - j$, $z_7 = 2e^{j3\pi/4}$, $z_8 = 3\underline{/-510°}$.
2. Express each of the following eight numbers in (a) exponential form and (b) polar form. (Write the argument of each number, θ, so that $-\pi < \theta \leq \pi$.) $z_1 = 1$, $z_2 = j$, $z_3 = -1$, $z_4 = -j$, $z_5 = 1 + j$, $z_6 = 1 - j$, $z_7 = -1 - j$, $z_8 = -1 + j$.
3. Express each of the following seven numbers in rectangular form: $z_1 = 1\underline{/0°}$, $z_2 = 2e^{j\pi}$, $z_3 = 3e^{-j\pi/2}$, $z_4 = 10\underline{/-450°}$, $z_5 = -3e^{j\pi/4}$, $z_6 = 7\underline{/30°}$, $z_7 = -(2\underline{/60°})$.
4. (a) Illustrate $z_1 + z_2 + z_3$ graphically as the sum of vectors in the complex plane, where $z_1 = 1$, $z_2 = j$, and $z_3 = 2\underline{/135°}$.

   (b) Find $|z_1 + z_2 + z_3|$ and $\mathscr{arg}\{z_1 + z_2 + z_3\}$.
5. Draw z_1, z_2, and $z_1 z_2$ as vectors in the complex plane, where $z_1 = 2\underline{/30°}$ and $z_2 = -4 + 4j$. Specify both the polar and the rectangular coordinates of $z_1 z_2$.
6. Draw z as an arbitrary vector in the complex plane, showing its length $|z|$ and angle θ and its rectangular coordinates x and y. Draw $1/z$ on the same plane and indicate its length and angle and its rectangular coordinates.
7. Evaluate each of the following expressions, putting your answers in (i) rectangular form, (ii) exponential form, and (iii) polar form.

   (a) $\dfrac{3\underline{/60°}}{1 + 3j}$ (b) $\dfrac{1 + j}{-1 - 3j}$

   (c) $\dfrac{e^{2j} + 1}{e^{j} - 1}$ (d) $\dfrac{e^{2j} + 1}{e^{j} + 1}$

   (e) $\dfrac{(1 - j)(4e^{j45°})}{1 + 3j + 3\underline{/60°}}$

   (f) $\dfrac{2\underline{/45°} + 8\underline{/135°}}{(4\underline{/90°})(2\underline{/20°})}$

   (g) $(5\underline{/22°})^{-1} + (6\underline{/185°})^{-1}$

   (h) $\dfrac{4\underline{/65°} + (4 + j)^{-1}}{17 + e^{j60°}}$
8. Verify Eqs. (B.25) and (B.26).
9. Verify Eq. (B.27).
10. Verify Eqs. (B.28), (B.29), and (B.30).
11. Find the magnitudes and angles of the following expressions. (All quantities are real except j.)

(a) $\dfrac{1}{1 + j\omega RC}$ (b) $\dfrac{V_m\angle\phi}{j\omega RC + 1}$

(c) $\dfrac{1 + j\omega RC}{R - \omega^2 RLC + j\omega L}$ (d) $\dfrac{(a + bj)(c + dj)}{(e + fj)(h + ij)}$

12. Find the real and the imaginary parts of each of the following expressions. Put your answers in the simplest form possible. (All quantities are real except j.)

    (a) $\dfrac{V_m\angle\phi}{1 + j\omega RC}\, e^{j\omega t}$

    (b) $\dfrac{(1 + j\omega RC)V_m e^{j\phi}}{R - \omega^2 RLC + j\omega L}\, e^{j\omega t}$

    (c) $\dfrac{j\omega_1 L(A_1\angle\phi_1)}{2j\omega_1 L + R}\, e^{j\omega_1 t} + \dfrac{j\omega_2 L(A_2\angle\phi_2)}{2j\omega_2 L + R}\, e^{j\omega_2 t}$

13. Find the roots of the following equations and plot them as points in the complex plane.

    (a) $s^2 + 4s + 8 = 0$

    (b) $s^2 + 6s + 8 = 0$

    (c) $s^8 = 1$

    (d) $s^7 = 1$

14. Shade the regions in the complex (z) plane that satisfy

    (a) $\mathcal{Re}\{z\} \geq 1$

    (b) $\mathcal{Im}\{z\} \geq 1$

    (c) $|z| \leq 1$

    (d) $|z - 1| \leq 3$

15. Under what conditions does $|z_1 + z_2| = |z_1| + |z_2|$?

16. Under what conditions does $|z_1 + z_2| = |z_1| - |z_2|$?

17. (a) Take the real part of $e^{j(\theta+\phi)}$ to show that $\cos(\theta + \phi) = \cos\theta\cos\phi - \sin\theta\sin\phi$.

    (b) Take the imaginary part of $e^{j(\theta+\phi)}$ to show that $\sin(\theta + \phi) = \cos\theta \sin\phi + \sin\theta\cos\phi$.

18. Show that $z_1 z_2 = 0$ if and only if at least one of the numbers z_1, z_2 is zero.

19. Show that

$$|z_1 + z_2 + \cdots + z_n| \leq |z_1| + |z_2| + \cdots + |z_n|$$

When does the equality hold?

20. Let n be any integer. Express the nth roots of unity in (a) exponential form and (b) rectangular form.

21. Evaluate each of the following expressions, putting your answers in (i) rectangular form, (ii) exponential form, and (iii) polar form.

    (a) $j^{\sqrt{j}}$

    (b) j^e

    (c) $\cos(3 + j2)$

    (d) $\sin(3 + j2)$

    (e) $1 + z + z^2 + \cdots$, where $|z| < 1$. (*Hint*: Let S equal the infinite series and subtract zS from S to get a closed form for S.)

    (f) $1 + z + z^2 + \cdots + z^N$, where N is an integer.

C

The Second-Order Differential Equation

In this appendix we use integration to derive the solution to the second-order LTI differential equation of Chapter 9.

C.1 The Source-Free Equation

In Chapter 9 we used an intuitive approach to solve the differential equation that describes a source-free circuit with two energy-storage elements (*second-order circuits*). We assumed an exponential solution, substituted this into the source-free *second-order* differential equation

$$\frac{d^2}{dt^2}v + a_1\frac{d}{dt}v + a_0v = 0 \tag{C.1}$$

and found that the solution to this equation is

$$v_n = A_1e^{s_1t} + A_2e^{s_2t} \qquad s_1 \neq s_2 \tag{C.2}$$

where s_1 and s_2 are the roots of the characteristic equation,

$$\mathcal{A}(s) = s^2 + a_1s + a_0 = 0 \tag{C.3}$$

and A_1 and A_2 are the constants of integration. This completely solved the problem, except for the *unusual* case where $s_1 = s_2$.

We will now solve Eq. (C.1) by integration. We do this not just to solve for the special case of $s_1 = s_2$, but to also establish a procedure to solve the differential equation that describes a second-order circuit with sources. We write Eq. (C.1) as

$$\frac{d^2}{dt^2}v - (s_1 + s_2)\frac{d}{dt}v + s_1s_2v = 0 \tag{C.4}$$

where s_1 and s_2 are the roots of the characteristic equation. We use

$$\frac{d}{dt}\left(\frac{d}{dt}v\right) = \frac{d^2}{dt^2}v \tag{C.5}$$

to write Eq. (C.4) in the form

$$\frac{d}{dt}\left(\frac{d}{dt}v - s_2v\right) - s_1\left(\frac{d}{dt}v - s_2v\right) = 0 \tag{C.6}$$

and define the variable

$$x = \frac{d}{dt}v - s_2v \tag{C.7}$$

so that our second-order equation in terms of voltage v, Eq. (C.6), becomes a first-order equation in terms of the function x:

$$\frac{d}{dt}x - s_1x = 0 \tag{C.8}$$

As in Chapter 8 [Eqs. (8.3) to (8.6)], we multiply this first-order equation by the integrating factor e^{-s_1t}, use the rule for the derivative of the product of two functions, and integrate with respect to t to obtain

$$x = Ae^{s_1t} \tag{C.9}$$

where A is a constant of integration.

We substitute Eq. (C.9) into Eq. (C.7) to obtain a first-order differential equation in terms of v:

$$\frac{d}{dt}v - s_2v = Ae^{s_1t} \tag{C.10}$$

We multiply this equation by the integrating factor $e^{-s_2 t}$, use the rule for the derivative of the product of two functions, and integration with respect to t to obtain

$$e^{-s_2 t}v - A_0 = \begin{cases} \dfrac{A}{s_1 - s_2} e^{(s_1 - s_2)t} & s_1 \neq s_2 \quad \text{(C.11a)} \\ At & s_1 = s_2 \quad \text{(C.11b)} \end{cases}$$

We now multiply both sides of Eqs. (C.11a) and (C.11b) by $e^{s_2 t}$, rearrange them, and define the constant of integration $A_1 = A/(s_1 - s_2)$ and $A_2 = A_0$ for $s_1 \neq s_2$, and $A_1 = A_0$ and $A_2 = A$ for $s_1 = s_2$ to obtain:

The Natural Response of a Second-Order Circuit

$$v_n = \begin{cases} A_1 e^{s_1 t} + A_2 e^{s_2 t} & s_1 \neq s_2 \quad \text{(C.12a)} \\ (A_1 + A_2 t)e^{s_1 t} & s_1 = s_2 \quad \text{(C.12b)} \end{cases}$$

where s_1 and s_2 are the roots of the characteristic equation, and A_1 and A_2 are constants of integration.

In the next section we use the same procedure to solve for the response of a differential equation that describes a circuit with sources.

C.2 The Forced Equation

We can easily extend our preceding analysis to solve second-order differential equations

$$\frac{d^2}{dt^2} v + a_1 \frac{d}{dt} v + a_0 v = b_2 \frac{d^2}{dt^2} i + b_1 \frac{d}{dt} i + b_0 i$$
$$= g(t) \quad \text{(C.13)}$$

that describe second-order circuits with sources. We see that the left-hand side of this equation is identical to the left-hand side of Eq. (C.1), so we write

$$\frac{d}{dt}\left(\frac{d}{dt} v - s_2 v\right) - s_1 \left(\frac{d}{dt} v - s_2 v\right) = g(t) \quad \text{(C.14)}$$

as we did in Eq. (C.6). We again use the variable

$$x = \frac{d}{dt} v - s_2 v \quad \text{(C.15)}$$

to reduce Eq. (C.14) to the first-order equation

$$\frac{d}{dt} x - s_1 x = g(t) \quad \text{(C.16)}$$

and use the integrating factor, $e^{-s_1 t}$, to obtain

$$x = g_1(t) + g_2(t) \quad \text{(C.17)}$$

where

$$g_1(t) = Ae^{s_1 t} \tag{C.18}$$

with A a constant of integration, and

$$g_2(t) = e^{s_1 t} \int e^{-s_1 t} g(t)\, dt \tag{C.19}$$

Substitution of Eq. (C.17) into Eq. (C.15) yields a first-order differential equation in terms of ν:

$$\frac{d}{dt}\nu - s_2 \nu = g_1(t) + g_2(t) \tag{C.20}$$

where $g_1(t)$ and $g_2(t)$ are given by Eqs. (C.18) and (C.19).

We use the integrating factor $e^{-s_2 t}$ to integrate Eq. (C.20). The result is

$$\nu = A_0 e^{s_2 t} + e^{s_1 t} \int e^{-s_1 t}[g_1(t) + g_2(t)]\, dt \tag{C.21a}$$

$$= \left[A_0 e^{s_2 t} + Ae^{s_2 t} \int e^{-s_2 t} g_1(t)\, dt \right] + e^{s_2 t} \int e^{-s_2 t} g_2(t)\, dt \tag{C.21b}$$

$$= \nu_n + \nu_p \tag{C.21c}$$

We see that, as for the first-order equation, the complete response is the sum of a natural response component ν_n and a particular response component ν_p. The two terms enclosed in the brackets in Eq. (C.21b) yield the natural response ν_n as given in Eqs. (C.12a) or (C.12b), and the third term is the particular response

$$\nu_p = e^{s_2 t} \int e^{-s_2 t} g_2(t)\, dt \tag{C.22}$$

Although we can combine Eqs. (C.19) and (C.22) into a single equation by substitution, the result is more confusing than illuminating. We find it easier to solve for ν_p in steps:

To Calculate the Particular Response by Integration

First use the right side of Eq. (C.13) to calculate $g(t)$ from i. Next, substitute $g(t)$ into Eq. (C.19) to determine $g_2(t)$. Now substitute $g_2(t)$ into Eq. (C.22) to obtain the particular response ν_p.

Evaluation of the integrals in Eqs. (C.19) and (C.22) is generally difficult, but we only need to perform the integrations once to obtain our most useful result. We are very interested in exponential inputs, because this includes constant and sinusoidal sources as special cases. For the exponential input $i = Ke^{s_p t}$, Eq. (C.13) gives us

$$g(t) = b_2 \frac{d^2}{dt^2} i + b_1 \frac{d}{dt} i + b_0 i$$

$$= (b_2 s_p^2 + b_1 s_p + b_0)Ke^{s_p t} = \mathcal{B}(s_p)Ke^{s_p t} \tag{C.23}$$

and Eqs. (C.19) and (C.22) yield

$$v_p = \begin{cases} \dfrac{\mathcal{B}(s_p)}{\mathcal{A}(s_p)} Ke^{s_p t} & \mathcal{A}(s_p) \neq 0 \quad \text{(C.24a)} \\ \dfrac{\mathcal{B}(s_p)}{s_p - s_1} tKe^{s_p t} & s_1 \neq s_p = s_2 \quad \text{(C.24b)} \\ \dfrac{1}{2} \mathcal{B}(s_p)\, t^2 Ke^{s_p t} & s_p = s_1 = s_2 \quad \text{(C.24c)} \end{cases}$$

Equation (C.24a) represents the usual situation. We rewrite this equation as

$$\mathcal{A}(s_p) v_p = \mathcal{B}(s_p) Ke^{s_p t} \qquad \mathcal{A}(s_p) \neq 0 \tag{C.25}$$

and conclude:

> To obtain the particular response to an exponential input $Ke^{s_p t}$, replace differentiation with multiplication by s_p for the usual case where $\mathcal{A}(s_p) \neq 0$

Although we will not go through the details, a similar development shows that the particular response to an exponential source is given by Eq. (C.25) for a linear time-invariant (LTI) circuit of any order.

D

Elementary Mathematical Formulas

The standard formulas in this appendix may be helpful in solving certain circuit problems.

D.1 Quadratic Formula

The roots of $ax^2 + bx + c = 0$ are given by

$$x = \frac{-b \pm \sqrt{b^2 - 4ac}}{2a} = -\frac{b}{2a} \pm \sqrt{\left(\frac{b}{2a}\right)^2 - \frac{c}{a}}$$

D.2 Trigonometric Formulas

$$\sin(-x) = -\sin x$$

$$\cos(-x) = \cos x$$

$$\sin x = \cos(x - 90°)$$

$$\cos x = \sin(x + 90°)$$

$$\cos^2 x + \sin^2 x = 1$$

$$\cos^2 x - \sin^2 x = \cos 2x$$

$$2 \sin x \cos x = \sin 2x$$

$$\sin(x + y) = \sin x \cos y + \cos x \sin y$$

$$\cos(x + y) = \cos x \cos y - \sin x \sin y$$

$$A \cos x + B \sin x = \sqrt{A^2 + B^2} \cos\left(x - \arctan\frac{B}{A}\right)$$

$$\sin x \sin y = \frac{\cos(x - y) - \cos(x + y)}{2}$$

$$\sin x \cos y = \frac{\sin(x + y) + \sin(x - y)}{2}$$

$$\cos x \cos y = \frac{\cos(x + y) + \cos(x - y)}{2}$$

D.3 Plane Triangle Formulas

Let A, B, C denote the angles of any plane triangle and a, b, c denote the corresponding opposite sides. The following apply

Law of sines

$$\frac{a}{\sin A} = \frac{b}{\sin B} = \frac{c}{\sin C}$$

Law of cosines

$$a^2 = b^2 + c^2 - 2bc \cos A$$

Law of tangents

$$\frac{b - c}{b + c} = \frac{\tan \frac{1}{2}(B - C)}{\tan \frac{1}{2}(B + C)}$$

D.4 Taylor Series

$$e^x = 1 + x + \frac{1}{2!}x^2 + \frac{1}{3!}x^3 + \cdots$$

$$\sin x = x - \frac{1}{3!}x^3 + \frac{1}{5!}x^5 - \cdots$$

$$\cos x = 1 - \frac{1}{2!}x^2 + \frac{1}{4!}x^4 - \cdots$$

D.5 Euler's Relations

$$e^{jx} = \cos x + j \sin x \qquad \text{where } j^2 = -1$$

$$\cos x = \frac{e^{jx} + e^{-jx}}{2}$$

$$\sin x = \frac{e^{jx} - e^{-jx}}{2j}$$

D.6 Miscellaneous Series and Integrals

$$\frac{1}{1-z} = 1 + z + z^2 + z^3 + \cdots \qquad \text{for } |z| < 1$$

$$\frac{1 - z^{n+1}}{1 - z} = 1 + z + z^2 + \cdots + z^n \qquad \text{for any } z \neq 1 \text{, where } n \text{ is a positive integer}$$

$$\int_{-\infty}^{\infty} e^{-a^2x^2}\, dx = \frac{\sqrt{\pi}}{a} \qquad \text{for } a > 0$$

D.7 L'Hôpital's Rule

If $f(a) = g(a) = 0$, then

$$\lim_{x \to a} \frac{f(x)}{g(x)} = \lim_{x \to a} \frac{f'(x)}{g'(x)}$$

provided that $g'(a) \neq 0$. The primes denote derivatives with respect to x.

Answers to Problems

Chapter 1

1. (a) $4t$ C $t > 0$ **2. (d)** $20\pi^2 \cos 2\pi t$ A $t > 0$ **5. (a)** 2 V
6. (e) $200 \cos^2 2\pi t = 100 + 100 \cos 4\pi t$ W $t > 0$, $100t + (25/\pi) \sin 4\pi t$ J $t > 0$
9. (a) 30 W, **(g)** −74 W **12. (a)** 324 kC **15. (d)** 34.8 A
19. (b) −894 C/g **24. (a)** −4 C

Chapter 2

1. $i_{05} = 6$ A **5.** $v_{12} = -1$ V, $v_{64} = 2$ V
8. $i_x = 4$ A, 39 V, 108 A, −1620 W **12.** −450 W **15. (a)** 9 V, **(e)** $R = 2\,\Omega$
17. (b) $-0.24e^{-4000t}$ A $t > 0$ **18. (a)** $6 + 2 \times 10^6 t$ V $t > 0$
22. (b) $-13e^{-200t}$ V $t > 0$ **23. (a)** $2 + 4 \times 10^3 t$ A $t > 0$
27. $v_1 = 8$ V, $v_2 = -240$ V **32.** $i_x = 1$ A
36. $i_{45} = -2t^2$ A, $i_{23} = -3t^2$ A, $v_{10} = 195t^2$ V, $v_{20} = 5 + 195t^2 - (13/21)t^3$ V, (all for $t > 0$)
40. 5 Ω in series with 10 H **44.** $L_{eq} = L/(1 + A)$ **49.** 1.8 μA **53.** 2 mH

Chapter 3

1. (a) $v_{ab} = 60$ V, $v_{oc} = 60$ V, $i_a = 10$ A, $i_b = 6$ A, $i_c = 4$ A, $p_s = -600$ W, $p_{10} = 360$ W, $p_{15} = 240$ W, $p_6 = 0$ **(b)** $v_{ab} = 60$ V, $i_{cd} = 10$ A
5. $p_s = -45 - 75 \cos(4t - 53.13^0)$ W, $p_R = 45 + 45 \cos 4t$ W, $p_s = -60 \sin 4t$ W, No
10: (8/3) Ω, 2 H, and 11 F in parallel. **15.** 18 V
16. (a) $i = 6$ A, $v_{ab} = 60$ V, $v_{bc} = 90$ V, $v_{ac} = 150$ V, $v_{oc} = 90$ V, $p_s = -900$ W, $p_{10} = 360$ W, $p_{15} = 540$ W, $p_{R10} = 0$ **(b)** $i = 6$ A, $v_{ac} = 96$ V, $i_{de} = (18/5)$ A
20. $p_{10} = 12500 + 12500 \cos 4t$ W, $p_L = -100 \sin 4t$ W, $p_s = -12500 + 12500.4 \times \cos(4t + 179.54°)$W, No **23.** $v_1 = 48$ V, $v_2 = -7{,}200$ V, $v_3 = -10{,}080$ V
27. 15 Ω, 6 H, and (12/5) F in series. **28.** $v_{ab} = 60$ V, $v_{cb} = -90$ V
32. (a) −3000 W **(b)** −600 W **37. (a)** (12/5) Ω **42. (a)** 59 kΩ

45. **(a)** 20 kΩ **(b)** $i = vR_1/[R_1R_2 + R_x(R_1 + R_2)]$
(c) $R_x = [R_1/(R_1 + R_2)][(v/i) - R_2]$
(d) $R_1 = 30.02\ \Omega$, $R_2 = 60\ \text{k}\Omega$, $R_x = 40\ \text{k}\Omega$ **(e)** $R_1 = 30/\{1 - [30/(20{,}000v)]\}$
(f) No, but for R_s a few ohms or less the error will be small.
49. **(a)** a 40-V source with the (+) reference toward terminal *a*
(b) a 40-A source with the reference arrow pointing toward terminal *a*
52. **(a)** $6 - 0.5 \times 10^{-3}n$ A, $6 \times 10^{-3}n - 0.5 \times 10^{-6}n^2$ W **62.** **(c)** $5R/6$

Chapter 4

1. $v_1 = 12e^t$ V, $v_2 = -36e^t$ V, $i_1 = 96e^t$ A, $i_2 = -5 + 105e^t$ A, $i_3 = -15 + 96e^t$ A, $i_4 = 15 - 100e^t$ A
3. 1 V **6.** $v_1 = 2$ V, $v_2 = 9$ V, $i_x = -2$ A
10. **(b)** $v_1 = 8.63$ V, $v_2 = 23.77$ V, $v_3 = 30.16$ V, $v_4 = 40.93$ V, $v_5 = 50$ V
15. **(a)** $v_L/v_b = -(\beta_1\beta_2R_L)/[R_b(1 + \beta_2)]$, $v_b/i_b = R_b$
18. $i_1 = 12e^t$ A, $i_2 = -36e^t$ A, $v_1 = 96e^t$ V, $v_2 = -5 + 105e^t$ V, $v_3 = 15 - 100e^t$ V,
23. $i_1 = 2$ A, $i_2 = 9$ A, $v_x = -2$ V **26.** 67.13 A, −147.92 A, 66.34 A, −40 A
28. $i_1 = -0.96i_s$, $i_2 = 0.98i_s$, $i_3 = i_s$, $i_y/i_s = -0.96$

32.
$$\begin{bmatrix} 4 - 135d/dt + 11\int_{-\infty}^{t} d\lambda & -9d/dt - 11\int_{-\infty}^{t} d\lambda \\ 70 + 135d/dt - 11\int_{-\infty}^{t} d\lambda & 12 + 9d/dt + 11\int_{-\infty}^{t} d\lambda \end{bmatrix} \begin{bmatrix} i_1 \\ i_2 \end{bmatrix} = \begin{bmatrix} 38 \\ 1040 \end{bmatrix}$$

36. 1 A, −100 V **40.** $38 < \beta < \infty$, $R_E > 1000\ \Omega$ **45.** 7.692 A
48. 2010/121 = 16.6116 A, 2295/121 = 18.9669 A, 3305/121 = 27.3140 A, 40 A, 50 A, −2409/242 = −9.9545 V

Chapter 5

1. **(a)** $i_y = 5$ A **3.** $v_o = -(18/5)(v_a + v_b)$
5. **(a)** A voltage source of value $v_s/4$ [with the (+) mark toward terminal *a*] in series with a resistance of value *R*.
8. A 12-A current source [with the reference arrow pointing toward terminal *a*] in parallel with a 6-Ω resistance.
9. **(c)** A 410-V voltage source [with the (+) mark toward terminal *a*] in series with a 266-Ω resistance. A (205/133)-A current source [with the reference arrow pointing toward terminal *a*] in parallel with a 266-Ω resistance.
10. **(a)** (−14/5) V, (1/20) Ω **14.** **(a)** 4 Ω **16.** R = ∞ (an open circuit)
19. A current source of value i_s is connected in parallel with a (1/2)-H inductance and in parallel with a series connection of a 3-S conductance and a 4-F capacitance.
22. A 3-V voltage source [with the (+) mark toward terminal *a*] in series with a −10-Ω resistance.
26. $v_o/v_s = -(40/3)$, $R_o = 2000\ \Omega$ **31.** **(a)** 45 V, 15 Ω

Chapter 6

2. **(a)** 2 V **(b)** 10 V **(c)** ≈15 V (saturated) **(d)** ≈15 V (saturated)
6. $2\ v_s^2$ mW **9.** **(a)** $|v_o| \approx 15$ V, $|i_{sc}| = \infty$ **(b)** $|v_o| \approx 10$ V, $|i_{sc}| < 150$ mA

13. $R_a = 2\ \text{k}\Omega$ and $R_b = 8\ \text{k}\Omega$ **18.** $v_o = [1 + (R_b/R_a) + 2(R_b/R_c)](v_{s1} - v_{s2})$
24. (a) $G = -8.057$, $R_i' = 1193.3\ \Omega$, $R_o = 10.704\ \Omega$, no **28.** $100\ \text{k}\Omega$
32. $100\ \text{k}\Omega$

Chapter 7

2. (a) $10 \sin(2\pi t)[u(t) - u(t - T)]$ A
3. (a) $[2tu(t) - 2(t - 1)u(t - 1) - 2u(t - 2)$ A
5. (a) $20tu(t)$ A **(d)** $10(1 - e^{-2(t-1)})u(t - 1)$ A
6. (a) 1 **(d)** 9 **9. (a)** 0.6020 s **12. (a)** $-72°$, 7.7254
13. (a) $\mathcal{R}e\{10e^{j12\pi t}\}$ **14. (a)** $10e^{j3t} + 10e^{-j3t}$ **15. (a)** 20.086
16. (a) $10 \cos 2t + j10 \sin 2t$ **19. (a)** 0 **20. (a)** V_m **21. (b)** 0
22. (b) $5/\sqrt{2}$ A **23. (a)** $\sqrt{17}$ **25. (a)** $e^{-4t}(4 \cos 5t - 6 \sin 5t)$ **29. (b)** $A\tau/T$
30. 2.5285 A

Chapter 8

3. (a) $10e^{-20t}$ V $t > 0$ **6.** $8e^{-20t}$ A $t > 0$ **10.** $10e^{-t/2}$ A $t > 0$
13. (a) $48 - 24e^{-5t}$ V $t > 0$ **(b)** $88e^{-5t} - 40e^{-8t}$ V $t > 0$
18. (a) $12e^{-12t}$ V $t > 0$ **(b)** $12 - 12e^{-10t}$ V $t > 0$ **19. (c)** $12te^{-12t}u(t)$ A
23. (b) $6 - 6e^{-6t}$ A $t > 0$
25. 200 V $t < 0$, $120 + 80e^{-12t}$ V $0 < t < 0.1$ s, $200 - 55.9e^{-8(t-0.1)}$ V $t > 0.1$ s
31. $10 + 40e^{-2t} - 40e^{-t/5}$ V $t > 0$ **35.** Stable for $\beta > -1$
37. (a) e^{6t} V $t < 0$, $(3/2) - (1/2)e^{-3t}$ V $t > 0$ **40.** $-100e^{-100t}u(t)$ V
44. (c) $6 - 6e^{t/2}$ A $t > 0$, $30e^{t/2}$ V $t > 0$ **49.** $4e^{-4t}$ V $t > 0$
53. (a) $100e^{-100t} - 100$ V $t > 0$ **57.** $20(1 - te^{-t} - e^{-t})u(t)$ V

Chapter 9

1. $-14e^{-t} + 20e^{-2t}$ V $t > 0$ **4.** $20e^{-4t} \cos 2t$ V $t > 0$
9. $e^{-2t}(8 \cos 4t - 10 \sin 4t)$ A $t > 0$ **11.** $(2 + 13t)e^{-3t}$ A $t > 0$
15. (c) $15 + 1500te^{-5t}$ V $t > 0$ **19.** $75e^{-3t} - (60 + 125t)e^{-5t}$ A $t > 0$
20. $16 - 5e^{-4t} + 5e^{-36t}$ A $t > 0$
22. $20 \cos(2t + 36.87°) + (-16 + 160t)e^{-4t}$ A $t > 0$
24. $-9 + (12 + 22t)e^{-2t}$ V $t > 0$
29. $v_1 = 3 + e^{-2t}(4 \cos \sqrt{2}\, t - 2\sqrt{2} \sin \sqrt{2}\, t)$ V $t > 0$
34. $11e^{-t} - 5e^{10t}$ A $t > 0$ **38.** $200 - (100 + 500t)e^{-5t}$ V $t > 0$
43. $12 + 0.9841 \cos(3t + 85.56°) + 1.897 \cos(6t + 71.57°)$ V
48. $10 \cos 10t$ V $t > 0$

Chapter 10

4. $v_c(t) = Ae^{s_1 t}$ and $s_1 = -\dfrac{1}{(1 - \beta)RC}$. For $\beta < 1$, $v_c(t) \to 0$ as $t \to \infty$. For $\beta = 1$, $v_c(t) = 0$ for all t. For $\beta > 1$, $v_c(t) \to \infty$ as $t \to \infty$

6. $\mathcal{A}(D)e^{j\omega t}$ denotes that $\mathcal{A}(D)$ *operates* on $e^{j\omega t}$. $\mathcal{A}(j\omega)e^{j\omega t}$ is the *result* of the operation. Consider for example, $\mathcal{A}(D) = D$. Then $De^{j\omega t} = j\omega e^{j\omega t}$ but $D \neq j\omega$.

11. **(a)** $\mathbf{Z} = \dfrac{\omega^2 RL^2}{R^2 + (\omega L)^2} + j\dfrac{\omega R^2 L}{R^2 + (\omega L)^2}$ **(b)** 1 Mrad/s
14. **(a)** $3\underline{/15^\circ}$ A **(b)** $-j$ MΩ **(c)** $3\underline{/-75^\circ}$ MV
(d) $3\cos(1000t - 75^\circ)$ MV
(e) $3 \times 10^6 \cos(1000t - 75^\circ) + v_C(0)$ V This circuit is not stable. The answers to parts (d) and (e) differ by a constant because the complementary response does not decay to zero as t increases.
17. **(a)**
$$\underbrace{\left(3\frac{d}{dt} + 2\right)(v_R(t) + jv_I(t))}_{\mathcal{A}(D)\mathbf{v}(t)} = \underbrace{3\frac{d}{dt}v_R(t) + 2v_R(t)}_{\mathcal{A}(D)v_R(t)} + \underbrace{j\left[3\frac{d}{dt}v_I(t) + 2v_I(t)\right]}_{j\mathcal{A}(D)v_I(t)}$$
The real part of the left-hand side is $\mathcal{Re}\{\mathcal{A}(D)\mathbf{v}(t)\}$, and the real part of the right-hand side is $\mathcal{A}(D)v_R(t)$. Therefore, $\mathcal{Re}\{\mathcal{A}(D)\mathbf{v}(t)\} = \mathcal{A}(D)\,\mathcal{Re}\{\mathbf{v}(t)\}$ for $\mathcal{A}(D) = 3D + 2$
(b)
$$\underbrace{\left(\sum_{k=0}^{n} a_k D^k\right)(v_R(t) + jv_I(t))}_{\mathcal{A}(D)\mathbf{v}(t)} = \underbrace{\sum_{k=0}^{n} a_k D^k v_R(t)}_{\mathcal{A}(D)v_R(t)} + \underbrace{j\sum_{k=0}^{n} a_k D^k v_I(t)}_{j\mathcal{A}(D)v_I(t)},$$
so $\mathcal{Re}\{\mathcal{A}(D)\mathbf{v}(t)\} = \mathcal{A}(D)\,\mathcal{Re}\{\mathbf{v}(t)\}$ for every real $\mathcal{A}(D)$. **(c)** Same derivation as in (b). **(d)** We can take the real part of the equation $\mathcal{A}(D)\mathbf{V}(t) = \mathcal{B}(D)\mathbf{i}(t)$ to obtain $\mathcal{Re}\{\mathcal{A}(D)\mathbf{v}(t)\} = \mathcal{Re}\{\mathcal{B}(D)\mathbf{i}(t)\}$. By (b) and (c) the above becomes $\mathcal{A}(D)\,\mathcal{Re}\{\mathbf{v}(t)\} = \mathcal{B}(D)\,\mathcal{Re}\{\mathbf{i}(t)\}$. To obtain Eq. (10.29) let $\mathbf{i}(t) = \mathbf{I}e^{j\omega t}$ and $\mathbf{v}(t) = \mathbf{V}e^{j\omega t}$.
22. From Table 10.1, $R(\omega)G(\omega) = R^2(\omega)/[R^2(\omega) + X^2(\omega)]$. This equals 1 if and only if $X(\omega) = 0$.
26. **(a)** $1/(1 + j\omega RC)$, which can be written as $\dfrac{1/j\omega RC}{R + (1/j\omega RC)}$, has voltage divider form involving the impedances of R and C.
(b) $\dfrac{1}{\sqrt{1 + (\omega RC)^2}}\cos(\omega t + \phi_{\mathbf{V}} - \arctan(\omega RC))$ **32.** 5 V

Chapter 11

2. $i(t) = \sqrt{2}\cos(\omega t + 45^\circ)$ **7.** $2.77\underline{/4.98^\circ}$ A
10. $RLDi = Rv + LDv + RLCD^2 v$ **14.** $\mathbf{I}_b = 1\underline{/90^\circ}$, $\mathbf{V}_a = 1.58\underline{/71.57^\circ}$
18. **(a)** $2/(3 + j2\omega RC)$ **(b)** 1/2 **(c)** $1/(3 + j2\omega RC)$
(c) $2RCDv_3 + 3v_3 = v_1$
22.
$$\left[\begin{array}{c|c} j\omega C + \dfrac{1}{j\omega L} + \dfrac{1}{R} & -\dfrac{1}{R} \\ \hline -\dfrac{1}{R} & j\omega C + \dfrac{1}{j\omega L} + \dfrac{1}{R} \end{array}\right] = \left[\begin{array}{c} \dfrac{\mathbf{V}_s}{j\omega L} \\ \hline j\omega C\mathbf{V}_s \end{array}\right]$$
26.
$$\begin{bmatrix} -j & 3 \\ 2j & 1 \end{bmatrix}\begin{bmatrix} \mathbf{V}_1 \\ \mathbf{V}_2 \end{bmatrix} = \begin{bmatrix} 18 - j5 \\ 5 + j10 \end{bmatrix}, \begin{bmatrix} \mathbf{V}_1 \\ \mathbf{V}_2 \end{bmatrix} = \begin{bmatrix} 5.02\underline{/4.9^\circ} \\ 5.86\underline{/0^\circ} \end{bmatrix}$$
29. **(a)**
$$\left[\begin{array}{c|c|c} \mathbf{Z}_a^{-1} + \mathbf{Z}_b^{-1} & 0 & 0 \\ \hline 0 & \mathbf{Z}_1^{-1} + \mathbf{Z}_2^{-1} & -\mathbf{Z}_2^{-1} \\ \hline -A & A & 1 \end{array}\right]\left[\begin{array}{c} \mathbf{V}_{(+)} \\ \hline \mathbf{V}_{(-)} \\ \hline \mathbf{V}_0 \end{array}\right] = \left[\begin{array}{c} \mathbf{Z}_a^{-1}\mathbf{V}_{s2} \\ \hline \mathbf{Z}_1^{-1}\mathbf{V}_{s1} \\ \hline 0 \end{array}\right]$$

(b) $\mathbf{V}_0 = \dfrac{A}{1 + A\beta}\left[-\dfrac{\mathbf{Z}_2}{\mathbf{Z}_1 + \mathbf{Z}_2}\mathbf{V}_{s1} + \dfrac{\mathbf{Z}_b}{\mathbf{Z}_a + \mathbf{Z}_b}\mathbf{V}_{s2}\right]$, where $\beta = \mathbf{Z}_1/(\mathbf{Z}_1 + \mathbf{Z}_2)$

(c) $\mathbf{V}_0 = -\dfrac{\mathbf{Z}_2}{\mathbf{Z}_1}\mathbf{V}_{s1} + \dfrac{\mathbf{Z}_b(\mathbf{Z}_1 + \mathbf{Z}_2)}{\mathbf{Z}_1(\mathbf{Z}_a + \mathbf{Z}_b)}\mathbf{V}_{s2}$

35. $$\begin{bmatrix} \dfrac{1}{j\omega C} + j\omega L & -j\omega L & 0 \\ 0 & -1 & 1 \\ \dfrac{1}{j\omega C} & R_2 & R_1 \end{bmatrix}\begin{bmatrix} \mathbf{I}_1 \\ \mathbf{I}_2 \\ \mathbf{I}_3 \end{bmatrix} = \begin{bmatrix} -\mathbf{V}_s \\ \mathbf{I}_s \\ 0 \end{bmatrix}$$

39. $$\begin{bmatrix} R_1 & j\omega L & \dfrac{1}{j\omega C} \\ -1 & 1 & 0 \\ -\beta & -1 & \beta + 1 \end{bmatrix}\begin{bmatrix} \mathbf{I}_1 \\ \mathbf{I}_2 \\ \mathbf{I}_3 \end{bmatrix} = \begin{bmatrix} -\mathbf{V}_s \\ \mathbf{I}_s \\ 0 \end{bmatrix}$$

44. $i(t) = 4.17 \sin(2t + 33.7°) - 2.5$ A
47. $v_o(t) = 5 \cos(t + 10°) + 2.5 \sin(2t + 33.7°) + 10$ V
52. A one ohm resistance.
57. If $\mathbf{V}_s \neq 0$, we obtain an impossible theoretical circuit when we short the input terminals. Thus $\mathbf{I}_{sc}$ is not defined for this circuit.

Chapter 12

5. **(a)** 30 mW **(b)** 20 mW **(c)** 17.32 mVAR **10.** 21.18 W
15. **(a)** 0.7071 lagging **(b)** 2 W **(c)** 2.83 VA
(d) 3.885 volt rms **(e)** 0.728 amperes rms
20. $P = \omega^2 C^2 R V_{rms}^2/\{[1 - \omega^2 LC]^2 + (\omega RC)^2\}$,
$Q = j(\omega^2 LC - 1)\omega C V_{rms}^2/\{[1 - \omega^2 LC]^2 + (\omega RC)^2\}$
24. **(a)** $\mathbf{S}_R = I_m^2 R/2$, $\mathbf{S}_L = j\omega L I_m^2/2$, $\mathbf{S}_C = -jI_m^2/2\omega C$
(c) $Q = 0$ when $\omega = 1/\sqrt{LC}$
(d) $P = 0$ when $R = 0$ and $\omega \neq 1/\sqrt{LC}$
30. **(a)** 100 V **(b)** $100\angle 70°$ VA **(c)** $58.5\angle -35°$ VA
(d) 1.44 amperes rms **(e)** $-j82.75\ \Omega$
35. **(a)** $R + j/\omega C$ **(b)** a 10 Ω resistance in series with a 1 μH inductance
(c) 1.25 W 125 mVAR
40. **(a)** $4 - j\ \Omega$ **(b)** 31.25 mW

Chapter 13

1. **(a)** $R/(R + R_s + j\omega RR_sC)$ **(b)** $R/(R + R_s)$, 0
(c) capacitance acts like an open circuit for $\omega = 0$ and a short circuit for $\omega = \infty$.
(d) $R/\sqrt{(R + R_s)^2 + (\omega RR_sC)^2}\ \angle -\arctan[\omega RR_sC/(R + R_s)]$
5. $20 + 35 \sin(10t - 90°) + 4 \cos(15t - 15°)$ V
10. **(a)** 1900 Ω, 10 μF **(c)** $120 + 38.3 \cos(300t - 16.7°) + 62.4 \cos(800t - 38.6°) + 228 \sin(1100t - 2.73°) - 44.7 \cos(2000t - 63.4°)$ V
15. **(a)** $\dfrac{1}{j\omega C} \| j\omega L = \infty$ when $\omega = 1/\sqrt{LC}$ **(b)** 1 kΩ

(c) $\mathbf{I}_s = \mathbf{I}_R = 5\angle 0°$ mA, $\mathbf{I}_L + \mathbf{I}_C = 0$, $\mathbf{I}_L = 5\angle -90°$ A, $i_L(t)$ and $i_C(t)$ have amplitudes 1000 times greater than $i_s(t)$!
20. $2Dv + v = 4Di + i$

25. **(a)** $25 \dfrac{s^2 + 4}{s^3 + 2s^2 + 17s}$ **(b)** 0

31. $\pm j1/\sqrt{LC}$ **35.** $-\dfrac{R}{2L} \pm j\sqrt{\dfrac{1}{LC} - \left(\dfrac{R}{2L}\right)^2}$] $r < R_2$

45. **(a)** 0, $KV_{1m} \cos [\omega_1(t - \tau) + \phi_{v1}]$, 0
(b) $4 \sin (800t - 45.8°) + 17 \cos (1100t - 18°)$

50. **(a)** $\sqrt{\dfrac{1}{LC} - \left(\dfrac{1}{RC}\right)^2}$ **(b)** $\mathbf{Z}(j\omega_{upf}) = \dfrac{L}{RC}\ \Omega$

56. **(a)** $\omega_d \approx 130$ rad/s **(b)** peak $|H| \approx \dfrac{9990}{41.2}$

59. $$\frac{Ks^2\left(\frac{s}{10} + 1\right)}{\left(\frac{s}{0.7} + 1\right)^2\left(\frac{s}{100} + 1\right)\left(\frac{s}{500} + 1\right)^2}$$

66. **(a)** $\dfrac{1 - \tau_1\tau_2 s^2}{\tau_1\tau_2 s^2 + s(\tau_1 + \tau_2) + 1}$ where $\tau_1 = L/R$ *and* $\tau_2 = RC$

Chapter 14

1. $\dfrac{1}{s + 3}$, $\mathcal{Re}\{s\} > -3$ **5.** $\dfrac{1}{s} + \dfrac{1}{s - 3} + \dfrac{1}{s + 3}$ for $\mathcal{Re}\{s\} > 3$

9. $e^{-6t}u(t)$ **14.** $\dfrac{s + 3}{(s + 3)^2 + 16}$ **17.** $\dfrac{6\angle 45°}{s + 3 - j5} + \dfrac{6\angle -45°}{s + 3 + j5}$

22. $e^{-2(t-1)}u(t - 1)$ **26.** $e^{-4t}\cos (5t)u(t)$

31. $\lim_{s\to 0} \dfrac{sA}{s^2 + \omega_0^2} = 0$. This is incorrect because $A \cos (\omega_0 t)u(t)$ has no final value.
35. Since $D(s) = (s - s_1)Q(s)$
then $D'(s) = \dfrac{d}{ds}[(s - s_1)Q(s)] = Q(s) + (s - s_1)\dfrac{d}{ds}Q(s)$.

Therefore $D'(s_1) = Q(s_1)$ and $k_{11} = (s - s_1)\left.\dfrac{N(s)}{D(s)}\right|_{s=s_1} = \dfrac{N(s_1)}{Q(s_1)} = \dfrac{N(s_1)}{D'(s_1)}$

40. $(2 - 4e^{-t} + 2e^{-2t})u(t)$ V
44. $v_C(t) = 10[1 - \cos (t)]u(t)$, $v_L(t) = v_1(t) - v_C(t) = 10 \cos (t)u(t)$
48. $3.161u(t - 0.001) - 3.161e^{-2000(t-0.001)}u(t - 0.001)$ V
51. 10 V **59.** $\delta(t) - e^{-t}u(t)$ **63.** $\delta(t - t_d)$, e^{-st_d}

Chapter 15

3. The mean square value of $v(t)$ is equal to the squared length of the vector $\mathbf{v}$.
7. **(b)** No
(c) The only difference between $v(t)$ and $v'(t)$ is that $v'(t)$ contains a nonzero coefficient a_0.
10. $A_0 = A/T$, $A_n = 2A/T$, $\phi_n = -2\pi nt_d/T$ for $n = 1, 2, \ldots$

13. **(a)** $a_0 = A$, $a_n = b_n = 0$ for $n = 1, 2, \ldots$
(b) $a_0 = 1/T$, $a_n = 2/T$ and $b_n = 0$ for $n = 1, 2, \ldots$

18. **(a)** $\frac{A^2\tau}{2RT}(1 - e^{-2T/\tau})$ **(b)** The average power dissipated by the dc component (0th harmonic) is $\frac{1}{R}\left(\frac{A\tau}{T}\right)^2(1 - e^{-T/\tau})^2$ **(c)** The average power dissipated by the nth harmonic is $\frac{2}{R}\left(\frac{A\tau}{T}\right)^2 \frac{(1 - e^{-T/\tau})^2}{1 + (2\pi n f_1 \tau)^2}$ for $n = 1, 2, \ldots$

30. 13.7 μF **33.** $\frac{1}{\pi}$ V

37. **(a)** $\frac{\tau}{1 + j2\pi f\tau}(1 - e^{-j2\pi fT})$

(b) $|\mathbf{V}(f)| = \frac{\tau\sqrt{[1 - e^{-T/\tau}\cos 2\pi fT]^2 + [e^{-T/\tau}\sin 2\pi fT]^2}}{\sqrt{1 + (2\pi f\tau)^2}}$,

$\underline{/\mathbf{V}(f)} = \arctan\left\{\frac{e^{-T/\tau}\sin 2\pi fT}{1 - e^{-T/\tau}\cos 2\pi fT}\right\} - \arctan(2\pi f\tau)$

Chapter 16

3. $\mathbf{Z} = \begin{bmatrix} \mathbf{Z}_A \| (\mathbf{Z}_A + \mathbf{Z}_B) + \mathbf{Z}_C & \frac{\mathbf{Z}_A^2}{2\mathbf{Z}_A + \mathbf{Z}_B} + \mathbf{Z}_C \\ \frac{\mathbf{Z}_A^2}{2\mathbf{Z}_A + \mathbf{Z}_B} + \mathbf{Z}_C & \mathbf{Z}_A \| (\mathbf{Z}_A + \mathbf{Z}_B) + \mathbf{Z}_C \end{bmatrix}$, $\mathbf{V}_{oc} = \mathbf{0}$

4. $\mathbf{V}_{oc} = \mathbf{V}_{2oc} + \mathbf{z}_{21}\frac{\mathbf{V}_s - \mathbf{V}_{1oc}}{\mathbf{Z}_s + \mathbf{z}_{11}}$, $\mathbf{Z}_{Th} = \mathbf{z}_{22} - \frac{\mathbf{z}_{12}\mathbf{z}_{21}}{\mathbf{Z}_s + \mathbf{z}_{11}}$ **7.** No

10. $\mathbf{y}_{11} = \mathbf{y}_{22} = \frac{1}{\mathbf{Z}_1 + \mathbf{Z}_1 \| \mathbf{Z}_2} + \frac{1}{\mathbf{Z}_A + \mathbf{Z}_A \| \mathbf{Z}_B}$,

$\mathbf{y}_{12} = \mathbf{y}_{21} = -\left(\frac{\mathbf{Z}_2}{\mathbf{Z}_1 + \mathbf{Z}_1}\right)\frac{1}{\mathbf{Z}_1 + \mathbf{Z}_1 \| \mathbf{Z}_2} - \left(\frac{\mathbf{Z}_B}{\mathbf{Z}_A + \mathbf{Z}_B}\right)\frac{1}{\mathbf{Z}_A + \mathbf{Z}_A \| \mathbf{Z}_B}$

13. $\mathbf{I}_{sc} = \begin{bmatrix} 7.26\underline{/65.6^\circ} \\ 4.28\underline{/24.5^\circ} \end{bmatrix}$ A **19.** $i_2 = -\beta_2\beta_1 i_1$

24. $\begin{bmatrix} \mathbf{V}_1 \\ \mathbf{V}_2 \end{bmatrix} = \begin{bmatrix} R & 0 \\ 0 & R \end{bmatrix}\begin{bmatrix} \mathbf{I}_1 \\ \mathbf{I}_2 \end{bmatrix} + \begin{bmatrix} 0 \\ R\mathbf{I}_o \end{bmatrix}$

$\begin{bmatrix} \mathbf{I}_1 \\ \mathbf{I}_2 \end{bmatrix} = \begin{bmatrix} 1/R & 0 \\ 0 & 1/R \end{bmatrix}\begin{bmatrix} \mathbf{V}_1 \\ \mathbf{V}_2 \end{bmatrix} - \begin{bmatrix} 0 \\ \mathbf{I}_o \end{bmatrix}$

$\begin{bmatrix} \mathbf{V}_1 \\ \mathbf{I}_2 \end{bmatrix} = \begin{bmatrix} R & 0 \\ 0 & 1/R \end{bmatrix}\begin{bmatrix} \mathbf{I}_1 \\ \mathbf{V}_2 \end{bmatrix} + \begin{bmatrix} 0 \\ -\mathbf{I}_o \end{bmatrix}$

29. Replace each of networks A and B of the series connection illustrated in Problem 28 with their **z**-parameter equivalents of Fig. 17.36a. The terminal equations of the resulting connection are given by KVL as: $\mathbf{V}_1 = (\mathbf{z}_{11A} + \mathbf{z}_{11B})\mathbf{I}_1 + (\mathbf{z}_{12A} + \mathbf{z}_{12B})\mathbf{I}_2 + \mathbf{V}_{1ocA} + \mathbf{V}_{1ocB}$, $\mathbf{V}_2 = (\mathbf{z}_{21A} + \mathbf{z}_{21B})\mathbf{I}_1 + (\mathbf{z}_{22A} + \mathbf{z}_{22B})\mathbf{I}_2 + \mathbf{V}_{2ocA} + \mathbf{V}_{2ocB}$ and may be written as

$$\begin{bmatrix} \mathbf{V}_1 \\ \mathbf{V}_2 \end{bmatrix} = \begin{bmatrix} \mathbf{z}_{11A} + \mathbf{z}_{11B} & \mathbf{z}_{12A} + \mathbf{z}_{12B} \\ \mathbf{z}_{21A} + \mathbf{z}_{21B} & \mathbf{z}_{22A} + \mathbf{z}_{22B} \end{bmatrix}\begin{bmatrix} \mathbf{V}_1 \\ \mathbf{V}_2 \end{bmatrix} + \begin{bmatrix} \mathbf{V}_{1ocA} + \mathbf{V}_{1ocB} \\ \mathbf{V}_{2ocA} + \mathbf{V}_{2ocB} \end{bmatrix}$$

Let us define sums of matrices as follows

$$\begin{bmatrix} \mathbf{z}_{11A} & \mathbf{z}_{12A} \\ \mathbf{z}_{21A} & \mathbf{z}_{22B} \end{bmatrix} + \begin{bmatrix} \mathbf{z}_{11B} & \mathbf{z}_{12B} \\ \mathbf{z}_{21A} & \mathbf{z}_{22B} \end{bmatrix} = \begin{bmatrix} \mathbf{z}_{11A} + \mathbf{z}_{11B} & \mathbf{z}_{12A} + \mathbf{z}_{12B} \\ \mathbf{z}_{21A} + \mathbf{z}_{21B} & \mathbf{z}_{22A} + \mathbf{z}_{22B} \end{bmatrix}$$

and

$$\begin{bmatrix} \mathbf{V}_{1ocA} \\ \mathbf{V}_{2ocA} \end{bmatrix} + \begin{bmatrix} \mathbf{V}_{1ocB} \\ \mathbf{V}_{2ocB} \end{bmatrix} = \begin{bmatrix} \mathbf{V}_{1ocA} + \mathbf{V}_{1ocB} \\ \mathbf{V}_{2ocA} + \mathbf{V}_{2ocB} \end{bmatrix}$$

Then the **z**-parameter matrix of the series network equals the sum of the **z**-parameter matrices of networks *A* and *B*. Notice also that a similar result applies for the open-circuit voltage matrices.

32. $R_a = 660\ \Omega$, $R_b = 330\ \Omega$, $R_c = 220\ \Omega$ **36.** 1.2 A

39. $\dfrac{\mathbf{V}_o}{\mathbf{V}_s} = \dfrac{\mathbf{z}_{21}\mathbf{Z}_L}{(\mathbf{z}_{11} + \mathbf{Z}_s)(\mathbf{z}_{22} + \mathbf{Z}_L) - \mathbf{z}_{12}\mathbf{z}_{21}}$

41. $\dfrac{\mathbf{V}_2}{\mathbf{V}_s} = \dfrac{\mathbf{V}_o}{\mathbf{V}_s} = \dfrac{-\mathbf{h}_{21}\mathbf{Z}_L}{\mathbf{Z}_L(\mathbf{h}_{11} + \mathbf{Z}_s)(\mathbf{h}_{22} + \mathbf{Y}_L) + \mathbf{h}_{21}\mathbf{h}_{12}}$

Chapter 17

2. $v_1 = -1200 \sin 3t$ V, $v_1 = -600 \sin 3t$ V **4. (a)** $Leq = L_1 + L_2 + 2M$
6. (b) $10.706\underline{/-52.39^\circ}$ A
11. (a) 20:1 **(b)** 0.1 A **(c)** 12 W when doorbell energized
16. $\mathbf{I}_1 = 2\underline{/-23.13^\circ}$ A **19.** $\mathbf{I}_1 = 5$ A, $\mathbf{I}_2 = -10$ A
21. $\mathbf{V}_1 = 100$ V, $\mathbf{V}_2 = -50$ V
25. $L_1 = 100$ H, $L_2 = 1600$ H, $M = 380$ H, $k = 0.95$
30. $14.241 \cos(6t - 8.531^\circ) + 18.827 \cos(8t - 11.310^\circ)$ V
35. $(L_1L_2 - M^2)/(L_1 + L_2 - 2M)$ **40.** $\mathbf{I}_2/\mathbf{I}_1 = (1 - s)/(1 + s)$, $\mathbf{Z} = 2\ \Omega$
42. (a) $[n_2/(n_1 + n_2)]^2\mathbf{Z}_s$ **46.** $R_{cd} = (1/2)(2n_1/n_2)^2R_L + 2R_s$
52. $L_{\ell 1} = 50.2$ mH, $L_{\ell 2} = 0.1256$ mH, $L_{M1} = 4.95$ H

Chapter 18

2. (a) $\mathbf{S}_1 = 240\underline{/0^\circ}$ VA, $\mathbf{S}_2 = 120\underline{/0^\circ}$ VA
(b) $\mathbf{S}_1 = 228.6\underline{/0^\circ}$ VA, $\mathbf{S}_2 = 114.3\underline{/0^\circ}$ VA, $\mathbf{S}_3 = 5485\underline{/36.87^\circ}$ VA
3. (a) $\mathbf{V}_{an} = 480\underline{/0^\circ}$ V, $\mathbf{V}_{bn} = 480\underline{/-120^\circ}$ V, $\mathbf{V}_{cn} = 480\underline{/-240^\circ}$ V
(b) $\mathbf{V}_{ab} = 831\underline{/30^\circ}$ V, $\mathbf{V}_{bc} = 831\underline{/-90^\circ}$ V, $\mathbf{V}_{ca} = 831\underline{/-210^\circ}$ V
9. $I_L = 6.93$ A (rms), $\mathbf{S} = 5760\underline{/53.13^\circ}$ VA $= (3456 + j4608)$ VA
11. $I_L = 20.78$ A (rms), $\mathbf{S} = 17{,}280\underline{/53.13^\circ}$ VA $= (10{,}368) + j13{,}824)$ VA
13. (a) $57.6\underline{/-36.87^\circ}\ \Omega$ **(b)** 24.06 A **(c)** $24.06\underline{/6.87^\circ}$ A
(d) $172.8\underline{/-36.87^\circ}\ \Omega$ **(e)** $13.89\underline{/-36.87^\circ}$ A **(f)** 91.2 hp
17. (a) $81.46\underline{/45^\circ}$ kVA
19. $\mathbf{S} = (22.57 + j4.43)$ kVA $= 23\underline{/11.1^\circ}$ kVA, PF = 0.98, $I_L = 5.5$ A
22. $P = 12$ kW, PF = 0.655 leading, $\mathbf{S} = 18.33\underline{/-49.11^\circ}$ kVA
27. (a) $\mathbf{S} = 14.14\underline{/45^\circ}$ kVA $= 10 + j10$ kVA, $I_L = 17.0$ A
(b) $\mathbf{S} = 10 + j7.5$ kVA $= 12.5\underline{/36.87^\circ}$ kVA, $I_L = 15.04$ A
32. 564 kVA
36. (a) $\mathbf{I} = 0.667 + j0.333$ A, $\mathbf{V}_2 = 1.491\underline{/-26.57^\circ}$ V, $\mathbf{S}_1 = 0.7454\underline{/-26.57^\circ}$ VA, $\mathbf{S}_2 = 1.111\underline{/-53.13^\circ}$ VA

Appendix A

1. $\begin{bmatrix} 3 & 1 & 3 \\ 2 & -2 & -1 \\ 1 & 1 & 2 \end{bmatrix} \begin{bmatrix} \mathbf{v}_1 \\ \mathbf{v}_2 \\ \mathbf{v}_3 \end{bmatrix} = \begin{bmatrix} 14 \\ -5 \\ 9 \end{bmatrix}$

2. $\begin{bmatrix} \dfrac{d^2}{dt^2} & -2 \\ -3\dfrac{d}{dt} & 4\displaystyle\int dt \end{bmatrix} \begin{bmatrix} \mathbf{v}_1 \\ \mathbf{v}_2 \end{bmatrix} = \begin{bmatrix} 5\cos t \\ e^{-4t} \end{bmatrix}$

3. -2 for all parts **4.** 1 V, 2 V, 3 V **5.** 1 A, 2 A, 3 A

Appendix B

7. (a) (i) $0.929 - j0.190$ **(ii)** $0.948e^{-j11.6°}$ **(iii)** $0.948\underline{/-11.6°}$
(c) (i) $0.539 - j0.989$ **(ii)** $1.27e^{-j61.4°}$ **(iii)** $1.27\underline{/-61.4°}$
(e) (i) $0.376 - j0.842$ **(ii)** $0.922e^{-j65.9°}$ **(iii)** $0.922\underline{/-65.9°}$
(g) (i) $0.0193 - j0.0604$ **(ii)** $0.0634e^{-j72.2°}$ **(iii)** $0.0634\underline{/-72.2°}$

11. (a) $\dfrac{1}{\sqrt{1 + (\omega RC)^2}}$ and $-\arctan(\omega RC)$ **(b)** $\dfrac{V_m}{\sqrt{1 + (\omega RC)^2}}$ and $\phi - \arctan(\omega RC)$

(c) $\sqrt{\dfrac{1 + (\omega RC)^2}{(R - \omega^2 RLC)^2 + (\omega L)^2}}$ and $\arctan(\omega RC) - \arctan\left[\dfrac{\omega L}{R - \omega^2 RLC}\right]$

(d) $\sqrt{\dfrac{(a^2 + b^2)(c^2 + d^2)}{(e^2 + f^2)(h^2 + i^2)}}$ and $\arctan(b/a) + \arctan(d/c) - \arctan(f/e) - \arctan(i/h)$

13. (a) $-2 + j2$ and $-2 - j2$ **(b)** -2 and -4
(c) $e^{j2\pi n/8}$ for $n = 0, 1, 2, \ldots, 7$ **(d)** $e^{j2\pi n/7}$ for $n = 0, 1, 2, \ldots, 6$

15. $\underline{/z_2} = \underline{/z_1}$

19. We can demonstrate the inequality by sketching z_i, $i = 1, 2, \ldots, n$ as vectors and examining the vector sum $z_1 + z_2 + \cdots + z_n$. Inspection shows that the length $|z_1 + z_2 + \cdots + z_n|$ is less than or equal to the sum of the lengths $|z_1| + |z_2| + \cdots + |z_n|$, with equality if and only if $\underline{/z_1} = \underline{/z_2} = \cdots = \underline{/z_n}$.

21. (a) Because $\sqrt{j} = \pm(0.707 + j0.707)$, there are two answers.
(i) $0.1465 + j0.295$ and $1.3502 - j2.7198$ **(ii)** $0.3924e^{j1.11}$ and $3.0365e^{-j1.11}$
(iii) $0.3924\underline{/1.11 \text{ rad}}$ and $3.0365\underline{/-1.11 \text{ rad}}$ **(b) (i)** $\cos(e\pi/2) + j\sin(e\pi/2)$
(ii) $e^{je\pi/2}$ **(iii)** $1\underline{/e\pi/2}$
(c) (i) $-3.7245 - j0.5119$ **(ii)** $3.759e^{-j172.7°}$ **(iii)** $3.759\underline{/-172.2°}$

Index